The Evolution of Social Behavior
in Insects and Arachnids

In the broad array of human diversity, George Eickwort stood out as one of those who was both delightful and productive to be with. Whether as a student, a teacher, a co-researcher, or whatever else, he worked hard while always managing to be both stimulating and entertaining to those around him. I always felt I was fortunate to have him as a student and friend. He is much missed.

C. D. Michener

We dedicate this volume to the memory and inspiration of George Eickwort.
J. C. Choe & B. J. Crespi

The Evolution of

Social Behavior
in Insects and Arachnids

Edited by

JAE C. CHOE Museum of Zoology, University of Michigan, USA
Seoul National University, Korea

and BERNARD J. CRESPI Simon Fraser University, Canada

CAMBRIDGE
UNIVERSITY PRESS

PUBLISHED BY THE PRESS SYNDICATE OF THE UNIVERSITY OF CAMBRIDGE
The Pitt Building, Trumpington Street, Cambridge CB2 1RP, United Kingdom

CAMBRIDGE UNIVERSITY PRESS
The Edinburgh Building, Cambridge CB2 2RU, United Kingdom
40 West 20th Street, New York, NY 10011-4211, USA
10 Stamford Road, Oakleigh, Melbourne 3166, Australia

First published in 1997

Printed in the United Kingdom at the University Press, Cambridge

Typeset in MT Ehrhardt 9/12pt

A catalogue record for this book is available from the British Library

Library of Congress Cataloguing in Publication data

The evolution of social behavior in insects and arachnids/edited by
    Jae C. Choe and Bernard J. Crespi.
        p.   cm.
    Includes index.
    ISBN 0 521 58028 5 (hc). – ISBN 0 521 58977 0 (pbk.)
    1. Insects – Behavior. 2. Arachnida – Behavior. 3. Behavior
evolution. 4. Social evolution in animals. I. Choe, Jae C. II. Crespi,
Bernard J.
QL496.E95 1997
595.7051–dc20   96–2861 CIP

ISBN 0 521 58028 5 hardback
ISBN 0 521 58977 0 paperback

Contents

Contributors

LETICIA AVILÉS
Department of Ecology & Environmental Biology, University of
Arizona, Tuscon, AZ 85721, USA.

WILLIAM J. BELL
Haworth Hall, Department of Entomology, University of Kansas,
Lawrence, KS 66045, USA.

H. JANE BROCKMANN
Department of Zoology, University of Florida, Gainesville, FL
32611, USA.

JAE. C. CHOE
Department of Biology, Seoul National University, Kwanak-Gu,
Shillim-Dong San 56-1, Seoul, Korea; and Museum of Zoology,
University of Michigan, Ann Arbor, MI 48109-1079, USA.

JAMES T. COSTA
Museum of Comparative Zoology, Harvard University,
Cambridge, MA 02138, USA.

BERNARD J. CRESPI
Department of Biological Sciences, Simon Fraser University,
Burnaby, British Columbia, Canada V5A 1S6.

BRYAN A. DANFORTH
Department of Entomology, Comstock Hall, Cornell University,
Ithaca NY 14853, USA.

JANICE S. EDGERLY
Department of Biology, Santa Clara University, Santa Clara,
CA 95053, USA.

ANNE-KATRIN EGGERT
Zoologisches Institut der Albert-Ludwigs-Universität,
Albertstrasse 21a, D-79104, Freiburg, Germany.

GEORGE C. EICKWORT
Department of Entomology, Comstock Hall, Cornell University,
Ithaca NY 14853, USA.

WILLIAM A. FOSTER
Department of Zoology, University of Cambridge, Downing
Street, Cambridge CB2 3EJ, UK.

GONZALO HALFFTER
Instituto de Ecologia, AC km 2.5 Antigua Carretera a Coatepec,
Xalapa, 91000, Veracruz, Mexico.

CRAIG S. HIEBER
Department of Biology, Saint Anselm College, Manchester NH
03102-1323, USA.

P. S. HURST
School of Biological Sciences, The Flinders University of South
Australia, GPO Box 2100, Adelaide 5001 SA, Australia.

DEBORAH S. KENT
Research Division, State Forests of New South Wales, P.O. Box
100, Beecroft 2119 NSW, Australia.

LAWRENCE R. KIRKENDALL
University of Bergen, Zoological Institute, Allégaten 41, N-5007
Bergen, Norway.

LAURENCE A. MOUND
Department of Entomology, Natural History Museum, Cromwell
Road, London SW7 5BD, UK (present address: Department of
Entomology, CSIRO, Black Mountain, Canberra ACT 2601,
Australia).

JOSEF K. MÜLLER
Zoologisches Institut der Albert-Ludwigs-Universität,
Albertstrasse 21a, D-79104 Freiburg, Germany.

CHRISTINE A. NALEPA
Department of Entomology, Gardner Hall, North Carolina State
University, Raleigh, NC 27695-7613, USA.

CHRISTIAN PEETERS
CNRS (URA 667), Laboratoire d'Ethologie Expérimentale et
Comparée, Université Paris Nord, 93430 Villetaneuse, France.
(present address: Laboratoire d'Ecologie, CNRS URA 258,
Université Paris 6, Campus Jussieu, 7 Quai Saint Bernard,
F-75252 Paris 05, France).

DAN L. PERLMAN
Geological Museum, Harvard University, Cambridge, MA 02138,
USA.

NAOMI E. PIERCE
Museum of Comparative Zoology, Harvard University,
Cambridge, MA 02138, USA.

KENNETH A. RAFFA
Department of Entomology, University of Wisconsin, 237 Russell
Laboratory, 1630 Linden Drive, Madison, WI 53706, USA.

YUTAKA SAITO
Laboratory of Applied Zoology, Faculty of Agriculture, Hokkaido
University, Kita-Ku, Sapporo 060, Japan.

CARL SCHAEFER
Department of Ecology & Environmental Biology, University of
Connecticut, Storrs, CT 0629-3043.

JACK C. SCHUSTER
Departamento de Biologia, Universidad del Valle de Guatemala,
Apartado 82, Ciudad Guatemala, Guatemala.

LAURA B. SCHUSTER

Departamento de Biologia, Universidad del Valle de Guatemala, Apartado 82, Ciudad Guatemala, Guatemala.

M. P. SCHWARZ

School of Biological Sciences, The Flinders University of South Australia, GPO Box 2100, Adelaide, SA 5001, Australia.

JANET S. SHELLMAN-REEVE

Section of Neurobiology & Behavior, Seely G. Mudd Hall, Cornell University, Ithaca, NY 14853, USA.

L. X. SILBERBAUER

Department of Zoology, LaTrobe University, Bundoora, Victoria 3083, Australia.

ROBERT L. SMITH

Department of Entomology, University of Arizona, Tucson, AZ 85721, USA.

DAVID L. STERN

Department of Ecology & Evolutionary Biology, Princeton University, Princeton, NJ 08544, USA (present address: Wellcome/CRC Institute and Department of Genetics, Tennis Court Road, Cambridge CB2 1QR, UK).

DOUGLAS W. TALLAMY

Department of Entomology & Applied Ecology, Delaware Agricultural Experiment Station, College of Agriculture, University of Delaware, Newark, DE 19717-1303, USA.

GEORGE W. UETZ

Department of Biological Sciences, University of Cincinnati, Cincinnati, OH 45221-0006, USA.

WILLIAM T. WCISLO

Smithsonian Tropical Research Institute, Apartado 2072, Balboa, Panama (correspondence to: Smithsonian Tropical Research Institute, Unit 0948, APO AA 34002-0948, USA).

DOUGLAS YANEGA

Department of Entomology, University of Kansas, Lawrence, KS 66045, USA (present address: Illinois Natural History Service, 607 E. Peabody Drive, Champaign, IL 61820, USA).

Acknowledgements

For reviews of one or more chapters, we are grateful to A. Bourke, D. Bell, J. Borden, D. Edmonds, T. Fitzgerald, G. Gries, B. Heming, S. Lewis, P. Luykx, C. D. Michener, N. Moran, L. Packer, H. K. Reeve, M. Richards, I. Robertson, M. Scott, J. Strassmann, S. Trumbo, P. Ward, and the authors themselves. For inspiration, we thank R. D. Alexander, W. D. Hamilton, B. Hölldobler, C. D. Michener and E. O. Wilson. For a visit to Ann Arbor that was instrumental in organizing and producing this book, we are grateful to R. D. Alexander and the University of Michigan Museum of Zoology. For enthusiasm and dedication, we thank our editor, Tracey Sanderson. And for love, patience, and support we thank our families, friends, and cats; they taught us the true meanings of cooperation.

Introduction

BERNARD J. CRESPI AND JAE C. CHOE

The purpose of this book is to explore the causes of some of the most spectacular transitions in evolutionary biology, the shifts between cooperation and competition, commensalism, or parasitism. In particular, we seek to explain under what conditions such behavior as parental care, alloparental care, and other forms of altruism among adults have evolved from the individually selfish life histories so common in animals, and how such behavior, once evolved, can be lost. Insects and arachnids provide the most varied and numerous instances and forms of the evolution of cooperation in animals; hence, they represent the most useful, though often idiosyncratic, database for analyzing social evolution.

We approach our main question, understanding the causes of the diversity of animal social systems, from the viewpoint of behavioral ecology: explaining and predicting behavior from ecology. In the context of studying sociality 'ecology' encompasses resource distribution in time and space, interactions with other species, and aspects of demography such as voltinism, survivorship and fecundity. Social 'behavior' can be defined broadly as the presence of cooperation and some form of organization, whereby the actions of individuals are coincident and coordinated or communicative in some way (Wilson 1971, p. 469). Our unifying framework for analysis in behavioral ecology is the gene's eye view of inclusive fitness theory, coupled with optimization theory (see, for example, Oster and Wilson 1978). We define a strategy set of possible behaviors, assess the fitness consequences of the alternatives in terms of particular selective pressures, and ask what behavior leads to the highest inclusive fitness. From this perspective, social behavior is complex relative to other types of behavior because it involves both continual interaction with others (and thus contingency and frequency-dependence of the strategies) and lifetimes of behavioral decisions, from which the pattern that we call a 'society' emerges. As in reproductive interactions between males and females, which are analyzed in the companion to this book (Choe and Crespi 1997), social interactions often involve one set

of individuals more or less willingly providing a limiting resource to another set (here, workers providing labor to reproductives) (Queller 1994a). Because individuals of the two sets usually differ in their optima for resource allocation, their interests conflict and their behaviors reflect an intricate mix of cooperation and competition.

Our motivation for this volume follows from the history of research into the evolution of sociality. The modern era of integrating natural history with evolutionary theory began with the suggestion of Hamilton (1964) that the relatedness asymmetries caused by haplodiploidy might explain the high incidence of helping in Hymenoptera. Over the next ten years, three hypotheses for explaining the origin of insect sociality came to dominate the literature: (1) haplodiploidy and three-quarters relatedness; (2) parental manipulation (Alexander 1974; Michener and Brothers 1974): and (3) mutualism (Lin and Michener 1972). Although these hypotheses were often presented and analyzed as being in opposition (see, for example, Craig 1979), they reflect separate and necessary components of the inclusive fitness inequality $rb - c > 0$ (Grafen 1984), which determines whether or not an allele for helping will increase in frequency. Thus, relatedness (r) varies from zero to one (from an actor to a receiver of an act) and benefits (b) and costs (c) may be affected by selective agents including conspecifics (see, for example, Gamboa 1978), natural enemies (Lin 1964), and abiotic environmental factors (see, for example, Evans 1977).

Many of the early tests of the theory of social behavior involved direct measurement or inference of relatedness (see, for example, Noonan 1981; Lester and Selander 1981; Crozier *et al.* 1987), to assess the presence and strength of inclusive fitness effects, especially with respect to relatednesses over one-half. For most hymenopteran species without morphological castes or swarm-founding, Hamilton's haplodiploidy theory has been well supported (see, for example, Ross and Carpenter 1991). However, low relatednesses in many obligately eusocial species, the absence of positive associations between relatedness and degrees of helping (Hughes *et al.* 1993), and the recognition that high relatedness cannot cause the evolution or maintenance of eusociality in the absence of particular ecological conditions (Andersson 1984), have led to the realization that relatedness alone cannot accurately predict the distribution and forms of sociality.

Following Evans (1977), who stressed the importance of both intrinsic and extrinsic conditions for social evolution, researchers began to search for ecological correlates of eusociality (Brockmann 1984; Strassmann and Queller 1989). This search led directly to the nest, especially toward nest characteristics expected to increase the inclusive fitness gains from staying at home and helping, such as expandability and defensibility (Alexander *et al.* 1991). Abe (1991) and Crespi (1994) pointed out that nests and other domiciles such as galls that combine food and shelter provide ideal conditions for the evolution of eusociality. Indeed, the coincidence of three conditions, (1) a habitat combining food and shelter, (2) strong selection for defense, and (3) the ability to defend, may be sufficient to predict the phylogenetic distribution of eusociality for all taxa except Hymenoptera and Isoptera that forage outside of their nests (Crespi 1994). Thus, at least some cases of eusociality can be predicted from ecology and phenotype.

Genetic, ecological and phenotypic variables adjust the fitness costs and benefits for individual decision-making, but the decisions themselves, to leave or stay, for a while or for a lifetime, are demographic and life-historical. Queller (1989, 1994b) and Gadagkar (1990) have pointed out that staying and helping when young can increase inclusive fitness by allowing reproduction (via help to relatives) earlier in adult life and with a higher probability of success. Thus, high adult mortality rates can themselves favor the origin and maintenance of eusociality. Nepotism when young could, in turn, speed the onset of senescence and promote worker–queen divergence in lifespan (Alexander *et al.* 1991).

The presence and absence of eusociality, or any other social system, should be predictable from some set of the genetic, ecological, phenotypic, and demographic traits described above. However, whether or not the same particular set of traits causes eusociality in taxa as disparate as aphids, wasps, and mole rats is unclear (Crespi 1996). To the extent that different taxa exhibit different sufficient conditions for the evolution of eusociality, its causes should be sought more in the idiosyncrasies of natural histories than via tests of theory seeking universal explanations. We believe that Wilson's (1989) prediction of progress via a taxon-centered research is well met at this stage in the study of social biology, because the main questions may yield not so much to universal explanations as to overlapping mosaics of causes, each more or less specific to a particular clade or clades.

We have three specific reasons for organizing this volume. First, the comparative database used for analyzing hypotheses concerning the evolution of sociality has recently undergone surprising expansions. Coleoptera have finally joined the eusocial fold, in the form of a platypodid that breeds in living trees (Kent and Simpson 1992), gall-forming Thysanoptera have been discovered with soldiers (Crespi 1992; Mound and Crespi 1995), soldier aphids continue to expand their ranks (Itô 1989; Moran 1993; Stern 1994), social mole rats now comprise at least two species (Jarvis 1981; Jarvis and Bennett 1993; Jarvis *et al.* 1994; Solomon 1994), spiders and mites show new levels of complexity and social sophistication (Saito 1986; Rypstra 1993), and hope remains for eusociality in Diptera (Holloway 1976), scorpions (Polis and Lourenço 1986; Shivashankar 1994) and marine crustaceans (Spanier *et al.* 1993). These newcomers to eusociality increase the number of independent origins by a substantial, as yet unknown, degree; their ecological and genetic similarities to and distinctiveness from Hymenoptera and Isoptera should allow for easier identification of the forms and causes of behavioral convergence among taxa.

Second, new approaches for studying adaptation, known generically as 'the comparative method', have greatly expanded the empirical tool kit of behavioral ecologists (Felsenstein 1985; Ridley 1989; Harvey and Pagel 1991). These methods provide statistical tests for association among characters and between environments and characters, and create a new context in which hypotheses can be generated for testing by other means. We can divide comparative methods into two main types: those that seek convergences (see, for example, Felsenstein 1985; Losos and Miles 1994; Pagel 1994), and those that analyze transitions in detail (see, for example, Coddington 1988; Baum and Larson 1991). The usefulness and limitations of comparative methods remain to be explored, especially in the context of complex traits such as sociality (Crespi 1996). However, these methods provide the ability to combine the phylogenetic history of sociality with microevolutionary analyses based on optimality, functional design (Williams 1966) and measurement of selection (Lande and Arnold 1983; Crespi 1990). Time and testable evolutionary trajectories have, for the first time, explicitly entered the adaptationist program.

Table I-1. *A categorization of insect and arachnid social systems*

This table extends Crespi and Yanega (1995) in altering their 'communal' category and dividing it into two types of society (colonial and communal) based on the presence of cooperation in brood care (i.e. feeding, defense, and construction and maintenance of nests). Subsocial refers to parental or biparental care. Shared breeding sites are those that involve multiple females. Alloparental brood care refers to the presence of behaviorally distinct groups, with individuals specializing in either reproduction or helping the more reproductive individuals to reproduce; and castes are 'groups of individuals that become irreversibly behaviorally distinct at some point prior to reproductive maturity' (Crespi and Yanega 1995). This categorization scheme is designed for the broadest possible taxonomic scale of comparative tests, for explanation of the genetic, phenotypic and ecological causes of variation in social systems. See Table I-2 for examples of taxa in each category.

Type of society	Brood care	Shared breeding site	Cooperation in brood care	Alloparental brood care	Castes
Subsocial	+				
Colonial	–	+			
Communal	+	+	+		
Cooperatively breeding	+	+	+	+	
Eusocial	+	+	+	+	+

Third, new conceptualizations of social terminology may free us from historical constraints and from confusion due to the independent evolution of terms by researchers studying sociality in insects and vertebrates. The words symbolizing and categorizing insect social systems developed in the context of halictine bees (Batra 1966; Michener 1969, 1974), were borrowed for Isoptera and non-halictine Hymenoptera (Wilson 1971), and have more recently been applied to various other arthropods (see, for example, Kent and Simpson 1992) and naked mole rats (Jarvis 1981). Vertebrate social terminology for other mammals and birds has, however, developed independently, focusing on classifying various forms of communal and cooperative breeding (Stacey and Koenig 1990; Emlen 1992). Gadagkar (1994), Crespi and Yanega (1995) and Sherman *et al.* (1995) have pointed out several serious problems arising from this polyphyletic origin of social lexicons. First, students of invertebrate and vertebrate sociality may be studying similar alloparental social systems, but any convergences would be obscured because different terms are used to identify similar patterns. Second, of the three conditions in the definition of eusociality for insects, the most important, reproductive division of labor, is sufficiently imprecise as to allow different interpretations of its meaning. This imprecision has obscured difficulties in social terminology, because classification is as ambiguous as it is easy (Crespi and Yanega 1995).

Gadagkar (1994) and Sherman *et al.* (1995) suggest that the best solution to the problems of terminological

vagueness and divergence is to redefine eusociality as encompassing all alloparental social systems, from passalid beetles where offspring sometimes help their siblings build pupal cases, through scrub jays with helping over a time scale of years, to some insects with permanent divergence of morphology, behavior, and demography. Moreover, Sherman *et al.* (1995) suggest that their 'index of reproductive skew' serves as a useful indicator of 'degree' of eusociality in this great chain of helping (see also Keller and Perrin 1995). By contrast, Crespi and Yanega (1995; Table I-1) suggest that eusociality should be defined as the presence of alloparental care and permanent differences in phenotype or life history between castes, such that one or both castes is not totipotent (i.e. able to engage in the full behavioral repertoire of the population, even if not all expressed). Cooperative breeding is defined as a society containing distinct categories with respect to reproduction and alloparental behavior but with all individuals totipotent, and, as used for the broad-scale comparisons in this book, communal societies exhibit totipotency and cooperation in brood care, but not distinct categories with respect to alloparental care and reproduction (Table I-1). Determining the usefulness of categorization schemes for the study of social behavior requires empirical analysis of whether or not discretely different types of social system exist in nature, and determining how frequently the schemes lead to new questions and insights regarding convergences, especially between arthropods and vertebrates. For example, Crespi and Yanega (1995) classify various

Table I-2. *Social insects and arachnids, categorized by social system*

This table mainly includes taxa that are discussed in this volume. Social terminology is described in Table I-1. In Lepidoptera, cooperation in brood care involves larvae.

Taxon	Subsocial	Colonial	Communal	Cooperatively breeding	Eusocial
Embioptera	+		?	?	
Psocoptera	+	+	?		
Blattaria	+	+		+	
Isoptera				+	+
Hemiptera	+	+			
Aphididae				+	+
Thysanoptera	+	+	+		+
Coleoptera: Scolytidae and Platypodidae	+	+	?		+
Coleoptera: Silphidae	+		+		
Coleoptera: Scarabaeiinae	+		?		
Coleoptera: Passalidae	+		?	?	
Hymenoptera: Halictidae	+	+	+	+	+
Hymenoptera: Anthophoridae	+	+	+	+	+
Hymenoptera: Vespidae				+	+
Hymenoptera: Sphecidae	+	+	+	+	+
Hymenoptera: Stenogastrinae	+	?	?	+	+
Hymenoptera: Formicidae				+	+
Lepidoptera		+	+		
Acari	+	+	+		
Araneae	+	+	+	?	

insects, such as some ponerine ants and polistine wasps, as cooperative breeders, which suggests that tests of ecological constraints models developed for vertebrates might usefully be applied to such taxa.

Given these new social taxa, methods for analyzing adaptation, and conceptualizations, we are left with a host of the same old questions, born from the union of gene-level selection with ethology. Indeed, for some of the most important questions, specific hypotheses still require formalization. Most generally, we are searching for the necessary and sufficient genetic, phenotypic, ecological, and demographic conditions for the origin and maintenance of different social systems. We view two related problems as most immediate and important for empirical analysis. First, how do the agents of natural selection differ qualitatively or quantitatively for different social systems, such as parental care, cooperative breeding, and eusociality? Many authors have stressed the importance of three factors for the evolution of various forms of cooperation: (1) predator and parasite pressure;

(2) difficulties in founding a nest independently; and (3) the nature of the resources used to survive and rear offspring. However, specific, falsifiable hypotheses predicated upon these selective pressures have seldom been formulated. Second, can patterns of differences between social species and their non-social relatives indicate the causes of the origins and losses of social behavior? Thus far, the majority of research on social arthropods and vertebrates has focussed on social species exhibiting forms of behavior presumed to be similar to those present at its origin (see, for example, Litte 1977; Eickwort 1986; Sinha *et al.* 1993). However, to understand the evolution of traits we should know their origins, and the answers to these questions can only come from analyzing non-social sister taxa with the same enthusiasm as we analyze our favorite social taxa. The ultimate usefulness of this approach will depend upon the evolutionary lability of traits and environments with respect to speciation events, but such information on evolutionary variability should itself be of great interest.

This volume comprises chapters analyzing sociality in as wide a range of insects and arachnids as we could assemble (Table I-2), with limited representation of wasps and ants owing to the recent books by Ross and Matthews (1991) and Hölldobler and Wilson (1990) (see also Bourke and Franks 1995). We have compelled each of the authors to: (l) present a comprehensive review of sociality in their taxon of expertise; (2) analyze the causes of sociality in their arthropods from a comparative, phylogenetic perspective; and (3) discuss the most productive questions and taxa for future research. Our hope is that, from this approach, convergences will emerge within and among disparate taxonomic groups, and that insights from comparative, experimental, and theoretical viewpoints will likewise converge upon answers to the questions of sociality.

The main conclusion of this introduction is how little we truly know about the evolution of social behavior, even after 30 years' developing theory and tests involving hundreds of species. With this volume, we venture into the field, armed with well-developed theory, new analytic methods, and a range of taxa both old and new to explore. We hope that this book will serve as a field guide to the next generation of students gathering data from the burrows, nests, webs, and galls of insects and arachnids from Cape Breton Island to Alice Springs. As C. D. Michener remarked, it is time to 'watch the bees across the road'.

ACKNOWLEDGEMENTS

We are grateful to Jane Brockmann and Mark Winston for comments.

LITERATURE CITED

Abe, T. 1991. Ecological factors associated with the evolution of worker and soldier castes in termites. *Ann. Entomol.* **9**: 101–107.

Alexander, R. D. 1974. The evolution of social behavior. *Annu. Rev. Ecol. Syst.* **5**: 25–383.

Alexander, R. D., K. M. Noonan and B. J. Crespi. 1991. The evolution of eusociality. In *The Biology of the Naked Mole Rat*. P. W. Sherman, J. U. M. Jarvis and R. D. Alexander, eds., pp. 3–44. Princeton: Princeton University Press.

Andersson, M. 1984. The evolution of eusociality. *Annu. Rev. Ecol. Syst.* **15**: 165–189.

Batra, S. W. T. 1966. Nests and social behavior of halictine bees of India. *Indian J. Entomol.* **28**: 375–393.

Baum, D. A. and A. Larson. 1991. Adaptation reviewed: a phylogenetic methodology for studying character macroevolution. *Syst. Zool.* **40**: 1–18.

Bourke, A. F. G. and N. R. Franks. 1995. *Social Evolution in Ants*. Princeton: Princeton University Press.

Brockmann, J. 1984. The evolution of social behaviour in insects. In *Behavioural Ecology. An Evolutionary Approach*. J. R. Krebs and N. B. Davies, eds., pp. 340–361. Sunderland, Mass.: Sinauer Associates.

Choe, J. C. and B. J. Crespi. 1997. *The Evolution of Mating Systems in Insects and Arachnids*. Cambridge University Press.

Coddington, J. A. 1988. Cladistic tests of adaptational hypotheses. *Cladistics* **4**: 3–22.

Craig, R. 1979. Parental manipulation, kin selection, and the evolution of altruism. *Evolution* **33**: 319–334.

Crespi, B. J. 1990. The measurement of selection on phenotypic interaction systems. *Am. Nat.* **135**: 32–47.

–. 1992. Eusociality in Australian gall thrips. *Nature (Lond.)* **359**: 724–726.

–. 1994. Three conditions for the evolution of eusociality: are they sufficient? *Insectes Soc.* **41**: 395–400.

–. 1996. Comparative analysis of the origins and losses of eusociality: causal mosaics and historical uniqueness. In *Phylogenies and the Comparative Method in Animal Behavior*. E. Martins, ed., pp. 253–287. New York: Oxford University Press.

Crespi, B. J. and D. Yanega. 1995. The definition of eusociality. *Behav. Ecol.* **6**: 109–115.

Crozier, R. H., B. H. Smith and Y. C. Crozier. 1987. Relatedness and population structure of the primitively eusocial bee *Lasioglossum zephyrum* (Hymenoptera: Halictidae) in Kansas. *Evolution* **41**: 902–910.

Eickwort G. C. 1986. First steps into eusociality: the sweat bee *Dialictus lineatulus*. *Fla. Entomol.* **69**: 742–754.

Emlen, S. 1992. Evolution of cooperative breeding in birds and mammals. In *Behavioural Ecology. An Evolutionary Approach*. J. R. Krebs and N. B. Davies, eds., pp. 301–337. Oxford: Blackwell Scientific Publications.

Evans, H. E. 1977. Extrinsic versus intrinsic factors in the evolution of insect eusociality. *BioScience* **27**: 613–617.

Felsenstein, J. 1985. Phylogenies and the comparative method. *Am. Nat.* **125**: 1–15.

Gadagkar, R. 1990. Evolution of eusociality: the advantage of assured fitness returns. *Phil. Trans. R. Soc. Lond.* B329: 17–25.

–. 1994. Why the definition of eusociality is not helpful to understand its evolution and what we should do about it. *Oikos* **70**: 485–487.

Gamboa, G. J. 1978. Intraspecific defense: advantage of social cooperation among paper wasp foundresses. *Science (Wash. D.C.)* **199**: 1463–1465.

Grafen, A. 1984. Natural selection, kin selection and group selection. In *Behavioral Ecology: An Evolutionary Approach*. J. R. Krebs and N. B. Davies, eds., pp. 62–84. Oxford: Blackwell Scientific Publications.

Hamilton, W. D. 1964. The genetical evolution of social behaviour. *J. Theor. Biol.* 7: 1–52.

Harvey, P. H. and M. Pagel. 1991. *The Comparative Method in Evolutionary Biology.* Oxford University Press.

Holloway, B. A. 1976. A new bat-fly family from New Zealand (Diptera: Mystacinobiidae). *New Zeal. J. Zool.* 3: 279–301.

Hölldobler, B. and E. O. Wilson. 1990. *The Ants.* Cambridge, Mass.: Belknap Press of Harvard University Press.

Hughes, C. R., D. C. Queller, J. E. Strassmann and S. K. Davis. 1993. Relatedness and altruism in *Polistes* wasps. *Behav. Ecol.* 4: 128–137.

Itô, Y. 1989. The evolutionary biology of sterile soldiers in aphids. *Trends Ecol. Evol.* 4: 69–73.

Jarvis, J. U. M. 1981. Eusociality in a mammal: cooperative breeding in naked mole rat colonies. *Science (Wash. D.C)* 212: 571–573.

Jarvis, J. U. M. and N. C. Bennett. 1993. Eusociality has evolved independently in two genera of bathyergid mole-rats but occurs in no other subterranean mammal. *Behav. Ecol. Sociobiol.* 33: 253–260.

Jarvis, J. U. M., M. J. O'Riain, N. C. Bennett and P. W. Sherman. 1994. Mammalian eusociality: a family affair. *Trends Ecol. Evol.* 9: 47–51.

Keller, L. and Perrin. 1995. Quantifying the degree of eusociality. *Proc. R. Soc. Lond.* B260: 311–315.

Kent, D. S. and J. A. Simpson. 1992. Eusociality in the beetle *Austroplatypus incompertus* (Coleoptera: Curculionidae). *Naturwissenschaften* 79: 86–87.

Lande, R. and S. J. Arnold. 1983. The measurement of selection on correlated characters. *Evolution* 37: 1210–1226.

Lester, L. J. and R. K. Selander. 1981. Genetic relatedness and the social organization of *Polistes* colonies. *Am. Nat.* 117: 147–166.

Lin, N. 1964. Increased parasite pressure as a major factor in the evolution of social behavior in halictine bees. *Insectes Soc.* 11: 187–192.

Lin, N. and C. D. Michener. 1972. Evolution of sociality in insects. *Q. Rev. Biol.* 47: 131–159.

Litte, M. 1977. Behavioral ecology of the social wasp, *Mischocyttarus mexicanus. Behav. Ecol. Sociobiol.* 2: 229–246.

Losos, J. B. and Miles, D. B. 1994. Adaptation, constraint, and the comparative method: phylogenetic issues and methods. In *Ecological Morphology. Integrative Organismal Biology.* P. C. Wainwright and S. M,. Reilly, eds., pp. 60–98. Chicago: University of Chicago Press.

Michener, C. D. 1969. Comparative social behavior of bees. *Annu. Rev. Entomol.* 14: 299–342.

–. 1974. *The Social Behavior of the Bees.* Cambridge: Belknap Press of Harvard University Press.

Michener, C. D. and D. J. Brothers. 1974. Were workers of eusocial Hymenoptera initially oppressed or altruistic? *Proc. Natl. Acad. Sci. U.S.A.* 71: 671–674.

Moran, N. 1993. Defenders in the North American aphid *Pemphigus obesinymphae. Insectes Soc.* 40: 391–402.

Mound, L. A. and B. J. Crespi. 1995. Biosystematics of two new gall-inducing thrips with soldiers (Insecta: Thysanoptera) from *Acacia* trees in Australia. *J. Nat. Hist.* 29: 147–157.

Noonan, K. 1981. Individual strategies of inclusive-fitness-maximizing in *Polistes fuscatus* foundresses. In *Natural Selection and Social Behavior: Recent Research and New Theory.* R. D. Alexander and D. W. Tinkle, eds., pp. 18–44. New York: Chiron Press.

Oster, G. F. and E. O. Wilson. 1978. *Caste and Ecology in the Social Insects.* Princeton: Princeton University Press.

Pagel, M. 1994. The adaptationist wager. In *Phylogenetics and Ecology.* P. Eggleton and R. I. Vane-Wright, eds., pp. 29–51. New York: Academic Press.

Polis, G. A. and W. R. Lourenço. 1986. Sociality among scorpions. *Actas X Congr. Aracnol. Jaca/Espana* 1: 111–115.

Queller, D. C. 1989. The evolution of eusociality: reproductive head starts of workers. *Proc. Natl. Acad. Sci. U.S.A.* 86: 3224–3226.

–. 1994a. Male–female conflict and parent–offspring conflict. *Am. Nat.* 144: S84–S99.

–. 1994b. Extended parental care and the origin of eusociality. *Proc. R. Soc. Lond.* B256: 105–111.

Ridley, M. 1989. Why not to use species in comparative tests. *J. Theor. Biol.* 136: 361–364.

Ross, K. G. and J. M. Carpenter. 1991. Population genetic structure, relatedness, and breeding systems. In *The Social Biology of Wasps.* K. G. Ross and R. W. Matthews, eds., pp. 451–479. Ithaca: Comstock Publishing.

Ross, K. and R. Matthews. 1991. *The Social Biology of Wasps.* Ithaca: Comstock Publishing.

Rypstra, A. L. 1993. Prey size, social competition, and the development of reproductive division of labor in social spider groups. *Am. Nat.* 142: 868–880.

Saito, Y. 1986. Biparental defense in a spider mite (Acari: Tetranychidae) infesting *Sasa* bamboo. *Behav. Ecol. Sociobiol.* 18: 377–386.

Sherman, P. W., E. A. Lacey, H. K. Reeve and L. Keller. 1995. The eusociality continuum. *Behav. Ecol.* 6: 102–108.

Shivashankar, T. 1994. Advanced sub social behaviour in the scorpion *Heterometrus fulvipes* Brunner (Arachnida). *J. Biosci.* 19: 81–90.

Sinha, A., S. Premnath, K. Chandrashekara and R. Gadagkar. 1993. *Ropalidia rufoplagiata*: a polistine wasp society probably lacking permanent reproductive division of labour. *Insectes Soc.* 40: 69–86.

Solomon, N. G. 1994. Eusociality in a microtine rodent. *Trends Ecol. Evol.* 9: 264.

Spanier, E., J. S. Cobb and M.-J. James. 1993. Why are there no reports of eusocial marine crustaceans? *Oikos* 67: 573–576.

Stacey, P. B. and W. D. Koenig. 1990. *Cooperative Breeding in Birds: Long Term Studies of Ecology and Behavior.* Cambridge University Press.

Stern, D. L. 1994. A phylogenetic analysis of soldier evolution in the aphid family Hormaphididae. *Proc. R. Soc. Lond.* B256: 203–209.

Strassmann, J. E. and D. C. Queller. 1989. Ecological determinants of social evolution. In *The Genetics of Social Evolution*. M. D. Breed and R. E. Page, Jr., eds., pp. 81–101. Boulder, CO.: Westview Press.

Williams, G. C. 1966. *Adaptation and Natural Selection*. Princeton: Princeton University Press.

Wilson, E. O. 1971. *The Insect Societies*. Cambridge: Belknap Press of Harvard University Press.

–. 1989. The coming pluralization of biology and the stewardship of systematics. *BioScience* 39: 242–245.

1 · Are behavioral classifications blinders to studying natural variation?

WILLIAM T. WCISLO

ABSTRACT

Biologists frequently discuss behavioral attributes of species or more inclusive taxa, and organize these attributes into classifications for comparative studies. Such behavioral classifications may influence the kinds of questions biologists ask. Recently proposed revisions to a well-established definition of 'eusocial' behavior are intended to improve classifications of animal societies, and hence facilitate comparative studies. Crespi and Yanega (1995) propose to narrow the scope of eusociality, whereas Gadagkar (1994) and Sherman et al. (1995) aim to widen its scope. These new proposals, along with the traditional definition, are used here to discuss the important point that behavioral classifications are operational, and therefore different questions require different classifications. The revisions and the existing definition use arbitrary criteria, in the sense that *a priori* they do not specify 'meaningful' parameters. Moreover, social behavior can vary within taxa, so taxon-specific descriptions of social behavior may force a typological characterization that potentially masks interesting variability, especially for taxa that are not thoroughly studied. A natural classification would be advantageous in that it would take into account actual patterns of evolutionary history (phylogeny), and thus ensure the accurate assessment of convergent behavioral patterns. Without a natural classification, authors should *precisely* define terms in ways suited to their particular needs in a given study.

INTRODUCTION

In *The Forest and the Sea* Bates (1960) gave examples from so-called primitive people for whom the act of naming something gives them control over it. Names have a certain 'magical' power; Bates tried to caution biologists against this word-magic. There is, he argued, a pitfall in equating the ability to name entities or processes with an understanding of them.

This word-magic readily couples with a widespread tendency to categorize percepts (see references in Bernays and

Wcislo 1994), which modifies our view of variability. As Mayr (1982) and others have discussed, down-playing the importance of variability has been an impediment to many areas of evolutionary biology. Yet typological descriptions linger in comparative studies of behavior (for criticisms, see Slater 1986; McLennan and Brooks 1993; W. G. Eberhard, W. T. Wcislo and M. J. Ryan, in preparation). Numerous recent papers or chapters in this volume, for example, could be cited which discuss the behavior of species. These characterizations facilitate communication, and are accurate if there is no variation. They also fit well with a view of behavioral evolution as occurring concomitant with lineage splitting. In the context of social evolution, for instance, authors have hypothesized an 'ethocline' from a solitary species, to a species with group-living behavior, to one with simple castes, and on to highly eusocial species (summarized in Wilson 1971; criticized in Michener 1985; West-Eberhard 1986).

Three frequently *un*stated assumptions in such a view are that (1) the behavioral character (e.g. eusociality) is constant within the taxon of interest; (2) the described behaviors are 'natural units'; and (3) within-lineage evolution (anagenesis) can be safely ignored. To the extent that these assumptions are violated, current or proposed behavioral classifications can be misleading or overly restrictive. I use recently proposed revisions to social terminology as examples to illustrate how typological categorizations of behavior can divert attention from within-taxon variability, and potentially bias both the questions we ask, and the conclusions we draw. (See *Note added in proof*, p. 13.)

SOCIAL TERMINOLOGY AND CLASSIFICATIONS

The *Status Quo*

Historical accounts of attempts to characterize and classify animal societies are given by Wheeler (1928) and Le Masne (1952). The *status quo* is represented by Michener's scheme (reviewed in Michener 1974), which was popularized by

Wilson (1971). Michener used the term 'eusocial', which was introduced by Batra (1966a), to describe social groups of more than one generation in which some individuals care for offspring that are not their own, and do not themselves reproduce (see Batra 1995). In the Batra–Michener–Wilson (hereafter BMW) classification, three attributes are considered simultaneously in defining eusociality:

(1) Are individuals of more than one generation present?
(2) Is there reproductive division of labor into fertile and more or less sterile castes?
(3) Is there cooperative brood care?

Eusocial societies have been further subdivided into 'primitively' and 'advanced' eusocial groups, but these subdivisions will not be discussed further here (see Kukuk 1994), nor will the other kinds of social organizations (e.g. communal, semisocial, parasocial), which are defined traditionally by different combinations of the above-mentioned attributes (see Michener 1974).

Problems with the *Status Quo*

Gadagkar (1994) (hereafter G), Crespi and Yanega (1995) (hereafter CY), and Sherman *et al.* (1995) (hereafter SLRK) properly criticize the original BMW formulations because the criterion of 'reproductive division of labor' can have different meanings. The division can be permanent or temporary, and absolute or statistical. The original authors, however, needed a practical classification that would be useful for the limited data available (C. D. Michener personal communication). They opted for realism over precision (cf. Levins 1968). For many taxa, and for arthropods especially, there is still a dearth of information, and present extinction rates suggest that some briefly studied forms will remain so forever.

A second difficulty with the *status quo* definition is that the criterion of 'overlap of generations' is probably a bias of temperate-zone biologists, who are in the majority. G points out that there is probably no less 'altruism' in some tropical insects without overlap of generations than there is in their temperate-zone relatives with overlap of generations.

Lastly, CY, G and SLRK assert that the BMW classification hinders comparative studies of vertebrate and arthropod social behavior. The many obvious counter-examples (see, for example, Hamilton 1964; Wilson 1975; West-Eberhard 1983; Itô *et al.* 1987) make this criticism less persuasive. Indeed, SLRK themselves note that 'a number of authors have identified striking evolutionary parallels between the social systems of cooperatively breeding birds and mammals and those of social insects' (p. 102) (see also Brockmann, this volume).

Proposed changes to the *Status Quo*

Some of the shortfalls of the BMW definition led G, CY and SLRK to redefine 'eusociality'. G, CY and SLRK are unanimous in relaxing the criterion of 'overlap of generations' for defining eusociality. Both G and SLRK propose to expand the scope of 'eusociality' to include all organisms in which some individuals do not reproduce (at least temporarily at some point in their mature lives), and instead help nestmates in rearing offspring. SLRK believe that 'eusociality' is not a natural class, but represents the tail of a continuum of 'reproductive skew' among animals with alloparental care. In other words, they note that one individual lays *most* of the eggs ('the queen'), and *most* individuals lay *few* or no eggs ('the workers'). Since 'most' and 'few' are relative terms, the skew of reproductive success can be arranged along a continuum. This important fact by itself, however, is not a useful basis for defining social levels, because an unequal distribution of reproductive success usually occurs in all populations, even non-social ones (see, for example, Raw 1984), as pointed out by Darwin (1859). Additionally, reproductive success among social colonies can be skewed for reasons unrelated to social behavior, such as nest destruction due to flooding (see, for example, Batra 1966b; see also CY).

In contrast, CY propose to narrow the scope of 'eusociality'. They believe eusocial societies exist in nature as a distinct class, which they define as societies containing 'groups of individuals that become irreversibly behaviorally distinct ... prior to reproductive maturity' (p. 111). This definition suits their argument that understanding developmental differences associated with phenotypic determination of particular social roles ('castes') are essential to understanding the evolution of eusociality (see, for example, Yanega 1989). Acknowledging problems with typology (below), CY caution that their terminology applies to 'societies' and not to species. According to CY, criteria for a useful definition of eusociality are that it be *non-arbitrary* (categorizing societies by meaningful parameters), *universal* (applicable to all species), and *evolutionarily informative* in that it indicates differences between societies that differ in how natural selection affects behavioral interactions' (p. 110). The following section evaluates the G, CY and SLRK revisions in light of these criteria.

THE SUBJECTIVE BASIS OF BEHAVIORAL CLASSIFICATIONS

The above-mentioned classifications are useful for discussing questions associated with the evolution of reproductive division of labor or the problem of altruism, but adopting them *a priori* may bias the questions we ask. Delage-Darchen and Darchen (1985) even suggest that a neglect of questions other than those related to caste may be a bias from our own human concerns with hierarchical societies.

The definitions by G, CY and SLRK are based on overall similarity, but they use different criteria to assess similarity. Not surprisingly, they often disagree with one another. Disagreement arises because the definitions are based on criteria which *a priori* they deem to be important. Such arbitrary criteria are useful only in the context of particular questions (Panchen 1992). For example, if one is interested in testing the hypothesis that groups of individuals are more effective at deterring natural enemies than solitary individuals (see, for example, Lin and Michener 1972; Wcislo *et al.* 1988), then the appropriate social classification for a comparative study will categorize individuals simply as members of a group, or solitary. Indeed, definitions of the characters used in classifications are themselves elusive (Pogue and Mickevich 1990), and different character definitions are appropriate for different questions (see, for example, Sokal 1962; Tinbergen 1963; Golani 1976; Drummond 1981; Miller 1988; Wenzel 1992).

Humans are limited in their capability to acquire and process information, as are other animals (Bernays and Wcislo 1994); this limits our ability to describe variability. We generalize from incomplete information, and simplify complex processes in order to comprehend them. Artificial classifications are operationally useful for these purposes (Panchen 1992), but their usefulness is ephemeral: as new data become available, we assemble them in new ways to address different questions. At any time, the categories we use are meaningful descriptors if there is little variation. A danger, however, is to ignore variation as uninteresting 'noise', or not to acknowledge it.

Arguments relating to the use of skew in defining eusociality illustrate some potential pitfalls of 'hiding' variation. SLRK state that species are better represented as segments of the scale rather than as points, presumably to avoid typological comparisons. Nevertheless, sweat bees provide examples which show how the distribution of reproduction among entities conveys little information that is useful as a basis for comparative studies. SLRK list *Lasioglossum*

(Dialictus) figueresi in their 'intermediate skew' subinterval. Within a population (aggregation), however, most females are solitary (Wcislo *et al.* 1993), so any skew is unrelated to social behavior. A minority of nests (<20%) either have one major reproductive and one or two bees with undeveloped ovaries, or all the residents are reproductive, as inferred from ovarian dissections. A species-level classification of their social behavior would range from low (zero) to high skew. A similar range would be observed in many populations of other species, but would reflect different biological patterns. Skew in species such as *Lasioglossum (Evylaeus) calceatum* (Sakagami and Munakata 1972) or *Halictus rubicundus* (Eickwort *et al.* 1996), for example, would also range from low (zero) to high (for other examples see Wcislo, Chapter 15 of this volume). Most nests in most populations would be characterized by high skew since colonies contain a reproductive queen, and numerous workers with undeveloped ovaries. In contrast, females in high-altitude populations are solitary so there is no *within*-nest skew.

Analyses of social evolution require strict attention both to organizational levels (Eldridge and Grene 1992) and to logical levels (Sherman 1988). Social competition occurs within colonies, and thus skewness in reproductive success is a property of colonies (see, for example, West-Eberhard 1983), not of species or more inclusive taxonomic groupings. Indeed, Michener (1974, 1990) and others (see Wcislo, Chapter 15 of this volume) have cautioned that, at least for some bees, social classifications are applicable at the level of the colony, but not at higher taxonomic levels.

The reproductive skew metric, if applied at inappropriate organizational levels, masks substantial biological differences (e.g. frequency of occurrence, geographic distribution) and pools dissimilar phenomena (e.g. skew due to social and non-social factors). Such cases are not limited to arthropods. Eisenberg (1991), for example, noted that early attempts to classify the social behavior of primate species were deficient, because it had not been known that for some socially relevant traits the differences among *populations* were often as great as differences among *species*. Similarly, as suggested by Linares (1979), an improved understanding of human social evolution is more likely to come from a focus on behavioral and ecological processes, rather than from attempts to force cultural typologies on existing data. Recent studies support Linares' contention (see, for example, Piperno *et al.* 1991).

Both CY and SLRK claim that their respective definitions provide a better basis for an evolutionarily informative definition of eusociality. SLRK assert that reproductive skew 'can direct the evolution of key societal features' (p. 106) and that their measure unites 'all occurrences of alloparental helping of kin under a single theoretical umbrella (Hamilton's rule)' (p. 102). This argument, however, confounds the description of an historical pattern with its interpretation and explanation. CY, in turn, suggest that loss of 'totipotency' (i.e. loss of part of the behavioral repertoire needed to flexibly adopt alternative social roles) is 'probably the most evolutionarily-relevant event in social evolution' (p. 113). Alternatively, one could argue that the most evolutionarily relevant event was an initial association of conspecifics, because without its occurrence the evolutionary elaboration of societies would be impossible. Again, the important point is that different attributes are important for different questions.

Juxtaposing the papers by G, CY and SLRK, which new definition(s), if any, should we adopt? Is differential reproductive success, or the existence of irreversible castes, a more informative attribute for delineating eusocial societies? Both are important and interesting components, and it would probably be useful to simultaneously consider them and others. To reiterate, the criteria for social classifications proposed by those authors are artificial, so the bases for the classification will depend on its intended usage; there is nothing intrinsically better about one or another. Other social attributes are also important and interesting, and it is worthwhile to ask questions about them.

IS A NATURAL CLASSIFICATION OF SOCIAL BEHAVIOR POSSIBLE?

G, CY and SLRK assert that their modifications will lead to increased understanding of social evolution by enabling biologists to recognize evolutionary convergence. Convergence, by definition, is impossible to recognize without reference to historical patterns (phylogenies), yet none of these revisions incorporates this historical element. They will, therefore, combine social systems with totally different evolutionary patterns. For example, they would lump together all species of solitary bees or birds. 'Solitary' behavior, however, can be ancestral, or be derived following the evolutionary loss of social behavior, as hypothesized for some bees (see, for example, Richards 1994) and birds (Crook 1964). This distinction is important: factors which maintain solitary behavior may be different from those factors that select for the evolutionary loss of social behavior.

Michener (1974, pp. 236ff.) attempted to objectively classify social attributes of diverse bee taxa. He concluded that it is important to take into account historical patterns of ancestor–descendent relationships when classifying social behaviors. A multivariate analysis of 28 behavioral traits from 18 taxa produced clusters of historical groups (i.e. all Halictidae clustered together; all Apidae clustered together, etc.), rather than convergent behavioral clusters (i.e. all solitary bees together, all communal bees together, etc.). Michener (1990) re-emphasized that knowledge of history (phylogeny) is important when classifying social behaviors in order to understand their evolution, yet none of the revisions addressed this point. Others have made the same point for understanding the evolution of behavior in general (see, for example, Thorpe 1979; Wcislo 1989; Wenzel 1992).

DISCUSSION AND CONCLUSIONS

Biologists ask diverse questions about social behavior. Our focus changes as new information becomes available, compelling us to organize information in novel ways. Despite the promises of clarity, and of facilitating comparative studies, recent terminological revisions for social organizations may lead to confusion and further imprecision. Revisions proposed by G, CY and SLRK redefine a deeply entrenched term, and then require subsequent authors to remember differences among eusociality *sensu* Batra–Michener–Wilson, *sensu* Crespi–Yanega, *sensu* Gadagkar, *sensu* Sherman *et al.*, and so on, and explicitly state which definition they adopt. In effect, this demands a simpler alternative for communicating information about the diversity of social systems. This alternative is already practiced (see, for example, West-Eberhard 1978) and requires only that authors state in each paper precisely and explicitly how they operationally define the behavior in question for the specific hypothesis they wish to test. This option enables an author to bring to bear evidence from all relevant taxa, without psychological constraint from pre-existing (and possibly biased) definitions. Maher and Lott (1995) reached a similar conclusion after reviewing 136 publications which contained 48 definitions of vertebrate 'territoriality'.

Standardized behavioral classifications cannot be monolithic, unless they are natural (history-based) ones, and for many interesting questions natural classifications are not appropriate. A healthy skepticism of a standardized, artificial classification should not be construed as an anarchic attempt to impede communication about biological discoveries. Instead, it emphasizes that such classifications are risky in that they potentially 'hide' novel biological patterns from view because they are not part of the accepted scheme, or because the questions deviate from current fashions (see also Maher and Lott 1995). To some extent standardized characterizations are useful and necessary to facilitate communication about complex phenomena. Yet, because they fit well with particular views of behavioral evolution, we might lose sight of the fact that they may be oversimplifications. Social behavior is not a static construct, but changes at different organizational levels. As Bateson (1972) repeatedly emphasized, we need to be careful not to mistake the 'map' for the 'territory', to believe that our description of behavior is the same as the real thing.

ACKNOWLEDGEMENTS

This chapter began when Bernie Crespi and Doug Yanega, and Raghavendra Gadagkar sent me manuscript drafts for informal criticisms; B. Crespi later extended an offer to formally critique these publications. B. Crespi, Bill Eberhard, R. Gadagkar, Penny Kukuk, Paul Sherman, and D. Yanega also helped by providing preprints or manuscripts, as well as useful criticisms on this manuscript. Additional helpful comments came from Donna Conlon, Bryan Danforth, the late George Eickwort, Laurent Keller, Charles Michener, Laurence Packer, Ted Schultz, Mike Schwarz, Tom Seeley, John Wenzel and Mary Jane West-Eberhard. Financial support came from an U.S.A. National Science Foundation Environmental Biology Postdoctoral Fellowship (BSR-9103786), and general research funds from the Smithsonian Tropical Research Institute.

LITERATURE CITED

Bates, M. 1960. *The Forest and the Sea: A Look at the Economy of Nature and the Ecology of Man.* New York: Vintage Books.

Batra, S. W. T. 1966a. Nests and social behavior of halictine bees of India. *Indian J. Entomol.* 28: 375–393.

–. 1966b. Life cycle and behavior of the primitively social bee, *Lasioglossum zephyrum. Univ. Kans. Sci. Bull.* 46: 359–423.

—. 1995. The evolution of 'eusocial' and the origin of 'pollen bees'. *Maryland Nat.* 39: 1–4.

Bateson, G. 1972. *Steps to an Ecology of Mind.* San Francisco: Chandler Publ. Co.

Bernays, E. A. and W. T. Wcislo. 1994. Sensory capabilities, information processing, and resource specialization. *Q. Rev. Biol.* 69: 187–204.

Crespi, B. J. and D. Yanega. 1995. The definition of eusociality. *Behav. Ecol.* 6: 109–115.

Crook, J. H. 1964. The evolution of social organization and visual communication in the weaverbirds (Ploceidae). *Behaviour* (Suppl.) 10: 1–178.

Darwin, C. 1859. *On the Origin of Species* (1964 reprint Cambridge: Harvard University Press.)

Delage-Darchen, B. and R. Darchen. 1985. Niveaux et schémas d'évolution chez les insectes sociaux comparés a ceux des araignées sociales. *Ann. Biol.* 24: 69–87.

Drummond, H. 1981. The nature and description of behavior patterns. In *Perspectives in Ethology*, vol. 4. P. P. G. Bateson and P. H. Klopfer, eds., pp. 1–33. New York: Plenum Press.

Eickwort, G. C., J. M. Eickwort, J. Gordon and M. A. Eickwort. 1996. Solitary behavior in a high-altitude population of the social sweat bee *Halictus rubicundus* (Hymenoptera: Halictidae). *Behav. Ecol. Sociobiol.,* 38: 227–233.

Eisenberg, J. F. 1991. Mammalian social organization and the case of *Alouatta*. In *Man and Beast Revisited.* M. E. Robinson and L. Tiger, eds., pp. 127–138. Washington, D.C.: Smithsonian Institution Press.

Eldridge, N. and M. Grene. 1992. *Interactions: The Biological Context of Social Systems.* New York: Columbia University Press.

Gadagkar, R. 1994. Why the definition of eusociality is not helpful to understand its evolution and what we should do about it. *Oikos* 70: 485–488.

Golani, I. 1976. Homeostatic motor processes in mammalian interactions: a choreography of display. In *Perspectives in Ethology*, vol. 2. P. P. G. Bateson and P. H. Klopfer, eds., pp. 69–134. New York: Plenum Press.

Hamilton, W. D. 1964. The genetical evolution of social behaviour, Parts I and II. *J. Theor. Biol.* 7: 1–16; 17–52.

Itô, Y., J. L. Brown and J. Kikkawa, eds. 1987. *Animal Societies: Theories and Facts.* Tokyo: Japan Scientific Societies Press.

Kukuk, P. 1994. Replacing the terms 'primitive' and 'advanced': new modifiers for the term eusocial. *Anim. Behav.* 47: 1475–1478.

Le Masne, G. 1952. Classification et caractéristiques des principaux types de groupements sociaux réalisés chez les Invertébrés. In *Rapport au Colloque International sur la Structure et la Physiologie des Sociétiés animales (Paris, 1950)*, pp. 19–70. Paris: Colloque Intenrational.

Levins, R. 1968. *Evolution in Changing Environments.* Princeton: Princeton University Press.

Lin, N. and C. D. Michener. 1972. Evolution of sociality in insects. *Q. Rev. Biol.* 47: 131–159.

Linares, O. F. 1979. What is lower Central American archaeology? *Annu. Rev. Anthropol.* 8: 21–43.

Maher, C. R. and D. L. Lott. 1995. Definitions of territoriality used in the study of variation in vertebrate spacing systems. *Anim. Behav.* 49: 1581–1597.

Mayr, E. 1982. *The Growth of Biological Thought*. Cambridge: Harvard University Press.

McLennan, D. A. and D. R. Brooks. 1993. The phylogenetic component of cooperative breeding in perching birds: a commentary. *Am. Nat.* **141**: 790–795.

Michener, C. D. 1974. *The Social Behavior of the Bees*. Cambridge: Harvard University Press.

–. 1985. From solitary to eusocial: need there be a series of intervening species? *Fortschr. Zool.* **31**: 293–305.

–. 1990. Wasps and our knowledge of insect social behavior. In *Social Insects and the Environment*. G. K. Veeresh, B. Mallik and C. A. Viraktamath, eds., pp. 61–62. New Delhi: Oxford & IBH.

Miller, E. H. 1988. Description of bird behavior for comparative purposes. *Curr. Ornithol.* **5**: 347–394.

Panchen, A. L. 1992. *Classification, Evolution, and the Nature of Biology*. Cambridge University Press.

Piperno, D. R., M. B. Bush and P. A. Colinvaux. 1991. Paleoecological perspectives on human adaptation in central Panama. I. The Pleistocene; II. The Holocene. *Geoarchaeology.* **6**: 201–226; 227–250.

Pogue, M. G. and M. F. Mickevich. 1990. Character definitions and character state delineation: the bête noire of phylogenetic inference. *Cladistics* **6**: 319–361.

Raw, A. 1984. The nesting biology of 9 species of Jamaican West Indies bees (Hymenoptera). *Rev. Bras. Entomol.* **28**: 497–506.

Richards, M. H. 1994. Social evolution in the genus *Halictus*: a phylogenetic approach. *Insectes Soc.* **41**: 315–325.

Sakagami, S. F. and M. Munakata. 1972. Distribution and bionomics of a transpalearctic eusocial halictine bee, *Lasioglossum (Evylaeus) calceatum*, in northern Japan, with reference to its solitary life cycle at high altitude. *J. Fac. Sci., Hokkaido Univ.* (Ser. VI, Zool.) **18**: 411–439.

Sherman, P. W. 1988. The levels of analysis. *Anim. Behav.* **36**: 616–619.

Sherman. P. W., E. A. Lacey, H. K. Reeve and L. Keller. 1995. The eusociality continuum. *Behav. Ecol.* **6**: 102-108.

Slater, P. J. B. 1986. Individual differences in animal behaviour: a functional interpretation. *Accad. Nazion. Lincei* **259**: 159–170.

Sokal, R. R. 1962. Typology and empiricism in taxonomy. *J. Theor. Biol.* **3**: 230–267.

Thorpe, W. H. 1979. *The Origins and Rise of Ethology*. London: Heinemann.

Tinbergen, N. 1963. On aims and methods of ethology. *Z. Tierpsychol.* **20**: 410–433.

Wcislo, W. T. 1989. Behavioral environments and evolutionary change. *Annu. Rev. Ecol. Syst.* **20**: 137–169.

Wcislo, W. T., M. J. West-Eberhard and W. G. Eberhard. 1988. Natural history and behavior of a primitively social wasp, *Auplopus semialatus*, and its parasite, *Irenangelus eberhardi* (Hymenoptera: Pompilidae). *J. Insect Behav.* **1**: 247–260.

Wcislo, W. T., A. Wille and E. Orozco. 1993. Nesting biology of tropical solitary and social sweat bees, *Lasioglossum (Dialictus) figueresi* Wcislo and *L. (D.) aeneiventre* (Friese) (Hymenoptera: Halictidae). *Insectes Soc.* **40**: 21–40.

Wenzel, J. W. 1992. Behavioral homology and phylogeny. *Annu. Rev. Ecol. Syst.* **23**: 361-381.

West-Eberhard, M. J. 1978. Polygyny and the evolution of social behavior in wasps. *J. Kansas Entomol. Soc.* **51**: 832–856.

–. 1983. Sexual selection, social competition, and speciation. *Q. Rev. Biol.* **58**: 155–183.

–. 1986. Alternative adaptations, speciation, and phylogeny (a review). *Proc. Natl. Acad. Sci., U.S.A.* **83**: 1388–1392.

Wheeler, W. M. 1928. *The Social Insects*. London: Kegan Paul, Trench, Truber & Co., Ltd.

Wilson, E. O. 1971. *The Insect Societies*. Cambridge: Harvard University Press.

–. 1975. *Sociobiology*. Cambridge: Harvard University Press.

Yanega, D. 1989. Caste determination and differential diapause within the first brood of *Halictus rubicundus* (Hymenoptera: Halictidae). *Behav. Ecol. Sociobiol.* **24**: 97–107.

Note added in proof
The same phenomena are reviewed from a different perspective by J. T. Costa & T. D Fitzgerald (1996). *Trends Ecol. Evol.* **11**: 285–289.

2 · Life beneath silk walls: a review of the primitively social Embiidina

JANICE S. EDGERLY

ABSTRACT

I review and summarize the scattered information on embiids (Order Embiidina), with an emphasis on details of colony structure and maternal care. I summarize experimental and observational field results from a detailed study on parental and communal behavior of *Antipaluria urichi*, a Trinidadian webspinner. Topics discussed include the function of maternal behavior, interactions with egg parasitoids, antipredator attributes of communal living, and possible functions of silk. I also compare features of webspinner sociality to other communal insects and spiders. In addition, I discuss promising topics for future study, including male dimorphism, the possibility of higher sociality, and communication systems.

INTRODUCTION

Webspinners (Order Embiidina or Embioptera) construct a nest-like structure, exhibit parental care, and commonly live in the tropics where overlapping generations may occur. These attributes represent factors that allow for the evolution of complex social interactions in insects (Evans 1977; Eickwort 1981), making embiids an intriguing order with many research questions yet to be addressed. This cosmopolitan order, including 850 mostly tropical species in 14 families (E. S. Ross, personal communication), has been classified within the Orthopteroidea, which includes earwigs, cockroaches, walkingsticks, mantids, katydids, crickets, grasshoppers and termites (Hennig 1981). In a more recent phylogenetic treatment of hexapod orders (Minet and Bourgoin 1986), Embiidina and Zoraptera are sister-groups within the Polyneoptera, which includes all the orders mentioned above, plus Plecoptera. Boudreaux (1979) also proposed a close phylogenetic relationship between Plecoptera and Embiidina. Confined to warmer regions, webspinners are found only as far north as southern Virginia in the United States; the highest known altitudinal record is 3500 m, in the cloud forests of Cuenca,

Ecuador (Ross, personal communication). Based on extensive collecting throughout the world, Ross (1970) categorized embiids as subsocial (following Michener's (1969) terminology), exhibiting care of eggs and sometimes of nymphs. The term communal, traditionally used to designate social organization in Hymenoptera (*sensu* Wilson 1971), also seems appropriate for those embiids who share a nest-like structure wherein each female oviposits and tends her own offspring.

In this chapter, I summarize information from the scant literature on webspinner behavior and describe attributes relating to group-living and parental care. Despite the anecdotal nature of most reports, the available information provides an intriguing composite of behavior of these relatively rare and generally inconspicuous insects. After a brief literature review, I will summarize my work on social behavior in *Antipaluria urichi* (Clothodidae).

GENERAL REVIEW OF EMBIID BEHAVIOR

Living within silk

The common name, webspinners, refers to the order-wide characteristic of silk-spinning behavior. Nymphs, adult females (which are always wingless) and adult males spin silk issuing from swollen metatarsal glands in the forelegs (Barth 1954; Alberti and Storch 1976). They live in silken tunnels constructed on surfaces of trees and other objects in the more humid regions (Fig. 2-1) and under rocks, bark flaps, or logs in the drier parts of their range. Webspinners display numerous adaptations for life beneath silk walls, including a flexible, slender body allowing quick movements and U-turns in tight spaces, flexible antennae that resist tangling in silk, and wings that fold along a crease that runs perpendicular to the length of the wing. This latter feature prevents their wings from catching in the silk as the males dart backwards (Ross 1970). Webspinners graze on the outer bark of trees, leaf litter, mosses, algae, and lichens on bark, rocks, mounds and soil (Ross 1970).

Fig. 2-1. Silk tunnels of an *Antipaluria urichi* colony, approximately 1 m wide, on a mud bank along a road cut in Trinidad's Northern Range. Photo.: J. S. Edgerly.

Ross successfully reared species, collected throughout the range of the order, on a diet of dry oak leaves and lettuce, a testament to their generalist requirements.

Observational evidence suggests a multifaceted function for silk, including shielding from abiotic factors as well as from predators, as proposed below. The outer covering of silk provides protection from heavy rain in the tropical habitats common to most embiids. Torrential rains trigger stemflow on trees supporting embiid colonies. After one such inundation in Trinidad, I tore open silk of colonies of *A. urichi* and discovered virtually dry substrate beneath the silk. I further investigated the wetting properties of silk by placing a section of a silk wall across the mouth of a jar, pouring water onto it from above; water did not penetrate the silk (Edgerly 1986). Silk may also shield inhabitants from direct sunlight and desiccation, although thermoregulatory properties remain to be investigated in the field.

Subsocial behavior in embiids

Maternal care
Maternal care has been observed in all embiids that have been closely examined. For example, adult females position their bodies over or near clustered eggs in *Anisembia texana* (Anisembiidae) (Mills 1932; Choe 1994), *Antipaluria urichi* (Edgerly 1987a, 1988), *Embia major* (Embiidae) (Imms 1913), *Oligotoma ceylonica* (Oligotomidae) (Bradoo 1967), *O. greeniana* (Bradoo and Joseph 1970) and *O. humbertiana* (Ananthasubramanian 1957). Some females also produce a complex egg mass, covered with substrate materials, silk, and/or fecal

material (Imms 1913; Ananthasubramanian 1957; Bradoo 1967; Bradoo and Joseph 1970; Edgerly 1987a). In contrast, *O. ceylonica* produces a single egg per day for two to two and one-half months, laying them in linear rows in the silk (Bradoo 1967). After egg hatch, females stay with their nymphs, as reported for *Anisembia texana* (Mills 1932; Choe 1994), *Antipaluria urichi* (Edgerly 1987a, 1988), *O. ceylonica* (Bradoo 1967), *O. saundersii* (Ling 1935) and *E. major* (Imms 1913). Although maternal behavior appears ubiquitous throughout the order, its function has been investigated in the field only for *A.urichi* (Edgerly 1987a,b, 1988). In this case, females protect their eggs from egg parasitoids and provide silk for newly hatched nymphs (see below; Edgerly 1988). Bradoo (1967) noted that nymphs of *O. ceylonica* follow their mothers through silk tunnels, with newly hatched nymphs being especially tenacious. Laboratory observations of *E. ramburi* of the Mediterranean region revealed that females provide macerated food to nymphs (Denis 1949; LeDoux 1958), a relatively complex form of maternal behavior for an insect. Also striking was Ross' (1970) observation of an Afghan species, *Paedembia* n. sp., which harvests aromatic leaves of the shrub *Artemisia*, caching them in subterranean galleries. Providing food to young is a feature of other subsocial insects, including cockroaches (Seelinger and Seelinger 1983; Nalepa 1984), crickets (West and Alexander 1963), cydnids (Sites and McPherson 1982), dung beetles (see, for example, Halffter 1977), earwigs (Radl and Linsenmair 1991) and membracids (Wood *et al.* 1984) and may have promoted complex communication systems in some insects, as in passalids (Pearse *et al.* 1936; Schuster and Schuster, this volume) and *Nicrophorus* (Milne and Milne 1976). Embiids that provide food for offspring and that live in nest-like structures (whereby food is collected from the outside and brought into the silk tunnels) are of particular interest in a search for more complex societies within the order.

Males are not known to exhibit parental care, but because of limited observations such behavior cannot be ruled out. Of particular interest for future study are the many species with wingless males (e.g. all males in the family Australembiidae; Ross 1963) which may spend more time interacting with females, and perhaps with their young, within silk tunnels. For example, Melander (1903) frequently encountered wingless males of the dimorphic *Anisembia texana* living in galleries with females. However, because males of most species do not feed after reaching maturity and die soon after mating (Ross 1970), they appear to lack the longevity necessary for guarding eggs and nymphs.

Table 2-1. *Types of embiid colony*

Species	Solitary	Communal	Max. number of females	Number of colonies	Reference
Anisembia texana	×	×	6	—	Melander 1903[a]
Antipaluria urichi	×	×	24	143	Edgerly 1987, 1994
Embia major	×	—	1	130	Imms 1913
Oligotoma ceylonica ceylonica	—	×	39	—	Bradoo 1967[a]
O. humbertiana	×	—	1	—	Ananthasubramanian 1957[a]
O. saundersii	—	×	185	—	Bradoo 1967[a]

[a] Anecdotal report.

Tendency to live in groups
In addition to mother–offspring associations, field colonies of webspinners exhibit a variety of social groupings, varying from solitary individuals to females with young to combinations of solitary and communal groups to populations with apparently obligate communal colonies (Table 2-1). Embiid life cycles vary from one generation per two years to up to five generations per year (Table 2-2). In species with many generations per year, overlap of generations within a colony is possible and may promote complex social interactions as yet undiscovered in webspinners.

Sex ratio

Although embiid sex ratios have not been studied in detail, anecdotal reports suggest variability within the order. The males' short adult lifespan and ability to fly may contribute to their relative rarity in collections of inhabitants within silk. Bradoo (1967) found for *O. ceylonica* seven adult males

and 28 adult females in a total of 28 field colonies: most of the 203 embiids collected were nymphs. In an intriguing statement, Melander (1903) reported that a single male develops from a brood of nymphs of *A. texana*. Ananthasubramanian (1957) reared 40 males and 31 females of *O. humbertiana* in the laboratory. The sex ratio in this case was not significantly different from 50 : 50 (Edgerly 1986). Similarly, Bradoo and Joseph (1970) reported an overall sex ratio for *O. greeniana* of 57 males to 46 females from five laboratory cultures.

Parasites, predators and inquilines

A suite of parasitoids has evolved that specialize on embiids. In some cases, these parasitoids may have promoted group living and/or maternal care, although few studies have tested such hypotheses. Sclerogibbidae (Fig. 2-2), a family of aculeate Hymenoptera, are ectoparasitoids of embiid nymphs (Callan 1939; Krombein 1979; Sheltar 1973).

Table 2-2. *Generations per year for five species of embiid*

Species	Location	Generations per year				References
		0.5	1	2	4–5	
Anisembia texana	Texas	—	×	—	—	Mills 1932
						Melander 1903
Embia major[a]	Himalayas, India	—	×	—	—	Imms 1913
E. taurica	Crimea	×	—	—	—	Kusnezov 1904
Oligotoma ceylonica ceylonica[b]	India	—	—	—	×	Bradoo 1967
O. japonica	Japan	—	—	×	—	Okajima 1926

[a] Females lived up to 6.5 months after oviposition.
[b] Each generation was completed within 72 days.

Although I observed only two interactions, female *Antipaluria urichi* did not aid their nymphs when sclerogibbids entered their silk; the wasp's ability to sting may prevent the embiid from defending her offspring (Edgerly 1988). Indeed, one embiid female bolted from her silk covering in response to a lunging attack by a sclerogibbid. Scelionid wasps (*Embidobia* sp.) parasitize webspinner eggs (see, for example, Callan 1952) and may have played a significant role in promoting complex egg guarding as seen in *A. urichi* (see below). Other parasitoids attack mature embiids. For example, tachinid flies parasitize *Clothoda* in Peru (Arnaud 1963) and *Dictyoploca* in South Africa (Mesnil 1953). In addition, a newly discovered genus of braconid, *Sericobracon* Shaw, utilizes *Antipaluria* adults as hosts in Trinidad (Shaw and Edgerly 1985). Disease organisms such as gregarines (*Gregarina marteli*, *Diplocystis clarki*)

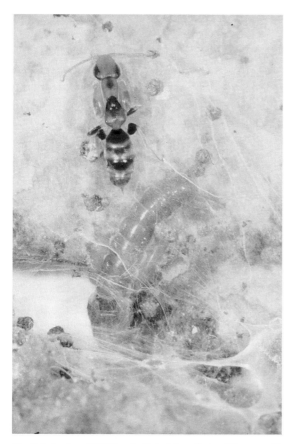

Fig. 2-2. A sclerogibbid female, *Probethylus callani*, walking on a silk wall of *Antipaluria urichi*. The abdomen of an embiid nymph, the wasp's host, is visible through the silk. Photo.: J. S. Edgerly.

(Stefani 1959, 1960) and coccidia (*Adelea transite*) (Denis 1949) also infect embiids. Interestingly, Stefani (1959) suggested that a protozoan, *Diplocystis clarki*, effectively sterilizes males in *Haploembia solieri*, leaving residual parthenogenetic populations in parts of Italy. The impact of these parasitoids and disease agents on the evolution of communal behavior in embiids has not been studied.

Predation is also a source of mortality for embiids, both within and outside of silk. In Trinidad, I observed ants, spiders, geckos, and a neuropteran larva (Ascalaphidae) killing *A. urichi* that were wandering outside of their silk (Edgerly 1988). In addition, Callan (1952) observed an asilid fly preying on this species. Other predators (ants) reach embiids by cutting through silk, pulling off sheets of silk (birds), or piercing through it (harvestmen, Edgerly 1988, 1993; reduviids, Denis 1949). Below I describe results of my investigation of the differential risk of predation for individuals in different-sized colonies, a possible selective agent maintaining facultative communal behavior in this species.

The silk of webspinners provides a habitat for embiophiles in the heteropteran family Plokiophilidae. The relationship between these bugs and their hosts is unclear: they may act as predators on embiid eggs or young nymphs (Callan 1952) or other insects and mites within the silk, or as scavengers on carcasses or scraps of material within the gallery (Carayon 1974). In *A. urichi*, plokiophilids occurred in 62 of 80 field colonies (Edgerly 1987a). Webspinners also live within silk of other embiid species; I observed *O. saundersii* inhabiting galleries of *A. urichi* in Trinidad (for similar examples from India, see Mukerji 1928). The outcome of such interactions is not understood. Perhaps even more surprising are the embiids (*O. ceylonica*, Bradoo 1967; *O. greeniana*, Bradoo and Joseph 1970) that live within colonies of the social spider *Stegodyphus sarasinorum*, where they appear to feed on algae growing on the spider silk. Bradoo suggested that webspinner silk protects embiids from being detected by spiders. In addition, I discovered two species of webspinners (*Diradius* n. sp. (Teratembiidae) and *O. saundersii*) living on the surface and within the outer fabric of termite nests (*Nasutitermes*) in Trinidad; the embiids apparently graze on algae and/or lichens on the termite nests. Soldier termites did not respond to webspinners that remained within their silk; however, when I opened the silk with forceps, the exposed embiids quickly elicited an alarm response from the termite soldiers (Edgerly 1986). Other embiid species have also been observed in close association with termites (references in Imms 1913).

CASE STUDY: RESEARCH ON *ANTIPALURIA URICHI*

Habitat and life cycle

Locally abundant, *A. urichi* is restricted to Trinidad (Ross 1987), an island situated approximately 11 km off the coast of Venezuela. I conducted my study in the tropical rain forest of the Northern Range at the Simla Research Station and Asa Wright Nature Center. Females are relatively elongate for a webspinner (1.6 ± 0.015 cm (SE)); the winged males are typically shorter in length (1.2 ± 0.02 cm) (Edgerly 1987a). Their silk galleries are generally found on tree surfaces and vertical banks along road cuts (Fig. 2-1), but also on posts, concrete walls and flower pots. Silk covers their resting sites and feeding zones at the colony's edge where fresh algae and other epiphytes are found. Once food is depleted, more silk is added, extending the forager's reach into unexploited sites. Resting sites remain fairly stable, often within a central location, as silk walls are expanded to enclose more feeding sites (Edgerly 1986). Ultimately, depletion of food may trigger dispersal from a colony site. Although webspinners typically do not venture outside their silk, one can see them by peering through new, thin silk as they graze on algae at night at the outer perimeters of the silk covering.

I concluded, by observing marked females and colonies, that the reproductive cycle of *A. urichi* is aseasonal; reproductive females were present every month I observed them, which included all months except April, May and June (Edgerly 1988). Embryonic development takes approximately 6 weeks; nymphal development requires 18 (Edgerly 1988) (Fig. 2-3). A key feature in more highly

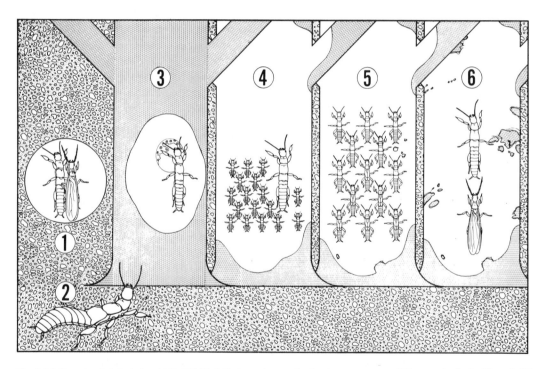

Fig. 2-3. Life cycle of *Antipaluria urichi* in Trinidad. Each number on the drawing represents a different point in the life cycle. Time is along the horizontal axis: each consecutive, diagrammatic tree to the right shows the same colony, only older. (1) Mating occurs within silk. (2) Dispersal often occurs in this species, from the natal colony to another tree or within the same tree, although the exact timing relative to reproduction is not known. (3) Egg-guarding by an adult female, requiring approximately six weeks. The white area represents silk spun by this solitary female; the circular object beneath her is her egg mass hidden under an egg mass covering. (4) The number of nymphs shown is the average number observed in field colonies. The female increases her silk-spinning activities as reflected in the larger expanse of silk. (5) The adult female generally disappears when the nymphs are approximately half grown. The nymphs also begin to disappear as they grow, as reflected in the fewer individuals in the colony. (6) Here only two mature embiids are left, 18 weeks after hatch. The silk shows many tears that have accumulated. For details used to produce this drawing see Edgerly (1986, 1987a,b, 1988). Drawing: Edward C. Rooks.

social insects, the overlap of generations within the nest, is apparently lacking in *A. urichi*. All adult females disappeared or died at varying times before their offspring matured (Edgerly 1988). Furthermore, only two of 15 closely monitored colonies produced female offspring that matured and reproduced in their natal colonies; all others disappeared. The structure of their life cycle lacks other factors that may favor more complex interactions, since individuals typically disperse prior to reproducing and do not interact with distant generations.

Maternal care

Egg-guarding behavior

As in many other subsocial insects (references in Wilson 1971; Eickwort 1981; Tallamy and Wood 1986), female embiids attend to clustered eggs until hatching, and subsequently remain with their nymphs. By employing focal sampling techniques, I determined that females invest approximately 85% of their time in egg attendance and care during the six-week developmental period (Edgerly 1987a). Such activities include spinning silk on the egg mass, positioning her body over the eggs while actively turning her head, and vigorously shaking and lunging at approaching plokiophilids, scelionid wasps, or other webspinners. These guarding behaviors persist day and night. Such activity is in sharp contrast to that displayed by pre-reproductive females and females with nymphs, who generally sit still during the day, and move, feed and spin silk at night. A cost of such maternal care is the inability of egg-guarding females to feed as often as other individuals. Their investment goes beyond behavioral protection, for they also provide physical protection by constructing an elaborate egg mass covering of macerated bark and other substrate materials, silk and possible salivary secretions that together produce a tough outer coating on each egg. After each egg is coated in this manner, the mother affixes it, with her other eggs, to the substrate, yielding an egg mass of 53 (± 2.7) eggs, on average (Edgerly 1987a). After oviposition, the mother adds macerated materials and silk to the top of the compact egg mass. The egg mass covering functions, at least in part, to prevent successful oviposition by scelionid wasps, which have difficulty penetrating the covering. The delay caused by the wasps having to dig and probe for eggs allows adult embiids time to repair their silk galleries and to feed at the perimeter of the silk covering without risking parasitism during their absence from the eggs.

Guarding females are effective in protecting most of their eggs from parasitoids, as revealed by experimental results where naturally occurring field eggs from which females were removed suffered significantly lower hatching rates ($12.9 \pm 8.7\%$) than did neighboring guarded eggs ($71.1 \pm 8.3\%$) (Edgerly 1987a). The lower hatch rate of unguarded eggs was due to parasitism by scelionid wasps and predation by ants.

The egg mass covering may protect eggs from desiccation in the dry season or fungal infestation in the rainy season. In order to test these ideas, I placed egg masses with approximately one-half of the covering removed in parasitoid-exclosure bags in the field or in Petri dishes in the laboratory. Even though approximately the same number of nymphs hatched from uncovered (treatment) and covered eggs (controls), 56% of those hatching from control eggs died entombed beneath the egg mass covering. After discovering this potentially lethal impact of the egg mass covering, I observed females more closely and discovered that at approximately six weeks after oviposition they remove the egg mass covering with their mandibles, thereby facilitating hatching. Similar behavior has been observed in spiders which open their egg sacs to facilitate spiderling emergence (references in Horel and Gunderman 1992). If a female embiid is killed prior to emergence of her nymphs, her eggs risk death by parasitism or emergence failure because of the egg mass covering. Parasitoids are ubiquitous (79% of 67 colonies housed egg parasitoids) and represent an almost constant threat to eggs; these scelionids are able to parasitize at any time during the egg's developmental period. The chance of eggs being abandoned, however, appears to be relatively low (18% of 44 females with eggs), and hence, the negative consequences of emergence failure because of the egg mass covering are rarely realized.

Maternal care of nymphs

After their eggs hatch, females invest significantly more time in behaviors displayed prior to their stint of egg guarding: spinning silk, feeding and moving within the galleries at night, and resting during the day (Edgerly 1987a). Females with nymphs may provide an important commodity for their developing young, as evidenced by an exponential increase in area of the silk covering after hatch. In contrast to *E. ramburi* (Denis 1949; LeDoux 1958), female *A. urichi* do not feed their nymphs, but, rather, feed alongside them at the periphery of the silk covering.

The mother's presence increases the development rate of the offspring (Edgerly 1988). The cause of the deficit in developmental rate when the mother is absent has not been clarified experimentally, but may be partly due to an increase in silk spinning by nymphs to counter the lack of maternal silk, which generally forms a dense outer covering. Choe (1994) discovered similar slow development in orphaned *Anisembia texana* nymphs in a laboratory experiment. Here, too, the nymph-produced silk covering was relatively thin.

Antipaluria urichi also exhibit hygienic behaviors, gathering fecal pellets into piles and silking them into the sides of their galleries. Other species have been observed gathering fecal pellets and pushing them through the silk wall to the outside (Ling 1935). Interestingly, E. S. Ross (personal communication) suggested that, in the tropics, feces pushed onto the silk surface serves as a substrate for microflora that may further enhance a colony's cover. Adult females may more effectively perform these activities, although the function of fecal handling remains to be tested. A similar observation was made on female subsocial crickets (*Anurogryllus muticus*) (West and Alexander 1963), and earwigs (see, for example, Radl and Linsenmair 1991), which remove excess waste material and fungi from underground burrows.

Comparison with other subsocial insects

Many species of insect guard their eggs against predators by utilizing chemical or other defensive behaviors. For example, lacebugs attack predators by fanning their wings and rushing at them (Tallamy and Denno 1981). *Antipaluria urichi* do not attack predators, other than intruding embiids, that threaten their eggs. Ants sometimes enter silk and destroy entire egg masses, while female webspinners retreat in response (Edgerly 1988). They apparently lack defenses against predators that easily sting and dismember an adult embiid, a soft-bodied insect. Embiids do, however, gain protection by producing silk walls, which, if intact, usually evade detection. Ants march along trails directly on webspinner silk, without attacking the occupants (E. S. Ross, personal communication; Myers 1928; Denis 1949; Edgerly 1988). If, however, the silk covering is breached, detection is afforded, and ants will enter and attack (*O. saundersii*, E. S. Ross, personal communication; *A. urichi*, Edgerly 1988).

Living within silk may protect *Antipaluria*, but it also concentrates their eggs, thus increasing the risk of annihilation if discovered by egg parasitoids or predators. Under these conditions, the resultant strong selective pressure exerted by egg parasitoids may have promoted the relatively elaborate behaviors associated with maternal care in *A. urichi*. For future work, a systematic comparison of embiids that vary in their tendency to construct egg masses (references in Edgerly 1987a) may reveal to what degree egg mass coverings and egg guarding are associated.

In their assessment of convergence in subsocial insects, Tallamy and Wood (1986) concluded that factors favoring the origin of parental care include reproduction confined to specific periods and places, extended adult longevity, and basic behavioral elements that can be molded into defensive posturing or nest building. The authors also suggested that an insect's association with a particular nutritional resource may promote parental care. According to their scheme, *A. urichi* resemble foliage feeders because of their surface-feeding tendencies. Hence, their exposure to predators as they feed may have promoted antipredator behaviors, such as production of the silk covering. Their relatively long adult life and pre-existing defensive behaviors as seen when they fight with intruding embiids – shaking, lunging and biting – may have further promoted the evolution of parental care. Egg-tending embiids use these same behaviors when encountering scelionid wasps or other webspinners near their eggs.

Insects in the Order Psocoptera exhibit behaviors observed in webspinners with some interesting differences. Psocids, commonly called bark lice, often spin conspicuous webs on bark where they feed in a manner similar to embiids: they add silk to cover new feeding sites, depleting food as the colony expands. Psocopteran silk shields against ants and other predators (references in New 1973), as suggested for webspinners. Colonies often form by convergence of groups derived from different founding females on the same tree, again reminiscent of webspinners. Unlike embiids, most psocids are not subsocial, however; only a few families contain species exhibiting maternal care and/or gregarious behaviors. According to Mockford (1957) group-living in psocids varies as follows: nymphs live in small groups under loose webs; nymphs live in large herds, which break up soon after they reach adulthood; females stay with eggs, but usually die prior to hatch; and, showing the highest degree of sociality, nymphs emerge in webs made by their mothers or by both parents plus their siblings. In this last type, found in the genus *Archipsocus*, nymphs stay in their natal colony and contribute to the webbing, as well as repairing it when torn. Some of the parental generation may be alive when the nymphs reach adulthood, so that generations may overlap within the colony. Species in the Family Archipso-cidae often have hundreds of psocids sharing a silk sheet that they spin together over bark and other substrates. The tendency to aggregate may be associated with silk-spinning, although one species, *Peripsocus nitens*, does not produce conspicuous webs, and yet appears to exhibit an intriguing phenomenon of communal oviposition. Female *P. nitens* stay with their eggs, often in clusters consisting of eggs numbering up to a few hundred produced by a number of females. They perhaps care for their eggs in some way (New 1985). The behavior is reminiscent of lace bugs, which add their eggs to other females' egg batches, thereby exploiting the care-giving of the receiving female. The egg-dumping lace bug leaves her eggs with another female, commences feeding, and ultimately produces more eggs (Tallamy 1985). The lace bug that receives the extra eggs may gain an antipredator advantage for her eggs via the dilution effect, or because her eggs gain safety by being in the center of the egg mass. It remains to be deter-mined what advantage a female *Peripsocus* gains by cluster-ing her eggs with hundreds of eggs produced by neighboring females. Protection against parasitism appears not be the function of maternal care here, however, because egg parasites seem rare for *P. nitens* (New 1985), a striking difference from what I observed for *Antipaluria*.

Given the similarities in silk-spinning and feeding behaviors of Psocoptera and Embiidina, it would be inter-esting to determine what factors contribute to the differ-ences in their tendency to exhibit subsocial behaviors. Although maternal behavior occurs in psocids, it appears rare, limited to few species within the order. This is in con-trast to the embiids, all of which appear to exhibit care of their eggs. Factors promoting such differences remain a puzzle but may include the high incidence of egg parasit-ism and perhaps egg cannibalism within webspinners. A closer comparison of the two orders, requiring more inten-sive field work, seems worthwhile.

To help complete the picture of evolution of embiid subsociality, the behavior of subterranean webspinners should be investigated further. These species may more closely resemble earwigs, burrowing crickets (references in Tallamy and Wood 1986), and others that suffer from threats not associated with surface feeding, such as infesta-tion by fungi or other soil contaminants. Behavior in these species may be more complex than in *Antipaluria*, espe-cially if females gather food from outside, returning it to offspring in a silk nest.

Communal behavior

Description of colonies

The facultative communal nature of *A. urichi* allows for the comparison of attributes associated with reproductive suc-cess for solitary females with their offspring, to those in colonies with as many as 72 individuals, including up to 24 adult females. Surveys of colony occupants in different areas in Trinidad in 1984 revealed variation from as many as 86% of the females being communal (of 138 females in 44 colonies), one colony with 9 egg masses, to 37% being communal (of 57 females in 64 colonies) (Edgerly 1986; 1987b). Individuals sharing silk hatch within the colony or enter as immigrants. Based on observations of marked individuals I inferred that relatedness of colony-mates varies from mother–offspring associations to groups of unrelated individuals that joined each other to combina-tions of the two types. No analysis exists determining relat-edness within a colony or on a tree, which may be high given the limited ability of wingless females to disperse over long distances (Edgerly 1987b). I documented a pat-tern of recruitment of new silk galleries on trees that sug-gests that dispersers remain on their host trees, establishing colonies close to their mothers' colonies, rather than crossing the forest floor to disperse to other trees (Edgerly 1987b). However, at least for most webspin-ner species, males fly in search of mates, potentially redu-cing relatedness of neighboring individuals. Relatedness within colonies needs to be examined in embiids; species of particular interest are those with more sedentary, nymph-like, wingless adult males (e.g. *Paedembia*) (Ross 1970).

Consequences of group-living

Impact of parasitoids. An advantage of group-living in *Anti-paluria* is enhanced egg production by communal females (Edgerly 1987b). However, egg parasitoids reduce this bonus, so that ultimately communal females produce the same number of hatched eggs as do solitary females. A key predictor of egg parasitism rate is how close egg masses are to each other on a tree; the closer they are, the greater the chance of parasitism. Furthermore, given egg masses of equal distance from one another, those produced in a communal colony suffer greater parasitism rates than do those produced by solitary females (Edgerly 1987b). Not only might egg parasitoids be differentially attracted to aggregated hosts (Hassell 1978), they may also hatch from eggs within a colony and search therein for nearby egg masses. The scelionid's ability to oviposit into eggs at any point in egg development heightens the risk of having parasitized eggs nearby. Ecological factors promoting differences in egg production for communal and solitary females remain unknown. They may include an increase in energy expenditure by solitary females that may spin more silk or walk further to distant sites when establishing a gallery. Such dispersal may be adaptive because females colonizing distant sites appear less likely to attract scelionid wasps (Edgerly 1987b), but may reduce their ability to produce as many eggs as they might have had they stayed close to their natal site. Furthermore, enhanced fecundity in communal females may not be due to benefits of social interactions *per se* but rather to site-specific resources, which elicit colonization by many females. In a laboratory study, LeDoux (1958) found that grouping by embiids may be partly in response to favored food resources. He observed that 15 *E. ramburi* initially fought when placed together, but over a two-week period showed a marked tendency to form silken tubes together in the vicinity of their food.

Impact of predation. Differential predation on colonies of varying sizes may maintain the facultative communal system of *Antipaluria*, whose colonies vary in expanse of silk covering, as well as in number of occupants. Recently (Edgerly 1993), I tested whether the probability of a predator detecting a colony increases proportionately with an increase in expanse of silk. Larger colonies would not gain an antipredator benefit from grouping if this were so, unless other attributes, such as group defense, enhancement of silk as a protective layer, or the dilution effect, served to protect them (see, for example, Foster and Treherne 1981; Turchin and Kareiva 1989). The predator avoidance effect hypothesis of Turner and Pitcher (1986; see also Dehn 1990; Inman and Krebs 1987) predicts that, in fact, larger groups are not proportionately more likely to be encountered by visually hunting predators. Using observational techniques, I determined that there was a less than one-to-one relationship between attack rate and increase in size of the colony. Holes cut in the silk increased during a three-week study at a rate of only 0.12 holes added per square centimeter of silk as the silk wall perimeter increased. Therefore, from the predator's point of view, less expansive silk walls are more conspicuous than their size would suggest; predators directed proportionally more attacks at smaller silk walls than would be predicted by their relatively small size. Webspinners, therefore, may be safer

when housed under larger sheets of silk because of the lower attack rate per unit area relative to that experienced, on average, by smaller colonies. The impact of conspicuousness of silk on predation rates may not fully explain differential predation risk; further analysis is required to determine whether crowding within a colony promotes safety in numbers via the dilution and/or confusion effects.

Although not yet investigated, vibratory signals through silk may enhance webspinner defensive responses by providing a warning of predator attacks elsewhere in the colony, as has been recently reported for spiders (Uetz and Hieber 1994). Reception of local vibratory signals in embiids seems likely because of the unusual stance adopted within their galleries: their first and third pairs of legs rest ventral to the body, while the middle pair is hooked dorsally into the silk. Any movement of silk could be perceived from above or below the animal. Furthermore, embiids appear to communicate by generating local vibrations by shaking rapidly, especially when an individual enters the gallery of another (Edgerly 1986; for *Anisembia texana*, Choe 1994).

In sum, costs and benefits of communal behavior may fluctuate depending on the age of the colony. Older colonies may suffer from heavy parasitoid loads (Edgerly 1986) and perhaps depleted food resources, but, on the other hand, may benefit from expansive silk walls. Furthermore, the larger number of individuals in older colonies may contribute to a potential dilution effect or other antipredator devices, and/or other attributes not yet determined.

Interactions between individuals. Aggressive interactions are common within *Antipaluria* colonies (Edgerly 1986), and have been reported for other webspinners as well (Ling 1935; LeDoux 1958; Bradoo and Joseph 1970; Choe 1994). Fights occur when dispersers cut into silk walls and enter established colonies. Webspinners respond to intruders by biting, pushing head-to-head, locking mandibles, and shaking vigorously. In experimental trials (Edgerly 1986), nymphs, adult females alone, and females with eggs were particularly responsive to intruding females. Females with nymphs, on the other hand, did not fight. Aggressive reactions generally did not cause the intruder to leave the silk; of 20 trials, 16 ended with the intruder staying. These experimental results paralleled field observations revealing that dispersers often join existing colonies, despite the sometimes aggressive reactions of residents (Edgerly 1987b). Fighting most likely results in spacing individuals out within the colony rather than pushing them out completely. Once established within a silk gallery, adult females rarely interact much beyond adding silk to, and sharing, contiguous silk walls.

Further studies are required in order to determine the function of fighting and why aggressive tendencies vary. One possibility is that females joining a colony are a threat to eggs (Edgerly 1986). In an preliminary attempt to determine whether females can recognize their own eggs, I moved marked females (maternal females; $n = 17$) from their own egg masses to other egg masses within their own or in neighboring colonies. In addition, I placed adult females without eggs (non-maternal; $n = 3$) onto egg masses to see if they would guard, abandon, or consume them. Ten of the maternal females relocated their own eggs and resumed guarding, one assumed a guarding posture over the adopted eggs, and the six others disappeared. The non-maternal females maintained their new positions, and consumed at least half of the eggs within two days of introduction to each site. Two of these egg masses were eventually completely consumed; the third was abandoned because of predation on the female. Other possible factors promoting the evolution of aggressive responses, yet to be examined, include risk of cannibalism for molting nymphs and risk of increased conspicuousness to parasitoids for closely grouped embiids.

CONCLUSION

Webspinners remain one of the least understood orders within the Class Insecta. Embiids appear to exhibit variation in parental behavior, but such variation has not been well documented. Furthermore, their tendency to live in groups also appears to vary; variability that is worth investigating. Of particular interest are species with dimorphic males, and especially those with nymph-like males. Males, in such species, may contribute to care of the young, or guard their mates. As revealed in this chapter, group living occurs within the order, but only one species, *A. urichi*, has been studied in detail in the field, and for that species many questions remain, such as: What role do the ubiquitous hemipteran inquilines (family Plokiophilidae) play? Does the dilution effect promote living in groups? and; Do females gain an advantage from sharing silk? In addition, the possibility of greater social complexity within the order is also an intriguing question. Subterranean species that store food in chambers, such as *Paedembia* n. sp. discovered by Ross in Afghanistan, may exhibit

more complex social interactions than seen in *Antipaluria*. Because the former collect food to stock their nests, generations may overlap within the colony, as they do within the Hymenoptera. In contrast, *Antipaluria* colonies slowly deplete algae from their natal colony sites perhaps triggering dispersal so commonly observed in this species. Nest-sharing in *Paedembia* and in other embiid species like them may facilitate more complex systems of communication and other social interactions. The harsh climate of Afghanistan, the difficulty of forming nests underground, and the unpalatability of their food (sagebrush leaves) resemble extrinsic factors proposed by Evans (1977) to promote the evolution of higher sociality. At the other end of the social spectrum, adult females in an African species discovered by E. S. Ross (personal communication) are intolerant of each other and extremely aggressive. Uncovering ecological correlates of such behaviors may yield answers concerning the evolution of communal behavior within the order.

The evolution of insect parental behavior also warrants closer inspection, for it appears in many orders of insects. If all webspinners are subsocial, as has been suggested by Ross (1970), then an examination of the order's sister group may help in discovering roots of parental care. A more difficult problem in analyzing the evolution of parental care in webspinners is the lack of a widely accepted phylogeny for orthopteroid insects. Because all embiids are subsocial, comparisons with asocial relatives would require identification of a sister group. Despite these difficulties, one could compare maternal investment for speices within the order with differing life history strategies (iteroparity vs. semelparity) or differing colony locations (arboreal vs. subterranean or beneath rocks) to determine what ecological factors promote heavy investment in a single brood. In addition, as suggested above, comparisons with insects in the Order Psocoptera may also help address questions of selective pressures that led to the evolution of subsocial behavior within the embiids, and more rarely within the psocids.

In conclusion, most important for future research will be basic field observations and quantitative analysis of behavior in as many webspinner species as possible, preferably from a variety of families and habitats. Such studies will add tremendously to the scant information that exists in the literature at the present time. Trends within the order cannot be identified without such information. Finally, phylogenetic analysis is still needed, and many newly discovered species remain to be described (E. S. Ross, personal communication).

ACKNOWLEDGEMENTS

I thank the editors for their efforts in coordinating this volume and for their constructive critiques of my chapter. I also thank D. W. Tallamy and anonymous reviewers for helpful comments on an earlier version of this chapter, the Director and staff of the Asa Wright Nature Center for providing field sites, and Edward C. Rooks for his meticulous drawing in Fig. 2-3. Aspects of this work were supported by grants from National Academy of Sciences (Research Grants in the Pure and Applied Sciences: no. 617 and no. 640), Sigma Xi, National Science Foundation (BSR-8312897), and Santa Clara University (Presidential Research Award: 5-28062). Finally, but certainly not least, I acknowledge the significant input of my mentor, George C. Eickwort, who suggested that I study the Order Embiidina, and supported me through much of my work on these insects.

LITERATURE CITED

Alberti, V. G. and V. Storch. 1976. Transmissions und rasterelektronenmikroskopische Untersuchung der Spinndrusen von Embien (Embioptera, Insecta). *Zool. Anz. (Jena)* **197**: 179–186.

Ananthasubramanian, K. S. 1957. Biology of *Oligotoma humbertiana* (Saussure)(Oligotoma, Embioptera). *Indian J. Entomol.* **18**: 226–232.

Arnaud, P.H. 1963. *Perumbyia embiaphaga*, a new genus and species of neotropical Tachinidae (Diptera) parasitic on Embioptera. *Am. Mus. Nov.* **2143**: 1–9.

Barth, R. 1954. Untersuchungen an den Tarsaldürsen von *Embilyntha batesi* MacLachlan, 1977 (Embioidea). *Zool. Jb., Anat. (Jena)* **74**: 172–188.

Boudreaux, H. B. 1979. *Arthropod Phylogeny with Special Reference to Insects.* New York; John Wiley & Sons.

Bradoo, B. L. 1967. Observations on the life history of *Oligotoma ceylonica ceylonica* Enderlein (Oligotomidae, Embioptera), commensal in the nest of social spider, *Stegodyphus sarasinorum* Karsch. *J. Bombay Nat. Hist. Soc.* **64**: 447–454.

Bradoo, B. L. and K. J. Joseph. 1970. Life history and habits of *Oligotoma greeniana* Enderlein (Oligotoma; Embioptera), commensal in the nest of social spider *Stegodyphus sarasinorum* Karsch. *Indian J. Entomol.* **32**: 16–21.

Callan, E. McC. 1939. A note on the breeding of *Probethylus callani* Richards (Hymenoptera, Bethylidae), an embiopteran parasite. *Proc. R. Entomol. Soc. Lond.* B**8**: 223–224.

–. 1952. Embioptera of Trinidad with notes on their parasites. In *Transactions of the 9th International Congress of Entomology*, vol. 1. J. de Wilde, ed., pp. 483–489. Amsterdam: International Congress of Entomology.

Carayon, J. 1974. Etude sur les Hemipteres Plokiophilidae. *Ann. Entomol. Soc. France* **10**: 499–525.

Choe, J. 1994. Communal nesting and subsociality in a webspinner, *Anisembia texana* (Insecta: Embiidina: Anisembiidae). *Anim. Behav.* **47**: 971–973.

Dehn, M. M. 1990. Vigilance for predators: detection and dilution effects. *Behav. Ecol. Sociobiol.* **26**: 336–342.

Denis, R. 1949. Ordre des Embiopteres. *Traite de Zoologie (Paris). Anatomie-Syst. Biologie.* **9**: 723–744.

Edgerly, J. S. 1986. Behavioral ecology of a primitively social webspinner (Embiidina: Clothodidae: *Clothoda urichi*). Ph. D. diss., Cornell University, Ithaca, NY.

–. 1987a. Maternal behaviour of a webspinner (Order Embiidina). *Ecol. Entomol.* **12**: 1–11.

–. 1987b. Colony composition and some costs and benefits of facultatively communal behavior in a Trinidadian webspinner, *Clothoda urichi* (Embiidina: Clothodidae). *Ann. Entomol. Soc. Am.* **80**: 29–34.

–. 1988. Maternal behaviour of a webspinner (Order Embiidina): mother–nymph associations. *Ecol. Entomol.* **13**: 263–272.

–. 1993. Is group living an antipredator defense in a facultatively communal webspinner? *J. Insect Behav.* **7**: 135–147.

Eickwort, G. C. 1981. Presocial insects. In *Social Insects*, vol. 2. H. R. Hermann, ed., pp. 199–226. New York: Academic Press.

Evans, H. E. 1977. Extrinsic versus intrinsic factors in the evolution of insect sociality. *BioScience* **27**: 613–617.

Foster, W. A. and J. E. Treherne. 1981. Evidence for the dilution effect in the selfish herd from fish predation on a marine insect. *Nature (Lond.)* **293**: 466–467.

Halffter, G. 1977. Evolution of nidification in the Scarabaeinae (Coleoptera: Scarabaeidae). *Quaest. Entomol.* **13**: 231–253.

Hassell, M. P. 1978. *The Dynamics of Arthropod Predator-Prey Systems.* Princeton, NJ: Princeton University Press.

Hennig, W. 1981. *Insect Phylogeny.* New York: John Wiley & Sons.

Horel, A. and J. L. Gunderman. 1992. Egg sac guarding by the funnel-web spider *Coelotes terrestris*: function and development. *Behav. Proc.* **27**: 85–93.

Imms, A. D. 1913. Contributions to a knowledge of the structure and biology of some Indian insects. II. On *Embia major*, sp. nov., from Himalayas. *Trans. Linn. Soc. Lond. (Zool.)* **2**: 167–195.

Inman, A. J. and J. Krebs. 1987. Predation and group living. *Trends Ecol. Evol.* **2**: 31–32.

Krombein, K. V. 1979. Biosystematic studies of Ceylonese wasps, VI. Notes on the Sclerogibbidae with descriptions of two new species (Hymenoptera: Chrysidoidea). *Proc. Entomol. Soc. Wash.* **81**: 465–474.

Kusnezov, N. J. 1904. Observations on *Embia taurica* Kusenov (1903) from the Southern Coast of the Crimea. *Horae Soc. Entomol.* **37**: 165–173.

LeDoux, A. 1958. Biologies et comportement de l'Embioptere *Monotylota ramburi* Rims.-Kors. *Ann. Sci. Nat. (Paris) (Zool.)* **20**(11): 515–532.

Ling, S. W. 1935. Notes on the biology and morphology of *Oligotoma* sp. *Peking Nat. Hist. Bull.* **9**: 133–139.

Melander, A. L. 1903. Notes on the structure and development of *Embia texana*. *Biol. Bull. Mar. Biol. Lab. Woods Hole* **4**: 99–118.

Mesnil, L. P. 1953. A new tachinid parasite of an Embiopteran. *Proc. R. Entomol. Soc. Lond.* B22: 145–146.

Michener, C. D. 1969. Comparative social behavior of bees. *Annu. Rev. Entomol.* **14**: 299–342.

Mills, H. B. 1932. The life history and thoracic development of *Oligotoma texana* (Mel.). *Ann. Entomol. Soc. Am.* **25**: 648–652.

Milne, L. J. and M. Milne. 1976. The social behavior of burying beetles. *Scient. Am.* **235**: 84–88.

Minet, J. and T. Bourgoin. 1986. Phylogenie et classification des hexapodes (Arthropoda). *Cahiers. Liaison* **20**: 23–29.

Mockford, E. L. 1957. Life history studies on some Florida insects of the genus *Archipsocus* (Psocoptera). *Bull. Florida State Mus., Biol. Sci.* **1**: 253–274.

Mukerji, S. 1928. On the morphology and bionomics of *Embia minor* sp. nov. with reference to its spinning organ. *Rec. Ind. Mus. Calcutta* **29**: 253–282.

Myers, J. G. 1928. The first known Embiophile, and a new Cuban embiid. *Bull. Brooklyn Entomol. Soc.* **23**: 87–90.

Nalepa, C. A. 1984. Colony composition, protozoan transfer and some life history characteristics of the woodroach *Cryptocercus punctulatus* Scudder (Dictyoptera; Cryptocercidae). *Behav. Ecol. Sociobiol.* **14**: 273–279.

New, T. R. 1973. The Archipsocidae of South America (Psocoptera). *Trans. R. Entomol. Soc. Lond.* **125**: 57–105.

–. 1985. Communal oviposition and egg-brooding in a psocid, *Peripsocus nitens* (Insecta: Psocoptera) in Chile. *J. Nat. Hist.* **19**: 419–423.

Okajima, G. 1926. Description of a new species of *Oligotoma* from Japan together with some notes on the family Oligotomidae (Embiidina). *J. Coll. Agric.* **7**: 411–434.

Pearse, A. S., M. T. Patterson, J. S. Rankin and G. W. Wharton. 1936. The ecology of *Passalus cornutus* Fabricius, a beetle which lives in rotting logs. *Ecol. Monogr.* **6**: 455–490.

Radl, R. C. and K. E. Linsenmair. 1991. Maternal behaviour and nest recognition in the subsocial earwig *Labidura riparia* Pallas (Dermaptera: Labiduridae). *Ethology* **89**: 287–296.

Ross, E. S. 1963. The families of Australian Embioptera, with descriptions of a new family, genus and species. *Wasmann J. Biol.* **21**: 121–136.

–. 1970. Biosystematics of the Embioptera. *Annu. Rev. Entomol.* **15**: 157–172.

–. 1987. Studies in the insect order Embiidina; a revision of the family Clothodidae. *Proc. Calif. Acad. Sci.* **45**: 9–34.

Seelinger, G. and U. Seelinger. 1983. On the social organization, alarm and fighting in the primitive cockroach *Cryptocercus punctulatus* Scudder. *Z. Tierpsychol.* **61**: 315–333.

Shaw, S. R. and J. S. Edgerly. 1985. A new braconid genus (Hymenoptera) parasitizing webspinners (Embiidina) in Trinidad. *Psyche* **92**: 505–511.

Sheltar, D. J. 1973. A redescription and biology of *Probethylus schwartzi* Ashmead (Hymenoptera: Sclerogibbidae) with notes on related species. *Entomol. News* **84**: 205–210.

Sites, R. W. and J. E. McPherson. 1982. Life history and laboratory rearing of *Sehirus cinctus cinctus* (Hemiptera: Cydnidae) with descriptions of immature stages. *Ann. Entomol. Soc. Am.* **75**: 210–215.

Stefani, R. 1959. Parassitosi da Gregarine e streilita negli Insetti. *Parassitol. (Ist. Parassit. Univ.) Univ. Roma* **2**: 311–313.

–. 1960. I rapporte tra parassitoisi, sterilita maschile e partenogenesi accidentale in popolazioni naturali di *Haploembia solieri* Ramb. anfigonica. *Riv. Parassitol. (Roma)* **21**: 277–287.

Tallamy, D. W. 1985. 'Egg dumping' in lace bugs (*Gargaphia solani*, Hemiptera: Tingidae). *Behav. Ecol. Sociobiol.* **17**: 357–362.

Tallamy, D. W. and R. F. Denno. 1981. Maternal care in *Gargaphia solani* (Hemiptera: Tingidae). *Anim. Behav.* **29**: 717–778.

Tallamy, D. W. and T. K. Wood. 1986. Convergence patterns in subsocial insects. *Annu. Rev. Entomol.* **31**: 369–390.

Turchin, P. and P. Kareiva. 1989. Aggregation in *Aphis varians*: an effective strategy for reducing predation risk. *Ecology* **70**: 1008–1016.

Turner, G. F. and T. J. Pitcher. 1986. Attack abatement: a model for group protection by combined avoidance and dilution. *Am. Nat.* **128**: 228–240.

Uetz, G. W. and C. S. Hieber. 1994. Group size and predation risk in colonial web-building spiders: analysis of attack-abatement mechanisms. *Behav. Ecol.* **5**: 326–333.

West, M. J. and R. D. Alexander. 1963. Sub-social behavior in a burrowing cricket *Anurogryllus muticus* (DeGeer) (Orthoptera: Gryllidae). *Ohio J. Sci.* **63**: 19–24.

Wilson, E. O. 1971. *The Insect Societies.* Cambridge, Mass.: Harvard University Press.

Wood, T. K., S. I. Guttman, and M. Taylor. 1984. Mating behavior of *Platycotis vittata*. *Am. Midl. Nat.* **112**: 305–313.

3 · Postovulation parental investment and parental care in cockroaches

CHRISTINE A. NALEPA AND WILLIAM J. BELL

ABSTRACT

Cockroaches show the entire range of reproductive modes: oviparous, ovoviviparous, viviparous, and intermediate stages. Postparturition parental care is likewise diverse, ranging from species in which females remain with neonates for a few hours, to biparental care that lasts several years and includes feeding the offspring on bodily fluids in a nest. Both ovoviviparity and parental care arose a number of times in the taxon. Evolution of reproductive mode seems most influenced by predators, parasites and cannibalism. Ovoviviparity, aggregation behavior of young nymphs, and diet are suggested as factors influential in the evolution of postparturition parental care. Females regulate parental investment via absorption of oocytes, abortion, cannibalism and brood reduction. The developmental status of cockroaches at hatching ranges along an altricial–precocial spectrum and is correlated with the presence and type of parental care. In several subsocial species neonates are blind, poorly sclerotized, and dependent for food, while in the sole viviparous cockroach nymphs hatch in an advanced state of development and require fewer molts to adulthood than any known cockroach. Association with microorganisms in both the digestive system and the fat body is suggested as one factor influential in the reproductive versatility of cockroaches. In particular, the endosymbiont flavobacteria which mediate the storage and recycling of nitrogenous waste products may allow for the variety of modes of postovulation provisioning of offspring.

INTRODUCTION

Cockroaches have achieved perhaps their widest range of adaptive diversity in the ways that females care for their eggs after fertilization (Roth and Willis 1954a; McKittrick 1964; Breed 1983). This variety is manifested in various combinations of the following components: production of egg cases (oothecae), preparation of burrows and oothecal deposition sites, concealment of oothecae, defense of oothecae, protection and provisioning of embryos with food, water, or both prior to parturition (ovovivipary, vivipary), protection and provisioning of young after parturition, care of offspring after nutritional independence, and care of young without provisioning (brooding, defense). Hinton (1981) considered cockroaches to be by far the largest group of insects that exhibit parental care, because he considered both ovoviviparity and viviparity as brood care, as did Hagan (in Roth and Willis 1955a). Their view is supported if parental care is defined in the broad sense, i.e. any form of parental behavior that promotes the survival, growth and development of immatures, including the care of eggs or young inside or outside the parent's body and the provisioning of young before or after birth (Tallamy and Wood 1986; Clutton-Brock 1991). A variety of cockroaches, however, also exhibit parental care in the narrow sense, i.e. care of juveniles after hatching by one or both parents.

Reproductive mode and parental care are so thoroughly enmeshed in cockroaches that both must be considered in analyzing parental investment. In the majority of cockroaches, reproduction is characterized by the formation of an ootheca: eggs are released from the ovaries, oriented into two rows by the ovipositor valves, then surrounded by a protective covering. Three general reproductive categories are recognized (based on Roth and Willis 1954a, 1957; Roth 1991).

Oviparous. This includes all families except Blaberidae and some Blattellidae. After ovulation, the ootheca is carried externally by the female for various periods of time (usually 24 hours or more) prior to deposition. Oviparous cockroaches are subdivided into two types: (a) those that drop the egg case shortly after formation (e.g., *Blatta orientalis*, *Eurycotis floridana*, *Periplaneta americana*, *Supella longipalpa*); and (b) those that carry the ootheca externally throughout embryonic gestation, then drop it immediately prior to hatch. Eggs may also hatch while the ootheca is attached to the mother (e.g., *Blattella germanica*, *B. vaga*). In these latter taxa, the anterior end of the egg case is permeable, allowing for transport of water and perhaps other materials from the female to the developing eggs.

Ovoviviparous. This category includes nearly all Blaberidae and two genera of Blattellidae (Roth 1989). In 'false' ovoviviparity, the ootheca is first extruded, as in oviparous taxa, but is retracted a short time later into a uterus (brood sac). The eggs develop inside the body of the mother and have sufficient yolk to complete development. Embryos receive water and possibly some nutrients from the female. When the nymphs are ready to hatch, the ootheca is fully extruded and the neonates emerge from their embryonic membranes and the ootheca (as in, for example, *Blaberus craniifer*, *Byrsotria fumigata*, *Rhyparobia* (=*Leucophaea*) *maderae*, *Nauphoeta cinerea*, *Sliferia lineaticollis*, *Stayella* spp.). In general, these cockroaches produce a thin, soft, lightly colored ootheca, which in some species only partly covers the eggs (Roth 1968a). In 'true' ovoviviparous species, the eggs are deposited directly from the oviducts into the brood sac. The ootheca is completely absent and the eggs occur in an unorganized mass, rather than in the two rows typical of oviparous and false ovoviviparous cockroach oothecae. True ovoviviparous cockroaches include the blaberid genera *Macropanesthia*, *Geoscapheus*, *Neogeoscapheus* and *Parapanesthia* (Rugg and Rose 1984a,b).

Viviparous. In viviparous forms, oviposition is similar to that of the false ovoviviparous cockroaches but the embryos are nourished within the brood sac on a proteinaceous fluid secreted by the mother. The only known example is the blaberid *Diploptera punctata*, but viviparity probably occurs in other species in this genus as well (Roth 1991).

Table 3-1 gives the taxonomic position of all cockroaches discussed in this chapter. Ovoviviparity arose at least three times within the Blaberoidea (Fig. 3-1); the major transitional stages are currently present within the group (Roth 1970, 1989). The Polyphagidae are all oviparous. In the Blattellidae, some species retain the egg case externally for the entire period of gestation, and ovoviviparity arose independently in two different subfamilies. With one exception, all Blaberidae are ovoviviparous, suggesting that they have radiated since an ancestor acquired the trait. The sole viviparous genus is in the Blaberidae. All Blattoidea are oviparous, dropping the egg case shortly after its formation.

We begin by briefly describing features of postovulation parental investment in each reproductive mode. Possible steps in the evolution of internal incubation of eggs are then suggested as effects of selection with respect to egg mortality, maternal foraging, and vulnerability to predation. Postparturition parental care is summarized and discussed in relation to predation pressure and relevant aspects of cockroach nutritional ecology and habitat. The distribution of maternal resources as reflected in the size and number of offspring is discussed, as well as mechanisms by which cockroaches adjust parental investment in relation to risks such as predation or starvation. We present a phylogenetic analysis of internal egg incubation and postparturition parental care, and describe patterns of postnatal development in relation to an precocial-altricial spectrum. Finally we address the role of microbial associations in the reproductive versatility of cockroaches. Preovulation parental contributions of males and females are reviewed by Bell and Adiyodi (1981), Schal *et al.* (1984) and Mullins *et al.* (1992).

REPRODUCTIVE MODES

Oviparity

Postoviposition parental investment in oviparous cockroaches includes production of a hardened egg case, selection of an appropriate microhabitat for deposition of the egg case, preparation of the site, concealment of the ootheca, agonistic behavior toward egg parasites and predators, and in species that carry their egg case until hatch, the provisioning of embryos with water and possibly nutrients. In most oviparous species, no parental care is shown after hatching. Females that deposit the egg case shortly after its formation depart before neonates emerge and may produce several more egg cases before the first hatches.

Oothecae

Oviparous species that abandon their egg cases produce a hard, dark brown ootheca that completely encloses the eggs (Stay and Roth 1962). During its formation, the ootheca presses against the valves of the ovipositor, producing a dorsal keel (crista); the nymphs force open the ootheca along this line of weakness (as in the opening of a handbag) at the time of hatching. The ootheca is structurally sophisticated (Lawson 1951; D. E. Mullins and J. Mullins, personal communication) and functions in gas exchange, water balance, and mechanical protection; it may also play a role in the suppression of pathogens and the deterrence of predators.

Table 3-1. *Classification of cockroaches (Blattaria) discussed in this chapter*

Family Cryptocercidae
    Subfamily Cryptocercinae
        Cryptocercus punctulatus
Family Blattidae
    Subfamily Blattinae
        Blatta orientalis, Periplaneta americana, P. brunnea
    Subfamily Polyzosteriinae
        Eurycotis floridana
Family Polyphagidae
    Subfamily Polyphaginae
        Therea (=Corydia) petiveriana
Family Blattellidae
    Subfamily Blattellinae
        Blattella germanica, B. vaga, Paratemnopteryx (=Shawella) couloniana, Parcoblatta virginica, Stayella spp.,
        Xestoblatta hamata
    Subfamily Ectobiinae
        Ectobius pallidus
    Subfamily Pseudophyllodromiinae (=Plecopterinae)
        Sliferia lineaticollis, Supella longipalpa (=supellectilium)
Family Blaberidae
    Subfamily Blaberinae
        Blaberus craniifer, Byrsotria fumigata
    Subfamily Diplopterinae
        Diploptera punctata
    Subfamily Epilamprinae
        Phlebonotus pallens, Pseudophoraspis nebulosa, Thorax porcellana
    Subfamily Geoscaphinae
        Geoscapheus spp., *Macropanesthia* spp., *Neogeoscapheus* spp., *Parapanesthia* spp.
    Subfamily Oxyhaloinae
        Gromphadorhina portentosa, G. laevigata, Nauphoeta cinerea, Rhyparobia (=Leucophaea) maderae
    Subfamily Panchlorinae
        Panchlora nivea, P. irrorata
    Subfamily Panesthiinae
        Panesthia angustipennis, P. australis, P. cribrata (=laevicollis), P. sloanei, Salganea esakii, S. gressiti, S. raggei,
        S. taiwanensis
    Subfamily Perisphaeriinae
        Aptera fusca (=cingulata), Perisphaerus spp., *Poeciloblatta* spp., *Trichoblatta sericea*
    Subfamily Pycnoscelinae
        Pycnoscelus surinamensis
    Subfamily Zetoborinae
        Thanatophyllum akinetum

Source: Based on McKittrick 1964; Rugg and Rose 1984b; L. M. Roth, personal communication; but see Grandcolas and Deleporte 1993.

The egg case consists primarily of quinone-tanned protein (Brunet and Kent 1955) and can represent a substantial protein sink for the female. The ootheca of *Periplaneta americana*, the most-studied species, consists of 20 mg of protein (Campbell 1929; Brunet 1951). Much of this investment, however, is recoverable. After hatch, either the neonates or the older nymphs and adults eat the embryonic membranes, inviable eggs and oothecal case (Roth and

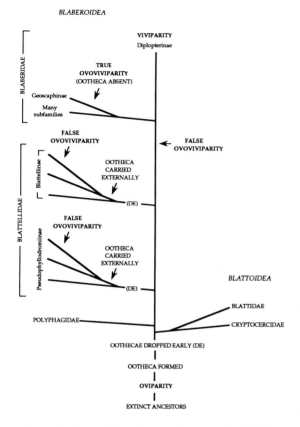

BLABEROIDEA

VIVIPARITY
Diplopterinae

TRUE
OVOVIVIPARITY
(OOTHECA ABSENT)

Geoscaphinae
Many
subfamilies

BLABERIDAE

FALSE
OVOVIVIPARITY

FALSE
OVOVIVIPARITY

Blattellinae

OOTHECA
CARRIED
EXTERNALLY

(DE)

BLATTELLIDAE

FALSE
OVOVIVIPARITY

Pseudophyllodromiinae

OOTHECA
CARRIED
EXTERNALLY

(DE)

BLATTOIDEA

BLATTIDAE

POLYPHAGIDAE

CRYPTOCERCIDAE

OOTHECAE DROPPED EARLY (DE)

OOTHECA FORMED

OVIPARITY

EXTINCT ANCESTORS

Fig. 3-1. Possible evolution of reproductive mode. Modified from Roth (1989).

Willis 1954a; Willis *et al.* 1958). It is estimated that female *Cryptocercus punctulatus* recover up to 59% of the nitrogen invested in a brood of eggs if they eat the empty egg cases (Nalepa and Mullins 1992). In several species of cockroach, oothecal predation by adults and the ingestion of the oothecal case after hatching by nymphs increase when other protein sources are lacking (W. J. Bell, unpublished). The degree to which this nitrogen is assimilated is unknown.

Deposition of oothecae
The time and energy invested in searching for or producing a patch of habitat suitable for oviposition is one form of parental effort (Lessells 1991). Although oviparous cockroaches differ in the manner in which they deposit their oothecae, the majority select and prepare the deposition site with some care (reviewed in Roth and Willis 1960; McKittrick *et al.* 1961; McKittrick 1964; Roth 1991). *Therea* (=*Corydia*) *petiveriana* simply deposits oothecae freely at

random in dry leaves (Ananthasubramanian and Ananthakrishnan 1959). Other species (e.g. *Supella longipalpa, Blatta orientalis*) 'glue' them (with saliva) to the substrate and many find or construct a crevice, glue the ootheca in a precise position inside it, then conceal the egg case with bits of debris, pieces of the substrate or excrement (e.g. *Eurycotis floridana, Ectobius pallidus, Periplaneta americana*). The material used to cover the egg case is often carefully selected; for example, in sand, small grains are discarded in favor of large grains (Rau 1943). The deposition and concealment process can take several hours; the stereotyped behavioral sequences involved may be of taxonomic value (McKittrick 1964).

The oothecae of oviparous cockroaches vary in their ability to prevent water loss from the eggs (Roth and Willis 1955b). In some species the ootheca and eggs at oviposition do not contain sufficient moisture for embryogenesis; the ootheca must be deposited in a humid or moist environment where the eggs absorb water (e.g. in *Ectobius pallidus, Parcoblatta virginica*). Alternatively, if the ootheca and eggs contain sufficient moisture for the needs of the embryos at the time of oviposition, the ootheca possesses a protective layer which retards water loss (e.g. in *Blatta orientalis, Periplaneta americana, Supella longipalpa*) (Roth and Willis 1955b, 1957).

External egg retention

Blattella germanica and some other Blattellidae may be the best models for an intermediate stage between oviparous and ovoviviparous forms (Roth and Willis 1957); egg cases are carried externally clasped in the genital cavity for the entire period of embryogenesis. The anterior end of the egg case is permeable, allowing for the transport of water from the vestibular tissue of the female to the developing eggs (Roth and Willis 1955b; Willis *et al.* 1958). Recent work suggests that the parent–offspring relationship in *B. germanica* is more elaborate than the unidirectional movement of water. Experiments with isotopes indicate that water movement may be bidirectional, and that small molecules may be transferred from the mother to the embryos. If a female is injected with radiolabeled leucine and glucose, the label can be detected in her nymphs after hatch (D. E. Mullins and K. R. Tignor, personal communication). There is also some evidence of communication between female *B. germanica* and the egg case she is carrying. In most instances, hatching of the egg case is initiated while it is still attached to the mother. The activity level of the female increases significantly prior to hatch, indicating either that she can detect impending hatch, or that her increased activity level initiates it (D. E. Mullins and K. R. Tignor, personal communication).

Except for retraction of the egg case into the body, *Blattella germanica* exhibits all of the characteristics of an ovoviviparous cockroach (Roth 1970; C. Schal, personal communication). The oothecal case is thinner and less darkly colored than in other oviparous cockroaches (Stay and Roth 1962), there is exchange of materials between mother and unhatched offspring, and oogenesis is suspended while females are carrying egg cases. The evolution of ovoviviparity would require only a minor transition from this starting point; the presence of the ovoviviparous genus *Stayella* in the same subfamily as *Blattella* supports this hypothesis.

Ovoviviparity

The final step to ovoviviparity from *Blattella germanica*-like ancestors was the internalization of the egg case. There are no known transitional stages among extant species, such as species with partial retraction of the ootheca into the brood sac. Retraction of the egg case into the body of the female required changes in the morphology and physiology of both the mother and the embryo. Modifications of the mother include the appearance of a brood sac and a decrease in the size of the colleterial glands in conjunction with the reduction of the egg case. The brood sac in ovoviviparous and viviparous cockroaches is large compared to the rest of the reproductive tract and is capable of enormous distension during gestation. It is, however, simply an elaboration of the membrane found below the laterosternal shelf in oviparous cockroaches (McKittrick 1964).

Among species, the colleterial glands of cockroaches vary in size in accordance with the size of the oothecal case (Stay and Roth 1962). During the evolution of ovoviviparity in the Blaberoidea, the ootheca changed from a hard rigid structure, completely enclosing the eggs, to a flexible, transparent, often incomplete membrane (Roth and Willis 1957; Roth 1968a). Although it has been suggested that an incomplete ootheca is important for respiration of the embryos (Nutting 1953), it is unclear whether there is gas exchange between the mother and embryos or whether ambient air drawn into the brood sac is sufficient for embryonic respiration. Embryos in both ovoviviparous and viviparous cockroaches have elongated pleuropodia (Roth and Willis 1957), which are long hollow appendages on the first abdominal segment extending outside the

serosal cuticle and lying beneath the chorion. These are thought to maintain ion and water balance between the embryo and the extraembryonic fluid (Stay 1977).

It is generally believed that ovoviviparous females provide only water and protection to the embryos they carry, the main source of energy and nutients being the yolk in the egg. In ovoviviparous *Rhyparobia maderae* and *Nauphoeta cinerea*, as well as oviparous *Periplaneta americana*, water content increases and dry mass decreases during embryogenesis (Roth and Willis 1955b). Given that transfer of materials may occur even in oviparous *Blattella germanica*, ovoviviparous cockroaches may be supplying more than water to their retained embryos, even though it is not reflected as mass gain. As pointed out by G. L. Holbrook (personal communication), ovoviviparous and *B. germanica* embryos may not lose as much dry mass as oviparous cockroaches during embryogenesis. Based on morphological evidence, Snart *et al.* (1984a,b) suggest that *Byrsotria fumigata* and *Gromphadorhina portentosa*, two Blaberidae commonly considered ovoviviparous, should in fact be classified as viviparous. The surface of the brood sac in these two cockroaches is covered with numerous closely-packed papillae. Pores in the apical region of each papilla exude material thought to result from secretory activity of the brood sac, and the brood sac wall has ultrastructural features characteristic of insect integumentary glands. These authors suggest that the brood sac in these two 'ovoviviparous' cockroaches is sufficiently similar to that of the viviparous *Diploptera punctata* to make it likely that the brood sacs of all three function in the same manner. Depriving *B. fumigata* and *G. portentosa* of food and water resulted in smaller nymphs (Snart *et al.* 1984a,b), but the relative effects of food and water deprivation are unknown.

Perhaps the main reason that the ootheca has not been selectively eliminated in most ovoviviparous cockroaches is because it determines the orderly arrangement of eggs and therefore assures contact and exchange of materials of each egg with the wall of the brood sac. A study of gestation, embryo growth and brood sac morphology in the Geoscaphinae, which do not form oothecae and whose eggs are incubated in a jumbled mass in the uterus (Rugg and Rose 1984a), would be especially instructive in this regard.

Viviparity

In the viviparous species *Diploptera punctata*, embryos receive nourishment directly from their mother in addition to that present in yolk. The brood sac is both a protective chamber and a 'mammary gland'; its epithelium is glandular and produces a proteinaceous secretion at a rate which parallels the growth rate of the embryos. The milk consists of approximately 45% protein, 5% free amino acids, 25% carbohydrate, and 16–22% lipids. Embryos begin orally ingesting milk just after closure of their dorsal body wall (at 19% of gestation) and continue until shortly before parturition. The ultimate source of nutrition for the embryos is the food intake of the mother; females normally double their body mass during gestation. At birth, first instars are 50 times heavier and contain 60 times more protein than the eggs at oviposition (Stay and Coop 1973, 1974; Ingram *et al.* 1977).

One consequence of viviparity is that at parturition, juveniles of *Diploptera punctata* are in an advanced state of development, requiring fewer molts to adulthood than any known cockroach. The female has three or four postembryonic instars compared with the usual 7–13 in a sample of 11 other species of Blattaria (Willis *et al.* 1958). The small number of instars in *D. punctata* is associated with their large size at parturition. *D. punctata* is at least twice the size of *Nauphoeta cinerea* at the time of hatching (see Fig. 3 in Roth and Hahn 1964), yet adults of *N. cinerea* are considerably larger than those of field-collected *D. punctata* (approximately 27 mm and 17 mm in length, respectively) (Cochran 1983a; W. J. Bell, unpublished data). Because there is little information available on *D. punctata* in its natural habitat (Hawaii and other Pacific Islands) (Roth and Willis 1955a), the selective pressures that resulted in the developmental 'jump-start' that viviparity confers on this species are unknown.

Evolution of reproductive mode

An initial step in the evolution of ovoviviparity was likely to be facultative transport of the egg case, as in the oviparous cockroaches that retain oothecae until a suitable microhabitat is found. For example, in *Therea petiveriana*, oothecae are generally deposited the same day following extrusion, but they may be retained for as long as 90 hours if a suitably moist substratum is not available (Livingstone and Ramani 1978; see Roth and Willis 1957 for additional examples). From that starting point, the trend toward ovoviviparity would be exemplified by cockroaches that retain the egg case for the entire period of embryogenesis, but provide no materials additional to those originally in the egg case; currently, there are no records of extant cockroaches that exhibit this pattern. A possible explanation for the lack

of this transitional stage is that the act of retaining the egg case may directly facilitate exchange of material between mother and embryos. Quinone tanning of the cockroach ootheca is an oxidation process (Sugumaran *et al.* 1991); if the ootheca is kept closely pressed to the female's vestibular tissues, the tanning process may be impeded by lack of oxygen.

The most information about possible origins of ovoviviparity in cockroaches would be gained by a study of the Blattellidae. This family contains oviparous species that drop the egg case shortly after its formation, species that externally retain the egg case throughout gestation, and two independent origins of internalization of eggs. For clues to the origins of viviparity, the Blaberidae are the logical focal group. As more data are gathered, we may find that there is a continuum between ovoviviparity and viviparity, ranging from water exchange at one end, to the extreme represented by *D. punctata* at the other. Worldwide, Blattellidae is the largest cockroach family with about 1740 species; there are approximately 1020 species of Blaberidae. The oviposition behavior of relatively few genera and species in these two families is known (Roth 1982).

Egg mortality

Most hypotheses that have been proposed to explain why live bearing has evolved invoke agents affecting embryonic viability as the selective forces for an evolutionary shift in reproductive mode. Factors affecting the female's costs of egg retention then either facilitate or constrain the transition (Shine 1985, 1989).

In cockroaches, ovoviviparity is thought to have first appeared in the Mesozoic as a evolutionary response to predators and parasites that first appeared during that era (Vishniakova 1968, in Roth 1989). The only time that the eggs of ovoviviparous cockroaches are exposed to mortality factors such as predators, pathogens, parasites and cannibals is when they are briefly extruded prior to retraction of the egg case. A number of Australian burrowing cockroaches have even eliminated that step by depositing eggs directly into the brood sac (Rugg and Rose 1984a). Once in this enemy-free space, the eggs will be eaten only if the mother herself is the prey item (Breed 1983). Nymphs of ovoviviparous cockroaches are, however, subject to cannibalism at the time of hatch. Attempts by conspecifics to eat the hatching nymphs as the female ejects the ootheca have been noted and may include pulling the still attached egg case away from the mother (Willis *et al.* 1958).

Besides protecting them from biotic factors, the body of an ovoviviparous female buffers her embryos from physical extremes, such as heat, cold, moisture, desiccation, anoxia and osmotic stress. Regardless of the sources of embryonic mortality, one potential advantage to egg retention is moderation of fluctuations in temperature or moisture that may affect embryonic development rate. Ovoviviparous females, as well as those that carry egg cases externally until hatching, may also behaviorally moderate the microenvironment of their embryonic cargo by seeking the most benign locations within their habitat. Nymphs of *B. germanica* are known to settle in microhabitats where temperatures are favorable to their development (Ross and Mullins 1995); it is probable that a female carrying an egg case acts similarly on behalf of her embryos.

Costs and benefits

Species in which the survivorship or food intake of the reproducing female would not be markedly affected if she retained eggs may be most likely to evolve internal incubation (Shine 1985). In reptiles, this includes species in which females are large or venomous, because gravid females of these forms are less vulnerable to predation. In small females, or in those that rely on mobility to escape predators or to forage widely for food, the physical burden of gestation might lower female survivorship (Shine 1985).

Although little comparative information on body size in cockroaches has been compiled, external or internal retention of eggs does not appear to be correlated with a large body size among the Blattellidae. Ovoviviparous *Sliferia* spp. are only 9–10 mm long (Roth 1989); adult *Blattella germanica* are 10–15 mm long (Cochran 1983a) and the egg case they carry throughout gestation is large (8 mm long) relative to their body size (Tanaka 1976). The use of noxious sprays is not a correlate of live bearing in cockroaches; defensive glands are found in a wide variety of both oviparous and ovoviviparous cockroaches (Roth and Alsop 1978).

It is unknown whether gravid ovoviviparous and egg-retaining cockroaches lose the running speed or agility that enables escape from predator attack, and there are no data on whether the physical burden of an egg clutch affects flying in those species that depend on it for evasion. Loss of mobility may not be an issue in cockroaches that rely on crypsis, immobility or shelter for escape, but in the latter the larger body dimensions of gravid females require a larger crevice (Wille 1920; Koehler *et al.* 1994).

Female cockroaches carrying eggs may be more impaired in their locomotory ability to forage than if they

had laid the eggs, but most feed little, if at all, during gestation, even when offered food *ad libitum* in the laboratory (e.g. *Blattella*, Cochran 1983b; Hamilton and Schal 1988; *Parcoblatta*, Cochran 1986; *Rhyparobia*, Engelmann and Rau 1965; *Trichoblatta*, Reuben 1988). The fat bodies of pregnant *Rhyparobia maderae* resemble those of starved animals (Engelmann and Rau 1965). Rather than exposing themselves to predators during gestation by foraging, they feed in cycles. An ootheca in the brood sac of *R. maderae* suppresses feeding, but when eggs are being produced, the female eats more than usual (Engelmann and Rau 1965). Unlike ovoviviparous females, gravid female *Diploptera punctata* feed; the nutrient secretion of the brood sac is normally derived from the maternal diet during the gestation period (Stay and Coop 1974). A female's behavioral response to her gravid condition may thus evolve in response to the costs and benefits of foraging. It might be argued that the evolution of viviparity should be favored by the presence of predictable, high-quality foods to support the considerable growth of embryos *in utero*. The diet of *D. punctata*, however, appears little different from that of many other cockroaches; it includes leaf litter, particularly leaves with mold or algae on the surface (W. J. Bell, personal observation), fruit, and the bark of living trees (Roth and Willis 1960; Roth 1979).

One benefit of ovoviviparity may be the loss of one or more of the costs of being oviparous (Shine 1985). In cockroaches, internal incubation permits females to dispense with producing a thick, protective oothecal case, and allows them to channel the protein that would have been required for its manufacture into present or future offspring or their own maintenance.

The retention of egg cases in cockroaches exacts a cost in terms of fecundity. Oviparous species that carry their ootheca externally for the entire embryogenetic period as well as ovoviviparous and viviparous females produce relatively few oothecae because the oocytes do not mature in the ovaries during the period an ootheca is being carried (Roth and Stay 1959, 1962). The number of egg cases per lifetime decreases and oviposition interval increases in the order oviparous, ovoviviparous, viviparous (see Fig. 21 in Roth 1970).

POSTEMBRYONIC CARE OF YOUNG

The first prerequisite for postembryonic parental care is that the female be in the vicinity of the hatching eggs. This requirement automatically excludes the majority of

Table 3-2. *Postparturition association of adults and nymphs*

Species	Brooding period (A) or description (B)	Source
(A) Short term maternal associations (brooding)		
Nauphoeta cinerea	an hour	Willis *et al.* 1958
Blaberus craniifer	an hour or more	Nutting 1953
Gromphadorhina portentosa	for some time	Roth and Willis 1960
G. laevigata		
Thanatophyllum akinetum	several hours	Grandcolas 1993a
(B) Parental care		
Byrsotria fumigata	with female first 15 days after birth	Liechti and Bell 1975
Panesthia australis? (see text)	in wood galleries	Matsumoto 1988; Shaw 1925
Salganea esakii	in wood galleries; biparental families	Matsumoto 1987
S. raggei	in wood galleries	T. Matsumoto, personal communication
S. gressiti	in wood galleries	
Poeciloblatta sp.	under stones	Scott 1929
Aptera fusca	under bark, stones	Skaife 1979
Perisphaerus armadillo	female surrounded by young	Karny 1924; in Roth and Willis 1960

oviparous cockroaches from subsociality, since females are no longer present when the nymphs hatch. The two exceptions are the genus *Cryptocercus*, which oviposits in a nest within rotting wood, and a brief parent–offspring association in *Blattella vaga*, which retains the ootheca externally until hatching. In ovoviviparous cockroaches, the egg cases are extruded from the female's body immediately prior to hatching; this provides a basis for the elaboration of subsocial behavior by putting the neonates in immediate contact with their mother.

Brooding

The simplest type of parental care shown by cockroaches after parturition is brooding, defined here as a short-term association of mother and neonates. In a number of blaberid species, the young cluster under, around and sometimes on the female for varying periods of time after hatching. Most associations last less than a day (Table 3-2A). Records of brooding behavior are based primarily on laboratory observations; Grandcolas (1993a), however, observed hatching of *Thanatophyllum akinetum* in the field. Nymphs aggregated beneath the mother's body for several hours before dispersing. Brooding after hatch may be more prevalent among the blaberids than the literature indicates, but because it is a brief association, the behavior may be easily overlooked.

The slight increase in parental investment required of a female that remains with first-instar nymphs may confer handsome benefits in relation to its cost. It takes several hours for the cuticle of neonates to harden, and soft, unpigmented nymphs are at risk from ants and cannibalism (Eickwort 1981).

Parental care

Table 3-2B lists the species in which there is evidence that females remain with offspring for more than a few hours. Many of these records are based on field observations, i.e. reports of females collected together with their young, often from sheltered situations such as under bark, within logs, or in soil pockets under stones. These cockroaches should probably be considered subsocial; nymphs in some cases were described as partly grown. There is, however, little information with which to evaluate their social status. Roth and Willis (1960) note a few anecdotal accounts additional to those in Table 3-2B.

Although there is no evidence of interaction beyond brooding in *Byrsotria fumigata*, we consider them subsocial because the parent–offspring association lasts longer than a few hours and because kin recognition has been demonstrated (Liechti and Bell 1975). After parturition, the female raises her body to allow nymphs access to the space

Table 3-3. *Postparturition parental care with a nutritional component.*

Genus	Subfamily	Offspring Location	Food	Source
(A) Feeding young on body fluids				
Perisphaerus sp.	Perisphaeriinae	Clings ventrally	Hemolymph?	1
Trichoblatta sericea	Perisphaeriinae	Clings ventrally	Sternal exudate	2
Pseudophoraspis nebulosa	Epilamprinae	Clings ventrally	?	3
Thorax porcellana	Epilamprinae	Under tegmina	Tergal exudate	2
Phlebonotus pallens	Epilamprinae	Under tegmina	?	3, 4
Salganea taiwanensis[a]	Panesthiinae	Wood galleries	Stomodeal fluids	5
Cryptocercus punctulatus[a,b]	Cryptocercinae	Wood galleries	Hindgut fluids	6, 7
Blattella vaga[b,c]	Blattellinae	Under tegmina	Tergal exudate	8
(B) Progressive provisioning or joint foraging				
Rhyparobia maderae	Oxyhaloinae	Under female	?	9, 10
Macropanesthia sp., *Parapanesthia* sp., *Neogeoscapheus* sp., *Geoscapheus* sp.	Geoscaphinae	Underground burrows	Leaf litter	11, 12

[a] Biparental families.
[b] Oviparous.
[c] Brief association.
Source: 1, Roth (1981); 2, Reuben (1988); 3. Shelford (1906); 4, Pruthi (1933); 5, T. Matsumoto and Y. Obata, personal communication; 6, Seelinger and Seelinger (1983); 7, Nalepa (1984); 8. Roth and Willis (1954a); 9, Séin (1923); 10, Wolcott (1948); 11, Rugg and Rose (1991); 12, H. A. Rose, personal communication.

beneath; first instars can recognize their own mother and prefer to aggregate beneath her for the first 15 days after hatch. Nymphs did not aggregate beneath their mother if they could not touch her, suggesting that contact chemoreception is involved. The female facilitates aggregation of the nymphs beneath her by remaining motionless.

Two different social structures have been reported for Australian wood-feeding panesthiines: family groups and aggregations that contain more than two adults. According to Shaw (1925), both *Panesthia australis* and *P. cribrata* (= *laevicollis*; Roth 1977) live in family groups consisting of a pair of adults and nymphs in various stages of development. Matsumoto's (1988) study of *P. australis* supports the notion of subsociality in this species. Of 29 social groups collected, 14 consisted of a female with nymphs, two were a male with nymphs, and two were an adult pair with nymphs. In his collections, the social group never contained more than a single adult of either sex, or an adult pair together with nymphs. Because the age of nymphs in

the group ranged widely, however, it is possible that these nymphs were aggregated individuals rather than a sibling group (T. Matsumoto, personal communication). The field studies of H. A. Rose (personal communication) indicate that neither *P. australis* nor any of the other wood-feeding Australian panesthiines are subsocial. Rugg and Rose (1984c) and O'Neill *et al.* (1987) found that although adult pairs with nymphs could be found in *P. cribrata* (12% of groups), social groups most commonly (29%) consisted of a number of adult females together with a single adult male and a number of nymphs. An intriguing possibility is that social structure in these cockroaches varies with habitat and population density. Multifemale groups seem to be common in areas of high population density, whereas family groups are generally found in marginal environments, or on the outer fringes of areas with high population density (D. Rugg, personal communication).

Table 3-3 lists species in which there is evidence for more elaborate forms of subsocial behavior. Females provide food to neonates, nymphs have been found clinging to females in the field, or there are morphological modifications of the female or nymphs that facilitate parental care. Maternal secretions from a variety of sources are the most prevalent form of food, but provisioning and possibly group foraging also occur. Some females have evolved external brood chambers under their wing covers, and others have the ability to roll into a ball, pill-bug-like, to protect ventrally clinging nymphs. Juveniles are highly active participants in the interactions, which include climbing under maternal wing covers, clinging to dorsal or ventral surfaces, and feeding on offered secretions. Morphological specializations of the juveniles include modifications of their appendages to aid in clinging to the female, and adaptations of their mouthparts to facilitate their unique feeding habits.

In the subfamily Perisphaeriinae there are two recorded cases of nymphs clinging to the ventral surface of the mother for both protection and nutrition. The subsocial relationship in *Perisphaerus* appears elaborate and intriguing but is known only from examinations of museum specimens (Roth 1981). Nymphs of this genus cling to the undersurface of the female for at least two instars. The first-instar nymph is blind and has an elongate head and specialized galeae, which suggest that it takes in liquid food from the mother. There are four distinct orifices on the ventral surface of the female, with one pair occurring between the coxae of both the middle and hind legs. Specimens have been collected with the mouthparts of a nymph inserted into one of these orifices of the mother, and the 'proboscis' of nymphs is 0.3 mm wide, about the same width as the intercoxal opening in the female. The food of the nymphs may be glandular secretions or possibly hemolymph (analogous to the 'larval hemolymph taps' employed by the ant *Leptanilla japonica*) (Masuko 1989). The female can roll up into a ball, protecting the clinging nymphs inside. Males are winged and do not show this behavior. At least nine nymphs are able to remain attached when the female assumes the defensive position (Roth 1981).

First-instar nymphs of *Trichoblatta sericea* cling to the underside of their mother for the first two to three days after hatching; adherence to her abdomen is facilitated by the enormously developed pulvilli and claws of the nymphs. During this period the female secretes a milky fluid from her ventral side. Although nymphs were not observed feeding on this secretion, no other food was taken during this period and nymphs isolated from their mother at hatching did not survive past the second instar (Reuben 1988).

In the subfamily Epilamprinae, females in two genera have an external brood chamber. Their wing covers (tegmina) are large and dome-shaped, and there is a shallow trough-like depression in the dorsal surface of the abdomen, forming a chamber in which the female carries the young. The hind wings of the female are small and rudimentary. In *Thorax porcellana*, oothecae contain 32–40 eggs; neonates scramble into the brood chamber immediately after hatching and remain there during the first and second instars. The protection is so perfect that the nymphs are not visible. If the nymphs are removed from the mother, they die. Nymphs do not feed on adult food, but they increase in size and molt. It is probable that nymphs feed on a pink material secreted from thin membranous areas on the dorsolateral regions of the 4th, 5th, 6th and 7th terga. The mouth parts of first instars are modified with dense setose arrangements on the maxillae and the labium, which may indicate that young nymphs feed on a liquid diet. Their midguts are filled with pink material rather than the leaf chips which they eat when older. The legs are well adapted for clinging, with large pulvilli and claws. Second-instar nymphs leave the protection of the mother for short periods of time to feed on dry leaves, but they do not venture far. Second-instar nymphs can survive, however, if they are removed from the mother. Maternal care lasts for about seven weeks (Reuben 1988).

In a manner similar to *Thorax porcellana*, *Phlebonotus pallens* carries about a dozen nymphs beneath its wing covers (Shelford 1906). The presence of young in the brood chamber does not interfere with the activities of the mother. The photograph in Pruthi (1933) suggests that incubated nymphs are large; the length of a nymph is more than 20% of the length of the adult female.

Biparental care in which adults feed their offspring has evolved independently in at least two distantly related genera of wood-feeding cockroaches, *Cryptocercus* and *Salganea*. *C. punctulatus* is an oviparous cockroach in the family Cryptocercidae that excavates galleries in rotted wood. Field evidence suggests that pairs of *C. punctulatus* have a single reproductive episode during which they produce a mean of 73 eggs, in up to four oothecae. An extended period of brood care, which can last three years or longer, follows and includes defense of the family, gallery excavation, sanitation of the nest and, in the early stages, trophallactic feeding of the young. Nymphs are born without the cellulolytic protozoan symbionts they require to digest their wood diet; consequently, neonates rely on adults for nourishment. Adults apparently provide all of the dietary needs of first-instar nymphs, and some degree of trophallactic feeding of offspring occurs until the hindgut symbiosis is established. Adults feed nymphs on hindgut fluids and specialized fecal pellets, but cuticular secretions may also play a role; nymphs can spend up to 20% of their time grooming adults (i.e. orally grazing their body surface). The young are capable of independence at the third instar, but the family structure is generally maintained until the death of the adults (Cleveland *et al.* 1934; Seelinger and Seelinger 1983; Nalepa 1984, 1990).

Salganea esakii, S. taiwanensis, S. raggei and *S. gressiti* are wood-feeding, ovoviviparous Blaberidae that live in family groups (Matsumoto 1987; T. Matsumoto, personal communication). In *S. taiwanensis*, nymphs cling to the mouthparts of their parents and may take liquids through stomodeal feeding (T. Matsumoto and Y. Obata, personal communication). Nymphs up to the fifth instar can distinguish their parents from conspecific pairs. High mortality results when neonates of *S. taiwanensis* are removed from their parents; removed nymphs that lived had a significantly longer duration of the first instar. These data indicate that although nymphs may not be totally dependent on parents, their survival and development are significantly improved by brood care.

Posthatch parent–offspring interactions have been reported in only one oviparous blattellid. Roth and Willis (1960) observed nymphs of *Blattella vaga* hatching out of the ootheca and then crawling over the body of the mother, who remained nearby. The female raised her wings, allowing nymphs to crawl under them and to feed on material covering her abdomen; after this feeding activity the nymphs scattered.

The remaining cases of parental care involve nymphs that are not fed on parental secretions, but are either taken to the feeding area by the adult, or are served adult food. Anecdotal evidence suggests that nymphs of *Rhyparobia maderae* follow their parent in coveys, like quail. In the words of Wolcott (1948), 'the mother broods over her young, and together they sally forth at night in search of food, until they are of such a size as to mingle with their elders'. First instars of *R. maderae* can recognize siblings; Evans and Breed (1984) suggest that this behavior enhances protection by the mother or the transferral of digestive microorganisms from mother to offspring.

Twenty species of Australian soil-burrowing cockroaches (genera *Macropanesthia, Parapanesthia, Neogeoscapheus* and *Geoscapheus*) live in underground burrows and feed on dead leaves, twigs and berries (H. A. Rose, personal communication). Females reproduce once per year, and provide their offspring with leaf litter collected from the soil surface and dragged down into the burrow. The best studied is the giant burrowing cockroach, *Macropanesthia rhinoceros*, which can reach a weight of over 30 g! The average number of eggs in the brood sac is 23; nymphs remain with females for five or six months and molt seven times before dispersal. Males are occasionally found in the family early in the nesting cycle. Females forage for leaf litter up to 50 cm from the entrance of their burrow (Rugg and Rose 1991; Matsumoto 1992). Because females drag leaves into their burrow whether or not they have offspring present, the progressive provisioning of offspring might be considered coincidental (H. A. Rose, personal communication). However, females with young are more aggressive (to adults of either sex) than those without young, and access to their mother's stash of food allows juveniles to delay the risks associated with dispersal and the construction of their own burrow.

The variety of subsocial interactions in the cockroaches detailed here is rare among arthropods. Unfortunately, because so much available information is incomplete or anecdotal, severe limits are imposed on our understanding of the selective pressures that resulted in this variation. The current state of knowledge, however, may also be viewed as opportunity; the taxon is rife with potential research projects. Specific topics that require attention range from the functional significance of brooding, to the characterization of social structure in Australian panesthiines in relation to habitat. It has been suggested that one alternative to paternal investment is maternal care characterized by physical transport of the young (Tallamy 1994); the idea may be explored using any of several cockroach species in which nymphs cling to the female. Studies of cockroaches that feed their offspring on bodily secretions (Table 3-3A) may increase our understanding of termite eusociality (Nalepa 1994). Surely the study of parental care in soil-burrowing cockroaches the size of small mammals is an appealing prospect regardless of the hypotheses being tested.

Benefits of parental care

Protection

There is an element of protection involved in all cases of postparturition parental care. Two features of parental defense deserve further treatment: mechanisms employed to protect eggs, and parental care that occurs within nests.

Protection of eggs

The tough layer that surrounds a clutch of cockroach eggs may protect them from both biotic threats and the effects of the physical environment. Schal *et al.* (1984) suggested that the calcium oxalate found in the egg cases of a variety of oviparous species deters predators, as these crystals do in tissues of aeroid plants and members of the genus *Oxalis*. Calcium oxalate occurs in the egg cases of most oviparous species; it is scarce in roaches that carry the egg case externally, and absent in those species that carry their eggs internally (Roth 1968a). These crystals can account for 15% of the dry mass of the ootheca in *Periplaneta americana*, but only 0.3–0.4% in *Blattella germanica*. Although it has been suggested that calcium oxalate in egg cases functions as an antibiotic (Stay *et al.* 1960), there is currently no experimental support for its role in the susceptibility of egg cases to pathogens. Fungi may be important sources of mortality in a damp environment such as the wood galleries of *Cryptocercus*, but oothecae of this cockroach contain no calcium oxalate (Roth 1968a). *C. punctulatus*, however, embeds its oothecae in the roof of galleries and tends to them by keeping the area actively cleared of the fungi and feces found in other parts of the gallery system (Nalepa 1987).

McKittrick (1964) suggested that destruction of oothecae by parasitism and cannibalism are the selective pressures favoring their concealment by oviparous cockroaches. Although few studies directly address this question, the evidence suggests that concealing oothecae attracts rather than discourages hymenopterous parasitoids. The body fluids that oviparous cockroaches use to attach egg cases to the substrate may act as kairomones, making them more vulnerable to attack (Narasimham 1984; Benson and Huber 1989). Parasitoids are even known to dig in sand to expose buried oothecae (Roth and Willis 1954b). Carrying the egg cases may be no better than concealment in conferring protection from parasitism; *Anastus floridanus* (Eupelmidae) will oviposit in egg cases that are still attached to female *Eurycotis floridana* (Fig. 9 in Roth and Willis 1954b). The cockroach can detect the presence of the wasp on the surface of the ootheca and tries to dislodge it with her hind legs (L. M. Roth, personal communication). *Blattella* spp. that carry their egg cases externally until hatching are also vulnerable to egg parasitoids (Roth 1985). Parasitism, however, has not been detected in the oothecae of ovoviparous blaberids (L. M. Roth, personal communication).

The value of concealing egg cases seems to lie in protecting them from predation and cannibalism. Concealment is almost 100% effective in saving oothecae from being devoured by other cockroaches (Rau 1940). McKittrick *et al.* (1961) found that the burying of oothecae by *Eurycotis floridana* prevented cannibalism by conspecifics and predation by ants, carabids, rodents and other predators. Exposed egg cases, and those still attached to a female, are subject to biting and cannibalism (Roth and Willis 1954a; Willis *et al.* 1958; Gorton 1979). These improprieties are countered with aggression on the part of the mother. For example, female *Periplaneta brunnea*, *P. americana* and *Paratemnopteryx* (=*Shawella*) *coulloniana* drive other females away from exposed oothecae (Haber 1920; Edmunds 1957; Gorton 1979). In *Blattella germanica*, two behavioral classes of female could be distinguished on the basis of the ovarian cycle (Breed *et al.* 1975): females carrying oothecae were more aggressive than females that had not yet formed them.

Often the protection of oothecae from sources of mortality may be an 'all or nothing' affair. If some of the eggs in an egg case are lost through inviability, parasitism or disease, the entire brood may fail to hatch. In most oviparous cockroaches, hatching requires a group effort by nymphs to force open the 'spring-loaded' oothecal keel at the time of hatch (Keil and Ross 1977). They swallow air, become inflated and force the separation of the two halves of the keel. Similarly, if the keel of an ootheca is merely nibbled by a predator or cannibal, the entire brood may be lost via desiccation (Roth and Willis 1955b).

Nests

In the wood-feeding cockroaches and the Australian soil-burrowers, the young are protected within excavated chambers. The structures offer some protection from natural enemies and act as buffers against temperature and moisture fluctuations, and one or both parents may actively defend the galleries. Because wood-feeding cockroaches nest within their food source, xylophagy offers a distinct advantage in that the young are never left untended. Adult soil-burrowing cockroaches, however, need only be away from their dwelling for brief periods of time because they nest only where their food source is ample.

There are costs associated with excavating nests and burrows, in that energy expended on nest construction can detract from a parent's capacity for subsequent investment in eggs and young (O'Connor 1984). Halffter (this volume), for example, describes a negative correlation in dung beetles (Scarabaeinae) between nest burrow complexity and fecundity. Insects that utilize nests may also invest time and energy in provisioning and hygienic activities (Tallamy and Wood 1986).

A heavy cuticle is associated with the burrowing habit and may be a substantial nitrogen sink in cockroaches. A sturdy exoskeleton is required for abrasion resistance and to provide insertions of considerable rigidity for the attachment of muscles, particularly the leg muscles (Day 1950). Burrowing cockroaches (Cryptocercidae, Geoscaphinae, Panesthiinae) have converged on a similar body type which is heavily armored, with spined legs and a thick, scoop-shaped pronotum to facilitate digging. Shed exoskeletons are a source of nitrogenous food; competition to feed on exuvia has been observed in both *Cryptocercus* (C. A. Nalepa, unpublished data) and *Macropanesthia* (M. Slaytor, personal communication).

Nutritional component of parental care

The maternal transfer of the physical necessities of life to offspring in cockroaches is a continuum that starts with the incorporation of vitellogenins into the oocytes. In a large number of cockroaches, parental investment continues through the embryonic and postembryonic stages, and may include gas and water exchange, small-molecule transfer, feeding embryos within a brood sac, a variety of modes of postembryonic ingestion of maternal secretions, and progressive provisioning. Fig. 3-2 classifies the timing of the nutritional component of parental care in cockroaches. The scheme is based primarily on Roth's (1991) analysis of reproductive mode, amended with postparturition parental provisioning of food and water. Although egg size, and consequently the amount of yolk nutrients in the egg at ovulation, is an important consideration, little comparative information has been compiled on the subject. Roth (1968b) presents data on the size of the terminal oocytes in a wide variety of cockroaches, but not in relation to the body size of the adult.

Nitrogenous food for the young.   One of the main factors selecting for the evolution of parental care may be the degree of difficulty in obtaining food for offspring (Lack 1954; Itô 1980; Tallamy and Wood 1986). In insects, the availability of nitrogenous food has a large potential impact on early instars because they have relatively small reserves, a high metabolism, and nutritional requirements that differ from those of adults (Slansky and Scriber 1985; Rollo 1986). Young cockroaches are described as inefficient foragers (*Blattella germanica* in a public building; Cloarec and Rivault 1991) and their hemimetabolous mode of development makes them particularly sensitive to their nutritional environment. Cuticle production in hemimetabolous insects has

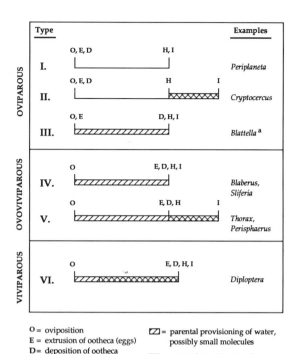

Fig. 3-2. Postoviposition provisioning in cockroaches. Oviposition refers to eggs being released from the ovaries. Extrusion is the permanent expulsion of the eggs from the body. Deposition is the disassociation of the egg case from the body. Independence is the ability of neonates to live apart from the parent(s).

[a] Neonates of *B. vaga* briefly feed on maternal secretions.

a negative effect on their nutritional efficiency (Bernays 1986) and may be the reason that, in comparison to holometabolous insects, they tend to grow less between molts and molt more often in the course of development (Cole 1980).

Cockroaches known to show parental care employ a variety of modes of postovulation provisioning of offspring; these parental investments in the form of brood milk, gut fluids and exudates may be analogous to lactation in that they are mechanisms for allowing young to grow rapidly and pass quickly through stages in which they are subject to cannibalism and predation or cannot satisfactorily process the adult diet. Parental tissues as food for the young are reliable, digestible and highly nutritious, and can support rapid growth even under suboptimal conditions (Pond 1977). Cockroaches employ a number of feeding behaviors associated with nitrogen conservation, including cannibalism, necrophagy, oophagy, feeding on exuviae, coprophagy, grooming, trophallaxis, and active and passive urate transfer (reviewed by Nalepa 1994). Regardless of whether these behaviors occur in the context of a family group or aggregation, food that originates from conspecifics may be especially important to those life stages with the highest nitrogen requirements: reproducing females and young nymphs.

Little is known of the natural diet of many of the cockroaches listed in Table 3-2. However, one category that is perceptible comprises of cockroaches adapted to a diet that is tough, high in fiber, and nutrient-poor: wood and detritus. These foods consist primarily of structural carbohydrate, from which it is difficult for juveniles to acquire sufficient protein reserves to ensure rapid postnatal growth (Nalepa 1994). Of these species, many live within or near their food source. For example, families of the wood-feeding *Cryptocercus* and *Salganea* live in galleries excavated within logs. The Australian burrowing cockroaches are fastidious feeders and nest only where sufficient food in the form of leaf litter occurs in the immediate vicinity of the burrow; *Geoscapheus dilatatus* thrives only on its normal diet of dried eucalyptus leaves (Slaytor 1992; H. A. Rose, personal communication). *Thorax porcellana* inhabits and feeds on aerial leaf litter in dry forests of India (Reuben 1988). *Perisphaerus* has been collected from the hollow, rotted-out core of vines (Roth 1981). In the field, *Blattella vaga* is most often associated with decaying vegetation, but when it invades houses, its feeding habits are similar to those of *B. germanica* (Flock 1941). *Trichoblatta sericea* is the anomaly in the group; it lives in the bark fissures of *Acacia* trees and feeds on the gum they exude.

How the nutritional needs of neonates are satisfied in the vast numbers of non-subsocial, non-viviparous cockroaches is yet unknown; many of these are also detritivores. Analysis of the stomach contents in 20 species of cockroach nymphs in Costa Rican rain forest indicates that they all feed on leaf litter (W. J. Bell, unpublished). The first meal for young nymphs of most ovoviparous cockroaches is the ootheca and embryonic membranes (Nutting 1953; Willis *et al.* 1958). In cockroaches that aggregate, sufficient high-nitrogen food may be found within the harborage in the form of feces, exuviae, corpses, etc. Support for this suggestion comes from Cochran (1979), who found that isolated nymphs of *Periplaneta americana* had less than half the stored uric acid of nymphs raised in groups. Species that disperse soon after hatching may rely on

yolk-derived nutrients that carry over into the postembronic stage. However, nitrogenous food for offspring may not be problematic for cockroaches with access to relatively rich food sources, such as cave dwellers that rely on carrion, guano and insect larvae (Gautier *et al.* 1988). The relationships between egg size, body size, reproductive mode, social behavior and nutritional ecology of cockroaches is an unexplored and potentially productive area of study.

Feces and microorganisms. The hindgut of cockroaches has a dense and varied microbiota (Cruden and Markovetz 1987). The advantage of these hindgut microorganisms is that they provide nutritional supplements that allow the insect to thrive with a much wider dietetic range than would otherwise be possible (Dadd 1985). Cockroaches are not, however, totally dependent on these microorganisms. The growth of nymphs is retarded when the anaerobic bacteria are killed, but adequately fed adults are not affected (Cruden and Markovetz 1987). It should be noted, however, that laboratory diets are poor approximations of the dietary regime that cockroaches experience in the field.

Newly hatched cockroaches acquire their hindgut microbiota by coprophagy; feces contain protozoan cysts and bacterial cells and spores (Hoyte 1961; Cruden and Markovetz 1984). Optimum growth of nymphs therefore depends on their consumption of feces from conspecifics. In ovoviviparous cockroaches that do not aggregate, short-term contact with the female may be necessary to ensure that the neonates partake of at least one fecal meal. Currently, however, there are no observations of the behavior of nymphs while beneath their mother, and, because so little is known of their association with gut bacteria, we cannot rule out the possibility that neonates acquire their gut biota from the substrate; for example, the dead leaves utilized by young cockroach nymphs in Costa Rican rain forest are saturated with a variety of microorganisms (W. J. Bell, unpublished). Cockroaches that aggregate are ensured a reliable supply of feces, and thus microbes, within the harborage. Little is known about feces as a source of either microorganisms or nutrients in cockroaches. In both *P. americana* and *B. germanica*, there are at least two distinct modes of defecation behavior, and the physical characteristics of the resultant feces also differ (Deleporte 1988; C. Schal, personal communication). In the latter species, the early-instar nymphs feed on the feces produced by other cockroaches when their yolk-derived nutrients are depleted (Silverman *et al.* 1991).

Some blattellids excrete urate-containing fecal pellets when in positive nitrogen balance, and these pellets are used as food by conspecifics (Cochran 1986; Lembke and Cochran 1990).

Cryptocercids are the only cockroaches known to have an obligate relationship with protozoan gut fauna; their hindgut is a fermentation chamber containing a community of interacting symbionts (reviewed by Cleveland *et al.* 1934; Breznak 1982; O'Brien and Slaytor 1982; Messer and Lee 1989). Neonates aquire their symbionts by proctodeal trophallaxis from parents. By the third instar, the symbiosis is fully established (Nalepa 1990); the young nymphs retain the protozoan symbionts for life (Cleveland *et al.* 1934). Because these wood-feeding cockroaches hatch devoid of gut fauna, they are clearly dependent on a food supply provided by their parents. Some degree of social behavior is thus concomitant with the obligate use of these protozoans for cellulose digestion (Nalepa 1991). The passage of symbionts, however, may not be the sole factor selecting for subsociality in wood-feeding cockroaches. *Salganea taiwanensis* and *S. esakii* are both subsocial wood-feeders (T. Matsumoto, personal communication), but their hindgut fauna is comprised of microbiota similar to those found in *Periplaneta* (Gijzen and Barugahare 1992; A. Messer, personal communication). The stomodeal feeding of offspring by adults in *S. taiwanensis* probably functions in nutritional supplementation, but it is possible that foregut bacteria are also transferred (A. Messer, personal communication). Cellulose digestion in the sole species of wood-feeding panesthiine studied to date (*Panesthia cribrata*) is thought to be independent of gut symbionts (Scrivener *et al.* 1989; Slaytor 1992; but see Martin 1991).

Evolution of parental care

The occurrence of parental care in cockroaches in relation to taxonomic position is illustrated in Fig. 3-3. Biparental care in a nest arose at least twice in wood-feeding cockroaches: in the oviparous Cryptocercidae (Blattoidea), and in the ovoviviparous Panesthiinae (Blaberoidea). Maternal care in a nest also occurs in the Geoscaphinae; the male may be present in the early stages of family life, but his role, if any, is unknown (Matsumoto 1992). Maternal care on the body evolved more than once, in related subfamilies.

Fig. 3-3 illustrates that ovoviviparity plays a role in the origins of parental care in cockroaches. With the exception of the Cryptocercidae, all instances of subsocial behavior are in the Blaberidae; retention of the egg case until

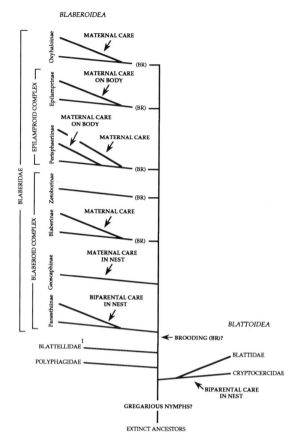

Fig 3-3. Possible evolution of parental care. In the Blaberidae, only those subfamilies with current records of post-parturition parental care are included. Brooding (BR) is assumed to be a characteristic of a subfamily if it was observed in any member species. Note that in the Blattellidae (labeled [1]) there is a brief parent–offspring association in *Blattella vaga*.

hatching allows proximity of adult and neonates in both time and space. We might therefore predict that we may eventually find postparturition parent–offspring associations among some blattellids that either externally carry the egg case until hatching or are ovoviviparous. The brief association of neonates with their mother in *Blattella vaga* supports this idea.

Young nymphs of both oviparous and ovoviviparous cockroaches have a tendency to aggregate (i.e. remain in groups) whether or not a parent is present (Roth and Cohen 1973). For example, newly hatched nymphs of *Blattella germanica* stay close to the hatched ootheca for several hours after emergence (Barson and Renn 1983), then aggregate during inactive periods. Aggregation behavior in nymphs facilitates the origin of parental care in that the female has ready access to offspring concentrated in space. The combination of ovoviviparity in adults and aggregation behavior in nymphs provides ample opportunity for brooding behavior. The multiple origins of parental care among the Blaberidae suggest that more elaborate forms of parent–offspring interactions evolved from that starting point.

Cryptocercus is the only known oviparous cockroach with well-developed parental care, and it is the only currently known oviparous genus to show nesting behavior.

If additional subsocial oviparous cockroaches are to be found, it will probably be among those that nest within or near a food source; nesting provides the same parental proximity to neonates as does egg retention. It is of note that biparental care arose twice in the cockroaches, in both cases among wood-feeders. Woodworking may warrant the cooperation of both parents; the colonization, manipulation and protection of wood has favored paternal investment not only in cryptocercids and panesthiines, but also in passalid and scolytid beetles (Tallamy and Wood 1986; Tallamy 1994; Kirkendall *et al.*, this volume). However, xylophagy also occurs in cockroaches that lack any form of parental care, most notably among the Australian wood-feeding panesthiines. There are additional reports of xylophagy in species thought to be solitary (Grandcolas 1993b) or with unknown social tendencies (Gurney 1937). Because the passage of hindgut symbionts clearly selects for parental care in *Cryptocercus*, the difference in social structures may represent a dichotomy between cockroaches that rely on vertical transmission of cellulose-digesting gut symbionts for the digestion of wood (i.e. direct transfer from parent to offspring) and those that (1) acquire them more casually via coprophagy, (2) have a complete complement of endogenous cellulases, or (3) rely on microorganisms acquired from the environment. A comparative study of the physiology of wood digestion in relation to social structure should help resolve the issue.

Many of the cockroaches that care for their young utilize dead wood or detritus as food. Detritus is low in food quality and can be abundant, but it may also be fragmented, dispersed and ephemeral; if so, detritivores must invest a lot of time and energy in locating sufficient quantities of food to meet their needs (Tallamy 1994). Resource location, abundance and distribution may thus determine whether parental care in cockroaches occurs in a nest, or whether the female carries her offspring on her body. *Cryptocercus*, panesthiines and geoscaphines are terrestrial, and protect their offspring in a nest in association with a local food source. The little we know of blaberids that carry their offspring on their body, however, indicates that they are arboreal. They feed on resources that may be more scattered, ephemeral, or seasonal (e.g. gum on the surface of trees, aerial litter, hollowed-out cores of vines), in a habitat that offers little raw material for building a structurally sound nest. In some tree-dwelling cockroaches, a mobile family unit has evolved to allow more wide-ranging foraging behavior in combination with offspring protection. Females that carry young on the body while foraging, however, are no doubt limited by the load they can transport. Excavated nests do not impose these restrictions on offspring number, size or mass.

REGULATION OF REPRODUCTIVE INVESTMENT

Termination of investment

It is clearly advantageous to abandon a reproductive attempt that has a high probability of failure if the investing parent thereby increases its likelihood of surviving to reproduce again (Parker 1977). If a female cockroach has initiated a reproductive episode that is threatened for lack of food or other reasons, she has several options for converting reproductive investment back into somatic tissue, thereby maintaining and redirecting her resources (Elgar and Crespi 1992). Termination of investment can occur at several points in the reproductive cycle. Prior to ovulation, starvation increases oocyte resorption in cockroaches (reviewed by Bell and Bohm 1975). In *P. americana*, most starved females produce one, sometimes two, oothecae in addition to the one being produced when starvation is initiated (Bell 1971). Large yolk-filled oocytes are retained in the ovaries of those females that do not deposit a second ootheca, and beginning on about the tenth day of starvation these oocytes are resorbed and the vitellogenins are stored. When feeding resumes, these stored yolk proteins are rapidly incorporated into developing oocytes. In *Xestoblatta hamata*, resorption of proximal oocytes and extension of the interval between oothecae are common in nature and result from unsuccessful foraging (Schal and Bell 1982; C. Schal, personal communication).

After ovulation, females have other mechanisms for terminating reproductive investment. Abortion will occur in laboratory cultures if gestating females are disturbed in *Pycnoscelus surinamensis*, *Panchlora irrorata* and *Blaberus craniifer* (Nutting 1953; Willis *et al.* 1958; Willis 1966). It is unknown if and under what circumstances ovoviviparous and viviparous cockroaches jettison egg cases in their natural habitat.

Cannibalism is a strong selection pressure that may have led to concealment of oothecae and brood care by the parents of potential victims (Crespi 1992). In the case of a female eating her own young, cannibalism is a means of recovering and recycling a threatened reproductive investment. For example, if disturbed when nymphs are freshly hatched, adults of *Cryptocercus punctulatus* may cannibalize their entire brood (C. A. Nalepa, unpublished). Other cockroach species are known to eat their young (Roth and Willis 1954a); starved females are often more likely to do so (Roth and Willis 1960; Rollo 1984; W. J. Bell, unpublished). Cannibalism, however, cannot always be ascribed to an obvious factor such as threat or starvation; in the case of young that are not their own, it may simply be a response to the presence of vulnerable individuals. Eggs and newborn animals are relatively defenseless unless guarded by a parent, and pose little or no risk to a cannibal (Fox 1975; Polis 1981). Cockroaches are generally opportunistic feeders and may take advantage of highly palatable morsels such as eggs and neonates.

Juvenile mortality

Overall, insects that exhibit postpaturition parental care may be expected to show low early mortality when compared with non-parental species (Itô 1980). This pattern, however, does not seem to apply to the few species of cockroach for which survivorship data are available. In *Macropanesthia*, mortality is about 35–40% by the time the nymphs disperse from the nest at the fifth to sixth instar (Rugg and Rose 1991; Matsumoto 1992). Both *Salganea esakii* and *S. taiwanensis* incubate an average of 15 eggs in the brood sac, but have only six nymphs (third instar) in field-collected families (T. Matsumoto and Y. Obata, personal communication). Family size of *Cryptocercus punctulatus* declines by about half during the initial stages; an average of 73 eggs is laid, but an average of only 36 nymphs is found in families prior to their first winter (Nalepa 1988, 1990). These data suggest that neonates may be subject to mortality factors such as disease or starvation despite the attendance of adults.

An alternative explanation for high neonate mortality in these species is that it represents an evolved strategy for adjusting parental investment after hatch. The eggs of *Cryptocercus punctulatus* are small for a cockroach of its size, and each neonate averages only 0.06% of its mother's post-oviposition dry mass (Nalepa and Mullins 1992). Young nymphs are dependent on the adults, and compete with each other for hindgut fluids, exuvia and some fecal pellets. Unlike other oviparous cockroaches, the hatching of nymphs from an egg case in *C. punctulatus* is not simultaneous, but extended in time; complete hatching of a single egg case may require up to four days. A brood most often consists of one to four egg cases laid in quick succession; the span of time between initiation of hatching of individual egg cases within a brood ranges from one to six days in the laboratory (C. A. Nalepa, unpublished data). The first-hatched nymphs of a brood therefore may be older than the last-hatched nymphs by two weeks or more (Nalepa 1988; C. A. Nalepa, unpublished data). Hatching asynchrony results in variation in competitive ability within a brood, a condition particularly conducive to the consumption of younger offspring by older siblings (Polis 1984). Nymphs of *C. punctulatus* 12 days old have been observed feeding on dead siblings; attacks by nymphs on moribund siblings have also been noted. Age differentials within broods may allow older nymphs to monopolize available food, leading to the selective mortality of the younger or weaker juveniles; necrophagy or cannibalism by adults or older juveniles then recycles the somatic nitrogen of the lower-quality offspring back into the family (C. A. Nalepa, unpublished manuscript). The production of expendable offspring to be eaten by siblings can be viewed as an alternative to producing fewer eggs, each containing more nutrients (Eickwort 1981; Polis 1981; Elgar and Crespi 1992).

Size vs. number

In birds, a negative effect of brood size upon nestling growth has been widely reported (see Klomp 1970) and is generally accepted as evidence that parents have a limited capacity to provide food for their offspring (Ricklefs 1983). In cockroaches, trade-offs associated with juvenile competition for adult-provided food are evident in both viviparous *Diploptera punctata* and oviparous *Cryptocercus punctulatus*. Although the former species provisions offspring before hatching and the latter exhibits postparturition provisioning, families in each species show trade-offs between size and number of offspring at independence. In both species, more offspring in the family means less food per nymph.

Roth and Hahn (1964) reduced the size of the litter in *Diploptera punctata* by surgically removing one of the ovaries. Neonates in these broods were larger than in the control families, presumably because of reduced competition and the greater amount of nutritive material made available to the fewer developing embryos. In ovoviviparous *Nauphoeta cinerea* and *Rhyparobia maderae*, however, the size of the nymphs remained constant regardless of the number incubated in the uterus. In the latter two species, embryos obtain water from their mother, but there is currently no evidence that they receive nutritive material. Nymphs within the same brood of *D. punctata* can also differ considerably in size depending on their position in the brood sac; embryos that have poor contact with the uterine wall have less ready access to the nutritive secretion provided by the mother (Roth and Hahn 1964).

In *Cryptocercus punctulatus*, parental investment is most intense immediately after hatching. Many neonates are present in families, they are all dependent for food, and individual dry mass increases substantially during the first few instars. By the time they reach nutritional independence, before their first winter, nymphs from small families are more advanced developmentally, on average (i.e. in a later instar), than nymphs from large families (C. A. Nalepa, unpublished manuscript).

Trade-offs between size and number of offspring are likely to occur in most of the cockroaches listed in Table 3-3A. *Perisphaerus* would be an especially interesting candidate for a study of this sort; females possess four intercoxal openings from which nymphs feed, but nine nymphs were associated with one of the museum specimens studied by Roth (1981).

Altriciality vs. precociality

In both birds and mammals, there are obvious gradients in the maturity and relative size of the young at birth, ranging from immature and helpless (altricial) to complete independence from adult support (precocial). This broad continuum of developmental patterns parallels a gradient in neonate independence of parental care and provides a useful basis for discussing patterns of postnatal growth (Nice 1962; Case 1978; Ricklefs 1983). In birds, Nice (1962) subdivided the precocial–altricial spectrum into categories based on the following criteria, determined at the time of hatching: open or closed eyes, presence or absence of down, mobility, and the nature of parental care.

Alexander (1990) suggested that the concepts of altricial and precocial development can also be applied to insects.

In the crickets (Gryllidae), for example, most juveniles have hard exoskeletons, are agile, seek their own food, and are independent at the time of hatch. In *Anurogryllus*, however, females tend their offspring in underground nests and feed them on unfertilized trophic eggs, and the offspring are 'soft and fat' (West and Alexander 1963; Alexander 1990). The degree of development at parturition also varies among cockroach taxa; with only minor adjustments, the categories used for birds by Nice (1962) may be used to classify developmental status of Blattaria at hatch: eyes (present or absent), cuticle (thin, unpigmented or thick, pigmented), mobility, and the presence and nature of parental care.

Eyes

Pigmentation of the eyes is one indicator used in timetables of embryonic development in hemimetabolous insects (Anderson 1972); it begins at 58% of the embryonic development period in *Blattella germanica* (Tanaka 1976). Eye pigmentation, however, is a postembryonic event in two cockroach species that show postparturition feeding of offspring. First-instar nymphs of *Cryptocercus punctulatus* lack compound eyes (Huber 1976); pigmentation first appears in the second instar. Huber proposed that the eyes may have degenerated because of the 'troglobiont' lifestyle of this species, in a manner similar to the loss of eyes in some cave-dwelling cockroaches (Roth 1980). However, young nymphs of *Perisphaerus*, a cockroach that lives in dead vegetation, also lack eyes; in one species at least the first two instars are blind (Roth 1981). This observation suggests that early blindness in these two cockroaches is more likely due to changes in developmental timing associated with the presence of parental care than to the degree of light in their environment. In a third species that feeds its young after hatching, *Trichoblatta sericea*, pigmentation of the eyes occurs during the embryonic stage, but late, relative to *B. germanica*: at 78% of gestation (Reuben 1988). These data are consistent with the idea that in some subsocial cockroaches, nymphs do not reach an advanced state of development while *in ovo*.

Cuticle

In birds, the heaviness of the integument (skin plus feathers) varies along the altricial–precocial spectrum; precocial birds generally have heavier integuments at hatching (Ricklefs 1983). Although quantitative data are lacking, there is some evidence to suggest that the thickness and pigmentation of the cuticle in first instar cockroaches may be correlated with the presence or absence of parental care. In most species of cockroach, the cuticle of first instars becomes pigmented and relatively sturdy a few hours after hatch. In *Cryptocercus punctulatus*, however, first instars have a pale, thin cuticle and the internal organs are clearly visible through the dorsal surface of the abdomen. T. Matsumoto (unpublished data) found that among wood-feeding panesthiine cockroaches, neonates of those species that are solitary or gregarious have strong, dark cuticles. Among these are *Panesthia sloanei*, *P. angustipennis* and *P. cribrata*. First instars of species that exhibit parental care, however, have thinner cuticles, varying in color from light brown to white (*Salganea raggei*, *S. taiwanensis*, *S. esakii* and *S. gressiti*).

Mobility

There are no published data allowing analysis of the relative mobility of neonate cockroaches. T. Matsumoto and Y. Obata (personal communication), however, have observed that first instars of the gregarious cockroach *Panesthia angustipennis* are more mobile and robust than those of the subsocial *Salganea esakii* and *S. taiwanensis*. First instars of species in which the female carries her offspring under the tegmina or attached to the ventral surface obviously rely on the locomotory abilities of the parent, and those confined to a nest have little need of well-developed locomotory capabilities for foraging or escape from predators.

Egg size

The small size of hatchlings in altricial birds is associated with the production of small eggs by the mother (reviewed by Dunbrack and Ramsay 1989). Once parental care becomes established, laying small eggs reduces the investment required for the production of large, well-provisioned eggs, and spreads initial parental costs over a longer interval. Monogamous pair bonds provide further opportunity for the evolution of small egg size because this postponement of parental investment allows some of the cost to be shared with the father (Zeveloff and Boyce 1982; Blackburn and Evans 1986; Alexander 1990).

Cryptocercus punctulatus appears monogamous, and exhibits biparental care; analysis of the limited data in the literature indicates that its eggs are small for a cockroach of its size. In six species of oviparous Blattidae, the length of the terminal oocyte prior to oviposition ranges from 9 to 16% of adult female length. In *C. punctulatus*, however, the terminal oocyte is 5% of adult length (Nalepa 1987). In this wood-feeding cockroach, a decrease in egg size and an

increase in dependence of neonates no doubt occurred after they became dependent on gut symbionts for cellulose digestion; it was then that parental care became obligate because of the lack of permanent establishment of the symbionts until the third instar. It is likely that an increased reliance on hindgut fluids to nourish offspring became coupled with the emergence of progressively more altricial young as investment in the egg was curtailed.

Advantages of altricial development

Rapid growth may be the adaptive function of altriciality in a wide variety of species (Alexander 1990); altricial birds grow three to four times more rapidly than precocial ones (reviewed by Ricklefs 1983). When external sources of mortality are removed, juveniles are relieved of the necessity of evolving protection from extrinsic hostile forces of nature and are freed to devote a greater proportion of their resources to becoming better adults (Alexander 1990).

In insects, the production of cuticle requires the commitment of assimilated nitrogen to chitin and proteins, and the commitment can be substantial. More than half the dry mass of a cockroach nymph may consist of the exoskeleton (Bernays 1986). Parental behavior in cockroaches may allow first instars to use for growth dietary nitrogen that might otherwise be diverted into a protective carapace. This diversion may be especially significant in subsocial burrowing cockroaches, which otherwise require a substantial cuticle.

Young nymphs of *Cryptocercus punctulatus* gain considerable mass and go through a relatively quick series of molts during the period of parental feeding after hatching. The dry mass of a second-instar nymph averages eleven times that of a newly hatched first instar (C. A. Nalepa and D. E. Mullins, unpublished data). The relative growth rates of other cockroach species that exhibit parental care have not been studied. A comparison of growth rates in gregarious and subsocial wood-feeding panesthiine cockroaches, however, would provide a good test of the rapid growth hypothesis in insects, after controlling for adult body size, mortality factors, quality of the diet and competition among siblings.

Precocial development

The altricial extreme of the developmental spectrum in cockroaches is currently represented by *Cryptocercus punctulatus*: this species hatches eyeless, with a pale, thin cuticle, is defended by parents in a nest, and is dependent on parents for symbionts and nourishment. The precocial end of the range includes the majority of cockroaches, which are born with eyes, develop a pigmented cuticle a few hours after hatch, and forage independently from their parents. The most extreme state of precociality in cockroaches, however, is represented by viviparous *Diploptera punctata*. Roth and Willis (1955a) described first-instar nymphs of *D. punctata* as unduly large in comparison with the nymphs of other species of cockroaches, even those with adults twice the size of adult *D. punctata*. There are other indications that *D. punctata* develops as an embryo longer than other cockroaches. During embryogenesis, closure of the dorsal body wall occurs at 19% of gestation time, after which they begin feeding on brood milk (Stay and Coop 1973). Dorsal closure occurs at 46% of gestation time in *Rhyparobia maderae* (Aiouaz 1974), at 50% of gestation in *Nauphoeta cinerea* (Imboden *et al.* 1978) and at 56% of gestation in *Periplaneta americana* (Lenoir-Rousseaux and Lender 1970). These data are consistent with the suggestion that nymphs of *D. punctata* complete a substantial portion of their development as embryos, feeding on the brood milk provided by the mother, with a corresponding decrease in the duration of postembryonic development. Gestation of *D. punctata* embryos takes 63 days at 27 °C (Stay and Coop 1973); nymphs require just 43–52 days to become adults (Willis *et al.* 1958).

Ecological factors

Two ecological factors – food availability and predation intensity – are important in determining the altricial–precocial spectrum (Case 1978; O'Conner 1984). If food is readily available, both to females for egg production and to independent young for growth, precocial development is feasible; most precocial birds subsist on food that is readily procured by the young (Ricklefs 1983). If, however, food is difficult to obtain, then the small eggs of altricial development are favored, and the young are dependent on others for the transduction of environmental resources. In cockroaches, species in which there is some indication that the offspring are altricial are those in which parental care includes the feeding of offspring (Table 3-3A). In a sense, these cockroaches are convergent with vertebrates that feed their young on bodily secretions (Pond 1977; Shine 1989). The mechanical and physiological problems encountered by poorly developed offspring attempting to feed on solid food are avoided, thus allowing significant size reduction (Pond 1983).

Most cockroaches in which parents are known to feed dependent young utilize low-quality resources. The diet of viviparous *Diploptera punctata*, however, appears unremarkable relative to other cockroaches, which suggests that precocial development in this species may have been an evolutionary response to predation pressure. The role of predation in structuring the life history of *Diploptera* is, however, unknown.

SYMBIONTS AND PARENTAL INVESTMENT

Cockroaches are a unique group in that they show nearly a complete range of possible parental investment, manifested in shifts in the distribution of resources in eggs, protection and nourishment of developing embryos, and the care and provisioning of nymphs following hatch. One source of this reproductive versatility may lie in their nutritional ecology. The classic explanation for the success of cockroaches as a group is that 'they can eat anything' (L. M. Roth, personal communication); their diets are as diverse as their habitats and their lifestyles. In general, however, cockroaches are closely associated with detritus (Mullins and Cochran 1987), a diet that is low in nitrogen and rich in structural polysaccharides. The success of cockroaches within this type of nutritional environment results in part from their relationship with microorganisms: their gut fauna and the intracellular bacteria in their fat body (Mullins and Cochran 1987).

The hindgut biota of cockroaches resembles that of the termite and the rumen in complexity (Cruden and Markovetz 1984); the gut is anaerobic and harbors methanogens and bacteria that digest carboxymethylcellulose (Bignell 1977; Bracke *et al.* 1979; Cruden and Markovetz 1979). The lining of the hindgut is covered with long cuticular spines that project into the lumen. This morphological adaptation provides a much larger surface area relative to gut volume than the gut wall of other animals, and indicates that the association with microorganisms is not a casual one. The gut lining and spines are coated with a thick layer of firmly attached microbes (Bracke *et al.* 1979; Cruden and Markovetz 1981; Hackstein and Stumm 1994).

The endosymbionts in the fat body of cockroaches are members of the flavobacteria group (Bandi *et al.* 1994) and have been found in all Blattaria studied to date (Cochran 1985). It has been known for a number of years that cockroaches deprived of these symbionts show poor growth and reproductive capacity (Richards and Brooks 1958) and that the symbiotic association is expressed in many metabolic pathways (Bignell 1981; Sacchi and Grigolo 1989). The best-known role of intracellular symbionts, however, is as mediators of uric acid degradation (Wren and Cochran 1987). Nitrogen excretion in cockroaches is a complex phenomenon that differs from the expected terrestrial insect pattern (Cochran and Mullins 1982; Mullins and Cochran 1987). The majority of cockroaches build up internal deposits of uric acid when their diet is high in nitrogen; later, via their fat body endosymbionts, they mobilize and use their urate stores when their diet is deficient in nitrogen or when nitrogen requirements increase.

Nourishment in an individual cockroach is thus a cooperative effort by the insect and all of its microbial associates. As a result, cockroaches are able to make metabolic sense of a wide variety of foods and their dependence on extrinsic sources of specific nutrients is reduced. The symbionts smooth out the valleys between infrequent peaks of nutritional input, and as a result cockroaches can withstand periods of food shortage, including 'self-imposed' periods of non-foraging while carrying egg cases. These microorganisms may also have a more direct effect on parent–offspring interactions in cockroaches. Transmission of the populations of hindgut microbes may require behavioral adaptations for each generation to acquire microflora from the previous one, and consequently select for association of neonates with older conspecifics. Cockroaches are a very social taxon, and, in addition to showing parental care, can be gregarious, territorial and hierarchical, as well as relatively solitary (Gautier *et al.* 1988).

Posthatching adaptations are not required for the acquisition of fat-body symbionts in cockroaches, since they are passed by transovarian transmission (reviewed by Sacchi and Grigolo 1985). These symbionts do, however, enable different aspects of parental investment in cockroaches. To reproduce, females must either find food, or withdraw materials and energy from reserves stored earlier. Morphology and physiology limit feeding rates, the level of reserves stored, and how fast reserves can be mobilized (Stearns 1992). The recycling of nitrogenous waste via fat-body endosymbionts may therefore be influential in the reproductive versatility of cockroaches. All of the nitrogen-scavenging activities of cockroaches (reviewed by Nalepa 1994) are facilitated by their storage–mobilization physiology, including cannibalism. Cannibalism, in turn, can be a selective pressure for the protective element of parental care, including oothecal concealment and female aggression in oviparous cockroaches and the retraction of the ootheca into the body in ovoviviparous cockroaches. Cannibalism also allows females to fine-tune their own

reproductive investment relative to risk of starvation, or as a strategy for assuring fewer, healthier offspring in the family. A storage–mobilization physiology furthermore allows an individual to share nitrogenous materials with conspecifics via social behavior (reviewed by Nalepa 1994). The diversity of modes of postovulation provisioning of offspring (brood milk, gut fluids, exudates) in cockroaches may be based in the physiological ability of a parent to mobilize and transfer stored reserves of nitrogen.

This chapter is a first step in making sense of the diversity in reproductive investment exhibited by cockroaches with the limited information at hand. Many of the conclusions we have reached are, without a doubt, premature, but at the very least, we have shown that the few species that forage in kitchens are not representative of the group as a whole. The most we can hope for is that the next generation of scientists recognizes the value of studying cockroaches in their natural habitat and accepts the challenge of gathering data that may integrate patterns of reproductive mode and parental care into the larger picture of ecology, phylogeny and life-history strategy.

ACKNOWLEDGEMENTS

We thank G. L. Holbrook, T. Matsumoto, A. Messer, D. E. Mullins, J. Mullins, Y. Obata, H. A. Rose, L. M. Roth, D. Rugg, C. Schal, M. Slaytor and K. R. Tignor for unpublished observations and data and for patiently answering our questions. The comments of B. Crespi, D. E. Mullins, L. M. Roth, D. W. Tallamy and an anonymous reviewer significantly improved the manuscript. Coby Schal's laboratory discussion group was a valuable source of feedback.

LITERATURE CITED

Aiouaz, M. 1974. Chronologie du développement embryonnaire de *Leucophaea maderae* Fabr. (Insecte, Dictyoptère). *Arch. Zool. Exp. Gén.* 115: 343–358.

Alexander, R. D. 1990. How did humans evolve? Reflections on the uniquely unique species. *Univ. Mich. Spec. Publ.* 1: 1–38.

Ananthasubramanian, K. S. and T. N. Ananthakrishnan. 1959. The structure of the ootheca and egg laying habits of *Corydia petiveriana* L. *Indian J. Entomol.* 21: 59–64.

Anderson, D. T. 1972. The development of hemimetabolous insects. In *Developmental Systems: Insects*, vol. 1. S. J. Counce and C. H. Waddington, eds., pp. 95–163. London: Academic Press.

Bandi, C., G. Damiani, L. Magrassi, A. Grigolo, R. Fani and L. Sacchi. 1994. Flavobacteria as intracellular symbionts in cockroaches. *Proc. R. Soc. Lond.* B 257: 43–48.

Barson, G. and N. Renn. 1983. Hatching from oothecae of the German cockroach (*Blattella germanica*) under laboratory culture conditions and after premature removal. *Entomol. Exp. Appl.* 34: 179–185.

Bell, W. J. 1971. Starvation-induced oocyte resorption and yolk protein salvage in *Periplaneta americana*. *J. Insect Physiol.* 17: 1099–1111.

Bell, W. J. and K. G. Adiyodi. 1981. Reproduction. In *The American Cockroach*. W. J. Bell and K. G. Adiyodi, eds., pp. 343–370. London: Chapman and Hall.

Bell, W. J. and M. K. Bohm. 1975. Oosorption in insects. *Biol. Rev.* 50: 373–396.

Benson, E. P. and I. Huber. 1989. Oviposition behavior and site preference of the brownbanded cockroach *Supella longipalpa* (F.) (Dictyoptera: Blattellidae). *J. Entomol. Sci.* 24: 84–91.

Bernays, E. A. 1986. Evolutionary contrasts in insects: nutritional advantages of holometabolous development. *Physiol. Ecol.* 11: 377–382.

Bignell, D. E. 1977. An experimental study of cellulose and hemicellulose degradation in the alimentary canal of the American cockroach. *Can. J. Zool.* 55: 579–589.

–. 1981. Nutrition and digestion. In *The American Cockroach*. W. J. Bell and K. G. Adiyodi, eds., pp. 57–86. London: Chapman and Hall.

Blackburn, D. G. and H. E. Evans. 1986. Why are there no viviparous birds? *Am. Nat.* 128: 165–190.

Bracke, J. W., D. L. Cruden and A. J. Markovetz. 1979. Intestinal microbial flora of the American cockroach, *Periplaneta americana* L. *Appl. Environ. Microbiol.* 38: 945–955.

Breed, M. D. 1983. Cockroach mating systems. In *Orthoptera Mating Systems*. D. T. Gwynne and G. K. Morris, eds., pp. 268–284. Boulder: Westview.

Breed, M. D., C. M. Hinkle and W. J. Bell. 1975. Agonistic behavior in the German cockroach, *Blattella germanica*. *Z. Tierpsychol.* 39: 24–32.

Breznak, J. A. 1982. Biochemical aspects of symbiosis between termites and their intestinal microbiota. In *Invertebrate-Microbial Interactions*. J. M. Anderson, A. D. M. Rayner and D. W. H. Walton, eds., pp. 171–203. Cambridge University Press.

Brunet, P. C. J. 1951. The formation of the ootheca by *Periplaneta americana*. I. The microanatomy and histology of the posterior part of the abdomen. *Q. J. Microsc. Sci.* 92: 113–127.

Brunet, P. C. and P. W. Kent. 1955. Observations on the mechanism of tanning reaction in *Periplaneta* and *Blatta*. *Proc. R. Soc. Lond.* B 144: 259–274.

Campbell, F. L. 1929. The detection and estimation of insect chitin, and the irrelation of "chitinisation" to hardness and pigmentation of the cuticla of the American cockroach, *Periplaneta americana*. *Ann. Entomol. Soc. Am.* 22: 401–426.

Case, T. J. 1978. On the evolution and adaptive significance of postnatal growth rates in the terrestrial vertebrates. *Q. Rev. Biol.* 53: 243–282.

Cleveland, L. R., S. R. Hall, E. P. Sanders and J. Collier. 1934. The wood-feeding roach *Cryptocercus*, its protozoa, and the symbiosis between protozoa and roach. *Mem. Am. Acad. Arts Sci.* 17: 185–342.

Cloarec, A. and C. Rivault. 1991. Age-related changes in foraging in the German cockroach (Dictyoptera: Blattellidae). *J. Insect Behav.* 4: 661–673.

Clutton-Brock, T. H. 1991. *The Evolution of Parental Care*. Princeton: Princeton University Press.

Cochran, D. G. 1979. Uric acid accumulation in young American cockroach nymphs. *Entomol. Exp. Appl.* 25: 153–157.

–. 1983a. *Cockroaches – Biology and Control.* World Health Organization Vector Biology and Control Series 82.856, pp. 1–53. WHO, Geneva.

–. 1983b. Food and water consumption during the reproductive cycle of female German cockroaches. *Entomol. Exp. Appl.* 34: 51–57.

–. 1985. Nitrogen excretion in cockroaches. *Annu. Rev. Entomol.* 30: 29–49.

–. 1986. Biological parameters of reproduction in *Parcoblatta* cockroaches (Dictyoptera: Blattellidae). *Ann. Entomol. Soc. Am.* 79: 861–864.

Cochran, D. G. and D. E. Mullins. 1982. Physiological processes related to nitrogen excretion in cockroaches. *J. Exp. Zool.* 222: 277–285.

Cole, B. 1980. Growth ratios in holometabolous and hemimetabolous insects. *Ann. Entomol. Soc. Am.* 73: 489–491.

Crespi, B. J. 1992. Cannibalism and trophic eggs in subsocial and eusocial insects. In *Cannibalism: Ecology and Evolution Among Diverse Taxa*. M. A. Elgar and B. J. Crespi, eds., pp. 176–213. Oxford University Press.

Cruden, D. L. and A. J. Markovetz. 1979. Carboxymethyl cellulose decomposition by intestinal bacteria of cockroaches. *Appl. Environ. Microbiol.* 38: 369–372.

–. 1981. Relative numbers of selected bacterial forms in different regions of the cockroach hindgut. *Arch. Microbiol.* 129: 129–134.

–. 1984. Microbial aspects of the cockroach hindgut. *Arch. Microbiol.* 138: 131–139.

–. 1987. Microbial ecology of the cockroach gut. *Annu. Rev. Microbiol.* 41: 617–43.

Dadd, R. H. 1985. Nutrition: Organisms. *Comp. Insect Phys. Biochem. Pharm.* 4: 313–390.

Day, M. F. 1950. The histology of a very large insect, *Macropanesthia rhinocerus* Sauss. (Blattidae). *Austr. J. Sci. Res.* B 3: 61–75.

Deleporte, P. 1988. Etude eco-ethologique et evolutive de *P. americana* et d'autre blattes sociales. These, L'Universite de Rennes.

Dunbrack, R. L. and M. A. Ramsay. 1989. The evolution of viviparity in amniote vertebrates: egg retention vs. egg size reduction. *Am. Nat.* 133: 138–148.

Edmunds, L. R. 1957. Observations on the biology and life history of the brown cockroach *Periplaneta brunnea* Burmeister. *Proc. Entomol. Soc. Wash.* **59**: 283–286.

Eickwort, G. C. 1981. Presocial insects. In *Social Insects*, vol. 2. H. R. Hermann, ed., pp. 199–280. New York: Academic Press.

Elgar, M. A. and B. J. Crespi. 1992. Ecology and evolution of cannibalism. In *Cannibalism: Ecology and Evolution Among Diverse Taxa.* M. A. Elgar and B. J. Crespi, eds., pp. 1–12. Oxford University Press.

Engelmann, F. and I. Rau. 1965. A correlation between the feeding and the sexual cycle in *Leucophaea maderae*. *J. Insect Physiol.* **11**: 53–64.

Evans, L. D. and M. D. Breed. 1984. Segregation of cockroach nymphs into sibling groups. *Ann. Entomol. Soc. Am.* **77**: 574–577.

Flock, R. A. 1941. The field roach *Blattella vaga*. *J. Econ. Entomol.* **34**: 121.

Fox, L. R. 1975. Cannibalism in natural populations. *Annu. Rev. Ecol. Syst.* **6**: 87–106.

Gautier, J. Y., P. Deleporte and C. Rivault. 1988. Relationships between ecology and social behavior in cockroaches. In *The Ecology of Social Behavior*. C. N. Slobodchikoff, ed., pp. 335–351. San Diego: Academic Press.

Gijzen, H. J. and M. Barugahare. 1992. Contribution of anaerobic protozoa and methanogens to hindgut metabolic activities of the American cockroach, *Periplaneta americana*. *Appl. Environ. Microbiol.* **58**: 2565–2570.

Gorton, R. E. 1979. Agonism as a function of relationship in a cockroach *Shawella couloniana* (Dictyoptera: Blattellidae). *J. Kansas Entomol. Soc.* **52**: 438–442.

Grandcolas, P. 1993a. Habitats of solitary and gregarious species in the Neotropical Zetoborinae (Insecta, Blattaria). *Stud. Neotrop. Fauna Environ.* **28**: 1–12.

–. 1993b. Le genre *Paramuzoa* Roth, 1973: sa répartition et un cas de xylophagie chez les Nyctiborinae (Dictyoptera, Blattaria). *Bull. Soc. Entomol. Fr.* **98**: 131–138.

Grandcolas, P. and P. Deleporte. 1993. La position systématique de *Cryptocercus* Scudder au sein des Blattes et ses implications évolutives. *C.R. Acad. Sci. Paris*, ser III, **315**: 317–322.

Gurney, A. B. 1937. Studies in certain genera of American Blattidae (Orthoptera). *Proc. Entomol. Soc. Wash.* **39**: 101–112.

Haber, V. R. 1920. Oviposition by a cockroach, *Periplaneta americana* Linn. (Orth.). *Entomol. News* **31**: 190–193.

Hackstein, J. H. P. and C. K. Stumm. 1994. Methane production in terrestrial arthropods. *Proc. Natl. Acad. Sci. U.S.A.* **91**: 5441–5445.

Hamilton, R. L. and C. Schal. 1988. Effects of dietary protein levels on reproduction and food consumption in the German cockroach (Dictyoptera: Blattellidae). *Ann. Entomol. Soc. Am.* **81**: 969–976.

Hinton, H. E. 1981. *The Biology of Insect Eggs*, vol. 1. New York: Pergamon Press.

Hoyte, H. M. D. 1961. The protozoa occuring in the hindgut of cockroaches. III. Factors affecting the dispersion of *Nyctotherus ovalis*. *Parasitology* **51**: 465–495.

Huber, I. 1976. Evolutionary trends in *Cryptocercus punctulatus* (Blattaria: Cryptocercidae). *J. N. Y. Entomol. Soc.* **84**: 166–168.

Imboden, H. B., J. Lanzrein, P. Delbecque and M. Luscher. 1978. Ecdysteroids and juvenile hormone during embryogenesis in the ovoviviparous cockroach *Nauphoeta cinerea*. *Gen. Comp. Endocrinol.* **36**: 628–635.

Ingram, M. J., B. Stay and G. D. Cain. 1977. Composition of milk from the viviparous cockroach *Diploptera punctata*. *Insect Biochemistry* **7**: 257–267.

Itô, Y. 1980. *Comparative Ecology*. Cambridge University Press.

Karny, H. H. 1924. Beiträge zur Malayischen Orthopterenfauna. V. Bemerkungen ueber einige Blattoiden. *Treubia* **5**: 3–19.

Keil, C. B. and M. H. Ross. 1977. An analysis of embryonic trapping in the German cockroach. *Entomol. Exp. Appl.* **22**: 220–226.

Klomp, H. 1970. The determination of clutch size in birds, a review. *Ardea* **58**: 1–24.

Koehler, P. G., C. A. Strong and R. S. Patterson. 1994. Harborage width preferences of German cockroach (Dictyoptera: Blattellidae) adults and nymphs. *J. Econ. Entomol.* **87**: 699–704.

Lack, D. 1954. *The Natural Regulation of Animal Numbers*. Oxford: Clarendon.

Lawson, F. A. 1951. Structural features of the oothecae of certain species of cockroaches (Orthoptera, Blattidae). *Ann. Entomol. Soc. Am.* **44**: 269–85.

Lembke, H. F. and D. G. Cochran. 1990. Diet selection by adult female *Parcoblatta fulvescens* cockroaches during the oothecal cycle. *Comp. Biochem. Physiol.* A **95**: 195–199.

Lenoir-Rousseaux, J. J. and T. Lender. 1970. Table de développement embryonnaire de *Periplaneta americana* (L.) Insecte, Dictyoptere. *Bull. Soc. Zool. Fr.* **95**: 737–751.

Lessells, C. M. 1991. The evolution of life histories. In *Behavioural Ecology. An Evolutionary Approach*, 3rd edn. J. R. Krebs and N. B. Davies, eds., pp. 32–68. Oxford: Blackwell Scientific Publications.

Livingstone, D. and R. Ramani. 1978. Studies on reproductive biology. *Proc. Indian Acad. Sci.* B **87**: 229–247.

Liechti, P. M. and W. J. Bell. 1975. Brooding behavior of the Cuban burrowing cockroach *Byrsotria fumigata* (Blaberidae, Blattaria). *Insectes Soc.* **22**: 35–46.

Martin, M. M. 1991. The evolution of cellulose digestion in insects. *Phil. Trans. Roy. Soc. Lond.* B **333**: 281–288.

Masuko, K. 1989. Larval haemolymph feeding in the ant *Leptanilla japonica* by use of a specialised duct organ the "larval haemolymph tap" (Hymenoptera: Formicidae). *Behav. Ecol. Sociobiol.* **24**: 127–132.

Matsumoto, T. 1987. Colony compositions of the subsocial wood-feeding cockroaches *Salganea taiwanensis* Roth and *S. easakii* Roth (Blattaria: Panesthiinae). In *Chemistry and Biology of Social Insects*. J. Eder and H. Rembold, eds., p. 394. München: Verlag J. Peperny.

–. 1988. Colony composition of the wood-feeding cockroach, *Panesthia australis* Brunner (Blattaria, Blaberidae, Panesthiinae) in Australia. *Zool. Sci.* **5**: 1145–1148.

–. 1992. Familial association, nymphal development and population density in the Australian giant burrowing cockroach, *Macropanesthia rhinocerus* (Blattaria: Blaberidae). *Zool. Sci.* **9**: 835–842.

McKittrick, F. A. 1964. *Evolutionary Studies of Cockroaches*. Memoir no. 389, Cornell University Agricultural Experiment Station.

McKittrick, F. A., T. Eisner and H. E. Evans. 1961. Mechanics of species survival. *Nat. Hist.* **70**: 46–51.

Messer, A. C. and M. J. Lee. 1989. Effect of chemical treatments on methane emission by the hindgut microbiota in the termite *Zootermopsis angusticollis*. *Microb. Ecol.* **18**: 275–284.

Mullins, D. E. and D. G. Cochran. 1987. Nutritional ecology of cockroaches. In *Nutritional Ecology of Insects, Mites, and Spiders*. F. Slansky Jr. and J. G. Rodriguez, eds., pp. 885–902. New York: John Wiley and Sons.

Mullins, D. E., C. B. Keil and R. H. White. 1992. Maternal and paternal nitrogen investment in *Blattella germanica* (L.) (Dictyoptera: Blattellidae). *J. Exp. Biol.* **162**: 55–72.

Nalepa, C. A. 1984. Colony composition, protozoan transfer and some life history characteristics of the woodroach *Cryptocercus punctulatus* Scudder. *Behav. Ecol. Sociobiol.* **14**: 273–279.

–. 1987. Life history studies of the woodroach *Cryptocercus punctulatus* Scudder (Dictyoptera: Cryptocercidae) and their implications for the evolution of termite eusociality. Ph.D. dissertation, North Carolina State University.

–. 1988. Reproduction in the woodroach *Cryptocercus punctulatus* Scudder (Dictyoptera: Cryptocercidae): mating, oviposition and hatch. *Ann. Entomol. Soc. Am.* **81**: 637–641.

–. 1990. Early development of nymphs and establishment of the hindgut symbiosis in *Cryptocercus punctulatus* Scudder (Dictyoptera: Cryptocercidae). *Ann. Entomol. Soc. Am.* **83**: 786–789.

–. 1991. Ancestral transfer of symbionts between cockroaches and termites: an unlikely scenario. *Proc. R. Soc. Lond.* B **246**: 185–189.

–. 1994. Nourishment and the evolution of termite eusociality. In *Nourishment and Evolution in Insect Societies*. J. H. Hunt and C. A. Nalepa, eds., pp. 57–104. Boulder: Westview Press.

Nalepa, C. A. and D. E. Mullins. 1992. Initial reproductive investment and parental body size in *Cryptocercus punctulatus* Scudder (Dictyoptera: Cryptocercidae). *Physiol. Entomol.* **17**: 255–259.

Narasimham, A. U. 1984. Comparative studies on *Tetrastichus hagenowii* (Ratzeburg) and *T. asthenogmus* (Waterson), two primary parasites of cockroach oothecae, and on their hyperparasite *Tetrastichus* sp. (*T. miser* (Nees) group) (Hymenoptera: Eulophidae). *Bull. Entomol. Res.* **74**: 175–189.

Nice, M. M. 1962. Development of behavior in precocial birds. *Trans. Linn. Soc. N.Y.* **8**: 1–211.

Nutting, W. L. 1953. Observations on the reproduction of the giant cockroach *Blaberus craniifer* Burm. *Psyche* **60**: 6–14.

O'Brien, R. W. and M. Slaytor. 1982. Role of microorganisms in the metabolism of termites. *Austr. J. Biol. Sci.* **35**: 239–62.

O'Connor, R. J. 1984. *The Growth and Development of Birds*. Chichester: John Wiley and Sons.

O'Neill, S. L., H. A. Rose and D. Rugg. 1987. Social behaviour and its relationship to field distribution in *Panesthia cribrata* Saussure (Blattodea: Blaberidae). *J. Austr. Entomol. Soc.* **26**: 313–321.

Parker, P. 1977. An ecological comparison of marsupial and placental patterns of reproduction. In *The Biology of Marsupials*. B. Stonehouse and D. Gilmore, eds., pp. 273–286. London: Macmillan.

Polis, G. A. 1981. The evolution and dynamics of intraspecific predation. *Annu. Rev. Ecol. Syst.* **12**: 225–251.

–. 1984. Intraspecific predation and "infant killing" among invertebrates. In *Infanticide. Comparative and Evolutionary Perspectives*. G. Hausfater and S. B. Hrdy, eds., pp. 87–104. New York: Aldine.

Pond, C. M. 1977. The significance of lactation in the evolution of mammals. *Evolution* **31**: 177–199.

–. 1983. Parental feeding as a determinant of ecological relationships in Mesozoic terrestrial ecosystems. *Acta Palaeontol. Pol.* **28**: 215–224.

Pruthi, H. S. 1933. An interesting case of maternal care in an aquatic cockroach, *Phlebonotus pallens* Serv. (Epilamprinae). *Curr. Sci. (Bangalore)* **1**: 273.

Rau, P. 1940. The life history of the american cockroach, *Periplaneta americana* Linn. (Orthop.: Blattidae). *Entomol. News* **51**: 186–189, 223–227.

–. 1943. How the cockroach deposits its egg case; a study in insect behavior. *Ann. Entomol. Soc. Am.* **36**: 221–226.

Reuben, L. V. 1988. Some aspects of the bionomics of *Trichoblatta sericea* (Saussure) and *Thorax porcellana* (Saravas) (Blattaria). Ph.D. dissertation, Loyola College, Madras, India.

Richards, A. G. and M. A. Brooks. 1958. Internal symbiosis in insects. *Annu. Rev. Entomol.* **3**: 37–56.

Ricklefs, R. E. 1983. Avian postnatal development. *Avian Biol.* **7**: 1–83.

Rollo, C. D. 1984. Resource allocation and time budgeting in adults of the cockroach *Periplaneta americana*: the interaction of behaviour and metabolic reserves. *Res. Pop. Ecol.* **26**: 150–187.

–. 1986. A test of the principle of allocation using two sympatric species of cockroaches. *Ecology* **67**: 616–628.

Ross, M. H. and D. E. Mullins. 1995. Biology. In *Understanding and Controlling the German Cockroach*. M. K. Rust, J. M. Owens and D. A. Rierson, eds. New York: Oxford University Press.

Roth, L. M. 1968a. Oothecae of Blattaria. *Ann. Entomol. Soc. Am.* **61**: 83–111.

–. 1968b. Ovarioles of Blattaria. *Ann. Entomol. Soc. Am.* **61**: 132–140.

–. 1970. The stimuli regulating reproduction in cockroaches. *Coll. Int. Cent. Nat. Res. Sci.* **189**: 267–286.

–. 1977. A taxonomic revision of the Panesthiinae of the world I. The Panesthiinae of Australia (Dictyoptera: Blattaria: Blaberidae). *Austr. J. Zool. Suppl. Ser.* **48**: 1–112.

–. 1979. Cockroaches and plants. *Horticulture*, August: 12–13.

–. 1980. Cave dwelling cockroaches from Sarawak, with one new species. *Syst. Entomol.* **5**: 97–104.

–. 1981. The mother–offspring relationship of some blaberid cockroaches (Dictyoptera: Blattaria: Blaberidae). *Proc. Entomol. Soc. Wash.* **83**: 390–398.

–. 1982. Ovoviviparity in the blattellid cockroach, *Symploce bimaculata* (Gerstaecker) (Dictyoptera: Blattaria: Blattellidae). *Proc. Entomol. Soc. Wash.* **84**: 277–280.

–. 1985. A taxonomic revision of the genus *Blattella* Caudell (Dictyoptera, Blattaria: Blattellidae). *Entomol. Scand. Suppl.* **22**: 1–221.

–. 1989. *Sliferia*, a new ovoviviparous cockroach genus (Blattellidae) and the evolution of ovoviviparity in Blattaria (Dictyoptera). *Proc. Entomol. Soc. Wash.* **91**: 441–451.

–. 1991. Blattodea, Blattaria (cockroaches). In *The Insects of Australia*. CSIRO, pp. 320–329. Ithaca: Cornell University Press.

Roth, L. M. and D. W. Alsop. 1978. Toxins of Blattaria. In *Arthropod Venoms*. S. Bettini, ed., *Handb. Exp. Pharmacol.* **48**: 465–487.

Roth, L. M. and S. Cohen. 1973. Aggregation in Blattaria. *Ann. Entomol. Soc. Am.* **66**: 1315–1323.

Roth, L. M. and W. Hahn 1964. Size of new-born larvae of cockroaches incubating eggs internally. *J. Insect Physiol.* **10**: 65–72.

Roth, L. M. and B. Stay. 1959. Control of oocyte development in cockroaches. *Science (Wash., D.C.)* **130**: 271–272.

–. 1962. A comparative study of oocyte development in false ovoviviparous cockroaches. *Psyche* **69**: 165–208.

Roth, L. M. and E. R. Willis. 1954a. The reproduction of cockroaches. *Smithsonian Misc. Coll.* **122**: 1–49.

–. 1954b. *Anastus floridanus* (Hymenoptera: Eupelmidae) a new parasite on the eggs of the cockroach *Eurycotis floridana*. *Trans. Am. Entomol. Soc.* **80**: 29–41.

–. 1955a. Intra-uterine nutrition of the "beetle-roach" *Diploptera dytiscoides* (Serv.) during embryogenesis, with notes on its biology in the laboratory (Blattaria: Diplopteridae). *Psyche* **62**: 55–68.

–. 1955b. Water content of cockroach eggs during embryogenesis in relation to oviposition behavior *J. Exp. Zool.* **128**: 489–509.

–. 1957. An analysis of oviparity and viviparity in the Blattaria. *Trans. Am. Entomol. Soc.* **83**: 221–238.

–. 1960. The biotic associations of cockroaches. *Smithsonian Misc. Coll.* **141**: 1–470.

Rugg, D. and H. A. Rose. 1984a. Reproductive biology of some Australian cockroaches (Blattodea: Blaberidae). *J. Austr. Entomol. Soc.* **23**: 113–117.

–. 1984b. The taxonomic significance of reproductive behaviour in some Australian cockroaches (Blattodea: Blaberidae). *J. Austr. Entomol. Soc.* **23**: 118.

–. 1984c. Intraspecies association in *Panesthia cribrata* (Sauss.) (Blattodea: Blaberidae). *Gen. Appl. Entomol.* **16**: 33–25.

–. 1991. Biology of *Macropanesthia rhinocerus* Saussure (Dictyoptera: Blaberidae). *Ann. Entomol. Soc. Am.* **84**: 575–582.

Sacchi, L. and A. Grigolo. 1985. Behavior of symbionts during oogenesis and early stages of development in the German cockroach *Blattella germanica* (Blattodea). *J. Invert. Path.* **46**: 139–152.

–. 1989. Endocytobiosis in *Blattella germanica* L. (Blattodea): recent acquisitions. *Endocytobiosis Cell Res.* **6**: 121–147.

Schal, C. and W. J. Bell. 1982. Ecological correlates of paternal investment of urates in a tropical cockroach. *Science (Wash., D.C.)* **281**: 171–173.

Schal, C., J.-Y. Gautier and W. J. Bell. 1984. Behavioral ecology of cockroaches. *Biol. Rev.* **59**: 209–254.

Scott, H. 1929. On some cases of maternal care displayed by cockroaches and their significance. *Entomol. Mon. Mag.* **65**: 218–222.

Scrivener, A. M., M. Slaytor and H. A. Rose. 1989. Symbiont-independent digestion of cellulose and starch in *Panesthia cribrata* Saussure, an Australian wood-eating cockroach. *J. Insect Physiol.* **35**: 935–941.

Seelinger, G. and U. Seelinger. 1983. On the social organization, alarm and fighting in the primitive cockroach *Cryptocercus punctulatus*. *Z. Tierpsychol.* **61**: 315–333.

Séin, F. Jr. 1923. *Cucarachas*. Puerto Rico Insular Exp. Sta. Circ. 64, 12pp.

Shaw, E. 1925. New genera and species (mostly Australasian) of Blattidae, with notes, and some remarks on Tepper's types. *Proc. Linn. Soc. N.S.W.* **1**: 171–213.

Shelford, R. 1906. Studies of the Blattidae. VI. Viviparity amongst the Blattidae. *Trans. Entomol. Soc. Lond.* **54**: 509–514.

Shine, R. 1985. The evolution of viviparity in reptiles: an ecological analysis. In *The Biology of the Reptilia*. C. Gans and F. Billett, eds., pp. 606–694. New York: John Wiley and Sons.

–. 1989. Ecological influences on the evolution of vertebrate viviparity. In *Complex Organismal Functions: Integration and Evolution in Vertebrates*. D. B. Wake and G. Roth, eds., pp. 263–278. New York: John Wiley and Sons.

Silverman, J., G. I. Vitale and T. J. Shapas. 1991. Hydramethylnon uptake by *Blattella germanica* (Orthoptera: Blattellidae) by coprophagy. *J. Econ. Entomol.* **84**: 176–180.

Skiafe, S. H. 1979. *African Insect Life*. Capetown: C. Struik Publishers.

Slansky, F. Jr. and J. M. Scriber. 1985. Food consumption and utilization. *Comp. Insect Physiol. Biochem. Pharm.* **4**: 87–163.

Slaytor, M. 1992. Cellulose digestion in termites and cockroaches: what role do symbionts play? *Comp. Biochem. Physiol. B* **103**: 775–784.

Snart, J. O. H., M. Greenwood, R. Beck and K. C. Highnam. 1984a. The functional morphology of the brood sac in two species of ovoviviparous cockroaches *Byrsotria fumigata* (Guerin) and *Gromphadorhina portentosa* (Schaum). 1. Scanning and light microscopy. *Int. J. Invert. Repro. Devel.* **7**: 345–355.

–. 1984b. The functional morphology of the brood sac in two species of ovoviviparous cockroaches *Byrsotria fumigata* (Guerin) and *Gromphadorhina portentosa* (Schaum). 2. Transmission electron microscopy. *Int. J. Invert. Repro. Devel.* **7**: 357–367.

Stay, B. 1977. Fine structure of two types of pleuropodia in *Diploptera punctata* (Dictyoptera: Blaberidae) with observations on their permeability. *Int. J. Insect Morphol. Embryol.* **6**: 67–95.

Stay, B. and A. C. Coop. 1973. Developmental stages and chemical composition in embryos of the cockroach, *Diploptera punctata*, with observations on the effect of diet. *J. Insect Physiol.* **19**: 147–171.

–. 1974. Milk secretion for embryogenesis in a viviparous cockroach. *Tissue Cell* **6**: 669–693.

Stay, B. and L. M. Roth. 1962. The colleterial glands of cock-roaches. *Ann. Entomol. Soc. Am.* **55**: 124–130.

Stay, B., A. King and L. M. Roth. 1960. Calcium oxylate in the oothecae of cockroaches. *Ann. Entomol. Soc. Am.* **53**: 79–86.

Stearns, S. C. 1992. *The Evolution of Life Histories.* Oxford University Press.

Sugumaran, M., V. Semensi, H. Dali and K. Nellaiappan. 1991. Oxidation of 3,4-dihydroxybenzyl alcohol: a sclerotizing precur-sor for cockroach ootheca. *Arch. Insect Biochem. Physiol.* **16**: 31–44.

Tallamy, D. W. 1994. Nourishment and the evolution of paternal investment in subsocial arthropods. In *Nourishment and Evolution in Insect Societies.* J. H. Hunt and C. A. Nalepa, eds., pp. 21–55. Boulder: Westview Press.

Tallamy, D. W. and T. K. Wood. 1986. Convergence patterns in social insects. *Annu. Rev. Entomol.* **31**: 369–390.

Tanaka, A. 1976. Stages in the embryonic development of the German cockroach, *Blattella germanica* Linné (Blattaria, Blattel-lidae). *Kontyû* **44**: 512–525.

Vishniakova, V. N. 1968. Mesozoic blattids with external ovipositor, and peculiarities of their reproduction. In *Jurassic Insects of Karatau.* B. B. Rohdendorf, ed., pp. 55–85. SSSR: Akad Nauk. (in Russian).

West, M. J. and R. D. Alexander. 1963. Sub-social behavior in a burrowing cricket *Anugryllus muticus* (De Geer). Orthoptera: Gryllidae. *Ohio J. Sci.* **63**: 19–24.

Wille, J. 1920. Biologie und bekampfung der deutschen schabe (*Phyllodromia germanica* L.). *Monogr. Angew. Entomol. (Suppl. I)* **7**: 1–140.

Willis, E. R. 1966. Biology and behavior of *Panchlora irrorata*, a cockroach adventive on bananas. *Ann. Entomol. Soc. Am.* **59**: 514–516.

Willis, E. R., G. R. Riser and L. M. Roth. 1958. Observations on reproduction and development in cockroaches. *Ann. Entomol. Soc. Am.* **51**: 53–69.

Wolcott, G. N. 1948. The insects of Puerto Rico. *J. Agric. Univ. Puerto Rico* **32**: 1–224.

Wren, H. N. and D. G. Cochran. 1987. Xanthine dehydrogenase activity in the cockroach endosymbiont *Blattabacterium cuenoti* (Mercier 1906) Hollande and Favre 1931 and in the cockroach fat body. *Comp. Biochem. Physiol.* **88**: 1023–1026.

Zeveloff, S. I. and M. S. Boyce. 1982. Parental investment and mating systems in mammals. *Evolution* **34**: 973–982

4 · The spectrum of eusociality in termites

JANET S. SHELLMAN-REEVE

ABSTRACT

I discuss phylogenetic relationships, nesting and feeding habits, kinship and ecological determinants of social behavior in the second largest taxon exhibiting eusociality, the termites. Contrary to previous hypotheses, cladistic analyses indicate that the woodroach *Cryptocercus* is probably not a sister taxon to the termites. Focusing on the reproductive consequences of termite societies, I place termite species along points on the spectrum of eusociality according to the reproductive potential of their offspring. The tendencies of workers to maintain their reproductive options in some species but not others are explained, in part, by termite nesting and feeding habits of four life types. Each life type is associated with a level of resource stability (nest and food resources) that sets the upper limit on the extent of worker altruism because resources influence maximum colony longevity. Cycles of inbreeding may yet be important in genera whose colonies are typically headed by sibling-mated supplemental reproductives. I discuss how reproductive-replacement strategies might minimize the effects of reduced genetic variability. Direct and indirect benefits explain why single-site and many multiple-site nesters provide alloparental care. Only indirect benefits, i.e. helping kin, appear to explain why central-site nesters and numerous multiple-site nesters stay and help. I suggest that the cumulative indirect fitness benefits (the potential benefits over time) earned per altruist may best explain individual decisions in termites throughout the social spectrum. In summary of key determinants, I show how eusociality is a 'mixed conditional strategy', where alternative reproductive options are frequency dependent and yield equal expected inclusive fitness pay-offs.

INTRODUCTION

All termite species exhibit some form of cooperative care among male and female reproductive pairs and non-reproductive workers and soldiers. This complex form of care is characterized by presence of reproductives and non-reproductive laborers for at least two overlapping generations (Wilson 1971; Sherman *et al.* 1995), and is the broad definition of eusociality I use to discuss the termites. At one extreme of the eusocial spectrum, some termites live in small colonies, in which a male and a female reproductive are helped by temporary workers and soldiers that are themselves future reproductives. At the other extreme are large, elaborate societies (highly complex societies) in which reproductives are assisted by permanent altruists, i.e. effectively sterile workers and soldiers that specialize to rear non-laboring future reproductives. Ecological constraints on independent reproduction and the fitness advantages associated with staying and working in a stable nest environment favor cooperative breeding and alloparental care by offspring in diverse vertebrate and arthropod taxa (see, for example, Hamilton 1964, 1972; Alexander 1974; Eickwort 1981; West-Eberhard 1981; Andersson 1984; Emlen 1984, 1991; Queller 1989, 1994; Alexander *et al.* 1991; Sherman *et al.* 1995), including termites. In the last section of this chapter, I show how ecological pressures, resource stability, and direct and indirect benefits shape social decisions of termites along the eusocial spectrum. First, however, I introduce the reader to this large taxon by summarizing termite phylogenetic relationships, by categorizing termite societies along the spectrum of eusociality by worker reproductive potential (*sensu* Sherman *et al.* 1995), and by describing the major life types of termites, i.e. their nesting and feeding habits.

PHYLOGENY AND EUSOCIAL CHARACTERISTICS OF TERMITES

Social origins via phylogenetic relationships

There are over 2291 fossil and living termite species that have been classified into seven families (sometimes more or less) in the order Isoptera (Fig. 4-1a, adapted from

(a)

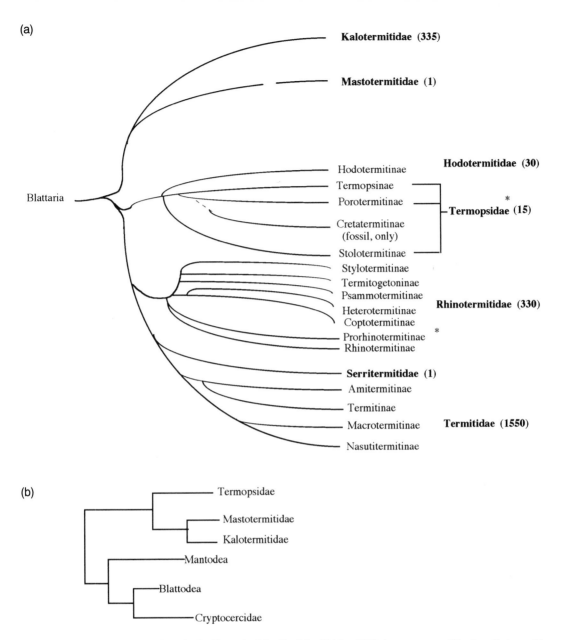

Figure 4-1(a) The phylogenetic tree of termite families and subfamilies (after Krishna 1970) shows recent modifications (designated by asterisks). One modification from Krishna (1970) is a split within the family Hodotermitidae: the three subfamilies (Termopsinae, Porotermitinae and Stolotermitinae), formerly of the Hodotermitidae, are considered to form in the family Termopsidae (see Grassé 1986). Grassé (1986) favored the split from the remaining subfamilies of the Hodotermitidae (Cretatermitinae and Hodotermitinae) because of habitat differences: species currently in the Hodotermitidae live in desert regions, whereas all species in the new family Termopsidae live in rotting logs. Roonwal and Chhotani (1989) have argued that only the Termopsinae should be placed in the Termopsidae based on the simplicity of wing venation of the Termopsinae relative to that found in other subfamilies making up the group. A second modification involved the genus *Prorhinotermes*: this genus was removed from the Rhinotermitinae and placed into its own subfamily, the Prorhinotermitinae, because the soldier caste of *Prorhinotermes* shows an absence of a 'daubing brush' of hairs on the labrum (Quennedey and Deligne 1975; Roonwal and Chhotani 1989). Estimates of the number of living species per family are indicated in parentheses. (b) The phylogenetic tree shown here is the result of Thorne and Carpenter's (1992) cladistic analysis of 70 characters in representative species in the Mantodea, Blattodea, Cryptocercidae, and three of seven termite families. Traits in the analysis include primarily morphological characters and some life-history traits, e.g. mode of foraging, soldier reproduction, soldier defense, soldier and worker morphological characters.

Krishna 1970; see also Emerson and Krishna 1975; Grassé 1986; Huang *et al.* 1987). The nearly cosmopolitan geographical distribution of more than 60 fossil termites and extant termites of most families has suggested to termitologists that most families evolved before the breakup of Pangaea in the Cretaceous period (see, for example, Emerson 1955; Krishna 1970; Emerson and Krishna 1975; Grassé 1986; Huang *et al.* 1987). Traditional systematic analyses of morphological characters (Emerson 1965; Krishna 1970; Emerson and Krishna 1975; Grassé 1986; Roonwal and Chhotani 1989), cladistic analysis of morphological, ecological, and social life-history characters (Thorne and Carpenter 1992), and cladistic molecular analyses (Vawter 1991; DeSalle *et al.* 1992; Kambhampati 1995; Bandi *et al.* 1995) provide evidence that the families Mastotermitidae and Termopsidae represent the two oldest living lineages in the Isoptera.

Much attention has focused on these two basal families in the Isoptera in an attempt to understand the ancestral characteristics of the first eusocial termites. The primitive morphological characters exhibited by the only extant species in the Mastotermitidae, such as the production of a membranous egg pod resembling a cockroach ootheca, the enlarged anojugal area of the hind wing, an expanded pronotum, the presence of intracellular bacteroids, and genitalia structures that are more complex than those of other termites, indicate that it shares more characters with cockroaches than do termites of other lineages (Emerson 1965; McKittrick 1965; Wilson 1971). The results of a cladistic analysis, however, led Thorne and Carpenter (1992) to emphasize derived characteristics of the Mastotermitidae. They also showed that the Termopsidae shares some primitive morphological characters with cockroaches that the Mastotermitidae do not, such as the presence of three mandibular teeth, the greater segmentation of the cerci, and the lack of curvature in the forewing. Indeed, in contrast to previous systematic analyses (Emerson 1965; Krishna 1970; Emerson and Krishna 1975), Thorne and Carpenter (1992) tentatively place the Termopsidae basal to the Mastotermitidae and Kalotermitidae (Fig. 4-1b). Their cladistic analysis, however, does not rule out the possibility that the Kalotermitidae is a sister group of the Mastotermitidae (Krishna 1970), which makes basal placement of one family over the other not yet possible.

To make broad comparisons between family groups in the Dictyopterodea complex, and where the order Isoptera is considered as a suborder within the Dictyoptera (McKittrick 1965; Vawter 1991; Thorne and Carpenter

1992; Kambhampati 1995), Thorne and Carpenter (1992) typically used a 'hypothetical' family member to characterize each family, i.e. characters for each family were compiled from several species. To address questions regarding which characters are more primitive and which lineage is more basal, the Mastotermitidae or the Termopsidae, an analysis at the species level, rather than at the family level, will likely provide greater resolution, using multiple species from all termite families, the Blattaria, other possible sister taxa, and outgroup taxa.

Nest, food, and social habits in conjunction with the developmental characteristics of the Mastotermitidae and the Termopsidae have also been used in reconstructions of termite eusociality. The nesting and feeding habits of the Mastotermitidae, which involve the ability to migrate through the soil to multiple food and nest sites, accompany a suite of characters that are considered highly derived, such as large colony size, inflexibility in development of offspring, and a functionally sterile soldier caste (see Watson and Sewell 1985; Thorne and Carpenter 1992). In contrast, species in the Termopsidae exhibit simpler nesting and feeding habits that do not involve the migration of a colony outside its single wood nest and food site. Termites of these 'single-site' wood-feeders also exhibit the simplest of social characteristics in comparison with termites exhibiting other life types (Noirot 1970, 1985a,b,c; Noirot and Pasteels 1987). Because the life habits of the Termopsidae are considered much less complex than those of the Mastotermitidae, termites like the Termopsidae, that live and feed entirely within a single wood site, are proposed to represent the 'primitive' life type, i.e. the life type of the first eusocial termites, whereas species that could migrate to multiple food and nest sites are considered 'advanced' life types (Noirot 1985a,b,c; Noirot and Pasteels 1987). Social and developmental characters, however, are not reliable indicators of 'primitive' versus 'advanced' conditions, because nesting and feeding habits may influence their expression, and hence, these may be 'convergent' characteristics (Watson and Sewell 1981, 1985). We still cannot rule out the possibility that the first eusocial ancestor, now extinct, located its nest in wood or soil, and gathered a complete and balanced diet not only from rotting wood, but also leaf and stem litter, seeds and other plant tissues (a diet similar to that of many cockroaches living today and, to some extent, *Mastotermes darwiniensis*) (Wilson 1971). Living within and feeding entirely off wood may be a derived trait; perhaps one that required greater coevolution between termite and specialized hindgut symbionts that currently sustain termites that feed entirely on wood (Bignell 1994).

To understand the possible events at the origins of termite eusociality, attention has also been directed at sister taxa that might resemble ancestral termites. The wood-roach genus (*Cryptocercus*), previously considered a basally branching member of the Blattaria (McKittrick 1965), has long been considered a sister taxon to the termites (Cleveland *et al.* 1934; Emerson 1938; McKittrick 1965; see Wilson 1971). Not only does *Cryptocercus* exhibit a similar nesting and feeding biology to wood-dwelling termites in the families Termopsidae and Kalotermitidae (see, for example, Emerson 1938; Noirot 1970; Nalepa 1988, 1994; Roisin 1993, 1994), it also harbors hindgut-protozoa symbionts similar to those found in termites in the families Mastotermitidae, Termopsidae, Kalotermitidae, Rhinotermitidae, Hodotermitidae and Serritermitidae (Cleveland *et al.* 1934; Honigberg 1970; Nalepa 1994; see also Bandi *et al.* 1995, for comparisons of endosymbionts). These protozoa are not yet found in other wood-feeding insects, and only some species of cockroaches are known to be hosts to closely related protozoa (Honigberg 1970). Previous studies documented that protozoa in extant termites must be transferred directly from one member to another or else protozoa would die from exposure to the air (see Cleveland *et al.* 1934; reviewed in Honigberg 1970). The necessity of direct transfaunation in extant termites has been used to argue that *Cryptocercus* and termites shared some common ancestry (Cleveland *et al.* 1934; Nalepa 1991) and that protozoa transfaunation favored the evolution of subsociality (Nalepa 1994) or even eusociality (Cleveland *et al.* 1934; Wilson 1971; Lin and Michener 1972).

The direct transfaunation hypothesis, however, provides weak support for a sister-taxon relationship and for the evolution of sociality, because convergent similarities in the nutritional ecology of *Cryptocercus* and wood-feeding species of termites may account for the common types of microbes that they share (Thorne 1990; Thorne and Carpenter 1992; Grandcolas and Deleporte 1992; Kambhampati 1995). Thorne (1990, 1991) demonstrated under laboratory conditions that the wood-dwelling termite *Zootermopsis* could have acquired these microbes by eating its heterospecific wood-feeding competitors, such as *Cryptocercus* (but cf. Nalepa 1991). Over the millions of years that symbiosis has evolved in termites, other modes of microbial acquisition also seem possible. Numerous protozoan species are capable of encysting, i.e. they form an outer sheath that offers protection from the aerobic environment

while outside the host's body (Hungate 1955; Honigberg 1970). Wood-feeding termites may have acquired encysted protozoa while they fed at contaminated feeding grounds that were visited by other herbivorous insects.

Subsocial or eusocial care need not have arisen solely out of the need to pass symbionts on to offspring. Indeed, Honigberg (1970) argued that protozoa eventually lost the benefits of encystment in termites because microbes were passed directly to colony members via the termite's anal exchange of nutrients: after termites formed a social life (see also Nalepa 1994). Under some circumstances, encystment can be induced in at least one protozoan species in the termite *Zootermopsis* (Messer 1990; Messer and Lee 1989). Thus, in a hypothetical ancestral termite that enjoyed the absence of ecological constraints that might have caused sociality, parents could have deposited encysted protozoa near an egg case to facilitate the subsequent inoculation of solitary-dwelling nymphs, since encysted protozoa are capable of surviving long periods outside the host. Offspring could also have picked up encysted symbionts directly by feeding on heterospecifics (*sensu* Thorne 1990, 1991), or from contaminated feeding grounds visited by heterospecific or conspecific adults, thus precluding direct transfaunation as a singular mechanism favoring eusocial or subsocial care.

Cladistic analyses of *Cryptocercus* with other family representatives in the Dictyoptera–Isoptera group (Thorne and Carpenter 1992; cf. Grandcolas 1994a), and of *Cryptocercus* with other cockroaches in the subfamily Polyphaginae (Grandcolas and Deleporte 1992; Grandcolas 1994a) also suggest that the woodroach has a more distant phylogenetic association with termites than was proposed by McKittrick (1965). Recent cladistic analyses by Grandcolas and Deleporte (1992) and Grandcolas (1994a) prompted them to remove *Cryptocercus* as a basal member of the Blattaria, and place it with a sister group *Therea*, as a highly derived member in the family Polyphagidae. Grandcolas and Deleporte (1992) argue that numerous derived morphological characters and its similarity to cockroaches in Polyphagidae make it an unlikely sister taxon to the termites (see also Grandcolas 1994a). Indeed, the relatively restricted biogeographical distribution of the genus *Cryptocercus* (Asia and North America) suggests to Grandcolas (1994b) that this woodroach may not have arisen until the Recent epoch, and hence was not in existence at the time most families of termites evolved. The cladistic data are also supported by the analysis of Thorne and Carpenter (1992), who concluded independently that, despite many similarities, *Cryptocercus* life history is not homologous with termite life history. Vawter's (1991) molecular phylogenetic analysis of *Cryptocercus*, a cockroach in the Blattidae, and termite representatives from Mastotermitidae and Rhinotermitidae also supports this view. Sequence data from the nuclear small subunit ribosomal RNA gene of *Mastotermes darwiniensis* in the Mastotermitidae suggests a close sister-taxon relationship with blattid roaches, excluding *Cryptocercus* (Vawter 1991). Indeed, comparisons of sequence data for *Blatta orientalis*, *Mastotermes darwiniensis*, and *Reticulitermes flavipes*, taken with other morphological characters, led Vawter to propose that the Mastotermitids belong with the Blattaria rather than with other Isoptera lineages. In contrast to these findings, Kambhampati's (1995) recent phylogenetic analysis of a mitochondrial ribosomal RNA gene of several blattarids, blaberids, and termite species led him to argue that *M. darwiniensis* does belong with the termite lineages. Both Kambhampati (1995) and Vawter (1991), however, find that termite taxa, with *M. darwiniensis* included, are more closely related to the Blattaria than they are to *Cryptocercus*. It can no longer be assumed that the most ancestral termites exhibited the wood-dwelling, wood-feeding characteristics of *Cryptocercus*. Feeding and nesting habits of the most ancient social termites, whether it was strictly wood-feeding and wood-dwelling (like *Zootermopsis* spp.) or a relatively more generalized feeding insect that lived in or adjacent to wood (like other blattid genera and *M. darwiniensis*), remains unknown.

The eusocial spectrum, life types, and corresponding social characteristics

Life types and social behaviors of termites have been categorized in various ways over the last several decades. The 'lower termites', which include Mastotermitidae, Kalotermitidae, Termopsidae, Hodotermitidae, Rhinotermitidae and Serritermitidae, are defined by their hindgut protozoa associations, while the 'higher termites', which include the Termitidae, are defined primarily by their external association with fungi and bacteria and loss of hindgut protozoa (see, for example, Honigberg 1970; Noirot 1970; Wood and Johnson 1986). Although this grouping is sometimes used to describe social differences among termites, it is not ideal, because highly complex social species (i.e. species with worker and soldier specialists) are present in both groups.

The phylogeny of termites is often referred to when the degrees of worker altruism in various termite societies are compared (Emerson 1955, 1965; Noirot 1985a,b,c, 1989; Abe 1987, 1991; Myles 1988; Noirot and Pasteels 1987; Roonwal and Chhotani 1989; Nalepa 1994; Roisin 1994). Terms such as 'primitive' social behavior (for example, with reference to wood-dwelling termites) or 'advanced' social behavior, (for example, with reference to soil-dwelling termites) imply that the social complexity of termites, i.e. its degrees of social integration and task specialization, coincides with their phylogeny. Phylogeny, however, does not necessarily correspond with social complexity (Watson and Sewell 1981, 1985; Noirot 1985a,b,c; Thorne and Carpenter 1992; Lenz 1994). For example, the single living termite *Mastotermes darwiniensis* in the relatively basal lineage Mastotermitidae exhibits a eusocial life history similar to that exhibited by species of more recently evolved termite lineages, e.g. species of *Reticulitermes* in the Rhinotermitidae (Watson and Sewell 1985). Similarly, the genus *Prorhinotermes*, in the recently evolved lineage Rhinotermitidae, exhibits a social life history similar to that exhibited by species in the relatively basal lineage Termopsidae (Miller 1942; Roisin 1988). Moreover, lineages that are extinct or nearly extinct (e.g. the Mastotermitidae) (Emerson 1965; Krishna 1970; Emerson and Krishna 1975) can obscure phylogenetic patterns of nesting and social behaviors. As Lenz (1994) points out, social behaviors of termites that do not support a phylogenetic scheme might best be understood by exploring the nesting and feeding habits of extant species of termites. Indeed, a phylogenetic analysis, although it might be useful, is not required to understand the functional advantages and maintenance of eusociality (Reeve and Sherman 1993).

In understanding the functional advantages and maintenance of eusociality, a fundamental correlate with eusocial complexity in termites is the reproductive potential of offspring produced in the colony. Offspring reproductive potential may be one of the best ways to characterize termite eusociality throughout its range. From Hamilton's (1964) rule, the extent to which there is worker altruism should be a function of the opportunity for personal reproduction. For altruism to be favored, $rb - c > 0$ (after Grafen 1984), where c is the net cost in terms of the loss of personal reproduction due to altruism, r is the genetic relatedness between worker and breeder, and b is the breeders' increase in reproduction due to altruism. When workers' direct reproductive opportunities are very limited, c will be very small, and altruism is thus more likely to occur (see also

West-Eberhard 1975, 1981). However, when indirect repro-ductive opportunities diminish, rb will be small, and direct reproduction will be favored at $rb - c < 0$, even with limited breeding opportunities.

Sherman *et al.* (1995) have quantified the degree of euso-ciality based on the degree of worker reproduction, and use it to classify social complexities of diverse vertebrate and arthropod taxa. The term they use, 'reproductive skew', measures the proportion of offspring that give up reproduc-tion. For purposes here, reproductive skew is calculated over the lifetime of the colony's members, not just while members reside within the group (see Reeve and Ratnieks 1993; Keller and Reeve 1994 for other measurements of reproductive skew in insects; and in vertebrates, Vehren-camp 1983; Emlen 1984). A low reproductive skew indicates that individuals produced in the colony become reproduc-tive at some point in their life; a high reproductive skew indicates that few individuals of a colony become reproduc-tive. Reproductive skew, in this broad sense, thus provides

an index of eusociality which varies from 0 to 1, where 0 is the case in which all colony members reproduce equally and 1 is the case in where each colony member foregoes direct reproduction, except for a single colony breeder (one from each sex). Thus, the degree of worker altruism is expected to increase as the index of eusociality increases. In termites, I focus on the reproductive potentials of a colony's workers and soldiers as an indicator of this index of eusociality. Termite workers and soldiers, collectively, which show a high degree of reproductive potential are placed on the left of the eusocial spectrum, near 0, whereas termite workers and soldiers which show a low degree of reproductive potential fall along the far right of the spec-trum, near 1 (Fig. 4-2; after Sherman *et al.* 1995). Thus, spe-cies of termites form a spectrum of eusociality according to worker and soldier reproductive potentials.

Position along the spectrum of eusociality is strongly correlated with termite food and nest habits. This is because food and nest resources associated with each life

0 to 0.25	0.2 to 0.50	0.40 to 0.75	0.70 to 1.0
in Termopsidae,	in Kalotermitidae,	Mastotermitidae;	in Rhinotermitidae
e.g. *Archotermopsis,*	e.g. *Postelectrotermes,*	in Kalotermitidae	e.g. many *Coptotermes,*
Hodotermopsis,	*Kalotermes,*	e.g. *Paraneotermes;*	*Schedorhinotermes,*
Zootermopsis;	*Incisitermes,*	in Rhinotermidae	in Hodotermitidae
Stolotermes,	*Pterotermes,*	e.g. *Heterotermes,*	e.g. *Hodotermes,*
Porotermes	*Cryptotermes,*	*Reticulitermes,* some	*Microhodotermes,*
	Neotermes,	*Coptotermes;*	*Anacanthotermes;*
	Bifiditermes,	Serritermitidae;	in Termitidae, e.g. *Amitermes,*
	Glyptotermes;	in Termitidae	*Microcerotermes, Cubitermes,*
	in Rhinotermitidae	e.g. *Termes,*	*Odontotermes, Macrotermes,*
	Prorhinotermes	*Prohamitermes*	*Nasutitermes, Cornitermes,*
			Trinervitermes

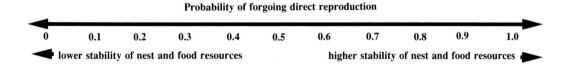

Figure 4-2. Termites are positioned along the spectrum of eusociality according to the estimated probability that alloparents (workers and soldiers) forgo direct reproduction. The index of eusociality varies from 0, representing species in which all alloparents attempt reproduction, to 1, representing species in which all alloparents are functionally sterile. Examples of termite genera that fall along the eusocial spectrum include the following: from 0 to 0.25, most workers and some soldiers attempt reproduction within their lifetime; from 0.20 to 0.50, most workers attempt reproduction and no soldiers reproduce; from 0.40 to 0.75, few workers reproduce and no soldiers reproduce; from 0.70 to 1.0, few to no workers reproduce, no soldiers reproduce, and task specializations based on morphology and sex are strongly expressed. The position of genera along the spectrum of eusociality is strongly correlated with termites' life types, which are defined by the stability of nest and food resources.

Table 4-1. *Distinguishing social characteristics associated with each nest type*

Characteristic	Single-site nesters: Termopsidae, most Kalotermitidae and the Prorhinotermitinae	Multiple-site nesters: Mastotermitidae, few Kalotermitidae, most Rhinotermitidae and some Termitidae	Central-site nesters: some Rhinotermitidae, most Termitidae and the Hodotermitidae	Inquiline-site nesters: some Rhinotermitidae, Serritermitidae and some Termitidae
Mode of colony formation	Colonies are started by a male and female dealate pair[a]	Colonies are started by a cooperative male and female dealate pair[a,b,c] or by a cohort of neotenics and workers[b,c]	Colonies are started by a cooperative male and female pair (sometimes multiple reproductives)[c,d,e,f] or by secondary (dealates) reproductives, brachypterous neotenics with workers, rarely ergatoid reproductives[d,e,f,g,z]	Colonies are started by a cooperative male and female pair or cohort of multiple neotenics with workers (Table 4-2)
Juvenile workers and non-workers	All juveniles perform work[h]; some workers exhibit laziness, i.e. they show reduced work compared with others[i,j]	In all species, all non-dispersing juveniles work[k,l,m]; degree of laziness not known; in *M. darwiniensis* dispersers' work status is not known[l]; in Rhinotermitidae and Termitidae dispersers do not work[m]	All non-dispersing juveniles work[k,m]; degree of laziness not known; dispersers do not work[m]	Work status not clearly defined (but see Table 4-2)
Task specializations	Soldier-morph specialists[h] and age polyethism[i]	Task specialization involves soldier-morph specialists[g] and age polyethism[k]	Task specialization involves soldier-morph specialists (except in Apicotermitinae)[k,n], age polyethism[k], other worker-morph specialists, sex specialists in most genera[k,m,n]	Task specialization involves soldier-morph specialists; other task characteristics not clearly defined (Table 4-2)

Age of primary reproductives	Field-observed reproductives can live 4–5 years in *Zootermopsis* spp.[o] and *Incisitermes schwarzi*[p]	*M. darwiniensis* lab-reared reproductives live up to 17 years[q]; *C. formosanus* at least 9 years[r]; *R. lucifugus* primaries and neotenics lived up to 7 years in the lab[a]	Primary reproductives estimated to live 20 years[s]; queens may live 20–50 years in Macrotermitidae spp. (estimates based on size of physogastric queens)[t]	Not known
Age of workers	*Zootermopsis* workers live up to 4 years prior to winged-reproductive age[a]; soldiers live 4–5 years[o]. *Neotermes insularis* soldiers and temporary workers lived more than 6 years in the lab[c]; *Incisitermes schwarzi* soldiers at least as similar to primary reproductives[p]	*R. lucifugus* soldiers lived up to 5 years in lab[a]; *R. hesperus*, 3–5 years as workers in the field[u]; *Coptotermes a.acinaciformis* workers and soliders survived for four years in the lab[c]	*Macrotermes* spp. workers and soldiers survive less than a year in the field and lab[z,w,x]; *Macrotermes natalensis*, up to 1.5 year in the lab[v]; *Cubitermes ugandensis*, mean worker life span, 6 months to 1 year[y]	Not known

Sources: general citations are included here, but see text also. [a] Nutting 1969; [b] Weesner 1970; [c] Gay and Calaby 1970; [d] Noirot 1970; [e] Thorne 1985; [f] Roisin 1993; [g] Thorne and Noirot 1982; [h] Miller 1969; [i] Howse 1968; [j] Springhetti 1990; [k] McMahan 1979; [l] Watson and Sewell 1985; [m] Noirot 1985c, 1989; [n] Noirot 1969; [o] Heath 1907; [p] Luykx 1993; [q] Watson and Abbey 1989; [r] Huang *et al.* 1987; [s] Roonwal 1970; [t] Grassé 1984; [u] Pickens 1934; [v] Josens 1982; [w] Darlington 1991; [x] Collins 1981; [y] Bouillon 1970; [z] Darlington 1993 type set an upper limit on how long colonies can persist, which in part, appears to determine the cost and benefits of becoming a dispersing reproductive versus remaining a worker (Table 4-1) (discussed in Watson and Sewell 1981, 1985; Abe 1987, 1991; Higashi *et al.* 1991; Lenz 1994). Abe (1987) presents a classification of termite life types based on their distinct nesting and feeding habits (after Noirot 1970). These categories of termites, sometimes grouped into three life types (Abe 1987), sometimes grouped broadly into two life types (Abe 1991; Higashi *et al.* 1991; Lenz 1994), help to clarify how resource-stability affects the reproductive options of workers. For purposes of characterizing termites along the spectrum of eusociality, I discuss Abe's (1987) three major distinct life types, which are referred to here as single-site nesters, multiple-site nesters, and central-site nesters. In addition, I add a fourth one, the termite-inquiline nesters, also emphasized by Noirot (1970). For each life type, I describe the key social characteristics, and in particular the worker's reproductive potential.

Single-site nesters

The single-site nesters, about 10–15% of all termite species, include all genera in the family Termopsidae, e.g. *Zootermopsis* and *Porotermes*, most genera in the Kalotermitidae, e.g. *Kalotermes* and *Incisitermes*, and one genus in the Rhinotermitidae, *Prorhinotermes*. Single-site wood nesters, also called the one-piece type, are termite species whose entire colony life is spent at a single wood dwelling that serves as both shelter and food. When wood nutrients becomes depleted or the shelter becomes uninhabitable, the workers molt into winged adults, which disperse, until finally a colony dies. Because neither nest nor wood nutrient is replenishable, this life type typically provides the least stable nesting and feeding habitat compared with that of termites exhibiting other life types (Abe 1987; see also Watson and Sewell 1981, 1985). Log diameter, mass (Abe 1987), volume, and the quality of nutrients (Lenz 1994) have all been used to estimate the longevity of wood resources available to single-site nesters. Colony longevity varies from 4 to 15 years in most species, and longer in larger pieces of wood (Lenz 1994; see also Abe 1987). Maximum colony sizes in species in the Termopsidae and Kalotermitidae vary from 600 to 8000 inhabitants, and sometimes more (reviewed by Lenz 1994). For the most part (with some minor exceptions) (Lenz 1994), single-site nesters are unable to migrate to new nest sites, perhaps because of the high probability of migrating to another log that is already occupied by termites.

In single-site nesters, the relative instability of nest and food resources is strongly associated with a number of social characteristics (Table 4-1). One characteristic is that single-site nesters, as a rule, exhibit one mode of nest-founding: colonies are initiated by a dispersing male and female reproductive pair (Nutting 1969; Lafage and Nutting 1978; Nalepa and Jones 1991). The reproductive pair are active care-givers (in *Zootermopsis*, Shellman-Reeve 1990, 1994a,b; Rosengaus and Traniello 1993a; other species, Nalepa and Jones 1991), although care-giving is eventually taken over by offspring in the third instar of development and older (e.g. in *Zootermopsis*, Howse 1968; Rosengaus and Traniello 1993a; *Kalotermes*, Springhetti and Sita 1986; see also McMahan 1969). Only scanty information is available on the lifespans of reproductives in the field. However, it is expected that longevity of reproductives will only correspond to the longevity of the non-replenishable nest and food resource. Heath's (1907) field study on reproductive longevity in 213 colonies suggested that the longevity of a male was similar to that of the female; the average reproductive lives four to five years after becoming adult. However, only 0.03% of the colonies sampled contained primary reproductives at the end of the fifth year. Luykx's (1993) genetic survey of *Incisitermes schwarzi* revealed that most primary reproductives are still living in most mature colonies; only one quarter of the colonies is replaced by neotenics, i.e. offspring that become precocial replacement reproductives. In studies of *Kalotermes minor*, Harvey (1934) tentatively estimated that one colony and its single primary female had lived 14 years. Most mature colonies and remaining primary females, however, were estimated to be between five and six years old, and primary male reproductives were not observed in colonies older than three years. Reproductives of species that are adapted to wood nests of larger diameter and larger volume (e.g. *Porotermes adamsoni* in *Eucalyptus* forests) (Lenz 1985, 1994; Abe 1987) may be expected to enjoy greater longevity.

Another effect of resource instability in single-site nesters is on offspring development: offspring of single-site nesters develop along a linear pathway, i.e. typically there is straight line of development from egg to young offspring, then worker to alate, with alates being the primary developmental endpoint (Fig. 4-3a) (Watson and Sewell 1981). Within this basic developmental scheme, offspring exhibit remarkable developmental plasticity. In the damp-wood termite *Zootermopsis*, for example, any juvenile might function as a temporary worker (sometimes referred to as a

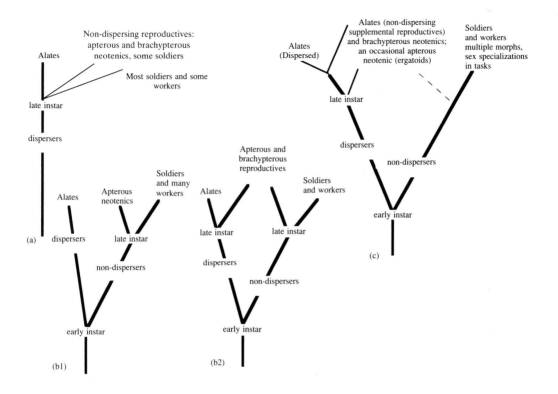

Figure 4-3. The endpoints of development can result in one of four reproductive forms and in one of two principal non-reproductive forms. The reproductive forms include: (1) *winged adult* (*alate*), which develops from a juvenile with wing pads; alates can disperse to a new nest site or can become a *secondary reproductive* as a *replacement reproductive* (i.e. when it replaces the parent in its natal nest) or as a *supplemental reproductive* (i.e. when it adds to reproduction of primary adult in the natal nest or starts a new colony by budding); (2) the *wing-padded neotenic* is a precocious reproductive that develops from a juvenile with wing pads, and can become a replacement or supplemental reproductive; (3) the *apterous neotenic* (sometimes called an ergatoid reproductive) is a precocious reproductive that develops from an apterous (wingless) worker, and can be a replacement or supplemental reproductive; (4) the *soldier reproductive* (fertile soldier) develops from a juvenile soldier morph and can become a replacement reproductive. Workers and soldiers can work in the colony as juveniles and if they do not become one of the reproductive forms described above, they remain as workers or soldiers at the endpoint of development, as an adult. Miller (1969), Noirot (1989), Noirot and Pasteels (1987) and Watson and Sewell (1985) may be consulted for the full array of worker and soldier developmental forms of termites in all families. The diagrams a, b₁, b₂ and c depict only the endpoints of development and the latest instar possible that shows development towards that endpoint; these diagrams do not show the multiple forms that juveniles might assume before reaching the endpoints. Bold lines indicate that a high proportion of the colony's members reach that endpoint; thin lines indicate the endpoint is frequently reached, and broken lines indicate the rarity of an endpoint. As a rule (a) **single-site nesters** follow one pathway that leads to the endpoint of winged reproductives, some wing-padded neotenics and apterous neotenics, some reproductive soldiers (in the Termopsidae) and some permanent workers and soldiers. (b) **Multiple-site nesters** follow one of two basic pathways producing three common endpoints. In b₁, one pathway leads to winged reproductives; the other leads to apterous neotenics, or permanent workers and effectively sterile soldiers (observed in *M. darwiniensis*). In b₂, one pathway leads to winged reproductives and wing-padded neotenics; the other leads to apterous neotenics or permanent workers and effectively sterile altruists, i.e. soldiers (observed in other multiple-site nesters). (c) **Central-site nesters** follow one of two pathways producing two common endpoints and one less common endpoint. One pathway leads to winged reproductives or wing-padded neotenics; the other leads to effectively sterile altruists, i.e. workers and soldiers. Apterous neoteny is very rare.

pseudergate), develop into a specialized fighting morph (called a soldier), or become a replacement reproductive (also called a neotenic) at any molt after the third developmental stage (instar), and before the penultimate molt (Castle 1934; Light and Illg 1945; Light and Weesner 1951; Miller 1969) (see Fig. 4-3 for definition of terms). Most offspring are dependent on care-givers up to the third instar, after which they serve as apterous workers until the fifth (or later) instar. After the fifth (or higher) instar, an individual may molt into a nymphal worker before becoming a winged adult (wing pads develop through at least two instars before an alate is formed). Nymphs can also undergo reversionary molts back to an apterous worker (as in *Kalotermes*) (Noirot 1985a; Watson and Sewell 1981, 1985). Offspring can delay, speed up, or even reverse their physiological and morphological development (*Zootermopsis*, Light and Illg 1945; Light and Weesner 1951; Miller 1969; *Kalotermes*, Watson and Sewell 1981; Noirot 1985a; *Prorhinotermes*, Miller 1942; Roisin 1988). Complete offspring development, from egg to winged adult, can take two to three years in *Kalotermes* and three to four years in *Zootermopsis* (see, for example, Nutting 1969; Lafage and Nutting 1978).

Substantial evidence suggests that many developmental changes are triggered by food and water availability (Termopsidae, Greenberg and Stuart 1982; Kalotermitidae, Nagin 1972; Lenz 1976, 1994) and reproductive vacancies (Termopsidae, Light and Illg 1945; Light and Weesner 1951; Greenberg and Stuart 1982; Mensa-Bonsu 1976; Kalotermitidae, Nagin 1972; see Noirot 1985c; Watson and Sewell 1981, 1985; Lenz 1985). Because single-site termites have a depletable food and nest resource, the selective advantage of a linear developmental schedule with great flexibility appears to lie in its ability to allow offspring to track changes in the relative benefits of personal reproduction versus altruistic helping (see section on genetic and ecological determinants; see also Higashi *et al.* 1991; Abe 1991).

Only a few individuals are permanent colony residents in single-site nesters: soldiers, some workers, and all secondary reproductives. Soldiers are workers that undergo at least two molts in development to obtain a large head morphology and elongated mandibles, suitable for nest defense (as in *Zootermopsis*, Castle 1934; Light and Weesner 1951; see also Miller 1969). At the point of change, soldiers are 'locked in' to being permanent residents: they become nutritionally dependent on nestmates and their large head size prevents them from dispersing by flight at the endpoint of development. Only 1–10% of the colony's

offspring are soldiers at any point in time (see Haverty 1977 for percentages of soldiers per colony in representatives of all Isoptera). Luykx's (1993) allozyme study on *Incisitermes schwarzi* indicates that soldiers are the oldest remaining colony residents, excluding the primary reproductive pair. Some workers may also remain as permanent colony residents; a small number remains in colonies that contain mostly alates (in *Zootermopsis*, Heath 1927; Heath and Wilbur 1927). However, the percentages of permanent workers have not yet been estimated. In *Incisitermes schwarzi*, there are apparently fewer older workers than soldiers in the colony; most are thought to have dispersed as reproductives (Luykx 1993). Because most workers will become reproductives each individual's total longevity, for example in *Zootermopsis*, is approximately seven to nine years, which includes three to four years as an immature (Nutting 1969) and four to five years as a reproductive (Heath 1907). In addition to the small pool of permanent workers and soldiers, some workers can become reproductives within their natal colony. If part of a colony is separated accidentally from the parents' portion, for example by branch breakage, or if one or both primary reproductives die, some workers accelerate their development to become secondary (replacement) reproductives (Weesner 1970; Lafage and Nutting 1978; Myles and Nutting 1988; Myles 1988). Additionally, male and female soldiers can become secondary reproductives, e.g. within one reversionary molt or one pre-soldier molt, under natural field conditions. These 'fertile soldiers' no longer function behaviorally as soldiers (Heath 1928). As far as is known, reproductive soldiers are observed only in species in the Termopsidae (Heath 1928; Myles 1986), although reproductive soldiers are physiologically possible in *Kalotermes* because soldiers implanted with active corpora alata from a roach become reproductive (Lebrun 1970).

In most species of the Termopsidae and in most other single-site nesters, replacement reproduction is assumed by apterous neotenics and wing-padded neotenics (Miller 1969; Weesner 1970, but c.f. Roisin and Pasteels 1991) (see Fig. 4-3 for definitions of terms). In orphaned colonies of *Zootermopsis*, many apterous workers of the fifth or later instars can molt quickly into active reproductive neotenics, while younger workers of the third instar have low egg production or delay reproduction for several weeks after becoming neotenic (Light and Weesner 1951). In laboratory cultures, late-instar wing-padded workers (nymphs) typically do not molt into replacement reproductives when primary adults are removed from the colony (in

Zootermopsis) (Light and Weesner 1951). In studies where wing-padded neotenics are produced, their numbers are always less than those of apterous neotenics (Greenberg and Stuart 1982). Nymphs of these single-site nesters appear to 'choose' dispersal over replacement reproduction in the nest. Perhaps nymphs are able to assess the availability of nutrients at their single wood site, dispersing when nutrient availability falls below some critical value.

Multiple-site nesters

The multiple-site nesters currently include the only living genus (*Mastotermes*) in the Mastotermitidae, the species *Paraneotermes simplicicornis* in the Kalotermitidae, many genera in the Rhinotermitidae, e.g. *Reticulitermes*, *Heterotermes*, and some *Coptotermes*, e.g. *C. formosanus* (Lenz 1994) and some genera in the Termitidae (Abe 1987). Multiple-site nesters, also called multiple-piece type and intermediate type, are species whose mature colony life is often spent in more than one nest. Each nest often interconnects with other nests of a single colony to form a large, loosely defined feeding territory. Wood serves as the primary nutrient, but seeds, other plant tissues, humus and fungi also supplement a wood diet (e.g. *Mastotermes*, Gay and Calaby 1970; species in Rhinotermitidae, Gay and Calaby 1970; Lenz 1994; and in the Termitidae, Noirot 1970). Living in or adjacent to its food resource further distinguishes the multiple-site nester from the single-site nester. *Mastotermes darwiniensis* (Mastotermitidae) and *Paraneotermes simplicicornis* (Kalotermitidae) nest primarily in logs or the stumps of trees (Hill 1942; Weesner 1970; Gay and Calaby 1970), whereas Rhinotermitidae and Termitidae of this life type nest either in or next to wood (e.g. species of Rhinotermitinae, Heterotermitinae) (Gay and Calaby 1970). In contrast to single-site nesters, colony members of all multiple-site nesters show some affiliation with the soil. Migration into the soil accompanies digging tunnels and galleries, and constructing covered foraging trails when outside the nest site (Noirot 1970). The ability to move outside of the nest structure, to forage or migrate to another location, allows a multiple-site nester to extend the colony's life beyond that possible for a single-site nester.

The average reproductive lifespan of multiple-site nesters is not well known. However, long lifespans have been recorded, e.g. one primary reproductive male of *M. darwiniensis* reaching least 17 years in the laboratory (Watson and Abbey 1989). Huang (1987) has followed the field production of primary reproductives to the ninth year of colonization in *Coptotermes formosanus*. Buchli (in Nutting 1969) determined that laboratory-kept primary reproductives (and neotenics) of *Reticulitermes lucifugus* were still present after seven years (Table 4-1). Perhaps the long lifespans of reproductives of multiple-site nesters, compared with single-site nesters, are attributable to greater resource stability. The lifespans of workers, however, are less well known.

Colony initiation in multiple-site nesters occurs by two principal modes. First, colonies may be founded by a male and female reproductive pair. As in single-site nesters, these reproductives are the initial care-givers. A second mode of colonization occurs when a group of colony members undergoes fission (budding) from the parents' colony. Typically, a cohort consisting of male and female neotenics, workers and soldiers migrates from the parents' nest to a new woody, nest site (*Mastotermes darwiniensis*, Gay and Calaby 1970; *Paraneotermes simplicicornis*, Nutting 1966; *Reticulitermes flavipes* and others, Weesner 1970). Propagation by budding is postulated by several termitologists to be at least as common as the release of alates in various circumstances (Banks and Snyder 1920; Weesner 1970; Hill 1942; Gay and Calaby 1970). Budding may be as successful a reproductive option as is dispersal, when success is measured in terms of the probability of survival and the number of offspring. Grassé (in Noirot 1970) once proposed that budding was the only successful mode by which colonies reproduce in one population of *Reticulitermes lucifugus*. Lenz and Barrett (1982) proposed that the invasion into new environments favors colony initiation by budding over colony founding by dispersers, because dispersers from the invading colony have difficulty locating other dispersers with which to mate. The possibility that reproduction by budding provides fitness benefits to non-alated offspring similar to those enjoyed by alated offspring must be further explored. Comparing the reproductive successes of budded versus dispersed (alated) reproductives may help to explain the developmental differences that are apparent between multiple-site nesters and single-site nesters.

Multiple-site nesters exhibit a forked pattern of development, in that offspring follow one of two distinct pathways early in development. In contrast to the linear development of the single-site nesters, the disperser and non-disperser pathways of the multiple-site nesters are present by the first molt (e.g. in *Mastotermes*) (Watson and Sewell 1981, 1985) or second molt (e.g. in species of *Reticulitermes*) (Noirot 1985a; Noirot and Pasteels 1987). Along the pathway of a disperser, an offspring progresses from egg to apterous larva (without wing pads) to nymph (with wing pads), and eventually develops into a winged adult or wing-padded neotenic (in all multiple-site nesters except *Masto-termes*, which has no wing-padded neotenics). In *Mastotermes darwiniensis*, it is not yet known whether offspring of the disperser pathway participate in colony work as juveniles (Watson and Sewell 1985). Offspring that become nymphs (and later alates) apparently do not work in the Rhinotermitidae (Noirot 1985b). Along the pathway of a non-disperser, a larva remains apterous; it may become an apterous neotenic (i.e. a reproductive adult retaining characteristics of a juvenile) or remain as a worker or soldier upon reaching the adult endpoint (Fig. 4-3b). Apparently, there is no permanent specialization of tasks of apterous workers in *M. darwiniensis*, and thus in this respect they are similar to single-site nesters (see, for example, Howse 1968; Springhetti 1990; Rosengaus and Traniello 1993a); workers are apparently capable of reaching adulthood in a shorter time, i.e. within 12 months (Watson and Abbey 1985). The divisions of labor may be more complex in *Reticulitermes*, as is certainly the case for species in some Rhinotermitidae (e.g. *Schedorhinotermes*) (Renoux 1985) and most Termitidae (see central-site nesters, below).

Much debate has centered upon why termites evolved apterous and disperser (winged) polymorphisms (a key characteristic of all multiple-site nesters) and whether species exhibiting these polymorphisms are historically derived from a eusocial termite with simpler developmental patterns (see, for example, Noirot and Pasteels 1987 versus Watson and Sewell 1981, 1985). On one side of the debate, it is hypothesized that species that first exhibited the forked-development type arose from species that exhibited linear development, after eusociality evolved. Under this view, it is proposed that developmental patterns of the first multiple-site nesters 'locked in' offspring into becoming workers, by forcing them early to develop along the non-disperser pathway (Noirot and Pasteels 1987). This step is said to be derived from one where temporary workers maintained the option to disperse, as in single-site nesters. However, the phylogenetic view that the non-disperser was 'locked in' as a worker, or was a 'reproductive loser' compared with the disperser (Taylor 1978; Roisin 1994) may be misleading. Developing as an apterous offspring and remaining a permanent worker are distinct events for some multiple-site nesters; for example, *Mastotermes* apterous workers often become neotenics (Watson and Abbey 1985). Like the single-site nesters, non-dispersers of *M. darwiniensis* have the ability to undergo numerous stationary and saltatory molts, and

hence delay or speed up their ability to reproduce directly (Watson and Sewell 1981, 1985). Presumably these juveniles, acting as temporary workers, can track conditions that favor personal reproduction. Thus, a forked developmental pathway, where offspring make the 'decision' to become apterous or winged early in development, need not be a product or a reflection of the degree of eusocial complexity *per se*, but rather, it may reflect changing reproductive opportunities (direct and indirect.) Also, the early fork in the developmental pathway may simply reflect the fact that the nest and food resource is replenishable and thus is more stable than for single-site nesters. Multiple-site offspring, unlike single-site offspring, may not have been selected to track changing resource conditions, and thus they make their developmental decisions earlier in development. Polymorphisms of this type are seen in other non-eusocial wood-dwelling insects, e.g. ptilid beetles (Taylor 1978; Watson and Sewell 1985). Because ecology may play a critical role in shaping developmental patterns, which life-type arose first in eusocial termites cannot be readily discerned with the available phylogenetic information.

In multiple-site nesters, the three adult endpoints – alates, neotenics (wing-padded and apterous ones), and altruists (workers and soldiers) (Fig. 4-3 b_1, b_2) – are commonly expressed in species alive today; the relative frequencies of these endpoints appear highly variable. Among non-dispersers, all soldiers are altruists at their endpoint; fertile soldiers are not known in this life-type. Many workers can become neotenics, and numerous workers remain as adult altruists. The percentage of colony members that become adult altruists can be inferred from studies on supplemental reproduction. To date, no field data are available for *Mastotermes*. However, in the Rhinotermitidae, Howard and Haverty (1980) determined that supplemental reproductives (neotenics) had a mean frequency of only 1.28% per colony in *R. flavipes*. In three populations of *R. lucifugus*, Fontana *et al.* (1982) determined that supplemental reproductives had a mean frequency of only 0.23% in colonies. Thus, in large colonies of the Rhinotermitidae most workers give up the option to reproduce, since only a fraction, about 1%, of the worker pool can replace the reproductives. Under these circumstances, the two most common endpoints for colony offspring are permanent altruists (workers and soldiers) and dispersive adults. However, the presence of budded colonies in a population can easily be overlooked (in *Reticulitermes*, Banks and Snyder 1920; Weesner 1970; also in *Mastotermes*, Hill 1942; Gay and Calaby 1970; Watson and Abbey 1989); budded colonies or mature colonies where resources are more limiting can account for colonies that have fewer inhabitants at maturity (Lenz 1994). Indeed, maximum colony sizes range widely in *Reticulitermes flavipes*, from 50 000 to 900 000 inhabitants (Howard *et al.* 1982; Grace 1990; reviewed by Lenz 1994). Similarly, in *Mastotermes darwiniensis* single colonies may exceed 1 million, while maximum colony sizes in other areas average only around 10 000 inhabitants (Hill 1942; Gay and Calaby 1970). The presence of numerous smaller colonies in *Mastotermes*, with maximum size of only 10 000 inhabitants, suggests that temporary workers in this species will have a relatively high probability of reproduction as apterous neotenics, compared with workers in larger colonies. Indeed, Watson and Sewell (1985) have described apterous neoteny as the 'natural' (typical) endpoint for *Mastotermes* workers. Watson and Abbey's (1985) laboratory studies on *M. darwiniensis* show that tens of neotenics can form in the company of a few workers (100–200 workers). Further effort is needed to assess the costs and benefits of neoteny in comparison with costs and benefits of worker and soldier altruism.

Central-site nesters

The central-site nesters include some genera in the Rhinotermitidae, e.g. some *Coptotermes*, some *Schedorhinotermes*, all genera in the Hodotermitidae, e.g. *Hodotermes*, and most genera in the Termitidae, e.g. *Amitermes*, *Microcerotermes*, *Nasutitermes*, *Trinervitermes*, *Cubitermes*, *Macrotermes* and *Odontotermes*. Central-site nesters and multiple-site nesters share numerous life-history traits; differences between these two life types are primarily quantitative with central-site nesters tending to exhibit traits at the upper extreme of a continuum (see Lenz 1994). Central-site nesters, also called separate-piece nesters, are species whose colony life is typically spent at one well-defined, centralized nest location, separate from their raw forage. Plants, grasses, wood, humus and other particulate matter in soil may serve as the raw forage and nutrients for diverse species. Food is gathered from outside their permanent nest structure, although initial colony growth may start in or near food site (Noirot 1970). In all central-site nesters (like multiple-site nesters) the construction of a nest (or portions of the nest) involves the use of wood pulp, plastered earth, excrement, or some mixture of feces, saliva, pulp and earth. The nest may be arboreal, subterranean, in an epigeous-mound, or partly subterranean and partly mound-like (see, for example, Noirot 1970; Darlington 1994; Lenz 1994). This lifestyle typically provides the greatest resource stability, in comparison

with other life types, since food is replenished continuously and brought back to a fixed, centralized nest site (Noirot 1970; Abe 1987). In some instances, nutrients are gained by processing raw resources at the nest site (as in fungus-growing termites) (Darlington 1994). A nest is under constant construction by the workers and has multiple functions. It buffers the inhabitants against predators, stabilizes temperature and moisture levels, and serves as a storage reservoir for nutrients that sustain a colony during the seasonal dearth of food (Bouillon 1970; Noirot 1970; see Darlington 1994 for details on fungus growers). Subterranean and mound nests can persist for several decades and can serve as permanent dwellings for several generations of reproductives and brood (e.g. some *Nasutitermes* species) (Hill 1942; Noirot 1970). Other nests (e.g. the arboreal nests of some *Nasutitermes* and *Microcerotermes* or the conical pillars of *Cubitermes*) (Noirot 1970) are not as robust; but according to Noirot *et al.* (1986) (for example, in *Cubitermes fungifaber*) the economy in design and material make construction of new nests timely, thereby preserving colony cohesion and stability.

There are two principal modes of colony initiation in species of central-site termites. First, colonies are started by alate reproductives, and sometimes multiple reproductives cofound a single nest (as in *Nasutitermes* spp., Thorne 1982, 1985; Roisin 1993; fungus termites, *Macrotermes* spp., Darlington 1985a, 1988). Second, colonies are started by budding, with primary (de-alates at the natal nest) or secondary (wing-padded neotenic) reproductives that leave the parents' nest with a worker and soldier cohort (reviewed by Thorne 1985; Darlington 1985a; Roisin 1993).

Like multiple-site nesters, offspring develop along two primary pathways that diverge early in development (Noirot 1985b; Noirot and Pasteels 1987). Unlike multiple-site nesters, offspring develop rapidly (within weeks, e.g. *Macrotermes* near *subhyalinus*, Okot-Kotber 1979; *Odontotermes* spp., Liu *et al.* 1981; *Trinervitermes geminatus*, Josens 1982; *Macrotermes michaelseni*, Darlington 1991) to become part of a permanent (effectively sterile) worker force that eventually numbers in the hundreds of thousands (reviewed in Lenz 1994). In some species, mature colonies may have one, two, or even five million colony members (*Macrotermes*, *Nasutitermes*) (Nutting 1969; Darlington 1979; Lenz 1994). Reproductives of central-site nesters may live two or more decades (Roonwal 1970; Grassé 1984) (Table 4-1). Workers and soldiers, however, are relatively short-lived individuals; in some *Trinervitermes* and *Macrotermes* species they live an average of three to six months (Josens 1982; Darlington 1991; see Table 4-1). This contrasts markedly with the greater longevity of offspring in the dispersive developmental pathway, i.e. the reproductive caste. It also contrasts sharply with the longevity of workers in single-site nesters and perhaps most multiple-site nesters: in these life types, the workers can live at least three to five years, and longer when they disperse as alates (single-site species) or become neotenics (also in multiple-site nesters). High risks of foraging and exposure to predation may explain why central-site workers live a briefer life in comparison with other life types (see, for example, Alexander *et al.* 1991); intense foraging pressures (e.g. predation by numerous soil arthropods) may also explain why brood production rates are much higher in central-site nesters than they are in single-site wood-nesters, which tend to live in less harsh environments (M. Lenz, personal communication). In addition, because individuals of central-site nesters appear to have almost no opportunities for direct reproduction, selection may have favored workers that take greater risks as the best means of increasing their inclusive fitness.

The complexity of cooperative interactions among workers and soldiers reaches a pinnacle in central-site nesters. Like ants and honey bees (their counterparts in the Hymenoptera) that also exhibit complex social integration, each colony functions, in part, like a superorganism (*sensu* Wilson 1975, 1990; Seeley 1985; Ratnieks and Reeve 1988, 1992). Depending on the species, task specialization is spatially defined by duties performed in the nest, on the nest periphery, and at forage sites, and also can depend on gender (e.g. in *Hodotermes*, Watson and Sewell 1985; numerous Termitidae, Noirot 1989), morphology (Traniello 1981; Noirot and Pasteels 1987; Lys and Leuthold 1991) and age (McMahan 1979; McMahan *et al.* 1983; Noirot 1985b; Gerber *et al.* 1988; Veeranna and Basalingappa 1984). Although laboratory studies show that any worker caste can perform diverse tasks (see Noirot 1985b), worker and soldier task specialization sharpens when it is integrated with that of other colony members (see, for example, McMahan 1979; Traniello 1981; Gerber *et al.* 1988; Veeranna and Basalingappa 1990; Lys and Leuthold 1991).

As a rule, only individuals in the nymphal pathway, i.e. those eventually capable of dispersal, can become reproductives (Fig. 4-3c) (species of Termitidae, Thorne and Noirot 1982; Noirot 1985b,c; *Coptotermes* in Rhinotermitidae, Lenz *et al.* 1986). Seasonally, nymphs in mature colonies develop into winged adults by the thousands to disperse and initiate new nests. Some wing-padded replacement individuals are present year round and are tolerated presumably because they represent insurance against loss of primary reproductives (Darlington 1986; *Amitermes*, Skaife 1955; *Nasutitermes princeps*, Roisin and Pasteels 1985; *Microcerotermes*, Weyer 1930; *Macrotermes* spp., Darlington 1986; *N. coxipoensis*, Lefeuve 1987; *Coptotermes*, Lenz and Runko 1993). Individuals in the nymphal pathway of large-colony species never serve as colony workers, except in rare instances (see, for example, Noirot 1985b; Darlington and Ritchie 1987) and in the vast majority of central-site termites, workers and soldiers never have an opportunity to reproduce (Grassé 1984; Noirot 1985b,c). Rare exceptions occur in undisturbed field colonies of a few species (Thorne and Noirot 1982; Noirot 1985c; Roisin and Pasteels 1987; Darlington *et al.* 1992); these same studies show that experimental manipulation can induce worker reproduction at higher levels than is found to occur naturally. Thus, in central-site termites there is typically a distinct non-worker class that functions only in the role of reproduction. Workers and soldiers perform all the colony maintenance and defense and have a reproductive potential that approaches zero. Thus, these taxa exhibit an index of eusociality that approaches 1.0 (Fig. 4-2).

Termite-inquiline nesters

The termite-inquiline nesters include the single genus *Serritermes* in the Serritermitidae, and some genera in the Termitidae, e.g. *Termes*, *Speculitermes* (see also Table 4-2). Termite-inquiline nesters are termite species that spend their lives in the nests of other termite species, and by doing so, apparently avoid the costs of colony maintenance (Noirot 1970; Roonwal 1970; Bouillon 1970; Araujo 1970; Basalingappa 1971; Collins 1980) (Table 4-2). Nests occupied by termite-inquilines are typically mound, subterranean, or arboreal nests; thus it appears that most hosts are central-site nesters. Collins (1980) distinguished two types of termite-inquilines: those that form an obligate or a close association with the host's nest while the host is still occupying it are called termitophils, and those that form a facultative association with the nest, but not necessarily during the host's occupation of the nest, are called termitariophils (see Collins 1980 for a list of termitariophilic inquilines). Very little is known about the social complexity of the termite-inquilines: the majority of workers may be either temporary residents or permanent, specialized altruists. Both types of termite inquilines gain numerous benefits from living in a nest built by the host termite: just as the host's nest is important to its colony members,

Table 4-2. *Termite-inquilines, their hosts, and nesting habits.*

Termite-inquiline	Termite host	Nesting habits of termite-inquiline
Termitinae: Termitidae		
Termes (= *Inquilinitermes*) *fur*	Termitidae: *Constrictotermes cyphergaster*	Feeds off host's storage reserves, shares dwelling but separate; hostile interaction when detected; small queen, small numbers, soldiers present[a]
Termes (= *Inquilinitermes*) *inquilinus*	*Constrictotermes cavifrons*	Feeds off host's storage reserves, shares dwelling, but inquiline and host have a distinct gallery system; interactions are hostile between inquiline and host[a]
T. winifredae	*Amitermes hastatus*	—
T. insitivus	*Nasutitermes magnus*[b]	
T. laticornis	*Macrotermes gilvus*	Facultative inquiline; it keeps separate galleries from host[d]
Dicuspiditermes incola	*Odontotermes ceylonicus, O. horni, O. redemanni* and *Hypotermes obscuripes*	Always associated with nest, makes its own galleries and chambers in host nest; host and inquiline are hostile when they meet[d]
Pericapritermes ceylonicus	*Odontotermes* species	Always associated with host; distinct activity from host, habits are similar to those of *Dicuspiditermes*[d]
Pericapritermes distinctus	*Odontotermes redemanni* and *Hypotermes obscuripes*	Lives in small colonies; makes its own galleries and chambers in host nest[d]
Microtermes obesi and *Speculitermes* spp	*Odontotermes wallonensis*	Lives in nest[e]
Speculitermes cyclops, Dicuspiditermes incola, Microtermes spp.	*Odontotermes assmuthi*	Inquiline excavates in the hosts' periphery walls; thought to derive shelter and protection; hostile interactions with host, except for *Microtermes*[f]
Protocapritermes kraepelinii	*Amitermes obeuntis*	Neotenics are unusual in that stumps resemble those of normal dealate; small colonies[g]
Nasutitermes inquilinus	*Anoplotermes*	Lives in abandoned nests of the host species[g]
Amitermitinae: Termitidae		
Ahamitermes hillii, A. nidicola, A. inclusus, Incolitermes pumilus	Rhinotermitidae: *Coptotermes*	Obligate inquiline; feeds off host's storage reserves, shares dwelling but separate; hostile interaction when detected, small colony number relative to host[g]
Ahamitermes hillii	*Coptotermes a. acinaciformis* and *C. a. raffrayi*	Obligate inquiline, feeds off host's storage reserves; alates of inquiline separate from host; brachypterous neotenics found[g]
Incolitermes pumilus	*Coptotermes a. acinaciformis*	Similar life to *Ahamitermes*; alates disperse with alates of the host. Inquiline found in nursery region of the host nest; inquiline eats the lamellae of the host nursery; both apterous and brachypterous neotenics are known[g]
Serritermitidae:		
Serritermes serrifer	Termitidae: *Cornitermes termitaria* or *Cornitermes cumulans*	Termite-inquiline lives on the periphery of the host's nests; it does not build a nest of its own[a,b]
Serritermes serrifer	*Cornitermes* spp. (*cumulans*)	Lives off the host's nutrient reserves; inquiline shows deformity in mid- and hindlegs in all colonies encountered[h,i]
Termitidae:		
Microcerotermes brachygnathus	*Cubitermes* spp.	Obligate association[c]
Euchilotermes umbraticola	*Cubitermes umbratus*	Cannot distinguish separate galleries; inquiline is rarely found in abandoned host nest; queens of the two species were found 3 inches apart[c]
Pseudacanthotermes spiniger	*Macrotermes natalensis*	Separate galleries from host[c]

Sources: [a] Araujo 1970; [b] Noirot 1970; [c] Bouillon 1970; [d] Roonwal 1970; [e] Rajagopal 1986; [f] Basalingappa 1971; [g] Gay and Calaby 1970; [h] Emerson and Krishna 1975; [i] Kitayama 1975.

termite-inquilines also obtain nest protection from predators, they are buffered against climatic extremes, and, as noted by most investigators, they obtain nutrients that the host has stored at the nest. Moreover, termite-inquilines that are strongly associated with the termite host (termitophilic) have access to nutrients replenished by the host's foragers, and probably gain from the host's active defense of the nest as well (Table 4-2). Thus, the inquiline workers probably experience lowered predation and reduced wear of body parts from foraging. These benefits may be important if workers have a high probability of reproduction, as in some multiple-site nesting species.

Basalingappa (1976) found that one inquiline, *Speculitermes cyclops*, initiates new nests by alate swarming as well as by budding. Little is known about when or how a termite-inquiline joins the nest of its host. Inquiline alates may disperse preferentially to large-sized, highly populated colonies of the host species, and thus escape detection by the host while initiating a nest within the host's nest wall. Nest-founding dealated pairs of different species are sometimes found settled in nests of fungus-growing termites (Noirot 1970). It is possible, too, that budded termite-inquilines are capable of entering large nests, undetected by their hosts. Perhaps in some species, colony initiation within a host's nest is a favored mode of new nest initiation because it avoids higher costs of independent nest-founding, after which budded colonies function independently.

Termitophils appear to exhibit a parasitic relationship with the host. The location of the inquiline's nest in that of its host differs in various species (Table 4-2). Some species live in the peripheral walls of the host's nest, sometimes in the host's nursery, and sometimes in the host's food storage chambers. Most often, termite-inquiline and host maintain separate galleries and nesting chambers within the host's nest; many studies note that when termite-inquiline and host do meet, the interactions are hostile (Noirot 1970; Roonwal 1970; Bouillon 1970; Araujo 1970; Basalingappa 1971). Such hostile eruptions suggest that termite-inquilines assume some risk by living with an enemy. The host has the distinct advantage of greater colony number which typically provides an advantage in competition for resources (see, for example, Franks and Partridge 1993). Bandeira's (1989) survey of termite fauna in Brazil led him to suggest that termite-inquilines avoid association with hosts that have well-developed chemical defenses. It is not known, however, whether termitophilic-inquilines avoid some termite species only because of their chemical

defenses. Typically, termite species with good chemical defenses (e.g. *Nasutitermes* spp.) spend less effort on the construction of a nest (Coles and Howse 1983). Thus, termite-inquilines may choose hosts based on nest preference rather than on the basis of the quality of the host's defenses. Selective factors that are involved in an inquiline's decision to invade its host must still be explored.

Because the life-histories of termite-inquilines are not well understood, it is not possible to draw comparisons of eusocial life history between inquilines and termites of other nest types. However among termitophilic inquilines, the high risk of host discovery should force colonies to remain small. Moreover, the relatively lower costs associated with nest maintenance for termitophilic termites, or the reduced cost of moving into an already functioning nest for termitariophilic species, suggests that the social system of the termite-inquilines should be similar to those of the multiple-site nesters rather than to those of the central-site nesters, but not to that of single-site nesters, since they are capable of migration (Basalingappa 1976) (see also Table 4-2). Both of the above considerations lead to the prediction that termite-inquiline workers should show a higher reproductive potential than other species of central-site nesters.

GENETIC AND ECOLOGICAL DETERMINANTS OF TERMITE EUSOCIALITY

Termites along a spectrum of eusociality range from genera whose workers (and sometimes soldiers) eventually become reproductives as adults, to genera whose workers and soldiers are permanent altruists as adults and perform specialized tasks. Correlations between colony life types and social behavior show how a colony's resource availability and stability affects the benefits for different worker–reproductive options and developmental decisions (described in the previous section, see also Watson and Sewell 1981, 1985; Abe 1987; Higashi *et al.* 1991; Lenz 1994). Understanding variation in the potential for altruism, however, must be accompanied by an explanation for why eusociality is or was favored at all. In the sections to follow, I address this question by examining the selective maintenance of eusociality in single-site, multiple-site, and central-site nesting termites, from a functional level of analysis, and by examining how the selective factors might relate to the origins of eusociality, from a historical level of analysis (*sensu* Sherman 1988; Alcock and Sherman 1994; after Tinbergen 1963).

Despite the consensus that Hamilton's rule is central to understanding the evolution and maintenance of eusociality (see, for example, Emlen 1984, 1991; Andersson 1984; Queller 1989, 1994), marked disagreements have occurred over the relative importance of its components. Hypotheses emphasizing the importance of genetic relatedness have been particularly controversial.

Kinship in termite societies

For helping to evolve by kin selection, helping must increase the reproduction of the recipients who are related to the helper (Hamilton 1964). Recognition and preferential care of kin over non-kin or less related kin are widespread in diverse eusocial taxa (see, for example, Gamboa *et al.* 1986; Breed and Bennett 1987). Where studied, nest-mate recognition and preferential cooperation with related nestmates are known to be important in termites as well (see, for example, Howick and Creffield 1980; Binder 1988; Adams 1991; Su and Haverty 1991). Intercolony conflicts to the death (Binder 1988; Su and Haverty 1991) arising as part of territory defense (see, for example, Adams and Levings 1987; Jones 1993) suggest one of many contexts where preferentially associating with nestmates, presumably kin, can have a selective advantage. There is little doubt that kin selection, in conjunction with ecological pressures, has played an important role in the evolution of cooperative societies in termites (as I discuss later). However, some genetic hypotheses have proposed that a relatedness asymmetry, with higher relatedness between alloparent and brood than between parent and brood, accounted for the origins and maintenance of eusociality (in termites, Hamilton 1972; Bartz 1976; Luykx and Syren 1979; Lacy 1980). These studies, involving inbreeding, cycles of inbreeding, and chromosome-translocation mechanisms, focused primarily on relatedness to explain eusociality.

Relatedness asymmetries favoring eusociality
Higher relatedness to non-descendent kin (such as full siblings) than to descendent kin is theoretically possible under various genetic mechanisms. For example, in eusocial colonies with a female-biased sex ratio and headed by singly mated, monogynous queens (in Hymenoptera, Hamilton 1964; Thysanoptera, Crespi 1992a), a relatedness asymmetry exists for females as a result of the haplodiploid genetic system. Higher relatedness between a worker and the brood it rears relative to its own offspring is also possible by continuous inbreeding (as, for example, in populations of naked mole rats) (Reeve *et al.* 1990). These studies show that relatedness asymmetry can be positively associated with alloparental care. However, no empirical evidence shows that a relatedness asymmetry is necessary or sufficient to explain eusociality in any taxon. To the contrary, many if not most social taxa exhibit intracolony relatedness between worker and recipient of aid that is no greater than (and often less than) relatedness between parent and offspring (in Hymenoptera, Strassmann *et al.* 1989; Gadagkar 1990; Hölldobler and Wilson 1991; in cooperatively breeding birds, Emlen and Wrege 1989).

In termites, genetic mechanisms proposed to explain eusociality have been extensively debated: among single-site nesters in the Kalotermitidae, these mechanisms have received extensive study by Peter Luykx and coworkers. Initially, Luykx and Syren (1979) and Lacy (1980) proposed that translocation complexes of sex-linked chromosomes (discovered by Syren and Luykx 1977) might favor eusociality through higher relatedness because, at loci on these chromosomes, same-sexed siblings are more highly related to each other than to opposite-sex siblings or to their own offspring ($r = 0.625$ versus $r = 0.375$ or $r = 0.50$, respectively; see Luykx 1985). However, Leinaas (1983) argued that preferential worker care of same-sexed brood is not likely to occur because most colony tasks, such as foraging and colony defense, do not differentially benefit brood of any particular sex, but yield benefits for the whole colony indiscriminately. Later studies showed that workers exhibit no overall preference towards same-sexed siblings (Luykx 1985; Luykx *et al.* 1986; Hahn and Stuart 1987). Moreover, numerous species of termite (e.g. species in the Mastotermitidae and Termopsidae) do not possess translocation sex chromosomes, even within the family of small-colony eusocial termites (the Kalotermitidae) that show a high frequency of sex-chromosome linkage (Crozier and Luykx 1985; Luykx 1990). Thus, higher relatedness between alloparents and non-descendent kin that might arise from large complexes of sex-linked chromosomes is neither a necessary nor a sufficient cause of alloparental cooperative care (in termites, Luykx 1985; Crozier and Luykx 1985).

Continuous inbreeding in termites can also cause relatedness asymmetries promoting eusociality (Hamilton 1972, 1978; Syren and Luykx 1977). Unfortunately, genetic studies that might test this hypothesis have not been conducted on the majority of termite groups. Thus far, *Incisitermes schwarzi* (Kalotermitidae) is the only species for which the degree of inbreeding of dealated nest founding pairs has been estimated: Luykx's (1986, 1993) data suggest that primary reproductives are typically *outbred* (matings are not between close relatives) and that approximately three-quarters of the colonies in the field are headed by primary reproductive pairs. Indirect evidence in numerous species, such as from patterns of sex dispersal (in multiple-site nesters and central-site nesters, Vincke and Tilquin 1978; Jones *et al.* 1988; McMahan *et al.* 1983; Lenz and Runko 1993; and single-site nesters, Luykx *et al.* 1986; Jones *et al.* 1981; Shellman-Reeve 1994b, 1996) and differences in sexual development (Greenberg and Stuart 1982; Luykx *et al.* 1986; Jones *et al.* 1988), suggest that outbreeding by primary reproductives occurs frequently. A laboratory study on nest-founding pairs of *Zootermopsis angusticollis* showed that the probability of mortality was higher for an 'outbred' pair (where an adult was paired with a mate that originated from a different log) than for an 'inbred' pair (where an adult was paired with a mate that originated from the same log) (Rosengaus and Traniello 1993b). However, the high mortality of 'outbred pairs' may be an effect of outbreeding depression (Bateson 1983) because 'different logs' were sometimes different species of log, and sometimes they were collected from different townships several kilometers away. In addition, 'inbred pairs' may not have been close kin because multiple colonies of *Z. angusticollis* typically reside in close proximity on a single log (Shellman-Reeve 1994a,b). Although inbreeding does occur through incestuous matings following the death of one or more primary reproductives (in *I. schwarzi*) (Luykx 1993), extensive population variability in the extent of inbreeding will weaken the role of inbreeding as a dominant factor favoring eusociality in termites of all life types.

Cycles of inbreeding, in which outbreeding primary-nest founding reproductives were themselves the offspring of inbreeding by replacement reproductives (Bartz 1979; see also Hamilton 1972), is yet another mechanism that can generate relatedness asymmetry favoring eusociality: the primary reproductives produce heterozygous workers that have greater genetic similarity to each other than to their own (outbred) offspring. Although plausible, the 'cycles of inbreeding' hypothesis for the maintenance of eusociality does not appear to apply to the single-site nesters, i.e. species whose workers are mostly temporary and have high reproductive potential (Table 4-1, Fig. 4-3a). Myles and Nutting (1988) point out that colony longevity for the great majority of these species is typically too brief for cycles of inbreeding to be effective. As is evident in populations

of *Incisitermes schwarzi* (and probably other single-site nesters), outbred primary reproductives probably produce most of the dispersing reproductives (Luykx 1985, 1993).

Cycles of inbreeding, however, might be of considerable importance among some multiple-site nesters and central-site nesters. In these species, there is greater opportunity for sibling-replacement reproduction because colonies persist for longer periods and because offspring, presumably close siblings, bud off from a parent's nest. The high degree of inbreeding that Reilly (1987) has discovered in *Reticulitermes flavipes* may be widespread in termites of this nest type. Myles (1988) proposes that only a few species are candidates for cyclic inbreeding, but he may have underestimated the importance of budding and replacement reproduction. Recent studies in species of central-site nester show that replacement and supplemental reproduction occurs by alates (not necessarily wing-padded neotenics) (Thorne and Noirot 1982; Noirot and Pasteels 1987; Roisin 1993). Thus, colonies headed by 'dealated' reproductives may look like founding primaries, but might be dealated siblings that decided to bud with a cohort of workers. Alate-replacement reproductives have also been reported in single-site nesters (*Neotermes* spp., Roisin and Pasteels 1991; M. Lenz, personal communication). Regardless of whether the parent colony dies out without replacement, colonies of numerous species may bud several times (see Weesner 1970; Roisin 1993). Frequent budding and frequent replacement reproduction can lead to a high frequency of homozygous alate offspring, that initiate new colonies by outbreeding, thereby renewing conditions that favor a relatedness asymmetry.

The cycles of inbreeding mechanism, however, poses potential difficulties for termites insofar as inbred colonies suffer reduced genetic variability. Reduced genetic variability can make colonies susceptible to parasites and pathogens (see, for example, Sherman *et al.* 1988; Hamilton 1990, 1993, Shykoff and Schmidt-Hempel 1991; Keller and Reeve 1994). It is possible that budding or replacement reproduction involves outbred matings (see Lenz and Runko 1993); in the Argentine ant, for example, dispersing males can enter into a mature colony to mate with unrelated females that then become replacement reproductives (Passera and Keller 1994). However, problems that can arise in offspring from incestuous replacement-reproductives, such as greater susceptibility to disease, might also be lessened in colonies that have a sufficiently large pool of neotenics. Lenz and co-workers (Lenz 1985; Lenz *et al.* 1982, 1985, 1986) have discovered two reproductive-replacement strategies in termites whose colonies are headed by supplemental reproductives: either numerous neotenics compete for reproduction with only those of highest quality surviving the ensuing conflict over the reproductive spots, or only a few neotenics attempt reproduction and these are replaced when they fail. In either case, neotenics of a high average quality usually tend to become the replacement reproductives (Lenz et al. 1985, 1986). Hence, perhaps the potentially deleterious effects of reduced genetic variability can be alleviated, as long as a colony is large enough that at least a few of the many neotenics that attempt reproduction do not have the deleterious complement of genes that might arise in offspring of inbred mates. Thus, cycles of inbreeding remains a potentially important genetic mechanism contributing to the maintenance of permanent, extreme worker altruism observed in central-site nesters and some multiple-site nesters.

Finally, it might be argued that relatedness asymmetries caused by one of the aforementioned genetic mechanisms were important in the origins, if not in the maintenance, of eusociality; for example, once higher relatedness favored eusociality, subsequent selection for worker specialization may have prevented evolutionary reversions to solitary living even if relatedness asymmetries were lost. Because historical-origins arguments like these are not easily tested (Frumhoff and Reeve 1994), we may never know for certain. However, eusociality and extensive cooperative group living in some vertebrates and arthropods is repeatedly lost in some lineages because of fluctuating ecological pressures altering costs and benefits of cooperation, for example in birds (Stacey and Koenig 1990), and in bees (Danforth and Eickwort, this volume). These examples throw some doubt on the notion that eusociality is permanently 'locked' in place once it evolves. This may be the case in termites as well; workers of all single-site nesters and many multiple-site nesters (Miller 1969; Watson and Sewell 1981), and under some circumstances even central-site nesters (Thorne and Noirot 1982), show remarkable plasticity in response to changing reproductive opportunities.

Ecological causes of eusociality

Recent discussions of cooperatively breeding vertebrates and eusocial invertebrates have emphasized ecological pressures that limit independent reproduction (e.g. in birds, Emlen 1984; other vertebrates and invertebrates,

Alexander 1974; Andersson 1984) and provide inclusive fitness advantages for altruistic helping (e.g. in insects: Hamilton 1972; West-Eberhard 1975, 1981; Queller 1989, 1994; Reeve and Ratnieks 1993; in birds, Emlen and Wrege 1989; Emlen 1991). In effect, these studies have emphasized the role of benefit and cost variables in satisfying Hamilton's rule for the evolution of altruism, keeping in mind that relatedness may be crucial in the formation of certain associations rather than others.

Ecological constraints limiting reproduction

Nest initiation and early colony growth. Perhaps the most potent selective pressures facilitating and maintaining eusocial life in termites of current ecological time operate during initial nest founding and early colony growth. From the perspective of a developing offspring, choosing to stay in a functioning nest might be safer than initiating a new one, especially if the natal nest offers a stable, well-protected shelter (Queller 1989, 1994). Ecological pressures on nest-founding adults are hypothesized to be especially harsh, as breeders must survive dispersal and then establish and maintain a nest for their young offspring in an environment in which longer-established neighbors probably have a competitive edge. Indeed, the severity of ecological constraints likely contributes to the extensive cooperation exhibited by nest-founding pairs and young workers.

Dispersal and migration. Dispersing alates are visible targets to a multitude of arthropods (e.g. arachnids, centipedes, and insects) and vertebrates (e.g. reptiles, amphibians, mammals, and birds) (Nutting 1969, 1979). Numerous accounts of predation during dispersal indicate that risks are high. Unfortunately, quantitative measures of dispersal-related mortality have not yet been gathered for any termite species (except under artificial laboratory situations) (see, for example, Basalingappa 1970). Typically, multiple colonies of varying sizes release alates that disperse varying distances, which tends to complicate any assessment of mortality based only on postflight densities at nest settlement sites versus preflight densities at the natal nest sites. Mortality per colony might be estimated by examining the percentage of colony-coded marked survivors departing from each colony.

Nutting (1979) pointed out that alates may reduce their dispersal risks by various predator-avoidance strategies: as a rule, when large numbers of alates disperse from a colony, the departure from a nest occurs in a narrow interval of time. Alates of this dispersal strategy may benefit from a selfish herd (Hamilton 1971) or satiation effect as predators confront an overwhelming number of prey, reducing the probability that any one alate will be caught. When only a small number of alates disperse from a nest, Nutting (1979) observed that single individuals depart from the nest over a long interval of time (seasonal, several months). Alates of this dispersal strategy apparently minimize their detection by predators that have restricted foraging times.

The migration process involving the fission of a new colony from that of the parents' colony has not been studied extensively. Only descriptive accounts of budding in progress are available (Banks and Snyder 1920; Weesner 1970; Noirot 1970; Noirot et al. 1986). Because the process of budding is rarely observed directly, almost nothing is known regarding the migration-related mortality of a budded colony. Presumably, detection by numerous soil-foraging predators makes budding as risky a process as alate dispersal, especially since a migrating cohort of workers and neotenics might be conspicuous. However, migration risks may be minimized by movement to new nest sites via the galleries constructed by foragers. Indeed, hopeful secondary breeders may receive information about good nesting sites from foragers that return with quality forage nutrients. Hence, budding may coincide with a return foraging trip back to the quality forage site.

Limited availability of quality nutrients favor cooperation among breeders and young colony workers. One likely pressure shaping initial colony founding is the limited availability of quality nutrients (Lafage and Nutting 1978; Prestwich et al. 1980; Breznak 1982; Nalepa 1984, 1988, 1994; Shellman-Reeve 1990, 1994a,b; Nalepa and Jones 1991; Lenz 1994; for other taxa, see White 1993). One quality nutrient, nitrogen, may be particularly important as its availability in wood xylem is typically low, between 0.03 and 0.10% (Lafage and Nutting 1978; Haack and Slansky 1987). The cambium layer of the log, a soft powdery layer just beneath the bark, contains the richest source of nitrogen on a log (up to 1% by dry mass in logs inhabited by *Zootermopsis*); but it diminishes quickly, as young colonies feed on it (Shellman-Reeve 1994a). Despite the severe limitation and patchy availability of nitrogenous nutrients in wood (Lafage and Nutting 1978; Collins 1983; Shellman-Reeve 1994a), termites, like other arthropods, require significant amounts of nitrogen for basic body needs (e.g. for DNA, muscle and exoskeleton). Nitrogen requirements

are especially high in egg development and larval growth (in all arthropods, Dadd 1985; other taxa, White 1993). Indeed, the percentage of nitrogen found in termites (approximately 7–15%) is similar to that found in other arthropods (Lafage and Nutting 1978; Prestwich et al. 1980; Scriber 1984), indicating that termites do obtain sufficient amounts of these critical nutrients. There is mounting evidence that termites of all colony ages compensate for their poor-quality diet through a variety of non-social and social mechanisms (for example, by storing and recycling nitrogen and other critical nutrients such as lipids, by fixing atmospheric nitrogen, by cooperatively sharing proteins, fats and sugars, and by cooperatively nesting and foraging at sites with high-quality nutrients) (reviewed in Amburgey 1979; Prestwich et al. 1980; Breznak 1982; MacKay et al. 1985; Shellman-Reeve 1990, 1994a,b, 1996; Nalepa and Jones 1991; Nalepa 1994; see also Nalepa and Bell, this volume, on cockroaches; White 1993, on diverse taxa).

One adaptive response to low external availability of quality nutrients is that new-nest founders disperse from their natal nest with fat-body reserves of high-quality nutrients (e.g. nitrogen, Basalingappa and Hegde 1974; Breznak 1982; Mugali and Basalingappa 1984; Shellman-Reeve 1990; and lipids, Cmlick 1969a,b; Mishra and Sen-Sarma 1981; Mugali and Basalingappa 1982; Han and Noirot 1983; van der Westhuizen and Hewitt 1984; Basalingappa and Gandhi 1985; Gandhi and Basalingappa 1992; Shellman-Reeve 1994b, 1996). There is direct and indirect evidence that alates are fed these nutrient reserves by worker nestmates prior to their leaving the natal nest (Howse 1968; McMahan 1969; Nutting 1969; Nutting and Lafage 1978; Darlington 1986). Of course, alates may use their flight muscles as an additional source of nitrogen (Basalingappa 1982; van der Westhuizen et al. 1987).

Nutrients of the fat body (such as nitrogenous compounds, minerals, lipids and sugars) appear especially important for successful production of initial offspring, as Lenz's (1987) experimental work on the single-site nester *Cryptotermes brevis* has shown. He compared the offspring production of numerous primary reproductive pairs (typically with high nutrient reserves) to that of numerous neotenic reproductive pairs (typically with low nutrient reserves) and found that the average primary reproductive pair produced more offspring than the average neotenic pair when neither had workers to assist them. Indeed, most neotenics perished during the attempt to reproduce (see also Ahmad et al. 1980). In contrast, when workers were present in the nest, neotenics had similar or even higher rates of brood production in comparison to primary reproductives (see also McMahan in Lenz 1987; Steward 1983a). Steward (1983b) also documented differences in egg-hatching time and success between primary pairs and neotenics which can be interpreted to reflect differences in fat body reserves and their subsequent allocation of nutrients in the eggs: in experimental groups of *C. brevis* and *C. dudleyi*, embryo development to egg-hatching took approximately eight weeks for eggs produced by primary reproductives, compared with approximately ten weeks for eggs produced by neotenics.

Despite the availability of fat reserves, many termites must feed before producing the first worker brood, examples are the single-site nester *Zootermopsis* (Cook and Scott 1933; Shellman-Reeve 1990), the multiple-site nester *Coptotermes formosanus* (Huang and Chen 1981), and perhaps also many central-site nesters (Noirot 1970). Several studies show that termite reproduction is lowered or halted when nitrogen availability is low (*Zootermopsis*, Hendee 1935; Hungate 1941; Shellman-Reeve 1990). The incidence of cannibalism is also higher under low nitrogen availability (in *Zootermopsis*, Cook and Scott 1933).

Body masses of both male and female reproductives decline significantly within the first year of nest settlement in single-site and central-site termites; for example, adult body mass declines by 25% in *Zootermopsis nevadensis* (Shellman-Reeve 1994a), by 70% in *Microtermes* spp. (Wood and Johnson 1986) and by 30% in *Macrotermes* spp. (Darlington 1986; Han and Bordereau 1992; Tahiri and Han 1993; see also Nalepa and Jones 1991). These declines in body mass often coincide with a high percentage of mortality during nest settlement. The percentage mortality in young colonies of *Zootermopsis nevadensis* is 60% (Shellman-Reeve 1994a) of *Macrotermes subhyalinus*, 58% (Han and Bordereau 1992) and 50% (Tahiri and Han 1993) of *M. michaelseni*, 75%; see also Table 1 and 2 in Pomeroy (1989). In a field study on one-year-old colony reproductives of *Z. nevadensis*, Shellman-Reeve (1994a) documented a positive correlation in dry mass between a male and his mate among reproductives of several one-year-old colonies (but not among newly paired reproductives) that cohabited a log. The lower masses of older pairs compared with those of new nest settling pairs indicate that, over time, a pair undergoes nutritional stress that varies according to locally available resources on a log (see similar result in the subsocial woodroach *Cryptocercus punctulatus*) (Nalepa and Mullins 1992). The reproductives' mass and size only increase after

workers are established in the colony (e.g. single-site nesters, Harvey 1934; central-site nesters, Darlington 1986).

The poor availability of quality nutrients and consequent high mortality during the first year of nest initiation suggests a selective advantage for cooperation between nest-founding reproductives. Indeed, laboratory experiments revealed that the behavioral activities of male and female nest-founding reproductives can shift in ways that conserve the body use of nitrogen of the female (*Zootermopsis*, Shellman-Reeve 1990). Nutrient assistance by the male (either by the procurement of external nutrients or by sharing of fat-body nutrients) is expected to provide a crucial contribution to egg production and offspring care in all termites, particularly in dispersing reproductives, i.e. reproductives that initiate nests without workers. In *Zootermopsis nevadensis*, a male transfers protein-rich nutrients to his mate nearly three times more often than the female transfers to the male during the time before egg laying (Shellman-Reeve 1990). Once eggs and larvae have been produced, the transfer of nutrients is evenly distributed between a male and female pair, and both contribute equally to the feeding of young (in *Z. nevadensis*, Shellman-Reeve 1990; and in *Z. angusticollis*, Rosengaus and Traniello 1991). Indeed, a male's involvement in all nesting activities is as great as the female's (in *Zootermopsis*, Shellman-Reeve 1990; Rosengaus and Traniello 1991).

The expected pressures of nutrient availability in the pre-worker period might also explain why multiple reproductives (many females and males; polygyny and polyandry) cooperate during nest initiation in some colonies of some central-site termite species (Thorne 1985; Darlington 1985a; Roisin 1993). Multiple nest-founding male and female reproductives are observed in the fungus-growing termites *Macrotermes michaelseni* and *M. herus* (Darlington 1985a, 1988). Alates of these termites typically initiate nests in soils devoid of an external source of nutrients; hence, reproductives are deprived of external nutrients prior to the production of first worker brood. Multiple breeders (multiple males with multiple females) that cooperate in a new colony may gain an advantage by producing a worker force of size similar to that produced by a single pair of breeders, but without each breeder severely depleting its fat-body reserves during the period when external nutrients cannot yet be gathered. Other advantages to multiple breeding are also apparent.

Thorne (1985) showed that young established colonies of *Nasutitermes corniger* headed by multiple queens and kings had faster colony growth and were of larger size compared with monogynous colonies in the same population. She proposed that nest-founding associations of multiple males with multiple females (and perhaps also among reproductives of newly budded colonies) (cf. Roisin 1993) benefit by attaining a higher rate of brood production than similar colonies headed by a single pair of reproductives.

Data gathered by Darlington (1985a) also suggest that multiple nest founders can enjoy a higher rate of growth. Darlington, however, proposes that the faster rate of colony growth is important in mature colonies. Fast-growing colonies may have an advantage in densely-populated habitats where competition for woody material is especially severe and where predation pressures are high (Thorne 1985; Darlington 1985a). Darlington (1985a) also proposes that multiple-nest founding might serve as insurance against loss of a reproductive: a single nest is expected to have greater probability of survival if headed by multiple reproductives than if headed by single reproductives (see also Roisin and Pasteels 1986; Roisin 1993).

Multiple nest-founding is facultative in all cases (in incipient colony founding, Darlington 1985a, 1988; and in satellite nests, Thorne 1985; Roisin 1993). Indeed, most incipient colonies are headed by a single male and female pair. Thus, further studies are needed to disentangle the costs and benefits associated with initiating a colony as a single reproductive pair compared with multiple reproductives. For example, such studies would allow us to answer the question: when does the advantage of reduced risk of starvation or accelerated rate of colony growth outweigh the disadvantages associated with sharing a single nest with multiple reproductives whose genetic interests are likely to differ? Investigations are also needed to examine how relatedness (Thorne 1985; Roisin 1993) and relative size or competitive differences among same-sex individuals (Roisin 1993; examined in *Zootermopsis*, Shellman-Reeve 1994b) influence decisions to form groups, and interactions within a cooperative group of reproductives. Data are also needed on how multiple reproductives cooperate in groups that vary in the number of active reproducers (as has been discussed in social species of other taxonomic groups) (see, for example, Keller and Reeve 1994).

Conspecific competition for quality nutrients favors cooperative defense. Competition for food and nesting resources is known to favor cooperative defense of resources in many taxonomic groups (see, for example, Wilson 1975, 1990; Wrangham and Rubenstein 1986). As might be expected,

there is growing evidence that termites exhibit cooperative and non-cooperative behaviors which permit them to gain access to and compete for quality nutrients and nest resources. In many single-site nesters (e.g. *Archotermopsis*, Imms 1919; *Pterotermes*, Jones et al. 1981; *Incisitermes*, Luykx 1986; *Zootermopsis*, Shellman-Reeve 1994a; and others, Weesner 1970; Lenz 1994), a reproductive pair must establish its nest on the same log where numerous nest-founding reproductives are attempting to settle. Because in single-site nesters resources are depleted and not replenishable, severe conflict between colonies is expected to ensue.

In the single-site nester *Zootermopsis nevadensis*, alates preferentially seek and most vigorously defend nests on regions of a log that are rich in cambium (a thin layer on the log that is high in nitrogen and other quality nutrients). As expected under competition for critical and limiting resources, nest settlement occurs in an ideal free distribution with respect to the availability of cambium nitrogen (Shellman-Reeve 1994a). Size of the log does not appear to be as important in nest site preference as the availability of nitrogen-rich cambium, at least within the range of branch and log diameters that were examined. Data on internest distances between nearest neighbors indicate that pairs obtain similar amounts of the cambium resource throughout a wide range of cambium availabilities that exist on logs. Although interpair competition is apparent under a wide range of cambium availabilities, the highest densities of nest-founding pairs are found on logs where the cambium layer is thickest. Under these conditions, direct competition for nutrient and nest resources is observed.

Fierce fights occur among newly flown pairs and unpaired reproductives over settlement of their nest sites. Nest-founding males initiate fights with as many females as they do males, and females do the same. Paired reproductives have a greater probability of holding onto a nest site than do single individuals, especially during times of aggression. These data indicate that cooperative nest defense is likely to be an important driving force favoring male–female association in *Z. nevadensis* (Shellman-Reeve 1994a). Cooperative nest defense is probably an important driving force facilitating male–female associations in termites of other life types as well. There is intense fighting among nest-founding adults of the central-site nester *Hodotermes mossambicus*. Nel's (1968) measure of aggression among single and paired reproductives revealed that the most aggression per termite was exhibited by individuals that were paired with the opposite sex, rather than when they were alone. Males and females also exhibit as many intersexual fights as intrasexual ones, indicating that fighting is not merely to gain access to mates (see also Shellman-Reeve 1994a). These fights ensued over access to nesting soil (see also van der Linde *et al.* 1989).

Even beyond the first year of colony-founding, the risks to reproductives and their young brood are extraordinarily high. The extreme pressures caused by conspecific (and interspecific) competition over access to the same resources is expected to favor cooperation among nestmates, in an effort to secure resources from threatening competitors. Levings and Adams (1984) and Adams and Levings (1987) studied intercolony competition in two species of *Nasutitermes* and found that intraspecific (and interspecific) aggression was very high, because colonies defended territories that were often overlapping. Fierce, deadly battles can occur at boundaries where two foraging parties meet. Both species are limited by the physical structure of their foraging arena (foraging trails are entirely arboreal, and can only be forged between trees that have physical links, e.g. branches). As a result, young colonies are particularly susceptible to failure since their small territories are often within the larger territories defended by more mature colonies. Nel (1968) and Van der Linde *et al.* (1989) have determined that territorial size varies widely in the harvester termite, *Hodotermes* (from 0.5 ha to 3.1 ha within a given area); their data suggest that foragers and soldiers of mature colonies present a similar threat to young colonies, since young colonies typically reside within their larger competitors' territory.

Even in more mature colonies competition from conspecifics persists, at is especially apparent at the territorial boundaries. Darlington's (1982) observations on the elaborate underground trail-forming habits of *Macrotermes michaelseni* revealed that mature colonies forage up to 50 m from the nest, with the main foraging area approximately 10–35 m, radiating from the nest. Foragers of *M. michaelseni* dig deep pits, which are used to temporarily store food at intermittent points along the forage trails. It appears that the number of pits roughly corresponds to the quantity of forage gathered per night. At portions of the underground trail that intersected near trails of other colonies, Darlington discovered that storage pits and passages contained dead termite carcasses and head capsules, indicating that intercolony fights at territory boundaries can be severe. Although nothing is known about the frequency of such conflicts, the occurrence of fights at storage pits suggests that intercolony thievery is a food-gathering strategy. Such thievery may

represent another selective pressure fueling intercolony competition, and which favors cooperation among workers and soldiers of a colony.

In the multiple-site nester *Heterotermes aureus*, battles between conspecific colonies are observed under both field conditions (Jones 1993) and controlled laboratory conditions (Binder 1988). Intercolony aggression in the laboratory is seen in *Coptotermes formosanus* (Su and Haverty 1991) and numerous other species (reviewed in Thorne and Haverty 1991). These studies suggest that competition between colonies for scarce resources is common under natural circumstances. Although fewer studies document the effects of exploitative competition on colony failure or success (but see Pomeroy 1989), exploitative competition may have severe negative effects on small colonies in circumstances in which small and large colonies overlap (e.g. *Macrotermes*, Abe and Darlington 1985; *Coptotermes lacteus*, M. Lenz, personal communication).

Predation costs of nest founding likely favor cooperation of breeders and young workers.   Predation is known to favor cooperative defense in numerous taxa (see, for example, Alexander 1974; Wilson 1975). In termites, slow offspring development (caused by a poor-quality wood diet) (Lafage and Nutting 1978; Breznak 1982; Shellman-Reeve 1994a; Nalepa 1994) likely increases the probability that offspring will be detected, injured or killed by a predator before reaching adulthood (maturation of a reproductive adult takes two to four years, or longer in most termites).

The best available evidence for the importance of predation as a selective pressure favoring worker and soldier cooperation is inferred from the variety of termite defenses against predators in mature colonies (see Stuart 1969; Noirot 1970; Prestwich 1983, 1984; Howse 1984a,b). Predation by ants and other soil arthropods is hypothesized to be especially harsh in species that forage outside their nest, i.e. in multiple-site and central-site nesters (see, for example, Wilson 1971, 1975; Howse 1984a,b). Indeed, predation from diverse soil arthropods may account for the diverse chemical weaponry that has evolved in the Mastotermitidae, Rhinotermitidae and Termitidae (Prestwich 1983; Howse 1984a,b). It is also termites of these life-types that exhibit one the most bizarre defensive adaptation known as autothysis. Autothysis is a form of self-sacrifice in species in the Apicotermitinae, Termitinae, and *Serritermes serrifer* of the Serritermitidae in which the abdominal wall of a worker (of soldierless species) or soldier is ruptured with its muscular contraction as it forces out a chemical toxin or immobilizing secretion onto an attacker (Wilson 1971; Sands 1981; Mill 1984).

Early colony failure due to predation is perhaps the most difficult ecological pressure to quantify. Descriptive accounts, however, can offer some hint of its importance. Among single-site wood-dwelling species, centipedes may be one of the more important predators. Adult centipedes (*Scolipendra* species) nest and reproduce on and adjacent to logs occupied by the single-site nesters *Zootermopsis nevadensis* and *Z. angusticollis*. The adult centipedes feed on termite reproductives and workers, while the centipedes' offspring feed on the termite eggs and early-instar termite larvae. Because centipedes continue to produce offspring as long as there is a continuous supply of nourishment (termites), these arthropods pose a continuous threat to developing colonies (Shellman-Reeve 1994b). The construction of a claustral nest by termite reproductive adults and nest barricades constructed by young colony members appear to be strategies that thwart or guard against predation (see, for example, Springhetti and Sita 1986; Nalepa and Jones 1991). Indeed, a thin wall of debris and fecal material is frequently all that separates a reproductive pair or young termite colony from foraging centipedes. Presumably the cooperative fighting observed against conspecific intruders (see, for example, Shellman-Reeve 1994a) would be effective against such predators. However, no studies have yet examined the cooperative nest defense of breeders in the context of predation.

Darlington *et al.* (1977) provide the only quantitative measure of initial nest failure due to predation (probably caused by ants). They determined that 42% of the colonies of the harvester termite *Hodotermes mossambicus* failed within one month of settlement. Ants were suspected of killing and absconding with nest-founding termites. Upon excavating nests to determine survivorship, the researchers found that all failed nests were completely devoid of termite carcasses. Attacks by doryline ants caused 80% of the observed mortality of maturing *Macrotermes* nests; foragers were at great risk (Darlington 1985b). Collins (1981) proposed that ant predation on foragers can account for most of the failures of young colonies of *Macrotermes bellicosus*: because there are no replacement foragers in young colonies, such predation can wipe out the colony's entire foraging force. In addition, such predation events would come at a time when reproductives are under highest nutritional stress (Darlington 1986).

Direct and indirect advantages favoring alloparental care

Assuming that limited reproductive opportunities favor some offspring to stay at home, then it must be asked: why do offspring help?

Why do offspring help? Choice or manipulation

Inherent in models in which individuals maximize their inclusive fitness is the notion that selection favors individuals that choose the best fitness option available to them. Some authors suggest that in many eusocial species, the offspring's options to become a worker or independent reproducer are severely limited by the individuals that raise it, i.e. the offspring is manipulated either by the parents or by the workers that rear it (Wilson 1971; Alexander 1974; Alexander *et al.* 1991). For example, parents or workers might force an offspring to become a worker by restricting the nutrients it receives as a juvenile. Manipulation may be especially important in eusocial Hymenoptera, because the holometabolous larvae are in a grub-like state and completely dependent on adults for their nutritive and hygienic needs. Colony offspring of hymenopteran species obtain independence from worker care only upon reaching adulthood. However, some studies that claim to have evidence for the parental or worker manipulation of offspring do not consider the alternative hypotheses (for example, in honey bees, Winston and Slessor 1992; in termites, Zimmerman 1983; Myles 1988; see also Roisin 1994). For example, offspring may develop along a pathway that serves the genetic interests of both putative manipulator and 'manipulated' individual (Seeley 1985; Keller and Nonacs 1993). This hypothesis is particularly likely if individuals are at least as related to siblings as to their own offspring (Stubblefield and Charnov 1986). In such cases, workers do not have to be manipulated to be in genetic agreement with the manipulators (i.e. reproductives, workers, etc.) over what the workers should do.

Single-site and multiple-site termites are an important group in which to examine the question of choice versus manipulation, since offspring attain a high degree of independence early in development. Termites undergo hemimetabolous development, resulting in juvenile offspring that are active participants in the nest. Offspring of single-site and many multiple-site nesters are only completely dependent on parents or workers for a brief period as a juvenile (up to the third instar of development). After this period, immatures can gather food for themselves and are free to exchange nutrients or microbes with any nestmate, and thus seem able to escape manipulation. The ability to choose a developmental pathway should be important in termites, since there are numerous alternative developmental endpoints; for example, an individual that starts off helping in the nest can become a soldier, a worker, a neotenic, and in single-site species, an alate. On this view, a termite offspring should be able to choose its best fitness option by altering its own development in response to changing social or resource conditions. For example, an offspring should disperse when wood becomes depleted of its nutrients, but should stay and rear non-descendent kin if resources at the natal nest are plentiful. Contrary to this view, there are numerous studies that emphasize that offspring are being manipulated (1) by older siblings that inflict injuries on them, thereby forcing them to stay in the nest or (2) by primary reproductives and soldiers that feed them pheromone-produced substances that 'control', 'suppress' or 'inhibit' their development to a particular end (in Watson *et al.* 1985; but see Keller and Nonacs 1993).

Data that have been used in support of parental or worker manipulation have not ruled out the possible importance of offspring choice in development. Under the sibling manipulation hypothesis (Zimmerman 1983; Myles 1988), some siblings manipulate the development of nymphs by injuring the wing buds of nymphs, thereby forcing them to stay at home and help, rather than allowing them to leave. Although nymphs of numerous single-site nesting species are found with apparently injured wing buds (Roisin 1994), no behavioral observations have been conducted that demonstrate how or why the injuries typically arise: they may be damaged by conflict over food (Roisin 1994) or by intercolony conflicts on logs cohabited by multiple colonies (as seen among new pairs) (Shellman-Reeve 1994a). A key prediction of the sibling manipulation hypothesis is that damaged nymphs are being manipulated for the purpose of rearing additional brood (Zimmerman 1983; Myles 1988). Since each worker produces, on average, fewer offspring at the end of the colony cycle, compared with early phases of the colony cycle (Table 4-3, Fig. 4-4), wing-bud damage should occur most frequently at the early phases of colony growth, when more food is available on which to rear additional brood, and less damage should occur in mature colonies, when little food is left to rear additional brood. Presently, there is little evidence that nymphs are damaged in young colonies, but nymphs with wing-pad damage are very common in mature colonies where nest and food resources have been depleted (Roisin 1988).

These findings are more consistent with the notion that wing-pad damage arises as a result of severe competition for food (Roisin 1994) and space. For example, if reproductively destined offspring receive the bulk of their nutrient reserves (i.e. their preflight dowry) after they develop into winged adults (see feeding of reproductives) (Howse 1968; Nutting 1969; Lafage and Nutting 1978), it would benefit nymphs and alates (but not younger workers) to damage the wing buds of developing nymphs, thereby preventing these would-be competitors from sharing food reserves. Younger workers and mature reproductives would benefit by permitting their dispersal rather than keeping them around as food competitors for a more prolonged period.

The pheromone manipulation hypothesis must also be critically examined. Pheromone-like substances produced by reproductives and soldiers of a colony often appear to channel the development of the receiver along a pathway that is adaptive to the sender (see, for example, Watson and Sewell 1985). Evidence indicates that soldiers appear to 'inhibit' workers from developing into soldiers and that primary reproductives appear to 'prevent' workers from developing into reproductives (in Lüscher 1976; McMahan 1969; in Watson *et al.* 1985). In these studies, the observations are of 'how' changes of development may be externally influenced. Terminology such as 'suppress', 'inhibit' and 'control' is used to describe the observed effect at the mechanistic level of analysis, not at the functional level of analysis (Sherman *et al.* 1988; Alcock and Sherman 1994). When the observations on how developmental changes occur are used to interpret 'why' developmental changes occur, then adopting the same terminology is not appropriate, in the absence of new kinds of evidence. In a functional analysis, the reproductive interests of both sender and receiver must be examined to see if manipulation is occurring or if genetic interests between the receiver and sender coincide. In other words, although a pheromone may cause the inhibition or redirection of development, it may be that the pheromone provides a cue or signal that contains information about the colony's status (health of a reproductive, food availability, injuries, etc.) which the offspring then uses to choose among its reproductive options, i.e. be either a worker, a soldier, a neotenic or an alate (see also Crespi 1992b). Under this scenario, the pheromone produced by the sender produces a stimulus for the receiver that provides information that is adaptive to the receiver (the stimulus may either be adaptive for the sender (a true signal) or only an incidental byproduct of the sender's actions (a cue)).

Table 4-3. *Cumulative numbers of offspring produced per alloparent as a function of colony age in* Zootermopsis nevadensis

The median (and range) number of residents per colony age were obtained from Heath (1927). The estimated offspring number per allo-parent (either soldier or worker morph) is obtained by the following formula: Offspring earned, $\text{year}_1 = \text{residents}_1/2$ parents; Offspring earned, $\text{year}_2 = (\text{residents}_2 - \text{residents}_1)/(\text{residents}_1 - 2 \text{ parents})$; Offspring earned, $\text{year}_3 = (\text{residents}_3 - \text{residents}_2)/(\text{residents}_2 - \text{residents}_1)$; Off-spring earned, $\text{year}_4 = (\text{residents}_4 - \text{residents}_3)/(\text{residents}_3 - \text{residents}_2)$, and so on. The cumulative number of earned offspring for an allo-parent that begins to help in years 2–7 is determined by summing the earned number of offspring per year for all remaining years a colony is in production. Because offspring work as immatures and most attempt to breed as adults, this cumulative number of offspring earned per allo-parent is a rough measure of the indirect component of an alloparent's inclusive fitness.

Year of production	Median numbers (range) of residents	Offspring gained per alloparent per year	Cumulative offspring earned per alloparent
1	10 (0–20)	5.00	
2	40	3.75	16.41
3	100	2.14	12.66
4	400	5.17	10.52
5	953 (786–1120)	1.86	5.35
6	2547 (2213–2881)	2.89	3.49
7	3500	0.60	0.60

Thus, 'soldier manipulation' that prevents a worker from becoming a soldier may actually be 'soldier signal-ing', which informs a worker that soldiers are present, causing the worker to choose a developmental pathway other than that leading to a soldier. This decision is potentially adaptive to both sender and receiver. Su and Lafage (1986) found that the soldiers were the first to starve under low food availability. Soldiers are nutri-tionally dependent on workers for food, so if food is scarce and higher soldier defense is not advantageous to the colony, then an excess of soldiers can lower colony survivorship.

The same hypothesis applies to pheromones produced by primary reproductives. Much evidence suggests that workers do not become neotenics in the presence of pri-mary reproductives (Greenberg 1980, 1982; Greenberg and Stuart 1982; in Watson *et al.* 1985), but evolutionary inhibi-tion or suppression need not be invoked to explain these observations. An alternative interpretation is that phero-mones signal the presence of a healthy king and queen, thus informing the receiver that its probability of success as a reproductive might be low (see also Alexander *et al.* 1991). Indeed, neotenics must have workers to become functional reproductives (Lenz 1987; Ahmad *et al.* 1980). Too many breeders and not enough workers may make it more advantageous for a worker to remain a worker, whereas not enough breeders and too many workers may make it advantageous for a worker to become a neotenic.

A pheromonal signal keeps a worker informed of these potentially changing fitness consequences.

Why might workers choose to help?
Brown's (1987) scheme for examining helper (worker and soldier) inclusive fitness is useful in understanding the fit-ness benefits for termite helpers. Inclusive fitness benefits are described along two axes: direct–indirect and present–future. Direct fitness gains are those accruing to the helper through its own offspring; indirect benefits accrue through the increased production of a helper's non-descendent kin. Present or immediate benefits are those achieved during the current breeding season; future benefits are realized in subsequent breeding seasons.

Selective advantages of helping are discussed by Emlen (1991) for cooperatively breeding vertebrates, i.e. those with parents and alloparents that provide care, and fall into four basic categories. By acting as a worker, an indivi-dual may (1) increase the production of non-descendent kin (immediate, indirect benefit); (2) improve the survival of its reproductive relatives (delayed, indirect benefits); (3) enhance its probability of filling an immediate reproduc-tive vacancy (immediate, direct benefit); (4) improve its later chances of becoming a successful reproductive (delayed, direct benefits).

In termites, workers of central-site nesters enhance their inclusive fitness apparently only through the produc-tion of non-descendent kin (immediate and future, indirect

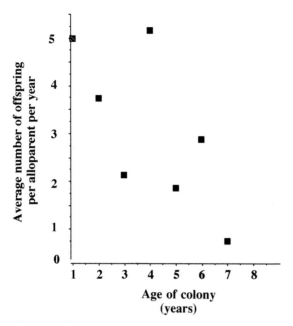

Figure 4-4. The average number of offspring per alloparent per year is regressed against the age of a colony in the single-site nester *Zootermopsis nevadensis*. The average number of offspring per alloparent (term includes both worker and soldier) declines with colony age ($y = -0.54x + 4.7$, $r = -0.695$; Pearson moment correlation, one tailed t-test $p < 0.05$). Each parent's offspring production (largely without alloparent help) is represented by the hatched square; all other years correspond to the number of offspring reared by alloparents. Median number of residents per colony age are taken from Heath (1927). The sharp decline in the sixth year is due to the depletion of food and nest resources in the log (Heath 1927, 1931). Knowing other parameters, such as the annual rate of mortality of alloparent, is desirable for more accurate measures of reproductives reared per alloparent. The average number of offspring reared per alloparent per year and the cumulative indirect benefits earned per alloparent (Table 4-3) provide only rough estimates of the relative indirect fitness benefits for workers produced in different years, since year-by-year growth data have not been collected for a single colony.

benefits). However, workers of single-site and multiple-site termite species can enjoy both indirect and direct benefits, as described below.

Indirect benefits in all nest types: immediate and future. The indirect component of a worker or soldier's inclusive fitness is the number of offspring successfully reared above that which the primary reproductives would have gained without its help, devalued by the helper's relatedness to

those offspring (Hamilton 1964; Grafen 1984). I examine two explanations for worker care through indirect reproduction.

The first explanation, put forward by Roisin (1994), is that workers 'make the best of a bad job'. Using Hamilton's rule to examine the origins of termite eusociality, Roisin (1994) proposed that intracolony competition for food results in intracolony conflicts that lead to the evolution of sterile worker and soldier castes in termites of all life types. Under this hypothesis, the permanent altruists are the phenotypic 'losers' in a colony. Because losers are likely to fail as reproductives, and because they are capable of perceiving themselves as losers, they stay at home to help, as part of a fitness-salvaging strategy (West-Eberhard 1981, for Hymenoptera). Losers in the contests over food are eventually favored to become colony-defenders or colony-workers in the attempt to improve their indirect fitness benefits (see also Nalepa 1994). Intracolony conflict is thus said to result in a type of stable, conditional polymorphism whereby offspring are expected to attempt direct reproduction if they perceive themselves as winners and become indirect reproductives (soldiers or workers) if they perceive themselves as losers.

Some life-history characteristics of termites, however, suggest that the 'loser strategy' may not account for the evolution of altruism. Contrary to Roisin's (1994) hypothesis, soldiers in some genera of the Termopsidae, e.g. *Zootermopsis*, can still reproduce, for example, in orphaned colonies (Heath 1928). The morphological transformation into a soldier apparently does not reduce the female or male soldier's ability to be an effective replacement reproductive under field conditions (Heath 1928). Indeed, according to Myles (1986), the enlarged mandibles of soldiers and their large size may have evolved to enhance their chances of outcompeting other siblings that are also attempting to become replacement reproductives in the nest. Since hypotheses regarding the origins of soldier evolution are not easily tested, we may never know why the soldier morphology was first favored. However, the 'making the best of a bad job' hypothesis may be tested in the context of selective maintenance. For example, under controlled contests over access to food, juvenile offspring that are losers of the contests over food are expected to be more likely to become altruistic soldiers than offspring that are winners.

Questions also arise over the proposed mechanism for permanent-worker castes. Roisin (1994) points out that nymphs that receive wing-pad damage (for example, as a result of intracolony competition over food) must remain in the colony while undergoing reversionary molts to repair the damage. He proposes that these damaged nymphs will 'make the best of a bad job' by helping out in the nest during the repair period. Because these damaged nymphs perceive themselves as losers, they work to improve their inclusive fitness. To explain the evolution of permanent worker castes, Roisin (1994) argues that juveniles, such as those that are susceptible to wing-bud damage, should be selected to perceive themselves as losers or winners of conflicts early in their development. That is because if an offspring is a loser, then its early development into a worker can improve its inclusive fitness. Indeed, permanent workers appear to channel all their efforts into indirect reproduction, and can even be specialists (or super-donors, *sensu* West-Eberhard 1975, for eusocial Hymenoptera). However, losers of some contests can be winners in others. For example, it is not known that damaged nymphs were always losers; perhaps previous 'winner' nymphs are as likely to obtain damage as previous 'losers', for example, when ambushed by another competitor. As long as an offspring is uncertain about its phenotypic status as a breeder, then that offspring should delay its decision to be a permanent altruist. A key prediction of the 'making the best of a bad job' hypothesis is that losers should be less effective reproductives than are winners. This prediction may be tested in the context of selective maintenance, for example, by comparing the direct reproductive efforts of offspring that are identified as 'losers' and 'winners' of contests over food. Losers (e.g. those with previous wing-pad damage) should do worse than winners (e.g. those with no previous damage) when total reproductive effort (e.g. dispersal survival, mate acquisition, cooperative nest formation, and offspring production) is compared.

The second explanation, I suggest, is that *workers gain cumulative indirect-fitness benefits*. Earlier discussions on ecological pressures at initial colony founding and initial colony growth provided examples of how biparental, multiple-parental and alloparental care may enhance a colony's survival and reproductive output. With regards to alloparents, these studies show how workers and soldiers can derive indirect fitness benefits: by obtaining limiting, critical nutrients, by acquiring resources through successful interference or exploitation competition, and by defending a colony against predators.

Not all potential workers and soldiers, however, should be expected to obtain equal indirect fitness benefits. In early and mid-phases of colony growth, typically when the availability of resources is great, each worker is expected to have ample opportunity to accumulate indirect fitness benefits. In contrast, in late colony phases, typically when the availability of resources levels off or diminishes, each worker is expected to have less opportunity to accumulate indirect fitness benefits. Since opportunities to accumulate indirect fitness benefits are expected to vary with colony age and the availability of food and nest resources, offspring should be selected to assess the potential cumulative indirect benefits for staying in the nest, then choose to stay in the nest either as a temporary helper or as a permanent worker or soldier, based on this assessment.

Each worker's assessment of his or her potential cumulative indirect fitness benefits may explain the varying degrees of altruism that are observed in termites of different life types, and within each life type, among workers present in different phases of colony growth. The cumulative indirect fitness assessment hypothesis resembles Queller's (1989, 1994) head-start hypothesis for adult Hymenoptera. In Hymenoptera, offspring are favored to help at home rather than attempt reproduction at a new nest site because of the benefits of remaining in a stable, functional colony. Workers have a head start on indirect reproduction, rearing non-descendent kin.

To illustrate the relationship between workers' indirect fitness benefits and colony age, I estimated the number of offspring reared per worker and soldier in colonies of different ages in the single-site nester *Zootermopsis nevadensis* (data from Heath 1927; Heath and Wilbur 1927). For all offspring that were censused, I assumed that genetic relatedness was constant in each colony's brood sampled, which is tantamount to the reasonable assumption that primary reproductives live until the production of alates. I first estimated the number of offspring reared per worker or soldier per year (Fig. 4-4). The cumulative indirect fitness benefits are the sum of the average number of offspring per helper per year determined for all years a helper might help (Table 4-3). Because all offspring of single-site nesters have the potential to reproduce, and most apparently do (Abe 1987; Luykx 1993; Lenz 1994) except for a small proportion of non-reproductive soldiers and workers (see Haverty 1977, for soldier estimate), it is assumed that the cumulative offspring reared per helper (including both temporary workers and soldiers) roughly reflects the number of reproductives reared per helper.

The estimates of the average offspring per helper per year for colonies of increasing colony age showed that the potential indirect fitness of a helper decreases with increasing colony age in the single-site species *Z. nevadensis* (Fig. 4-4). The first helpers produced by parents in the incipient colony (year 1) and helpers subsequently produced in two- to four-year old colonies exhibit the highest per capita brood production per year. In contrast, helpers that were produced in older colonies, i.e. five to seven years old in *Zootermopsis*, exhibited the lowest per capita brood production per year (Fig. 4-4). These data show a decline in annual offspring production per helper in colonies of older age; this decline is probably a reflection of decreased nutrient availability at the nest–food site (see, for example, Heath 1927, 1931; Maki in Abe 1987; Watson and Sewell 1981, 1985; Lenz 1994) and the resulting increase in competition from neighboring colonies that coinhabit a log (Shellman-Reeve 1994a).

Offspring produced early in colony ontogeny have the potential for greater cumulative indirect fitness returns; offspring produced in older colonies have fewer opportunities to accumulate indirect fitness (Table 4-3). Because offspring of single-site termites are in the nest for a minimum of three years as immatures (Castle 1934; Luykx 1993), and live four to five years as adults or longer (Heath 1907), any helper produced in the first years of colony ontogeny can earn from 11 to 17 offspring if it decides to delay dispersal, or forgo dispersal entirely (as a soldier), whereas offspring produced in the latter phases of colony growth would obtain about half that indirect benefit (Table 4-3). These data suggest there are strong selective advantages to delaying dispersal (if a worker) or forgoing dispersal (by becoming a soldier), if an individual develops early in colony ontogeny. These benefits decline for the average helper produced in older colonies, i.e. colonies with shortened expected lifetimes and fewer resources. Soldiers are expected to be the oldest members of the helper force in single-site termites, i.e. they should be among the brood produced early in colony ontogeny. Moreover, these findings suggest that temporary workers are expected to exhibit less altruism when they are in older colonies, especially in nests that are becoming depleted in nutrients.

Unfortunately, there are few data examining worker and soldier altruism as a function of colony age. In the single-site nesters, Heath (1931) reported that all the brood produced in first-year colonies of *Zootermopsis* express the soldier potential, i.e. all of this first cohort eventually became soldiers in colony ontogeny. Apparently he based his argument on direct observation, but no further studies were conducted to confirm the tendency (cf. Howse 1968).

Luykx (1993) determined that soldiers in colonies of *Incisitermes* also tended to be the oldest members of a colony; soldiers in colonies with replacement reproductives typically had the genotype of the primary parents and not the supplemental parents.

The cumulative indirect benefits hypothesis may also be applied to central-site species. Since in central-site nesters worker and soldier life spans are typically less than a year (Table 4-1), most benefits to a worker and soldier must be determined by its future indirect benefits (as calculated by Mumme *et al.* 1989). The high proportion of permanent altruists versus alate-brood in the juvenile and mid-phases of colony growth is consistent with the notion that workers have an abundance of food resources, and a high potential for cumulative indirect fitness returns. The high proportion of alate-broods produced in 'mature' colonies may reflect a decline in growth, due to a leveling-off or reduction of resources. Many authors suggest that once a colony reaches maturity, the proportion of workers and alates becomes constant year after year (for *Macrotermes*, Collins 1981; Darlington 1986; see Lenz 1994). This observation is compatible with the cumulative indirect fitness benefits hypothesis if resources are being replenished at a constant rate in mature colonies. The hypothesis also predicts that an increased proportion of altruists will reflect an increase in food resources, whereas an increased proportion of alates will reflect a decrease in food resources, relative to the number of workers gathering these resources. Thus, this hypothesis may be tested in all life types by manipulating resource availability, workers, and breeders (e.g. last-instar nymphs).

Direct benefits of single-site and multiple-site nesters: immediate and future. The direct benefits gained by helping are expected to be high in single-site nesters and some multiple-site nesters, i.e. *Mastotermes* and some *Reticulitermes* species; few workers of central-site nesters, however, attempt reproduction. For the single-site and multiple-site termites, I review four explanations to show how direct benefits can favor worker care (see Myles 1988 for another discussion of direct benefits).

The first explanation is that *helpers improve their chances of becoming successful reproductives.* One way in which helping can increase a worker's probability of survival is that, by helping, more offspring will be present to enhance colony vigilance, defend against predators and competitors, acquire and defend valued resources, or any combination of the latter. Two critical predictions of this hypothesis are

(1) that helping augments group size and (2) that survival of the helper is higher in larger than in smaller groups (Emlen 1991). The first prediction is met for most workers (for example, in early and mid-phases of colony growth); the numbers of offspring in mature colonies (Lenz 1994) far exceed numbers of offspring that a reproductive pair might produce on its own. The exponential growth typical of most colonies is directly attributable to worker participation (see Fig. 4-4 and Table 4-3).

It is likely that the second prediction, increased survival in larger groups, is also met. For a given stage of development, offspring in smaller colonies are morphologically smaller and take longer to mature (Castle 1934; Nagin 1972; Roisin 1988). Body sizes of same-age workers increase in colonies of larger size, as does their rate of development (Nutting 1969; Lafage and Nutting 1978). An immediate benefit of helping is that more offspring are available to gather nutrients; a high number of foragers increases the probability of finding rich pockets of patchily distributed nutrients. Springhetti (1990) also found that individuals in groups of larger size (but not smaller size) exhibit a reduced rate of activity per individual. Thus, by helping to rear offspring a worker may ultimately reduce its own foraging effort, as well increase its chances of obtaining quality nutrients, at least in early and mid-phases of colony growth, where resources can be found in abundance. Greater colony size is also expected to improve a helper's access to nutrient resources through territory expansion and maintenance (Shellman-Reeve 1994a).

Greater group size might also permit a worker to hide more easily from predators, by the selfish herd effect (Hamilton 1971); greater numbers of offspring in the nest likely increase defense capabilities (see, for example, Franks and Partridge 1993). Having more workers present also increases the diversity of methods for deterring predators. Some colony mates defend by dumping copious amounts of feces onto the attacker or into the tunnel system to block attack, while others produce a predator alarm, e.g. by head-tapping (Howse 1984a; Stuart 1988) or by pheromones (Stuart 1969).

The second explanation is that *workers gain a dowry.* Former workers that leave the nest by dispersal must initiate a new nest without the help of a worker force. Nest initiation by this mode favors male and female pairing and biparental care (Nel 1968; Wood and Johnson 1986; Darlington 1986; Han and Bordereau 1992; Tahiri and Han 1993; Shellman-Reeve 1990, 1994a; Nalepa and Jones 1991). Even with two individuals participating fully in food acquisition

and nest defense at nest founding, dispersing adults take with them stores of nutrients that are probably essential in colony initiation. The resources available per individual are likely greater in larger colonies (for a given external resource level) because of the greater ability of such colonies to secure critical resources. In the 'dowry' hypothesis, offspring reared in the current season can be used by workers to increase their chances of successful dispersal and colony initiation, as observed in cooperatively breeding birds (Emlen 1991). In single-site nesters, workers are likely to benefit by helping to rear offspring who, in turn, will make available the critical nutrients that workers require for dispersal and new nest initiation. However, this hypothesis is probably not applicable in multiple-site nesters, because most multiple-site workers do not disperse and cannot benefit from a dowry.

The third explanation is that *workers gain future helpers.* Workers that become replacement reproductives in the natal nest or original reproductives in budded fragments of the natal colony may benefit by rearing non-descendant offspring, which subsequently help them. In multiple-site nesters, field colonies of many species are more frequently found to be headed by neotenics than by primary reproductives (Gay and Calaby 1970; Weesner 1970); these findings indicate that budding or replacement reproduction at the natal nest is a common direct reproductive avenue by apterous offspring. Colonies of single-site nesters are, however, less likely to be headed by neotenics, except when nests are located in large logs (see Lenz 1994). When neotenic reproduction does occur, neotenics must acquire a small helper force before attempting to reproduce; apparently neotenics do not have the fat reserves needed for independent founding (for example, in *Cryptotermes brevis,* Lenz 1987; *Heterotermes indicola,* Ahmad *et al.* 1980). Thus, it is possible that workers provide help, not to gain a dowry for dispersal, but to gain future helpers as they attempt to bud or become replacement reproductives at the nest. Because the vast majority of workers do not become neotenics in most termite species, this hypothesis by itself is not sufficient to explain worker helping.

Almost nothing is known about the factors underlying the decision to bud rather than to stay at the nest. It may be that most budded groups leave the parent's nest after the primary reproductive pair has died (Weesner 1970), in which case workers no longer have a choice to stay and preferentially help to rear parents' brood rather than sibs' brood. In may also be that diminishing resources at nests headed by primary reproductives trigger budding in multiple-site nesters. Roisin and Pasteels (1987) suggest that patchily distributed resources (in central-site nesters *Nasutitermes princeps* and *N. polygynus*) make colony cohesion difficult to maintain because resource location would require foraging groups to be spread out too far from each other; thus this condition may trigger budding. Neotenics may rear their own brood before budding; these offspring, rather than siblings, may accompany the neotenics during budding.

The fourth explanation is that *workers benefit through nest inheritance*. By staying in the nest, workers may improve their chances of becoming reproductives by inheriting all or part of their parents' nest (see discussions in Myles 1988). As explained by Myles (1988), nest inheritance can favor offspring that stay close to home. However, nest inheritance does not explain why workers would also *help* to rear offspring (which are future competitors). Indeed, to the extent that nest inheritance is an important pathway for direct reproduction, selection should favor some level of worker laziness (*sensu* Reeve 1992). In termites, Howse (1968) provides evidence of worker laziness in *Zootermopsis nevadensis*. His observations of marked apterous and nymphal workers revealed that about one-half of the nymphs consistently performed little or no work, whereas the other half exhibited high levels of activity. Springhetti's (1990)

studies on *Kalotermes* also showed that in larger colonies, workers performed, on average, less work than workers in smaller colonies. These findings suggest that some individuals may be more lazy than others, perhaps in the attempt to enhance their chances for direct reproduction.

CONCLUSIONS

To summarize the key factors that may influence termite social evolution, I present a model that incorporates the major ecological and genetic determinants affecting the potential fitness gains for worker altruism (Fig. 4-5). The ecological determinants involve the costs of independent founding, and the availability and stability of resources; these affect the potential for both direct and cumulative indirect benefits. When integrated together, as intended under Hamilton's rule, these determinants can explain why offspring delay or even forego direct reproduction, and conversely, why offspring attempt direct reproduction.

Under this view of eusociality, I suggest that indirect reproduction (altruism) and direct reproduction in termites are 'mixed strategies' (*sensu* Maynard Smith 1982). In other words, staying at the nest and being altruistic vs. leaving the nest and attempting direct reproduction are events that lead to equal expected inclusive fitness pay-offs

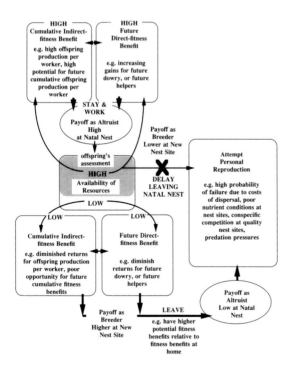

Figure 4-5. Altruists and breeders enjoy equal pay-offs under this frequency-dependent model for the evolution of termite eusociality. Indirect reproduction (being an altruist) and direct reproduction (being a breeder) are 'mixed conditional strategies' (*sensu* Maynard Smith 1982). The model incorporates kinship and ecological determinants to show how changes in these determinants influence offspring either to become altruistic or breeders. The expected pay-offs as an altruist or a breeder are sensitively dependent on the following changing conditions: (1) costs of dispersal, independent founding or budding, (2) the availability and stability of resources at the natal nest site, (3) the potential for direct fitness benefits per worker, and (4) the potential for cumulative indirect fitness benefits per worker. The pay-off for the altruist is high when high cumulative indirect benefits per worker or high future direct benefits per worker offset the high costs associated with direct reproduction at a new nest site. An abundance of resource is a strong predictor that cumulative indirect fitness benefits or future direct fitness benefits will give increasing returns. The pay-off for the breeder is high when the expect returns on cumulative indirect fitness benefits per worker or on future direct benefits per worker diminish at the natal nest, and when direct benefits such as a dowry or helpers make direct reproduction a viable option. The decline of resources at the natal nest is a strong predictor of diminishing returns as an altruist.

for offspring at the time caste decisions are made. One necessary criterion of the mixed strategies (whether they are fixed or conditional) (Maynard Smith 1982) is that the two pay-offs have equal fitness probabilities at the time an offspring makes a decision. I assume that fitness equality is a function of survival, i.e. $1 -$ (probability of failure in dispersal, nest founding, or nest establishment) and reproduction. Another necessary criterion of the mixed strategy is that it be frequency-dependent: the pay-off for one strategy increases as the other increases in frequency. This appears reasonable because, for example, as the frequency of workers increases, the pay-offs for attempted dispersal should increase since a greater worker force increases colony resources per disperser.

In the model for termites (Fig. 4-5), I show how the pay-off for indirect reproduction (altruism) is higher when abundant resources allow for higher cumulative indirect benefits and/or future direct benefits (e.g. gaining a dowry or helpers), given that the probability of failure for dispersers, new nest-founders, and budders is high; the pay-off

for direct reproduction (breeding) is higher when resources decline and the potential for cumulative indirect benefits or future direct benefits yield diminishing returns per worker, or when a dowry or helpers have been gained. The predicted result is that the evolutionarily stable proportion of workers should decrease as resources become depleted, as is observed (discussed earlier).

A mixed strategy based on cumulative indirect fitness benefits of workers vs. breeders is contrasted with alternative hypotheses for termite evolution, which propose that eusociality emerges from a 'pure conditional strategy' (*sensu* Maynard Smith 1982). Here, altruism evolved because individuals are making the best of a bad job (Roisin 1994) or are manipulated into helping (Myles 1988). Under this view, the pay-offs are not equal and offspring become altruistic because they assess themselves as inferior breeders or are forced into becoming altruists.

For the case of 'making the best of a bad job', 'losers' are predicted to be ineffectual reproductives, since not being able to succeed as a reproductive is proposed as the

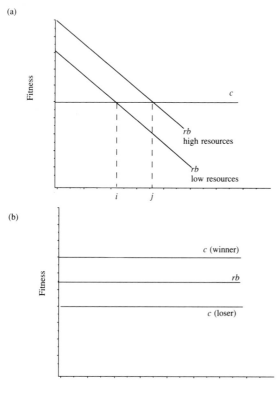

Figure 4-6. Components of Hamilton's rule are depicted on fitness graphs to distinguish differences between the mixed conditional-strategy and pure conditional strategy hypotheses. (a) Under the mixed strategy, worker altruism and direct reproduction have equal pay-offs to offspring at the decision point, hence $rb - c = 0$. The c (costs to the individual in terms of future offspring as a breeder) are held constant. In other words, costs of dispersal, nest-founding and colonization do not differ for each offspring, on average. In contrast, rb (coefficient of relatedness times benefit of working) is a frequency-dependent function; becoming an altruist rather than a breeder is dependent on the frequency of workers. A low frequency of worker corresponds to high worker rb, and thus favors offspring to become workers; a high frequency of workers corresponds to a low worker rb, and thus favors offspring to become breeders. At the equilibrium points i and j is the decision switch point where $rb - c = 0$ for low and high resources, respectively. The equilibrium frequency of workers is expected to be higher when resources are higher than when resources are lower. (b) Under the pure strategy, rb is held constant for both losers (offspring of poor phenotypic status) and winners (offspring of high phenotypic status). The c of losers (costs to the individual in terms of future offspring as a breeder) is expected to be low, because the individual of poorer phenotypic status gives up fewer future offspring than it would if it were of better phenotypic status. In contrast, the c of winners is higher, because a winner is expected to lose more future offspring by being a worker. Hence, the loser is always at $rb > c$.

reason why an offspring becomes a worker. If an individual of poorer status can be an effective reproductive, the reproducing 'loser' type will spread at the expense of the loser type that is expected to act as a worker (see Maynard Smith 1982, p. 82). If altruism is expressed despite evidence that even 'losers' can make effective reproductives, then a mixed strategy, where altruist and the breeder enjoy equal probabilities of fitness success, must be considered (Maynard Smith 1982). I examine the fitness components of Hamilton's rule (Fig. 4-6) to show the key differences between the two alternative hypotheses. Testing predictions of the two alternative strategy promises to set a challenging course for future studies in termite social evolution, as in other taxonomic groups.

ACKNOWLEDGMENTS

B. Crespi, G. C. Eickwort, M. Engel, M. Lenz, H. K. Reeve, L. Vawter, M. J. West-Eberhard and two anonymous readers were most helpful in providing invaluable comments and suggestions on various drafts of the manuscript. I give special thanks to J. Choe for inviting me to be a contributor in this book, and to B. Crespi for his extensive and valuable editorial efforts. This chapter is dedicated to the late George Eickwort.

LITERATURE CITED

Abe, T. 1987. Evolution of life types in termites. In *Evolution and coadaptation in biotic communities*. S. Kawano, J. J. H. Connell and T. Hidaka, eds., pp. 125–148. Tokyo: University of Tokyo Press.

–. 1991. Ecological factors associated with the evolution of worker and soldier castes in termites. *Ann. Entomol.* 9: 101–107.

Abe, T. and J. P. E. C. Darlington. 1985. Distribution and abundance of a mound-building termite *Macrotermes michaelseni* with special reference to its subterranean colonies and ant predators. *Physiol. Ecol. Japan.* 22: 59–74.

Adams, E. S. 1991. Nest-mate recognition based on heritable odors in the termite, *Microcerotermes arboreus*. *Proc. Natl. Acad. Sci. U.S.A.* 88: 2031–2034.

Adams, E. S. and S. C. Levings. 1987. Territory size and population limits in mangrove termites. *J. Anim. Ecol.* 56: 1069–1081.

Ahmad, M., M. Afzal and Z. Salihah. 1980. Laboratory study of post-flight behavior and colony formation in *Heterotermes indicola* (Wasmann). (Isoptera: Rhinotermitidae: Heterotermitinae). *Pakistan J. Zool.* 12: 227–237.

Alcock, J. and P. W. Sherman. 1994. The utility of the proximate-ultimate dichotomy in ethology. *Ethology* 96: 58–62.

Alexander, R. D. 1974. The evolution of social behavior. *Annu. Rev. Ecol. Syst.* 5: 325–383.

Alexander, R. D., K. M. Noonan and B. J. Crespi. 1991. The evolution of eusociality. In *The Biology of the Naked Mole Rat*. P. W. Sherman, J. U. M. Jarvis and R. D. Alexander, eds., pp. 3–44. Princeton: Princeton University Press.

Amburgey, T. L. 1979. Review and checklist of the literature on interactions between wood-inhabiting fungi and subterranean termites: 1960–1978. *Sociobiology* 4: 279–295.

Andersson, M. 1984. The evolution of eusociality. *Annu. Rev. Ecol. Syst.* 15: 165–189.

Araujo, R. L. 1970. Termites of the neotropical region. In *Biology of Termites*, vol. 2. K. Krishna and F. M. Weesner, eds., pp. 527–576. New York: Academic Press.

Bandeira, A. G. 1989. Analysis of termite fauna insecta: Isoptera of a primary forest and a pasture in eastern Amazonia Brazil. *Bol. Mus. Para. Emil. Goel. Ser.* 5: 225–242.

Bandi, C., M. Sironi, G. Damiani, L. Magrassi, C. A. Nalepa, U. Laudani and L. Sacchi. 1995. The establishment of intracellular symbiosis in an ancestor of cockroaches and termites. *Proc. R. Soc. Lond.* B 259: 293–299.

Banks, N. and T. E. Snyder. 1920. A revision of the Nearctic termites (Banks) with notes on their biology and geographical distribution (Snyder). *U.S. Nat. Mus. Bull.* 108: 1–228.

Bartz, S. J. 1979. Evolution of eusociality in termites. *Proc. Natl. Acad. Sci. U.S.A.* 76: 5764–5768.

Basalingappa, S. 1970. Environmental hazards to reproductives of *Odontotermes assmuthi* Holmgren. *Ind. Zool.* 1: 45–50.

–. 1971. Association of *Speculitermes cyclops*, *Dicuspiditermes (Capritermes) incola*, and an undetermined species of *Microtermes* with *Odontotermes assmuthi*. *Ind. Zool.* 2: 67–72.

–. 1976. Sociotomy, a process of colony formation in the termites *Odontotermes assmuthi* (Holmgren) and *Speculitermes cyclops* (Roonwal & Sen-Sarma). *J. Karnatak University* 21: 251–253.

–. 1982. Degeneration of thoracic muscles in sexual forms of *Odontotermes assmuthi* Holmgren (Termitidae: Isoptera) following swarming and colony establishment. *Entomon* 7: 63–64.

Basalingappa, S. and M. R. Gandhi. 1985. Determination of per cent free sugars from potential reproductives and the royal couple of the termite *Odontotermes obesus* (Termitidae: Isoptera). *Ind. Zool.* 9: 37–38.

Basalingappa, S. and S. N. Hegde. 1974. The probable significance of the differential occurrence of protein in various castes of the termite *Odontotermes assmuthi* (Termitidae: Isoptera). *Experientia* 30: 758.

Bateson, P. P. G. 1983. Optimal outbreeding. In *Mate Choice*. P. P. G. Bateson, ed., pp. 257–277. Cambridge University Press.

Bignell, D. E. 1994. Soil-feeding and gut morphology in higher termites. In *Nourishment and Evolution in Insect Societies*. J. H. Hunt and C. A. Nalepa, eds., pp. 131–158. Boulder: Westview Press.

Binder, B. F. 1988. Intercolonial aggression in the subterranean termite, *Heterotermes aureus* Isoptera: Rhinotermitidae. *Psyche* 95: 123–137.

Bouillon, A. 1970. Termites of the Ethiopia region. In *Biology of Termites*. vol. 2. K. Krishna and F. M. Weesner, eds., pp. 153–280. New York: Academic Press.

Breed, M. D. and B. Bennett. 1987. Kin recognition in highly eusocial insects. In *Kin Recognition in Animals*. D. J. C. Fletcher and C. D. Michener, eds., pp. 243–285. New York: John Wiley and Sons.

Breznak, J. A. 1982. Intestinal microbiota of termites and other xylophagous insects. *Annu. Rev. Microbiol.* 36: 323–343.

Brown, J. L. 1987. *Helping and Communal Breeding in Birds*. Princeton: Princeton University Press.

Castle, G. B. 1934. The damp-wood termites of western United States, genus *Zootermopsis* (formerly, *Termopsis*). In *Termites and Termite Control*. C. A. Kofoid, ed., pp. 273–310. Berkeley: University of California Press.

Cleveland, L. R., S. R. Hall, E. P. Sanders and J. Collier. 1934. The wood-feeding roach *Cryptocercus*, its protozoa, and the symbiosis between protozoa and roach. *Mem. Am. Acad. Arts Sci.* 17: 185–342.

Cmlick, S. H. W. 1969a. The neutral lipids from various organs of the termite *Macrotermes goliath*. *J. Insect Physiol.* 15: 839–849.

–. 1969b. Composition of the neutral lipids from termite queens. *J. Insect Physiol.* 15: 1481–1487.

Coles, H. R. and P. E. Howse. 1983. Chemical defense in termites: Ecological aspects. In *Social Insects in the Tropics*, vol. 2. P. Jaisson, ed., pp. 21–29. Paris: Université Paris Nord.

Collins, N. M. 1980. Inhabitation of epigeal termite (Isoptera) nests by secondary termites in Cameroun rain forest. *Sociobiology* 5: 47–54.

–. 1981. Populations, age structure and survivorship of colonies of *Macrotermes bellicosus* (Isoptera: Macrotermitinae). *J. Anim. Ecol.* 50: 293–311.

–. 1983. The utilisation of nitrogen resources by termites (Isoptera). In *Nitrogen as an Ecological Factor*. J. A. Lee, S. McNeill and I. H. Rorison, eds., pp. 381–412. Oxford: Blackwell Scientific Publications.

Cook, S. F. and K. G. Scott. 1933. The nutritional requirements of *Zootermopsis angusticollis*. *Cell Comp. Physiol.* 4: 95–111.

Crespi, B. J. 1992a. Eusociality in Australian gall thrips. *Nature (Lond.)* 359: 724–726.

–. 1992b. Cannibalism and trophic eggs in subsocial and eusocial insects. In *Cannibalism – Ecology and Evolution in Diverse Taxa*. M. Elgar and B. Crespi, eds., pp. 176–213. Oxford University Press.

Crozier, R. H. and P. Luykx. 1985. The evolution of termite eusociality is unlikely to have been based on a male-haploid analogy. *Am. Nat.* 126: 867–869.

Dadd, R. H. 1985. Nutrition: Organisms. In *Comprehensive Insect Physiology, Biochemistry and Pharmacology*, vol. 4–8. G. A. Kerkut and L. I. Gilbert, eds., pp. 313–390. Oxford: Pergamon Press.

Darlington, J. E. C. P. 1979. Populations of nests of *Macrotermes* species in Kajiado and Bissell. *Annu. Report, Int. Centre. Insect Physiol. Ecol.* 6: 22–23.

–. 1982. The underground passages and storage pits used in foraging by a nest of the termite *Macrotermes michaelseni* in Kajiado, Kenya. *J. Zool. (Lond.)* 198: 237–247.

–. 1985a. Multiple primary reproductives in the termites, *Macrotermes michaelseni* (Sjöstedt). In *Caste Differentiation in Social Insects*. J. A. L. Watson, B. M. Okot-Kotber and Ch. Noirot, eds., pp. 187–200. Oxford: Pergamon Press.

–. 1985b. Attacks by doryline ants and the termite nest defences (Hymenoptera: Formicidae; Isoptera: Termitidae) *Sociobiology* 11: 184–200.

–. 1986. Seasonality in mature nests of the termite *Macrotermes michaelseni* in Kenya. *Insectes Soc.* 33: 168–189.

–. 1988. Multiple reproductives in nests of *Macrotermes herus* (Isoptera: Termitidae). *Sociobiology* 2: 347–351.

–. 1991. Turnover in the populations within mature nests of the termite, *Macrotermes michaelseni* in Kenya. *Insectes Soc.* 38: 251–262.

–. 1994 Nutrition and evolution in fungus-growing termites. In *Nourishment and Evolution in Insect Societies*. J. H. Hunt and C. A. Nalepa, eds., pp. 105–130. Boulder: Westview Press.

Darlington, J. E. C. P. and J. M. Ritchie. 1987. Intercastes between steriles and nymphs in *Macrotermes michaelseni* (Isoptera: Termitidae) *Sociobiology* 13: 67–73.

Darlington, J. E. C. P., E. M. Cancello and O. F. F. de Souza. 1992. Ergatoid reproductives in termites of the genus *Dolichorhinotermes* (Isoptera, Rhinotermitidae). *Sociobiology* 20: 41–47.

Darlington, J. E. C. P., W. A. Sands and D. E. Pomeroy. 1977. Distribution and post-settlement survival in the field by reproductive pairs of *Hodotermes mossambicus* Hagen (Isoptera: Hodotermitidae). *Insectes Soc.* 24: 353–358.

DeSalle, R., J. Gatesy, W. Wheeler and D. Grimaldi. 1992. DNA sequences from a fossil termite in oligo-miocene amber and their phylogenetic implications. *Science, (Wash., D.C.)* 257: 1933–1936.

Eickwort, G. C. 1981. Presocial insects. In *Social Insects*, vol. 2. H. R. Hermann, ed., pp. 199–279. New York: Academic Press.

Emerson, A. 1938. Termite nests – A study of the phylogeny of behavior. *Ecol. Monogr.* 8: 247–284.

–. 1955. Geographical origin and dispersion of termite genera. *Field. Zool.* 37: 465–521.

–. 1965. A review of the Mastotermitidae (Isoptera), including a new fossil genus from Brazil. *Am. Mus. Nov.* 2236: 1–46.

Emerson, A. and K. Krishna. 1975. The termite family Serritermitidae (Isoptera). *Am. Mus. Nov.* 2570: 1–31.

Emlen, S. T. 1984. Cooperative breeding in birds and mammals. In *Behavioural Ecology. An Evolutionary Approach*. J. R. Krebs and N. B. Davies, eds., pp. 305–339. Oxford: Blackwell Scientific Publications.

–. 1991. Evolution of cooperative breeding in birds and mammals. In: *Behavioural Ecology. An Evolutionary Approach*. J. R. Krebs and N. B. Davies, eds., pp. 301–337. Oxford: Blackwell Scientific Publications.

Emlen, S. T. and L. W. Oring. 1977. Ecology, sexual selection, and the evolution of mating systems. *Science, (Wash., D. C.)* **197**: 215–223.

Emlen, S. T. and P. H. Wrege. 1989. A test of alternate hypotheses for helping behavior in white-fronted bee-eaters of Kenya. *Behav. Ecol. Sociobiol.* **25**: 303–320.

Fontana, F., A. M. Gravioli and M. Amorelli. 1982. Osservazioni sulla composizione castale di tre populazioni toscane di *Reticulitermes lucifugus* Rossi. (Isoptera: Rhinotermitidae). *Frust. Entom.* **2**: 121–124.

Frumhoff, P. C. and H. K. Reeve. 1994. Using phylogenies to test hypotheses of adaptation: a critique of some current proposals. *Evolution* **48**: 172–180.

Franks, N. and L. W. Partridge. 1993. Lancaster battles and the evolution of combat in ants. *Anim. Behav.* **45**: 197–199.

Gadagkar, R. 1990. The haplodiploidy theshold and social evolution. *Curr. Sci.* **59**: 374–376.

Gamboa, G. J., H. K. Reeve and D. W. Pfennig. 1986. The evolution and ontogeny of nestmate recognition in social wasps *Annu. Rev. Entomol.* **31**: 431–454.

Gandhi, M. R. and S. Basalingappa. 1992. Analysis of fatty acids from the alates of the termite, *Odontotermes obesus* (Termitidae: Isoptera). *Ind. J. Comp. Anim. Physiol.* **10**: 73–76.

Gay, F. J. and J. H. Calaby. 1970. Termites of the Australian region. In *Biology of Termites*, vol. 2. K. Krishna and F. M. Weesner, eds., pp. 393–448. New York: Academic Press.

Gerber, C., S. Badertscher and R. H. Leuthold. 1988. Polyethism in *Macrotermes bellicosus* (Isoptera). *Insectes Soc.* **35**: 226–240.

Grace, J. K. 1990. Mark–recapture sutdies with *Reticulitermes flavipes* (Isoptera, Rhinotermtidae). *Sociobiology* **16**: 297–303.

Grafen, A. 1984. Natural selection, kin selection, and group selection. In *Behavioural Ecology. An Evolutionary Approach*. J. R. Krebs and N. B. Davies, eds., pp. 62–86. Oxford: Blackwell Scientific Publications.

Grandcolas, P. 1994a. Phylogenetic systematics of the subfamily Polyphaginae, with the assignment of *Cryptocercus* Scudder, 1862 to this taxon (Blattaria, Blaberoidea, Polyphagidae). *Syst. Entomol.* **19**: 145–158.

–. 1994b. When did *Cryptocercus* cockroaches get their protoza symbionts from termites? In *Les Insectes Sociaux*. A. Lenior, G. Arnold and M LaPage, eds., p. 57. Villetaneuse: Publications Université Paris Nord.

Grandcolas, P. and P. Deleporte. 1992. The systematic position of *Cryptocercus* Scudder in cockroaches and its evolutionary implications. *C.R. Acad. Sci. Paris* **315**: 317–322.

Grassé , P-P. 1984. *Foundation des Sociétés – Construction. Termitologia*, vol. 2. Paris: Masson.

–. 1986. *Comportement – Socialité – Ecologie – Evolution – Systématique. Termitologia*, vol. 3. Paris: Masson.

Greenberg, S. L. W. 1980. Pheromonal inhibition of neotenic reproductive development in a primitive termite. *Am. Zool.* **20**: 905.

–. 1982. Studies on neotenic reproductive development in a primitive termite. In *The Biology of Social Insects*. M. D. Breed, C. Michener and H. E. Evans, eds., pp. 218–219. (Proc. Ninth Cong. Inter. Union Soc. Soc. Insects.) Boulder: Westview Press.

Greenberg, S. L. W. and A. M. Stuart. 1982. Precocious reproductive development (neoteny) by larvae of a primitive termite, *Zootermopsis angusticollis* (Hagen). *Insectes Soc.* **29**: 535–547.

Haack, R. A. and F. Slansky. 1987. Nutritional ecology of wood feeding Coleoptera, Lepidoptera, and Hymenoptera. In *Nutritional Ecology of Insects, Mites, and Spiders*. F. Slansky and J. G. Rodriguez, eds., pp. 449–486. New York: Wiley and Sons.

Hahn, P. D. and A. M. Stuart. 1987. Sibling interactions in two species of termites: a test of the haplodiploid analogy (Isoptera: Kalotermidae; Rhinotermitidae). *Sociobiology* **13**: 83–93.

Hamilton, W. D. 1964. The genetical evolution of social behavior. *J. Theor. Biol.* **7**: 1–52.

–. 1971. Geometry for the selfish herd. *J. Theor. Biol.* **31**: 295–311.

–. 1972. Altruism and related phenomena, mainly in social insects. *Annu. Rev. Ecol. Syst.* **3**: 193–232.

–. 1978. Evolution and diversity under bark. In *Diversity of Insect Faunas*. L. A. Mound and N. Waloff, eds., pp. 154–175. (Symposia of the Royal Entomological Society of London 9.) Oxford: Blackwell Scientific Publications.

–. 1990. Memes of Haldane and Jayakar in a theory of sex. *J. Genet.* **69**: 17–32.

–. 1993. Inbreeding in Egypt and in this book: A childish perspective. In *Natural History of Inbreeding and Outbreeding*. N. W. Thornhill, ed., pp. 429–450. Chicago: University of Chicago Press.

Han, S. H. and C. Bordereau. 1992. From colony foundation to dispersal flight in a higher fungus-growing termite, *Macrotermes subhyalinus* (Isoptera, Macrotermitinae). *Sociobiology* **20**: 219–231.

Han, S. H. and C. Noirot, 1983. Développement de la jeune colonie chez *Cubitermes fungifaber* (Sjöstedt) (Isoptera: Termitidae). *Ann. Soc. Entomol. France*. **19**: 413–420.

Harvey, P. 1934. Life history of *Kalotermes minor*. In *Termites and Termite Control*. C. A. Kofoid, ed., pp. 217–233. Berkeley: University of California Press.

Haverty, M. I. 1977. The proportion of soldiers in termite colonies: a list and a bibliography (Isoptera). *Sociobiology* **2**: 199–216.

Heath, H. 1907. The longevity of members of different castes of *Termopsis angusticollis*. *Biol. Bull.* **13**: 161–164.

–. 1927. Caste formation in the termite genus *Termopsis*. *J. Morph. Physiol.* **43**: 387–425.

–. 1928. Fertile termite soldiers. *Biol. Bull.* **54**: 324–326.

–. 1931. Experiments in termite caste development. *Science (Wash., D.C.)*. **30**: 421.

Heath, H. and B. C. Wilbur. 1927. The development of the soldier caste in the termite genus *Termopsis*. *Biol. Bull.* **53**: 145–156.

Hendee, E. C. 1935. The role of fungi in the diet of the common damp-wood termite, *Zootermopsis*. *Hilgardia* **9**: 499–525.

Higashi, M., N. Yamamura, T. Abe and T. P. Burns. 1991. Why don't all termite species have a sterile worker caste? *Proc. R. Soc. Lond.* B **246**: 25–30.

Hill, G. F. 1942. *Termites (Isoptera) from the Australian Region.* Melbourne: Council Sci. Ind. Res.

Hölldobler, B. and E. O. Wilson. 1991. *The Ants.* Cambridge: Harvard University Press.

Honigberg, B. 1970. Protozoa associated with termites and their role in digestion. In *Biology of Termites*, vol. 2. K. M. Krishna and F. M. Weesner, eds., pp. 1–36. New York: Academic Press.

Howard, R. W. and M. I. Haverty. 1980. Reproductives in mature colonies of *Reticulitermes flavipes*: abundance, sex ratio, and association with soldiers. *Environ. Entomol.* **9**: 458–460.

Howard, R. W., S. C. Jones, J. K. Mauldin and R. H. Beal. 1982. Abundance distribution, and colony size estimates for *Reticulitermes* spp. (Isoptera: Rhinotermitidae) in southern Mississippi. *Environ. Entomol.* **11**: 1290–1293.

Howick, C. D. and J. W. Creffield. 1980. Intraspecific antagonism in *Coptotermes acinaciformis* (Froggatt) (Isoptera: Rhinotermitidae). *Bull. Entomol. Res.* **70**: 17–23.

Howse, P. E. 1968. On the division of labour in the primitive termite *Zootermopsis nevadensis* (Hagen). *Insectes Soc.* **15**: 45–50.

–. 1984a. Sociochemicals of termites. In *Chemical Ecology of Insects.* W. J. Bell and R. T. Cardé, eds., pp. 475–519. Sunderland: Sinauer Associates.

–. 1984b. Alarm, defence and chemical ecology of social insects. In *Insect Communication.* T. Lewis, ed., pp. 151–167. (Twelveth Symp. R. Entomol. Soc. Lond.) London: Academic Press.

Huang, F. and L. L. Chen. 1981. Influence of food factors on colony formation by *Coptotermes formosanus* Shiraki. [Based on summary.] *Acta Entomol. Sinic.* **24**: 147–151.

Huang, F., S. Zhu and G. Li. 1987. Effect of continental drift on phylogeny of termites. *Zool. Res.* **8**: 55–60.

Huang, L. W. 1987. The growth of colony size in relation to increasing reproductive capacity of *Coptotermes formosanus* Shiraki. *Acta Entomol. Sinica* **30**: 393–396. (English summary).

Hungate, R. E. 1941. Experiments on the nitrogen economy of termites. *Ann. Entomol. Soc. Am.* **34**: 457–489.

–. 1955. Mutualistic intestinal protozoa. In *Biochemistry and Physiology of Protozoa.* S. H. Hutner and A. Lwoff, eds., pp. 159–199. New York: Academic Press.

Imms, A. D. 1919. On the structure and biology of *Archotermopsis*, together with descriptions of new species of intestinal protozoa and general observations on the Isoptera. *Phil. Trans. R. Soc. Lond.* B **209**: 75–189.

Jones, S. C. 1993. Field observations of intercolony aggression and territory changes in *Heterotermes aureus* Isoptera: Rhinotermitidae. *J. Insect Behav.* **6**: 225–236.

Jones, S. C., J. P. Lafage and R. W. Howard. 1988. Isopteran sex ratios: phylogenetic trends. *Sociobiology* **14**: 89–156.

Jones, S. C., J. P. Lafage and V. L. Wright. 1981. Studies of dispersal, colony caste and sexual composition, and incipient colony development of *Pterotermes occidentis* (Walker) (Isoptera: Kalotermitidae) *Sociobiology* **6**: 221–242.

Josens, G. 1982. Le bilan énérgétique de *Trinervitermes geminatus* Wasmann (Termitidae: Nasutitermitinae):1. Mesure de biomasses, d'equivalents énérgétique, de longévité et de production en laboratoire. *Insectes Soc.* **29**: 295–307.

Kambhampati, S. 1995. Phylogeny of cockroaches and related insects based on DNA sequence of mitochondrial ribosomal RNA genes. *Proc. Natl Acad. Sci. U.S.A.* **92**: 2017–2020.

Keller, L. ed. 1993. *Queen Number and Sociality in Insects.* Oxford: Oxford University Press.

Keller, L. and P. Nonacs. 1993. The role of queen pheromones in social insects: queen control or queen signal? *Anim. Behav.* **45**: 787–794.

Keller, L. and H. K. Reeve. 1994. Genetic variability, queen number, and polyandry in social Hymenoptera. *Evolution* **48**: 694–704.

Kitayama, K. 1975. Nota preliminar sobre a distribuiçao, o regime alimentar e teratologia de *Serritermes serrifer* (Hagen) (Isoptera). *Stud. Entomol.* **18**: 614–618.

Krishna, K. 1970. Taxonomy, phylogeny, and distribution of termites. In *Biology of Termites.* vol. 2. K. Krishna and F. M. Weesner, eds., pp. 127–152. New York: Academic Press.

Lacy, R. C. 1980. The evolution of eusociality in termites: a haplo-diploid analogy? *Am. Nat.* **116**: 449–451.

Lafage, J. P. and W. L. Nutting. 1978. Nutrient dynamics of termites. In *Production Ecology of Ants and Termites.* M. V. Brian, ed., pp. 165–232. London: Cambridge University Press.

Lebrun, D. 1970. Intercastes expérimentaux de *Calotermes flavicollis* Fabr. *Insectes Soc.* **17**: 159–176.

Lefeuve, P. 1987. Replacement queens in the neotropical termite *Nasutitermes coxipoensis. Insectes Soc.* **34**: 10–19.

Leinaas, H. P. 1983. A haplodiploid analogy in the evolution of termite eusociality? *Am. Nat.* **121**: 302–304.

Lenz, M. 1976. The dependence of hormone effects in termite caste determination of external factors. In *Phase and Caste Determination in Insects. Endocrine Aspects.* M. Lüscher, ed., pp. 73–89. Oxford: Pergamon Press.

–. 1985. Is inter- and intraspecific variability of lower termite neotenic numbers due to adaptive thresholds for neotenic elimination? Considerations from studies on *Porotermes adamsoni* (Froggatt) Isoptera: Termopsidae. In *Caste Differentiation in Social Insects.* J. A. L. Watson, B. M. Okot-Kotber and Ch. Noirot, eds., pp. 125–145. Oxford: Pergamon Press.

–. 1987. Brood production by imaginal and neotenic pairs of *Cryptotermes brevis* (Walker): The significance of helpers (Isoptera: Kalotermitidae). *Sociobiology* **13**: 59–66.

–. 1994. Food resources, colony growth and caste development in wood-feeding termites. In *Nourishment and Evolution in Insect Societies.* J. H. Hunt and C. A. Nalepa, eds., pp. 159–209. Boulder: Westview Press.

Lenz, M. and R. A. Barrett. 1982. Neotenic formation in field colonies of *Coptotermes lacteus* (Froggatt) in Australia, with comments on the roles of neotenics in the genus *Coptotermes* (Isoptera: Rhinotermitidae). *Sociobiology* 7: 47–60.

Lenz, M. and S. Runko. 1993 Long-term impact of orphaning on field colonies of *Coptotermes lacteus* (Froggatt) (Isoptera: Rhinotermitidae). *Insectes Soc.* 40: 439–456.

Lenz, M., R. A. Barrett and L. R. Miller. 1986. The capacity of colonies of *Coptotermes acinaciformis acinaciformis* from Australia to produce neotenics (Isoptera; Rhinotermitidae). *Sociobiology* 11: 237–244.

Lenz, M., R. A. Barrett and E. R. Williams. 1985. Reproductive strategies in *Cryptotermes*: neotenic production in indigenous and "tramp" species in Australia (Isoptera: Kalotermitidae). In *Caste Differentiation in Social Insects*. J. A. L. Watson, B. M. Okot-Kotber and Ch. Noirot, eds., pp. 147–163. Oxford: Pergamon Press.

Lenz, M., E. McMahan and E. R. Williams. 1982. Neotenic production in *Cryptotermes brevis* (Walker): influence of geographical origin, group composition, and maintenance conditions (Isoptera: Kalotermitidae). *Insectes Soc.* 29: 148–163.

Lepage, M. 1990. Développement au laboratorie des jeune colonies de *Macrotermes michaelseni* (Sjöstedt) (Isoptera: Macrotermitinae). *Ann. Soc. Entomol. France.* 26: 39–50.

Levings, S. C. and E. S. Adams. 1984. Intra- and inter-specific territoriality in *Nasutitermes* (Isoptera: Termitidae) in a Panamanian mangrove forest. *J. Anim. Ecol.* 53: 705–714.

Light, S. F. and P. L. Illg. 1945. Rate and extent of development of neotenic reproductives in groups of nymphs of the termite genus *Zootermopsis*. *University Calif. Pub. Zool.* 53: 1–40.

Light, S.F and F. M. Weesner. 1951. Further studies on the production of supplementary reproductives in *Zootermopsis* (Isoptera). *J. Exp. Zool.* 117: 397–414.

Lin, N. and C. D. Michener. 1972. Evolution of sociality in insects. *Q. Rev. Biol.* 47: 131–159.

Liu, Y. Z., G. Q. Tang, Y. Z. Pang, L. D. Chen and Y. Z. He. 1981. Observations on the construction of the unilocular nest of *Odontotermes formosanus* (Shiraki). *Acta Entomol. Sinica* 24: 361–366.

Lüscher, M., ed. 1976. *Phase and Caste Determination in Insects. Endocrine Aspects*. Oxford: Pergamon Press.

Luykx, P. 1985. Genetic relations among castes in lower termites. In *Caste Differentiation in Social Insects*. J. A. L. Watson, B. M. Okot-Kotber and C. Noirot, eds., pp. 17–25. Oxford: Pergamon Press.

–. 1986. Termite colony dynamics as revealed by the sex- and caste-ratios of whole colonies of *Incisitermes schwarzi* Banks (Isoptera: Kalotermitidae). *Insectes Soc.* 33: 221–248.

–. 1990. A cytogenetic survey of 25 species of lower termites from Australia. *Genome* 33: 80–88.

–. 1993. Turnover in termite colonies: A genetic study of colonies of *Incisitermes schwarzi* headed by replacement reproductives. *Insectes Soc.* 40: 191–205.

Luykx, P. and R. M. Syren. 1979. The cytogenetics of *Incisitermes schwarzi* and other Florida Termites. *Sociobiology* 1: 91–209.

Luykx, P., J. Michel and J. Luykx. 1986. The spatial distribution of the sexes in colonies of the termite *Incisitermes schwarzi* Banks (Isoptera: Kalotermitidae). *Insectes Soc.* 33: 406–421.

Lys, J. A. and R. H. Leuthold. 1991. Task-specific distribution of the two worker castes in extranidal activities in *Macrotermes bellicosus* Smeathman: Observation of behavior during food acquisition. *Insectes Soc.* 38: 161–170.

MacKay, W. P., J. H. Blizzard, J. J. Miller and W. G. Whitford. 1985. Analysis of above-ground gallery construction by the subterranean termite *Gnathamitermes tubiformans* (Isoptera: Termitidae). *Environ. Entomol.* 14: 470–474.

Maynard Smith, J. 1982. *Evolution and the Theory of Games*. Cambridge University Press.

McKittrick, F. A. 1965. A contribution to the understanding of cockroach-termite affinities. *Ann. Entomol. Soc. Am.* 58: 18–22.

McMahan, E. A. 1969. Feeding relationships and radioisotopes. In *Biology of Termites*, vol. 1. K. Krishna and F. M. Weesner, eds., pp. 387–406. New York: Academic Press.

–. 1979. Temporal polyethism in termites. *Sociobiology* 4: 153–168.

McMahan, E. A., P. K. Sen-Sarma and S. Kumar. 1983. Biometric, polyethism, and sex ratio studies of *Nasutitermes dunensis* Chatterjee and Thakur (Isoptera: Termitidae). *Ann. Entomol. Soc. Am.* 1: 15–25.

Mensa-Bonsu, A. 1976. The production and elimination of supplementary reproductives in *Porotermes adamsoni* (Froggatt) (Isoptera: Hodotermitidae). *Insectes Soc.* 23: 133–154.

Messer, A. C. 1990. Chemical ecology in an Indonesian context. Ph. D. dissertation, Cornell University, Ithaca, N. Y.

Messer, A. C. and M. J. Lee. 1989. Effect of chemical treatments on methane emission by the hindgut microbiota in the termite, *Zootermopsis angusticollis*. *Microbiol. Ecol.* 1989: 275–284.

Mill, A. E. 1984. Exploding termites: An unusual defensive behaviour. *Entomol. Mon. Mag.* 120: 179–183.

Miller, E. M. 1942. The problem of castes and caste differentiation in *Prorhinotermes simplex* (Hagen) *Bull. Univ. Miami* 15: 3–27.

–. 1969 Caste differentiation in the lower termites. In *Biology of Termites*, vol. 1. K. Krishna and F. M. Weesner, eds., pp. 283–310. New York: Academic Press.

Mishra, S. C. and P. K. Sen-Sarma. 1981. Neutral lipids in different castes in termites. *Indian J. Entomol.* 43: 80–82.

Mugali, R. N. and S. Basalingappa. 1982. Total body lipid from different castes of termites, *Odontotermes wallonensis and O. feae* (Isoptera: Termitidae). *Comp. Physiol. Ecol.* 7: 148–150.

–. 1984. Total body protein from different castes of termites, *Odontotermes wallonensis* and *O. feae* (Isoptera: Termitidae). *Life Sci Adv.* 3: 95–97.

Mumme, R. L., W. D. Koenig and F. L. W. Ratnieks. 1989. Helping behavior reproductive value and the future component of indirect fitness. *Anim. Behav.* 38 : 331–343.

Myles, T. G. 1986. Reproductive soldiers in Termopsidae (Isoptera). *Pan Pac. Entomol.* **62**: 293–299.

–. 1988. Resource inheritance in social evolution from termites to man. In *The Ecology of Social Behavior.* C. N. Sloboidchikoff, ed., pp. 379–423. New York: Academic Press.

Myles, T. G. and W. L. Nutting. 1988. Termite eusocial evolution: a re-examination of Bartz's hypothesis and assumptions. *Q. Rev. Biol.* **63**: 1–24.

Nagin, R. 1972. Caste determination in *Neotermes jouteli* (Banks). *Insectes Soc.* **19**: 39–61.

Nalepa, C. A. 1984. Colony composition, protozoan transfer and some life history characteristics of the woodroach *Cryptocercus punctulatus* Scudder (Dictyoptera: Cryptocercidae). *Behav. Ecol. Sociobiology* **14**: 273–279.

–. 1988. Cost of parental care in the woodroach *Cryptocercus punctulatus* Scudder (Dictyoptera: Cryptocercidae). *Behav. Ecol. Sociobiology* **23**: 135–140.

–. 1991. Ancestral transfer of symbionts between cockroaches and termites: an unlikely scenario. *Proc. R. Soc. Lond.* B **246**: 185–189.

–. 1994. Nourishment and the origin of termite eusociality. In *Nourishment and Evolution in Insect Societies.* J. H. Hunt and C. A. Nalepa, eds., pp. 57–104. Boulder: Westview Press.

Nalepa, C. A. and S. C. Jones. 1991. Evolution of monogamy in termites. *Bioscience* **66**: 83–97.

Nalepa, C. A. and D. E. Mullins. 1992. Initial reproductive investment and parental body size in *Cryptocercus punctulatus* Scudder (Dictyoptera: Cryptocercidae). *Physiol. Entomol.* **17**: 255–259.

Nel, J. J. C. 1968. Aggressive behaviour of the harvester termites *Hodotermes mossambicus* (Hagen) and *Trinervitermes trinervoides* (Sjöstedt). *Insectes Soc.* **15**: 145–156.

Noirot, C. 1970. The nests of termites. In *Biology of Termites*, K. Krishna and F. M. Weesner, eds., pp. 73–126. New York: Academic Press.

–. 1985a. Pathways of caste development in the lower termites. In *Caste Differentiation in Social Insects.* J. A. L. Watson, B. M. Okot-Kotber and Ch. Noirot, eds., pp. 41–58. Oxford: Pergamon Press.

–. 1985b. The caste system in higher termites. In *Caste Differentiation in Social Insects.* J. A. L. Watson, B. M. Okot-Kotber and Ch. Noirot, eds., pp. 75–86. Oxford: Pergamon Press.

–. 1985c. Differentiation of reproductives in higher termites. In *Caste Differentiation in Social Insects.* J. A. L. Watson, B. M. Okot-Kotber and Ch. Noirot, eds., pp 177–186. Oxford: Pergamon Press.

–. 1989. Social structure in termite societies. *Ethol. Ecol. Evol.* **1**: 1–18.

Noirot, C., C. Noirot-Timothé and S. H. Han. 1986. Migration and nest building in *Cubitermes fungifaber* (Isoptera, Termitidae). *Insectes Soc.* **33**: 361–374.

Noirot, C. and J. Pasteels. 1987. Ontogenetic development and evolution of the worker caste in termites. *Experientia* **43**: 851–952.

Nutting, W. L. 1966. Colonizing flights and associated activities of termites. I. The desert damp-wood termite *Paraneotermes simplicicornis* (Kalotermitidae). *Psyche* **73**: 131–149.

–. 1969. Flight and colony foundation. In *Biology of Termites.* vol. 1. K. Krishna and F. M. Weesner, eds., pp. 233–282. New York: Academic Press.

–. 1979. Termite flight periods: strategies for predator avoidance? *Sociobiology* **4**: 141–151.

Okot-Kotber, B. M. 1979. Recent findings on mechanisms of caste differentiation in *Macrotermes* species near *subhyalinus.* *Annu. Rep. Int. Centre Insect Physiol. Ecol.* **6**: 27–30.

Passera, L. and L. Keller. 1994. Mate availability and male dispersal in the Argentine ant *Linepithema humile* (Mayr) (=*Iridomyrmex humilis*). *Anim. Behav.* **48**: 361–369.

Pickens, A. L. 1934. The biology and economic significance of the western subterranean termite, *Reticulitermes hesperus.* In *Termites and Termite Control.* C. A. Kofoid, ed., pp. 157–183. Berkeley: University of California Press.

Pomeroy, D. E. 1989. Studies on a two species population of termites in Kenya (Isoptera). *Sociobiology* **15**: 219–236.

Potrikus, C. J. and J. A. Breznak. 1980. Uric acid in wood-eating termites. *Insect Biochem.* **10**: 19–27.

Prestwich, G. D. 1983. The chemical defenses of termites. *Scient. Am.* **249**: 68–75.

–. 1984. Defense mechanisms of termites. *Annu. Rev. Entomol.* **29**: 201–232.

Prestwich, G. D., B. L. Bentley and E. J. Carpenter. 1980. Nitrogen sources for neotropical nasute termites: fixation and selective foraging. *Oecologia (Berl.)* **46**: 397–401.

Queller, D. C. 1989. The evolution of eusociality: Reproductive head starts of workers. *Proc. Natl. Acad. Sci. U.S.A.* **86**: 3224–3226.

–. 1994. Extended parental care and the origin of eusociality. *Proc. R. Soc. Lond.* B**256**: 105–111.

Quennedey, A. and J. Deligne. 1975. L'arme frontel des soldats de termites, I. Rhinotermitidae. *Insectes Soc.* **22**: 243–267.

Rajagopal, D. 1986. Biological activities of the mound building termite, *Odontotermes wallonensis* (Wasmann) (Isoptera: Termitidae) in Karnataka. *J. Soil Biol Ecol* **6**: 42–52.

Reeve, H. K. 1992. Queen activation of lazy workers in colonies of the eusocial naked mole-rat. *Nature (Lond.)* **358**: 147–149.

Reeve, H. K. and F. L. W. Ratnieks. 1993. Queen–queen conflicts in polygynous societies: Mutual tolerance and reproductive skew. In *Queen Number and Sociality in Insects.* L. Keller, ed., pp. 45–85. Oxford University Press.

Reeve, H. K. and P. W. Sherman. 1993. Adaptation and the goals of evolutionary research. *Q. Rev. Biol.* **68**: 1–32.

Reeve, H. K., D. F. Westneat, W. A. Noon, P. W. Sherman and C. F. Aquadro. 1990. DNA fingerprinting reveals high levels of inbreeding in colonies of the eusocial naked mole-rat. *Proc. Natl. Acad. Sci. U.S.A.* **87**: 2496–2500.

Reilly, L. M. 1987. Measurements of inbreeding and average relatedness in a termite population. *Am. Nat.* **130**: 339–349.

Renoux, J. 1985. Dynamic study of polymorphism in *Schedorhinotermes lamanianus* (Rhinotermitidae). In *Caste Differentiation in Social Insects.* J. A. L. Watson, B. M. Okot-Kotber and C. Noirot, eds., pp. 59–74. Oxford: Pergamon Press.

Ratnieks, F. L. W. and H. K. Reeve. 1988. Reproductive harmony via worker-policing in eusocial Hymenoptera. *Am. Nat.* **132**: 217–236.

–. 1992. Conflict in single queen hymenopteran societies: the structure of conflict and processes that reduce conflict in advanced eusociality species. *J. Theor. Biol.* **158**: 33–65.

Roisin, Y. 1988. Morphology, development, and evolutionary significance of the working stages in the caste system of *Prorhinotermes* (Insecta, Isoptera). *Zoomorphology* **107**: 339–348.

–. 1993. Selective pressures on pleometrosis and secondary polygyny: a comparison of termites and ants. In *Queen Number and Sociality in Insects.* L. Keller, ed., pp. 402–421. Oxford University Press

–. 1994. Intragroup conflicts and the evolution of sterile castes in termites. *Am. Nat.* **143**: 751–765.

Roisin, Y. and J. M. Pasteels. 1985. Imaginal polymorphism and polygyny in the Neo-Guinean termite *Nasutitermes princeps* (Desneux). *Insectes Soc.* **32**: 140–157.

–. 1986. Reproductive mechanisms in termites: polycalism and polygyny in *Nasutitermes polygynus* and *Nasutitermes costali. Insectes Soc.* **33**: 149–167.

–. 1987. Caste development potentialities in the termite *Nasutitermes novarumhebridarum. Entomol. Exp. Appl.* **44**: 277–287.

–. 1991. Sex ratio and asymmetry between the sexes in the production of replacement reproductives in the termite, *Neotermes papua* Desneux. *Ethol. Ecol. Evol.* **3**: 327–336.

Roonwal, M. L. 1970. Termites of the oriental region. In *Biology of Termites.* K. Krishna and F. M. Weesner, eds., pp. 315–391. New York: Academic Press.

Roonwal, M. L. and O. B. Chhotani. 1989. *Fauna of India and the Adjacent Countries.* Calcutta: Zoological Survey of India.

Rosengaus, R. B. and J. F. A. Traniello. 1991. Biparental care in incipient colonies of the dampwood termite *Zootermopsis angusticollis* Hagen (Isoptera: Termopsidae). *J. Insect Behav.* **4**: 633–648.

–. 1993a. Temporal polyethism in incipient colonies of the primitive termite, *Zootermopsis angusticollis.* A single multiage caste. *J. Insect Behav.* **6**: 237–252.

–. 1993b. Disease risk as a cost of outbreeding in the termite *Zootermopsis angusticollis. Proc. Natl. Acad. Sci. U.S.A.* **90**: 6641–6645.

Sands, W. A. 1981. Agonistic behavior of African soldierless Apicotermitinae (Isoptera: Termitidae). *Sociobiology* **7**: 61–72.

Scriber, J. M. 1984. Host-plant suitability. In *Chemical Ecology of Insects.* W. J. Bell and R. T. Cardé, eds., pp. 159–204. Sunderland: Sinauer Associates.

Seeley, T. D. 1985. *Honeybee Ecology. A Study of Adaptation in Social Life.* Princeton: Princeton University Press.

Shellman-Reeve, J. S. 1990. Dynamics of biparental care in the dampwood termite, *Zootermopsis nevadensis* (Hagen): response to nitrogen availability. *Behav. Ecol. Sociobiology* **26**: 389–397.

–. 1994a. Limiting nutrient availability: Nest preference, competition, cooperative nest defence. *J. Anim. Ecol.* **63**: 921–932.

–. 1994b. Mating ecology of the dampwood termite, *Zootermopsis nevadensis.* Ph. D. dissertation, Cornell University, Ithaca, N. Y.

–. 1996. Operational sex ratios and lipid reserves in the dampwood termite, *Zootermopsis nevadensis* Hagen (Isoptera: Termopsidae). *J. Kan. Entomol. Soc.,* in press.

Sherman, P. W. 1988. The levels of analysis. *Anim. Behav.* **36**: 616–619.

Sherman, P. W., H. K. Reeve and T. D. Seeley. 1988. Parasites, pathogens, and polyandry in social Hymenoptera. *Am. Nat.* **131**: 602–610.

Sherman, P. W., E. A. Lacey, H. K. Reeve and L. Keller. 1995. The eusociality continuum. *Behav. Ecol.* **6**: 102–108.

Shykoff, J. A. and P. Schmid-Hempel. 1991. Genetic relatedness and eusociality parasite-mediated selection on the genetic composition of groups. *Behav. Ecol. Sociobiology* **28**: 371–376.

Skaife, S. H. 1955. *Dwellers in Darkness.* New York: Longmans.

Springhetti, A. 1990. Nest digging of *Kalotermes flavicollis* Fabr. (Isoptera: Kalotermitidae) by groups of different numbers of pseudergates. *Ethol., Ecol. Evol.* **2** : 165–174.

Springhetti, A. and E. Sita. 1986. Prime fasi della ricostruzione del nido di *Kalotermes flavicollis* Far. (Isoptera: Kalotermitidae). *Redia* **69**: 11–23.

Stacey, P. B. and W. D. Koenig, eds. 1990. *Cooperative Breeding in Birds: Long-Term Studies of Ecology and Behavior.* London: Cambridge University Press.

Steward, R. C. 1983a. The effects of humidity, temperature and acclimation on the feeding, water balance, and reproduction of dry-wood termites (*Cryptotermes*). *Entomol. Exp. Appl.* **33**: 135–144.

–. 1983b. Microclimate and colony foundation by imago and neotenic reproductives of dry-wood termite species (*Cryptotermes* spp.) (Isoptera: Kalotermitidae). *Sociobiology* **7**: 311–332.

Strassmann, J. E., C. R. Hughes, D. C. Queller, S. R. Turillazi, R. S. Cervo, S. K. Davis and K. F. Goodnight. 1989. Genetic relatedness in primitively eusocial wasps. *Nature (Lond.)* **342**: 268–269.

Stuart, A. M. 1969. Social behavior and communication. In *Biology of Termites.* vol. 1. K. Krishna and F. M. Weesner, eds., pp. 193–232. New York: Academic Press.

–. 1979. The determination and regulation of the neotenic reproductive caste in the lower termites (Isoptera) with special reference to the genus *Zootermopsis* (Hagen). *Sociobiology* **4**: 223–237.

–. 1988. Preliminary studies on the significance of head-banging movements with special reference to *Zootermopsis angusticollis* (Hagen) (Isoptera: Hodotermitidae). *Sociobiology* **14**: 49–60.

Stubblefield, J. W. and E. L. Charnov. 1986. Some conceptual issues in the evolution of eusociality. *Heredity* **57**: 181–187.

Su, N. Y. and M. I. Haverty. 1991. Agonistic behavior among colonies of the Formosan subterranean termite, *Coptotermes formosanus* Shiraki (Isoptera: Rhinotermitidae) from Florida, USA and Hawaii: lack of correlation with cuticular hydrocarbon. *J. Insect Behav.* **4**: 115–128.

Su, N. Y. and J. P. Lafage. 1986. Effects of starvation on survival and maintenance of soldier proportion in laboratory groups of the Formosan subterranean termites, *Coptotermes formosanus* Isoptera: Rhinotermitidae. *Ann. Entomol. Soc. Am.* **79**: 312–316.

Syren, R. M. and P. Luykx. 1977. Permanent segmental interchange complex in the termite *Incisitermes schwarzi*. *Nature (Lond.)* **266**: 167–168.

Tahiri, A. and S. H. Han. 1993. Ovarian evolution of the queen during colony development in *Macrotermes subhyalinus* (Isoptera: Termitidae). *Ann. Soc. Entomol. France* **29**: 321–327.

Taylor, V. A. 1978. A winged élite in a subcortical beetle as a model for a prototermite. *Nature (Lond.)* **276**: 73–75.

Tinbergen, N. 1963. On aims and method of ethology. *Z. Tierpsychol.* **20**: 410–429.

Thorne, B. L. 1982. Reproductive plasticity in the neotropical termite *Nasutitermes corniger*. In *Social Insects in the Tropics*. P. Jaisson, ed., pp. 21–29. Paris: Université Paris Nord.

–. 1985. Termite polygyny: the ecological dynamics of queen mutualism. In *Experimental Behavioral Ecology and Sociobiology*. B. Hölldobler and M. Lindauer, eds., pp. 325–341. Sunderland: Sinauer Associates.

–. 1990. A case for ancestral transfer of symbionts between cockroaches and termites. *Proc. R. Soc. Lond.* B **241**: 37–41.

–. 1991. Ancestral transfer of symbionts between cockroaches and termites: an alternative hypothesis. *Proc. R. Soc. Lond.* B **246**: 191–196.

Thorne, B. L. and J. M. Carpenter. 1992. Phylogeny of the Dictyoptera. *Syst. Entomol.* **17**: 253–268.

Thorne, B. L. and M. L. Haverty. 1991. A review of intracolony, intraspecific, and interspecific agonism in termites. *Sociobiology* **19**: 115–145.

Thorne, B. L. and C. Noirot. 1982. Ergatoid reproductives in *Nasutitermes corniger* (Motschulsky) (Isoptera: Termitidae). *Int. J. Insect Morph. Embryol.* **11**: 213–226.

Traniello, J. F. A. 1981. Enemy deterrence in the recruitment strategy of a termite: Soldier-organized foraging in *Nasutitermes costalis*. *Proc. Natl. Acad. Sci. U.S.A.* **78**: 1976–1979.

Trivers, R. L. 1972. Parental investment and sexual selection. In *Sexual Selection and the Descent of Man*. B. Campbell, ed., pp. 136–179. Chicago: Aldine.

van der Linde, T. C. D. K, P. H. Hewitt, M. C. van der Westhuizen and J. Mitchell. 1989. The use of iodine-131 and iodine-125 and aggressive behavior to determine the foraging area of *Hodotermes mossambicus* Hagen (Isoptera: Hodotermitidae). *Bull. Entomol. Res.* **79**: 537–544.

van der Westhuizen, M. C. and P. H. Hewitt. 1984. The utilisation of energy reserves by founding pairs of the harvester termite, *Hodotermes mossambicus* (Hagen). *Int. Congr. Entomol.* **17**: 495.

van der Westhuizen, M. C., P. H. Hewitt and T. C. D. K. van der Linde. 1987. Physiological changes during colony establishment in the termite *Hodotermes mossambicus* (Hagen): energy reserves – neutral lipids and glycogen. *Insect Biochem.* **17**: 793–797.

Vawter, L. 1991. Evolution of blattoid insects and of the small subunit ribosomal RNA gene. Ph. D. dissertation, University of Michigan, Ann Arbor, MI. USA.

Veeranna, G. and S. Basalingappa. 1984. Distribution of various castes in different parts of the mound of the termite, *Odontotermes wallonensis* Wasmann (Isoptera: Termitidae). *Entomon* **9**: 217–220.

–. 1990. Population density in different parts of the mound nests of the termite *Odontotermes obsesus* Rambur and their functional behavior. *Entomon* **15**: 59–62.

Vehrencamp, S. L. 1983. Optimal degree of skew in cooperative societies. *Am. Zool.* **23**: 327–335.

Vincke, P. P. and J. P. Tilquin. 1978. A sex-linked ring quadrivalent in termitidae (Isoptera). *Chromosoma* **67**: 151–156.

Watson, J. A. L. and H. M. Abbey. 1985. Development of neotenics in *Mastotermes darwiniensis* Froggatt: an alternative strategy. In: *Caste Differentiation in Social Insects*. J. A. L. Watson, B. M. Okot-Kotber and C. Noirot, eds., pp. 107–124. Oxford: Pergamon Press.

–. 1989. A 17-year old primary reproductive of *Mastotermes darwiniensis* (Isoptera). *Sociobiology* **15**: 279–284.

Watson, J. A. L. and J. J. Sewell 1981. The origin and evolution of caste systems in termites. *Sociobiology* **6**: 106–118.

–. 1985. Caste development in *Mastotermes* and *Kalotermes*, which is primitive? In *Caste Differentiation in Social Insects*. J. A. L. Watson, B. M. Okot-Kotber and C. Noirot, eds., pp. 27–40. Oxford: Pergamon Press.

Watson, J. A. L., J. J. C. Nel and P. H. Hewitt. 1972. Behavioural changes in founding pairs of the termite, *Hodotermes mossambicus*. *J. Insect Physiol.* **18**: 373–387.

Weesner, F. M. 1970. Termites of the Nearctic region. In *Biology of Termites*. K. Krishna and F. M. Weesner, eds., vol. 2, pp. 477–525. New York: Academic Press.

West-Eberhard, M. J. 1975. The evolution of social behavior by kin selection. *Q. Rev. Biol.* **50**: 1–33.

–. 1981. Intragroup selection and the evolution of insect societies. In *Natural Selection and Social Behavior: Recent Research and New Theory*. R. A. Alexander and D. W. Tinkle, eds. pp. 3–17. New York: Chiron.

Weyer, F. 1930. Ueber Ersatzgeschlechtstiere bei Termiten. *Z. Morphol. Oekol. Tiere* **19**: 364–380.

White, T. C. R. 1993. *The Inadequate Environment: Nitrogen and the Abundance of Animals*. Berlin: Springer-Verlag.

Wilson, E. O. 1971. *The Insect Societies*. Cambridge: Belknap Press of Harvard University Press.

–. 1975. *Sociobiology: the New Synthesis*. Cambridge: Belknap Press of Harvard University Press.

–. 1990. *Success and Dominance in Ecosystems: The Case of the Social Insects*. Oldendorf: Ecology Institute.

Winston, M. and K. N. Slessor. 1992. The essence of royalty: honey bee queen pheromone. *Sci. Res. Soc.* **80**: 374–385.

Wood, T. G. and R. A. Johnson. 1986. The biology, physiology, and ecology of termites. In *Economic Impact and Control of Social Insects*. S. B. Vinson, ed., pp. 1–63. New York: Praeger Publishing.

Wrangham, R. W. and D. I. Rubenstein. 1986. Social evolution in birds and mammals. In *Ecological Aspects of Social Evolution*. D. I. Rubenstein and R. W. Wrangham, eds., pp. 452–470. Princeton: Princeton University Press.

Zimmerman, R. B. 1983. Sibling manipulation and indirect fitness in termites. *Behav. Ecol. Sociobiology* **12**: 143–145.

5 · Maternal care in the Hemiptera: ancestry, alternatives, and current adaptive value

DOUGLAS W. TALLAMY AND CARL SCHAEFER

ABSTRACT

Using the Hemiptera as a model, this chapter develops an alternative hypothesis for the evolution of subsocial behavior, one that challenges the traditional view that maternal behavior is an exceptional and relatively recent evolutionary leap forward. Data are presented that support the argument that maternal care is not a recent behavioral innovation of the Arthropoda; indeed, it is a common phenomenon in phyla as primitive as the Cnidaria. Evidence that parental behavior has been a constant trait throughout the evolution of the Hemiptera is weak, but there is solid support for a claim of plesiomorphy in the Membracoidea, the Cimicomorpha, and the Pentatomoidea. Hemipteran maternal care is not a behavior restricted to occupants of unusually harsh environments and it does not appear to provide taxa that express it with superior survivorship or with an unusual capacity to radiate. Instead, maternal behavior is a trait fraught with ecological costs. When compared with females of related asocial taxa, mothers and the young they seek to protect are subject to increased exposure to predators, reduced fecundity, and lower intrinsic rates of natural increase. Maternal costs are so high that subsocial species have developed cost-reducing mechanisms such as egg-dumping, ant mutualism, and the avoidance of maternal risks during periods of high reproductive value. Through various permutations of clutch placement and size, most hemipterans have abandoned the maternal option in favor of dozens of alternatives that protect eggs from environmental dangers without loss of life or fecundity. These oviposition methods are apparently so successful that at least 95% of heteropteran taxa exhibit them. In fact, it is only when host-plant seasonality or other unidentified phenomena prevent the distribution of eggs over space and time that relict maternal behavior persists.

INTRODUCTION

More than two centuries ago, Modeer described what appeared to be maternal behavior in the acanthosomatid shield bug *Elasmucha grisea*. He noted that the female remained steadfast over her compact egg mass and tilted her body towards intruding objects instead of taking flight (Modeer 1764). Similar behavior went unnoticed among homopterans until 1887, when it was observed in *Entylia sinuata* treehoppers (Murtfeldt 1887). Today new records of parental care in Hemiptera are hardly noteworthy; such behavior has already been confirmed in 74 genera (31 Homoptera; 43 Heteroptera) of 19 families and the list continues to grow (Bequaert 1935; Melber and Schmidt 1977; Tachikawa 1991). The acknowledgment of parental behavior in insects has not always been universal, however. As late as 1971, the noted heteropteran scholar N. C. E. Miller declared that 'interpreting these actions as parental behaviour is surely an example of the loose manner in which some observers endow organisms which are very low on the scale of development with the sentiments of animals much higher in the scale' (Miller 1971, p. 56).

Such anthropocentric attitudes thwarted advances in our understanding of insect behavior until Wilson (1971) laid them to rest in his classic *Insect Societies*. His synthesis of the numerous reports of parental activities forever dispelled the notion that insects are too 'primitive' to care for their young. Yet the implication remained that maternal care was an exceptional advancement that relatively few arthropods had managed to achieve. Wilson believed that insect parental behavior is a polyphyletic behavioral solution to exceptional environmental challenges. Only when physical conditions are particularly rigorous, nutritional resources are exceptionally rich or ephemeral, or predation is unusually intense, is selection powerful enough to provide insects with parental abilities. When parental care is achieved, however, it permits the use of environments or resources that are too challenging for related, solitary species (Wilson 1971, 1975). These ideas have been developed and expanded by many researchers (Melber and Schmidt 1977; Eickwort 1981; Hinton 1981; Wood 1982; Tallamy 1984; Wood 1984; Tallamy and Wood 1986; Tallamy 1994), but never tested. In this chapter, we use the Hemiptera to provide evidence for an alternative hypothesis in which maternal behavior is not viewed as an apomorphic advancement, but rather as a costly, plesiomorphic relic that only rarely functions as well as alternative life-history adaptations.

ANTHROPOCENTRIC ASSUMPTIONS

The anthropocentric assumption that maternal care is an apomorphic advancement over not providing care occurs almost instinctively to many workers. It is intuitively natural to think that if vertebrates, particularly the highly intelligent mammals, provide extended periods of care to their young, any insect that behaves similarly must be evolutionarily precocious and therefore more 'successful' than non-caregivers. Fortunately, these assumptions need not be accepted or rejected on faith; they can be empirically tested by examining the accuracy of the following predictions. If the evolution of maternal behavior were truly an adaptive leap forward, its appearance must have been the result of some extraordinary change in phenotype that exceeded the phylogenetic capacities of asocial taxa. Consequently, maternal care should predominate in the most derived members of a particular group. It also follows that the acquisition of parental abilities should have improved juvenile survivorship and thus fitness in those taxa that had made this adaptive leap. Parental abilities, in turn, should have enhanced the competitive abilities of those taxa, increasing population sizes in relation to non-parental taxa, and possibly permitting the invasion of new adaptive zones and subsequent radiation. Quite simply, if maternal care worked better, taxa that cared for their young should have dominated and eventually replaced those that did not, particularly in the rigorous, rich, or threatening habitats described by Wilson (1971). In the following sections we will examine the evidence for and against these predictions.

WAS MATERNAL CARE AN ADAPTIVE LEAP?

The early stages of an evolutionary event may be reconstructed by examining the simplest forms of its extant expression. In the case of hemipteran maternal care, the simplest and most common form of the behavior is egg-guarding. In all subsocial homopterans, and in most heteropterans that exhibit maternal behavior, females physically shield their eggs from parasitoids and predators. To accomplish this, only the slightest change in behavior need occur; rather than leaving eggs after oviposition, females simply need to remain on or near the egg mass for some period following oviposition. No morphological innovations are necessary; indeed, the morphology of related subsocial and asocial taxa is typically indistinguishable. Of course, mothers will be more effective guardians if they actively defend their eggs; but, again, no special defensive abilities have evolved in parental hemipterans that are not also present in asocial species. Wing-fanning, for example, is a typical maternal response to the approach of arthropod predators (Bequaert 1935; Wood 1974; Melber and Schmidt 1975; Tachikawa 1980, 1991; Tallamy and Denno 1981b), but asocial species readily fan their wings when threatened or to discourage unwanted suitors (Tallamy 1984). Even the most complex parental behaviors exhibited by subsocial insects are simply an amalgamation of typical feeding and defensive behaviors exhibited by asocial relatives. Thus, the evolution of maternal care in its simplest form required no adaptive change in morphology and the modification of only one behavior: the urge to leave the oviposition site. In fact, rather than a complex evolutionary advancement, maternal protection may have been the most parsimonious (though hardly the most efficient) means of decreasing egg mortality.

When considering the evolutionary differences between subsocial and asocial taxa, it is essential to establish whether subsocial taxa evolved maternal care from disparate traits that existed previously in asocial ancestors, or whether asocial taxa retained the morphological and behavioral traits of their subsocial forebears. In some lineages, evidence supports the latter. For example, it may not be coincidental that subsocial – but also asocial – pentatomoid shield bugs typically look like medieval shields. Maternal pentatomoids use their broad, flat bodies to physically shield their eggs from predators and especially parasitoids (Bequaert 1935; Eberhard 1975; Melber and Schmidt 1975; Tachikawa 1980). Mappes and Kaitala (1994) have shown that acanthosomatid body size and shape are critical determinants of maternal success. Because any eggs not shielded are inevitably lost to parasitoids, females adjust their clutch sizes according to the surface area of their bodies. Since asocial pentatomoids also have shield-like bodies, it is not for the lack of this trait that members of these taxa have taken a different route to egg protection. Instead, we submit the following: maternal behavior was pervasive in cydnoid-like protopentatomoids that guarded egg clusters in small depressions on the ground. As these taxa moved from the ground onto plants to feed and oviposit, shield-like morphology was favored because it improved maternal success. For reasons discussed later, most pentatomoids subsequently abandoned maternal care for more effective, less costly means of protecting eggs. Clutch sizes shrank and eggs were safely dispersed through time and space, leaving only the shield-shaped body as a morphological remnant of past maternal activities.

This scenario suggests that morphology and behavior common in asocial hemipteran taxa may, in some cases, be remnant traits of ancestral groups with subsocial behaviors, rather than vice versa. But the strength of this argument depends upon establishing the antiquity of maternal care in our example taxon, the Pentatomoidea. In the following section we review evidence that maternal care is a trait characteristic of not only ancestral pentatomoids; it is a primitive solution to egg-care problems in most arthropod relatives and in numerous, more plesiomorphic, taxa as well.

THE QUESTION OF PLESIOMORPHY

If the advent of maternal behavior was not dependent upon major morphological or behavioral change, care must have been an evolutionary option even for ancestral groups. But relative to asocial behavior, is care-giving typically ancestral (plesiomorphic) rather than derived (apomorphic) among arthropods? The determination of character polarity is often difficult to establish; techniques for doing so have been rigorously debated for decades (reviewed by Nelson 1978). Several methods have been proposed (Nelson 1978) but evidence of character plesiomorphy is most commonly attained by comparing in closely related groups, or sister groups, the expression of the trait in question (Wiley 1981). If both groups share the same character-state, it is assumed that this condition was inherited from a common ancestor. Alternatively, if the character-state differs in the two groups, it is assumed that the character has recently changed in one or both of the groups. Of course, this method is only as accurate as the phylogenies on which it is based (Nelson 1973). Furthermore, if only some members of both groups

possess the same character-state, the polarity of the trait can be difficult to resolve by this approach.

Let us first apply this approach to a phylogeny of arthropod classes. It is immediately obvious from Fig. 5-1 that parental care is the rule rather than the exception in most extant relatives of insects. Among the Chelicerata, paternal behavior occurs in all Pycnogonida, or sea spiders (Jarvis and King 1972), while maternal care is common throughout the Arachnida: universally present in scorpions (Scorpiones) (Vachon 1952), pseudoscorpions (Pseudoscorpiones) (Weygoldt 1969), windscorpions (Solifugae), whipscorpions (Uropygi) and short-tailed whipscorpions (Schizomida) (Weygoldt 1972), and the dominant life-history option in nine families of spiders (Araneae) (Buskirk 1981). Maternal behavior is also extremely common in four of the five classes of Crustacea (Schöne 1961) (Lutz 1986), occurs in all centipedes (Chilopoda) (Eason 1964) and symphylans (Jones 1935), and has been recorded in pauropods (Pauropoda) (Harrison 1914) and millipedes (Diplopoda) (Schubart 1934). Even arthropod

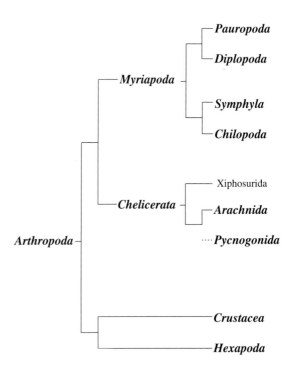

Fig. 5-1. Phylogenetic classification of the extant Arthropoda (modified from Boudreaux 1979; Friedrich and Tautz 1995). At least some members of the taxa in italics exhibit parental behavior.

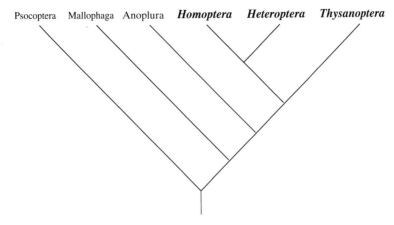

Hypothetical rhynchotoid ancestor

Fig. 5-2. Phylogeny of rhynchotoid orders based on characters of the spermatozoa (from Baccetti 1979). Maternal behavior occurs in italicized taxa.

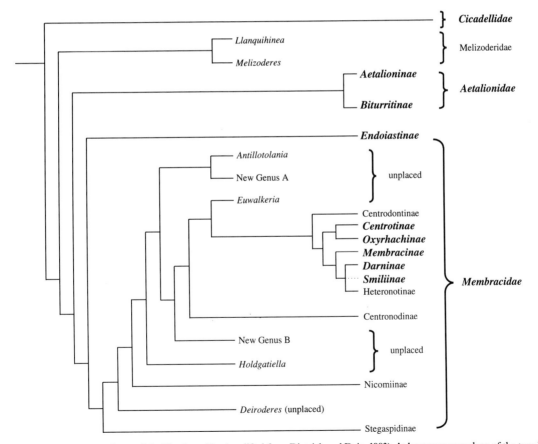

Fig. 5-3. Phylogeny of selected taxa of the Membracoidea (modified from Dietrich and Deitz 1993). At least some members of the taxa in italics exhibit maternal care.

relatives and other primitive invertebrates exhibit parental care: the behavior is common in cephalopods (Mollusca) (Van Heukelem 1979), polychaetes (Annelida) (Schroeder and Hermans 1975), Rotifera (Thane 1974), Platyhelminthes (Henley 1974), Ctenophora (Pianka 1974) and Cnidaria (Campbell 1974). There is little agreement about the inter-relationships of the Hexapoda, Myriapoda, Crustacea and Chelicerata (Boudreaux 1979; Weygoldt 1979, 1986; Manton 1979; Anderson 1979; Boore et al. 1995; Friedrich and Tautz 1995); recent evidence from ribosomal DNA sequences suggests that the Crustacea (Friedrich and Tautz 1995) and not the Myriapoda (Kristensen 1975; Boudreaux 1979; Manton 1979; Anderson 1979) is the sister group of the Hexapoda. However, since both the Crustacea and the Myriapoda are groups in which maternal behavior is a firmly entrenched life-history trait, it is clear that insects were not the first arthropods to protect their young from a hostile environment.

We can extend the investigation of plesiomorphy in hemipteran maternal behavior to the Rhynchota (Fig. 5-2). Most authors agree that the Thysanoptera represent the sister group of the Hemiptera (Baccetti 1979; Ananthakrish-nan 1984; Boudreaux 1979; von Dohlen and Moran 1995). Maternal care is a frequent, but not pervasive, feature of this outgroup (Hean 1943; Crespi 1990, 1992), so polarity at this level cannot be determined. Within the Homoptera proper, a recent phylogeny of the Membracoidea proposed by Dietrich and Deitz (1993) reveals the same problem (Fig. 5-3). Dietrich and McKamey's (1990) recent discovery of maternal behavior among idiocerine leafhoppers satisfies the inclusion of this trait in the outgroup Cidadel-lidae, but it is a rare exception. Despite our ignorance of the life histories of rare groups such as the Melizoderidae, Nicomiinae and Stegaspidinae (L. Deitz and T. Wood, personal communication), however, there is good evidence that maternal care arose early in the membracoid lineage. All females in the Aetalionidae, the sister group of the Membracidae, guard their eggs (Haviland 1925; Colthurst 1930; Brown 1976), as do members of the Endoiastinae (Wood 1984), a sister group of the remaining membracid subfamilies.

In contrast, the evidence that heteropteran maternal behavior descended from a common ancestor is weak (Fig. 5-4). Although there is little agreement regarding the composition and phylogeny of heteropteran infraorders (Leston et al. 1954; Stys and Kerzhner 1975; Cobben 1978; Sweet 1979; Schuh 1979; Stys 1985; Wheeler et al. 1993), most cladists believe that the Pentatomomorpha, which

contains 10 of the 15 subsocial heteropteran families (Table 5-1), is the most derived group in the Heteroptera (reviewed by Wheeler et al. 1993). Arguments for the primi-tive nature of the Pentatomomorpha have been made on the basis of the structure and orientation of eggs during ovipo-sition (Leston 1954; Dupuis 1948; Southwood 1956; Cobben 1968) and the presence of a salivary feeding cone (Sweet 1979). However, these arguments have not been made in a cladistic context with other heteropteran infraorders.

The ancestral status of maternal care is less controversial within infraorders. No cladogram of the Pentatomomorpha exists yet (Schaefer 1993) but within the Pentatomoidea there is general agreement that the Cydnidae form the base of the superfamily, with Acanthosomatidae and Scu-telleridae arising as early offshoots (Cobben 1978; South-wood 1956; Dupuis 1948; Leston 1954; Schaefer 1984, 1993). Maternal care is present in all three of these families, but is absent in many of the more derived pentatomoids (Table 5-1). Similarly, Schuh and Stys (1991) place the Reduviidae, a family in which parental care appears in at least six genera, at the base of their Cimicomorpha clado-gram. With the exception of the tingid genera Gargaphia (Tallamy and Denno 1981b) and Leptobyrsa (Melksham 1984), the remaining 13 cimicomorphan families are com-pletely asocial.

At least two hypotheses can be erected to explain the patterns described above. The commonly held belief is

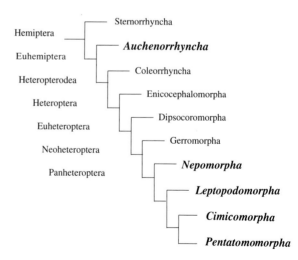

Fig. 5-4. Phylogeny of higher Heteroptera proposed by Wheeler et al. (1993). Maternal care is present in at least some members of the taxa in italics.

Table 5-1. *Oviposition diversity in the heteroptera*

Taxon	Mode of oviposition	Protection index	Reference
Enicocephalomorpha			
Enicocephalidae	eggs glued singly (?) to substrate roots	5	Myers 1926
Dipsocoromorpha			
Dipsocoridae	eggs hidden singly (?) in crevices	20	Cobben 1968
Gerromorpha			
Gerridae			
Gerrinae	eggs glued in small clutches to the underside of floating debris	6	Cobben 1968
Rhagadotarsinae	eggs inserted into plants in small (?) groups	18	Cobben 1968
Mesoveliidae	small (?) groups of eggs inserted into stems	18	Miller 1971
Veliidae	small (?) groups of eggs glued to floating debris or rocks	6	Cobben 1968
Hebridae			
Hebrus	eggs inserted into plants in small (?) groups	18	Cobben 1968
Hyrcanus	small (?) groups of eggs glued to floating debris	6	Cobben 1968
Hydrometridae	small clutches elevated on stalks on water plants	9	Cobben 1968
Nepomorpha			
Notonectidae			
Notonecta	small groups of eggs inserted into submerged plants	18	Cobben 1968
Notonecta maculata	eggs glued in small (?) clusters to submerged objects	6	Cobben 1968
Pleidae	eggs inserted into submerged plants in small groups	18	Cobben 1968
Helotrephidae	eggs glued to submerged plants in small (?) groups	6	Cobben 1968
Ochteridae	eggs hidden within damp soil singly	20	Hinton 1981
Gelastocoridae			
Gelastocorinae	small clutches buried in sand	12	Hungerford 1922
Nethra martini	♀ guards large clutch within a soil depression under stones	4	Usinger 1956
Naucoridae			
Ambrysinae, Potamocorinae, Cryphocricinae	eggs glued to submerged objects in small (?) groups	6	Cobben 1968
Naucorinae	eggs inserted into plants in small (?) groups	18	Cobben 1968
Belostomatidae			
Lethocerinae	♂ guards large clutches glued to emergent vegetation	5	Ichikawa 1988
Belostomatinae	♂ aerates large clutches glued to its dorsum	5	Miller 1971
Nepidae	eggs inserted into submerged plants	18	Cobben 1968
Corixidae			
Micronectinae (*Diaprepocoris*)	eggs glued to submerged objects in small (?) groups	6	Cobben 1968, Miller 1971
Cymatiinae, Corixinae, Heterocorixinae, Micronectinae (*Tenagobia*)	eggs elevated on stalks in small (?) groups on submerged objects	9	Hungerford 1948
Leptopodomorpha			
Leptopodidae–Omaniidae	small (?) groups of eggs buried in sand	12	Cobben 1968
Saldidae			
Chiloxanithinae, some Saldinae	small (?) groups of eggs inserted into plants	18	Cobben 1968
Aepophilus	♀ guards many (?) eggs within a soil depression	4	Butler 1923
most Saldinae	small (?) groups of eggs glued to substrate	3	Cobben 1968
Cimicomorpha			
Reduviidae			Miller 1971

Table 5-1. *(cont.)*

Taxon	Mode of oviposition	Protection index	Reference
Phymatinae, some Apiomerinae some Harpactorinae (*Panthous, Sycanus, Isyndus*), some Ectrichodiinae (*Scadra*)	small clutches covered with glutinous froth	6	Cobben 1968, Southwood 1956
Macrocephalinae, most Emesinae, Phonolibinae, some Harpactorinae (*Coranus*), Rhaphidosomatinae, some Apiomerinae	eggs glued singly to substrate	5	Miller 1971, Cobben 1968, Southwood 1956
Cetherinae, Triatominae	eggs dropped singly to substrate	5	Miller 1971
Holoptilinae	small clutches concealed in crevices and covered with protective exudate	24	Miller 1971
some Harpactorinae (*Pisilus, Endochus*), some *Rhinocoris*, some Emesinae (*Ghinallelia*)	♀ guards large (?) clutches glued to vegetation	1	Bequaert 1935, Parker 1965, Edwards 1962
some Harpactorinae (*Vestula, Nagusta*)	small clutches glued to substrate	3	Miller 1971
Piratinae, Stenopodinae, most Ectrichodiinae	small clutches buried in soil	12	Miller 1971
Reduviinae	eggs hidden singly in crevices and covered with glutinous material	40	Southwood 1956
some Harpactorinae (*Rhinocoris, Zelus*)	♂ guards large clutches glued to substrate	5	Odhiambo 1959, Ralston 1977
some Emesinae (*Stenolemus arachniphagus*)	♂ carries small (?) clutch on dorsum	15	Cobben 1968
Nabidae			
Nabinae, Prostemminae, Gorpinae, Carthasinae	eggs inserted in small clutches into plants	18	Cobben 1968
Arachnocorinae	eggs glued in small numbers to spider webs	15	Cobben 1968
Anthocoridae			
Xylocoris	small clutches hidden in crevices	12	Southwood 1956
Physopleurella	pseudoplacentary viviparity	15	Miller 1971
Triphleps	eggs inserted in small groups into plants	18	Miller 1971
Cimicidae	eggs hidden singly in crevices	20	Miller 1971
Microphysidae			
Myrmedobia	eggs hidden singly in crevices	20	Miller 1971
Loricula	oviparity (viviparity)	15	Miller 1971
Tingidae			
Gargaphia, Leptobyrsa	♀ guards large clutches cemented to leaf surface	1	Tallamy and Denno 1981b, Melksham 1984
Corythucha	eggs partly inserted into leaves in small clutches	3	Tallamy and Denno 1981a
Stephanitis	eggs completely buried in leaves and covered with a shellac-like secretion	36	D. W. Tallamy, personal observations
most Tinginae, Agramminae and Cantacaderinae	eggs inserted into plants in small groups	18	Southwood 1956

Table 5-1. *(cont.)*

Taxon	Mode of oviposition	Protection index	Reference
Miridae			
some Orthotylinae	eggs partly inserted into plants in small clutches	3	Southwood 1956
some subfamilies	eggs deeply buried in plants in small clutches	18	Cobben 1968
most species	eggs inserted singly into plants	30	Hinton 1971
Thaumastocoridae	eggs hidden singly in crevices	20	Miller 1971
Pentatomomorpha			
Lygaeidae			
some *Nysius*	eggs attached singly to underside of leaves	5	Michalk 1935
most species	eggs inserted singly into crevices	20	Cobben 1968
some Rhyparochrominae	eggs buried in flower heads singly	30	Miller 1971
Aradidae			
Aradus	eggs glued singly under bark	20	Miller 1971
Aneurinae	viviparity	15	Miller 1971
Mezira, Neuroctenus, Ctenoneurus, Brachyrhynchus	♀ guards large clutches of eggs	1	Taylor 1988, Takahashi 1934
Piesmatidae	eggs glued singly to leaf surface	5	Miller 1971, D. W. Tallamy, personal observations
Berytidae			
Metacanthinae	single eggs elevated over plant surface on stalks	15	Miller 1971
Berytinae	single eggs hidden in plant crevices	20	Miller 1971
Colobathristidae	eggs buried singly in soil	20	Miller 1971
Pyrrhocoridae	small clutches buried in soil	12	Southwood 1956
Largidae	small clutches buried in soil	12	Miller 1971
Coreidae			
most Coreinae	eggs deposited singly on food plant	5	Miller 1934
Leptoglossus, Anoplocnemus	small clutches laid in chains on plant	3	Southwood 1956
some Coreinae (*Plunentis, Phyllomorpha*)	eggs glued to dorsum of conspecifics in small groups	15	Miller 1971, Lima 1940
some Coreinae (*Mictis*)	small groups of eggs are covered with wax-like substance	6	Miller 1971
Physomerus	♀ guards large clutches glued to plants	1	Nakamura 1990
Alydidae	small groups of eggs glued in rows to substrate	3	Cobben 1968
Rhopalidae	eggs laid singly on substrate	5	Wheeler and Hoebeke 1988
Cydnidae			
Sehirinae (*Sehirus, Adomerus, Canthophorus, Tritomegas*), *Brachypelta, Legnotus,* Parastrachinae (*Parastrachia*)	♀ guards unstructured mass of eggs within a soil crevice	4	Miller 1971, Schorr 1957, Sites and McPherson 1982, Tachikawa and Schaefer 1985
Cydninae	eggs buried singly in soil	20	Ayyar 1930
Thyreocoridae	eggs glued singly to plants	5	Southwood 1956, Biehler and McPherson 1982
Acanthosomatidae			
some Acanthosomatinae (*Anaxandra, Elasmucha, Sastragala*)	♀ guards large clutches glued to leaves	1	Tachikawa 1991
other Acanthosomatinae	small (?) clutches glued to leaves	3	Cobben 1968

Table 5-1. (cont.)

Taxon	Mode of oviposition	Protection index	Reference
Scutelleridae			
some Scutellerinae (*Cantao*, *Augocoris*), some Pachycorinae (*Pachycoris*, *Tectocoris*)	♀ guards large clutches glued to leaves	1	Tachikawa 1991, Hussey 1934
Eurygastrinae	small (?) clutches glued to leaves	3	Cobben 1968
Plataspidae	small clutches covered with frothy secretion	6	Southwood 1956
Urostylidae	small clutches hidden in bark crevices and covered with dark mucus	24	Yamada 1914, 1915
Phloeidae (*Phloea*, *Phloeophana*)	♀ guards large clutches hidden in bark tissues	12	Miller 1971, Bequaert 1935
Dinidoridae			
Cyclopelta	♀ guards large clutches glued to substrate	1	Tachikawa 1991
Aspongopus	small clutches glued to substrate	3	Cobben 1968
Tessaratomidae			
Tessarotoma	eggs laid in small clusters on plants	3	Southwood 1956
Pygoplatys, *Erga*	♀ guards large clutches glued to plants	1	Tachikawa 1991
Pentatomidae			
Birketsmithia	eggs inserted into plant tissue in small groups	18	Leston 1954
some Discocephalinae (*Antiteuchus, Mecistorhinus,* (*Dinocoris*)), Pentatominae (*Chlorocoris, Eumecopus*), Podopinae (*Scotinophora*), Edessinae (*Edessa*)	guards large egg masses glued to leaves	1	Eberhard 1975, Corbett and Yusope 1924, Tachikawa 1991, Rau 1918
most pentatomids	small clutches attached to leaf surface	3	Cobben 1968

[a] The protection index is a value describing the degree to which various aspects of a particular mode of oviposition are likely to confer to the embryo protection from predators and parasites. It is computed by multiplying the artificially assigned values of each of the protective attributes listed in Table 5-4 that the taxon in question exhibits.

that parental behavior arose independently (and irreversibly) hundreds of times in a remarkable display of convergence throughout the Arthropoda (Bequaert 1935; Wilson 1971; Eickwort 1981; Tallamy and Wood 1986). The cladograms above generally support this hypothesis, with the inclusion of an important caveat: though not a trait common to all hemipteroid ancestors, care arose very early in a number of hemipteroid lineages that subsequently became largely or completely asocial. Alternatively, maternal care may be a plastic trait that has appeared *and* disappeared repeatedly as taxa moved through major evolutionary shifts. Unfortunately, distinguishing between these alternatives may be impossible.

EVIDENCE FROM RADIATION PATTERNS

If maternal care is a successful evolutionary innovation that provides better protection to juveniles than do asocial life-history traits, then taxa whose members have achieved this behavior should have radiated more extensively in species richness than taxa whose members provide no care. To avoid untestable assumptions and confounding effects from phylogenetic inertia, comparisons of this nature should be restricted to sister groups (Slowinski and Guyer 1993). Unfortunately, there are few cases in the Hemiptera where subsocial behavior occurs as a generic trait in sample sizes large enough to make statistically valid comparisons, and there are no such cases where phylogenies establishing sister relationships have been completed. Perhaps the earliest opportunity to test this hypothesis, however, will come from the membracid Smiliinae (Appendix 1). Our knowledge of Smiliinae life history patterns is far from complete, but it appears that maternal care is a tribal trait in the Polyglyptini, is common in the Tragopini, is present in only one genus of both the Smiliini and Amastrini, and is completely absent from the Acutalini, Ceresini, Micrutalini, Quadrinareini and Thuridini (reviewed by Wood 1984). In spite of the fact that maternal care is more common in the Smiliini than in other membraciid subfamilies, both conservative (only genera *known* to exhibit maternal behavior; $n = 14$) and liberal (genera *suspected* of having maternal species; $n = 27$) comparisons of the number of subsocial vs. asocial genera ($n = 59$) within the Smiliinae reveal the extent to which asocial genera currently dominate the subfamily. We will have to await the completion of generic phylogenies before we can determine whether this domination is reflected in species radiation patterns as well.

A SOLUTION TO UNCOMMON ENVIRONMENTAL PROBLEMS?

Wilson's proposal (1971) that parental behavior arose under conditions that were too extreme for insects that provide no care finds little support in the Hemiptera. There is nothing uncommonly harsh or threatening about the niches occupied by maternal hemipterans. The vast majority of subsocial hemipterans (all homopterans, the maternal pentatomids, tessaratomids, scutellerids, phloeids, dinidorids, acanthosomatids, tingids and coreids) are herbivores that feed on the exposed parts of living plants (Table 5-1). Subsocial cydnids specialize on fruits and seeds that have fallen to the ground, while aradids feed on fungal hyphae under bark. Though rare, maternal care also occurs among heteropteran predators such as *Nethra* gelastocorids, *Aepophilus* saldids, and four reduviid genera, *Rhinocoris*, *Endochus*, *Ghinallelia* and *Pisilus*. Eggs of herbivores and predators alike are threatened by predators and parasites. Certainly these threats are serious for subsocial hemipterans, but they are hardly less so for asocial species. Furthermore, subsocial hemipterans are not better able to exploit plant sap, fruits, seeds, or invertebrate prey by virtue of their maternal behavior; many more species of hemipteran develop successfully on these same resources without the aid of maternal protection. Rather than a means of exploiting uncommon resources, maternal behavior is a tactic used by relatively few of the species dependent upon the most abundant resources available to the Insecta.

THE COSTS OF CARING

If maternal behavior is not an evolutionarily recent innovation, and if it is not necessarily the product of unique environmental challenges, why is it not a life-history option common to most or all hemipterans? Only in the homopteran family Aetalionidae is the behavior exhibited by all species; there are no heteropteran families in which the number of subsocial species is more than a tiny fraction of the total. It is likely that the extraordinary costs associated with maternal care have much to do with its scarcity (Tallamy 1984, 1994; Tallamy and Wood 1986). One such cost comes from the risk involved in defending offspring from predators. For example, in the lace bug *Gargaphia solani*, females that guard their nymphs are three times as likely to be killed as females without brood (Tallamy and Horton 1990).

A less obvious problem is caused by clustering eggs in one spot, an essential component of parental care. Insect eggs or prey items of any sort that are aggregated in one area may be more attractive to predators than isolated eggs (Hassell 1978). Females that do not hide their eggs or distribute them over space and time are caught in a maternal 'bind' (to borrow the terminology of Trivers 1974). If they do not physically guard their eggs, they risk the loss of much or all of their reproductive effort to predators. If females do stand guard, there may be selection for increased clutch size to justify the large time commitment allocated to a single clutch. But as the size of a clutch increases, so does its vulnerability. Once caught in this bind, it may be difficult to escape it without a complete restructuring of the life history.

Yet another problem with the maternal option comes from the timing of reproduction. Mothers that invest from one to several weeks in their first clutch of offspring are postponing subsequent oviposition by at least that much. As Cole (1954) clearly demonstrated, any delay in reproduction has a profound negative effect on the intrinsic rate of natural increase. This, delay, in turn, creates another selective advantage for a large initial reproductive effort, even to the point of semelparity, in species committed to maternal behavior. Table 5-2 demonstrates the

Table 5-2. *Relative size of the first clutch in subsocial and asocial Pentatomoidea*

Taxon	Total fecundity	Mean clutch size	Percentage of total fecundity	Reference
Subsocial Species				
Cydinidae				
Parastrachia japonensis	80	80	100	Tachikawa and Schaefer 1985
Sehirus cinctus cinctus	270	135	*ca.* 50[a]	Sites and McPherson 1982
S. cinctus albonotatus	88	44	*ca.* 50[a]	McDonald 1968
S. cinctus bicolor	60	30	*ca.* 50[a]	Southwood and Hine 1950
S. sexmaculatus	250	125	*ca.* 50[a]	Southwood and Hine 1950
Scutelleridae				
Tectocoris diophthalmus	218	109	*ca.* 50[a]	Ballard and Holdaway 1926
Acanthosomatidae				
Elasmucha putoni	106	53	50	Honbo and Nakamura 1985
Pentatomidae				
Antiteuchus tripterus	56	28	50	Eberhard 1975
Asocial species				
Pentatomidae				
Carbula humerigera	602	14	2.3	Kiritani 1985
Eysarcoris inconspicous	101.6	12	11.8	Shalaby *et al.* 1983
Halyomorpha mista	456.9	28	6.5	Kawada and Kitamura 1982
Nezara viridula	825	72	8.73	McLain *et al.* 1990
Oebalus pognax	946	21.4	2.26	Nilakhe 1976
Perillus bioculatus	118.9	17.8	5.6	Tamaki and Butt 1978
Picromerus bidens	224	38	16.9	Javahery 1986
Piezodorus guildinii	204.8	16.8	8.2	Panizzi and Slansky 1985
Podisus modestus	132.7	16.7	5.9	Tostowaryk 1971
Thyanta perditor	534.4	35.6	6.6	Panizzi and Herzog 1984

[a] No information could be found regarding reproductive attempts beyond the first clutch. These species may be semelparous, but to be conservative, it was assumed that two clutches of equal size are produced. The mean proportion of total fecundity committed to the first clutch by subsocial species is 56.3% whereas that committed by asocial species is 7.5% (t-test adjusted for unequal variances: $t = 6.88$; $p = 0.0004$). The mean fecundity of subsocial species is 141.0 eggs; asocial species produce on average 414.6 eggs (t-test: $t = 2.68$; $p = 0.0216$).

significantly greater proportion of total fecundity that sub-social pentatomoids commit to their first (or only) clutch when compared with their asocial relatives.

Maternal care is also costly because of its detrimental effects on subsequent, and thus total, fecundity (Tallamy 1994). Mothers that are standing guard are not foraging for the nutrients required to produce a new batch of eggs. For example, when the lifetime fecundity of subsocial pentatomoids is compared with that of asocial species (Table 5-2), subsocial females are significantly less fecund than species without maternal care. Similarly, a study of the life-history traits of four species of lace bug found

that maternal *G. solani* females laid significantly fewer eggs than did two of three asocial *Corythucha* species (Table 5-3) (Tallamy and Denno 1981a). However, *G. solani* females that were not permitted to care for eggs laid more than twice as many eggs as females that guarded eggs and nymphs (Tallamy and Denno 1982). Of course, lace bugs and all subsocial pentatomids are herbivorous, but the trade-off between caring and eating is greatest among predators because of the rigorous foraging demands on this guild of animals (Tallamy and Wood 1986; Tallamy 1994). Unless they are 'sit and wait' hunters (as are most maternal spiders), female predators that guard eggs typically forgo

Table 5-3. *Selected life history traits of four species of lace bugs*

Means with different letters are significantly different (Duncan's Multiple Range; $p < 0.05$). Bugs were reared from eclosion until death at 27 °C and a 15 : 9 (L : D) photoperiod.

Trait	Gargaphia solani		Corythucha marmorata	Corythucha ciliata	Corythucha pruni
Oviposition mode	eggs exposed on leaf surface		eggs embedded in leaves	eggs hidden in leaf trichomes	eggs exposed on leaf surface
Maternal care	present	absent	absent	absent	absent
Clutch size	71.0a	71.0a	7.6b	5.0b	11.7c
Number of clutches	1.8a	5.9b	18.1d	33.2e	13.7c
Fecundity	127.8a	294.9c	136.7a	166.0b	160.7b
Survivorship at egg hatch (%)	68.1a	44.2b	68.0a	84.3a	71.2a
Survivorship at adult eclosion (%)	23.1a	3.4b	20.0a	31.0a	32.6a

Source: From Tallamy and Denno (1981a, 1982).

eating altogether. It is no surprise that, of the 65 hemipteran genera that exhibit maternal care, only six are predators (Table 5-1).

Reduced fecundity among parental species is often thought to be offset by the increase in juvenile survivorship resulting from extended parental investment (Wilson 1975). This may be the case in vertebrates, but among insects there is growing evidence to the contrary. In fact, the most ironic aspect of hemipteran maternal behavior is that it may be no more effective in protecting offspring than hiding eggs or spreading them through time and space. A large group of eggs exposed on the substrate is a rich resource for persistent predators and parasitoids, and maternal efforts to thwart these enemies are frequently ineffective. Acanthosomatid females typically lose 50% of their eggs to parasitoids (Tachikawa 1980). The pentatomid *Antiteuchus tripterus* suffers 70% mortality from egg parasitoids alone (Eberhard 1975). How do these losses compare with those of asocial species? Life-table data on such species are rare, but in our lace bug comparisons above (Table 5-3), maternal care of eggs and nymphs provided *G. solani* with no advantage in juvenile survivorship over three species of solitary *Corythucha*, either at egg hatch or adult eclosion.

A convincing demonstration of the costs associated with maternal behavior comes from the evolution within subsocial species of mechanisms to avoid or reduce them. In the lace bug *G. solani* (Tallamy 1985) and the treehopper

Polyglypta dispar (Eberhard 1986), females dodge the risks from, and time commitment to, guarding their young by attempting to oviposit in the egg masses of conspecifics. If they succeed, these 'egg dumpers' are free to lay a second clutch almost immediately, whereas egg recipients cannot resume oviposition until the eggs hatch (*Polyglypta*) or the nymphs reach adulthood (*Gargaphia*). If they have no opportunity to dump their eggs, *G. solani* females reduce the risks associated with care by defending aggressively only when their own reproductive value is low (older females) and the reproductive value of their young is high (third–fifth instars) (Tallamy 1986). *Pubilia* treehoppers limit maternal costs in a different way (Bristow 1983). Mothers remain with young until ants discover the group and begin to feed on honeydew produced by the nymphal aggregation. Then the mother abandons them, apparently transferring care of her young to the very capable ant defenders. Added to these fascinating examples is the most convincing evidence that maternal care is a costly evolutionary option: the proliferation of non-parental mechanisms for protecting eggs without risking death or reducing fecundity.

ALTERNATIVE LIFE-HISTORY STRATEGIES

We have argued that it is not evolutionarily difficult for an arthropod to include maternal care in its suite of life-history traits. Rather, the difficulty may have been to escape the ecological bind that results from doing so. Selection to replace maternal behavior with more efficient and less costly means of egg care must have been intense for early hemipteroids, particularly predatory groups; but a series (or combination) of behavioral, physiological or morphological changes was necessary before a taxon could effectively avoid the costs of caring without suffering unacceptable juvenile losses to enemies.

Several lines of evidence suggest that some 260 million years ago (Kukalová-Peck 1991) hemipteran lineages were derived from scavengers that lived, fed and bred on the ground (Ananthakrishnan 1984; reviewed by Schaefer 1981, 1988). These ground-dwellers soon diverged into root-feeding homopterans (Schaefer 1988) and predaceous heteropterans, many of which developed herbivorous habits secondarily (Schaefer 1981). Because desiccation and predation threatened the eggs of terrestrial arthropods (Cobben 1968), it is possible that these early stocks laid their eggs in large clutches (large in relation to total fecundity) in damp soil and then guarded them from abundant soil predators in much the same way that sehirine cydnids do today. Of course, those heteropterans that invaded aquatic habitats and developed eggs that could survive long periods of submergence (most Gerromorpha and Nepomorpha; Table 5-1) successfully avoided terrestrial predators and parasitoids as well as the threat of egg desiccation.

Perhaps the first change among early ground-dwellers was to slow or pulse the rate of egg maturation, allowing females to distribute their ova in smaller batches throughout the resources their nymphs required. Evidence from extant herbivores suggests that this departure alone may have provided enough protection to eggs to ecologically encourage non-caring females. If *Gargaphia* lace bugs are prevented from guarding their large clutch of aggregated eggs, 56% of those eggs are lost to predators before they hatch (Table 5-3). *Corythucha pruni*, in contrast, oviposits on exposed leaf surfaces in exactly the same way as *Gargaphia*; however, instead of producing one large clutch and guarding it, this species lays 14 small clutches on many different leaves over the course of its adult life. By ovipositing in this way, they lose only 29% of their eggs to predation (Tallamy and Denno 1981a).

In addition to losses from predation, the efficient exploitation of plants, either as food for the phytophagous Homoptera, or as hunting sites by the predatory Heteroptera, may have been hampered by problems with egg desiccation (Cobben 1968). This challenge was met in at least two ways. As evidenced by some of the oldest hemipteroid fossils

(Becker-Migdisova 1960), the piercing ovipositor appeared very early in the hemipteran lineage as a solution to problems of both desiccation and protection. Taxa whose members were endowed with this structure could bury their eggs deep within juicy plant tissue, out of harm's way. So great were the advantages of the piercing ovipositor that all Auchenrrhyncha, many Nepomorpha and most Cimicomorpha have retained this mode of oviposition.

The piercing ovipositor was not cost-free, however. Eggs of hemipterans that oviposit through such structures are typically 2–3 times smaller in volume than those of hemipterans that have developed a short, flattened ovipositor (calculation based on Southwood 1956). Pentatomoids, the most speciose hemipterans to have adopted the flattened ovipositor, lay huge, barrel-shaped eggs that give rise to exceptionally large hatchlings. Apparently these bugs have traded the advantages of the piercing ovipositor for the advantages of large body size. Because their eggs are no longer hidden within the humid environment of plant tissues, those pentatomoids that deposit large eggs on plant surfaces avoid egg desiccation through the development of thicker chorions with fewer but longer micropylar processes (Cobben 1968). But such eggs are again vulnerable to enemies and those taxa that do not hide or distribute their eggs through time and space are obliged to guard them.

Piercing ovipositors may be an unusually successful solution to the problems of egg care and desiccation, but the need for maternal protection has been avoided in many other ways as well, particularly in the Heteroptera (Table 5-1). Groups including the Piesmatidae and Berytidae as well as many Aradidae strategically glue single eggs to the substrate at a rate of two or three per day; Triatomine reduviids simply drop them as they walk (Southwood 1956; Cobben 1968; Miller 1971; D. W. Tallamy, unpublished data). It is difficult for enemies to track or specialize on eggs dispersed in this manner. A reduction in clutch size also permits the use of natural cavities as oviposition sites. This approach to egg protection appears throughout the Heteroptera, but it is most common in groups that do not have piercing ovipositors, such as lygaeids, dipsocorids and thaumastocorids. Urostylids and holoptiline reduviids enhance the effectiveness of hiding eggs in cracks and crevices by covering these openings after oviposition with a protective secretion.

Although dispersing numerous small clutches through time and space is the most common means of avoiding maternal care, interesting alternatives to clutch size reduction have evolved in the Hemiptera. Treehoppers ovipositing into woody stems often cannot insert their eggs deeply enough to completely hide their large clutches from predators or, more particularly, parasites. This problem is exacerbated by defensive responses of the host plant: callous tissue proliferates at the oviposition site, literally forcing the eggs back out of the plant (Wood *et al.* 1990). This vulnerability of eggs despite the efforts of the female to bury them may help explain the presence of maternal care in so many membracid taxa. Groups such as the Membracini, however, avoid care by covering each clutch with a dense, wax-like secretion that apparently deters at least some egg consumers (Haviland 1925; Hinton 1977; Wood 1984). Heteropterans such as the Plataspidae and many harpactorine reduviids also hide entire clutches in shellac-like material excreted from abdominal glands (Miller 1971).

Perhaps evolution's most innovative tack to avoid post-ovipositional maternal costs has been to transfer the responsibilities of egg care from females to males (Tallamy 1994). Exclusive paternal care is rare among animals for several reasons, not the least of which are problems with paternity assurance and the potential loss of male promiscuity (Ridley 1978). It is difficult for males of most taxa to be certain that the offspring they might nurture are carrying their own genes. And even if males can be certain of their paternity, long periods of care usually prevent them from achieving additional matings. Both of these impediments to paternal contributions can be eliminated, however, when the time, energy, and risk associated with acquiring nutritional resources are great (Tallamy 1994). Under these conditions, females typically cannot guard young without sacrificing foraging activities and thus subsequent egg production. Males, in contrast, can free females from this conflict by trading care for mating opportunities. When sperm and associated seminal fluids can be produced without large energy expenditures, males can forgo eating during periods of care without sacrificing their future reproductive potential. In fact, typical sex roles become reversed: by offering paternal protection to offspring, males become the limiting sex and are actively sought by females, i.e. paternal care becomes an epigamically selected trait. Females benefit from this arrangement because the safety of their current clutch is enhanced without reducing future fecundity. Males also gain because promiscuity is encouraged, their young are protected, and males are certain that they have inseminated the eggs they guard because they have demanded to mate before accepting eggs.

When are maternal costs great enough to select for this type of paternal investment? As discussed above, the physical restrictions of maternal behavior usually impact female predators the most, because of their wide-ranging foraging requirements. Thus, it is no surprise that, in the Insecta, exclusive paternal care has evolved only in heteropteran belostomatid and reduviid predators (Table 5-1) (Tallamy 1994).

Paternal egg-guarding has been confirmed in *Rhinocoris* and *Zelus* assassin bugs (Odhiambo 1959; Ralston 1977; Thomas 1995) and is suspected in *Stenolemus arachniphagus*, which was observed by Van Doesburg with two upright rows of eggs glued in the dorsal concavity of its abdomen (Cobben 1968). *Rhinocoris albopilosus* males simultaneously guard the eggs of several females, but only after mating with each (Odhiambo 1959, 1960). The genus *Rhinocoris* provides an extraordinary opportunity to examine the evolution of egg-protection mechanisms because it includes maternal species (*R. carmelita*) (Edwards 1962), paternal species (*R. albopilosus, R. tristis* and *R. albopunctatus*) (Odhiambo 1959; Nyiira 1970; Thomas 1995) and species that protect their eggs without care (*R. bicolor* and *R. tropicus*) (Parker 1969). A preliminary phylogenetic analysis of this genus suggests that paternal care may indeed have been derived from maternal species (L. Thomas, personal communication).

Paternal care is universal in the Belostomatidae, but exists in two manifestations. Primitive giant water bugs in the Lethocerinae oviposit on emergent vegetation. The primary role of male lethocerines is to hydrate their huge eggs by dripping on them and also to guard against egg predation by unrelated females (Ichikawa 1988, 1995). In the Belostominae, females secure their eggs to the backs of their mates, which then proceed to aerate them until hatching (Smith 1976, this volume). As advanced as these behaviors may appear, their evolution required neither a reduction in clutch size nor any morphological change on the part of males or females.

Other heteropterans have been cited as exhibiting paternal behavior, but an examination of the evidence suggests otherwise (Tallamy 1994). Males of the coreid *Phyllomorpha laciniata* have been observed with eggs neatly inserted among the spines on their abdominal tergites (Bolivar 1894). There is no evidence, however, that these males modify their behavior on behalf of their charges, as do male belostomatids and reduviids. It is probably true that no behavior modification in male *Phyllomorpha* is necessary to impart some level of antipredator protection to eggs glued to their abdomens. However, reports of female *Phyllomorpha* also carrying eggs on their dorsum suggest that females rather than males deserve parental credit for selecting safe oviposition sites. Apparently Bolivar's observations represent a female strategy to protect eggs without providing care themselves by ovipositing on the backs of other (male *or* female) conspecifics.

The most derived method of reducing maternal costs actually increases maternal commitment. By internalizing care through viviparity (some Anthocoridae, some Aradidae, many Aphidoidea) and ovoviviparity (some Microphysidae), females can guarantee the well-being of their young without sacrificing foraging activities. Fecundity is reduced by the physical constraints of internalized development, but the benefits of increased survivorship may counter this disadvantage.

The effectiveness in reducing egg mortality of the oviposition options discussed above is difficult to evaluate without the benefit of detailed life-table analyses. However, by making logical (but highly subjective) assumptions about the protective value of each component of these options (Table 5-4), we can for purposes of comparison compute an index that describes the degree to which each option is likely to protect eggs from predators and parasites (Table 5-1). For example, certain Reduviinae hide single eggs (protective value = 5) in natural crevices (protective value = 4) and then cover the crevice with a chemical

Table 5-4. *Relative protective value of hemipteran oviposition behaviors*

Clutch size	Protection index
eggs laid singly	5
eggs in small groups	3
eggs in large groups	1
Oviposition site	
on terrestrial substrate	1
on aquatic substrate	2
elevated on stalks	3
hidden in natural crevices	4
glued to conspecific	5
glued to spider web	5
given to male for care	5
eggs retained until hatching	5
inserted into plant tissue	6
Protective secretion	
eggs covered with protective secretion	2

secretion (protective value = 2). By multiplying these values we can assign an overall protection index of 40 (5 × 4 × 2) to this mode of oviposition.

This exercise is useful for two reasons. First, it reveals clearly the relationship between oviposition mode and the retention of maternal care. The protective value of the oviposition modes of heteropteran females that offer no care to young ($\bar{x} = 12.7$; $n = 80$) is significantly greater than that of subsocial females ($\bar{x} = 2.4$; $n = 14$; t-test: $t = -8.42$; $p = 0.0001$, taxa from Table 5-1). There are no heteropteran taxa with members that lay large, exposed clutches – the most vulnerable oviposition state – without providing care. The available data suggest that the converse is also true: all subsocial species lay large clutches that are either exposed on the substrate or, less often, in ground cavities (Tachikawa 1991). The single exception to this generalization occurs in the family Phloeidae, in which females hide eggs in bark crevices and then place their flattened bodies over the crevice until egg-hatch (Bequaert 1935). When the tiny nymphs emerge from the bark, it is thought that they feed directly from their mother's body because their stylets are too short to penetrate the thick bark of their host plant. Thus, maternal behavior in this species may reflect selection for feeding facilitation rather than egg protection.

The use of the protective index also emphasizes the degree to which alternative oviposition habits, e.g. those that do not rely on maternal protection, dominate heteropteran life histories. The deposition of small numbers of eggs through time and space is known or suspected in all or most members of 95% of heteropteran families (Table 5-1). With such a clear demonstration of the success of alternatives, the question to be pondered is not 'why don't more hemipterans care for their young?' but rather 'why is maternal care retained or developed anew as often as it is?'

THE PERSISTENCE OF MATERNAL CARE

Maternal behavior may have persisted or even arisen *de novo* in some hemipteran taxa for several reasons. If, as the data suggest, care is retained only when a large proportion of total egg production is simultaneously exposed to enemies, we must consider why certain groups have not distributed their reproductive output more evenly over their adult life. Herbivores, such as long-lived membracids that imbed eggs into leaf midribs or branches, may be constrained by the narrow window during which young host-plant growth remains soft and vulnerable. If ovipositors are only able to penetrate tissue that is soft, all eggs must be deposited during this short period. In many cases, hardening plant tissues force buried eggs partly or completely out of the twig (Wood *et al.* 1990), nullifying the effectiveness of this mode of oviposition. Maternal care, therefore, may persist in these treehoppers as a necessary, if inefficient, evil.

Just as seasonality in host-tissue vulnerability may maintain maternal behavior in membracids, highly seasonal food resources may also favor egg care by selecting for semelparous reproduction and thus eliminating the most serious cost of maternal behavior: reduced fecundity. The cydnid *Parastrachia japonensis*, for example, feeds as a nymph only on fallen drupes of *Schoepfia jasminodora*. All reproduction must coincide with the brief period during which host drupes are abundant. Females, therefore, can produce one large clutch and then guard it and the ensuing nymphs for weeks without sacrificing reproduction in the future (Tachikawa and Schaefer 1985; Tsukamoto and Tojo 1992).

Even if highly seasonal host-tissue or food resources prove to be important in the persistence of maternal care, these factors do not explain the expression of maternal care in many heteropteran species. Why do all 66 species of *Gargaphia* (as far as we know) and at least one species of *Leptobyrsa* lace bugs guard their eggs and nymphs when the other 1740 species of lace bug do not (numbers from Drake and Ruhoff 1965)? Why do females of a few herbivorous acanthosomatids, dinidorids and pentatomids care for their young when the overwhelming majority protect their eggs in other ways? Why, when most *Rhinocoris* assassin bugs provide no care for eggs, does this duty fall to females of *R. carmelita* (Edwards 1962) and to males of *R. albopilosus* (Odhiambo 1959)? These exceptional taxa will provide rigorous tests of the hypotheses proposed in this chapter and hold the key to our understanding of the evolution of subsocial behavior in general.

ACKNOWLEDGEMENTS

We thank Toby Schuh, Tom Wood and Lew Deitz for their invaluable assistance during this chapter's writing and for helpful reviews upon its completion; we also are grateful to Bernie Crespi for his editorial skills and thoughtful comments and Debbie Hughes for compiling the references. This work was supported, in part, by NSF grant BSR-9020014 to D. Tallamy.

LITERATURE CITED

Ananthakrishnan, T. N. 1984. *Bioecology of Thrips.* Oak Park, **MI**: Indira Publishing House.

Anderson, D. T. 1979. Embryos, fate maps, and the phylogeny of arthropods. In *Arthropod Phylogeny.* A. P. Gupta, ed., pp. 59–98. New York: Van Nostrand.

Ayyar, P. N. K. 1930. A note on *Stibaropus tabulatus*, Schiö (Hemiptera: Pentatomidae), a new pest of tobacco in South India. *Bull. Entomol. Res.* **22**: 29–31.

Baccetti, B. 1979. Ultrastructure of sperm and its bearing on arthropod phylogeny. In *Arthropod Phylogeny.* A. P. Gupta, ed., pp. 609–644. New York: Van Nostrand.

Ballard, E. and F. G. Holdaway. 1926. The life-history of *Tectacoris lineola* F, and its connection with internal boll rots in Queensland. *Bull. Entomol. Res.* **16**: 329–346.

Bequaert, J. 1935. Presocial behavior among the Hemiptera. *Bull. Brooklyn Entomol. Soc.* **30**: 177–191.

Becker-Migdisova, E. E. 1960. Die Archescytinidae als vermutliche Vorfahren der Blattläuse. In *Proceedings XI Internationaler Kongress für Entomologie* (Wien, August 17–25, 1960), vol. 1. H. Strouhal and M. Beir, eds., pp. 298–301. Vienna: Organisation-skomitee d. XI Internationales Kongress für Entomologie.

Biehler, J. A. and J. E. McPherson. 1982. Life history and laboratory rearing of *Galgupha ovalis* (Hemiptera: Corimelaenidae), descriptions of immature stages. *Ann. Entomol. Soc. Am.* **75**: 465–470.

Bolivar, I. 1894. Observations sur la *Phyllomorpha laciniata* de Villiers. *Feuille. Jeun. Nat.* **24**: 43.

Boore, J. L., T. M. Collins, D. Stanton, L. L. Daehler and W. M. Brown. 1995. Deducing the pattern of arthropod phylogeny from mitochondrial DNA rearrangements. *Nature (Lond)* **376**: 163–165.

Boudreaux, H. B. 1979. Arthropod phylogeny with special reference to insects. In *Arthropod Phylogeny.* A. P. Gupta, ed., pp. 551–584. New York: Van Nostrand.

Bristow, C. M. 1983. Treehoppers transfer parental care to ants: a new benefit of mutualism. *Science (Wash., D.C.)* **220**: 532–533.

Broomfield, P. S. 1976. A revision of the genus *Amastris* (Homoptera: Membracidae). *Bull. Br. Mus. Nat. Hist. Entomol.* **33**: 347–460.

Brown, R. L. 1976. Behavioral observations on *Aethalion reticulatum* and associated ants. *Insectes Soc.* **23**: 99–108.

Buskirk, R. E. 1981. Sociality in the Arachnida. In *Social Insects*, vol. 2. H. R. Hermann, ed., pp. 281–367. New York: Academic Press.

Butler, E. A. 1923. *A Biology of the British Hemiptera-Heteroptera.* London: Witherby.

Campbell, R. D. 1974. Cnidaria. In *Reproduction of Marine Invertebrates*, vol. 1. A. C. Giese and J. S. Pearse, eds., pp. 133–199. New York: Academic Press.

Cobben, R. H. 1968. *Evolutionary Trends in Heteroptera*, part 1. *Eggs, Architecture of the Shell, Gross Embryology and Eclosion*. Wageningen: Centre for Agricultural Publishing and Documentation.

–. 1978. Evolutionary trends in Heteroptera. Part II. Mouthpart structures and feeding strategies. *Syst. Zool.* 28: 653–656.

Cole, L. C. 1954. The population consequences of life history phenomena. *Q. Rev. Biol.* 29: 103–137.

Colthurst, I. 1930. A Himalayan hillside. *Darjeeling Nat. Hist. Soc. J.* 4: 68–72.

Corbett, G. H. and M. Yusope. 1924. *Scotinophara coarctata* F. (the black bug of padi). *Malay. Agric. J.* 12: 91–106.

Crespi, B. J. 1990. Subsociality and female reproductive success in a mycophagous thrips: an observational and experimental analysis. *J. Insect Behav.* 3: 61–74.

–. 1992. Eusociality in Australian gall thrips. *Nature (Lond.)* 359: 724–726.

Deitz, L. L. 1975. Classification of the higher categories of the New World treehoppers (Homoptera: Membracidae). *North Carolina Agric. Exp. Sta. Tech. Bull.* No. 225.

Dietrich, C. H. and L. L. Deitz. 1993. Superfamily Membracoidea (Homoptera: Auchenorrhyncha) II. Cladistic analysis and conclusions. *Syst. Entomol.* 18: 297–311.

Dietrich, C. H. and S. H. McKamey. 1990. Three new idiocerine leafhoppers (Homoptera: Cicadellidae) from Guyana with notes on ant-mutualism and subsociality. *Proc. Entomol. Soc. Wash.* 92: 214–223.

Drake, C. J. and F. A. Ruhoff. 1965. *Lacebugs of the World: A Catalog (Hemiptera: Tingidae)*. Washington: Smithsonian Institute.

Dupuis, C. 1948. Nouvelles données biologiques et morphologiques sur les diptères Phasiinae parasites d'Hemiptères Hétéroptères (Contributions III et IV á l'étude des Phasiinae cimicophages). *Ann. Parasitol. Hum. Comp.* 22: 201–232; 397–441.

Eason, E. H. 1964. *Centipedes of the British Isles*. London: Frederick Warne and Co., Ltd.

Eberhard, W. G. 1975. The ecology and behavior of a subsocial pentatomid bug and two scelionid wasps: Strategy and counterstrategy in a host and its parasitoids. *Smithson. Contrib. Zool.* 205: 1–39.

–. 1986. Possible mutualism between females of the subsocial membracid *Polyglypta dispar* (Homoptera). *Behav. Ecol. Sociobiol.* 19: 447–453.

Edwards, J. S. 1962. Observation on the development and predatory habit of two reduviid Heteroptera, *Rhinocoris carmelita* Stal and *Platymeris rhadamanthus* Gerst. *Proc. R. Entomol. Soc. Lond.* 37: 89–98.

Eickwort, G. C. 1981. Presocial insects. In *Social Insects*. H. R. Hermann ed., pp. 199–280. New York: Academic Press.

Friedrich, M. and D. Tautz. 1995. Ribosomal DNA phylogeny of the major extant arthropod classes and the evolution of myriapods. *Nature (Lond.)* 376: 165–167.

Harrison, L. 1914. On some Pauropoda from New South Wales. *Proc. Linn. Soc. N.S.W.* 39: 615–634.

Hassell, M. P. 1978. *The Dynamics of Arthropod Predator-Prey Systems*. Princeton: Princeton University Press.

Haviland, M. D. 1925. The Membracidae of Kartabo. *Zoologica* 6: 229–290.

Hean, A. F. 1943. Notes on maternal care in thrips. *J. Entomol. Soc. Southern A.* 6: 81–83.

Henley, C. 1974. Platyhelminthes (Turbellaria). In *Reproduction of Marine Invertebrates*, vol. 1. A. C. Giese and J. S. Pearse, eds., pp. 267–343. New York: Academic Press.

Hinton, H. E. 1977. Subsocial behavior and biology of some Mexican membracid bugs. *Ecol. Entomol.* 2: 61–79.

–. 1981. *Biology of Insect Eggs*, vol. 2. New York: Pergamon Press.

Honbo, Y. and K. Nakamura. 1985. Effectiveness of parental care in the bug *Elasmucha putoni* (Hemiptera: Acanthosomidae). *Jap. J. Appl. Entomol. Zool.* 29: 223–229.

Hungerford, H. B. 1922. The life history of the toad bug (Heteroptera). *Univ. Kansas Sci. Bull.* 14: 145–171

–. 1948. The eggs of Corixidae (Hemiptera). *J. Kansas Entomol. Soc.* 21: 141–147.

Hussey, R. F. 1934. Observations on *Pachycoris torridus* (Scop.), with remarks on parental care in other Hemiptera. *Bull. Brooklyn Entomol. Soc.* 29: 133–145.

Ichikawa, N. 1988. Male brooding behavior of the giant water bug *Lethocerus deyrollei*. *J. Ethol.* 6: 121–127.

–. 1995. Male counter-strategy against infanticide of the giant water bug *Lethocerus deyrollei*. *J. Insect Behav.* 8: 181–188.

Jarvis, J. H. and P. E. King. 1972. Reproduction and development in the pycnogonoid *Pycnogonum littorale*. *Mar. Biol.* 13: 146–154.

Javahery, M. 1986. Biology and ecology of *Picromerus bidens* (Hemiptera: Pentatomidae) in southeastern Canada. *Entomol. News* 97: 87–98.

Jones, S. 1935. A note on the distribution, oviposition and parental care of *Scutigerella unguiculata* Hansen var. *indica* Gravely. *J. Bombay Nat. Hist. Soc.* 38: 209–211.

Kawada, H. and C. Kitamura. 1982. The reproductive behavior of the brown marmorated stink bug, *Halyomorpha mista* Uhler (Heteroptera: Pentatomidae). *Appl. Entomol. Zool.* 18: 234–242.

Kiritani, Y. 1985. Timing of oviposition and nymphal diapause under the natural daylengths in *Carbula humerigera* (Heteroptera: Pentatomidae). *Appl. Entomol. Zool.* 20: 252–256.

Kopp, D. 1979. A taxonomic review of the tribe Ceresini (Homoptera: Membracidae). *Misc. Pub. Entomol. Soc. Am.* 11: 1–98.

Kopp, D. D. and J. H. Tsai. 1983. Systematics of the genus *Idioderma* Van Duzee (Homoptera: Membracidae) and biology of *I. virescens* Van Duzee. *Ann. Entomol. Soc. Am.* 76: 149–157.

Kristensen, N. P. 1975. The phylogeny of hexapod 'orders.' *Z. Zool. Evol. Forsch* 13: 1–44.

Kukalová-Peck, J. 1991. Fossil history and the evolution of hexapod structures. In *The Insects of Australia*. T. P. Naumann, ed., pp. 141–179. Melbourne: Melbourne University Press.

Leston, D., J. D. Pendergrast and T. R. E. Southwood. 1954. Classification of the terrestrial Heteroptera (Geocorisae). *Nature (Lond.)* **174**: 91–92.

Lima, A. da Costa. 1940. *Insectos do Brasil*, vol. 2. *Hemipteros*. Rio de Janeiro: Escola Nacional de Agronomia.

Lutz, P. E. 1986. *Invertebrate Zoology*. Reading: Addison-Wesley Publishing Company.

Manton, S. M. 1979. Functional morphology and the evolution of the hexapod classes. In *Arthropod Phylogeny*. A. P. Gupta, ed., pp. 387–466. New York: Van Nostrand.

Mappes, J. and A. Kaitala. 1994. Does a female parent bug lay as many eggs as she can defend; experiments with *Elasmucha grisea* L. (Heteroptera; Acanthosomatidae). *Behav. Ecol. Sociobiol.* **35**: 314–317.

McDonald, F. J. D. 1968. Some observations on *Sehirus cinctus* (Palisot de Beauvois) (Heteroptera: Cydnidae). *Can. J. Zool.* **46**: 855–858.

McLain, D. K., D. L. Lanier and N. B. Marsh. 1990. Effects of female size, mate size, and number of copulations on fecundity, fertility, and longevity of *Nezara viridula* (Hemiptera: Pentatomidae). *Ann. Entomol. Soc. Am.* **83**: 1130–1136.

Melber, A. and G. H. Schmidt. 1975. Sozialverhalten zweier *Elasmucha*-Arten (Heteroptera: Insecta). *Z. Tierpsychol.* **39**: 403–414.

–. 1977. Sozialphänomene bei Heteropteren. *Zoologica* **127**: 19–53.

Melksham, J. A. 1984. Colonial oviposition and maternal care in two strains of *Leptobyrsa decora* Drake (Hemiptera: Tingidae). *J. Austr. Entomol. Soc.* **23**: 205–210.

Metcalf, Z. P. and V. Wade. 1965. *General catalogue of the Homoptera: Membracoidea, Section I.* Baltimore: Waverly Press.

Michalk, O. 1935. Zur Morphologie und Ablage der Eier bei den Heteropteran, sowie über ein System der Eiablagetypen. *Deut. Entomol. Z.* **1935**:148–157.

Miller, N. C. E. 1934. The developmental stages of some Malayan Rhynchota. *J. Fed. Malay States* **17**: 502–525.

–. 1971. *The Biology of the Heteroptera*. Hampton: E.W. Classey, Ltd.

Modeer, A. 1764. Några märkvärdigheter hos insectet *Cimex ovatus* pallide griseus, abdominis lateribus albo nigrogre variis alib albis basi scutelli nigricante. *Vetensk. Akad. Handl., Stockholm* **25**: 41–57.

Murtfeldt, M. E. 1887. Traces of maternal affection in *Entilia sinuata* Fabr. *Entomol. Am.* **3**: 177–178.

Myers, J. G. 1926. Biological notes on New Zealand Heteroptera. *Trans. New Zealand Inst.* **56**: 449–511.

Nakamura, K. 1990. Ecology of Sumatran bug *Physomerus grossipes*, with reference to egg guarding behavior of females. *Insectarium* **25**: 18–27 (in Japanese).

Nelson, G. J. 1973. The higher level phylogeny of vertebrates. *Syst. Zool.* **22**: 87–91.

–. 1978. Ontogeny, phylogeny, paleontology, and the biogenetic law. *Syst. Zool.* **27**: 324–345.

Nilakhe, S. S. 1976. Overwintering, survival, fecundity, and mating behavior of the rice stink bug. *Ann. Entomol. Soc. Am.* **69**: 717–720.

Nyiira, Z. M. 1970. The biology and behavior of *Rhinocoris albopunctatus* (Hemiptera: Reduviidae). *Ann. Entomol. Soc. Am.* **63**: 1224–1227.

Odhiambo, T. R. 1959. An account of parental care in *Rhinocoris albopilosus. Proc. R. Entomol. Soc. Lond.* **34**: 175–187.

–. 1960. Parental care in bugs and nonsocial insects. *New Scientist* **8**: 449–451.

Panizzi, A. R. and D. C. Herzog. 1984. Biology of *Thyanta perditor* (Hemiptera: Pentatomidae). *Ann. Entomol. Soc. Am.* **77**: 646–650.

Panizzi, A. R. and F. Slansky. 1985. Legume host impact on performance of adult *Piezodorus guildinii* (Westwood) (Hemiptera: Pentatomidae). *Environ. Entomol.* **14**: 237–242.

Parker, A. H. 1965. The maternal behaviour of *Pisilus tipuliformis* Fabricius (Hemiptera: Reduviidae). *Entomol. Exp. Appl.* **8**: 13–19.

–. 1969. The predatory and reproductive behaviour of *Rhinocoris bicolor* and *R. tropicus* (Hemiptera: Reduviidae). *Entomol. Exp. Appl.* **12**: 107–117.

Pianka, H. D. 1974. Ctenophora. In *Reproduction of Marine Invertebrates*. vol. 1. A. C. Giese and J. S. Pearse eds., pp. 201–265. New York: Academic Press.

Ralston, J. S. 1977. Egg guarding by male assassin bugs of the genus *Zelus* (Hemiptera: Reduviidae). *Psyche* **84**: 103–107.

Rau, P. 1918. Maternal care in *Dinocoris tripterus*. *Entomol. News* **29**: 75–76.

Ridley, M. 1978. Paternal care. *Anim. Behav.* **26**: 904–932.

Schaefer, C. W. 1981. The land bugs and their adaptive zones. *Rostria* **33**: 67–83.

–. 1984. Non-homology of hypopygidial appendages in the Podopinae (Pentatomidae) and Scutellerinae (Scutelleridae) (Hemiptera: Pentatomoidea). *J. Kansas Entomol. Soc.* **57**: 532–533.

–. 1988. The food plants of some 'primitive' Pentatomoidea (Hemiptera: Heteroptera). *Phytophaga (Madras)* **2**: 19–45.

–. 1993. The Pentatomomorpha (Hemiptera: Heteroptera): an annotated outline of its systematic history. *Eur. J. Entomol.* **90**: 105–122.

Schöne, H. 1961. Complex behavior. In *The Physiology of Crustacea*, vol. 2. T. H. Waterman, ed., pp. 465–520. New York: Academic Press.

Schorr, H. 1957. Zur verhaltensbiologie und Symbiose von *Brachypelta aterrima* Forst (Cydnidae, Heteroptera). *Z. Morphol. Oekol. Tiere* **45**: 561–602.

Schroeder, P. C. and C. O. Hermans. 1975. Annelida: Polychaeta. In *Reproduction in Marine Invertebrates*, vol. 3. A. C. Giese and J. S. Pearse, eds., pp. 1–213. New York: Academic Press.

Schubart, O. 1934. Tausend füssler oder Myriapoda. I. Diplopoda. *Tierwelt Deutschlands* **28**: 301.

Schuh, R. T. 1979. Review of evolutionary trends in Heteroptera. Part 2. Mouthpart structures and feeding strategies. *Syst. Zool.* **28**: 653–656.

Schuh, R. T. and P. Stys. 1991. Phylogenetic analysis of cimicomorphan family relationships. *J. N.Y. Entomol. Soc.* **99**: 298–351.

Shalaby, F. F., M. M. Assar and A. M. Tantawi. 1983. Biological studies on *Eysarcoris inconspicous* H. Sc. (Hemiptera-Heteroptera: Pentatomidae). *Ann. Agric. Soc. Moshtohor* **19**: 557–572.

Sites, R. W. and J. E. McPherson. 1982. Life history and laboratory rearing of *Sehirus cinctus cinctus* (Hemiptera: Cydnidae) with descriptions of immature stages. *Ann. Entomol. Soc. Am.* **75**: 210–215.

Slowinski, J. B. and C. Guyer. 1993. Testing whether certain traits have caused amplified diversification: an improved method based on a model of random speciation and extinction. *Am. Nat.* **142**: 1019–1024.

Smith, R. L. 1976. Brooding behavior of a male water bug *Belostoma flumineum* (Hemiptera: Belostomatidae). *J. Kansas Entomol. Soc.* **49**: 333–43.

Southwood, T. R. E. 1956. The structure of the eggs of the terrestrial Heteroptera and its relationship to the classification of the group. *Trans. R. Entomol. Soc. Lond.* **108**: 163–221.

Southwood, T. R. E. and D. J. Hine. 1950. Further notes on the biology of *Sehirus bicolor*. *Entomol. Mon. Mag.* **86**: 299–301.

Stys, P. 1985. Soucasny stav beta-taxonomic radu Heteroptera. *Pr. Slov. Ent. Spol.* **4**: 205–235.

Stys, P. and I. M. Kerzhner. 1975. The rank and nomenclature of higher taxa in recent Heteroptera. *Acta Entomol. Bohem.* **86**: 1–32.

Sweet, M. H. 1979. On the original feeding habits of the Hemiptera. *Ann. Entomol. Soc. Am.* **72**: 575–579.

Tachikawa, S. 1980. Parental care of Pentatomoidea (Heteroptera) in Japan. *XVI Int. Congr. Entomol.* Abstract No. 16P, p. 436. Kyoto.

–. 1991. *Studies on the Subsociality of Japanese Heteroptera.* Tokyo: College of Tokyo Agriculture Press.

Tachikawa, S. and C. W. Schaefer. 1985. Biology of *Parastrachia japonensis* (Hemiptera: Pentatomoidea: ?-idae). *Ann. Entomol. Soc. Am.* **78**: 387–397.

Takahashi, R. 1934. Parental care of *Mezira membranacea* Fabr. (Hemiptera: Aradidae). *Trans. Nat. Hist. Soc. Formosa* **24**: 315–316.

Tallamy, D. W. 1984. Insect parental care. *BioScience* **34**: 20–24.

–. 1985. 'Egg dumping' in lace bugs (*Gargaphia solani*, Hemiptera: Tingidae). *Behav. Ecol. Sociobiol.* **17**: 357–362.

–. 1986. Age specificity of 'egg dumping' in *Gargaphia solani* (Hemiptera: Tingidae). *Anim. Behav.* **34**: 599–603.

–. 1994. Nourishment and the evolution of paternal investment in subsocial arthropods. In *Nourishment and Evolution in Insect Societies.* J. H. Hunt and C. A. Nalepa, eds., pp. 21–56. Boulder: Westview Press.

Tallamy, D. W. and R. F. Denno. 1981a. Alternative life history patterns in risky environments: an example from lace bugs. In *Insect Life History Patterns: Geographic and Habitat Variation.* R. F. Denno and H. Dingle, eds., pp. 129–148. New York: Springer-Verlag.

–. 1981b. Maternal care in *Gargaphia solani. Anim. Behav.* **29**: 771–778.

–. 1982. Life history tradeoffs in *Gargaphia solani* (Hemiptera: Tingidae): the cost of reproduction. *Ecology* **63**: 616–620.

Tallamy, D. W. and L. A. Horton. 1990. Costs and benefits of the egg-dumping alternative in *Gargaphia* lace bugs (Hemiptera: Tingidae). *Anim. Behav.* **39**: 352–359.

Tallamy, D. W. and T. K. Wood. 1986. Convergence patterns in subsocial insects. *Annu. Rev. Entomol.* **31**: 369–390.

Tamaki, G. and B. A. Butt. 1978. Impact of *Perillus bioculatus* on the Colorado potato beetle and plant damage. *U.S. Dept. Agric. Tech. Bull.* 1581:1–11.

Taylor, S. J. 1988. Observations on parental care in the family Aradidae. *Great Lakes Entomol.* **21**: 159–161.

Thane, A. 1974. Rotifera. In *Reproduction of Marine Invertebrates,* vol. 1. A. C. Giese and J. S. Pearse, eds., pp. 471–484. New York: Academic Press.

Thomas, L. 1995. Parental care in assassin bugs (Reduviidae). Ph.D. dissertation, University of Cambridge.

Tode, W. D. 1966. Taxionomische Untersuchungen an der Sudamerikanischen Membracidengattung *Tragopa* Latreillei 1829, und deren Neugliederung. *Mitt. Hamburg Zool. Mus. Inst.* **63**: 265–328.

Tostowaryk, W. 1971. Life history of *Podisus modestus* (Hemiptera: Pentatomidae) in boreal forest in Quebec. *Can. Entomol.* **103**: 662–674.

Tsukamoto, L. and S. Tojo. 1992. A report of progressive provisioning in a stink bug, *Parastrachia japonensis* (Hemiptera: Cydnidae). *J. Ethol.* **10**: 21–29.

Trivers, R. L. 1974. Parent-offspring conflict. *Am. Zool.* **14**: 249–264.

Usinger, R. L. 1956. A revised classification of the Reduvioidea with a new subfamily from South America. *Ann. Entomol. Soc. Am.* **36**: 602–618.

Vachon, M. 1952. *Etudes sur les Scorpions.* Algiers: Institut Pasteur D'Algerie.

Van Heukelem, W. F. 1979. Environmental control of reproduction and lifespan in *Octopus:* an hypothesis. In *Reproductive Ecology of Marine Invertebrates,* vol. 9. S. Stanyck, ed., pp. 123–133. Columbia: University of South Carolina Press.

von Dohlen, C. D. and N. A. Moran. 1995. Molecular phylogeny of the Homoptera: a paraphyletic taxon. *J. Molec. Evol.* **41**: 211–223.

Weygoldt, P. 1969. *The Biology of Pseudoscorpions.* Cambridge: Harvard University Press.

–. 1972. Geisselskorpione und Geisselspinnen (*Uropygi* und *Amblypygi*). *Z. Kölner Zoo* **15**: 95–107.

–. 1979. Significance of later embryonic stages and head development in arthropod phylogeny. In *Arthropod Phylogeny.* A. P. Gupta, ed., pp. 107–136. New York: Van Nostrand.

–. 1986. Arthropod interrelationships – the phylogenetic-systematic approach. *Z. Zool. Syst. Evol. Forsch.* **24**: 19–35.

Wheeler, A. G. and E. R. Hoebeke. 1988. Biology and seasonal history of *Rhopalus* (*Brachycarenus*) *tigrinus*, with descriptions of immature stages (Heteroptera: Rhopalidae). *J. N.Y. Entomol. Soc.* **96**: 381–389.

Wheeler, E. C., R. T. Schuh and R. Bang. 1993. Cladistic relationships among higher groups of Heteroptera: congruence between morphological and molecular data sets. *Entomol. Scand.* **24**: 121–137.

Wiley, E. O. 1981. *Phylogenetics.* New York: Wiley-Interscience.

Wilson, E. O. 1971. *The Insect Societies.* Cambridge: Harvard University Press.

–. 1975. *Sociobiology.* Cambridge: Harvard University Press.

Wood, T. K. 1974. Aggregating behavior of *Umbonia crassicornis. Can. Entomol.* **106**: 169–173.

–. 1982. Selective factors associated with the evolution of membracid sociality. In *The Biology of Social Insects.* M. D. Breed, C. D. Michener and H. E. Evans, eds. pp. 175–179. Boulder: Westview Press.

–. 1984. Life history patterns of tropical membracids (Homoptera: Membracidae). *Sociobiology* **8**: 299–344.

–. 1993. Diversity in the New World Membracidae. *Annu. Rev. Entomol.* **38**: 409–435.

Wood, T. K., K. L. Olmstead and S. I. Guttman. 1990. Insect phenology mediated by host-plant water relations. *Evolution* **44**: 619–628.

Yamada, Y. 1914. On *Urostylus westwoodii* Scott. *Insect World* **18**: 138–142.

–. 1915. On *Urostylus striicornis* Scott. *Insect World* **19**: 313–316.

Appendix I. Radiation in asocial and subsocial Smiliinae (Membracidae)*

Tribe	Genus	Species	Asocial	Subsocial	Reference
Acutalini	*Acutalis*	11	X		
	Euritea	5	X		
	Thrasymedes	7	X		
Amastrini	*Amastris*	49	X		Broomfield 1976
	Bajulata	1	X		
	Erosne	1	X		
	Harmonides	3		X	
	Iidioderma	3	X		Kopp and Tsai 1983
	Lallemandia	1	X		
	Tynelia	11	X		
	Vanduzeea	12	X		
Ceresini	*Amblyophallus*	5	X		Kopp 1979 (all Ceresini)
	Anisostylus	4	X		
	Antonae	22	X		
	Clepsydrius	20	X		
	Cyphonia	26	X		
	Eucyphonia	8	X		
	Hadrophallus	3	X		
	Melusinella	2	X		
	Paraceresa	7	X		
	Parantonae	4	X		
	Penichrophorus	14	X		
	Poppea	23	X		
	Proxolonia	1	X		
	Spissistilus	7	X		
	Stictocephala	15	X		
	Stictolobus	4	X		
	Tapinolobus	2	X		
	Tortistilus	10	X		
	Trichaetipyga	3	X		
	Vestistilus	6	X		

Appendix I. (*continued*)

Tribe	Genus	Species	Asocial	Subsocial	Reference
Micrutalini	*Micrutalis*	31	X		
	Trachytalis	2	X		
Polyglyptini	*Adippe*	7		X	
	Aphetea	6		X	
	Bilimekia	2		X	
	Bryantopsis	2		?	
	Dioclophara	2		?	
	Ecuatoriana	3		?	
	Ennya	12		X	
	Entylia	8		X	
	Hemiptycha	1		?	
	Heranice	3		?	
	Hille	7		?	
	Hygris	1		?	
	Incolea	2		?	
	Matunaria	7		?	
	Membracidoidea	1		?	
	Mendicea	1		?	
	Metheisa	2		X	
	Phormophora	4		?	
	Polyglypta	4		X	
	Polyglyptodes	4		X	
	Polyrhyssa	1		?	
	Publilia	6		X	
Quadrinareini	*Quadrinarea*	1	X		
Smiliini	*Antianthe*	7		X	
	Ashmeadea	1	X		
	Atymna	13	X		
	Atymnina	1	X		
	Carynota	5	X		
	Cyrtolobus	43	X		
	Glossonotus	5	X		
	Godingia	1	X		
	Grandolobus	3	X		
	Heliria	1	X		
	Helonica	12	X		
	Hemicardiacus	1	X		
	Ophiderma	16	X		
	Palonica	6	X		
	Smilia	3	X		
	Telamona	2	X		
	Telamonanthe	2	X		
	Telonaca	3	X		
	Thelia	2	X		
	Tropidarnis	1	X		
	Xantholobus	12	X		

Appendix I. (*continued*)

Tribe	Genus	Species	Asocial	Subsocial	Reference
Tragopini	*Anobilia*	15	X		Tode 1966
	Chelyoidea	5	X		
	Horiola	3		X	
	Richteria	5		X	
	Stilbophora	3	X		
	Tragopa	4		X	Wood 1984
	Tropidolomia	3	X		
	Walkeria	2	X		
Thuridini	*Thuris*	1	X		

* Higher classification by Deitz (1975). Species numbers unless otherwise noted by Metcalf and Wade (1965). Wood (1984, 1993) reviewed the distribution of maternal care in the Smiliinae.

6 · Evolution of paternal care in the giant water bugs (Heteroptera: Belostomatidae)

ROBERT L. SMITH

ABSTRACT

Unilateral postzygotic paternal care is extremely rare among animals. The giant water bug family Belostomatidae contains most of the arthropod species known to exhibit this unusual behavior. In the subfamily Lethocerinae, males brood eggs laid on emergent vegetation. Brooding in this group involves watering eggs, shading them, and defending them against predation. In the subfamily Belostomatinae, males employ a variety of behavior patterns to aerate eggs attached to their backs by their mates. Brooding is obligatory in all belostomatid species studied; unattended eggs invariably die if left in the open air or submersed.

This chapter explores the biology, phylogeny and fossil record of the Belostomatidae and related taxa in an attempt to discern the selection forces, the constraints, and the sequence of historical events responsible for the evolution of this unusual behavior and its subsequent diversification. Selection for large bug size, in order to take advantage of vertebrate prey, together with the dual phylogenetic constraints of Dyar's Law and the apparent inability of heteropterans to add molts, coupled egg size to body size. Thus selection for large bugs also produced large eggs: too large to develop unattended submersed in water. A past history of eggs being laid in water left these larger eggs lacking the necessary adaptations to survive desiccation when laid unattended in the open air. Consequently, large eggs created selection for an innovation to lift egg-size limitations on imago size. Ergo, emergent brooding evolved in the lethocerine lineage.

The initial costs of emergent brooding were minimal for males, who required less food than females, possessed a perfected anti-cuckoldry mechanism, and were able to obtain multiple mates and brood multiple clutches of eggs. Females in contrast, would have sacrificed substantial future fitness in order to brood eggs and were therefore selected to abandon. Transitions from non-brooding to emergent-brooding and back-brooding are inferred under different assumptions about the behavior of members of the enigmatic genus *Horvathinia*. The evolution of back-brooding was followed by an adaptive radiation. Water bugs in the fossil genus *Mesobelostomum* from the Upper Jurassic of the Solnhofen, Bavaria, probably brooded their eggs more than 150 million years before present.

Giant water bug brooding, although complex as expressed in extant species, may have represented a modest innovation at the time of its origin. However, this behavioral innovation lifted constraints, and thus permitted the evolution of larger bugs able to prey on larger aquatic vertebrates, shifting the lineage to new adaptive peaks. Although existing general theory on parental care could not have predicted brooding in the Belostomatidae, much of the theory on unilateral postzygotic paternal care fits giant water bugs very well.

The term 'ancillary selection' may usefully distinguish natural selection that produces characters supportive of primary traits. Brooding in the Belostomatidae is deemed to be a product of ancillary selection, which supports primary selection for large bug size to facilitate the use of vertebrate prey.

INTRODUCTION

The ultimate, if never fully achievable, goal of evolutionary biology is to understand the environmental circumstances, the phylogenetic constraints, and the sequence of historical events that produced the characteristics of organisms (Darwin 1872). Reconstructing the history of patterns of animal behavior can be particularly daunting, especially if the behavior of interest requires no special morphological structures for its expression and is restricted to only one or a few extant taxa that exhibit no antecedent patterns. Despite such difficulties, the undertaking may be justified in that even a partial comprehension of how 'unusual' behavior has evolved may provide insights into the history of more common patterns (Williams 1966). In addition, attempts to reconstruct the development of rare (vs. common) patterns are more likely to advance our understanding of evolutionary processes. Unilateral postzygotic paternal care is a highly unusual behavioral pattern, which requires no special structures, and for which no intermediate patterns are to be found. This chapter represents an attempt to account for the evolutionary history of exclusive paternal care in a single insect clade, the giant water bugs (Belostomatidae).

Delivery of nurture by fathers alone is exceptional among animals. By far the largest concentration of species that exhibit this unusual behavior is found in the fish, while a few anurans, birds, and arthropods provide the handful of additional examples (Ridley 1978). Among insects, parental care of any kind is uncommon, and unilateral paternal care is extremely rare. Of the nearly 1.5 million described insect species, unassisted paternal care is found in fewer than 150 species (Smith 1980). Of these, more than 99% occur in a group of aquatic insects known as the giant water bugs (Belostomatidae). Giant water bug males in the subfamily Lethocerinae brood eggs attached to emergent vegetation above the surface of the water by irrigating and shading them to prevent their desiccation and by defending them against predators. Females in another subfamily, the Belostomatinae, attach their eggs to the backs of males, who then periodically expose the clutch at the air–water interface and employ various behavior patterns to aerate them when under water. Brooding in both the Lethocerinae and the Belostomatinae is obligatory: embryos in unattended eggs submersed or left in open air always die.

The accumulated theory (see below) on unilateral postzygotic paternal care predicts that this behavior should be associated with: (1) external fertilization, (2) eggs or young vulnerable to the physical and/or biotic environment, (3) sparse distribution in a harsh environment, and (4) dependence on patchy, scarce, or abundant but ephemeral resources. Giant water bugs are antithetical to this profile. They are hemimetabolous insects whose precocial larvae swim adeptly soon after hatching. The belostomatids have no known egg predators or parasites (see Hoffman 1932; Masner 1972; Smith and Larsen 1993) and live in typically resource-rich aquatic habitats where they are buffered against temperature fluxes, abrasion and desiccation. They have internal fertilization, and belong to an infraorder (Nepomorpha) of aquatic and semiaquatic insects which contains over 1000 species in nine families all of whose eggs develop perfectly well unattended by parents. So,

why is paternal care so extensively expressed in the giant water bug clade and what is the history of its initial cause and the subsequent diversification of male brooding patterns?

Efforts to reconstruct this history are handicapped by our imperfect knowledge of the biology of extant species of giant water bugs, and by an almost total ignorance of certain key taxa in the family and its sister group, the Nepidae. In addition, some pivotal studies needed to resolve especially troublesome issues have not been performed on any species. Despite these difficulties, enough is now known to discern a clear path to paternal brooding of eggs. When the data run thin, I conjecture freely in the hope of inspiring future work. To avoid confusion, I ignore the diversity of behavioral terms applied by different authors, and have selected a single term for what I deem to be homologous behavior patterns in different taxa. The work is generously laced with whatever unpublished observations are known to me that bear on the subject. Some of these are my own and some have been provided as personal communications from other workers. I attempt to distinguish facts from preliminary observations, anecdotes, and my own expansive supposition.

The chapter is organized as follows. I first review parental care theory with emphasis on the singular case of exclusive postzygotic paternal care. This is followed by a summary of what is known of the phylogenetics and biology of the Belostomatidae. I then explore several 'special topics' that bear on belostomatid evolution and ultimately on paternal brooding. I present an evolutionary scenario, which attempts to account for why belostomatid eggs require brooding and why fathers rather than mothers evolved to perform the task. It also examines the transitions among non-brooding, emergent-brooding and back-brooding patterns. Next, the nepid solution to large eggs is compared to the belostomatid solution. I infer ancient brooding from the belostomatid fossil record; finally, my summary revisits applicable theory to determine its fit in this case and I discuss the subordinate role of egg-brooding behavior to the product of primary selection for large bugs.

PARENTAL CARE THEORY

Parental investment may be defined as any contribution to current offspring which could cause a measurable cost in production of future progeny (Trivers 1972). To be selected, parental investment must benefit offspring by enhancing their survival or improving their competitiveness. Selection cannot favor delivery of care that exceeds the return in benefit for offspring. Therefore, if a single parent can defend offspring against some form of predation, then only one parent will be selected to invest in such defense. Females or males or both sexes may contribute to offspring either before (Thornhill 1976) or after the onset of embryogenesis, but when offspring can obtain full benefit of care from one parent alone, it is typically the mother who provides. Unilateral maternal care is the usual pattern in mammals. Biparental care may evolve when offspring require, or can substantially benefit from, more than a single parent's care. This pattern is often found in birds.

It is highly unusual for males to engage in unilateral postzygotic parental care. Clutton-Brock (1991) noted, following Bateman (1948), Williams (1966) and Trivers (1972), that the nature of parental contribution to offspring, in whatever form, determines the degree of competition between members of one sex for mates of the other, which in turn affects all aspects of mating behavior. It is the usual pattern, because of anisogamy (Parker et al. 1972; Parker 1984), that eggs and hence females limit reproduction of males, thereby creating mate competition among them. Male–male combat, sexual aggressiveness and elaborate display are conspicuous manifestations of competition for females or the resources necessary to attract them. Williams (1966) first predicted that, in those rare cases where males make a large postzygotic contribution, their investment could limit female reproduction, causing females to compete for mates and thereby induce a reversal of reproductive roles in courtship. This prediction inspired behavioral ecologists to study the few species of fish, birds, frogs, and arthropods that exhibit unilateral postzygotic paternal investment (Berglund et al. 1986, 1989; Jenni 1974; Jenni and Collier 1972; Oring and Knudsen 1972; Smith 1979a,b; Mora 1990; Rosenqvist 1990; Summers 1990, 1992); several of these studies have confirmed the prediction (but see Summers 1992 and Vincent et al. 1992).

A substantial body of theory attempts to discern the circumstances under which parental care is likely to evolve and to predict which gender will provide, if only one parent is selected to invest in care. Wilson (1971) conceived of several environmental circumstances which he believed would favor parental care. Tallamy and Wood (1986) amplify on this and review the literature to test its predictions (see also Tallamy and Schaeffer, this volume). They find clear patterns of behavioral convergence when resources are viewed in terms of their spatial, temporal, physical and broader ecological characteristics. Trivers

added dimensions to the theoretical analysis of parental care behavior in his 1972 paper. Maynard Smith (1977) developed a game-theory model, which evaluates the relative costs and benefits of care for each sex. This model has been modified by inclusion of a term for certainty of parentage (see Zeh and Smith 1985) which, under different circumstances, may retard or facilitate the evolution of male care. The heuristic worth of Maynard Smith's model is undisputed, but it may lack utility because patterns of investment by the sexes and the requirements of offspring are likely to have evolved together following the historical initiation of parental care. Such coevolution may have dramatically altered, from the starting condition, both the degree of dependency of offspring, and the costs and the benefits of care for adults (Clutton-Brock and Godfray 1991).

Several hypotheses attempt to explain the rarity of exclusive male care. First, the *past investment hypothesis* (Trivers 1972) proposes the idea that, because of their large differential investment in ova over male investment in sperm, females are under much stronger selection than males not to abandon their offspring. The notion that future contributions of nurture should be influenced by past investment was disputed by Dawkins and Carlisle (1976) who labeled this concept 'the Concorde fallacy'. Second, the *cruel bind hypothesis* (Trivers 1972; Dawkins and Carlisle 1976; Maynard Smith 1977; Grafen and Sibly 1978; Carlisle 1982) posits a bias toward maternal care in internal-fertilizing species and toward paternal care in species with external fertilization because the first sex to release its gametes has an opportunity to escape before zygotes are produced, thus leaving the other sex to care for offspring alone. Third, the *confidence of paternity hypothesis* (Trivers 1972; Ridley 1978; Blumer 1979; Perrone and Zaret 1979; see also Maynard Smith 1977; Grafen 1980; Werren et al. 1980; Wittenberger 1981; Zeh and Smith 1985; Scott 1990; Xia 1991; Whittingham et al. 1992; Møller and Birkhead 1993a,b) likewise involves mode of fertilization. When fertilization is internal, and females mate with more than one male, nurturing fathers can be cuckolded by previously stored ejaculates or sperm remaining in the female tract from prior matings. Low confidence of paternity can promote cheating strategies (see Smith 1980; Zeh and Smith 1985; Whittingham et al. 1992) or it can increase variance in fitness for caring vs. deserting fathers in small populations (Xia 1991), both of which can undermine selection for paternal care. Fourth, the *unequal pay-off hypothesis* (Williams 1975; Gross and Shine 1981) proposes that internal fertilization, and delayed oviposition or birth, creates a male cost bias that favors female care. Males may forfeit additional matings while waiting for the current mate to produce her off-spring, making them available to him for his investment.

The latter three hypotheses implicate external fertilization as a condition that would favor the evolution of exclusive paternal care (Ridley 1978). If correct, external fertilization may account for the preponderance of examples of independently evolved unilateral paternal care among fish. Further support for these hypotheses may be found in the absence or rarity of exclusive male care in the mammals, birds, and terrestrial arthropods, all of which have internal fertilization.

Parental care provides benefits that can be assigned to three broad non-mutually-exclusive categories; one is density-independent and the other two are density-dependent. First, care may mitigate the rigors of the physical environment. For example, birds must incubate their eggs through the fluxes of ambient temperature to assure embryonic development. Second, resources may be provided to offspring. Young mammals, altricial birds, and the larvae of social insects lack the ability to garner resources from their environments and must be fed by parents. Third, parents may protect their offspring against biotic threats. Care that defends eggs and young against disease, parasitism and predation has evolved independently many times in diverse taxa.

Eggs of terrestrial arthropods other than insects are wholly unprotected against the rigors of life on land and are almost universally nurtured by their mothers (Zeh *et al.* 1988) or very rarely their fathers (see, for example, Mora 1990). Eggs of the terrestrial pterygote insects, however, are superbly well protected against both desiccation and drowning by an elaborate chorionic architecture (Hinton 1961a, 1969, 1970, 1981; Zeh *et al.* 1988), and of course poikilothermic embryos require no parental thermoregulation. As a result, most insect eggs develop perfectly well unattended even when fully exposed to the elements (Zeh *et al.* 1988). Consequently, parental care of any kind is unusual in all but the social insects (Smith 1980), and where it occurs, it usually involves provisioning offspring or protecting them against biotic threats (see Eickwort 1981 for examples).

WATER BUG PHYLOGENETICS

The superfamily Nepoidea is monophyletic, consisting of the sister groups Nepidae and Belostomatidae. Mahner (1993) reviewed past efforts to reconstruct a phylogeny for the Nepomorpha (=Cryptocerata) and has assembled characters from the literature to produce a traditional Hennigian analysis of this Heteroptera infraorder. The Nepomorpha contains all the aquatic (subsurface dwelling) true bugs, and the semiaquatic families Octeridae and Gelastocoridae. Mahner's cladogram shows that the

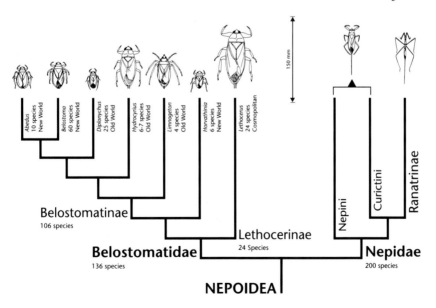

Fig. 6-1. Pictorial cladogram of the Nepoidae. Illustrations are scaled to the largest known representatives of each genus or higher taxon. Species numbers are approximate. (After Lauck and Menke 1961; Mahner 1993.)

remaining taxa in the infraorder form the sister group to the Nepoidea. Mahner's tree for the Belostomatidae does not deviate substantially from an intuitive phylogeny produced by Lauck and Menke (1961) over 30 years ago, but both trees were constructed using relatively few characters.

Figure 6-1 portrays the Nepoidea with scaled illustrations of the largest species representing each lineage. At least two subfamilies are recognized in the Belostomatidae. One of these, the Lethocerinae, is monobasic, containing only the genus *Lethocerus*. This is the earliest derivative lineage in the family and contains about 24 species having a pantropical distribution with a few temperate representatives. All of the remaining taxa in the family make up the sister group to the Lethocerinae. Five genera, three Old World and two New World, are contained in the much larger subfamily, Belostomatinae (>100 species). Of the five, only the three Old World genera are fully resolved. *Limnogeton* is considered the most primitive of the belostomatines (Lauck and Menke 1961; Mahner 1993). Members of this genus are the only giant water bugs that lack middle and hind leg modifications for swimming and a strong prey-grasping function of the front legs. Members of this genus also express a variety of other, possibly plesiomorphic, characters. *Limnogeton* spp. are ecologically unique in their highly specialized diet (see below). Despite having included *Limnogeton* in the Belostomatinae, Lauck and Menke (1961) noted that this genus: '...may, in fact, be the most primitive member of the family'.

Mahner (1993) could find no autapomorphy for the New World genus *Belostoma*, and could not resolve the genus with respect to the other New World belostomatine genus *Abedus*, which does possess at least two autapomorphies. Schnack *et al.* (1990) believe *Abedus* should be a subgenus of *Belostoma*. There may be several distinct clades within the genus *Belostoma*, and it should be considered unresolved. The nominal New World genus *Weberiella* De Carlo (1965) is known from only a few specimens. Menke (1965) placed the single described species in *Belostoma*, but Mahner (1993) holds the opinion that there are insufficient characters for its assignment.

Finally, *Horvathinia*, an enigmatic South American genus with a strangely limited distribution, is thought to occupy a position between Lethocerinae and Belostomatinae. Lauck and Menke (1961) accorded this genus monobasic subfamilial status as the Horvathiniinae. *Horvathinia* exhibits many autapomorphies which contribute nothing to our understanding of its relationship to other genera in the family. No member of this genus has ever been collected

in water, and nothing has been published on its ecology or reproductive behavior. The few specimens in collections have all been taken at lights. *Horvathinia*'s putative phylogenetic position between the emergent-brooders (Lethocerinae) and the back-brooders (Belostomatinae) portends its premier importance for understanding transitions among non-brooding, emergent-brooding and back-brooding states.

The Nepidae (water scorpions), sister group to the Belostomatidae, contains the Nepini, which together with the 'Curictini' (a paraphyletic group; see Fig. 6-1) share common features with the giant water bugs. The Ranatrinae is a specialized subfamily that has diverged from the general broad-bodied form of the Nepoidea *bauplan* (Mahner 1993). Nepids have, I believe, evolved under selection and constraints similar to those of the belostomatids but they have not evolved parental care. Accordingly, it is appropriate to visit the Nepidae in an outgroup comparison with the Belostomatidae. The stem of the Nepomorpha is also the stem of the Nepoidea because of the latter's basal position in the infraorder. It would be a great advantage in this present enterprise to know something about this ancestor of all of the aquatic bugs, but unfortunately its characteristics are obscure.

GIANT WATER BUG BIOLOGY

Belostomatids are medium-sized (9 mm) to very large (>110 mm), brown, dorsoventrally flattened, aquatic bugs that resemble dead leaves. They have stout beaks and hidden antennae, and adults have a pair of retractable, straplike, posterior appendages which function snorkellike to obtain atmospheric air for an otherwise submersed bug. These insects feed by inserting their piercing mouthparts into prey, injecting saliva that contains proteolytic enzymes, then imbibing the liquified prey tissues. All of the giant water bugs except members of the genus *Limnogeton* have the middle and hind legs broadened, flattened, and fringed with hairs to facilitate swimming, and the femora of the forelegs dramatically expanded to contain a massive musculature whose function it is to close the tibiotarsi on prey. All belostomatids for which there are data brood their eggs, which may double in size and mass during embryonic development by imbibing water (Madhavan 1974). What follows in this section is a review of the biology and reproductive behavior of species representing all genera in the three subfamilies of Lauck and Menke (1961) and Mahner (1993).

Lethocerinae

General

The single genus in this subfamily contains large to giant, elongate bugs ranging from 40 to over 110 mm in length. In fact, the genus contains an insect superlative: *Lethocerus maximus* De Carlo, which is by far the largest true bug and certainly among the largest insects. *Lethocerus* species prefer ponds or lakes, but are sometimes found in slow waters of streams and rivers (Menke 1979a; DuBois and Rackouski 1992). It has long been known that these insects lay their eggs on emergent vegetation and other objects out of the water, but only recently has it been discovered that males attend and brood these eggs. Ichikawa (1988) and Smith and Larsen (1993) have published accounts of egg attendance and brooding by males of *Lethocerus deyrollei* Vuillefroy in Japan and *Lethocerus medius* (Guerin) in Arizona, respectively. Additionally, N. Ichikawa has witnessed brooding by a single male *Lethocerus indicus* (Le Peletier & Serville) from Thailand, and Rogelio Macas-Ordóñez observed male care by a few individuals of both *Lethocerus colossicus* (Stål) in Mexico and *Lethocerus americanus* (Leidy) collected from Wisconsin (personal communications.). All observations except those on *L. medius* were conducted in the laboratory. *Lethocerus medius* has been studied exclusively in the field (Smith and Larsen 1993; R. L. Smith, unpublished). Lethocerine bugs are ambush predators that feed on virtually all aquatic animals, but they clearly prefer (and may require) vertebrate prey. Use of vertebrate prey, a crucial element in the scenario for the evolution of emergent-brooding, is discussed as a special topic later in the chapter.

Courtship, mating and oviposition

Tawfik (1969) was first to observe courtship and repeated copulation in a lethocerine, *Lethocerus niloticus* Stål. I observed courtship and repeated copulation, but no oviposition, by *L. medius* in the laboratory (Smith 1975). Hirayama (1977) published an account of the reproductive behavior of *L. deyrollei* in Japanese, but unfortunately this paper was never abstracted or listed in any of the biological databases. Ichikawa (1989a) has published the only detailed English account of mating and oviposition behavior in any lethocerine. More recently, I have studied courtship, mating, and oviposition by *L. medius* in the field, and Rogelio Macas-Oróñez has obtained preliminary observations on *L. colossicus* and *L. americanus* mating in the laboratory. What is known of these behaviors for *Lethocerus* spp. seems consistently exhibited among all of them. Males always take up a position on emergent vegetation (or other emergent substrate) with head down and the posterior at, or protruding slightly from the surface of the water. This position permits the submersed bug to breathe atmospheric air via its air straps. In response to any movement in the vicinity, and sometimes spontaneously, males begin to do display pumping or 'push-ups' by flexing the middle and hind legs at a rate of up to 3 pumps per second. Movement of the male's body produces wave motion on the surface of, and through, the water column. These waves attract receptive females. I have preliminary evidence that an acoustical signal accompanies display pumping in *L. medius*; the sounds may also be an important component in courtship by lethocerines, as they have been shown to be in other belostomatids (Smith 1982; W. F. Kraus and R. L. Smith, unpublished). After a receptive female arrives, copulation takes place from one to ten times in the water (Ichikawa 1989a; R. L. Smith, unpublished). These copulations are interrupted by one or the other or both sexes climbing out of the water onto the emergent substrate then returning to the water. Finally, the female climbs out of the water to a position where she will lay her eggs, and the male follows. Copulation then alternates with oviposition for a period of hours. Ichikawa (1989a) found that from 3 to 10 eggs were laid after each bout of copulation in *L. deyrollei*.

Females deposit their eggs embedded in a foamy substance, which bubbles up around the ova as they are released one at a time from the female reproductive tract (Ichikawa 1989a, R. L. Smith, unpublished) (Fig. 6-2a). The origin and nature of the foam has not been investigated but it is assumed to be an accessory gland product, into which air from the tracheal system is somehow injected. The material may have some historically important function in the hydration of eggs. Mr Hidehiro Hashizume of Osaka, Japan, has discovered that the foam contains numerous sperm (N. Ichikawa, personal communication). During oviposition, the female only rarely returns to the water, but the male does so fairly regularly, especially as the clutch of eggs is enlarged. Oviposition alternates with copulation and egg-watering (see below) by the male at intervals which permit only a few eggs to be laid without the female being interrupted. Bouts of oviposition and copulation alternate until the female has laid all of her eggs. When the female has finished her last bout of egg laying, she allows the male to assume the 'watering' position (see below) over the eggs. She then jumps off the substrate directly into the water, leaving the male in charge of brooding. The abandoning mother never returns to her clutch.

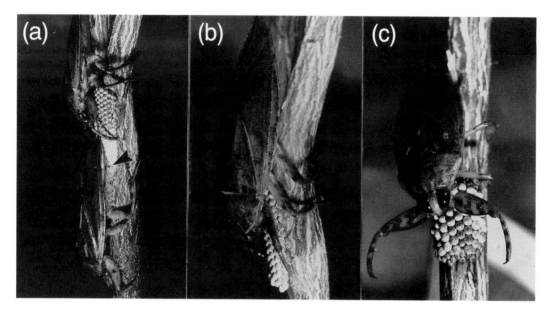

Fig. 6-2. (a) Reproductive pair of *Lethocerus medius*. Male above eggs in the 'watering position', female below ovipositing. Arrow points to oviposition foam produced by the female. (b) Brooding male *Lethocerus medius* in watering position, watering eggs. (c) Brooding male *Lethocerus medius* in defensive display directed toward the photographer.

Ichikawa (1989a) performed a series of 'male removal' experiments that reveal much about male and female reciprocal benefits in the evolution of emergent-brooding. When he removed males any time during the alternating bouts of copulation and oviposition that take place out of the water, the females ($n = 5$) always stopped laying eggs within five minutes and each abandoned her partial clutch if a male was not returned to her within about an hour. This observation suggests that females of *L. deyrollei*, at least, do not brood their own eggs and have no conditional tactic to do so in the event of their abandonment by a mate. This important finding suggests that historically, females may never have brooded (see below).

Emergent-brooding

Prior to the published work of Ichikawa (1988) there were only hints in the literature from which one might infer lethocerine brooding. It was known only that *Lethocerus* spp. eggs are laid in large compact, single-layered clusters on stems of emergent vegetation (e.g. cattails, papyrus, rushes, rice and other aquatic graminoids) or tree branches (dead or alive) that project into aquatic habitats (Hoffman 1924; Hungerford 1925; Rankin 1935; Tawfik 1969; Cullen 1969; Menke 1979a). Rarely, eggs have been deposited on other objects out of the water (De Carlo 1962; Smith and

Larsen 1993). Weed and Parker (in Hoffman 1924) in separate anecdotes reported finding adult *L. americanus* associated with, and apparently 'defending', their eggs. In addition, a single specimen of *L. americanus* in the insect collection of the University of Michigan Biological Station at Douglas Lake bears the note: 'guarding eggs, female?'. This specimen was a male (R. L. Smith, personal observation).

Ichikawa (1988) published the first detailed account of emergent-brooding in a lethocerine bug in the laboratory. He found that *L. deyrollei* males remain on the emergent substrate to which eggs are attached until they hatch. Brooding consisted of the male bug irregularly ascending from the water to the eggs, then positioning his beak in the interstices between an upper row of eggs such that water carried on the bug's body flows from the body down the beak and into the clutch of eggs (Fig. 6-2b). N. Ichikawa (personal communication) and I both now believe that *Lethocerus* spp. males also imbibe water, which is regurgitated onto the eggs. This is an especially important mode of irrigation for *L. medius* in the Sonoran Desert where temperatures are high and relative humidities are low such that, by the time a bug reaches his eggs, his body is almost dry. During some bouts of watering, the bug may change the position of his beak one or more times, presumably to assure that all eggs are wetted. When the bug is not watering eggs, he typically rests on the emergent substrate below the water's surface, head down, with his posterior near the surface so that the air straps can be extended for breathing atmospheric air. Smith and Larsen (1993) studied *L. medius* in ponds of the Altar Valley in southern Arizona near Tucson and found that males always attended wooden laths upon which eggs had been laid. During survey work on *L. medius* I occasionally observed males delivering water to their eggs in a manner identical to that of *L. deyrollei*. Later, extended observations on *L. medius* revealed that males of this species water eggs throughout the day at frequencies up to once every 0.5 h. Males spent three to five minutes watering eggs before descending into the water. At night *L. medius* males watered eggs less frequently than during the daylight hours, but remained on the eggs for much longer (often >30 min) periods of time. This pattern of watering differs from that reported by Ichikawa (1988), who observed that *L. deyrollei* males remained below the surface during the daylight hours and brooded most regularly at night. However, Dr Ichikawa has occasionally observed *L. deyrollei* males brooding in rice fields during the day and now believes the infrequent daylight brooding he initially reported for this species may be a laboratory artifact (N. Ichikawa, personal communication). Rogelio Macas-Ordóñez observed one male *L. colossicus* brood in a pattern very different from either of the other two species. His observations were conducted in a laboratory near Cancun, Mexico, where the insects were collected. He observed that the male remained out of the water on the eggs for most of the day, descending into the water only occasionally and remaining under the water for only 2–35 min. Ichikawa found that *L. deyrollei* eggs required from five to nine days for development to hatching at temperatures ranging from 23 to 31 °C, and I have observed that those of *L. medius* require from three to four days to develop under temperatures ranging from 30 to 38 °C. All *L. medius* males ($n = 11$) always continued to brood their eggs until all nymphs had hatched. Hatching always took place at night (R. L. Smith, unpublished). N. Ichikawa (personal commication) believes brooding males of *L. deyrollei* may remain in the water to defend their young hatchlings until they have dispersed. The complexity and variety in brooding patterns shown among the few *Lethocerus* spp. studied clearly suggests the need for more intensive work on previously studied species as well as a survey of brooding patterns in all members of this genus.

Ichikawa (1988) performed a series of experiments to determine whether eggs of *L. deyrollei* could hatch unattended by a male bug. He submersed some clutches at a depth of 10–15 cm below the surface of the water, and placed some eggs 10 cm above the surface of the water; all failed to hatch. Similar experiments were performed on *L. medius* eggs with identical results. Smith and Larsen (1993) also found that unattended eggs in the field always failed to hatch. If field-laid eggs of *L. medius* were submersed in the pond after rainstorms they were always abandoned by the brooding male and none ever hatched. Cullen (1969) provides data on the height above water at which *L. maximus* laid in the dry season and the wet season in Trinidad. From Cullen's data I calculate a mean of 4.63 (± 1.25 SD) cm during the dry season and a mean of 29.42 (± 5.16 SD) cm during the wet season. Elevating eggs during the wet season is apparently an adaptation to protect the clutch against its being submersed and drowned. Ichikawa has used a gauze wick to keep eggs of *L. deyrollei* moistened but not submersed, and I have dripped water on eggs of *L. medius* suspended on a wire screen, to keep them hydrated without their being submersed. Both of these procedures resulted in a high percentage hatch of each species' eggs.

Smith and Larsen (1993) observed that *L. medius* males resting on their eggs sometimes defended them by displaying aggressively towards a threatening approach, in this case the extended hand of the investigator. The aggressive display consisted of the bug elevating itself slightly above the substrate and widely spreading the raptorial front legs in the plane of the body (Fig. 6-2c). Threatened bugs did not hesitate to grab anything that was brought within their reach. N. Ichikawa, and R. Macías-Ordóñez have observed similar threat displays in *L. deyrollei* and *L. colossicus*, respectively (personal communication).

In the summer of 1993, I found that 'wiggling' an egg-bearing stick attended by a submersed male often caused the male to ascend the stick and defend the eggs as described above. This response was less predictable in males who had been in attendance of eggs for less than two days, but increased with males who had brooded longer. Later in the season, I discovered that tapping the egg-bearing stick (to simulate a pecking bird) was a much more reliable stimulus. In fact, tapping never failed to cause six brooding bugs to ascend their sticks in more than 30 trials of the procedure (R. L. Smith, unpublished). Thus it seems that defense of eggs against predation (probably by birds) is a component of lethocerine paternal care behavior (R. L. Smith, unpublished) Also, *L. medius* males sometimes appear to shade their eggs by covering them with their bodies during the daylight hours (Smith and Larsen 1993).

Horvathiniinae

Horvathinia is truly an enigma. Members of this genus are morphologically quite different from all others in the family; no immatures of any species have ever been seen, their natural habitat is unknown, and the only observations of their behavior are those disclosed below. Specimens are known from the northeastern provinces of Argentina, from Uruguay, Paraguay, Bolivia, and from the southeastern provinces of Brazil. The genoholotype, *Horvathinia pelocoroides* Montandon, was described in 1911 and the type and one paratype reside in the Hungarian National Museum. Prior to 1981, only nine other specimens in this genus were held in museums (all in Argentina) and all nine specimens had been described, each as a different species, based on a few superficial characters (De Carlo 1958; Schnack 1976). Apparently all nine of the specimens in Argentina museums had been taken at lights.

In March 1981, Dr Juan Alberto Schnack collected 55 specimens of a single *Horvathinia* species, during one week, at street lights in Torres, Rio Grande do Sul, Brazil. Thirty-five of these specimens survived to be taken alive to the Instituto de Limnologa 'Dr Raúl A. Ringuelet', at the Universidad Nacional de La Plata, Argentina. At my urging, and following some of my suggestions, a student in the Instituto, Sr Eduardo Domici, set up 8 male–female pairs in aquaria (28 cm × 15 cm × 17 cm) containing sand, stones, water and *Elodea*. The sand was fashioned to produce a 'shore' at one end of the aquarium and a depression filled with water at the other end. Stones were placed on the shore, and stones and *Elodea* were placed in the water. The bugs were offered small fish (*Cnesterodon decemmaculatus*), but none was ever observed feeding on a fish. Sr Domici casually observed the bugs from early February through the middle of March 1981. Domici (*in litt.* 1981) observed that the *Horvathinia* swam in the water, but with less facility than is generally expected of belostomatids. He also noted that, for much of the time, individuals remained hidden among stones in the water, or out of the water among the stones on the experimental shore. One individual was also observed to burrow into the sand. The most significant of Sr Domici's findings was that females laid eggs half-buried in moist sand out of the water. Usually eggs were laid in groups of 3–5, but occasionally larger groups ranging from 15 to 22 were laid. Unfortunately, none of these eggs hatched, suggesting either that they had not been fertilized (no coupling of the pairs was observed) or that moist sand was not a suitable substrate for embryonic development in this species. Freshly laid eggs had mean lengths of 2.74 ± 0.06 mm and mean widths of 1.32 ± 0.06 mm in width ($n = 20$). The eggs were yellow-greenish with grayish markings. Such markings are found on the eggs of *Lethocerus* spp., but never on those of the Belostomatinae. *Horvathinia* eggs had no unusual chorionic structures such as the respiratory horns found in the Nepidae (see below). Significantly, none of the *Horvathinia* females ever attached eggs to the backs of males. Also, no female of any other belostomatid species has been observed to deposit her eggs in sand or soil out of the water. However, many species of water scorpions, for which there are data, always lay their eggs in mud on the shores of their aquatic habitats (Menke 1979b). Our dearth of knowledge about the ecology and behavior of *Horvathinia* represents a major impediment to reconstructing the history of paternal brooding-state changes in the Belostomatidae.

Belostomatinae

General

This subfamily is worldwide in distribution but none of its genera are cosmopolitan. Considerable variation in size and external morphology exists among genera and among species in *Belostoma* and *Diplonychus*. The smallest belostomatines likewise belong to the New World genus *Belostoma* (*ca.* 9 mm) and the Old World genus *Diplonychus* (*ca.* 12 mm). The largest bugs in the subfamily belong to the genus *Hydrocyrius* whose species range in length from 42 to 70 mm. All members of the New World genus *Abedus* are found in streams (Menke 1960); Poisson (1949) reports *Hydrocyrius* from African rivers. The rest of the Belostomatines are limited to lentic waters, varying in size from small puddles to the margins of large lakes. *Limnogeton*, with its non-swimming middle and hind legs, and non-raptorial but highly dexterous front legs, is the most divergent genus in this subfamily. Females of all belostomatine bugs lay eggs on the backs of their mates, where they remain and are brooded until eclosion. With the exception of *Limnogeton*, all species of belostomatines are generalist predators on aquatic animals of all kinds. The smallest species feed on small prey including midge larvae and amphipods (McPherson and Packauskas 1986) and dragonfly larvae (Ichikawa 1989b; Okada and Nakasuji 1993), but all species take vertebrate prey of appropriate size. Members of the genus *Limnogeton* are snail specialists.

Courtship, mating and oviposition

Courtship, mating and oviposition in belostomatine genera are functionally identical to those of *Lethocerus* spp., the differences being that, in the belostomatines, the whole sequence takes place under water and eggs are laid on the male's back rather than on emergent vegetation. Courtship begins with display-pumping or push-ups. This is apparently homologous to the lethocerine pattern, and likewise involves longitudinal up-and-down body movement produced by flexure of the mesothoracic legs while the hind legs anchor the bug. Display pumping is usually performed at or near the surface of the water such that it produces both surface and subsurface wave motion. The pumping display occurs in response to the presence of another bug, but is sometimes produced spontaneously. This pattern was not observed in *Limnogeton fieberi* Mayr, but Voelker (1968) may not have witnessed a full courtship and mating sequence and may therefore have missed seeing this behavior. Kopelke (1982) observed the pattern in *Hydrocyrius columbiae* Spinola, and Lee *et al.* (1970) and Böttger (1974) report display-pumping in *Diplonychus esakii* Miyamoto and Lee and *Diplonychus grassei* Poisson, respectively. I have seen it performed in *Belostoma flumineum* Say (R. L. Smith, unpublished). Schnack *et al.* (1990) observed display-pumping in *Belostoma oxyurum* (Dufour) and the pattern has been studied extensively in *Abedus herberti* Hidalgo (Smith 1979a) and in *Abedus indentatus* (Haldeman) (Kraus 1989a). Females are attracted from up to 30 cm (in *A. indentatus* (Kraus 1989a)) to the wave motion produced by display-pumping. An acoustical signal of unknown origin accompanies display-pumping in *A. herberti* (Smith 1982) and *A. indentatus* (W. F. Kraus and R. L. Smith, unpublished data). When the female reaches the male, there is often some form of sparring with the front legs of both sexes. This is followed by repeated female attempts to oviposit on the male's back, but the male always resists and attempts to initiate copulation. Only after one or more (usually several) couplings does the male stand still for the deposition of the first egg. Oviposition begins at the apex of the hemelytra and proceeds forward, uniformly covering the entire dorsal surface of the folded hemelytra. As in *Lethocerus* spp., egg-laying is regularly interrupted by the male for additional copulations. These interruptions take the form of the male moving repeatedly, display-pumping, or both, to deprive the female of a stable platform upon which to oviposit. Pairs are reported to remain in copula for from 20 s to 20 min depending on species and conditions (Voelker 1968; Böttger 1974; Smith 1979a; Kopelke 1982; Schnack *et* *al.* 1990). Uncoupling is a cooperative effort. When the male is ready to uncouple, he signals the female by scrubbing her back with his hind leg, then both sexes move to disengage (Smith 1979a,b; Kopelke 1982). Alternating copulation and oviposition has been observed in *Limnogeton fieberi* (Voelker 1968), *H. columbiae* (Kopelke 1982), *D. grassei* (Böttger 1974), *Diplonychus annulatum* (Fabricius) (Saha and Raut 1992), *Diplonychus major* Esaki (Ichikawa 1989b), *B. flumineum* (Kruse 1990), *Belostoma thomasi* Lauck and *Belostoma ellipticum* Latreille (Kopelke 1980), *A. herberti* (Smith 1979a,b) and *A. indentatus* (Kraus 1989b). The presence of this important pattern among species in all of the belostomatine genera suggest that it is probably ubiquitous. The time permitted for oviposition between bouts of copulation allows between one and four eggs to be laid in the species for which there are data (Böttger 1974; Kopelke 1982; Saha and Raut 1992; Smith 1979a,b). The time taken for females of various species to deposit their full complement of eggs ranged from 6 to 36 h (Böttger 1974; Saha and Raut 1992; Kopelke 1982; Smith 1979b).

Back-brooding

Brooding behavior by male belostomatine bugs consists of three basic patterns; one is found in all species, one is widely expressed, and one may be restricted to two genera. All of these patterns function to aerate eggs attached to the males' backs. A simple and universal pattern, 'surface-brooding', consists of exposing eggs horizontally at the air–water interface for protracted periods. This keeps water in the interstices among the eggs while exposing their apices to air. This behavior seems ideal for embryonic respiration, but places both brooder and eggs at high risk of predation. Representatives of all belostomatine genera perform surface-brooding (Voelker 1968; Kopelke 1982; Venkatesan 1983; Smith 1976a; Jawale and Ranade 1988). Females and unencumbered males never assume this position. K. C. Kruse (personal communication) has observed that brooding *B. flumineum* males are differentially captured by fishing spiders, and Scott Sakaluk and I (unpublished) have accumulated field data which suggest that brooding males of *A. herberti* experience differentially higher predation than females and unencumbered males in mountain streams and rock plunge-basin pools in Arizona.

The second most common brooding pattern in the belostomatines is 'brood-pumping'. This active behavior is performed below the surface of the water and is identical to display-pumping except that brood-pumping is expressed at a rate of less than one pump per second, much slower than the display. Species representing all genera but *Belostoma* have been observed to brood-pump (Voelker 1968; Kopelke 1982; Venkatesan 1983; Smith 1976a). This pattern is apparently a modification of display-pumping, which is plesiomorphic for the family; thus emergent-brooders may have been exapted for this form of subsurface back-brooding.

'Brood-stroking', a pattern that consists of the bug using its hind legs to rhythmically brush and circulate water over the egg clutch while the brooder is below the surface, has been observed in *Diplonychus indicus* Venk. and Rao and *B. flumineum* (Venkatesan 1983; Smith 1976b), but not in any of the other genera. *Abedus herberti* and *A. indentatus* definitely do not brood-stroke, but these species and *B. flumineum* among others occasionally pat their egg clutch with the hind legs. Egg-patting is assumed to provide tactile feedback to the brooder that his egg pad is still attached and intact. I have shown (though not conclusively) that this pattern serves to maintain males in 'broody' condition (Smith 1976a,b). It seems that brood-pumping and perhaps brood-stroking may aid eclosing nymphs in freeing themselves from their egg-shells. After all nymphs have hatched from the clutch, it is typical for a male to kick the spent egg pad off of his back (Smith 1976a,b). The ability to kick off an egg pad gives brooding males the option to abort, which they routinely do when stressed under laboratory conditions (Smith 1976b).

Brooded eggs in all belostomatine species that have been studied have high survival to hatching (typically >95%), but as in the case of *Lethocerus* eggs, belostomatine eggs cannot survive if removed from the male's back and left unattended in open air or completely submersed in water (Cullen 1969; Kopelke 1982; Venkatesan 1983; Smith 1976a,b). To the extent that brooding males succeed in avoiding their predators, so too do they defend the eggs that are attached to their backs. Surface-brooding males of many *Belostoma* and *Abedus* spp. always dive below the surface in response to movement of objects above them (R. L. Smith and S. K. Sakaluk, unpublished).

SPECIAL TOPICS

The following 'special topics' are assembled here, set apart from the preceding overview of giant water bug biology, because these issues have special import in reconstructing the historical evolution of paternal brooding.

Snails and vertebrates as prey

A key element in my scenario for the evolution of paternal brooding is the large size of giant water bugs compared with other aquatic insects. I argue that the evolution of great size occurred under selection to facilitate the handling of progressively larger and larger aquatic vertebrate prey. However, the acquisition of dexterous grasping front legs may have evolved earlier to permit the use of snails as prey by small bugs deep in the belostomatid lineage. A review of the literature on the predatory habits of modern belostomatids is offered below in support of these hypotheses.

Giant water bugs are the only insects that, as both immatures and adults, routinely prey on freshwater gastropods (the larvae of marsh flies (Sciomyzidae) are predatory and parasitic on snails). *Limnogeton* spp. are obligate snail predators. Tawfik *et al.* (1978) challenged *L. fieberi* with a variety of potential prey including various aquatic insects, tadpoles, and fish, all of which are readily taken by belostomatid species in the other genera (see below). However, this array of items was rejected by *Limnogeton* spp. in favor of a monophagous diet of snails. *Limnogeton* seem to recognize their prey when the mollusks are creeping very slowly and even when they are still. *Limnogeton fieberi* approaches snails with its beak extended. The stylets are then inserted into the body of the snail, causing it to release its grip if attached to a plant or other substrate. The bug then grasps the shell with its forelegs, and rotates it until it is held cup-like so the beak can be inserted. The bug then liquifies and imbibes the snail's tissues (Tawfik *et al.* 1978). Voelker (1968) provided a detailed quantitative account of predation by *L. fieberi* on snails belonging to nine genera. *Limnogeton fieberi* was also fed members of three snail genera by Miller (1961).

Other belostomatids, including those in the genera *Abedus* and *Belostoma* (R. L. Smith, unpublished), have occasionally been observed to eat snails; and *D. annulatum* routinely ate them in the laboratory (Saha and Raut 1992). Other *Diplonychus* spp., including *Diplonychus urniator* Dufour (Tawfik *et al.* 1978) and *Diplonychus japonicus* Vuillefroy, were observed to prey on snails in the field. *Diplonychus japonicus* preyed primarily on Lymnaeidae and Physidae snails in small ponds (Okada and Nakasuji 1993). Cullen (1969) observed that *Belostoma malkini* Lauck 'thrived particularly well upon the gastropod *Marisa* sp.' I have observed *A. herberti* nymphs feeding on *Physa* sp. snails in Arizona streams and Harvey (1907) notes that

Table 6-1. *Categories of prey taken by giant water bug genera*

Cells containing the word 'yes' indicate published record or reliable unpublished data that prey category is taken by a representative(s) of the indicated genus. Cells containing the word 'no' indicate that members of the genus refused prey of this category. Cells containing a question mark indicate that no data exist for the genus and prey category. Cells containing the word 'no' enclosed in parentheses indicate no data, but a low probability that the item would be taken by members of the genus.

Genus	Snails	Crustacea	Insects	Fish	Larval Amphib.	Adult Amphib.	Snakes	Birds
Lethocerus	?[a]	?	yes	yes	yes	yes	yes	yes
Horvathinia	?	?	?	(no)[b]	?	?	(no)[c]	(no)[c]
Limnogeton	yes	no	no	no	no	no	(no)[d]	(no)[d]
Hydrocyrius	?	?	yes	yes	yes	yes	?	yes
Diplonychus	yes	yes	yes	yes	yes	yes	(no)[c]	(no)[c]
Belostoma	yes	yes	yes	yes	yes	yes	?	?
Abedus	yes	yes	yes	yes	yes	yes	(no)[c]	(no)[c]

[a] One unpublished report states that *L. indicus* takes snails, but N. Ichikawa (personal communication) notes that *L. deyrollei* do not favor snails.
[b] Fish offered to *Horvathinia* sp. in laboratory, but not taken (E. Domici, personal communication)
[c] All members of the genus are deemed too small to take the indicated prey.
[d] *Limnogeton* spp. are snail specialists; they reject all other categories of prey except snails.
Sources: Matheson 1907; Brockelman 1969; Heyer and Belin 1973; Heyer *et al.* 1975; Smith and Larsen 1993; De Carlo 1962; Cullen 1969; Demmock (in Rankin 1935); Ichikawa 1989a,b; Miller 1961; Saha and Raut 1992; Jawale and Ranade 1990; Venkatesan 1983; Schnack *et al.* 1990; Harvey 1907.

A. indentatus fed on snails in an aquarium. N. Ichikawa (personal communication) finds that *L. deyrollei* does not favor snails; however, I am aware of an unconfirmed report that *L. indicus* consumes snails of medical importance in Vietnam.

The existence of a belostomatid specialist on snails among a group of aggressive polyphagous predators seems remarkable, but apparently most if not all belostomatids, perhaps including members of the earliest derivative genus, are able to recognize and use gastropods as prey. It seems that recognition is by general form, since gastropods lack the sudden movements that normally release predatory attacks by belostomatids. In addition the uniqueness of snail-feeding by belostomatids hints that this may be an old habit retained through the radiation of the family. If so, snail predation may have selected for the extraordinary range of foreleg movements exhibited by the giant water bugs, which clearly contribute to their proficiency in handling vertebrate prey.

The giant water bugs, some of the water scorpions, dragonfly larvae and, rarely, tabanid larvae (Jackman *et al.* 1983) are the only insects that routinely kill vertebrates for food. Prey capture in *Lethocerus* bugs is by ambush; attack is released by movement of the prey. A dead fish (or other prey normally taken alive by belostomatids) presented to *Lethocerus* spp., if gradually introduced to the bug, elicits no reaction; however, when moved abruptly, the dead offering is instantly seized (Cullen 1969; Smith 1975, R. L. Smith, unpublished). Three studies have focussed on belostomatine predatory behavior. Victor and Ugwoke (1987) explored the relationship of prey size to detection, handling, and feeding times in *Diplonychus nepoides* Fabricius. This study demonstrated that *D. nepoides* will attack prey twice its own size. Kehr and Schnack (1991) examined tadpole (*Bufo arenarum*) prey size and density on rates of predation by *Belostoma oxyurum*. Although ambush is clearly the primary predatory strategy employed by belostomatines, Cloarec (1991, 1992) has shown that *D. indicus* may actively pursue, as well as ambush, prey.

Table 6-1 provides a checklist of belostomatid genera and their prey. Representatives of all belostomatid genera except *Limnogeton* and *Horvathinia* have been reported to take vertebrates, but the largest giant water bugs in the genera *Lethocerus* and *Hydrocyrius* are vertebrate specialists (Fig. 6-3); the predatory feats of these huge insects are legend. Published anecdotes record giant water bugs

Fig. 6-3. A *Lethocerus deyrollei* feeding on a frog that it has captured and immobilized by injecting it with potent saliva by means of its beak (at arrow). Note that the frog is substantially larger than the insect. (Photo. by N. Ichikawa.)

and *Hydrocyrius*), the dexterity and vice-like grip of their raptorial front legs, and the lightning speed with which they grasp their prey. Also remarkable is the fearless tenacity of these bugs once they have grabbed a prey of any size, including the occasional careless biologist. It is accurate to say that lethocerine bugs have adopted the obdurate policy of not voluntarily releasing seized prey of any size. No amount of jumping, thrashing, writhing, shaking, swimming or even flying by a victim will cause the bug to release its grasp. Nor can any maneuver long prevent puncture of the victim by the bug's stylets, and the injection of toxic saliva. Saliva in *Lethocerus* and other belostomatids is known to contain powerful proteolytic and hemolytic enzymes and heart-stopping neurotoxins (Dan *et al.* 1993; Rees and Offord 1968; De Carlo 1959; Picado 1937, 1939). Large prey are quickly subdued by this venom. Although some work has been done on the pharmacological effects of giant water bug saliva, most fractions have yet to be characterized and their physiological modes of action on vertebrate nervous and other tissues are in need of further elucidation. It seems possible that belostomatid saliva may contain more powerful natural anesthetic and neurotoxic agents than have yet been discovered; as will be disclosed later, lethocerine saliva was probably refined on the tissues of fish and amphibians beginning at least 150 million years before present.

Clearly, giant water bugs have evolved a variety of adaptations to capture, subdue, and consume large vertebrate prey, and they are nearly unique in this habit among insects. Large size, among all of the attributes that facilitate vertebrate prey capture, indirectly created the circumstances that led to selection for parental care of eggs.

Nutrition, enhanced fecundity and who broods

The 'enhanced fecundity hypothesis', originally proposed for birds (Graul 1973; Nethersole-Thompson 1973) and fish (Emlen 1973; Perrone and Zaret 1979) and accounted for in Maynard Smith's (1977) model, proposes that paternal care has evolved as a sexually selected male trait that frees females for maximal nutrient accrual and egg production. Put another way, if one of the costs of brooding for females is a significant diminution in rate of nutrient accrual, with a consequent reduction in lifetime production of eggs, then females may be selected to prefer males willing to invest in care. Implied in this hypothesis is that males, who require only sufficient nutrition for maintenance,

having taken prey many times the bugs' size and mass, including 'foot long' trout and water snakes, bullfrogs, wading birds (see Menke 1979a), and even a hapless woodpecker that was apparently grabbed by a bug when it waded (on too short legs) into a pond for a drink (Matheson 1907). Lethocerines are known to be the most important predators on frog larvae (Brockelman 1969; Heyer and Belin 1973; Heyer *et al.* 1975; see also Brodie *et al.* 1978), and tadpoles of desert toads are believed to be a critical item in the diet of *L. medius* (Smith and Larsen 1993).

Among the attributes of giant water bugs that permit them to take vertebrate prey, one is instantly impressed with their great size (especially in the genera *Lethocerus* may invest at less cost than females. Tallamy (1995) examined the arthropod literature to test this hypothesis and found that belostomatids may provide an exceptionally good example of the enhanced fecundity hypothesis. This hypothesis would seem to be operative if males are shown to require substantially less food than females, if predatory efficiency is reduced by brooding, and if number of eggs produced increases with enhanced nutrition.

What is the evidence that this theory applies to giant water bugs? Böttger (1974) fed adult *D. grassei* chironomid larvae exclusively and observed that after oviposition females gorged, consuming up to 18 larvae per day compared with the adult male consumption of 3–5 larvae per day. Voelker (1968) found that adult female *L. fieberi* consumed from 2 to 3.5 times as many same-sized snails as unencumbered males of that species. He also noted that males encumbered with eggs substantially reduced feeding, or ceased to feed altogether because of the incompatibility of brooding with prey acquisition and feeding. In addition, Smith (1976b) observed that brooding *A. herberti* males were inhibited from feeding on prey in the size class of first-instar nymphs. This inhibition may minimize the risk of males cannibalizing their own hatchlings. Ichikawa (1988) set up a laboratory experiment where brooding males and their recent mates were given equal opportunity to prey on goldfish. He found that females killed from five to nine times as many fish as males over a seven to nine day period; in the extreme case, one female killed and consumed 30 fish in nine days. Crowl and Alexander (1989) showed that back-brooding behavior significantly decreased the efficiency of prey capture in *B. flumineum* males. In predation experiments, *B. oxyurum* adult females consumed four times as many tadpoles of the same size class as did adult males (Kehr and Schnack 1991).

Finally, R. L. Smith and D. W. Tallamy (unpublished) designed an experiment to compare the effects of food limitation on fitness of male and female *B. flumineum*. We fed one group of males and females daily to simulate bugs free to hunt at will; the other treatment group was fed at 10 day intervals, simulating the feeding rates established for males encumbered with eggs. Each group received one prey item at its designated treatment interval. Treatment males and females were paired with well-fed mates every other day until eggs were laid. We kept an accounting of the fitness of both sexes in the two treatments, with the following dramatic results. There was no significant difference between the number of offspring sired by males fed daily and by those fed only once every ten days. However,

the well-fed females laid 3.5 times as many viable eggs as did females fed once every ten days. We believe this experiment clearly demonstrates the disparate impact in loss of fitness that would be experienced by females if they, instead of males, provided care in the Belostomatidae. The female cost could be even higher in the emergent-brooders (Lethocerinae) who would have to service clutches of eggs out of the water, away from prey, for much of the time while brooding and who could not change locations if prey availability were poor in the vicinity of the brooded eggs. In addition, lethocerine brooders cannot hunt actively very far from the clutch, whereas back-brooders are nearly always in the water and by comparison are quite portable.

Paternity assurance

Belostomatid mating behavior always involves short bouts of oviposition repeatedly interrupted by the male for coupling such that no more than four eggs are laid without an intervening copulation. I have shown that male-controlled alternation of copulation and oviposition in *A. herberti* assures brooders almost perfect certainty of paternity of eggs that are placed on their backs despite most females in natural populations having been previously mated such that they usually contain stored sperm from a previous mate. I also demonstrated that stored sperm from a previous mate poses a real threat of cuckoldry to a current mate (Smith 1979b). An ejaculatory 'ductectomized' male was mated to a female previously mated to a male homozygous for a dominant genetic marker. The offspring in this experiment, although hatched from eggs brooded by the current male, all exhibited the genetic label from the female's previous mate (Fig. 6-4). It is assumed, but has not yet been demonstrated, that each copulation in the alternating series displaces sperm from previous mate(s) to the blind end of the female's spermatheca, making it unavailable to fertilize eggs. Alternatively, the multiple ejaculates may simply swamp the female reproductive tract with an overwhelming numerical advantage for the current male's sperm.

Whatever the actual mechanism of sperm precedence, it is clear that alternating bouts of copulation with oviposition is the behavioral adaptation that assures paternity in *A. herberti*. Every species for which there are data, including representatives of all genera in the back-brooders and three species of emergent-brooders, exhibit alternating copulation–oviposition. Apparently this male behavior is

Fig. 6-4. A brooding male *A. herberti*. This male was ejaculatory-ductectomized to prevent him from transferring sperm during alternating bouts of copulation and oviposition. His mate had previously mated with a genetically marked male; the stripe on the first hatched nymph's back identifies its father as the marked male, not the brooder. Thus, the brooder has been made an experimental cuckold. (Photo. reprinted with permission from: Smith, R.L. 1979. Repeated copulation and sperm precedence: Paternity assurance for a male brooding water bug. *Science* **205**: 1029–1031. Copyright © 1979 American Association for the Advancement of Science).

ancient, perhaps having originated before emergent-brooding as a mate-guarding pattern (see Alcock 1994). The high confidence of paternity accorded by alternating oviposition and copulation may well have provided an ancestral exaptation to male care.

Multi-clutch brooding and the oocide paradox

In some species of lethocerine water bug, two or more females may mate sequentially with a single male and deposit multiple adjacent clutches to be brooded by the polygynous father. In one and perhaps two other species, it has been observed that gravid females who encounter a

brooding male may fight with him, cannibalize his eggs, then mate with him and substitute their own eggs for the ravaged clutch. Accounts of these paradoxically antithetical patterns are presented below, followed by comments on what additional research is needed to more fully understand the phenomenon and its implications for the evolution of paternal brooding.

In 1935, Rankin published a life history and some field observations on *L. americanus*. He worked on this species at the University of Michigan, Douglas Lake Biological Field Station, where he collected and photographed eggs of *L. americanus* from a shoreline pond on the lake and from sandpit ponds around the nearby town of Pellston, Michigan. Rankin collected nine groups of eggs, mostly from *Typha* stalks. Of these, only two contained a single clutch of eggs. All others contained from two to eight clutches of eggs. Rankin did not observe brooding, probably because he was too eager to collect the egg masses for his laboratory studies, but it is apparent (to me) from his photographs (contained in the 1935 publication) that the eggs in multiple clutches were viable. This would not have been the case if they had been abandoned or were inadequately cared for by their fathers.

In the laboratory, Rogelio Macas-Ordóñez (personal communication) provided an *L. americanus* male with two gravid females in succession. The first female mated and deposited a clutch of eggs. This was followed by the second female, who likewise mated and deposited her eggs below the existing clutch. The male then brooded both clutches of eggs until they hatched. Smith and Larsen (1993) discovered a double clutch of *L. medius* eggs attended by a single male in an Arizona pond (Fig. 6-5a). The double clutch was taken from the field and brooded artificially in the laboratory where most eggs (>75%) in both clutches hatched. The onset of hatching of one clutch lagged by about one day behind, the first hatching from the other clutch, suggesting that the two clutches had been deposited on separate nights. These accounts lead me to conclude that at least two *Lethocerus* spp. have the capability to successfully acquire and brood multiple clutches.

In contrast with the preceding accounts of benign female behavior and multiclutch brooding are dramatic patterns of female cannibalism and destruction of non-filial eggs in other species. Cullen (1969) studied the life history of *L. maximus* in Trinidad. He kept bugs in large outdoor cage-covered concrete tanks that contained sedges. On one occasion, Cullen observed a female climbing a sedge that contained a clutch of eggs. When she

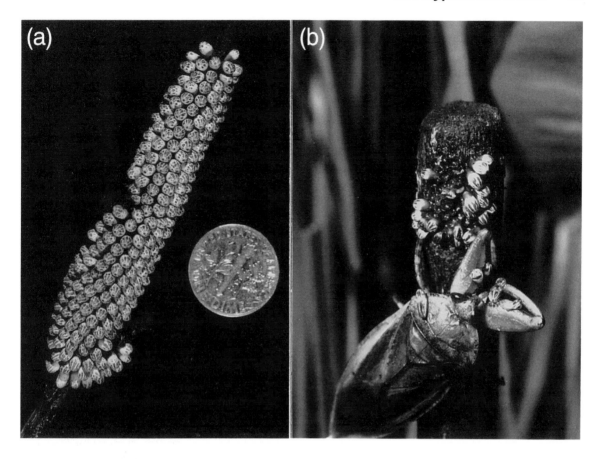

Fig. 6-5. The paradox of multiclutch shared care and oocide behavior. (a) This double clutch of eggs of *Lethocerus medius* was brooded by a single male. (b) *Lethocerus deyrollei* female attacking and cannibalizing a clutch of eggs from another female. Sometimes a brooding male is successful in defending against an oocidal female, but if not, the male mates with the marauder, then broods her eggs, which she substitutes for the clutch she has destroyed. (Photo. 5(b) by N. Ichikawa.)

reached the eggs, she began sucking them and continued until all had been consumed. Cullen returned the next day to find that the same female had replaced the cannibalized clutch with her own eggs. It should be noted that lethocerine paternal care had not been discovered at the time of Cullen's observations and so he was unaware of emergent-brooding patterns.

Ichikawa (1990) provides enthralling accounts of egg-clutch destruction in the Japanese water bug *L. deyrollei*. Upon introduction of new females to tanks containing brooding males previously mated to other females, Ichikawa occasionally observed the results of egg cannibalism as Cullen had done over 20 years before. These initial observations prompted Ichikawa to make continuous observations of the phenomenon and to arrange some simple

experiments to determine the condition of females that engaged in this carnage. A generalized summary of his findings follows.

Ichikawa found that only gravid females attacked brooded eggs. When a gravid female first approached a brooding male, the male would attempt to drive her away. However, under laboratory conditions, in all six cases Ichikawa observed that the female managed to circumvent male attempts at defense in order to ascend the guarded stick to his eggs. When an aggressor female reached the eggs, she would begin immediately to suck them and then, using her front legs, she would very deliberately pry and tear the eggs from the substrate (Fig. 6-5b). Bouts of cannibalism alternated with bouts of tearing eggs out of their mucilage so that the expended eggs dropped into the water. After several minutes of attacking eggs, the female typically descended into the water for variable periods, after which she would climb to the clutch again and renew the onslaught until few if any eggs remained attached to the substrate. Male behavior was amazingly ambivalent. Males display-pumped (courted) and copulated with marauding females while attempting to defend the existing clutch against female attacks. Following destruction of the brooded clutch, aggressor females always mated repeatedly with the resident male and replaced the demolished clutch with their own eggs in the manner earlier described and then abandoned them. Males always dutifully brooded the new clutch.

In Ichikawa's aquarium studies, defensive attempts by the resident brooding male always failed to dissuade the aggressor females, who inevitably destroyed the existing clutch. However, in an outdoor study with bugs housed in a large concrete pond containing sedges, Ichikawa found that in three of nine cases males accosted by gravid females were successful in fending them off and saving the brooded eggs; however, in one unusual instance, a female chased a brooding male some distance away from the clutch he was attending. She then mated with the displaced male and laid eggs at the new site, where the male remained to brood the new clutch and hence abandoned his old clutch, which failed to hatch.

The potential for multiclutch brooding eliminated the ultimate cost and principal impediment to selection favoring male investment (i.e. loss of future matings). Thus, as I discuss later, multiclutch brooding and the polygyny it permitted may have been a key element in the origin of male care in *Lethocerus*. If, as suggested from the meager data assembled above, brooding by a single male can assure survival of multiple egg clutches, then males should always be eager for polygynous matings. But why should females ever tolerate multiclutch brooding if they can prevent it? Multiple clutches could result in dilution of care, nymphal competition, and interclutch nymphal cannibalism with a disparate negative impact on the offspring of the newcomer because her eggs will always hatch later than her competitors' eggs. In addition, why should the new arrival forgo the nutritional benefit of eating the previous clutch of eggs if it is possible to do so?

Our current state of knowledge suggests only two explanations for multiclutch brooding in the species that exhibit this behavior. Because egg-clutch destruction is certainly derived, some *Lethocerus* species may simply not have evolved the patterns. Alternatively, males in some species may possess perfected defenses against female attack.

A species survey for multiclutch brooding and oocide in *Lethocerus* together with a phylogenetic analysis of species in the genus could reveal whether egg destruction has had a single or multiple origins. Finally, some species of *Lethocerus* are dimorphic for size, females being larger than males. This raises the possibility that oocidal behavior will be found in those species whose females are substantially larger than the males, but not in species monomorphic for size.

The scent gland riddle

The metathoracic scent gland system (MSGS) is a basic feature of the order Heteroptera (Carayon 1971) where it commonly functions in chemical defense (see Staddon 1986 and references therein). Members of most of the Nepomorpha families possess these glands (Staddon and Thorne 1979), but they are mysteriously absent from all of the back-brooders, and no function has ever been discerned for the exocrine product in emergent-brooders. This section reviews the literature, presents some preliminary data on the MSGS in *L. medius*, and posits a function for the glands and their product that bears on the evolution of emergent-brooding.

Paired MSGS glands were first described in a belostomatid, *L. griseus*, by Leidy (1847). Buntenandt and Tâm (1957) found the glands in *L. indicus* and identified their product and stored contents as *trans*-hex-2-enyl acetate. The existence of this pleasantly aromatic substance had been known and appreciated for over two millennia by the gourmet chefs of Southeast Asia, where it is an essential ingredient in special fare of Thailand, Cambodia, Vietnam and southern China (Tieu 1928; Pemberton 1988). Pattenden & Staddon (1970) found the MSGS in *L. medius* and *Lethocerus cordofanus* Mayr, and analyzed the contents of the glands of *L. cordofanus*. This they found to be the same as that in *L. indicus*. Male *L. maximus* and *L. annulipes* Herrich-Schäffer were dissected by Cullen (1969), who found that both species had the MSGS. R. L. Smith *et al.* (unpublished) identified the principal compound in *L. medius* and *L. americanus*, which we likewise determined to be *trans*-hex-2-enyl acetate. The MSGS is absent in all genera of the back-brooding subfamily Belostomatinae (Staddon 1971) and does not occur in any members of the giant water bugs' sister group, the Nepidae (Staddon and Thorne 1979). As noted in the phylogenetics section of this chapter, presence or absence of the MSGS in *Horvathinia* represents a crucial anatomical clue to the phylogenetic position and reproductive behavior of members of this genus.

Butenandt and Tâm (1957) apparently failed to find glands in females of *L. indicus* and they advanced their belief that the MSGS served a sexual attractant function in this species. Pattenden and Staddon (1970) dissected males and females of *L. indicus*, *L. medius* and *L. cordofanus* and found that females do indeed have the MSGS, but female glands are much smaller than those of males. Male glands in *L. cordofanus* were estimated to hold about 25 times the contents of the female glands by Pattenden and Staddon (1970), but we (R. L. Smith *et al.*, unpublished) used gas chromatography to quantify gland contents in *L. medius* and found that male glands contained just over 10 times more *trans*-hex-2-enyl acetate than the glands of females (mean $9.95\,mg \pm 0.72\,SD$ male; mean $= 0.96\,mg \pm 0.88\,SD$ female; $n = 5$ males, 5 females).

To summarize, seven species of *Lethocerus* definitely have metathoracic glands, and sexual dimorphism for gland size has been shown in five species. Males have glands 10–25 times larger than those of females. Finally, the content of these glands is the same isomer of a single compound in both sexes of four species of *Lethocerus*. Pattenden and Staddon (1970) deduce that sexual dimorphism in glands is characteristic of all members of the genus *Lethocerus*, and that the secretions from these glands are probably neither sex-specific nor species-specific. If this is true, then the glands' exocrine product may not be a sexual pheromone in *Lethocerus* spp. and could not function in species isolation.

So what is the primary function of these glands and their product? Staddon (1971) prophetically observed: '...*a difference in behavior* [emphasis mine] is to be sought between the members of the subfamilies Lethocerinae and Belostomatinae, a difference associated with the presence of metathoracic scent glands in the former and their absence in the latter'. At the time of Staddon's prediction, it was not known that *Lethocerus* spp. males brood eggs attached to emergent vegetation, and emergent-brooding behavior is of course not found in either the Belostomatinae or the Nepidae. Could this be the behavioral difference divined by Staddon?

I posit that the product of the MSGS in lethocerines functions primarily as a labeling and trail pheromone that marks the substrate to which eggs are attached and a path from the water's surface to the clutch. The chemical marks reliably identify egg-bearing substrates and efficiently lead parents to their eggs. The opening of these glands, between the coxae of the hind legs, would be ideal for marking emergent substrate while on the move. The female could

initially mark a trail from water to eggs on her preliminary trips to the oviposition site, and the male could repeatedly mark the trail during his many ascents and descents through several hours of mating and oviposition and while ascending and descending the substrate over the several days he will spend carrying water to the clutch. Thus, the abandoning mother would not require the males' much larger stores of chemical marker. The hypothesis would explain why females possess the MSGS but have glands much smaller than those of males. Loss of the MSGS in the Belostomatinae is likewise accommodated by the trail-marking hypothesis, in that emergent-brooding (and hence any trail-marking function) is supplanted by back-brooding in this lineage.

Is there evidence for chemical marking by *Lethocerus*? My field observations and the results of a series of field experiments to be published elsewhere (R. L. Smith *et al.*, unpublished) provide preliminary support for this hypothesis. I have repeatedly observed that *L. medius* males leave emergent substrate containing eggs, to swim from several centimeters up to 1.5 m away (R. L. Smith, unpublished). After these excursions, males always ($n = 50$ observations on nine brooding males) succeeded in relocating egg-bearing substrate. Although *Lethocerus* spp. eggs are often laid on uncomplicated substrate such as cattails, rice, or papyrus stalks, in Arizona *L. medius* frequently deposits eggs on plants of very complex architecture such as desert broom (*Baccharis* sp.), cocklebur (*Xanthium* sp.) or branches of desert legumes (e.g. *Prosopis* sp.). The male depicted brooding in Fig. 6-6 repeatedly ascended this complicated plant with many branch choices and always found his eggs with only an occasional error. In addition, I observed nine brooding males for over two hours each at night during the 1992 and 1993 seasons. All nine of these males regularly ascended from the water to their eggs with agility and without significant error in the dark when it seems unlikely that they could have been using visual cues to find the clutches.

Ichikawa (1988) found that *L. deyrollei* males were able to find, but unable to distinguish, others' eggs and inevitably brooded alien clutches in laboratory experiments. My field experiments produced the same result for *L. medius* (R. L. Smith *et al.*, unpublished). Japanese workers refer to the MSGA exudate as the 'good smell'. (There is a 'bad smell': contents of the gut sprayed from the anus in a defensive context.) Mr Hashizume (personal communication, through N. Ichikawa) has observed that there is an 'explosion' of good smell during the initiation of courtship by *L. deyrollei*. He further observed that females, but not

Fig. 6-6. A *Lethocerus medius* male brooding eggs (at arrow) on a complicated plant structure, dead cocklebur (*Xanthium* sp.). Despite the complexity of the substrate, the brooding male has no difficulty finding his eggs when he ascends to bring them water. The circle indicates the location of a previously hatched clutch of eggs.

males, seem to respond to the good smell. Additional research on the MSGS may reveal the range of activity of the gland product and is needed to evaluate the trail-marking hypothesis.

ELEMENTS IN AN EVOLUTIONARY SCENARIO

The ancestor

Here, I review characters apparently plesiomorphic for giant water bugs or for aquatic Heteroptera in an effort to reconstruct the common ancestor to the Belostomatidae or the Nepoidea.

All aquatic bugs (excluding some corixids) in the Nepo-morpha are predatory, but all other bugs are much smaller than the smallest extant species in the basal giant water bug genus *Lethocerus*. Furthermore, members of the Belostomatidae and the Nepidae are larger than the other bugs, and members of the largest *Lethocerus* spp. are among the largest of living insects. Thus, it seems reasonable to

conclude that 'small' is plesiomorphic and 'big' is derived in Nepoidea. Representatives of most belostomatine genera feed on snails, which suggests that recognition and use of snails for food may be an old habit with its origin deep in the belostomatid phylogeny. Early nepoid snail predators must have possessed the ability to hold and manipulate snail shells to access the contained animal, and selection for this ability may have exapted the front legs for capture of more challenging prey. All extant species of aquatic bugs other than Nepoidea lay their eggs on or in substrate under water where embryogenesis occurs perfectly with no parental care and without special chorionic respiratory adaptations such as exist in the Nepidae (see below). Adults of most species representing all groups of aquatic Heteroptera possess the ability to disperse by flight and quickly colonize shallow-water habitats including temporary ponds. Most other aquatic Heteroptera have legs adapted for swimming, but some are crawlers.

I infer from the foregoing that the common ancestor of the giant water bugs was a relatively small aquatic insect with dexterous front legs that fed primarily on snails and small arthropods, but had begun to develop a taste for fish and amphibian larvae. Its ability to capture vertebrates was enabled by grasping legs, perhaps originated to hold and manipulate snail shells or other small invertebrates, and toxic (proteolytic) saliva, perhaps initially selected to liquify invertebrate tissues. Very small vertebrate prey could have been grasped dexterously while pierced, injected with saliva, and quickly debilitated by proteolytic enzymes. However this ancestral bug was constrained by its small size to capture only the smallest, youngest aquatic vertebrate prey. This ancestor possessed the ability to disperse by flight and to colonize new aquatic habitats including temporary ponds rich in amphibian larvae, where it deposited its eggs below the surface of the water and abandoned them to develop on their own. It was either a crawler or a swimmer.

Cobben (1968) cites embryological evidence to support an opinion at variance with the infraordinal phylogenetic schemes of China, Parsons, Popov, Rieger and Mahner (see Menke 1979a for references and discussion). He believes the predecessors of the Nepomorpha and hence the Nepoidea were naucorid-like. I cannot evaluate the basis of this view, but naucorids are broad-bodied, have raptorial front legs and metasternal scent glands, are reported to prey on mollusks (Uhler 1884), and appear abundantly as fossils of the Upper Jurassic (Popov 1971; Carpenter 1992). In addition, some extant naucorids from the Austro-Malayan and Indo-Malayan subregions are exceptionally large for non-Nepoidea water bugs (John Polhemus, personal communication). Specimens of an undescribed species from the Truong Son mountain range of Vietnam measure *ca.* 17 mm (Doug Currie, personal communication), almost twice the length of the smallest belostomatids. Virtually no data are available on the biology of these huge creeping water bugs, except that they inhabit swiftly flowing mountain rivers. It seems that cold, rapidly moving water supersaturated with oxygen may sustain development of the relatively large static eggs of these bugs. I cannot ignore the possibility that the belostomatid ancestor might have been a relatively large naucorid-like heteropteran that lived in an oxygen-rich cold-water stream.

The opportunity

The earliest unequivocal belostomatids in the fossil record are from the Upper Jurassic with an undescribed nepoid-like creature from the Upper Triassic (see below). Thus the family probably evolved its many remarkable attributes in the Upper Triassic through Lower Jurassic.

Fresh waters during this period contained ray-finned fish and larval amphibians including those of anurans (which originated in the Triassic) and salamanders (which evolved in the Jurassic). Temporary ponds, shallow backwaters, and the littoral zones of larger and more permanent lentic environments were the primary habitats for amphibian larvae. Here they exploited rich blooms of algae and an array of phyllopod crustacean and aquatic insect prey. Shallow and temporary waters were free from the large predaceous fish that inhabited deeper more permanent aquatic habitats. Significantly, the first undisputed freshwater gastropods are found in the early Jurassic, but their thin-shelled forbears were almost certainly present in the Triassic.

Until the late Jurassic, there were no birds to swoop down upon and swiftly harvest the abundant resources to be found in shallow transient aquatic habitats. However, insects at that time possessed flight, and both dragonflies and the giant water bug ancestor were able to quickly disperse to these habitats and to avail themselves of this ecological opportunity.

Getting bigger

Imagine that a small, say 5–7 mm long, water bug species in the belostomatid lineage has begun to perfect predation on small vertebrate prey. Selection has already strengthened its front femora and the toxic fractions of its saliva. Our hypothetical species' only competitors are dragonfly nymphs who, lacking venom, must chew their prey to death while restraining it with delicate prehensile labial structures. The intricate hydraulic mechanics of prehensile labial prey capture drastically limit the size of vertebrate prey dragonfly nymphs can handle (Westfall and Tennessen 1996). Competition among water bugs (and dragonfly nymphs) belonging to the ancestral population was most intense for the smallest amphibian and fish larvae, so any slightly larger than average individuals enjoyed a selective advantage because of their ability to handle larger, less contested size classes of vertebrate prey. As we move through early giant water bug history, we can envisage repeated allopatric speciation events followed by geographic reunion of parental and descendent species such that approximately equal-sized members of new species were placed in competition for same-sized amphibian and fish larvae. However, there was always 'room at the top', and strong directional selection favored progressively larger individual bugs and thus species, until eventually the process slowed to a stop when evolution collided with a physiological constraint: eggs that were too large to develop unattended, submersed in static water and unprotected against desiccation if laid and abandoned in the open air.

Big bugs, big eggs

Insect development is a step function. Significant increases in the dimensions of developing larvae or nymphs occur only in the moments following each molt. Dyar (1890) observed that the size of the head capsule of lepidopteran larvae increased by a factor of 1.6 per molt. This discovery became known as Dyar's Law. 'Insect growth shows regular geometrical progression in successive instars' is a more general statement of this 'Law'. Cole (1980) reviewed evidence from studies of 105 hemi- and holometabolous insect species and found that the median growth ratios for insects ranged from 1.27 to 1.52. Whatever the ratio of growth for a particular species or structure thereof, Dyar's Law decrees that there are only two ways an insect can become significantly larger than its ancestors: (1) it can add molts (=instars), or (2) it can retain the ancestral number of molts and start life as a larger hatchling (i.e. come from a larger egg). Mayflies hatch from tiny eggs and molt through 12–45 nymphal instars before their fleeting aerial existence (Edmunds and Waltz 1996). Odonata attain large size by hatching from very small eggs and passing though from 10 to 15 instars (Westfall and Tennessen 1996). By contrast, the overwhelming majority of Heteroptera (true bugs) pass through five nymphal instars.

Štys and Davidová-Vilmová (1989) exhaustively reviewed the literature for all 'other-than-five' instar heteropteran ontogenies. Their review validated very few exceptions to the five-instar pattern of development. Most of the verified exceptions involved reductions to four instars. The addition of one instar to produce six has been confirmed in only two species. Five instars is clearly the plesiomorphic condition in Nepomorpha, with only a few reductions to four instars having occurred in the nepid clade. All giant water bugs have only five instars. Thus selection for large bugs inadvertently produced big eggs.

To review the logic, insect growth is a step function governed by Dyar's Law. A collolary to Dyar's Law is that descendant insects can be larger than their ancestors only by adding instars or by starting as a larger first instar, which would necessitate hatching from an egg larger than the ancestral egg. Members of the Nepomorpha including the Belostomatidae were apparently unable to add instars; therefore the only developmental route to larger bugs involved production of larger first instars (eggs). Thus in the Belostomatidae, adult bug size and egg size are inevitably and inextricably coupled: every incremental increase in adult bug size produced a proportional enlargement in the egg.

Any mutation (had it occurred) that added a sixth instar to a giant water bug's ontogeny would have enjoyed immediate success because any adult bug that possessed the mutant gene would have grown (according to Dyar's Law; see Cole 1980) to be *ca.* 30–50% larger than its parents and could therefore have taken much larger (and previously unexploitable) vertebrate prey. However, I calculate that to attain the size of the largest extant giant water bug, starting with a one-millimeter egg (the size of the eggs of some smaller extant belostomatids), (see, for example, Tawfik *et al.* 1978; Böttger 1974) would have required from four to seven additional molts for a total of 9–12 nymphal instars. Apparently, such variation in ontogeny was not available at the critical juncture in belostomatid history and the lack of additional instars upon which selection could have acted surely qualifies as a phylogenetic constraint in this historical context.

Big eggs, big problems

Oxygen diffuses about 324 000 times less rapidly in water than it does in air and its concentration in water is at maximum 15 ppm compared with 200 000 ppm in air (Eriksen *et al.* 1996). These parameters represent a critical ceiling for embryogenesis of any kind of submersed, sessile, unattended egg in static water (R. L. Smith, unpublished). An egg's chorion is the embryo's window on its world. Embryos must acquire oxygen and discharge carbon dioxide across the chorion. As the size of a sphere is increased, the surface area is expanded by about 12.6 times the *square* of its diameter while the volume is increased by four times the diameter *cubed*. Thus the ratio of surface to volume is diminished with every increment of increase in diameter of the sphere. Therefore, for each increase in the size of an embryo (egg) there occurs a concomitant proportionate reduction in the size of its window for embryonic gas exchange. At some point, the window will attain the minimum size required for embryonic development, and no further enlargement of submersed eggs can evolve.

Of course, larger eggs could be laid out of water in the much richer concentration of oxygen and improved diffusion dynamics provided by the atmosphere. Eggs laid in open air suffer no oxygen transfer deficiency but are challenged by the problem of water loss. Since oxygen (O_2) molecules are larger than water (H_2O) molecules, membranes permeable to O_2 permit the rapid outward diffusion of H_2O. Desiccation could be especially acute for insect eggs because of their relatively small size and consequent high ratio of surface to volume compared with eggs of terrestrial vertebrates. Eggs of pterygote insects with a long terrestrial history, including all terrestrial Heteroptera (Cimicomorpha and Pentatomomorpha), are protected against desiccation by a wax layer, which restricts water loss by evaporation through multilaminate eggshells, and by an elaborate chorionic architecture (Hinton 1961a, 1969, 1970, 1981; Margaritis 1985; Zeh *et al.* 1988). Although these chorionic adaptations are certainly plesiomorphic for Heteroptera, some were apparently lost owing to relaxed selection or by counteractive selection for enhanced gas flow as eggs enlarged under water in the belostomatid lineage. Also, giant water bug eggs are apparently provisioned with more yolk proteins (and less lipids) than are the eggs of terrestrial bugs (Indira *et al.* 1965; Madhavan 1974). The metabolism of proteins to fuel embryogenesis requires water whereas the use of fats produce metabolic water. Consequently, giant water bug embryogenesis requires water and belostomatid eggs are protected inadequately, if at all, against desiccation when laid in open air.

The dawn of emergent-brooding

Emergent-brooding is a complex of behavior patterns that clearly could not have emerged full-blown. Below, I surmise how the transition from non-brooding to emergent-brooding might have progressed incrementally, in small steps.

Once eggs were so large that their diminished surface to volume ratio restricted oxygen uptake to the limits of survival for developing embryos, further enlargement of the egg (and imago) would have stopped. However, the process could have continued with an adjustment in maternal oviposition habits. One option would have been to deposit eggs at the splash zone or air–water interface: on rocks, on emergent vegetation, or on the shore. Alternatively, eggs could have been partly inserted into mud or wet sand. So positioned, they would be moistened but not usually submersed for extended periods. The mucilage by which eggs of modern belostomatids are attached to substrate in both the emergent-brooders and the back-brooders appears to

be a hydrophilic mucopolysaccharide. When moist, this material is dense and rubbery, but when dried it is crisp and scaly. Water stored in this material may act as a buffer by slowly releasing moisture to the embryos as they develop. The foam placed around eggs by *Lethocerus* spp. females during their deposition forms a sort of 'bubble nest' (Fig. 6-2b) that may have had some transitional role in egg hydration. Finally, climates or seasons that provide high humidity and daily rains may have allowed occasional development of large eggs laid unattended out of water. Any one or a combination of the foregoing changes in oviposition behavior must have permitted the evolution of giant water bug eggs beyond a critical upper size limit such that they could no longer develop when submersed for a protracted period of time.

Large unattended eggs laid out of water were vulnerable to both drowning and desiccation. Loss of full clutches of eggs by these causes must have been a common occurrence and any trait that reduced these risks could have been selected. The frequency of egg-drowning could have been decreased by depositing eggs higher on emergent vegetation, but elevated eggs would lose the benefit of hydration by waves or spray from the water's surface. Brooding probably evolved in a physically benign environment where emergent, unattended eggs, laid high enough to escape drowning, regularly succeed in development, but not always. Eggs in some clutches desiccated and the embryos died during dry periods.

Now let us suppose that one or the other or both parents spent some time, after mating and oviposition, back in the water clinging to the emergent vegetation upon which their eggs had been laid. Giant water bugs periodically leave the water to dry themselves (R. L. Smith, unpublished; W. F. Kraus, R. Macas-Ordóñez, and N. Ichikawa, personal communications). The function of this behavior is unknown, but it may discourage periphyton growth on the bugs' integument. *Lethocerus* spp. climb emergent vegetation to dry themselves. I posit that each time a bug ascended the substrate to dry itself, water that dripped from the bug's body down the substrate effectively irrigated the eggs. Having come this far, it is not unreasonable to imagine that selection would favor those individuals who focussed their drip-drying in the vicinity of egg clutches, and especially those who positioned themselves over the eggs to provide optimal irrigation of ova. Later, the more active behavior of imbibing water to be regurgitated onto the eggs could be selected. Much later, large, formidably armed bugs in attendance of eggs could be selected to defend them against avian predators. And so, through an unremarkable series, there could have evolved a truly remarkable pattern of behaviors: emergent-brooding.

Adult bugs of either sex needed only: to remain in the vicinity of their hatching eggs, to water them, to have been inhibited from nymphal cannibalism, and to have captured potential predators to protect teneral first-instar nymphs until they became sclerotized and dispersed. Defense of eggs must have lagged behind these patterns by scores of millions of years until passerine birds first appeared and began to feed on lethocerine ova.

Why paternal care?

Here I review evidence in support of the proposition that male giant water bugs were selected to do emergent-brooding because the cost to males was negligible in contrast to profound female costs, which selected for abandonment by mothers.

Postinsemination associations (i.e. males remaining with mates after sperm transfer) (Alcock 1994) occur in Nepidae (Steve Keffer, personal communication) and may have existed in the belostomatid ancestor. These associations may initially have evolved to protect sperm delivered by a guarding male against preemption or dilution from subsequent female mating. This may be the case for water scorpions. All giant water bugs exhibit alternation of copulation and oviposition (Alcock's 1994 'additional mating gain'). This behavior has been shown to accomplish nearly 100% confidence of paternity in a back-brooding giant water bug (Smith 1979b) and is assumed to do so for emergent-brooders as well. If alternating copulation and oviposition behavior pre-dated paternal care, it would have removed cuckoldry as an impediment to paternal care evolution, while keeping males nearby during oviposition. Thus the sexes would have had equal opportunity to contribute care, or abandon, despite internal fertilization.

The ultimate cost of care for males of all kinds is their forfeiture of future mating opportunities during the time of caring, and loss of calories on caring that could have been expended searching for and courting additional mates. However, we have seen that in some *Lethocerus* spp., males are able to mate polygynously and brood several clutches simultaneously. If this were the transitional condition, it would have removed the key deterrent to male investment. As previously discussed, non-filial oocide by females, which may impose a degree of serial monogamy on males, is clearly a derived trait found only in some

Lethocerus spp. Therefore, at the transition to emergent-brooding, males who remained to brood eggs probably stood as good a chance of attracting additional mates by wave motion from their display-pumping (and possible acoustical and olfactory signals) as they would if they changed locations. In fact, brooding males could have been more attractive to females than males without eggs, as is the case for some fish having paternal care of eggs (Sikkel 1989; Petersen and Marchetti 1989; Hoelzer 1990).

Finally, it has been shown that the amount of food required to maintain males is much less than what females require for maintenance and egg production. Brooding has been shown to interfere with prey acquisition, and food deprivation has been demonstrated to reduce female (but not male) fitness, in a giant water bug species. If these gender-specific cost patterns are applicable to giant water bugs and their ancestors, then the maternal cost of brooding alone would have been a potent force compelling females to abandon their eggs. Consequently, it is extremely doubtful that females ever brooded singly, or as biparental providers.

Transitions

Because brooding is not expressed in the belostomatid's sister group, the Nepidae (nor elsewhere in Nepomorpha) we can confidently infer that parental care in Belostomatidae evolved from a non-brooding ancestor somewhere in the stem. However, within Belostomatidae, transitions among behavioral states are by no means clear. Deficiency in our knowledge of *Horvathinia*'s reproductive behavior especially fogs the picture.

Fig. 6-7 displays eight most parsimonious reconstructions of brooding behavioral evolution for the three subfamilies of the Belostomatidae, given that all belostomatines are back-brooders (BB) and all lethocerines are emergent-brooders (EB), but varying assumptions about the behavior of *Horvathinia*. Characters were unordered for these reconstructions. Only one most parsimonious reconstruction

Fig. 6-7. MacClade (Maddison and Maddison 1992) generated these most parsimonious reconstructions of brooding behavior. NB, non-brooding; EB, emergent-brooding; BB, back-brooding. Reconstructions assume different states for *Horvathinia*: (a) assumes *Horvathinia* is an emergent-brooder; (b) and (c) assume *Horvathinia* is a back-brooder; (d–h) assume *Horvathinia* is non-brooding.

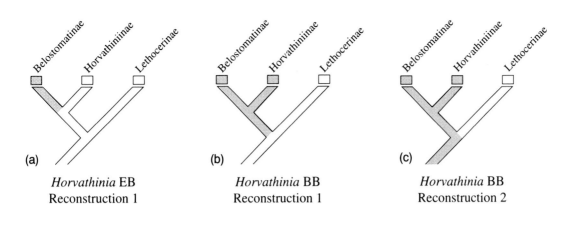

(a)

Horvathinia EB
Reconstruction 1

(b)

Horvathinia BB
Reconstruction 1

(c)

Horvathinia BB
Reconstruction 2

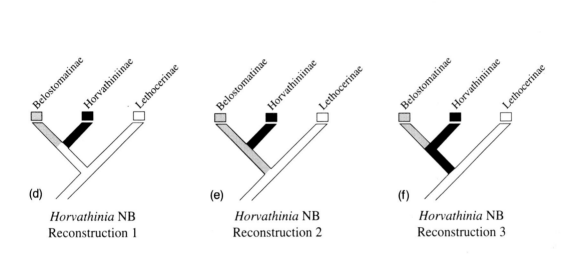

(d)

Horvathinia NB
Reconstruction 1

(e)

Horvathinia NB
Reconstruction 2

(f)

Horvathinia NB
Reconstruction 3

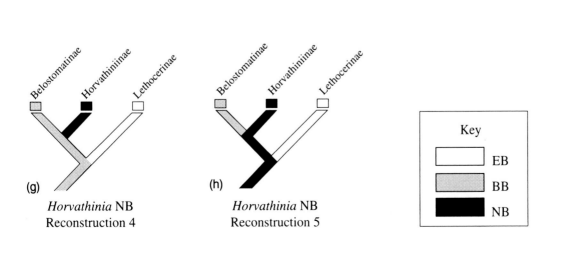

(g)

Horvathinia NB
Reconstruction 4

(h)

Horvathinia NB
Reconstruction 5

Key

EB

BB

NB

(one state change) exists if *Horvathinia* is an emergent-brooder (Fig. 6-7a). Two equally parsimonious reconstructions (each with a single state-change) result if *Horvathinia* is held to be a back-brooder (Fig. 6-7b,c), and there are five equally possible reconstructions, all with two state-changes, if *Horvathinia* turns out to be non-brooding (NB, Fig. 6-7d–h). Although we don't know very much about *Horvathinia*, we do know that it has eggs too large to develop unattended if submersed or if left in open air, and we also know these eggs have unelaborated chorions (see below). Sr Domici's observations on pairs of *Horvathinia* in the laboratory (see 'Water bug biology', above) suggest that they are not back-brooders. Based on this fragmentary information, my best guess is that *Horvathinia* is an emergent-brooder, a view that would favor reconstruction 7a. Of course, *Horvathinia* may be found to do something entirely different from all others in the family.

Putting named taxa aside for the moment, let us consider the likelihood of all of the possible transitions among NB, EB, and BB. There are four general transitions, BB from EB (Fig. 6-7a,b,d,e), EB from BB (Fig. 6-7c,g), both BB and EB from NB (Fig. 6-7f,h), or NB from BB or EB (Fig. 6-7d,e,f,g). Again, I hold the opinion that NB is not likely to have arisen from either of the brooding states, because I believe the eggs of even the smallest belostomatids are too large to develop without their being brooded. If this is true, then reconstructions 6-7d,e,f,g can be disregarded. Both brooding states may have evolved independently from NB, or one brooding state may have evolved from the other. Although both of these possibilities involve two state-changes, the former posits two independent inventions of brooding *per se*, whereas the latter proposes one invention followed by a change in kind of brooding. If we accept that the latter is more parsimonious, we may then ask which brooding state evolved from the other. I favor BB from EB and believe there is considerable support for this position and not the reverse.

Emergent-brooding requires emergent vegetation, which is absent from many aquatic habitats, such that emergent-brooders are limited in their ecological distribution, whereas back-brooders are not. This ecological limitation may be reflected in the small number of species of extant emergent-brooders relative to the back-brooders (more than five times as many of the latter; see Fig. 6-1); thus, it is difficult to imagine an advantage that would have favored the evolution of emergent-brooding from back-brooding. Moreover, there is evidence that the survivorship of back-brooded eggs is on average significantly

higher than that of emergent-brooded eggs (Smith and Larsen 1993). Finally, there are no accounts of a back-brooder having ever laid eggs on any substrate out of the water, and those dumped on static substrate beneath the water never hatch (Smith 1976a,b).

In addition to the points made above, anecdotal evidence (reviewed below) supports an easy transition of BB from EB, but not of EB from BB. Hungerford (1925) paired several emergent-brooders, *Lethocerus griseus* (Say), in aquaria without emergent substrate. From these pairs he obtained egg masses which were laid on (aquarium) supports above the water's surface. However, in one case, Hungerford found that a female had laid 17 eggs on the back of a male that he discovered to be '...resting high and dry above the water on the screening of the cage'. Hungerford does not comment on the fate of these eggs, but Noritaka Ichikawa (personal communication) provides a similar story with a fascinating outcome. Dr Ichikawa had placed several *L. deyrollei* bugs in an aquarium containing water to a depth of about 10 cm. He visited this aquarium the next morning, and while observing a female ascending a stick to dry herself, was startled to find that she had a clutch of eggs attached to her back. These eggs had apparently been laid by another female while the egg-bearer was out of the water drying herself on the previous night. Ichikawa left the eggs attached to the egg-bearing female's back and returned several days later to find that most of the eggs had hatched. Apparently, the periodic surfacing on emergent substrate that *Lethocerus* spp. bugs all perform had in this case been sufficient to keep the eggs from drowning or desiccating, and thus promoted embryonic development to hatching.

All accounts of oviposition errors by back-brooders demonstrate remarkable fidelity to the back-brooding habit, and no evidence of any predilection to emergent-brooding as an alternative. There are several accounts of females laying eggs on the backs of females. Bequaert (1935) stated that female belostomatids usually deposit their eggs on the backs of males but occasionally oviposit on the back of another female. In the course of studies on a natural pond population of *B. flumineum* in Coles County, Illinois, Kruse and Leffler (1984) found two females with eggs attached to their backs. One of these individuals retained four eggs attached for 17 days before they fell off. In another case, while conducting an exhaustive mark and recapture study of an *A. indentatus* population in Deep Canyon Creek, Palm Desert, California, Kraus (1985) found that 3 of 706 adult females had eggs attached to their backs. These three females, one each captured in mid-May, late May, and early June, contained 13, 2, and 52 eggs, respectively. The female carrying 52 eggs was recaptured approximately one month later, in early July, at which time she carried only 27 eggs of the original 52, but 14 of the remaining 27 were hatched. During a two-year survey of a population of *A. herberti* at Cave Creek, Maricopa Co., Arizona, I occasionally observed females carrying eggs during the summer months (R. L. Smith, unpublished), but I have no data on the viability of female-borne eggs in this species. In a similar vein, Menke (1960) found that *A. indentatus* females will oviposit on the backs of males of a different species if conspecific males are unavailable.

The preceding suggest directly and by analogy that the transition to back-brooding from emergent-brooding could have begun as mistakes or as 'acts of desperation' by gravid females lacking appropriate oviposition substrate. Such circumstances might arise for emergent-brooders who found themselves in a habitat lacking emergent vegetation. In the evolutionary transition to back-brooding from some other state, stringency for oviposition substrate choice could have been relaxed if the initial desperate acts or errors resulted in no penalty on developmental rates or embryonic survivorship. If embryos in back-borne eggs developed faster, or survived better on average than those laid elsewhere, selection would begin to favor a female preference for laying eggs on other bugs' backs. Mates would likely have provided the most consistently accessible 'other bug's back', especially if mating already involved the alternation of copulation and oviposition. The unmodified respiratory and drying behaviors in these bugs exapted males to provide for embryonic needs. Emergent-brooding males could initially have tolerated having eggs laid on their backs by some females while others continued to lay their eggs on emergent vegetation. Emergent-brooding patterns would have insured high survivorship of back-laid eggs. The female preference for back-brooding males could have fixed by sexual selection, reduced male opportunity for polygyny, and thus created intense pressure on males to refine their back-brooding efficiency.

The only other documented insect case of back-carried eggs is that of *Phyllomorpha laciniata* de Villers, a coreid bug of very unusual appearance and habits having a circum-Mediterranean distribution. Females of this species lay their eggs into the concavity formed by the recurved thorax and abdomen of conspecifics, usually males. However, both males and females serve as egg-carriers, especially late in the reproductive season, and some eggs are laid on the host plant (Arja Kaitala, personal communication). Eggs are glued to the bugs' backs and are contained by long spines that project toward the mid-dorsal line and thus form a sort of basket or cage around them (Weber 1930). Eggs attached to conspecific backs are apparently protected against predation by ants and possibly also against cannibalism (Arja Kaitala, personal communication). Thus the deposition of eggs on conspecific backs in *Phyllomorpha* reveals by analogy how some immediate benefit to having eggs carried rather than laid on inanimate substrate could have selected the female oviposition behavior.

The possibility that water bug females or both sexes initially brooded is a complication in the transition from non-brooding to paternal emergent-brooding. The only data bearing on this matter are those provided by Ichikawa (1989a, reviewed earlier) concerning removal of males during the course of oviposition. When males were removed in experiments that simulated male abandonment, females ceased laying eggs and never remained with the clutch to brood. This observation, coupled with the gender based cost–benefit analysis in the previous section, reinforces my view that females probably never brooded. In addition, it is doubtful whether male brooding could be evolutionarily stable if females retained a contingency to brood in the event of paternal abandonment. The evolution of back-brooding suffers no complications from the possibility of transitions from shared or maternal care because female giant water bugs cannot lay eggs on their own backs.

THE WATER SCORPION SOLUTION

The water scorpions, sister group to the Belostomatidae, are large predatory aquatic bugs with big eggs, but water scorpions do not brood. Here I compare the characteristics of the water scorpions and the giant water bugs in support of my thesis that the two taxa independently evolved large eggs under similar selection, that both experienced similar constraints, and that each evolved a different solution to the problems created by large eggs in water.

Nepids, like belostomatids, are large aquatic bugs, generally larger than any of the other non-nepoid Nepomorpha. However, the largest nepids are considerably smaller than the largest belostomatids and all water scorpions are less robust than giant water bugs. In fact, members of the derived nepid lineage Ranatrinae are gracile, stick-like insects. None of the nepids have legs well developed for swimming. The water scorpions, like the belostomatids, are all predators with raptorial front legs for grasping prey. The earliest derivative and most robust nepids capture vertebrate prey. For example, *Laccotrephes brachialis* Gerstracher, a *Lethocerus*-like species, is a top predator in papyrus swamps of the Kibale Forest, Uganda, East Africa, where it feeds almost entirely on fish (Lauren Chapman, personal communication). I find nothing on nepid snail-feeding. Typically, nepids develop through five instars, although a few species have four. All water scorpions produce relatively large eggs that are never brooded. Water-scorpion eggs are inserted into both living and dead plant tissue under water, and are deposited in mud (Menke 1979a,c). Eggs laid in mud survive even when inundated (Jay McPherson, personal communication).

So how are large unbrooded nepid eggs able to develop unattended? The chorion at the apex of the eggs of all nepids is produced into long respiratory processes called respiratory horns (Cobben 1968; Hinton 1961b). Eggs of the earliest nepids are characterized by a crown of many short horns, which has been reduced to two long horns in the derived Ranatrinae (Fig. 6-8). These structures greatly enlarge the surface area available for gas exchange and permit these relatively large eggs to develop unassisted in water.

That the largest nepids are considerably smaller than the largest belostomatids (Fig. 6-1) suggests that the respiratory-horn solution to the large egg problem may have been a less efficient adaptation than brooding. There are no vestiges of brooding in the nepids nor of chorionic respiratory horns in the Belostomatidae, indicating that each solution arose independently within its respective clade. This pattern is consistent with the view that if either solution had evolved in an ancestor common to both clades, it would have extinguished selection for the alternative solution.

FOSSIL BUGS AND ANCIENT BROOD CARE

Fossils can be used to infer ancient behavior, but their utility for this purpose is highly variable (Boucot 1990). Here I briefly review the fossil record of giant water bugs in an attempt to discern when brooding may first have evolved. I present data on the size of a fossil belostomatid which belonged to the *Lethocerus* clade from the Upper Jurassic (Popov 1971; R. L. Smith, unpublished). These data satisfy Boucot's (1990) 'category 2B' criteria for fossil evidence of behavior (in this case, emergent-brooding) in that bug size is morphological evidence of brooding in a taxon having extant representatives that perform the behavior.

Working backwards through geologic time, many specimens of 'modern' *Lethocerus* have been recovered from the La Brea tar pits of the Late Quaternary (Miller 1983). Giant

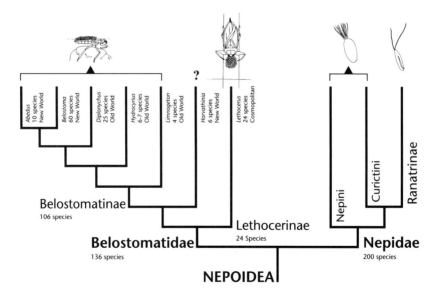

Fig. 6-8. Nepoidea cladogram with adaptations for large eggs in aquatic environments. Left to right: the Belostomatinae are back-brooders. *Horvathinia* is an unknown. Lethocerinae are emergent-brooders. The Nepini together with the Curictini have eggs with multiple respiratory horns, while the Ranatrinae have eggs with a pair of longer respiratory horns.

water bugs are represented in the Santana formation of the Lower Cretaceous of Brazil. These fossils are mostly nymphs of an undescribed species, two of which are depicted in Grimaldi and Maisey (1990). They are broadly ovoid in body shape, and therefore conclusively belong to the back-brooding subfamily Belostomatinae. One of the nymphs is slightly over 10 mm long, so the imago would have to be at least 14 mm in length if the figured nymph was a fifth instar, or larger if the depicted fossil was of an earlier instar.

The Solnhofen limestone of southern Germany contains a remarkable record of Nepoidea. Included are two described species, *Mesobelostomum deperditum* (Germ.) and *Mesonepa primordialis* (Germ.), which have swimming legs, making them Belostomatidae. *Laccotrephes incertus* Popov, nepoid bugs without swimming legs, can be ascribed to the Nepidae (Popov 1971; R. L. Smith, unpublished). Another nepomorph, *Stygeonepa foersteri* Popov, although clearly in Nepoidea, is unlike any extant species (Popov 1971; R. L. Smith, unpublished). The co-occurrence of representatives of these sister groups in the Solnhofen places a minimum time for their divergence at >150 million years before present. The two belostomatids from this age may each belong to a different belostomatid subfamily. If this classification can be confirmed, it places divergence of the Belostomatinae from the Lethocerinae before the Late Jurassic or earlier than 150 million years before present. This dating is reinforced by belostomatine fossils from the Lower Lias of England, including a small *Mesonepa* and an undescribed species that is *ca.* 30 mm in length (Wootton 1988; Whalley 1985).

The remarkable nepoid record of Solnhofen and of Dorset, England, is spectacularly pre-dated by an undescribed nepoid fossil from Triassic (Carnian) lacustrian deposits of the Eastern United States (Olsen *et al.* 1978; C. Remington, personal communication). This material represents the only substantial Triassic insect assemblage in the world other than the Issyk-Kul of Russia and Central Asia. Of the 300+ insect specimens recovered from these sites, more than 200 have been tentatively identified as aquatic Heteroptera, and among those depicted in Olsen *et al.* (1978) is an adult bug, which based on general morphology, size, and probable swimming legs is a belostomatid. This specimen bears a closer superficial resemblance to modern *Lethocerus* than it does to any of the belostomatine genera.

I have studied and measured numerous specimens (>40) of adult *Mesobelostomum deperditus* Germar fossils at the Jura-Museum Naturwissenschaftliche Sammlung in Eichstätt, Bayrische Staatssammlung für Paläontologie

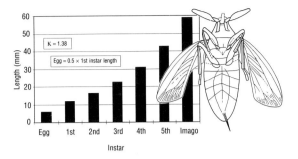

Fig. 6-9. Stadia and egg lengths of the fossil of an adult giant water bug, *Mesobelostomum deperditum* (Germar), calculated back through five instars using a growth constant of 1.38 for nymphal growth and 0.5 × 1st instar to estimate egg size. An egg of the size estimated could not survive unattended submersed in water or exposed to air. (Fossil reconstruction from Popov 1971.)

und Historische Geologie in München, and Museum für Naturkunde der Humboldt Universität in Berlin. Based on these investigations, I concur with Popov (1971) that *Mesobelostomum* is a lethocerine bug.

Did *Mesobelostomum* brood its eggs? Fossils of this species included one specimen that measured 59 mm in length. This individual is assumed to represent one of the largest of this species. I have used nymphal measurements of *L. americanus* from Rankin (1935) and R. Macas-Ordóñez (personal communication) and my own data on *L. medius* to calculate an average developmental growth constant for bug length ($K = 1.38$). Also from Rankin (1935) and other sources, I estimate the size of a *Lethocerus* egg to be about half the size of the first-instar nymph. When these parameters are applied through five instars, working back from a 59 mm imago, I extrapolate an egg size of *ca.* 5.0 mm (Fig. 6-9). Eggs of this size are much larger than those of many extant belostomatid species that have obligatory brooding, and are much larger than what might be considered a transitional size. From this extrapolation, I infer with a high degree of certainty that *Mesobelostomum* eggs were brooded: they were simply too large to have developed submersed in water, and the terrestrial environs surrounding the lagoons where the limestones of Solnhofen occur were generally arid with occasional rains that produced the freshwater ponds (Barthel *et al.* 1990; Gunter Viohl, personal communication) in which *Mesobelostomum* lived and reproduced. Thus there is scant possibility that eggs laid out of water could have received sufficient hydration from rain to have developed unattended, and they would certainly have quickly drowned if submersed.

DISCUSSION

At the beginning of the chapter I reviewed general theory on evolution of parental care. We may now ask: could brooding have been predicted in the giant water bugs based on what is known, or could be discovered, about the ecology of the group? It seems to me highly improbable that the real adaptive significance or evolutionary history of parental brooding in giant water bugs could have yielded to any general theory concerning parental care, because brooding in this group evolved as an adaptation to mitigate constraints created or exposed by selection for another (primary) trait. If there is doubt about the obscurity of parental care in the giant water bug clade, consider that emergent-brooding in *Lethocerus*, sister group to the highly conspicuous and well known back-brooding belostomatines, was first discovered less than two decades ago (Hirayama 1977) and has been generally known for only a very few years (Ichikawa 1988; Smith and Larsen 1993).

An obvious benefit of emergent-brooding is the prevention of embryonic death by desiccation, but this is not the ultimate function of the behavior. Brooding functions to permit development of large eggs that in turn produce large bugs able to exploit vertebrate prey. Conspicuous behavioral features of other animals that provide obvious benefits may likewise have obscure functions similarly evolved to permit progress by some primary selection force. If this is true, it may be useful to distinguish between 'primary selection' which favors traits that directly enhance fitness and historically always precede and create 'ancillary selection' by exposing constraints. Ancillary selection may be defined as a force that overrides constraints and thus supports rectilinear evolution by primary selection. To test for a suspected ancillary trait, we may ask: what forces maintain the trait? Giant water bug brooding is maintained by large eggs, and large eggs are maintained by primary selection for large bugs. If eggs could be reduced in size by relaxed selection for large bugs, then ova could develop unattended in water, and the necessity for brooding would be obviated. Thus, selection for large bugs is the force that ultimately produced and is responsible for maintaining parental care, and selection for large bugs should be recognized as primary selection. To further illustrate the point, suppose that brooding in the Lethocerinae had been shown to effectively defend eggs against some pervasive predator such as passerine birds. If one looked no further, then the behavior might readily be attributed to selection for its obvious value, when in fact

defense of eggs is twice removed from the primary selection that produced large bugs, which caused large eggs to be placed out of water on emergent vegetation subjecting them first to desiccation, and historically millions of years later, to attack by birds.

Parental care has been correlated with the use of temporally abundant, but fleeting resources. Under this circumstance, care may function to accelerate offspring development to maximize their use of the resource while it is available. Lethocerines use fish and amphibian larvae, upon which they are probably dependent for reproduction. These resources are seasonal and often occur in one or at most a few annual pulses. Temporary ponds and the amphibian larvae that they contain represent an especially transitory ecological system. It could be argued that selection resisted the addition of instars (if more than five instars ever occurred in this clade) because added molts would have slowed development, and inhabitants of temporary ponds must develop rapidly in the race not only to harvest transitory resources, but also to acquire wings for dispersal from an ephemeral habitat. Thus selection for rapid development may also have limited the ultimate size of imagos in the Belostomatidae.

If parental care theory would have failed to predict or explain the evolution of brooding in the Belostomatidae, that portion of theory applicable to the evolution of paternal care seems very close to the mark. Three factors that depress male cost of care figure prominently in explaining why unilateral and paternal, rather than maternal or shared, brooding evolved in the belostomatids. These factors are (1) the perfection of paternity assurance by males; (2) the ability of males to simultaneously brood and therefore fertilize multiple clutches of eggs from different females; and (3) the ability of females (but not males) to enhance their fitness by abandoning eggs and capturing more prey to make additional clutches of eggs. The first two of these factors essentially removed the primary impediment to the evolution of male brooding, and the third accounts for an overriding female expense which effectively blocked the evolution of maternal care. Female costs diminish the probability that maternal or biparental brooding was ever expressed as a transient state in the evolution of unilateral paternal brooding.

Back-brooding and emergent-brooding may have evolved independently from the non-brooding state, or one brooding state may have evolved from the other. The former requires that brooding *per se* be twice invented; the latter posits a single invention with subsequent modification. I favor, and the evidence supports, back-brooders diverging from an emergent-brooder, which evolved from a non-brooder. Back-brooders perform both surface-brooding, which involves exposing eggs at the air–water interface, and subsurface brooding, which circulates water over the eggs for aeration. The unmodified respiratory behavior of back-brooders was sufficient to promote embryonic development of back-laid eggs, but general and specific patterns of surface and subsurface brooding have evolved. The evolution of back-brooding was followed by an adaptive radiation of species, due in large part to the fact that back-brooders require no emergent vegetation as a critical feature of their habitats and were thus able to utilize a variety of habitats unsuitable for emergent-brooders. Our better understanding of transitions among giant water bug brooding behavioral states awaits improved phylogenetic information and our enlightenment on the anatomy, ecology, and behavior of the conundrum bugs in the genus *Horvathinia*.

ACKNOWLEDGEMENTS

First, I thank our gentle editors, for splendid editing of course, but more importantly, for their persistence and clever tactics, always charmingly deployed, that finally brought this manuscript to the light of day. Dr Noritaka Ichikawa has been extremely generous with discussion, unpublished data, photographs, and an important eleventh-hour review. Dr Ichikawa also called my attention to interesting investigations published in Japanese by other workers. I have benefitted from useful discussions with Rogelio Macas-Ordóñez, who contributed preliminary observations on two species of *Lethocerus*. I also thank Juan Schnack, Eduardo Domici, Kip Kurse, Bill Kraus, Lauren Chapman, Doug Currie, Jay McPherson, John Polhemus, Steve Keffer and Arja Kitala for their unpublished observations and data, and for beneficial discussion. Robin Wootton, Paul Olsen and Charles Remington encouraged my late-budding interest in insect paleontology. Katja Schulz dissected scent glands from *L. medius* and *L. americanus*, and Philip Evans and William Bowers analyzed scent gland contents. A German Academic Exchange (DAAD) grant made possible my two-month visit in Germany to study the Solnhofen water-bug fossils. Dr Gunter Viohl generously provided fossils for study, and amenities, during my stay at the Jura Museum in Eichstätt, Bavaria. Finally, I am indebted to both Katja Schulz and David Maddison for tutorials in modern cladistics. They helped me to learn what I don't know about giant water bug evolution – which seems an excellent point of departure for my future research on the group.

LITERATURE CITED

Alcock, J. 1994. Postinsemination associations between males and females in insects: The male-guarding hypothesis. *Annu. Rev. Entomol.* **39**: 1–21.

Barthel, K. W., N. H. M. Swinburne and S. Conway Morris. 1990. *Solnhofen, A Study in Mesozoic Palaeontology*. Cambridge University Press.

Bateman, A. J. 1948. Intra-sexual selection in *Drosophila. Heredity* **2**: 349–368.

Bequaert, J. 1935. Presocial behaviour among the Hemiptera. *Bull. Brooklyn Entomol.* Soc. **30**: 177–191.

Berglund, A., G. Rosenqvist and I. Svensson. 1986. Reversed sex roles and parental energy investment in zygotes of two pipefish (Syngnathidae) species. *Mar. Ecol. Prog. Ser.* **29**: 209–215.

–. 1989. Reproductive success of females limited by males in two pipefish species. *Am. Nat.* **133**: 506–516.

Blumer, L. S. 1979. Male parental care in bony fishes. *Q. Rev. Biol.* **54**: 149–161.

Boucot, A. J. 1990. *Evolutionary Paleobiology of Behavior and Coevolution*. Amsterdam: Elsevier.

Böttger, K. 1974. Zur Biologie von *Sphaerodema grassei ghesquierei. Arch. Hydrobiol.* **71**: 100–122.

Brockelman, W. Y. 1969. An analysis of density effects and predation on *Bufo americanus* tadpoles. *Ecology* **50**: 632–644.

Brodie, E. D., Jr., D. R. Formanowicz, Jr. and E. D. Brodie, III. 1978. The development of noxiousness of *Bufo americanus* tadpoles to aquatic insect predators. *Herpetologica* **34**: 302–306.

Butenandt, A. and N. Tâm. 1957. Ueber einen geschlechtsspezifischen Duftstoff der Wasserwanze *Belostoma indica* Vitalis (*Lethocerus indicus* Lep.). *Hoppe-Seyler's Z. Physiol. Chem.* **308**: 277–83.

Carayon, J. 1971. Notes et documents sur l'appareil odorant métathoracique des Hémiptères. *Ann. Soc. Entomol. Fr.* (N.S.) **7**: 737–770.

Carlisle, T. S. 1982. Brood success in variable environments: Implications for parental care allocation. *Anim. Behav.* **30**: 824–836.

Carpenter, F. M. 1992. *Treatise on Invertebrate Paleontology*, Part R, *Arthropoda 4. Hexapoda*, vols. 3, 4. Boulder: Geological Society of America; Lawrence: University of Kansas.

Cloarec, A. 1991. Predatory versatility in the water bug *Diplonychus indicus. Behav. Proc.* **23**: 231–242.

–. 1992. The influence of feeding on predatory tactics in a water bug. *Physiol. Entomol.* **17**: 25–32.

Clutton-Brock, T. H. 1991. *The Evolution of Parental Care*. Princeton: Princeton University Press.

Clutton-Brock, T. H. and C. Godfray. 1991. Parental investment. In *Behavioural Ecology*, 3rd edn. J. R. Krebs and N. B. Davies, eds., pp. 234–262. Oxford: Blackwell Scientific Publications.

Cobben, R. H. 1968. *Evolutionary Trends in Heteroptera. Part I. Eggs, Architecture of the Shell, Gross Embryology and Eclosion*. Wageningen: Cent. Agric. Publ. Doc.

Cole, B. J. 1980. Growth ratios in holometabolous and hemimetabolous insects. *Ann. Entomol. Soc. Am.* **73**: 489–491.

Crowl, T. A. and J. E. Alexander, Jr. 1989. Parental care and foraging ability in male water bugs (*Belostoma flumineum*). *Can. J. Zool.* **67**: 513–515.

Cullen, M. J. 1969. The biology of the giant water bugs (Hemiptera: Belostomatidae) in Trinidad. *Proc. R. Entomol. Soc. Lond.* **44**: 123–137.

Dan, A., M. H. Pereira, A. L. Melo, A. D. Azevedo and L. Freire-Maia. 1993. Effects induced by saliva of the aquatic hemepterian *Belostoma anurum* on isolated Guinea-pig heart. *Comp. Biochem. Physiol.* **106**: 221–228.

Darwin, C. 1872. *The Expression of Emotions in Man and Animals.* London: Murry.

Dawkins, R. and T. R. Carlisle. 1976. Parental investment, mate desertion and a fallacy. *Nature (Lond.)* **262**: 131–133.

De Carlo, J. A. 1958. Identificaión de las especies del género *Horvathtinia* Montandon. Descripción de tres especies nuevas (Hemiptera-Belostomatidae) *Rev. Soc. Entomol. Arg.* **20**: 45–52.

–. 1959. Hemipteros Cryptocerata. Efectos de sus picaduras. *Prim. J. Entomoepid. Argent. Sexta Sex. Cient.*, pp. 715–719.

De Carlo, J. M. 1962. Consideraciones sobre la biologia de *Lethocerus mazzai* De Carlo (Hem. Belostomatidae). *Physis* **23**: 143–151.

–. 1965. Un nuevo género, nuevas especies y referencias de otras poco conocidas de la familia Belostomatidae (Hemiptera). *Rev. Soc. Entomol. Arg.* **28**: 97–109.

DuBois, R. B. and M. L. Rackouski. 1992. Seasonal drift of *Lethocerus americanus* (Hemiptera: Belostomatidae) in a Lake Superior Tributary. *Great Lakes Entomol.* **25**: 85–89.

Dufour, L. 1863. Essai monographique sur les Bélostomides. *Ann. Entomol. Soc. France* **32**: 373–400.

Dyar, H. G. 1890. The number of molts of lepidopterous larvae. *Psyche* **5**: 420–422.

Edmunds, G. F. and R. D. Waltz. 1996. Ephemeroptera. In *An Introduction to the Aquatic Insects.* R . W. Merritt and K. W. Cummins, eds., pp. 126–163. Dubuque: Kendall/Hunt Publishing Company.

Eickwort, G. C. 1981. Presocial insects. In *Social Insects.* H. R. Hermann, ed., pp. 199–280. New York: Academic Press.

Emlen, J. M. 1973. *Ecology: An Evolutionary Approach.* Reading: Addison-Wesley.

Eriksen, C. H., V. H. Resh and G. A. Lamberti. 1996. Aquatic Insect Respiration. In *An Introduction to the Aquatic Insects.* R. W. Merritt and K. W. Cummins, eds., pp. 29–40. Dubuque: Kendall/Hunt Publishing Company.

Grafen, A. 1980. Opportunity cost, benefit and the degree of relatedness. *Anim. Behav.* **28**: 967–968.

Grafen, A. and R. Sibly. 1978. A model of mate desertion. *Anim. Behav.* **26**: 645–652.

Graul, W. D. 1973. Breeding biology of the Mountain Plover. *Wilson Bull.* **7**: 2–31.

Grimaldi, D. and J. Maisey. 1990. Introduction. In *Insects from the Santana Formation, Lower Cretaceous of Brazil.* D. A. Grimaldi, ed., pp. 5–14. (*Bull. Am. Mus. Nat. Hist.* **195**.) New York: American Museum of Natural History.

Gross, M. R. and R. Shine. 1981. Parental care and mode of fertilization in ectothermic vertebrates. *Evolution* **35**: 775–793.

Harvey, G. W. 1907. A ferocious water-bug. *Proc. Entomol. Soc. Wash.* **8**: 72–75.

Heyer, W. R. and M. S. Belin. 1973. Ecological notes on five sympatric *Leptodactylus* (Amphibia, Leptodactylidae) from Equador. *Herpetologica* **29**: 66–72.

Heyer, W. R., R. W. McDiarmid and D. L. Weigmann. 1975. Tadpoles, predation, and pond habits in the tropics. *Biotropica* **7**: 100–111.

Hinton, H. E. 1961a. How some insects, especially the egg stages, avoid drowning when it rains. *Proc. S. Lond. Entomol. Nat. Hist. Soc.* 1960:138–154.

–. 1961b. The structure and function of the eggshell in the Nepidae (Hemiptera). *J. Insect Physiol.* **7**: 224–257.

–. 1969. Respiratory systems of insect egg shells. *Annu. Rev. Entomol.* **14**: 343–368.

–. 1970. Insect eggshells. *Scient. Am.* **223**: 84–91.

–. 1981. *Biology of Insect Eggs*, 3 vols. Oxford: Pergamon Press.

Hirayama, T. 1977. The reproductive behavior of *Lethocerus deyrollei* Vuillefroy. *Nature Study* **23**: 50–54. (in Japanese)

Hoelzer, G. A. 1990. Male-male competition and female choice in the Cortez damselfish *Stegastes rectifraenum*. *Anim. Behav.* **40**: 339–349.

Hoffman, W. E. 1924. Biological notes on *Lethocerus americanus*. *Psyche* **31**: 175–183.

–. 1932. Hymenopterous parasites from the eggs of aquatic and semiaquatic insects. *J. Kansas Entomol. Soc.* **5**: 33–37.

Hungerford, H. B. 1925. Notes on the giant water bugs. *Psyche* **32**: 88–91.

Ichikawa, N. 1988. Male brooding behavior of the giant water bug *Lethocerus deyrollei* Vuillefroy (Hemiptera: Belostomatidae). *J. Ethol.* **6**: 121–127.

–. 1989a. Repeated copulations benefit of the female in *Lethocerus deyrollei* Vuillefroy (Heteroptera: Belostomatidae). *J. Ethol.* **7**: 113–117.

–. 1989b. Breeding strategy of the male brooding water bug, *Diplonychus major* Esaki (Heteroptera: Belostomatidae): Is male back space limiting? *J. Ethol.* **7**: 133–140.

–. 1990. Egg mass destroying behavior of the female giant water bug *Lethocerus deyrollei* Vuillefroy (Heteroptera: Belostomatidae). *J. Ethol.* **8**: 5–11.

Indira, T., P. Govindan and V. Sriramulu. 1965. Utilization of yolk proteins during embryogenesis in *Sphaerodema molestum* (Duf.). *J. Anim. Morphol. Physiol.* **12**: 69–75.

Jackman, R., S. Norwicki, D. J. Aneshansley and T. Eisner. 1983. Predatory capture of toads by fly larvae. *Science (Wash. D.C.)* **222**: 515–516.

Jawale, S. M. and D. R. Ranade. 1988. Observations on the parental care in *Sphaerodema* (=*Diplonychus*) *rusticum* Fabr. *Geobios*. **15**: 44–46.

–. 1990. Morphology of the ovaries of *Sphaerodema* (*Diplonychus*) *rusticum* (Heteroptera, Belostomatidae). *J. Morphol.* **205**: 183–192.

Jenni, D. A. 1974. Evolution of polyandry in birds. *Am. Zool.* **14**: 129–144.

Jenni, D. A. and G. Collier. 1972. Polyandry in the American Jacana (*Jacana spinosa*). *Auk* **89**: 743–765.

Kehr, A. I. and J. A. Schnack. 1991. Predator-prey relationship between giant water bugs (*Belostoma oxyurum)* and larval anurans (*Bufo arenarum*). *Alytes.* **9**: 61–69.

Kopelke, J.-P. 1980. Morphologische und biologische Studien an Belostomatiden am Beispiel der mittelamerikanischen Arten *Belostoma ellipticum* und *B. thomasi. Entomol. Abh. Mus. Tierk. Dresden* **44**: 59–80.

–. 1982. Brutpflegende Räuber – die Belostomatidae. *Natur Mus.* **112**: 1–14.

Kraus, B. 1985. Oviposition on the backs of female giant water bugs, *Abedus indentatus*: The consequence of a shortage in male back space? (Heteroptera: Belostomatidae). *Pan Pac. Entomol.* **61**: 54–57.

Kraus, W. F. 1989a. Surface wave communication during courtship in the giant water bug, *Abedus indentatus. J. Kansas Entomol. Soc.* **62**: 316–328.

–. 1989b. Is male back space limiting? An investigation into the reproductive demography of the giant water bug, *Abedus indentatus* (Heteroptera: Belostomatidae) *J. Insect Behav.* **2**: 623–648.

Kruse, K. C. 1990. Male backspace availability in the giant water-bug (*Belostoma flumineum* Say). *Behav. Ecol. Sociobiol.* **26**: 281–289.

Kruse, K. C. and T. R. Leffler. 1984. Females of the giant water bug, *Belostoma flumeneum* (Hemiptera: Belostomatidae). *Ann. Entomol. Soc. Am.* **77**: 20.

Lauck, D. and A. S. Menke. 1961. The higher classification of the Belostomatidae (Hemiptera). *Ann. Entomol. Soc. Am.* **54**: 664–657.

Lee, C. E., H. M. Cho and S. O. Park. 1970. The bionomics of the water-bug *Diplonychus esakii* in Korea (Het., Belostomatidae). *Nature and Life* **1**: 1–11 (in Korean).

Leidy, J. 1847. History and anatomy of the hemipterous genus *Belostoma. J. Acad. Nat. Sci. Philadelphia* **1**: 57–67.

Maddison, W. P. and D. R. Maddison. 1992. *MacClade: Analysis of Phylogeny and Character Evolution*, version 3. 0. Sunderland, MA: Sinauer Associates.

Madhavan, M. M. 1974. Structure and function of the hydropyle of the egg of the bug, *Sphaerodema molestum. J. Insect Physiol.* **20**: 1341–1349.

Mahner, M. 1993. Systema Cryptoceratorum Phylogeneticum (Insecta, Heteroptera). *Zoologica* **143**: 1–302.

Margaritis, L. H. 1985. Structure and physiology of the eggshell. In *Comprehensive Insect Physiology, Biochemistry and Pharmacology*. G. A. Kerkut and L. I. Gilbert, eds., pp. 153–230. Oxford: Pergamon Press.

Masner, L. 1972. The classification and interrelationships of Thoronini (Hym.: Proctotrupoidae, Scelionidae). *Can. Entomol.* **104**: 833–849.

Matheson, R. 1907. *Belostoma* eating a bird. *Entomol. News* **18**: 452.

Maynard Smith, J. 1977. Parental investment: A prospective analysis. *Anim. Behav.* **25**: 1–9.

McPherson, J. E. and R. J. Packauskas. 1986. Life history and laboratory rearing of *Belostoma lutarium* (Heteroptera: Belostomatidae) with description of immature stages. *J. N. Y. Entomol. Soc.* **94**: 154–162.

Menke, A. S. 1960. A taxonomic study of the genus *Abedus* Stål. (Hemiptera, Belostomatidae) *Univ. Calif. Publ. Entomol.* **16**: 393–440.

–. 1965. A new South American toe biter (Hemiptera: Belostomatidae). *L. A. County Mus. Contrib. Sci.* **89**: 1–4.

–. 1979a. Introduction. In *The Semiaquatic and Aquatic Hemiptera of California* (Heteroptera: Hemiptera). A. S. Menke, ed., pp. 1–15. (*Bull. Calif. Insect Surv.*, vol. 21.) Berkeley, CA: University of California Press.

–. 1979b. Family Belostomatidae – giant water bugs, electric light bugs, toe biters. In *The Semiaquatic and Aquatic Hemiptera of California (Heteroptera: Hemiptera)*. A. S. Menke, ed., pp. 76–86. (*Bull. Calif. Insect Surv.*, vol. 21.) Berkeley, CA: University of California Press.

–. 1979c. Family Nepidae – water scorpions. In *The Semiaquatic and Aquatic Hemiptera of California (Heteroptera: Hemiptera)*. A. S. Menke, ed., pp. 70–75. (*Bull. Calif. Insect Surv.*, vol. 21.) Berkeley, CA: University of California Press.

Miller, P. L. 1961. Some features of the respiratory system of *Hydrocyrius columbiae* Spin. (Belostomatidae, Hemiptera). *J. Insect Physiol.* **6**: 243–271.

Miller, S. E. 1983. Late Quaternary insects of Rancho La Brea. *Quart. Res.* **20**: 90–104.

Mora, G. 1990. Paternal care in a neotropical harvestman, *Zygopachylus albomarginis* (Arachinda, Opiliones: Gonyleptidae). *Anim. Behav.* **39**: 582–593.

Møller, A. P. and T. R. Birkhead. 1993a. Cuckoldry and sociality: A comparative study in birds. *Am. Nat.* **142**: 118–140.

–. 1993b. Certainty of paternity covaries with paternal care in birds. *Behav. Ecol. Sociobiol.* **33**: 261–268.

Nethersole-Thompson, D. 1973. *The Dotterel*. London: Collins.

Okada, H. and F. Nakasuji. 1993. Comparative studies on the seasonal occurrence, nymphal development, and food menu in two giant water bugs, *Diplonychus japonicus* Vuillefroy and *Diplonychus major* Esaki (Hemiptera: Belostomatidae). *Res. Pop. Ecol.* **35**: 15–22.

Olsen, P. E., C. L. Remington, B. Cornet, and K. S. Thompson. 1978. Cyclic changes in Late Triassic lacustrine communities. *Science (Wash., D.C.).* **201**: 729–733.

Oring, L. W. and M. L. Knudsen. 1972. Monogamy and polyandry in the spotted sandpiper. *Living Bird* **11**: 59–73.

Parker, G. A. 1984. Sperm competition and the evolution of animal mating strategies. In *Sperm Competition and the Evolution of Animal Mating Systems*. R. L. Smith, ed., pp. 1–60. New York: Academic Press.

Parker, G. A., R. R. Baker and V. G. F. Smith. 1972. The origin and evolution of gamete dimorphism and the male-female phenomenon. *J. Theor. Biol.* **36**: 529–553.

Pattenden, G. and B. W. Staddon. 1970. Observations on the metasternal scent gland of *Lethocerus* spp. (Heteroptera: Belostomatidae). *Ann. Entomol. Soc. Am.* **63**: 900–901.

Pemberton, R. W. 1988 The use of the Thai giant water bug, *Lethocerus indicus* (Hemiptera: Belostomaticae), as human food in California. *Pan Pacific Entomol.* **64**: 81–82

Perrone, M., Jr. and T. M. Zaret. 1979. Parental care patterns in fish. *Am. Nat.* **113**: 351–361.

Petersen, C. W. and K. Marchetti. 1989. Filial cannibalism in the Cortez damselfish *Stegastes rectifraenum*. *Evolution* **43**: 158–168.

Picado, C. 1937. Estudo experimental sobre o veneno de *Lethocerus delpontei* De Carlo. *Mem. Inst. Butantan (Sao Paulo)* **10**: 303–310.

–. 1939. Etude experimentale du venin de *Lethocerus delpontei* De Carlo. *Trav. Stn. Zool. Wimereux* **13**: 553–562.

Poisson, R. 1949. Hémiptères aquatiques. *Explor. Parc Natl. Albert*, **58**: 1–94.

Popov, Y. A. 1971. The historical development of bugs of the infraorder Nepomorpha. *Trans. Paleontol. Inst. Akad. Nauk USSR* **129**: 1–228 (in Russian).

Rankin, K. P. 1935. Life history of *Lethocerus americanus* (Leidy) (Hemiptera-Belostomatidae). *Univ. Kans. Sci. Bull.* **22**: 479–491.

Rees, A. R. and R. E. Offord. 1968. Studies on the protease and other enzymes from the venom of *Lethocerus cordofanus*. *Nature (Lond.)* **221**: 675–677.

Ridley, M. W. 1978. Paternal care. *Anim. Behav.* **26**: 904–932.

Rosenqvist, G. 1990. Male mate choice and female-female competition for mates in the pipefish *Nerophis ophidion*. *Anim. Behav.* **39**: 1110–1115.

Saha, T. C. and S. K. Raut. 1992. Bioecology of the water-bug *Sphaerodema annulatum* Fabricius (Heteroptera: Belostomatidae). *Arch. Hydrobiol.* **124**: 239–253.

Schnack, J. A. 1976. Los Belostomatidae de la Republica Argentina. *Fauna de Agua Dulce de la República Argentina* **35**: 1–86.

Schnack, J. A., E. A. Domizi and A. L. Estévez. 1990. Comportamiento reproductivo del *Belostoma oxyurum* (Hemiptera, Belostomatidae) *Rev. Soc. Entomol. Arg.* **48**: 121–128.

Scott, M. P. 1990. Brood guarding and the evolution of male parental care in burying beetles. *Behav. Ecol. Sociobiol.* **26**: 31–39.

Sikkel, P. C. 1989. Egg presence and developmental stage influence spawning site choice by female garibaldi. *Anim. Behav.* **38**: 447–456.

Smith, R. L. 1975. Bionomics and behavior of *Abedus herberti* with comparative observations on *Belostoma flumineum* and *Lethocerus medius* (Hemiptera: Belostomatidae). Ph. D. dissertation, Arizona State University, Tempe, AZ.

–. 1976a. Male brooding behavior of the water bug *Abedus herberti* (Heteroptera: Belostomatidae). *Ann. Entomol. Soc. Am.* **69**: 740–747.

–. 1976b. Brooding behavior of a male water bug *Belostoma flumineum* (Hemiptera: Belostomatidae). *J. Kansas Entomol. Soc.* **49**: 333–343.

–. 1979a. Paternity assurance and altered roles in the mating behaviour of a giant water bug *Abedus herberti* (Heteroptera: Belostomatidae). *Anim. Behav.* **27**: 716–725.

–. 1979b. Repeated copulation and sperm precedence: Paternity assurance for a male brooding water bug. *Science (Wash., D.C.)* **205**: 1029–1031.

–. 1980. Evolution of exclusive postcopulatory paternal care in the insects. *Fla. Entomol.* **63**: 65–78.

–. 1982. Reproductive behavior of giant water bugs. In *Insect Behavior, A Sourcebook of Laboratory and Field Exercises*. J. R. Matthews and R. W. Matthews, eds., pp. 149–153. Boulder: Westview Press.

Smith, R. L. and E. Larsen. 1993. Egg attendance and brooding by males of the giant water bug *Lethocerus medius* (Guerin) in the field (Heteroptera: Belostomatidae). *J. Insect Behav.* **6**: 93–106.

Staddon, B. W. 1971. Metasternal scent glands in Belostomatidae (Heteroptera). *J. Entomol.* **46**: 69–71.

–. 1986. Biology of scent glands in the Hemiptera-Heteroptera. *Ann. Soc. Entomol. Fr.* **22**: 183–190.

Staddon, B. W. and M. J. Thorne. 1979. The metathoracic scent gland system in Hydrocorisae (Heteroptera: Nepomorpha). *Syst. Entomol.* **4**: 239–250.

Štys, P. and J. Davidová-Vilímová. 1989. Unusual numbers of instars in Heteroptera: A review. *Acta Entomol. Bohem.* **86**: 1–32.

Summers, K. 1990. Paternal care and the cost of polygyny in the green dart-poison frog, *Dendrobates auratus*. *Behav. Ecol. Sociobiol.* **27**: 307–313.

–. 1992. Dart-poison frogs and the control of sexual selection. *Ethology.* **91**: 89–107.

Tallamy, D. W. 1995. Nurishment and the evolution of paternal investment in subsocial arthropods. In *Nourishment and Evolution in Insect Societies*. J. H. Hunt and C. A. Nalepa, eds., pp. 21–55. Boulder: Westview Press.

Tallamy, D. W. and T. K. Wood. 1986. Convergence patterns in subsocial insects. *Annu. Rev. Entomol.* **31**: 369–390.

Tawfik, M. F. S. 1969. The life history of the giant water-bug *Lethocerus niloticus* Stael (Hemiptera: Belostomatidae). *Bull. Soc. Ent. Egypte* **53**: 299–310.

Tawfik, M. F. S., S. I. El-Sherif and A. F. Lutfallah. 1978. On the life-history of the giant water-bug *Limnogeton fieberi* Mayr (Hemiptera: Belostomatidae), predatory on some harmful snails. *Zool. Anz. (Jena)* **86**: 138–145.

Thornhill, R. 1976. Sexual selection and paternal investment in insects. *Am. Nat.* **110**: 153–163.

Tieu, N. C. 1928. Notes sur les insectes comestibles au Tonkin. *Bull. Econ. Indocine.* **8**: 738–741.

Trivers, R. L. 1972. Parental investment and sexual selection. In *Sexual Selection and the Descent of Man 1871–1971*. B. G. Campbell, ed., pp. 136–179. Chicago: Aldine.

Uhler, P. R. 1884. Order IV. Hemiptera. In *The Standard Natural History*, vol. 2. J. Kingsley, ed., pp. 204–296. Boston: Boston Cassino & Co.

Venkatesan, P. 1983. Male brooding behaviour of *Diplonychus indicus* Venk. & Rao (Hemiptera: Belostomatidae). *J. Kansas Entomol. Soc.* **56**: 80–87.

Victor, R. and L. I. Ugwoke. 1987. Preliminary studies on predation by *Sphaerodema nepoides* Fabricius (Heteroptera: Belostomatidae). *Hydrobiologia.* **154**: 25–32.

Vincent, A., I Ahnesjö, A. Berglund, G. Rosenqvist. 1992. Pipefish and seahorses: Are they sex role reversed? *Trends Ecol. Evol.* **7**: 237–241.

Voelker, J. 1966. Wasserwanzen als obligatorische Schneckenfresser im Nildelta (*Limnogeton fieberi* Mayr: Belostomatidae, Hemiptera). *Z. Tropenmed. Parasitol.* **17**: 155–165.

–. 1968. Untersuchungen zu Ernährung, Fortpflanzungsbiologie und Entwicklung von *Limnogeton fieberi* Mayr (Belostomatidae, Hemiptera) als Beitrag zur Kenntnis von natürlichen Feinden tropischer Süßwasserschnecken. *Entomol. Mitt. Zool. Staatsinst. Zool. Mus. Hamb.* **3**: 1–24.

Whalley, P. E. S. 1985. The systematics and palaeogeography of the Lower Jurassic insects of Dorset, England. *Bull. Br. Mus. Nat. Hist. Geol.* **39**: 107–189.

Weber, H. 1930. *Biologie der Hemipteren. Eine Naturgeschichte der Schnabelkerfe.* Berlin: Springer.

Westfall, M. J. and K. J. Tennessen. 1996. Odonata. In *An Introduction to the Aquatic Insects of North America.* R. W. Merritt and K. W. Cumming, eds., pp. 164–211. Dubuque: Kendall/Hunt Publishing Co.

Werren, J. H., M. R. Gross and R. Shine. 1980. Paternity and the evolution of male parental care. *J. Theor. Biol.* **82**: 619–631.

Whittingham, L. A., P. D. Taylor and R. J. Robertson. 1992. Confidence of paternity and male parental care. *Am. Nat.* **139**: 1115–1125.

Williams, G. C. 1966. *Adaptation and Natural Selection.* Princeton: Princeton University Press.

–. 1975. *Sex and Evolution.* Princeton: Princeton University Press.

Wilson, E. O. 1971. *The Insect Societies.* Cambridge: Belknap Press of Harvard University Press.

Wittenberger, J. F. 1981. *Animal Social Behavior.* Boston: Duxbury Press.

Wootton, R. J. 1988. The historical ecology of aquatic insects: An overview. *Palaeogeogr. Palaeoclimatol. Palaeoecol.* **62**: 477–492.

Xia, X. 1991. Uncertainty of paternity can select against paternal care. *Am. Nat.* **139**: 1126–1129.

Zeh, D. W. and R. L. Smith. 1985. Paternal investment by terrestrial arthropods. *Am. Zool.* **25**: 785–805.

Zeh, D. W., J. A. Zeh and R. L. Smith. 1988. Of oviposition, amnions and eggshell architecture: A role for the egg stage in the diversification of the terrestrial arthropods. *Q. Rev. Biol.* **64**: 147–168.

7 · The evolution of sociality in aphids: a clone's-eye view

DAVID L. STERN AND WILLIAM A. FOSTER

ABSTRACT

A number of aphid species produce individuals, termed soldiers, that defend the colony by attacking predators. Soldiers have either reduced or zero direct reproductive fitness. Their behavior is therefore altruistic in the classical sense: an individual is behaving in a way that incurs reproductive costs on itself and confers reproductive benefits on another. However, comparison with the better-known eusocial insects (Hymenoptera, Isoptera) indicates that there are important differences between clonal and sexual social animals.

Here we take a clone's-eye view and conclude that many facets of aphid sociality are best thought of in terms of resource allocation: for example, the choice between investment in defense and reproduction. This view considerably simplifies some aspects of the problem and highlights the qualitatively different nature of genetic heterogeneity in colonies of aphids and of other social insects. In sexually reproducing social insects, each individual usually has a different genome, which leads to genetic conflicts of interest between individuals. In social aphids, all members of a clone have identical genomes, barring new mutations, and there should be no disagreement among clonemates about investment decisions. Genetic heterogeneity within colonies can arise, but principally through clonal mixing, and this means that investment decisions will vary between different clones rather than among all individuals.

Since the discovery of soldiers in aphids almost 20 years ago, approximately 50 species have been found to produce some type of defensive morph. All known species producing soldiers are members of the two closely related families Pemphigidae and Hormaphididae. None the less, soldier-producing species are taxonomically widely distributed within these two families, and even with our very crude understanding of aphid phylogeny it is clear that a minimum of six evolutionary events (gains and losses) are necessary to explain the distribution of known soldier-producing species. Thus, although soldier-producing species are relatively rare, soldier production is evolutionarily

quite plastic. This evolutionary plasticity makes the aphids an ideal group for comparative studies, and we discuss several directions for future research aimed at elucidating the causes of the evolution of soldiers.

INTRODUCTION

Aphids hold a special place in the study of the evolution of highly social behavior. They show a range of social behavior culminating in the existence of sterile castes and they reproduce parthenogenetically, which means that all members of a colony descended from a single individual are genetically identical. Despite this possible genetic predisposition toward highly social behavior (Hamilton 1987), the most extreme displays of social behavior are limited to approximately one percent of aphid species and sociality in these species does not come close to rivaling the division of labor found in social Hymenoptera and Isoptera. The best-studied social behavior in aphids is the phenomenon of soldiers and is the subject of this chapter.

We approach the problem of soldier evolution at two levels: historical and ecological. These two levels are not naturally distinct; indeed, the past ecological adventures of a lineage are its history, but it is useful to employ different methods to study each. We first examine the biology of the soldier-producing aphids in some detail and compare their biology with that of other social insects to illustrate the special place that aphids hold. Next we examine what is currently known about the phylogenetic distribution of soldier-producing aphids and conclude that soldier evolution has been extremely evolutionarily labile. With this historical framework we then explore various features of the natural history that might have predisposed some aphid species to soldier production.

APHID BIOLOGY

Aphids are small (adults usually under 7 mm long), soft-bodied insects (Suborder Homoptera: Order Hemiptera) that feed on phloem fluid. Most generations are parthenogenetic and bear live young. Freed from the requirements of fertilization, mothers can develop daughter and granddaughter embryos within themselves. This telescoping of generations, together with a short generation time (approximately 10 days), allows aphid clones to grow very quickly (Dixon 1985). Most species are cyclically parthenogenetic; that is, a series of parthenogenetic generations alternates with a single generation of sexual reproduction. About 10% of aphid species alternate between two host plant species: the primary host, where sex occurs and several parthenogenetic generations are produced, and the secondary host, where successive entirely parthenogenetic generations of females develop. In temperate regions, aphids spend autumn, winter and spring on the primary host and summer on the secondary host. In the tropics, strict seasonality of host alternation has been lost in at least some species (Stern *et al.* 1994). The primary hosts are usually woody and tend to be highly taxonomically conserved for each of the major aphid groups, whereas the secondary hosts, which are not usually closely related to the primary hosts, tend to be much more diverse (for a review see Moran 1992).

Aphids form aggregations that sometimes include many tens of thousands of individuals. These aggregations form primarily because larvae tend to remain close to their birthplace. In addition, migrant individuals, both winged and unwinged, may settle on the same leaves.

A distinctive feature of aphids is their high degree of polymorphism (Miyazaki 1987). Within one particular clone there may be as many as eight distinct morphological forms (Dixon 1985). The clones therefore have the potential to develop a flexible division of labor, with different morphs specialized for the often conflicting requirements of, for example, rapid growth, dispersal, sexual reproduction or survival. The elaboration of different morphs is particularly noticeable in host-alternating families, such as the Pemphigidae and Hormaphididae (Moran 1988).

A CLONE'S-EYE VIEW

Aphids are unique among the social insects because they reproduce by thelytokous parthenogenesis. Therefore, all members of a colony are, potentially, genetically identical. The only source of genetic heterogeneity, barring mutation (which would probably not provide significant variation within the life of a colony), is mixing of members from different clones. This genetic purity should lead to a clarity of purpose among individuals of a clone. This is the primary

reason that we hesitate to categorize soldier-producing aphid species as eusocial. Recently, two new definitions of eusociality have been proposed in an attempt to unite studies of social behavior (Crespi and Yanega 1995; Sherman *et al.* 1995). Because aphids are clonal, we feel that they cannot be accomodated either by these two new definitions or by the original definition of eusociality, which was constructed primarily to aid study of social bees (Michener 1969). Implicit in all these definitions is potential conflict among colony mates arising from genetic heterogeneity. None of these definitions aids our understanding of the factors leading to the evolution of soldiers in aphids. Although we admit the utility of attempts to unite studies of social behavior among all animals, we agree with Wcislo (Chapter 1, this volume) that such definitions, in their quest for all-encompassing explanations, will tend to distract attention from the natural history unique to each lineage that may be crucial to understanding the evolution of highly social behavior.

Since aphids are clonal, we suggest that soldier production should be viewed purely as a problem of resource allocation. All the individuals of a clone should have identical interests and they should therefore all agree on the optimal allocation of colony resources. Most importantly, we expect no conflict between a soldier and her mother. We call this perspective a clone's-eye view, since the collective genomes of an uninvaded colony can be thought of as a clone. This perspective is easily extended to situations where several clones mix to form a single colony. All the members of clone A should agree on optimal resource allocation, as should all the members of clone B, but clones A and B might disagree on this allocation. A conflict might arise between clones, but never within clones.

Despite this clear advantage, reproductive altruism is seemingly rare in clonal animals, and the lack of extreme forms of reproductive altruism among clonal insects has provided considerable concern to theorists (Hamilton 1987). For example, only one percent of aphid species are known to produce soldiers. Why should altruism be so rare in clonal animals?

This question is slightly misleading. Reproductive altruism is fairly common among colonial marine invertebrates (see, for example, Wilson 1975; Hughes 1989) and is also found in polyembryonic parasitoid wasps (Cruz 1981, 1986). Moreover, from the genome's point of view, the cells of a metazoan body or a plant might be considered highly reproductively altruistic individuals. There seems to be another factor besides relatedness influencing the extent of reproductive altruism between organisms. This additional factor has to do with the extent and permanence of spatial proximity (Hamilton 1987). That is, identical genomes that spend considerable time in a definite or predictable spatial relationship (cells of a body, zooids of a Portuguese man-of-war, etc.) tend to participate in extreme forms of reproductive altruism. Perhaps a useful way to think of this problem is that genomes are essentially individuals of a population with extremely low levels of migration between subpopulations (i.e. bodies). These are ideal conditions for group selection to act and we should not be surprised to see 'group'-level adaptations (Leigh 1983; Wilson and Sober 1989).

The evolution of altruism immediately becomes more difficult as soon as genomes or individuals are not permanently connected, as in aphids. In aphids, because they are clonal but unconnected, altruistic behavior is available for stealing by unrelated clones but there is no question as to the genetic benefits if the altruistic act reaches individuals of the same clone. In other social insects, not only are individuals unconnected but individuals are genetically heterogeneous (owing to sexual reproduction) within colonies. The aphids, therefore, offer an ideal opportunity for studying the determinants of relatively simple forms of reproductive altruism without the confounding factors of genetic heterogeneity among individuals.

NATURAL HISTORY OF SOLDIERS

Definitions of soldiers: the role of sterility

Soldiers can be thought of as part of the system of morph specialization that is common to all aphids, their particular role being the defense of the clone. What has singled them out for the attention of evolutionary biologists is the fact that they may exhibit the ultimate in reproductive altruism: sterility. However, we think it is unhelpful to restrict the term 'soldier' to sterile morphs. This is partly because it is, in fact, difficult to define sterility, but also because to define soldiers in this way directs attention away from the central issue, which is the cost imposed on the clone by investment in defensive morphs. This cost is the reduced growth rate of the clone and can be made up of many factors, including increased investment in weaponry and defender body size, increased risk of death and injury to the defenders, increased developmental time required for producing defenders, extension of the duration of a larval instar, and sterility. It seems inappropriate

to have regard for only one of these costs (sterility) and we will therefore refer to all the different types of defenders as 'soldiers'.

Aphid soldiers provide an excellent illustration of Alexander *et al.*'s (1991) remark that 'sterility is not an all-or-nothing phenomenon' (p. 26). In some species, the defenders are apparently unable to develop and reproduce (although it is not clear that any species has been studied in sufficient detail to be sure; see Kurosu and Aoki 1991a). These soldiers could perhaps be described as 'obligately sterile'. All other soldiers might be described as 'facultatively sterile': that is, they may develop into a reproductive if they do not die before maturity while defending the clone, either by being killed or injured by predators, by falling from the colony or being permanently excluded from it, or by being born too late in the colony cycle.

A common feature of most soldiers, whether or not they are sterile, is that their development is delayed, such that the defensive instar persists for a relatively long time. A striking example of this is provided by the defenders of *Pemphigus obesinymphae*, which remain in the first instar for two to three months (Moran 1993). Akimoto (1992) suggested that the prolongation of the defensive stage might be a crucial first step in the evolution of soldiers. Obligate sterility may be a later acquisition.

The diversity of soldiers

Aphid soldiers are surprisingly diverse. At least four different morphological types can be defined on the basis of taxonomic divisions and differences in behavior and associated weaponry (see Foster and Northcott 1994). It is probable that this diversity will increase as more aphid species are examined.

Pemphigus type (Pemphigini)

These soldiers are always restricted to the gall generations on the primary host, they are never sterile, and they are almost always first instars. Their weapons are their stylets and legs, usually just the hind-legs (e.g. *Pemphigus spyrothecae* (Foster 1990)) but in some cases all six legs are used (e.g. *P. obesinymphae* (Moran 1993)). The legs are enlarged and heavily sclerotized, with well-developed claws (Lampel 1968–9). They attack predators by stabbing with their stylets and by squeezing with their legs, attempting to rupture the predator's cuticle (Aoki and Kurosu 1986; Foster 1990; Moran 1993). Soldiers produced towards the end of the season may be functionally sterile, because they will not have time to develop to adulthood (Foster and Northcott 1994).

Colophina type (Eriosomatini)

These soldiers occur on both the primary and secondary hosts. Their weapons are their stylets and their fore and mid-legs. They hold predators firmly with their enlarged hardened fore and mid-legs, and pierce the predator's cuticle with their stylets, which are supported by a short, powerful rostrum (Aoki 1977a). The soldiers on the primary host are either first instars (*Hemipodaphis* (Akimoto 1992)) or second instars (*Colophina* (Aoki 1978, 1980)) and they are never sterile. Those on the secondary hosts are first instars and may be either sterile (e.g. *Colophina arma, C. clematis, C. monstrifica* (Aoki 1977a,b, 1980, 1983) or not

sterile (e.g. *C. clematicola* (Kurosu and Aoki 1988a)). The most specialized sterile soldiers are morphologically distinct from the other non-defenders of the same instar.

Styrax-gall type (Cerataphidini)

These soldiers occur on the primary host (*Styrax*) and an example from one species, *Pseudoregma bambucicola*, is illustrated in Fig. 7-1a. Their only weapons are their stylets and they are usually second instars, obligately sterile and morphologically distinct from non-soldiers (see, for example, Aoki *et al.* 1977; Aoki 1982a; Aoki and Kurosu 1989a, 1993; Aoki and Usuba 1989). All species of the tribe Cerataphidini found on the primary host produce these soldiers within galls. At least three cerataphidines, *Ceratovacuna nekoashi, Astegopteryx bambucifoliae* and *Ceratoglyphina*

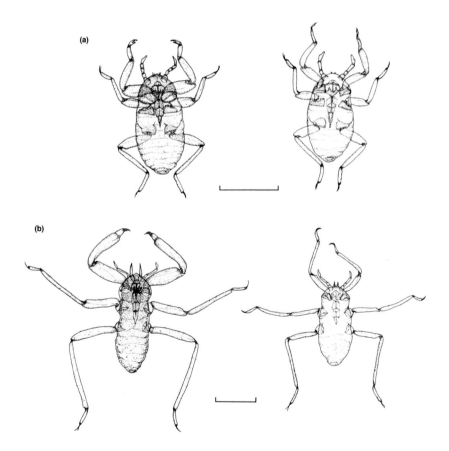

Fig. 7-1. Two types of soldiers produced by the cerataphidine aphid *Pseudoregma bambucicola*. (a) Primary-host soldier, on left, is a second-instar nymph and has enlarged forelegs and heavy sclerotization compared with normal second-instar nymph, on the right. Scale = 0.5 mm. (b) Secondary-host soldier, on left, is a first-instar nymph with longer legs, enlarged forearms and frontal horns, and heavier sclerotization, compared with normal first-instar nymph, on right. Scale = 0.5 mm. (From Stern and Foster (1996), with permission.)

styracicola, produce an interesting variant on these soldiers, the so-called outsiders: these are first instars, which are usually monomorphic and effectively sterile because they are born too late to have access to an open incipient subgall. They patrol the surface of the developing subgalls, apparently warding off predators that might otherwise eat into the galls (Kurosu and Aoki 1988b, 1991b,c).

Horned type (Cerataphidini)

These are first-instar soldiers produced on the secondary hosts; an example from *Pseudoregma bambucicola* is shown in Fig. 7-1b. These soldiers have strongly sclerotized tergites and enlarged, greatly thickened forelegs armed with strong claws, and they attack predators with a novel weapon: horns on the front of their heads. They grasp predators with their forelegs and attempt to rip holes in them with their frontal horns (see, for example, Aoki and Miyazaki 1978). They range from the relatively unspecialized soldiers of *Ceratovacuna lanigera*, which are monomorphic and not sterile and primarily attack predator's eggs (Aoki *et al.* 1984; Aoki and Kurosu 1987), to the highly morphologically specialized soldiers of *Pseudoregma alexanderi*, which are obligately sterile and distinct from non-soldiers, perhaps having soldier subcastes, and which aggressively attack a range of predators (Aoki and Miyazaki 1978; Aoki *et al.* 1981).

Although some species (e.g. *Pseudoregma* spp.) have both *Styrax*-gall type soldiers on the primary host and horned-type soldiers on the secondary host (see Fig. 7-1), there are, as we will discuss later, good reasons for thinking that these two types of soldier are not homologous but represent separate origins of soldier behavior (Aoki 1987; Aoki and Kurosu 1989b; Stern 1994).

Other types of soldier

Soldiers have been reported in two other tribes, the Prociphilini (*Pachypappa marsupialis* (Aoki 1979b)) and the Nipponaphidini (*Nipponaphis distyliicola* (Kurosu *et al.* 1995); *Distylaphis foliorum* (Noordam 1991)). The small amount of information available for these soldiers suggests that they might represent further independent origins of soldier behavior, but clearly more field data on these and related species are required. There is an urgent need, also, to look for possible defensive behavior and morphology in aphids from the three tribes of the Hormaphididae and Pemphigidae (Hormaphidini, Fordini and Melaphidini) from which soldiers have so far not been described.

How to be a soldier

Broadly speaking, there are three ways to be a larval soldier (Fig. 7-2). Soldiers can be a monomorphic caste, but then they may at some point molt and reproduce. Alternatively, soldiers can be distinct, either morphologically or behaviorally, from normal larvae of the same instar. In this case soldiers have the option of molting and reproducing or remaining in their larval instar indefinitely and being sterile. Different species of aphid produce all three types of soldiers. More importantly, many aphid species produce different kinds of soldiers during different generations or different phases of the life cycle (see Kurosu and Aoki 1988b, 1991b; Foster and Northcott 1994; Stern and Foster 1996). What are the relative costs and benefits of producing these different types of soldier? Are certain larval-instars better for being a soldier and are certain morphologies and weapons more appropriate for different larval-instar soldiers? These problems invite a theoretical analysis explicitly examining the costs and benefits, and possible trade-offs, of soldier size, larval instar, and efficiency.

Although it might seem surprising that all soldiers are larvae, there may be good reasons for this. Hamilton (1987) argued that aphid soldiers are larvae because aphid larvae have a more generalized homopteran morphology than adults which might allow them to more easily evolve new morphologies. It is not clear to us that the morphology of larvae is significantly less specialized than adults; in fact we argue the opposite below. More importantly, we do not believe that soldier evolution initially involved any morphological specialization (see below and Table 7-1). There are several facts of aphid biology that might explain the predominance of larval soldiers. First, the first instars of both aphids and coccids are frequently the dispersal morph and they frequently have morphological specializations for crawling or otherwise moving to new locations (Foster 1978; Foster and Treherne 1978). Second, the primary weapon of most soldiers is their stylets. The size of the stylets does not increase much during development and slightly larger stylets may not offer a significantly better weapon. Third, the smaller size of larvae may allow many soldiers to attack a predator simultaneously, thereby inflicting more stylet punctures per unit area of the predator. Finally, early-instar larvae are available for defense sooner than late-instar larvae and adults.

Several facts suggest that there may be more to the problem of why soldiers are always larvae. When larvae

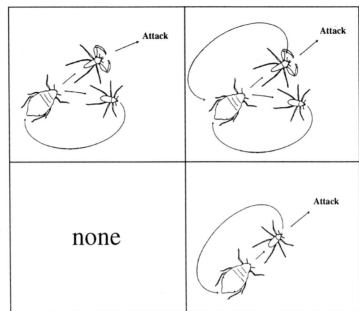

Fig. 7-2. The three different ways to be a soldier. Soldiers can represent either a morphologically or a behaviorally specialized caste (dimorphic larvae) or not (monomorphic larvae). In addition, morphologically specialized soldiers can be either facultatively sterile or obligately sterile. Facultatively sterile soldiers have the ability to molt, continue development and reproduce, but they may not always do so.

are dimorphic, soldiers are inevitably the larger morph and some aphid soldiers display a wide size range, even intraspecifically (e.g. *Pseudoregma alexanderi* (Aoki and Miyazaki 1978)). This suggests that there might sometimes be an advantage to relatively large body size for soldiers. However, for soldiers within galls there might be distinct advantages to being small. A small soldier can easily exit through and defend a small gall opening that might deny larger predators access to galls. A larger soldier, for example an adult, would require a larger gall opening that would allow access to more, and larger, predators. It is noteworthy that gall openings increase in size, or if galls are usually closed they open for the first time, late in the season as winged adults leave the gall. It would be interesting to know how patterns of predation vary with changes in the size of the gall entrance.

Recruitment and communication

All social animals require efficient systems of communication between the individual members of the society. Aphids have sensory capabilities that are mainly concerned with mediating the relationship between the aphid and its host plant (Anderson and Bromley 1987). Nevertheless, aphids are known to have both chemical (see, for example, Nault and Montgomery 1977) and mechanical (see Kennedy *et al.* 1967) means of communication with each other. Have the soldier-producing species developed communication systems in any way comparable to those known from the social Hymenoptera (see, for example, Hölldobler and Wilson 1990)?

Aphids do not forage for food. They have to *find* suitable host plants, and chemical signals between aphids may play some role in this, but because they are not central-place foragers, there is no requirement for the elaboration of trails and other food-locating mechanisms. Communication in soldier-producing aphids is therefore largely restricted to two areas: information about predator attack and information about clone size and development in relation to soldier-production itself. Almost nothing is known about this second area, although it is clearly of great interest to know how, if at all, the aphid clones are able to monitor their own development to produce the right amount, and perhaps the right type, of soldier at the appropriate time (Stern *et al.* 1994).

Aphids are well known to produce alarm pheromones from their cornicles in response to predator attack. These pheromones either cause nearby aphids to walk, jump or fly from the area, or they may simply heighten the aphids' responses to stimuli associated with natural enemies (Dixon 1985). The only detailed account of an alarm pheromone in a soldier-producing species is that of Arakaki (1989) on *Ceratovacuna lanigera*. Arakaki showed experimentally that a cornicular secretion, which was produced by all aphid instars, had a dual effect on the other aphids: the first-instar soldiers approached the secretion and attacked whatever it was adhered to (either paper or a predator) and the later instars and adults all moved away from the secretion. The soldiers were observed to have a special behavior of lifting and bending their abdomens to smear droplets onto a predator.

Schütze and Maschwitz (1991) reported a different mechanism of recruitment of soldiers to a predator. Soldiers of *Pseudoregma sundanica*, which have no alarm pheromones (Schütze and Maschwitz 1992), probably because they are obligately ant-tended (see Nault *et al.* 1976), do not respond to tactile stimulation but do respond to contact with insect hemolymph. Paper balls soaked in hemolymph evoked a permanent clinging response in 70–85% of trials: the soldiers responded approximately equally to aphid, noctuid and grasshopper hemolymph. It would be interesting to try this experiment on other soldier aphids, and to analyze what components of the hemolymph are responsible for the effect. An important consequence of this behavior is that there will be strong selection for predators to develop clean eating habits: those that can siphon off aphid hemolymph without causing any drips will be able to feed undisturbed.

Kin discrimination

There are two distinct reasons why we might expect soldier-producing aphids to discriminate against aphids

from another clone. First, invasion of aphids from other clones would reduce the benefit of soldier production to the host clone. Second, the alien aphids might impose direct costs on the host clone, for example by taking up valuable feeding space or by fouling the local environment. This second effect will perhaps be important only in gall-living aphids and will occur whether or not the clone produces soldiers.

The only reported example of discrimination in a soldier-producing species is that shown by *Ceratoglyphina styracicola*: soldiers in the gall-forming generations always attacked conspecific non-soldiers, regardless of kinship, provided that they encountered them outside the gall (Aoki *et al.* 1991). Other soldiers found outside the gall and apparently aphids within the gall were never attacked. The effect of this behavior is to exclude unrelated non-soldiers from the gall. Unrelated soldiers are presumably not much of a threat to the clone, since they will not be able to reproduce (although there is some evidence to the contrary (Kurosu and Aoki 1991a)); indeed, they might provide a bonus to the clone, as unpaid mercenaries. This mode of soldier attack raises a paradox: if soldiers on the outside of the gall attack anything that is not a conspecific soldier, then how do the winged migrants leave the gall without being attacked by soldiers? It turns out that soldiers change their behavior late in the gall cycle when alates start leaving the gall. At this time, soldiers no longer attack any conspecifics, although they continue to attack other aphid species and predators (Aoki and Kurosu 1992).

There is no evidence that any kind of aphid is able to discriminate kin from non-kin, either in the context of mating (Foster and Benton 1992) or fighting (Foster 1990; Aoki *et al.* 1991; Sakata and Itô 1991). However, the observations of Aoki *et al.* (1991) indicate that these effects might be relatively subtle, and it would be well worth looking at other species in greater detail. For example, Foster (1990) showed that soldiers of *Pemphigus spyrothecae* did not attack alien conspecifics when confined with them in experimental chambers, but this does not rule out the possibility of more complex context-specific types of discrimination.

A SCENARIO FOR SOLDIER EVOLUTION

By considering the diversity of soldiers reviewed above we have generated a general hypothesis for the steps involved in the origin of different soldier types (Table 7-1). Perhaps the first step in the evolution of soldiers was the development of attacking behavior by the appropriate instar.

Table 7-1. *A scenario for the evolution of sterile dimorphic soldiers*

It is easy to imagine that stages 2 and 3 might be switched so that instar prolongation preceded the evolution of morphological specializations. A question mark after an example species indicates that not enough data have been published to determine whether these species truly fit the stage in which they have been placed.

Stage	Description	Examples
1. Behavioral specialization	The instar defends the clone but is not morphologically specialized compared with the ancestral state	*Pemphigus dorocola* (?)[1]
2. Morphological specialization	The instar defends the clone and is morphologically specialized compared with the ancestral state	*Pemphigus gairi* (?)[2]
3. Prolongation of defensive instar	Molting to the next instar of defenders is delayed	*Pemphigus obesinymphae*[3], *Hemipodaphis persimilis*[4], *Ceratovacuna lanigera*[5]
4. Dimorphism	The instar becomes differentiated into two morphological castes: soldier: slow-growing, defensive non-soldier: rapid-growing, non-defensive	*Pemphigus spyrothecae*[6]
5. Sterility	The soldier caste does not molt, becomes 'deterministically' sterile	Many cerataphidine aphids on the 1° and 2° hosts[7]

Sources: [1]Aoki (1978); [2]W. A. Foster (personal observations); [3]Moran (1993); [4]Akimoto (1992); [5]Aoki *et al.* (1984), Takano (1941); [6]Aoki and Kurosu (1986), Foster (1990); [7]Aoki (1987).

These instars would not have been morphologically specialized or physically different from the ancestral state. These defender instars might then have developed morphological specializations. At this point, the duration of the defensive instars might have become prolonged as an easy way to increase defense investment; however, it is also easy to imagine that instar prolongation preceded morphological specialization. The instar might then have split into two developmental paths, producing two sorts of aphid within the same instar: soldiers and non-soldiers. Part of the reason for this might have been selection for a fast-growing route to the production of adults. Finally, soldier development might have become so delayed that soldiers became obligately sterile.

Further elaboration of the soldier caste might have taken place. For example, there is some evidence for morphological soldier subcastes (Aoki and Miyazaki 1978). Another strong possibility is the development of temporal polyethism within the defensive instar. For example, within the Pemphigidae, it is conceivable that the young defenders specialize in gall-cleaning (see below), whereas the older ones are specialized warriors. It would be well worth explicitly looking for temporal polyethism in the soldier castes.

It might be possible to test this scenario phylogenetically, but we must stress that this scenario was constructed by considering all the known soldier-producing species. Particular taxonomic groups possess species representative of some, but not all, stages. In addition, we do not advocate thinking about soldier evolution as a one-way, linear process. In fact, we might expect lineages to move quickly (in both directions) among stages or to skip some stages altogether (cf. Michener 1985). There is already evidence that some species have lost soldiers (Stern 1994). It will be extremely useful to determine the relative costs and benefits of the different stages in the proper historical context to fully understand why this diversity of soldier types exists.

EXPLAINING THE PHYLOGENETIC DISTRIBUTION OF SOLDIERS

About 50 of the approximately 4400 species of aphid have been found to produce soldiers. These soldier-producing species are restricted to two of the ten aphid families, the Pemphigidae and the Hormaphididae (Fig. 7-3). Within these two closely related families, soldiers have been found in six of the nine tribes (Fig. 7-4). This taxonomic distribution suggests that soldiers have been gained and lost many times. These multiple evolutionary events will eventually allow comparative tests of hypotheses for the causes of the evolution of soldiers. At the moment,

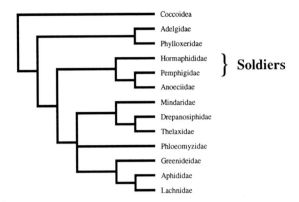

Fig. 7-3. Phylogeny of Aphidoidea, redrawn from Heie (1987), indicating soldier-producing families.

however, there is little ecological information with which we can test ideas for the evolution of soldiers.

In the remaining sections we begin to address the causes for the observed distribution of soldiers with the data that are available. But perhaps the most useful role of the discussion will be to pinpoint the areas that we think are most important for future research on soldier aphid biology.

Historical aspects of soldier production

The current phylogenetic distribution of soldier-producing aphids presents us with two major problems. First, why have soldier aphids been found only in the two closely related families, Pemphigidae and Hormaphididae, and, second, what has determined the distribution of soldier production within these two families?

It is sometimes assumed that helping behavior can only evolve in insects if they are mandibulate, because these are the only species that have the requisite, generalized cutlery for the delicate tasks of nest construction and the manipulation of food, eggs and young larvae (see, for example, Wilson 1971; Eickwort 1981). On this view, aphids would be considered too specialized to provide help for each other (Hamilton 1987). However, because aphids are viviparous, exopterygote, and can readily obtain their own food, the requirement for elaborate parental care is greatly relaxed and therefore the absence of mandibulate mouthparts does not provide a major barrier to the evolution of sociality. In any case, aphids do have excellent, ready-made weapons – their stylets – with which they might defend the clone. Long before

Aoki's (1977a) demonstration of a soldier caste, there were reports of aphids attacking humans, predators' eggs, and each other with their stylets (summarized in Banks et al. 1968). In addition, aphids (Stern 1995) and also thrips (Crespi 1992), by the subtle use of their stylets, are able to provoke plants into producing the equivalent of a nest for their offspring; a nest, moreover, produced with an economy of means unparalleled in the other social insects and providing for their offspring food and shelter from the environment, from competitors and from natural enemies. These nests are, of course, plant galls; those induced by aphids can be of wonderful, baroque complexity (see, for example, Aoki and Kurosu 1989a, 1993; Aoki and Usuba 1989). We believe that the galling habit has played a vital role in soldier evolution and we discuss this in the next section.

Galls and the evolution of soldiers

The two soldier-producing families share the trait of inducing galls on their primary hosts; the ancestral condition for the two families is probably gall production. Foster and Northcott (1994) have discussed how gall-living might have facilitated the evolution of sociality in aphids. Aphid galls, being defensible, long-lasting, initially small, expansible, food-rich 'nest-sites', fulfill all the requirements suggested by Alexander et al. (1991) to be critical for facilitating social evolution. Because aphids do not need to exit the gall to forage, aphid galls are more similar to the nests of mole rats and termites than to those of the Hymenoptera. Alexander et al. (1991) consider this to be an important predisposing factor in the evolution of eusociality in the former groups. Aphid galls differ considerably in the extent to which they are long-lasting; we will consider this variation in a later section.

There are two ways in which gall-living might have influenced the evolution of soldiers. (1) Gall-living and associated morphological and behavioral specializations might have provided useful traits that were later adopted and modified by the soldier morph. We consider these specializations to be preadaptations to soldier production. (2) Gall-living and associated traits might have directly selected for soldier production. In addition, some traits may have acted as preadaptations as well as directly selecting for soldiers.

Gall-associated preadaptations. Aphids do not normally fight each other, but an important exception to this is the first-instar foundresses which have been shown to fight each other for access to good gall-initiation sites. Such fighting has been observed in the pemphigids *Pemphigus betae* (Whitham 1979) and *Epipemphigus niisimae* (Aoki and Makino 1982) and in the hormaphidid *Ceratovacuna nekoashi* (Kurosu and Aoki 1990a). In these and other examples, it is interesting that the first-instar foundresses seem to be well prepared for fighting, in terms of both morphology and behavior, whereas the later instars are not specialized in this way. The limited amount of data available suggests that fighting by first-instar foundresses is widespread in the Hormaphididae and Pemphigidae. Therefore, in these two families there already existed a fighting stage whose specializations could be adopted by later gall generations in defending the clone against predators.

There might be considerable selective advantage for clones to invest in individuals that dispersed to seek out and colonize other galls (Aoki 1982b). These aphids might acquire behaviors and morphology, such as well-developed walking abilities, increased sclerotization to combat desiccation, and general breaking-and-entering capabilities, that would be useful to them not only as migrants but also as soldiers. A convincing example of this is provided by *Pachypappa marsupialis* (Aoki 1979a), whose first instars are migratory: their cuticle is well sclerotized, they are able to enter conspecific galls, and they attack syrphid eggs. There is suggestive evidence from several studies that intergall migration is common (Setzer 1980; Williams and Whitham 1986; Kurosu and Aoki 1990b).

Galling traits selecting for soldiers. First, being enclosed in a gall ought to make the clone more readily defendable. The significance of this factor will depend strongly on the details of gall morphology (for example, see Stern *et al.* 1994), in particular on the number and extent of the openings to the gall.

Second, living in a gall will tend to preserve the genetic integrity of the colony, simply by reducing the chances that alien conspecifics will enter the colony. This reduced genetic mixing would increase the benefit that soldiers would provide for the clone and therefore increase the likelihood that the clone will invest in them (Stern and Foster 1996). However, there is little empirical data on the genetics of gall-aphid populations. In some species, for example *Pemphigus populitransversus* and *P. populicaulis* (Setzer 1980), genetic mixing occurs at a high level, whereas in other gall aphids, for example *Melaphis rhois* (Hebert *et al.* 1991), mixing appears to occur at a very low level. There is an urgent need for further observations and measurements on genetic mixing in social aphid colonies.

Finally, the ability to defend the gall might be an evolutionary offshoot of the need to keep the gall clean. All gall aphids face the potentially severe problem of contamination with their own honeydew. A common solution to this is to wrap the honeydew in wax and push the wax-covered material, along with cast skins and dead aphids, out of the gall. Benton and Foster (1992) established in *Pemphigus spyrothecae* that this housekeeping behavior was costly to the soldiers and essential for the survival of the clone. It is interesting that it is the soldiers or, if there are no defenders (e.g. *Hormaphis betulae* (Kurosu and Aoki 1991d)), the first instars that clean the gall. In addition, gall-cleaning seems to be phylogenetically more widespread than defensive behavior (for example, both *H. betulae* and *Pemphigus bursarius* (W. A. Foster and T. G. Benton, unpublished observations) have gall-cleaners but not defenders). It is plausible that some first or second instars might have become specialized as cleaners, perhaps because they were small enough to work effectively at the gall opening, and in pushing things out of the gall they might have come into contact with predators trying to get in. In other words, the waste orifice of the gall might have become its 'Achilles heel', allowing easy access for predators and strong selection for the cleaning morphs, already in the appropriate location, to develop an additional role as defensive warriors. Thus, gall-cleaning might have served as both a preadaptation as well as directly selecting for soldiers.

The origins of soldiers within the Hormaphididae and Pemphigidae

Although gall-living might provide the key to a general understanding of why soldiers are restricted to aphids belonging to the Hormaphididae and Pemphigidae, it is not a complete explanation and it does not go very far in explaining the pattern of distribution of soldiers within these two families. Many gall-living species, in these and other aphid families, do not have soldiers, and within these families soldiers occur on the secondary hosts where galls are usually not produced. Understanding the details of soldier evolution within these two families is dependent on reliable phylogenies; phylogenies for the relevant groups are beginning to appear (Stern 1994). These phylogenies, along with consideration of the taxonomic distribution of soldiers, indicate that soldiers have been lost and gained many times within these two families (Fig. 7-4).

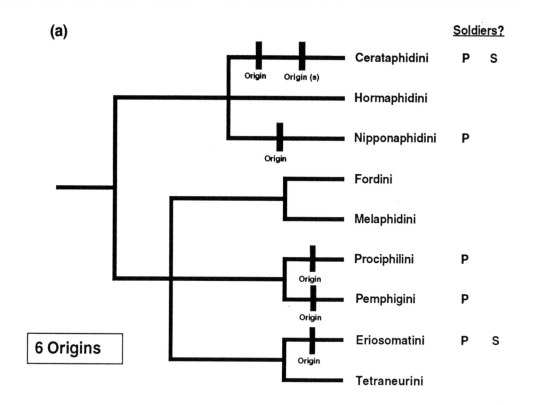

(a)

Soldiers?

Cerataphidini   P   S

Origin   Origin (s)

Hormaphidini

Nipponaphidini   P

Origin

Fordini

Melaphidini

Prociphilini   P

Origin

Pemphigini   P

Origin

Eriosomatini   P   S

Origin

Tetraneurini

6 Origins

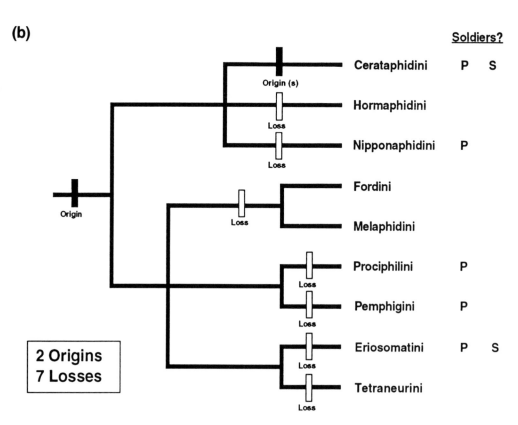

(b)

Soldiers?

Cerataphidini   P   S

Origin (s)

Hormaphidini

Loss

Nipponaphidini   P

Loss

Fordini

Loss

Melaphidini

Prociphilini   P

Loss

Pemphigini   P

Loss

Eriosomatini   P   S

Loss

Tetraneurini

Loss

Origin

2 Origins
7 Losses

These patterns of soldier evolution will provide the setting for studies addressing how natural selection has been acting on soldier production in these lineages. We discuss elsewhere the potential role of these ecological factors (Stern and Foster 1996). Below we discuss two topics of special interest: the evolution of secondary-host soldiers and the importance of the peculiarities of aphid life cycles to the evolution of soldiers.

The origins of soldiers on the secondary hosts
Soldiers occur on the secondary hosts in two tribes: the Eriosomatini and the Cerataphidini (Fig. 7-4). These soldier-producing species also have soldiers on the primary host in all species where the complete life cycle is known. Clearly, the most parsimonious explanation for this distribution is that the secondary-host soldiers are simply a carry-over from those on the primary host. Once a set of behaviors, in this case attacking rather than running from predators, became established in one part of the life cycle, it might easily have been adopted in another part of the life cycle. For the soldiers within the Eriosomatini, this is a plausible argument; the soldiers on the primary and secondary hosts are morphologically similar and use similar weapons, although the primary-host soldiers are usually second instars whereas those on the secondary host are first instars (Aoki and Miyazaki 1985).

For the soldiers within the Cerataphidini, however, there are convincing arguments (Aoki 1987; Aoki and Kurosu 1989b) that the primary- and secondary-host soldier are not homologous and represent at least two separate origins. The

Fig. 7-4. Phylogeny of Hormaphididae and Pemphigidae illustrating two possible scenarios for the origins and losses of soldiers. This phylogeny was constructed based on the classification of Eastop (1977) assuming all taxonomic categories represent monophyletic groups. The type of soldiers produced, primary-host (P) or secondary-host (S), is indicated next to the name of each tribe. In all tribes, except primary host soldiers in the Cerataphidini, only some species produce soldiers. Therefore, each tribe with soldier-producing species requires a minimum of one event to account for the presence of soldiers in only some species. In addition, primary and secondary host soldiers of the Cerataphidini are likely to represent independent evolutionary events (see text). (a) If origins and losses are assumed to be equally likely then the most parsimonious reconstruction requires a minimum of six independent origins. (b) If losses are assumed to be more likely than origins then the most parsimonious reconstruction requires a minimum of two origins and seven losses.

two types of soldier are very different in morphology and behavior: those on the primary host fight with their stylets, whereas those on the secondary host have large offensive forelegs and use entirely novel weapons – their frontal horns – to tear open predators (see Fig. 7-1). Primary-host soldiers are taxonomically more widespread than secondary-host soldiers. Finally, phylogenetic evidence strongly suggests that the secondary-host soldiers represent a separate and relatively recent innovation (Stern 1994).

We find it somewhat surprising that secondary-host soldiers of the Cerataphidini have evolved a novel defensive weapon (horns) when primary-host soldiers of the same species, which evolved first, use their stylets for defense. This is even more surprising when it is noted that one species, which has horns, has been found to use its stylets to attack predators' eggs on the secondary host (Aoki and Kurosu 1989c) and a second horned species might also use its stylets for attack (Agarwala et al. 1984).

If the horned secondary-host soldiers of the Cerataphidini are not derived from the gall-inhabiting soldiers, what might be their evolutionary antecedents? An early function of the frontal horns was probably as a weapon with which to fight conspecific aphids for feeding sites on the secondary host; this head-butting behavior seems to occur in all the species of the Cerataphidini that have horns (Aoki and Kurosu 1985; Aoki 1987; W. A. Foster, S. Aoki, U. Kurosu and D. L. Stern, unpublished data). Aphids with butting behavior might then have attacked predators, perhaps initially piercing predator's eggs (see, for example, Aoki and Kurosu 1987). However, the only way to begin to understand why the horned soldiers are restricted to the genera *Pseudoregma* and *Ceratovacuna* will be to find out more about the ecological context of social behavior in these taxa and their close relatives lacking soldiers.

It is worth noting that secondary-host soldiers may serve as better model systems for some ecological studies of soldiers. These colonies are typically exposed on leaves and very easy to observe, contrasting with the major difficulty of observing soldier behavior in galls. In addition, these colonies are much easier to manipulate experimentally and they often occupy the host plants for a long period of time.

The importance of gall duration
The only obvious life-history correlate of soldier production within the Pemphigidae and Hormaphididae is the duration of colonies within galls. Soldier-producing species tend to inhabit long-lasting galls. For example, the

most aggresive soldiers in the genus *Pemphigus* are produced by *P. obesinymphae* (Moran 1993), which produces galls that last several months longer than those of other *Pemphigus* species. In addition, galls of the Cerataphidini, which possess some of the most aggressive soldiers (Aoki 1979b), can last up to two years (Aoki and Kurosu 1990, 1992). Further quantitative data are still needed to verify this correlation, but unpublished observations from a number of workers seem to support this trend (S. Aoki, U. Kurosu and S. Akimoto, personal communication).

There are two possible reasons for the correlation between gall duration and soldier production. First, a gall may be a valuable, potentially long-lasting, defendable resource (cf. Alexander *et al.* 1991) and this may have selected for defense. That is, defense allowed colonies to occupy the gall for longer than predation might otherwise allow. Alternatively, aphid colonies may be forced, because of constraints imposed by other parts of their life cycle, to remain in galls for an extended period, perhaps longer than is otherwise optimal. In such a case, they may have no recourse but to defend the colony until they are able to move to the next stage of their life cycle.

Many aphid species migrate between two host plant species, and the timing of migration between hosts may be dependent on a number of factors including host plant phenology. Certain plants may simply be unavailable to aphids at certain times of year or their nutritional status may make migration disadvantageous. In such cases, aphids may be better off remaining on one host, for example in a gall, even though the benefits of staying are not particularly high. In addition, galls provide a relatively beneficial microclimate and predictable adverse environmental conditions, for example a long hot summer, may make it beneficial for colonies to remain in galls.

These two hypotheses are very similar and it may be difficult to differentiate between them. The first suggests that soldiers are produced to allow the utilization of a gall for an extended period because the gall is a favorable resource even though the alternate host may be available as a suitable resource. The second suggests that aphids are stuck in their gall for an extended period for some reason and they resort to defense to survive this difficult period. It will also be difficult to discriminate between these hypotheses and more general life-history-based hypotheses (see Stern and Foster 1996). For example, colonies in long-lasting galls probably have lower growth rates than colonies in short-lived galls (cf. Moran 1993), and this may also select for soldier production (Stern and Foster 1996).

Finally, it must be remembered that soldiers are not always produced within galls (for example on the secondary hosts in the Cerataphidini and Eriosomatini) and it seems unlikely that soldiers are produced to defend a valuable resource in these cases.

CONCLUSIONS AND PROSPECTS

The three major points we have sought to establish in this review are: (1) the clonality of aphids sets them apart from all other social insects, in particular by expunging all traces of conflict from colony members descended from a single mother; (2) aphid soldiers are diverse in morphology and behavior; and (3) there have been multiple origins of soldiers within the aphids.

To what extent are we now able to account for the restricted and patchy taxonomic distribution of soldier-producing aphids? At the gross level, gall-living is of enormous importance in the evolution of sociality: the crucial ingredients provided by living in a gall are increased genetic integrity and defendability of the clone. There are parallels here with the social gall-living thrips: Crespi and Mound (this volume) suggest that the important role played by the gall in the evolution of sociality in thrips is as a home and feeding site. Although the gall will provide a good feeding site and physical protection for the aphids, we suggest that its role in the evolution of sociality is that it encourages a mother to invest in soldiers by increasing the likelihood that they are defending clonemates, rather than aliens, and by making the clone more readily defendable.

Although gall-living may be necessary, it is certainly not sufficient for the evolution of social behavior. Most gall-living species do not produce soldiers. Explaining the pattern of soldier-producing species within the gall-living aphids remains the main challenge for the future. The first task is to provide a theoretical framework, including a general model of optimal resource allocation to defense and growth, and simulation models, applicable to particular gall-living species, that use measured values of the critical variables (growth rate, predation rate, etc.). Second, we need to understand how aphid life cycles impose constraints on the evolution of sociality and select for soldiers on either the primary or the secondary host. Third, we need to measure the costs and benefits of producing soldiers: this will be crucial not only in helping to explain why a particular species has soldiers, but also in understanding why particular types of soldiers (size, instar, weaponry) are produced. Finally, we need to elucidate the importance of clonal mixing: this will require both the development of an appropriate theoretical framework and reliable biochemical methods for measuring mixing under natural conditions.

Soldier aphids prompt several other fascinating questions that are not directly related to the evolution of sociality. In particular, we know nothing about the proximate mechanisms controlling caste. Developmental decisions about morph production must be made by the mother, probably at an early stage of embryogenesis. How does she induce soldier production in embryos and how is this influenced by external cues? The social aphids are a rich vein of natural history that we have only just begun to mine: many soldier-producing species remain undiscovered, many extraordinary phenomena await description.

ACKNOWLEDGEMENTS

We are very grateful to S. Aoki for intitiating the study of social aphids. D. L. S. is grateful to S. Aoki, U. Kurosu, S. Akimoto, Y. Yamaguchi, N. Pierce and C. Martinez del Rio for many hours of helpful discussion and advice on many of the ideas presented here. Thanks to S. Aoki, J. Bonner, J. Choe, B. Crespi and N. Moran for helpful comments on the manuscript. We offer special thanks to S. Aoki for permission to adapt his Fig. 4-1 from Aoki (1987) for use in Fig. 7-4. Kristina Thalia Grant graciously illustrated Fig. 7-1 within a significant time constraint.

LITERATURE CITED

Agarwala, B. K., S. Saha and A. K. Ghosh. 1984. Studies on *Ceratovacuna silvestrii* (Takahashi) (Homoptera: Aphididae) and its predator *Anisolemnia dilatata* (Fab.) on *Bambusa arundinacea*. *Rec. Zool. Surv. India* 81: 23–42.

Akimoto, S. 1992. Shift in life-history strategy from reproduction to defense with colony age in the galling aphid *Hemipodaphis persimilis* producing defensive first-instar larvae. *Res. Pop. Ecol.* 34: 359–372.

Alexander, R. D., K. M. Noonan and B. J. Crespi. 1991. The evolution of eusociality. In *The Biology of the Naked Mole-Rat*. P. W. Sherman, J. U. M. Jarvis and R. D. Alexander, eds., pp. 3–44. Princeton: Princeton University Press.

Anderson, M. and A. K. Bromley. 1987. Sensory system. In *Aphids and Their Biology, Natural Enemies and Control*. A. K. Minks and P. Harrewijn, eds., pp. 153–162. Amsterdam: Elsevier.

Aoki, S. 1977a. *Colophina clematis* (Homoptera, Pemphigidae), an aphid species with 'soldiers'. *Kontyû* 45: 276–282.

–. 1977b. A new species of *Colophina* (Homoptera, Aphidoidea) with soldiers. *Kontyû* 45: 333–337.

–. 1978. Two pemphigids with first instar larvae attacking predatory intruders (Homoptera, Aphidoidea). *New Entomol.* 27: 7–12.

–. 1979a. Dimorphic first instar larvae produced by the fundatrix of *Pachypappa marsupialis* (Homoptera: Aphidoidea). *Kontyû* 47: 390–398.

–. 1979b. Further observations on *Astegopteryx styracicola* (Homoptera: Pemphigidae), an aphid species with soldiers biting man. *Kontyû* 47: 99–104.

–. 1980. Life cycles of two *Colophina* aphids (Homoptera, Pemphigidae) producing soldiers. *Kontyû* 48: 464–476.

–. 1982a. Pseudoscorpion-like second instar larvae of *Pseudoregma shitosanensis* (Homoptera, Aphidoidea) found on its primary host. *Kontyû* 50: 445–453.

–. 1982b. Soldiers and altruistic dispersal in aphids. In *The Biology of Social Insects*. M. D. Breed, C. D. Michener, and H. E. Evans, eds., pp. 154–158. Boulder, Colorado: Westview Press.

–. 1983. A new Taiwanese species of *Colophina* (Homoptera, Aphidoidea) producing large soldiers. *Kontyû* 51: 282–288.

–. 1987. Evolution of sterile soldiers in aphids. In *Animal Societies: Theories and Facts*. Y. Itô, J. L. Brown and J. Kikkawa, eds., pp. 53–65. Tokyo: Japan Sci. Soc. Press.

Aoki, S., S. Akimoto and S. Yamane. 1981. Observations on *Pseudoregma alexanderi* (Homoptera, Pemphigidae), an aphid species producing pseudoscorpion-like soldiers on bamboos. *Kontyû* 49: 355–366.

Aoki, S. and U. Kurosu. 1985. An aphid species doing a headstand: butting behavior of *Astegopteryx bambucifoliae* (Homoptera: Aphidoidea). *J. Ethol.* 3: 83–87.

–. 1986. Soldiers of a European gall aphid, *Pemphigus spyrothecae* (Homoptera: Aphidoidea): Why do they molt? *J. Ethol.* 4: 97–104.

–. 1987. Is aphid attack really effective against predators? A case study of *Ceratovacuna lanigera*. In *Population Structure, Genetics and Taxonomy of Aphids and Thysanoptera*. J. Holman, J. Pelikan, A. F. G. Dixon and L. Weisman, eds., pp. 224–232. The Hague: SPB Academic Publishers.

–. 1989a. Soldiers of *Astegopteryx styraci* (Homoptera, Aphidoidea) clean their gall. *Jap. J. Entomol.* 57: 407–416.

–. 1989b. Two kinds of soldiers in the tribe Cerataphidini (Homoptera: Aphidoidea). *J. Aphidol.* 3: 1–7.

–. 1989c. A bamboo horned aphid attacking other insects with its stylets. *Jap. J. Entomol.* 57: 663–665.

–. 1990. Biennial galls of the aphid *Astegopteryx styraci* on a temperate deciduous tree, *Styrax obassia*. *Acta Phytopathol. Entomol. Hung.* 25: 57–65.

–. 1992. No attack on conspecifics by soldiers of the gall aphid *Ceratoglyphina bambusae* (Homoptera) late in the season. *Jap. J. Entomol.* 60: 707–713.

–. 1993. The gall, soldiers and taxonomic position of the aphid *Tuberaphis taiwana* (Homoptera). *Jap. J. Entomol.* 61: 361–369.

Aoki, S., U. Kurosu and D. L. Stern. 1991. Aphid soldiers discriminate between soldiers and non-soldiers, rather than between kin and non-kin, in *Ceratoglyphina bambusae*. *Anim. Behav.* **42**: 865–866.

Aoki, S., U. Kurosu and S. Usuba. 1984. First instar larvae of the sugar-cane wooly aphid, *Ceratovacuna lanigera* (Homoptera, Pemphigidae), attack its predators. *Kontyû* **52**: 458–460.

Aoki, S. and S. Makino. 1982. Gall usurpation and lethal fighting among fundatrices of the aphid *Epipemphigus niisimae* (Homoptera, Pemphigidae). *Kontyû* **50**: 365–376.

Aoki, S. and M. Miyazaki. 1978. Notes on the pseudoscorpion-like larvae of *Pseudoregma alexanderi* (Homoptera, Aphidoidea). *Kontyû* **46**: 433–438.

–. 1985. Larval dimorphism in Hormaphidinae and Pemphiginae aphids in relation to the function of defending their colonies. In *Evolution and Biosystematics of Aphids, Proceedings of the International Aphidological Symposium at Jablonna, 1981.* H. Szelegiewicz, ed., pp. 337–338. Warszawa: Polska Akademia Nauk.

Aoki, S. and S. Usuba. 1989. Rediscovery of '*Astegopteryx*' *takenouchii* (Homoptera, Aphidoidea), with notes on its soldiers and hornless exules. *Jap. J. Entomol.* **57**: 497–503.

Aoki, S., S. Yamane and M. Kiuchi. 1977. On the biters of *Astegopteryx styracicola* (Homoptera, Aphidoidea). *Kontyû* **45**: 563–570.

Arakaki, N. 1989. Alarm pheromone eliciting attack and escape responses in the sugar cane wooly aphid, *Ceratovacuna lanigera* (Homoptera, Pemphigidae). *J. Ethol.* **7**: 83–90.

Banks, C. J., E. D. M. Macaulay and J. Holman. 1968. Cannibalism and predation by aphids. *Nature (Lond.)* **218**: 491.

Benton, T. G. and W. A. Foster. 1992. Altruistic housekeeping in a social aphid. *Proc. R. Soc. Lond.* B**247**: 199–202.

Crespi, B. J. 1992. Eusociality in Australian gall thrips. *Nature (Lond.)* **359**: 724–726.

Crespi, B. J. and D. Yanega. 1995. The definition of eusociality. *Behav. Ecol.* **6**: 109–115.

Cruz, Y. P. 1981. A sterile defender morph in a polyembryonic hymenopterous parasite. *Nature (Lond.)* **294**: 446–447

–. 1986. The defender role of the precocious larvae of *Copidosomopsis tanytmemus* Caltagirone (Encyrtidae, Hymenoptera). *J. Exp. Zool.* **237**: 309–318.

Dixon, A. F. G. 1985. *Aphid Ecology.* Glasgow: Blackie & Son Limited.

Eastop, V. F. 1977. Worldwide importance of aphids as virus vectors. In *Aphids as Virus Vectors.* K. F. Harris and K. Maramorosch, eds., pp. 3–62. New York: Academic Press.

Eickwort, G. C. 1981. Presocial insects. In *Social Insects,* vol. 2. H. R. Hermann, ed., pp. 199–280. New York: Academic Press.

Foster, W. A. 1978. Dispersal behaviour of an intertidal aphid. *J. Anim. Ecol.* **47**: 653–659.

–. 1990. Experimental evidence for effective and altruistic colony defence against natural predators by soldiers of the gall-forming aphid *Pemphigus spyrothecae* (Hemiptera: Pemphigidae). *Behav. Ecol. Sociobiol.* **27**: 421–430.

Foster, W. A. and T. G. Benton. 1992. Sex ratio, local mate competition and mating behaviour in the aphid *Pemphigus spyrothecae*. *Behav. Ecol. Sociobiol.* **30**: 297–307.

Foster, W. A. and P. A. Northcott. 1994. Galls and the evolution of social behaviour in aphids. In *Plant Galls: Organisms, Interactions, Populations.* M. A. J. Williams, ed., pp. 161–182. Oxford University Press.

Foster, W. A. and J. E. Treherne. 1978. Dispersal mechanisms in an intertidal aphid. *J. Anim. Ecol.* **47**: 205–217.

Hamilton, W. D. 1987. Kinship, recognition, disease, and intelligence: Constraints of social evolution. In *Animal Societies: Theories and Facts.* Y. Itô, J. L. Brown and J. Kikkawa, eds., pp. 81–102. Tokyo: Japan Sci. Soc. Press.

Hebert, P. D. N., T. L. Finston and R. Foottit. 1991. Patterns of genetic diversity in the sumac gall aphid, *Melaphis rhois.* *Genome* **34**: 757–762.

Heie, O. E. 1987. Paleontology and phylogeny. In *Aphids: Their Biology, Natural Enemies and Control,* vol A. A. K. Minks and P. Harrewijn, eds., pp. 367–391. Amsterdam: Elsevier.

Hölldobler, B. and E. O. Wilson. 1990. *The Ants.* Berlin: Springer-Verlag.

Hughes, R. N. 1989. *A Functional Biology of Clonal Animals.* London: Chapman and Hall.

Kennedy, J. S., L. Crawley and A. D. MacLaren. 1967. Spaced-out gregariousness in sycamore aphids *Drepanosiphum platanoides* (Schrank) (Hemiptera, Callaphididae). *J. Anim. Ecol.* **36**: 147–170.

Kurosu, U. and S. Aoki. 1988a. Monomorphic first instar larvae of *Colophina clematicola* (Homoptera, Aphidoidea) attack predators. *Kontyû* **56**: 867–871.

–. 1988b. First-instar aphids produced late by the fundatrix of *Ceratovacuna nekoashi* (Homoptera) defend their closed gall outside. *J. Ethol.* **6**: 99–104.

–. 1990a. Formation of a 'Cat's-paw' gall by the aphid *Ceratovacuna nekoashi* (Homoptera). *Jap. J. Entomol.* **58**: 155–166.

–. 1990b. Transformation of the galls of *Astegopteryx bambucifoliae* by another aphid, *Ceratoglyphina bambusae. Acta phytopathol. Entomol. Hung.* **25**: 113–122.

–. 1991a. Molting soldiers of the gall aphid *Ceratoglyphina bambusae* (Homoptera). *Jap. J. Entomol.* **59**: 576.

–. 1991b. The gall formation, defenders and life cycle of the subtropical aphid *Astegopteryx bambucifoliae* (Homoptera). *Jap. J. Entomol.* **59**: 375–388.

–. 1991c. Incipient galls of the soldier-producing aphid *Ceratoglyphina bambusae* (Homoptera). *Jap. J. Entomol.* **59**: 663–669.

–. 1991d. Gall cleaning by the aphid *Hormaphis betulae. J. Ethol.* **9**: 51–55.

Kurosu, U., I. Nishitani, Y. Itô and S. Aoki. 1995. Defenders of the aphid *Nipponaphis distyliicola* (Homoptera) in its completely closed gall. *J. Ethol.* **13**: 133–136.

Lampel, G. 1968–9. Untersuchungen zur Morphenfolge von *Pemphigus spirothecae* Pass. 1860 (Homoptera, Aphidoidea). *Bull. Naturforsch. Ges. Freiburg* **58**: 56–72.

Leigh, E. G., Jr. 1983. When does the good of the group override the advantage of the individual? *Proc. Natl. Acad Sci. U.S.A.* **80**: 2985–2989.

Michener, C. D. 1969. Comparative social behavior of bees. *Annu. Rev. Entomol.* **14**: 299–342.

–. 1985. From solitary to eusocial: need there be a series of intervening species? *Fortschr. Zool.* **31**: 293–305.

Miyazaki, M. 1987. Forms and morphs of aphids. In *Aphids: Their Biology, Natural Enemies and Control*, vol A. A. K. Minks and P. Harrewijn, eds., pp. 27–50. Amsterdam: Elsevier.

Moran, N. A. 1988. The evolution of host-plant alternation in aphids: evidence for specialization as a dead end. *Am. Nat.* **132**: 681–706.

–. 1992. The evolution of aphid life cycles. *Annu. Rev. Entomol.* **37**: 321–348.

–. 1993. Defenders in the North American aphid *Pemphigus obesinymphae*. *Insectes Soc.* **40**: 391–402.

Nault, L. R., M. E. Montgomery and W. S. Bowers. 1976. Ant-aphid association: role of aphid alarm pheromones. *Science (Wash., D.C.)* **192**: 1349–1351.

Nault, L. R. and M. E. Montgomery. 1977. Aphid pheromones. In *Aphids as Virus Vectors*. K. F. Harris and K. Maramorosch, eds., pp. 527–545. New York: Academic Press.

Noordam, D. 1991. Hormaphidinae from Java (Homoptera: Aphididae). *Zool. Verhand. (Leiden)* **270**: 1–525.

Sakata, K. and Y. Itô. 1991. Life history characteristics and behaviour of the bamboo aphid, *Pseudoregma bambucicola* (Hemiptera: Pemphigidae), having sterile soldiers. *Insectes Soc.* **38**: 317–326.

Schütze, M. and U. Maschwitz. 1991. Enemy recognition and defence within trophobiotic associations with ants by the soldier caste of *Pseudoregma sundanica* (Homoptera: Aphidoidea). *Entomol. Gen.* **16**: 1–12.

–. 1992. Investigations on the trophobiosis of *Pseudoregma sundanica* (Homoptera: Aphidoidea: Hormaphididae), a Southeast-Asian aphid with sterile soldiers. *Zool. Beitr.* **34**: 337–347.

Setzer, R. W. 1980. Intergall migration in the aphid genus *Pemphigus*. *Ann. Entomol. Soc. Am.* **73**: 327–331.

Sherman, P. W., E. A. Lacey, H. K. Reeve and L. Keller. 1995. The eusociality continuum. *Behav. Ecol.* **6**: 102–108.

Stern, D. L. 1994. A phylogenetic analysis of soldier evolution in the aphid family Hormaphididae. *Proc. R. Soc. Lond.* B**256**: 203–209.

–. 1995. Phylogenetic evidence that aphids, rather than plants, determine gall morphology. *Proc. R. Soc. Lond.* B**260**: 85–89.

Stern, D. L., S. Aoki and U. Kurosu. (1994) A test of geometric hypotheses for soldier investment patterns in the gall producing tropical aphid *Cerataphis fransseni* (Homoptera, Hormaphididae). *Insectes Soc.* **41**: 457–460.

–. (1995) The life cycle and natural history of the tropical aphid *Cerataphis fransseni* (Homoptera, Hormaphididae), with reference to the evolution of host alternation. *J. Nat. Hist.* **29**: 231–242.

Stern, D. L. and W. A. Foster. (1996) The evolution of soldiers in aphids. *Biol. Rev.* **71**: 27–79.

Takano, S. 1941. The sugar cane woolly aphis, *Ceratovacuna lanigera* Zehntner, in Formosa. *Rep. Govt. Exp. Stn, Tainan, Formosa* **9**: 1–76.

Whitham, T. G. 1979. Territorial behaviour of *Pemphigus* gall aphids. *Nature (Lond.)* **279**: 324–325.

Williams, A. G. and T. G. Whitham. 1986. Premature leaf abscission: An induced plant defence against gall aphids. *Ecology* **67**: 1619–1627.

Wilson, D. S. and E. Sober. 1989. Reviving the superorganism. *J. Theor. Biol.* **136**: 337–356.

Wilson, E. O. 1971. *The Insect Societies*. Cambridge: Belknap Press of Harvard University Press.

–. 1975. *Sociobiology: The New Synthesis*. Cambridge: Belknap Press of Harvard University Press.

8 · Ecology and evolution of social behavior among Australian gall thrips and their allies

BERNARD J. CRESPI AND LAURENCE A. MOUND

My walls give me food,
And protect me from foes,
I eat at my leisure,
In safety repose.

A. B. COMSTOCK (1911) *Handbook of Nature-Study*. Ithaca: Comstock Publishing.

ABSTRACT

Australian gall-forming thrips and their allies comprise a group of several dozen species with a remarkable range of life histories and social systems. These species can be categorized into six ecological modes: (1) gall-formers on *Acacia*, which include species of *Oncothrips*, *Kladothrips* and *Onychothrips* in which galls are initiated by a single female, or a male and a female; (2) kleptoparasites in the genus *Koptothrips*, which usurp galls, kill the gall-formers, and breed inside; (3) opportunistic gall-inhabitants, such as some *Warithrips*, *Grypothrips* and *Csirothrips*, which breed in abandoned galls after the original inhabitants have left; (4) domicile-formers in the genera *Lichanothrips*, *Panoplothrips*, *Dunatothrips* and *Carcinothrips*, which either glue phyllodes (petioles modified as leaves) together and live inside, or use a cellophane-like material to create an enclosed space containing multiple apical *Acacia* phyllodes; (5) lepidopteran leaf-tie inhabitants, *Warithrips*, which live in phyllodes tied by lepidopteran silk; and (6) pre-existing hole inhabitatants, species of *Dactylothrips* and *Katothrips* that live in old Hymenoptera galls, abandoned leaf mines, or holes in split-stem galls. Social behavior of these Australian thrips include soldier castes in four species of *Oncothrips* and two species of *Kladothrips*, a wingless, apparently non-soldier morph in *Oncothrips sterni*, pleometrotic (multiple-adult) colony founding in species of *Katothrips*, *Dunatothrips*, and *Lichanothrips*, and group foraging in a *Lichanothrips*. Evidence from adult morphology, gall morphology and mitochondrial DNA sequencing suggests that soldiers have evolved twice in *Acacia* gall thrips. Invasion by *Koptothrips* may be the main selective force for the evolution of soldiers, and selection for rapid domicile formation may favor pleometrosis.

INTRODUCTION

The order Thysanoptera, or thrips, include about 5000 described species, distributed worldwide (Stannard 1968; Lewis 1973; Mound and Teulon 1994). Thrips have asymmetrical piercing–sucking mouthparts, which are used, in all non-predatory species, to feed on plant cell contents, fungal hyphae, or fungal spores (Heming 1993). All bisexual thrips species are apparently haplodiploid (Crespi 1991); all thrips have a form of development that involves two mobile larval stages, two or three quiescent pupal stages, and mobile adults (Davies 1969; Heming 1973). In most species, adults are winged, with the fringed wings characteristic of tiny insects, but wing polymorphism and winglessness are common (Crespi 1986a).

Approximately 300 species of thrips in the suborder Tubulifera form galls on plant leaves, stems, or buds (Mound 1994). Most gall-formers are recorded from the tropical or subtropical Old World, and make simple leaf-rolls on dicots. In Australia, however, some species form rounded, more or less closed galls on phyllodes (petioles modified to the structure and function of leaves) of *Acacia* (Mound 1971) (Fig. 8-1a). These gall-formers are found primarily in the arid and semiarid interior of Australia, and associated with them phylogenetically and ecologically are several dozen thrips species that do not form galls, but have found diverse other ways of inhabiting enclosed spaces and feeding on *Acacia* plant tissue. The purpose of this chapter is to: (1) describe recent discoveries of social behavior among these Australian thrips on *Acacia*, which include new species with soldiers and multiple new modes of gall thrips life-histories and social systems; and (2) present and evaluate hypotheses for explaining the forms and distribution of thysanopteran social behavior.

Fig. 8-1. (a) A gall of *Oncothrips tepperi*. (b) Micropterous soldiers (top row) and female macropterae (bottom row) of *Oncothrips tepperi*. (c) On the right, a domicile of *Lichanothrips* sp. on *A. lysiphloia*, formed from glueing apical phyllodes together with a cellophane-like substance; on the left is a normal set of apical phyllodes, without thrips. (d) An opened domicile of a *Dunatothrips* sp. female, formed from glueing together two phyllodes of *Acacia torulosa*. (e) A foraging group of *Lichanothrips* on *A. tephrina*. (f) An old dried Hymenoptera gall of the type inhabited by a species of *Dactylothrips*.

Gall-forming thrips and their allies exhibit a suite of traits suggestive of the presence of complex social behavior, which motivated the studies described here. First, many species show extreme foreleg allometry among females (Mound 1970, 1971). Among male thrips, such morphological variation is indicative of fighting (Crespi 1986a,b, 1988); among female insects, fighting is found most commonly in social species (Crespi 1992a). Second, foreleg size variation is associated in some species with wing polymorphism: micropterous females and males have large forelegs, and macropterous adults have much smaller ones (Mound 1970; Crespi 1992a,b). Dimorphism in females is usually suggestive of behavioral differences, such as fighter–flier variation or division of labor. Third, sex ratios of gall thrips are often female-biased, sometimes extremely so (Crespi 1993). Female-biased sex ratios in haplodiploid species may favor the evolution of helping under inclusive fitness theory (Hamilton 1967, 1972). Fourth, females of some gall thrips develop extreme physogastry, and lay hundreds of eggs (Crespi 1992a). High fecundity may entail high variation in fecundity, which could favor the evolution of reproductive division of labor (West-Eberhard 1978). Finally, in many species the life of gall foundresses overlaps with that of their adult offspring, and such families inhabit a valuable, long-lived resource. Species of *Koptothrips*, and various lepidopteran, dipteran and hymenopteran invaders, specialize in gall usurpation, and defense against invaders may be an important selective pressure for the evolution of social behavior (Lin 1964; Lin and Michener 1972).

Previous work on thrips sociality includes analysis of group foraging in the spore-feeding species *Anactinothrips gustaviae* in Panama (Kiester and Strates 1986), description of the life cycles, morphology, fighting behavior, and sex ratios of 12 Australian gall thrips taxa (Crespi 1992a), and the discovery of soldier castes in the gall-formers *Oncothrips tepperi*, *O. habrus*, *O. waterhousei* and *Kladothrips hamiltoni* (Crespi 1992b; Mound and Crespi 1995). These studies have recently been augmented by additional collections and observations of *Acacia* gall thrips from throughout most of their range in the Australian interior. Because the study of gall thrips social behavior has literally just begun, the life-cycles of many of the taxa discussed are not yet known in sufficient detail to answer many simple and fundamental questions. However, we believe that the outline of life history and sociality in Australian gall thrips can now be sketched in sufficient detail to serve as an accurate guide for future studies, and that this remarkable group adds great depth and range to the taxa available for addressing the outstanding problems in the theory of social evolution.

COMPARATIVE SOCIAL BIOLOGY OF AUSTRALIAN GALL THRIPS AND THEIR ALLIES

Australian thrips on *Acacia* can be categorized into at least six distinct ecological modes (Table 8-1). For the species in

each mode, we will discuss their morphology, life cycle, and behavior, with emphasis on evidence and intimations of social life.

Gall-forming species

Australian thrips species that form galls on *Acacia* include at least six species of *Kladothrips*, three species of *Onychothrips*, and eight species of *Oncothrips* (Mound 1971; Mound and Crespi 1995; L. A. Mound, B. J. Crespi and B. Kranz, unpublished manuscript). In *Oncothrips* and *Onychothrips*, galls are founded by single females, but in *Kladothrips rugosus*, *K. ellobus*, *K. harpophyllae*, and *K. acaciae*, newly formed galls contain either a single female, or a female and a male together (Crespi 1992a; B. J. Crespi, unpublished data) Females of *Kladothrips*, *Onychothrips*, *Oncothrips tepperi*, *O. habrus*, and *O. waterhousei* have markedly enlarged forelegs; fighting between foundresses for possession of incipient galls has been described in *Kladothrips rugosus*, *Onychothrips arotrum*, *O. tepperi* and *Oncothrips tepperi* (Crespi 1992a), and recently was observed in *O. antennatus* (B. J. Crespi and T. Chapman, unpublished data). In *Kladothrips rugosus*, males are also known to fight one another during gall formation, and newly-formed galls contain either a single female or a male–female pair.

In *Onychothrips tepperi*, *O. arotrum*, *Oncothrips antennatus* and *O. waterhousei* and *Kladothrips hamiltoni*, which inhabit highly arid areas, galls are formed by single females after rainfall, when their host plants produce new phyllodes. The *Onychothrips* species and *Oncothrips antennatus* breed within their galls for a single generation, producing a large brood of adults which remain within the gall for some unknown period of time before dispersal. By contrast, galls of *O. waterhousei* and *K. hamiltoni* apparently persist for multiple generations of breeding, with the micropterous offspring of the foundress reproducing after her death.

Kladothrips rugosus, *Oncothrips tepperi* and *O. habrus*, three species that live in semiarid areas, have annual life cycles. Galls are formed in spring when new phyllodes are produced, larvae develop during summer, and newly eclosed adults apparently disperse from the galls in fall or winter, perhaps when the gall-containing phyllodes drop from the plant. In *O. tepperi*, however, collections from one site suggest that breeding sometimes continues within galls for a second generation and year (Crespi 1992a); the same could also be true of *O. habrus*, but collections are currently insufficient to test this hypothesis.

Table 8-1. *Ecology and sociality in Australian gall thrips and their allies*

Collection data for each of these species are available from BJC. Usage of the terms for social systems follows Crespi and Yanega (1995). *Kladothrips rugosus, K. ellobus, K. harpophyllae* and *K. acaciae* galls are founded by either one female, or a male and a female together. As described in Fig. 8-2, *Oncothrips habrus* is also found on *A. pendula*, *O. antennatus* is also found on *A. adsurgens*, and *K. rugosus* is also found on *A. tephrina* and *A. melvillei*.

Mode of life	Taxon	Host	Founding	Social system
Gall-forming	*Oncothrips antennatus*	*Acacia aneura*	1 female	solitary
	O. rodwayi	*A. melanoxylon*	1 female	solitary
	O. tepperi	*A. oswaldii*	1 female	eusocial
	O. habrus	*A. melvillei*	1 female	eusocial
	O. waterhousei	*A. tephrina*	1 female	eusocial
	O. sterni	*A. aneura*	1 female	?
	O. torus	*A. citrinoviridis*	1 female	solitary
	Kladothrips ellobus	*A. cambagei*	1 female or 1 each sex	solitary
	K. rugosus	*A. pendula*	1 female or 1 each sex	solitary
	K. acaciae	*A. harpophylla*	1 female or 1 each sex	solitary
	K. hamiltoni	*A. cambagei*	1 female	eusocial
	K. harpophyllae	*A. harpophylla*	1 female or 1 each sex	eusocial?
	Onychothrips arotrum	*A. aneura*	1 female	solitary
	O. tepperi	*A. aneura*	1 female	solitary
	O. pilbara	*A. citrinoviridis*	1 female	solitary
Gall-usurping	*Koptothrips* spp.	Various *Acacia* galls	1 female	solitary
Opportunistic gall-inhabiting	*Warithrips sp.*	Old *O. antennatus* galls	?	communal?
	Grypothrips sp.	Old *K. ellobus* galls	1 female	solitary
	Csirothrips watsoni	Old *O. arotrum* galls	1 female	solitary or subsocial
	Xaniothrips xantes	Old *Lichanothrips* nests	1 female	solitary or subsocial
Phyllode nest-forming	*Dunatothrips aneurae*	*A. aneura*	1–many females	communal?
	New genus and species	*A. lysiphloia*	1 female	solitary or subsocial
	Lichanothrips sp. n.	*A. tephrina*	1–many adults	communal?
	Lichanothrips spp.	*A. harpophylla*	1 female	solitary or subsocial
	Lichanothrips spp.	*A. cambagei*	1 female	solitary or subsocial
	Panoplothrips australiensis	*A. shirleyi*	1 female	solitary or subsocial
	Carcinothrips leai	*A. torulosa*	1 female	solitary or subsocial
	Dunatothrips sp. n.	*A. torulosa*	1 female	solitary or subsocial
Lepidopteran tie-inhabiting	*Warithrips maelzeri*	*A. tumida*	Many adults	communal?
Pre-existing hole-inhabiting	*Katothrips sp.*	*A. acradenia*	1–many adults	?
	Dactylothrips sp.	*A. adsurgens*	?	solitary or subsocial
	Dactylothrips sp.	*A. stenophylla*	?	solitary

Oncothrips rodwayi is the only gall-former found in non-arid habitats, on *Acacia melanoxylon* in temperate regions of the Great Dividing Range. At Upper Fern Tree Gully, Victoria, and Canberra, A. C. T., this species initiates galls in the spring, and complete broods can be collected during winter.

In all of the gall-forming species, gall initiation is usually highly synchronous in any one location, coinciding with the production of new phyllodes. In the seasonal species

Oncothrips tepperi and *O. habrus*, collections during June of galls containing broods of adults suggest that they overwinter as adults within the gall. By contrast, in the seasonal *Kladothrips*, such as *K. rugosus* and *K. acaciae*, we have found second-instar larvae, but never broods of adults, or pupae, within the galls in May and June. These data suggest that the larvae of these species leave the gall to pupate, or leave soon after eclosion, but no direct evidence supports this hypothesis.

Social biology of *Oncothrips* and *Kladothrips* with soldiers
In *Oncothrips tepperi*, *O. habrus*, *O. waterhousei* and *Kladothrips hamiltoni*, the first adults to develop within the brood of a foundress have reduced wings and enlarged forelegs, and they use their forelegs to defend the gall against invaders (Fig. 8-1b)(Crespi 1992b; Mound and Crespi 1995). These morphologically distinct females and males will attack insects other than members of their own species, including *Koptothrips*, lepidopteran larvae and ants (Crespi 1992b; Mound and Crespi 1995). In *O. habrus*, micropterous adults have a higher propensity than foundresses to emerge from a small hole cut in the gall and to attack invaders artificially held at the hole; a non-significant trend for such a division of labor is found in experiments with *O. tepperi* (Crespi 1992b). Dissections indicate that foundresses of both *O. tepperi* and *O. habrus* have greater ovarian development than do micropterous females from the same gall, although micropterous females apparently do lay eggs (Crespi 1992b).

In *O. tepperi* and *O. habrus*, most attacks on *Koptothrips* are by female soldiers, but micropterous males of both species have been observed to attack on several occasions. Data collected while testing other hypotheses suggests that males are substantially less likely to attack than are females; for example, in one experiment with *O. habrus*, 33% of 270 females but only 9% of 23 males emerged from their galls and attacked *Koptothrips*, and in *O. tepperi*, 6% of 16 males and 36% of 56 females emerged from small holes cut in their galls. However, a sex difference in soldier behavior remains to be tested rigorously. A micropterous male of *O. tepperi* was once seen mating with a micropterous female, in an experimental observation gall. However, in many galls of both *O. tepperi* and *O. habrus* micropterous males are not present, so reproducing micropterous females are presumably laying only unfertilized, male eggs.

In *Oncothrips waterhousei* and *K. hamiltoni*, most non-kleptoparasitized galls that we have collected contained several to several dozen micropterous adults, a live or dead foundress, and many larvae and eggs. Micropterous adults in these colonies will, as in *O. tepperi* and *O. habrus*, spontaneously attack *Koptothrips* and lepidopteran larvae held in front of them or loose in their gall (Mound and Crespi 1995). In both of these species, micropterous adults apparently breed within the galls for an extended period after the foundress has died (Mound and Crespi 1995), and dispersing macropterous brood may be primarily the offspring of micropterae.

Oncothrips antennatus and *O. rodwayi* broods do not contain micropterous adults, and macropterous first-brood adults do not attack invaders held up to them at a hole in the gall. Instead, these adults will simply crawl out of the hole, apparently trying to disperse.

We have recently discovered two new species for which behavioral observations have not yet been conducted, but which are of considerable interest for understanding the origin of soldiers. *Kladothrips harpophyllae*, a species collected from *A. harpophylla* near Rollston, Queensland, exhibits gall-founding by single females and female–male pairs, and has micropterae with soldier morphology developing early in the first brood (L. A. Mound, B. J. Crespi and B. Kranz, unpublished manuscript). This is the first species with apparent soldiers and frequent gall-founding by female–male pairs, a habit likely to ensure relatedness of three-quarters among females in a brood. *Oncothrips sterni* from *A. aneura* near Wittenoom, Western Australia, has gall-founding by single females, and wingless, weakly sclerotized, and completely unarmed offspring developing early in the first brood. These individuals appear to be non-soldier, within-gall reproductives, and their presence raises the possibility that soldiers evolved from neotenic within-gall reproductives rather than from formerly dispersing macropterae.

Phylogeny of gall-forming thrips
We have inferred the phylogeny of gall-forming thrips on *Acacia* using a combination of nucleotide sequence data, from the mitochondrial genes cytochrome oxidase 1 and 16S rDNA, and data on adult and gall morphology (Fig. 8-2). The analysis shows that: (1) a species of *Gynaikothrips* is basal to the clade of gall-formers on *Acacia*; (2) soldiers have apparently evolved at least twice, once in *Oncothrips* and once in *Kladothrips*; and (3) *Oncothrips sterni*, the species with a wingless within-gall non-soldier, is apparently not the sister taxon to either of the clades with soldiers. As discussed below, this phylogenetic analysis also serves as a basis for assessing the necessary and sufficient conditions for the origin or loss of soldiers, and the presence of differences in ecology, demography and genetics between species with and without a soldier caste.

Gall kleptoparasites: the genus *Koptothrips*

The genus *Koptothrips* contains four described species, but morphological variation within most species is pronounced and at least several species are as yet undescribed. Females

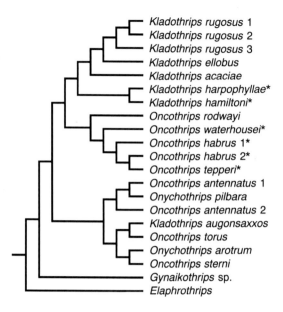

Figure 8-2. A phylogeny of Australian gall-forming thrips, inferred from nucleotide sequence of cytochrome oxidase 1 (477 characters) and 16S rDNA (264 characters), adult morphology (16 characters) and gall morphology (10 characters). Asterisks indicate that the species has a soldier morph. *Kladothrips rugosus* 1 is from *Acacia melvillei*, *K. rugosus* 2 is from *A. tephrina*, and *K. rugosus* 3 is from *A. pendula*; *Oncothrips habrus* 1 is from *A. pendula* and *O. habrus* 2 is from *A. melvillei*; *Oncothrips antennatus* 1 is from *A. adsurgens* and *O. antennatus* 2 is from *A. aneura*. Within each of these three sets of operational taxonomic units, the taxa are morphologically indistinguishable. However, mtDNA genetic distances for cytochrome oxidase 1 of between 8% and 16% between pairs of morphologically-indistinguishable forms on the different host plants suggest that each represents a different host race or species. Maximum parsimony analysis of the data sets, combined such that each character has equal weight, in PAUP 3.1.1 (Swofford 1993) yielded this one shortest tree of length 962. The data set is available from BJC; details of data and analysis will be presented elsewhere.

of these thrips invade the galls of *Kladothrips*, *Oncothrips* and *Onychothrips*, kill the gall-formers, and produce broods of offspring (Mound 1971; Crespi 1992a). Some *Koptothrips* species are host specific, such as *K. zelus* attacking *Kladothrips acaciae* and *Koptothrips xenus* attacking *Kladothrips ellobus*. Specimens that key to *Koptothrips flavicornis* and *K. dyskritus* in Mound (1971) are found attacking multiple gall thrips host species on multiple host plants, and recent DNA evidence (P. Abbot, B. J. Crespi and D. Carmean, unpublished data) suggests that each of these two taxa comprises multiple sibling species, each host-specific.

Koptothrips females kill the host thrips by grasping them with their forelegs. However, in contrast to gall-formers, most of which have enlarged forelegs, *Koptothrips* species, with the exception of *K. xenus*, do not show extreme foreleg allometry. Despite this lack of armament, gall-forming thrips grasped even momentarily by *Koptothrips* usually die within several minutes, usually without any obvious sign of injury. These observations, and the observation that defending gall thrips must clearly puncture and hold a *Koptothrips* for many minutes to kill it, suggest that *Koptothrips* kill their enemies with a venom. Moreover, some *Koptothrips* species are heavily sclerotized, and have a shiny cuticle that may be slippery and cause the tarsi of their attackers to slip off. The finding of multiple galls of *Oncothrips tepperi* containing one live *Koptothrips flavicornis* female and one or more dead ones suggests that they kill competing conspecifics as well as their hosts.

The life cycle of *Koptothrips* appears accelerated in relation to that of their hosts. In any given collection, most gall-forming foundresses might have a few larvae and mostly unhatched eggs, while the *Koptothrips* foundresses in other galls on the same plants have broods of teneral adults, pupae, and larvae.

Galls with dead *Koptothrips* and live soldiers have been found in *Oncothrips habrus* and *O. tepperi*, and in some galls of these two species, and in galls of *O. waterhousei*, dead *Koptothrips* have been found squashed within the gall lips, apparently having died while attempting to enter the gall. Galls containing live *Koptothrips* and dead soldiers have been found in all three species of *Oncothrips* with soldiers, and in *Kladothrips hamiltoni*. These collections are consistent with behavioral observations that *Koptothrips* kill soldiers and vice versa (Crespi 1992b; Mound and Crespi 1995), such that the outcome of attempted usurpation varies in natural populations.

Adult females of *Koptothrips xenus* use a cellophane-like material to enclose small areas within *Kladothrips ellobus* galls whose walls have been chewed through by other insects or birds, and which are otherwise open to the outside. With this alternative method of gall utilization, *Koptothrips xenus* females need not kill a *Kladothrips ellobus* host to produce a brood of offspring. *Koptothrips zelus* uses this material in a third way: multiple usurpers of this species invade galls of *Kladothrips acaciae*, and each invader uses the cellophane-like material to form a partition enclosing about one square centimeter inside the gall, which separates it from the others. After the young of these *Koptothrips zelus* foundresses mature, the partitions are broken down, such that the offspring of the females mix after eclosion. Galls of *Kladothrips acaciae* that have been invaded by *Koptothrips zelus* are much thinner laterally than uninvaded galls, such that much less material is needed to form the partitions than would otherwise be necessary.

Opportunistic gall-inhabiting species

Some species of thrips on *Acacia* neither form nor usurp galls, but instead they breed in galls after the original inhabitants have left. This mode of life is found in *Warithrips*, which breed in old *Oncothrips* galls on *Acacia adsurgens*, in *Grypothrips*, which breed in old, dried galls and in other enclosed spaces such as abandoned glued phyllodes or lepidopteran leaf ties, and in *Csirothrips watsoni*, which breeds in old galls of *Onychothrips arotrum* (Crespi 1992a).

In both *Warithrips* and *Csirothrips*, adults have enlarged forelegs, with several spikes on the inner margin of the fore femur; by contrast, *Grypothrips* have the forelegs enlarged but the fore femora smooth. A third species that invades galls, *Thaumatothrips froggatti*, also has spikes on the inner margin of the fore femur (Mound and Crespi 1992). The presence of this trait in three species that invade galls before or after host abandonment suggests common functional significance, perhaps relating to the manner in which they kill hosts, competitors or other invaders.

Warithrips live in groups of many micropterous and macropterous adults and larvae, in old, yellowish galls which are sometimes clearly open to the outside. *Grypothrips* are commonly found in dry, brown galls of *Kladothrips ellobus* on *Acacia cambagei*, where single females lay a clutch of eggs. These old *Kladothrips ellobus* galls are often fragmentary, offering only a few square centimeters of living space, 1–2 mm thick, for eggs, larvae and female; the galls have no green tissue and are open to the outside. In *Csirothrips watsoni*, single females move into galls of *Onychothrips* after the gall lips have opened (Crespi 1992a). In the laboratory, these foundresses will attack live *Onychothrips*, but whether any *Onychothrips* normally remain in the galls after they open is unknown. Females of *Csirothrips* apparently plug the gall lips with cast *Onychothrips* exuviae, and they probably breed within the galls for a single generation, since old *Onychothrips* galls tend to fall easily off the plant.

Cooperative behavior has not been observed in these opportunistic gall inhabitants, but their morphology and ecology suggest several possibilities. First, the presence of micropterous individuals in *Warithrips* indicates that multiple generations live within the galls, and the foreleg armature found in this species suggests that they defend the gall against intruders. Second, *Grypothrips* adults and larvae living in old dry *Kladothrips ellobus* galls must leave their domicile in order to feed on green plant tissue. Presumably they do so at night, but whether foraging is coordinated, and how they orient to the domicile, are unknown. Finally, the closure of gall entrances by *Csirothrips* is probably conducted by a single foundress, but could also involve larvae, and the enlarged forelegs of females suggests that they defend their gall against intruders, unless the forelegs are used only to kill any *Onychothrips* left within the gall when it is invaded. If *Csirothrips* females do defend against non-conspecifics after egg-laying, then they could be regarded as subsocial.

Phyllode domicile-formers

Some species of thrips on *Acacia* do not inhabit galls, but instead use a cellophane-like material, probably some sort of glandular secretion, to either: (1) enclose multiple apical *Acacia* phyllodes, creating a space that in different species is either clearly open to the outside or apparently closed (Fig. 8-1c); or (2) glue flat, overlapping, mature, non-apical *Acacia* phyllodes together, forming oval or rectangular spaces one-half to several centimeters on a side and one to several millimeters thick (Fig. 8-1d).

Two species of apical phyllode domicile-formers have thus far been discovered. In *Dunatothrips aneurae*, which inhabits *Acacia aneura*, eleven recently founded colonies contained one to five dealate females (mean 3) with unhatched eggs, and, in two colonies, one or two macropterous males. Older colonies contained one to five dealate females (mean 2.3), one to eight macropterous adults, and many pupae, larvae, and hatched and unhatched eggs. This species thus exhibits both single- and multiple-female colony founding, and multiple foundresses apparently breed in established groups. In an undescribed species on *Acacia lysiphloia*, collections at a single site yielded seven domiciles containing single dealate females with eggs or both larvae and eggs, and three domiciles with a single dealate female, four to seven macropterous adults, and multiple larvae and pupae. This species therefore apparently has colony-founding by single females.

In both of the species that glue apical phyllodes, life cycles are asynchronous. Usurpation of the domiciles by other thrips species or other insects had not occurred in any of the colonies that we collected.

Mature, non-apical phyllodes are glued together by species in four genera: *Lichanothrips*, *Panoplothrips*, *Carcinothrips* and *Dunatothrips*. These domiciles are always initiated and formed by single females, which rear a single brood that disperses after reaching adulthood. Most of these species induce discoloration of the phyllode at the domicile site, which suggests that they feed mainly or exclusively there. At any given collection site, these species are usually not synchronized in their life history, and newly formed glued phyllodes can be found at the same time and on the same trees as groups consisting of a foundress and her adult offspring.

In *Lichanothrips*, *Panoplothrips* and *Carcinothrips*, females have massively enlarged forelegs, which they might use to hold pairs of phyllodes together while glueing them; indeed, these species have exceptionally well-developed tarsal pads, used for holding tightly to the substrate. Alternatively, females may fight one another with their forelegs over ownership of their domiciles, as do most gall-forming species. However, attempts to elicit fighting among *Lichanothrips* foundresses have failed.

In the apical-phyllode and mature-phyllode domicile-forming species described above, the thrips apparently do not leave their domicile before dispersal by macropterous adults. However, in a *Lichanothrips* species on *A. tephrina*, adults and larvae leave their domicile to forage in groups. Ten recently founded colonies of this species contained 2–16 macropterous adults (mean 6.8) with eggs or eggs and larvae, whereas older domiciles at the same site contained 1–30 macropterous adults, dozens of larvae and pupae, and dozens of hatched and unhatched eggs. During early morning, and when their plants were shaded from the sun, adults and larvae of this species were observed walking slowly, in groups of several to a few dozen, in the open on phyllode surfaces near their domicile (Fig. 8-1e). These groups usually moved rapidly back into the domicile after a nearby human hand or head movement. Larvae of this species are brilliant red; casual observations suggested that larvae tended to move in groups with other larvae, whereas adults moved in groups with other adults. The domiciles of this species comprise two to several phyllodes about 2 cm wide and 8 cm long, glued together irregularly at the edges to form enclosed spaces. Colonies are clearly founded by multiple adults, but the sex of these adults remains to be determined.

In *Panoplothrips*, *Carcinothrips* and *Dunatothrips*, unhatched eggs and recently born larvae are often still present after the offspring of a foundress have eclosed. These data indicate that generation overlap occurs between a breeding female and her adult offspring. Whether or not foundresses or broods defend themselves against invaders is unknown; in *Lichanothrips*, *Carcinothrips* and *Dunatothrips* the glued areas completely enclose the thrips, and adults and brood apparently do not leave to feed. By contrast, the glued areas of *Panoplothrips australiensis* are open on one side and oriented downwards, which suggests that females, and perhaps their offspring, feed away from the glue site, and that foundresses may be selected to defend against predators or parasites. Usurpers, such as *Koptothrips* species or lepidopteran larvae, have not been found in any of these species that glue phyllode pairs. Although species in the genus *Xaniothrips* (Mound 1971) use the domiciles formed by *Lichanothrips*, they apparently do so only after the *Lichanothrips* have finished breeding and left (Crespi 1992a).

Warithrips colonies in lepidopteran leaf-ties

Acacia tumida dominates many valley floors in the Hamersley Range, Western Australia; *Warithrips maelzeri* forms large colonies on this plant, in the spaces formed between pairs of phyllodes tied together by lepidopteran larvae. Females of this species have greatly enlarged forelegs with two spikes on the inner margin of the fore femur, whereas males show little foreleg enlargement or armature (Mound 1971).

Colonies of *W. maelzeri* are initiated by multiple adults of both sexes: 13 colonies were collected that contained one or more adults of both sexes but no larvae or pupae, and unhatched eggs were found in four of these 13 groups. Of the other 18 colonies collected, most contained multiple adults and dozens of pupae, larvae, and eggs, with most individuals massed together in the enclosed space between phyllodes. Lepidopteran larvae were still present in only two of the leaf ties collected, both of which contained only three or four apparent founders and no eggs or larvae.

Adults and larvae were sometimes seen out in the open on phyllode surfaces, apparently feeding. Such individuals did not seem to be coordinated in their movements, but they did tend to move back into the enclosed area between phyllodes after being disturbed.

Colonies of *W. maelzeri* are founded by multiple adults, of both sexes, and they apparently persist for a single generation on a pair of phyllodes. The enlarged forelegs in females suggest fighting, perhaps against conspecifics or to rid the leaf-tie of its lepidopteran inhabitatant.

Dactylothrips and *Katothrips* inhabiting pre-existing cavities

We have collected species of *Dactylothrips* and *Katothrips* living in three habitats: abandoned weevil leaf mines on *Acacia acradenia*, exit holes of dry, brown Hymenoptera galls on *A. adsurgens* (Fig. 8-1f), and shallow cracks, apparently fungal 'split stem' galls, on *A. stenophylla*. Females of the *Dactylothrips* species in the old Hymenoptera galls have highly modified abdominal sternites and tergites, with a pair of large, converging tubercles on the posterior margin of the ninth tergite, which can be opposed against abdominal segment ten (the tube) to form pincers. By contrast, the *Katothrips* species in leaf mines and the *Dactylothrips* in split stem galls lack the tubercles.

The *Dactylothrips* species in old Hymenoptera galls were found in exit holes 1–2 mm in diameter and 5–10 mm deep. Each hole with thrips contained a single female, with her abdomen pointed outward. Disturbance of such a female, using the hairs of a paintbrush, caused her to raise her tenth abdominal segment against the two tubercles on the ninth tergite, forming pincers which grasped and held the paintbrush hairs. Galls often contained multiple exit holes, in close proximity, all inhabited by *Dactylothrips*. Because the galls are brown and dry, these *Dactylothrips* may leave the gall to feed on phyllode tissue on adjacent stems, but how they orient to their holes where multiple inhabited holes are present in close proximity is unclear.

The *Katothrips* in abandoned weevil leaf mines gained access to the mines via their single exit holes, about 0.4 mm in diameter. Mines were one to a phyllode, about 10–15 mm × 10–15 mm, more or less circular or rectangular, and several millimeters thick. Observation of these exit holes under the microscope showed that, in eight of ten undisturbed mines (mines removed from the plant but unopened), the abdomens of one to three thrips projected slightly out of the exit hole (four exit holes with one abdomen, one with two abdomens, and three with three abdomens). When these exit holes were held to the nose, a distinctly unpleasant smell, similar to that produced by pentatomid bugs, was noticeable; this smell could not be detected from leaf mines without thrips. Three adults pulled from the exit holes were females; two were dealate and one macropterous.

Of twelve leaf mines with thrips, eight mines contained several dozen to several hundred adults, pupae, larvae and eggs. When these mines were opened, the same strong smell was detected as in recently founded colonies. The other four mines had apparently been invaded only recently, and these contained 1, 5, 6, and 8 adults, as well as 0–15 eggs and up to several larvae. Sexing of the eight adults from one of the mines showed that it contained six dealate females and two macropterous males.

To determine whether or not the presence of individuals with their abdomens in exit holes would prevent other *Katothrips* from entering, we placed two adults from other colonies onto a leaf mine and gently prodded them towards an exit hole from which two abdomens projected. These two adults, one male and one female, each antennated the thrips in the exit hole, which moved their abdomens sharply, after which the adults outside of the hole left. However, in a third trial, a female from another colony did enter this same exit hole while two abdomens were projecting from it. Once, in the field, we saw an ant investigating an exit hole, which it did not enter.

On *Acacia stenophylla*, single female *Dactylothrips* were found in small crevices associated with split stems. They kept their abdomens pointed outward, and flicked them sharply when disturbed. Three times, a male was found in a crevice near to a crevice containing a female.

DISCUSSION

We view the behavior found in a given species of thrips as evolving in concert with its morphology, habitat and life cycle. Thus, morphology and behavior allow, direct, bias or limit the evolution of each other; spatial and temporal aspects of the habitat select for particular ways of utilizing and protecting resources for breeding; and the life cycle forms a labile framework within which social interactions take place. A further influence on thrips life cycles and behavior stems from haplodiploidy, which creates relatedness asymmetries, allows sex-ratio manipulation, permits virgin females to reproduce, and may engender relatedness above one-half between females.

Morphology

Morphological aspects of thrips that may influence the evolution of their behavior patterns include: (1) forelegs that can become enlarged and modified to serve as weapons or structures to manipulate the habitat; (2) abdomens that can be modified to serve as defensive structures; (3) wing polymorphism, which engenders morphological specialization in characters other than wings; (4) hyperfecundity of macropterous females, which can produce enough offspring to completely fill a cavity such as a gall; (5) piercing–sucking mouthparts, which preclude direct food transfer between parents and offspring; (6) mobile larvae, with the same food requirements as adults, which are less morphologically suited for defense than adults because they are much less sclerotized (but see Pelikán 1990). The main effects of these traits on behavior are that adults can build and defend, but cannot feed offspring, and that behavioral and reproductive specialization can evolve in conjunction with morphological divergence of macropterous and micropterous morphs, wherever resources otherwise allocated to dispersal structures can better increase inclusive fitness in other ways.

Morphological variation among gall thrips species and their allies mainly involves overall body size, the presence of wing polymorphism, and the presence, degree and form of foreleg enlargement. The functional significance of interspecific variation in the form of foreleg armature remains unclear, but wing polymorphism is indicative of multiple generations of breeding within the same domicile.

Domiciles, holes and other enclosed habitats

Spatial and temporal aspects of the habitat are perhaps the most important determinants of behavior in gall thrips and their allies, and the most variable. The most critical features of such habitats, for the presence and form of social interactions, are: (1) habitat initiation, and whether or not initiation involves building; (2) habitat size, duration, expandability and defensibility, which shape the life cycle; and (3) whether or not the habitat can be replaced should it be usurped or otherwise lost, which affects habitat value and the strength of selection for defense.

Colonies are initiated by multiple adults in the apical phyllode gluer *Dunatothrips aneurae*, *Lichanothrips* on *A. tephrina*, the leaf-mine inhabiting *Katothrips*, and *Warithrips maelzeri*. Possible selective pressures for such pleometrosis include: (1) fast, efficient building of the habitat, which could be important because of high predation on adults in incomplete structures or limitations on ability to produce glue quickly (in *Dunatothrips* and *Lichanothrips* on *A. tephrina*); or (2) benefits of cooperative habitat defense (in *Katothrips*). These hypotheses require testing using experiments or natural variation in foundress number. Single founding may be favored in all other species because one female can easily create a gall, or create and monopolize a glued space or other domicile, alone,

and thereby avoid any reproductive costs resulting from sharing the domicile with others. For example, the *Lichanothrips*, *Panoplothrips*, *Carcinothrips* and *Dunatothrips* that glue pairs of phyllodes together all create enclosed spaces that are quite small and appear to require relatively little glue.

The domiciles of thrips that use glue appear not to be expanded after the initial building. With the possible exception of *Lichanothrips* on *A. tephrina*, the domiciles that contain mature offspring are not noticeably larger, nor do they comprise more glue, than newly founded ones. Similarly, in gall-formers on *Acacia*, the galls apparently reach their final, mature size while the foundress is still ovipositing. The general lack of habitat expandability in thrips means that an offspring worker–builder force, so vital in social Hymenoptera, is not important here, and that the space available for offspring development is often strictly limited.

The defensibility of gall thrips habitats can be inferred mainly from evidence of attempted and successful invasion. In gall-formers, invasion of *Oncothrips* and *Kladothrips* galls by *Koptothrips* is common, although its frequency varies tremendously between species and sites (B. J. Crespi, unpublished data). For thrips species on *Acacia* that do not form galls, evidence of defensibility comes from defensive morphology and behavior (in *Dactylothrips* and *Katothrips*) and from the nature of the habitat itself. For example, in *Panoplothrips australiensis* the glued area between two phyllodes is U-shaped and open at one end, whereas in *Lichanothrips*, *Dunatothrips* and *Carcinothrips* the glued area appears to completely enclose the group, which may make the thrips more or less invulnerable to arthropod predation. We have never collected domiciles of phyllode-gluing thrips that had been usurped by other insects. However, we expect that an important threat to such species, and to many other *Acacia* thrips, would come from ants, which are abundant on most of their host plants. The ability of most thrips species, when exposed, to physically defend against ant predation is doubtful; we expect that only protective enclosure, unpalatability, or chemical defenses could be effective against such large, persistent threats.

The effectiveness of the diverse forms of *Acacia* gall against invasion by *Koptothrips* and other insects is difficult to assess. Galls of almost all species of *Oncothrips*, *Kladothrips*, and *Onychothrips* appear closed to the outside, but *Koptothrips* have dorsoventrally flattened heads, which are probably an adaptation for squeezing into galls through their lips. The collection of some *Oncothrips* galls with *Koptothrips* dead within the lips, and the obvious requirement for macropterous adults of the gall formers to be able to exit the gall to disperse, suggest that *Koptothrips* normally invade galls in this manner. Invasion times and routes of lepidopteran, dipteran and hymenopteran invaders are generally unknown, although the presence of some lepidopterans and dipterans in newly-founded galls suggests that they normally invade before closure.

Whether or not *Acacia* thrips habitats can be replaced, if usurped or otherwise disturbed, depends upon how the habitat was formed. Species that form galls do so on young, growing phyllodes, which are available for only a narrow time window of several days or weeks at yearly or otherwise infrequent intervals. Thus, if a colony is invaded, the host thrips cannot leave and form another gall. In species that form domiciles from glue, young phyllodes appear to be less essential to habitat formation, and domiciles can probably be replaced at a cost of glue production and possible increased mortality while exposed. For species that use pre-existing habitats, replacing a disturbed habitat is probably easiest, although it depends upon availability, which may vary from high in species such as *Warithrips*, which use lepidopteran leaf-ties, to low in *Katothrips*, which require abandoned leaf mines of a particular age.

Life cycles

The nature of the habitat structures the life cycle with respect to synchrony of colony initiation, the presence and extent of generation overlap between parents and offspring, and number of generations in a colony. In gall-forming species, gall initiation is highly synchronous. By contrast, in thrips species that glue phyllodes, and those that inhabit other cavities, life cycles are asynchronous: recently founded colonies can be collected at the same site and time as colonies with mature offspring. Some of these species, such as the *Lichanothrips* on *A. tephrina*, may be dependent on relatively young (although not necessarily newly produced) phyllodes for colony initiation, but such phyllodes are apparently available, in this *Acacia* species, for periods of at least a few weeks, such that breeding is somewhat asynchronous.

Some degree of generation overlap between parents and adult offspring is usual in *Acacia* thrips. In species that glue phyllodes, the common coincidence of foundresses, unhatched eggs and macropterous adult offspring indicates that foundresses are still breeding after their offspring have eclosed. In *Oncothrips waterhousei*, some galls

contain live foundresses and their micropterous offspring, but in most mature galls that we have collected the foundress is dead. For *O. habrus*, *O. tepperi* and *O. waterhousei*, the collection of many galls with live foundresses and many micropterous adults indicates that generations overlap for weeks, if not several months. The amount of generation overlap in other taxa, such as *O. rodwayi*, *O. antennatus*, and *Dactylothrips*, is unknown.

The three main forms of cooperative behavior found in *Acacia* thrips are building, defense and, possibly, coordinated foraging. Patterns of among-species variation in these behaviors are predictable, to some extent, from aspects of the morphology, habitat and life cycle. Thus, species that live on phyllodes in the open use glue to create a domicile, species whose domiciles are invaded by other insects have means of defense, and at least one species that forms large colonies forages away from the domicile in groups. The presence of soldiers in some species of gall-forming thrips, but not in thrips in other habitats, may be related to three characteristics of galls: (1) high life-cycle synchrony, which creates a temporally superabundant and predictable resource for invaders such as *Koptothrips* to exploit (Wcislo 1987); (2) founding by single females, which, if singly inseminated would produce broods of mainly females, related by three-quarters; and (3) high value of the habitat, because it is irreplaceable and provides food and shelter for a lifetime.

Comparative biology of gall-forming thrips with and without soldiers

Six observations indicate that thrips species with soldiers can be considered eusocial, having castes and alloparental care (Crespi and Yanega 1995): (1) the discrete morphological differences between soldiers and macropterous individuals; (2) the lack of dispersal and gall-forming by soldiers, which indicates that they have lost behavioral totipotency; (3) the higher propensity of soldiers than foundresses to attack *Koptothrips*, when both morphs are present, in at least one species (Crespi 1992b); (4) the spontaneity of attack by thrips soldiers on *Koptothrips* in all species (Crespi 1992b; Mound and Crespi 1995); (5) the observation that attack often leads directly to the attacking soldier being killed; and (6) the dissection data suggesting that soldiers have lower fecundity than foundresses. Despite these data, the amount of generation overlap between foundresses and soldiers may be rather low in *K. hamiltoni* and *O. waterhousei*, such that soldiers may often be defending themselves and their offspring, siblings, and the offspring of siblings, but not their mother.

The most parsimonious explanation for the origin of soldier morphs is that they evolved from normal macropterous offspring, such that the first adults to eclose within a gall were selected to engage in defensive behavior against insects of other species, using their enlarged forelegs that had evolved in the context of intraspecific fights (see Crespi 1996). Selection for defense would, by this scenario, have led to both morphological dimorphism and phenological divergence, such that soldiers eclose first in a discrete group, as in *Onychothrips tepperi*. This hypothesis can be tested by searching for species with monomorphic offspring and soldiering, as are found in some soldiers aphids (Aoki 1987; Itô 1989; Stern and Foster this volume), and for species with soldiers but without fighting morphology and behavior among foundresses.

Why do some species of gall-forming thrips have soldiers, but others do not? Differences between the *Oncothrips* species with soldiers, and their closest relative, *O. rodwayi*, without them (Fig. 8-2) include: (1) enlarged forelegs, and fighting over ownership of incipient galls, in macropterous forms only of species with soldiers; (2) distributions that span semiarid regions in species with soldiers, whereas *O. rodwayi* lives in relatively wet, temperate areas; (3) high rates of successful invasion by *Koptothrips* in *Oncothrips* species with soldiers, and apparent low rates of invasion in *O. rodwayi* (B. J. Crespi, unpublished data). Of these differences, the first may involve a morphological preadaptation to the evolution of soldiers, if small forelegs, in the ancestors of extant species without soldiers, were sufficiently useless for defense that no variation in defensive ability was available for selection. However, among-species variation in invasion rates by *Koptothrips* offers a more telling explanation for the comparative pattern: in lineages with soldiers, selection for defense has been, and apparently remains, strong and perhaps stronger than in lineages without soldiers. The success of *Koptothrips* in invading galls of species with soldiers may stem from two causes: they often invade before soldiers have eclosed, and they are apparently often able to usurp the gall despite the presence of soldiers.

Differences between the two *Kladothrips* species with soldiers, and the congenerics without them (Fig. 8-2), are more difficult to identify, especially because a solitary species, *Kladothrips ellobus*, inhabits the same host plant, *A. cambagei*, as *K. hamiltoni*. However, three factors that may be involved include: (1) the life-cycle duration of

K. hamiltoni, which appears to be substantially greater than that of other *Kladothrips*; (2) the lack of pupae in galls of *Kladothrips* species other than *K. hamiltoni* and *K. harpophyllae*, which suggests that second-instar larvae of these other species leave the gall to pupate or that adults leave the gall soon after eclosion, and (3) high levels of sucessful *Koptothrips* invasion in the two species with soldiers (though also in *K. ellobus* and *K. acaciae*). All *Kladothrips* have fighting among foundresses, or morphological indications that such fighting takes place.

Given these comparative differences and similarities in *Oncothrips* and *Kladothrips*, the presence of soldiers in a species is accurately predictable from three conditions: (1) fighting in foundresses; (2) pupation within the galls or remaining within the gall for a long time after eclosion; and (3) high rates of kleptoparasitism (Crespi 1996). The causes of the comparative distribution of these three conditions requires further study.

What factors might be involved in the form of eusociality in gall thrips? The apparent lack of reproductive dominance among female soldiers, when the foundress is dead, may result from a high level of compatibility between defense and reproduction. Thus, behavioral specialization among soldiers may have few benefits to the dominant individual in terms of ability to reproduce more (since food need not be sought), and subordinate individuals would suffer large costs, and gain few if any inclusive fitness benefits, were their reproduction curtailed. Alternatively, if competition for living space for offspring limits the reproduction of soldiers, then dominance could be advantageous. However, it is unclear how a dominant individual could reduce the reproduction of subordinates through altering their behavior, since altruism is only required in rare emergencies and food should be freely available to all.

Sex ratios of macropterous offspring in gall thrips are highly female-biassed in all species thus far investigated, which include *Oncothrips tepperi*, *O. antennatus*, *O. rodwayi*, and *Onychothrips arotrum* (Crespi 1992a; B. J. Crespi, unpublished data). These female biases are probably due to local mate competition and inbreeding (Hamilton 1967; Herre 1985), but macroptery in males of all species, and observations of mating between macropterous adults of *Oncothrips tepperi* on phyllodes in the open, suggest some degree of outbreeding. Genetic relatedness within broods has not yet been estimated for any gall-forming species, but since founding by single females is the rule, mating more than twice by foundresses is the only possible cause of average relatednesses of less than one-half among daughters.

Comparison with other insect taxa

The closest parallels to gall thrips come from gall-forming aphids, many of which have soldier first- or second-instar larvae (Aoki 1987; Itô 1989; Stern and Foster, this volume). All aphid species with soldiers form galls at some point in their life cycle, and larvae have horns, forelegs, or mouthparts serving as weapons to defend the colony. As in gall thrips, some aphid species are specialized to invade and usurp the galls of others (Aoki 1979; Akimoto 1989). However, the main selective force for the evolution of aphid soldiers appears to be predation from other arthropods, and possibly vertebrates in some cases, rather than usurpation by other species of aphids (Foster 1990; Kurosu *et al.* 1990; Aoki and Kurosu 1987, 1991; however, see Aoki *et al.* 1991).

The main similarity between gall thrips and many other eusocial taxa is the presence of food within the relative safety of the nest. This type of habitat is also found in termites (Wilson 1971; Abe 1991), mole rats (Jarvis 1981; Sherman *et al.* 1991; Jarvis and Bennett 1993), a eusocial ambrosia beetle (Kent and Simpson 1992) and most of the soldier-producing aphids discussed above. Coupled with selection for and ability to defend, the presence of such food–shelter coincidence appears sufficient to explain most or all origins of eusociality outside of the Hymenoptera (Abe 1991; Alexander *et al.* 1991; Crespi 1994).

A final parallel between *Acacia* thrips and social insects in other taxa is pleometrosis, the colony-founding by multiple adults found in many social wasps, bees and ants (Wilson 1971; Keller 1993). Among polybiine wasps, predation by ants appears to be an important cause of pleometrosis (Chadab 1980); the same may be true for *Dunatothrips* and *Katothrips*. Among other polistine wasps, and halictine bees, pleometrosis usually entails dominance of one foundress over the others (Reeve 1991; Packer 1993). We doubt that thrips foundresses exhibit reproductive division of labor, because the benefits of such behavior, in terms of any increased reproduction of dominant individuals, increased inclusive fitness of subordinates via defending kin, or more efficient defense of the incipient colony, are likely to be few in relation to the cost of lower fecundity to subordinates. However, we have no data that bear directly on this important question.

CONCLUSIONS AND SUGGESTIONS FOR FUTURE RESEARCH

The primary motivation for seeking social behavior in gall thrips was the presence of haplodiploidy, and new opportunities to test whether or not relatednesses above one-half favors the evolution of helping. We have yet to measure relatednesses in any gall-former, but regardless of relatedness values, two aspects of ecology, (1) the irreplacable, defensible nature of the gall habitat, and (2) the common presence of kleptoparasites, appear crucial for the origin and evolution of soldier castes. Thus, we believe that alleles for becoming a soldier have spread where the costs of being a soldier, which include inability to disperse, lower maximum potential reproduction, and the possibility of being killed by a *Koptothrips* while defending the gall, are offset by the benefits of a high probability of some reproduction within the gall and inclusive fitness gained via reproduction of the kin saved through gall defense.

The main unanswered questions in the study of thrips sociality include: (1) the nature of the later stages of the life cycle in species such as *O. waterhousei* and *K. hamiltoni*, and whether soldiers produce the macropterae that disperse from the galls; (2) whether or not female soldiers are indeed more likely to defend than are male soldiers; (3) whether one or both sexes, or only certain individuals of one sex, engage in building and defense in pleometrotic species that glue phyllodes; (4) levels of genetic relatedness between soldiers and the individuals who benefit from their defense, and between cofoundresses in species that form colonies pleometrotically; and (5) the presence, extent and selective basis of coordinated foraging. Gall thrips and their allies provide striking parallels to other social insects. We believe that elucidation of the genetic, phenotypic, and ecological causes of the diverse social systems in this taxon will considerably advance our understanding of how cooperation evolves.

ACKNOWLEDGEMENTS

We are grateful to Greg Leach, Brendan Lepschi, Bruce Maslin and Les Pedley for plant identifications, and to Geoff Clarke, Ross Crozier, Susan Lawler, Roger Lowe and Lisa Vawter for field assistance and logistical help. For helpful comments and discussions, we thank Penny Kukuk, David Stern and Joan Strassmann. For permission to collect, we thank the Australian National Parks and Wildlife Service, the Department of Conservation and Land Management of Western Australia, the Department of Conservation and Environment of Victoria, and the National Parks and Wildlife Services of New South Wales, Queensland, and South Australia. We are especially grateful to the National Geographic Society for financial support.

LITERATURE CITED

Abe, T. 1991. Ecological factors associated with the evolution of worker and soldier castes in termites. *Ann. Entomol.* **9**: 101–107.

Akimoto, S. 1989. Gall-invading behavior of *Eriosoma* aphids (Homoptera, Pemphigidae) and its significance. *Jap. J. Entomol.* **57**: 210–220.

Alexander, R. D., K. M. Noonan and B. J. Crespi. 1991. The evolution of eusociality. In *The Biology of the Naked Mole Rat.* P. W. Sherman, J. U. M. Jarvis and R. D. Alexander, eds., pp. 3–44. Princeton: Princeton University Press.

Aoki, S. 1987. Evolution of sterile soldiers in aphids. In *Animal Societies: Theories and Facts.* Y. Itô, J. L. Brown and J. Kikkawa, eds., pp. 53– 65. Tokyo: Japan Sci. Soc. Press.

Aoki, S. & U. Kurosu. 1987. Is aphid attack really effective against predators? A case study of *Ceratovacuna lanigera.* In *Population Structure, Genetics and Taxonomy of Aphids and Thysanoptera.* J. Holman, J. Pelikán, A. F. G. Dixon and L. Weisman, eds., pp. 224–232. The Hague: SPB Academic Publications.

–. 1991. Galls of the soldier-producing aphid *Ceratoglyphina bambusae* broken by vertebrates (Homoptera, Aphidoidea). *Jap. J. Entomol.* **59**: 743–746.

Aoki, S., U. Kurosu and D. L. Stern. 1991. Aphid soldiers discriminate between soldiers and non-soldiers, rather than between kin and non-kin in *Ceratoglyphina bambusae. Anim. Behav.* **42**: 865– 866.

Chadab, R. 1980. Army-ant predation on social wasps. Ph. D. dissertation, University of Connecticutt.

Crespi, B. J. 1986a. Territoriality and fighting in a colonial thrips, *Hoplothrips pedicularius,* and sexual dimorphism in Thysanoptera. *Ecol. Entomol.* **11**: 119–130.

–. 1986b. Size assessment and alternative fighting tactics in *Elaphrothrips tuberculatus* (Insecta: Thysanoptera). *Anim. Behav.* **34**: 1324–1335.

–. 1988. Risks and benefits of lethal male fighting in the polygynous, colonial thrips *Hoplothrips karnyi. Behav. Ecol. Sociobiol.* **22**: 293–301.

–. 1991. Heterozygosity in the haplodiploid Thysanoptera. *Evolution* **45**: 458–464.

–. 1992a. The behavioral ecology of Australian gall thrips. *J. Nat. Hist.* **26**: 769–809.

–. 1992b. Eusociality in Australian gall thrips. *Nature (Lond.)* **359**: 724–726.

–. 1993. Sex ratio selection in Thysanoptera. In *Evolution and Diversity of Sex Ratio in Insects and Mites.* D. L. Wrensch and M. Ebbert eds., pp. 214–234. New York: Chapman and Hall.

–. 1994. Three conditions for the evolution of eusociality: are they sufficient? *Insectes Soc.* **41**: 395–400.

–. 1996. Comparative analysis of the origins and losses of eusociality: causal mosaics and historical uniqueness. In *Phylogenies and the Comparative Method in Animal Behavior.* E. Martins, ed., pp. 253– 287. Oxford University Press.

Crespi, B. J. and D. Yanega. 1995. The definition of eusociality. *Behav. Ecol.* **6**: 109–115..

Davies, R. G. 1969. The skeletal musculature and its metamorphosis in *Limothrips cerealium* Haliday (Thysanoptera: Thripidae). *Trans. R. Entomol. Soc. Lond.* **121**: 167–233.

Foster, W. A. 1990. Experimental evidence for effective and altruistic colony defence against natural predators by soldiers of the gall-forming aphid *Pemphigus spyrothecae* (Hemiptera: Pemphigidae). *Behav. Ecol. Sociobiol.* **27**: 421–430.

Hamilton, W. D. 1967. Extraordinary sex ratios. *Science (Wash., D.C.)* **156**: 477–488.

–. 1972. Altruism and related phenomena, mainly in social insects. *Annu. Rev. Ecol. Syst.* **3**: 193–232.

Heming, B. S. 1973. Metamorphosis of the pretarsus in *Frankliniella fusca* (Hinds) (Thripidae) and *Haplothrips verbasci* (Osborn) (Phlaeothripidae) (Thysanoptera). *Can J. Zool.* **51**: 1211–1234.

–. 1993. The structure, function, ontogeny, and evolution of feeding in thrips (Thysanoptera). In *Thomas Say Publications in Entomology: Proceedings.* C. W. Schaefer and R. A. B. Leschen, eds., pp. 3–41. Lanham, Maryland: Entomological Society of America.

Herre, E. A. 1985. Sex ratio adjustment on fig wasps. *Nature (Lond.)* **228**: 896–898.

Itô, Y. 1989. The evolutionary biology of sterile soldiers in aphids. *Trends Ecol. Evol.* **4**: 69–73.

Jarvis, J. U. M. 1981. Eusociality in a mammal: cooperative breeding in naked mole-rat colonies. *Science (Wash., D.C.)* **212**: 571–573.

Jarvis, J. U. M. and N. C. Bennett. 1993. Eusociality has evolved independently in two genera of bathyergid mole-rats-but occurs in no other subterranean mammal. *Behav. Ecol. Sociobiol.* **33**: 253– 260.

Keller, L., ed. 1993. *Queen Number and Sociality in Insects.* Oxford University Press.

Kent, D. S. and J. A. Simpson. 1992. Eusociality in the beetle *Austroplatypus incompertus* (Coleoptera: Curculionidae). *Naturwissenschaften* **79**: 86–87.

Kiester, A. R. and E. Strates. 1986. Social behavior in a thrips from Panama. *J. Nat. Hist.* **18**: 303–314.

Kurosu, U., D. L. Stern, and S. Aoki. 1990. Agonistic interactions between ants and gall-living soldier aphids. *J. Ethol.* **8**: 139–141.

Lewis, T. 1973. *Thrips, their Biology, Ecology and Economic Importance.* London: Academic Press.

Lin, N. 1964. Increased parasite pressure as a major factor in the evolution of social behavior in halictine bees. *Insectes Soc.* **11**: 187–192.

Lin, N. and C. D. Michener 1972. Evolution of sociality in insects. *Q. Rev. Biol.* **47**: 131–159.

Mound, L. A. 1970. Intragall variation in *Brithothrips fuscus* Moulton with notes on other Thysanoptera induced galls on *Acacia* phyllodes in Australia. *Entomol. Mon. Mag.* **105**: 159–162.

–. 1971. Gall-forming thrips and allied species (Thysanoptera: Phlaeothripinae) from *Acacia* trees in Australia. *Bull. Br. Mus. Nat. Hist. Entomol.* **25**: 389–466.

–. 1994. Thrips and gall induction: a search for patterns. In *Plant Galls*. M. A. J. Williams, ed., pp. 131–149. Oxford: Systematics Association and Oxford University Press.

Mound, L. A. and B. J. Crespi. 1992. The complex of phlaeothripine thrips (Insecta: Thysanoptera) in woody stem galls on *Casuarina* in Australia. *J. Nat. Hist.* **26**: 395–406.

–. 1995. Biosystematics of two new gall-inducing thrips with soldiers (Insecta: Thysanoptera) from *Acacia* trees in Australia. *J. Nat. Hist.* **29**: 147–157.

Mound, L. A. and D. A. J. Teulon. 1994. Thysanoptera as phytophagous opportunists. In *Thrips Biology and Management*. B. L. Parker, M. Skinner, and T. Lewis., eds., pp. 3–19. New York: Plenum Press.

Packer, L. 1993. Multiple-foundress associations in sweat bees. In *Queen Number and Sociality in Insects*. L. Keller, ed., pp. 215–233. Oxford University Press.

Pelikán, J. 1990. Butting in Phlaeothripid larvae. In *Proc. 3rd. International Symposium on Thysanoptera*, Kazimierz Dolny, Poland. pp. 51–55. Warsaw: Warsaw Agricultural University Press.

Reeve, H. K. 1991. Polistes. In *The Social Biology of Wasps*. K. G. Ross and R. W. Matthews,. eds., pp. 99–148. Ithaca: Cornell University Press.

Sherman, P. W., J. U. M. Jarvis and R. D. Alexander, eds. 1991. *The Biology of the Naked Mole Rat*. Princeton: Princeton University Press.

Stannard, L. J. 1968. The thrips or Thysanoptera of Illinois. *Ill. Biol. Monogr.* **29**: 215–552.

Swofford, D. 1993. *PAUP: Phylogenetic Analysis Using Parsimony*, version 3.1.1. Computer Program distributed by Illinois Natural History Survey, Champaign, Illinois.

Wcislo, W. 1987. The roles of seasonality, host synchrony, and behavior in the evolutions and distributions of nest parasites in Hymenoptera (Insecta), with special reference to bees (Apoidea). *Biol. Rev.* **62**: 515–543.

West-Eberhard, M. J. 1978. Polygyny and the evolution of social behavior in wasps. *J. Kansas Entomol. Soc.* **51**: 832–856.

Wilson, E. O. 1971. *The Insect Societies*. Cambridge: Harvard University Press.

9 · Interactions among males, females and offspring in bark and ambrosia beetles: the significance of living in tunnels for the evolution of social behavior

LAWRENCE R. KIRKENDALL, DEBORAH S. KENT AND KENNETH F. RAFFA

ABSTRACT

Parental care and colonial breeding are both widespread in two related groups of weevils known traditionally as Scolytidae and Platypodidae. Within-family cooperative breeding and eusociality also occur; in at least one platypodid ambrosia beetle, non-reproductive females help presumed relatives to raise offspring.

Although they breed in a wide variety of woody tissues, the majority of species fall into two ecological categories, those reproducing under the bark and feeding directly on inner bark ('bark beetles', most scolytids) and diverse taxa feeding upon microbial ectosymbionts they have introduced to the walls of their tunnel systems ('ambrosia beetles', many scolytids and all true platypodids). In both scolytids and platypodids, females lay eggs over an extended period of time in long tunnels. Males usually remain with females in these tunnel systems, controlling and expelling refuse. We examine hypotheses for prolonged male residence, and find that the most likely explanations for long stays in burrows are either blocking out natural enemies or increasing the reproductive rate of the resident females.

Species reproducing in bark or wood usually breed in large aggregations. In most species, these colonies are an incidental effect of mutual attraction to odors emanating either from the resource itself or from the beetles (pheromones). We suggest that the key feature predisposing these beetles to the evolution of breeding in aggregations is the presence of mate-attracting pheromones coupled with the utilization of resource patches that cannot be monopolized by single families. In most species, juvenile survivorship is reduced by breeding in dense aggregations; however, in species that kill trees, rapid recruitment of conspecifics is essential to the success of the first beetles to attack a tree. We discuss the trade-offs tree-killing species must make when recruiting additional colonists, which are at the same time helpers and competitors.

Although colonial behavior has been studied intensively (particularly in regards to tree killing), there has been little direct study of other forms of social behavior in scolytids and platypodids. Behaviors that increase offspring or sibling survivorship have evolved repeatedly in these taxa. Living inside a well-protected food source such as an ambrosia beetle tunnel system, we argue, is especially conducive to the evolution of parental care. We summarize observations on direct and indirect forms of parental care as well as possible instances of siblings helping siblings. We suggest that key factors in the evolution of the social behaviors we discuss are: (1) dead woody tissue is ephemeral, divisible, patchily distributed, and relatively infrequent; (2) bark beetles breed in tunnel systems, which are relatively safe from predation, provide both food and relatively favorable temperature and humidity conditions, and are easily barricaded; (3) finding and establishing a new tunnel system is difficult and dangerous.

INTRODUCTION AND NATURAL HISTORY

It is not generally known that complex social behaviors are well developed in bark beetles and ambrosia beetles. Unfortunately, social behavior has seldom been a research focus in these insects, and consequently we have only a superficial knowledge of the details of interactions within and between families.

Social behavior can be so broadly defined that it covers virtually all interactions between two organisms (see, for example, Trivers 1985). Here, we will concentrate on within-species, regular interactions between individuals of bark and ambrosia beetles. We begin this chapter by trying to place these beetles in a phylogenetic perspective, in an attempt to deduce the ecology and behavior of ancestral taxa. Next, we discuss: (1) prolonged contact between females and offspring; (2) prolonged male residence in gallery systems; (3) interactions among offspring; (4) colonial breeding (breeding aggregations), and (5) eusociality. We conclude with a general discussion and suggestions for future research.

Phylogenetic background

Forest entomologists have traditionally treated scolytids ('bark beetles') and platypodids ('pinhole borers') together (usually as two families), because of their morphological, ecological and behavioral similarity; they co-occur in the same logs, have similar tunnel systems, and create similar economic problems for the timber industry. Systematists disagree as to whether these taxa should be considered as families (see, for example, Browne 1961a; S. L. Wood 1973, 1978, 1982, 1986, 1993; Morimoto 1976; Beaver 1989) or subfamilies of Curculionidae (Crowson 1968; Kuschel 1990, 1995; Lawrence and Newton 1995). Until recently, there has been a consensus that, together, the two groups form a monophyletic lineage of highly specialized weevils, either inside or outside the Curculionidae. However, four recent studies force us to re-examine the hypotheses that Scolytidae and Platypodidae (1) have a separate origin outside the Curculionidae and can thus be considered separate families (see Wood 1978–1993), and (2) are sister taxa (Thompson 1992; Kuschel 1995; Lyal 1995; Lyal and King 1996). There is a consensus among these four studies that at least scolytids can be readily placed within the Curculionidae, near the Cossoninae; there is no support in these analyses for considering them as a separate family. The phylogenetic placement of Platypodidae is much less clear; none of these studies present unique, shared derived character states which link platypodids unequivocally with scolytids, or with any other single weevil group. Although the exact relationships among curculionids, scolytids and platypodids are thus uncertain, we will follow a long tradition in forest entomology and discuss scolytids and platypodids together in this chapter; we will also continue to use 'scolytids' and 'platypodids' to refer to these groups, while we await the results of more detailed morphological and biochemical data on their phylogenetic relationship.

Feeding ecology

Scolytid and platypodid beetles exhibit considerable variability in mating systems and social behavior. Almost all scolytid and platypodid species breed in tissues of woody plants, mostly in the inner bark ('bark beetles', in the strict sense) or wood. Most species breed in tunnel systems in dead inner bark, but wood, seeds, woody fruits, pith, leaf petioles, and immature pine cones are also used. The various breeding habits have each evolved independently on multiple occasions (see, for example, S. L. Wood 1982; Kirkendall 1983; Beaver 1989). The current, widely used ecological designations 'bark beetle' and 'ambrosia beetle' refer broadly to larval feeding habits. In the former, larvae usually feed directly on the plant tissues in which they are embedded; in the latter, they feed exclusively on microorganisms (predominantly fungi, including yeasts) growing on tunnel walls within plant tissue.

The *ambrosia beetles* depend upon their highly nutritious symbionts and do not feed directly upon the wood (see, for example, Norris and Baker 1967; Norris 1975; Beaver 1988). For the sake of simplicity, we will refer to the food of ambrosia beetles as 'the' ambrosia fungus. However, the presence of several microbial species may be essential for the successful reproduction of a given beetle species, as there is usually an entire microbial community (fungi and bacteria) with its own succession associated with ambrosia beetles (Haanstad and Norris 1985; Beaver 1989). The fungus-growing habit has evolved at least eight times, once in the ancestor to the platypodids and at least seven times in the scolytids (S. L. Wood 1982; Kirkendall 1983; Beaver 1986, 1989). Most taxa designated as ambrosia beetles are true fungus-growing insects, and possess a wide array of pouches, pores, and cuticular invaginations (mycangia) for carrying, protecting, and nourishing their 'starter cultures' (Batra 1963; Francke-Grosmann 1956, 1966, 1967; Kok 1979; Norris 1975, 1979; Nakashima 1975; reviewed in Beaver 1989).

Those wood-tunneling taxa that do not 'cultivate' microorganism communities are usually lumped with bark beetles, although the more technical term 'xylophage' is also used. In more general discussions, species with other habits, such as breeding in pine cones, pith, seeds and woody fruits, or herbaceous plants, are also commonly lumped with bark beetles. Most bark beetle species breed in the inner bark or outer sapwood tissues of dead trees. We will refer to these phloeophagous species as 'true bark beetles', when we wish to distinguish them from species breeding in sapwood and heartwood, pith, seeds and fruits, or other plant tissues.

For the 1339 scolytid species from North and Central America whose larval feeding habits could be categorized (using primarily information in S. L. Wood 1982), roughly one-half breed in inner bark, one-third breed as ambrosia beetles, and the remaining 19% breed in wood, pith, seeds/fruits, or in herbaceous tissue (Kirkendall 1993). The proportion of species breeding as ambrosia beetles is highest in tropical regions (Beaver 1977, 1979; Kirkendall 1993, fig. 7.6); to give one example, for 216 Indonesian species, only one-fifth breed in inner bark while three-fourths are ambrosia beetles (Kalshoven 1958, 1960).

Mating systems and gallery systems

An important feature of scolytid and platypodid biology is a variety of mating systems rarely found in other insects (reviewed in Kirkendall 1983, 1993; see also Atkinson and Equihua1986a,b). It is usual for males to remain with ovipositing females during most or all of the oviposition period (see, for example, Kirkendall 1983, Table 2). Although egg tunnel systems usually are constructed by one female ('monogynous' species), in a variety of unrelated lineages 'harem polygyny' has evolved, in which several to many females are associated with each successful male (Kirkendall 1983, Table 3). Brother–sister inbreeding has evolved numerous times, and is perhaps the dominant mating system in tropical regions (Kirkendall 1993, Table 7.1, Fig. 7.1).

The mating system terminology used here differs from usage in the social insect literature. 'Monogynous' denotes a breeding unit (for an outbreeding species) with one female, with or without an accompanying male. 'Polygynous' refers to a breeding unit with two or more females; no polygynous species are known in which the male does not stay for at least several days, and usually males remain with their mates for several weeks or more. Species in which breeding units include a variable number of females have been termed 'harem polygynous', while those with regular brother–sister mating, where one or a few males inseminate their many sisters, have been called 'inbreeding polygynous' (Kirkendall 1983).

Most bark and ambrosia beetles construct tunnel systems in the breeding material. The egg tunnel or tunnels made by single females are usually referred to as 'galleries', 'egg galleries', 'burrows', or 'nests'; in harem polygynous species, the collective work of the females mated to a single male is referred to as a 'gallery system', 'tunnel system', or 'nest'. It is important to emphasize that these nests are normally isolated from the external environment, accessible only through the entrance tunnel or, in some species, through secondary holes often termed 'ventilation holes'.

Mate location

Mate location in most scolytids and platypodids is accomplished by long-range pheromonal attraction. Although the (long-range) pheromone-producing sex is constant within a species (and usually within a genus), it varies at higher taxonomic levels. Females are the pheromone producers in most monogynous species, and males the producing sex in all harem polygynous groups (Kirkendall 1983); Kirkendall (1983) has argued that male pheromone production in monogynous groups is usually found in taxa derived from harem polygynous lineages.

In those species not using long-distance pheromonal attraction, males and females are attracted by host odors (see, for example, Rudinsky 1966; Lyttyniemi et al. 1988; Ytsma 1989; Eidmann et al. 1991). In both pheromone-producing and non-producing species, copulation may take place at overwintering sites or, more commonly (in those taxa in which maturation feeding in fresh tissue is a part of the life cycle), at maturation feeding sites, prior to the normal breeding activities (see, for example, Spessivtseff 1921; Fisher 1931; Bartels and Lanier 1974; Rudinsky et al. 1978; Mendel 1983; Janin and Lieutier 1988).

Mating systems

Promiscuous male sexual behavior is the exception rather than the rule in bark and ambrosia beetles. We attribute this to the ease with which males can defend females, and the high costs to searching for mates. Bark and ambrosia beetles are noteworthy for the variety of mating systems they exhibit, in most of which the males remain with females during most or all of the oviposition period (reviewed in Kirkendall 1983). Sibling-mating inbreeding has also evolved numerous times in Scolytidae, and is characterized by highly female-biased sex ratios and strong sexual dimorphism (Hamilton 1967; Beaver 1976; Kirkendall 1993).

PROLONGED CONTACT BETWEEN FEMALES AND OFFSPRING

Overlap between parents and their brood is unusual in insects, although it has evolved in many orders (Wilson 1971). However, it is the norm in bark and ambrosia beetles, arising naturally from their egg-laying habits: females, and in many cases males, reside in the gallery system. Neither sex need leave the nest for food.

Table 9-1. *Care given eggs or larvae by bark and ambrosia beetles*

Note: in most cases, the effects postulated here have not been demonstrated experimentally. For references, see text.

Type of care	Possible effects	Examples
Eggs placed in special niches	Spaces out larvae, gives them a starting orientation, keeps them out of way of adults	Most bark beetles, corthyline and xyloterine ambrosia beetles
Egg niche covered over by frass, or single or clumped eggs (not in niches) covered by frass	Protection against mites? Protection against other predators or parasites? Protection against desiccation?	Most bark beetles, corthyline and xyloterine ambrosia beetles
Eggs not in niches, 'tended' by female, moved if female disturbed	Protection from microorganisms	Observed in *Xyleborus*
Eggs or larvae rescued from ejection / falling out of tunnel	Accidental loss of offspring averted	Platypodids, *Xyleborus*
Tending of larvae	Pushing larvae onto fresh food source; renewing food, removing refuse	*Xyleborus, Trypodendron, Monarthrum*

Female care is difficult to observe directly. True bark beetles can be observed through a Plexiglas or plastic 'roof' (see, for example, Reid 1962, Figs. 1, 2), but behavior of ambrosia beetles can only be observed by allowing the beetles to tunnel in artificial medium in glass tubes or jars (see, for example, Norris 1975). As with bark beetles, brief glimpses are afforded when gallery systems are carefully and laboriously exposed, but these are in effect one-time 'snapshots'. None the less, most of what we know about behavior of ambrosia beetles (and everything about behavior of platypodids) has been pieced together from such observations.

We discuss first the effect of type of breeding substrate on contact between parents and offspring. We then discuss, separately for bark and ambrosia beetles, what parent females might or might not be able to do for offspring during the oviposition period (types of care given to offspring are summarized in Table 9-1) and why they might remain in the gallery system after egg-laying is completed. Some of these arguments apply equally well to male presence, but males will be treated explicitly in the next section.

Breeding habits and family cohesion

Differences in breeding habits have important consequences for the evolution of social behavior. Larvae of most true bark beetles, for example, tunnel independently through the thin layer of inner bark/outer sapwood, and the larval tunnels gradually diverge. These larval mines are packed tightly with frass. Consequently, there can be no physical contact between such larvae and their parents. Furthermore, most bark beetles usually breed in aggregations, and larvae may be in closer proximity to unrelated larvae than to siblings (see Kirkendall 1989).

Conversely, ambrosia beetle larvae do not tunnel as they feed, though in some groups they construct their own so-called 'cradles' (feeding chambers) and in most platypodids (see below) older larvae extend the parental tunnel systems. The result is that siblings are potentially in close contact with each other, either all the larvae of a single family or (in more extensive tunnel systems) contemporaneous batches of larvae. The entire set of siblings is potentially directly affected by parental actions. Consequently, the potential for the evolution of certain forms of social behavior is much greater for ambrosia than bark beetles.

Spatial clustering of larvae also commonly results from breeding in seeds or in pith. Sibling larvae remain in relatively close contact, although they may tunnel individually, and many will be close to their parents. As with ambrosia beetles, they nest deeply enough in the resource to be largely protected from the community of natural enemies preying upon true bark beetles. In pith, extensive movement within the nest may or may not be possible, depending on the shape of the gallery with respect to beetle diameter; regardless, the actions of a given individual will potentially affect one or more family members. Thus, with respect to factors important for the evolution of social behavior, pith- and seed-breeding species are more similar to ambrosia beetles than to the true bark beetles from which they are derived.

Prolonged female–offspring contact in bark beetles

Females of most bark beetles lay many or all of their eggs in one egg gallery. Eggs are laid singly or in small batches, and are relatively large with respect to the female's body size: browsing through arbitrarily selected reprints revealed ratios of egg length to female body length varying from 1/8 (the smallest we could find, for *Scolytus rugulosus*: Kemner 1916) to 1/4 or larger in a number of species (e.g. *Trypophloeus populi* and *Procryphalus mucronatus*, Petty 1977; *Xyleborus affinis*, Roeper *et al.* 1980). Producing a clutch of such large eggs requires feeding during the extended oviposition period, leading to temporal overlap with early-produced offspring. While parents are present, what behaviors increase egg and larval survivorship, and are the effects on offspring fitness due to adaptive behaviors or incidental effects of 'normal' oviposition behavior? Which, if any, female bark beetle behaviors qualify as parental investment, i.e. do certain behaviors help current offspring at the cost to future reproduction?

In those species with independent larval mines, care of eggs or larvae is restricted to any general protection afforded by the female's presence. Ovipositing females of most species keep the egg tunnels clean, and presumably offer some defense against intruding predators and parasitoids (though not against parasitoids ovipositing through the bark, or predators which do not come through the entrance to a defending female's egg tunnel). Diseased or dead adults are entombed, which may prevent further spread of pathogens within the gallery system (though it is not clear whether entombing is the work of the female or the male).

When males are absent, females may also block entrances before oviposition is finished, either with their bodies (*Phloeotribus demessus*, Kirkendall 1984; *Hypothenemus curtipennis*, Beaver 1986) or with a hardened plug (*Cryptocarenus heveae*, Beaver 1974; *Pityophthorus puberulus*, Deyrup and Kirkendall 1983). Hard plugs are also found in a few species *with* male residence, as in *Hylesinus varius* (Løyning and L. R. Kirkendall unpublished observation) and *Camptocerus niger* (Beaver 1972), and some *Dendroctonus* pairs block the gallery entrance with resin-laden frass (see, for example, Blackman 1931). Possible benefits to blocking

during oviposition are illustrated by two *Scolytus* species in which males do not stay with females after copulation. Females of *Entodon leucogramma* (Hymenoptera: Eulophidae) wait until *S. scolytus* females are at the far end of the egg tunnel, then dash in and oviposit into *S. scolytus* eggs (Beaver 1967); *Entodon ergias* apparently attacks *S. intricatus* eggs in a similar fashion (Yates 1984).

Female presence after oviposition could represent maternal care if females can block out potential predators or parasitoids whose offspring follow larval mines away from egg tunnels (e.g. *Roptrocerus xylophagorum*, Dix and Franklin 1981). However, female residence after cessation of oviposition could have two other explanations: (1) females remain to feed and regenerate flight muscles; (2) females use the egg tunnel as a safe place to overwinter. Both behaviors are commonly reported for scolytids.

Blocking the entrance during or after oviposition has been reported for a number of species, and dying in the entrance has been reported for a number of species. For a number of outbreeding monogynous species, for example, males leave after oviposition but females stay (e.g. *Chramesus exul*, *Chaetophloeus minimus*, T. H. Atkinson *in litt.* 1986). Males do not stay at all, in tunnel systems of *Scolytus intricatus*: females usually die in the entrance (92% of 130 galleries had a dead female (Yates 1984)).

Prolonged female–offspring contact in ambrosia beetles

Adult presence in an ambrosia-beetle gallery normally ought to confer a wider variety of benefits than the presence of a parent in a bark-beetle gallery. Maternal activity is critical for maintaining an appropriate rate of food production: cropping the fungal or yeast mats on tunnel walls keeps the tunnel systems from being overgrown, as has been shown experimentally for *Xyleborus affinis* (Roeper *et al.* 1980) and *X. ferrugineus* (Kingsolver and Norris 1977) and has often been inferred from field or laboratory observations (see, for example, Hubbard 1897; Batra 1963). In addition, *Xyleborus* females have been observed to assist newly emerging larvae from the eggs, to push eggs and young larvae onto food, and to 'tend' both young and older larvae (French and Roeper 1975; Kingsolver and Norris 1977; Roeper *et al.* 1980). *Trypodendron lineatum* eggs are laid in tunnel-wall niches, and the larvae gradually expand these niches into 'cradles'; females keep the egg galleries open by cropping the ambrosia growth, and remove larval excreta; the absence of females leads to death of the

Table 9-2. *Mate fidelity in* Chramesus hicoriae[a]

Season, stage[b]	Both present	Female only	Male only	Neither present
Summer, no eggs	7	2	0	0
Summer, some eggs	7	0	0	0
Fall, end of breeding	7	2	1	0
Spring, parents dead	23	2	1	1

[a] Population breeding in *Carya ovata* and *C. glabra* on the E.S. George Reserve, Washtenaw Co, Michigan. These previously unpublished observations were made between 1979 and 1980 by LRK. For the fall data, only the two females that were alone were dead; these were the two shortest galleries.

[b] 'Stage' = stage of reproductive cycle: recently started, no eggs laid; some eggs laid; all eggs laid (end of breeding); all laid, now the next spring.

offspring (reviewed for this species in Borden 1988). Finally, *Xyleborus* females either eat dead and weakened progeny, or entomb them in a short branch tunnel (Gadd 1941; Norris 1979).

The most elaborate care thus far detailed remains that in the scolytid genus *Monarthrum*, detailed by Hubbard (1897) and Doane and Gilliland (1929). According to Hubbard, females of *M. fasciatum* first prepare a bed of wood chips which they inoculate with the ambrosia fungus; chips are later placed in the niches ('cradles') along with each egg. Females seal each cradle with a plug of ambrosia fungus. After hatching, the larvae crop the ambrosia from the chips and from the plug, and periodically perforate the plug in order to eject waste. The mother beetles remove refuse to the fungus beds, and renew and reseal the food plug. According to Hubbard (1897, p. 27), 'The mother beetle is in constant attendance upon her young during the period of their development, and guards them with jealous care'. Doane and Gilliland report similar habits, for two California species of *Monarthrum*, as well as for a species of the closely related genus *Gnathotrichus*. For both genera, the fungi are said to occur in patches in the tunnel system, where they are tended zealously.

Female blocking after the conclusion of oviposition should be especially effective in those taxa breeding as ambrosia beetles in sapwood, or breeding in pith or seeds. First, in these taxa, offspring remain in close proximity to their mother. Second, access of natural enemies is limited. When the tunnel systems are in solid sapwood (ambrosia beetles) or in twigs or seeds, where breeding units are usually isolated and the one entrance hole will be the only hole leading under the surface for the entire resource unit, they cannot move readily from tunnel system to tunnel system as they can in inner bark. Blocking the tunnel entrance could exclude predators and possibly help regulate microclimate, and might have a much greater effect on offspring survivorship in such species than for species which breed in large aggregations. It would be interesting to know if the incidence of postovipositional entrance blocking is higher for pith- and seed-boring species and for ambrosia beetles than for species breeding in inner bark.

Postovipositional blocking may be widespread in ambrosia beetles; in some cases, this includes blocking both during and after oviposition. It has been mentioned for *Sueus niisimai* (Beaver 1984), *Xyleborinus saxeseni* (Bremmer 1907; Hosking 1973), *Euwallacea destruens* (Kalshoven 1962), *Trypodendron betulae*, *Xyleborus obesus*, and *X. sayi* (Kirkendall 1984). Female postovipositional blocking has also been observed in pith-boring species, such as *Hypothenemus dissimilis* (Kirkendall 1984) and the ambrosia beetles *Sueus niisimai* (Beaver 1984) and *Hypothenemus curtipennis* (Beaver 1986).

PROLONGED MALE RESIDENCY

Males remain in tunnel systems with females for at least some days or weeks, in most Scolytidae and Platypodidae (Kirkendall 1983). In at least 12 genera, both sexes degenerate their wing muscles to a non-functional condition at the start of breeding, which restricts male mating behavior options (see, for example, Roberts 1961; references in Borden 1974 and Langor 1987). Males die with females in some taxa (e.g. *Chramesus hicoriae*) (Table 9-2; see Kirkendall 1983), and in other species they may even remain in their breeding system after females have left (discussed below, and see Table 9-2 and Kirkendall 1983). In platypodids, the commitment of males to their original tunnel systems may often be total: it is apparently common that both parents break off tarsal segments and claws, thereby rendering them incapable of walking outside their gallery systems (Beeson 1917; Browne 1961a,b; Roberts 1961; Harris *et al.* 1976; Milligan 1979; Kent and Simpson 1992).

Such mate fidelity is extremely unusual in insects: male association with females seldom extends beyond a few hours, or perhaps a few days, pre- or postcopulatory guarding (Thornhill and Alcock 1983; Zeh and Smith 1985). In this section, we will discuss the primary hypotheses for prolonged male postcopulatory residence – mate attraction, mate guarding, increased offspring survivorship, increased offspring number – followed by a summary of experimental evidence, and finally a discussion of the alternatives to male residency.

Male residency and mate attraction

Males may stay with the first female because of the potential of attracting additional mates. Most scolytids and all typical platypodids are monogynous, and normally only one female or egg tunnel is found in these species; the possibility that later-arriving females copulate with established males but do not enter their tunnel systems has not been carefully tested, but it is unlikely that this can be a significant benefit to male residency. Even in harem polygynous species, mate attraction may be a factor in male residency for only a brief period. At least in temperate, outbreeding species, colonization of a given resource unit is frequently highly synchronous, weather permitting (Berryman 1982; Coulson 1979; Reid and Roitberg 1994; for exceptions, see Kirkendall 1983; Bustamente and Atkinson 1984; Yates 1984). The result is that few new females arrive after the first few days of gallery initiation, in such species.

Male residency as mate-guarding

Competition between males for access to females is often fierce in bark beetles (Kirkendall 1983). Males that have burrowed mostly or entirely under the bark can usually resist challenges from other males (see, for example, Oester and Rudinsky 1975; Vernoff and Rudinsky 1980). Mate-guarding can only be important as long as rival males are still on the bark; this seems to be the case for only a relatively short period, however, at least in those bark and ambrosia beetles for which the colonization process has been carefully studied. However, although pheromone production by females of *Trypodendron lineatum* is drastically reduced by pairing, if males are removed, production begins again (Klimetzek *et al.* 1981). If this capability is widespread in female-colonizing, monogynous species (which category includes the vast majority of scolytid phylogenetic lineages), then male residency may be favored as mate-guarding, as long as there is a good probability that an abandoned female can still produce pheromone and attract a new male. It is likely that this capability is lost as females begin ovipositing; at least a low frequency of solitary-breeding females are usually found in such species, and this frequency increases over time (see, for example, McMullen and Atkins 1959, 1962; Borden 1969; Salonen 1973).

As mentioned above, colonization of a given resource unit is frequently highly synchronous. Consequently, males are attracted to an active breeding site for only a short period of time, under favorable weather conditions for only a week or less. Where local colonization is synchronous, defense against cuckoldry cannot be a benefit for more than the first few days (see, for example, Reid and Roitberg 1994).

Male residence and offspring survivorship

Blocking out predators, parasites, inquilines, and other species

An obvious possible function of burrow-blocking is excluding predators, parasites, inquilines, or competing species. The efficacy of prolonged burrow-blocking in this regard has not yet been investigated in detail for any bark beetle. However, Reid and Roitberg (1994) found that removal of *Ips pini* males for only three or four days led to a major increase in the occurrence in their tunnel systems of two species of beetle that prey on eggs and larvae. The effects of blocking on predators or burrow-usurpers ('cleptoinquilines': Naumann-Etienne 1978) of ambrosia beetles has never been measured; there are few natural enemies of ambrosia beetles, but most of those that do occur are highly specialized for the task, often with elaborate extensions of the pronotum, forehead (frons), or mandibles, which look as though they would be especially useful for prying and puncturing (see drawings of *Amphicranus* in Blandford 1905, and of *Gnathotrupes* (as *Gnathocortus* and *Gnathomimus*) in Naumann-Etienne 1978), and specially shaped, toothed elytral declivities (the hind-ends) which would appear to aid in evicting burrow residents with backward shoving movements (Naumann-Etienne 1978 and drawings cited above). These specialized beetles are most often from the families Brenthidae, Colydiidae and Curculionidae (Schedl 1961, 1962a,b, 1972; Browne 1963; Roberts 1969; Crowson 1981), but nest-usurpers also include several unrelated lineages of scolytid ambrosia beetles (Naumann-Etienne 1978; S. L. Wood 1982).

Entrances under the bark are apparently an extremely valuable resource for subcortical and xylophagous insects. Excavating the entrance hole is time-consuming and risky: a tunneling individual is vulnerable to host defenses and natural enemies. Blocking the tunnel entrance might therefore also serve to keep out other species seeking easy entry under the bark. There are numerous observations of one species either occasionally or regularly using the entrances (or even egg tunnels) of another as a starting point for their own egg galleries (see, for example, Schwartz 1894; Chararas 1962; Stewart 1965; S. L. Wood 1971; Rudnew and Kosak 1974; Wood 1982); two studies reported that the intruding *Crypturgus* reduced breeding success of the host species (of *Tomicus*, Stark 1926; of *Ips* species, Karpinski 1935) through interspecific larval competition and egg destruction. In addition, monogynous pairs are occasionally found sharing an entrance with pairs of the same species (e.g. *Trypodendron retusum*, Hubbard 1897; *Hylurgops pinifex*, Blackman 1919a; *Ips avulsus*, Cook et al. 1983). Although it is plausible to think that this intraspecific entrance-sharing would lead to increased competition (which would imply a benefit for burrow-blocking), this aspect has not been studied.

Blocking out unwanted females (control of harem size)
Burrow-blocking allows males to control the number of females gaining access to the tunnel system. It remains to be established, however, whether monogynous systems remain so because males exclude second females, or because subsequent females choose not to enter systems in which one female is already present. For harem polygynous species, studies have shown that males in harem *Ips paraconfusus* and *I. pini* refuse entry to additional females, once they have acquired their normal complement (Borden 1967; Swaby and Rudinsky 1976), and that *I. pini* males with three or four females are much less attractive to females. Males of *Polygraphus rufipennis* with large harems produce less pheromone than the same individuals when unmated (Bowers et al. 1991); *Ips calligraphus* males

with three or more females cease synthesizing the pheromone component ipsdienol (Vité et al. 1972). All of these data suggest the tantalizing possibility that males of at least some harem polygynous species attempt to limit harem size because there is a cost to accepting too many females (see Kirkendall 1983): an extremely unusual and interesting phenomenon which deserves more research.

Controlling microclimate
Several researchers have suggested that blocking by males regulates humidity (Jover 1952; Husson 1955; Cachan 1957; Kalshoven 1959; Beaver 1986). Ambrosia fungi require a high moisture content in the wood and in the atmosphere to thrive (Cachan 1957; Browne 1961b; Francke-Grosmann 1967), and the tunneling activity and attack density of ambrosia beetles is highest at high sapwood moisture contents (Kabir and Giese 1966a,b; Saunders and Knoke 1967). In several platypodids, during the main period of larval development, males plug the entrance with long cylinders of wood fibers; these are rapidly replaced if they are experimentally removed (Jover 1952; Husson 1955). In *Platypus caviceps*, the male uses his body in conjunction with a cone of fibers stuck together with sap (Holloway 1973). Males of the ambrosia beetle genus *Scolytoplatypus* were observed to make rapid pumping movements with the abdomen, interpreted by Kalshoven (1959) as possibly aerating the tunnel.

Keeping larvae and eggs in gallery systems
Neither Xyleborini nor the typical platypodids place eggs in special niches or cradles. ('Typical platypodids' refers to the monogamous ambrosia beetles that make up the bulk of this family, groups placed by Wood (1993) in Tesserocerinae and Platypodinae, by Thompson (1992) in Platypodinae.) Since adult females frequently move about in the tunnel system, and shove frass and wood chips backwards towards the entrance, they risk removing offspring with the refuse. Males of two platypodids (*Doliopygus* spp.) have been seen to rescue eggs in danger of being ejected along with wood fragments (Cachan 1957; Browne 1963); Chapman (1870) states that larvae fall out, if male *Platypus cylindrus* are removed. Referring to his observations on three *Platypus* species in New Zealand, Milligan (1979) remarks similarly that 'if the male is lost, live larvae may fall from the entrance'. In culture tubes with families of *Xyleborus affinis* (an inbreeding species, with no adult male present in the nest), larvae were occasionally found on the surface of the culture medium (Roeper et al. 1980). Thus, for bark or ambrosia beetle species in which eggs and larvae are

Table 9-3. *The effects of male presence on offspring production and survivorship, in bark and ambrosia beetles*

For most data sets, only these means were published (no measures of variation, or raw data, were given).

Species	Treatment	n	No. of eggs	Tunnel length (mm)	Eggs mm^{-1}	Breeding situation, source
Hylesinus californicus	Male present	49	—	25.4	—	From logs colonized during a pheromone study; the egg tunnels were dissected and analyzed 8.5 days after initial colonization of the logs (SE = 0.8 for both; Rudinsky and Vernoff 1979).
	No male present	52	—	15.3	—	
Scolytus multistriatus	Male present	179	68.5	27.8	2.5	Introduced population, Connecticut. Pairs or females alone were allowed to colonize freshly cut log sections (Wallace 1940).[a]
	No male present	7	23.9	19.7	1.2	
Scolytus ventralis	Both present	66	58.0	61.0	0.95	From naturally occuring tunnel systems in three colonized fir trees (Ashraf and Berryman 1969, Table 1).
	Male absent	105	28.6	61.0	0.47	
Trypodendron lineatum	With male	17	17.0	—	—	Females were confined on logs, singly or with males (Chapman 1959).
	Female alone	50	9.5	—	—	

[a] No difference in egg hatch. Bartels and Lanier (1974) seldom found males in dissected galleries, in New York. Chapman (1869) never found males after oviposition had commenced (England: no data).

Table 9-4. *The effect of male presence on reproduction, in the ambrosia beetle genus* Camptocerus

Data are for several species combined (Beaver 1972, Table III). The larvae, although feeding primarily on ambrosia fungus, do enlarge their egg niches into 'cradles', one larva per cradle. Data are from natural colonization.

	No. of galleries	No. of cradles	No. of cradles per gallery	No. of dead dead immatures	Mortality (%)
Both present	7	123	18	3	2
Male absent	5	41	8	8	19
Both absenty	8	41	5	29	71

not confined in some way, entrance-blocking may prevent accidental loss of offspring.

In one genus of Platypodinae, *Crossotarsus*, females have evolved an intricate solution to this problem, summarized by Browne (1961a, p. 221):

The head is deeply hollowed in front, and the antennal scapes are enlarged and compressed so that they can be used as a pair of forceps. As the mother beetle lays her eggs, she picks them up with her antennae and places them in the hollow of her head, where they adhere, and the adult is thus able to move freely in the tunnel without danger to them.

Male residence and number of offspring produced

Available data on female reproductive success with and without males present reveal a consistent pattern: among monogynous and at least some polygynous species, females lay more eggs or produce more adult offspring with males present than with males absent (Tables 9-3, 9-4). However, experimental manipulation is necessary, since male absence could be related to low female quality; perhaps males do not stay long with sick or feeble females. Data in Tables 9-3 and 9-4 for *Scolytus ventralis* and *Camptocerus* species could be confounded in this way.

There is also a tendency in bigynous and harem polygynous species for females to lay fewer eggs, in proportion to the number of other females with which they must share the services of a male (Fig. 9-1). For the two species in which this is not true, *Pityophthorus lautus* and *Pityogenes chalcographus*, it is interesting to note that they have higher average harem sizes than the other species in this figure, perhaps because in these species male help can be more readily shared than in others.

It is not known why fewer eggs are laid when males are absent (when such is the case), or why in some bigynous or harem polygynous species fewer eggs are laid in larger than in smaller harems. These data support the hypothesis, however, that the help males give females during egg tunnel excavation, by keeping the egg tunnels free of refuse, is a significant factor in determining the rate at which females can deposit eggs (see, for example, Reid and Roitberg 1994; G. Helland and L. R. Kirkendall, manuscript in preparation).

General considerations of male residency

Experimental evidence

Two studies of harem polygynous species in the Ipini have specifically addressed the question of long male residence (i.e. longer than a few days). *Ips pini* breeds in weakened and recently dead pines; males stay with females for several weeks. Reid and Roitberg (1994) tested the hypotheses that male residence (1) attracts more females; (2) is a form of mate-guarding; (3) leads to a higher reproductive rate of females; and (4) reduces mortality of offspring from predation or parasitism. The mate attraction and mate-guarding hypotheses were clearly rejected; prolonged male residence did not increase the number of mates they acquired, and experimentally removed males were not replaced after the short colonization period had ended. After three or four days, the females had laid 11% fewer eggs in systems from which the male had been removed. In addition, two-thirds of these systems had at least one predator present, whereas only a few systems with males present had predators.

G. Helland and L. R. Kirkendall (manuscript in preparation) tested hypotheses 1–3 for a common European harem polygynous bark beetle breeding in dead spruce, *Pityogenes chalcographus*. In a field experiment, males were removed from every other gallery; in laboratory experiments, harems of one, three, and five females were established, after which the males were removed from half the systems. As in *Ips pini*, hypotheses 1 and 2 were rejected.

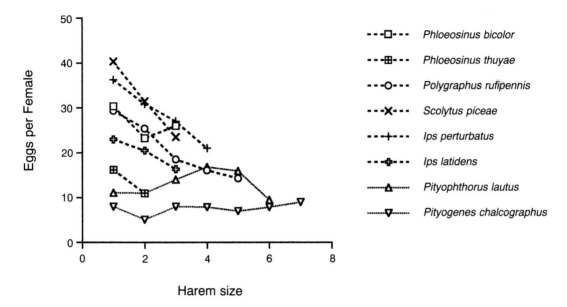

Fig. 9-1. The relationship between mean egg production per female and number of females per male, for harem polygynous scolytids. All data sets are from field collections. Sources for data: *Phloeosinus*, Chararas 1962: *Polygraphus* and *Scolytus*, Blackman and Stage 1918; *Ips perturbatus*, Gobeil 1936; *I. latidens*, Blackman 1991a; *Pityophthorus lautus*, Kirkendall 1983 (Table 4, 'dead pieces'); *Pityogenes chalcographus*, Chararas 1960.

Females with males present laid significantly more eggs than females in systems from which males had been removed; the difference arose both because females left systems earlier when a male was not present, and because they laid slightly but significantly more eggs when males were present.

Cheap, multifunctional defense
Remaining in the tunnel system entrance requires neither exceptional morphological modifications nor major changes in pre-existing behaviors, and can be done with little or no energetic expenditure beyond that of normal metabolism. Feeding, however, presents a problem, and presumably usually requires leaving the entrance for a short period of time. There is a certain risk of mortality to guarding individuals from predators or parasitoids (see, for example, Beaver 1986), but often this same mortality might have been incurred by males *not* blocking the entrance, if the predator or parasitoid entered the tunnel system. Burrow defense is also multifunctional; regardless of the context in which it may have first evolved, most or all of the possible functions discussed above may be served simultaneously.

Offspring in tunnel systems are much easier to protect than exposed batches of eggs or larvae. The entryway is only slighter larger than the beetle: this close fit, coupled with males' ability to wedge themselves firmly in place, makes pushing or pulling males from their guarding position difficult. Blocking out intruders, then, does not require complex behaviors.

The value of entrance-blocking will be reduced or negated, however, if the entrance is not the only hole to the outside. Holes from egg tunnels to the surface occur in a variety of species, although their taxonomic distribution has not been reviewed. These holes are additional points of entry for intruders, including clerid larvae (Berryman 1966) and torymid parasitoids (Dix and Franklin 1981).

Alternatives to fidelity
Males of perhaps most scolytids and platypodids may be committed to the local resource unit, and platypodids may be confined to their original tunnel system, by slowly reversible (wing-muscle degeneration) or irreversible (loss of tarsi) morphological changes, which occur shortly after mating. At least theoretically, though, or before these

changes, have taken place, the alternative to staying with a female (or group of females) is to seek further mating opportunities. In species with relatively synchronized breeding, this alternative may disappear after only a few days, at least in the current resource unit and perhaps in neighboring units as well. Finding other mates will then require leaving the immediate area, if the period of colonization has not ended. Males do not stay more than minutes or hours with females in a few species; in other species, males may continue searching out females at first, and eventually stay with one female (Kirkendall 1983). In most monogynous species, at least some mated but unpaired females are found; the extent to which this is due to (a) male departure, (b) inseminated, unpaired females establishing a new egg tunnel, or (c) predation, has not been quantified.

The general costs and benefits of searching vs. staying have most recently been discussed by Zeh and Smith (1985), Yamamura (1986) and Alcock (1994). Important parameters include the operational sex ratio (OSR), female receptivity to other males, search time and mortality cost to searching for further matings, the capacity for resisting takeovers, sperm precedence patterns, and the strategies pursued by rival males (see also Alcock 1994, Table 2).

The most important problem faced by males is finding additional breeding sites. Bark and ambrosia beetle populations are generally thought to be limited by availability of suitable breeding material (see, for example, Berryman 1982), and mortality during the search for new hosts is thought to be over 50% for bark beetles (Klein *et al.* 1978; Pope *et al.* 1980; Garraway and Freeman 1981; Anderbrant 1989). (It has been estimated that the average life expectancy of a bark beetle is only one or a few days when dispersing (Pope *et al.* 1980).) Freshly dead woody material is usually uncommon in natural forests, occurs unpredictably in space and time, and is only of an appropriate quality for a relatively short period. Species breeding in dead inner bark usually have specific requirements with respect to one or more factors such as host taxa, material diameter, bark moisture content, temperature conditions, presence or absence of certain fungi, etc.: demands which further reduce the spectrum of available breeding material. Species cultivating fungus in wood are much less specialized with respect to host species (Browne 1958; Beaver 1979; Atkinson and Equihua 1986a), but may still have preferences with respect to host material size, moisture content, and other physical and biotic factors.

Summary: why do males stay?

The available evidence leads us to the following conclusions. Mate-guarding is probably the current primary function of male presence in only a few scolytid species, particularly those in which males leave before oviposition has stated. However, it may well have been the initial advantage to remaining with females. Reid and Roitberg (1994) point out that mate-guarding seems to have preceded offspring care by males in other insects (citing in particular Smith 1980; Brockmann and Grafen 1989) and that parental defense of offspring is often associated with some sort of brood chamber. The additional benefits to prolonging the mate-guarding period, either one particular benefit or the summed effects of several factors, then selected for longer residency, and usually minor changes in male behaviors and morphology as well.

A few scolytid species are characterized by a mating system in which males pair only briefly with females, and do not enter the gallery system ('tachygamous' mating systems) (Kirkendall 1983). The breeding behavior in these species has been described only superficially. The species in which males do not stay are scattered among four unrelated tribes, Hylesinini, Diamerini, Scolytini and Corthylini-Pityophthorina (Kirkendall 1983, Table 2). The relevant behavioral and ecological differences between them, and those taxa in which males remain, are in no way obvious. A careful comparative study of, for example, *Scolytus rugulosus* (males leave, Gossard 1913) and *S. mali* (males stay, Chapman 1869; Rudinsky *et al.* 1978), two common, widespread species often found breeding in the same host material, could potentially prove enlightening as to the adaptive significance of male postcopulatory residence in such monogynous breeding systems. Similar comparisons could be made among *Scolytus* species in elm (Chapman 1869; Kirkendall 1983), or certain *Pityophthorus* species breeding in twigs (Blackman 1919b; Hedlin and Ruth 1970; Amman *et al.* 1974; Pfeffer 1976).

Experimental tests of the consequences of male presence for scolytid reproduction are now providing tentative answers for harem polygynous ipines. However, the 'primitive' (plesiomorphic) mating system for both scolytids and platypodids is almost certainly monogyny, so the ultimate test of ideas concerning the evolution of male residency in scolytids and platypodids will come from the thorough investigation of tachygamous mating systems or by experimental analysis of systems with naturally occurring variation in male residence, variation which probably occurs within populations in many monogynous species. The genera *Scolytus* and *Scolytodes*, and *Pityophthorus* plus *Conophthorus*, would seem to be the best scolytid clades for detailed experimental and comparative studies of the evolutionary ecology of male residency; each contains a wide range of male behaviors and of mating systems (see, for example, Bovey 1976; S. L. Wood 1982; Kirkendall 1983, 1984; L. R. Kirkendall, personal observations; Atkinson and Equihua 1985, 1986a).

INTERACTIONS AMONG OFFSPRING

Interactions among larvae and parents

Bark and ambrosia beetle larvae can move through tunnels or chew through wood, but are otherwise limited in the possibilities for physical interactions with each other. No special adaptations for larval helping have been found in these beetles, and no dimorphisms (within an instar) outside of slight differences between the sexes are known for larvae of any scolytids or platypodids.

Last- (fifth-) instar platypodid larvae make their own pupal chambers in all species studied, and in most species they also extend the branch tunnels which they used as larvae (see, for example, Table 9-7) prior to pupating. Fifth-instar larvae are equipped with mandibles capable of chewing wood, unlike the earlier instars, which can only crop fungi (Browne 1961a). Apparently, there is little or no such tunneling when the wood has already been invaded by competing fungi (Milligan 1979). This larval tunneling leads to a rapid expansion of the nest (see, for example, Chapman 1870; Holloway 1973). Some xyleborine ambrosia beetles make large brood chambers (as opposed to only systems of cylindrical tunnels); their larvae 'appear, at least during the later instars, to assist in the widening of the tunnels, and for this purpose they have relatively powerful mandibles...' (Browne 1961a, p. 40; Hubbard 1897 describes the same, for *Xyleborinus saxeseni*).

It is not clear whether such larval tunneling in ambrosia beetles contributes to the success of the family as a whole, or is primarily advantageous to the individual larvae doing the tunneling. However, in *Doliopygus chapuisi*, these larvae form a 'bucket brigade', passing along wood fibers from larva to larva, out to the female and finally the male, who shoves them out of the entrance tunnel (Jover 1952). This participation in nest hygiene presumably benefits siblings and parents, possibly at some small cost to the larvae. Roberts (1960) reported that last-instar larvae and young

adults of *Trachyostus ghanaensis* perform 'sanitary duties' in older gallery systems, including packing frass and feces into old branch tunnels.

Holloway (1973) suggested that the new branch tunnels contributed by the larvae of *Platypus caviceps* provide more oviposition sites than would be available in the original nest, and that the additional tunnel length also provides future ambrosia for later brood. *Trachyostus aterrimus* females oviposit in such vertical branch tunnels (Roberts 1962a). Browne (1961b) observed that the younger larvae of *Doliopygus dubius* quickly move into these new tunnels, to feed on the fresh fungal mycelia growing in them. In such instances, the branch tunnels may or may not directly benefit the larvae making them, but they clearly benefit parents or siblings.

Hubbard, writing generally about North American *Platypus* species, stated that the older larvae assist in excavating the galleries and furthermore that they 'show evident regard for the eggs and very tender young...'. If disturbed, the older larvae 'will frequently stop at the nearest intersecting passage way to let the small fry pass, and show fight to cover their retreat' (Hubbard 1897, p. 15).

In bark beetles, gregarious larvae occur in solitary breeding *Dendroctonus terebrans*, *D. valens*, *D. punctatus* and *D. micans*. Larval behavior has only been closely studied in the latter, European species, in which larvae feed side by side in groups of up to 50 individuals. Larvae actively aggregate in response to larval pheromones, and larvae reared in groups have higher survivorship and gain mass faster than isolated larvae (Grégoire 1985, 1988).

Interactions involving adult offspring

Adult offspring helping parents and siblings
The most likely 'helping' behavior to be commonly found among ambrosia beetle offspring is in aiding parent beetles with keeping the tunnels from being overgrown by ambrosia fungi, when these fungi have continued to grow late in the breeding cycle (Hadorn 1933). As useful as this behavior might be for the last-developing offspring, particularly if the parents have died or deserted, it involves little or no cost to the young adults: the 'helping' apparently involves little more than feeding on the ambrosial growth on the tunnel walls (see discussion of reproductive altruism, below). Nonetheless, this feeding behavior is essential for maintaining the normal gallery environment, because in the absence of continual cropping galleries are quickly overgrown by the ambrosial fungus or by contaminants (Batra 1966).

It is occasionally reported that daughters have replaced their (missing) mothers, in blocking a gallery system (see, for example, Beaver 1986), or that they assist in nest hygiene (Roberts 1960). It cannot be determined whether these behaviors are a regular occurrence. Both platypodid species in which helping by progeny has been reported breed in living trees. In older burrows of *Trachyostus ghanaensis*, young adults and last instar larvae participate in 'refuse disposal', which involves packing all waste, frass and feces into old branch tunnels that have become vacant; at this time, the 'parent male has frequently died' (Roberts 1960, p. 7). In the second species, *Austroplatypus incompertus*, young adult females (but not males) from earlier broods are long-term helpers, which defend the nest and keep it clean (up to 11 helpers have been found in a single nest). In particular, they keep the outer tunnel portions free of the polyphenol-rich exudate known as kino, which flows continuously from the surrounding live sapwood (Kent and Simpson 1992; D. S. Kent, unpublished observations).

Two other behaviors of offspring adults are occasionally found: extension of parental tunnels, and blocking of the entrance tunnel when a parent beetle cannot do so. The former makes more tunnel-wall surface available for ambrosial growth, which can benefit younger siblings in those species in which larvae move freely through the tunnel systems. The latter provides protection from intruders, and possibly helps in regulating tunnel climate. It is not clear to what extent either involves a significant sacrifice of personal fitness for siblings.

Do adult offspring breed together?

During his investigation of *T. ghanaensis* – the species most similar ecologically to the eusocial *A. incompertus* (discussed below) – Roberts (1961) occasionally found instances of more than one older female in a gallery system. Initially, based on their mature ovaries, he considered them to be mated females (but see Roberts 1968). However, D. S. Kent (unpublished observations) found that the unmated 'worker' (daughter) females of *A. incompertus* do show ovary development, ranging from expanded ovaries with developing oocytes to the presence of well-developed eggs. Roberts thought it unlikely that the 'extra' females in *T. ghanaensis* nests had entered the gallery systems after the original pair had become established (Roberts 1961, p. 35).

Unfortunately, there are no quantitative data on how often Roberts encountered more than one adult female in a *T. ghanaensis* gallery system. Nor can we be certain if they are mated or not, or, if mated, whether they can be early-maturing offspring which have somehow been inseminated, or females that have cofounded or secondarily invaded a nest. If they are not offspring, then these instances might represent groups of females breeding together, and the species could be considered quasisocial (if sisters; otherwise, communal). If they are offspring, then eusociality becomes a possibility. Further study of this species would thus be of great interest with respect to the evolution of complex forms of social behavior.

Multiple older females in one tunnel system have only been found in three platypodid species, all of which breed in living trees. The second species, *A. incompertus*, is discussed below, as eusocial. In the third species, *Platypus tuberculosus*, D. S. Kent (unpublished observation) found six females within a single large, well-established gallery system that had ramified throughout the host tree both vertically and horizontally. Five of the females were dissected; all had spermatozoa in the spermatheca. Four of these five females had laid eggs. It is an intriguing possibility, then, that the original tunnel system was being extended by the descendants of the initial colonizing pair; regardless of the actual source of these mated females, this observation falsifies the old picture of platypodid mating behavior as universally lifelong monogamous (see, for example, Browne 1961a; Kirkendall 1983).

ADULT–ADULT INTERACTIONS: COLONIAL BREEDING

One of the striking features of most outbreeding species is the rapid and simultaneous colonization of suitable host material. In a few species, this results in tree-killing. Breeding in aggregations almost certainly evolved early in the history of non-aggressive (not tree-killing) bark and ambrosia beetles: in no higher taxon is it conspicuously absent, with the notable exception of groups of species breeding in non-shareable resources such as seeds and small twigs. Although mutual attraction to odors emanating from the host (including odors from microorganisms) (see, for example, Brand *et al.* 1975; Brand and Barras 1977; Leufvén *et al.* 1984) mediates aggregation in a few outbreeding and many inbreeding species (detailed below), in most species sex pheromones attract both sexes and lead to rapid, dense colonization.

Table 9-5. *Primary factors in the evolution of colonial breeding in bark and ambrosia beetles*

Factor	Allowing solitary breeding	Encouraging colonial breeding
Resource availability	abundant	rare, unpredictable
Resource unit size	small	large
Resource condition	dead or degraded	alive
Breeding system	inbreeding, or outbreeding without sex pheromones	outbreeding with sex pheromones

We will use the term 'colonial' in this chapter only to refer to *aggregations* of breeding females, pairs, or harems. 'Colony' has been used to refer to the gallery system belonging to a single family, but we will not use 'colony' or 'colonial' in this sense. 'Solitary', then, denotes a breeding situation characterized normally by one family unit (a single female, a male–female pair, or a harem) per resource unit (i.e. per twig, per seed, per tree trunk, etc.).

Solitary breeding appears to be the derived character state in scolytids and (if it occurs at all in the group) in platypodids. Its phylogenetic distribution is very scattered, and it occurs most often in tribes (Dryocoetini, Cryphalini, Corthylini) thought to be highly derived (Wood 1978). Regular solitary breeding is associated with adopting the use of breeding material such as living or dead small twigs, seeds, or pine cones, or (in the case of a few *Dendroctonus* species) with breeding in living trees without killing them. Many solitary breeders are species with regular inbreeding (Kirkendall 1993).

Colonial breeding characterizes most scolytid and platypodid species. Colonial behavior in most species is a result of requiring a breeding habitat that is of limited and patchy occurrence, but which cannot be monopolized by a single individual. A variety of chemical and acoustic signals have evolved at least in part as a result of the competition engendered by group living; one effect is that entrance tunnels are usually found to be more evenly spaced than would be expected from random settling (reviewed by Byers 1989a,b), which is hypothesized to reduce competition between larvae from different broods.

In the following discussion, we emphasize species regularly attacking living trees ('aggressive' species). These few species are one end of a spectrum; the vast majority exclusively breed in completely dead woody tissue, but it is important to emphasize that many species fall between these two extremes (see, for example, Bletchley 1961; Browne 1961a; Rudinsky 1962). Aggressive species are of

particular interest in the context of this chapter because they have evolved a strong dependence upon colonial breeding, and thus provide the most logical candidates in a search for behaviors that might be classified as cooperative. Why some colonial species regularly kill trees but most do not is an interesting question in itself, but beyond the purview of this chapter.

The ecology of solitary vs. colonial breeding

Role of the host plant

The host plant plays a critical role in the radiation of scolytid behavior along solitary or colonial lines (Table 9-5). Two aspects of the host, its quality as a breeding substrate, and the mechanisms it possesses to defend itself, often exert opposing selective pressures. Fresh inner bark, for example, has more favorable carbon-to-nitrogen ratio and mineral content than older tissues, which enables females to breed faster and produce more eggs in fresher plant tissues (Shrimpton 1973; Kirkendall 1983; Haack and Slansky 1987; Popp *et al.* 1989; Schowalter *et al.* 1992). However, living inner bark and the outer sapwood may also be actively defended chemically by toxic terpenes and phenols as well as physically by gums, latexes or resins (see, for example, Chapman 1974; Raffa and Berryman 1983a; Nebeker *et al.* 1993; Raffa *et al.* 1993; Klepzig *et al.* 1995), and recently killed tissue will at least initially have dangerous residues of defensive compounds.

The more time has elapsed since host tissue death, the more available the tissue is to competitors. Thus, a species that can in some way avoid or overcome the defensive system of a living tree will benefit from higher resource quality and lower competition for that resource, than one breeding in dead plant tissues (Raffa and Berryman 1987).

Analogous arguments apply also to some extent to ambrosia beetles, even though they are usually confined to dead sapwood or heartwood where they may experience considerable competition from fungi and other insects (see, for example, Schowalter *et al.* 1992). Their requirement for sustained high humidity is well met by the dead wood within the constantly moist, living tissues. This habitat is not itself actively defended, but tunnels may become blocked by the flow of defensive chemicals from the living tissues they have successfully penetrated. Species must be able to prevent this burrow-flooding to be able to avail themselves of the unexploited, undegraded resources within a living tree.

Exploiting living tissue

There are three ways of securing suitable breeding material: (1) kill a large host plant; (2) kill only a small part of a large host plant, or an entire small plant; (3) avoid host defenses, and breed in dying or already dead tissue. The first requires cooperation, in the sense that successful breeding for an individual can only result from group activity. The latter two do not require multiple colonization. Few species possess the ability to kill healthy trees; many more can kill trees weakened by drought or pathogen attack (see, for example, Blackman 1924; Browne 1958; Moore 1959; Bletchly 1961; Rudinsky 1962; Howard 1973; Goheen and Cobb 1980; Mendel and Halperin 1982; Benz 1985).

Large units of living plant material can only be killed by dense, concerted colonization. However, a single breeding unit (a female, a pair, or a harem, depending on the species) can kill a small plant part. *Corthylus punctatissimus* regularly kills (by girdling) and breeds in small saplings (Robert 1947; Finnegan 1967; Roeper *et al.* 1987). *Conophthorus* species breed in girdled pine cones, tops or branch tips (S. L. Wood 1982; Flores and Bright 1987). *Hypothenemus hampei* (the coffee berry borer) attacks green coffee berries and destroys the seed (see, for example, LePelley 1968). In such cases, colonization by groups of individuals is not necessary for the success of individuals.

Single family units in inner bark can also neutralize or avoid local defenses (for example, *Dendroctonus micans*, Grégoire 1988), as can those of ambrosia beetles breeding in wood (*Corthylus columbianus*, and the platypodids *Doliopygus conradti, D. dubius, Trachyostus ghanaensis, Dendroplatypus impar, Platypus mutatus, A. incompertus, P. tuberculosus*). *Notoplatypus elongatus* avoids defenses by directly entering the heartwood of living trees through scars (D. S. Kent, personal observations). In these cases, too, it is possible to breed in living trees without being part of a large group.

Aggregation

Aggregated breeding (multiple colonization) occurs in some groups without beetle-produced long-distance attractants. For example, careful research has not revealed any long-distance beetle-produced attractants in such aggregating bark beetles as *Hylastes cunicularius* (Hylastini: Eidmann *et al.* 1991), *Tomicus piniperda* (Tomicini: Byers *et al.* 1985; Lanne *et al.* 1987; Löyttyniemi *et al.* 1988), *Hylurgopinus rufipes* (Tomicini: Swedenborg *et al.* 1988) or *Pseudohylesinus nebulosus* (LeConte) (Tomicini: Walters and McMullen 1956; Rudinsky 1966) or in the ambrosia beetle *Platypus caviceps* Broun (Ytsma 1989). Mate location in these taxa is apparently mediated by common responses to host volatiles (see also Rudinsky 1962; Kangas *et al.* 1970; Moeck *et al.* 1981).

Entire dead trees, broken branches or tops, and dead patches of bark are seldom abundant under natural circumstances. However, such resource patches can support up to thousands of adults and their resulting brood (Knight *et al.* 1956; Thalenhorst 1958; Berryman 1974; Raffa and Berryman 1983a; Knight 1969; DeMars *et al.* 1986; Werner 1986; Anderbrant 1990). Hence, long-range attractants which initially evolved in the context of mate location can be exploited by eavesdropping members of the same sex in the context of resource location (Raffa *et al.* 1993).

Group colonization does not necessarily indicate cooperation among senders and responders (Raffa and Berryman 1980, 1983a, 1987; Alcock 1982; Schlyter and Birgersson 1989; Raffa *et al.* 1993). The behavioral literature provides numerous examples from other types of insects in which an individual that succeeds in locating suitable host material is perceived by, and attractive to, conspecifics (see, for example, Thornhill and Alcock 1983; Dickens and Wiygul 1987; Eggert and Müller 1989; White and Chambers 1989). Chemicals associated with host colonization, such as frass, plant-wounding, mate-signaling and microbial development, are often used as cues by conspecifics seeking host material. The 'sender' that deliberately or incidentally releases such chemicals may derive no benefit, and may have its share of the resource reduced by subsequent arrivals. By contrast, senders may benefit from signaling if their ability to subdue prey (or avoid predation) is increased by the arrival of conspecifics. Under these latter conditions, cooperative behavior is more likely to evolve.

The cooperative features of aggregation differentiate tree-killing species from their less aggressive, non-tree-killing relatives. If aggregation results simply from incoming beetles exploiting the efforts of earlier, risk-taking colonists (see, for example, Alcock 1982), we would expect: (1) the strongest and most localized response to be by the opposite (mate-seeking) sex; and (2) the calling sex to cease producing pheromone when mated (at least, for monogamous species). To the extent that the colonizing, pheromone-producing sex is actively eliciting further colonization, we would expect that sex to exhibit the strongest response to the first colonizers, and we would expect that, especially in monogamous species, cessation of pheromone production by individual beetles would be at least partly related to the exhaustion of host-tree resistance. If early-colonizing females experience further local colonization solely as competition, then females should stop releasing attractive substances immediately after having secured a mate.

Among non-aggressive aggregating species, there is normally a much stronger opposite- than same-sex response to pioneer beetles (Raffa et al. 1993); however, in tree-killing species of Dendroctonus, the same-sex (female–female) response dominates in the early phases of colonization, and only later does the sex ratio of arriving beetles become male-biassed (Renwick and Vité 1970; Raffa and Berryman 1983a). Furthermore, in the tree-killing southern pine beetle (D. frontalis), mating does not lead to an abrupt cessation of attraction as found in the less aggressive D. pseudotsugae (Rudinsky 1969), and there is relatively little difference in attractiveness of unmated and mated females (Coster and Vité 1972). Thus, there is substantial support for the view that some bark-beetle behaviors involved with colonization and killing of living trees include an important component of cooperation.

Cooperation and competition during colonization: the costs and benefits of pioneering

Aggregating bark beetles include representatives along the entire continuum of host-material freshness. Those species that colonize dead plants or plant parts do not encounter active host resistance mechanisms, and the amount of residues of defensive chemicals will depend on time since death. High beetle density in such species only increases larval mortality due to competition and resource deterioration. Individuals and species that colonize the stems of living trees, however, enter an environment that will soon become lethal if additional recruits do not arrive (Raffa and Berryman 1983a; Raffa 1991a; Raffa and Smalley 1995). During rapid aggregation, each beetle contributes by physically draining resin, severing resin canals, and inoculating

the tissue with phytopathogenic fungi (Berryman 1972; Raffa and Berryman 1983a; Christiansen and Ericsson 1986; Christiansen et al. 1987).

The evolution of intraspecific cooperation within members of the same sex requires several proximate mechanisms (Raffa and Berryman 1983a). In scolytids, these include: (1) a complex communication system that facilitates sufficiently rapid aggregation to outpace the dynamic wound responses of host trees (Wood 1972; Raffa and Berryman 1983a,b; Schlyter and Birgersson 1989); (2) association with phytopathogenic fungi that augment the virulence of each beetle (Whitney 1982); (3) mechanisms for minimizing intraspecific and interspecific exploitation of communication signals (Rudinsky 1969; Birch et al. 1980a,b; Lanier et al. 1980; Payne et al. 1984; Raffa and Klepzig 1989; Herms et al. 1991; Raffa 1991b; Raffa and Dahlsten 1995); and (4) mechanisms for facilitating synchronous brood emergence (Raffa and Berryman 1987; Bentz et al. 1991).

Cooperative attack always includes elements of competition, because the thin phloem layer in which bark beetles breed can support only a fixed number of progeny. Hence, each arriving beetle diminishes the available substrate for all other individuals. Scolytids have evolved an array of mechanisms for reducing intraspecific competition, such as the emission of 'anti-aggregation' pheromones (usually terpene ketones) and agonistic acoustic signaling in the form of stridulation (Rudinsky 1969; Ryker 1984; Byers 1989a). Adjustment of egg clutch size by ovipositing females can also reduce the adverse effects of competition. At high colonization densities, reproductive success is higher for females that deposit only part of the egg load, re-emerge from trees and seek new hosts (Coulson 1979). This behavior is more common among species that colonize healthy trees, and hence rely more heavily on conspecifics for food procurement, than among species that only colonize highly stressed or dying trees (Raffa and Berryman 1987).

Pheromonal communication among tree-killing scolytids (see D. L. Wood, 1982; Borden 1985; Payne and Coulson 1985; Raffa et al. 1993) can lead to optimal colonization density (i.e. maximal brood per parent) through the physiological linkage of beetle communication with host defense physiology (Berryman et al. 1985, 1989; Zhang et al. 1992). That is, beetles are most attractive while trees resist attack, which both increases the likelihood of successful colonization and protects against excessive intraspecific competition (Renwick and Vité 1970; Raffa and Berryman 1983a; Birgersson 1989). The multicompound, context-dependent pheromone systems of tree-killing species include incorporation of trees' defensive chemicals into pheromone blends as synergists and/or precursors (D. L. Wood 1982), the linkage of 'antiaggregation' pheromone synthesis to mating (Rudinsky 1969), host terpene oxidation (Borden *et al.* 1986a) and the successful establishment of 'antiaggregation' pheromone-producing fungal symbionts (Brand *et al.* 1976). Responding beetles react differently to the same pheromone, depending upon the ratio of the pheromone to host chemicals; the ratio reflects the current state of the colonization attempt. For example, the compound MCH, the 'antiaggregation' pheromone of *D. pseudotsugae*, attracts beetles when the ratio of MCH to resin compounds is low, and repels them when the concentration of resin in the mixture is low (Rudinsky 1973). Similarly, the effects of the *D. ponderosae* pheromone components *exo*-brevicomin and frontalin at least in some circumstances are reversed with concentration, acting to increase attraction at low doses but repelling incoming beetles at high doses (Borden *et al.* 1986b). The evolutionary stability of these communication systems is reinforced by the gain to receivers from recognizing spacing signals: (1) their likely brood production is low under crowded conditions; and (2) the late-arriving beetles have a high likelihood of success if they simply reorient their attacks to adjacent uncolonized trees (Gara 1967; Gara and Coster 1968; Geiszler *et al.* 1980; Raffa and Berryman 1983a; Raffa 1988).

Rapid termination of group attack might also reduce the possibility of persistent 'cheating' among some members of scolytid populations. That is, might some individuals respond to pheromones but not signal, and thereby avoid the costs of pheromonal production and host plant selection, yet benefit from group attack? Most authors believe that such behavior, at least as a genetically fixed strategy, would be maladaptive, particularly during periods of low population densities (Birgersson *et al.* 1988; Schlyter and Birgersson 1989). However, a flexible system in which optimal strategies vary with changing conditions might favor some degree of cheating (Schlyter and Birgersson 1989). We currently lack the information to evaluate this possibility, but several studies have demonstrated high intraspecific variation in pheromone synthesis (Miller *et al.* 1989; Seybold *et al.* 1992; Seybold 1993; Teale *et al.* 1994).

Dependence upon group attack to acquire favorable breeding sites also requires relatively synchronous emergence of adult beetles, particularly in univoltine species (Raffa and Berryman 1987). Consistent with this view, some species, such as *D. ponderosae*, appear to have inherent

stage-specific temperature thresholds that preclude prolonged emergence periods regardless of seasonal conditions (Bentz *et al.* 1991). Interestingly, there are no mass-attacking, tree-killing scolytids in tropical dipterocarp forests with monospecific dominance (see, for example, Browne 1961a). Dipterocarpaceae is one of the families used most by Southeast Asian scolytids and platypodids (Browne 1958), and, unlike most other tropical forests, in these forests the potential hosts (the dominant dipterocarp species) occur at relatively high densities (Hart 1990).

In summary, research into the consequences of intraspecific competition and cooperation for the population dynamics of scolytids has given us fascinating insights into the evolution of their colonial lifestyle. Breeding in (and killing) living trees seems to require a level of cooperation that is unusual for insects. Many details are known of the mechanisms of communication, but since they have been gathered largely through observations of and experiments on populations rather than individuals, we cannot be sure of the extent and nature of true cooperation in these beetles, nor do we know to what degree flexibility in colonizing behavior is conditional as opposed to being polymorphic. Research focussed on the costs and benefits of individual options would tell us much about the forces that have sculpted these complex signaling systems. Specifically, we need a better understanding of the details of the behavior of colonizing-sex individuals which come early vs. late in the colonization process, and of the effects of colonization density and mating status (i.e. mated or not) of these individuals on their pheromone and stridulatory behavior. Much could also be learned by studying these behaviors in species that breed in both living and dead trees (facultative tree-killers, which includes a much wider range of taxa). Finally, the evolution of cooperation in colonial scolytids or platypodids is intimately related to the evolution of breeding in living or freshly killed trees; comparative research into why so few species have evolved cooperation will thus have to address the interplay between the ecology, physiology and architecture of the host plant, as well as associations with microorganisms and the impacts of competitors and natural enemies.

THE EVOLUTION OF REPRODUCTIVE ALTRUISM IN AMBROSIA BEETLES

Eusociality has traditionally been characterized by overlap of generations, cooperative brood care, and the presence of non-reproductive helpers (see, for example, Michener

1969; Wilson 1971). Crespi and Yanega (1995) suggest that the essential characteristic of eusociality is the presence of irreversible, behaviorally distinct castes, and that eusocial societies are characterized by two traits: (1) caste-associated helping behavior; and (2) division of labor, in the sense that no caste, or only the reproductive caste, is capable of performing all behaviors (totipotent). Species with helping behavior and division of labor but without permanent castes (all individuals start out totipotent) are then classified as 'cooperative' rather than 'eusocial'. We will review what is known about the one example of possible eusociality in platypodids, then discuss in a more general way the nature of reproductive altruism and conditions favoring the evolution of complex social behavior (whether eusociality or cooperative breeding) in ambrosia beetles.

Possible eusociality in *Austroplatypus incompertus*

Austroplatypus incompertus is unusual among ambrosia beetles for breeding in living trees, *Eucalyptus* spp., in southeastern Australia. The hosts are not seriously affected by these galleries. Tree-ring scarring reveals that some family tunnel systems have persisted for up to 37 years. The primary evidence for eusociality is that uninseminated females, apparently daughters of the reproducing female, clean, defend, and maintain the tunnel system (Kent and Simpson 1992). All older gallery systems examined so far contain such females. These uninseminated females cannot survive outside their natal galleries, as they have broken off the last four tarsal segments and can no longer cling to surfaces (D. S. Kent, unpublished observations). What is not yet known, however, is what possibilities they have of future reproduction within their home tunnel system, or whether the helpers constitute a permanent caste. It is possible that at least some of them become reproductives, as it is unlikely that founding females survive to the age of the older colonies. Of 42 young females present in nests that contained an older, reproductive female, none had spermatozoa, suggesting that within-nest matings are infrequent at best. Male progeny are not found in nests after the usual two- to three-month period of maturation feeding.

Austroplatypus incompertus is the only known typical platypodid in which males do not stay and help the females with which they have mated (Kent and Simpson 1992). Because *A. incompertus* males do nothing to help, females (and eventually her female offspring) must do all the tunneling and housekeeping, and in addition must continually remove the build-up of kino, flowing from the living sapwood. Lack of male help no doubt explains at least in part the unusually long delay between the start of tunneling and the commencement of egg-laying; it takes females 4–5 months to reach the heartwood, and another 6–7 months pass before the first branch tunnel is completed and the first batch of eggs has been produced.

No other species were placed in the genus *Austroplatypus* Browne (Browne 1971). The sister taxon of *A. incompertus* is unknown; the larvae are most similar to *Dendroplatypus impar* from Malaya (Browne 1972). *Austroplatypus* and *Dendroplatypus* were both synonymized without comment by Wood (1993), in the only recent generic revision of platypodids. In light of the unique morphological and behavioral characteristics shown by larvae and adults of *Austroplatypus* (Browne 1971, 1972; Kent and Simpson 1992), the relationship of this monotypic genus to *Platypus* (*sensu* Wood 1993) should be re-examined before accepting the synonym proposed by Wood.

The nature of reproductive altruism in ambrosia beetles

For extreme self-sacrificing helping behavior to evolve, there must be either overlap of generations or grouping of sisters, and there must be some way potential helpers can behave that significantly increases the production or survivorship of related offspring. These are necessary, but not sufficient, conditions for the evolution of eusociality. Most of what can be accomplished by helpful offspring can be done with behaviors that benefit most or all offspring simultaneously, and which involve only slight (if any) changes in pre-existing reproductive behavior patterns. Thus, keeping tunnels clean is largely accomplished simply by grazing, and removal of frass, dead bodies, etc. from the gallery system requires little work. It is not clear if young adults that help in this way entail any cost to their future reproduction: the nest is a safe place for young adults to remain until ready for breeding. Furthermore, female offspring of the inbreeding species can often breed by extending the parental tunnels (see, for example, Zehntner 1900; Beeson 1930; Kalshoven 1962; Saunders and Knoke 1967; Gagne and Kearby 1979).

Even in the one investigated case of possible eusociality, *A. incompertus*, the reproductive sacrifice, while significant, is probably not absolute, as some females remaining in the nest might become new breeders in long-lasting family colonies. However, beginning reproduction without first dispersing, although an easily achieved alternative for

Table 9-6. *Major factors influencing the evolution of helping behavior in bark and ambrosia beetles*

(a) *Overlap of generations*

Factor	Favoring overlap	Not favoring overlap
Larval feeding	In tunnel system	Mine away, in bark or wood
Length of oviposition period	Long	Short
Development, egg to adult	Rapid	Slow
Maturation feeding by young adults	Prolonged	Over short period
Resource stability	Stable for several generations	Rapid loss of quality

(b) *Females helping their own offspring*

Factor	Favoring helping	Not favoring helping
Offspring dispersion	Clumped	Dispersed
Costs of helping	Low	High
Larval feeding	In tunnel system	Mine away, in bark or wood
Potential future reproduction	Low	High

(c) *Females helping to rear siblings or siblings' offspring*

Factor	Favoring helping	Not favoring helping
Progeny mortality if parent departs	High, or highly variable	Low
Effect on offspring survivorship	Potentially major	Little or none
Offspring nutritional requirements	Same as parent	Different from parent
Overlap of generations	High degree of overlap	Low degree overlap
Expandability of living quarters	Expandable for ≥1 generation	Little or no possibility
Relatedness to recipients	Inbreeding, high relatedness	Outbreeding, low relatedness

females of sibling-inbreeding or mother–son-inbreeding species, would entail a sequence of changes in the mating behavior for species (such as monogamous platypodids) with complex mate attraction and courtship behaviors in which neither sex is normally sexually mature before leaving the nest (see, for example, Roberts 1961). It is not known whether any platypodids accomplish within-nest matings. For example, it cannot be ruled out that intruding, unrelated males are the source of matings for non-dispersing females of *A. incompertus* that then go on to become new queens. (Mated daughters have been found in natal nests of *Trachyostus ghanaensis* and *Platypus tuberculosus*, both of which breed in living trees.)

A costlier form of helping is blocking the entrance. Feeding and blocking are incompatible, and blocking individuals are also vulnerable to parasitoids specialized on sessile adults (see Bushing 1965; Mills 1983). In contrast to parasitism, the risk of predation may not be significantly different for blocking vs. non-blocking individuals; some of the most common predators frequently kill most or all

of the brood (Browne 1961b, 1963; Roberts 1962a). If this is the case, then there may be no opportunities for the evolution of physical self-sacrifice (heroic sacrifice: Alexander *et al.* 1991) as a form of altruism in ambrosia beetles.

Conditions favoring reproductive altruism

Overlap of generations

Overlap of generations – contact between parents and adult offspring – is common in ambrosia beetles. Spatial overlap between parents and adult offspring will normally only be likely in species in which larvae do not tunnel independently away from the egg gallery: ambrosia beetles, and many species breeding in seeds or pith. Temporal overlap will most likely occur when oviposition extends over a long period, when the resource degrades relatively slowly, and when larval development is relatively rapid (Table 9-6a).

There is generational overlap in many platypodids (Table 9-7: only one is known not to have overlap), because oviposition extends over a long period. As a consequence, emergence from single families can take weeks or months to complete, and females may still be laying eggs when the first tenerals appear, depending on the food productivity of the tunnels (the stability of the resource). All stages of xyleborine offspring are occasionally found in a single nest (see, for example, Zehntner 1900; Beeson 1930; Kalshoven 1959; Schneider 1987). However, the data are too fragmentary to allow any conclusions about how common overlap of generations actually is in this group.

Stability of the host. In the warm, moist climate of tropical rain forests, decay is much faster than in temperate forests (Odum 1970; Lang and Knight 1979). Consequently the dead tree – especially compared with a live tree – is an ephemeral habitat, capable usually of supporting a single generation of bark or ambrosia beetles (Hubbard 1897; Browne 1961a; Beaver 1977, 1984). According to Browne (1961a, p. 41), for example, development in *Diapus pusillimus* in Malaysia takes only a few weeks, while for the same species in Queensland it requires one year. In Malaysia, the length of the life cycle for most ambrosia beetles is 4–6 weeks (Browne 1961a); 3–6 weeks is considered usual in the tropical forests of Ghana (Jones *et al.* 1959) (6–12 weeks for platypodids colonizing trunks: Roberts 1961). Temperate ambrosia beetles frequently require one or more years to develop to maturity.

Thus, tropical platypodids have shorter life cycles than temperate representatives, and in the tropical platypodids, exceptionally long life cycles occur only in the three species that occasionally (*D. conradti*: Browne 1963) or regularly complete their life cycle in living trees (see Table 9-7). There are too few data to draw any definite conclusions, but it is interesting to note from Table 9-7 that as life-cycle length increases there seems to be a corresponding increase in the difference between time of first progeny emergence and overall duration of the nest. This pattern suggests that the ratio of larval development time to oviposition period is much larger in temperate species. Thus, with respect to host stability, reproductive altruism is more likely to evolve in species breeding in temperate than tropical regions, and in living trees than dead.

For ambrosia beetles of temperate forests, large logs occasionally support multiple generations if not too densely colonized (Hubbard 1897; Chamberlin 1918; Beeson 1930). In such cases, populations are dense, and tunnels merge. Schneider (1987) cites an example of a wind-thrown *Quercus megalocarpa* from Kansas, which was gradually colonized downwards by *Xyleborus affinis* over three years; in the fourth spring the network of tunnels was more than 6 m in extent.

Resource stability
Resource stability allows prolongation of adult life span relative to the developmental period and hence overlap of generations, and increases survivorship of potential helpers that remain in their natal nest area (see, for example, Alexander *et al.* 1991). In addition, resource stability can lead to a difference in reproductive rates between: (1) reproduction *via* offspring produced by parents; and (2) offspring produced by individuals choosing to disperse rather than to help, a difference that will be proportional to the difficulty of finding a mate and of establishing a new nest site.

As in wood-feeding termites and ants, and similar to the situation in naked mole rats, ambrosia beetles live in the safety of extensive tunnel systems, which are surrounded by food. The tunnel systems are potentially long-lived and expandable. However, colonization by ambrosia beetles normally initiates an irreversible succession of wood-feeding organisms (primarily insects, fungi and yeasts, and bacteria) which leads to alterations in the physical and chemical properties of the woody tissue. Locally, the ambrosia fungi exhaust the wood in the area around the initial tunnels, and the mycelia on the tunnels walls gradually cease growing (see, for example, Nord 1972). Consequently, if a gallery system cannot be expanded, it must be abandoned.

Living vs. dead trees. Overlap of generations in ambrosia beetles requires extended tunnel-system productivity, which will depend upon the densities of both conspecifics and the other organisms exploiting the wood. Finding unexploited wood will frequently be difficult or impossible if colonization density is high. Boring into living rather than dead trees, however, may lead to favorable, low within-tree densities for two reasons: first, the much greater availability of potential hosts should act to spread out populations and lead to lower recruitment during colonization; second, colonization of living trees involves high mortality and a high rate of nest failure, further reducing within-tree density.

Once colonized, living trees provide a much more stable environment than is possible in dead trees. Although the tissue being utilized is dead, it is protected from insect and microbial attack by the surrounding living sapwood and inner bark (as well as the dead outer bark). Thus,

Table 9-7. *Important life-history features related to the evolution of social behavior in platypodid ambrosia beetles.*

Only *Austroplatypus incompertus* is known to have non-breeding female helpers. nm, Not mentioned in the cited literature; *, the species regularly breeds in living trees.

Tribe	Species	Region	First emergence	Duration of nest	Longevity of adults	Generations overlap?	Larvae tunnel?	Reference(s)
(a) Tropical species:								
DIAP:	*Geniocerus furtivus*	India	10–11 weeks	7–9 months	nm	yes	no	Beeson 1917, 1941
DIAP:	*Diapus pusillimus*	Australia	1 year	2–3 years	nm	yes	yes	Smith 1935
PLAT:	*Doliopygus conradti*	Malaysia	a few weeks	nm	same as nest	nm	nm	Browne 1961a
PLAT:	*D. dubius*	Africa	*ca.* 2 months	4–5 months	same as nest; only one brood	yes	yes	Browne 1962, 1963
PLAT:	*Trachyostus aterrimus*	Africa	13 weeks	14–16 weeks	as nest, male dies before female	no	yes	Browne 1961b; Roberts 1968
PLAT:	*T. ghanaensis**	Africa	avg. 27 wks (18–43); 6 months (1960); 12 months (1962b); avg 105 wks (60–147) (1968)	avg. 39 weeks (range 21–74 weeks); avg. 9–12 months but some up to 2 yrs (1960); avg 139 weeks (61–242) (1968)	as nest, male dies before female	yes	yes	Roberts 1962b, 1968; Roberts 1960, 1962b, 1968
PLAT:	*T. schaufussi*	Africa	avg. 25 wks (15–29)	avg. 36 weeks (16–54 weeks)	nm	nm	yes	Roberts 1968
PLAT:	*Dendroplatypus impar**	Malaysia	nm	1 year	both sexes 1 yr	yes	nm	Browne 1961a
(b) Temperate species:								
TESS:	*Notoplatypus elongatus**	Australia	unknown	>3 years	both sexes ≥3 years	unknown	yes	D. S. Kent (unpublished)
PLAT:	*Platypus mutatus* (*)[b]	S. America	8–9 months	1 year	same as nest	nm	no	Santoro 1957
PLAT:	*P. subgranosus*	Australia	10–12 months	avg. 2–3 yrs but can extend to 4–5 yrs	nm	yes	yes	Hogan 1948; Candy 1990
PLAT:	*P. caviceps*	New Zealand	2 years	up to 4 yrs	nm	yes	yes	Holloway 1973; Milligan 1979
PLAT:	*P. cylindrus*	Europe	*ca.* 8 months	3–4 yrs under favorable conditions	nm	nm	yes	Husson 1955; Baker 1956
PLAT:	*P. apicalis*	New Zealand	2 years	up to 4 yrs	nm	yes	yes	Milligan 1979
PLAT:	*P. gracilis*	New Zealand	2 years	6 yrs (when observations were stopped)	nm	yes	yes	Milligan 1979
PLAT:	*Austroplatypus incompertus**	Australia	4 years	4–37 years	female > 4 yrs, male 12–20 weeks	yes	yes	Harris *et al.* 1976; Kent and Simpson 1992; D. S. Kent (unpublished)
PLAT:	*P. tuberculosus**	Australia	unknown	up to 30 years	unknown	yes	yes	D. S. Kent (unpublished)
PLAT:	*Crossotarsus armipennis*	Australia	5–6 months	8–12 months	unknown	unknown	yes	D. S. Kent (unpublished)

[a] TESS = Tesserocerinae, Tesserocerini; DIAP = Tesserocerinae, Diaporini; PLAT = Platypodinae, Platypodini (Wood 1993).

[b] The article discusses attacks on living exotic trees, mostly in plantations; it is not clear to us whether this species regularly attacks living, native hosts.

those families successful in invading a living tree will frequently be able to expand their tunnel systems without encountering competition, and the tissue contacted will usually be fresh.

Breeding exclusively in living trees is known for only a handful of species: the scolytids *Corthylus columbianus* (see, for example, Kabir and Giese 1966a; Milne and Giese 1969; Nord 1972) and *C. fuscus* (Bustamente and Atkinson 1984), and the platypodids *Trachyostus ghanaensis* (Roberts 1960), *Dendroplatypus impar* (Browne 1961a), *Platypus mutatus* (Santoro 1957), *Austroplatypus incompertus* (Kent and Simpson 1992), *P. tuberculosus* (D. S. Kent, unpublished) and *Notoplatypus elongatus* (D. S. Kent, unpublished observations). Eusocial behavior or cooperative breeding has apparently evolved in at least one of these six species. In addition, *Doliopygus dubius* regularly breeds in living *Terminalia superba* in Ghana (Browne 1961b) and Nigeria (Roberts 1960) but also in logs of this and other species (Browne 1961b). Published observations are detailed enough to conclude that cooperative or eusocial behavior does not regularly occur in *C. columbianus*, *D. dubius*, *Dendroplatypus impar* or *P. mutatus*.

Living trees simultaneously provide two of the factors considered most important in the evolution of reproductive self-sacrifice: they provide an expandable, defensible nest site, which is capable of persisting for many generations, but at the same time they are difficult to colonize and cause high rates of mortality among colonizing beetles. The combination of a stable breeding site combined with a high cost to establishing a new nest site clearly favors remaining 'at home', if direct (self) or indirect (through relatives) reproduction is possible.

In outbreeding species in which males (fathers or brothers) are present, the choice for those females not dispersing is between (1) breeding in the natal nest, which entails close inbreeding (presumably leading to fewer and lower quality offspring due to inbreeding depression), and (2) helping related females raise offspring. The indirect gains from helping a mother or sister coupled with the advantage of immediate (if indirect) reproduction over the delay necessary to attain sexual maturity, may outweigh the genetic pay-off to delayed, direct production of inbred, low-quality offspring. The helping alternative would be further favored in circumstances where mate location is difficult or costly.

For inbreeding species, it is difficult to imagine similar advantages to indirect reproduction over extending the nest and reproducing directly. Presumably, when further reproduction is indeed possible in the natal nest, non-dispersing

females will not suffer a cost due to inbreeding depression, nor would there be a time loss due to mate location.

Given the advantages, why have so few species adopted breeding in living trees? Trees have evolved quite effective active and passive defenses against insects and fungi that attempt to penetrate them (see, for example, Mattson *et al.* 1988; Blanchette and Biggs 1992). The extreme longevity of these plants (up to several hundred or even a thousand or more years) attests to the success of these defense systems. Trees must attempt to seal off open holes, to prevent infection by microorganisms. Perhaps future research will reveal whether the few beetle species capable of breeding in living trees share particular behavioral or physiological characteristics which make this lifestyle possible.

ANCESTRAL BIOLOGY

We now turn to the question of the ancestral biology for the scolytid–platypodid lineage. We will briefly present the biology of what are considered to be the most 'primitive' (plesiomorphic) platypodids, and (even more briefly) consider the biology of cossonine weevils. We do this because those groups closest to an ancestral type in morphology might also have retained ancestral behavioral traits, but we acknowledge that there need be no such correlation.

'Primitive' platypodids

The platypodids discussed in this chapter have been placed in the two main subfamilies, Tesserocerinae and Platypodinae (Wood 1993). A third subfamily, Wood's Coptonotinae (six genera), may be polyphyletic (Thompson 1992; Kuschel 1995). Kuschel disperses the four genera that were available to him for study among the Curculioninae (*Protohylastes*, biology unknown), Cossoninae (*Protoplatypus*), Scolytinae (*Coptonotus*, biology unknown) and Platypodinae (*Schedlarius*). Neither author discusses relationships of *Scolytotarsus* Schedl, and nothing has been published on its habits; Thompson agrees with Wood that *Mecopelmus* is a platypodid.

Of the aberrant groups placed with the platypodids – *Mecopelmus*, *Schedlarius* and (according to Thompson) *Carphodicticus* and relatives – none are ambrosia beetles, but all are monogynous and lay eggs in subcortical tunnels (Wood 1971). As they are rarely collected, details of their social behavior are not known. They are clearly quite different from normal Platypodidae, however, in both egg-tunnel construction and feeding habits. *Mecopelmus* (one species from Panama) is monogynous and breeds in dead inner bark, where the female constructs a cave-type chamber and deposits eggs in clusters; at least the later larval instars feed singly in the inner bark (Wood 1966). The sole *Schedlarius* species, morphologically the most primitive platypodid, places its eggs in niches in an irregular egg tunnel, and larvae mine singly in the wood of dead *Bursera* (Wood 1957a; Atkinson *et al.* 1986). The mined wood is darkly stained, possibly by fungus, and the larval mines are much shorter than in bark- or wood-feeding species not associated with fungi. *Carphodicticus*, too, appears to be relatively 'primitive'. It was placed by Wood in his Scolytidae–Carphodicticini along with *Craniodicticus* Blandford (two species breeding in wood) and *Dendrodicticus* Schedl (one species, habits not known) (Wood 1986). The sole *Carphodicticus* species is monogynous and breeds in inner bark apparently like a typical scolytid bark beetle (Wood 1971).

Curculionidae–Cossoninae

Cossonine weevils are generally believed to be closely related to bark and ambrosia beetles, as discussed above; adults usually live and feed inside plant tissues, mainly in dead and dying parts (Kuschel 1995). An important difference from scolytids and platypodids, however, is that adults periodically return 'to browse temporarily on the surface'. However, both *Araucarius* and *Protoplatypus* have bark-beetle-like breeding habits (Kuschel 1966; Rühm 1977; Wood 1993). *Protoplatypus* parent adults form radiate tunnel systems similar to those of harem polygynous scolytids, complete with egg niches and individual larval mines (Wood 1993).

Conclusion

Regardless of the true relationships of these primitive curculionoid taxa, all resemble typical bark beetles or certain cossonines in their biology, and none are ambrosia beetles. The typical platypodids and scolytids, with their complex social behavior, apparently arose from a non-ambrosial, tunnel-breeding ancestor in which parental care and male residence may or may not have been especially sophisticated. Originally, parent beetles probably moved in and out of tunnels, as apparently do certain cossonines (Rühm, in Kuschel 1966); the next step was prolonged residence in the oviposition tunnel. Presumably there was overlap between at least females and juveniles,

as seen today in scolytids and the aforementioned cossonines. Male residence was presumably favored very early in the evolutionary history of both scolytids and platypodids, for reasons discussed later. Adoption of regular associations with fungi and other microorganisms that enriched the food source would have provided further opportunities for slight changes in behavior that would increase offspring survivorship at little cost to the parents, and could have led repeatedly to the evolution of 'ambrosia' mutualisms. Moving deep into sound wood provided a relatively safe and stable environment, and possibly a relatively underexploited resource; the dead-wood habitat in tropical forests is now dominated by ambrosia beetles, in both numbers of individuals and numbers of species.

DISCUSSION AND SUGGESTIONS FOR FUTURE RESEARCH

Discussion

The unusually rich complexity of bark and ambrosia beetle social behavior may stem largely from their ovipositional habits. Rather than simply laying eggs on plant surfaces or in cracks or excavations in plant tissues, as do so many other insects (including most weevils), bark and ambrosia beetles bore into the larval food medium, laying eggs over a protracted period in the same tunnel system. This behavior has a number of important consequences for the development of complex social interactions: (1) the beetles are *well hidden* from potential predators, parasitoids, or inquilines, and since the tunnel systems normally have only one entrance, they are *easily defended*; (2) they work in an atmosphere of constant *high humidity*; and (3) they live surrounded by their *food*. As a consequence of (2) and (3), they can remain in the tunnel system for long periods of time, in some instances for their entire reproduction.

Thus, in the context of evolution of social behavior, extensive tunneling into the food resource was the key innovation for bark and ambrosia beetles (as well as for the wood-feeding blattoid ancestors of termites); tunneling into food favors laying eggs in batches, prolonged presence of the mother, and reduced vulnerability to generalist predators for both adults and offspring. Tunnels by their very nature are more easily defended than open surfaces, providing an opportunity for inexpensive parental or sibling care (tunnel-blocking). This defensibility also can readily favor the evolution of prolonged male residence, probably initially as a form for mate-guarding, as argued above.

The relative safety of tunnel systems, and probably the more constant microclimate as well, increases the benefits to parents and offspring of remaining in the burrows as long as the resource can still be exploited, which again promotes within-family contact. Burrow-blocking and burrow maintenance are almost universal in bark and ambrosia beetles; even if these behaviors evolved early in these lineages, perhaps only once or twice, they could easily be lost if current ecological conditions were not actively favoring their maintenance.

For many lineages, selection on both sexes has favored increased investment in the current breeding situation (as long as the resource quality remains high enough) rather than leaving to breed again elsewhere. For the vast majority of species, host material is relatively scarce, ephemeral, and unpredictably available, but abundant when found; finding a suitable breeding site is difficult, and the searching process exposes the beetles to a wide variety of predators and to the high risk of dying of either starvation or desiccation. It seems likely that many species experience conditions in which the costs of helping are low and the benefits to relatives high (Table 9-6b); furthermore, the life cycles of many species lead to overlap of generations (Table 9-6a), which produces a situation in which adult offspring are present at a time when their mother is still reproducing and in which there are eggs and small larvae that can benefit from helping.

The nutritional enhancement afforded by fungus cultivation has led to loss of larval mining in all true ambrosia beetles; larval involvement in the production or extension of cradles in certain lineages can be seen as vestiges of larval tunneling, and even this behavior has been lost in at least the platypodids and xyleborine scolytids. Much more direct contact between parents or older siblings and eggs or young siblings was the result, opening up new potential pathways for social evolution. Ambrosia beetles have apparently become completely dependent upon their microbial ectosymbionts, in effect committing them to the initial tunnel system for at least the duration of the current brood; abandoning a brood (or dying) leads to almost certain death of the offspring. The constant tending of fungus 'gardens' and, in some cases, the occasional extension of tunnels, may reduce the number of eggs a female can lay; it remains to be seen whether there are gradations in maternal reproductive investment among ambrosia beetles which entail trade-offs with quantity of offspring. We simply know too little about the details of most ambrosia-beetle social behavior.

Ambrosia beetles in particular have evolved social behaviors of increasing complexity, including cooperative breeding or eusociality. As far as is known, none live in large, socially integrated 'societies'. Ambrosia-beetles differ from the other ecologically successful fungus-growing social insects, the ants and termites (see Shellman-Reeve, this volume), in one important feature: the ants and termites bring food (leaves) to the fungus gardens, while the ambrosia beetles bring the fungus to the food (fresh wood). The ant and termite gardens are housed in a structure that is long-lasting, and their food input is renewable. The wood used by a fungus-growing beetle, on the other hand, is locally exhaustible. For species breeding in dead trees, there is also considerable competition with other organisms. This difference may explain why the immense colonies with complex social organization of the former have not evolved in the latter. Fungus-growing termites and ants can expand their gardens physically to an impressive extent, and the size of a garden (and hence of the primary food source) is presumably correlated with the numbers of nest laborers and foragers the colony can produce. By contrast, the primary fungus of an ambrosia beetle nest is limited to the wood of the one tree in which it is growing, and in most cases to only a small portion of that tree (owing to competition with other ambrosia beetles of the same or different species). Consequently the beetles' tunnel systems must be constantly expanded if they are to continue to supply food. Apparently, these habitat units seldom last for more than one generation; although there are numerous anecdotal reports of species apparently successfully expanding a nest for several generations, for most species this is probably not a regular occurrence. An additional factor limiting fungal growth and hence the number of ambrosia beetles that can be supported in single nest must be the very low nutritional quality of wood, compared with plant materials such as leaves (see, for example, Haack and Slansky 1987; White 1993). As a consequence, the rate of food production for the beetles must be considerably lower than that for fungus-growing ants and termites, and must be a severely limiting factor for colony size in the beetles.

The distinction between true bark beetles and ambrosia beetles (plus pith- and seed-breeders) would seem to parallel that between precocial and altricial birds. The comparison is best between altricial birds and ambrosia beetles, as ambrosia-beetle offspring are apparently dependent upon the actions of their parents for a continual supply of food. Precocial bird offspring can be 'herded' and led to good feeding sites and can be actively protected by parents, but bark-beetle offspring can only be defended, since they are in their food already and the larvae of most species tunnel further and further away from their parents with time. Helping at the nest has evolved almost exclusively in altricial species (Andersson 1984) and similarly, as we have shown, only in 'altricial' scolytids and platypodids: that is, those species in which larvae are sessile or remain in contact with their parents.

In this discussion, we have so far focussed upon familial interactions. Bark and ambrosia beetles have attracted the interests of researchers, however, largely because of another class of social interaction: their ability to kill large stands of living trees through the coordinated attack of tens of thousands of individuals. Tree-killing is mediated by complex pheromone and stridulatory communication systems, and, we argue, is only possible if unrelated beetles cooperate, at least initially, in overcoming tree defenses. Our use of the term 'cooperation' will be met with a degree of skepticism, owing to earlier interpetations that did not consider selection at the level of the individual. We believe, however, that it might be justified in discussing tree-killing. In fact, a careful re-evaluation of the pheromone and acoustic behaviors of both sexes and of the context-dependent responses to these signals might reveal that there is indeed a significant 'trait-group' selection component (Wilson 1980) to natural selection on some of these behaviors: if it is largely true, for example, that the next generation of colonizers comes only from trees that have been successfully killed by group action, and this group action has included elements of cooperative behavior.

In his landmark review of insect social behavior, Wilson (1971) was able to cite only two studies dealing directly with scolytid or platypodid behavior (Hubbard 1897; Barr 1969), and he relied solely upon Hubbard's classic paper for his discussion of ambrosia-beetle social behavior. However, earlier works (including several books) (Chamberlin 1939; Browne 1961a; Chararas 1962) had already made significant contributions in this area. Unfortunately, most scolytid and platypodid treatises are in the forestry or applied entomology literature, much of which is not normally encountered by general biologists.

In many respects, bark and ambrosia beetles make excellent model organisms for investigating a wide range of phenomena in population biology and ecology, and for simultaneously serving both 'basic' and 'applied' research interests (Mitton and Sturgeon 1982; S. L. Wood 1982; Raffa and Berryman 1983a, 1987; Berryman et al. 1985, 1989;

Kirkendall and Stenseth 1989; Schlyter and Birgersson 1989). Nonetheless, evolutionary biologists and behavioral ecologists have only just begun to explore the rich diversity of social systems found in these beetles. We hope that the tantalizingly incomplete picture we have presented here we will awaken greater interest in the biology of scolytids and platypodids. Possible eusociality in platypodids was first described only a few years ago: surely, many similarly exciting discoveries will reward future research on bark and ambrosia beetles.

Suggestions for future research

The social behavior of bark and ambrosia beetles seems to have afforded them with some degree of evolutionary success; over 5800 species of scolytid are recognized, plus almost 1500 species of platypodid (Wood and Bright 1992). Nonetheless, they have received insufficient attention from evolutionary ecologists, and hence many critical details of social behavior have been overlooked or misinterpreted. We are sure that virtually every in-depth study of social behavior will reveal new, often surprising features. In particular, we expect to find further examples of more complex ('advanced') forms of social behavior in platypodids and ambrosia beetles, especially in situations where the resource supply for a family unit can be exploited for several to many generations. New twists on known techniques may be required, however, given the cryptic lifestyles of these insects.

Research in phylogenetic systematics would contribute much to our understanding of the evolution of the various forms for social behavior in these insects. Bright (1993) has reviewed the current situation with respect to current research in taxonomy and systematics, and discussed primarily lower-level phylogenetic analyses; we will focus here on the need for higher-level phylogenetic research. Decades of research and synthesis by Stephen Wood and his student Don Bright have culminated in a series of catalogues and systematic publications treating the higher relationships of scolytid and platypodid beetles (see, for example, Bright 1968, 1976, 1993; Wood 1954, 1957b, 1973, 1978, 1982, 1986, 1993; Wood and Bright 1992). The next step will be more formal, cladistic phylogenetic analyses completing and extending the morphological character set implicit in Wood's verbal analyses (Wood 1978 and later), but also including larval characters (Thomas 1960; Lekander 1968; Browne 1972), the proventriculus (Nobuchi 1969), stridulatory structures (Barr 1969; Menier 1976), behavior (Wood 1980; Kirkendall 1983; Bright 1993) and biochemical data such as nuclear and mtDNA sequences (Bright 1993). Many of the critical phylogenetic junctures within the scolytids, for example, have not been unambiguously resolved (see, for example, S. L. Wood 1982; Kirkendall 1993). The evolutionary lability of social behavior cannot be assessed until we have more exact, formal phylogenetic hypotheses. The questions we see as outstanding with respect to higher systematics are the relationship of scolytids and platypodids to each other and to other cuculionoid weevils; the placement of the primitive, aberrant taxa placed currently in Coptonotinae by Wood; which scolytid taxa are closest the base of the scolytid tree; and, the resolution of that section of the tree including the two haplodip-loid lineages, *Coccotrypes* and the Xyleborini (discussed in Kirkendall 1993).

Obviously, comparative analysis of social behaviors at a lower level (within-tribe, or within-genus, for example) will also depend upon further phylogenetic study; no scolytid or platypodid tribe or genus has been subjected to a cladistic analysis, not even those most thoroughly investigated taxonomically. Unfortunately, there are few younger researchers or students doing systematic or taxonomic research on these economically important and biologically fascinating beetles (Bright 1993). Robust answers to our questions about the evolution of male residence, female postovipositional blocking, cooperative tree colonization, and the evolution of cooperative or eusocial reproduction will not be possible before this systematic work has been realized.

ACKNOWLEDGMENTS

We thank Tom Phillips, USDA-ARS, for critiquing an earlier version of this manuscript. Comments by two conscientious reviewers, Ian Robertson and John Borden, and our editor Bernard Crespi (all three, Simon Fraser University) were extremely helpful. This circumglobal collaboration would not have been feasible without the Internet, for which we are enormously grateful.

KFR was supported by the University Of Wisconsin College of Agricultural and Life Sciences, National Science Foundation grant DEB-9408264, and USDA Regional Project W-187. Logistic support for LRK came from the University of Bergen Zoological Institute. Research by DSK was supported by State Forests of New South Wales.

LITERATURE CITED

Alcock, J. 1982. Natural selection and communication among bark beetles. *Fla. Entomol.* 65: 17–32.

–. 1994. Postinsemination associations between males and females in insects: the mate-guarding hypothesis. *Annu. Rev. Entomol.* 39: 1–21.

Alexander, R. D., K. M. Noonan and B. J. Crespi. 1991. The evolution of eusociality. In *The Biology of the Naked Mole-Rat.* P. W. Sherman, J. U. M. Jarvis and R. D. Alexander, eds., pp. 3–44. Princeton, N. J.: Princeton University Press.

Amman, A. G., S. L. Amman and G. D. Amman. 1974. Development of *Pityophthorus confertus. Environ. Entomol.* 3: 562–563.

Anderbrant, O. 1989. Reemergence and second brood in the bark beetle *Ips typographus. Holarct. Ecol.* 12: 494–500.

–. 1990. Gallery construction and oviposition of the bark beetle *Ips typographus* (Coleoptera: Scolytidae) at different breeding densities. *Ecol. Entomol.* 15: 1–8.

Andersson, M. 1984. The evolution of eusociality. *Annu. Rev. Ecol. Syst.* 15: 165–189.

Ashraf, M. and A. A. Berryman. 1969. Biology of *Scolytus ventralis* (Coleoptera: Scolytidae) attacking *Abies grandis* in northern Idaho. *Melanderia* 2: 1–23.

Atkinson, T. H. and A. Equihua. 1985. Lista comentada de los coleopteros Scolytidae y Platypodidae del Valle de Mexico. *Folia Entomol. Mex.* 65: 63–108.

–. 1986a. Biology of bark and ambrosia beetles (Coleoptera: Scolytidae and Platypodidae) of a tropical rain forest in southeastern Mexico with an annotated checklist of species. *Ann. Entomol. Soc. Am.* 79: 414–423.

–. 1986b. Biology of the Scolytidae and Platypodidae (Coleoptera) in a tropical deciduous forest at Chamela, Jalisco, Mexico. *Fla. Entomol.* 69: 303–310.

Atkinson, T. H., E. Martínez F., E. Saucedo C. and A. Burgos S. 1986. Scolytidae y Platypodidae (Coleoptera) asociados a selva baja y comunidades derivadas en el estado de Morelos, Mexico. *Folia Entomol. Mex.* 69: 41–82.

Baker, J. M. 1956. Investigations on the oak pinhole borer, *Platypus cylindrus* Fab. *Rec. Conv. Br. Wood Pres. Ass.* 1956: 92–111.

Barr, B. A. 1969. Sound production in Scolytidae (Coleoptera) with emphasis on the genus *Ips. Can. Entomol.* 101: 636–672.

Bartels, J. M. and G. N. Lanier. 1974. Emergence and mating in *Scolytus multistriatus* (Coleoptera: Scolytidae). *Ann. Entomol. Soc. Am.* 67: 365–370.

Batra, L. R. 1963. Ecology of ambrosia fungi and their dissemination by beetles. *Trans. Kans. Acad. Sci.* 66: 213–236.

–. 1966. Ambrosia fungi: extent of specificity to ambrosia beetles. *Science (Wash., D.C.)* 153: 193–195.

Beaver, R. A. 1966 (1967). Notes on the biology of the bark beetles attacking elm in Wytham Wood, Berks. *Entomol. Mon. Mag.* 102: 156–170.

–. 1972. Biological studies of Brazilian Scolytidae and Platypodidae (Coleoptera). I. *Camptocerus* Dejean. *Bull. Entomol. Res.* 62: 247–256.

–. 1974. Biological studies of Brazilian Scolytidae and Platypodidae (Coleoptera). IV. The tribe Cryphalini. *Stud. Neotrop. Fauna* 9: 171–178.

–. 1976. Biological studies of Brazilian Scolytidae and Platypodidae (Coleoptera) V. The tribe Xyleborini. *Z. Angew. Entomol.* **80**: 15–30.

–. 1977. Bark and ambrosia beetles in tropical forests. *Proc. Sympos. For. Pests and Diseases in Southeast Asia*, Bogor, Indonesia, April 1976. BIOTROP Sp. Publ. No. 2, pp. 133–149.

–. 1979. Host specificity of temperate and tropical animals. *Nature (Lond.)* **281**: 139–141.

–. 1984. The biology of the ambrosia beetle, *Sueus niisimai* (Eggers) (Col., Scolytidae), in Fiji. *Entomol. Mon. Mag.* **120**: 99–102.

–. 1986. The taxonomy, mycangia and biology of *Hypothenemus curtipennis* (Schedl), the first known cryphaline ambrosia beetle (Coleoptera: Scolytidae). *Entomol. Scand.* **17**: 131–135.

–. 1989. Insect-fungus relationships in the bark and ambrosia beetles. In *Insect-Fungus Interactions. 14th symposium of the Royal Entomological Society of London in collaboration with the British Mycological Society*. N. Wilding, N. M. Collins, P. M. Hammond and J. F. Webber, eds., pp. 121–143. London, UK: Academic Press.

Beeson, C. F. C. 1917. The life-history of *Diapus furtivus* Sampson. *Indian For. Rec.* **6**: 1–29.

–. 1930. The biology of *Xyleborus*, with more new species. *Indian For. Rec. (Entomol.)* **14**: 209–272.

–. 1941. *The Ecology and Control of the Forest Insects of India and the Neighboring Countries*. Dehra Dun, India: publ. by the author. 253pp.

Bentz, B. J., J. A. Logan and G. D. Amman. 1991. Temperature-dependent development of the mountain pine beetle (Coleoptera: Scolytidae) and simulation of its phenology. *Can. Entomol.* **123**: 1083–1094.

Benz, G. 1985. *Cryphalus abietis* (Ratz.) and *Ips typographus* (L.) new for Turkey and a note on the tree killing capacity of *Pityophthorus pityographus* (Ratz.). *Bull. Soc. Entomol. Suisse* **58**: 275.

Berryman, A. A. 1966. Factors influencing oviposition and the effect of temperature on development and survival of *Enoclerus lecontei* (Wolcott) eggs. *Can. Entomol.* **98**: 579–585.

–. 1972. Resistance of conifers to invasion by bark beetle-fungus associations. *BioScience* **22**: 598–602.

–. 1974. Dynamics of bark beetle populations: towards a general productivity model. *Environ. Entomol.* **3**: 579–585.

–. 1982. Population dynamics of bark beetles. In *Bark Beetles in North American Conifers*. J. B. Mitton and K. B. Sturgeon, eds., pp. 264–314. Austin, TX: University of Texas Press.

Berryman, A. A., B. Dennis, K. F. Raffa and N. C. Stenseth. 1985. Evolution of optimal group attack, with particular reference to bark beetles (Coleoptera: Scolytidae). *Ecology* **66**: 898–903.

Berryman, A. A., K. F. Raffa, J. A. Millstein and N. C. Stenseth. 1989. Interaction dynamics of bark beetle aggregation and conifer defense rates. *Oikos* **56**: 256–263.

Birch, M. C., D. M. Light, D. L. Wood, L. E. Browne, R. M. Silverstein, B. J. Bergot, G. Ohloff, J. R. West and J. C. Young. 1980a. Pheromonal attraction and allomonal interruption of *Ips pini* in California by the two enantiomers of ipsdienol. *J. Chem. Ecol.* **6**: 703–717.

Birch, M. C., P. Svihra, T. D. Paine and J. C. Miller. 1980b. Influence of chemically mediated behavior on host tree colonization

by four cohabiting species of bark beetles. *J. Chem. Ecol.* **6**: 395–414.

Birgersson, G. 1989. Host tree resistance influencing pheromone production in *Ips typographus* (Coleoptera: Scolytidae). *Holarct. Ecol.* **12**: 451–456.

Birgersson, G., F. Schlyter, G. Bergström and J. Löfqvist. 1988. Individual variation in the aggregation pheromone content of the spruce bark beetle *Ips typographus*. *J. Chem. Ecol.* **14**: 1737–1761.

Blackman, M. W. 1919a. Notes on forest insects. I. On two bark-beetles attacking the trunks of white pine trees. *Psyche* **26**: 85–96, pl. IV.

–. 1919b. Notes on forest insects. II. On several species of *Pityophthorus* breeding in the limbs and twigs of white pine. *Psyche* **26**: 134–142.

–. 1924. The effect of deficiency and excess in rainfall upon the hickory bark beetle (*Eccoptogaster quadrispinosus* Say). *J. Econ. Entomol.* **17**: 460–470.

–. 1931. The black hills beetle (*Dendroctonus ponderosae* Hopk.). *N.Y. St. Coll. For., Syracuse*, Tech. Pub. **4**: 1–97.

Blackman, M. W. and H. H. Stage. 1918. Notes on insects bred from the bark and wood of the American larch. *N.Y. State Coll. For. Syracuse Tech. Pub.* **10**: 1–115.

Blanchette, R. A. and A. R. Biggs, eds. 1992. *Defense Mechanisms of Woody Plants Against Fungi*. New York: Springer-Verlag.

Blandford, W. F. H. 1905. Fam. Scolytidae. *Biol. Centrali-Americana, Insecta, Coleoptera.*, vol. 4, part 6, pp. 281–298.

Bletchly, J. D. 1961. A review of factors affecting ambrosia beetle attack in trees and felled logs. *Empire For. Rev.* **40**(103):13–18.

Borden, J. H. 1967. Factors influencing the response of *Ips confusus* (Coleoptera: Scolytidae) to male attractant. *Can. Entomol.* **99**: 1164–1193.

–. 1969. Observations on the life history and habits of *Alniphagus aspericollis* (Coleoptera: Scolytidae) in southwestern British Columbia. *Can. Entomol.* **101**: 870–878.

–. 1974. Aggregation pheromones in the Scolytidae. In *Pheromones*. M. C. Birch, ed., pp. 135–160. Amsterdam: North-Holland Publ. Co.

–. 1985. Aggregation pheromones. In *Comparative Insect Physiology, Biochemistry and Pharmacology*, G. A. Kerkut and L. I. Gilbert, eds. vol. 9. *Behaviour*, G. A. Kerkut, ed., pp. 257–285. Oxford: Pergamon Press.

–. 1988. The striped ambrosia beetle. In *Dynamics of Forest Insect Populations*. A. A. Berryman, ed., pp. 579–596. New York: Plenum.

Borden, J. H., D. W. A. Hunt, D. R. Miller and K. N. Slessor. 1986a. Orientation in forest Coleoptera: an uncertain outcome to responses by individual beetles to variable stimuli. In *Mechanisms in Insect Olfaction*. T. L. Payne, M. C. Birch and C. E. J. Kennedy, eds., pp. 97–109. Oxford University Press.

Borden, J. H., L. C. Ryker, L. J. Chong, H. D. Pierce, B. D. Johnston and A. C. Oehlschlager. 1986b. Response of the mountain pine beetle, *Dendroctonus ponderosae* Hopkins (Coleoptera: Scolytidae), to five semiochemicals in British Columbia lodgepole pine forests. *Can. J. For. Res.* **17**: 118–128.

Bovey, P. 1976. Sur une capture intéressante de *Pityophthorus car-niolicus* Wichmann (Col. Scolytidae). *Bull. Soc. Entomol. Suisse* **49**: 73–78.

Bowers, W. W., G. Gries, J. H. Borden and H. D. Pierce. 1991. 3-methy-3-buten-1-ol: an aggregation pheromone of the four-eyed spruce bark beetle, *Polygraphus rufipennis* (Kirby) (Coleoptera: Scolytidae). *J. Chem. Ecol.* **17**: 1989–2002.

Brand, J. M. and S. J. Barras. 1977. The major volatile constituents of a basidiomycete associated with the southern pine beetle. *Lloydia* **40**: 398–400.

Brand, J. M., J. W. Bracke, A. J. Markovetz, D. L. Wood and L. E. Browne. 1975. Production of verbenol pheromone by a bacterium isolated from bark beetles. *Nature (Lond.)* **254**: 136–137.

Brand, J. M., J. W. Bracke, L. N. Britton, A. J. Markovetz and S. J. Barras. 1976. Bark beetle pheromones: production of verbenone by a mycangial fungus of *Dendroctonus frontalis*. *J. Chem. Ecol.* **2**: 195–199.

Bremmer, O. E. 1907. The ambrosia beetle (*Xyleborus xylographus* Say), as an orchard pest. *Can. Entomol.* **39**: 195–196.

Bright, D. E. 1968. Review of the tribe Xyleborini in America north of Mexico (Coleoptera: Scolytidae). *Can. Entomol.* **100**: 1288–1323.

–. 1976. *The insects and arachnids of Canada*, Part 2. *The bark beetles of Canada and Alaska, Coleoptera: Scolytidae*. Ottawa: Canada Dep. Agric., Res. Branch, Biosystematics Res. Inst.

–. 1993. Systematics of bark beetles. In *Beetle-Pathogens Interactions in Conifer Forests*. T. D. Schowalter and G. M. Filip, eds., pp. 23–36. London: Academic Press.

Brockmann, H. J. and A. Grafen. 1989. Mate conflict and male behaviour in a solitary wasp, *Trypoxylon (Trypargilum) politum* (Hymenoptera, Sphecidae). *Anim. Behav.* **37**: 232–255.

Browne, F. G. 1958. Some aspects of host selection among ambrosia beetles in the humid tropics of south-east Asia. *Malay. For.* **21**: 164–182.

–. 1961a. The biology of Malayan Scolytidae and Platypodidae. *Malayan For. Rec.* **22**: xi + 255pp.

–. 1961b. Preliminary observations on *Doliopygus dubius* (Samps.) (Coleopt: Platypodidae). *Fourth Rep. West Afr. Timber Borer Res. Unit*, pp. 15–30. London: Crown Agents.

–. 1962. The emergence, flight and mating behaviour of *Doliopygus conradti* (Strohm.), (Coleoptera, Platypodidae). *Fifth Rep., West Afr. Timber Borer Res. Unit 1961-62*, pp. 21–27. London: Crown Agents.

–. 1963. Notes on the habits and distribution of some Ghanaian bark beetles and ambrosia beetles (Coleoptera: Scolytidae and Platypodidae). *Bull. Entomol. Res.* **54**: 229–266.

–. 1971. *Austroplatypus*, a new genus of Platypodidae (Coleoptera), infesting living *Eucalyptus* trees in Australia. *Commonw. For. Rev.* **50**: 49–50.

–. 1972. Larvae of the principal Old World genera of Platypodinae (Coleoptera: Platypodidae). *Trans. R. Entomol. Soc. Lond.* **124**: 167–190.

Bushing, R. W. 1965. A synoptic list of the parasites of Scolytidae (Coleoptera) in North America north of Mexica. *Can. Entomol.* **97**: 449–492.

Bustamente O. F. and T. H. Atkinson. 1984. Biología del barrenador de las ramas del peral *Corthylus fuscus* Blandford (Coleoptera: Scolytidae), en el norte del estado de Morelos. *Folia Entomol. Mex.* **60**: 83–101.

Byers, J. A. 1989a. Behavioral mechanisms involved in reducing competition in bark beetles. *Holarct. Ecol.* **12**: 466–476.

–. 1989b. Chemical ecology of bark beetles. *Experientia* **45**: 271–283.

Byers, J. A., J. Lanne, J. Löfqvist, F. Schlyter and G. Bergström. 1985. Olfactory recognition of host-tree susceptibility by pine shoot beetles. *Naturwissenschaften* **72**: 324–326.

Cachan, P. 1957. Les Scolytoidea mycétophages des forêts de Basse Côte d'Ivoire. *Rev. Pathol. Veg. Agr. Fr.* **36**: 1–126.

Candy, S. G. 1990. Biology of the mountain pinhole borer, *Platypus subgranosus* Schedl, in Tasmania. M. Sc. (Agr.) thesis, University of Tasmania, Australia.

Chamberlin, W. J. 1918. Bark beetles infesting the Douglas fir. *Bull. Oregon Agric. Coll. Exp. Sta.* **147**: 1–40.

–. 1939. *The Bark and Timber Beetles of North America*. Corvallis, Oregon: Oregon State College Cooperative Association.

Chapman, J. A. 1959. Forced attacks by the ambrosia beetle, *Trypodendron*. *Can. Dep. Agric. For. Biol. Div. Bi-monthly Prog. Rep.* **15**(5): 3.

Chapman, R. F. 1974. The chemical inhibion of feeding by phytophagous insects: a review. *Bull. Entomol. Res.* **64**: 339–363.

Chapman, T. A. 1869. Observations on the oeconomy of the British species of *Scolytus*. *Entomol. Mon. Mag.* **6**: 126–131.

–. 1870. On the habits of *Platypus cylindrus*. *Entomol. Mon. Mag.* **7**: 103–107; 132–135.

Chararas, C. 1960. Recherches sur la biologie de *Pityogenes chalcographus* L. *J. For. Suisse* **111**: 82–97.

–. 1962. *Étude Biologique des Scolytidae des Conifères*. Paris: Lechevalier.

Christiansen, E. and A. Ericsson. 1986. Starch reserves in *Picea abies* in relation to defence reaction against a bark beetle transmitted blue-stain fungus, *Ceratocystis polonica*. *Can. J. For. Res.* **16**: 78–83.

Christiansen, E., R. H. Waring and A. A. Berryman. 1987. Resistance of conifers to bark beetle attack: searching for general relationships. *For. Ecol. Mgmt.* **22**: 89–106.

Cook, S. P., T. L. Wagner, R. O. Flamm, J. C. Dickens and R. N. Coulson. 1983. Examination of sex ratios and mating habits of *Ips avulsus* and *I. calligraphus* (Coleoptera: Scolytidae). *Ann. Entomol. Soc. Am.* **76**: 56–60.

Coster, J. E. and J. P. Vité. 1972. Effects of feeding and mating on the pheromone release in the southern pine beetle. *Ann. Entomol. Soc. Am.* **65**: 263–266.

Coulson, R. N. 1979. Population dynamics of bark beetles. *Annu. Rev. Entomol.* **24**: 417–447.

Crespi, B. J. 1992. Eusociality in Australian gall thrips. *Nature (Lond.)* **359**: 724–726.

Crespi, B. J. and D. Yanega. 1995. The definition of eusociality. *Behav. Ecol.* **6**: 109–115.

Crowson, R. A. 1968. *A Natural Classification of the Families of Coleoptera*. Hampton, England: E. W. Classey, Ltd.

–. 1981. *The Biology of the Coleoptera*. London: Academic Press.

DeMars, C. J., Jr., D. L. Dahlsten, N. X. Sharpnack and D. L. Rowney. 1986. Tree utilization and density of attacking and emerging populations of the western pine beetle (Coleoptera: Scolytidae) and its natural enemies, Bass Lake, California, 1970–1971. *Can. Entomol.* **118**: 881–900.

Deyrup, M. and L. R. Kirkendall. 1983. Apparent parthenogenesis in *Pityophthorus puberulus* (Coleoptera: Scolytidae). *Ann. Entomol. Soc. Am.* **76**: 400–402.

Dickens, J. C. and G. Wiygul. 1987. Conspecific effects on pheromone production by the boll weevil, *Anthonomus grandis* Boh. (Col., Curculionidae). *J. Appl. Entomol.* **104**: 318–326.

Dix, M. E. and R. T. Franklin. 1981. Observations on the behavior of the southern pine beetle parasite *Roptrocerus eccoptogastri* Ratz. (Hymenoptera: Torymidae). *J. Georgia Entomol. Soc.* **16**: 239–248.

Doane, R. W. and O. J. Gilliland. 1929. Three California ambrosia beetles. *J. Econ. Entomol.* **22**: 915–921.

Eggert, A. and J. K. Müller. 1989. Mating success of pheromone-emitting *Necrophorus* males: do attracted females discriminate against resource owners? *Behaviour* **110**: 248–257.

Eidmann, H. H., E. Kula and Å. Lindelöw. 1991. Host recognition and aggregation behaviour of *Hylastes cunicularius* Erichson (Col., Scolytidae) in the laboratory. *J. Appl. Entomol.* **112**: 11–18.

Finnegan, R. J. 1967. Note on the biology of the pitted ambrosia beetle, *Corthylus puctatissimus* (Coleoptera: Scolytidae) in Ontario and Quebec. *Can. Entomol.* **99**: 49–54.

Fisher, R. C. 1931. Notes on the biology of the large elm bark-beetle, *Scolytus destructor* Ol. *Forestry* **5**: 120–131.

Flores L. J. and D. E. Bright. 1987. A new species of *Conophthorus* from Mexico: descriptions and biological notes. *Coleopt. Bull.* **41**: 181–184.

Francke-Grosmann, H. 1956. Hautdrüsen aus Träger der Pilz-symbiose bei Ambrosa-Käfern. *Z. Morphol. Oekol. Tiere* **45**: 275–308.

–. 1966. Ueber Symbiosen von xylo-mycetophagen und phloeophagen Scolytoidea mit holzbewohnenden Pilzen. *Holz. Org. Int. Symp. Berlin-Dahlem* (1965) **1**: 503–522.

–. 1967. Ectosymbiosis in wood-inhabiting beetles. In *Symbiosis*. S. M. Henry, ed., pp. 141–205. New York: Academic Press.

French, J. R. J. and R. A. Roeper. 1975. Studies on the biology of the ambrosia beetle *Xyleborus dispar* (F.) (Coleoptera: Scolytidae). *Z. Angew. Entomol.* **78**: 241–247.

Gadd, C. H. 1941. The life history of the shot-hole borer of tea. *Tea Quart.* **14**: 5–22.

Gagne, J. A. and W. H. Kearby. 1979. Life history, development and insect-host relationships of *Xyleborus celsus*. *Can. Entomol.* **111**: 295–304.

Gara, R. I. 1967. Studies on the attack behavior of the southern pine beetle. I. The spreading and collapse of outbreaks. *Contrib. Boyce Thompson Inst.* **23**: 349–353.

Gara, R. I. and J. E. Coster. 1968. Studies on the attack behavior of the southern pine beetle, III. Sequence of tree infestation within stands. *Contrib. Boyce Thompson Inst.* **24**: 69–79.

Garraway, E. and B. E. Freeman. 1981. Population dynamics of the juniper bark beetle *Phloeosinus neotropicus* in Jamaica. *Oikos* **37**: 363–368.

Geiszler, D. R., V. F. Gallucci and R. I. Gara. 1980. Modeling the dynamics of mountain pine beetle aggregation in a lodgepole pine stand. *Oecologia (Berl.)* **46**: 244–253.

Gobeil, A. R. 1936. The biology of *Ips perturbatus* Eichhoff. *Can. J. Res.* D **14**: 181–204.

Goheen, D. J. and F. W. Cobb Jr. 1980. Infestation of *Ceratocystis wagneri*-infected ponderosa pines by bark beetles (Coleoptera: Scolytidae) in the central Sierra Nevada. *Can. Entomol.* **112**: 725–730.

Gossard, H. A. 1913. Orchard bark beetles and pin-hole borers. *Ohio Agric. Exp. Sta. Bull.* **264**: 1–68.

Grégoire, J.-C. 1985. Host colonization strategies in *Dendroctonus*: larval gregariousness or mass attack by adults? In *Proceedings of the Meeting of the IUFRO Working Group Parties 52.07-05 and -06*, Banff, Canada, Sept. 1983. L. Safranyik, ed., pp. 147–154. Banff.: Can. Dept. Environ., Can. For. Serv. and USDA For. Serv.

–. 1988. The greater European spruce bark beetle. In *Dynamics of Forest Insect Populations*. A. A. Berryman, ed., pp. 455–478. New York: Plenum.

Haack, R. A. and F. Slansky. 1987. Nutritional ecology of wood-feeding Coleoptera, Lepidoptera and Hymenoptera. In *The Nutritional Ecology of Insects, Mites, Spiders, and Related Invertebrates*. F. Slansky and J. G. Rodriguez, eds., pp. 449–486. New York: Wiley.

Haanstad, J. O. and D. M. Norris. 1985. Microbial symbionts of the ambrosia beetle, *Xyloterinus politus*. *Microb. Ecol.* **11**: 267–276.

Hadorn, C. 1933. Recherches sur la morphologie, les stades évolutifs et l'hivernage du bostryche liseré (*Xyloterus lineatus* Oliv.). *Z. Schweiz. Forstvereins (Suppl.)* **11**: 1–120.

Hamilton, W. D. 1967. Extraordinary sex ratios. *Science (Wash., D.C.)* **156**: 477–488.

Harris, J. A., K. G. Campbell and G. M. Wright. 1976. Ecological studies on the horizontal borer *Austroplatypus incompertus* (Schedl) (Coleoptera: Platypodidae). *J. Entomol. Soc. Austr. (N.S.W.)* **9**: 11–21.

Hart, T. B. 1990. Monospecific dominance in tropical rain forests. *Trends Ecol. Evol.* **5**: 6–11.

Hedlin, A. F. and D. S. Ruth. 1970. A Douglas-fir twig mining beetle, *Pityophthorus orarius* (Coleoptera: Scolytidae). *Can. Entomol.* **102**: 105–108.

Herms, D. A., R. A. Haack and B. D. Ayres. 1991. Variation in semiochemical-mediated prey-predator interaction: *Ips pini* (Scolytidae) and *Thanasimus dubius* (Cleridae). *J. Chem. Ecol.* **17**: 515–524.

Hogan, T. W. 1948. Pin-hole borers of fire-killed mountain ash. The biology of the pinhole borer, *Platypus subgranosus* S. *J. Dep. Agric. Vict.* **46**: 373–380.

Holloway, W. A. 1973. A Study of *Platypus caviceps* in Felled Hard Beech. New Zeal. For. Serv., For. Res. Inst., For. Entomol. Report no. 39, 14 pp. (Unpublished; available from L. R. Kirkendall or D. S. Kent.)

Hosking, G. P. 1973. *Xyleborus saxeseni*, its life-history and flight behaviour in New Zealand. *New Zeal. J. For. Sci.* **3**: 37–53.

Howard, T. M. 1973. Accelerated tree death in mature *Nothofagus cunninghamii* Oerst forests in Tasmania. *Victorian Nat.* **90**: 343–345.

Hubbard, H. G. 1897. The ambrosia beetles of the United States. (Pages 9–30 in Howard, L. O., ed., Some miscellaneous results of the work of the Division of Entomology.) *U.S. Dept. Agric. Bur. Entomol., Bull.* (N.S.) 7: 1–87.

Husson, R. 1955. Sur la biologie du Coléoptère xylophage *Platypus cylindrus* Fabr. *Ann. Univ. Sarav.–Sci.* **4**: 348–356.

Janin, J. and F. Lieutier. 1988. Existence de fécondations précoces dans le cycle biologique de *Tomicus piniperda* L. (Coleoptera Scolytidae) en forêt d'Orléans. *Agronomie* **8**: 169–172.

Jones, T., H. Roberts and J. M. Baker. 1959. Report of the West African Timber Borer Research Unit, 1955–1958. *W. Afr. Timber Borer Res. Unit Report* 1959. 61pp.

Jover, H. 1952. Note préliminaire su la biologie des Platypodidae de basse Côte d'Ivoire. *Rev. Pathol. Vég. Entomol. Agric. Fr.* **31**: 73–81.

Kabir, A. K. M. and R. L. Giese. 1966a. The Columbian timber beetle, *Corthylus columbianus* (Coleoptera: Scolytidae). I. Biology of the beetle. *Ann. Entomol. Soc. Am.* **59**: 883–894.

–. 1966b. The Columbian timber beetle, *Corthylus columbianus* (Coleoptera: Scolytidae). II. Fungi and staining associated with the beetle in soft maple. *Ann. Entomol. Soc. Am.* **59**: 894–902.

Kalshoven, L. G. E. 1958. Studies on the biology of Indonesian Scolytoidea. I. *Xyleborus fornicatus* Eichh. as a primary and secondary shot-hole borer in Java and Sumatra. *Entomol. Ber. Amst.* **18**: 147–160.

–. 1959. Studies on the biology of Indonesian Scolytoidea. 4. Data on the habits of Scolytidae. Second part. *Tijdschr. Entomol.* **102**: 135–173, pls. 15–22.

–. 1960. Studies on the biology of Indonesian Scolytoidea. 7. Data on the habits of Platypodidae. *Tijdschr. Entomol.* **103**: 31–50.

–. 1962. Notes on the habits of *Xyleborus destruens* Blfd., the near-primary borer of teak trees on Java. *Entomol. Ber. Amst.* **22**: 7–18.

Kangas, E., H. Oksanen and V. Perttunen. 1970. Responses of *Blastophagus piniperda* L. (Col., Scolytidae) to trans-verbenol, cis-verbenol and verbenone, known to be population pheromones of some American bark beetles. *Ann. Entomol. Fenn.* **36**: 75–83.

Karpinski, J. J. 1935. (Les causes qui limitent la reproduction de *Bostryches typographes* (*Ips typographus* L. et *Ips duplicatus* Sahlb.) dans la forêt primitive.) *Inst. Recherches For. Doman. Pologne* (Ser. A), no. 15. 86 pp. (Czech with long French summary.)

Kemner, N. A. 1916. Några nya eller mindre kända skadedjur på frukträd. [Some new or lesser-known pests of fruit trees.] Report no. 133 från Centralanstalten för Försöksväsendet på Jordbruksområdet, Entomol. avd. nr. 25. 21pp.

Kent, D. S. and J. A. Simpson. 1992. Eusociality in the beetle *Austroplatypus incompertus* (Coleoptera: Platypodidae). *Naturwissenschaften* **79**: 86–87.

Kingsolver, J. G. and D. M. Norris. 1977. External morphology of *Xyleborus ferrugineus* (Fabr.) (Coleoptera: Scolytidae) I. Head and prothorax of adult males and females. *J. Morphol.* **154**: 147–156.

Kirkendall, L. R. 1983. The evolution of mating systems in bark and ambrosia beetles (Coleoptera: Scolytidae and Platypodidae). *Zool. J. Linn. Soc. Lond.* **77**: 293–352.

–. 1984. Notes on the breeding biology of some bigynous and monogynous Mexican bark beetles (Scolytidae: *Scolytus*, *Thysanoes*, *Phloeotribus*) and records for associated Scolytidae (*Hylocurus*, *Hypothenemus*, *Araptus*) and Platypodidae (*Platypus*). *Z. Angew. Entomol.* **97**: 234–244.

–. 1989. Within-harem competition among *Ips* females, an overlooked component of density-dependent larval mortality. *Holarct. Ecol.* **12**: 477–487.

–. 1993. Ecology and evolution of biased sex ratios in bark and ambrosia beetles (Scolytidae). In *Evolution and Diversity of Sex Ratio: Insects and Mites*. D. L. Wrensch and M. A. Ebbert, eds., pp. 235–345. New York: Chapman and Hall.

Kirkendall, L. R. and N. C. Stenseth. 1989. Population dynamics of bark beetles, with special reference to *Ips typographus*: contributions of applied bark beetle studies to basic research in ecology and population biology. *Holarct. Ecol.* **12**: 526–527.

Klein, W. H., D. L. Parker and C. E. Jenson. 1978. Attack, emergence and stand depletion of the mountain pine beetle, in a lodgepole pine stand during an outbreak. *Environ. Entomol.* **7**: 732–737.

Klepzig, K. D., E. L. Kruger, E. B. Smalley and K. F. Raffa. 1995. Effects of biotic and abiotic stress on the induced accumulation of terpenes and phenolics in red pines inoculated with a bark beetle fungus. *J. Chem. Ecol.* **21**: 601–626.

Klimetzek, D., K. Kiesel, C. Mhring and A. Bakke. 1981. *Trypodendron lineatum*: Reduction of pheromone response by male beetles. *Naturwissenschaften* **68**: 149–150.

Knight, F. B. 1969. Egg production by the Engelmann spruce beetle, *Dendroctonus obesus*, in relation to status of infestation. *Ann. Entomol. Soc. Am.* **62**: 448.

Knight, F. B., W. F. McCambridge and B. H. Wilford. 1956. Estimating Engelmann spruce beetle infestations in the central Rocky Mountains. *USDA For. Serv. Rocky Mtn. For and Range Expt. Stn. Pap.* **25**: 1–12.

Kok, L. T. 1979. Lipids of ambrosia fungi and the life of mutualistic beetles. In *Insect–Fungus Symbiosis: Nutrition, Mutualism and Commensalism*. L. R. Batra, ed., pp. 33–52. New York: John Wiley and Sons.

Kuschel, G. 1966. A cossonine genus with bark-beetle habits, with remarks on relationships and biogeography (Coleoptera Curculionidae). *New Zeal. J. Sci.* 9: 3–29.

–. 1990. *Beetles in a suburban environment: a New Zealand case study. The identity and status of Coleoptera in the natural and modified habitats of Lynfield, Auckland (1974–1989).* DSIR *Plant Protection Report* no. 3, pp. 1–118.

–. 1995. A phylogenetic classification of Curculionoidea to families and subfamilies. *Mem. Entomol. Soc. Wash.* 14: 5–35.

Lang, G. E. and D. H. Knight. 1979. Decay rates for boles of tropical trees in Panama. *Biotropica* 11: 316–317.

Langor, D. W. 1987. Flight muscle changes in the eastern larch beetle, *Dendroctonus simplex* LeConte. *Coleopt. Bull.* 41: 351–357.

Lanier, G. N., A. Claesson, T. Stewart, J. Piston and R. M. Silverstein. 1980. *Ips pini:* The basis for interpopulational differences in pheromone biology. *J. Chem. Ecol.* 6: 667–687.

Lanne, B. S., F. Schlyter, J. A. Byers, J. Löfqvist, A. Leufvén, G. Bergström, J. N. C. van der Pers, R. Unelius, P. Baeckström and T. Norin. 1987. Differences in attraction to semiochemicals present in sympatric pine shoot beetles, *Tomicus minor* and *T. piniperda. J. Chem. Ecol.* 13: 1045–1067.

Lawrence, J. F. and A. F. Newton. 1995. Families and subfamilies of Coleoptera (with selected genera, notes, references and data on family-group names). In *Biology, Phylogeny and Classification of Coleoptera. Papers Celebrating the 80th Birthday of Roy A. Crowson.* J. Pakaluk and S. A. Slipinski, eds., pp. 779–1006. Warsaw: Museum i Instytut Zoologii PAN.

Lekander, B. 1968. *Scandinavian bark beetle larvae, descriptions and classification.* Inst. Skogszoologi, Skogshögskolan, Rapporter och Uppsatser, no. 4, 186pp.

LePelley, R. H. 1968. *Pests of Coffee.* London: Longmans, Green and Co.

Leufvén, A., G. Bergström and E. Falsen. 1984. Interconversion of verbenols and verbenone by identified yeasts isolated from the spruce bark beetle *Ips typographus. J. Chem. Ecol.* 10: 1349–1361.

Löyttyniemi, K., K. Heliovaara and S. Repo. 1988. No evidence of a population pheromone in *Tomicus piniperda* (Coleoptera: Scolytidae): a field experiment. *Ann. Entomol. Fenn.* 54: 93–95.

Lyal, C. H. 1995. The ventral structures of the weevil head (Coleoptera: Curculionoidea). *Mem. Entomol. Soc. Wash.* 14: 37–54.

Lyal, C. H. C. and T. King. 1996. Elytro-tergal stridulation in weevils (Insecta: Coleoptera: Curculionoidea). *J. Nat Hist.* 30: 703–773.

Mattson, W. J., J. Levieux and C. Bernard-Dagan, eds. 1988. *Mechanisms of Woody Plant Defenses against Insects, Search for Pattern.* New York: Springer-Verlag.

McMullen, L. H. and M. D. Atkins. 1959. Life-history and habits of *Scolytus tsugae* (Swaine) (Coleoptera: Scolytidae) in the interior of British Columbia. *Can. Entomol.* 91: 416–426.

–. 1962. The life history and habits of *Scolytus unispinosus* LeConte (Coleoptera: Scolytidae) in the interior of British Columbia. *Can. Entomol.* 94: 17–25.

Mendel, Z. 1983. Seasonal history of *Orthotomicus erosus* (Coleoptera: Scolytidae) in Israel. *Phytoparasitica* 11: 13–24.

Mendel, Z. and J. Halperin. 1982. The biology and behaviour of *Orthotomicus erosus* in Israel. *Phytoparasitica* 10: 169–181.

Menier, J. J. 1976. Existence d'appareils stridulatoires chez les Platypodidae (Coleoptera). *Ann. Soc. Entomol. Fr.* (N.S.) 12: 347–353.

Michener, C. D. 1969. Comparative social behavior of bees. *Annu. Rev. Entomol.* 14: 299–342.

Miller, D. R., J. H. Borden and K. N. Slessor. 1989. Inter- and intrapopulation variation of the pheromone, ipsdienol produced by male pine engravers, *Ips pini* (Say) (Coleoptera: Scolytidae). *J. Chem. Ecol.* 15: 233–247.

Milligan, R. H. 1979. *Platypus apicalis* White, *Platypus caviceps* Broun, *Platypus gracilis* Broun (Coleoptera: Platypodidae). The native pinhole borers. *Forest Res. Inst., New Zealand For. Serv. Forest and Timber Insects of New Zealand,* no. 37. 16pp.

Mills, N. J. 1983. The natural enemies of scolytids infesting conifer bark in Europe in relation to the biological control of *Dendroctonus* spp. in Canada. *Biocontrol News Info.* 4: 305–328.

Milne, D. H. and R. L. Giese. 1969. The Columbian timber beetle, *Corthylus columbianus* (Coleoptera: Scolytidae). IX, Population biology and gallery characteristics. *Entomol. News* 80: 225–237.

Mitton, J. B. and K. B. Sturgeon. 1982. *Bark Beetles in North American Conifers.* Austin: University of Texas Press.

Moeck, H. A., D. L. Wood and K. Q. Lindahl. 1981. Host selection behavior of bark beetles attacking *Pinus ponderosae,* with special emphasis on the western pine beetle, *Dendroctonus ponderosae. J. Chem. Ecol.* 7: 49–83.

Moore, K. M. 1959. Observations on some Australian forest insects. 4. *Xyleborus truncatus* Erichson 1842 (Coleoptera: Scolytidae) associated with dying *Eucalyptus saligna* Smith (Sydney blue-gum). *Proc. Linn. Soc. N. S. W.* 84: 186–193, pl. vii.

Morimoto, K. 1976. Notes on the family characters of Apionidae and Brentidae (Coleoptera) with key to the related families. *Kontyû* 44: 469–476.

Nakashima, T. 1975. Several types of the mycetangia found in platypodid ambrosia beetles (Coleoptera: Platypodidae). *Insecta Matsumurana,* (N.S.) 7: 1–69.

Naumann-Etienne, K. 1978. Morphological, zoogeographical and biological aspects of the Scolytidae from *Nothofagus dombeyi* in Argentina. *Stud. Neotrop. Fauna Environ.* 13: 51–62.

Nebeker, T. E., J. D. Hodges and C. A. Blanche. 1993. Host response to bark beetle and pathogen colonization. In *Beetle-Pathogen Interactions in Conifer Forests.* R. D. Schowalter and G. M. Filip, eds., pp. 157–178. New York: Academic Press.

Nobuchi, A. 1969. A comparative morphological study of the proventriculus in the adult of the superfamily Scolytoidea (Coleoptera). *Bull. Govt. For. Exp. Sta.* 224: 39–110, pls. 1–17.

Nord, J. C. 1972. Biology of the Columbian timber beetle, *Corthylus columbianus* (Coleoptera: Scolytidae). *Ann. Entomol. Soc. Am.* 65: 350–358.

Norris, D. M. 1975. Chemical interdependencies among *Xyleborus* spp. ambrosia beetles and their symbiotic microbes. *Material Organismen* 3: 479–488.

–. 1979. The mutualistic fungi of Xyleborini beetles. In *Insect-fungus Symbiosis*. L. R. Batra, ed., pp. 53–63. Montclair, New Jersey: Allanheld, Osmun and Co.

Norris, D. M. and J. K. Baker. 1967. Symbiosis: effects of a mutualistic fungus (*Fusarium solani*) upon the growth and reproduction of *Xyleborus ferrugineus*. *Science (Wash., D.C.)* 156: 1120–1122.

Odum, H. T. 1970. Summary: an emerging view of the ecological system at El Verde. In *A Tropical Rain Forest*. H. T. Odum and R. F. Pigeon, eds., pp. I-191 – I-281. Washington, D.C.: U.S. Atomic Energy Commission Tech. Info. Ctr. Oak Ridge, Tennessee.

Oester, P. T. and J. A. Rudinsky. 1975. Sound production in Scolytidae: stridulation by 'silent' *Ips* bark beetles. *Z. Angew. Entomol.* 79: 421–427.

Payne, T. L. and R. N. Coulson. 1985. Role of visual and olfactory stimuli in host selection and aggregation behavior by *Dendroctonus frontalis*. In *The Role of Hosts on the Population Dynamics of Forest Insects*. L. Safranyik, ed., pp. 73–82. (Proc. IUFRO Conf., Alberta.) Banff: Can. Dept. Environ., Can. For. Serv. and USDA For. Serv.

Payne, T. L., J. C. Dickens and J. V. Richerson. 1984. Insect predator-prey coevolution via enantiomeric specificity in a kairomone-pheromone system. *J. Chem. Ecol.* 10: 487–491.

Petty, J. L. 1977. Bionomics of two aspen bark beetles, *Trypophloeus populi* and *Procryphalus mucronatus* (Coleoptera: Scolytidae). *Great Basin Nat.* 37: 105–127.

Pfeffer, A. 1976. Revision der paläarktischen Arten der Gattung *Pityophthorus* Eichhoff (Coleoptera, Scolytidae). *Acta Entomol. Bohem.* 73: 324–342.

Pope, D. N., R. N. Coulson, W. S. Fargo, J. A. Gagne and C. W. Kelly. 1980. The allocation process and between-tree survival probabilities in *Dendroctonus frontalis* infestations. *Res. Popul. Ecol.* 22: 197–210.

Popp, M. P., R. C. Wilkinson, E. J. Jokela, B. R. Harding and T. W. Phillips. 1989. Effects of slash pine phloem nutrition on the reproductive performance of *Ips calligraphus* (Coleoptera: Scolytidae). *Environ. Entomol.* 18: 795–799.

Raffa, K. F. 1988. The mountain pine beetle, *Dendroctonus ponderosae* in western North America. In *Population Dynamics of Insects: Patterns, Causes, and Management Practices*. A. A. Berryman, ed., pp. 505–530. New York: Plenum Press.

–. 1991a. Induced defensive reactions in conifer-bark beetle systems. In *Phytochemical Induction by Herbivores*. D. W. Tallamy and M. J. Raupp, eds., pp. 245–276. New York: Academic Press.

–. 1991b. Temporal and spatial disparities among bark beetles, predators and associates responding to synthetic bark beetle pheromones: implications for coevolution and pest management. *Environ. Entomol.* 20: 1665–1679.

Raffa, K. F. and A. A. Berryman. 1980. Flight responses and host selection by bark beetles. In *Proceedings of the Second IUFRO Conference on the Dispersal of Foret Insects: Evaluation, Theory, and Management Implications*. A. A. Berryman and L. Safranyik, eds., pp. 213–233. Pullman, WA: Conference Office, Cooperative Extension Service, Washington State University.

–. 1983a. The role of host plant resistance in the colonization behavior and ecology of bark beetles (Coleoptera: Scolytidae). *Ecol. Monogr.* 53: 27–49.

–. 1983b. Physiological aspects of lodgepole pine wound responses to a fungal symbiont of the mountain pine beetle, *Dendroctonus ponderosae* (Coleoptera: Scolytidae). *Can. Entomol.* 115: 723–734.

–. 1987. Interacting selective pressures in conifer-bark beetle systems: A basis for reciprocal adaptation? *Am. Nat.* 129: 234–262.

Raffa, K. F. and D. L. Dahlsten. 1995. Differential reponses among natural enemies and prey to bark beetle pheromones. *Oecologia (Berl.)* 102: 17–23.

Raffa, K. F. and K. D. Klepzig. 1989. Chiral escape of bark beetles from predators responding to a bark beetle pheromone. *Oecologia (Berl.)* 80: 556–569.

Raffa, K. F., T. W. Phillips and S. M. Salom. 1993. Strategies and mechanisms of host colonization by bark beetles. In *Beetle-Pathogens Interactions in Conifer Forests*. T. D. Schowalter and G. M. Filip, eds., pp. 103–128. London: Academic Press.

Raffa, K. F. and E. B. Smalley. 1995. Interaction between pre-attack and induced monoterpene concentrations in conifer defense against bark beetle-microbial complexes. *Oecologia (Berl.)* 102: 285–295.

Reid, M. L. and B. D. Roitberg. 1994. Benefits of prolonged male residence with mates and brood in pine engravers (Coleoptera: Scolytidae). *Oikos* 70: 140–148.

Reid, R. W. 1962. Biology of the mountain pine beetle, *Dendroctonus monticolae* Hopkins, in the East Kootenay region of British Columbia. II. Behaviour in the host, fecundity and internal changes in the female. *Can. Entomol.* 94: 605–613.

Renwick, J. A. A. and J. P. Vité. 1970. Systems of chemical communication in *Dendroctonus*. *Contrib. Boyce Thompson Inst.* 24: 283–292.

Robert, A. 1947. Etude préliminaire du *Corthylus punctatissimus* Zimm. le Scolyte de l'érable. *Travaux de l'Inst. biol. générale et de zool. de l'Univ. Montréal* no. 37, pp. 1–5.

Roberts, H. 1960. *Trachyostus ghanaensis* Schedl (Col., Platypodidae) an ambrosia beetle attacking Wawa, *Triplochiton scleroxylon* K. Schum. *West Afr. Timber Borer Res. Unit Tech. Bull.* No. 3. 17pp.

–. 1961. The adult anatomy of *Trachyostus ghanaensis* Schedl (Platypodidae), a W. African beetle and its relationship to changes in adult behaviour. *Fourth Rep. West Afr. Timber Borer Res. Unit*, pp. 31–38.

–. 1962a. A description of the developmental stages of *Trachyostus aterrimus* (Schauf.), a West African platypodid and some remarks on its biology. *Fifth Rep. West Afr. Timber Borer Res. Unit*, pp. 29–46.

–. 1962b. An examination of the biology of *Trachyostus ghanaensis* Schedl (Platypodidae), an ambrosia beetle attacking living trees of *Triplochiton scleroxylon* K. Schum, in West Africa. In *Proceedings XI Int. Colgr. Entomol.*, Wien, 17-25 Aug. 1960, vol. 2, pp. 241–244.

–. 1968. Notes on the biology of ambrosia beetles of the genus *Trachyostus* Schedl (Coleoptera: Platypodidae) in West Africa. *Bull. Entomol. Res.* **58**: 325–352.

–. 1969. A note on the Nigerian species of the genus *Sosylus* Erichson (Col., Fam. Colydiidae) parasites and predators of ambrosia beetles. *J. Nat. Hist.* **3**: 85–91.

Roeper, R. A., L. M. Treeful, K. M. O'Brien, R. A. Foote and M. A. Bunce. 1980. Life history of the ambrosia beetle *Xyleborus affinis* (Coleoptera: Scolytidae) from in vitro culture. *Great Lakes Entomol.* **13**: 141–144.

Roeper, R. A., D. V. Zestos, B. J. Palik and L. R. Kirkendall. 1987. Distribution and host plants of *Corthylus punctatissimus* (Coleoptera: Scolytidae) in the lower peninsula of Michigan. *Great Lakes Entomol.* **20**: 69–70.

Rudinsky, J. A. 1962. Ecology of Scolytidae. *Annu. Rev. Entomol.* **7**: 327–348.

–. 1966. Scolytid beetles associated with Douglas-fir response to terpenes. *Science (Wash., D.C.)* **152**: 218–219.

–. 1969. Masking of the aggregation pheromone in *Dendroctonus pseudotsugae* Hopk. *Science (Wash., D.C.)* **166**: 884–885.

–. 1973. Multiple functions of the Douglas-fir beetle pheromone 3-methyl-2-cyclohexen-1-one. *Environ. Entomol.* **2**: 579–585.

Rudinsky, J. A., V. Vallo and L. C. Ryker. 1978. Sound production in Scolytidae: attraction and stridulation of *Scolytus mali* (Col., Scolytidae). *Z. Angew. Entomol.* **86**: 381–391.

Rudinsky, J. A. and S. Vernoff. 1979. Evidence of a female-produced aggregative pheromone in *Leperisinus californicus* Swaine (Coleoptera: Scolytidae). *Pan Pac. Entomol.* **55**: 299–303.

Rudnew, D. F. and W. T. Kosak. 1974. Ueber die Variabilität der Brutbilder von Kiefernborkenkäern (Coleoptera, Ipidae). *Anz. Schälingskde. Pflanzen-Umweltschutz* **47**: 155–158.

Rühm, W. 1977. Rüsselkäfer (Arucariini, Cossoninae, Col.) mit einer Borkenkäfern (Scolytoidea) änlichen Brutbiologie an der *Araucaria araucana* (Mol.) Koch in Chile. *Z. Angew. Entomol.* **84**: 283–295.

Ryker, L. C. 1984. Acoustic and chemical signals in the life cycle of a beetle. *Scient. Am.* **250**: 112–123.

Salonen, K. 1973. On the life cycle, especially on the reproduction biology of *Blastophagus piniperda* L. (Col., Scolytidae). *Acta For. Fenn.* **127**: 1–72.

Santoro, F. H. 1957. Contribución al conocimento de la biología de *Platypus sulcatus* Chapuis. *Rev. Inv. For.* **1**(3): 7–19.

Saunders, J. L. and J. K. Knoke. 1967. Diets for rearing the ambrosia beetle *Xyleborus ferrugineus* (Fabricius) in vitro. *Science (Wash., D.C.)* **157**: 460–463.

Schedl, K. E. 1961. Forstentomologische Beiträge aus Belgisch-Kongo. Familie Brenthidae. *Mitt. Forstl. Bundes-Versuchsanst. Mariabrunn* **61**: 1–95.

–. 1962a. Forstentomologische Beitrge aus dem Kongo, Ruber und Kommensalen. *Entomol. Abh. Ber.* **28**: 37–84.

–. 1962b. Scolytidae und Platypodidae Afrikas. Band 3. Familie Platypodidae. *Rev. Entomol. Moçmbique* **5**: 595–1352.

–. 1972. *Monographie der Familie Platypodidae Coleoptera*. Den Haag: W. Junk.

Schlyter, F. and G. Birgersson. 1989. Individual variation of pheromone in bark beetles and moths – a comparison and an evolutionary background. *Holarct. Ecol.* **12**: 457–465.

Schneider, I. 1987. Verbreitung, Pilzubertragung und Brutsystem des Ambrosiakäers *Xyleborus affinis* im Verlich mit *X. mascarensis* (Col., Scolytidae). *Entomol. German.* **12**: 267–275.

Schowalter, T. D., B. A. Caldwell, S. E. Carpenter, R. P. Griffiths, M. E. Harmon, E. R. Ingham, R. G. Kelsey, J. D. Lattin and A. R. Moldenke. 1992. Decomposition of fallen trees: initial conditions and heterotroph colonization rates. In *Tropical Ecosystems: Ecology and Management*. K. P. Singh and J. S. Singh, eds., pp. 373–383. New Delhi: Wiley Eastern Ltd.

Schwarz, E. A. 1894. A 'parasitic' scolytid. *Proc. Entomol. Soc. Wash.* **3**: 15–17.

Seybold, S. J. 1993. Role of chirality in olfactory-directed behavior – aggregation of pine engraver beetles in the genus *Ips* (Coleoptera: Scolytidae). *J. Chem. Ecol.* **19**: 1809–1831.

Seybold, S. J., S. A. Teale, D. L. Wood, A. Zhang, F. X. Webster, K. Q. Lindahl and I. Kubo. 1992. The role of lanierone in the chemical ecology of *Ips pini* (Coleoptera: Scolytidae) in California. *J. Chem. Ecol.* **18**: 2305–2329.

Shrimpton, D. M. 1973. Extractives associated with the wound response of lodgepole pine attacked by the mountain pine beetle and associated organisms. *Can. J. Bot.* **51**: 527–534.

Smith, J. H. 1935. The pinhole borer of North Queensland cabinet woods. *Queensland Dept. Agric. Div. Entomol. Bull.* (N.S.) **12**: 38pp.

Smith, R. L. 1980. Evolution of exclusive postcopulatory paternal care in the insects. *Fla. Entomol.* **63**: 65–78.

Spessivtseff, P. 1921. Bidrag till känedomen om splintborrarnas näringsgnag. *Medd. Stat. Skogsf.* **18**: 315–326 (in Swedish with German summary).

Stark, V. 1926. Influence du développement de *Crypturgus cinereus* et sur celui de *Blastophagus minor* Hart. dans les conditions du gouvernement de Brjansk. *Défense Plantes* **1926**: 164–167.

Stewart, K. W. 1965. Observations on the life history and habits of *Scierus annectans* (Coleoptera: Scolytidae). *Ann. Entomol. Soc. Am.* **58**: 924–927.

Swaby, J. A. and J. A. Rudinsky. 1976. Acoustic and olfactory behavior of *Ips pini* (Say) (Coleoptera: Scolytidae) during host invasion and colonisation. *Z. Angew. Entomol.* **81**: 421–243.

Swedenborg, P. D., R. L. Jones, M. E. Ascerno and V. R. Landwehr. 1988. *Hylurgopinus rufipes* (Eichhoff) (Coleoptera: Scolytidae): attraction to broodwood, host colonization behavior and seasonal activity in central Minnesota. *Can. Entomol.* **120**: 1041–1050.

Teale, S. A., B. J. Hager and F. X. Webster. 1994. Pheromone-based assortative mating in a bark beetle. *Anim. Behav.* **48**: 569–578.

Thalenhorst, W. 1958. Grundzüge der Populationsdynamik des grossen Fichtenborkenkäfers *Ips typographus* L. *Schrift. Forstl. Fak. Univ. Göttingen* 21: 1–126.

Thomas, J. B. 1960. The immature stages of Scolytidae: the tribe Xyloterini. *Can. Entomol.* 92: 410–419.

Thompson, R. T. 1992. Observations on the morphology and classification of weevils (Coleoptera, Curculionoidea) with a key to major groups. *J. Nat. Hist.* 26: 835–891.

Thornhill, R. and J. Alcock. 1983. *The Evolution of Insect Mating Systems.* Cambridge, Mass.: Harvard University Press.

Trivers, R. 1985. *Social Evolution.* Menlo Park, CA: Benjamin/ Cummings.

Vernoff, S. and J. A. Rudinsky. 1980. Sound production and pairing behavior of *Leperisinus californicus* Swaine and *L. oreganus* Blackman (Coleoptera: Scolytidae) attacking Oregon ash. *Z. Angew. Entomol.* 90: 58–74.

Vité, J. P., A. Bakke and J. A. A. Renwick. 1972. Pheromones in *Ips* (Coleoptera: Scolytidae): occurrence and production. *Can. Entomol.* 104: 1967–1975.

Wallace, P. 1940. Notes on the smaller European elm bark beetle *Scolytus multistriatus* Marsham. *Conn. Exp. Stat. Bull.* 434: 293–311.

Walters, J. and L. H. McMullen. 1956. Life history and habits of *Pseudohylesinus nebulosus* (LeConte) (Coleoptera: Scolytidae) in the interior of British Columbia. *Can. Entomol.* 88: 197–202.

Werner, R. A. 1986. The eastern larch beetle in Alaska. *U.S.D.A. For. Serv. Res. Pap. PNW-357.* 13pp.

White, P. R. and J. Chambers. 1989. Saw-toothed grain beetle *Oryzaephilus surinamensis* (L.) (Coleoptera: Silvanidae): antennal and behavioural responses to individual components and blends of aggregation pheromone. *J. Chem. Ecol.* 15: 1015–1031.

White, T. C. R. 1993. *The Inadequate Environment. Nitrogen and the Abundance of Animals.* Berlin: Springer-Verlag.

Whitney, H. S. 1982. Relationships between bark beetles and symbiotic organisms. In *Bark Beetles in North American Conifers.* J. B. Mitton and K. B. Sturgeon, eds., pp. 183–211. Austin: University of Texas Press.

Wilson, D. S. 1980. *The Natural Selection of Populations and Communities.* Menlo Park, CA: Benjamin/Cummings.

Wilson, E. O. 1971. *The Insect Societies.* Cambridge, MA: Belknap Press of Harvard University Press.

Wood, D. L. 1972. Selection and colonization of ponderosa pine by bark beetles. *Symp. R. Entomol. Soc. Lond.* 6: 110–117.

–. 1982. The role of pheromones, kairomones and allomones in the host selection and colonization behavior of bark beetles. *Annu. Rev. Entomol.* 27: 411–446.

Wood, S. L. 1954. A revision of North American Cryphalini (Scolytidae, Coleoptera). *Univ. Kan. Sci. Bull.* 36: 959–1089.

–. 1957a. A new generic name for and some biological data on an unusual Central American beetle (Coleoptera: Platypodidae). *Great Basin Nat.* 17: 103–104.

–. 1957b. Ambrosia beetles of the tribe Xyloterini (Coleoptera: Scolytidae) in North America. *Can. Entomol.* 89: 337–354.

–. 1966. New records and species of neotropical Platypodidae (Coleoptera). *Great Basin Nat.* 26: 45–70.

–. 1971. New records and species of neotropical bark beetles (Scolytidae: Coleoptera), Part V. *Brigham Young Univ. Sci. Bull., Biol. Ser.* 15(3). 54pp.

–. 1973. On the taxonomic status of Platypodidae and Scolytidae (Coleoptera). *Great Basin Nat.* 33: 77–90.

–. 1978. A reclassification of the subfamilies and tribes of Scolytidae (Coleoptera). *Ann. Soc. Ent. Fr.* (N. S.) 14: 95–122.

–. 1980. The correlation of biology and behavior with classification and phylogeny in Scolytidae. *Proc. XVI Int. Cong. Entomol. (Kyoto, Japan)* p. 286.

–. 1982. The bark and ambrosia beetles of North and Central America (Coleoptera: Scolytidae), a taxonomic monograph. *Great Basin Nat. Mem.* 6: 1–1359.

–. 1986. A reclassification of the genera of Scolytidae (Coleoptera). *Great Basin Nat. Mem.* 10: 1–126.

–. 1993. Revision of the genera of Platypodidae (Coleoptera). *Great Basin Nat.* 53: 259–281.

Wood, S. L. and D. E. Bright. 1992. A Catalog of Scolytidae and Platypodidae. (Coleoptera), Part 2: Taxonomic Index. *Great Basin Nat. Mem.* 13: 1–1553.

Yamamura, N. 1986. An evolutionarily stable strategy (ESS) model of postcopulatory guarding in insects. *Theor. Popul. Biol.* 29: 438–455.

Yates, M. G. 1984. The biology of the oak bark beetle, *Scolytus intricatus* (Ratzeburg) (Coleoptera: Scolytidae), in southern England. *Bull. Entomol. Res.* 74: 569–579.

Ytsma, G. 1989. Colonization of southern beech by *Platypus caviceps* (Coleoptera: Platypodidae). *J. Chem. Ecol.* 15: 1171–1176.

Zeh, D. W. and R. L. Smith. 1985. Paternal investment by terrestrial arthropods. *Am. Zool.* 25: 785–805.

Zehntner, L. 1900. De Riet-schorskever, *Xyleborus perforans* Wollaston. *Meded. Proefst. Suikerriet West Java, Kagok-Tegal* 44: 1–21, 1 plate.

Zhang, Q. H., J. A. Byers and F. Schlyter. 1992. Optimal attack density in the larch bark beetle, *Ips cembrae* (Coleoptera, Scolytidae). *J. Appl. Ecol.* 29: 672–678.

10 · Biparental care and social evolution in burying beetles: lessons from the larder

ANNE-KATRIN EGGERT AND JOSEF KARL MÜLLER

ABSTRACT

Burying beetles (Coleoptera: Silphidae: *Nicrophorus*) exhibit elaborate biparental care. Males and females independently search for small vertebrate carcasses, which serve as the sole food source for developing young. Both parents prepare the carcass for burial, excavate a cavity in the carcass within which the young feed, provision the young during the early stages of their development, and protect them from conspecific and interspecific predators and competitors. Carrion is an extremely nutrient-rich resource, but it can vary greatly in quantity and quality, and owing to its rarity and ephemeral nature, its occurrence is highly unpredictable. We propose that many of the sexual and parental behaviors of *Nicrophorus* can be regarded as adaptations to the unique problems posed by these resource features.

Competition for carrion is intense; consequently, traits that help reduce or eliminate competition, such as carcass burial, should increase reproductive success. Competition among burying beetles is manifest in inter- and intraspecific aggressive interactions that can escalate into damaging fights. However, losers of contests over carcasses can adopt alternative reproductive tactics: subordinate females can leave some young to be cared for by a dominant female, and subordinate males can sire offspring by surreptitiously mating with the resident female. Males can also inseminate females without having found a carcass, but the number of offspring resulting from these matings is small relative to that of parental males on carcasses.

Carcass size affects not only the number of larvae that can be reared in a reproductive attempt, but also the kinds of social interactions that occur between adult beetles. On large carcasses, several females may jointly care for a brood of mixed maternity. Breeding associations are variable (from single females and monogamous pairs to polygynandrous groups on a carcass) and are affected by carcass size. On large carcasses, different male and female optima for the number of reproducing females on the carcass lead to sexual conflict, manifest in female interference with male attempts to attract additional females to the carcass. Recent research on burying-beetle breeding associations has replaced the view of *Nicrophorus* as a model of parental solicitude with a more complex view encompassing a whole array of more or less reluctant alliances that are rife with conflict and competing interests, and has raised a variety of interesting problems waiting to be solved in the future.

INTRODUCTION

The peculiar habits of burying beetles (genus *Nicrophorus* Fabricius; for a discussion of the correct spelling of the genus name see Herman 1964) have fascinated naturalists for centuries. The earliest descriptions of their behaviors can be found in the works of eighteenth- and nineteenth-century entomologists (Gleditsch 1752; Rösel vom Rosenhof 1761; Herbst 1792; Lacordaire 1834–38). The French entomologist Jean Henri Fabre (1899) carried out extensive experimental studies of burying-beetle behavior near the turn of the century. In 1933, German biologist Erna Pukowski published a detailed report on the behavior of European *Nicrophorus*, still the most comprehensive account of burying-beetle reproductive behavior. Half a century later, behavioral ecologists and evolutionary biologists rediscovered *Nicrophorus* as a suitable study organism for questions related to competition, cooperation and parental care (Wilson *et al.* 1984; Bartlett 1987; Müller and Eggert 1987; Trumbo 1987; Otronen 1988a; Scott 1989). In this chapter, we review the social biology of *Nicrophorus* beetles, focussing on the ecological causes of variation in behavior. After a brief description of the natural history of the genus, we present evidence of severe competition for carrion, and illustrate the reproductive options that are open to individuals that have not been able to secure a carcass. We then discuss the evolutionary origin of parental care, the current benefits of care to care-giving females and males, the ultimate causes of sex roles, and the kinds of adjustments that care-giving individuals show in response to variation in resource and brood size. We discuss possible causes of tolerance and joint care of offspring by several females on large carcasses and depict a sexual conflict that arises in this context. Finally, we compare different forms of parental care in the Coleoptera and attempt to identify the selective pressures that have shaped the social system of *Nicrophorus*.

NATURAL HISTORY OF BURYING BEETLES

The genus *Nicrophorus* belongs to the coleopteran family Silphidae (carrion beetles). Although most other silphids utilize vertebrate carrion as an adult food source or oviposition site, only species in the genus *Nicrophorus* bury small vertebrate corpses in the soil as food for their larvae. Carcasses are buried by the repeated passage of the beetles under the carcass, by which soil is dug up from beneath it. The carcass either falls into the cavity that the beetles have created (Fabre 1899) or is dragged into the hole (Pukowski 1933). The beetles are able to bury corpses that weigh 100–300 times as much as the insects themselves. When they find a corpse resting on very hard ground, they can move it several meters away from its original location to a spot with softer soil that allows burial (Milne and Milne 1976). They inter the carcass until it is covered with soil, or at least with leaf litter or moss; there is considerable species-specific variation in the depth at which the carcass is buried (Pukowski 1933; Wilson and Fudge 1984; Wilson and Knollenberg 1987).

Before burial is complete there may be violent fights for possession of the carcass (Pukowski 1933). Fights occur between conspecifics of the same sex as well as between members of different species (Pukowski 1933), and they may result in one conspecific pair of beetles residing on the buried carcass (Pukowski 1933), the monogamous situation that is described as typical of *Nicrophorus* in numerous texts (see, for example, Wilson 1975; Thornhill and Alcock 1983). However, other kinds of breeding associations are possible: almost every imaginable situation from single females (Scott and Traniello 1990; Eggert 1992) to groups with multiple males and females (Peck 1986; Müller *et al.* 1990a; Scott and Traniello 1990; Trumbo 1992) has been found on buried carcasses in nature. Lone males on carcasses emit pheromones that attract mainly conspecific females, whereas lone females bury carcasses by themselves.

Once underground, the carcass is rolled up into a ball, and fur or small feathers covering the outer surface of the corpse are removed (Pukowski 1933). The beetles' continual movement around the carrion ball results in the construction of a crypt, a small brood chamber with stable walls of compressed soil. Oviposition begins eight hours to several days after the female has detected the carcass, depending on the species. The eggs are deposited singly in the soil near the carrion ball. In *N. vespillo*, eggs are deposited along the sides of a larger longitudinal cavity in the soil, the 'Muttergang' (Pukowski 1933), which may facilitate larval access to the carcass. In many other species (e.g. *N. vespilloides*) eggs are scattered in the soil. Before and during the oviposition period, the pair mate frequently, with individual matings lasting only seconds to a few minutes (Müller and Eggert 1989).

While the embryos develop in the soil, the parent beetles remain on the carcass. The adults remove any soil particles that fall on the surface of the carrion ball, continually moisten it with anal and oral secretions (Pukowski 1933), and remove any fungal hyphae, such that the growth of mold is strongly reduced relative to carcasses that are not so treated (Halffter *et al.* 1983; Wilson and Fudge 1984).

After the larvae have hatched, they crawl to the carcass. There is evidence that attraction of first instars to the carcass is olfactory (Pukowski 1933). Additionally, larvae may be attracted to stridulatory sounds produced by adult beetles on the carcass (Pukowski 1933; Niemitz and Krampe 1971, 1972). The larvae eventually aggregate in or near a small patch of exposed flesh on top of the carrion ball that the parent beetles create by repeatedly piercing the skin of the corpse with their mandibles (Pukowski 1933). At this time the feeding of larvae begins: the adults feed on the carcass, and later they regurgitate predigested carrion to the larvae. The larvae exhibit a specialized begging behavior in which they rear up, sometimes climbing the parent's forelegs, bringing their mouthparts closer to the parent's ones. As the larvae grow older, they spend less time begging and more time feeding on their own. The depression at the top of the carrion ball grows into an increasingly larger cavity inside the carcass, since it is the primary location on the carcass where feeding takes place (Pukowski 1933). The larvae continue feeding for a total of six days to two or three weeks, depending on the species and ambient temperature. The female stays on the carcass until larval development is complete; the male usually leaves a few days before the female (Bartlett 1988; Reinking 1988; Scott and Traniello 1990; Trumbo 1991; Schwarz and Müller 1992). The larvae typically consume the carrion ball almost entirely and leave the crypt when nothing but pieces of skin and bones remain. The female flies off in search of another carcass, and the larvae crawl into the soil to build separate pupal chambers (Pukowski 1933). In most species, pupation occurs several days later and adult emergence takes place a few weeks after the carcass was first buried; in some species, such as *N. investigator*, *N. fossor* and *N. tomentosus*, larvae stay in the soil and overwinter as prepupae, and adults do not emerge until the following summer.

Although adult beetles do not require vertebrate carrion for their own subsistence, they will feed on it, especially when the carcass is unsuitable for reproduction (Wilson and Knollenberg 1984). At the end of the reproductive season, groups of *N. vespilloides* can frequently be found feeding on carcasses that are incompletely buried, and aggressive interactions are noticeably absent (A.-K. Eggert, unpublished data). Adults also feed on dead insects and prey on carrion fly larvae; adult *N. germanicus* are specialized predators of dung beetles (Pukowski 1933). Reproduction, however, requires the availability of a suitable corpse. Females do not undergo the final stages of ovarian development until they have located a small carcass and found it acceptable for use in a reproductive attempt (Wilson and Knollenberg 1984; Scott and Traniello 1987), a feature that likely reduces wing loading during search flight, and delays the allocation of resources to ovarian development until there is an opportunity for successful reproduction.

CARRION: A RARE AND VALUABLE RESOURCE?

Recent analyses of the reproductive behavior of burying beetles (see, for example, Bartlett 1987; Müller 1987; Trumbo 1990a; Eggert 1992; Scott and Gladstein 1993) suggest that securing a corpse suitable for offspring production is extremely difficult. However, an empirical assessment of the availability of carrion to burying beetles appears to be virtually impossible. One would need to know the rate at which carcasses of a certain size become available within an area, the period over which they remain in a condition suitable for reproduction, the frequency of competitive preemption by flies or vertebrate scavengers, and the area that beetles can effectively search within a given period. Not surprisingly, reliable data on the number of carcasses that become available to burying beetles in a breeding season are entirely lacking.

The best evidence supporting the view that the availability of carrion is extremely low comes from the beetles' own behavior. Adult longevity and the duration of the reproductive season would easily allow for more than two successive breeding attempts, and females in the laboratory can produce and rear up to seven successive broods (Bartlett and Ashworth 1988; J. K. Müller, unpublished data; I. C. Robertson, personal communication). Nonethe- less, burying beetles reproducing on a carcass often behave as if they have no chance of finding another carcass in their lifetime.

Damaging or lethal fights should occur only when the contested resource is extremely valuable (Thornhill and Alcock 1983), i.e. when the bulk of an individual's reproductive success is at stake (Enquist and Leimar 1990). In natural populations of burying beetles, many individuals lack antennal or tarsal segments, injuries that are probably due to fights over the possession of carcasses (Pukowski 1933; Trumbo 1990b). The killing of subordinate contestants has been observed under conditions that denied these individuals the opportunity to abandon the carcass (Scott 1990; Trumbo 1990c; Robertson 1993). The behavior of subordinate individuals is also consistent with the rarity of reproductive opportunities. Subordinate males and females stay near the carcass and return to it repeatedly, risking severe injuries for a benefit that is small relative to that of the dominant individuals (Bartlett 1988; Müller *et al.* 1990a, 1991). Males regularly employ another tactic with low expected benefits, pheromone emission in the absence of carrion (Eggert 1992), which suggests that the alternative activity, searching for carcasses, is unlikely to be successful. Females accept carcasses of marginal size that allow for the rearing of one or two offspring only (Müller *et al.* 1990b), although 30 or more offspring can be reared on a carcass under optimal conditions (Eggert and Müller 1992; Trumbo and Wilson 1993). Females also provide full-length parental care to a single larva on larger carcasses (Müller 1987). Even after several brood failures, burying beetles often do not give up a carcass to search for a new reproductive opportunity (Müller 1987).

COMPETITION FOR CARRION

The high nutritive value of carrion makes it a highly sought-after resource (a 'bonanza' resource: Wilson 1971) for a variety of terrestrial scavengers. Various species in a number of insect orders utilize this protein-rich resource as adult or larval food: a host of coleopterans (carabids, silphids, nitidulids, scarabaeid dung beetles, etc.) feed or oviposit on carcasses, hymenopterans (wasps, ants) carry off pieces of carrion for their own provisioning or as food for their larvae, and numerous dipterans (sarcophagid and calliphorid flesh flies, phorids, sphaerocerids) oviposit or larviposit on carcasses (see, for example, Hennig 1950; Walker 1957; Lundt 1964; Payne 1965). Hence, competition for this spatially unpredictable, ephemeral resource is intense. Competition for carrion among insects is mostly of the 'scramble' type (Nicholson 1954), whereas 'contest'-type competition (Nicholson 1954), the exclusive utilization of the carcass by a small number of individuals as it occurs in *Nicrophorus*, appears to be fairly rare.

Their burying habit and relatively large and heavily sclerotized bodies enable *Nicrophorus* to exclude most other insect competitors, but congeners and conspecifics are another matter. On small carcasses, which are arguably the most abundant ones, there is just enough carrion for one female's brood to survive (Pukowski 1933; Müller *et al.* 1990a; Trumbo and Fiore 1994), and aggressive interactions are the rule (Pukowski 1933).

Many of the rules governing the structure and the outcome of these altercations as delineated by Pukowski (1933) have been supported by more recent studies. The studies of Wilson *et al.* (1984) and Trumbo (1990a) have confirmed the competitive advantage of larger species in interspecific competition, and several studies corroborate Pukowski's finding that larger individuals win fights on unburied carcasses (Bartlett and Ashworth 1988; Otronen 1988a, 1990; Müller *et al.* 1990a). Experiments by Dressel (1987), Bartlett (1988), and Bartlett and Ashworth (1988) confirm that conspecific males do not fight vigorously on carcasses unless a female is present. For males, the arrival of a female changes the carcass from a potential to an actual oviposition site. In the presence of a female close to oviposition, the exclusion of rivals from the carcass becomes essential to the maximization of a male's reproductive success (Bartlett 1988; Müller *et al.* 1991).

Conspecific and congeneric burying beetles can sometimes locate buried carcasses (Trumbo 1990a, 1994; Robertson 1993). Such intruders subsequently attempt to take over the carcass and secure as large an amount of carrion as possible for their own reproduction; when successful, they kill the larvae present on the carcass (Scott 1990; Trumbo 1990c) and destroy eggs present in the soil (Robertson 1993). The resident beetles attempt to defend their brood and the carcass by attacking the intruder vigorously, regardless of the intruder's sex, size, or species. Even

intruders that are larger than the residents are not always successful in this situation, suggesting that factors other than body size, such as the differential value of the carcass to the contestants, can affect the outcome of fights. Prior experience could be another factor influencing contestants' behavior: *N. humator* individuals that were raised as 'winners' by having them fight weaker individuals at regular intervals tended to win over conspecifics of equal size that had been raised as 'losers' (Otronen 1990).

ALTERNATIVE TACTICS: MAKING THE BEST OF A BAD JOB?

Even when competition is intense, smaller and competitively inferior individuals retain the opportunity to reproduce on contested carcasses. After only one or two fights, it becomes obvious which of two contestants will secure control of a carcass. Nonetheless, defeated individuals rarely abandon the corpse at this point, but their behavior is quite different from the behavior of winners. Losers typically remain in the soil a short distance from the carcass for long periods of time, but return to the carcass for brief visits (Bartlett 1988; Dressel and Müller 1988; Müller *et al.* 1990a). These truncated visits, during which subordinates are often encountered and attacked by the dominant consexual, can have beneficial effects on the reproductive success of losers of both sexes.

Satellite males

The co-occurrence of two or more males near a buried carcass has been observed in several field studies (during burial, Fabre 1899; the morning after discovery: 21–31%, Wilson and Fudge 1984; 24 h after burial is complete: 16%, Eggert 1992). Dominant females spend most of their time on the carcass (Müller *et al.* 1990a) and do not reject mating attempts by vanquished males. Hence, subordinate males have a high probability of inseminating resident females during their visits to the carcass. However, such visits are rare relative to the extended periods of time dominant males spend on the carcass (Müller *et al.* 1991), and thus, the dominant male probably achieves a much larger number of matings than the subordinate. A large number of repeated matings is required for males to achieve a high paternity (Müller and Eggert 1989). Studies using the irradiated-male technique (Bartlett 1988) and genetic markers (Eggert 1992; Scott and Williams 1993) demonstrate that the first male to leave a brood (the loser) sires a much smaller portion of the brood than the winner (roughly 7%, Bartlett 1988; 0–35% of the brood, median = 10%, Eggert 1992; 0–38% of the brood, median = 10.5%, Scott and Williams 1993). In many territorial animal species (see, for example, Arak 1984), surreptitious matings constitute a less profitable reproductive tactic, but the only one available to physically inferior males.

Intraspecific brood parasites

Vanquished females also retain the opportunity to produce offspring on a contested carcass. Despite frequent attacks by the winner, such females often stay for several days and lay eggs near the lost carcass. The first female to desert a carcass (the loser) frequently has some surviving offspring (Müller *et al.* 1990a; Scott and Williams 1993). However, even when the loser succeeds in producing offspring, her share of the brood is smaller than that of the winner (17–18%) on small and medium-sized carcasses (Müller *et al.* 1990a; Scott and Williams 1993), the exact proportion increasing with the size of the carcass (Eggert and Müller 1992).

On small carcasses, the loser female's offspring are raised almost exclusively by the winner (Müller *et al.* 1990a). The winner, typically the larger female, remains with the brood for a much longer time than the loser, such that relative body size and time spent on the carcass are correlated (Eggert and Müller 1992; Scott and Williams 1993). The winner clearly provides most of the care to a brood that often (60%, Müller *et al.* 1990a; 67%, Scott and Williams 1993) includes some of the loser's offspring. Additionally, the reproduction of subordinate females on small and medium-sized carcasses reduces the dominant's reproductive success (Müller *et al.* 1990a; Eggert and Müller 1992; Scott 1994), and thus constitutes a case of intraspecific brood parasitism (Müller *et al.* 1990a). The primary benefit to a brood-parasitic *Nicrophorus* female is not the dominant female's parental care, but rather the opportunity for her and her offspring to secure access to the resource. Defeated females do provide parental care to their own offspring when experimentally provided with a carcass by themselves (Müller *et al.* 1990a).

Possible mechanisms causing the lower reproductive success of losers on small carcasses include: (1) the loser's inability to produce as large an egg clutch as the winner (Müller *et al.* 1990a); (2) the selective destruction of the loser's eggs by the winner (oophagy has been described for intruder males and females, Robertson 1993); and (3) the

selective killing of the loser's larvae by the winner (Müller and Eggert 1990). We consider the first of these hypothetical mechanisms the one most likely responsible for the differential reproductive success of winners and losers. Both females should attempt to produce the maximum number of eggs, thereby contributing as many offspring as possible to the surviving brood. We have suggested earlier (Müller *et al.* 1990a) that losers may be unable to produce as large a clutch in this situation because they have less opportunity to feed on the carcass than do winners. In the absence of rivals, females typically feed on a carcass extensively; they oviposit but still gain mass during the interval from carcass discovery to the hatching of larvae (Reinking 1988). Ingestion of considerable amounts of protein from the carcass may be a necessary prerequisite for the maturation of a large number of eggs, such that a female's fecundity probably increases with the amount of carrion she consumes. Based on these considerations, we suggest that defeated females benefit from their trips to the carcass because feeding on the carcass increases their fecundity.

Selective oophagy of the competitor's eggs requires females to discriminate between their own eggs and those of other females. In takeover situations, the intruder female likely destroys eggs (Robertson 1993) before she herself starts to oviposit. When two females oviposit more or less simultaneously, a female would have to use chemical cues or the specific location of her own eggs for their recognition; it is not known whether females possess this capability.

Infanticide may contribute to the differential in offspring number between winner and loser females on small carcasses, but is unlikely to be its sole cause. Although an effective mechanism of parent–offspring recognition would be highly advantageous in this situation, burying beetles do not appear to possess such a mechanism. So long as unrelated larvae appear on the carcass at approximately the same time as the caring individual's own larvae, they are accepted and cared for (Müller and Eggert 1989, 1990), even when they belong to a different species of burying beetle (Trumbo 1994). Larvae appearing on the carcass outside a certain time window are killed and eaten by the adults (Müller and Eggert 1990). Unrelated larvae are killed when they appear on the carcass eight hours or more before the caring female's own offspring hatch, but the adult's response to the late arrival of larvae is even less exact. A large differential in the onset of oviposition, as would be required for this mechanism to benefit the winner, appears unlikely. The victorious female has unlimited access to the carcass and should therefore be able to mature her ovaries sooner than her subordinate rival. The temporal recognition mechanism may not be particularly beneficial in the context of brood parasitism, but may instead play a more important role during or after take-overs, when oviposition periods differ greatly and usurpers benefit from the infanticide of premature larvae that cannot be their own (Trumbo 1990c). Although it is feasible that a dominant female might kill the offspring of a brood parasite by selectively culling late larvae in the context of brood reduction (Bartlett 1987), no evidence exists for such selective infanticide (Bartlett 1987).

Emission of sex pheromones by males

Although the production of offspring requires carrion, matings nonetheless regularly occur in the absence of carcasses and on carcasses unsuitable for reproduction. When no carcasses are available, sexual encounters are made possible by the emission of a sex pheromone by males (Müller and Eggert 1987; Eggert and Müller 1989a,b). Males of *N. vespilloides* (Eggert 1990, 1992) and other species (Müller and Eggert 1987) regularly emit the pheromone at the end of the species-typical daily activity period. Significant additive genetic variation in the extent to which males exhibit this behavior exists in at least one population (Eggert 1992). Females attracted to phero-mone-emitting males in the field leave after a single mating (Eggert and Müller 1989b). They benefit from the acquisition of fresh sperm: sperm in the spermatheca become inviable after two to three weeks (Eggert 1992), and females depend on a supply of fertile sperm for successful reproduction in the absence of males, which frequently occurs (Wilson and Fudge 1984; Scott and Traniello 1990; Eggert 1992).

The benefit of single matings to males is small (Eggert 1992). If the female carries the sperm of other males (as do nearly all females in the field) (Müller and Eggert 1989), and reproduces in the absence of other males on a carcass immediately after a single mating, only about 10% of her brood is sired by the last male (Müller and Eggert 1989; Eggert 1992). When the female reproduces with the assistance of another male, the pheromone-emitter's paternity is close to zero (Eggert 1992). However, this tactic affords males a large number of mates (Eggert and Müller 1989b), which increases the probability that males leave some offspring should they fail to find a carcass. The search for carrion is likely to be less successful at the end of the activity period because many of the available carcasses have already

been detected by conspecifics (Eggert 1992). Pheromone emission does not appear to be merely a best-of-a-bad-job tactic employed by smaller or weaker individuals: neither body size nor nutritional status influences phero-mone emission (Eggert 1990).

PARENTAL CARE

Elaborate parental care as exhibited by *Nicrophorus* is rare among insect species outside the social Hymenoptera and Isoptera (Eickwort 1979; Wilson 1971, 1975). Hypotheses about the evolutionary origin of such parental care can be based on phylogenetic information and on present-day analogies of the presumed historical situation. The present function or maintenance of such care can be assessed directly by using manipulative experiments. Aside from questions pertaining to its origin and maintenance, several unique aspects of care in *Nicrophorus* call for an evolutionary explanation: the co-occurrence of uni- and biparental care, the shift from biparental to uniparental care prior to the dispersal of larvae from the carcass, and sex differences in the duration of care.

The origin of parental care

The lack of phylogenetic information regarding the Silphidae greatly impairs the development of hypotheses concerning the evolution of parental care in this family. The family consists of two subfamilies, the Silphinae and the Nicrophorinae; the latter subfamily includes *Nicrophorus* and its presumed sister genus from southeast Asia, *Ptomascopus* (Peck 1982; Peck and Anderson 1985). Within the Silphinae, a few species have become specialized predators or phytophages (Heymons *et al.* 1927, 1928, 1929; Anderson and Peck 1984), but the adults and larvae of most species feed on carcasses left exposed on the soil surface (Peck and Anderson 1985). The use of carrion as a reproductive resource also occurs in *Ptomascopus* and is thus most parsimoniously interpreted as an ancestral character-state in *Nicrophorus*.

Parental care, however, is likely to be a synapomorphic character of the species in the genus *Nicrophorus*. Phylogenetic relations among different *Nicrophorus* species are largely obscure (but see Peck and Anderson 1985), but all *Nicrophorus* species that have been studied bury and maintain carcasses and feed and defend larvae. Neither *Ptomascopus* nor any of the silphine species studied to date engage in either of these activities (Peck 1982; Peck and Anderson

1985; M. Kon, personal communication; A.-K. Eggert and J. K. Müller, personal observation). *Ptomascopus morio* females oviposit on carrion made available to them (Peck 1982; M. Kon, personal communication; A.-K. Eggert and J. K. Müller, personal observation), but *P. morio* larvae are also frequently found in nests of *N. concolor*, indicating that this species is a facultative kleptoparasite of *Nicrophorus* (M. Kon, personal communication). Kleptoparasitism could only evolve after the origin of care in *Nicrophorus*, which suggests that some of the current reproductive habits of *Ptomascopus* are not representative of the situation in the common ancestor of the two genera.

We hypothesize that a prolonged residence with the brood and the carcass evolved subsequent to the evolution of the carcass-burying habit. Although burying beetles occasionally care for broods on exposed carcasses (Peck 1986), no extant silphid species outside the genus *Nicrophorus* is known to exhibit guarding of broods or carcasses, suggesting that such defense is not effective unless carcasses are buried.

The burying of carcasses may have reduced desiccation and decelerated decomposition of the corpse owing to the lower temperature and higher humidity of the subterranean environment. More importantly, however, it would have made the carcass more difficult to locate for competitors, and would thus have enabled parental adults to defend the carcass as long as it remained attractive to competitors. We suggest that parental preparation of larval food and direct regurgitation were selectively favored once such parental defense of the carcass was established. Direct feeding may have accelerated consumption of the carcass, thereby decreasing the duration of care and reducing the vulnerability of the brood to intruders. The silphine species *Oeceoptoma thoracica* and *Thanatophilus sinuatus* have developmental times very similar to that of *N. vespilloides* (30, 28, and 31 days, respectively, from egg to adult at 20°C) (Reinking 1988), but the silphine larvae spend a much larger portion of this time on the carcass (64 and 62%, respectively) than do larvae of *N. vespilloides* (19%) (Reinking 1988). In *Ptomascopus*, however, the period of larval feeding on the carcass is short despite the absence of parental care, suggesting an independent evolution of the two characters.

Benefits of parental care

One of the main current benefits of parental care is the improved survival of the brood through parental defense against predators such as staphylinid beetles (Pukowski 1933; Scott 1990) as well as congeneric intruders (Scott 1990). When other burying beetles attempt to take over a buried carcass (Wilson *et al.* 1984), parents attack and pursue such intruders, thereby reducing the risk that intruders will destroy eggs and kill larvae (Scott 1990; Trumbo 1990a,c, 1991; Robertson 1993). Increased amounts of food are available to the brood because adults repel or kill many other carrion consumers, such as blowflies (Pukowski 1933). Parental regurgitation and carrion maintenance aid larval survival and growth even in the absence of competitors and predators (Pukowski 1933; Wilson and Fudge 1984; Trumbo 1990d, 1992; Reinking and Müller 1990).

In *N. vespilloides*, larval dependency on parental presence is especially pronounced in the first 12 h after hatching (Fig. 10-1; see also Reinking and Müller 1990). During this time, the parent chews a small hole in the surface of the carrion ball, typically at the top of the ball. When the carcass is protected from predators, the mere existence of such an opening in the carcass, whether it has been created by the beetles themselves or artificially, by a human experimenter, leads to larval survival rates that are not significantly different from those observed when parents provide care (Fig. 10-1). Continued parental feeding of larvae results in an increase in the mass larvae attain (Reinking

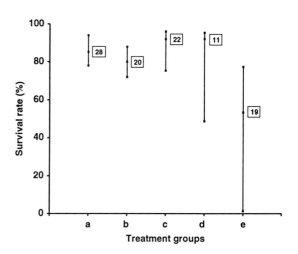

Fig. 10-1. Survival rates of broods of 15 larvae on 15 g carcasses (median and interquartile range). Treatments: (a) broods receiving full-length care, (b) receiving 12 hours of care, (c) receiving no care on carcasses opened by the parents, (d) receiving no care on carcasses opened artificially (using scissors), (e) receiving no care on closed carcasses. Numbers in boxes are sample sizes.

and Müller 1990). The larvae of many smaller species, including *N. vespilloides*, can be reared on chopped liver in the absence of parents (Trumbo 1992; J. K. Müller, unpublished data).

The extended presence of the female on the carcass also affords her the opportunity of producing new clutches should some or all larvae from her first clutch be lost, allowing for some reproductive success so long as sufficient carrion remains (Müller 1987; Wussler and Müller 1994).

Uni- versus biparental care

The participation of males in parental care is an especially intriguing aspect of burying beetle reproduction. In burying beetles, biparental care coexists with uniparental (maternal) care. Uniparental care occurs whenever a female discovers a carcass first and buries it before any conspecific males detect it, but if a male detects an unburied carcass to which a female is also attracted, he remains with the brood and the female for several days (Bartlett 1988; Scott and Traniello 1990; Trumbo 1991; Schwarz and Müller 1992), except on very small carcasses (Bartlett 1988).

For many of the benefits of parental care described above, uniparental care suffices as well as biparental care. Indeed, the presence of a male with the female typically has no significant effect on the number and size of surviving young when predators and competitors are excluded (Bartlett 1988; Reinking and Müller 1990; Trumbo 1991). Benefits arising through male assistance before and during burial cannot explain the prolonged stay of males on carcasses once these tasks have been completed. Male presence at this early time does improve the chances of successfully producing a brood in the face of severe fly competition (Trumbo 1994). An effect of male presence on burial speed, as suggested by Pukowski (1933), was not found in the only field study to investigate this effect (Scott 1990).

Currently, the improved defense that biparental care affords against intruding conspecifics and congeners is considered the most important factor maintaining male participation in care (Scott 1990; Trumbo 1990a,c, 1991; Robertson 1993). Especially on large carcasses, pairs have significantly higher success rates than single females at retaining buried carcasses in the face of intense conspecific and congeneric competition. This was found at natural densities (Trumbo 1990a, 1991; Robertson 1993) and when intruders were artificially introduced (Scott 1990; Trumbo 1990c; Scott and Gladstein 1993). A male's assistance in

repelling intruders after larvae become present on the carcass (Trumbo 1990c), and his contribution to carcass maintenance, which makes the carcass less easy to locate for competitors (Trumbo 1994), may both contribute to the observed effect. Whether carcass maintenance and control of microbial activity by two adults also have a positive effect on reproductive success when intrusions do not occur (Bartlett 1988) has yet to be investigated in the field.

Replacement of the female as the care-giver has also been suggested as a possible function of extended male residency on carcasses (Trumbo 1991; Fetherston *et al.* 1994). When the female is absent, males extend their residency over the entire period of larval development (Trumbo 1991) and feed the brood as effectively as a pair or a female would (Bartlett 1988; Scott 1989; Reinking and Müller 1990; Trumbo 1991; Schwarz and Müller 1992; Fetherston *et al.* 1994). However, this is unlikely to be the primary benefit the male gains from his extended presence on the carcass, because female death or disappearance during a breeding attempt is probably a rare event.

Another benefit to an extended male residency may arise when the female produces a replacement clutch. When a female is about to produce a second clutch, males resume mating at a rate similar to that occurring during carcass burial and preparation (Müller and Eggert 1989; J. K. Müller and A.-K. Eggert, unpublished data), presumably as a means of paternity assurance. Again, replacement clutches may not occur frequently enough for this to provide a compelling explanation for an extended male presence on the carcass. The opportunity for copulations with expelled intruder females (I. C. Robertson, personal communication) also cannot explain male care, because males could probably attract additional females using the pheromone emission tactic if they chose to do so.

The finding that a male's presence up until the larvae disperse can lead to a reduced brood size (Scott 1989; Scott and Gladstein 1993) may be an artifact of the inclusion of smaller replacement broods in comparisons between treatments, and of a particular experimental design (see also Trumbo 1991). Forcing males to remain with the brood beyond the time at which they otherwise would have abandoned the carcass, as was done in the above studies, does not induce prolonged paternal care (Bartlett 1988; Trumbo 1991). In a normal biparental situation, males abandon the carcass days before the larvae disperse from the carcass. When forced to remain beyond this time, males attempt to fly in their containers (Bartlett 1988) and frequently begin to kill and eat their own offspring

(S. T. Trumbo, personal communication; A.-K. Eggert and J. K. Müller, personal observation). Such paternal infanticide probably contributed to the observed reduction in larval numbers. When males are given the opportunity to desert the carcass, or when they are removed three days after larvae hatch, such a reduction is not observed (Trumbo 1991; Scott and Gladstein 1993).

Although Scott and Gladstein (1993) concede that 'pathological infanticide' may have contributed to the observed effect, they ascribe it to 'social instability' and sexual conflict over the duration of male care. The consumption of carrion by paternal males is interpreted as the underlying cause of this conflict, and it is also assumed to contribute to reduced larval numbers. However, males providing full-length care on a carcass typically do not gain mass (Reinking 1988; Scott and Gladstein 1993). To account for the observed negative effect of prolonged male presence, male feeding and mass gain must occur primarily at the end of parental care, but the data show that males gain mass before the larvae hatch and experience considerable mass loss towards the end of larval development (Scott and Gladstein 1993; see also Bartlett 1988; Reinking 1988).

Why does one parent leave early ?

Even in broods that initially receive biparental care, it never continues until the end of larval development, because males leave the brood two to five days earlier than females (Bartlett 1988; Scott and Traniello 1990; Trumbo 1991; Schwarz and Müller 1992; but see Scott and Traniello 1990 for *N. orbicollis*). The most likely explanation for this transition from biparental to uniparental care in the course of a breeding attempt is related to the rapidly progressing consumption of the carcass and its concomitantly decreasing value to potential intruders. Even before the amount of carrion remaining is insufficient for rearing a brood, the remaining carrion may become extremely difficult to locate chemically (Robertson 1993). Biparental care should become unnecessary once the carcass is no longer attractive or valuable to potential competitors, because a single parent is sufficient to feed the brood. Recent studies on natural (Robertson 1993) and experimentally induced (Scott and Gladstein 1993) takeover attempts demonstrate a reduced risk of intrusion towards the end of larval development in *N. orbicollis*. Males probably remain beyond the period during which most takeovers occur (12 vs. 8 days: Robertson 1993), because their contribution to brood defense, even when intrusions are rare, yields greater reproductive benefits than does searching for another carcass.

Males abandon very small carcasses at about the time the larvae hatch (Bartlett 1988), but male residence times increase on larger carcasses (Bartlett 1988; Scott and Traniello 1990; Trumbo 1991), which take longer to be consumed. Both observations are consistent with the hypothesis that male residence times depend on the attractiveness or value of the remaining carrion to potential competitors. However, considerable variation in the duration of male residency occurs even on carcasses of the same size (Robertson 1993). The hypothesis that the value of the remaining carrion to an intruder is the most important variable in this context (Robertson 1993; Scott and Gladstein 1993) awaits testing in manipulative experiments.

Sex differences

The above scenario explains why only one adult should stay until the larvae disperse from the carcass, but does not explain why this is typically the female. It may not be the male's decision to leave first; Pukowski (1933), Bartlett (1988) and Fetherston *et al.* (1990) observed females attacking males and forcing them away from carcasses, especially from small carcasses. Nonetheless, the ultimate factors underlying female attacks on the male, or male decisions to leave, remain obscure.

For each sex, the time of desertion should represent a point at which the benefits of desertion start to outweigh the cost, irrespective of costs and benefits to the other sex. For this equilibrium point to be reached earlier in males than in females, the cost of desertion must be lower in males than in females, or the benefits higher, or both. It has been suggested that the potential cost of desertion (the destruction of the brood by an intruder) may not be as high for the male as for the female: when the female produces a replacement clutch after a takeover by a male, some of this second clutch will still be fathered by the first male (Scott and Traniello 1990). However, if our hypothesis about the transition from biparental to uniparental care is correct, any such cost (and therefore the concurrent asymmetry) is moot, because after the male has left, the probability that takeovers occur and replacement clutches are produced is negligible.

Trumbo (1991) suggested that a female may gain lower benefits from leaving early than her mate because the brood she produces on a subsequent carcass will be smaller than her first, whereas a male can expect a full-size brood in his second breeding attempt if it is his mate's first one. Females' second clutches were found to be smaller than their first clutches in two American species (Scott and Traniello 1990; Trumbo 1990d). In *N. vespilloides*, however, females' second broods are no smaller than their first (Fig. 10-2), so this explanation cannot account for sex differences in the duration of care in this species.

Females may also gain higher benefits from staying with the brood. When the male leaves, there is still some carrion available for feeding, which enables the female to replenish her energy reserves during her longer presence with the brood. Replenishment of energy reserves may be more important for females than males because of greater physiological stress imposed by egg production.

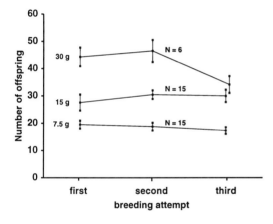

Fig. 10-2. Number of offspring produced in the first, second and third breeding attempts of female *N. vespilloides* (mean and standard error). Different curves represent broods on carcasses of different size. Single inseminated females, or male–female pairs, were given a carcass of 7.5 g, 15 g, or 30 g in a container that permitted the beetles to abandon the carcass as soon as it had been buried. Details on experimental conditions are described for the 'black-box experiment' in Müller *et al.* (1990a). Females were given a new carcass of the same mass two days after they had finished caring for the previous brood. On 7.5 g and 15 g carcasses, offspring numbers did not differ significantly between the three breeding attempts (7.5 g: $\chi^2 = 3.17$, $p = 0.21$; 15 g: $\chi^2 = 0.93$, $p = 0.63$, Friedman test); on 30 g carcasses, however, clutch size in the third breeding attempt was significantly lower than that of the first two attempts ($\chi^2 = 6.33$, $p = 0.04$, Friedman test; multiple comparisons after Wilcoxon–Wilcox demonstrate that the only significant difference is the one between second and third broods: $D = 8.0$, $p < 0.10$).

There is yet another asymmetry that may account for a sex difference in residence times. Females have a higher probability of finding a carcass than males, because they can either locate a carcass themselves or be attracted by a male that has located one. Males, in turn, secure some reproductive success by emitting pheromones and inseminating females attracted to them (Müller and Eggert 1987; Eggert and Müller 1989a,b; Eggert 1992). A female that has been inseminated by a pheromone-emitting male will have a certain probability of finding a carcass and producing this male's offspring not only on the day she is inseminated, but also on each day following the mating. The reproductive success a male can expect from this female decreases over time and eventually goes down to zero, because his sperm will age and face increasing numbers of competitors' sperm transferred in subsequent matings. Nonetheless, a male loses non-parental matings with each day he spends with his present brood, and these losses carry over into the future. It is this loss of additional future reproductive benefits that causes the loss of potential offspring through care, or the cost of staying with the present brood, to be slightly higher for male than for female burying beetles. However, when the remaining breeding season is shorter than the time over which females can use sperm from a particular mating, the cost of care to females may exceed the cost to males (unpublished model by J. K. Müller and A.-K. Eggert).

RESOURCE UTILIZATION: OPTIMAL UNDER THE CIRCUMSTANCES

The upper limit to the range of carcass sizes utilized for reproduction by a *Nicrophorus* species is likely set by the beetles' ability to bury and maintain large carcasses, whereas the lower limit to this range probably depends on the efficiency of carrion utilization (i.e. the total mass of a brood produced from a given carrion mass), the portion of the carcass consumed by adults, and the minimum larval mass at dispersal that is required for the successful completion of development. The size of carcasses utilized, and thus the amount of larval food available, varies within and between species (4 to over 120 g in *N. defodiens*, 8 to over 300 g in *N. orbicollis*, Trumbo 1992; 2–75 g in *N. vespilloides*, Bartlett 1987; Bartlett and Ashworth 1988; Müller *et al.* 1990b). Hence, we would expect females to show some degree of adjustment of the size of their egg clutches and broods to the amount of carrion available if they are physiologically capable of such adjustments.

Clutch size regulation

Several studies investigating the regulation of egg numbers strongly suggest that clutches on medium to large carcasses always contain the maximum number of eggs that females can lay within a short period of time. Clutch size is reduced only on carcasses that are very small for the respective species. In *N. vespilloides*, a significant correlation between egg clutch size and carcass mass was found on carcasses of mass less than 10 g (Müller *et al.* 1990b). On larger carcasses, the slope of the regression line obtained was not significantly different from zero (Müller *et al.* 1990b), corroborating earlier findings that carcass size (>10 g) has no effect on the number of eggs laid (Easton 1979; Bartlett 1987). In the slightly larger species *N. investigator*, clutches were smaller on mouse carcasses (mean mass 21.5 g) than on rats (mean mass 227 g) (Easton 1979), indicating that mouse carcasses were too small to trigger the production of maximum clutches. In *N. orbicollis*, no significant difference existed in the size of clutches laid around carcasses of 15–20 g and 30–35 g mass, respectively (Wilson and Fudge 1984), but reduced clutches were observed on carcasses weighing less than 10 g (Trumbo 1990d). In their assessment of carcass size, *N. vespilloides* appear to be using the volume or surface area of the corpse rather than its mass (J. K. Müller, unpublished data).

Overproduction of eggs and infanticide

The adjustment of brood size at the egg stage is far from perfect. The number of eggs laid per gram of carrion differs vastly between small and large carcasses (Easton 1979; Müller *et al.* 1990b). In *N. vespilloides*, the average clutch laid on 3 g mice contained 17 eggs, equivalent to 5.6 eggs per gram carrion, whereas the average clutch on 30 g mice consisted of 37 eggs, i.e. only 1.2 eggs per gram carrion (Müller *et al.* 1990b). Hatching success does not vary systematically with carcass size, and thus the initial number of larvae competing for a given amount of carrion is higher on small carcasses, where it regularly exceeds the carrying capacity of the corpse (Bartlett 1987). Wilson and Fudge (1984) observed that despite similar clutch sizes, fewer larvae survived on small than on large carcasses, and the larvae produced were of similar size on both kinds of carcasses. Their suggestion that brood size is regulated by the parents has been supported by Bartlett's (1987) and Trumbo's (1990b) observations of systematic infanticide of early larval instars by parental *N. vespilloides* and *N. tomentosus* on small carcasses. Probably all burying beetles lay excess eggs on small carcasses, and cull some of the larvae arriving on the carcass when hatching success is high. On large carcasses that provide sufficient food for an average clutch, culling is reduced or absent (Bartlett 1987). One of the proximate cues used by the adult beetles in determining the number of larvae allowed to survive appears to be a variable related to carcass volume rather than mass (C. Creighton, unpublished data); however, the cues used in choosing the larvae that are culled remain obscure.

What are the benefits to the overproduction of eggs and subsequent infanticide on small carcasses? The availability of large quantities of protein facilitates the production of excess eggs, but does not explain the phenomenon *per se* (Trumbo 1990d). Unpredictable variation in the exact amount of carrion left after oviposition, owing to the varying activity of fungi and bacteria (Wilson and Fudge 1984), occurs in the field, but it cannot explain the production of clutches that are so large that massive culling is required even when the activity of fungi and bacteria is minimized (Trumbo 1990d). Bartlett's (1987) insurance hypothesis, that the production of excess eggs is an insurance against unpredictable mortality at this stage, provides a more compelling explanation for the overproduction of eggs. Even if such mortality is rare, it could still select for an overproduction of eggs if future reproductive opportunities are limited (Trumbo 1990d). The overproduction of eggs can also be interpreted as a mechanism by which females increase their genetic representation in broods of mixed maternity (Müller *et al.* 1990a; Trumbo 1990d). Females may try to swamp the carcass with their own larvae to reduce the amount of carrion utilized by rival larvae. When two females oviposit on the same carcass, each female's genetic representation in the surviving brood will largely be determined by the proportion of all eggs laid on the carcass that are her own. The fact that females lay excess eggs even when no rival females are present does not necessarily falsify this hypothesis. Females may not always be able to detect a rival female near a carcass, or they may decide on the size of their clutch at a time after which competitors may still arrive; therefore, the benefit of producing excess eggs even when a rival is not detected may be greater than the cost.

Responses to brood loss

Occasionally, the mortality of eggs and larvae can be high, such that there are too few larvae for the amount of carrion

Table 10-1. *Offspring number and mass, and relative mass of broods produced by* N. vespilloides *on carcasses of different size*

All values are given as mean ± standard error of mean. Inseminated females or male–female pairs were each provided with a mouse carcass and allowed to reproduce without disturbance in breeding containers filled with moist peat as described in Müller *et al.* (1990a). Newly emerged beetles were collected and weighed to the nearest milligram within twelve hours upon leaving the breeding containers. Carcass mass treatments differed significantly in the number of offspring ($H = 33.08$, $p < 0.0001$, Kruskal–Wallis test), the mass of individual offspring ($H = 38.09$, $p < 0.0001$), and the relative mass of broods ($H = 22.14$, $p < 0.0001$).

Carcass mass	7.5 g	15 g	25 g	35 g
n	16	16	19	9
Number of offspring	19.2 ± 1.4	30.4 ± 1.5	37.1 ± 1.8	38.1 ± 2.9
Mean mass of individual offspring (mg)	120.2 ± 6.1	144.6 ± 5.01	181.1 ± 4.6	189.4 ± 6.3
Relative brood mass (% of initial carcass mass)	29.2 ± 1.2	28.6 ± 0.7	26.4 ± 0.91	20.2 ± 1.2

that the adults have secured. Female burying beetles remedy this situation by producing replacement clutches (Müller 1987), as they also do when male intruders kill their initial broods (Trumbo 1990c). *N. vespilloides* females can produce up to five successive clutches on the same carcass when larvae are prevented from gaining access to it; the production of new clutches ceases only when the carcass is no longer suitable for the production of offspring (Müller 1987).

When only a single larva is present on a large carcass, it is sometimes found dead inside the carcass even after it has already reached a considerable size (Wussler and Müller 1994). It appears that the female closes up the opening to the cavity in the carrion ball within which the larva develops, resulting in the suffocation of the larva. Although a single larva reduces the quality and quantity of carrion for any offspring arising from a replacement clutch, such infanticide appears to be an accidental consequence of the female's attempts to maintain the carcass. Not all females show this behavior, and the factors governing individual differences are obscure.

Larval numbers vs. size: optimality and limits

The optimal number of larvae that an adult beetle should rear on a given amount of carrion is not easily determined. The more offspring that are raised, the higher the probability that some of them will reproduce; however, there is a trade-off between offspring number and size. Very small burying beetles are disadvantaged in aggressive interactions with conspecifics, and possibly in other contexts as well (see, for example, Bartlett and Ashworth 1988). Therefore, reproducing adults should not merely maximize larval numbers, but rather opt for a compromise between the number and size of offspring that yields the highest number of second-generation offspring. Such an optimal compromise should be valid irrespective of carcass size, and the average size of larvae produced should be the same on carcasses of different sizes. It is clear that optimal body size will depend on the degree of intraspecific competition that exists at the time the offspring become reproductive, which may be unpredictable. In this case, fluctuating selection could lead to the maintenance of different parental tactics with respect to offspring numbers and size. The differences in the number and size of offspring reared on small carcasses may be due to such variation in parental tactics.

Contrary to the prediction that larval sizes should be constant across different carcass weights, size and mass of individual offspring in undisturbed breeding attempts of *N. vespilloides* were greater on large than on small carcasses (Table 10-1). In addition, large carcasses are not used as efficiently as are small ones. The likely cause of these effects is physiological limits on the production of eggs within a short period of time (Müller *et al.* 1990b; Trumbo 1992). Females do rear more and smaller larvae when the initial number of larvae is experimentally increased. In experimentally enlarged laboratory broods of *N. vespilloides*, a median of 60 out of 75 larvae survived to dispersal on 25 g carcasses (n = 15, interquartile range = 14.5), compared with 37 adults produced in undisturbed broods (Table 10-1). Mortality between larval dispersal and adult emergence is very low in *N. vespilloides* and hence cannot explain why females in undisturbed broods produced fewer offspring.

JOINT BREEDING ON LARGE CARCASSES

True monopolization of the carcass by a male and female, as reported by early researchers (Fabre 1899; Pukowski 1933; Milne and Milne 1944), appears to be rare when conspecifics of the same sex are present. Not only can subordinate individuals reproduce through the use of alternative tactics, but on large carcasses two females (Eggert and Müller 1992; Trumbo 1992), or two males (Trumbo 1992), may also tolerate each other and feed the brood simultaneously.

Tolerance among sexually mature consexuals occurs regularly in non-reproductive contexts, i.e. on carcasses that are not suitable for offspring production and serve as an adult food source only (Trumbo and Wilson 1993; A.-K. Eggert, unpublished data). On carcasses used for reproduction, tolerance is contingent on carcass size. In *N. vespilloides*, there is a transition from intense competition on small to medium-sized carcasses (15 g, Müller *et al.* 1990a), to greater tolerance on relatively large carcasses (35 g, Eggert and Müller 1992). On small and medium-sized carcasses, the smaller female is typically subordinate, leaves first, and has far fewer offspring than the dominant one (Müller *et al.* 1990a), but with increasing carcass size, the smaller female remains on the carcass for longer periods of time, and produces a larger proportion of the offspring. On 35 g carcasses, the difference in residence time and reproductive output between smaller and larger female vanishes, such that there is no longer a correlation between a female's relative body size and her genetic contribution to the brood (Eggert and Müller 1992). Injury rates on large carcasses (35 g) are unaffected by the females' difference in body size (A.-K. Eggert and J. K. Müller, unpublished data), but are generally lower than on smaller (15 g) carcasses, indicating a lower frequency or intensity of fights (Eggert and Müller 1992). Two females have even been observed to feed the larvae side by side (Eggert and Müller 1992). In *N. defodiens*, *N. orbicollis* and *N. tomentosus*, joint brood care by several females is also tied to carcasses that are large for the respective species (Trumbo 1992;

Trumbo and Wilson 1993). In *N. orbicollis*, and likely in the other species as well, cohabiting females exhibit alloparental care, i.e. they feed their own and the other female's offspring indiscriminately (Trumbo and Wilson 1993).

Intraspecific brood parasitism and joint breeding describe the extreme ends of a behavioral continuum; the relative frequency of these extremes, and of intermediate situations, is determined by resource size (Müller *et al.* 1990a; Eggert and Müller 1992; Trumbo 1992; Trumbo and Wilson 1993). Even on carcasses of small and intermediate size, a small proportion of subordinate females remain long enough to provide parental care to broods of mixed maternity (Müller *et al.* 1990a; Eggert and Müller 1992; Trumbo 1992; Scott and Williams 1993; Trumbo and Wilson 1993). However, on small carcasses most subordinate females provide minimal or no care for their own young. Reproduction by the subordinate reduces the dominant female's reproductive success on small and on large carcasses (Eggert and Müller 1992; Trumbo and Fiore 1994). It is unlikely that jointly breeding individuals are close genetic relatives, or that kin-selected altruism can explain the tolerant behavior of females on large carcasses, because siblings must disperse from the parental nest if they are to survive, and at the end of the two- to three-week maturation period, siblings are likely to be separated by vast distances. At this point, they have yet to cover long distances before they eventually locate a carcass suitable for reproduction.

To summarize, mutual care of offspring occurs in pairs of female burying beetles, and the evidence available to date suggests that dominants do not derive an overall benefit from an increase in the number of individuals reproducing on the carcass. Both females contribute more or less equally to the shared brood, and regurgitate to larvae indiscriminately.

In traditional classifications of animal societies, jointly breeding *Nicrophorus* would be considered 'quasisocial' (Michener 1969), 'communal' (Brown 1987) or 'cooperative' (Emlen 1991) breeders. Burying beetles exhibit 'joint nesting – plural breeding' (Brown 1987) or 'plurimatry' (Choe 1995). However, they are hard to place in some recent classification schemes attempting to more clearly define eusociality (Crespi and Yanega 1995; Sherman *et al.* 1995; Crespi and Choe, this volume). In Sherman *et al.*'s (1995) 'eusociality continuum', burying beetles do not have a place because the individuals breeding together are not kin. In Crespi and Yanega's (1995) classification scheme, they fall between the 'communal' ('nest-sharing but no alloparental care') and the 'cooperatively-breeding' category ('behaviorally distinct groups, with individuals of one group specializing in reproduction and individuals of the other group specializing in helping'). In Crespi and Choe's introduction to this volume, the latter condition is labelled 'alloparental care', which places burying beetles in the 'communal' category; however, it appears counterintuitive to claim that they do not exhibit alloparental care in view of the fact that jointly nesting females show no indication of discrimination against unrelated young when they feed a mixed brood (see Trumbo and Wilson 1993). In effect, we think the various classification schemes do not contribute to our understanding of the evolution of either joint breeding or eusociality (see also Gadagkar 1994).

More research is needed to better assess costs and benefits of joint breeding to the dominant and subordinate invidual on a large carcass. In males, little is known about the extent to which joint care exists; in females, two separate aspects of joint breeding call for evolutionary explanations: (1) the greater tolerance of females, especially the larger one; and (2) the contribution of both females to parental care beyond the period during which their own larvae appear on the carcass.

Tolerance of consexuals

When two or more females locate the same large carcass, it would not benefit either of them to abandon it and search for a new one because suitable corpses are rare. However, it is not clear why the larger female attacks smaller conspecific females less vigorously on large carcasses than on small and medium-sized ones. As detailed below, greater tolerance on large carcass may be due to: (1) reproductive benefits that the larger female gains from the smaller female's presence (Eggert and Müller 1992; Trumbo 1992; Scott 1994); or (2) decreased benefits, or increased costs, of aggressive behaviors directed at a smaller rival on large carcasses (Eggert and Müller 1992; Trumbo 1992, 1995; Trumbo and Wilson 1993).

First, several authors have considered possible mutualistic effects of joint reproduction on large carcasses arising from the improved effectiveness of care, such as improved regurgitation, carcass maintenance, brood defense in larger groups, or the improved extermination of fly eggs and maggots present on the carcass (Eggert and Müller 1992; Trumbo 1992; Scott 1994). However, when the smaller and the larger female contribute equally to the mixed brood (Eggert and

Müller 1992), groups of two females should produce at least twice as many offspring as single females for the relationship to be called mutualistic. This prediction has not been met in the laboratory (Eggert and Müller 1992) or in the field (Trumbo and Wilson 1993; Scott 1994; Trumbo and Fiore 1994). On the contrary, in *N. vespilloides* it appears that the larger female loses a greater number of offspring through another female's presence on large (35 g) than on medium-sized (15 g) carcasses (Eggert and Müller 1992), because there is no longer a reproductive skew in her favor. The larger female gains some search time when she is the first to leave the brood, but the expected benefits of this search are low. Joint care might reduce the physiological cost of care (Eggert and Müller 1992), but this cost appears trivial (Bartlett 1988; Reinking 1988). None of these presumed benefits provide a convincing explanation of the difference in aggression on small and large carcasses, rendering mutualism an unlikely cause of females' greater toleration of rivals on large carcasses.

Secondly, decreased benefits or increased costs of aggression on large carcass may result from four types of selective pressure, described here as four hypotheses that are not mutually exclusive. The superabundant-resource hypothesis (Trumbo 1992) is based on the observation that single females cannot fully utilize large carcasses, because the number of eggs they can lay in any one reproductive bout is limited (Müller *et al.* 1990b; Trumbo 1992). As carcass size increases, the increase in available carrion may cause a decrease in the cost of tolerating another female, as postulated by Scott (1994) and Trumbo and Fiore (1994). However, there is a reproductive skew in favor of the larger female on small and medium-sized but not on large carcasses, such that in *N. vespilloides*, the cost of another female's presence to the larger female may be even greater on large (35 g) than on medium-sized (15 g) carcasses (Eggert and Müller 1992). Regardless of this difference in reproductive skew, there is still a cost of tolerance even on very large carcasses (Trumbo and Fiore 1994).

The nesting-failure hypothesis states that burying beetles delay fights on carcasses until they have obtained sufficient information that an attempt to reproduce is likely to succeed (Trumbo and Wilson 1993; Trumbo 1995). On large carcasses, high nesting failure rates greatly decrease the reproductive output expected from a breeding attempt, such that the benefit of fighting may not exceed the cost (Trumbo 1995). By the time larvae are established on the carcass, there is no longer a conflict of interest between females (Eggert and Müller 1992) and thus no benefits can be gained from aggressive interactions (Trumbo and Wilson 1993).

The resource-defensibility hypothesis also attributes greater female tolerance on large carcasses to reduced benefits of aggression (Eggert and Müller 1992). In contrast to the previous hypothesis, however, benefits are decreased owing to the ineffectiveness of the dominant female's aggressive behavior in excluding her smaller rival from a large carcass. The increased surface area of a large carcass probably affords the smaller female more time to feed undetected by the larger female than does a small carcass. Feeding on the carcass promotes ovarian development (J. K. Müller and A.-K. Eggert, unpublished data), and likely increases the size of the clutch that a female can produce. Once the smaller female's intake of carrion enables her to produce a full clutch, the dominant female gains nothing further by attacking her. This hypothesis might explain the difference in reproductive bias and in aggressiveness on smaller and larger carcasses.

Finally, groups of tolerant beetles may bury a carcass faster than mutually aggressive ones (the burial-speed hypothesis), which does not necessarily imply that two beetles bury the carcass faster than one. The risk of losing the entire carcass to larger congeners or dipteran competitors is high for larger carcasses (Trumbo 1992, 1995) and is likely to be increased by time lost in fights, because time spent fighting is lost for carcass burial. Thus, the beetles may experience higher reproductive success when they tolerate the presence of consexual adults on a large carcass, despite the cost to their own reproductive success.

Tests of these hypothesis have yet to be carried out. The superabundant-resource hypothesis requires field data on the individual reproductive success of females reproducing alone and in the presence of a smaller consexual, respectively. Comparisons of different geographical populations of a species would clarify whether tolerant behavior is less common in areas where nesting attempts are more likely to be successful, as predicted by the nesting-failure hypothesis. If the burial-speed hypothesis is correct, pairs of females should tolerate each other from the very beginning of a breeding attempt on a large carcass; in addition, tolerant pairs should bury a carcass faster than pairs exhibiting aggressive behaviors. The resource-defensibility hypothesis predicts that an aggressive phase during the initial stages of a reproductive attempt is followed by a tolerant phase; even during the initial phase, subordinates are predicted to spend longer periods of time on the carcass than subordinates on small carcasses do.

Joint parental care

Although the scarcity of reproductive opportunities provides a ready explanation for the fact that all the females attracted to a large carcass remain until they have oviposited, the observation that all of them stay well beyond this time is not as easily explained. On small carcasses, the smaller female leaves early and provides little if any care to the larvae (Müller et al. 1990a).

Residence times for all-female pairs on large carcasses are similar to those of male–female pairs (Eggert and Müller 1992), suggesting that additional females help to decrease the likelihood of a takeover by conspecific or congeneric competitors, as does the male in heterosexual pairs (Scott 1990; Trumbo 1990a, 1991; Robertson 1993). The presence of a male in this situation would be expected to reduce the tendency of one of the two females, or the male, to provide lengthy care, a prediction that has not been tested to date. The second female's contribution to regurgitation is not required: a single female can feed at least 60 larvae adequately (N. vespilloides: J. K. Müller, unpublished data). The second female may benefit from remaining on the carcass for reasons that are unrelated to parental care; for example through continued access to a virtually unlimited food supply.

Scott and Williams (1993) and Scott (1994) have suggested that the care provided by females in the presence of another female may be 'unselected', i.e. constrained physiologically, and therefore not adaptive. According to their hypothesis, the process of burying causes endocrine changes that in turn produce brood-care behavior; owing to such hormonal constraints, a beetle is unable to avoid care-giving once it has participated in the burial of a carcass.

A female's readiness to provide care is likely influenced by oviposition (Müller and Eggert 1990) and may be mediated by hormones (Trumbo et al. 1995). It is entirely possible that subordinate females base their 'decision' to stay or leave on the number of eggs they have laid or an associated hormonal level. However, neither mechanism provides a suitable explanation beyond the proximate level of analysis. Non-adaptive extended residence times would have been selected against because of the concomitant loss of time to search for new reproductive opportunities, and thus modifiers uncoupling hormones and residence times should have been selectively favored. Moreover, the parental behaviors of beetles that have oviposited or participated in burial are not as stereotypic as required for the constraints argument to hold. On carcasses of intermediate size, the smaller of two females remains from one to ten days in *N. tomentosus* (Trumbo and Wilson 1993); in *N. vespilloides*, a smaller female that has oviposited remains from one to eight days, covering the entire range from brood parasitism to joint care (Müller *et al.* 1990a). Even if both females provide care, they do not both remain on the carcass until the larvae disperse. Instead, one of them leaves at about the same time a care-giving male would leave, or even earlier (Eggert and Müller 1992; Scott and Williams 1993). Males are equally flexible in their parental behaviors: if the female leaves early or dies, they extend their residence on the carcass (Reinking 1988; Trumbo 1991) and increase the frequency of feeding behaviors (Fetherston *et al.* 1994). The supposed physiological constraint is clearly subject to much situation-dependent modification, which could only have been shaped by natural selection.

Invoking physiological constraints as an explanation of the participation of two females in parental care appears premature at best. To date, not a single test of adaptive hypotheses of the phenomenon has been carried out. Some experiments have compared the reproductive success of females reproducing in the absence of consexuals with that of individuals within multiple-female groups (i.e., Eggert and Müller 1992; Scott 1994; Trumbo and Fiore 1994; Trumbo and Eggert 1994). However, a critical test of the adaptiveness of multifemale parental care would require a comparison of polygynous groups of two or three females, all of which contributed to care for the joint brood, with groups in which second and third females were prevented from providing care by removing them after oviposition. If the additional females' participation in care resulted in an increase in the number of surviving offspring, then the behavior must be interpreted as adaptive, whether or not the mechanism governing the behavior is a hormonal one (see Reeve and Sherman (1993) for a discussion of 'unselected' hypotheses for the evolution of alloparental care in birds).

Jointly breeding females provide indiscriminate alloparental care, but there is no reason to assume that their behavior is not aimed at the maximization of their own reproductive success. Asynchronous oviposition by two females reproducing on the same carcass might result in conflicts at the time larvae hatch. Late oviposition might enable a female to selectively kill some unrelated larvae reaching the carcass early, but it would also put her own larvae at a competitive disadvantage relative to the other female's offspring. More detailed studies are urgently needed to better understand conflict and cooperation in multiple-female groups.

SEXUAL CONFLICT

In burying beetles, the reproductive success of both sexes is limited by the availability of carrion, and male and female contributions to pre- and posthatching care of their offspring are similar in many respects. However, the interests of the sexes can differ radically, even when they are reproducing together on a carcass. For example, females drive their mates away from the carcass (Pukowski 1933) if it is very small (Bartlett 1988).

The most impressive and obvious sexual conflict occurs on large carcasses. On large carcasses, two or more females produce a brood larger than either alone could produce (Eggert and Müller 1992; Trumbo and Eggert 1994; Trumbo and Fiore 1994). When only one male is present, he fathers most of the offspring reared on a carcass. Thus, male reproductive success is higher when there are two or more females on the carcass (polygyny) than when only one female is present (monogamy). A female, in turn, will always lose offspring when another female is present, even on large carcasses, because the two females' larvae compete for the available carrion (Eggert and Müller 1992; Trumbo and Fiore 1994). A conflict of interest thus occurs when a male and a female are the only individuals that have located a large carcass. The male would gain offspring, but the female would lose offspring, if another female were attracted to the carcass. Both sexes behave according to their own interests in this situation. Males emit pheromones in the presence of their mates, and females attempt to interfere with this behavior: they push the male, crawl under him, mount him and sometimes even attempt to bite the tip of the male's abdomen with their mandibles (Trumbo and Eggert 1994). The interference behavior shown by the female, in turn, reduces the time the male can emit the pheromone (Eggert and Sakaluk 1995). The female's behavior can cause the male to interrupt pheromone emission, or cease it altogether. On small carcasses, none of these behaviors are observed (Trumbo and Eggert 1994; Eggert and Sakaluk 1995), presumably because male and female interests concur. Small carcasses can only support one female's brood, such that males gain nothing by attracting additional females.

Another conflict between the sexes seems to occur between males that emit pheromone on carcasses and the females they attract. Males attempt to mate with these females as soon as possible, but the females are reluctant to do so at first. They run away from males that try to mount, bend the tip of their abdomen away from the male's aedeagus, or crawl under rocks or sticks in an attempt to dislodge mounted males (Eggert and Müller 1989b). The female's reluctance gradually recedes, and the duration of copulations increases with time since arrival (Eggert 1990). Pheromone-emitting males that have not located a carcass are rarely rejected by females (Eggert and Müller 1989b), and the same is true for satellite males (Dressel 1987). We can only speculate about possible benefits to the females' coy behavior in the situation described above. On a large carcass, the male should attempt to attract additional females; the female's reluctance may serve to delay the onset of this behavior by keeping the male occupied with mating attempts. If this is true, then females should allow copulations if they determine that a carcass is small, or when it is so late in the day that no other females are likely to be attracted.

DISCUSSION

Burying beetles utilize a resource with many unusual features, and it appears that their entire biology has been shaped by this resource, including some physiological adaptations. Adult *Nicrophorus* are good, persistent fliers that can cover large distances in a short period of time, and they possess highly sensitive chemical receptors that aid them in locating relatively fresh carcasses (Boeckh 1962; Ernst 1969; Waldow 1973). The completion of ovarian development is postponed until the female detects a suitable carcass, thereby freeing up energy for consumption in sustained flight and reducing wing loading during flights. We have argued that most aspects of burying-beetle behavior have also been influenced by the qualities of the larval food resource. Various phenomena indicate that the beetles attempt to maximize individual reproductive success in each reproductive attempt, even when reproducing in multiple-female groups. Burial and defense of the carcass, feeding of young, and joint defense of the brood yield important fitness benefits primarily because of the extreme difficulty associated with securing the resources needed for reproduction. Defense of the carcass after burial and during larval development also yields benefits because the carcass still remains attractive to competitors. Nonetheless, it is important to keep in mind that it was not the resource alone that affected the evolution of care; after all, there are numerous species that utilize carrion without

exhibiting any parental behavior. Morphological and behavioral preadaptations were also important requirements for the evolution of carcass burial, maintenance, and defense, as well as feeding of offspring.

Some dung beetles (Scarabaeidae) resemble *Nicrophorus* in the degree of specialization of parental-care behaviors, possibly because of similar qualities of the larval food resources the two groups utilize (Tallamy and Wood 1986), as well as similar preadaptations. Carrion and vertebrate dung are valuable and ephemeral resources, and some scarabaeid beetles also conceal dung from competitors by interring it in the soil (Lengerken 1952; Halffter 1977, this volume; Cambefort and Hanski 1991). In the tropical horned beetle *Coprophanaeus ensifer*, which buries pieces of carrion severed from large corpses, inter- and intrasexual fights are common, but nothing is known about parental care (Otronen 1988b). African dung beetles of the genus *Kheper*, which reproduce on elephant dung, also resemble burying beetles in several respects. In *K. platynotus*, pairs or single females roll balls of elephant dung and bury them to provision their larvae; intrasexual fights are frequent during burial (Sato and Hiramatsu 1993). Male *K. nigroaeneus* and *K. aegyptiorum* release sex pheromones near reproductive resources (Tribe 1975; Sato and Imamori 1986, 1988; Edwards and Aschenborn 1988); *K. nigroaeneus* males also emit the sex pheromone in the absence of a dung pat or brood ball (Edwards and Aschenborn 1988). Unlike *Nicrophorus* males, *Kheper* males do not contribute to parental care beyond the construction of brood balls (Sato and Hiramatsu 1993).

In a few dung-beetle species, parental care is comparable to that of burying beetles in complexity and exceeds it in duration (e.g. *Copris hispanus*, Lengerken 1952). The necrophagous tropical *Canthon cyanellus* exhibits biparental care throughout larval development up until adult emergence (Halffter 1977; Favila 1993). Parental feeding of larvae, a regular component of burying-beetle parental care, is not found in dung beetles, perhaps because of the different nature of dung and carrion. The brood balls of most dung beetles consist primarily of undigested plant particles, which require microbial symbionts for their breakdown. Such resources are most efficiently utilized by repeated passage through the digestive tract (Cambefort 1991), such that parental digestive enzymes can do little to speed up larval development. In addition, the evolutionary history of certain patterns of oviposition may play a role. Dung beetles typically lay their eggs into balls of larval food, and larvae develop inside these balls; this pattern is maintained in the necrophagous *C. cyanellus*, in which parental care involves control of fungal growth on the surface of the carrion ball but no direct feeding of young.

The level of sociality reached in *Nicrophorus* has also been limited by resource features: (1) carcasses are ephemeral, which restricts the duration of parental associations to a few weeks and precludes the development of associations with overlapping generations; (2) they are not large enough to support large colonies, because interring imposes severe size constraints; and (3) they are non-renewable, rendering each additional adult a competitor for the limited amount of carrion available. As pointed out before, the distribution of the resource also requires the dispersal of parents and adult offspring from the parental nest, which leads to a rapid separation of related individuals. For these reasons, *Nicrophorus* social behavior is not likely to be the precursor of a eusocial system, although they exhibit parental care and joint breeding, which have been interpreted as transitory states preceding eusocial behavior (Michener 1969; Wilson 1975). Ambrosia beetles in the curculionid subfamily Platypodinae have apparently reached higher levels of sociality (Kent and Simpson 1992; Kirkendall *et al.*, this volume). The resource used by these ambrosia beetles is hard to obtain because of difficulties associated with establishing tunnel systems in trees and founding a colony (Kent and Simpson 1992); it is also durable and large, and helping by additional individuals increases the amount of food available to the colony. Passalidae live in rotting logs, and for various species, colonies have been described that contain sexually mature beetles and teneral adults along with larvae, pupae and eggs (see, for example, Heymons 1929; Schuster and Schuster 1985, this volume; Kon and Johki 1992); apparently, larvae are dependent on the presence of adult beetles for their nourishment, and immature adults participate in the construction and repair of pupal cells in the colony (Schuster and Schuster 1985; Valenzuela-Gonzalez 1993).

Burying beetles have experienced a recent surge in popularity among students of insect behavior and ecology. After the publication of Pukowski's (1933) impressive work, it must have appeared to many researchers that most or all of the puzzles around burying beetles were solved; 60 years later, most students of *Nicrophorus* biology would likely agree that we are still a long way from a complete understanding of their biology, and many unresolved problems in burying-beetle biology remain to be identified, pondered, and tackled in the future. Most of our knowledge of burying beetle natural history is based on a handful of species, although about 70 species have been described worldwide (Peck and Kaulbars 1987). Species differences with respect to reproductive biology appear to be mostly of a quantitative nature, and are often related to differences in body size or phenology, but undetected qualitative differences may also exist. For example, *N. pustulatus* is not found on small carcasses in the field, although it can reproduce on them like other species (Robertson 1992), and female *N. pustulatus* produce vast numbers of eggs on very large carcasses, up to 200 compared with the 20–40 typical of other species (Trumbo 1992). Studies on additional species are required to ascertain whether parental care is basically similar in all species of the genus; in addition, more work on the sister genus *Ptomascopus* might contribute to a better understanding of the selective forces affecting the evolution of parental care. In addition, costs and benefits associated with joint reproduction by several (presumably unrelated) males or females are not well understood; investigations on this subject are highly desirable. Reproductive associations between non-relatives appear rare and could prove to be of special interest to students of cooperative behavior. Last, but not least, evolutionary interpretations of studies on the ecology of individual species would be greatly facilitated by an established phylogeny of the family Silphidae, or even of the subfamily Nicrophorinae or the genus *Nicrophorus*, none of which has been worked out to date.

Unlike the genus *Nicrophorus*, many genera in the Scarabaeidae and the Passalidae that exhibit elaborate parental care appear to have been largely neglected by behavioral ecologists. The literature on parental care and social behavior in both taxa is full of contradictions, ambiguous or unfounded statements and interpretations that are in clear contradiction to current evolutionary paradigms. It appears to us that species in either group would also be highly rewarding objects for studies in behavioral ecology.

ACKNOWLEDGMENTS

Thanks are due to all the colleagues who provided us with copies of their work, and those who helped generate and develop the ideas presented in this chapter in discussions of our favorite animal. We thank Stewart Peck, Ian Robertson, Scott Sakaluk, Michelle Scott, Doug Tallamy and Steve Trumbo for many helpful comments on earlier versions or parts of this chapter, and Scott Sakaluk for linguistic corrections. Many thanks to Masahiro Kon for providing information on *Ptomascopus morio* and shipping beetles to Freiburg, to Curtis Creighton for allowing us to cite his unpublished results on *N. orbicollis*, and to Steve Trumbo for providing copies of papers that were still in the press. During preparation of this chapter, A.-K. Eggert was supported by a postdoctoral fellowship from the DFG, while Prof. K. Peschke provided working facilities at the Institut für Zoologie in Freiburg.

LITERATURE CITED

Anderson, R. S. and S. B. Peck. 1984. Bionomics of Nearctic species of *Aclypea* Reitter: phytophagous 'carrion' beetles (Coleoptera: Silphidae). *Pan-Pac. Entomol.* **60**: 248–255.

Arak, A. 1984. Sneaky breeders. In *Producers and Scroungers: Strategies of Exploitation and Parasitism.* C. J. Barnard, ed., pp. 154–194. London: Croom Helm.

Bartlett, J. 1987. Filial cannibalism in burying beetles. *Behav. Ecol. Sociobiol.* **21**: 179–183.

–. 1988. Male mating success and paternal care in *Nicrophorus vespilloides* (Coleoptera: Silphidae). *Behav. Ecol. Sociobiol.* **23**: 297–303.

Bartlett, J. and C. M. Ashworth. 1988. Brood size and fitness in *Nicrophorus vespilloides* (Coleoptera: Silphidae). *Behav. Ecol. Sociobiol.* **22**: 429–434.

Boeckh, J. 1962. Elektrophysiologische Untersuchungen an einzelnen Geruchsrezeptoren auf den Antennen des Totengräbers (*Necrophorus*, Coleoptera). *Z. Vergl. Physiol.* **46**: 212–248.

Brown, J. L. 1987. *Helping and Communal Breeding in Birds.* Princeton, New Jersey: Princeton University Press.

Cambefort, Y. 1991. From saprophagy to coprophagy. In *Dung Beetle Ecology.* I. Hanski and Y. Cambefort, eds., pp. 22–35. Princeton, New Jersey: Princeton University Press.

Cambefort, Y. and I. Hanski. 1991. Dung beetle population biology. In *Dung Beetle Ecology.* I. Hanski and Y. Cambefort, eds., pp. 36–50. Princeton, New Jersey: Princeton University Press.

Choe, J. C. 1995. Plurimatry – new terminology for multiple reproductives. *J. Insect Behav.* **8**: 133–137.

Crespi, B. J. and D. Yanega. 1995. The definition of eusociality. *Behav. Ecol.* **6**: 109–115.

Dressel, J. 1987. The influence of body size and presence of females on intraspecific contests of males in the carrion beetle *Necrophorus vespilloides* (Coleoptera, Silphidae). *Verh. Dt. Zool. Ges.* **80**: 307.

Dressel, J. and J. K. Müller. 1988. Ways of increasing the fitness of small and contest-losing individuals in burying beetles (Silphidae, Coleoptera). *Verh. Dt. Zool. Ges.* **81**: 342.

Easton, C. 1979. The ecology of burying beetles (*Necrophorus*: Coleoptera, Silphidae). Ph. D. dissertation, University of Glasgow.

Edwards, P. B. and H. H. Aschenborn. 1988. Male reproductive behaviour of the African ball-rolling dung beetle, *Kheper nigroaeneus* (Coleoptera: Scarabaeidae). *Coleopt. Bull.* **42**: 17–27.

Eggert, A.-K. 1990. Chemische Kommunikation beim Totengräber *Necrophorus vespilloides* Herbst (Coleoptera: Silphidae): Pheromonabgabe als alternative Fortpflanzungstaktik der Männchen. Ph. D. dissertation, Universität Bielefeld.

–. 1992. Alternative male mate-finding tactics in burying beetles. *Behav. Ecol.* **3**: 243–256.

Eggert, A.-K. and J. K. Müller. 1989a. Pheromone-mediated attraction in burying beetles. *Ecol. Entomol.* **14**: 235–237.

–. 1989b. Mating success of pheromone-emitting *Necrophorus* males: do attracted females discriminate against resource owners? *Behaviour* **110**: 248–257.

–. 1992. Joint breeding in female burying beetles. *Behav. Ecol. Sociobiol.* **31**: 237–242.

Eggert, A.-K. and S. K. Sakaluk. 1995. Female-coerced monogamy in burying beetles. *Behav. Ecol. Sociobiol.* **37**: 147–153.

Eickwort, G. C. 1979. Presocial insects. In *Social Insects*, vol. 2. H. R. Herman, ed., pp. 199–280. New York: Academic Press.

Emlen, S. T. 1991. Evolution of cooperative breeding in birds and mammals. In *Behavioural Ecology: An Evolutionary Approach*, 3rd edn. J. R. Krebs and N. B. Davies, eds., pp. 301–337. Oxford: Blackwell.

Enquist, M. and O. Leimar. 1990. The evolution of fatal fighting. *Anim. Behav.* **39**: 1–9.

Ernst, K.-D. 1969. Die Feinstruktur der Riechsensillen auf der Antenne des Aaskäfers *Necrophorus* (Coleoptera). *Z. Zellforsch.* **94**: 72–102.

Fabre, J. H. 1899. *Bilder aus der Insektenwelt* (authorized translation from *Souvenirs Entomologiques*). Stuttgart, Germany: Franck'h'sche Verlagshandlung.

Favila, M. A. 1993. Some ethological factors affecting the lifestyle of *Canthon cyanellus* (Coleoptera Scarabaeidae): An experimental approach. *Ethol. Ecol. Evol.* **5**: 319–328.

Fetherston, I. A., M. P. Scott and J. F. A. Traniello. 1990. Parental care in burying beetles: the organization of male and female brood-care behavior. *Ethology* **85**: 177–190.

–. 1994. Behavioural compensation for mate loss in the burying beetle *Nicrophorus orbicollis*. *Anim. Behav.* **47**: 777–785.

Gadagkar, R. 1994. Why the definition of eusociality is not helpful to understand its evolution and what should we do about it. *Oikos* **70**: 485–488.

Gleditsch. 1752. *Acta reg. soc.* (Berolin) (cited in Pukowski 1933).

Halffter, G. 1977. Evolution of nidification in the Scarabaeinae (Coleoptera, Scarabaeidae). *Quaest. Entomol.* **13**: 231–253.

Halffter, G., S. Anduaga and C. Huerta. 1983. Nidification des *Nicrophorus* (Col. Silphidae). *Bull. Soc. Entomol. Fr.* **88**: 648–666.

Hennig, W. 1950. Entomologische Betrachtungen an kleinen Wirbeltierleichen. *Z. Hyg. Zool.* **38**: 33–88.

Herbst, J. F. W. 1792. *Natursystem aller bekannten in- und ausländischen Insekten, als eine Fortsetzung der von Buffonschen Naturgeschichte.* Berlin: Joachim Pauli.

Herman, L. H. Jr. 1964. Nomenclatural consideration of *Nicrophorus* (Coleoptera: Silphidae). *Coleopt. Bull.* **18**: 5–6.

Heymons, R. 1929. Ueber die Biologie der Passaluskäfer. *Z. Morphol. Oekol. Tiere* **16**: 74–100.

Heymons, R., H. von Lengerken and M. Bayer. 1927. Studien über die Lebenserscheinungen der Silphini (Coleopt.). II. *Phosphuga atrata* L. *Z. Morphol. Oekol. Tiere* **9**: 271–312.

–. 1928. Studien über die Lebenserscheinungen der Silphini (Coleopt.). III. *Xylodrepa quadripunctata* L. *Z. Morphol. Oekol. Tiere* **10**: 330–352.

–. 1929. Studien über die Lebenserscheinungen der Silphini (Coleopt.). IV. *Blithophaga opaca* L. (Glattstreifiger Rübenaaskäfer). *Z. Morphol. Oekol. Tiere* **14**: 234–260.

Kent, D. S. and J. A. Simpson. 1992. Eusociality in the beetle *Austroplatypus incompertus* (Coleoptera: Curculionidae). *Naturwissenschaften* **79**: 86–87.

Kon, M. and Y. Johki. 1992. Passalid beetles (Coleoptera, Passalidae) collected from Sabah, Borneo, with special reference to their colony composition and habitats. *Elytra, Tokyo* **20**: 207–216.

Lacordaire, M. T. 1834–38. *Introduction à l'Entomologie*. Paris (cited in Pukowski 1933).

Lengerken, H. von. 1952. *Der Mondhornkäfer und seine Verwandten*. Leipzig: Akademische Verlagsgesellschaft Geest & Portig K.-G.

Lundt, H. 1964. Oekologische Untersuchungen über die tierische Besiedlung von Aas im Boden. *Pedobiologia* **4**: 158–180.

Michener, C. D. 1969. Comparative social behavior of bees. *Annu. Rev. Entomol.* **14**: 299–342.

Milne, L. J. and M. J. Milne. 1944. Notes on the behavior of burying beetles (*Nicrophorus* spp.). *J. N. Y. Entomol. Soc.* **52**: 311–327.

–. 1976. The social behavior of burying beetles. *Scient. Am.* **8**: 84–89.

Mosebach, E. 1936. Aus dem Leben des Totengräbers (*Necrophorus*). *Natur und Volk* **66**: 222–231.

Müller, J. K. 1987. Replacement of a lost clutch: a strategy for optimal resource utilization in *Necrophorus vespilloides* (Coleoptera: Silphidae). *Ethology* **76**: 74–80.

Müller, J. K. and A.-K. Eggert. 1987. Effects of carrion-independent pheromone emission by male burying beetles (Silphidae: *Necrophorus*). *Ethology* **76**: 297–304.

–. 1989. Paternity assurance by "helpful" males: adaptations to sperm competition in burying beetles. *Behav. Ecol. Sociobiol.* **24**: 245–249.

–. 1990. Time-dependent shifts between infanticidal and parental behavior in female burying beetles: a mechanism of indirect mother-offspring recognition. *Behav. Ecol. Sociobiol.* **27**: 11–16.

Müller, J. K., A.-K. Eggert and J. Dressel. 1990a. Intraspecific brood parasitism in the burying beetle, *Necrophorus vespilloides* (Coleoptera: Silphidae). *Anim. Behav.* **40**: 491–499.

Müller, J. K., A.-K. Eggert and E. Furlkröger. 1990b. Clutch size regulation in the burying beetle *Necrophorus vespilloides* Herbst (Coleoptera: Silphidae). *J. Insect Behav.* **3**: 265–270.

Müller, J. K., A.-K. Eggert and H. H. Schwarz. 1991. Reproductive success of male burying beetles competing for a carcass (*Necrophorus vespilloides*: Coleoptera, Silphidae). *Verh. Dt. Zool. Ges.* **84**: 322.

Nicholson, A. S. 1954. An outline of the dynamics of animal populations. *Austr. J. Zool.* **2**: 9–65.

Niemitz, C. and A. Krampe. 1971. Gehörsinn bei polyphagen Käfern nachgewiesen. *Naturwissenschaften* **58**: 368–369.

–. 1972. Untersuchungen zum Orientierungsverhalten der Larven von *Necrophorus vespillo* F. (Silphidae Coleoptera). *Z. Tierpsychol.* **30**: 456–463.

Otronen, M. 1988a. The effect of body size on the outcome of fights in burying beetles (*Nicrophorus*). *Ann. Zool. Fenn.* **25**: 191–201.

–. 1988b. Intra- and intersexual interactions at breeding burrows in the horned beetle, *Coprophanaeus ensifer*. *Anim. Behav.* **36**: 741–748.

–. 1990. The effect of prior experience on the outcome of fights in the burying beetle, *Nicrophorus humator*. *Anim. Behav.* **40**: 980–982.

Payne, J. A. 1965. A summer carrion study of the baby pig *Sus scrofa* Linnaeus. *Ecology* **46**: 592–602.

Peck, S. B. 1982. The life history of the Japanese carrion beetle *Ptomascopus morio* and the origins of parental care in *Nicrophorus* (Coleoptera, Silphidae, Nicrophorini). *Psyche* **89**: 107–111.

–. 1986. *Nicrophorus* (Silphidae) can use large carcasses for reproduction (Coleoptera). *Coleopt. Bull.* **40**: 44.

Peck, S. B. and R. S. Anderson. 1985. Taxonomy, phylogeny, and biogeography of the carrion beetles of Latin America (Coleoptera: Silphidae). *Quaest. Entomol.* **21**: 247–317.

Peck, S. B. and M. M. Kaulbars. 1987. A synopsis of the distribution and bionomics of the carrion beetles (Coleoptera: Silphidae) of the conterminous United States. *Proc. Entomol. Soc. Ont.* **118**: 47–81.

Pukowski, E. 1933. Oekologische Untersuchungen an *Necrophorus* F. *Z. Morphol. Oekol. Tiere* **27**: 518–586.

Reeve, H. K. and P. W. Sherman. 1993. Adaptation and the goals of evolutionary research. *Q. Rev. Biol.* **68**: 1–32.

Reinking, M. 1988. Aufwand und Erfolg der Brutpflege bei *Necrophorus vespilloides* (Coleoptera: Silphidae). Staatsexamensarbeit, Universität Bielefeld, Germany.

Reinking, M. and J. K. Müller. 1990. The benefit of parental care in the burying beetle, *Necrophorus vespilloides*. *Verh. Dt. Zool. Ges.* **83**: 655–656.

Robertson, I. C. 1992. Relative abundance of *Nicrophorus pustulatus* (Coleoptera: Silphidae) in a burying beetle community, with notes on its reproductive behavior. *Psyche* **99**: 189–197.

–. 1993. Nest intrusions, infanticide, and parental care in the burying beetle, *Nicrophorus orbicollis* (Coleoptera: Silphidae). *J. Zool. (Lond.)* **231**: 583–593.

Rösel vom Rosenhof, A. J. 1761. *Insektenbelustigung*, vol. 4 (cited in Pukowski 1933).

Sato, H. and K. Hiramatsu. 1993. Mating behaviour and sexual selection in the African ball-rolling scarab *Khepher platynotus* (Bates) (Coleoptera: Scarabaeidae). *J. Nat. Hist.* **27**: 657–668.

Sato, H. and M. Imamori. 1986. Production of two brood pears from one dung ball in an African ball-roller, *Scarabaeus aegyptiorum* (Coleoptera, Scarabaeidae). *Kontyû* **54**: 381–385.

–. 1988. Further observations on the nesting behaviour of a subsocial ball-rolling scarab, *Kheper aegyptiorum*. *Kontyû.* **56**: 873–878.

Schuster, J. C. and L. B. Schuster. 1985. Social behavior in passalid beetles (Coleoptera: Passalidae): cooperative brood care. *Fla. Entomol.* **68**: 266–272.

Schwarz, H. H. and J. K. Müller. 1992. The dispersal behaviour of the phoretic mite *Poecilochirus carabi* (Mesostigmata, Parasitidae): adaptation to the breeding biology of its carrier *Necrophorus vespilloides* (Coleoptera, Silphidae). *Oecologia (Berl.)* **89**: 487–493.

Scott, M. P. 1989. Male parental care and reproductive success in the burying beetle, *Nicrophorus orbicollis*. *J. Insect Behav.* **2**: 133–137.

–. 1990. Brood guarding and the evolution of male parental care in burying beetles. *Behav. Ecol. Sociobiol.* **26**: 31–39.

–. 1994. Competition with flies promotes communal breeding in the burying beetle, *Nicrophorus tomentosus*. *Behav. Ecol. Sociobiol.* **34**: 367–373.

Scott, M. P. and D. S. Gladstein. 1993. Calculating males? An empirical and theoretical examination of the duration of paternal care in burying beetles. *Evol. Ecol.* **7**: 362–378.

Scott, M. P. and J. F. A. Traniello. 1987. Behavioural cues trigger ovarian development in the burying beetle, *Nicrophorus tomentosus*. *J. Insect Physiol.* **33**: 693–696.

–. 1990. Behavioural and ecological correlates of male and female parental care and reproductive success in burying beetles (*Nicrophorus* spp.). *Anim. Behav.* **39**: 274–283.

Scott, M. P. and S. M. Williams. 1993. Comparative reproductive success of communally breeding burying beetles as assessed by PCR with randomly amplified polymorphic DNA. *Proc. Natl. Acad. Sci. U.S.A.* **90**: 2242–2245.

Sherman, P. W., E. A. Lacey, H. K. Reeve and L. Keller. 1995. The eusociality continuum. *Behav. Ecol.* **6**: 102–108.

Tallamy, D. W. and T. K. Wood. 1986. Convergence patterns in subsocial insects. *Annu. Rev. Entomol.* **31**: 369–390.

Thornhill, R. and J. Alcock. 1983. *The Evolution of Insect Mating Systems*. Cambridge, Mass.: Harvard University Press.

Tribe, G. D. 1975. Pheromone release by dung beetles (Coleoptera: Scarabaeidae). *S. Afr. J. Sci.* **71**: 277–278.

Trumbo, S. T. 1987. The ecology of parental care in burying beetles (Silphidae: *Nicrophorus*). Ph. D. dissertation, University of North Carolina, Chapel Hill.

–. 1990a. Interference competition among burying beetles (Silphidae, *Nicrophorus*). *Ecol. Entomol.* **15**: 347–355.

–. 1990b. Reproductive success, phenology and biogeography of burying beetles (Silphidae, *Nicrophorus*). *Am. Midl. Nat.* **124**: 1–11.

–. 1990c. Reproductive benefits of infanticide in a biparental burying beetle *Nicrophorus orbicollis*. *Behav. Ecol. Sociobiol.* **27**: 269–273.

–. 1990d. Regulation of brood size in a burying beetle, *Nicrophorus tomentosus* (Silphidae). *J. Insect Behav.* **3**: 491–500.

–. 1991. Reproductive benefits and the duration of paternal care in a biparental burying beetle, *Nicrophorus orbicollis*. *Behaviour* **117**: 82–105.

–. 1992. Monogamy to communal breeding: exploitation of a broad resource base by burying beetles (*Nicrophorus*). *Ecol. Entomol.* **17**: 289–298.

–. 1994. Interspecific competition, brood parasitism, and the evolution of biparental cooperation in burying beetles. *Oikos* **69**: 241–249.

–. 1995. Nesting failure in burying beetles and the origin of communal associations. *Evol. Ecol.* **9**: 125–130.

Trumbo, S. T. and A.-K. Eggert. 1994. Beyond monogamy: territory quality influences sexual advertisement in male burying beetles. *Anim. Behav.* **48**: 1043–1047.

Trumbo, S. T. and A. J. Fiore. 1994. Interspecific competition and the evolution of communal breeding in burying beetles. *Am. Midl. Nat.* **131**: 169–174.

Trumbo, S. T. and D. S. Wilson. 1993. Brood discrimination, nest mate discrimination, and determinants of social behavior in facultatively quasisocial beetles (*Nicrophorus* spp.). *Behav. Ecol.* **4**: 332–339.

Trumbo, S. T., D. W. Borst and G. E. Robinson. 1995. Rapid elevation of juvenile hormone titer during behavioral assessment of the breeding resource by the burying beetle, *Nicrophorus orbicollis*. *J. Insect Physiol.* **41**: 535–543.

Valenzuela-Gonzalez, J. 1993. Pupal cell-building behavior in passalid beetles (Coleoptera: Passalidae). *J. Insect Behav.* **6**: 33–41.

Waldow, U. 1973. Elektrophysiologie eines neuen Aasgeruchrezeptors und seine Bedeutung für das Verhalten des Totengräbers (*Necrophorus*). *J. Comp. Physiol.* **83**: 415–424.

Walker, T. J. 1957. Ecological studies of the arthropods associated with certain decaying materials in four habitats. *Ecology* **38**: 262–276.

Wilson, D. S. and J. Fudge. 1984. Burying beetles: intraspecific interactions and reproductive success in the field. *Ecol. Entomol.* **9**: 195–203.

Wilson, D. S. and W. G. Knollenberg. 1984. Food discrimination and ovarian development in burying beetles (Coleoptera: Silphidae: *Nicrophorus*). *Ann. Entomol. Soc. Am.* **77**: 165–170.

–. 1987. Adaptive indirect effects: the fitness of burying beetles with and without their phoretic mites. *Evol. Ecol.* **1**: 139–159.

Wilson, D. S., W. G. Knollenberg and J. Fudge. 1984. Species packing and temperature dependent competition among burying beetles (Silphidae, *Nicrophorus*). *Ecol. Entomol.* **9**: 205–216.

Wilson, E. O. 1971. *The Insect Societies*. Cambridge, Mass.: Belknap Press of Harvard University Press.

–. 1975. *Sociobiology – the New Synthesis*. Cambridge, Mass.: Belknap Press of Harvard University Press.

Wussler, G. L. and J. K. Müller. 1994. Optimal family planning through infanticide? *Verh. Dt. Zool. Ges.* **87**: 66.

11 · Subsocial behavior in Scarabaeinae beetles

GONZALO HALFFTER

ABSTRACT

The nidification (nesting) process of the Scarabaeinae beetles exhibits a range from relatively primitive levels, similar
to those seen in other subfamilies of the Scarabaeidae, to
highly complex nests. It is possible to follow the successive
stages of this evolution in different suprageneric groups
and even among species of the same genus.

Evolved nidification in Scarabaeinae has two main
forms, the adaptive values of which appear to be equivalent. In the first of these behaviors, the mother covers the
food destined for the larva (the brood ball) with a thick
layer of soil. Immediately after laying the egg, the mother
abandons the nest. The other behavior includes postoviposition care by the mother, with variable participation
(depending on the species) by the male. These types of subsocial care have been identified in 67 species belonging to
13 genera. These genera belong to distinct phylogenetic
lines, with no relationship other than that they are all Scarabaeinae (a monophyletic group).

Subsocial care in Scarabaeinae cannot be related to a
specific morphological characteristic or to a concrete physiological function common to all known cases. From an
ecological perspective, subsocial care appears to be a
response, according to known cases, to different selective
pressures.

In *Oniticellus* and *Tragiscus* and perhaps also in *Eurysternus*, *Cyptochirus* and *Paraphytus*, the main pressure is predation on the nest by other animals (on the food accumulated
in the brood balls and on the larvae). The next strongest
pressure is the threat of the development of fungus on the
brood balls.

Cephalodesmius exhibits the most complex subsocial
behavior within the Scarabaeinae, with the greatest participation by the male. In this genus, care is directly related to
the very special nature of the food and to the continuous
provisioning of the brood balls by the mother during
larval development. The food (leaves and other plant fragments) is collected by the male from the surface and incorporated by the female into the mass that acts as the external
rumen. The intense microbial fermentation that occurs
within this mass produces a material which has the consistency of dung. This material is continually added to the
outside of each of the brood balls by the female.

In the other cases of subsocial care, there is well defined
bisexual cooperation in the process of nidification and this
is extremely important for the relocation and burial of the
food, considering that dung pats are a highly prized
resource. Postoviposition care by the female (with variable
participation by the male) prevents the invasion of the
brood balls by fungus, as well as preventing ruptures
which can cause desiccation and attacks by parasites.

Although subsocial behavior in Scarabaeinae does not
reach the level of complexity seen in Passalidae or the
close parent–offspring relationship exhibited by *Nicrophorus* in the form of trophallaxis, for the study of the evolution of behavior the Scarabaeinae have the advantage of
representing various independent responses within the
general process of evolution of nidification. An additional
advantage is that this process can be studied in its different
stages, whereas in Passalidae and in *Nicrophorus* there are
few differences between species in their subsocial behavior.

INTRODUCTION

The dung beetles of the subfamily Scarabaeinae (Coleoptera: Scarabaeidae) provide exceptional material for the
study of the evolution of behavior associated with reproduction. Within these insects it is possible to find different
levels of complexity in nesting which correspond to increasing offspring protection and a decreasing number of progeny. Subsocial care is found in various genera of
Scarabaeinae that are not phylogenetically related. To date
subsocial care has been identified in 13 genera and in 67
species, although the true number of species must be
much higher since for some genera (e.g. *Copris*) this type
of care is probably present in all species.

This chapter is divided into three sections. First, an
overview of the relationship between the environment

(food in particular) and behavior in Scarabaeinae is presented, with an emphasis on nidification. The second section includes a synthesis of reproductive behavior for all cases of dung beetles with subsocial care. The third section is a discussion of the aspects of subsocial behavior in Scarabaeinae which I believe are the most important. A taxonomic list of the tribes and genera of Scarabaeinae exhibiting subsocial behavior is included at the end of the chapter (Appendix 11-1).

ECOLOGY AND BEHAVIOR IN DUNG BEETLES: AN OVERVIEW

The name 'dung beetle' is applied to the members of three subfamilies of Scarabaeidae: Aphodiinae, Geotrupinae (only Geotrupini) and Scarabaeinae. Subsocial behavior is found only in the Scarabaeinae, and the scope of this chapter is limited to this subfamily, which includes 200 genera with 4500 species. Halffter and Matthews (1966) presented a review of the natural history of Scarabaeinae, with respect to ecology, behavior and distribution in different ecosystems. In 1982 Halffter and Edmonds compiled current information on nesting behavior in this subfamily, analyzing the relationships between ecological conditions and morphological and behavioral responses. I rely on this review for many of the comments included in this overview. Recently, Hanski and Cambefort (1991) edited a book on the ecology of dung beetles. There are, therefore, current bibliographic sources for an analysis of the relationship between environmental conditions and behavior in the Scarabaeinae beetles.

Halffter and Edmonds (1982) consider that the evolution of Scarabaeinae has been strongly linked to the ecological characteristics of the excrement of large mammals. Exposed excrement on the surface of the soil is subject to strong physical and biotic alteration, which can rapidly render it useless to dung beetles. Among these detrimental elements are rapid desiccation (the mouthparts of adult dung beetles are modified for manipulating soft, highly pliable, semi-liquid food) and intense competition with other coprophagous insects, particularly flies and other dung beetles, as well as nematodes and fungi (see Hanski and Cambefort 1991). Excrement must be quickly taken advantage of because it is ephemeral, patchy in its spatial distribution and in high demand. This requirement has resulted in one of the most characteristic traits of Scarabaeinae behavior: the relocation of food.

Food relocation is a key feature of adult behavior and is accomplished by one of two principal methods (unusual modifications do exist): by packing pieces into the blind end of a tunnel dug beneath or near the food source beforehand; or by forming a ball at the food source, which is then rolled for some distance and buried intact at a shallow depth (Fig. 11-1). The fact that food relocation is effected in different ways (burrowing at different depths, rolling various distances) results in spatial partitioning of a resource originally occupying a small area and, thereby, in a reduction of competition. From the beetles' perspective, the crucial advantage of relocation is protection from competitors, unfavorable climatic conditions, and predators.

The separation between the two forms of relocation occurred early in the evolutionary history of the Scarabaeinae and resulted in two phylogenetic branches: ball-rollers and burrowers (see Halffter and Halffter 1989; Halffter and Edmonds 1982; Halffter 1977). All Scarabaeinae except the neotropical tribe Eurysternini belong to these two branches. In some other cases there are changes in the form of relocation (or it does not occur) but a clear evolutionary derivation may be established for the two great branches of Scarabaeinae (for an analysis of these cases see Halffter and Halffter 1989; Halffter and Edmonds 1982).

Fossil evidence suggests that the origin and basic divergence (rollers–burrowers) in nidification are ancient. There are Oligocene nests which are similar to present-day nests (Halffter 1959; Halffter and Edmonds 1982). The main event in the evolution of nidification would seem to be the

(a)

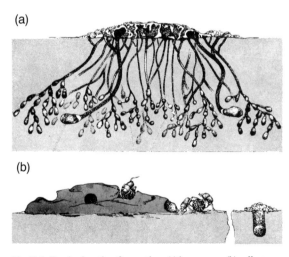

(b)

Fig. 11-1. Food relocation for nesting: (a) burrowers; (b) rollers (from Bornemissza 1976).

appearance of the brood ball. Its importance with respect to the parents' behavior and offspring survival is examined in the section 'Evolution of subsocial behavior in Scarabaeinae', below. The brood ball is not found in the burrowers with the most primitive nidification behavior. In evolutionary terms, it must have appeared relatively early, before the separation of the rollers and burrowers, since it is present and has the same basic characteristics in both groups, as well as in the Eurysternini.

The different procedures used by Scarabaeinae beetles for food relocation correspond not only to different behavior, but also to marked morphological differences: long and slender legs in the rollers and shorter, thicker legs in the burrowers. Burrowers have a more compact body and the male has horns on its head and prothoracic protuberances, neither of which are found on rollers. In spite of these differences, Scarabaeinae have a series of characteristics in common which clearly delimit them as a monophyletic group (Halffter and Edmonds (1982) review the major adaptive features of the subfamily in comparison with other dung beetles).

In addition to food relocation, the most important characteristics of Scarabaeinae associated with reproduction are as follows. (1) Subterranean nesting is a direct consequence of food-relocation behavior. (2) Within a phylogenetic sequence (burrowers or rollers) there is an increase in male–female cooperation. This cooperation is favored by the ephemeral and concentrated nature of the dung (two beetles can dig, stock or roll better and more quickly than one). (3) Fecundity is extremely low. All Scarabaeinae that have been studied have a single ovary, with a single ovariole. In insects, such an extreme reduction is only encountered in certain aphids (Pemphigidae). The number of eggs laid per female is low (usually less than 20) and inversely proportional to the complexity of nesting behavior. Moreover, Scarabaeinae eggs are enormous relative to the size of the female. (4) Larvae and pupae exhibit adaptations for development within an enclosed space.

The same ecological pressures that favor food relocation by the adult influence food relocation for offspring, i.e. the nesting process. The development of feeding and nesting behavior likely proceeded in tandem in both ball-rollers and burrowers. However, in nidification some new characteristics arise. The food for the larva must be protected from desiccation for a prolonged period. The adult, with highly modified mouthparts (soft incisive parts, shredders at the base) eats microorganisms and liquids (Halffter and Matthews 1966, 1971; Halffter and Edmonds

1982). Larvae eat much drier food with a high vegetable fiber content. The intestine of the adult is long and adapted to the digestion of microorganisms, whereas in the larva – as in other Scarabaeoidea – there is a fermentation chamber (Halffter and Matthews 1966). Furthermore, food accumulated by the larva is processed in an external rumen, inoculated with the mother's and the larva's own excrement, which is mixed with the accumulated food.

Although the evolution of nidification occurred independently in rollers and burrowers, there is a common tendency towards an increase in parental investment in each of their offspring and a decrease in fecundity. As nidification becomes more complex there is an increase in bisexual cooperation, as well as an increased investment in nest construction, with the appearance of postoviposition care in several unrelated genera (Fig. 11-2).

The role of competition in the evolution of behavior in Scarabaeinae is very important (see Hanski and Cambefort 1991). Adults are only active during a certain period of the year in many ecosystems. In addition, the spatial distribution of excrement, particularly the droppings of large herbivores, is not uniform simply because the animals that produce it are not evenly spaced. The droppings are not only clumped, but their distribution is clearly discontinuous and this increases aggregation of the beetles. Under some conditions (e.g. in African savannas) thousands of individuals may gather at a dung pat. Added to this is the short useful life of excrement from the beetles' perspective. All these conditions promote fierce competition among adults.

At one extreme of food-use mechanisms are the large burrowing beetles, a pair of which removes the entire resource (just as occurs with *Nicrophorus* and a small carcass; see Eggert and Müller, this volume). At the other extreme, a group of species may assemble at the same resource patch. Hanski and Cambefort (1991) refer to the first case as a dominance, determined by lottery dynamics. In the second case they define the community as structured by variance–covariance dynamics. There is a continuum of variation between these two extremes.

Competition takes the form of intraspecific combat (interference competition), which is much more common in the rollers. It also takes the form of preemptive resource competition, both between different roller species as well as between burrowers and rollers. Competition between burrowers is for food and space (including nidification space). Between rollers, competition is for food. Competition is limited to the adults. Each one of the larvae is

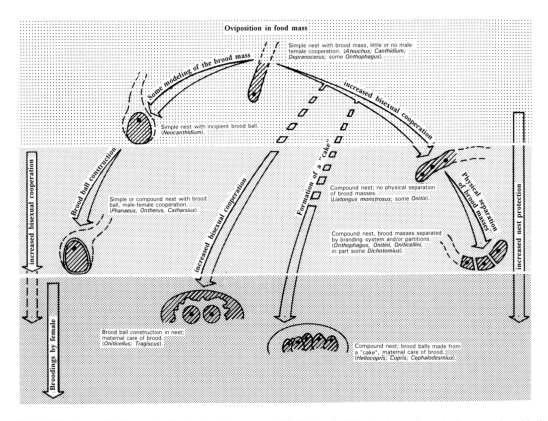

Fig. 11-2. Evolution of nidification in Scarabaeinae burrowers and *Cephalodesmius*. The simple nest has a single mass or brood ball; the compound nest has more than one (modified from Halffter and Edmonds 1982).

confined to an excrement mass or brood ball which has been prepared by the parents, thus there is no competition between them.

Hanski and Cambefort (1991, pp. 327–8) indicate that the large rollers and fast burrowers are superior competitors because of their ability to relocate excrement.

Large size is an advantage in digging deep tunnels fast; but as the size of the beetles becomes large compared to the size of the dropping, the number of offspring that may be produced becomes necessarily small, creating a selection pressure toward a high level of parental care ... good competitive ability, low fecundity, and high level of parental care constitute a single syndrome of large dung beetles.

The ideas put forth by Hanski and Cambefort are true in general terms, if within the 'high level of parental care' we include not only species with subsocial care, but also those that prepare a brood ball with a thick, well-made external covering of soil (*Phanaeus* and *Ontherus* spp.). In

these cases the preparation of each one of the brood balls requires the mother to make a substantial investment of time and energy, and this is associated with a reduced number of offspring.

Nevertheless, the association of factors proposed by Hanski and Cambefort does not explain all cases of subsocial behavior in Scarabaeinae (see below, 'Scarabaeinae with postoviposition care'). As I indicate on various occasions in this chapter, subsocial behavior in Scarabaeinae appears in genera that are separate phylogenetically and as a response to pressures that are not always the same. For example, within the burrowers, subsocial *Copris*, *Synapsis* and *Heliocopris* are large beetles, which dig deep tunnels quickly and, as such, have good competitive ability. However, other beetles of the same size and with the same resource management dynamic have primitive nidification (e.g. *Dichotomius*). Subsocial species of *Oniticellus*, which are small to medium-sized, do not build nesting galleries and do not relocate food; their fecundity is not especially low. The appearance of postoviposition care in *Oniticellus* might be associated with the use of dung pats that are much drier that those used by the majority of Scarabaeinae. In these dung pats, which have diminished pliability, the mother rapidly prepares various brood balls. In the days following oviposition, the female must add food to the ball's exterior. Under these nest conditions, i.e. inside a dry dung pat, defense against predators becomes a very important reason for the mother to stay in the nest.

Within the rollers, *Kheper* and *Scarabaeus* are among the large beetles with good competitive ability. The medium- to small-sized *Canthon cyanellus* uses small cadavers for nesting. In *Cephalodesmius*, subsocial behavior (with the greatest participation by the male) is clearly derived from the necessity, during the postoviposition period, to continue collecting small leaves and fruit, which are transformed into the external rumen of the nesting chamber (see below). The female adds the transformed material to the exterior of the brood balls, each of which contains a developing larva.

Based on the previous comments, one can conclude that large size and good competitive ability are characteristic of the subsocial beetles that attack large, fresh dung pats that are subject to much competition. The other cases of subsocial behavior correspond to species which, owing to different conditions, could be considered as marginal to the guild that attacks large excrement deposits.

SCARABAEINAE WITH POSTOVIPOSITION CARE

Burrowing beetles: *Copris*, *Synapsis* and *Heliocopris*

Subsocial behavior occurs in three genera of typical burrowers: *Copris*, *Synapsis* and *Heliocopris*, all of which belong to the Coprini tribe. All the processes of nidification described for these genera (22 species in *Copris*, 2 in *Synapsis* and 12 in *Heliocopris*) include care of the nest by the mother.

In this synopsis, wherever there is a recent review I use it as a point of reference, indicating the source for details, differences between species and earlier bibliography (generally omitted here; I concentrate on recent contributions). Halffter and Edmonds (1982) present a detailed review of knowledge on the nidification of these dung beetles. Later publications with information on subsocial behavior in the beetles include Klemperer (1982a,b, 1986), Anduaga and Huerta (1983), Tyndale-Biscoe (1983,1984),

Cambefort and Lumaret (1986), Anduaga *et al.* (1987) and Anduaga and Halffter (1991).

Species of *Copris*, *Synapsis* and *Heliocopris* prepare a subterranean nest with several brood balls. The first step in the process of nidification is the excavation of a gallery just beside or below the dung pat. The majority of these beetles – from medium to large, even to very large – take advantage of the sizable pats of large herbivorous mammals, which are pasty in texture. The gallery ends in a spacious nidification chamber.

The mating pair forms when the excavation of the gallery begins, and stays together throughout nest preparation and during part of the period of offspring care. The excavation of the gallery and the chamber is primarily done by the female. The male often helps remove the soil that the female excavates (the female may help with soil removal). Then the chamber is stocked with dung. The dung is taken in small 'armfuls' by the male from the outside to the interior of the gallery, where the female compacts it in the subterranean chamber to form a 'cake'. This 'cake', which is round or oval and has a flat base, in compacted by the female, usually with the help of the male, and the excrement of the parents is incorporated (in addition to references found in Halffter and Edmonds 1982, also see Anduaga *et al.* 1987). Under these conditions (i.e. high microbial content of the original material, the incorporation of the parents' excrement, compaction, and even the addition of a layer of soil to the 'cake' by some species) the 'cake' functions as an external rumen, causing a change in the dung and microbial enrichment which will continue later in each of the brood balls (Fig. 11.3).

The female makes several brood balls from the 'cake'. Each one contains an egg in an upper chamber which is located immediately below a porous aeration duct. In most species, at different times of larval development, each of the brood balls is covered with an external layer of soil by the female, increasing its insulation and protection (Anduaga *et al.* 1987).

Once the brood balls have been made the female stays in the chamber, continually attending to their outer surface and making slight adjustments in their position. It is during this period of time that the female applies the outer layer of soil. Care occurs until pupation of the offspring or until their emergence. In *Copris diversus* Waterhouse, Tyndale-Biscoe (1984) mentions an exceptional phenomenon: during the first days of larval development, the dung of brood balls in which the egg or larva had died was relocated to the outer surface of other brood balls in

(a) (b)

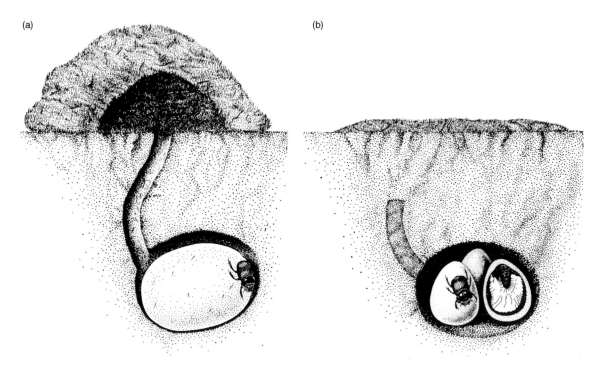

Fig. 11-3. (a) Initial stage (dung 'cake') and (b) a stage close to the end of care (larvae III) in the nest of *Copris armatus* Harold (from Huerta *et al*. 1981).

which the progeny was still alive. Thus, the remaining brood balls were larger and produced larger offspring.

The time that the male spends in the nest varies according to species. In some, the male leaves when the 'cake' has been finished, or even before. In many species the male leaves when the female is preparing the brood balls. In some cases the male stays in the nest chamber until the larvae are in the second stage (see Anduaga *et al*. 1987).

The large amount of time that the female spends preparing and looking after the nest results in the preparation of only one nest per year in species that live in environments with marked seasonality. For those species in which development is more rapid, or where seasonality is less marked (as occurs in various species of the humid and sub-humid tropics) each female can prepare more than one nest per year (Klemperer 1986; Anduaga *et al*. 1987).

There is a single exception known with respect to the efficiency and continuity of maternal care in these three genera, and it occurs in *Heliocopris*. In many species of this genus the female normally spends several days taking care of the nest.

If conditions become very dry, the female may remain in the nest much longer than usual until another good rain, upon which she emerges to construct another nest. During this period the female ceases brood ball construction and actively moistens completed ones. These are coated with what appears to be moist maternal faeces (not soil). If the dung 'cake' is exhausted, or after its remains are consumed by the female, she will begin to feed on brood balls to the extent of destroying her progeny (observed frequently in *H. neptunus* and *H. andersoni*).

(Halffter and Edmonds 1982, p. 107.)

This infanticide under special ecological conditions is the only known instance of this phenomenon within subsocial Scarabaeinae, with the notable exception of *Eurysternus* (see below).

Nesting in *Litocopris* (a subgenus of *Copris*, considered as a genus by some authors) has been described for *C. (L.) punctiventris* Waterhouse. According to Cambefort and Lumaret (1986) it appears less elaborate when compared with the characteristic nesting of *Copris*. Instead of resting on the soil in a hemispherical chamber, the brood balls occupy the entire volume of the spherical nest chamber.

Table 11-1. *Postoviposition care in* Oniticellus *and* Tragiscus.

Species	Immature stage when female departs	External layer of soil applied to brood ball	References
O. planatus Laporte	Adult (emerged)	No	Davis 1977, 1989
O. cinctus (F.)	Pupa or adult	No	Klemperer 1983; Davis 1989
O. formosus Chevrolat	Pupa	No	Davis 1977, 1989
O. pictus Hausman	—	No	Davis 1977, 1989
O. pseudoplanatus Balthasar	—	No	Cambefort 1982
O. egregius Klug	Second instar-larva	Yes	Davis 1977, 1989
O. rhadamistus (F.)	First-instar larva	Yes	Lumaret and Moretto 1983
Tragiscus dimidiatus Klug	Third-instar larva	No	Davis 1977, 1989

This is a result of the brood balls not being made from a 'cake' located on the soil in the chamber, but rather their being separated from a mass which fills the entire spherical nest. The brood balls themselves are not as well finished as those of *Copris* and there are more of them (6–8, compared with an average of 5 in *Copris*).

Dweller beetles: Oniticellini and Coprini

As a reference point for these genera I use Cambefort (1982) and Cambefort and Lumaret (1983). Later information about subsocial behavior is found in Klemperer (1983) and Davis (1989). *Oniticellus* and *Tragiscus* belong to a tribe of burrowing beetles, Oniticellini, in which some species may prepare the nest (with various brood balls) in the interior of a dung pat or at the dung–soil interface (hence the name dweller beetles, coined by Hanski and Cambefort 1991). The brood balls are built *in situ* using nearby dung. There is no vertical relocation before nesting as is the case with *Copris*, nor is there horizontal relocation as is seen in *Canthon cyanellus* or *Kheper*. Dung relocation is the phase of nesting in which cooperation by the male is the most striking. When relocation is unnecessary, as in *Oniticellus* and *Tragiscus*, cooperation of the male is also absent and the nest is built exclusively by the female. Davis (1989) indicates the presence of the male in the nest of *O. planatus* and *O. formosus* for half or less of the time that the female spends in the nest.

Not all species of these two genera exhibit subsocial behavior. In some, postoviposition care is replaced by a layer of soil deposited on the exterior of the brood balls by the female when she has finished preparing them, or during the first days of larval development (Table 11-1).

One of the most interesting adaptive aspects of the Oniticellini with prolonged maternal care (*O. cinctus*, *O. planatus*, *O. formosus*, *O. pictus* and *Tragiscus dimidiatus*) is that the mother adds material to the exterior of the brood balls and makes them bigger after ovipositing and during the first stages of larval development. Hence, the nest may have up to 20 brood balls, made one after the other and later made larger. This mechanism makes these the nests with the most balls within subsocial Scarabaeinae.

Belonging to a different subtribe of Oniticellini from *Oniticellus* and *Tragiscus*, *Cyptochirus distinctus* Janssens prepares a nest which (Cambefort 1981, 1982; Cambefort and Lumaret 1983) consists of a mass of dung placed in a cavity below the dung pat and then covered with a layer of soil. Each mass has several submasses (two to five), each of which contains an egg and later a larva, until its pupation. For pupation the larva was observed to construct a capsule from its own excrement. When the entire mass was moved, each one of the submasses separated as if they had been built in succession and later pressed together and covered with a layer of soil. At least during the first days of larval development the female remained at the excavation, outside the covering layer of soil.

This nest could either be interpreted as a case in which several submasses (equivalent to balls) are arranged together in a nest with a common soil covering, or as an example of subsocial care prior to the appearance of individual brood balls. A similar nest was found for *Eurysternus foedus* (Halffter *et al.* 1980) which differs considerably from the nests described for other *Eurysternus* spp. (see below).

Table 11-2. *Nidification in* Eurysternus

Species	Nuptial feast: maximum number of balls made (duration in days)	Percentage of cases in which experimental nests occur (number of successive experimental nests)	Definitive nest: number of brood balls (duration in days)
E. magnus	17 (12–28)	0%	3 (25–26)
E. balachowskyi	55 (60–90)	100% (±3)	2 (40)
E. caribaeus	94 (7–69)	65%	2–6 (38–53)
E. mexicanus	14	(4)	3–9

The genus *Paraphytus* belongs to a different group: Coprini, Dichotomiina. Several nests of *P. aphodioides* Boucomont have been described by Cambefort and Walter (1985). Each nest was found in a poorly defined cavity formed between debris and the excrement of xylophagous insects (primarily passalids) in a rotten trunk. One brood ball made from the same material and containing a larva was found in this cavity. The female was present and stayed in the nest until pupation. Independent of the material used, this nest appears to correspond in a general sense to that of *Oniticellus*, i.e. a brood ball made from the feeding material, which is not relocated.

In all the cases presented here the fundamental role of postoviposition care appears to be the protection of the nest against predators and parasites, for which the frequency and probability of attacks is greater than in subterranean nests (Fig. 11-4). This protection has special importance because the nest is not located within the soil, as it is in *Copris*, *Kheper*, *Scarabaeus*, and almost all Scarabaeinae that do not exhibit subsocial care.

Fig. 11-4. Female *Oniticellus* rejecting an invading larva (*Aphodius*). Original photograph provided by Dr. Hugh G. Klemperer.

Eurysternus

Although nidification in *Eurysternus* appears to be similar to that of *Oniticellus* and *Tragiscus* (i.e. taking advantage of a resource without its relocation) it has some unexpected characteristics. *Eurysternus* is the only genus in the tribe Eurysternini, occupying a taxonomically isolated position within the Scarabaeinae. The nidification process of four species, *E. magnus* Laporte, *E. balachowskyi* Halffter and Halffter, *E. caribaeus* (Herbst) and *E. mexicanus* Harold, is well known. For all four species nidification has been observed only under laboratory conditions (Halffter 1977; Halffter *et al.* 1980), except for a nest of *E. magnus*, which was observed in the field (Halffter and Matthews 1966). The typical nidification pattern has three stages: (1) the nuptial feast; (2) 'experimental' nests and (3) the definitive nest.

The beginning of nidification by *E. balachowskyi* (some of the more notable differences among the four species studied are presented in Table 11-2) is marked by the nuptial feast, which corresponds to the maturation of the ovary. It is characterized by the rapid and massive construction of balls. These balls are made exclusively by the female and only at the beginning of nidification. *Eurysternus* cannot roll the balls as do the roller beetles (Scarabaeini), so the balls remain in the vicinity of the dung pat. From nuptial feast (up to now only the female has participated in the actual nesting procedure) the mating pair is established and does not separate until the process of nidification is well advanced. Both the male and the female are found in the experimental nest, although only the female occupies the definitive nest. Some of the balls made during the nuptial feast are eaten by the female or the male a few hours or days later; others are retouched and pushed around a little, and others are abandoned. Copulation occurs at the end of the nuptial feast or in the first stages of 'experimental' nesting and may be repeated several times.

'Experimental' nesting begins when the beetles excavate a very superficial crater under several brood balls. Other balls are pushed into this crater. Among all these balls, the female lays an egg in some (the others are destroyed) and then partly covers them with soil. This nest is cared for by the female in the presence and with the participation of the male. At a given moment, both the female and the male may eat one part of the balls containing eggs, while continuing to take care of the others. A few days later, the female or the male destroys the remaining balls and both abandon the nest. At random, some brood balls can survive this destruction and the juvenile can continue developing. This process is what we refer to as 'experimental' nesting.

After the female eats directly from the dung pat she begins making brood balls again (31–50 additional balls) and 'experimental' nidification is repeated several times (±3), resulting in abandoned craters containing partly eaten balls. The definitive nest is begun when the female selects two brood balls which already contain eggs and moves them to a flat surface, one in contact with the other, forming an figure eight. Afterwards, the female excavates a furrow around the balls and covers them with soil. The furrow is made deeper and wider until the balls are perched on a minimal column of supporting soil. The female cares for the nest over the course of about 40 days, retouching the surface of the brood balls and covering them with soil toward the end of the period of larval development, without any attempt to destroy them. The definitive nest is very similar in *E. caribaeus* (Fig. 11-5). In *E. magnus* and *E. mexicanus* the brood balls have an external covering of soil, but are not joined together. The preparation of the definitive nest is linked to the maturation of the last ovocytes of a cycle.

When the new imagos finish their development and emerge, after a month of feeding directly on the dung, the female begins the nidification process again, with a new nuptial feast.

Nidification in *Eurysternus* presents a strong contrast with that of other Scarabaeinae. *Eurysternus* appears to waste a notable quantity of food, time and energy on the nuptial feast, infanticide and the deliberate abandonment of the nests during the 'experimental' stage. It takes the female of *E. balachowskyi* about six months to produce two brood balls (plus whichever ones escape the parents' destruction). It would appear that *Eurysternus* lives in a world with limitless resources and no competition. However, this is not so. The four species studied are

Fig. 11-5. *Eurysternus caribaeus* Hbst. female caring for a definitive nest (from Halffter *et al.* 1980).

coprophagous and live in American tropical forests. Competition by Scarabaeinae in these ecosystems and over this food is fierce. To explain its survival, in addition to the effect of elements with which we are not familiar, there are two possibilities. First, all descriptions of the nidification process of *Eurysternus* were made from studies carried out in the laboratory (with the observation of one nest in the field). Observation under field conditions is difficult, owing to the length of the process, and has never been done. In the laboratory the dung pat is made available from the beginning of the observation period and fresh dung is added to it every two or three days. There are no competitors. It is possible that laboratory conditions favor the nuptial feast and 'experimental' nests. Nevertheless, even within a group with such behavioral plasticity as the Scarabaeinae, one must ask whether it is possible for *Eurysternus* to develop such a complicated behavior in the laboratory. The second possibility is that nidification is carried out by *Eurysternus* in the field during the season when the Scarabaeinae guild is less active. Even supposing that the nuptial feast and 'experimental' nests are less wasteful in the field than they are in the laboratory, this process would only be viable if pressure by other Scarabaeinae were limited. It is important to note that in American tropical forests there is no season in which the Scarabaeinae are completely inactive, although activity does decrease significantly during the driest season.

As far as intrinsic determinants are concerned, there could be different explanations for the behavior of the male and of the female. The male is attracted by the female during the massive formation of the balls (nuptial feast). When this happens the ovary of the female is mature. During this process copulation occurs, followed by the 'experimental' nests (copulation may also occur during the period of experimental nidification). The destruction of the brood balls by the male in the 'experimental' nests could be interpreted as an effort on the male's part to ensure that it was his sperm that fertilized the egg. In Scarabaeinae, sperm remain viable in the sperm receptacle for a long time. By destroying the brood balls, the male eliminates the possibility that the first brood balls might have been fertilized by sperm from a copulation previous to his arrival.

To explain the destruction of the brood balls by the female, Halffter et al. (1980) suggest a lack of coordination between the processes of ovogenesis and nest care. In Scarabaeinae where subsocial care has been extensively studied, maternal displays are accompanied by a cessation in vitellogenesis and by reabsorption of the ovary. In Eurysternus (Halffter et al. 1980) vitellogenesis continues, and this provokes the attack and later abandonment of the brood balls after several days of looking after them. The definitive nest would begin with the maturation of the last ovocytes in a series.

Cephalodesmius

Nidification in the Australian genus Cephalodesmius was described in 1981 by G. H. Monteith and R. I. Storey. Nidification by C. armiger Westwood is well known; field observations by the previously mentioned authors allow us to recognize that two other species nest similarly.

The adults of C. armiger emerge at the end of summer and live in individual feeding galleries. In autumn they form bisexual pairs, each of which occupies a feeding gallery finished into a chamber. This pair will continue to occupy the gallery (the chamber of which is gradually made larger) from the beginning of the reproductive process until death, for a period of at least 12 months. The pair winters in the feeding gallery, where it survives on provisions collected during forays on the surface. With the onset of spring, active foraging is resumed. The process of foraging consists of carrying partly decomposed leaves, small flowers, parts of seeds, small fruit and occasionally other material such as excrement to the gallery. Although foraging is initially done by both sexes, it soon becomes a task exclusively carried out by the male. The female works with the provisions, to which she adds her own feces as well as those of the male, making a compressed mass. The microbiological action (principally fungal) that occurs in the mass results in its taking on the appearance and consistency of fresh dung. This rumen–mass is the source of food for the adults and the material from which the brood balls are made. New material is continually added to the mass and it is never allowed to dwindle during the life of the nest (Fig. 11-6).

Construction of the brood balls is begun when the female separates a portion of the mass and molds a small ball from it. Four to ten brood balls are constructed. If any ball is not used for oviposition it is incorporated into the other balls or into the rumen–mass. Copulation occurs several times while the balls are being constructed and also on the surface in the short period before the preparation of the nest. Once the brood balls have been made the female rarely leaves the nest. Each female goes through the process of nidification once in her life.

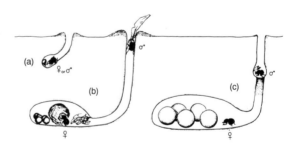

Fig. 11-6. Advances in *Cephalodesmius armiger* Westwood nest construction. (a) Short feeding burrow made by a newly emerged adult. (b) The nest in its first stages. The male is dragging leaves into the nest and the female is attending the rumen–mass she has made from the leaves. Five small brood balls containing developing larvae are shown at left. (c) The nest in the last stage. (From Monteith and Storey 1981.)

On emerging, the larvae eat in the small initial brood ball. As the food contained in each ball decreases, the walls becoming thinner, the female adds a new layer to the outer surface of the ball. The combination of consumption by the larva and the addition of new food to the outside by the female results in a constant brood-ball wall thickness of approximately 2 mm. During the first stages of larval development the material added to the outside of the brood-balls has always been completely processed by the rumen-mass. Later on the provisions added may contain pieces of plants that have not completely decomposed and the female may even incorporate material that the male brings without first processing it in the rumen-mass. In this case fermentation occurs in the walls of the brood ball.

Developing larvae stridulate in an audible manner using a gula–abdomen mechanism which is thought to be unique (Monteith and Storey 1981; Paulian *et al.* 1983). Monteith and Storey suggest that this sound indicates to the female the need to add more food to the walls of the brood ball, by communicating the thickness of the wall. The adults do not stridulate.

When the larva completes its development, the female covers the outer surface of each brood ball with a thin layer which is a mixture of adult and larval excrement, the latter having been ejected by the larva through fissures in the wall of the brood ball. When this layer dries it becomes very hard. Once pupation occurs the female continues tending the brood balls. The male no longer enters the nest chamber and is excluded by a soil plug at the upper part of the gallery. At the time of emergence both parents are usually dead.

Cephalodesmius exhibits the most complex subsocial behavior of all the Scarabaeinae. Its behavior is strongly associated with the external rumen. With respect to the latter, and to the type of material transformed in it, the following features stand out: the prolonged occupation of the nest, constant foraging and the progressive addition of food to the outside of the brood balls.

How are the fungi which are responsible for the transformation of the rumen-mass transmitted? Monteith and Storey (1981) found fungal fruiting bodies projecting from the interior of the pupal cell. A few days before the teneral adult emerges from the pupal cell the fungal bodies have disappeared or are broken. The hypothesis of the previously mentioned authors is that the first food of the adult might be these fungal bodies. This hypothesis is strengthened by the presence of the fungal bodies in the digestive tube of the new imagos. From the digestive tube via excrement, the fungi form part of the new 'cake' which the adults establish as soon as they begin their free life and which will soon become the rumen-mass of the new breeding pair.

The rollers: Scarabaeini

Roller Scarabaeinae (Scarabaeini tribe) make a ball from the source of food and the male and female roll it together for a certain distance. The male then buries it and the female transforms it into the brood ball. Cooperation between the male and female during rolling is noteworthy. Each is in a different position and their efforts are complementary. The investment made in the nest by the male is greatest within the Scarabaeinae. This behavior includes initial formation of the ball, attraction of the female, the most important effort during rolling, and the burial of the ball.

There are two patterns of nidification in roller beetles, one with and the other without subsocial care. In species without postoviposition care (the majority), the ball is rolled, the egg is laid in it, and it is transformed into the brood ball, which is then abandoned. In contrast, when there is care, the ball is divided into more than one brood ball, or the nest (a superficial crater) is stocked with more than one ball rolled in successive stages one after the other. This nest receives intensive care by the female, who is accompanied part of the time by the male. This subsocial behavior in roller beetles was first described for *Canthon cyanellus cyanellus* Le Conte by Halffter (1977). This description was later broadened by Halffter *et al.* (1983).

Canthon cyanellus

The behavior of *C. cyanellus cyanellus* Le Conte has become the most studied for Scarabaeinae both in the laboratory and in the field. In the past ten years 28 articles and theses have been published on *C. cyanellus*, including experimental analyses of its reproductive behavior, studies of the ovary and testicle and mechanisms of neuroendocrine regulation, and the description of tegumental glands associated with behavior. The most relevant contributions with respect to subsocial behavior include Halffter (1977), Halffter *et al.* (1983), Bellés and Favila (1983), Martínez and Caussanel (1984), Favila (1988a,b, 1992, 1993), Pluot-Sigwalt (1991), Martínez (1991, 1992a,b), Martínez and Cruz (1992), Cruz (1993), and Favila and Díaz (1996).

The following is a description of nidification and behavior of the parents given by Halffter *et al.* (1983) completed with later contributions. The pair forms as a response to the long-distance pheromonal call made by the male from the food source, after his rolling a ball alone, or without a pheromonal call during the formation of the ball. Copulation occurring at the food source is the exception. It occurs with much greater frequency in the nest, always close to the ball. Favila (1992) and Cruz (1993) indicate the presence of spermatophores and spermatozoids in females dissected during the rolling process, which are from previous copulations with different males than that with which she is rolling the ball. Favila (1992) estimates that the spermatozoids remain viable in the female's spermatheca for 16 days. This retention of viable sperm would explain the repetition of copulation during the beginning of the process of nidification, as an attempt of the nesting male to assure the priority of his sperm, displacing that of males with which the female had previously copulated.

Although both the male and the female may prepare balls for feeding, the initial preparation of the ball destined for nidification (which is larger than the feeding balls) is the task of the male. The male and female finish preparing the future brood ball together and roll it. The male is in the pushing position (in 85% of observed cases according to Favila 1988b), which requires a greater energy expenditure. The pushing position facilitates continuous contact between the glands on the ventral portion of the male's abdomen and the ball. Rolling is followed by one or more attempts at burying the ball (very superficially), until definitive burial is achieved. These activities are carried out by the male. At the definitive nesting site, i.e. a small, shallow crater, the female covers the ball with a thick layer of soil, molds it, lays the egg in the upper chamber, and seals the ball (now a brood ball). Once this has been done the construction and transportation of other balls continue one to five times until the compound nest of two to six brood balls has been completed. The male forms and rolls the new balls, although the female may momentarily abandon the nest and help with the rolling. Conversely, the male may stay at the nest and the female may leave to make a new ball, but this rarely occurs. Stocking of the nest by continuous rolling was described (Halffter 1977; Halffter et al. 1983) based on laboratory observations under conditions in which carrion (ground meat) was continually available. Under field conditions, Favila (1992) and Favila and Díaz (1996) have observed cooperative rolling of a ball from which the female makes several (1–4) brood balls (Fig. 11-7).

The female takes care of the nest until the larvae reach the third stage or even until eclosion. The male stays in the nest for 5–10 of the 25 ± 3 days required for the complete

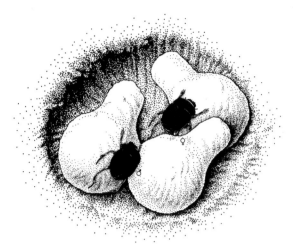

Fig 11-7. Male and female *Canthon cyanellus* LeConte taking care of the nest. Original.

development process (Favila 1993). In *C. cyanellus*, care does not include continuous recovering and attention to the exterior of the ball, although it is kept free from fungus. From the first larval stage small balls of excrement expelled by the larva appear on the surface of the ball. These are left on its surface by the female (Halffter et al. 1983; Favila 1992).

Each female prepares several nests during her lifetime. One inseminated female is capable of making a nest alone, but under these circumstances the nest contains no more than two brood balls.

There is evidence that allows us to surmise that other South American Canthonina such as *Canthon bispinus* Germar, *C. muticus* Harold, *C. edentulus* Harold, *Malagoniella bicolor* (Guerin) and *M. puncticollis tubericeps* (Gillet) (see Halffter 1977; Halffter and Halffter 1989) exhibit nesting behavior that includes the building of compound nests and possibly social care. This may also be true of the African *Anachalcos cupreus* (F.) (Walter 1978). In *Canthon edentulus* this behavior has been confirmed under field conditions by Mario Zunino (personal communication), who found a nest with a chamber containing six brood balls with larvae immediately beneath a dung pat. These were in advanced stages of development and were cared for by the female.

Kheper and *Scarabaeus*

The discovery of subsocial care in roller beetles not belonging to the Canthonina subtribe can be credited to two groups of researchers. These are Penelope B. Edwards and

Table 11-3. *Nidification in* Kheper *and* Scarabaeus *with postoviposition care*

A Initial preparation of the brood ball. B Number of brood balls per nest. C The brood ball(s) are covered with the female's excrement during the period of care. D Presence or absence of the male in the nest. E The female cares for the brood balls until offspring are at the stage indicated.

Species	A	B	C	D	E	References
K. nigroaeneus (Boheman)	Male	1 (always)	Yes	Leaves before ovipositing	Adult	Edwards and Aschenborn 1988, 1989; Edwards 1988a,b.
K. platynotus (Bates)	Female	1–4	Yes	Leaves before ovipositing	Pupa	Sato and Imamori 1986a, 1987; Sato and Hiramatsu 1993
K. aegyptiorum (Latreille)	Male	1–2	Yes	Leaves before ovipositing	Pupa	Sato and Imamori 1986b, 1988
K. lamarcki (MacLeay)	—	2	Yes	Leaves before ovipositing	Adult	P. B. Edwards (personal communication)
K. subaeneus (Harold)	—	2-3	—	Leaves before ovipositing	Adult	P. B. Edwards (personal communication)
K. aeratus (Gerstaecker)	—	1	—	Leaves before ovipositing	Pupa	Palestrini *et al.* 1992
K. cupreus (Laporte)	—	1–2	—	Leaves before ovipositing	Not confirmed before pupa	P. B. Edwards (personal communication)
K. clericus (Boheman)	—	1	—	Leaves before ovipositing	Not confirmed before pupa	P. B. Edwards (personal comminication)
S. funebris (Boheman)	—	1–2 (usually 1)	—	Leaves before ovipositing	Pupa	P. B. Edwards (personal communication)
S. galenus (Westwood)	—	1–4	—	Leaves after ovipositing	Adult	P. B. Edwards (personal communication)

H. H. Aschenborn (*Kheper nigroaeneus*) and Hiroaki Sato, Mitsuhiko Imamori and K. Hiramatsu (*K. aegyptiorum* and *K. platynotus*). Findings seem to indicate that many species in the African genus *Kheper* (Scarabaeinae: Scarabaeini), if not all, exhibit subsocial care (Table 11-3).

There is an aspect of nidification in *Kheper* that is not found in the subsocial Scarabaeinae previously presented: the prenuptial offering. This offering is very well developed in *Scarabaeus* (a related genus) in species that do not exhibit subsocial care. This offering involves the male presenting the female with a quantity of food, which allows the maturation of the ovary to be completed. In *Scarabaeus* a complete ball is offered by the male and consumed by the female in a few days, after which new rolling begins with a different male, this time for nesting (the mechanism of the offering was described by Fabre 1897; see synthesis in Halffter and Matthews 1966). In *K. aegyptiorum* a prenuptial offering is made (Sato and Imamori 1988). The offering may be less obvious (but equal in energetic value) when the female eats a large part of the ball prepared and rolled by the male, before transforming it into the brood ball (*K. nigroaeneus*).

In general terms the nidification process in *Kheper* and *Scarabaeus* is similar to that of *C. cyanellus*, except that there is no indication of progressive provisioning of the nest with balls. Differences between species of *Kheper* and *Scarabaeus* with subsocial care are included in Table 11-3.

EVOLUTION OF SUBSOCIAL BEHAVIOR IN SCARABAEINAE

In this section I review all aspects of the behavior of the Scarabaeinae that have influenced the appearance of postoviposition care.

The brood ball

All Scarabaeinae beetles store and protect food for their offspring. In the most primitive forms of nidification (Fig. 11-2) this is done by packing a quantity of food at the back of a gallery. This is the brood mass of Halffter and Edmonds (1982). The brood ball (following the terminology of the authors just cited) represents a very important stage in the evolution of nidification. This brood ball, which may be spherical or piriform, is always prepared by the female. It contains a nucleus of food manipulated by the mother and an upper chamber which contains the egg. Between the egg chamber and the surface there is a less compact area, more or less distinguishable for different genera, which may even have fibers conveniently located in it and which facilitates gas exchange. The brood ball represents an excellent example of how parental behavior changes the physical environment in which the progeny develop (see Wcislo 1989). This elaboration is a very important evolutionary step in the protection of both the food for the larva and the larva itself.

The larva will develop inside the brood ball, eating the accumulated food, defecating and eating the resulting mixture repeatedly. The larva lives in the brood ball throughout its life and pupates in it. The integrity of the brood ball is indispensable for the survival of the juvenile.

All subsocial Scarabaeinae make brood balls (but see *Cyptochirus* and *Paraphytus*). In order for the brood ball to appear in the evolutionary process of the Scarabaeinae it is necessary for the larva to have acquired an efficient system of maintenance and repair of the ball from within, using its own excrement (see Halffter and Edmonds 1982; Klemperer 1983). These mechanisms do not exist or are not very efficient in the Scarabaeinae that make brood masses or in the subfamily Geotrupinae, which belong to the same family (Scarabaeidae) and also make brood masses (Klemperer 1978; G. Halffter, unpublished observations).

The preparation of the brood ball has been associated with the deposit of substances secreted by the sternal glands of the female (see Pluot-Sigwalt 1991). These glands are found in both rollers and those burrowers that make brood balls; the substances that they produce appear to have an antifungal effect. In preliminary tests with *C. cyanellus* in which the layer of soil is removed from the ball or in which the ball is bathed in organic solvents, the ball rapidly becomes covered with filamentous fungi (M. E. Favila, personal communication).

Two activities of the mother, i.e. depositing a layer of soil on the outside of the brood ball and the care of the ball, may appear to be redundant since they both provide similar protection against external elements. However, unlike Scarabaeinae in which there is no parental care and where the application of the layer of soil is the last step in the preparation of the brood ball before oviposition, in subsocial burrowing Scarabaeinae the layer of soil is applied in the advanced stages of larval development (especially the third larval stage and before pupation) when the mother might die or abandon the nest and when the ability of the juvenile to repair the ball from the inside disappears.

Bisexual cooperation

In nidification of all Scarabaeinae there is some degree of bisexual cooperation. In terms of interspecific competition, the occurrence of more or less participation by the male determines the different capacity in which the dung pat is used for nidification. Cooperation by the male is especially important in relation to the speed with which the nest is built (burrowers), relocation distance achieved (rollers) and the quantity of food gathered for the offspring (burrowers and rollers). Trumbo (1994) analyzes the importance of bisexual cooperation in relation to interspecific competition for *Nicrophorus*. In this aspect, as in others (relocation, external rumen) convergence in behavior between *Nicrophorus* and Scarabaeinae is a result of ecological similarities in the food used for nidification. Bisexual cooperation in nesting has different manifestations in the two principal phylogenetic groups. In the burrowers the female excavates, prepares and mostly provisions the nest. The male helps with provisioning. In the rollers (with the exception of *Kheper platynotus*) the male takes the initiative in preparing and rolling the ball, as well as in attracting the female, and makes the greatest energy expenditure during rolling; the male also buries the ball. In all Scarabaeinae the female molds the brood ball and is responsible for most or all of the subsocial care (although the male may participate in nest care). In *Cephalodesmius*, where stocking the nest is continuous, cooperation lasts throughout the period of care. In the rollers (with or without subsocial care), although the female might roll a ball and nest alone, the participation of the male assures a larger ball (balls rolled together are always bigger than those rolled alone) and therefore more food for the larva. In several species of rollers with subsocial care the collaboration of the male assures a ball that is large enough to make two or more brood balls. In *Oniticellus* and *Tragiscus*, genera in which food is not relocated (the nest is within the dung pat), cooperation by the male is minimal or non-existent.

Male behavior

In Scarabaeinae, increasingly as nidification evolves, the male invests a considerable quantity of time and energy helping the female in the preparation of the nest. This investment is even greater in subsocial species that also protect the nest and participate in the care of the brood balls. This behavior does not pay in terms of fitness unless the male is assured that the offspring is his. In Scarabaeinae, confidence of paternity is a serious problem. In the period prior to nidification, the female may have copulated with various different males. Furthermore, these copulations seem to be a prerequisite for the conclusion of ovocyte maturation (Halffter and López-Guerrero 1977; Halffter *et al.* 1980; Huerta *et al.* 1981; Edwards and Aschenborn 1989; Martínez and Cruz 1990, 1992; Favila 1992; Cruz 1993).

The male has certain behaviors to ensure paternity. First, copulation may be repeated more than once during the stages of nidification. This repetition is more evident in rollers than in burrowers and may represent an effort to ensure sperm priority. A similar pattern is observed in *Nicrophorus* (Müller and Eggert 1989). Second, the male stays in the nest until the egg or eggs are laid, preventing insemination of the female by other males. A similar phenomenon explains the presence of the male in the first days of *Nicrophorus* nests (Scott 1990; Trumbo 1990, 1991; Robertson 1993; Eggert and Müller, this volume). Third, as discussed above, a male of *Eurysternus* will attack and kill the first offspring, which might be the result of a fertilization previous to his arrival and the process of nidification.

Repeated copulation may ensure sperm priority in spite of the longevity of spermatozoids. After copulation the spermatophore disintegrates on the *bursa copulatrix* and after a few hours the spermatozoids are stored in the spermatheca (Martínez and Cruz 1990; Cruz 1993). The spermatheca has the form of a blister, which connects at one end to the *ductus receptaculi*. A strong muscle distends (suction) or folds (expulsion of spermatozoids) the blister. Based on anatomical observation I believe that it is possible that when suctioned, the spermatozoids from a second copulation push spermatozoids from previous copulations towards the bottom. When the spermatheca is folded by the contraction of the muscle, evidently the spermatozoids of the most recent copulation will be expelled first. Even considering the longevity of the sperm in the spermatheca, this mechanism would give sperm priority to the last male to copulate with the female and, if copulation is repeated more than once, greater assurance of sperm priority.

Male parental care in insects is a relatively rare phenomenon (reviewed in Tallamy 1994). In Coleoptera, in addition to the Scarabaeinae, it occurs in *Nicrophorus* (Eggert and Müller, this volume) and Passalidae (Schuster and Schuster,

this volume). We cannot speak of subsocial care by the male for species other than those in which the male stays in the nest after oviposition occurs. This happens, to a different extent depending on the species, in *Copris* and, in a more highly defined manner, in *Canthon cyanellus*. It is the key element to successful nesting in *Cephalodesmius*. In *Canthon cyanellus*, the only species in which this has been studied, the male reabsorbs the testicle during the period of offspring care: equivalent to the reabsorption of the ovary by the female (Martínez 1991; Martínez and Cruz 1992; Cruz 1993).

It is not evident why the male participates in postoviposition care in *Nicrophorus*, Passalidae and in some Scarabaeinae, but not in other Coleoptera. The cases mentioned have in common the duration of offspring care by the mother at a fixed site: the nest. This results in fewer offspring, but also in fewer females available to be fertilized. In terms of fitness, it is more advantageous for the male to care for his offspring than to leave in search of another female with which to copulate. The presence of the male is not only beneficial to the offspring (to the point of his being able to replace the female in *C. cyanellus*), but also prevents intrusions by other males.

The external rumen

In all Scarabaeinae the brood mass or brood ball acts as an external rumen. The larva defecates and its excrement mixes with the food accumulated by the mother. This mixture is eaten and defecated again several times, which produces a progressive increase in its microbial content. In subsocial species, the rumen begins to function in the 'cake' (*Copris, Cephalodesmius*) or in the ball that will be transformed into the brood ball (*Kheper*), both of which are inoculated with the female's excrement. In several species of *Kheper* (Table 11-3) the mother adds her own excrement to the outside of the brood ball during larval development. The inoculation of the juvenile's food with the parents' excrement is also found in *Nicrophorus* and Passalidae (Halffter 1982, 1991), making it one of the causes of the parents' staying in the nest after ovipositing.

In Scarabaeinae the external rumen is associated with the undigested fibers found in the excrement of large herbivorous mammals, and with essentially microbial feeding (Halffter and Matthews 1971). The rumen is most useful in those species that nest with plant remains (*Cephalodesmius*) or excrement, such as that of the elephant, with a high fiber content (*Kheper*). The importance of the care taken by the mother in the preparation of the external rumen is accentuated throughout the fundamental evolutionary sequence in the nidification of Scarabaeinae: increase in its complexity and protection of the nest vs. a reduction in the food accumulated for the offspring. In terms of energy expenditure, less food requires more efficient use. This efficiency is achieved with the external rumen.

Fecundity

In Scarabaeinae species that exhibit subsocial care, the number of eggs per nest is very low: two to five per nest, and in extreme cases only one (*Kheper*; see Table 11-3). The exception within subsocial species is *Oniticellus*. The number of eggs laid by species with less complex nidification is much higher per season (in some cases, e.g. *Digitonthophagus gazella* (F.), more than 20 times higher). Although the reduction in fecundity is extreme in subsocial species, it is also found in some species with no postoviposition care, but with complex nidification (e.g. *Phanaeus* and *Ontherus* spp.).

Although the proportion of medium-sized and large beetles in the subsocial species and, in general, in those with complex nidification is greater than in those species with primitive nidification, this is not always the rule. There are large species with primitive nidification and relatively high fecundity.

Advantages of subsocial care in terms of survival

We have quantitative information that demonstrates higher offspring survivorship when there is parental care. In *Copris fricator* (F.) only 59% of the progeny survive to the adult stage when the mother is absent, compared with 93% when the mother takes care of the nest (Halffter and Matthews 1966). In *C. diversus* the results were 32% and 76% (Tyndale-Biscoe 1984), and in *C. incertus*, 22% and 76%, respectively (G. Halffter and C. Huerta, unpublished data).

In *Canthon cyanellus* survivorship of offspring that received no care was 59% whereas in those that received care until the third larval stage survivorship was 93% (Favila 1992, 1993). For *Kheper nigroaeneus*, Edwards (1988a) found significant differences in the survivorship of offspring that did and did not receive care, under laboratory conditions.

Only in *C. cyanellus* has the value of care by the male been experimentally studied. Nests that were cared for by the male only never filled with fungi and had high survivorship: 80% versus 93% with biparental care (Favila 1992). At least under experimental conditions, care by the father can be a substitute for care by the mother. With the exception of *Cephalodesmius*, *Canthon cyanellus* is the Scarabaeinae beetle in which the male is present for the greatest length of time. *Cephalodesmius* represents an equilibrium in which the participation of both sexes (for different tasks) is indispensable for offspring survival.

In all Scarabaeinae, care is principally the responsibility of the mother, which, in most species exhibiting subsocial care, stays in the nest until pupation is well advanced or even until the new adults emerge. The mother repairs the ball from outside, closing fissures. This is a complementary activity to what the larva is doing from the inside, placing and spreading its excrement on the inner wall of the nest ball. The continuous activity of the female prevents the growth of fungi. In *Canthon cyanellus*, balls that are not cared for become covered with filamentous fungi (Fig. 11-8). When *Copris incertus* parents are removed from their nests, 100% of the balls are invaded by fungi.

Another very important task of the mother during the postoviposition period, in which the male participates, is protecting the nest against intruders. These include histerids and quilopods, which prey on the larva, kleptoparasites in search of the food stocked for the larva, such as dipteran larvae and beetles from the Aphodiinae subfamily, as well as predators on the 'cake' and the nest balls, such as earthworms (Carmen Huerta and G. Halffter, unpublished observations). In *Copris* and *Oniticellus* the mother has been observed fending off and even killing these interlopers (Klemperer 1983, 1986; Tyndale-Biscoe 1984). The scarce number of observations (and complete absence of experimental analysis) do not do justice to the importance of this task, considering the fact that laboratory work to date has always been done in the absence of intruders. On the other hand, in the field one finds many brood balls that have been destroyed in the absence of the mother (*Copris*) (Carmen Huerta and G. Halffter, unpublished observations).

With respect to survival, the brood balls and the offspring in a dry pat (*Oniticellus*) are much more exposed to the action of other insects than those in a subterranean nest. A series of parasites and predators has been reported in *Oniticellus*'s nests: *Oniticellobia sublaevis* Boucek (Hymenoptera-Pteromalidae) (Davis 1977); *Bombylius* sp. and *B. ornatus* Wiedeman (Diptera-Bombyllidae) (Davis 1977; Cambefort 1983); *Adelopygus decorsei* Desbrochers

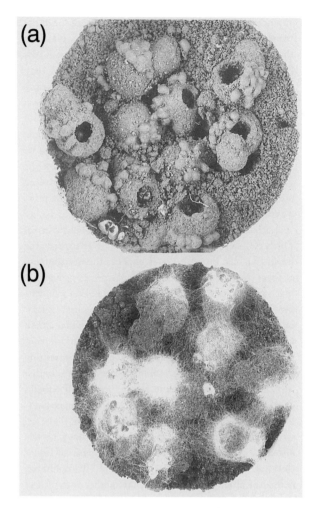

Fig 11-8. Brood balls of *Canthon cyanellus*. (a) With parental care, showing the projections of larval excrement and the circular shaped hole from which the young adult emerges. (b) Without parental care and attacked by fungi (from Favila 1993).

(Coleoptera: Histeridae) (Rougon and Rougon 1982); as well as termites, which may attack the dung content of the brood ball (Lumaret and Moretto 1983). Although it has not been experimentally demonstrated to date, there are several indications (abundance of intruders, aggressive reactions by the mother) that the defense of the brood balls against intruders is the main reason for postoviposition care in *Oniticellus*.

In some species, another role of the mother is the application of additional food (during the development of the larva) to the brood balls initially formed, before covering them with a layer of soil. This mechanism, an integral part of care in *Cephalodesmius*, is also found in *Oniticellus*, in some species of *Kheper* and as an exception in one species of *Copris*. The application of additional food to the exterior of each brood ball during postoviposition care allows these beetles to overcome one of the principle restrictions for nidification in Scarabaeinae: the need to prepare the nest rapidly and the limited amount of food that the mother can provide for each brood ball, as well as the number of brood balls that she can prepare in a short period of time.

Semiochemicals

In Scarabaeinae, there are two different types of chemical communication related to sexual behavior and nidification: long-distance and short-distance. Long-distance communication is found in both rollers (Scarabaeini) that do and those that do not care for their offspring. First described by Tribe (1975), this call is made by the male in order to attract the female. Depending on the species, the call can occur at the source of food, beside the ball, during the formation or rolling of the ball, or immediately after burial of the ball. Differences in the place where emission occurs (and when it occurs during the process of nidification) depend on where copulation occurs. The objective of this call is to attract the female to copulate and subsequently nest. The behavior is always the same in all the species in which it has been observed: the male puts his head down, supporting himself on his front and middle legs, while the hind legs pass over the abdomen and stretch alternately (*Canthon cyanellus*) or simultaneously (*Kheper nigroaeneus*). The tibiae sweep over the outer surface of the abdomen where glandular pores are grouped in 'beaches'. In *C. cyanellus* they are located on sternite VII (Pluot-Sigwalt 1983, 1988a,b; 1991); in *Kheper nigroaeneus* on sternite I (Tribe 1975). A secretion has been observed leaving these pores; this secretion is spread by the movement of the legs. This behavior has been described for *Kheper* (Tribe 1975; Edwards and Aschenborn 1988) and for *Canthon cyanellus* (Bellés and Favila 1983; Favila 1992; G. Halffter, unpublished observations) as well as in numerous Scarabaeini that do not exhibit subsocial behavior. The compound emitted has been analyzed for three species of *Kheper* (Burger *et al.* 1983, 1990; B. V. Burger, personal communication) and is made up of a complex mixture of volatile compounds: mostly aliphatic acids, methyl and ethyl esters, and alkanes. In *C. cyanellus* the compound emitted is a mixture of alcohols, aliphatic acids and methyl esthers of hexadecanoic and octodecanoic acids (M. E. Favila, and K. Jaffe, personal communication).

The other form of chemical communication is short-distance communication, which has a longer-lasting effect. In rollers this communication occurs, for the most part, with the ball as an intermediary. The sternal glands, much more common in rollers and with a different distribution between sexes (Pluot-Sigwalt 1988a,b, 1991), must be associated with this communication. The activity of these sternal glands is greater during the reproductive stage of the beetles. Females of burrower species that make brood balls also have external glands. In these cases, the activity of these glands appears to be linked to preparation of the brood ball by the mother.

We cannot ignore the possibility of secretions from other parts of the body coming into contact with the ball during its preparation or rolling. Both front and back legs have exocrine glands (Pluot-Sigwalt 1988a,b). There are two large glands, believed to be used for defense, which are located on the pigidium.

We can associate several behavioral responses with short-distance communication. Although it might be possible to generalize these responses to the rollers, they have been thus far described only for *Canthon cyanellus*. Favila (1988a, 1992) shows that sexually mature males, but not old or sexually immature beetles (which is related to the limited activity of the glandular systems of the old and young), impregnate the ball with semiochemicals during rolling. The chemically labeled ball has a short-distance effect and is capable of luring the female to roll the ball and of keeping her in the nest. It also appears that there is a chemical label, perhaps the same one, which allows the male to find the ball when he loses it while rolling (Favila 1992). Finally, there are 'cuticle odors' (Favila 1992), independent of the impregnation of the ball, which play a part in the sexes' recognition of each other. These odors do not necessarily originate from the sternal glands (although it is difficult to distinguish how big a part abdominal secretions play). Houston (1986) describes a group of glands on the ventral surface of the anterior coxae, the activity of which appears to be associated with recognition between the sexes. Houston states that he has found these glands in all Scarabaeinae studied and in other Scarabaeoidea with bisexual cooperation whose adults engage in below-ground activities.

Evolution

Subsocial behavior in Scarabaeinae is not associated with a specific phyletic line. It appears at least eight times and can appear in groups of related genera or in one genus, without the occurrence of similar behavior in close taxa. It is clear that subsocial behavior completely abandons taxonomic limits and affinities. I believe that this indicates that, although there are ecological and ethological conditions which can lead to the development of subsocial care, such care is not the only evolutionary response to the same set of circumstances. Subsociality has a high survival value, but we cannot say that it is the most efficient one until we know more about the balance between survivorship and the number of offspring in closely related genera with evolved nidification without subsocial care, and with parental post-oviposition care.

Subsocial care fulfills a function, the importance of which is different for different genera: it guarantees the integrity of the brood ball and impedes the development of fungi (*Copris, Synapsis, Heliocopris, Eurysternus* rollers); it maintains the external rumen in the stages after ovipositing (*Cephalodesmius, Kheper*); it protects the nest against intruders (in all species, but especially in *Oniticellus* and in *Canthon cyanellus*); and it permits the rapid addition of more food to the brood balls at the moment of ovipositing (*Oniticellus, Cephalodesmius*, some *Kheper*). Male *Canthon cyanellus* deposit an allomone on the brood balls (Bellés and Favila 1983) and this repels flies of the genus *Calliphora* (recall that the food in the balls is carrion).

What similarities are there between the dung beetles discussed here and subsocial care exhibited by other Coleoptera? In both Chrysomelidae and Hydrophilidae, neither of which nest, the role of the mother is to protect the offspring (Goidanich 1956; Wilson 1971; Hinton 1981; Eickwort 1981). In Carabidae (Brandmayr and Zetto-Brandmayr 1974; Brandmayr 1977, 1992; Thiele 1977; Brandmayr and Zetto-Brandmayr 1979; Eickwort 1981) the principal role of postoviposition maternal care is to protect the eggs and, later, the offspring from predators and fungi. This also appears to be the case in Staphylinidae (Hinton 1981; Ashe 1986). In *Bledius spectabilis* Kraate, the female also provisions the nest during larval development (Larsen 1953) in a process similar to that of *Cephalodesmius* and which is not seen in other Coleoptera exhibiting subsocial care. In the Tenebrionidae, adult *Phrenapates bennetti* Kirby prepare wood to facilitate its consumption by the larvae, similarly to the preparation of food that occurs in passalids (Wilson 1971). *Nicrophorus* (Silphidae) and subsocial Scarabaeinae converge in various aspects of nidification, and these are derived from ecological similarities in the condition of the food used for nesting (see Eggert and Müller, this volume).

In *Nicrophorus* the small cadaver is relocated and, once buried and prepared, acts like an external rumen (Halffter 1991). It is also defended against intruders (Robertson 1993). In Passalidae the food is stable, but needs to be fragmented by the parents and inoculated with their excrement (Reyes-Castillo and Halffter 1983; Valenzuela-González 1992; Schuster and Schuster, this volume). Male–female cooperation reaches the highest level attained in Coleoptera (as do acoustic communication systems). Passalidae exhibit behavior not found in any other Coleoptera: the juvenile imagos collaborate in the protection (preparation of the pupal cocoon) of their younger siblings. In the fungus-growing beetles (Platypodidae and Scolytidae) the main form of care by the adults is the preparation of the substrate for the development and control of the fungus on which their larvae feed (Norris 1979; Hinton 1981; Morales 1984; Tallamy and Wood 1986; see Kirkendall *et al.*, this volume).

SUGGESTIONS FOR FUTURE RESEARCH

Over the past 20 years we have undoubtedly advanced a great deal in our knowledge of subsocial behavior in the Scarabaeinae. We have progressed from only knowing this process for *Copris* (and because of similarities in the nest, *Synapsis*) to currently having descriptions of postoviposition care for 13 genera. Nevertheless, many unknowns still remain to be explored, especially those related to environmental elements that favor the appearance of this type of care in some genera, but not in others that are taxonomically close.

Until now we have focussed our attention on the description of the process of nidification and on demonstrating, under laboratory conditions, the advantages of care in terms of offspring survival. The results are clear but provide an incomplete picture of the adaptive value of subsocial care. In the laboratory, the comparison of offspring survival in the presence and absence of an intruder species provides interesting results (G. Halffter and C. Huerta, in preparation), but also provides a reductionist view of the complex and variable effects that environmental elements, especially predators and kleptoparasites, have on the nest. A very different approximation would result from the comparison of offspring survival in the field for species with an evolved brood ball but without postoviposition care and subsocial species that live in the same area. In Mexico and Central America this study could be done by comparing *Phanaeus* and *Ontherus* with *Copris*. In Africa, species of *Oniticellus* with postoviposition care could be compared with those

not exhibiting this behavior. In Africa and southeast Asia, *Catharsius* could be compared with *Copris* and *Heliocopris*. These comparative studies would elucidate whether, in terms of the fitness of the subsocial female, greater offspring survival justifies her caring for a single nest each season. Just as has been found for vertebrates (see Rubenstein and Wrangham 1986), research under field conditions is necessary for the proposal of a hypothesis that might explain the advantages of sociality.

In order to understand why male roller beetles invest the greatest amount of time and energy further research on mechanisms that ensure paternity are required. M. E. Favila (in preparation) has selected a genetic marker for *Canthon cyanellus* which manifests in phenotype as variations in color, thus allowing the paternity of the offspring to be identified. Studies such as this, or the use of DNA fingerprinting, will allow us to broaden our knowledge of sperm priority mechanisms.

In all Scarabaeinae beetles (with the exception of *Cephalodesmius*) a fertilized female is capable of nesting without the help of the male. As such, under laboratory conditions it is possible to study the effects of the male's participation, or lack thereof, in the nidification process and offspring care on the number and survivorship of the offspring. *Canthon cyanellus*, *Copris* and *Eurysternus* are taxa that are especially appropriate for this type of study.

One field of research that is practically unexplored is the study of possible subsocial behavior in Central and South American Canthonina other than *Canthon cyanellus*. There are indications that subsocial care does occur in some species (see above 'Scarabaeinae with postoviposition care') but we truly know little about nesting of the numerous and varied group of Neotropical Canthonina.

Given the difficulties of working with *Eurysternus* under field conditions, I feel that it would be interesting to undertake new laboratory studies to determine the effects of different feeding regimes, the absence of the male and competition between several males.

Another area of research that has just begun to produce results is the study of the role of semiochemicals in the process of bisexual cooperation and nest care. The rollers are most appropriate for this type of study, and in particular behavior that has the brood ball as the center of activity.

ACKNOWLEDGEMENTS

Various colleagues have kindly provided information on subsocial behavior in Scarabaeinae and related activities which is unpublished or in press: Dr Penelope B. Edwards, Division of Entomology, CSIRO, Australia; Prof. Mario Zunino, Università di Palermo; Prof. B. V. Burger, University of Stellenbosch, South Africa; Dr Jean-Pierre Lumaret, Université Paul Valéry, Montpellier. My colleagues at the Instituto de Ecología, A. C., Dr Mario Enrique Favila, Carmen Huerta, M.Sc., and Violeta Halffter were kind enough to review the text and provide me with unpublished observations.

Bianca Delfosse, M.Sc. (Instituto de Ecología, A. C.) translated the text into English from the original in Spanish. I greatly appreciate her patient and careful work.

Dr W. David Edmonds (California State University, Pomona, California) reviewed the manuscript. I am indebted to Prof. Bernard Crespi for his interesting and generous comments. I am grateful for his efforts.

LITERATURE CITED

Anduaga, S. and G. Halffter. 1991. Escarabajos asociados a madrigueras de roedores (Coleoptera: Scarabaeidae: Scarabaeinae). *Folia Entomol. Mex.* 81: 185–197.

Anduaga, S., G. Halffter and C. Huerta. 1987. Adaptaciones ecológicas de la reproducción en *Copris* (Coleoptera: Scarabaeidae: Scarabaeinae). *Boll. Mus. Reg. Sci. Nat. Torino* 5: 45–65.

Anduaga, S. and C. Huerta. 1983. Factores que inducen la reabsorción ovárica en *Copris armatus* Harold (Coleoptera, Scarabaeidae, Scarabaeinae). *Folia Entomol. Mex.* 56: 53–73.

Ashe, J. S. 1986. Subsocial behavior among Gyrophaenine Staphylinids (Coleoptera: Staphylinidae, Aleocharinae). *Sociobiology* 12: 315–320.

Bellés, X. and M. E. Favila. 1983. Protection chimique du nid chez *Canthon cyanellus cyanellus* Le Conte (Coleoptera, Scarabaeidae). *Bull. Soc. Entomol. Fr.* 88: 602–607.

Bornemissza, G. F. 1976. The Australian dung beetle project 1965–1975. *Aust. Meat Res. Cttee Rev.* 30: 1–32.

Brandmayr, P. 1977. Ricerche etologiche e morfofunzionali sulle cure parentali in Carabidi Pterostichini (Coleoptera: Carabidae, Pterostichinae). *Redia* 60: 275–316.

–. 1992. Short review of the presocial evolution in Coleoptera. *Ethol. Ecol. Evol.* (Spec. Issue) 2: 7–16.

Brandmayr, P. and T. Zetto-Brandmayr. 1974. Sulle cure parentali e su altri aspetti della biologia di *Carterus* (*Sabienus*) *calydonius* Rossi, con alcune considerazioni sui fenomeni di cura della prole sino ad oggi riscontrati in Carabidi (Coleoptera, Carabidae). *Redia* 55: 143–175.

–. 1979. The evolution of parental care phenomena in Pterostichini, with particular reference to the genera *Abax* and *Molops*. *Misc. Pap. Landbouwhogescoll. Wageningen* 18: 35–49.

Burger, B. V., Z. Munro, M. Röth, H. S. C. Spies, V. Truter, G. D. Tribe and R. M. Crewe. 1983. Composition of heterogeneous sex attracting secretion of the dung beetle, *Kheper lamarcki. Z. Naturforsch.* C 38: 848–855.

Burger, B. V., Z. Munro and W. F. Brant. 1990. Pheromones of the Scarabaeinae, II. Composition of the pheromone disseminating carrier material secreted by male dung beetles of the genus *Kheper. Z. Naturforsch.* C 45: 863–872.

Cambefort, Y. 1981. La nidificacion du genre *Cyptochirus* (Coleoptera, Scarabaeidae). *C. R. Acad. Sci. Paris* (III) 292: 379–381.

–. 1982. Nidification behavior of Old World Oniticellini (Coleoptera: Scarabaeidae). In *The Nesting Behavior of Dung Beetles (Scarabaeinae): An Ecological and Evolutionary Approach.* G. Halffter and W. D. Edmonds, eds., pp. 141–145. México: Instituto de Ecología.

–. 1983. Étude écologique des coléoptères Scarabaeidae de Côte d'Ivoire. Thèse Doctorat, Université Paris VI.

–. 1986. Nidification et larve du genre *Litocopris* Waterhouse (Coleoptera, Scarabaeidae). *Nouv. Rev. Entomol.* (N.S.) 3: 251–256.

Cambefort, Y. and J. P. Lumaret. 1983. Nidification et larves des Oniticellini afro-tropicaux. *Bull. Soc. Entomol. Fr.* 88: 542–569.

Cambefort, Y. and P. Walter. 1985. Description du nid et de la larve de *Paraphytus aphodioides* Boucomont et notes sur l'origine de la coprophagie et l'evolution des coléoptères Scarabaeidae s. str. *Ann. Soc. Entomol. Fr.* (N. S.) 21: 351–356.

Cruz, M. 1993. Actividad reproductora de los machos de *Canthon indigaceus chevrolati* Harold y *Canthon cyanellus cyanellus* Le Conte y su influencia en el comportamiento reproductor de las hembras (Coleoptera, Scarabaeinae). Tesis Maestría Ciencias, Escuela Nacional de Ciencias Biológicas, IPN, México.

Davis, A. L. V. 1977. The endocoprid dung beetles of Southern Africa (Coleoptera: Scarabaeidae). M. Sc. thesis, Rhodes University, Grahamstown, South Africa.

–. 1989. Nesting of afrotropical *Oniticellus* (Coleoptera: Scarabaeidae) and its evolutionary trend from soil to dung. *Ecol. Entomol.* 14: 11–21.

Edwards, P. B. 1988a. Field ecology of a brood-caring dung beetle *Kheper nigroaeneus.* Habitat predictability and life history strategy. *Oecologia* (Berl.) 75: 527–534.

–. 1988b. Contribution of the female parent to survival of laboratory-reared offspring in the dung beetle *Kheper nigroaeneus* (Boheman) (Coleoptera: Scarabaeidae). *J. Austr. Entomol. Soc.* 27: 233–237.

Edwards, P. B. and H. H. Aschenborn. 1988. Male reproductive behaviour of the African ball-rolling dung beetle, *Kheper nigroaeneus* (Coleoptera: Scarabaeidae). *Coleopt. Bull.* 42: 17–27.

–. 1989. Maternal care of a single offspring in the dung beetle *Kheper nigroaeneus*: The consequences of extreme parental investment. *J. Nat. Hist.* 23: 17–27.

Eickwort, G. C. 1981. Presocial insects. In *Social Insects*, vol. 2. H. R. Hermann, ed., pp.169–280. New York: Academic Press.

Fabre, J. H. 1897. *Souvenirs Entomologiques* vol. V. Paris: Delagrave.

Favila, M. E. 1988a. Chemical labelling of the food ball during rolling by males of the subsocial Coleopteran *Canthon cyanellus cyanellus* LeConte (Scarabaeidae). *Insectes Soc.* **35**: 125–129.

–. 1988b. Comportamiento durante el período de maduración gonádica en un escarabajo rodador (Coleoptera: Scarabaeidae: Scarabaeinae). *Folia Entomol. Mex.* **76**: 55–64.

–. 1992. Análisis del comportamiento subsocial de *Canthon cyanellus cyanellus* LeConte (Coleoptera: Scarabaeidae). Tesis Doctoral, Escuela Nacional de Ciencias Biológicas, IPN, México.

–. 1993. Some ecological factors affecting the life-style of *Canthon cyanellus cyanellus* (Coleoptera: Scarabaeidae): An experimental approach. *Ethol. Ecol. Evol.* **5**: 319–328.

Favila, M. E. and A. Díaz. 1996. *Canthon cyanellus cyanellus* Le Conte (Coleoptera: Scarabaeidae) makes a nest in the field with several brood balls. *Coleopt. Bull.* **50**: (1): 52–60.

Goidanich, A. 1956. Gregarismi od individualismi larvali e cure materne nei crisomelidi (Col., Chrysomelidae). Centro di Entomologia alpina e forestale - Consiglio Nazionale delle Ricerche, Pub. no. 11, pp.151–182.

Halffter, G. 1959. Etología y paleontología de Scarabaeinae. *Ciencia (Mex.)* **19**: 165–178.

–. 1977. Evolution of nidification in the Scarabaeinae (Coleoptera, Scarabaeidae) *Quaest. Entomol.* **13**: 231–253.

–. 1982. Evolved relations between reproductive and subsocial behaviors in Coleoptera. In *The Biology of Social Insects*. M. D. Breed, C. D. Michener and H. E. Evans, eds., pp.164–170. Boulder, Colorado: Westview Press.

–. 1991. Feeding, bisexual cooperation and subsocial behavior in three groups of Coleoptera. In *Advances in Coleopterology*. M. Zunino, X. Bellés and M. Blas, eds., pp.281–296. Barcelona: Asociación Europea de Coleopterología.

Halffter, G. and W. D. Edmonds. 1982. *The Nesting Behavior of Dung Beetles (Scarabaeinae). An Ecological and Evolutive Approach.* México: Instituto de Ecología.

Halffter, G. and V. Halffter. 1989. Behavioral evolution of the non-rolling roller beetles (Coleoptera: Scarabaeidae: Scarabaeinae). *Acta Zool. Mex.* (N.S.) **32**: 1–53.

Halffter, G., V. Halffter and C. Huerta. 1980. Mating and nesting behavior of *Eurysternus* (Coleoptera: Scarabaeinae). *Quaest. Entomol.* **16**: 599–620.

–. 1983. Comportement sexuel et nidification chez *Canthon cyanellus cyanellus* Le Conte (Col. Scarabaeidae). *Bull. Soc. Entomol. Fr.* **88**: 586–594.

Halffter, G. and Y. López-Guerrero. 1977. Development of the ovary and mating behavior in *Phaneus*. *Ann. Entomol. Soc. Am.* **70**: 203–213.

Halffter, G. and E. G. Matthews. 1966. The natural history of dung beetles of the subfamily Scarabaeinae (Coleoptera, Scarabaeidae). *Folia Entomol. Mex.* 12–14: 1–312.

–. 1971. The natural history of dung beetles: A supplement on associated biota. *Rev. Lat. Amer. Microbiol.* **13**: 147–164.

Hanski, I. and Y. Cambefort. 1991. Competition in dung beetles. In *Dung Beetle Ecology*. I. Hanski and Y. Cambefort, eds., pp. 305–329. Princeton: Princeton University Press.

Hinton, H. E. 1981. *Biology of Insect Eggs*, 3 vols. Oxford: Pergamon Press.

Houston, W. W. K. 1986. Exocrine glands in the forelegs of dung beetles in the genus *Onitis* F. (Coleoptera: Scarabaeidae). *J. Austr. Entomol. Soc.* **25**: 161–169.

Huerta, C., S. Anduaga and G. Halffter. 1981. Relaciones entre nidificación y ovario en *Copris* (Coleoptera: Scarabaeidae: Scarabaeinae). *Folia Entomol. Mex.* **47**: 139–170.

Klemperer, H. G. 1978. The repair of larval cells and other larval activities in *Geotrupes spiniger* Marsham and other species (Coleoptera, Scarabaeidae). *Ecol. Entomol.* **3**: 119–131.

–. 1982a. Normal and atypical nesting behaviour of *Copris lunaris* (L.): Comparison with related species (Coleoptera, Scarabaeidae). *Ecol. Entomol.* **7**: 69–83.

–. 1982b. Parental behaviour in *Copris lunaris* (Coleoptera: Scarabaeidae): Care and defence of brood balls and nest. *Ecol. Entomol.* **7**: 155–167.

–. 1983. Subsocial behaviour in *Oniticellus cinctus* (Coleoptera: Scarabaeidae): Effect of the brood on parental care and oviposition *Physiol. Entomol.* **8**: 393–402.

–. 1986. Life history and parental behaviour of a dung beetle from neotropical rainforest, *Copris laeviceps* (Coleoptera, Scarabaeidae). *J. Zool. (Lond.)* **209**: 319–326.

Larsen, E. B. 1953. Studies on the soil fauna of Skallingen. Qualitative and quantitative studies on alterations in the beetle fauna during five years natural development of some sand and salt-marsh biotopes. *Oikos* **3**: 166–192.

Lumaret, J. P. and P. Moretto. 1983. Contribution a l'étude des Oniticellini. Nidification et morphologie larvair d'*Oniticellus rhadamistus* (F.) (Coleoptera: Scarabaeidae) et considérations sur la position taxonomique de cette espàce. *Ann. Soc. Entomol. Fr.* (N.S.) **19**: 311–316.

Martínez, I. 1991. Activité reproductrice et ses controles chez le male et la femelle de *Canthon indigaceus chevrolati* et *C. cyanellus cyanellus* (Coléoptères: Scarabaeinae). These Doctorat, Université Paris, VI.

–. 1992a. L'activité ovarienne pendant la vie imaginale chez deux espèces de *Canthon* (Coleoptera: Scarabaeidae). *Boll. Mus. Reg. Sci. Nat. Torino* **10**: 367–386.

–. 1992b. Données comparatives sur l'activité reproductive de *Canthon indigaceus chevrolati* Harold et *Canthon cyanellus cyanellus* LeConte (Coleoptera: Scarabaeidae). *Ann. Soc. Entomol. Fr.* (N.S.) **28**: 397–408.

Martínez, I. and C. Caussanel. 1984. Modifications de la pars intercerebralis des corpora allata, des gonades et comportement reproducteur chez *Canthon cyanellus cyanellus* Le Conte (Coleoptera: Scarabaeidae: Scarabaeinae). *C.R. Acad Sci. Paris* (III) **299**: 597–602.

Martínez, I. and M. Cruz. 1990. Cópula, funcion ovárica y nidificación en dos especies del género *Canthon* Hoffmannsegg (Coleoptera: Scarabaeidae). *Elytron* 4: 161–169.

–. 1992. L'activité de l'appareil reproducteur mâle pendant la vie imaginale chez deux espèces de *Canthon* (Coleoptera: Scarabaeidae). *Acta Zool. Mex.* (N.S.) 49: 1–22.

Monteith, G. B. and R. I. Storey. 1981. The biology of *Cephalodesmius*, a genus of dung beetles which synthesizes "dung" from plant material (Coleoptera: Scarabaeidae: Scarabaeinae). *Mem. Queensl. Mus.* 20: 253–277.

Morales, J. A. 1984. Estructura de los nidos y comportamiento subsocial de *Xyleborus volvulus* (Fabricias) (Coleoptera, Scolytidae). *Folia Entomol. Mex.* 61: 35–47.

Müller, J. K. and A.-K. Eggert. 1989. Paternity assurance by "helpful" males: adaptations to sperm competition in burying beetles. *Behav. Ecol. Sociobiol.* 24: 245–249.

Norris, D. M. 1979. The mutualistic fungi of xyleborini beetles. In *Insect Fungus Symbiosis*. L. R. Batra, ed., pp. 53–63. New York: Allanheld, OSMUN Co.

Palestrini, C., E. Barbero and M. Zunino. 1992. The reproductive behaviour of *Kheper aeratus* (Gerstaecker) and the evolution of subsociality in Scarabaeidae (Coleoptera). *Ethol. Ecol. Evol.* (Spec. Issue) 2: 27–31.

Paulian, R., J. P. Lumaret and G. B. Monteith. 1983. La larve du genre *Cephalodesmius* Westwood. *Bull. Soc. Entomol. Fr.* 88: 635–648.

Pluot-Sigwalt, D. 1981. Le syste[ac]me glandulaire abdominal des Coléoptères coprophages Scarabaeidae: Ses tendances évolutives et ses relations avec la nidification. *Ann. Soc. Entomol. Fr.* (N.S.) 27: 205–229.

–. 1983. Les glandes tégumentaires des Coléoptères Scarabaeidae: répartition des glandes sternals et pygidiales dans la famille. *Bull. Soc. Entomol. Fr.* 88: 597–602.

–. 1988a. Le système des glandes tégumentaires des Scarabaeidae rouleurs, particulièrement chez deux espèces de *Canthon* (Coleoptera). *Folia Entomol. Mex.* 74: 79–108.

–. 1988b. Données sur l'activité et le rôle de quelques glandes tégumentaires, sternales, pygidiales et autres, chez deux espèces de *Canthon* (Col. Scarabaeidae). *Bull. Soc. Entomol. Fr.* 93: 89–98.

–. 1991. Le systéme glandulaire abdominal des Coléoptères coprophages Scarabaeidae: Ses tendances évolutives et ses relations avec la nidification. *Ann. Soc. Entomol. Fr.* (N.S.) 27: 205–229.

Reyes-Castillo, P. and G. Halffter. 1983. La structure sociale chez les Passalidae (Col.). *Bull. Soc. Entomol. Fr.* 88: 619–635.

Robertson, I. C. 1993. Nest intrusions, infanticide, and parental care in the burying beetle, *Nicrophorus orbicollis* (Coleoptera: Silphidae). *J. Zool. (Lond.)* 231: 583–593.

Rougon, D. and C. Rougon. 1982. Le comportement nidificateur des Coléoptères Scarabaeinae Oniticellini en zone sahélienne. *Bull. Soc. Entomol. Fr.* 87: 272–279.

Rubenstein, D. I. and R. W. Wrangham, eds. 1986. *Ecological Aspects of Social Evolution*. Princeton: Princeton University Press.

Sato, H. and M. Imamori. 1986a. Nidification of an African ball-rolling scarab, *Scarabaeus platynotus* Bates (Coleoptera: Scarabaeidae). *Kontyû* 54: 203–207.

–. 1986b. Production of two brood pears from one dung ball in an African ball-roller, *Scarabaeus aegyptiorum* (Coleoptera, Scarabaeidae). *Kontyû* 54: 381–385.

–. 1987. Nesting behaviour of subsocial African ball-roller *Kheper platynotus* (Coleoptera: Scarabaeidae). *Ecol. Entomol.* 12: 415–425.

–. 1988. Further observations on the nesting behaviour of a subsocial ball-rolling scarab, *Kheper aegyptiorum*. *Kontyû* 56: 873–878.

Sato, H. and K. Hiramatsu. 1993. Mating behaviour and sexual selection in the African ball-rolling scarab *Kheper platynotus* (Bates) (Coleoptera: Scarabaeidae). *J. Nat. Hist.* 27: 657–668.

Scholtz, C. H. 1990. Phylogenetic trends in the Scarabaeoidea (Coleoptera). *J. Nat. Hist.* 24: 1027–1066.

Scott, M. P. 1990. Brood guarding and the evolution of male parental care in burying beetles. *Behav. Ecol. Sociobiol.* 26: 31–39.

Tallamy, D. W. 1994. Nourishment and the evolution of paternal investment in subsocial arthropods. In *Nourishment and Evolution in Insect Societies*. J. H. Hunt and C. A. Nalepa, eds., pp. 21–55. Boulder, Co: Westview Press.

Tallamy, D. W. and T. K. Wood. 1986. Convergence patterns in subsocial insects. *Annu. Rev. Entomol.* 31: 369–390.

Thiele, H. U. 1977. *Carabid Beetles in their Environments*. Berlin/Heidelberg/New York: Springer-Verlag.

Tribe, G. D. 1975. Pheromone release by dung beetles (Coleoptera: Scarabaeidae). *S. Afr. J. Sci.* 71: 277–278.

Trumbo, S. T. 1990. Interference competition among burying beetles (Silphidae, *Nicrophorus*). *Ecol. Entomol.* 15: 347–355.

–. 1991. Reproductive benefits and the duration of parental care in a biparental burying beetle, *Nicrophorus orbicollis*. *Behaviour* 177: 82–105.

–. 1994. Interspecific competition, brood parasitism, and the evolution, of biparental cooperation in burying beetles. *Oikos* 69: 241–249.

Tyndale-Biscoe, M. 1983. Effects of ovarian condition on nesting behaviour in a brood-caring dung beetle, *Copris diversus* Waterhouse (Coleoptera: Scarabaeidae). *Bull. Entomol. Res.* 73: 45–72.

–. 1984. Adaptative significance of brood care of *Copris diversus* Waterhouse (Coleoptera: Scarabaeidae). *Bull. Entomol. Res.* 74: 453–461.

Valenzuela-González, J. E. 1992. Adult-juvenile alimentary relationship in Passalidae (Coleoptera). *Folia Entomol. Mex.* 85: 25–37.

Walter, P. 1978. Recherches écologiques et biologiques sur les Scarabéides Coprophages d'une savane du Zaïre. Sc. D. thesis, Université de Montpellier, France.

Wcislo, W. T. 1989. Behavioral environments and evolutionary change. *Annu. Rev. Ecol. Syst.* 20: 137–169.

Wilson, E. O. 1971. *The Insect Societies*. Cambridge: Belknap Press.

Zunino, M. 1983. Essai préliminaire sur l'évolution des armures génitales des Scarabaeinae, par rapport à la taxonomie du groupe et à l'évolution du comportement de nidification. *Bull. Soc. Entomol. Fr.* **88**: 531–542.

–. 1985. Las relaciones taxonómicas de los Phanaeina (Coleoptera, Scarabaeinae) y sus implicaciones biogeográficas. *Folia Entomol. Mex.* **64**: 101–115.

Appendix 11-1 Taxonomic scheme

The subfamily Scarabaeinae is divided into its respective tribes, showing the genera mentioned in the text and in Fig. 11-2. Subsocial care is present in some species, but not all, for the genera marked with an asterisk (*). Subsocial care is present in all species whose nidification has been studied, in those genera marked **. Letters in parentheses after tribes indicate: b, burrowers; d, dwellers; r, rollers; x, other form of horizontal relocation. Taxonomic structure after Halffter and Edmonds (1982), Zunino (1983, 1985) and Scholtz (1990).

Order	COLEOPTERA
Superfamily	SCARABAEOIDEA
	(or LAMELLICORNIA)
Family	SCARABAEIDAE
Subfamily	SCARABAEINAE (dung beetles)

ONTHOPHAGINI (b)
    Onthophagus
    Digitonthophagus
ONITINI (b)
    Phanaeus
    Onitis
ONITICELLINI (b, some d)
    *Oniticellus**
    *Tragiscus**
    Liatongus
    Drepanocerus
    *Cyptochirus**
COPRINI (b)
    *Copris***
    *Synapsis***
    *Helicopris***
    Catharsius
    Ontherus
    Dichotomius
    *Paraphytus**
    Ateuchus
    Canthidium
    Neocanthidium
EURYSTERNINI (d)
    *Eurysternus***
EUCRANIINI (x)
SCARABAEINI (r)
    *Canthon**
    *Kheper***
    *Scarabaeus**

12 · The evolution of social behavior in Passalidae (Coleoptera)

JACK C. SCHUSTER AND LAURA B. SCHUSTER

ABSTRACT

Most members of the Passalidae live in rotting wood. They occur in family groups including male and female parents, eggs, larvae, pupae, and teneral and mature offspring. All stages must eat the feces of the mature adults. Feces are comprised of wood that is fragmented, digested, inoculated with bacteria and fungi from the digestive tract of the adults, and further decomposed after being excreted, an example of an external rumen. Larvae and adults cooperate in pupal case construction and teneral adults repair pupal cases of siblings.

The selective advantages of adults and offspring staying together may include: (1) gaining protection and food from the log in which egg-laying occurred; and (2) adults supplying the larvae with shredded wood and feces, which speeds juvenile development and consequently increases the degree of overlap of generations. The selective value of cooperative pupal-case construction is apparently protection of the easily damaged pupa. The value of emergence of the adult offspring while their exoskeleton is still soft may be that they become inoculated with the wood-decomposing bacteria or fungi and help in pupal case repair before migrating. Hypotheses for the selective advantage of the male staying with the family include: (1) contribution to feeding his progeny; (2) his sperm, delivered in repeated copulations, stimulates egg production and may nourish the female; and (3) defense of his mate against other males. Both males and females defend the tunnel system against passalids of the same or opposite sex, which probably prevents their progeny or food from being eaten by an intruder.

Passalids show little interspecific variation in social behavior among the three species studied in detail and over 100 species examined in some respect. To infer stages in their social evolution, therefore, it is necessary to look at the most closely related families, the Lucanidae and the Scarabaeidae. Little is known of the lucanids. Significantly, in contrast to the burrowing scarab *Cephalodesmius*

armiger, passalid parents are alive when their adult offspring emerge and passalid teneral adults repair pupal cases. In contrast with the dung–soil interface scarabs *Oniticellus* and *Tragiscus*, the passalid male stays with the family.

INTRODUCTION

Passalids are large (14–80 mm long), pantropical beetles related to lucanids and scarabs. Most of the approximately 600 species (see Reyes-Castillo 1970 for a synopsis of passalid taxonomy and Schuster 1992 for larval taxonomy) live in family groups in rotting wood (usually at least two years dead), although a few species live in decaying plant material among the rhizomes of epiphytic ferns (Kon and Araya 1992) or in the detritus chambers of leaf-cutter ants (Hendrichs and Reyes-Castillo 1963; Schuster 1984). Their social behavior has been known since the works of Ohaus (1900, 1909). Although sound communication has been studied in 57 species (Schuster 1983) and life cycle and ecology has been studied in over 100 species, detailed studies of social behavior have been conducted on only a few species, all from the New World, notably *Odontotaenius disjunctus* (Illiger) (Miller 1932; Schuster 1975; Schuster and Schuster 1985), *Popilius haagi* (Kaup) (Schuster and Schuster 1985) and *Heliscus tropicus* (Percheron) (Valenzuela 1984). Passalids apparently have similar behavior and life cycles, which involve overlap of generations, cooperative brood care, and an external rumen. In this chapter, we describe the life cycle of passalids, discuss how their social behavior might have evolved and compare passalid social behavior to that of other insects in similar habitats.

BEHAVIOR AND LIFE CYCLE

A newly begun tunnel system usually contains one black (fully sclerotized) adult, either male or female. In 112 cases of nine species in newly colonized logs in Los Tuxtlas,

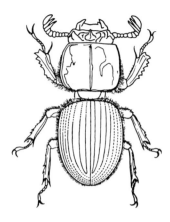

Fig. 12.1. Dorsal view of the giant, flightless passalid *Proculus goryi* Melly, endemic to the volcanos of the Pacific coast of Guatemala and Chiapas, Mexico.

Mexico, 61 were single males and 51 single females (M. L. Castillo, personal communication, 1994). The first adult is joined later by another of the opposite sex (see Schuster 1975 for log colonization experiments with *O. disjunctus*). Both sexes show territoriality, forcing intruders of the same and opposite sex out of the tunnel system, during which the defenders produce characteristic stridulations (Schuster 1975, 1983). Copulation occurs in the log (Castillo and Reyes-Castillo 1989, and personal observation) after an elaborate courtship ritual involving various stridulatory signals by both sexes (Schuster 1975, 1983).

Eggs are placed in a nest of finely chewed wood (Gray 1946). Females of *O. disjunctus* produce two to four, rarely six, eggs per day, and their nests usually have fewer than 35 eggs (Gray 1946). Adults have been observed carrying eggs in their mandibles, so eggs may be laid elsewhere in the tunnel and carried to the nest (Gray 1946; Schuster 1975). Isolated females can produce clutches months apart, either using sperm stored in the spermatheca or possibly by parthenogenesis (Schuster and Schuster 1985). When a male and female are together, repeated copulation occurs (Schuster and Schuster 1985). Although more sperm may not be essential for fertilization, repeated copulation may increase egg production; Fonseca (1988) found that females with little sperm in their spermathecae showed little ovarian activity, whereas those with full spermathecae had growing ovarian follicles.

Larvae feed on wood triturated (shredded and chewed) by the mature adults and on feces of the mature adults. The feces are richer in protein than the wood itself (Valenzuela

1984). Larval jaws are apparently not strong enough to excavate much wood directly from the log (Reyes-Castillo and Jarman 1981). Adults also feed on their own feces (Valenzuela 1984). Teneral and mature adults and larvae cannot survive on wood alone (Valenzuela 1984). Food passing through the digestive tract of the mature adult is contaminated with bacteria and fungi (Heymons and Heymons 1934), which continue digestion in the excreted feces in the manner of an external rumen (Mason and Odum 1969). Larvae are frequently found close to the adults. They possess specialized stridulatory organs, which produce weak sounds of unknown function. The possibility that the stridulations stimulate the adults to produce food for the larvae is untested.

When larvae are ready to pupate, they cooperate with adults in the construction of a frass (shredded wood and feces) case, the larva working from the inside, the adult from the outside, as described by Miller (1932). If the case is damaged, adults will repair it. Adults have been seen to build cases for conspecific pupae from other families in the laboratory; however, they killed pupae of other species (Valenzuela-Gonzalez 1993). Of particular interest is the observation that teneral-adult siblings of the pupa will also repair the case, though not as efficiently as the parents (Schuster and Schuster 1985; Valenzuela-Gonzalez 1993).

Passalid pupae are delicate and easily damaged. We suspect that the case protects the pupa from accidental mechanical damage, which may result in cannibalism, as well as from predators and parasites that might enter the tunnels (e.g. Reduviidae and Tachinidae) (Schuster 1975). A larva cannot construct a case without adult help, but it can pupate successfully without a case (Gray 1946 and personal observations).

The time from egg to pupal ecdysis is short, only two and one-half months in some species (Gray 1946; Schuster 1975). Adult *O. disjunctus* live for two or more years in nature (Schuster 1975). One adult *Proculus burmeisteri* Kuwert lived for five years in captivity (W. Dix, personal communication, 1986).

Just after ecdysis, adults are orange with white elytra. They are completely orange when they leave the pupal case, slowly darkening to red and then to black as they mature. This color change takes eight to ten weeks in some species, including *O. disjunctus* (Reyes-Castillo 1970; Schuster 1975), and four to over eight months in *H. tropicus* (Valenzuela-Gonzalez and Castillo 1984; Valenzuela-Gonzalez 1986a).

Teneral adults are not yet sexually mature (Virkki 1965; Virkki and Reyes-Castillo 1973); however, we have found eggs in the ovaries of dark red individuals. Teneral adults do not copulate, as demonstrated by their empty spermathecae (Fonseca 1988). We presume that black adults are sexually mature and capable of reproduction. Adults do not migrate until they are fully blackened. As they blacken, their exoskeleton becomes more sclerotized. Fonseca (1988) observed that some *Passalus* teneral adults excavate a gallery off the main system. We have found teneral adults of other species both alone in a portion of the tunnel system and in contact with black adults.

What is the number of reproductives (black adults) in a tunnel system? Data for occupants of fully explored tunnel systems are given in Table 12-1. Data for tunnel systems not necessarily fully explored, yet containing more than two black adults, are given in Table 12-2. In a single tunnel system at one time, it is possible to find black (mature) adults, red (teneral) adults, pupae, larvae and eggs, presumably of the same genetic lineage (Table 12-1). In 56 fully explored Neotropical and Philippine tunnel systems of multiple species, five contained a single black female without other black adults, three contained a single black male, 12 held a black male–female pair, two held two black females and a black male, one contained three unsexed black adults and one had four unsexed black adults (Table 12-1). Of the remaining 32 tunnel systems, 11 contained a single unsexed black adult and 21 held an unsexed black pair. Of 358 tunnel systems we investigated in the Neotropics and the Philippines, 59 had more than two black adults, representing 31 species (Table 12-2) of the 59 species involved.

The most common situation is one female and one male in a tunnel system (Table 12-1). In the case of a single female or a single male, we may have either the initiation of a new tunnel system or, if offspring are present, the loss of one of the partners. No evidence exists for females starting colonies alone after mating. If there are three or more black adults in a tunnel system (Tables 12-1 and 12-2), we likely have progeny with parents. Age, and thus generation, can be inferred by examining the wear on the mandibles. For example, in the Philippines, we found three black adults of *Leptaulax bicolor* (Fabricius) under bark in the same tunnel system. The male and one of the females had sharp mandibular teeth and were probably relatively young, whereas the second female had worn teeth, and was probably their mother.

Is there more than one reproducing female in a tunnel system? Gray (1946), working with *O. disjunctus*, determined that the maximum number of eggs laid in 24 h was six and that 16 days were required at 27 °C before ecdysis. Therefore, the maximum number of unhatched eggs in the nest of a given female should be no more than 96. Under bark in Chiapas, 430 eggs were found in a single nest of *Passalus interstitialis* Eschscholtz (Moron *et al.* 1988) and four nests of *Paxillus leachi* MacLeay each contained from 264 to 532 eggs (Valenzuela 1984). If the rates of egg-laying and development are the same as in *O. disjunctus*, then more than one female would have had to contribute to each of these nests. Nobody has dissected females from the same system to determine whether more than one female is producing eggs at the same time.

Reyes-Castillo and Halffter (1984) suggested that the dissolution of the colony occurs after adult progeny mature, expulsion occurring from sexual conflict and aggression; however, we have not seen good evidence for this. On the contrary, on several occasions, we have found more than two mature adults in a tunnel system (Tables 12-1 and 12-2). Mature adults may peacefully coexist until the beginning of the migration season. In tropical areas, where most species of passalid live, peak migration occurs at the beginning of the wet season, although adults may occasionally be found outside of logs during most of the year. In Guatemala, half (25 of 50) migrating beetles were collected in the last half of May, when the rainy season begins.

When passalids colonize new logs, they migrate by either flying or walking. Even beetles capable of flight will walk for long distances. We have observed flight only at night or at dusk. In Mexico and Guatemala, of 31 sexed dispersing passalids we found, 17 were female and 14 were male.

SELECTIVE VALUE OF SOCIAL BEHAVIOR

What factors appear to be important in the origin of complex subsociality in passalids? Haplodiploidy has been implicated in the evolution of social behavior in Hymenoptera; however, passalids have an XO or XY sex determination system, depending on the species (Virkki and Reyes-Castillo 1973). We propose that habitation of logs by adults and by larvae appears to be of key importance for the evolution of sociality in Passalidae.

First, what brings adults and juveniles together? Dispersing adult passalids tunnel into and take up abode in logs, where they lay eggs. The log supplies the adults with food and protection during and after copulation and oviposition. In contrast, many Cerambycidae and Scarabaeidae deposit their eggs in logs but only the larva lives in the wood.

Table 12-1. *Contents of 64 completely explored passalid tunnel systems*

L1, L2, L3, first, second, third larval instar; ( ), unknown larval instar; +, many; m, male; f, female.

Species and collection site	eggs	L1	L2	L3	Pre pupa	Pupa	Red adult	Intermediate adult	Black adult
PROCULINI:									
Oileus sargi	—	—	—	1	2	1	4	—	1
Guatemala	—	—	—	—	—	—	—	—	2
	—	—	—	—	1	—	—	mf	—
Chondrocephalus debilis	—	—	—	—	—	—	—	—	1
Guatemala	—	—	—	2	2	2	—	—	2
	—	—	—	—	—	—	1	—	2
Ch. purulensis	—	5	5	1	—	—	—	8	—
Guatemala									
Ch. granulifrons	1	—	—	—	—	—	—	—	2
Guatemala	—	—	—	—	—	—	—	—	2
	—	—	—	—	—	—	—	—	m
	—	—	—	—	—	—	—	—	mf
	—	—	—	—	—	—	—	—	mf
Ch. n. sp.	12	2	—	—	—	—	—	—	2
Guatemala									
Vindex sp.	1	—	—	1	—	—	—	—	1
Guatemala	1	—	—	2	1	—	1	1	4
Petrejoides reyesi	—	—	—	1	—	2	—	—	mf
Honduras	—	—	—	—	—	—	—	—	mf
	—	—	1	3	—	—	—	—	2
	—	—	1	2	—	—	—	—	mf
P. guatemalae	—	—	—	3	—	—	—	—	mf
Guatemala	—	—	—	—	—	—	—	—	1
Odontotaenius	—	—	—	—	—	—	—	1	—
striatopunctatus	1	—	—	—	—	—	ff	—	—
	—	—	—	—	—	—	—	m	—
	—	—	—	—	—	—	—	mf	—
O. zodiacus	5	—	—	—	—	—	—	—	f
Sierra Madre Oriental, Mexico									
Popilus haagi	—	—	—	—	—	—	—	—	2
Guatemala									
Proculejus brevis	—	—	—	—	—	2	2	—	1
SMO, Mex									
Spurius bicornis	—	—	—	—	—	—	—	—	2
Ogyges championi	—	—	—	—	—	—	—	2	*1*
Guatemala	2	—	—	—	—	—	—	—	2
	—	—	—	3	1	—	—	m	1
	—	—	—	—	—	2	1	—	—
O. crassulus	—	—	—	—	—	—	—	—	f

Table 12-1. (*cont.*)

Species and collection site	eggs	L1	L2	L3	Pre pupa	Pupa	Red adult	Intermediate adult	Black adult
O. adamsi Honduras	2	+	+	—	—	—	—	—	2
O. hondurensis Honduras	—	—	—	—	—	—	—	—	2
O. furcillatus Guatemala	—	—	—	—	—	—	—	—	ffm
Proculus mniszechi Guatemala	—	—	—	—	3	—	—	—	2
	—	—	—	—	—	—	—	—	2
P. burmeisteri Honduras	—	—	—	—	—	—	—	—	f
Publius n.sp. Costa Rica	—	—	—	2	—	—	—	—	mf
	—	—	—	—	—	—	—	—	2
Pseudacanthus subopacus Guatemala	—	—	—	—	—	—	—	—	2
Veturius boliviae Peru	—	—	—	—	—	2	1	—	1
	—	—	—	—	—	—	—	—	1
Verres hageni	—	—	—	—	—	—	—	—	1
	—	—	—	—	—	—	—	—	2
V. corticicola PASSALINI:	—	—	—	—	—	—	—	—	2
Passalus n.sp. Guatemala	—	—	—	—	—	—	—	—	mf
P. aff. *morio*	—	—	(2)	—	—	2	1	—	3
P. punctatostriatus	—	—	—	—	—	—	—	—	m
	—	—	—	—	—	2	7	3	2
	—	—	—	2	1	—	—	—	mf
	—	—	—	—	—	—	—	—	1
	—	—	—	—	—	—	—	—	m
	11	13	7	—	—	—	—	—	2
	—	—	—	—	—	—	—	—	ffm
	—	—	—	—	—	—	—	—	2
	—	—	—	—	—	—	—	—	mf
	—	—	—	—	—	—	—	mff	mf
	—	—	—	—	—	—	mmmff	—	f
P. punctiger	—	—	(3)	—	—	—	—	—	mf
P. coniferus Peru	—	—	—	—	—	1	—	—	f

Table 12-2. *Passalid tunnel systems with more than two black adults*

The whole tunnel system was not necessarily traced; therefore, it might have contained even more than the number of black adults found. Abbreviations: m, one male; f, one female.

Species	No. of tunnel systems: occupant sex not determined	No. of tunnel systems: occupant sex determined
PROCULINI		
Chondrocephalus debilis	2	—
Ch. purulensis	3	—
Ch. granulifrons	1	—
Ch. sp.	1	—
Petrejoides tenuis	1	—
P. sp. (Mexico)	—	1 : ffm
Spurius bicornis	2	—
Popilius haagi	2	—
P. eclipticus	4	—
P. sp.	1	—
Pseudoarrox karreni	1	1 : ffmmm
Vindex sp.	2	—
Oileus sargi	4	1 : mmm
Ogyges adamsi	1	—
O. crassulus	1	1 : ffm
O. furcillatus	—	1 : ffm
Pseudacanthus grannulipennis	1	—
Odontotaenius striatopunctatus	2	1 : fffm
Proculus mniszechi	1	—
P. opacipennis	1	—
Verres hageni	1	—
V. cavicollis	—	1 : fff
V. furcilabris	1	—
PASSALINI		
Passalus caelatus	2	—
P. aff. *morio*	1	—
P. spiniger	4	—
P. guatemalensis	1	—
P. punctatostriatus	4	1 : ffm
P. punctiger	2	1 : mmf
P. unicornis	2	—
Leptaulax bicolor	—	1 : ffm

Second, what favors extended cohabitation by parents and offspring? Tallamy and Wood (1986) suggest that insects can utilize wood, which is hard to fragment, low in nitrogenous nutrients and difficult to digest and assimilate, in one of three ways: '(a) through obligate symbiotic relationships with cellulose-digesting protozoans, (b) by feeding indirectly on cultivated or naturally occurring fungi, and/or (c) if the organism can tolerate slow growth and development'. No protozoans have yet been found in passalid digestive tracts; however, fungi have been found (Heymons and Heymons 1934). Dispersing passalid adults bring the wood-digesting microorganisms in their digestive tracts. The adults shred the wood, eat it, and pass inoculated feces, which continue digestion on the floor of the tunnel. These feces are an essential food for all stages (Valenzuela 1984). The predigested wood and less investment in larval jaws allow the larvae to develop more rapidly, two and one-half to three months in medium-sized passalids, compared with several years for log-inhabiting larvae of Cerambycidae and Scarabaeidae that are found without adults. This rapid development gives rise to the presence of older, even adult, progeny in a tunnel system with reproductive parents. Dependence on adults for treated food (feces) may select for longer-lived adults, again increasing interaction between parents, immatures and adult progeny.

Third, what selects for cooperative pupal-case construction? The soft and delicate prepupae and pupae, being sluggish or immobile, are vulnerable to accidental (or deliberate) mechanical damage, which may result in cannibalism or attack by predators or parasites. Piling tunnel waste against prepupae and pupae could help them to avoid detection or damage, which could lead to the more formal construction of pupal cases.

Fourth, what selects for the adult progeny staying in the colony after emergence from the pupal case? (1) The teneral adults can eat and thereby become innoculated with the wood-digesting microorganisms and build food reserves; or (2) they can work in pupal case building and repair, behavior that augments the inclusive fitness of the teneral adults. Before leaving the log, Cerambycidae and Scarabaeidae remain in pupal cells until relatively hard. Passalids emerge from the pupal case in a soft, teneral state. In this condition they are easily injured and it is advantageous for them to stay in the log until the exoskeleton is harder. (The early emergence itself may have been selected for by factors (1) and (2) above).

Fifth, what is the advantage to the male of staying with the female in a log? Tallamy and Wood (1986) point out that cooperation between male and female is frequent when wood is the food source because of the difficulty in fragmenting, digesting and enriching it. In addition, multiple copulations may increase the availability of sperm for fertilization, stimulate egg production, and/or provide nutrients for use in egg construction. In addition, the male may protect the female from other males.

Sixth, why should both sexes defend the tunnel system against intruders of either sex (Schuster and Schuster 1985)? Apparently, aggression occurs only when immatures are present in the colony, and more frequently against the same sex (Valenzuela-Gonzalez 1986b). It would be disadvantageous for parents and their progeny to compete for food with other individuals of the same or opposite sex. A new individual of the opposite sex, however, could provide more opportunities for reproduction. Because cannibalism of larvae is frequent in passalids, at least in the laboratory, we hypothesize that it would be advantageous for an intruder to eat any larvae present not its own. Ejecting members of the opposite sex thus becomes advantageous because a new spouse might eat larvae by an earlier spouse. It would be interesting to test this hypothesis by introduction experiments with established laboratory colonies.

STAGES IN THE EVOLUTION OF SOCIAL BEHAVIOR

Studies to date have not revealed any substantive or qualitative differences in social behavior among species of Passalidae. Therefore, we look outside the family for possible stages in their social evolution. The closest relatives to Passalidae are the Scarabaeidae and Lucanidae. We lack data on the Lucanidae; however, *Cephalodesmius armiger* Westwood, a scarab that burrows in the ground (Halffter, this volume), possesses certain behavioral and ecological similarities with, for example, *O. disjunctus*, a passalid that tunnels in wood.

A comparison of these two species, with the scarab data taken from Monteith and Storey (1981) (see also Halffter, this volume), helps in understanding the evolution of sociality in scarab and passalid beetles. (1) Both *C. armiger* and *O. disjunctus* make tunnels, the former in soil, the latter in wood. (2) Both excavate a larger chamber in the tunnel; the eggs are located in this chamber. *C. armiger* lays the eggs in this chamber. It is not known whether *O. disjunctus* lays the eggs elsewhere and then carries them to the chamber. (3) Larval and adult food is partly decayed plant material mixed with gut microorganisms of male and female and allowed to decay further. In *C. armiger*, leaves, flowers, seed parts, fruit and excrement of other animals are brought to the tunnel, at first by both sexes, later only by the male. The female forms

these materials into a mass and adds the feces of both parents. Larvae and adults feed on this mass. In *O. disjunctus*, wood bordering the tunnels is shredded and eaten by male and female black adults, and their feces are eaten by all stages. (4) The members of the pair copulate repeatedly with each other in both species. (5) Larvae are confined with their food. *C. armiger* forms a ball from the decaying mass and lays an egg in this; the larva develops within and is thus prevented from straying in the earth chamber. *O. disjunctus* forms tunnels littered with feces, which restrain the larva with its food. (6) Food must be continually supplied to the larvae, and the larvae stridulate. The *C. armiger* female continually adds decaying material to the outside of the brood balls. The larvae inside stridulate by a gula–abdomen mechanism. Adults do not stridulate. *O. disjunctus* adults continually produce feces, which the larvae eat. Larvae stridulate by rubbing the third leg on the second coxa. Adults do stridulate. (7) The female forms the outer surface of the pupal case. The *C. armiger* female applies a mixture of adult and larval excrement to the outer surface of the brood ball. At this stage, the male is excluded from the nest chamber by a soil plug. *O. disjunctus* parents and adult siblings form the outer surface of the pupal case. (8) At pupal ecdysis, *C. armiger* parents are usually dead but *O. disjunctus* parents are usually alive. At this time food is probably scarce for *C. armiger*, yet usually still abundant for *O. disjunctus*. (9) Teneral adults ingest fungi associated with the larval food source. In *C. armiger*, fungal fruiting bodies found on the inside of the pupal case and in the digestive tubes of tenerals were similar. In *O. disjunctus*, teneral adults eat the feces in the tunnels.

If a passalid ancestor with the characteristics of *C. armiger* began to tunnel in a log, many of the differences of the passalid from the burrowing scarab should appear. First, foraging is not necessary, as the beetles are living in their food. Second, because there is no danger that the feces will become mixed with soil, there is no selection to form them into a mass. In fact, decomposition is faster with greater dispersion and hence better aeration. Third, the tunnel keeps the larva with its food, so there is no gain from the construction of a ball around the larva. Fourth, the food source is exhausted slowly, so there is selection for long parental life; the parents are therefore alive when the offspring emerge and there is selection for delayed dispersal of offspring, which opens up the opportunity for adult siblings to aid in pupal-case construction and repair. Presumably, the food gathered by *C. armiger* becomes scarce owing to its seasonality in Australia.

An alternative comparison may be made between (1) two genera of scarabs, *Oniticellus* and *Tragiscus*, which nest at the dung–soil interface (or within the dung) and do little burrowing (Halffter, this volume) and (2) the passalid *Taeniocerus platypus* (Kaup), which lives at the log–soil interface and does not tunnel in wood (Kon and Araya 1992).

A comparison of these species follows, with the scarab data based on literature summarized by Halffter (this volume) and the data for *T. platypus* from Kon and Araya (1992). (1) *Oniticellus* and *Tragiscus* and *Taeniocerus platypus* live at the interface between ground and food; they tunnel little or not at all. The scarabs make the nest at the dung soil interface or in the interior of the dung pat. The *T. platypus* family lives on the ground under the log, not tunneling into the log, but shredding wood from the log's lower surface. (2) Decaying plant material is the food source for all, dung for the scarabs and wood for the passalid. (3) Little or no transport of food occurs; the scarabs make dung balls next to, or within, the dung pat. No transport is reported for the passalid. (4) In *Oniticellus* and *Tragiscus*, the female alone makes the nest and brood balls. (The male's principal role in nesting in most Scarabaeinae is relocation of dung, which is not carried out by these genera.) In *Taeniocerus platypus*, a male and two females (color not reported) were found with pupae and larvae. (5) The female's excrement is mixed with food for the larvae. In *Oniticellus* and *Tragiscus* no mention is made of this by Halffter, but it is probably unavoidable. In *Taeniocerus platypus*, male and female (parents?) were present with larvae, which, therefore, could eat adult excrement. (6) The larvae are confined within their food. *Oniticellus* and *Tragiscus* make dung balls in which an egg is laid and a larva develops. A large amount of wood frass and feces on the floor of the family area is reported for *Taeniocerus bicanthatus* (Perch.) (Kon and Johki 1987), and thus can be expected with *T. platypus*. The *T. platypus* larvae would, therefore, have processed food below them and wood above them. (7) Food is continually supplied to larvae during development. *Oniticellus cinctus*, *O. formosus* and *O. planatus* mothers add material to the outside of brood balls after ovipositing and during the first stages of larval development. In *T. platypus*, the adult male and female apparently supply shredded wood and feces to offspring into the adult stage. (8) The larvae may be encased with soil. *O. egregius* mothers cover the brood balls with soil and leave soon after ovipositing. *T. platypus* pupal cases were made of soil. This implies that passalid adults excavated soil. (9) The female takes care of the nest until the

teneral adults appear. This is true of *Oniticellus planatus* and *O. formosus*. In *T. platypus*, the presence of a male and two females with pupae suggests that one female is a daughter and, therefore, that both father and mother are present with adult offspring.

If a passalid ancestor with the characteristics of *Oniticellus* and *Tragiscus* began to eat wood at the soil interface, certain differences between the passalid and the interface scarab should appear. First, males would be selected to remain with the offspring to help the female fragment and digest the wood (Tallamy and Wood 1986). Alternatively, the help of the male in feeding the offspring, which occurs in various dung beetles (Halffter, this volume), could be ancestral and the non-participation derived in *Oniticellus* and *Tragiscus*. Second, it would be disadvantageous for passalids to form balls of triturated wood and adult excrement around individual larvae, since this would force the larvae to repeatedly eat their own protein-depleted feces accumulating inside the ball rather than the richer adult feces. Larval excrement from a wood diet would not be expected to be as rich in nitrogen as larval excrement from a dung diet. The latter might act as nitrogenous fertilizer for the decomposers on the inside of the dung ball. Also, larval excrement in Passalidae is solid and disc-shaped and would not pass through fissures in the walls of the brood ball as does the apparently semiliquid excrement of *Cephalodesmius* larvae (Halffter, this volume).

SIMILAR WOOD-INHABITING INSECTS

Additional associations of complex social behavior with adults and immatures living and feeding in wood are found in termites and *Cryptocercus* cockroaches (Wilson 1971; Nalepa and Bell, this volume; Shellman-Reeve, this volume) and in beetles in the families Scolytidae, Platypodidae (see, for example, Wheeler 1923; Kirkendall *et al.*, this volume), Tenebrionidae (*Phrenapates*) (Arrow 1951), Staphylinidae (*Leptochirus*) and probably Lucanidae (*Nigidius*). *Leptochirus* lives under bark of rotting logs in colonies that include adults, larvae and pupae. Pupal cases of these staphylinids, similar to those of passalids, were found (J. C. and L. B. Schuster, personal observations) in cloud and rain forests of Guatemala and Peru. More than four adults of *Nigidius* were found together in the same tunnel systems in lowland forests in southernmost Taiwan and near Subic Bay in the Philippines in December. Morphologically, these lucanids are very similar to some passalids of the Oriental subfamily Aulacocyclinae, especially in possessing vertical horns on the mandibles. Little is known concerning the social behavior of the latter three genera. In contrast, both adults and larvae live and feed in logs in some true bark beetles (Scolytidae) but little social behavior is displayed (Kirkendall *et al.*, this volume). Why? In most species, the larvae tunnel off alone and pack the tunnels behind themselves with frass. It thus is impossible for the adult to be supplying fragmented and digested wood or microorganisms to these larvae as do passalids. Parental feeding of larvae as well as a habitat in common between parents and offspring is therefore important for the development of social behavior.

Additional associations of high levels of social behavior with symbiotic microorganisms decomposing wood occur in termites, *Cryptocercus* roaches (Nalepa and Bell, this volume), Scolytidae and Platypodidae (Tallamy and Wood 1986; Kirkendall *et al.*, this volume). Halffter (1991) suggests that the development of the external rumen is the major factor in the development of complex social behavior in beetles, especially Passalidae, Scarabaeidae, Silphidae and perhaps *Phrenapates*. Therefore, we suggest that high levels of social behavior are promoted both by log cohabitation by adults and juveniles and trophic relationships with symbiotic microorganisms.

SUGGESTIONS FOR FUTURE WORK

Much more needs to be learned concerning passalid ecology and social behavior. The following questions should be investigated. Since five or more species may live in the same logs, how do different species divide the resources? Is there competition between species? Do those living under bark have a different social structure from those invading the heartwood? How do Old World and New World passalids compare with respect to ecology and social structure? What similarities and differences exist between passalid and lucanid ecology and behavior?

Concerning juvenile social behavior and life cycle: what is the function of the larval sounds? Do teneral adults help defend a tunnel system against other passalid invaders? How long do adult progeny remain in the natal tunnel system? Do energy reserves accumulate in the teneral adult after emergence from the pupa?

Concerning colony composition and adult behavior: Do colonies last for multiple generations, and if so, are these communal groups of parents and offspring, all breeding? Why is there (apparently) no reproductive dominance? Interesting data could be obtained by dissecting beetles to

determine the degree of egg and sperm presence in various females from the same system. To what degree does biparental and communal care of juveniles exist? Are introduced (new) potential spouses accepted if the previous spouse is removed, yet larvae are still present? Will a new spouse kill larvae of a previous spouse?

ACKNOWLEDGEMENTS

We thank Gonzalo Halffter and Jae Choe for stimulating us to write this chapter. Charles MacVean, Kalara Schuster, Bernie Crespi and two anonymous reviewers provided very helpful criticisms. Pedro Reyes-Castillo and Maria Luisa Castillo have provided information and stimulating discussion over the years.

LITERATURE CITED

Arrow, G. J. 1951. *Horned Beetles*. The Hague: W. Junk.

Castillo, C. and P. Reyes-Castillo. 1989. Copulation in natura of passalid beetles (Coleoptera: Passalidae). *Coleopt. Bull.* **43**: 162–164.

Fonseca, C. R. V. 1988. Contribucao conhecimento da bionomia de *Passalus convexus* Dalman, 1817 e *Passalus latifrons* Percheron, 1841 (Coleoptera: Passalidae). *Acta Amazon.* **18**: 197–222.

Gray, I. E. 1946. Observations on the life history of the horned passalus. *Am. Midl. Nat.* **35**: 728–746.

Halffter, G. 1991. Feeding, bisexual cooperation and subsocial behavior in three groups of Coleoptera. In *Advances in Coleopterology*. M. Zunino, X. Belles and M. Blas, eds, pp. 281–296. Barcelona: AEC.

Hendrichs, J. and P. Reyes-Castillo. 1963. Asociación entre coleopteros de la familia Passalidae y hormigas. *Ciencia Mex.* **22**: 101–104.

Heymons, R. and H. Heymons. 1934. Passalus und seine intestinale Flora. *Biol. Zentr.* **54**: 40–51.

Johki, Y. and M. Kon. 1987. Morpho-ecological analysis on the relationship between habitat and body shape in adult passalid beetles (Coleoptera: Passalidae). *Mem. Fac. Sci., Kyoto Univ., Biol. Ser.* **12**: 119–128.

Kon, M. and K. Araya. 1992. On the microhabitat of the Bornean passalid beetle, *Taeniocerus platypus* (Coleoptera, Passalidae). *Elytra* **20**: 129–130.

Kon, M. and Y. Johki. 1987. A new type of microhabitat, the interface between the log and the ground, observed in the passalid beetle of Borneo *Taeniocerus bicanthatus* (Coleoptera: Passalidae). *J. Ethol.* **5**: 197–198.

—. 1992. Passalid beetles (Coleoptera, Passalidae) collected from Sabah, Borneo, with special reference to their colony composition and habitats. *Elytra* **20**: 207–216.

Mason, W. and E. Odum. 1969. The effect of coprophagy on retention and bioelimination of radionuclides by detritus-feeding animals. *Proc. Second Nat. Symp. Radioecology, Michigan*, pp. 721–724.

Monteith, G. B. and R. I. Storey. 1981. The biology of *Cephalodesmius*, a genus of dung beetles which synthesizes "dung" from plant material (Coleoptera: Scarabaeidae: Scarabaeirae). *Mem. Queensl. Mus.* **20**: 253–277.

Moron, M., J. Valenzuela and R. Terron. 1988. La macro-coleopterofauna saproxilofila del Soconusco, Chiapas, Mexico. *Folia Entomol. Mex.* **74**: 145–158.

Miller, W. C. 1932. The pupa-case building activities of *Passalus cornutus* Fab. (Lamellicornia). *Ann. Entomol. Soc. Am.* **25**: 709–712.

Ohaus, F. 1900. Bericht ber eine entomologische Reise nach Centralbrasilien. *Stett. Entomol. Z.* **61**: 164–173.

–. 1909. Bericht ber eine entomologische Studienreise in S damerica. *Stett. Entomol. Z.* **70**: 1–39.

Reyes-Castillo, P. 1970. Coleoptera, passalidae: morfologia y division en grandes grupos; generos americanos. *Folia Entomol. Mex.* **20–22**: 1–240.

Reyes-Castillo, P. and G. Halffter. 1984. La estructura social de los Passalidae (Coleoptera: Lamellicornia). *Folia Entomol. Mex.* **61**: 49–72.

Reyes-Castillo, P. and M. Jarman. 1981. Estudio comparativo de la fuerza ejercida por las mandibulas de larva y adulto de Passalidae (Coleoptera, Lamellicornia). *Folia Entomol. Mex.* **48**: 97–99.

Schuster, J. 1975. Comparative behavior, acoustical signals, and ecology of New World Passalidae (Coleoptera). Ph. D. dissertation, University of Florida.

–. 1983. Acoustical signals of passalid beetles: complex repertoires. *Fla. Entomol.* **66**: 486–496.

–. 1984. Passalid beetle (Coleoptera: Passalidae) inhabitants of leafcutter ant (Hymenoptera: Formicidae) detritus. *Fla. Entomol.* **67**: 176–177.

–. 1992. Passalidae: State of larval taxonomy with description of new world species. *Fla. Entomol.* **75**: 357–369.

Schuster, J. and L. Schuster. 1985. Social behavior in passalid beetles (Coleoptera: Passalidae): cooperative brood care. *Fla. Entomol.* **68**: 266–272.

Tallamy, D. and T. Wood. 1986. Convergence patterns in subsocial insects. *Annu. Rev. Entomol.* **31**: 369–390.

Valenzuela, J. 1984. Contribution a l'étude du comportement des Passalidae (Coleoptera, Passalidae). Thesis, Univ. Paris XIII. 112pp.

Valenzuela-Gonzalez, J. 1986a. Life cycle of the subsocial beetle *Heliscus tropicus* (Coleoptera: Passalidae) in a tropical locality in Southern Mexico. *Folia Entomol. Mex.* **68**: 41–51.

–. 1986b. Territorial behavior of the subsocial beetle *Heliscus tropicus* under laboratory conditions (Coleoptera: Passalidae). *Folia Entomol. Mex.* **70**: 53–63.

–. 1993. Pupal cell-building behavior in passalid beetles (Coleoptera: Passalidae). *J. Insect Behav.* **6**: 33–41.

Valenzuela-Gonzalez, J. and M. L. Castillo. 1984. El comportamiento de cortejo y copula en *Heliscus tropicus* (Coleoptera, Passalidae). *Folia Entomol. Mex.* **61**: 73–92.

Virkki, N. 1965. Insect gametogenesis as a target? *Agr. Sci. Rev.* **3**: 24–37.

Virkki, N. and P. Reyes-Castillo. 1973. Cytotaxonomy of Passalidae (Coleoptera). *An. Esc. Nac. Cienc. Biol. Mex.* **19**: 49–83.

Wheeler, W. 1923. *Social Life Among the Insects*. New York: Harcourt, Brace and Co.

Wilson, E. O. 1971. *The Insect Societies*. Cambridge, Mass.: Belknap Press.

13 · The evolution of social behavior in the augochlorine sweat bees (Hymenoptera: Halictidae) based on a phylogenetic analysis of the genera

BRYAN. N. DANFORTH AND GEORGE. C. EICKWORT

ABSTRACT

In this chapter we review the published literature on the social behavior of the halictid tribe Augochlorini, and present a generic-level phylogeny. The augochlorine genera show a wide variety of social behaviors ranging from solitary nesting to eusociality, and there is considerable within-species variation. Most species can be characterized as facultatively solitary to semisocial; three genera, *Augochlorella*, *Pereirapis* and *Augochlora (Oxystoglossella)*, contain primitively eusocial species. In order to determine the pattern of social evolution in the Augochlorini we performed a phylogenetic analysis using eighty-one morphological characters derived from Eickwort's (1969b) generic revision of the tribe. These phylogenetic results have some significance for interpreting patterns of social evolution within the Augochlorini. First, although eusociality occurs in three genera, our phylogeny indicates that eusociality arose once within the tribe, in the common ancestor of *Augochlorella*, *Ceratalictus*, *Pereirapis* and *Augochlora* sensu lato. Second, these results also suggest that eusociality in the Augochlorini arose independently of its origin in the Halictini. Third, the existence of solitary behavior in at least one species of *Augochlora* sensu stricto can be most parsimoniously explained as a reversal to solitary nesting from a eusocial ancestor. Finally, the sole kleptoparasitic genus in the Augochlorini, *Temnosoma*, appears closely related to its presumed host, *Augochloropsis* sensu lato. We discuss the importance of these results for other studies of social evolution and indicate future directions for research on this tribe of bees.

INTRODUCTION

A common and implicit assumption made by students of social insects is that eusocial behavior is always derived relative to solitary behavior (see, for example, Wilson 1971). In many cases this is clearly true (e.g., Isoptera relative to Dictyoptera, Formicidae relative to the Scolioidea). But in the superfamily Apoidea (the bees), and the Halictidae (the sweat bees) in particular, there are many taxa containing both solitary and eusocial members, including Allodapini (Schwarz *et al.*, this volume), Ceratinini, Halictini (Wcislo, this volume), and the Augochlorini. Given only knowledge of the behavior and biology of the solitary and eusocial members of a group, it is impossible to interpret the direction of evolutionary change in behavior. Such an interpretation can only be based on a phylogenetic hypothesis. Cladistic reasoning provides the best way to identify the direction of character state change in social evolution (cf. Carpenter 1989; Sillén-Tullberg 1988). Ridley (1983), Pagel and Harvey (1988), and Brooks and McLennan (1991) provide reviews of the use of phylogenetic analysis in reconstructing character evolution.

The tribe Augochlorini is an excellent group with which to illustrate this application of cladistic reasoning to the study of social evolution. Within this group there are solitary, communal, semisocial and primitively eusocial species. Three of the 33 genera and subgenera recognized by Eickwort (1969b) and Moure and Hurd (1987) (as modified in Table 13-1) are known to contain primitively eusocial species: *Augochlorella* (Ordway 1964, 1965, 1966), *Pereirapis* (Oliveira Campos 1980) and *Augochlora* sensu lato (Eickwort and Eickwort 1972). The genus *Augochlora* contains subgenera that include both solitary (*Augochlora* sensu stricto, Stockhammer 1966) and eusocial (*A.* [*Oxystoglossella*], Eickwort and Eickwort 1972) species (the third subgenus, *A.* [*Mycterochlora*], is not known behaviorally). *Augochlora* sensu lato and *Augochlorella* have been considered as close relatives by some authors (Vachal 1911; Moure 1943; Eickwort 1969b). It is therefore possible that the solitary behavior seen in some species of *Augochlora* sensu stricto has arisen from a eusocial ancestor

Table 13-1. *Augochlorine genera and subgenera*

Taxa that were not included in the phylogenetic analysis are indicated with an asterisk. Numbers in parentheses are estimated number of species (based on Moure and Hurd 1987).

Genus	Typical Social behaviors observed
Subgenus	
Corynura Spinola (23)	
Corynura Spinola (18)	Semisocial or communal ?
Callochlora Moure (4)	Semisocial or communal ?
Halictillus Moure (1)	Semisocial or communal ?
Rhinocorynura Schrottky (5)	Semisocial
Corynurella Eickwort (1)	—
Rhectomia Moure * (1)	—
Neocorynura Schrottky (65)	
Neocorynura Schrottky (64)	Usually solitary (rarely semisocial)
Neocorynuroides Eickwort (1)	—
Paroxystoglossa Moure (9)	Usually solitary (rarely semisocial)
Andinaugochlora Eickwort (1)	—
Chlerogas Vachal * (2)	—
Augochloropsis Cockerell (138)	
Augochloropsis Cockerell (14)	Solitary and semisocial (rarely communal)
Paraugochloropsis Schrottky (32)	Solitary and semisocial
Augochlorodes Moure (1)	Semisocial
Thectochlora Moure (1)	—
Augochlora Smith (131)	
Augochlora Smith (82)	Solitary and semisocial
Oxystoglossella Eickwort (28)	Eusocial
Mycterochlora Eickwort (3)	—
Augochlorella Sandhouse (16)	Eusocial (rarely semisocial)
Ceratalictus Moure (5)	—
Pereirapis Moure (6)	Eusocial
Pseudaugochloropsis Schrottky (7)	Solitary and semisocial
Caenaugochlora Michener (14)	
Caenaugochlora Michener (13)	Solitary and semisocial
Ctenaugochlora Eickwort (1)	—
Megalopta Smith (28)	Semisocial or communal ?
Megommation Moure (4)	
Megommation Moure (1)	Semisocial
Megaloptina Eickwort (2)	—
Megaloptilla Moure and Hurd *(1)	—
Megaloptidia Cockerell (4)	—
Micrommation Moure *(1)	—
Ariphanarthra Moure (1)	—
Chlerogella Michener* (3)	—
Temnosoma Smith (7)	Kleptoparasitic

Source: Eickwort 1969b; Moure & Hurd 1987.

(if *Augochlora* sensu lato is monophyletic and is the true sister group to *Augochlorella*), as was suggested by Michener (1990), or that eusociality has arisen independently in *Augochlorella* and *Oxystoglossella* (if both are found to arise from separate clades of predominantly solitary species). Following our review of the biology of the Augochlorini, we present a phylogeny based on morphological characters to distinguish these alternative hypotheses.

AUGOCHLORINE LIFE HISTORY

The tribe Augochlorini is restricted to the western hemisphere; these bees have their greatest diversity in the tropics and subtropics, where they may represent the majority of halictid species and individuals (see, for example, Heithaus 1979b; Laroca *et al.* 1982). Over 75% of the 470 described species occur in South America, about 20% in North America, and about 5% in the West Indies (Moure and Hurd 1987; Eickwort 1988). Two genera (*Corynura* and *Halictillus*) occur in south temperate Chile and are restricted to that region. Five genera (*Pseudaugochloropsis*, *Temnosoma*, *Augochloropsis*, *Augochlora* and *Augochlorella*) range into the USA, but only the latter three genera occur north of the Mexican border. In contrast to the Chilean augochlorines, these are primarily tropical genera, and only *Augochlorella* contains more than one species restricted to the nearctic region, with one species (*A. striata*) ranging as far north as Cape Breton Island, Canada (Packer *et al.* 1989b; Packer 1990). Thus the social biology of the augochlorine bees largely reflects long-term occupation of tropical and subtropical environments, although there are a few genera in temperate climatic regimes. None of the species occurs so far north as to have adapted to the shortened season of the subarctic, and none successfully crossed the Bering land bridge during the Tertiary or Pleistocene; thus the Augochlorini are absent from Eurasia and Africa. Similarly, no augochlorines have adapted to high altitudes in the North Temperate Zone, as have some halictine species (Eickwort *et al.* 1996), although at least two genera (*Andinaugochlora* and *Neocorynura*) have adapted to high altitudes in the Andes (Eickwort 1969b). Unfortunately, nothing is known about the biology of the high Andean sweat bees.

The common features of augochlorine nesting biology are as follows. Nests are usually initiated by a single, inseminated female, although multifoundress associations are known (Ordway 1966; Michener and Kerfoot 1967). In localities with a pronounced seasonal cycle caused by the intervention of winter or alternation of dry and wet seasons,

females may enter diapause before the beginning of the unfavorable season and spend it as inseminated gynes. In many tropical localities, especially those without discrete and prolonged dry seasons, there are no discrete periods of inactivity (Heithaus 1979c). In these areas nests may be founded throughout the year and in any one locality be developmentally asynchronous (see, for example, Gimenes *et al.* 1991).

Most species construct their nests in the soil, but all *A. (Augochlora)* (Stockhammer 1966; Sakagami and Moure 1967; Eickwort and Eickwort 1973a) and *Megalopta* (Sakagami 1964; Sakagami and Moure 1967; Janzen 1968) and some (but not most) *N. (Neocorynura)* (Lüderwaldt 1911; Schremmer 1979; Eickwort 1979) construct nests in rotting or standing wood. The abundance and diversity of *Augochlora* sensu stricto in forested habitats points to the importance of fallen logs and standing wood as nest sites. Largely unexplored is the importance of canopy twigs and branches for nests of *Megalopta* and other wood-nesting augochlorines (Wolda and Roubik 1986).

Nest architecture varies considerably among genera of Augochlorini, although patterns are typically consistent within genera (Eickwort and Sakagami 1979). Eickwort and Sakagami hypothesized that the primitive arrangement is to excavate horizontal cells adjacent to the main burrow, as in *Corynura (Callochlora)*, a pattern secondarily evolved in *Augochlora (Oxystoglossella)*. In the majority of species, however, cells are constructed in a clustered, comb-like arrangement, usually in a cavity in the soil, a pattern independently evolved in some Halictini (e.g. *Lasioglossum* [*Evylaeus*], a few *Halictus*) and a few Nomiinae. Cells may either be oriented primarily vertically, with openings pointed upwards (e.g. *Megommation*, *Rhinocorynura*, *Augochloropsis*), or oriented primarily horizontally (e.g. *Corynura*, *Paroxystoglossa*, *Augochlora*, *Augochlorella*), or oriented irregularly (e.g. *Caenaugochlora*, *Pseudaugochloropsis*, *Neocorynura*) (Eickwort and Sakagami 1979).

Nests may be excavated widely scattered from each other or in loose or dense aggregations, as in other Halictinae. The features that select for aggregation are so far unknown, but restrictions due to edaphic limitations do not appear to play a role (see Wcislo, this volume). In the North and South Temperate Zones and among those tropical species that live in disturbed habitats, nest sites are similar to those of other Halictinae and indeed nests are often intermingled with those of such genera as *Lasioglossum* (*Dialictus* and *Evylaeus*), *Halictus*, *Agapostemon* and *Pseudagapostemon*. In general, halictine bees in open habitats nest in well-drained soils with abundant exposure to sun. At the northernmost extent of its range on Cape Breton Island, *Augochlorella striata* preferentially builds its nests under stones, the solar warming of which presumably assists in development of brood (Packer *et al.* 1989a), as occurs with ant nests under similar conditions.

Augochlorines are among the most diverse and abundant taxa of bees in the tropical forest habitat. There augochlorine nests may be built in heavily shaded and very moist locations, such as under deep, overhanging banks (e.g. *Pseudaugochloropsis graminea*, Michener and Kerfoot 1967; G. C. Eickwort, personal observations) and in very wet soil (e.g. *Megommation insigne*, Jörgensen 1912; Michener and Lange 1958b; Sakagami and Moure 1967). Nests of the latter species even survive flooding. The excavation of a cavity around clustered cells, as well as the excavation of vertical burrows surrounding clustered cells of *Augochloropsis* and *Augochlora (Oxystoglossella)*, probably assists in drainage and aeration of cells in saturated soils (Packer *et al.* 1989a; see Wcislo, this volume). Augochlorine nests are shallow, with cells located only a few centimeters below the surface. This may facilitate warming through insolation (Eickwort and Eickwort 1972) and also help avoid submergence in saturated soils after rains. However, under other conditions (especially in the temperate zone), shallow nests make the brood more susceptible to dehydration or overheating during drought (Ordway 1966; Packer *et al.* 1989a; Packer 1990).

Nest cells are lined with a shiny, hydrophobic coating produced by the Dufour's gland (reviewed in Duffield *et al.* 1984), which waterproofs the cell and apparently provides some protection from microbial contamination of provisions. These same chemicals may be incorporated into the provision mass (e.g. that of *Augochlora pura*). A single egg is laid on the top of the provision mass and the cell is then closed with a soil partition. Although these closures are thin in comparison to those of many other Halictinae, there is no evidence of extensive interaction between the adults in the nest and the brood once the cell has been closed (see Michener and Kerfoot (1967) for photographs of cell and nest structure). Thus, development of augochlorine bees is very similar to that of other Halictinae.

NEST ASSOCIATES AND MORTALITY FACTORS

Augochlorine bees appear to maintain basically the same set of nest associates, parasites and predators that afflict all halictine bees, and to suffer similar sources of mortality (see Wcislo, this volume). Adult bees suffer mortality principally when foraging, from birds, reptiles, amphibians, and such arthropods as crab spiders (Thomisidae), ambush bugs (Reduviidae) and sphecid wasps (*Philanthus*, *Trachypus*) (Eickwort and Eickwort 1972; Evans and Matthews 1973; Menke 1980; Evans and O'Neill 1988). Foragers are also subject to parasitism by conopid flies (*Zodion*) (Michener and Lange 1958b; Ordway 1964; Eickwort and Eickwort 1972; Packer 1990; Mueller 1993), and to acquiring first-instar larvae of Strepsiptera (*Halictoxenos*: Stylopidae) (Michener and Lange 1958b; Ordway 1964) and of rhipiphorid beetles (*Rhipiphorus*) (Eickwort and Eickwort 1972; Packer *et al.* 1989a), which are then transmitted back to the nest to parasitize the brood. Carabid beetles (*Scarites*) and cicindelid beetles and their larvae are predators on adults and brood (Ordway 1964; Mueller 1993). Nests are subject to destruction by ants and by mammals such as skunks and moles (Ordway 1964; Packer *et al.* 1989a). The nest cells are hosts to a variety of mites (Acari), which are predominantly commensal or even mutualistic, feeding on microbial contaminants. A few mites are deleterious, feeding on the provisions and perhaps even killing the brood (Eickwort 1993). Nests also harbor springtails (Collembola), which do not harm the brood (Eickwort and Eickwort 1972, 1973b). Numerous fungi are found in the cells and on the provisions and feces (Batra *et al.* 1973). Diplogasterid nematodes (*Aduncospiculum halicti*) develop in the cell linings, provisions, and feces, and their 'dauer' juveniles are phoretic in the Dufour's glands of the adult females and genital capsules of the males (Giblin-Davis *et al.* 1990). Mutillid and tiphiid wasps (*Pseudomethoca*, *Paramutilla*, *Myrmilloides*, *Myrmosula*) (Claude-Joseph 1926; Michener and Lange 1959; Ordway 1964; Eickwort and Eickwort 1973b; Mickel 1973) invade nests and parasitize larvae and pupae. The kleptoparasitic halictine bee *Sphecodes pimpinellae* attacks nearctic *Augochlorella* (Ordway 1964; Packer *et al.* 1989a); the highly derived augochlorine genus *Temnosoma* contains kleptoparasites, which presumably attack other augochlorine bees (Eickwort 1969b; Michener 1978).

Taken together, the various biotic factors cause surprisingly low mortality rates in augochlorine nests. The single most important mortality factor appears to be conopid parasitism of adults, ranging from 2.4 to 4.0% in foundresses and from 1.1 to 2.5% in workers of *Augochlorella striata* in New York (Mueller 1993) and apparently higher in foundresses of the same species in Nova Scotia (Packer 1990), and in 7% in females of *Augochlora nominata* in

Costa Rica (Eickwort and Eickwort 1972). *Sphecodes pimpinellae* attacked 10% of nests of *A. striata* in Kansas (Ordway 1964) although it was absent or nearly so in studies of the same species in New York and Nova Scotia.

Extremes of temperature and soil humidity undoubtedly can play a major role in mortality. Drought has been noted to strongly reduce brood production in *A. striata* in both Nova Scotia and Kansas (Packer *et al.* 1989a; Ordway 1966). In Kansas, 80% of the nests succumbed during drought.

Overall nest mortality has been carefully recorded only in the nearctic *Augochlorella striata*. Mueller (1993) estimated that 10–40% of foundresses in upstate New York died of unknown causes before completing production of the first brood, and 10–13% of the second brood nests failed without producing adults. Packer *et al.* (1989a) recorded 7% cell mortality in Nova Scotian *A. striata*, the bulk of it due to unknown causes or apparent microbial infestation.

Based on admittedly poor data, the tropical environment does not contribute to high rates of nest mortality, especially not due to parasitism. No studies mention significant mortality due to biotic factors, nor stress nest failures due to any cause. There have been no studies comparing mortality in isolated vs. aggregated nests or in solitary vs. social colonies.

FORAGING BEHAVIOR

Those augochlorines that have adapted to disturbed habitats are typically polylectic, using a wide variety of plant genera and families for pollen and nectar, as in the better-studied Halictinae (e.g. *Lasioglossum* [*Dialictus* and *Evylaeus*] and *Halictus*) that occur in the same habitats. Such polylecty is obvious in *Augochlorella* and *Augochlora* in the North Temperate Zone and *Corynura* (*Callochlora*) in the South Temperate Zone (see flower records in Moure and Hurd 1987 and also Claude-Joseph 1926). Polylecty also characterizes such inhabitants of disturbed tropical habitats as species of *Pereirapis*, *A. (Oxystoglossella)*, *Paroxystoglossa*, and *Neocorynura pubescens* (Michener *et al.* 1966; Eickwort and Eickwort 1972; Michener 1977; Heithaus 1979b; G. C. Eickwort, personal observations). Polylecty appears to be plesiomorphic in the Halictinae (Wcislo and Cane 1996) and it is reasonable to assume that it is also plesiomorphic in the Augochlorini. Polylecty is usually considered an important prerequisite to eusociality, which requires two or more successive broods in a year, and thus seasonal activity that extends beyond the blooming period of any one taxon of plants.

Little is really known about the foraging behavior of those augochlorines that inhabit tropical forests and indeed most other habitats. Our limited data suggest higher degrees of floral specialization in some genera than characterize the Halictini. *Caenaugochlora costaricensis* and other members of its species group are apparently oligoleges on squashes (*Cucurbita*) (Michener and Kerfoot 1967). Both temperate and tropical species of *Augochloropsis* frequently vibrate the poricidal anthers of Ericaceae, Fabaceae, Malpighiaceae and Melastomataceae (Wille 1963; Laroca 1970; Heithaus 1979a; Cane *et al.* 1985 and references therein; Renner 1986; Rêgo and Albuquerque 1990; G. C. Eickwort, personal observations); although they are probably not true oligoleges, in the same habitats they are rarely encountered on the flowers used by other polylectic augochlorines. Species of *Pseudaugochloropsis* also buzz-pollinate poricidal-anther flowers of Fabaceae and Solanaceae (Wille 1963; Michener and Kerfoot 1967; Rick *et al.* 1978; G. C. Eickwort, personal observations). *Pseudaugochloropsis* also has pointed, sclerotized maxillae (Eickwort 1969b) which are used to pierce tubular corollas and 'steal' nectar (Sakagami and Moure 1967; G. C. Eickwort, personal observations). The elongate heads of *Chlerogas* and *Chlerogella*, the elongate glossa of many augochlorines (such as *Augochlora* sensu stricto), and the narrow proboscis of the *Megaloptidia* group (Eickwort 1969b) imply nectar-foraging on plants not utilized by typical short-tongued halictines. The extraordinarily elongate maxillary palpus of *Ariphanarthra palpalis* certainly indicates specialized foraging on unknown plants (Eickwort 1969b).

Specialization is also exhibited in the diel foraging period. Some species can fly very early in the morning, like *Megommation* (*Megaloptina*) in Panama, visiting a 'blue-flowered monocot' (Janzen 1968) and *A. (Augochlora) magnifica* in the Dominican Republic, visiting *Ipomoea* (G. C. Eickwort, personal observations; see also Schlising 1970). *Megalopta* is well known as a strictly nocturnal genus of bees (Wolda and Roubik 1986), which Janzen (1968) characterizes as flying very slowly from flower to flower and collecting pollen from *Solanum*. Other augochlorines with large ocelli are apparently also nocturnal foragers, like *M. (Megommation)* (Jörgensen 1912; Michener and Lange 1958b) and *Megaloptidia*, although even less is known about their foraging habits.

Both solitary and parasocial augochlorines exhibit specialized foraging habits. This is not surprising, because parasocial societies do not have two broods and thus can complete their development within the blooming period of single species of plants. Nevertheless, all of the observed parasocial augochlorines are multiple-brooded; they occur in tropical regimes in which the blooming period of their host plants also is extended (Heithaus 1979c). Given our admittedly limited data, there is at present no evidence that solitary behavior is coincident with specialized foraging, as occurs in the Halictini (e.g. *Lasioglossum* [*Sphecodogastra*] with *Onagraceae* and *L.* [*Hemihalictus*] with *Pyrrhopappus*).

SOCIAL BEHAVIOR

Table 13-2 presents our review of studies of augochlorine bees which provide indications of their social behavior. In it, we follow the terminology for social behavior established by Michener (1974, 1990):

Solitary: a single female excavates and occupies each nest. She constructs and provisions her own cells.

Parasocial: More than one female of the same generation occupies a nest.

Parasocial behavior is subdivided into three categories:

Communal: All the females are reproductively active but they do not cooperate in cell construction or provisioning. Communal nesting in augochlorines is usually a facultative alternative to solitary nesting, both occurring within the same species and typically in the same nest aggregation.

Quasisocial: All the females are reproductively active but they cooperate in cell construction or provisioning. This form of social behavior may be rare among bees and is not positively known in the Augochlorini.

Semisocial: There is reproductive division of labor among the females, and some evidence of cooperative provisioning. The reproductive division of labor is typically indicated by variation among nestmates in ovarian development and often by absence of sperm in the spermathecae of some females. Semisocial behavior typically occurs as a facultative alternative to solitary behavior, with both kinds of nest often occurring in the same aggregation. It also can occur facultatively in the first brood of colonies that will later become eusocial, and as a later development in second-brood eusocial colonies in which the maternal queen dies and is replaced by a daughter queen, a sister of the workers.

Eusocial: Females of more than one generation occupy a nest, with evidence of cooperative provisioning and reproductive division of labor. The first-generation females (queens) are reproductively active, as indicated by developed ovaries and presence of sperm in their spermathecae, whereas some of the second-generation females (workers) have no or less reproductive activity, as indicated by undeveloped ovaries and frequently by absence of sperm in their spermathecae.

Why have we chosen the above classification over other, newly proposed, terminologies (Crespi & Yanega 1995; Sherman *et al.* 1995)? First, the majority of literature on halictid social behavior has used this terminology; rather than reinterpret the results of literally hundreds of studies, we chose simply to adopt the original terminology used in those studies. Second, this terminology was originally developed by Michener and Batra specifically to describe social systems in halictines (Crespi & Yanega 1995), and it remains generally applicable across halictine bees. And finally, we do not believe that the newer terminologies provide significantly less arbitrary or more practical criteria for differentiating among social states. The greatest problem inherent in any system of classifying social systems is the tremendous range of intraspecific variability. Most species of augochlorine bees are best characterized by a *range* of social behaviors (whichever system is chosen), typically involving a mixture of solitary, communal and/or semisocial nests within the same species (see Table 13-2).

In practice it is not easy to determine the status of a given species. As mentioned above, few or no augochlorine species are entirely communal or semisocial; at least some nests of each well-studied species that otherwise fits into these categories are solitary. Even in species that are otherwise characterized as eusocial, there are some nests that are semisocial at points in their development, and of course all monogynous nests of eusocial species are solitary during the production of the first brood. Even in many species that are primarily solitary, there is a small percentage of nests that contain more than one female, thus being either communal or semisocial.

Moreover, studies of augochlorine species throughout an entire seasonal cycle are very limited. The vast majority of studies cited in Table 13-2 are based on observations of only a few (often one!) nests over a very limited period of time, thus giving an imperfect and perhaps highly misleading picture of the social status of these species. In addition, there can be important differences in social behavior within a species when it is studied throughout its range and under different environmental conditions, as emphasized by Wcislo (this volume). Only two species (*Augochlorella striata* and *Pseudaugochloropsis graminea*) have been

Table 13-2. *Review of social behavior in the Augochlorini*

Species	Social behavior	Locality	Reference
Augochlora (Augochlora) pura[a]	Solitary	Lawrence, Kansas	Stockhammer 1966
A. (Oxystoglossella) cordiaefloris	Eusocial?	Turrialba, Costa Rica	Eickwort and Eickwort 1972
A. (O.) nominata	Eusocial	Turrialba, Costa Rica	Eickwort and Eickwort 1972
A. (O.) semiramis	Semisocial or eusocial	Curitiba, Brazil	Michener and Lange 1958b; Sakagami and Moure 1967
Augochlorella edentata	Solitary and semisocial?	Turrialba, Costa Rica	Eickwort and Eickwort 1973b
A. michaelis	Semisocial or eusocial	Curitiba, Brazil	Michener and Lange 1958b; Sakagami and Moure 1967
A. persimilis	Eusocial	Lawrence, Kansas	Ordway 1966
A. striata	Eusocial	Lawrence, Kansas	Ordway 1964, 1965, 1966
A. striata	Eusocial	Cape Breton Is., Canada	Packer *et al.* 1989a,b; Packer 1990
A. striata	Eusocial	Ithaca, New York	Mueller 1991, 1993
Augochlorodes turrifasciens	Solitary and semisocial	Rio de Janeiro, Brazil	Michener and Campos Seabra 1959
Augochloropsis (Augochloropsis) brachycephala	Solitary (rarely semisoc?)	Rio de Janeiro, Brazil	Michener and Campos Seabra 1959
A. (A.) diversipennis	Solitary and communal	Curitiba, Brazil	Michener and Lange 1959
A. (A.) ignita	Semisocial or eusocial	Filadelfia, Costa Rica	Michener and Lange 1959
A. (A.) notophos	Communal or semisocial	Sao Paulo, Brazil	Gimenes *et al.* 1991
A. (Paraugochloropsis) sparsilis	Semisocial (rarely solitary)	Curitiba, Brazil	Michener and Lange 1958c, 1959
A. (P.) sumptuosa	Communal or semisocial	New Jersey, USA	Smith 1901
A. (P.) sumptuosa	?	Lawrence, Kansas	Michener and Lange 1959 (as *Augochloropsis humeralis*)
A. iris	Solitary and semisocial	Curitiba, Brazil	Michener and Lange 1959
Caenaugochlora (Caenaugochlora) costaricensis	Solitary and semisocial	San José, Costa Rica	Michener and Kerfoot 1967 (as *Pseudaugochloropsis costaricensis*)
Corynura (Corynura) apicata	Solitary?	Temuco, Chile	Claude-Joseph 1926 (as *Halictus apicatus*)
C. (Callochlora) chloris	Communal or semisocial?	Santiago, Chile	Claude-Joseph 1926 (as *Halictus chloris*)
C. (Corynura) cristata	Communal or semisocial?	Santiago, Chile	Claude-Joseph 1926 (as *Halictus cristatus*)
C. (Corynura) herbsti	Communal or Solitary?	Manquehue, Chile	Claude-Joseph 1926 (as *Halictus herbsti*)
Halictillus 'glabriventris'	Communal or semisocial?	Santiago, Chile	Claude-Joseph 1926 (as *Halictus glabriventris*)
Megalopta centralis[a]	Semisocial or communal?	Rincon, Costa Rica	Janzen 1968
M. sp.[a]	Semisocial or communal	Manaus, Brazil	Sakagami 1964
Megommation (Megommation) insigne	Semisocial	Curitiba, Brazil	Michener and Lange 1958b; Sakagami and Moure 1967
Neocorynura (Neocorynura) colombiana[a]	Semisocial?	Santa Marta, Colombia	Schremmer 1979; Eickwort 1979
N. (N.) errinys[a]	Solitary and semisocial?	Raiz da Serra, Brazil	Lüderwaldt 1911

Table 13-2. (*cont.*)

Species	Social behavior	Locality	Reference
N. (N.) fumipennis	Solitary (rarely semisocial?)	San José, Costa Rica	Michener *et al.* 1966
N. (N.) polybioides	Solitary and communal	Curitiba, Brazil	Michener and Lange 1958b; Sakagami and Moure 1967
N. (N.) pubescens	Solitary?	San José, Costa Rica	Michener *et al.* 1966
N. (N.) pubescens	Solitary	Rio Negro, Colombia	Michener 1977
N. sp. 1	Solitary?	Cerro las Vueltas, Costa Rica	Michener *et al.* 1966
Paroxystoglossa andromache	Solitary (rarely semisocial)	Curitiba, Brazil	Michener and Lange 1958b
P. jocasta	Solitary (rarely semisocial)	Curitiba, Brazil	Michener and Lange 1958a
P. seabrai	Solitary (rarely semisocial)	Rio de Janeiro, Brazil	Michener and Campos Seabra 1959
P. spiloptera	Solitary (rarely semisocial)	Guaruva, Brazil	Michener and Campos Seabra 1959
Pereirapis semiauratus	Eusocial	Ribeirao Preto, Brazil	Oliveira Campos 1980
Pseudaugochloropsis graminea	Solitary (rarely semisocial)	San José, Costa Rica	Michener and Kerfoot 1967
P. graminea	Solitary and semisocial	Curitiba, Brazil	Michener and Lange 1958b (as *Pseudaugochlora nigromarginata*)
P. sordicutis	Solitary and semisocial	San José, Costa Rica	Michener and Kerfoot 1967 (as *Pseudaugochloropsis nigerrima*)
Rhinocorynura inflaticeps	Semisocial	Curitiba, Brazil	Eickwort and Sakagami 1979
Temnosoma spp.	Kleptoparasitic		Michener 1978

[a] Species nests in wood.

studied more than cursorily in one location. We consider only *Augochloropsis diversipennis*, *A. sparsilis*, *Augochlora (A.) pura*, *A. (Oxystoglossella) nominata*, *Augochlorella striata*, *A. persimilis*, *Caenaugochlora costaricensis*, *Neocorynura fumipennis*, *Paroxystoglossa jocasta*, *P. seabrai*, *Pereirapis semiaurata*, *Pseudaugochloropsis graminea* and *P. sordicutis* to have been studied in sufficient depth to provide a good picture of their social biology. Some other studies that have followed numerous nests throughout a season have not involved dissections of females to determine castes. The early and extensive study by Claude-Joseph (1926) provides fascinating hints about social interactions among nestmates, but does not provide any substantive information about their reproductive capacities or even their generations, and is further hampered by Claude-Joseph's belief that he was observing alternating bisexual and parthenogenetic broods. Thus these Chilean genera (*Corynura* and *Halictillus*), which exhibit some level of parasocial or even eusocial colony development, await modern study.

Few of the studied augochlorine species appear to be strictly solitary, the only well-documented case being that of the nearctic species *Augochlora (A.) pura* (Stockhammer 1966). Solitary behavior also appears to characterize *Corynura apicata* (Claude-Joseph 1926).

The subgenus *Augochlora* sensu stricto is often characterized as being solitary (based on *A. pura*), but the other, tropical species are known from only one nest apiece, which could readily have been first-brood colonies that would later become social. Dissections of foraging *Augochlora (A.) magnifica* in the West Indies indicated that many were worker-like, and the strong size variation of many other tropical species (e.g. *Augochlora [A.] nigrocyanea* and *A. smaragdina*) implies the presence of castes (G. C. Eickwort, personal observations).

On the other hand, numerous augochlorine species are predominantly solitary, with a small proportion of the nests being plurimatrous (having more than one adult female (Choe 1995)). This condition characterizes most *Neocorynura* and *Paroxystoglossa*, *Augochloropsis brachycephala*, and Costa Rican *Pseudaugochloropsis graminea* (Table 13-2). In most cases, dissections indicated that the co-nesting females differed in reproductive capacity, implying a semisocial association. The one plurimatrous nest of *Neo-corynura polybioides* found by Michener and Lange (1958b) had both females reproductively capable, indicating a communal association.

Communal behavior in bees typically occurs as an evolutionary alternative to caste-based sociality (semisociality and eusociality) and in different phyletic lines (Kukuk and Eickwort 1987; Wcislo, this volume; Packer 1993). Communal associations, in contrast to caste-based societies, frequently involve unrelated bees (Kukuk and Sage 1994), and discrimination by nest guards against unrelated but conspecific bees is absent. Communal halictine nests often contain large numbers of bees and cells. The only well-documented, frequently communal species of Augochlorini is *Augochloropsis diversipennis* (Michener and Lange 1959) in which second-brood nests may contain two or three reproductively active females, each with her own cell cluster. *Augochloropsis* is a large genus, and it is not known whether the communal species is closely related to the semisocial species or instead belongs to a separate clade. Communal behavior may also occur in *Corynura herbsti*, in which second-brood females first start working together on the old comb of cells, then each builds her own comb (Claude-Joseph 1926).

Semisociality as a frequent, terminal stage of colony development is rare among bees and is best known in the Augochlorini. The studies of *Augochloropsis sparsilis* by Michener and Lange (1958c, 1959) and of *Caenaugochlora costaricensis* and *Pseudaugochloropsis sordicutis* by Michener and Kerfoot (1967) provide the best picture of semisociality anywhere among bees. In the latter two species about 60–80% of the females live in semisocial associations of up to seven adults, with the nests containing up to 50 or 67 cells. Semisociality is even more advanced in *A. sparsilis*, in which an average of 1.5–3.7 females cooperate in guarding the nest and provisioning cells.

The distinction between semisocial and eusocial colonies depends upon the generation of the queen. If she is the mother or aunt of the workers, the colony is eusocial; if she is a sister or cousin, the colony is semisocial. Because semisocial colonies are assumed to be initiated by sisters reusing a maternal nest (their mother having died or left), practical distinction of a eusocial queen requires either that the queen and workers be very different in wear (indicating differences in age) or that the queen be marked while producing the first brood. Slight size differences are not definitive, because semisocial queens are typically the largest siblings. Because none of the studies of semisocial

augochlorines have involved marked queens, the possibility that some (or even most!) of the presumably semisocial colonies were actually eusocial cannot be eliminated.

With this in mind, presumed semisocial behavior also appears to be characteristic of many or most nests (at least of the second brood) of *Augochlorodes turrifasciens*, many *Augochloropsis*, *Megalopta*, *Megommation (M.) insigne*, *Rhinocorynura*, Brazilian *Pseudaugochloropsis graminea*, and perhaps *Augochlorella edentata* (Table 13-2). The descriptions of females working together in nests of *Corynura (Callochlora) chloris*, *C. (Corynura) cristata*, and *Halictillus 'glabriventris'* by Claude-Joseph (1926) also suggest semisociality, although without dissections, communal (or even eusocial) behavior cannot be ruled out. Similarly, the studies of *Megalopta* do not exclude the possibility of communal (or even eusocial) colonies.

Eusociality characterizes all studied *Augochlora (Oxystoglossella)*, *Augochlorella* (except perhaps *A. edentata*), and *Pereirapis* (Table 13-2). Colonies of *Augochlorella* in both the tropics and the temperate zone are rather small (a maximum of 6–8 adults with a mean of 4–5 during the gyne-production phase), with only slight differences in queen and worker size (Ordway 1965, 1966; Mueller 1993). Similar colony development appears to occur in *A. (Oxystoglossella)*, although in the only well-studied species (*A. nominata*) colony size in the gyne-producing phase is even smaller (maximum of 6, mean of 3.3 adults) (Eickwort and Eickwort 1972). In contrast to *Augochlorella*, queens and workers are very different in size, and indeed in the Jamaican *A. (Oxystoglossella) decorata*, the caste size difference is the greatest to be found in the Halictidae (G. C. Eickwort, personal observations). *Pereirapis semiauratus* is similar in colony development to *A. (Oxystoglossella)* (Oliveira Campos 1980). In all studied eusocial species, colonies are predominantly haplometrotic (founded by single females) and replacement of foundress queens by sisters of the workers may occur.

For the purposes of tracing the evolution of social behavior on the phylogeny presented below, we will accept the current view that only *Augochlorella*, *Augochlora (Oxystoglossella)* and *Pereirapis* contain eusocial species. Data presented by species in Table 13-2 are summarized by genera in Table 13-1.

PHYLOGENETIC ANALYSIS

There is little doubt that the Augochlorini is a monophyletic group. Moure (1943) erected the tribe and Eickwort

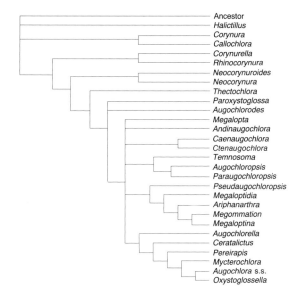

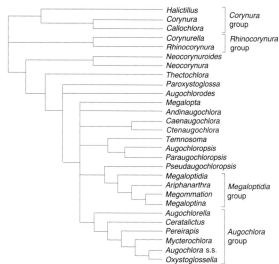

Fig. 13-1. Nelson consensus tree of 15 equally parsimonious trees based on outgroup rooting. Sixteen of 32 multistate characters were treated as ordered. Length=349, ci = 0.36, ri = 0.59.

Fig. 13-2. Nelson consensus tree of five equally parsimonious trees based on using *Halictillus*, *Corynura*, *Callochlora*, *Corynurella* and *Rhinocorynura* as outgroups for the remaining taxa (ingroup rooting). Sixteen of 32 multistate characters were treated as ordered. Length = 341, ci = 0.37, ri = 0.60.

(1969a) expanded its scope to include 24 genera united by a number of apomorphic features including the absence of a pygidial plate and the presence of a spiculum on metasomal sternum VIII in males, and the presence of a median slit in tergum V in females. In an analysis of halictine tribal relationships, Pesenko (1983) listed five characters that support the monophyly of the Augochlorini. We used Eickwort's (1969b) generic descriptions to construct a data matrix for a phylogenetic analysis of the augochlorine genera (see Appendixes 13-1 and 13-2 for character descriptions and data matrix). The character-states listed in the data matrix are representative of all the species in each genus. Characters that were variable within genera were coded as missing (= 9 in the data matrix). We did not include social behavior as a character in the data matrix because to do so would bias our results in favor of finding a single origin of eusociality. Hennig 86, version 1.5 (Farris 1988, 1989; Fitzhugh 1989; Platnick 1989) was used to find the most parsimonious phylogenetic hypothesis for the relationships of twenty-eight of the thirty-one subgenera and genera recognized by Eickwort (1969b) (three taxa, *Chlerogas*, *Chlerogella* and *Rhectomia*, were excluded because one sex was unknown at the time of Eickwort's study). We used the commands mhennig* followed by the branch and bound algorithm (bb*) for all analyses that follow. We also

ran the analyses multiple times with the taxa arranged in different orders to increase the likelihood of finding different 'islands' of most parsimonious trees (Maddison 1991).

An initial analysis was performed with forty of the characters polarized based on outgroup comparisons with other halictid taxa (Table 1 in Eickwort 1969b) and 'no data' (= 9 in the data matrix, Appendix 13-1) for the remaining characters (see Lundberg 1972 and Maddison *et al.* 1984 for an explanation of outgroup analysis). All multistate characters were initially treated as unordered (non-additive) unless the character-states could be arranged in a logical series. Sixteen of the 32 multistate characters could thus be ordered either because they included relative lengths (e.g. character 6) or degree of emargination (e.g. character 10). This analysis resulted in 15 equally parsimonious trees with consistency indices of 0.36. Fig. 13-1 shows the Nelson consensus tree of these fifteen trees. In all trees, five genera, *Corynura*, *Callochlora*, *Halictillus*, *Rhinocorynura* and *Corynurella*, form a paraphyletic basal grade relative to a monophyletic group containing the remaining genera. Characters 10[1], 21[1], 76[1], 77[2] and 78[1] are synapomorphies of the remaining genera.

In order to increase the number of polarized characters and to improve tree resolution, we performed a second analysis in which the basal five genera were treated as

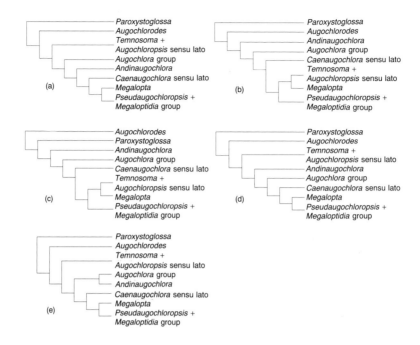

Fig. 13-3. Five equally parsimonious resolutions of the consensus tree shown in Fig. 13-2. Length = 341, ci = 0.37, ri = 0.60.

outgroups for the remaining (ingroup) genera (see Watrous and Wheeler (1981) for an explanation of the rationale underlying this approach). We considered this justified because of the strong evidence that the remaining genera form a monophyletic group, and because ingroup rooting gives more accurate information on character polarity.

In the second analysis we obtained five equally parsimonious trees. Fig. 13-2 shows the Nelson consensus tree of these five trees and Fig. 13-3a–e shows the alternative resolutions. As one would expect, ingroup rooting gave a slightly higher consistency index (ci = 0.37) and fewer steps (seven fewer steps) than rooting based on the original outgroup, but the overall topologies are similar and the unresolved polytomy in the first analysis is not resolved with ingroup rooting. The consensus trees shown in Figs. 13-1 and 13-2 are identical except for the relationships among the outgroup genera in the *Corynura* and *Rhinocorynura* groups (Fig. 13-2).

In a third analysis we applied successive approximations character weighting (Farris 1969; Carpenter 1988) to the analysis based on ingroup rooting and obtained three equally parsimonious trees. The consensus tree of the successive approximations analysis is shown in Fig. 13-4. The three trees differ only in the relationships of *Thectochlora*, *Neocorynura* senso lato and a monophyletic group

including the remaining genera, giving rise to a single unresolved trichotomy at the base of the consensus tree (Fig. 13-4). Our successive approximations tree closely

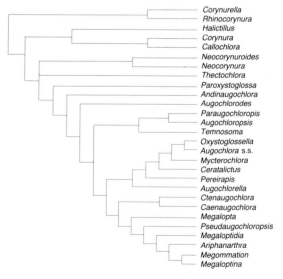

Fig. 13-4. Nelson consensus tree of three equally parsimonious trees resulting from successive approximations character weighting. Length = 605, ci = 0.58, ri = 0.77.

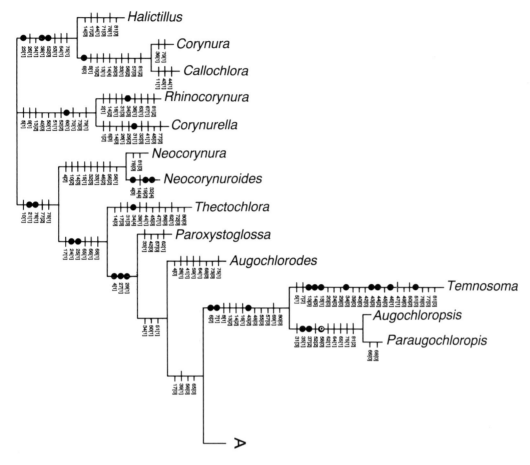

Fig. 13-5. Same tree as in Fig. 13-3d with character state changes mapped on. Circles indicate unique and unreversed character state changes; complete hatch marks indicate parallelisms; half hatch marks indicate reversals.

resembles the trees based on our unweighted analysis, especially the trees shown in Fig. 13-3a,d,e. The only differences involve the placement of the problematic genus *Andinaugochlora* and the relative positions of *Pereirapis* and *Ceratalictus*.

None of our trees (Figs. 13-3a–e and Fig. 13-4) bears much resemblance at a higher level to either Eickwort's (1969b) tree based on the distance matrix (his Fig. 417) or his tree based on the correlation matrix (his Fig. 418). Eickwort's Fig. 417 requires 39 additional steps ($ci = 0.33$; retention index, $ri = 0.52$) and Eickwort's Fig. 418 requires 25 additional steps ($ci = 0.35$; $ri = 0.55$) to explain the morphological data matrix as opposed to our minimum-length trees (Fig. 13-3a–e; $ci = 0.37$; $ri = 0.60$; length = 341). At a lower level, our trees show some of the same monophyletic groups as Eickwort's. Eickwort recognized seven generic

groups, of which our trees support four: the *Augochlora*, *Corynura*, *Rhinocorynura*, and *Megaloptidia* groups (Fig. 13-2).

We conclude from the positions of the four nocturnal genera (*Megalopta*, *Megaloptidia*, *Ariphanarthra* and *Megommation* sensu lato) in Fig. 13-3a–e that nocturnal foraging either arose twice independently, in *Megalopta* and in the monophyletic group *Megaloptidia* + (*Ariphanarthra* + *Megommation* sensu lato) (e.g. Fig. 13-3b), or that nocturnal foraging arose in the common ancestor of the nocturnal genera and reversed to diurnal foraging in the common ancestor of *Pseudaugochloropsis* (e.g. Fig. 13-3a). Finding multiple origins of nocturnal foraging is surprising but plausible, because many of the characters that have led to the assumption that the nocturnal species are monophyletic, such as pallid coloration and enlarged ocelli, are

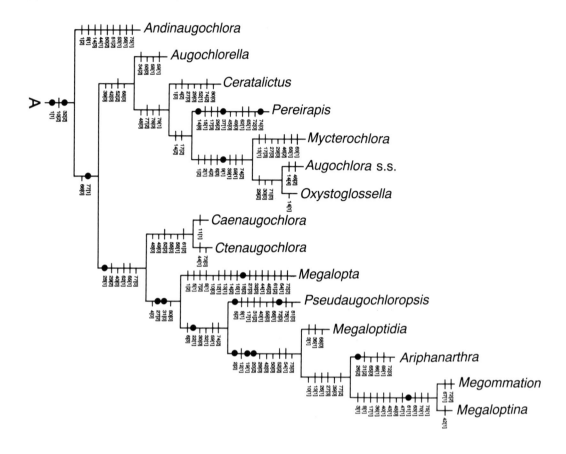

Fig. 13-5. (*cont.*)

associated with nocturnal foraging in other, clearly unrelated, nocturnal bees (Kerfoot 1967a,b,c). The genus *Megalopta*, which appears either as the sister group to *Temnosoma + Augochloropsis* (e.g. Fig. 13-3d) or as the sister group to *Pseudaugochloropsis* + the *Megaloptidia* group (Figs. 13-3e and 13-4), could not be placed with certainty in any species group by Eickwort (1969b).

We believe that the tree based on ingroup rooting and several ordered characters is the best estimate of the phylogeny given the current data set. Nevertheless, in our unweighted analyses (Fig. 13-3) the consistency index is relatively low (0.37), and a number of nodes are weakly supported (Fig. 13-5). The consistency index gives a rough indication of the robustness of phylogenetic trees but is highly correlated with the number of taxa included (Sanderson and Donoghue 1989). Based on Sanderson and

Donoghue's empirical analysis of consistency indices, for our data set we should expect a consistency index of roughly 0.44. Our observed consistency index is slightly below that value, indicating that our data are relatively noisy in comparison to those studies included in Sanderson and Donoghue's survey.

The major source of ambiguity in our analysis based on ordered characters involves the relationships among six monophyletic groups: *Megalopta*, *Andinaugochlora*, *Caenaugochlora* sensu lato, *Temnosoma + Augochloropsis* sensu lato, *Pseudaugochloropsis + Megaloptidia* group, and the *Augochlora* group (Fig. 13-2). Although the consensus tree shown in Fig. 13-2 shows an unresolved polytomy, inspection of the five equally parsimonious trees in Fig. 13-3a–e indicates that the major source of ambiguity results from the placement of only a few taxa. First, the monotypic genus

Andinaugochlora moves around considerably in these trees, appearing as the sister group to the eusocial genera in Fig. 13-3e, as a relatively basal group in Fig. 13-3b–d, and as the sister group to *Caenaugochlora* sensu lato, *Pseudaugochloropsis*, and the nocturnal genera in Fig. 13-3a. Secondly, the monophyletic group *Temnosoma* + *Augochloropsis* sensu lato varies in placement among the various trees in Fig. 13-3 from being placed relatively basally (Fig. 13-3a,d,e) to being placed as the sister group to *Megalopta* (Fig. 13-3b,c). The placement of *Andinaugochlora* as the sister group to the eusocial genera is supported by a single homoplasious character (65[0]), making this hypothesis particularly doubtful. Similarly, the placement of *Megalopta* as the sister group to *Temnosoma* + *Augochloropsis* sensu lato is weakly supported. For these reasons we believe Fig. 13-3d shows the most likely phylogeny of the Augochlorini based on our unweighted data set.

That our successive approximations tree closely resembles Fig. 13-3d lends support to this hypothesis. Successive approximations may provide a method of evaluating the robustness of phylogenetic hypotheses (Carpenter *et al.* 1993; Wheeler *et al.* 1993). The close agreement between our successive approximations tree (Fig. 13-4) and our preferred tree based on the unweighted analysis (Fig. 13-3d) indicates that our phylogenetic hypothesis for the relationships among the augochlorine genera is robust.

None of the alternative topologies shown in Figs. 13-3 or 13-4 alters our interpretation of how social behavior has evolved in the augochlorines, which we discuss below.

EVOLUTION OF SOCIALITY IN THE AUGOCHLORINI

Although eusociality occurs in three genera within the Augochlorini, our phylogenetic results suggest that eusociality has arisen only once within the tribe. In all our analyses (Figs. 13-1, 13-2 and 13-4), *Augochlorella* belongs to a monophyletic group including *Ceratalictus*, *Pereirapis* and *Augochlora* sensu lato (the *Augochlora* group) (Fig. 13-2) indicating a single origin of eusociality in the Augochlorini (Fig. 13-6). In the analysis based on ingroup rooting, the monophyly of the eusocial genera is supported by from one to four characters (39[0], 42[0], 52[2], 68[0]), none of which is unique and unreversed (Fig. 13-5). The one character uniting the eusocial genera in all trees is the presence

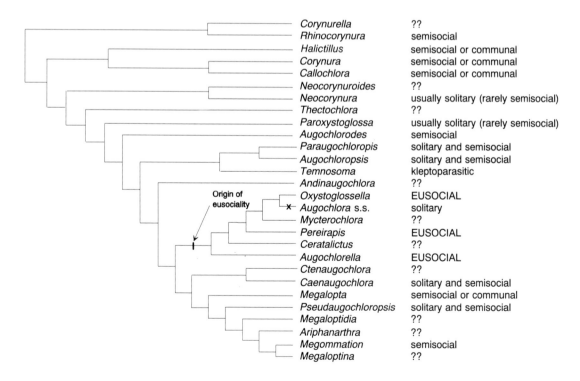

Fig. 13-6. Same tree as in Fig. 13-3d with social behavior mapped on.

of serrate inner hind tibial spurs (42[0]). This is a rather variable character that arises elsewhere in the tree: in *Temnosoma*, *Paroxystoglossa* and the ancestor of *Megaloptidia* and its sister group. We therefore cannot place great confidence in this character. Although our best estimate of the phylogeny based on our morphological data set indicates a single origin of eusociality, more work is clearly needed before this hypothesis can be accepted with any degree of confidence.

Our phylogenetic results support a hypothesis, proposed by Michener (1990), that the solitary behavior in *Augochlora* sensu stricto is derived from a eusocial ancestor. The clade *Augochlorella* + (*Ceratalictus* + (*Pereirapis* + (*Mycterochlora* + (*Augochlora* sensu stricto + *Oxystoglossella*)))) is supported in all our trees. Because *Augochlorella*, *Pereirapis* and *Oxystoglossella* are eusocial, the most parsimonious explanation of solitary behavior in *Augochlora* sensu stricto is that it results from a reversal to solitary behavior (Figs. 13-6 and 13-7). Michener (1990) suggested that such a reversal may have resulted from reduced predation in

woodland habitats, where *Augochlora* sensu stricto are found. This hypothesis could be overturned if both *Mycterochlora* and *Ceratalictus* are found to be solitary in future studies. If this were the case, then the hypothesis of three independent losses of sociality (in *Mycterochlora*, *Ceratalictus* and *Augochlora* sensu stricto) would be equally parsimonious as the hypothesis of three independent derivations of eusociality (in *Augochlorella*, *Pereirapis* and *Oxystoglossella*).

Our phylogenetic analyses also indicate that eusociality in the Augochlorini, whether derived multiply or singly, is evolutionarily distinct from any origins of eusociality in its sister group, the Halictini (Pesenko 1983). Because none of the basal lineages of Augochlorini shows eusociality, the ancestral state for the augochlorines is presumably a mix of solitary, communal or semisocial behavior. The presence of semisocial behavior in some of the most basal augochlorine genera (e.g. *Rhinocorynura* in Table 13-2) indicates that a mix of solitary, communal and semisocial behavior may be plesiomorphic for the tribe.

Finally, our results are satisfying because they suggest one hypothesis for the origin of kleptoparasitism in the Augochlorini. *Temnosoma* is a kleptoparasitic genus of augochlorines (Michener 1978). Although they are clearly kleptoparasites, based on their heavily armored cuticle and lack of pollen-collecting structures, their host association is not known. Moure and Hurd (1987) list *Augochlora* sensu stricto, *Augochloropsis* and *Pseudaugochloropsis* as potential hosts of *Temnosoma*, presumably because each of these genera encompasses the entire geographic range of *Temnosoma*. Based on our analysis, *Temnosoma* is the sister group to *Pseudaugochloropsis* sensu lato. The monophyly of this group is well supported by three unique and unreversed characters and nine additional characters that show some degree of homoplasy (Fig. 13-5). That *Temnosoma* appears to be the sister group to one of its presumed hosts suggests that interspecific kleptoparasitism may have arisen from intraspecific kleptoparasitism. Evidence of intraspecific kleptoparasitism has been found in *Halictus ligatus* (Packer 1986), based on behavioral observations, and in *Augochlorella striata* (Mueller *et al.* 1994), based on DNA fingerprinting.

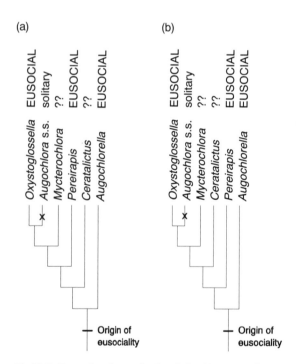

Fig. 13-7. Alternative schemes for the relationships among the eusocial genera. Part (a) shows the relationships based on unweighted data and (b) shows the relationships based on successive approximations character weighting. Reversal to solitary nesting is indicated by x.

DISCUSSION

The major conclusion of our study, that eusociality has arisen once in the common ancestor of the *Augochlora* group and then reversed to solitary behavior in *A. (Augochlora) pura*, is founded on a number of assumptions, which should be made explicit. First, we are assuming that our phylogenetic hypotheses are accurate. Second, we are assuming that we have an adequate understanding of the range and limits of social behavior. For example, we are assuming that all members of the eusocial genera are indeed eusocial based on studies of, in some cases, a single species (e.g. *Pereirapis*).

Third, and most importantly, we assume that eusociality is a discrete and heritable trait, which evolves by descent with modification. As Michener (1985), Wcislo (this volume), and others have pointed out, the high degree of intraspecific variation in bee social behavior could indicate that social behavior is such a phenotypically plastic trait that one cannot map it as a character on a cladogram. Perhaps the potential to evolve eusociality is a heritable trait but the expression of eusociality is simply a function of local environmental conditions. We believe that eusociality *is* a discrete behavioral phenotype that can be distinguished from alternative social systems involving alloparental care (such as semisociality) (but see Sherman *et al.* (1995) for an alternative view). Characteristics of eusociality include the existence of overlapping generations (Michener–Batra system) or the existence of castes: 'discretely different, irreversible life history trajectories' (Crespi–Yanega system). Our phylogenetic results bolster the view that the evolution of eusociality is a discrete and historically recognizable event. Eusociality in the Augochlorini arose a single time from a common ancestor that probably exhibited a mix of solitary and semisocial nesting habits, and has been retained in almost all of the extant descendants of that common ancestor. Furthermore, eusociality occurs in both tropical and north-temperate species, suggesting that it is not simply a phenotypic response to local environmental conditions.

It is worth pointing out that the traits that are used to characterize eusociality in the Michener–Batra system and in the Crespi–Yanega system are functionally correlated. Overlap of generations means that mothers are capable, through their control of offspring provisions, of manipulating their offspring into alternative developmental phenotypes. Overlap of generations may provide the means by which mothers direct their offspring down different developmental trajectories (Alexander 1974; Michener 1990). Such selfish manipulation of offspring phenotype provides no direct parental benefits in species which lack overlapping generations.

One of the major questions in insect sociobiology is 'Why has eusociality arisen in some taxa and not others?'

One approach to this question is to analyze the selective factors that lead to social or solitary nesting within socially polymorphic species (e.g. *Augochlorella striata*, Mueller 1991). Although studies of intraspecific variation may give us indirect information about the evolutionary origins of eusociality, they may not give us a clear indication of why eusociality has arisen in one clade rather than another. Our phylogenetic information provides us with a unique opportunity to analyze why eusociality has arisen in the common ancestor of the *Augochlora* group and in no other genera of augochlorines. Adaptive hypotheses for the origins of eusociality hinge on extrinsic factors, such as the relative costs and benefits of solitary vs. social life, and intrinsic factors, such as the relatedness among interacting individuals (Hamilton 1964a,b; Andersson 1984). Presumably, for eusociality to have arisen in the common ancestor of the *Augochlora* group, a change must have occurred, either in the ecological constraints associated with solitary nesting (extrinsic) or in the relatedness asymmetries within nests (intrinsic).

That eusociality occurs in cool-temperate as well as tropical habitats suggests that climatic or macrogeographic factors that affect the costs associated with solitary or social nesting may not be particularly important in the origin of eusociality in the Augochlorini. However, microgeographic factors may partly explain the origins of eusociality in the augochlorines because all eusocial genera inhabit disturbed, open areas. The eusocial *Augochlochlora (Oxystoglossella)*, *Augochlorella* and *Pereirapis* are among the most abundant of all bees in disturbed, open areas in temperate and tropical habitats. Nests of all these genera are built in exposed, sunny areas, subject to discovery by nest parasites, and in which brood develop quickly and foundresses readily survive to become queens during the production of the second brood. Strong ecological constraints could simultaneously favor both worker-like behavior and parental manipulation. To the extent that founding a new nest is a risky endeavor, potential workers may find that remaining in the natal nest as a worker is their best reproductive option. To the extent that nest defense is required, foundress females may be under selection to manipulate their daughters to stay within the natal nest and serve as foragers or guards.

Alternatively, changes in intrinsic factors, such as the frequency of mating by foundresses and offspring sex ratio, may have altered the relatedness asymmetries sufficiently to favor eusociality in the common ancestor of the *Augochlora* group. Both decreased mating frequency by foundress females and increasingly female-biassed broods would elevate intra-nest relatedness, and hence the benefits associated with worker-like behavior. If eusociality has arisen in the *Augochlora* group from semisocial ancestors, as our phylogenetic results indicate, it is worth considering the role of relatedness in the transition from semisociality to eusociality. In semisocial colonies, assuming females mate once, the *maximum* relatedness among workers and the brood they rear is 0.375 (nieces). In eusocial colonies the *maximum* relatedness among workers and the brood they rear is 0.75 (full-sib sisters). Hence the advantage to potential workers of cooperating in eusocial colonies is potentially twice that in semisocial colonies, all other things being equal. To the extent that females in semisocial colonies can produce a male-biassed offspring sex ratio (as was found in *Augochlorella striata*) (Mueller 1991), the benefits associated with behaving as a worker in such colonies increases. The important point is that both evolutionary change in queen longevity and changes in sex ratio might be sufficient to favor the evolution of eusociality. Given these relatedness considerations, why do semisocial colonies persist at all? We suspect that environmental constraints to solitary nest founding must be pronounced in the semisocial augochlorines, such that worker behavior is selected even under low relatedness.

Our results unambiguously support the hypothesis that solitary behavior in *Augochlora* sensu stricto is derived from a eusocial ancestor. Our study demonstrates such a phenomenon in social insect evolution and challenges the widely held, but rarely stated, assumption that solitary nesting is always primitive with respect to eusociality. Packer (1991) found reversals to solitary behavior based on phylogenetic studies in the subgenus *Evylaeus* using electrophoretic characters; Richards (1994) found similar reversals in *Halictus*. Both these studies and ours should serve as cautions to students of social behavior; eusociality should not be assumed *a priori* to be derived. Future phylogenetic analyses will be needed to address these questions in the Halictini, where evolutionary changes in social behavior are seemingly frequent (cf. Wcislo, this volume).

FUTURE DIRECTIONS FOR RESEARCH ON THE AUGOCHLORINI

The augochlorine bees are unique among insects in exhibiting a predominance of semisocial behavior as a terminal strategy. The benefits of such behavior are obscure, as there should be conflicting selective pressures to evolve towards solitary or eusocial behavior. Future studies should center on a tropical, preferably forest, population and follow it through the year. Bees should be marked to ascertain whether there is generational overlap, dissected to determine caste, and analyzed genetically to determine the intracolony relatedness. Solitary and semisocial nests within the population should be compared as to sex ratio, productivity, and survival. Manipulation of colonies, such as selective removal of individuals, would allow one to determine whether the alternative behavioral phenotypes (foragers vs. queens) are discrete and irreversible (and thus indicative of eusociality *sensu* Crespi and Yanega (1995)) or interchangable.

From a comparative viewpoint, there are tantalizing taxa awaiting study. High degrees of cephalic polymorphism are exhibited by species of *Augochlora (Augochlora* and *Oxystoglossella), Augochloropsis, Rhinocorynura* and *Megalopta* (Sakagami and Moure 1965), implying morphological castes and advanced types of semisociality or eusociality. Members of the *Augochlora* group especially deserve attention, as it is within this clade that eusociality has evolved. Nothing is known about the nesting biology of *Ceratalictus* and *Augochlora (Mycterochlora)* within that clade. The tropical species of *Augochlora* sensu stricto need investigation, as dissections of foragers and presence of polymorphism suggests the presence of castes, in contrast to the only well-studied species in the subgenus (*A. pura*), which is solitary. Investigations of these bees will help determine the number of times that eusociality has evolved and been lost in the Augochlorini.

From a phylogenetic viewpoint, the Chilean Augochlorini merit the most biological attention. The basal augochlorine genera occur there (*Corynura* and *Halictillus*) and the tantalizing studies of Claude-Joseph (1926) indicate that most exhibit some form of sociality. At least some species nest in conspicuous aggregations (e.g. *Corynura [Callochlora] chloris*) which should facilitate future studies of social biology. From these studies we may ascertain the plesiomorphic state of augochlorine social biology, and therefore understand the base from which the diversity of social behaviors exhibited by these beautiful tropical bees has evolved.

ACKNOWLEDGEMENTS

We are grateful to Rex Cocroft, Nancy Jacobson, Ulrich Mueller, Ted Schulz, Bill Wcislo and Doug Yanega for discussions of this paper and their extensive comments on earlier versions of the manuscript, and to the two reviewers for their comments. BND is grateful to Charles Michener and James Hunt for initial discussions of this paper. Preparation of this work was made possible by a National Science Foundation Post-doctoral Fellowship in Environmental Biology to BND (DEB-9201921) and a National Science Foundation Research Grant to GCE (BSR-8413229). I (BND) wish to thank George Eickwort for his guidance, help and encouragement during my tenure at Cornell as a Post-Doctoral Fellow. George was the wisest advisor and most spirited collaborator one could ever hope for. He had a profound influence on me personally and professionally and I am grateful for having known him and worked with him.

LITERATURE CITED

Alexander, R. D. 1974. The evolution of social behavior. *Annu. Rev. Ecol. Syst.* **5**: 325–383.

Andersson, M. 1984. The evolution of eusociality. *Annu. Rev. Ecol. Syst.* **15**: 165–189.

Batra, L. R., S. W. T. Batra and G. E. Bohart. 1973. The mycoflora of domesticated and wild bees (Apoidea). *Mycopathol. Mycol. Appl.* **49**: 13–44.

Brooks, D. R. and D. A. McLennan. 1991. *Phylogeny, Ecology, and Behavior: A Research Program in Comparative Biology.* Chicago: University of Chicago Press.

Cane, J. H., G. C. Eickwort, F. R. Wesley and J. Spielholz. 1985. Pollination ecology of *Vaccinium stamineum* (Ericaceae: Vaccinioideae). *Am. J. Bot.* **72**: 135–142.

Carpenter, J. M. 1988. Choosing among multiple equally parsimonious cladograms. *Cladistics* **4**: 291–296.

–. 1989. Testing scenarios: wasp social behavior. *Cladistics* **5**: 131–144.

Carpenter, J. M., J. E. Strassmann, S. Turillazzi, C. R. Hughes, C. R. Solís and R. Cervo. 1993. Phylogenetic relationships among paper wasp social parasites and their hosts (Hymenoptera: Vespidae; Polistinae). *Cladistics* **9**: 129–146.

Choe, J. C. 1995. Plurimatry: new terminology for multiple reproductives. *J. Insect Behav.* **8**: 133–137

Claude-Joseph, F. 1926. Recherches biologiques sur les Hyménoptères du Chili (Mellifères). *Ann. Sci. Nat. (Paris) Zool.* (Ser. 10) **9**: 113–268.

Crespi, B. J. and D. Yanega. 1995. The definition of eusociality. *Behav. Ecol.* **6**: 109–115.

Duffield, R. M., J. W. Wheeler and G. C. Eickwort. 1984. Sociochemicals of bees. In *Chemical Ecology of Insects.* W. J. Bell and R. T. Cardé, eds., pp. 387–428. New York: Chapman and Hall.

Eickwort, G. C. 1969a. Tribal positions of western hemisphere green sweat bees, with comments on their nest architecture (Hymenoptera: Halictidae). *Ann. Entomol. Soc. Am.* **62**: 652–660.

–. 1969b. A comparative morphological study and generic revision of the augochlorine bees (Hymenoptera: Halictidae). *Univ. Kans. Sci. Bull.* **48**: 325–524.

–. 1979. A new species of wood-dwelling sweat bee in the genus *Neocorynura*, with description of its larva and pupa (Hymenoptera: Halictidae). *Entomol. Gen.* **5**: 143–148.

–. 1988. Distribution patterns and biology of West Indian sweat bees (Hymenoptera: Halictidae). In *Zoogeography of Caribbean Insects.* J. K. Liebherr, ed., pp. 231–253. Ithaca, N. Y.: Cornell University Press.

–. 1993. Evolution and life-history patterns of mites associated with bees. In *Mites. Ecological and Evolutionary Analyses of Life-History Patterns.* M. A. Houck, ed. pp. 218–251. New York: Chapman & Hall.

Eickwort, G. C. and K. R. Eickwort. 1972. Aspects of the biology of Costa Rican halictine bees, IV. *Augochlora (Oxystoglossella)* (Hymenoptera: Halictidae). *J. Kansas Entomol. Soc.* **45**: 18–45.

–. 1973a. Notes on the nests of three wood-dwelling species of *Augochlora* from Costa Rica (Hymenoptera: Halictidae). *J. Kansas Entomol. Soc.* **46**: 17–22.

–. 1973b. Aspects of the biology of Costa Rican halictine bees, V. *Augochlorella edentata* (Hymenoptera: Halictidae). *J. Kansas Entomol. Soc.* **46**: 3–16.

Eickwort, G. C. and S. F. Sakagami. 1979. A classification of nest architecture of bees in the tribe Augochlorini (Hymenoptera: Halictidae; Halictinae), with description of a Brazilian nest of *Rhinocorynura inflaticeps*. *Biotropica* **11**: 28–37.

Eickwort, G. C., J. M. Eickwort, J. Gordon and M. A. Eickwort. 1996. Solitary behavior in a high-altitude population of the social sweat bee *Halictus rubicundus* (Hymenoptera: Halictidae). *Behav. Ecol. Sociobiol.*, in press.

Evans, H. E. and R. W. Matthews. 1973. Observations on the nesting behavior of *Trachypus petiolatus* (Spinola) in Colombia and Argentina (Hymenoptera: Sphecidae: Philanthini). *J. Kansas Entomol. Soc.* **46**: 165–175.

Evans, H. E. and K. M. O'Neill. 1988. *The Natural History and Behavior of North American Beewolves.* Ithaca N. Y.: Cornell University Press.

Farris, J. S. 1969. A successive approximations approach to character weighting. *Syst. Zool.* **18**: 374–385.

–. 1988. *Hennig86 reference.* Documentation for version 1. 5. (Distributed by the author, Port Jefferson Station, New York.)

–. 1989. The retention index and the rescaled consistency index. *Cladistics* **5**: 417–419.

Fitzhugh, K. 1989. Cladistics in the fast lane. *J. N.Y. Entomol. Soc.* **97**: 234–241.

Giblin-Davis, R. M., B. B. Norden, S. W. T. Batra and G. C. Eickwort. 1990. Commensal nematodes in the glands, genitalia, and brood cells of bees (Apoidea). *J. Nematol.* **22**: 150–161.

Gimenes, M., C. K. Kajiwara, F. A. do Carmo and L. R. Bego. 1991. Seasonal cycle and nest architecture of *Augochloropsis notophos* Vachal (Hymenoptera, Halictidae, Halictinae). *R. Bras. Entomol.* **35**: 767–772.

Hamilton, W. D. 1964a. The genetical evolution of social behavior, I. *J. Theor. Biol.* **7**: 1–16.

–. 1964b. The genetical evolution of social behavior, II. *J. Theor. Biol.* **7**: 17–52.

Heithaus, E. R. 1979a. Flower-feeding specialization in wild bee and wasp communities in seasonal neotropical habitats. *Oecologia* **42**: 179–194.

–. 1979b. Flower visitation records and resource overlap of bees and wasps in northwest Costa Rica. *Brenesia* **16**: 9–52.

–. 1979c. Community structure of neotropical flower visiting bees and wasps: Diversity and phenology. *Ecology* **60**: 190–202.

Janzen, D. H. 1968. Notes on nesting and foraging behavior of *Megalopta* (Hymenoptera: Halictidae) in Costa Rica. *J. Kansas Entomol. Soc.* **41**: 342–350.

Jörgensen, P. 1912. Beitrag zur Biologie einiger südamerikanischer Bienen. *Z. Wiss. Insektenbiol.* **8**: 268–272.

Kerfoot, W. B. 1967a. Nest architecture and associated behavior of the nocturnal bee, *Sphecodogastra texana* (Hymenoptera: Halictidae). *J. Kansas Entomol. Soc.* **40**: 84–93.

–. 1967b. Correlation between ocellar size and the foraging activities of bees (Hymenoptera: Halictidae). *Am. Nat.* **101**: 65–70.

–. 1967c. The lunar periodicity of *Sphecodogastra texana*, a nocturnal bee (Hymenoptera; Apoidea). *Anim. Behav.* **15**: 479–486.

Kukuk, P. F. and G. C. Eickwort. 1987. Alternative social structures in halictine bees. In *Chemistry and Biology of Social Insects*. J. Eder and H. Rembold, eds., pp. 555–556. Munich: J. Peperny.

Kukuk, P. F. and G. K. Sage. 1994. Reproductivity and relatedness in a communal halictine bee *Lasioglossum (Chilalictus) hemichalceum. Insectes Soc.* **41**: 443–455.

Laroca, S. 1970. Contribuição para o conhecimento das relações entre abelhas e flôres: Coleta de pólen das anteras tubulares de certas Melostomataceae. *Rev. Floresta* **2**: 69–74.

Laroca, S., J. R. Cure and C. de Bortoli. 1982. A associação de abelhas silvestres (Hymenoptera, Apoidea) de uma área restrita no interior da cicade de Curitiba (Brasil): Uma abordagem biocenótica. *Dusenia* **13**: 93–117.

Lüderwaldt, H. 1911. Nestbau von *Neocorynura erinnys* Schrottky. *Z. Wiss. Insektenbiol.* **7**: 94–96.

Lundberg, J. G. 1972. Wagner networks and ancestors. *Syst. Zool.* **21**: 398–413.

Maddison, W. P. 1991. The discovery and importance of multiple islands of most-parsimonious trees. *Syst. Zool.* **40**: 315–328.

Maddison, W. P., M. J. Donoghue and D. R. Maddison. 1984. Outgroup analysis and parsimony. *Syst. Zool.* **33**: 83–103.

Menke, A. S. 1980. Biological notes on *Trachypus mexicanus* Saussure and *T. petiolatus* (Spinola) (Hymenoptera: Sphecidae). *J. Kansas Entomol. Soc.* **53**: 235–236.

Mickel, C. E. 1973. *Paramutilla halicta* n. genus, n. species, a parasite of the halictine bee *Augochlorella edentata* (Hymenoptera: Mutillidae). *J. Kansas Entomol. Soc.* **46**: 1–3.

Michener, C. D. 1974. *The Social Behavior of the Bees.* Cambridge, Mass.: Harvard University Press.

–. 1977. Nests and seasonal cycle of *Neocorynura pubescens* in Colombia (Hymenoptera: Halictidae). *Rev. Biol. Trop.* **25**: 39–41.

–. 1978. The parasitic groups of Halictidae (Hymenoptera, Apoidea). *Univ. Kans. Sci. Bull.* **51**: 291–339.

–. 1985. From solitary to eusocial: need there be a series of intervening species? In *Fortschritte der Zoologie, vol. 31. Experimental Behavioral Ecology.* B. Hölldobler and M. Lindauer, eds., pp. 293–305. Stuttgart: Gustav Fischer Verlag.

–. 1990. Reproduction and castes in social halictine bees. In *Social Insects. An Evolutionary Approach to Castes and Reproduction.* W. Engels, ed., pp. 77–121. Berlin: Springer-Verlag.

Michener, C. D. and C. A. Campos Seabra. 1959. Observations on the behavior of Brasilian halictid bees, VI, Tropical species. *J. Kansas Entomol. Soc.* **32**: 19–28.

Michener, C. D. and W. B. Kerfoot. 1967. Nests and social behavior of three species of *Pseudaugochloropsis. J. Kansas Entomol. Soc.* **40**: 214–232.

Michener, C. D. and R. B. Lange. 1958a. Observations on the behavior of Brazilian halictid bees II: *Paroxystoglossa jocasta. J. Kansas Entomol. Soc.* **31**: 129–138.

–. 1958b. Observations on the behavior of Brasilian halictid bees, III. *Univ. Kans. Sci. Bull.* **39**: 473–505.

–. 1958c. Distinctive type of primitive social behavior among bees. *Science (Wash., D.C.)* **127**: 1046–1047.

–. 1959. Observations on the behavior of Brazilian halictid bees (Hymenoptera, Apoidea) IV. *Augochloropsis*, with notes on extralimital forms. *Am. Mus. Nov.* **1924**: 1–41.

Michener, C. D., W. B. Kerfoot and W. Ramírez B. 1966. Nests of *Neocorynura* in Costa Rica (Hymenoptera: Halictidae). *J. Kansas Entomol. Soc.* **39**: 245–258.

Moure, J. S. 1943. Notas sôbre abelhas da coleção Zikán (Hym. Apoidea). *Rev. Entomol. (Rio)* **14**: 447–484.

Moure, J. S. and P. D. Hurd, Jr. 1987. *An Annotated Catalog of the Halictid Bees of the Western Hemisphere (Hymenoptera: Halictidae).* Washington, D. C.: Smithsonian Institution Press.

Mueller, U. G. 1991. Haplodiploidy and the evolution of facultative sex ratios in a primitively eusocial bee. *Science (Wash., D.C.)* **254**: 442–444.

–. 1993. Haplodiploidy and the evolution of facultative sex ratios in a primitively eusocial bee. Ph. D. dissertation, Cornell University, Ithaca, N. Y.

Mueller, U. G. , G. C. Eickwort and C. F. Aquadro. 1994. DNA fingerprinting analysis of parent-offspring conflict in a primitively eusocial bee. *Proc. Natl. Acad. Sci. U.S.A.* **91**: 5143–5147.

Oliveira Campos, M. J. 1980. Aspectos da sociologia e fenologia de *Pereirapis semiauratus* (Hymenoptera, Halictidae, Augochlorini). M.S. thesis, Universidade Federal de São Carlos, Brazil.

Ordway, E. 1964. *Sphecodes pimpinellae* and other enemies of *Augochlorella* (Hymenoptera: Halictidae). *J. Kansas Entomol. Soc.* **37**: 139–152.

–. 1965. Caste differentiation in *Augochlorella* (Hymenoptera, Halictidae). *Insectes Soc.* **12**: 291–308.

–. 1966. The bionomics of *Augochlorella striata* and *A. persimilis* in eastern Kansas (Hymenoptera: Halictidae). *J. Kansas Entomol. Soc.* **39**: 270–313.

Packer, L. 1986. The biology of a subtropical population of *Halictus ligatus* IV: A cuckoo-like caste. *J. N.Y. Entomol. Soc.* **94**: 458–466.

–. 1990. Solitary and eusocial nests in a population of *Augochlorella striata* (Provancher) (Hymenoptera; Halictidae) at the northern edge of its range. *Behav. Ecol. Sociobiol.* **27**: 339–344.

–. 1991. The evolution of social behavior and nest architecture in sweat bees of the subgenus *Evylaeus* (Hymenoptera: Halictidae): a phylogenetic approach. *Behav. Ecol. Sociobiol.* **29**: 153–160.

–. 1993. Multiple-foundress associations in sweat bees. In *Queen Number and Sociality in Insects*. L. Keller, ed., pp. 215–233. Oxford University Press.

Packer, L., B. Sampson, C. Lockerbie and V. Jessome. 1989a. Nest architecture and brood mortality in four species of sweat bee (Hymenoptera; Halictidae) from Cape Breton Island. *Can. J. Zool.* **67**: 2864–2870.

Packer, L., V. Jessome, C. Lockerbie and B. Sampson. 1989b. The phenology and social biology of four sweat bees in a marginal environment: Cape Breton Island. *Can. J. Zool.* **67**: 2871–2877.

Pagel, M. D. and P. H. Harvey. 1988. Recent developments in the analysis of comparative data. *Q. Rev. Biol.* **63**: 413–440.

Pesenko, Y. 1983. Phylogenetic relationships between the tribes of the subfamily Halictinae and within the tribe Nomioidini. *Fauna of the U.S.S.R., Hymenopteran Insects.* vol. 17, no. 1. *Halictid bees (Halictidae), subfamily Halictinae, tribe Nomioidini (in the Palearctic fauna).* Moscow: Nauka (in Russian).

Platnick, N. I. 1989. An empirical comparison of microcomputer parsimony programs, II. *Cladistics* **5**: 145–161.

Rêgo, M. M. C. and P. M. C. D. Albuquerque. (1989) 1990. Comportamento das abelhas visitantes de murici, *Byrsonima crassifolia* (L.) Kunth, Malpighiaceae. *Bol. Mus. Para. Emil. Goel. Ser. Zool.* **5**: 179–194.

Renner, S. S. 1986. Reproductive biology of *Bellucia* (Melastomataceae). *Acta Amazon.* **16/17**: 197–208.

Richards, M. 1994. Social evolution in the genus *Halictus*: a phylogenetic approach. *Insectes Soc.* **41**: 315–325.

Rick, C. M., M. Holle and R. W. Thorp. 1978. Rate of cross-pollination in *Lycopersicon pimpinellifolium*: Impact of genetic variation in floral characters. *Pl. Syst. Evol.* **129**: 31–44.

Ridley, M. 1983. *The Explanation of Organic Diversity. The Comparative Method and Adaptations for Mating.* Oxford: Clarendon Press.

Sakagami, S. F. 1964. Wiederentdeckung des Nestes einer Nachtfurchenbiene, *Megalopta* sp. am Amazonas (Hymenoptera, Halictidae). *Kontyû* **32**: 457–463.

Sakagami, S. F. and J. S. Moure. 1965. Cephalic polymorphism in some neotropical halictine bees (Hymenoptera – Apoidea). *An. Acad. Bras. Ciênc.* **37**: 303–313.

–. 1967. Additional observations on the nesting habits of some Brazilian halictine bees (Hymenoptera, Apoidea). *Mushi* **40**: 119–138.

Sanderson, M. J. and M. J. Donoghue. 1989. Patterns of variation in levels of homoplasy. *Evolution* **43**: 1781–1795.

Schlising, R. A. 1970. Sequence and timing of bee foraging in flowers of *Ipomoea* and *Aniseia* (Convolvulaceae). *Ecology* **51**: 1061–1067.

Schremmer, F. 1979. Zum Nest-Aufbau der neuen neotropischen Furchenbienen-Art *Neocorynura colombiana* (Hymenoptera: Halictidae). *Entomol. Gen.* **5**: 149–154.

Sherman, P. W., E. A. Lacey, H. K. Reeve and L. Keller. 1995. The eusociality continuum. *Behav. Ecol.* **6**: 102–108.

Sillén-Tullberg, B. 1988. Evolution of gregariousness in aposematic butterfly larvae: phylogenetic analysis. *Evolution* **42**: 293–305.

Smith, J. B. 1901. Notes on some digger bees. II. *J. N.Y. Entomol. Soc* **9**: 52–72.

Stockhammer, K. A. 1966. Nesting habits and life cycle of a sweat bee, *Augochlora pura* (Hymenoptera: Halictidae). *J. Kansas Entomol. Soc.* **39**: 157–192.

Vachal, J. 1911. Etude sur les *Halictus* d'Amérique (Hym.). *Misc. Entomol. (Narbonne)* **19**: 9–24, 41–56, 107–116.

Watrous, L. E. and Q. D. Wheeler. 1981. The out-group comparison method of character analysis. *Syst. Zool.* **30**: 1–11.

Wcislo, W. T., and J. H. Cane. 1996. Floral resource utilization by solitary bees (Hymenoptera: Apoidea) and exploitation of their stored foods by natural enemies. *Annu. Rev. Entomol.* **41**: 257–286.

Wheeler, W. C., P. Cartwright and C. Y. Hayashi. 1993. Arthropod phylogeny: a combined approach. *Cladistics* **9**: 1–39.

Wille, A. 1963. Behavioral adaptations of bees for pollen collecting from *Cassia* flowers. *Rev. Biol. Trop.* **11**: 205–210.

Wilson, E. O. 1971. *Insect Societies.* Cambridge: Harvard University Press.

Wolda, H. and D. W. Roubik. 1986. Nocturnal bee abundance and seasonal bee activity in a Panamanian forest. *Ecology* **67**: 426–433.

Appendix 13-1. Character list

Eighty-one characters were included. Character numbers in parentheses are those of Eickwort (1969b). Ordered characters are indicated by a + next to the number. Since polarity decisions were based in some analyses on ingroup rooting, character state 0 does not always correspond to the plesiomorphic condition. See Eickwort (1969b) for illustrations and complete descriptions of characters.

Female characters

1(1)+   angle of epistomal sulcus: 0 = obtuse; 1 = right; 2 = acute

2(2)   lateral profile of clypeus: 0 = normal, bevelled; 1 = flat; 2 = protuberant

3(2)   clypeal tooth: 0 = normal; 1 = prolonged

4(4)   preoccipital ridge: 0 = round; 1 = angulate; 2 = carinate; 3 = lamellate

5(4)   vertex: 0 = normal; 1 = swollen; 2 = pointed

6(5)+   hypostomal relative length: 0 = very short; 1 = short; 2 = medium; 3 = long

7(6)+   length of closed hypostomal bridge suture: 0 = short; 1 = medium; 2 = long

8(7)   hypostomal carina posterior flange: 0 = normal; 1 = projecting

9(7)   anterior angle of the hypostomal carina: 0 = normal; 1 = sharp

10(8)+   eye emargination: 0 = weakly emarginate; 1 = moderately emarginate; 2 = strongly emarginate

11(9)   eye hair: 0 = short, visible only under microscope; 1 = long

12(10)   size of ocelli: 0 = normal, small; 1 = enlarged

13(11)   labral distal process: 0 = narrowly triangular; 1 = broadly triangular; 2 = expanded; 3 = quadrate

14(11)   labral basal elevation (shape): 0 = orbiculate; 1 = suborbiculate; 2 = bituberculate; 3 = longitudinal; 4 = transverse; 5 = four-lobed; 6 = triangular

15(11)   labral basal elevation (profile): 0 = protuberant; 1 = low

16(11)   labral fimbria: 0 = on surface; 1 = on edge

17(11)+   labral teeth: 0 = absent; 1 = weak; 2 = moderate; 3 = strong

18(12)   mandibular teeth: 0 = bidentate; 1 = monodentate; 2 = supplementary teeth

19(12)   mandibular width: 0 = narrow; 1 = broad

20(13)+   relative length of distal portion of maxilla: 0 = short/broad; 1 = medium; 2 = long/narrow

21(13)   base of galea vs. base of stipes: 0 = distal; 1 = equal

22(13)   apex of galea: 0 = lobed; 1 = pointed

23(13)   inner strip of galea: 0 = setae present; 1 = setae absent

24(13)   markings on inner strip of galea: 0 = broad; 1 = narrow

25(13)   galeal comb: 0 = present; 1 = absent

26(14)+   maxillary palp length: 0 = short; 1 = long; 2 = extremely long

27(15)+   prementum width: 0 = extremely wide; 1 = wide; 2 = medium; 3 = narrow

28(16)   V-shaped brace of salivary plate: 0 = present; 1 = absent

29(17)+   glossa length/width: 0 = short; 1 = medium; 2 = long

30(18)   length of labial palpi 2+3 vs. 1: 0 = shorter; 1 = longer

31(19)   pronotal dorsal ridge: 0 = rounded; 1 = angled; 2 = carinate; 3 = lamellate

32(19)   pronotal lateral ridge: 0 = absent; 1 = rounded; 2 = sharply angled; 3 = carinate; 4 = lamellate

33(20)   mesoscutal shape in dorsal view: 0 = normal, broadly rounded anteriorly; 1 = narrowed anteriorly

34(20)   mesoscutum lip: 0 = absent; 1 = rounded; 2 = angled; 3 = carinate; 4 = lamellate

35(21)   tegula shape: 0 = semioval in outline; 1 = inner angle produced

36(22)   basal area of propodeum: 0 = pitted or striate; 1 = smooth

37(23)   posterior surface of propodeum: 0 = normal, lateral carina diverging; 1 = narrow; 2 = wide

38(23)   propodeal pit: 0 = narrow; 1 = enclosed within V-shaped notch

39(24)   apex of marginal cell: 0 = acute; 1 = truncate

40(25)   marginal cell length/width: 0 = long; 1 = short

41(25)   stigma size: 0 = normal; 1 = enlarged

42(27)   inner hind tibial spur: 0 = serrate; 1 = pectinate

43(29)+   basitibial plate shape: 0 = normal, rounded; 1 = = short; 2 = very short; 3 = absent

44(29)+   rim basitibial plate: 0 = strong; 1 = weak anteriorly; 2 = absent

45(30)   anterior basitarsal brush: 0 = absent; 1 = present

46(31)   setae on pseudopygidial area: 0 = fine; 1 = absent; 2 = scaly

47(32)   gradulus tergum VI: 0 = present; 1 = absent

48(34)   gradulus sternum IV: 0 = present; 1 = absent

49(34)   gradulus sternum V: 0 = present; 1 = absent

Male characters

50(35)+   antennal flagellomere 2 vs. 1: 0 = short; 1 = medium; 2 = long

51(35)+   antennal length: 0 = short; 1 = moderate; 2 = long

52(35)+   scape length: 0 = short; 1 = medium; 2 = long

53(35)   antennal sensory plate areas: 0 = absent; 1 = present

54(36)   labral distal process: 0 = present; 1 = absent

55(36)   labral basal elevated notch: 0 = present; 1 = absent

56(38)   metasoma shape: 0 = oval; 1 = elongate; 2 = petiolate

57(39)   tergum VII: 0 = gradually convex; 1 = abruptly convex

58(39)+   gradulus of tergum VII: 0 = present; 1 = weak; 2 = absent

59(40)   posterior anal filaments: 0 = absent; 1 = present

60(40)   microtrichia on proctiger: 0 = absent; 1 = present

61(41)   apex of sternum III: 0 = unmodified; 1 = median point; 2 = bilobed

62(42)   sternum IV postgradular setae: 0 = setae present but widely scattered; 1 = setae appearing in discrete clumps

63(42)   sternum IV gradulus: 0 = unmodified; 1 = median interruption

64(42)   sternum IV gradulus: 0 = unmodified; 1 = touches antecosta anteriorly

65(42)   posterior margin of sternum IV: 0 = unmodified; 1 = modified, typically with median emargination

66(43)   sternum V postgradular setae: 0 = setae present but widely scattered; 1 = setae appearing in discrete clumps

67(43)   sternum V gradulus: 0 = unmodified; 1 = modified

68(43)   posterior margin of sternum V: 0 = unmodified; 1 = modified

69(44)   sternum VI postgradular setae: 0 = setae present but widely scattered; 1 = setae appearing in discrete clumps

70(44)   gradulus of sternum VI: 0 = unmodified; 1 = median interruption

71(44)   posterior margin of sternum VI: 0 = entire; 1 = notched

72(45)   posterior margin of sternum VII: 0 = unmodified; 1 = median projection; 2 = bilobed; 3 = lateral projection

73(45)   posterior margin of sternum VIII: 0 = unmodified; 1 = median projection

74(45)   junction of sterna VII & VIII: 0 = before apodemes; 1 = at apodemes; 2 = bilobed

75(45)   spiculum of sternum VIII: 0 = narrow; 1 = broad

76(46)   ventral bridge of gonobase: 0 = broad; 1 = narrow

77(46)+   dorsal lobes of gonobase: 0 = small; 1 = moderate; 2 = strong

78(47)   basal process of gonostylus: 0 = present; 1 = absent

79(47)   setae on basal process of gonostylus: 0 = absent; 1 = present

80(47)   parapenial lobes: 0 = present; 1 = absent

81(48)   venter of penis valve: 0 = unmodified; 1 = prong; 2 = keel

Appendix 13-2. Data matrix (81 characters)

```
                  000000000111111111122222222223333333333344444444445555555555566666666666777777777788
                  123456789012345678901234567890123456789012345678901234567890123456789012345678901

ANCESTOR    00900099900099900099009900090999090999990009900009991990909900090090000999990991
TEMNOSOM    0001122101003501010010001010012100000000003201100201000002000000001091019110001000
CHLEROGL    20000390000092109009999999999900900100000100100119999999999999999999999999999999999
ARIPHANR    120003900101110000121109123100210100000000000100110120910121001000101101002012 1001
MEGALODI    12100390020101000012110910210001010100100000010011092911099999100100000019999999999
MEGOMAT     12100391010111000121109113100010101000100100011101291901210111010110112021121001
MEGALOPI    12100391010111000121109113100010101000101100011101291901210111090001011102112 1091
MEGALOPA    20001121000112010201100110312103010000100101120111110010129021011001001211010 1001
CAENAUGO    10010100021001000001001101121220100000100100100001120000110021001001001111010 1011
CTENAUGO    10010100020001000001001101121220100001001011000011200001190210010010011010101011
PSEUDAUG    10002301020001001001101102120210100001101001001111190001910010011010013121101000
PEREIRAP    10010100020006103001100110102122010010010000000011200101900010010000010100120111
CERATALI    00020100020001000001100110000122010000010000100011110010120000000000011120120101
AUGOELLA    10010100020001000001100110101122020000010000100110120010111000000000011110111011
AUGOCHLO    21020300120004002001100110102022010000011000012001112001012100000000000011201 20111
OXYSTOGL    21020300120001002001100110102022010000110000100011120010121000000000000112012 0111
MYCTEROC    21020300120012003001100110000122010000110000120011200101210000000011011120120111
CORYNURA    00000010200014000000001001000121110101010900100112201112020000000000001111110 00112
CALLOCHL    00000010210140000000001001000121110001010111001122011120200000000000111110 00012
HALICTIL    00000100000000002001001001000121010001010101100112201111120000000000001111001010
RHINOCOR    01001101000022010001000000001310301000001001001111200111201001000100111010000192
CORYNURL    20001301000020000001000001002113020000001100000111120911120100990090011101 0020111
RHECTOMI    0000139000001200000100000100113010100010100000011999999999999999999999999999999999
NEOCORYA    00020100020003100001100990000123120090010100120112219012110000000000009911012 0010
NEOCROYD    00030100020004100021100990000124120090010100120112210012110000000000009911012 1011
PAROXYST    00010100010010010011001101011211200009100001001122100110900010011010011110 121011
ANDINAUG    20010101020003000001100190101122010000110101100122110111900000001010011111121011
CHLEROGA    10010999910099999999909999999911010010010999999912119110000009901010011111 111010
AUGOPSIS    00010211010022010001100110101131011020110120100011200000090010111011011111121092
PARAUGOP    00010211010022010001100110101131011020110120100011200000090010110001011111121092
AUGOODES    00000100010001001001100111101121010000011100100111110011121000011100001101 1121091
THECTOCH    00000100010000003001100110000131040001101010011221001012000100110100101101 21001
```

14 · Demography and sociality in halictine bees (Hymenoptera: Halictidae)

DOUGLAS YANEGA

ABSTRACT

I propose two hypotheses regarding the relationships between halictine bee demography and social behavior and the environment: the 'Mating Limitation Hypothesis' (MLH), that a female's social role ('caste') is dependent on whether she mates while young (and thus dependent on male demography); and the 'Environmental Control Hypothesis' (ECH), that the decision to lay eggs of one sex or another (thus defining male demography) is dependent upon the temperature and/or photoperiodic conditions experienced at the time the egg is laid, in temperate species. Published demographic and behavioral data typically lack the precision needed for truly conclusive analysis and review, and there are some important recently discovered phenomena that were historically overlooked, and extremely difficult to document. Nonetheless, a review of the available data reveals many notable features that can help evaluate the proposed hypotheses, and offers some guidance for future research. An older hypothesis that is also addressed here is that sexual selection is expected to promote protandry, for many reasons, yet the data on halictine bees suggests that protandry is absent in social populations, although it is characteristic of solitary populations.

Geographic variation in demography and behavior is common, and this variation appears to correlate with the seasonality of the habitat, in a manner consistent with the predictions of the two proposed hypotheses. The overall pattern in temperate halictines is that those populations that begin provisioning earliest in the year are generally those that are most clearly eusocial, and I propose that the proximate explanation is that foundresses begin laying eggs at a time when both photoperiod and temperature favor production of female brood alone; there are thus no males for these females to mate with when they emerge, and they become non-gynes as a result. Tropical and subtropical halictines exhibit a wide range of behavioral types, but there is a general negative association between male abundance during female emergence periods and the degree to which sociality is developed. Populations that are either active year-round or have males and females in diapause together are either solitary/communal or have weakly differentiated 'castes' (apparently related to unmatedness) and a scarcity of mated non-reproductive females. These patterns are consistent with the MLH. Colony structure in aseasonal halictine populations suggests that what have been called 'castes' may be facultative behavioral classes, and thus represent semisociality rather than eusociality, a different syndrome from seasonal populations, with different evolutionary implications, although both might be viewed as expressions of a shared form of maternal manipulation based on unmatedness.

Developmental periods can vary considerably, with larger-bodied species tending to take longer, but temperature effects on the rate of development can be substantial, and potentially of great demographic importance. The evidence suggests that sex ratio in some halictines is affected by temperature, but that this effect may be more subtle in some species than in others; thus the data are equivocal regarding this aspect of the ECH. Studies of temperate halictines suggest that photoperiod variation has profound effects upon sex ratio of eggs laid, supporting this aspect of the ECH, and it may be a major factor in determining demography and social structure.

The conclusions of this review suggest, at the very least, a need for new alternative theories and models to approach the links between environment, demography, and sociality, as well as critical empirical tests of these hypotheses. Most halictines studied in detail do not appear to be limited strictly to one mode of social behavior, even within populations, and species-level characterizations of sociality should be abandoned in favor of more precise and quantitative descriptions of social structure. I view this as a form of phenotypic flexibility, which has evolved to allow halictines to adjust their behavior to match the characteristics of their environment.

INTRODUCTION

The production of reproductive individuals in the Hymenoptera is a complex phenomenon, especially owing to the haplodiploid mechanism of sex determination and the control of egg fertilization by reproductive females. In social Hymenoptera, there are also various behavioral complexities related to the presence of subfertile individuals, such as the queen–worker conflict over reproduction. Variation in nest-founding habits, colony size, dominance interactions, degree of worker sterility, caste determination, mating systems, response to environmental cues (and their variation), and other factors, may lead to considerable variation in the social patterns expressed, even between related species or conspecific populations. The relationships among these factors are not well understood for any taxa, and different 'rules' may well apply to different taxa, as various forms of sociality have arisen independently in several hymenopteran lineages (Wilson 1971). The halictine bees are especially noteworthy, as the range of forms of sociality expressed and the variation within and between genera, species, and even populations is essentially unparalleled in any other social animal group (with the possible exception of the allodapine Apidae; see Schwarz *et al.*, this volume). Other chapters on halictines (Danforth and Eickwort, this volume; Wcislo, this volume) have discussed some conceptual and evolutionary issues surrounding this diversity of social behaviors, and I would like to propose two specific, testable hypotheses as to proximate mechanisms – perhaps unique to halictines – that may underlie this variation.

I have hypothesized that caste in the facultatively eusocial bee *Halictus rubicundus* is determined by whether a female mates soon after emergence, and that the likelihood of mating is a function of relative male abundance on the day a female first emerges from her natal nest (Yanega 1989, 1992). If males are scarce relative to newly emerged females, then most females will fail to mate and become workers, and if males are relatively abundant, then most females will mate and become gynes, which enter diapause. Apparently, the presence of males in the population is what results in the 'production' of gynes, meaning that analysis of the pattern of production of reproductives can be reduced largely to analysis of *male* production. The mechanisms controlling the timing and quantity of male production seem to be of primary importance in determining the demography and social structure of the population (Yanega 1993). Replacement queens in halictines are members of the same caste as workers, by definition (Yanega 1989; Crespi and Yanega 1995); their differentiation is a phenomenon independent of caste determination. There are a number of patterns visible in halictines in general

that could be explained by a mechanism similar to that suggested above: i.e. that a female's behavior is determined by whether or not (and when) she mates. I formalize this as a specific hypothesis of behavioral determination below.

As a corollary to this hypothesis, if male production affects sociality, then it is clearly quite important to understand factors which influence male production. For halictine bees, temperature and photoperiod effects on sex ratio have been documented in the laboratory and are discussed below. Long photoperiod correlates with an increase in the percentage of male eggs laid by *H. rubicundus* in the field (with a peak at the summer solstice), and warm temperatures appear to increase this male bias (Yanega 1993). I propose a second formal hypothesis below as to abiotic environmental factors influencing male production, and this hypothesis (though not the first) may have application outside the halictines.

At a functional level, the behavioral patterns that would result from the mechanisms I propose here for halictines could conceivably be viewed as adaptive in several ways.

(a) The production of gynes in a foundress' first brood in a seasonal habitat may be a form of bet-hedging against environmental uncertainty (see, for example, Yanega 1992).

(b) An unmated female can presumably maximize her inclusive fitness by remaining and helping her mother produce female siblings, and if the decision to act as a worker *is* irreversible, then this decision should be made as soon as possible (after some time interval without mating), as every day not working may be a potential sibling (or more) lost. If foundresses can influence this process in their favor, they should certainly be expected to do so, but only when a female offspring is worth more as a worker than as a potential gyne. In contrast, a mated female has the capacity to produce her own female offspring, and should pursue the course that maximizes her own fitness; in seasonal populations, this will generally entail diapause and subsequent nest initiation. In aseasonal populations, a female may have the option to stay in the natal nest, but it should be as a dominant or equal (not a subordinate) to other females in the nest; this will depend on costs and benefits of independent nesting.

(c) For a female in a temperate population, photoperiod is one of the more reliable seasonal cues available to help her allocate reproductive resources in an appropriate manner, especially if caste determination is dependent upon male production by the other females in the population. Reliance on photoperiod will also ensure that production of reproductives and cessation of worker production are at the appropriate time relative to the end of the season (similarly, in experiments with some polistine wasps, photoperiod has been shown to influence a late-season switch to male egg production) (Suzuki 1981, 1982).

PREDICTIVE HYPOTHESES

If the above arguments apply to many halictine species, certain patterns should exist in both male production and social structure within the Halictinae. Given knowledge of the photoperiod and temperature regimes experienced by any given temperate halictine population at the start of its active season, for example, it should be possible to predict general patterns of demography, and thus social behavior. Further details (on longevity and developmental rates, especially) would allow for more precise predictions on a case-by-case basis. It might also be possible, if the hypothesis regarding mating is more generally applicable, to make predictions regarding populations which do not exhibit diapause. My formal hypotheses and their predictions can be analyzed separately, as they are not mutually dependent.

The 'mating limitation hypothesis' (MLH)

If a female mates promptly after emerging, she then becomes a member of the maximally reproductive behavioral class (whether this class is permanent (a caste) or facultative (a subcaste): for example, a diapausing gyne in most annual-cycle species). Females that fail to mate promptly should remain in the natal nest, as helpers when possible, as long as this is a fitness-maximizing strategy. If prompt mating determines females as gynes, the recruitment of new workers should diminish in proportion to increasing male abundance during the season, and new gynes should never be found well before males. Clear exceptions, such as high percentages of workers emerging at a time when males are also relatively numerous, would indicate that there is some other factor influencing caste. By extension, any population with clear protandry (i.e. numbers of males present before *any* females appear) should have few or no workers, and demonstrate solitary or communal behavior, even if there is overlap of generations. The existence of strictly solitary, partly bivoltine populations is not expected under the MLH: females that mate promptly should enter diapause rather than remain active (assuming overwintering *is* the superior reproductive strategy), and any unmated females are expected to form eusocial groups with their mothers. For example, if foundresses are typically dead by the time

their daughters emerge, and the season is long enough for production of a second generation, mating should not necessarily induce diapause, because there may be no advantages to overwintering in this case; such a population would be solitary and strictly bivoltine.

Populations in which both males and unmated females diapause (any life stage) would not be expected to develop eusocial behavior under the MLH, unless there is some aspect of the life cycle, such as different male and female emergence times or forms of maternal control (e.g. inhibition of receptivity to mating), that can regularly cause some some females to remain unmated; this circumstance should also be necessary for eusociality to exist in populations lacking diapause altogether. Therefore, the most general prediction of the MLH is that eusociality is unlikely to occur in halictines unless males are rare or absent for part of the season during which young females are emerging, so some females will be unmated, at least initially. Note also that I specify that the MLH should apply in cases with subcastes rather than true (permanent) castes, so the same restrictions on male production may be expected in semisocial halictine populations.

The MLH also offers a mechanism for precisely the sort of association of decreasing male bias in the first brood with 'increasing sociality' as has long been noted in temperate halictines (Knerer and Plateaux-Quénu 1967; Breed 1976; Packer and Knerer 1985). The MLH states that relative male abundances during first brood emergence determine the proportion of first-brood females that become workers; this should lead to a general inverse correlation of first-brood males and features associated with sociality, between and within temperate species and/ or populations. Increased male abundance will lead to decreased numbers of eusocial nests (Yanega 1993) and decreased morphological distinctiveness of castes (first-brood females are smaller, and if large numbers become gynes rather than workers, the size distribution of foundresses will more broadly overlap that of workers; see also Yanega 1989). The opposite hypothesis, that females are born already determined as gynes or workers, and that males are produced to mate with emerging gynes, is discussed below.

The 'environmental control (of sex ratio) hypothesis' (ECH)

Above some threshold, increasing photoperiod will result in an increased proportion of unfertilized (male) eggs laid by mated queens, and higher temperatures will accentuate this effect. If long photoperiods serve to increase the proportion of unfertilized eggs a mated female lays, then males should first appear at roughly similar times in all temperate halictine populations experiencing similar photoperiod and/or temperature conditions, and reach their peak of production (as eggs) at the summer solstice. Of course, there may well be some minimal photoperiodic variation over the year needed for this effect, and at lower latitudes the ECH is unlikely to apply. Evidence for a temperature effect on male bias would include finding that populations with higher thermal regimes in given years produce a relatively high proportion of unfertilized eggs at any specific photoperiod in those years (see, for example, Yanega 1993).

If the thermal peak is shifted relative to the solstice, then the peak in male frequency may be correspondingly shifted and/or flattened. Variation in developmental times between species, between sexes, or due to temperature should also lead to variation in the actual emergence patterns observed. Indeed, the interactions of the three major parameters (photoperiod, temperature, and developmental characteristics) may make interpretation or prediction of the patterns found in nature difficult, unless sex ratio data can be obtained directly at time of egg-laying. Finding two halictine populations which differ significantly in only *one* of these parameters seems rather unlikely; studies of one population over several years, or laboratory data (see below), are required to test critically whether this hypothesis applies to species other than *H. rubicundus*, and to what degree. Similarly, the patterns of male emergence in the second brood will also reflect production (if any) of males by workers and replacement queens; this could further skew or flatten the seasonal peak in male production, and detailed data would be required to quantify this effect.

General predictions

Taken together, the MLH and ECH predict that annual temperate populations that begin provisioning early in the season (under cool temperatures and shorter daylengths) should be eusocial, and the later provisioning begins (regardless of how long provisioning could continue), the lower the likelihood that (and degree to which) eusociality will be expressed. Fertilized (female) eggs laid after the solstice, for example, should rarely if ever lead to workers. Although the ECH does not predict male production regimes in the tropics, the MLH predicts that a correlation between mating and reproductivity should still exist, such

that different patterns of male production and seasonal activity will lead to different social structures. An important corollary, in terms of analysis of both past and future studies, pertains to laboratory rearings: (1) they sometimes utilize unnatural photoperiods and temperatures; (2) they always have artificial spatial structure (which may profoundly affect male–female encounter dynamics); (3) they do not allow females to enter mid-season diapause away from the natal nest. As all three of these features violate the conditions under which the MLH and ECH operate in nature, such studies are expected to yield unnatural results, and caution must be taken when attempting to extrapolate such results to behavior in the natural environment.

TERMINOLOGY

The different forms of social behavior seen in halictines are as defined elsewhere (Danforth and Eickwort, this volume), with a few definitions modified following Crespi and Yanega (1995).

Quasisocial: several females share a nest, with a unimodal distribution of female lifetime reproductive output, but females exhibit cooperative brood care (i.e. shared provisioning duties).

Semisocial: as in quasisocial, but some females reproduce more (and typically provision less) than others, resulting in bimodality of lifetime reproduction. Females retain the capacity, even if not expressed in all individuals, to switch from the less reproductive to the more reproductive behavioral class (subcastes) after they attain reproductive maturity.

Eusocial: females belong to two permanent behavioral classes (castes), which are determined at some point prior to reproductive maturity. Differences between these classes may include (but are not limited to) physiology (e.g. diapause), mean lifetime reproduction, or morphology.

Such terms are best applied to colonies, although if the behavior is uniform they may be applied to populations or possibly even species (see also Wcislo, this volume). If, however, a population with mostly eusocial colonies has some parasocial colonies or solitary individuals (for example, as a result of foundress death), it is appropriate to term the population as *primarily* eusocial, because the population as a whole does contain members of two distinct castes, even if some nests do not. These terms make no reference to overlap or non-overlap of generations; there is no evidence that behavioral classes in parasocial halictines (traditionally defined by their lack of generational overlap) are permanent, and some suggestion that they are not. Until demonstrated otherwise, all halictines considered communal, quasi-, or semisocial in the past remain in these categories at present (Crespi and Yanega 1995). The possibility exists, however, that some 'eusocial' populations lack true castes, and are thus semisocial (see below).

For brevity, I refer to subgenera of *Lasioglossum* (including *Dialictus*, *Evylaeus* and *Sphecodogastra*) as distinct genera throughout.

ALTERNATIVE VIEWS

Alternative hypotheses to the MLH

Maternal dominance over adult daughters may influence the differentiation of non-gyne subcastes (see, for example, Buckle 1982), but if mothers determine caste through behavioral inhibition, then females emerging into nests lacking a queen should always become gynes. However, this pattern has never been observed. Existing data simply do not support maternal inhibition as a *caste-*determining mechanism itself (Yanega 1989), but inhibition could easily work in concert with the MLH (if maternal inhibition affects receptivity to mating, as in Greenberg and Buckle 1981), and as such would not be an alternative hypothesis *per se*. In fact, if the MLH is valid, then mothers should benefit from being able to actively inhibit mating, as they would thus increase the likelihood of recruiting daughters as workers, and in such a case inhibition would become a corollary aspect of the MLH. If mating has no effect on caste, then the existence of mating inhibition becomes more difficult to explain.

A few of the same patterns as predicted by the MLH might be expected if caste were determined before eclosion, but there is no evidence for such a mechanism in halictines, and there are other arguments against it (Yanega 1989, 1992; see below). However, as caste predetermination has never been explicitly tested, it has not yet been disproven. The argument might be phrased as 'males are produced only when gynes are produced because they are intended to mate with gynes', or something similar, but this viewpoint is in clear contrast to the MLH (and ECH) as to the significance of worker matings, and whether protandry is expected to occur in seasonal social populations or not.

(a) If caste – at the very least meaning whether a female will diapause or not – is predetermined, Bulmer's (1983) model should apply (see below) and there should be strong selection for protandry. Given the short lifespans of male halictines studied so far (Barrows 1976; Yanega 1990), and the short time during which gynes may be available for mating (Yanega 1989, 1992), males produced just before gyne emergence would have the greatest opportunities for mating. However, this reasoning assumes that there is zero benefit from mating with non-gynes, and this is unlikely given that replacement queens (and sometimes laying workers) occur in virtually all eusocial halictines. If non-gyne status is predetermined, then non-gynes should be free to mate immediately upon emergence, and males *should* mate with them at the earliest possible opportunity; those non-gynes that are first to emerge into nests in which the foundress died before brood emergence are likely to become replacement queens, yet they typically do not leave the nest after assuming queen status (see, for example, Yanega 1989), so they are only available for mating for a few days at most. Thus, not only should males emerge before gynes, but some males should precede non-gynes as well. Furthermore, the male bias in sex ratio should peak early and diminish over time (males produced late in the first brood would have much less mating success), until just prior to the emergence of the second brood. Protogyny in the strict sense would also at least be a possibility (as in some polistines) if caste is predetermined; the timing of male and gyne production in polistines is uncoupled (both protandrous and protogynous species exist (Suzuki 1986)), as might be expected since mating is clearly not the determinant of caste in polistines (see, for example, Gadagkar *et al.* 1988; Mead and Gabouriaut 1993). In short, if mating has no effect on caste, male and gyne production need not be synchronous.

(b) The MLH, however, predicts that males and new gynes will *always* occur together. As male abundance increases, increasing numbers of females are expected to mate and become gynes, and therefore males cannot emerge together with or in advance of non-gynes only (some gynes would result), and gynes cannot emerge in advance of males (if mating must occur promptly after emergence (Yanega 1989, 1992)). Workers and replacement queens should not be found mated soon after emergence. The ECH further predicts that the sex ratio will be most male-biased at the end of the first brood, rather than at the beginning.

The occurrence of either clearly protandrous (with males preceding females by several days, rather than mean emergence date alone) or protogynous (in the strict sense, with *gynes* first) eusocial populations would therefore constitute evidence against the MLH. However, most past studies have not even considered the possibility of 'worker brood' gyne production, and therefore cannot be considered evidence for protandry. That is, although protandry in the 'reproductive brood' is universal, this is irrelevant if gynes occur in the 'worker brood' (*H. rubicundus* exemplifies both features (Yanega 1988)), because then there is essentially one period of gyne emergence, extended over both broods, and it is extremely difficult to detect gynes emerging within a 'worker brood'. If one insists on analyzing broods separately, then the population cycle consists of a foundress phase, a eusocial phase ('worker brood[s]'), and a final solitary phase ('reproductive brood'), and as long as the eusocial phase is not protandrous there is no refutation of the MLH. Some recent studies that have looked specifically for gynes in the 'worker broods' have indeed found direct or suggestive evidence (Packer 1986a, 1990; Packer *et al.* 1989; G. C. Eickwort, personal communication).

Alternative interpretation of data

Some of the phenomena seen in *H. rubicundus* are almost undetectable without multiyear study of marked individuals; even the most sophisticated indirect analyses may be inadequate and yield a false negative result. For example, in Packer's (1992) detailed study of *Dialictus laevissimus* in Alberta, he suggested three criteria to detect first-brood gynes: (1) a reduction in numbers of workers found in nests, relative to numbers of pupae found earlier; (2) a difference in size distributions between pupae and those females later found as workers; and (3) a skew in the size distribution of foundresses owing to the inclusion of small first-brood females. He found no evidence in support of these predictions. However, these criteria all fail (at least in some years) for *H. rubicundus* as well, even though some 40–70% of the first-brood females are gynes. First, pupal counts are drawn from the pool of nests with foundresses, but worker counts are drawn only from those nests which are active later, when anywhere from 20–50% of the nests may have produced all-gyne broods and been abandoned as a result. For example, in 1984, when 43.2% of the first-brood females were gynes, *H. rubicundus* pupal counts would have estimated 5.2 workers per nest, and actual numbers were 4.1 workers per active nest (data from Yanega 1993). If there had been a limited sampling of nests within the population (and allowing for sampling error), then such a difference might not be significant, and would at most suggest only a 20% gyne frequency. Second, the size distributions of first-brood gynes

and workers overlap almost completely, and mean size of gynes is not always greater than workers (workers were larger on average in 1986, for example (Yanega 1992)). The removal of first-brood gynes from the overall distribution will thus rarely have a detectable effect. Third, there is no reason to expect normality in size distributions of second-brood females (given that offspring size is based on resource availability, number of workers per nest, social structure, temperature, and other non-normally distributed factors; see, for example, Kamm 1974), and a departure from normality in queen sizes may be meaningless, or reflect a bias in overwintering success. Moreover, large numbers of first-brood gynes may be included *without* creating a skewed or bimodal distribution (see, for example, Yanega 1988).

Of the three criteria, the first is probably the most likely to be detectable, though unreliable (in addition to reasons above, high levels of worker mortality in a given population prior to a worker census could obscure the phenomenon of gyne disappearance). However, given knowledge of the size distribution of the second brood in one season, it might be possible to detect the presence of excess small females in the foundress pool the following season (Yanega 1988, 1989). Ultimately, none of the features of the biology of *H. rubicundus* for which truly comparable data are available are idiosyncratic in any way. The problem at present is that this is the only halictine population that has been studied in a manner sufficient to detect and document many of these novel phenomena; in every other aspect, the behavior is unremarkable, and the possibility yet exists that all of these formerly unknown phenomena are widespread in temperate halictines, including species that have already been relatively well studied.

SURVEY OF THE LITERATURE

There is a great deal of literature pertaining to the natural history of halictines, much of it anecdotal, but sufficient detailed studies now exist to attempt an evaluation of large-scale patterns of demography and behavior. The accuracy of much of the data from the literature discussed in this chapter is limited in various degrees (see also Wcislo, this volume), and it is important to bear several points in mind.

First, a critical limitation of published data is the frequency and thoroughness of observations; with one exception, no studies have involved daily observations throughout the season, or for more than a few hours each day, and the precision of parameters such as dates of first male emergence or first provisioning activity often depend upon the visibility of the phenomenon and the intensity of the observer's scrutiny. In some cases, phenologies are given based on individuals sampled on flowers, rather than nest-site observations, or vice versa; the margin for error is large, as in a study of *Dialictus rohweri* (Breed 1975), where males were seen in the field a month before male pupae were found in nest samples. Stöckhert (1933) reports when overwintering females emerge without specifying the date of first provisioning; for many of his records, it seems likely that there is a long gap between emergence of over-wintering females and the start of provisioning (for example, compare *Lasioglossum costulatum* and *L. nitidum* in Table 14-1; there is a three-month difference in apparent initiation dates, yet brood emergence is synchronous), or perhaps the populations may not be solitary as he reported (for example, *H. tumulorum* may be eusocial (Sakagami and Ebmer 1979)). Some authors present data that are self-contradictory, such as when males first appear (see, for example, Litte 1977; Knerer 1983; in the former case, for example, some 6% of the workers were inseminated in July, although males reportedly did not begin emerging until the very end of the month). Both the social structure and the sex ratio reported for *Evylaeus villosulus* (Plateaux-Quénu *et al.* 1989) are based largely on laboratory rather than field observations, and may be subject to laboratory artifacts. Also, as mentioned above, virtually all published studies have assumed that workers and gynes emerge in separate broods, so published statements as to first dates of gyne emergence are essentially meaningless.

Second, there must be variation from year to year in many of the demographic parameters given, and one can only assume for present purposes that the various studies that present data from one or a few years are representative of each population's seasonal cycle.

Third, with few exceptions, data on sex ratio have been based on counts of pupae and teneral adults, and involve grouping of data from large (typically 10–15 days) time intervals. This results in considerable overlap of data from one observation period to the next, owing to the large variation in age (and therefore the emergence dates) within each sample. Thus, the resolution of precise demographic data, such as the peak in male bias, is difficult even where such analysis has been attempted (see below). This lack of resolution also makes analysis of protandry or protogyny more difficult, because there may not be a simple relationship between which sex was seen first and the relative abundances of the two sexes at given points in time (which may actually be more important).

Finally, it is unfortunate that much past work has assigned the most 'advanced' social structure expressed in a single population (usually eusociality) to species as a whole, rather than quantifying the frequencies of occurrence of various social structures (Danforth and Eickwort, this volume; Wcislo, this volume). For the purpose of this review, I must primarily follow the original authors, unless there is evidence for variation that might be relevant to the MLH. There are 'eusocial' populations in which less than half the nests may contain eusocial colonies in some years (see, for example, *H. rubicundus* (Yanega 1993)), or that are largely solitary with only occasional eusocial nests (see, for example, *Augochlorella striata* (Packer *et al.* 1989; Packer 1990)). At the other extreme are species such as *H. hesperus*: a 'eusocial species' with completely discrete female morphs (group E of Wille and Orozco 1970), but for which there is also evidence of occasional semisocial colonies (Wille and Michener 1971; Packer 1985). There are many such halictines that appear to contain *both* eusocial and parasocial colonies (or solitary and parasocial colonies; see Danforth and Eickwort, this volume) in various proportions within their populations. This variation should be acknowledged and quantified whenever possible, as so few species appear monotypic between or even within populations upon careful analysis.

Although there are varying degrees of uncertainty associated with virtually all studies of halictines to date, I will comment in greater detail on uncertainty in those studies that might disprove the MLH or ECH. Given the difficulty of detecting some of the crucial phenomena involved, it is essential to evaluate possible exceptions critically.

TESTS OF THE MLH

Occurrence of protandry in halictines

A general demographic prediction has been made that in social Hymenoptera there should often be protandry in those groups, hereafter referred to as 'seasonal', which produce offspring in distinct annual or semiannual broods (Bulmer 1983). Protandry is known to occur in many other types of organism, but it should be especially facilitated in Hymenoptera (all else being equal) owing to their precise control over sex determination. Thus far, numerous examples of protandrous social Hymenoptera exist to support this hypothesis, although contrary examples have also been reported (see, for example, Suzuki 1986).

Table 14-1. *Demography and phenology of annual-cycle halictines*

Abbreviations: S.L., social level (characteristic form reported for the first seasonal brood in population; some species are polygynous (plurimatrous) in foundress phase, but this is not indicated): S = solitary, C = communal, S = semisocial, E = eusocial. LOC, location of population(s) studied. F.P., first provisioning activity: E = early in month (days 1–10), M = middle of month (days 11–20), L = late in month. Months are indicated by numerals (e.g. E4 = early April). 1ST F/M, dates of emergence of first female and male offspring, respectively. E.O., emergence order: F = females clearly first, M = males clearly first, S = clearly simultaneous, SF = simultaneous, but females predominant initially, SM = simultaneous, but males predominant initially (if in parentheses, the predominance is either my own interpretation of the data, or there is some uncertainty in the author's estimation), S? = apparently simultaneous, uncertain due to imprecision of the original data. S.R., sex ratio (given as % males) in the first brood of offspring; M-B = Male-biassed, F-B = Female-biassed.

Species	S.L.	LOC	F.P.	1ST F/M	E.O.	S.R.	Reference
Agapostemon texanus	S	Kansas	E4	E6/E6	S	F-B	Roberts 1973
A. virescens	C	New York	E6	E8/L7	M	67%	Abrams and Eickwort 1980
Augochlora nominata	E	Costa Rica	M6	E8/E8	S	33%	Eickwort and Eickwort 1972
A. pura	S	Kansas	M5	E6/E6	SM	25%[a]	Stockhammer 1966
Augochlorella striata	E	Kansas	E5	E6/E6	S	19%[a]	Ordway 1966
A. striata	E	Nova Scotia	M6	E8/E8	S(F)	ca. 80%[a]	Packer 1990
Augochloropsis diversipennis	S?[b]	S. Brazil	M11	E2/E2	S?	ca. 44%	Michener and Lange 1959
A. sparsilis	S?[b]	S. Brazil	E10	E1/E1	S(M)	F-B	Michener and Lange 1959
Dialictus coeruleus	E	Kansas	L4	E6/E6	S(F)	—	Stockhammer 1967
D. figueresi	S	Costa Rica	M12	M4/M4	S	ca. 50%	Wcislo et al. 1993
D. imitatus	E	Kansas	E4	E6/E6	S(F)	>10%[a]	Michener and Wille 1961
D. nr. laevissimus	E	Maryland	M4	L5/M6	F	0%	Batra 1987
D. lineatulus	E	New York	E5	M6/M6	S?	ca. 25%	Eickwort 1986
D. rhytidophorus	E	S. Brazil	E9L	10/L10	S	41%[a]	Michener and Lange 1958d
D. rohweri	E	Kansas	E4	L5/L5	SF	<40%[a]	Breed 1975
D. umbripennis	E	Costa Rica	L12	L1/L1	S?	8%	Wille and Orozco 1970
D. versatus	E	Kansas	M4	M6/M6	S(F)	5%[a]	Michener 1966
D. zephyrus	E	Kansas	M4L	5/L5	SF	15%[a]	Batra 1966
Evylaeus affinis	E	Honshu	E6	L7/E8	SF	F-B	Sakagami et al. 1982b
E. calceatus	E	Germany	L3?	L6/E7	F	–	Stöckhert 1933
E. calceatus	E	Hokkaido	E5	M7/L7	F	F-B	Sakagami and Munakata 1972
E. calceatus	S	Hokkaido[c]	M6E	8/E8	SM	52%	Sakagami and Munakata 1972
E. calceatus	E	Netherlands	M4E	7/M7	F	F-B	Vleugel 1973
E. comagenensis	S	New York	L5	E7/E7	S?	ca. 50%	Batra 1990
E. cooleyi	E	Vancouver	5	6/6	S?	F-B?	Packer and Owen 1989
E. duplex	E	Hokkaido	L4	E7/E7	SF	4%[a]	Sakagami and Hayashida 1958, 1961, 1968
E. fulvicornis	S	Germany	L3?	7/L6	M	—	Stöckhert 1933
E. intermedius	S?	Germany	4	M7/E7	M	—	Stöckhert 1933
E. laticeps	E	England	L4	E6/E6	S?	24%	Packer 1983
E. linearis	E	France	E4	L5/E7[d]	F	1%	Knerer 1983
E. malachurus	E	Germany	M4	M6/M7	F	0%	Stöckhert 1923
E. minutus	S?	Germany	L3?	M7/E7	M	—	Stöckhert 1933
E. nigripes	E	France	E5	M7/M7	S?	4%	Knerer and Plateaux-Quénu 1970
E. nitidiusculus	S	Germany	E4?	M7/E7	M	—	Stöckhert 1933
E. nupricolus	S	Hokkaido	M6	M8/M8	S(F)	M-B	Sakagami 1988
E. ohei	C	Hokkaido	M6	M8/M8	SM	76%[a]	Sakagami et al. 1966

Table 14-1 (*cont.*)

Species	S.L.	LOC	F.P.	1ST F/M	E.O.	S.R.	Reference
E. pauxillus	E	Germany	L4	M6/M6	Sf	—	Stöckhert 1923
E. pauxillus	E	Germany	L3?	M6/L6	F	—	Stöckhert 1933
E. punctatissimus	S?	Germany	L3?	L7/M7	M	—	Stöckhert 1933
E. sakagamii	S	Honshu	E6	L7/L7	SM	—	Sakagami *et al.* 1982a
E. smeathmanellus	S?	Germany	4	7/M6	M	—	Stöckhert 1933
E. villosulus	S?	Germany	E5	E7/E7	S(M)	—	Stöckhert 1923
E. villosulus	C?[b]	France	5	M6/M6?	S?	21%	Plateaux-Quénu *et al.* 1989
Halictus confusus	S?	Nova Scotia	E6	E7/E7	S(M)	—	Atwood 1933
H. confusus	E	Indiana	M4L	5/L5	S?	F-B	Dolphin 1966
H. farinosus	E	Utah	L5	L6/L7	F	F-B	Nye 1980
H. farinosus	E	California	E4	E6/E6	S(F)	F-B	Eickwort 1985
H. ligatus	E	Indiana	E5	L6/L7	F	0%	Chandler 1955
H. ligatus	E	New York	L5	E7/L7[d]	F	F-B	Litte 1977
H. ligatus	E	Ontario	M5L	6/L6	S	—	Michener and Bennett 1977
H. ligatus	E	Ontario[e]	E6	E7/E7	S	15%	Packer 1986b
H. quadricinctus	S	Udmurt SSR	E5	L7/M7	M	M-B?	Sitdikov 1988
H. rubicundus	E	Nova Scotia	M5M	7/L7	F	F-B	Atwood 1933
H. rubicundus	E	Germany	E4?	L7/L7	S	—	Westrich 1989
H. rubicundus	E	New York	L4[f]	E6/E6	SF	25%[a]	Yanega 1988, 1989
H. rubicundus	E	Kansas	E4	M5/M6	F	0%	D. Yanega, unpublished
H. sexcinctus	S	Germany	M4?	L7/M7	M	—	Stöckhert 1933
H. sexcinctus	S	France	M5M	7/M7	S(M)	—	Knerer 1980
H. tsingtouensis	S	Hokkaido	E6	E8/L7	M	F-B?	Sakagami 1980
H. tumulorum	S?	Germany	E4?	L7/M7	M	—	Stöckhert 1933
Lasioglossum arcuatum	E	Nova Scotia	M5	L6/M7	F	F-B	Atwood 1933
L. costulatum	S	Germany	E6	L7/M7	M	—	Stöckhert 1933
L. esoense	S	Hokkaido	M6	L8/L8	S	—	Sakagami *et al.* 1966
L. laevigatum	S	Germany	L3?	L7/M7	M	—	Stöckhert 1933
L. leucozonium	S	Nova Scotia	E6	L7/L7	S	—	Atwood 1933
L. leucozonium	S	Germany	E5	E7/E7	SM	—	Stöckhert 1933
L. nitidum	S?	Germany	E3?	L7/M7	M	—	Stöckhert 1933
L. occidens	S	Hokkaido	E6	M8/E8	M	—	Sakagami and Hayashida 1968
L. pallidulum	S	Hokkaido	E5	E8/E8	S	M-B?	Sakagami and Munakata 1966
L. zonulum	S	Germany	E5	L7/M7	M	—	Stöckhert 1933
Pseudagapostemon divariticus	C?	S. Brazil	E10	E2/E2	SM	62%	Michener and Lange 1958d
Sphecodogastra galpinsiae	S	Utah	E6	M7/M7	S	*ca.* 50%	Bohart and Youssef 1976
S. oenotherae	S	Toronto	M6L	7/L7	S	M-B	Knerer and MacKay 1969

[a] This is only the initial sex ratio reported; the author listed those of later periods, as well.

[b] Uncertain: my interpretation and the author's differ (see text).

[c] This site is different from the preceding site, and at a much greater altitude.

[d] Equivocal: the author's own data suggests males may emerge earlier (see text).

[e] This site is different from the preceding site, and is farther north.

[f] This and subsequent parameters are averaged results from several years.

The literature on seasonal eusocial halictine bees would superficially appear to offer evidence of protandry, but as discussed above, gynes may emerge far sooner than has been assumed, so as to raise doubt as to whether true protandry is common in social halictines, or even if it occurs at all. The summary in Table 14-1 includes only studies of seasonal halictines in which sufficiently detailed comparative data on male and female production are presented (in particular, dates of offspring emergence and first provisioning). Species in which males and females are both present at all times are excluded from Table 14-1, as are studies that present only gross phenological data.

For the two seasonal social halictines in which the timing of gyne production is definitely known (*Evylaeus marginatus*, Plateaux-Quénu 1962; *Halictus rubicundus*, Yanega 1989, 1992), males and gynes are produced synchronously. All of the known protandrous populations are reported as solitary or communal (with one possible exception, *Augochloropsis sparsilis*; Table 14-1), and all of the 'protogynous' populations are eusocial. There is thus no solid evidence for protandry among social halictines, contrary to Bulmer's (1983) hypothesis. It is unlikely that cause and effect are reversed here; that is, whereas I propose that male abundance determines gynehood, it might conversely be suggested that gynehood is predetermined and males are produced along with gynes. Referring back to the predetermination hypothesis, however, Bulmer's model still applies, as it implicitly assumes that males do not influence caste determination, so sexual selection should still lead to protandry in some populations. Again, despite the nearly universal occurrence of replacement queens and mated workers in eusocial halictine populations, which suggests an additional benefit to protandry, no such populations are protandrous. Under a predetermination model, social behavior and demography are uncoupled, leaving no explanation as to why solitary populations are always protandrous whereas social populations are not.

An important related question is why foundresses do not produce male offspring earlier than they do; protandry (as a form of 'cheating') would seem to have a potentially large fitness pay-off for females practicing it, and thus be favored by selection. However, such behavior could be favored only if the average early male has a number of matings exceeding the number of siblings an early worker female will help to produce. Nothing is presently known about reproductive values of male halictines, and if early workers contribute to the production of many additional offspring, and/or if a male can only mate a few times (there

is some evidence for this; D. Yanega, personal observations), then 'cheating' might well *not* be a superior strategy. A female must choose, based on the time of egg-laying, whether to produce a son or a daughter, and if the average daughter at time X yields a higher pay-off as a worker than the average son at that time, a female should produce a daughter. As the value of the average worker diminishes towards the end of the first brood (owing to cessation of foraging prior to second brood emergence; see, for example, Packer 1986b), the relative value of the average son will increase, and male production should increase, which will in turn decrease the likelihood of a daughter becoming a worker. As these are frequency-dependent strategies, there should be a gradual shift in sex bias, rather than a sudden change (if all foundresses switched to all-male production, there would be no virgin females with which to mate!). Females that produce daughters first therefore have the highest average overall fitness, and this behavior should be evolutionarily stable; detailed modeling of this process should yield further insights.

As foundresses are not all synchronized in their provisioning, and cannot have knowledge of how far ahead or behind the remainder of the population they are, the most reliable temporal cue for modulating male production would be photoperiod. This is indeed what appears to happen in *H. rubicundus* (Yanega 1989), where nest initiation is staggered and male production begins as early as possible while still allowing for a number of workers. There are a few days of daughters only, then males gradually begin to emerge, and by the last week of the first brood, all emerging daughters mate and become gynes rather than workers or replacement queens. Photoperiodic triggering of male production as suggested in the ECH may then be viewed as the proximate mechanism that yields the highest fitness pay-off for the average foundress.

Low-latitude populations

Packer and Knerer (1987) report that around latitude 27°N, populations of *Halictus ligatus* switch from an annual cycle to one with essentially year-round activity and at most a quiescent phase experienced by individuals of all ages and both sexes. Males and reproductive females are produced throughout the season in the southern populations, and it is clear that photoperiod is not acting to limit male production there. The longest daylength experienced at the latitude in question (25–26°N) is roughly 13.75 h, which is the equivalent of mid–late April at the latitudes of most of the temperate halictines studied (40–45°), and this is roughly equal to or less than the shortest daylength to which active adults of many temperate populations are ever exposed. If photoperiod alone determined male production, then males might almost never be produced at lower latitudes. There must be some adaptation of sex-ratio control to local conditions, and the ECH will clearly not apply at all latitudes. There are no populations of halictines known from latitudes above 27° that exhibit year-round activity. Given these observations, I will consider the biology of halictine populations of the lower latitudes (tropics and subtropics) separately from temperate ones, to investigate the predictions of the MLH independent of the ECH. Some relatively clear patterns in behavior emerge.

First, there is an apparent association of seasonality with being 'highly eusocial'. Those populations that exhibit distinct caste differences and a predominance of eusocial colonies (groups D and E of Wille and Orozco 1970) all exhibit diapause as mated females: *Augochlora nominata* (see Table 14-1), *Dialictus exiguus* (Packer 1985), *D. umbripennis* (in Damitas; see Table 14-1), *Halictus aerarius* (Sakagami and Fukushima 1961), *H. hesperus* (Wille and Michener 1971; Brooks and Roubik 1983; Packer 1985), *H. lutescens* (Wille and Michener 1971) and *H. latisignatus* (Sakagami and Wain 1966). Populations of some other species are not as clearly eusocial, but again exhibit diapause as mated females: *Augochlora semiramis* (Michener and Lange 1958a), *Augochlorella michaelis* (Michener and Lange 1958a; Sakagami and Moure 1967), *D. rhytidophorus* (Sakagami and Moure 1967; see also Table 14-1), *D. seabrai* (Michener and Seabra 1959; Sakagami and Moure 1967) and *Megommation insigne* (Michener and Lange 1958a).

If one relaxes the restrictions on caste distinctiveness, however, exceptions appear. For example, *D. aeneiventris* (Wcislo *et al.* 1993) and *Pereirapis semiauratus* (Campos 1980) both have a quiescent phase rather than true diapause, exhibit a fair degree of caste overlap, and have all mated females reproductive (males of *P. semiauratus* are apparently rare, and produced at the same time that new females begin new nests). The Turrialba population of *D. umbripennis* (Eickwort and Eickwort 1971) contained numerous queens (roughly 18% of all females, often more than one to a nest), essentially the only mated females in the population, and considerable overlap in caste sizes. There are insufficient data to tell whether there is a quiescent phase in this population. Two other similar cases are the Colombian populations of *D. seabrai* and *D. breedi* (Michener *et al.* 1979), in which the castes are not at all distinct; at least half

the foragers are reproductive, many are mated, and virtually all mated females reproduce, regardless of 'caste' (additionally, there is no size difference between castes in *D. breedi*, and males of this species are uncommon). These populations, together with some cases discussed below, may represent a social 'syndrome' different from the traditional categorization of eusociality.

Second, the few seasonal populations in which males are known to diapause along with the females (only known from low latitudes) do not exhibit eusociality: *Caenohalictus curticeps* (Michener and Lange 1958a), *Dialictus opacus* (Michener and Lange 1958b), *Habralictus canaliculatus* and *Paroxystoglossa andromache* (Michener and Lange 1958a), *P. jocasta* (Michener and Lange 1958c), and *Ruizantheda mutabilis* (Claude-Joseph 1926). Despite a potential overlap of generations (only *D. opacus* is univoltine), communal behavior is the most complex form of sociality recorded for any of these populations.

Third, the remaining seasonal populations for which data exist are also not known to be eusocial: *Augochloropsis diversipennis* and *A. sparsilis* are reported to be semisocial, although matrifilial colonies may exist (Michener and Lange 1959; Danforth and Eickwort, this volume). There are often several reproductives in each nest, many of which apparently also act as workers (in *A. diversipennis*, at least, this is true to such an extent that the authors consider it 'ordinarily' communal, although it might also be semi- or quasisocial), and the few unmated females found are distinctly more 'worker-like' (increased foraging, little or no reproduction) than mated females (Michener and Lange 1959). *Neocorynura polybioides* is normally solitary (rarely parasocial (Michener and Lange 1958a)). *Pseudagapostemon divariticus* and *P. perzonatus* are both univoltine and solitary or communal (Michener and Lange 1958d).

Fourth, most of the reportedly aseasonal populations are solitary or communal: *Agapostemon nasutus* (Eickwort and Eickwort 1969), *Caenohalictus eberhardorum* and *Habralictus bimaculatus* (Michener *et al.* 1979), *Lasioglossum albescens* (Sakagami 1968), and *Neocorynura fumipennis* and *N. pubescens* (Michener *et al.* 1966; Michener 1977). Possible exceptions are *Caenaugochlora costaricensis*, and *Pseudaugochlora graminea* and *P. sordicutis* in Costa Rica (Michener and Kerfoot 1967), which are reportedly semisocial; there is some evidence for year-round activity in *P. graminea*, but multifemale nests are quite uncommon (as are males) and all mated females are reproductive. In the other two species multifemale nests and males are both common, but the evidence for aseasonality is weak.

Fifth, populations that appear to have a quiescent period (rather than diapause) during the year vary in behavior. *Augochloropsis brachycephala* and *Paroxystoglossa seabrai* typically have one female per nest, and the social relationships in the rare multifemale nests are unknown (Michener and Seabra 1959). *Dialictus figueresi* is typically solitary (rarely parasocial), with sporadic bivoltinism; it lacks overlap of generations, all mated females are reproductive, and males survive the quiescent period (Wcislo *et al.* 1993). *D. aeneiventris* and *Pereirapis semiauratus* are eusocial, as discussed above; *Halictus ligatus* in several localities (Michener and Bennett 1977; Packer and Knerer 1986) and *Pseudaugochlora graminea* in southern Brazil (Michener and Lange 1958a) both have large proportions of mated and reproductive females of various ages, many of them foraging (in addition to more typical 'workers'), all in the same nests. The social structures of these populations thus appear to be an unusual form of either semisociality or eusociality, depending on whether the behavioral classes seen are permanent or facultative. The 'semisocial' *Augochloropsis* species discussed above are rather similar (and are probably quiescent rather than in diapause, as females were found flying in midwinter (Michener and Lange 1959)). The data for *H. ligatus* are especially noteworthy: in Trinidad, the phenology is suggestive of a quiescent period, all mated females (58% of the population) have developed ovaries, less than 20% are unmated non-reproductive foragers, 'caste' sizes vary widely but are not bimodal, and the authors suggest that 'females mate when young or not at all' (Michener and Bennett 1977). In the Florida Keys, 57% of all *H. ligatus* workers had mated (mated ones were larger on average, interpreted as resulting from male mating preference), 68% of all females were reproductive, some 30% of the population was classified as gynes (based on pupal counts), an apparently kleptoparasitic class of large mated females existed, and colonies were found to increase male production as the colony grew larger (Packer 1986c,d; Packer and Knerer 1986). These populations demonstrate distinct similarities to some of the aseasonal populations in previous categories (such as *D. umbripennis* in Turrialba, *D. seabrai* in Colombia, or *P. sordicutis*) in the poor differentiation of 'castes' and the reproductive status of essentially all the mated females; the variation in social structure between these populations may correspond to periodicity (or lack thereof) of male abundance, suggested by the presence of unmated females in all of these populations.

Summarizing the diversity of social and demographic patterns above, it is noteworthy that: (1) no species active year-round, or with a quiescent phase only, is also highly eusocial (groups D or E of Wille and Orozco 1970); (2) no species in which males diapause is eusocial; (3) no study reported males to be numerous at the time of worker emergence; (4) mated, non-reproductive females are rare in aseasonal populations (at higher latitudes, such females occur regularly). These patterns match the general predictions of the MLH: (1) presence of males when females are emerging (as might occur in an aseasonal population) is predicted to make the formation of a worker caste difficult (if mating leads to gynehood); (2) in the absence of diapause, mated females are expected to become maximally reproductive. Species such as *Pereirapis semiauratus* are especially instructive in this regard, as castes are fairly well-defined, and the phenology is virtually aseasonal (though there is still some demographic periodicity), but males are produced in low numbers and most females remain unmated. There may yet be aseasonal, clearly eusocial populations to be discovered, but the MLH predicts they would have a similar periodic or chronic limitation on male abundance.

TESTS OF THE ECH

Developmental periods

Given that the ECH relates to environmental influences at the time of egg-laying, it is unfortunate that the subsequent pattern of adult emergence is often the only evidence available to evaluate the sex ratio produced at a given time. It is therefore important to know whether there are possible confounding factors, related to development, which would invalidate the use of adult emergence patterns to test the ECH. There are some direct data on developmental periods available for various species (Table 14-2). Data for *D. zephyrus* (not in table) are the most detailed: Batra (1966) reports 21–24 days at 28–33 °C, and 38 days at 18.5 °C (in the field); Kamm (1974) reports that laboratory-reared females mature in 23.8, 22.4, and 17.9 days at 21, 26, and 30 °C, respectively, and males mature in 23.4, 20.9, and 18.3 days under the same conditions. There appears to be a general association of larger body size with longer developmental times both within (males are smaller and mature marginally faster than females) and between species (the *Halictus* and *Agapostemon* are the largest; none of the listed studies included mass). Temperature changes can dramatically alter developmental times, although the range is roughly 20–35 days. However, there is clearly enough variation to limit the potential applicability of the

Table 14-2. *Developmental times of various halictines*

Egg–adult, average total egg-to-adult intervals, in days (ranges given in parentheses, temperature given in square brackets);
Pupa, days as pupa out of total; N.A., data not available.

Species	Egg–adult	Pupa	Reference
Agapostemon nasutus	*ca.* 37–38	21–22	Eickwort and Eickwort 1969
A. texanus[a]	*ca.* 32 [25–30 °C]	N.A.	Roberts 1969
Augochlora pura[a]	17–21 [24–32 °C]	N.A.	Stockhammer 1966
Augochlorella spp.	*ca.* 27 (20–31)	7–16	Ordway 1966
Dialictus aeneiventris	35	15	Wcislo *et al.* 1993
D. figueresi	*ca.* 80[b]	N.A.	Wcislo *et al.* 1993
D. imitatus	*ca.* 30/*ca.* 21[c]	6–19[d]	Michener and Wille 1961
D. nr. *laevissimus*	*ca.* 35 [24–29 °C]	11	Batra 1987
D. rohweri	*ca.* 21–30	*ca.* 14–15[d]	Breed 1975
D. umbripennis	22	11	Wille and Orozco 1970
Evylaeus duplex	*ca.* 38/*ca.* 31[c]	13–14	Sakagami *et al.* 1982a
E. sakagamii	*ca.* 35–40	13–14	Sakagami *et al.* 1982b
Halictus farinosus	*ca.* 28 (21–35)	7–10	Nye 1969
H. rubicundus	*ca.* 30	11–13	D. Yanega, unpublished
Sphecodogastra galpinsiae	(19–29)	7–10	Bohart and Youssef 1976

[a] Laboratory rearings.

[b] Mature adults may wait months before leaving their natal cells.

[c] The first figure is time required in spring, the second is for summer.

[d] The author(s) specify that the developmental times are strongly temperature-dependent.

ECH across species, and it is questionable as to whether the relative timing of male vs. female emergence is a reflection of egg-laying sequence alone. For those species that have not been studied in detail, it is not safe to assume that the time interval between first provisioning and first emergence is a good measure of developmental period, because mature individuals might not immediately leave their cells or reopen the nests, as in *D. figueresi* (Wcislo *et al.* 1993) and *E. marginatus* (Plateaux-Quénu 1962).

Laboratory studies, temperature and photoperiod

Several species have been reared in the laboratory, although year-round rearing has been achieved only for *Dialictus zephyrus* (Greenberg 1982); these studies offer some data relevant to temperature and photoperiod effects on sex ratio, to help evaluate the ECH, which predicts increasing male bias with increasing photoperiod and temperatures. Some reports are essentially anecdotal, but noteworthy; in *Evylaeus nigripes*, 'suitable' variation in temperature and photoperiod resulted in queen-sized workers, worker-sized queens, and mixed broods of males and females (Knerer and Plateaux-Quénu 1970), and in *E. cinctipes*, *E. linearis* and *E. malachurus*, 'summer photoperiods' resulted in 'queens' (presumably males as well (Knerer 1987)). Under summer-like temperatures (24–28 °C) and unnaturally short photoperiod (only 8 h of light!), the following patterns were observed (Knerer and Plateaux-Quénu 1967): (a) *E. malachurus* produced small females; (b) *E. cinctipes* produced 80% males, after which the colonies 'died out'; (c) *D. laevissimus* produced a brood of workers, then males and females; (d) *Lasioglossum forbesii* (a solitary species) produced a female-biased brood (9 females and 2 males), instead of a 50:50 ratio or mostly males; (e) *Halictus ligatus* produced mostly males in monogynous nests, and mostly females in polygynous nests; (f) in *H. scabiosae*, the results were almost identical to those for *H. ligatus*, with the additional observation of a female that laid male eggs when alone, but female eggs when another female visited the nest. In *Agapostemon* at >27° C and 16 hrs light (unnaturally long), 18 of 21 offspring were male (Roberts 1969). Work with *D. zephyrus* was well-documented: at 22-28° C and 13.5 h light (equivalent to late April or late August), 28.7% of the offspring were male

(a)

(b)

(c)

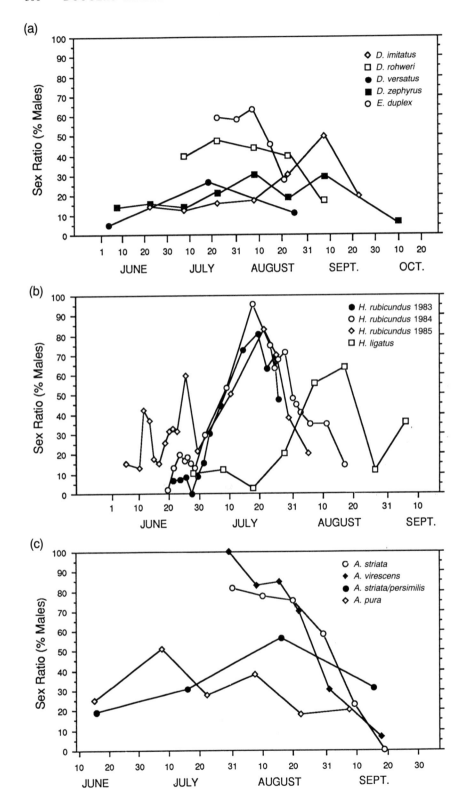

(Greenberg 1982); at 15 h light (early June conditions) there were increasing percentages of males with increasing temperatures (56% at 21° C, 78% at 26° C and 100% at 30° C (Kamm 1974)). Cell sizes tended to be smaller at longer photoperiods at 26° C (11 vs. 13 and 15 h), although the females used for this test were all unmated and thus unable to produce female offspring (smaller cells are presumed more likely to have male eggs laid in them (Kamm 1974)). Taken together, these observations clearly indicate strong effects of both photoperiod and temperature on sex ratio, in a manner consistent with the ECH, although when warm temperatures and short photoperiods were combined (the ECH predicts these influences would be in opposition), the results were mixed (Knerer and Plateaux-Quénu 1967), suggesting that different species may rely more heavily on one factor than the other.

Peaks of male production

Some studies present data on changes in sex ratio over the course of the season, and these are of particular interest in evaluating the ECH. Eleven temperate-zone studies are summarized in Fig. 14-1; note that the data for *Agapostemon virescens* and *Halictus rubicundus* are based on daily emergence censuses and thus are free of problems with pupal counts, but might conceivably reflect any post-maturity delays in emergence. The peaks in male production for all but two species appear to fall between mid-July and mid-August (the two exceptions being *Augochlora pura*, in early July, and *Dialictus imitatus*, in early September), although the patterns for some (*A. pura*, *D. versatus* and *D. zephyrus*) are difficult to resolve precisely. Only the data for *H. rubicundus* cover more than one year, and the results are consistent

Fig. 14-1. (a) Sex ratios as reported for various temperate populations of *Lasioglossum*, subgenera *Dialictus* and *Evylaeus*. References: *D. imitatus* in Kansas, Michener and Wille 1961; *D. rohweri* in Kansas, Breed 1975; *D. versatus* in Kansas, Michener 1966; *D. zephyrus* in Kansas, Batra 1966; *E. duplex* in Hokkaido, Sakagami and Hayashida 1968. (b) Sex ratios as reported for various temperate populations of *Halictus*. References: *H. rubicundus* in New York, Yanega 1993; *H. ligatus* in Ontario, Packer 1986b. (c) Sex ratios as reported for various temperate populations of Halictinae. References: *Augochlorella striata* in Nova Scotia, Packer 1990; *A. striata/persimilis* in Kansas, Ordway 1966; *Augochlora pura* in Kansas, Stockhammer 1966; *Agapostemon virescens* in New York, Abrams and Eickwort 1980.

among years, despite variation in dates of first provisioning (Yanega 1993). Packer (1992) graphs pupal sex ratio of a population of *D. laevissimus* in Alberta, but does not give percentage data; however, the sex ratio reaches 100% male roughly three weeks after the solstice, as expected under the ECH. The one subtropical species studied in detail, *D. rhytidophorus* (Michener and Lange 1958b; not figured), which experiences the least photoperiod variation (and therefore might not be a valid test of the ECH), shows a distinct drop in male production in the summer months, almost the exact opposite of the pattern in the temperate species. The profiles for *D. laevissimus*, *D. rohweri*, *D. versatus*, *E. duplex*, *H. rubicundus*, *A. striata* and *A. virescens* approximate a photoperiodic curve (with a peak in male eggs apparently in mid to late June), whereas *D. imitatus*, *D. zephyrus*, *H. ligatus*, and *A. striata/persimilis* better approximate thermal profiles (male egg peaks from July to August). These latter cases, however, may well reflect some degree of worker production of males; until more detailed data are available, these observations do not serve to refute the hypothesis. The early peak for *A. pura*, however, is the one clear temperate-zone exception, suggesting a potentially different mode of control over sex ratio, as also appears to be the case for the subtropical *D. rhytidophorus*.

Phenological studies

Some studies unfortunately lack first dates of provisioning and/or sex ratios, but contain some data suggesting that male eggs are typically not produced at short photoperiods: there were 17 halictine species in southern Ontario for which MacKay and Knerer (1979) captured more than five males, and for 16 of these the earliest record was between June 26 and July 22. Similarly, six of seven annual-cycle halictines studied in France by Poursin and Plateaux-Quénu (1982) produced their first males between mid-June and mid-July (the exception was *Evylaeus calceatus*, with males in mid-May, the earliest date reported for any temperate species in the literature); the variation in dates of first female emergence was greater than for males, from early May to late July. Packer *et al.* (1989) provide an example of a solitary population (*E. comagenensis*) initiating nests earlier than a weakly eusocial population (*Augochlorella striata*) in the same habitat, which again suggests interspecific variation in threshold conditions under which males are produced (the authors themselves suggest that *A. striata*, a member of a taxon of tropical origin, might exhibit unusual behavior under cold conditions).

GEOGRAPHIC VARIATION IN DEMOGRAPHY AND SOCIALITY

There are several species in Table 14-1 for which data from two or more populations are presented, and these data indicate that it takes relatively little geographic distance between populations to yield different phenologies and, in most cases, different social structures. This distinct geographic variation in demography and/or phenology holds for other localities not given in Table 14-1, such as *Dialictus rohweri* in Ontario (Breed 1975), *D. umbripennis* in Turrialba, Costa Rica (Table 14-1 lists parameters for Damitas; see above), *Evylaeus affinis* elsewhere in Japan (Sakagami *et al.* 1982b), *E. malachurus* elsewhere in Europe (Knerer 1987), *Halictus ligatus* in the tropics and subtropics (Michener and Bennett 1977; Packer 1986c,d; Packer and Knerer 1986), *H. rubicundus* in Colorado (Eickwort *et al.* 1996) and *H. tumulorum* in Hokkaido (Sakagami and Ebmer 1979). The most notable pattern that emerges is that several species are solitary in one locality and eusocial in others, and/or have extra numbers of broods. (a) On the island of Hokkaido, *E. calceatus* is solitary at 1167 m altitude but has two broods and is eusocial at 150 m (Sakagami and Munakata 1972). (b) *E. malachurus* has two broods in England, three in central Europe, and four in southern France (Knerer 1987). (c) *H. rubicundus* is solitary at 2850 m in Colorado (Eickwort *et al.* 1996), has two broods in New York, and three broods in Kansas (Yanega 1993), with nest initiation a month earlier in each successive locality (though the season ends at roughly the same time in all three). In Kansas, no males are present in the first brood, and all females become workers. (d) *Augochlorella striata* and *H. confusus* appear to be exclusively or largely solitary in Nova Scotia (L. Packer, personal communication; Atwood 1933) and eusocial farther south, although in long, warm seasons in Nova Scotia the former species exhibits a small number of eusocial nests (Packer *et al.* 1989; Packer 1990; L. Packer, personal communication).

Thus, the earlier provisioning begins, the longer the period of worker production, yet males and gynes appear (and the active season ends) at essentially the same time (or photoperiod) in all populations of the same species, in keeping with the combined predictions of the MLH and ECH.

SPECIAL CASES

A particularly interesting temperate species is *Evylaeus marginatus*, which is the only halictine known to have an obligate multiyear nesting cycle (with one long-lived queen and one brood per year); males are produced only by colonies in their final year (the fifth or sixth (Plateaux-Quénu 1962)). Also unusual is that all males and females are produced synchronously, but do not emerge synchronously; nests that contain males are opened in the fall (the females from these nests remain inside, mate with males that enter the nest, and become gynes), but nests containing only females remain closed and these females emerge the following spring, after all the males have died, and become workers. Manually opening such closed nests, Plateaux-Quénu (1960) demonstrated that these females will mate and become gynes, given the opportunity. That some nests remain closed when males are active seems essential for the worker caste to persist, and supports the MLH. It is not clear why the females in *all* colonies do not simply open the nests, and it would not be surprising to find that the queens control this in some manner.

The temperate *Sphecodogastra galpinsiae* is a specialist of *Oenothera* and demonstrates some unique behaviors (Bohart and Youssef 1976); females are active and provisioning whenever the host plant is in flower, but at all other times females are apparently in diapause and can remain alive for more than one season. Females seem to enter and exit diapause essentially at will, 'tracking' the *Oenothera* bloom; provisioning females at any time of year can stop, allow their ovaries to regress, and wait in diapause for another flowering period (both between and within seasons), and can repeat this up to three or more times during their lifespan, depending upon conditions. Male and female offspring are produced simultaneously, and young mated females dig and provision their own nests, often while their mothers are apparently still foraging, as long as host flowering continues. Diapause appears to be a behavioral strategy to enhance synchrony with the host plant, rather than being a one-time event associated with mating and overwintering, making this population more similar, in some respects, to aseasonal halictines.

Two of the populations in Table 14-1 (*Halictus quadricinctus* and *H. sexcinctus*) are noteworthy in that they have overlapping generations but are none the less solitary and univoltine; males emerge before females, and none of the females are recruited as workers by the surviving foundresses. The Nova Scotia population of *Augochlorella striata* might exhibit such behavior in some years, but in 1987 (and 1991; L. Packer, personal communication) a small number of early-emerging daughters were recruited as workers by their mothers (mean of less than one per nest; Packer *et al.*

1989; Packer 1990). It is also possible that males of *Evylaeus villosulus* in France emerge before females, given that all first-brood females examined were mated (despite the small number of males observed alive at that point in the season; Plateaux-Quénu *et al.* 1989). This population is also unusual in that it is described as solitary/communal and partly bivoltine, although the authors also report that a small number of third-brood offspring may be produced, vitiating the claim of partial bivoltinism. A third brood would also mean that the lifetime reproductive output of first-brood females and overwintering gynes may not differ much (as both appear to have a small chance to produce two broods of offspring), such that overwintering might not be a substantially better strategy. This is the only known halictine population that approaches solitary partial bivoltinism, a life-history pattern required by Seger's (1983) model of halictine social evolution as the ancestral (and presumably stable) condition, upon which selection could then act to produce eusociality via inclusive fitness effects; however, this particular life-history pattern appears to be rare or non-existent in halictines.

CONCLUSIONS: EVALUATION OF HYPOTHESES

Mating limitation hypothesis

The specific predictions of the MLH are supported, on the whole: worker recruitment diminishes as relative male abundance increases (Sakagami and Hayashida 1968; Yanega 1989, 1993); gynes are not found in advance of males; protandrous populations are solitary or communal (even with overlap of generations) (Knerer 1980; Sitdikov 1988); there is only one potentially partly bivoltine solitary/communal population known (Plateaux-Quénu *et al.* 1989; contrary to the scheme proposed by Seger 1983); there is evidence from many different aseasonal populations that females that mate typically become reproductive, even if there are already other reproductive females (sometimes of different generations) present, while unmated females typically become workers. This latter pattern is in contrast to temperate populations, in which mated non-reproductive females are much more common, and further suggests that the restriction on *prompt* mating stated in the MLH only applies to populations that exhibit diapause. There are no known halictines that possess well-defined castes and are active year-round (or in which both males and females diapause), whereas there are many apid

and vespid species of this nature. If caste were predetermined, then there should be nothing to prevent aseasonal halictine populations from having well-defined castes, and there should likewise be protandrous, eusocial populations.

Environmental control of sex ratio hypothesis

There is ample evidence, from both field and laboratory, that both long photoperiods and warm temperatures have male-biassing effects in temperate halictines. However, there is considerable variation in the responses of individual species to similar conditions (see, for example, Packer *et al.* 1989), as well as apparent differential reliance on one cue or the other. Considering the number of possible confounding factors, the reported phenological patterns are fairly consistent as to dates of first male emergence and peaks of male production: first male emergences are typically from early June to late July (one report of males in mid-May, three in late May, and all but one of the early August reports are cases where *no* brood emerged before that point), and peaks of male emergence are mostly from early or mid-July to mid-August. The latter observation roughly matches the expectation of a peak in male egg production near the solstice, taken together with a 20–35 day estimation of developmental times; if one allows for temperature effects and worker production of males, however, the prediction is better supported. The original hypothesis, based as it was on studies of *H. rubicundus*, would appear to place undue emphasis on photoperiod. The data clearly support the contention that laboratory studies performed under artificial conditions may give correspondingly artificial results; extrapolations from such studies to natural behavior should be viewed with skepticism, although such data can still be used to verify or reject mechanistic hypotheses such as the ECH.

Overall patterns

Most temperate eusocial populations begin provisioning between early April and mid-May, with four in late May or early June; the one population that began in mid-June had the least worker recruitment (*Augochlorella striata*) (Packer 1990). Most solitary or communal populations begin provisioning in early or mid-June, although there are some 'early' exceptions; three are reported as having female-biased first broods (*Augochlora pura*, *Agapostemon texanus* and *Evylaeus villosulus*), two have overlap of generations (*Halictus quadricinctus* and *H. sexcinctus*), and there are a number of populations of reportedly solitary species, all from Germany, with records of overwintering females emerging as early as the beginning of March (Stöckhert 1933; but these females apparently do not begin provisioning until much later, as discussed above). The overall pattern corresponds to the prediction of decreasing sociality as the date of first provisioning approaches the solstice. This is particularly evident in species demonstrating geographic variation; in *Augochlorella striata*, *E. calceatus*, *H. confusus*, *H. ligatus* and *H. rubicundus*, the populations that start the soonest are most clearly eusocial (with, for example, longer worker production phase, higher percentage of nests with workers, greater female size variation, greater relative scarcity of first-brood males), and in all but *H. ligatus*, the latest-starting population is characteristically solitary. Altogether, these data suggest that each species in a given habitat has its own characteristic thermal and photoperiodic response thresholds, such that under identical environmental conditions there may still be differences between species, but that any given species will demonstrate predictable variation between populations in different environments. In any event, it appears that eusocial behavior (i.e. the formation of a society containing two permanent female castes) may not be expressed unless some females emerge when males are rare or absent, which would explain why protandry and eusociality appear to be mutually exclusive in halictines. If this link between demography and caste is unique to halictines, so may be the route they have followed to eusociality.

CONCLUSIONS: THEORY AND PRACTICE

The MLH and ECH appear to offer some potential both to explain the patterns we already know of in halictines and to predict the patterns for cases that have not yet been investigated. A particular strength of these hypotheses is that they are both open to testing; laboratory work with halictines has already yielded important results, and with more sophisticated techniques under more natural conditions, clearer answers may be forthcoming. It will be especially valuable to develop either a laboratory procedure or a form of controlled field-rearing that allows individuals to forage, nest, mate, and (most importantly) enter diapause in a natural manner. I expect that under properly controlled conditions, or careful transplant ('common garden') experiments, the following should be possible: (a) to make bees from temperate social halictine populations act as if solitary, by exposing foundresses to long photoperiods and/or high temperatures (or by somehow introducing large numbers of males at the start of what is normally worker emergence); (b) to make bees from solitary populations exhibit sociality (by similar, but reverse, procedures).

If these hypotheses are valid, particularly the MLH, the elaboration of a new body of accompanying theory, both general and specific, will be required. No extant models of social evolution, sex ratio evolution, sexual selection, or even general demography in social insects allow for the possibility that the presence of reproductives of one sex (males) could be the factor responsible for the 'production' of functional reproductives of the opposite sex (gynes).

It is absolutely essential to recognize that sociality in halictines is based on two apparently independent phenomena: determination of caste (permanent behavioral alternatives) and the establishment of reproductive dominance hierarchies within and between castes. The latter has been the focus of virtually all past research on caste in halictines (see Michener 1990) and only one study has focused on the former (Yanega 1989, 1992). At present, the *only* feature known to define permanent behavioral classes in halictines is diapause; at some point in some halictines (including some with non-overlapping size distributions, e.g. *H. hesperus* (Packer 1985)), individuals functioning as workers can later, under proper circumstances, be found mating, establishing dominance, producing future reproductives of both sexes, or even founding nests independently, but there is no evidence that such individuals can ever enter diapause. This may appear a trivial issue, given that the traditional view of caste is based on reproductive differences, and a replacement queen, for example, seems quite capable of normal reproduction. However, in cases where diapausing foundresses can survive to produce more than one brood of offspring, this represents a substantial increase in *lifetime* reproductive success relative to females that cannot diapause. The crucial question, then, is how this demographic feature relates to the expression and evolution of sociality.

Both theory (Godfray and Grafen 1988) and empirical observations (Michener 1985) have suggested that unmatedness is a condition that may favor the expression of helping behavior in hymenopterans; the common premise is that there may be a fundamental difference in the costs and benefits of different behavioral and reproductive strategies between mated and unmated females that favors helping behavior (sib-rearing) in unmated females. What the preceding analyses suggest is that these cost–benefit differences *themselves* vary under seasonal vs. aseasonal life histories, resulting in characteristic patterns in the expression of sociality in halictines.

In aseasonal populations, virtually all mated females reproduce, and such females tend not to be subordinate to others; the less reproductive behavioral class consists of unmated females. Whether a mated female remains in the natal nest presumably depends on the likelihood of successful nest founding elsewhere and on the potential availablity of females to help rear brood and defend the natal nest. If mated females decide to stay, a society composed of numerous fully reproductive females of one or more generations, along with some less reproductive females, may form, as is sometimes seen. Whether the behavioral alternatives are permanent or facultative (eusociality vs. semisociality (Crespi and Yanega 1995)) is presently an open issue, but it may be possible to make some predictions. If a female that has been acting as a subordinate for some time can subsequently mate and attain reproductive status, then the behavioral strategies she employs prior to such a point should reflect this (i.e. she should not take great risks or make great energetic or reproductive sacrifices, because she may have the opportunity to produce her own offspring) and there should be relatively little morphological or behavioral differentiation of 'castes'. She should contribute to sib-rearing only so long as she remains unmated. If, however, the subordinate role is fixed after some time without mating, then she should more readily and completely exhibit 'worker-like' behavior such as vigorous foraging activity, minimal concern for personal nutrition or reproduction; moreover, it should then be possible to find mated, non-reproductive females. The observations are most suggestive of the former possibility; in those cases where 'castes' appear in aseasonal populations, they are always poorly defined (even in those cases where there is considerable size variation; Michener and Bennett 1977) and correlated with unmatedness. Populations such as Turrialba *D. umbripennis* or Colombian *D. seabrai* might therefore be categorized as exhibiting semisociality with an overlap of generations (a logical impossibility under traditional definitions of sociality; see, for example, Michener 1974).

In contrast, in seasonal populations, a mated, diapause-capable female should rarely if ever opt to remain active in the natal nest, as she will typically lose one brood's worth of potential reproductive output. A female that fails to mate is then in a position where sib-rearing, if possible, might be the best strategy to pursue, and societies should contain only one dominant female and some number of subordinates. As above, if the subordinate role is perma-nent, then mated, non-reproductive females might be found, and this *is* commonly seen in seasonal halictines. If females in aseasonal populations can mate and change status, then the original MLH restriction on 'prompt mating' may apply only to seasonal populations, in which diapause may be induced during a narrow physiological window. Such a difference in mating physiology should lead to different effects of *timing* of male production on social structures, as well, and possibly increase selection in aseasonal populations for other mechanisms, such as behavioral dominance of offspring, in order to increase the odds of unmatedness in the presence of males. Thus, the strategies adopted by both mated and unmated females, as well as mothers with their offspring, are expected to be quite different in seasonal vs. aseasonal habitats (i.e. when prompt mating is associated with diapause).

There are some fundamental similarities expected and observed, however, between seasonal and aseasonal social structures. First, under the MLH, the presence of subordinate, 'helper' females is a facultative property of the society that relies on some finite frequency of unmatedness to persist. As unmatedness is the factor determining 'castes' in both cases, it would not be expected that obligately sterile workers could ever occur in halictines, and there is no evidence that they do. Second, if there is overlap of generations, maternal behaviors that increase the likelihood of daughters remaining unmated can be considered 'maternal manipulation' (Brian 1983) which may favor the evolution and maintenance of helping behavior by making sib-rearing a daughter's best option (through inclusive fitness effects). Third, interactions between females living together are based on asymmetries; unmatedness may be the primary factor, but within groups of females of similar status, dominance hierarchies based on age or size differences may develop, such as is seen in the differentiation of replacement queens (such females typically mate *after* attaining 'queen' status if they mate at all; Packer 1986b; Yanega 1989). Fourth, some aspects of habitat preference, developmental profiles, longevity, and thermal or photoperiodic thresholds may be determined genetically, but the formation of colonies and differentiation of castes may be only facultatively expressed, the resulting social structure having little or no heritability in itself. The precise form of social behavior (or lack thereof) seen in a given population may only represent one possible result of a broad 'norm of reaction', and thus not necessarily correspond to phylogeny (see Wcislo, this volume); the patterns of sociality among species or populations may instead reflect their biogeography, for example.

Thus, the variability in social behavior in halictines is seen as not representing genetic variability alone, but as a reflection of the many combinations of environmental, demographic, and behavioral factors that can lead to associations of mated and unmated females. Again, the difference in characteristic social structures between aseasonal populations (apparently semisocial) and seasonal ones (eusocial) is thus seen as a function of the different costs and benefits faced by mated vs. unmated females in these environments, in the context set by demography. This review points out a need for much more detailed study of halictine populations, preferrably multiseasonal studies using marked individuals. Geographically, there is a need for studies in the warm temperate regions and subtropics (between 25 and 35° latitude), for different patterns are seen in aseasonal habitats, and we need to understand the transitional areas (see, for example, Packer and Knerer 1987). On the whole, however, there is presently ample evidence for a significant environmental influence on demography, which in turn influences the social structure expressed in any given population. As stated elsewhere (Packer *et al.* 1989; Yanega 1993), the combination of environmental control and caste flexibility apparent in halictines may represent a mechanism by which, from a wide range of possibilities, the forms of behavior expressed by a population are attuned to the specific environmental conditions experienced; a type of phenotypic plasticity that promotes ecological fine-tuning. The few apparent exceptions to the general rules noted in this review may eventually prove to be particularly instructive, for they may possess different mechanisms than those hypothesized here, and suggest further specific areas of inquiry that could shed light on the evolution of social behavior in this group.

ACKNOWLEDGEMENTS

I thank the many individuals who have commented upon and contributed to drafts of this manuscript over the last several years; B. J. Crespi, J. H. Hunt, R. L. Minckley, L. Packer, W. T. Wcislo, and especially the members of my Thesis Committee: C. D. Michener, B. A. Alexander, W. J. Bell and G. W. Byers. This is contribution no. 3069 from the Department of Entomology, University of Kansas.

LITERATURE CITED

Abrams, J. and G. C. Eickwort. 1980. Biology of the communal sweat bee *Agapostemon virescens* (Hymenoptera: Halictidae) in New York state. *Search: Agr.* 1: 1–19.

Atwood, C. E. 1933. Studies on the Apoidea of western Nova Scotia with special reference to visitors to apple bloom. *Can. J. Res.* 9: 443–457.

Barrows, E. M. 1976. Mating behavior in halictine bees (Hymenoptera: Halictidae): I, patrolling and age-specific behavior in males. *J. Kansas Entomol. Soc.* 49: 105–119.

Batra, S. W. T. 1966. The life cycle and behavior of the primitively social bee, *Lasioglossum zephyrum* (Halictidae). *Univ. Kans. Sci. Bull.* 46: 359–423.

–. 1987. Ethology of the vernal eusocial bee, *Dialictus laevissimus* (Hymenoptera: Halictidae). *J. Kansas Entomol. Soc.* 60: 100–108.

–. 1990. Bionomics of *Evylaeus comagenensis* (Knerer and Atwood) (Halictidae), a facultatively polygynous, univoltine, boreal halictine bee. *Proc. Wash. Entomol. Soc.* 92: 725–731.

Bohart, G. E and N. N. Youssef. 1976. The biology and behavior of *Evylaeus galpinsiae* Cockerell (Hymenoptera: Halictidae). *Wasmann J. Biol.* 34: 185–234.

Breed, M. D. 1975. Life cycle and behavior of a primitively social bee, *Lasioglossum rohweri* (Hymenoptera: Halictidae). *J. Kansas Entomol. Soc.* 48: 64–80.

–. 1976. The evolution of social behavior in primitively social bees: A multivariate analysis. *Evolution* 30: 234–240.

Brian, M. V. 1983. *Social Insects: Ecology and Behavioral Biology.* New York: Chapman and Hall.

Brooks, R. W. and D. W. Roubik. 1983. A halictine bee with distinct castes: *Halictus hesperus* (Hymenoptera: Halictidae) and its bionomics in Central Panamá. *Sociobiology* 7: 263–282.

Buckle, G. R. 1982. Differentiation of queens and nestmate interactions in newly established colonies of *Lasioglossum zephyrum* (Hymenoptera: Halictidae). *Sociobiology.* 7: 8–20.

Bulmer, M. G. 1983. The significance of protandry in social Hymenoptera. *Am. Nat.* 121: 540–551.

Campos, M. J. de O. 1980. Aspectos da sociologia e fenologia de *Pereirapis semiauratus.* Dissertação (Mestre) da Universidad Federal de São Carlos, SP, Brazil. 188pp.

Chandler, L. 1955. The ecological life history of *Halictus (H.) ligatus* Say with notes on related species. Ph. D. dissertation, Purdue University.

Claude-Joseph, F. 1926. Recherches biologique sur les Hyménoptères du Chili (Mellifères). *Ann. Sci. Nat., Zool.* (Ser 10) 9: 113–268.

Crespi, B. J. and D. Yanega. 1995. The definition of eusociality. *Behav. Ecol.* 6: 109–115.

Dolphin, R. E. 1966. The ecological life history of *Halictus (Seladonia) confusus* Smith (Hymenoptera, Halictidae). Ph. D. dissertation, Purdue University.

Eickwort, G. C. 1985. The nesting biology of the sweat bee *Halictus farinosus* in California, with notes on *H. ligatus* (Hymenoptera: Halictidae). *Pan Pac. Entomol.* 61: 122–137.

–. 1986. First steps into eusociality: the sweat bee *Dialictus lineatulus. Fla. Entomol.* 69: 742–754.

Eickwort, G. C. and K. R. Eickwort. 1969. Aspects of the biology of Costa Rican halictine bees, I. *Agapostemon nasutus* (Hymenoptera: Halictidae). *J. Kansas Entomol. Soc.* **42**: 421–452.

–. 1971. Aspects of the biology of Costa Rican halictine bees, II. *Dialictus umbripennis* and adaptations of its caste structure to different climates. *J. Kansas Entomol. Soc.* **44**: 343–373.

–. 1972. Aspects of the biology of Costa Rican halictine bees, IV. *Augochlora (Oxystoglossella)* (Hymenoptera: Halictidae). *J. Kansas Entomol. Soc.* **45**: 18–45.

Eickwort, G. C., J. M. Eickwort, J. Gordon and M. A. Eickwort. 1996. Solitary behavior in a high-altitude population of the social sweat bee *Halictus rubicundus* (Hymenoptera: Halictidae). *Behav. Ecol. Sociobiol.*, in press.

Gadagkar, R., C. Vinutha, A. Shanubhogue and A. P. Gore. 1988. Pre-imaginal biasing of caste in a primitively eusocial insect. *Proc. R. Soc. Lond.* B**233**: 175–189.

Godfray, H. C. J. and A. Grafen. 1988. Unmatedness and the evolution of eusociality. *Am. Nat.* **131**: 303–305.

Greenberg, L. 1982. Year-round culturing and productivity of a sweat bee, *Lasioglossum zephyrum* (Hymenoptera: Halictidae). *J. Kansas Entomol. Soc.* **55**: 13–22.

Greenberg, L. and G. R. Buckle. 1981. Inhibition of worker mating by queens in a sweat bee, *Lasioglossum zephyrum*. *Insectes Soc.* **28**: 347–352.

Kamm, D. R. 1974. Effects of temperature, day length and number of adults on the sizes of cells and offspring in a primitively social bee (Hymenoptera: Halictidae). *J. Kansas Entomol. Soc.* **47**: 8–18.

Knerer, G. 1980. Biologie und sozialverhalten von bienenarten der gattung *Halictus* Latreille (Hymenoptera, Halictidae). *Zool. J. (Jena), Abt. Syst., Oekol. Geog. Tiere* 107: 511–536.

–. 1983. The biology and social behavior of *Evylaeus linearis* (Schenck) (Apoidea; Halictinae). *Zool. Anz.* **211**: 177–186.

–. 1987. Photoperiod as cue for voltinism and caste regulation in halictine bees. In *Chemistry and Biology of Social Insects*. J. Eder and H. Rembold, eds., p. 305. Munich: Verlag J. Peperny.

Knerer, G. and P. MacKay. 1969. Bionomic notes on the solitary *Evylaeus oenotherae* (Stevens) (Hymenoptera: Halictinae), a matinal summer bee visiting cultivated Onagraceae. *Can. J. Zool.* **47**: 289–294.

Knerer, G. and C. Plateaux-Quénu. 1967. Sur la production de males chez les Halictinae (Insectes, Hymenoptères) sociaux. *C. R. Acad. Sci. Paris* 264: 1096–1099.

–. 1970. The life cycle and social level of *Evylaeus nigripes* (Hymenoptera: Halictinae), a Mediterranean halictine bee. *Can. Entomol.* 102: 185–196.

Litte, M. 1977. Aspects of the social biology of the bee *Halictus ligatus* in New York state (Hymenoptera, Halictidae). *Insectes Soc.* **24**: 9–36.

MacKay, P. A. and G. Knerer. 1979. Seasonal occurrence and abundance in a community of wild bees from an old field habitat in southern Ontario. *Can. Entomol.* **111**: 367–376.

Mead, F. and D. Gabouriaut. 1993. Post-eclosion sensitivity to social context in *Polistes dominulus* Christ females (Hymenoptera, Vespidae). *Insectes Soc.* **40**: 11–20.

Michener, C. D. 1966. The bionomics of a primitively social bee, *Lasioglossum versatum* (Hymenoptera: Halictidae). *J. Kansas Entomol. Soc.* **39**: 193–217.

–. 1974. *The Social Behavior of the Bees: A Comparative Approach.* Cambridge, Mass.: Harvard University Press.

–. 1977. Nests and seasonal cycle of *Neocorynura pubescens* in Colombia (Hymenoptera: Halictidae). *Rev. Biol. Trop.* **25**: 39–41.

–. 1985. From solitary to eusocial: need there be a series of intervening species? *Fortschr. Zool.* **31**: 293–305.

–. 1990. Reproduction and caste in social halictine bees. In *Social Insects: an Evolutionary Approach to Castes and Reproduction.* W. Engels, ed., pp. 75–119. Berlin: Springer-Verlag.

Michener, C. D. and F. D. Bennett. 1977. Geographical variation in nesting biology and social organization of *Halictus ligatus*. *Univ. Kans. Sci. Bull.* **51**: 233–260.

Michener, C. D., M. D. Breed and W. J. Bell. 1979. Seasonal cycles, nests and social behavior of some Colombian halictine bees (Hymenoptera; Apoidea). *Rev. Biol. Trop.* **27**: 13–34.

Michener, C. D. and W. B. Kerfoot. 1967. Nests and social behavior of three species of *Pseudaugochloropsis* (Hymenoptera: Halictidae). *J. Kansas Entomol. Soc.* **40**: 214–232.

Michener, C. D., W. B. Kerfoot and W. Ramírez B. 1966. Nests of *Neocorynura* in Costa Rica (Hymenoptera: Halictidae). *J. Kansas Entomol. Soc.* **39**: 245–258.

Michener, C. D. and R. B. Lange. 1958a. Observations on the behavior of Brasilian halictid bees, III. *Univ. Kans. Sci. Bull.* **39**: 473–505.

–. 1958b. Observations on the behavior of Brasilian halictid bees. V, *Chloralictus.* *Insectes Soc.* **5**: 379–407.

–. 1958c. Observations on the behavior of Brazilian halictid bees. II: *Paroxystoglossa jocasta.* *J. Kansas Entomol. Soc.* **31**: 129–138.

–. 1958d. Observations on the behavior of Brazilian halictid bees (Hymenoptera, Apoidea) I. *Pseudagapostemon.* *Ann. Entomol. Soc. Am.* **51**: 155–164.

–. 1959. Observations on the behavior of Brazilian halictid bees (Hymenoptera, Apoidea) IV. *Augochloropsis*, with notes on extralimital forms. *Am. Mus. Nov.* 1924: 1–41.

Michener, C. D. and C. A. C. Seabra. 1959. Observations on the behavior of Brasilian halictid bees, VI, Tropical species. *J. Kan. Entomol. Soc.* **32**: 19–28.

Michener, C. D. and A. Wille. 1961. The bionomics of a primitively social bee, *Lasioglossum inconspicuum.* *Univ. Kans. Sci. Bull.* **42**: 1123–1202.

Nye, W. P. 1980. Notes on the biology of *Halictus (Halictus) farinosus* Smith (Hymenoptera: Halictidae). *U.S.D.A. Sci. Ed. Admin. Agr. Res. Results* ARR-W-11: 1–29.

Ordway, E. 1966. The bionomics of *Augochlorella striata* and *A. persimilis* in eastern Kansas (Hymenoptera: Halictidae). *J. Kansas Entomol. Soc.* **39**: 270–313.

Packer, L. 1983. The nesting biology and social organisation of *Lasioglossum (Evylaeus) laticeps* (Hymenoptera, Halictidae) in England. *Insectes Soc.* **30**: 367–375.

–. 1985. Two social halictine bees from southern Mexico with a note on two bee hunting philanthine wasps (Hymenoptera: Halictidae and Sphecidae). *Pan Pac. Entomol.* **61**: 291–298.

–. 1986a. Multiple-foundress associations in a temperate population of *Halictus ligatus* (Hymenoptera; Halictidae). *Can. J. Zool.* **64**: 2335–2332.

–. 1986b. The social organisation of *Halictus ligatus* (Hymenoptera; Halictidae) in southern Ontario. *Can. J. Zool.* **64**: 2317–2324.

–. 1986c. The biology of a subtropical population of *Halictus ligatus* Say (Hymenoptera; Halictidae). II. Male behaviour. *Ethology* **72**: 287–298.

–. 1986d. The biology of a subtropical population of *Halictus ligatus* IV: A cuckoo-like caste. *J. N.Y. Entomol. Soc.* **94**: 458–466.

–. 1990. Solitary and eusocial nests in a population of *Augochlorella striata* (Provancher) (Hymenoptera; Halictidae) at the northern edge of its range. *Behav. Ecol. Sociobiol.* **27**: 339–344.

–. 1992. The social organisation of *Lasioglossum (Dialictus) laevissimum* (Smith) in southern Alberta. *Can. J. Zool.* **70**: 1767–1774.

Packer, L., V. Jessome, C. Lockerbie and B. Sampson. 1989. The phenology and social biology of four sweat bees in a marginal environment: Cape Breton Island. *Can. J. Zool.* **67**: 2871–2877.

Packer, L. and G. Knerer. 1985. Social evolution and its correlates in bees of the subgenus *Evylaeus* (Hymenoptera; Halictidae). *Behav. Ecol. Sociobiol.* **17**: 143–149.

–. 1986. The biology of a subtropical population of *Halictus ligatus* Say (Hymenoptera; Halictidae). I. Phenology and social organisation. *Behav. Ecol. Sociobiol.* **18**: 363–375.

–. 1987. The biology of a subtropical population of *Halictus ligatus* Say (Hymenoptera; Halictidae). III. The transition between annual and continuously brooded colony cycles. *J. Kansas Entomol. Soc.* **60**: 510–516.

Packer, L. and R. E. Owen. 1989. Notes on the biology of *Lasioglossum (Evylaeus) cooleyi* (Crawford), an eusocial halictine bee (Hymenoptera: Halictidae). *Can. Entomol.* **121**: 431–438.

Plateaux-Quénu, C. 1960. Nouvelle preuve d'un déterminisme imaginal des castes chez *Halictus marginatus* Brullé. *C. R. Acad. Sci. Paris* **250**: 4465–4466.

–. 1962. Biology of *Halictus marginatus* Brulle. *J. Apic. Res.* **1**: 41–51.

Plateaux-Quénu, C., L. Plateaux and L. Packer. 1989. Biological notes on *Evylaeus villosulus* (K.) (Hymenoptera, Halictidae), a bivoltine, largely solitary halictine bee. *Insectes Soc.* **36**: 245–263.

Poursin, J.-M. and C. Plateaux-Quénu. 1982. Niches écologiques de quelques Halictinae I. Comparaison des cycles annuels. *Apidologie* **13**: 215–226.

Roberts, R. B. 1969. Biology of the bee genus *Agapostemon* (Hymenoptera: Halictidae). *Univ. Kans. Sci. Bull.* **48**: 689–719.

–. 1973. Bees of northwestern America: *Agapostemon* (Hymenoptera: Halictidae). *Oregon St. Univ. Agr. Expt. Sta. Tech. Bull.* **125**: 1–23.

Sakagami, S. F. 1968. Nesting habits and other notes on an Indo-Malayan halictine bee, *Lasioglossum albescens*, with a description of *L. a. iwatai* ssp. nov. (Hymenoptera, Halictidae). *Malay. Nat. J.* **21**: 85–99.

–. 1980. Bionomics of the halictine bees in northern Japan. 1. *Halictus (Halictus) tsingtouensis* (Hymenoptera, Halictidae), with notes on the number of origins of eusociality. *Kontyû* **48**: 556–565.

–. 1988. Bionomics of the halictine bees in northern Japan. IV. *Lasioglossum (Evylaeus) nupricola* sp. nov., a climatic relic. *Kontyû* **56**: 337–353.

Sakagami, S. F. and A. W. Ebmer. 1979. *Halictus (Seladonia) tumulorum higashi* ssp. nov. from the Northeastern Palaearctic (Hymenoptera: Apoidea; Halictidae). *Kontyû* **47**: 543–549.

Sakagami, S. F., A. W. Ebmer, T. Matsumura and Y. Maeta. 1982a. Bionomics of the halictine bees in northern Japan. II. *Lasioglossum (Evylaeus) sakagamii* (Hymenoptera, Apoidea, Halictidae), with taxonomic notes on allied species. *Kontyû* **50**: 198–211.

Sakagami, S. F. and K. Fukushima. 1961. Female dimorphism in a social halictine bee, *Halictus (Seladonia) aerarius* (Smith) (Hymenoptera, Apoidea). *Jap. J. Ecol.* **11**: 118–124.

Sakagami, S. F. and K. Hayashida. 1958. Biology of the primitive social bee, *Halictus duplex* Dalla Torre I. Preliminary report on the general life history. *Ann. Zool. Jap.* **31**: 151–155.

–. 1961. Biology of the primitive social bee, *Halictus duplex* Dalla Torre, III. Activities in spring solitary phase. *J. Fac. Sci., Hokkaido Univ., Ser. Zool.* **14**: 639–682.

–. 1968. Bionomics and sociology of the summer matrifilial phase in the social halictine bee, *Lasioglossum duplex. J. Fac. Sci., Hokkaido Univ., Ser. Zool.* **16**: 413–513.

Sakagami, S. F., Y. Hirashima, Y. Maeta and T. Matsumura. 1982b. Bionomic notes on the social halictine bee *Lasioglossum affine* (Hymenoptera, Halictidae). *Esakia* **19**: 161–176.

Sakagami, S. F., Y. Hirashima and Y. Ohé. 1966. Bionomics of two new Japanese halictine bees (Hymenoptera, Apoidea). *J. Fac. Agr., Kyushu Univ.* **13**: 673–703.

Sakagami, S. F. and J. S. Moure. 1967. Additional observations on the nesting of some Brazilian halictine bees (Hymenoptera: Apoidea). *Mushi* **40**: 119–138.

Sakagami, S. F. and M. Munakata. 1966. Bionomics of a Japanese halictine bee, *Lasioglossum pallidulum* (Hymenoptera: Apoidea). *J. Kansas Entomol. Soc.* **39**: 370–379.

–. 1972. Distribution and bionomics of a transpalaearctic eusocial halictine bee, *Lasioglossum (Evylaeus) calceatum*, in northern Japan, with reference to its solitary life cycle at high altitude. *J. Fac. Sci., Hokkaido Univ., Ser. Zool.* **18**: 411–437.

Sakagami, S. F. and F. L. Wain. 1966. *Halictus latisignatus* Cameron: a polymorphic Indian halictine bee with caste differentiation (Hymenoptera, Halictidae). *J. Bombay Nat. Hist. Soc.* **63**: 57–73.

Seger, J. 1983. Partial bivoltinism may cause alternating sex-ratio biases that favour eusociality. *Nature (Lond.)* **301**: 59–62.

Sitdikov, A. A. 1988. Nesting of the bee *Halictus quadricinctus* (F.) (Hymenoptera, Halictidae) in the Udmurt ASSR. *Entomol. Rev.* **66**: 66–77.

Stockhammer, K. A. 1966. Nesting habits and life cycle of a sweat bee, *Augochlora pura* (Hymenoptera: Halictidae). *J. Kansas Entomol. Soc.* **39**: 157–192.

–. 1967. Some notes on the biology of the blue sweat bee, *Lasioglossum coeruleum* (Apoidea: Halictidae). *J. Kansas Entomol. Soc.* **40**: 177–189.

Stöckhert, E. 1923. Ueber Entwicklung und Lebensweise der Bienengattung *Halictus* Latr. und ihrer Schmarotzer (Hym.). *Konowia* **2**: 48–64, 146–165; 216–247.

Stöckhert, F. K. 1933. Die Bienen Frankens. *Dt. Entomolog. Z. (Suppl.)* 1932: 1–294.

Suzuki, T. 1981. Effect of photoperiod on male egg production by foundresses of *Polistes chinensis antennalis* Pérez (Hymenoptera, Vespidae). *Jap. J. Ecol.* **31**: 347–351.

–. 1982. Cessation and resumption of laying of female-producing eggs by foundresses of a Polistine wasp *Polistes chinensis antennalis* (Hymenoptera, Vespidae) under experimental conditions. *Kontyû* **50**: 652–655.

–. 1986. Production schedules of males and reproductive females, investment sex ratios and worker-queen conflict in paper wasps. *Am. Nat.* **128**: 366–378.

Vleugel, D. A. 1973. Observations on the behaviour of the primitively social bee, *Evylaeus* (=*Halictus*) *calceatus* Scop. I. Preliminary report on the general life history. *Entomol. Ber.* **33**: 121–127.

Wcislo, W. T., A. Wille and E. Orozco. 1993. Nesting biology of tropical solitary and social sweat bees, *Lasioglossum (Dialictus) figueresi* Wcislo and *L. (D.) aeneiventre* (Friese) (Hymenoptera: Halictidae). *Insectes Soc.* **40**: 21–40.

Westrich, P. 1989. *Die Wildbienen Baden-Württembergs*. Stuttgart: Verlag Eugen Ulmer.

Wille, A. and C. D. Michener. 1971. Observations on the nests of Costa Rican *Halictus* with taxonomic notes on Neotropical species (Hymenoptera: Halictidae). *Rev. Biol. Trop.* **18**: 17–31.

Wille, A. and E. Orozco. 1970. The life cycle and behavior of the social bee *Lasioglossum (Dialictus) umbripenne* (Hymenoptera: Halictidae). *Rev. Biol. Trop.* **17**: 199–245.

Wilson, E. O. 1971. *The Insect Societies*. Cambridge, Mass.: Harvard Univ. Press.

Yanega, D. 1988. Social plasticity and early-diapausing females in a primitively social bee. *Proc. Natl. Acad. Sci. U.S.A.* **85**: 4374–4377.

–. 1989. Caste determination and differential diapause within the first brood of *Halictus rubicundus* in New York (Hymenoptera: Halictidae). *Behav. Ecol. Sociobiol.* **24**: 97–107.

–. 1990. Philopatry and nest founding in a primitively social bee, *Halictus rubicundus*. *Behav. Ecol. Sociobiol.* **27**: 37–42.

–. 1992. Does mating determine caste in sweat bees? (Hymenoptera: Halictidae). *J. Kansas Entomol. Soc.* **65**: 231–237.

–. 1993. Environmental effects on male production and social structure in *Halictus rubicundus* (Hymenoptera: Halictidae). *Insectes Soc.* **40**: 169–180.

15 · Behavioral environments of sweat bees (Halictinae) in relation to variability in social organization

WILLIAM T. WCISLO

Do you know the Halicti? Perhaps not . . . [T]hese humble creatures with
no history can tell us some very singular things; and their acquaintance is
not to be disdained if we would enlarge our ideas upon the bewildering
swarm of this world.

Jean Henri Fabre (1915, p. 365)

ABSTRACT

An overview of variability in social behavior of sweat bees
shows why a consideration of environmental influences
helps us understand the current distribution of halictine
social behavior and how it evolved. The breadth of reaction
norms for social behavior apparently varies among species,
based on limited information. Conspecific attraction,
mutual tolerance and group-living occur within lineages
of solitary sweat bees, providing opportunities to study
weakly or undifferentiated social organizations, analogous
to early phases of social evolution. Perceptual capabilities
to manipulate social environments are reviewed, but there
is insufficient evidence to determine whether such capabilities constrain social organizations. Hypotheses relating to
the importance of natural enemies for the expression of
social behavior are also reviewed, but again pertinent data
are scarce.

INTRODUCTION

Hamilton (1964, 1972) enlarged our concept of reproductive
behavior by taking into account a kinship-weighted fraction of offspring produced by collateral relatives. This
gene's-eye view of reproduction clarified certain features
of animal societies and their evolution, as discussed elsewhere in this volume (see also Gadagkar 1991; Mueller
1991). Similarly, we need to enlarge our concept of 'the
environment'. We need to take into account modifications
induced by behavior, and other developmental changes
(see Wcislo 1989; Cairns et al. 1990; Bateson 1991; West-
Eberhard 1986, 1992a; Gottlieb 1992). Following Fabre's
(1915) lead, sweat bees ('Halicti') may show us how and
why this expansion is important.

An understanding of behavioral environments integrates knowledge of intrinsic (genetic) and extrinsic (environmental) factors in two ways (Evans 1977; see also
Alexander et al. 1991; Schwarz et al., this volume). First, phenotypes ('reaction norms') are determined by complex
interactions between genic and environmental information
(see, for example, Waddington 1940; Schmalhausen 1949;
Zuckerkandl and Villet 1988). We need to know the breadth
of these reaction norms with respect to sociality for diverse
lineages. To what extent do the breadths of the norm
evolve? A large gap in our knowledge of sweat bees concerns the physiological and developmental mechanisms
that generate these norms. Secondly, as emphasized in
other chapters, all evaluations of Hamilton's (1964, 1972)
inequality require information on the ecological benefits
and costs associated with social or solitary behavior, but
these extrinsic factors have undergone a period of neglect.

Solitary, social (communal, semisocial, eusocial) and
parasitic behaviors occur both among lineages of sweat
bees and sometimes among and within populations. This
variability provides excellent opportunities for studying
environmental influences on the expression and organization of social behavior. In this chapter I examine sweat-bee
biology from an environmental point of view. To describe
the social behavior of different colonies, I use Michener's
(1974, p. 38ff.) terminology, with modifications indicated
where appropriate. Following a brief overview of natural
history, and a review of intraspecific variation in social
behavior, I focus on the following areas.

(1) An animal's activity can modify its own environment, or those of its neighbors (including nestmates and
offspring), in passive, evocative, and active ways. Passive
biotic effects include those influenced by spatial proximity
to other bees, which are important for understanding

316

the earlier phases of social (and parasitic) evolution (Evans 1958; Hoogland and Sherman 1976; Wcislo 1984, 1987a, 1992b; for passive abiotic effects, see Yanega, this volume).

(2) Different individuals can evoke different responses from conspecifics, and thus influence their behavior. Conspecific attraction (including mates) and mutual tolerance are the simpler forms of socialized behavior (Allee 1938), and are likely important features for understanding how and under what conditions more integrated societies evolve.

(3) A behaving animal actively modifies its environment (cf. Wcislo 1989); the modifications can contribute to the genesis of social structures, and to their elaboration. These activities require perceptual capabilities to assess physical and social environmental conditions, including differences in ecological opportunities (cf. Stebbins and Hartl 1988).

(4) An abundance of natural enemies makes for hostile environments; depredations from enemies are hypothesized to favor social behavior for defensive reasons (see, for example, Lin and Michener 1972; Evans 1977; Alexander *et al.* 1991).

Much information on the biology and natural history of social halictine bees is available in the following reviews, from which readers can get more detailed information on a particular topic: Yanega (this volume) on male production; Packer (1993) on multifoundress associations; Michener (1990) on reproduction and caste determination; Kukuk (1989) on population genetics; Greenberg (1988) and Michener and Smith (1987) on kin discrimination and recognition; Sakagami (1974) on social polymorphism; Michener (1974) on general biology; and Sakagami and Michener (1962) on nest architecture.

OVERVIEW OF SWEAT-BEE NATURAL HISTORY

Taxonomy and systematics

The Halictidae consists of over 5000 species of mostly small bees in three subfamilies: Rophitinae (=Dufoureinae), Nomiinae and Halictinae. Halictids are among the more important elements of the bee fauna in most localities around the world (Michener 1979; Roubik 1989; Ayala *et al.* 1993), accounting for up to one-half of all individual bees collected on flowers at one site (see, for example, Sakagami *et al.* 1967; Kakutani *et al.* 1990).

The Rophitinae may be the sister taxon to [Nomiinae + Halictinae], but this relationship is not well supported (Eickwort *et al.* 1986). Known rophitines are solitary, and many have specialized morphologies for utilizing pollen from specific plants (see Rozen 1993). Pollen specialization is hypothesized to be ancestral for Halictidae (Eickwort 1992). This hypothesis is important for understanding halictine social evolution because resource specialization can limit evolutionary opportunities for mother–daughter social behavior, if a bee generation is longer than the flowering period of a host-plant taxon in an area. Advantages associated with sociality may facilitate an evolutionary escape from specialization (*sensu* Hardy 1954).

The Nomiinae is likely the sister taxon to the Halictinae. Group-living is known in nine of twenty nomiine bee species that have been studied, but the biology of most species is not well known (reviewed in Wcislo and Engel 1996).

The subfamily Halictinae consists of three tribes. The Nomioidini is the sister taxon to [Augochlorini + Halictini] (Pesenko 1983). Most nomioidines are solitary. Group-living is known in an Indian population of *Nomioides minutissimus* (Batra 1966b), even though *N. minutissimus* are solitary in the Ukraine (Radchenko 1980). The Augochlorini is treated by Danforth and Eickwort (this volume).

Unless indicated otherwise, taxa discussed in this chapter are in the Halictini. Taxonomic boundaries of the numerous (>50) genera and subgenera are uncertain, as are relationships among lineages, so halictine classification and nomenclature are not stable (Moure and Hurd 1987). In the past especially, different authors have used different nomenclature, so generic or subgeneric names used here may differ from the original publications.

Halictus, *Lasioglossum (Evylaeus)* [= *L. (E.)*], and *L. (Dialictus)* [= *L. (D.)*] are abundant taxa in many localities (Fig. 15-1); more is known about their biology than about that of other halictines. Capabilities to express caste-based social behavior may have been present in the ancestors of both *Halictus* (Michener 1978a, 1990; Richards 1994a) and *Lasioglossum (Evylaeus)* (Packer 1991), and evolved more than several times in *L. (Dialictus)* (G. C. Eickwort, personal communication). Brief reports indicate that social behavior is also known in two other lineages, each of which is probably not closely related to those above: the African *Lasioglossum rubricaudis* (Michener 1969) and the Mediterranean *L. aegyptiellum* (Knerer, in Ebmer 1976). As reviewed below, within taxa there is often considerable variability in the expression of social behavior.

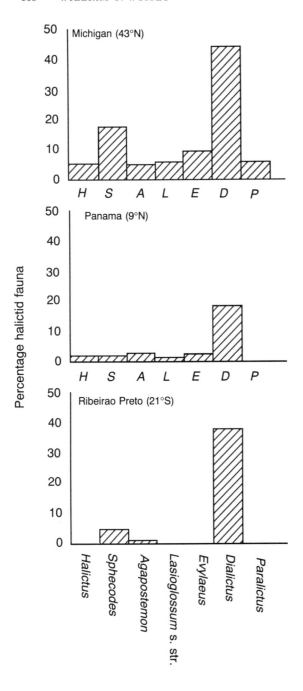

Fig. 15-1. Histograms showing relative abundance of major genera or subgenera of Halictini at several localities in the Western Hemisphere. *L. (Dialictus)* always dominates the fauna, as would *L. (Evylaeus)* in the Old World. Data are from Evans (1986) (Michigan), Michener (1954) (Panamá) and Sakagami *et al.* (1967) (Ribeirão Preto).

Regular cases of communal behavior in associations of apparent reproductive equals are known in Australian *L. (Chilalictus)* spp. (Knerer and Schwarz 1978; Kukuk and Crozier 1990; Kukuk 1992), new world agapostemonine bees (Eickwort and Eickwort 1969; Eickwort 1981) and at least one nomioidine bee (Batra 1966b). Based on behavioral observations of nest-switching in *Agapostemon* (Abrams and Eickwort 1981) and on biochemical studies of *L. (Chilalictus)* spp. (Kukuk and Sage 1994), it is likely that bees in a communal society are not close relatives.

Social parasitism or cleptoparasitic behavior (i.e. laying eggs in cells constructed and provisioned with pollen and nectar by bees of another species) has evolved in *Sphecodes*, which then diversified at the subgeneric level (*Ptilocleptis, Austrosphecodes, Microsphecodes*). Parasitic halictids are apparently more speciose in temperate regions than tropical ones (Fig. 15-1) (reviewed by Wcislo 1987a; Wcislo and Cane 1996). Parasitism also evolved once each in *Echthralictus, Paradialictus* and *Parathrincostoma*, and probably more than once in *Lasioglossum (Paralictus)* (Michener 1978b; Wcislo 1987a, 1996a). The biology of these parasites is not well known (Sick *et al.* 1994; Tengö *et al.* 1992; Wcislo 1987a, 1996a; Eickwort and Eickwort 1972).

Natural history of social attributes

Representative life histories for solitary, communal, semi-social and eusocial halictids are described by Danforth and Eickwort (this volume) and Yanega (this volume; see also Michener 1974).

Within a colony, a socially dominant ('queen') sweat bee usually has mated, has developed ovaries, lays eggs, eats others' eggs, dominates others, rarely forages for pollen and nectar, and is the oldest or largest bee in the group. A subordinate ('worker'), in contrast, usually is not mated, has underdeveloped ovaries, lays few eggs, has eggs eaten, is dominated, frequently forages or guards the nest, and is a younger or smaller bee in the group. These differences can occur even in associations of only two bees. A queen usually lives for a single season, although in *L. (E.) marginatum* queens live for more than one year in perennial colonies (Plateaux-Quénu 1959); in other halictines, queens sometimes live for more than one year in certain localities (Plateaux-Quénu 1992; Plateaux-Quénu and Plateaux 1994).

Social sweat bees can have weakly differentiated social roles. In single-generation laboratory colonies of

L. (D.) zephyrum with three bees, for example, the behavioral repertoires of colony members overlap, but multivariate statistical analyses discriminate a principal forager, egg-layer and guard (Brothers and Michener 1974; Michener 1974, 1990; Eickwort 1986). In these societies, the death of the queen can cause a worker to attain status as a replacement queen (which is different from a foundress queen; see Yanega, this volume).

Caveats for comparative studies of halictine bees

The majority of sweat bees are unknown behaviorally, and for others there is limited information from only one or a few populations, often based on a very limited number of nests. These circumstances encourage typological thinking about the social behavior of *species* (see Wcislo, Chapter 1, this volume). Generalizations to species from one or a few populations (see, for example, Breed 1976; Packer and Knerer 1985) are certainly invalid in at least some cases. Recent surprising findings about the occurrence of food-exchange among adults (Kukuk and Crozier 1990), and of facultative diapause by future foundresses (Yanega, this volume), show the value of caution in drawing broad generalizations about halictine life history.

In cases for which phylogenetic hypotheses are available, social behavior is sometimes unknown for the relevant taxa, or is known only from a single locality, forcing a typological approach; Itô (1993) makes a similar point with respect to vespid wasps. Such character-state delineations may or may not be valid (see Wcislo, this volume; Danforth and Eickwort, this volume). Species-level phylogenetic relationships are resolved for few halictines (see McGinley 1986; Packer 1991; Richards 1994a), so terms such as 'facultatively eusocial' may be misleading in an evolutionary sense. If the ancestor was eusocial (as hypothesized, for example, for *L. (E.) calceatum*) (Packer 1991), then populations with both eusocial and solitary individuals should properly be described as 'facultatively solitary,' to incorporate knowledge of evolutionary history.

Finally, most halictines that have been studied are in populations from the temperate zone, mostly in Europe, North and South America, and Japan (Sakagami 1974; Michener 1974, 1990). This bias towards temperate-zone fauna may be significant, because environmental correlates of seasonality (e.g. photoperiod, nesting and flowering synchrony) may influence social biology, as discussed below (see also Yanega, this volume).

'SOLITARY' BEES AND VARIABILITY IN NUMBERS OF BEES PER NEST

Solitary halictine bees are apparently common, yet behavioral information from different populations is scarce; and few studies have specifically looked for behavioral variability (see, for example, Claude-Joseph 1926; Michener and Lange 1958; Knerer and MacKay 1969; Sakagami 1980; Sakagami *et al.* 1982, 1985; McGinley 1986; Golubnichaya and Moskalenko 1992 (the *Lasioglossum* bees studied by the latter authors were misidentified as *Halictus*)). Based on extensive and careful observations of insect behavior, Rau (1933, p. 275) contended that

Anyone who has watched insects or animals living their lives freely in the open knows that they do not always do their everyday duties in the same way; that there are in each population conservatives and radicals, stupid, brilliant, mediocre, and probably insane individuals.

Even in so-called solitary species, multifemale nests occur, supporting Rau's contention. There is no evidence that these associations are themselves antecedents to more elaborate social behavior, but they demonstrate behavioral variability and provide opportunities to study situations analogous to early phases of social differentiation. In three populations of the bee, *L. (D.) figueresi*, for example, between 9 and 26% of the nests contained two females (Wcislo *et al.* 1993). Approximately one-half of these nests contained females that were equal in reproductive appearance (ovarian development and mating status). In the other two-female nests, one female had queen-like attributes and the other had worker-like attributes. Queens were not significantly larger than workers.

Female *Halictus quadricinctus* are solitary bees in some populations (Sitdikov 1988), although in others they sometimes nest in groups of several females, apparently of equal reproductive status (Vasić 1967; for other examples, see Plateaux-Quénu *et al.* 1989; Sakagami and Munakata 1966; Sakagami *et al.* 1966). The pollen specialist *L. (Sphecodogastra) galpinsiae* usually lives alone, but under changed environmental conditions associated with drought, about 30% of its nests were occupied by two or three females each (Bohart and Youssef 1976). Soil is especially hard during a drought, raising the cost of digging a nest, possibly favoring nest-sharing (cf. Evans and Hook 1986; McCorquodale 1989). Jarvis *et al.* (1994) also hypothesized that mole rat (Rodentia: Bathyergidae) sociality is related to the fact that solitary individuals are unable to dig sufficiently large burrow systems in hard-baked soil.

VARIABILITY IN
SOCIAL ORGANIZATION

Many of the better-studied 'eusocial species' also show variation in social behavior under varying environmental conditions. In southwest France, for example, *L. (E.) albipes* produces two generations per season and is social; the same species has only a single generation and is solitary in eastern France (Plateaux-Quénu 1993). At high altitudes (1167 m) in Japan, *L. (Evylaeus) calceatum* is solitary (Sakagami and Munakata 1972), yet lives in eusocial colonies at lower altitudes in Japan and elsewhere in Europe (Plateaux-Quénu 1992). In extreme temperate regions, and at high altitudes of temperate latitudes, the favorable season is so short that bees must be solitary because they would not have time to first produce a generation of workers. Close relatives of *L. (E.) calceatum* are eusocial, leading Packer (1991) to hypothesize that the solitary behavior represents an evolutionary loss of 'social behavior'. Alternatively, the apparently fixed solitary behavior is the phenotypic expression of individual condition-sensitive genetic systems, without evolutionary change; for a debate on this topic, compare Crozier (1992) with West-Eberhard (1992b). Experimental manipulations are needed to resolve these alternatives.

Another example similar to *L. (E.) calceatum* involves *Halictus rubicundus*, which is solitary at high altitudes (2850 m) in Colorado (USA), yet lives in social colonies at lower altitudes there (Eickwort *et al.* 1996) and elsewhere (Yanega, this volume). Bonelli (1967) presented some suggestive, but not conclusive (Sakagami 1980), evidence that *H. rubicundus* also is solitary at high altitudes in Italy.

Geographic variation in social behavior also occurs within tropical regions. In a wet/dry seasonal area of western Costa Rica, *L. (D.) umbripenne* was active only during the dry season (Wille and Orozco 1970). Colonies had many bees; each was headed by a queen, which was morphologically distinct from the workers. In contrast, in an less seasonal eastern area, bees were probably active nearly the year round; there were fewer bees per colony, and queens were only statistically larger than the workers (Eickwort and Eickwort 1971). It is not known whether the populations are genetically differentiated. Not all examples of solitary behavior in 'social species', however, can be due to seasonal limitations, since behavioral variation occurs in areas where bees are active the year round. In several tropical *L. (Dialictus)* spp., for example, active nests had one to several bees (Table 15-1). Even in two-bee societies,

frequently one bee had queen-like attributes, and one had worker attributes.

Temperate and tropical populations often differ in social organization (Table 15-1). *H. (Halictus) ligatus*, for example, is a well-studied species showing much geographic variation in social behavior (Table 15-2), including cuckoo-like behavior in some individuals from subtropical Florida (Packer 1986a). Michener and Bennett (1977) noted that populations from Trinidad do not neatly fit into any of the standard social classifications (for other examples compare Nye 1980 and Eickwort 1985; or Batra 1966c, Quénu 1957 and Knerer and Plateaux-Quénu 1967). In contrast, with respect to differences *among* species, statistical comparisons show no significant differences between temperate and tropical *L. (Dialictus)* with respect to the parameters listed in Table 15-1 (assuming these data are realistic for the species: Mann–Whitney U-tests, $p > 0.05$).

SOCIAL MODIFICATIONS OF
BEHAVIORAL ENVIRONMENTS

Scarr and McCartney (1983) outline environmental effects associated with children's behavior, which can be applied to behavioral environments of sweat bees.

(1) *Passive* effects occur independently of the actor.
 (i) Biotic factors (e.g. food availability, relative abundance of natural enemies).
 (ii) Abiotic factors (e.g. photoperiod, temperature) (Plateaux-Quénu 1988; Yanega, this volume).
(2) *Evocative* effects result from an actor eliciting different responses from different individuals (for example, a large female might be more attractive to males than a small one).
(3) *Active* effects result from an individual's activities and behavior (e.g. dominance behavior).

Collectively, these effects represent an organism's behavioral environment, emphasizing its importance for the organization and evocation of social roles.

Passive effects

Influence of nest-site selection and nest architecture on social behavior
The different facets of nest-site selection and nest architecture point to abiotic and biotic factors that passively modify social environments, creating novel opportunities for social or parasitic behavior by altering the potential frequency of social interactions.

Table 15-1. *A comparison of social attributes for temperate and tropical* L. (Dialictus) *sweat bees*

Data are presented as means; standard deviations are given in parentheses; n.a., not available.

Species and Region	Number of bees per nest	Difference in caste size (%)	Workers with developed ovaries (%)	Workers mated (%)
zephyrum (Kansas)	14.3	9.1	38.0	8.0
rohweri (Kansas)	4.9	10.0	9.0	37.9
versatum (Kansas)	29.6	11.9	25.0	3.0
imitatum (Kansas)	8.1	9.9	12.0	2.5
lineatulum (New York)	7.0	4.4	28.3	20.0
laevissimum (Alberta)	3.5	7.0	63.3	35.0
near *laevissimum* (Maryland)	6.7	9.0	53.6	n.a.
rhytidophorum (Brazil)	3.8	6.0	28.0	12.9
Temperate Mean[a]	9.7	8.4	32.1	17.0
	(8.7)	(2.4)	(18.9)	(14.6)
umbripenne (Damitas, Costa Rica)	75.6	16.9	32	2.5
umbripenne (Turrialba, Costa Rica)	16.7	9.8	25	rare
breedi (Colombia)	1.4	3	40	n.a.
seabrai (Colombia)	2.1	3	40	6
seabrai (Brazil)	1.6	5	n.a.	n.a.
aeneiventre (Costa Rica)	2.7	2	59	68.0
Tropical Mean (all)	16.7	6.6	39.2	25.5
	(29.5)	(5.8)	(12.7)	(36.9)
Tropical Mean	4.9	4.6	41.0	37.0
(excl. Damitas *umbripenne*)	(6.6)	(3.1)	(13.9)	(43.8)

Sources: Data taken from summaries by Breed (1976) and Packer (1992), as well as Eickwort and Eickwort (1971), Michener and Seabra (1959), Michener *et al.* (1979), Batra (1987) and Wcislo *et al.* (1993).

Nest-site selection. Most halictine bees nest in the soil, although there have been at least several independent origins of nesting in dead wood among Augochlorini (Danforth and Eickwort this volume) and some *L. (Dialictus)* spp. (Sakagami and Moure 1967). Sweat bees prefer certain edaphic conditions, but within an area of suitable habitat there are no obvious abiotic correlates with the frequently observed patchy distribution of nests (Michener *et al.* 1958; reviewed in Wcislo and Cane 1996).

Aggregations. Probably many halictine bees have nests isolated from one another, but these nests are hard to find and not much is known of their biology (e.g. *H. (Seladonia) lutescens*, Sakagami and Okazawa 1985). Tiny bees like *L. (D.) imitatum* and *L. (D.) opacum* have nest entrances less than 2 mm in diameter, and nest both in aggregations and isolated from conspecifics (Michener and Wille 1961; Michener and Lange 1958; W. T. Wcislo, personal observations), but usually only the aggregated nests are studied. There

may be differences among species in tendencies to aggregate nests, but this has not been tested (see Sakagami and Michener 1962).

A tendency to aggregate nests influences population viscosity (Wilson *et al.* 1992; Taylor 1992) and the extent of non-random spatial associations of genotypes (Crozier *et al.* 1987; Kukuk 1989; Kukuk and Decelles 1986). Aggregations create opportunities for positive or negative social interactions (Gadgil *et al.* 1983; Wcislo *et al.* 1985; Wcislo 1984, 1987a), which in turn may select for refined perceptual capabilities for recognition and assessment (Hamilton 1971b; Wcislo 1992b). Gregarious behavior may have the same automatic costs as other social behaviors (Alexander 1974; Hoogland and Sherman 1976; Wcislo 1984), creating environmental conditions beneficial to group-living. For example, overall cell mortality for solitary nests of *Halictus rubicundus* within an aggregation was 15.5%, whereas for isolated nests the rate was 8.1% (Eickwort *et al.* 1996). No data are available to assess the hypothesis that

Table 15-2. *A comparison of social attributes for temperate and tropical populations of* Halictus ligatus

Region	Number of bees per nest	Difference in caste size (%)	Workers with developed ovaries (%)	Workers mated (%)
North Temperate[a]	5	13.8	28.7	25
Florida[b]	12.5	16	68	57
Trinidad[c]	11.3	n.a.	53	55.6

Sources: [a]Data from populations in Ithaca, New York (Litte 1977); and Toronto and Victoria, Ontario (Packer 1986c,d); [b]Packer and Knerer (1986); [c]Michener and Bennett (1977).

sociality is more frequent in lineages with aggregated nests than in those with isolated nests.

Nesting aggregations can persist for at least 35 years (for example, in *L. (E.) malachurum*) (Stoeckhert 1954). Others, like those of a Kansas population of *L. (D.) rohweri* (Breed 1975), may last only a single season, although the same species has persistent aggregations in New York (G. C. Eickwort, personal communication). Theoretically, persistent aggregations permit the build-up of larger populations of parasites, which, in turn, potentially increases benefits associated with group-living. Long-term population studies of sweat bees and their parasites are needed in this context.

Nest densities of sweat bees can be as high as 150–200 nests per square meter (see, for example, Breed 1975). One evolutionary hypothesis proposed to explain tendencies to spatially aggregate nests is that the collective group, which Hamilton (1971a) termed a 'selfish herd', provides 'cover' to a particular individual in the aggregation from enemies (Wcislo 1984). This hypothesis has not been tested for sweat bees, and evidence from other taxa is mixed (reviewed in Rosenheim 1990; Wcislo and Cane 1996).

Anecdotal observations suggest that nests often may be more densely aggregated in temperate areas than tropical ones (Michener and Bennett 1977; Eickwort 1985; W. T. Wcislo, personal observations), but no quantitative data are available to test this hypothesis. A broadly-distributed species, *Halictus ligatus*, for example, has nests at high densities in the northern part of its range (Canada), whereas nests are at lower densities, or even isolated, in the southern part of its range in southern Florida, Trinidad, Costa Rica or Colombia (Michener and Bennett 1977; Packer and Knerer 1986; Richards 1994b; W. T. Wcislo, personal observations). Among tropical species, counter examples of dense aggregations are easy to find (see, for example, Michener *et al.* 1979; Wcislo *et al.* 1993; W. T. Wcislo, personal observations). Jeanne (1975) hypothesized that latitudinal

gradients in ant predation influenced wasp (Polistinae) social behavior, but the possible relevance of such predation to variation in sweat-bee nesting and social behavior has not been studied.

Architecture. For Halictini, nest architecture generally follows variations on several themes: (1) each cell is at the end of a long side tunnel, and the side tunnel itself is connected to the main tunnel from the soil surface, or cells are arranged serially along these side tunnels; (2) cells are attached directly to the main nest tunnel ('sessile cells'); and 3) there is a single main tunnel, and cells are built in a cluster that is often surrounded by a cavity, which might help drain water in moist soils. Respectively, each of these different architectures provides an environment with increasing opportunities for social contact among nestmates.

Within lineages, there is no direct relationship between increasing levels of sociality and nest architecture (Sakagami and Michener 1962; Packer and Knerer 1985; but see Knerer 1969). Kukuk and Eickwort (1987), however, hypothesized that nest architecture influenced the likelihood that either communal or eusocial behavior evolved among lineages. They assumed that it is more difficult for one bee to dominate egg-laying in nests with widely dispersed cells, but this assumption has not been tested. Nests of some communal bees (e.g. *L. (Chilalictus)* spp.) have cells widely dispersed in a branched tunnel system, but others do not (see Kukuk and Eickwort 1987; Wcislo 1993).

Evocative effects

Solitary bees and mutual tolerance

The majority of bees lead solitary lives and are usually intolerant of conspecifics of the same sex (see, for example, Wcislo 1987a; Field 1992). One of the earlier stages of social integration involves mutual tolerance among conspecifics, but information is scarce on individual variability in tolerance thresholds among solitary bees, or during the solitary phase for social bees. Likewise, factors that inhibit conspecific agonistic behavior are probably important for understanding how social behavior arises.

Intraspecific tolerance or aggression was assayed by using paired solitary females of *L. (Dialictus) figueresi* (*n* = 15 pairs) taken from different nests of reproductively active females (W. T. Wcislo, unpublished). Using a circular arena formed by a plastic tube (after Breed *et al.* 1978), two females could be forced to interact repeatedly. Females in seven pairs appeared indifferent to the other's presence. In four pairs, one female repeatedly passed the other, and *vice versa*. In the remaining four pairs, one female expressed some queen-like behavior and repeatedly nudged the other bee, then rapidly backed away (as do queens in a social colony), and never passed the other bee. The other bee expressed worker-like behavior such as frequently following the queen after being nudged, or sometimes passing her. These behaviors occur in social colonies, and are thought to relate to dominant-subordinate interactions (see 'Social competition', below); they show the potential for the expression of communicative behaviors in otherwise solitary bees (but these bees may be secondarily solitary; Wcislo *et al.* 1993).

Mate attraction
Passive (abiotic) environmental effects determined by demographic factors may influence social roles, as hypothesized for primates by Altmann and Altmann (1979) and for halictines by Yanega (this volume). An important social difference in some sweat bees is that foundress queens are typically mated (and lay all or most eggs) and workers are often unmated (and lay few or no eggs) (Tables 15-1 and 15-2; see also Michener 1974).

Evocative environmental effects may be important in determining whether a female mates or not. Individual females differ in their relative attractiveness to conspecific males, regardless of social level (Wcislo 1987b, 1992a; Packer 1986b). Ayasse *et al.* (1990, 1993) found that unmated females have different chemical profiles from mated females in *L. (E.) malachurum*. Males of this species are preferentially attracted to unmated bees (Smith and Ayasse 1987), but whether or not the preference is learned is unknown. In populations of another species, *L. (D.) zephyrum*, males are attracted to both mated and unmated

females, including females that repeatedly refuse to mate (Wcislo 1987b; Smith and Ayasse 1987), and even to black dots if female odors are present (Barrows *et al.* 1975).

The proportion of mated workers varies sometimes within and among populations (see, for example, Tables 15-1 and 15-2), and also can vary seasonally (see, for example, Packer 1986b,d; Eickwort 1985). In areas with year-round activity, a high proportion of the bees are frequently found to be mated, whereas in strongly seasonal areas there is much variability in this proportion. These differences may be associated with geographic differences in social behavior (see Yanega, this volume). Seasonal differences in frequency of mated workers may be due to changed environments from (1) an increasing abundance of males (which is not evolutionarily stable since earlier production of males would be favored); or (2) decreasing inhibitory control from an increasingly senescent queen (if queens somehow actively inhibit a daughter's mating). We do not understand the mechanisms by which queens inhibit sexual receptivity of other females. Ultimately, the payoffs for worker-mating are higher later in the season. Unmated laying workers can produce males early in the season, which is when males are relatively less abundant because queens are producing female workers; later in the season, queens produce both male and female reproductives, so each male produced then by unmated laying workers would be relatively less valuable.

Active effects

Waddington (1940) discussed hypothetical organizers for the expression of structure during development. We need to understand analogous environmental organizers for halictine social development, and identify the behavioral traits which directly influence behavior of conspecifics, complementing genetic studies. How might the behavior of a future potential queen promote its chances of becoming queen? And how might she promote worker-like behavior in a nest-mate?

Genesis of social structure
The evolutionary establishment of a 'queen–worker' relationship requires tolerance for repeated interactions among otherwise solitary individuals (see previous section). Repeated interactions themselves are important for two reasons, especially under circumstances in which reciprocal altruism is important (see Trivers 1985, pp. 361ff.). First, an environment in which repeated interactions

occur is favorable for the establishment and spread of coop-erative behavior in an otherwise ego-driven world (Axelrod 1984). Secondly, as Bateson (1935) realized from studies of the Indo-Pacific Iatmul people, tolerance for competitive interactions can spontaneously result in two kinds of social organization, one with undifferentiated interactants, and one with differentiated interactants (for discussion, see Geiger 1990).

In an undifferentiated social structure, competing indi-viduals escalate aggressive interactions (an 'eye for an eye' response), although at times the interactants may appear indifferent (see Staw and Ross 1989). Presumably, commu-nal bees behave in this way (for a communal spider wasp, see Wcislo *et al.* 1988). In contrast, differentiated structures *decrease* the possibility of physical violence (ritualized beha-vior). Dominant–subordinate relations characterize differ-entiated social structures. A submissive response to aggressive behavior can increase the likelihood of similar behavior by both individuals during subsequent interac-tions. These repeated interactions are known to lead to learned dominant–subordinate relationships in other ani-mals (Cheney and Seyfarth 1990; references in Wcislo 1989). These initially spontaneous interactions can be rein-forced by changes in hormonal titer (Leshner 1983) and by learning (Kagan *et al.* 1988; Shors *et al.* 1992). If bees can assess and recognize individuals (see 'Perceptual capabil-ities and social manipulation', below), then these features will be subject to social selection. In turn, if there is genetic variability, these interactions can serve as varia-tion for the origins of novel social organizations (West-Eberhard 1979). All these areas remain unstudied for sweat bees, yet are probably essential for understanding how behavioral flexibility relates to the origins of eusociality (West-Eberhard 1992a). Manipulations of a colony's social structure (see Mueller 1991) make it possible to experimen-tally study the costs and benefits of different kinds of social organizations, but such studies have not yet been made.

Social competition

For group-living animals, behavioral environments are important because of competition for socially limited resources (Alexander 1974). Competition is important for structuring social differences, and can be influenced by a variety of factors, such as relatedness, size or age asymme-tries, and nutritional differences (see, for example, Richards and Packer 1994).

In *L. (D.) zephyrum* colonies, for example, the degree of behavioral differentiation (as queen, worker, or guard)

varies with the number of bees in the colony; a queen's abil-ity to suppress worker reproduction (or worker acquies-cence to be suppressed) declines with increasing number of workers in the colony (Brothers and Michener 1974; Buckle 1985). Larger queens suppress worker reproduction more effectively than can smaller queens (Kukuk and May 1991). Moreover, workers rarely passed the queen in the artificial colonies of unrelated bees studied by Brothers and Michener (1974). This behavior is a likely example of widespread tendencies in social animals to avoid a domi-nant individual (Trivers 1985). Using colonies of genealogi-cally related individuals, however, Smith (1987) found that workers were much more likely to pass the queen than were the unrelated workers in the Brothers and Michener study (but see Kukuk and May 1988; Smith 1988).

Perceptual capabilities and social manipulation

Abilities to advantageously manipulate information (*sensu* Dusenbury 1992) concerning nestmates and offspring are important in various social contexts, especially concerning relatedness asymmetries and dominance status (Alexander 1979, pp. 112ff.).

Relatedness. The recognition and discrimination capabil-ities of female sweat bees have been studied in the context of nest guarding. Dufour's-gland chemicals function to line nest cells and in some cases to line nest entrances (see Wcislo 1992b). These compounds have species- and indivi-dual-specific components (Hefetz *et al.* 1978, 1986; Hefetz 1987; Smith *et al.* 1985; Smith and Wenzel 1988), and they may serve as 'recognition badges' to maintain colony integ-rity among familiar nestmates (Greenberg 1979). Using an individual recognition system, social group sizes of sweat bees may be limited by the storage capacities of the bees' memory, as Dunbar (1992) suggested for group size in pri-mates. Sweat bees such as *H. (Seladonia) lutescens* (Sakagami and Okazawa 1985) or *H. (S.) tripartitus* (G. C. Eickwort, per-sonal communication) have colonies with hundreds of bees. If they too recognize nestmates as individuals, then Dunbar's hypothesis might be rejected for sweat bees.

For *L. (D.) zephyrum* the chemical badges themselves are not strongly modified by environmental circumstances (Bell *et al.* 1974), but a bee's behavioral response to them is modified by learning, since she uses her nestmates' odors as bases to admit or reject bees at the nest entrance (Green-berg and Buckle 1981). In nest-related contexts other than guarding, much less is known about how perception of familiarity modifies behavior, although sex-ratio studies demonstrate its occurrence (Mueller 1991; Boomsma and Eickwort 1993).

Reproductive and dominance status. Little is known about the modalities by which individuals assess the reproductive capabilities and social status of nestmates. In *L. (D.) zephyrum*, socially dominant individuals had high levels of activity, which was associated with the inhibition of workers' ovarian development (Brothers and Michener 1974; Breed and Gamboa 1977; Buckle 1984). The dominant bee frequently 'nudges' the other bees in the colony. In *L. (Evylaeus) malachurum* larger females (gynes) were more aggressive, on average, than were smaller gynes in paired interactions (Smith and Weller 1989). Topical application of synthetic Dufour's-gland secretions to the smaller gyne decreased the aggressiveness of the larger individual such that there were no longer significant differences in aggression, implicating some pheromonal assessment of social status.

Origins of recognition capabilities. Hölldobler and Michener (1980) proposed that the two likely sources for origins of the perceptual capabilities for social recognition were either mate recognition by males, or nest recognition by females. There is still not enough evidence to eliminate either possibility (Wcislo 1992a,b); indeed, nothing is known about female perceptual capabilities in the context of sexual behavior. Food-plant recognition is yet another context for the origin of recognition capabilities (cf. Robinson 1985; Bernays and Wcislo 1994), but these perceptual processes are utilized in a context unrelated to the nest environment, so they may be less relevant to nest-related recognition capabilities.

NATURAL ENEMIES AND HOSTILE ENVIRONMENTS

Natural enemies of sweat bees range from bacteria to parasitic fungi, nematodes, insects, and mites, to predatory insects, and arachnids. Danforth and Eickwort (this volume) list the natural enemies of augochlorine bees; an equally formidable list could be made for halictines (see, for example, Batra 1965). Any given colony, however, is usually attacked by a subset of these enemies. Morbidity rates from these different enemies are not known. An understanding of the importance of natural enemies for sweat-bee social behavior requires detailed studies on morbidity from both microparasites (e.g. bacteria) and macroparasites (e.g. flies, mutillid wasps). Hypothetically, microparasites will select against eusociality (especially among kin-groups if queens are singly mated) (Hamilton 1987), whereas macroparasites will select for group-living (see also Crespi and Choe, this volume).

Lin and Michener (1972) reviewed mutual advantages from group-living in an environment sufficiently hostile to offset disadvantages from group-living (Alexander 1974). Since then, not much information has been gathered on the relative costs and benefits of group-living under varying parasite loads (see also Schwarz et al., this volume). We do not even know how nest site selection influences parasitism rates for bees (Michener 1985; Wcislo 1996b). Freeman (1982) found that mud-daubing wasps (*Sceliphron fistularum*) (Sphecidae) that built their nests on vines had 5% of their cells parasitized by *Melittobia* (Eulophidae) wasps, whereas 49% of cells in nests built on flat substrates at the same locality were parasitized.

Much qualitative evidence points to the importance of macroparasites for sweat-bee social behavior. Nest architecture itself suggests the importance of defense against enemies. An entrance to an underground burrow is typically constricted to a diameter just slightly larger than that of the guard's head (Sakagami and Michener 1962), which facilitates blocking the entrance. Among many taxa with group-living, entrances are guarded by bees which lunge at, bite or attempt to sting enemies that wander too close to an entrance. In some taxa, a guard that is challenged by an enemy will adopt a peculiar C-shaped posture, blocking the entrance with an abdominal (metasomal) tergite (Wcislo 1996a). This behavior is not observed in all taxa, and studies of its phyletic distribution may give important insights to relevance of nest defense for social evolution.

Nest survivorship is often quite low (*ca.* 25%) (see, for example, Michener and Wille 1961; Batra 1966a; Sakagami 1977; Sakagami and Fukuda 1989), although Richards (1994b) reported higher survivorship (from 51% to about 92%) for foundress nests of *Halictus ligatus*. Despite the array of enemies which decrease survivorship, few studies attribute significant morbidity to any one cause and, for those that do, epidemiological studies are not available (see, for example, Batra 1965; Michener 1966; Packer *et al.* 1989).

In sweat bee colonies with annual cycles the foundress usually behaves as a solitary bee at the nest-founding stage, and when she is foraging for food her nest is left untended. These unattended nests are attacked by the same brood parasites which successfully attack nests of solitary bees. Batra (1965), for example, lists a beefly (*Bombylius pulchellus*, Diptera: Bombyliidae) and another fly (*Leucophora johnsoni*, Diptera: Anthomyiidae) as parasites of the spring, solitary-phase, nests of *L. (D.) zephyrum*. After the emergence of workers (one of which acts as a guard), these parasites are usually unable to successfully enter nests. Some other parasites (e.g. Mutillidae, *Sphecodes* bees) attack both solitary and social nests.

Quantitative information on the efficacy of nest-guarding is limited, and experimental evidence is totally lacking. In three populations of *L. (D.) figueresi*, nests with one female and those with two females had similar rates of brood mortality (about 20%) due to attacks by parasitic flies, *Phalacrotophora halictorum* (Phoridae) (Wcislo *et al.* 1993; Wcislo 1990). Solitary (foundress) spring nests of *Halictus ligatus* are unguarded when a foundress forages; such nests had rates of brood survival similar to cofounded spring nests (in which the dominant bee guards) (Packer 1988). In contrast, cell mortality was 9.4% in nests of *H. rubicundus* where a foundress was present and 29.3% in nests that lacked a foundress at the time of excavation (Eickwort *et al.* 1996). In *L. (E.) duplex*, nest mortality from ants was about 17% in spring nests, which are unattended when females are foraging, yet was nearly 32% in the summer nests, which are always guarded (Sakagami and Fukuda 1989). In other taxa, guarding effectively keeps out enemies. A kleptoparasitic bee (*Nomada articulata*) never successfully entered communal multifemale nests of *Agapostemon virescens*, but successfully entered a rare nest ($n = 1$!) with only a single bee (Abrams and Eickwort 1981; Eickwort and Abrams 1980).

CONCLUSIONS AND FUTURE DIRECTIONS

Environmental bases for variability in social organization or behavior are not popular areas for investigation (for possible reasons, see Thomas 1925; Wcislo 1989). Our tendency to perceive only the external 'environment' generally fails to take into account modifications induced by activities of animals themselves. An organism-centered approach will broaden this perspective to focus on environmental organizers for social behavior. Many examples given above belabor the point that social behavior and organization is labile in sweat bees, yet typological thinking has impeded an appreciation of the essential links among genic and environmental sources of information, by setting the two sources in opposition.

The tremendous variability in halictine social behavior makes this group an important one for future studies of social evolution. The same behavioral variability, however, makes it difficult to draw general conclusions from limited field studies. Fortunately, many of the more common species are weed bees, which thrive in areas like gardens tended by universities, and deserve more long-term study.

We need to obtain to obtain behavioral data from *multiple* populations of the numerous species which have not been studied, in order to document breadths of reaction norms, and to suggest which environmental features are especially important in influencing the expression of social behavior. Secondly, we need information on historical relationships among taxa to determine how many times *a capability to express social behavior* has arisen. Since social behavior may not be fixed within a taxon, a major difficulty is that it is still not obvious how socially relevant characters should be defined for phylogenetic studies of these bees. To this end, experiments are needed to determine whether solitary bees can be manipulated to live in social groups, and *vice versa*. Natural enemies are frequently hypothesized to influence the expression of social behavior, but few pertinent data are available. Environmentally induced behavioral variability needs to be emphasized to counterbalance our tendency to place in opposition the intrinsic and extrinsic factors that shape social evolution.

ACKNOWLEDGEMENTS

I thank the following people for criticisms, helpful discussions, or the use of prepublication manuscripts: B. Crespi, B. Danforth, P. Kukuk, C. Michener, U. Mueller, L. Packer, M. Richards, M. Schwarz, P. Sherman and D. Yanega; I am especially grateful to the late George Eickwort for his help. Financial support was initially provided by a National Science Foundation (USA) Environmental Biology Postdoctoral Fellowship (BSR-9103786) and general research funds of the Smithsonian Tropical Research Institute.

LITERATURE CITED

Abrams, J. and G. C. Eickwort. 1981. Nest switching and guarding by the communal sweat bee *Agapostemon virescens* (Hymenoptera, Halictidae). *Insectes Soc.* 28: 105–116.

Alexander, R. D. 1974. The evolution of social behavior. *Annu. Rev. Ecol. Syst.* 5: 325–383.

–. 1979. *Darwinism and Human Affairs.* Seattle: University of Washington Press.

Alexander, R. D., K. M. Noonan and B. J. Crespi. 1991. The evolu-
tion of eusociality. In *The Biology of the Naked Mole Rat*. P. W.
Sherman, J. U. M. Jarvis and R. D. Alexander, eds., pp. 3–44.
Princeton: Princeton University Press.

Allee, W. C. 1938. *The Social Life of Animals*. New York: Beacon
Paperbacks.

Altmann, S. A. and J. Altmann. 1979. Demographic constraints on
behavior and social organization. In *Primate Ecology and Human
Origins*. I. S. Bernstein and E. O. Smith, eds., pp. 47–63. New
York: Garland Press.

Axelrod, R. 1984. *The Evolution of Cooperation*. New York: Basic
Books.

Ayala, R., T. L. Griswold and S. H. Bullock. 1993. The native bees
of México. In *Biological Diversity of México, Origins and Distribution*. T. P. Ramamoorthy, R. Bye, A. Lot and J. Fa, eds.,
pp. 180–227. New York: Oxford University Press.

Ayasse, M., W. Engels, A. Hefetz, G. Lubke and W. Francke. 1990.
Ontogenetic patterns in amounts and proportions of Dufour's
gland volatile secretions in virgin and nesting of *Lasioglossum
malachurum* (Hymenoptera: Halictidae). *Z. Naturforsch.* **45**:
709–714.

Ayasse, M., W. Engels, A. Hefetz, J. Tengö, G. Lubke, and
W. Francke, 1993. Ontogenetic patterns of volatiles identified in
Doufour's gland extracts from queens and workers of the primitively eusocial halictine bee, *Lasioglossum malachurum* (Hymenoptera: Halictidae). *Insectes Soc.* **40**: 41–58.

Barrows, E. M., W. J. Bell and C. D. Michener. 1975. Individual odor
differences and their social functions in insects. *Proc. Natl. Acad.
Sci. U.S.A.* **72**: 2824–2828.

Bateson, G. 1935. Culture contact and schismogenesis. *Man* **35**:
178–183.

Bateson, P., ed. 1991. *The Development and Integration of Behavior*.
Cambridge University Press.

Batra, S. W. T. 1965. Organisms associated with *Lasioglossum zephyrum* (Hymenoptera: Halictidae). *J. Kansas Entomol. Soc.* **38**:
367–389.

–. 1966a. The life cycle and behavior of the primitively social bee,
Lasioglossum zephyrum (Halictidae). *Univ. Kans. Sci. Bull.* **46**:
359–423.

–. 1966b. Nests and social behavior of halictine bees of India
(Hymenoptera: Halictidae). *Indian J. Entomol.* **28**: 375–393.

–. 1966c. Nesting behavior of *Halictus scabiosae* in Switzerland
(Hymenoptera, Halictidae). *Insectes Soc.* **13**: 87–92.

–. 1987. Ethology of the vernal eusocial bee, *Dialictus laevissimus*
(Hymenoptera: Halictidae). *J. Kansas Entomol. Soc.* **60**: 26–34.

Bell, W. J., M. D. Breed, K. W. Richards and C. D. Michener. 1974.
Social stimulatory and motivation factors involved in intraspeci-
fic nest defense of a primitively eusocial halictine bee. *J. Comp.
Physiol.* **93**: 173–181.

Bernays, E. A. and W. T. Wcislo. 1994. Sensory capabilities, information processing, and resource specialization. *Q. Rev. Biol.* **69**:
187–204.

Bohart, G. E., and N. N. Youssef. 1976. The biology and behavior
of *Evylaeus galpinsiae* Cockerell (Hymenoptera: Halictidae).
Wasmann J. Biol. **34**: 185–234.

Bonelli, B. 1967. Osservazioni biologiche sugli imenotteri melliferi
e predatori della Val di Fiemme. XXIII. Contributo *Halictus
rubicundus* Christ (Hymenoptera-Halictidae). *Stud. Trentini Sci.
Nat.*, Sez B **44**: 85–96.

Boomsma, J. J. and G. C. Eickwort. 1993. Colony structure, provisioning and sex allocation in the sweat bee *Halictus ligatus*
(Hymenoptera: Halictidae). *Biol. J. Linn. Soc.* **48**: 355–377.

Breed, M. D. 1975. Life cycle and behavior of a primitively social
bee, *Lasioglossum rohweri* (Hymenoptera: Halictidae). *J. Kansas
Entomol. Soc.* **48**: 64–80.

–. 1976. The evolution of social behavior in primitively social bees:
a multivariate analysis. *Evolution* **30**: 234–240.

Breed, M. D. and G. J. Gamboa. 1977. Control of worker activities
by queen behavior in a primitively eusocial bee. *Science (Wash.,
D.C.)* **195**: 694–696.

Breed, M. D., J. M. Silverman and W. J. Bell. 1978. Agonistic behavior, social interactions, and behavioral specialization in a primitively eusocial bee. *Insectes Soc.* **25**: 351–364.

Brothers, D. J. and C. D. Michener. 1974. Interactions in colonies of
primitively social bees III. Ethometry of division of labor in
Lasioglossum zephyrum (Hymenoptera: Halictidae). *J. Comp. Physiol.* **90**: 129–168.

Buckle, G. R. 1984. A second look at queen-forager interactions in
the primitively eusocial halictid, *Lasioglossum zephyrum*. *J. Kansas
Entomol. Soc.* **57**: 1–6.

–. 1985. Increased queen-like behavior of workers in large colonies
of the sweat bee *Lasioglossum zephyrum*. *Anim. Behav.* **33**: 1275–1280.

Cairns, R. B., J. L. Gariepy and K. E. Hood. 1990. Development,
microevolution, and social behavior. *Psychol. Rev.* **97**: 49–65.

Cheney, D. L. and R. M. Seyfarth. 1990. *How Monkeys See the World*.
Chicago: University of Chicago Press.

Claude-Joseph, F. 1926. Recherches biologiques sur les Hyménoptères du Chili (Melliferes). *Ann. Sci. Nat. (Paris) (Zool.)* (10) **9**:
113–268.

Crozier, R. H. 1992. The genetic evolution of flexible strategies.
Am. Nat. **139**: 218–223.

Crozier, R. H., B. H. Smith and Y. Crozier. 1987. Relatedness and
population structure of the primitively eusocial bee *Lasioglossum
zephyrum* (Hymenoptera: Halictidae) in Kansas. *Evolution* **41**:
902–910.

Dunbar, R. I. M. 1992. Neocortex size as a constraint on group size
in primates. *J. Human Evol.* **20**: 469–493.

Dusenbury, D. B. 1992. *Sensory Ecology*. New York: W. H. Freeman
& Co.

Ebmer, A. W. 1976. Liste der Mitteleuropaischen *Halictus-* und
Lasioglossum-arten. *Linzer Biol. Beitr.* **8**: 393–405.

Eickwort, G. C. 1981. Aspects of the nesting biology of five nearctic
species of *Agapostemon* (Hymenoptera: Halictidae). *J. Kansas
Entomol. Soc.* **54**: 337–351.

–. 1985. The nesting biology of the sweat bee *Halictus farinosus* in California, with notes on *Halictus ligatus*, Hymenoptera: Halictidae. *Pan Pac. Entomol.* **61**: 122–137.

–. 1986. First steps into eusociality: The sweat bee *Dialictus lineatulus. Fla. Entomol.* **69**: 742–754.

–. 1992. Foraging behavior of sweat bees: evolution, ecology, and future directions [abstract]. In *Internatl. Workshop on non-*Apis *Bees and their Role as Crop Pollinators* (USDA-ARS Bee Biology & Systematics Laboratory, Logan, UT), p. 16. Logan, UT: USDA.

Eickwort, G. C. and J. Abrams. 1980. Parasitism of sweat bees in the genus *Agapostemon* by cuckoo bees in the genus *Nomada* (Hymenoptera: Halictidae, Anthophoridae). *Pan Pac. Entomol.* **56**: 144–152.

Eickwort, G. C. and K. R. Eickwort. 1969. Aspects of the biology of Costa Rican halictine bees, I. *Agapostemon nasutus* (Hymenoptera: Halictidae). *J. Kansas Entomol. Soc.* **42**: 421–452.

–. 1971. Aspects of the biology of Costa Rican halictine bees, II. *Dialictus umbripennis* and adaptations of its caste structure to different climates. *J. Kansas Entomol. Soc.* **44**: 343–373.

–. 1972. Aspects of the biology of Costa Rican halictine bees, III. *Sphecodes kathleenae*, a social cleptoparasite of *Dialictus umbripennis* (Hymenoptera: Halictidae). *J. Kansas Entomol. Soc.* **45**: 529–541.

Eickwort, G. C., K. R. Eickwort, J. M. Eickwort, J. Gordon and A. Eickwort. 1996. Solitary behavior in a high-altitude population of the social sweat bee *Halictus rubicundus* (Hymenoptera: Halictidae). *Behav. Ecol. Sociobiol.* **38**: 227–233.

Eickwort, G. C., K. R. Eickwort, P. F. Kukuk and F. R. Wesley. 1986. The nesting biology of *Dufourea novaeangliae* (Hymenoptera: Halictidae) and the systematic position of the Dufoureinae based on behavior and development. *J. Kansas Entomol. Soc.* **59**: 103–120.

Evans, F. C. 1986. Bee-flower interactions on an old field in southeastern Michigan. In *The Prairie: Past, Present, and Future.* G. K. Clambey and R. H. Pemble, eds., pp. 103–109. Fargo, N. D.: Tri-College University Center for Environmental Studies.

Evans, H. E. 1958. The evolution of social life in wasps. *Proc. X Int. Congr. Entomol.* (Montreal) **2**: 449–457. Montreal: International Congress of Entomology.

–. 1977. Extrinsic versus intrinsic factors in the evolution of insect sociality. *BioScience* **27**: 613–617.

Evans, H. E. and A. W. Hook. 1986. Nesting behavior of Australian *Cerceris* digger wasps, with special reference to nest reutilization and nest sharing (Hymenoptera, Sphecidae). *Sociobiology.* **11**: 275–302.

Fabre, J. H. 1915. *Bramble Bees and Others.* New York: Dodd, Mead & Co.

Field, J. 1992. Intraspecific parasitism as an alternative reproductive tactic in nest-building wasps and bees. *Biol. Rev.* **67**: 79–126.

Freeman, B. E. 1982. The comparative distribution and population dynamics in Trinidad of *Sceliphron fistularium* (Dahlbom) and *S. asiaticum* (L.) (Hymenoptera: Sphecidae). *Biol. J. Linn. Soc.* **17**: 343–360.

Gadagkar, R. 1991. On testing the role of genetic asymmetries created by haplodiploidy in the evolution of eusociality in the Hymenoptera. *J. Genet.* **70**: 1–31.

Gadgil, M., N. V. Joshi and S. Gadgil. 1983. On the moulding of population viscosity by natural selection. *J. Theor. Biol.* **104**: 21–42.

Geiger, G. 1990. *Evolutionary Instability. Logical and Material Aspects of a Unified Theory of Biosocial Evolution.* New York: Springer-Verlag.

Gottlieb, G. 1992. *Individual Development and Evolution.* Oxford University Press.

Golubnichaya, L. V. and P. G. Moskalenko. 1992. Peculiarities of biology of *Halictus zonulus* Smith, Hymenoptera, Halictidae. *Entomol. Rev.* 71(2): 6–11.

Greenberg, L. 1979. Genetic component of bee odor in kin recognition. *Science (Wash., D. C.)* **206**: 1095–1097.

–. 1988. Kin recognition in the sweat bee, *Lasioglossum zephyrum. Behav. Genet.* **18**: 425–437.

Greenberg, L. and G. R. Buckle. 1981. Inhibition of worker mating by queens in a sweat bee, *Lasioglossum zephyrum. Insectes Soc.* **28**: 347–352.

Hamilton, W. D. 1964. The genetical evolution of social behaviour, Parts I and II. *J. Theor. Biol.* **7**: 1–16; 17–52.

–. 1971a. Geometry for the selfish herd. *J. Theor. Biol.* **31**: 295–311.

–. 1971b. Selection of selfish and altruistic behavior in some extreme models. In *Man and Beast: Comparative Social Behavior.* J. F. Eisenberg and W. S. Dillon, eds., pp. 59–91. Washington, D.C.: Smithsonian Institution Press.

–. 1972. Altruism and related phenomenon, mainly in social insects. *Annu. Rev. Ecol. Syst.* **3**: 193–232.

–. 1987. Kinship, recognition, disease, and intelligence: constraints of social evolution. In *Animal Societies: Theories and Facts.* Y. Itô, J. L. Brown and J. Kikkawa, eds., pp. 81–102. Tokyo: Japan Sci. Soc. Press.

Hardy, A. C. 1954. Escape from specialization. In *Evolution as a Process.* J. Huxley, A. C. Hardy and E. B. Ford, eds., pp. 122–142. London: George Allen & Unwin Ltd.

Hefetz, A. 1987. The role of Dufour's glands secretions in bees. *Physiol. Entomol.* **12**: 243–253.

Hefetz, A., G. Bergström and J. Tengö. 1986. Species, individual and kin specific blends in Dufour's gland secretions of Halictine bees chemical evidence. *J. Chem. Ecol.* **12**: 197–208.

Hefetz, A., M. S. Blum, G. C. Eickwort and J. W. Wheeler. 1978. Chemistry of the Dufour's gland secretion of halictine bees. *Comp. Biochem. Physiol.* **61**: 129–132.

Hölldobler, B. and C. D. Michener. 1980. Mechanisms of identification and discrimination in social Hymenoptera. In *Evolution of Social Behavior: Hypotheses and Empirical Tests.* H. Markl, ed., pp. 35–58. Weinheim: Verlag Chemie.

Hoogland, J. L. and P. W. Sherman. 1976. Advantages and disadvantages of bank swallow (*Riparia*) coloniality. *Ecol. Monogr.* **46**: 33–58.

Itô, Y. 1993. The evolution of polygyny in primitively eusocial polistine wasps with special reference to the genus *Ropalidia*. In *Queen Number and Sociality in Insects*. L. Keller, ed., pp. 171–187. Oxford: Blackwell Scientific Publications.

Jarvis, J. U. M., M. J. O'Riain, N. C. Bennett and P. W. Sherman. 1994. Mammalian eusociality: a family affair. *Trends Ecol. Evol.* **9**: 47–51.

Jeanne, R. L. 1975. The adaptiveness of social wasp nest architecture. *Q. Rev. Biol.* **50**: 267–287.

Kagan, J., J. S. Reznick and N. Snidman. 1988. Biological bases of childhood shyness. *Science (Wash., D.C.)* **240**: 167–171.

Kakutani, T., T. Inoue, M. Kato and H. Ichihashi. 1990. Insect-flower relationship in the campus of Kyoto University, Kyoto: an overview of the flowering phenology and the seasonal pattern of insect visits. *Contr. Biol. Lab., Kyoto Univ.* **27**: 465–521.

Knerer, G. 1969. Synergistic evolution of halictine nest architecture and social behavior. *Can. J. Zool.* **47**: 925–930.

Knerer, G. and P. MacKay. 1969. Bionomic notes on the solitary *Evylaeus oenotherae* (Stevens) (Hymenoptera: Halictinae), a matinal summer bee visiting cultivated Onagraceae. *Can. J. Zool.* **47**: 289–294.

Knerer, G. and C. Plateaux-Quénu. 1967. Usurpation de nids étrangers et parasitisme facultatif chez *Halictus scabiosae* (Rossi) (Insecte Hymenoptere). *Insectes Soc.* **14**: 47–50.

Knerer, G. and M. Schwarz. 1978. Beobachtungen an australischen Furchenbienen (Hymenoptera; Halictinae). *Zool. Anz.* **200**: 321–333.

Kukuk, P. F. 1989. Evolutionary genetics of a primitively eusocial halictine bee, *Dialictus zephyrus*. In *Genetics of Social Evolution*. R. E. Page and M. D. Breed, eds., pp. 183–202. Boulder: Westview Press.

–. 1992. Social interactions and familiarity in a communal halictine bee *Lasioglossum hemichalceum*. *Ethology* **91**: 291–300.

Kukuk, P. F. and R. H. Crozier. 1990. Trophallaxis in a communal halictine bee *Lasioglossum (Chilalictus) erythrurum*. *Proc. Natl. Acad. Sci. U.S.A* **87**: 5401–5404.

Kukuk, P. F. and P. Decelles. 1986. Behavioral evidence of population structure in *Lasioglossum zephyrum* (Hymenoptera: Halictidae). *Behav. Ecol. Sociobiol.* **19**: 233–239.

Kukuk, P. F. and G. C. Eickwort. 1987. Alternative social structures in halictine bees. In *Chemistry and Biology of Social Insects*. J. Eder and H. Rembold, eds., pp. 555–556. Munchen: Verlag J. Peperny.

Kukuk, P. F. and B. P. May. 1988. Dominance hierarchy in the primitively eusocial bee *Lasioglossum (Dialictus) zephyrum*: is genealogical relationship important? *Anim. Behav.* **36**: 1848–1850.

–. 1991. Colony dynamics in a primitively eusocial halictine bee *Lasioglossum (Dialictus) zephyrum* (Hymenoptera: Halictidae). *Insectes Soc.* **38**: 171–188.

Kukuk, P. F. and G. K. Sage. 1994. Reproductivity and relatedness in a communal halictine bee *Lasioglossum (Chilalictus) hemichalceum*. *Insectes Soc.* **41**: 443–455.

Leshner, A. J. 1983. The hormonal responses to competition and their behavioral significance. In *Hormones and Behavior*, B. B. Svare, ed., pp. 339–404. New York: Plenum Press.

Lin, N. and C. D. Michener. 1972. Evolution of sociality in insects. *Q. Rev. Biol.* **47**: 131–159.

Litte, M. 1977. Aspects of the social biology of the bee *Halictus ligatus* in New York state (Hymenoptera, Halictidae). *Insectes Soc.* **24**: 9–36.

McGinley, R. J. 1986. Studies of Halictinae (Apoidea: Halictidae), I: Revision of New World *Lasioglossum* Curtis. *Smithson. Contrib. Zool.* **429**: 1–294.

McCorquodale, D. B. 1989. Soil softness, nest initiation and nest sharing in the wasp *Cerceris antipodes* (Hymenoptera: Sphecidae). *Ecol. Entomol.* **14**: 191–196.

Michener, C. D. 1954. Bees of Panamá. *Bull. Am. Mus. Nat. Hist.* **104**: 1–175.

–. 1966. The bionomics of a primitively social bee, *Lasioglossum versatum* (Hymenoptera: Halictidae). *J. Kansas Entomol. Soc.* **39**: 193–217.

–. 1969. Notes on the nests and life histories of some African halictid bees with descriptions of a new species. *Trans. Am. Entomol. Soc.* **94**: 473–497.

–. 1974. *The Social Behavior of the Bees*. Cambridge: Harvard University Press.

–. 1978a. The classification of halictine bees: tribes and old world nonparasitic genera with strong venation. *Univ. Kans. Sci. Bull.* **51**: 501–538.

–. 1978b. The parasitic groups of Halictidae (Hymenoptera: Apoidea). *Univ. Kans. Sci. Bull.* **51**: 291–339.

–. 1979. Biogeography of the bees. *Ann. Missouri Bot. Gard.* **66**: 277–347.

–. 1985. From solitary to eusocial: Need there be a series of intervening species? *Fortschr. Zool.* **31**: 293–306.

–. 1990. Reproduction and castes in social halictine bees. In *Social Insects: an Evolutionary Approach to Castes and Reproduction*. W. Engels, ed., pp. 77–121. New York: Springer-Verlag.

Michener, C. D. and F. D. Bennett. 1977. Geographical variation in nesting biology and social organization of *Halictus ligatus*. *Univ. Kansas Sci. Bull.* **51**: 233–260.

Michener, C. D., M. D. Breed and W. J. Bell. 1979. Seasonal cycles, nests, and social behavior of some Colombian halictine bees (Hymenoptera; Apoidea). *Rev. Biol. Trop.* **27**: 13–34.

Michener, C. D. and R. B. Lange. 1958. Observations on the behavior of Brasilian halictid bees, III. *Univ. Kans. Sci. Bull.* **39**: 473–505.

Michener, C. D., R. B. Lange, J. J. Bigarella and R. Salamuni. 1958. Factors influencing the distribution of bees' nests in earth banks. *Ecology* **39**: 207–217.

Michener, C. D. and C. A. C. Seabra. 1959. Observations on the behavior of Brasilian halictid bees. VI, Tropical species. *J. Kansas Entomol. Soc.* **32**: 19–30.

Michener, C. D. and B. H. Smith. 1987. Kin recognition in primitively eusocial insects. In *Kin Recognition in Animals.* D. J. C. Fletcher and C. D. Michener, eds., pp. 209–242. Chichester: John Wiley & Sons.

Michener, C. D. and A. Wille. 1961. The bionomics of a primitively social bee, *Lasioglossum inconspicuum. Univ. Kans. Sci. Bull.* **42**: 1123–1202.

Moure, J. C. and P. D. Hurd, Jr. 1987. *An Annotated Catalog of the Halictid Bees of the Western Hemisphere (Hymenoptera: Halictidae).* Washington, D.C.: Smithsonian Institution Press.

Mueller, U. G. 1991. Haplodiploidy and the evolution of facultative sex ratios in a primitively eusocial bee. *Science (Wash., D.C.)* **254**: 442–444.

Nye, W. P. 1980. Notes on the biology of *Halictus (Halictus) farinosus* Smith (Hymenoptera: Halictidae). *Sci. Edu. Administr., Agric. Res. Results, West. Ser.* (U.S. Dep. Agric.) **11**. 29 pp.

Packer, L. 1986a. The biology of a subtropical population of *Halictus ligatus*: IV. A cuckoo-like caste. *J. N. Y. Entomol. Soc.* **94**: 458–466.

–. 1986b. The biology of a subtropical population of *Halictus ligatus* Say (Hymenoptera; Halictidae). II. Male behaviour. *Ethology* **72**: 287–298.

–. 1986c. Multiple-foundress associations in a temperature population of *Halictus ligatus* (Hymenoptera: Halictidae). *Can. J. Zool.* **64**: 2325–2332.

–. 1986d. The social organization of *Halictus ligatus* (Hymenoptera: Halictidae), in southern Ontario Canada. *Can. J. Zool.* **64**: 2317–2324.

–. 1988. The effect of *Bombylius pulchellus* (Diptera, Bombyliidae) and other mortality factors upon the biology of *Halictus ligatus* (Hymenoptera; Halictidae) in southern Ontario. *Can. J. Zool.* **66**: 611–616.

–. 1991. The evolution of social behavior and nest architecture in sweat bees of the subgenus *Evylaeus* (Hymenoptera: Halictidae): a phylogenetic approach. *Behav. Ecol. Sociobiol.* **29**: 153–160.

–. 1992. The social organisation of *Lasioglossum (Dialictus) laevissimum* (Smith) in southern Alberta. *Can. J. Zool.* **70**: 1767–1774.

–. 1993. Multiple foundress associations in sweat bees (Hymenoptera: Halictidae). In *Queen Number and Sociality in Insects.* L. Keller, ed., pp. 215–233. Oxford: Blackwell Scientific Publications.

Packer, L. and G. Knerer. 1985. Social evolution and its correlates in bees of the subgenus *Evylaeus* (Hymenoptera; Halictidae). *Behav. Ecol. Sociobiol.* **17**: 143–150.

–. 1986. The biology of a subtropical population of *Halictus ligatus* Say (Hymenoptera: Halictidae) I. Phenology and social organisation. *Behav. Ecol. Sociobiol.* **18**: 363–375.

Packer, L., B. Sampson, C. Lockerbie and V. Jessome. 1989. Nest architecture and brood mortality in four species of sweat bee (Hymenoptera, Halictidae) from Cape Breton Island Nova Scotia, Canada. *Can. J. Zool.* **67**: 2864–2870.

Pesenko, Y. A. 1983. Filogeneticheskie svyazi mezhdu tribami podsemeistva Halictinae i vnutri triby Nomiodini. *Fauna SSSR, Nasekomye Pereponchatokrylye* **17**: 199 pp.

Plateaux-Quénu, C. 1959. Un nouveau type de société d'insectes: *Halictus marginatus* Brullé (Hym., Apoidea). *Ann. Biol.* **35**: 325–455.

–. 1988. Réalisation expérimentale de grands individus de première couvée chez *Evylaeus calceatus* (Scop.) (Hym., Halictinae): biométrie et caste. *Ann. Sci. Nat., (Paris) (Zool.)* **9**: 263–270.

–. 1992. Comparative biological data in two closely related eusocial species: *Evylaeus calceatus*, Scop., and *Evylaeus albipes*, F., (Hym., Halictinae). *Insectes Soc.* **39**: 351–364.

–. 1993. Flexibilité sociale chez *Evylaeus albipes* (F.) (Hymenoptera, Halictinae). *Actes Coll. Ins. Soc.* **8**: 127–134.

Plateaux-Quénu and L. Plateaux. 1994. Polyphenism of *Halictus (Seladonia) tumulorum* (L.) (Hymenoptera, Halictinae). *Insectes Soc.* **41**: 219–222.

Plateaux-Quénu, C., L. Plateaux and L. Packer. 1989. Biological notes on *Evylaeus villosulus* (K.) (Hymenoptera, Halictidae), a bivoltine, largely solitary halictine bee. *Insectes Soc.* **36**: 245–263.

Quénu, C. 1957. Sur les femelles d'été de *Halictus scabiosae* (Rossi). *C. R. Acad. Sci Paris.* **244**: 1073–1076.

Radchenko, V. G. 1980. The nesting of *Nomioides minutissimus* (Rossi) (Hymenoptera, Halictidae). *Entomol. Rev.* **58**: 71–74.

Rau, P. 1933. *The Jungle Bees and Wasps of Barro Colorado Island.* St. Louis: Van Hoffmann.

Richards, M. H. 1994a. Social evolution in the genus *Halictus*: a phylogenetic approach. *Insectes Soc.* **41**: 315–325.

–. 1994b. Social responses to changing environments: reproductive behaviour and reproductive options in a primitively eusocial sweat bee. Ph. D dissertation, York University (Ontario).

Richards, M. H. and L. Packer. 1994. Trophic aspects of caste determination in *Halictus ligatus*, a primitively eusocial sweat bee. *Behav. Ecol. Sociobiol.* **34**: 385–391.

Robinson, M. H. 1985. Predator-prey interactions, informational complexity, and the origins of intelligence. *J. Wash. Acad. Sci.* **75**: 91–104.

Rosenheim, J. A. 1990. Density dependent parasitism and the evolution of aggregated nesting in the solitary Hymenoptera. *Ann. Entomol. Soc. Am.* **83**: 277–286.

Roubik, D. W. 1989. *Ecology and Natural History of Tropical Bees.* Cambridge: Cambridge University Press.

Rozen, J. G., Jr. 1993. Nesting biologies and immature stages of rophitine bees (Halictidae) with notes on the cleptoparasite *Biastes* (Anthophoridae) (Hymenoptera: Apoidea). *Am. Mus. Nov.* **3066**: 1–28.

Sakagami, S. F. 1974. Sozialstruktur und Polymorphismus bei Furchen- oder Schmalbienen (Halictidae). In *Sozialpolymorphismus bei Insekten.* G. H. Schmidt, ed., pp. 257–293. Stuttgart: Wissenschaftliche Verlagsgesellschaft MBH.

–. 1977. Seasonal change of nest survival and related aspects in an aggregation of *Lasioglossum duplex* (Dalla Torre), a eusocial halictine bee (Hymenoptera: Halictidae). *Res. Pop. Ecol.* **19**: 69–86.

–. 1980. Bionomics of the halictine bees in northern Japan. I. *Halictus (Halictus) tsingtouensis* (Hymenoptera, Halictidae), with notes on the number of origins of eusociality. *Kontyû* **44**: 525–536.

Sakagami, S. F., A. W. Ebmer, T. Matsumura and Y. Maeta. 1982. Bionomics of the halictine bees in northern Japan: 2. *Lasioglossum sakagamii*, Hymenoptera, Apoidea, Halictidae), with taxonomic notes on allied species. *Kontyû* **50**: 198–211.

Sakagami, S. F. and H. Fukuda. 1989. Nest founding and nest survival in a eusocial halictine bee, *Lasioglossum duplex*: additional observations. *Res. Popul. Ecol.* **31**: 139–151.

Sakagami, S. F., Y. Hirashima and Y. Ohe. 1966. Bionomics of two new Japanese halictine bees (Hymenoptera, Apoidea). *J. Fac. Agric., Kyushu Univ.* **13**: 673–703.

Sakagami, S. F., S. Laroca and J. S. Moure. 1967. Wild bee biocoenotics in São Jose dos Pinhais (RP), South Brazil, preliminary report. *J. Fac. Sci., Hokkaido Univ., Ser. VI, Zool.* **16**: 253–291.

Sakagami, S. F., T. Matsumura and Y. Maeta. 1985. Bionomics of the halictine bees in northern Japan: III. *Lasioglossum allodalum*, with remarks on the serially arranged cells in the halictine nests. *Kontyû* **53**: 409–419.

Sakagami, S. F. and C. D. Michener. 1962. *The Nest Architecture of the Sweat Bees*. Lawrence: University of Kansas Press.

Sakagami, S. F. and J. S. Moure. 1967. Additional observations on the nesting habits of some Brazilian halictine bees (Hymenoptera, Apoidea). *Mushi* **40**: 119–138.

Sakagami, S. F. and M. Munakata. 1966. Bionomics of a Japanese halictine bee, *Lasioglossum pallidum* (Hymenoptera: Apoidea). *J. Kansas Entomol. Soc.* **39**: 370–379.

–. 1972. Distribution and bionomics of a transpalearctic eusocial halictine bee, *Lasioglossum (Evylaeus) calceatum*, in northern Japan, with reference to its solitary life cycle at high altitude. *J. Fac. Sci., Hokkaido Univ., Ser. VI, Zool.* **18**: 411–439.

Sakagami, S. F. and T. Okazawa. 1985. A populous nest of the halictine bee *Halictus (Seladonia) lutescens* from Guatemala (Hymenoptera, Halictidae). *Kontyû* **53**: 645–651.

Scarr, S. and K. McCartney. 1983. How people make their own environments: a theory of genotype → environmental effects. *Child Devel.* **54**: 424–435.

Schmalhausen, I. I. 1949. *Factors of Evolution* (1986 reprint). Chicago: University of Chicago Press.

Shors, T. J., C. Weiss and R. F. Thompson. 1992. Stress-induced facilitation of classical conditioning. *Science (Wash., D.C.)* **257**: 537–539.

Sick, M., M. Ayasse, J. Tengö, W. Engels, G. Lubke and W. Francke. 1994. Host-parasite relationships in six species of *Sphecodes* bees and their halictid hosts: nest intrusion, intranidal behavior and Dufour's gland volatiles (Hymenoptera, Halictidae). *J. Insect Behav.* **7**: 101–118.

Sitdikov, A. A. 1988. Nesting of the bee *Halictus quadricinctus* (F.) (Hymenoptera, Halictidae) in the Udmurt ASSR. *Entomol. Rev.* **66**: 66–77.

Smith, B. H. 1987. Effects of genealogical relationship and colony age on the dominance hierarchy in the primitively eusocial bee *Lasioglossum zephyrum*. *Anim. Behav.* **35**: 211–217.

–. 1988. Genealogical relationship and social dominance in bees: a reply to Kukuk and May. *Anim. Behav.* **36**: 1850–1851.

Smith, B. H. and M. Ayasse. 1987. Kin-based male mating preferences in two species of halictine bee. *Behav. Ecol. Sociobiol.* **20**: 313–318.

Smith, B. H., R. G. Carlson and J. Frazier. 1985. Identification and bioassay of macrocyclic lactone sex pheromone of the halictine bee *Lasioglossum zephyrum*. *J. Chem. Ecol.* **11**: 1447–1456.

Smith, B. H. and C. Weller. 1989. Social competition among gynes in halictine bees: the influence of bee size and pheromones on behavior. *J. Insect Behav.* **2**: 397–411.

Smith, B. H. and J. W. Wenzel. 1988. Pheromonal covariation and kinship in social bee *Lasioglossum zephyrum* (Hymenoptera: Halictidae). *J. Chem. Ecol.* **14**: 87–94.

Staw, B. M., and J. Ross. 1989. Understanding behavior in escalation situations. *Science (Wash., D.C.)* **246**: 216–220.

Stebbins, G. L. and D. L. Hartl. 1988. Comparative evolution: latent potentials for anagenetic advance. *Proc. Natl. Acad. Sci. U.S.A.* **85**: 5141–5145.

Stoeckhert, F. K. 1954. Fauna Apoideorum Germaniae. *Abh. Bayer. Akad. Wiss.* **65**: 1–87.

Taylor, P. D. 1992. Altruism in viscous populations – an inclusive fitness model. *Evol. Ecol.* **6**: 352–356.

Tengö, J., M. Sick, M. Ayasse, W. Engels, B. G. Svensson, G. Lubke and W. Francke. 1992. Species specificity of Dufour's gland morphology and volatile secretions in kleptoparasitic *Sphecodes* bees (Hymenoptera: Halictidae). *Biochem. Syst. Ecol.* **20**: 351–362.

Thomas, F. 1925. *The Environmental Basis of Society*. New York: The Century Co.

Trivers, R. 1985. *Social Evolution*. Menlo Park: Benjamin/Cummings.

Vasić, Z. 1967. *Halictus quadricinctus* et le probleme de la polygynie. *Bull. Mus. Hist. Nat.* **22**: 181–187.

Waddington, C. H. 1940. *Organizers and Genes*. Cambridge University Press.

Wcislo, W. T. 1984. Gregarious nesting of a digger wasp as a 'selfish herd' response to a parasitic fly (Hymenoptera: Sphecidae; Diptera: Sacrophagidae [sic]). *Behav. Ecol. Sociobiol.* **15**: 157–160.

–. 1987a. The roles of seasonality, host synchrony, and behaviour in the evolutions and distributions of nest parasites in Hymenoptera (Insecta), with special reference to bees (Apoidea). *Biol. Rev.* **62**: 515–543.

–. 1987b. The role of learning in the mating biology of a sweat bee *Lasioglossum zephyrum*, Hymenoptera: Halictidae. *Behav. Ecol. Sociobiol.* **20**: 179–186.

–. 1989. Behavioral environments and evolutionary change. *Annu. Rev. Ecol. Syst.* **20**: 137–169.

–. 1990. Parasitic and courtship behavior of *Phalacrotophora halictorum* (Diptera: Phoridae) at a nesting site of *Lasioglossum figueresi* (Hymenoptera: Halictidae). *Rev. Biol. Trop.* **38**: 205–209.

–. 1992a. Attraction and learning in mate-finding by solitary bees, *Lasioglossum (Dialictus) figueresi* Wcislo and *Nomia triangulifera* Vachal (Hymenoptera: Halictidae). *Behav. Ecol. Sociobiol.* **31**: 139–148.

–. 1992b. Nest localization and recognition in a solitary bee, *Lasioglossum (Dialictus) figueresi* Wcislo, Hymenoptera: Halictidae), in relation to sociality. *Ethology* **92**: 108–123.

–. 1993. Communal nesting in a north American pearly-banded bee, *Nomia tetrazonata*, with notes on nesting behavior of *Dieunomia heteropoda* (Hymenoptera: Halictidae: Nomiinae). *Ann. Entomol. Soc. Amer.* **86**: 813–821.

–. 1996a. Invasion of nests of *Lasioglossum imitatum* by a social parasite, *Paralictus asteris* (Hymenoptera: Halictidae). *Ethology*, in press.

–. 1996b. Rates of parasitism in relation to nest site in bees and wasps (Hymenoptera: Apoidea). *J. Insect Behav.* **9**: 643–656.

Wcislo, W. T. and J. H. Cane. 1996. Resource utilization by solitary bees (Hymenoptera: Apoidea) and exploitation by their natural enemies. *Annu. Rev. Entomol.* **41**: 257–286.

Wcislo, W. T. and M. S. Engel. 1996. Social behavior and nest architecture of nomiine bees (Hymenoptera: Halictidae: Nomiinae). *J. Kansas Entomol. Soc.*, in press.

Wcislo, W. T., B. S. Low and C. J. Karr. 1985. Parasite pressure and repeated burrow use by different individuals of *Crabro* (Hymenoptera: Sphecidae; Diptera: Sarcophagidae). *Sociobiology.* **11**: 115–125.

Wcislo, W. T., M. J. West-Eberhard and W. G. Eberhard. 1988. Natural history and behavior of a primitively social wasp, *Auplopus semialatus*, and its parasite, *Irenangelus eberhardi* (Hymenoptera: Pompilidae). *J. Insect Behav.* **1**: 247–260.

Wcislo, W. T., A. Wille and E. Orozco. 1993. Nesting biology of tropical solitary and social sweat bees, *Lasioglossum (Dialictus) figueresi* Wcislo and *L. (D.) aeneiventre* (Friese) (Hymenoptera: Halictidae). *Insectes Soc.* **40**: 21–40.

West-Eberhard, M. J. 1979. Sexual selection, social competition, and evolution. *Proc. Am. Phil. Soc.* **123**: 222–234.

–. 1986. Alternative adaptations, speciation, and phylogeny (a review). *Proc. Natl. Acad. Sci. U.S.A.* **83**: 1388–1392.

–. 1992a. Behavior and evolution. In *Molds, Molecules and Metazoa.* P. R. Grant and H. S. Horn, eds., pp. 57–79. Princeton: Princeton University Press.

–. 1992b. Genetics, epigenetics, and flexibility: a reply to Crozier. *Am. Nat.* **139**: 224–226.

Wille, A. and E. Orozco. 1970. The life cycle and behavior of the social bee *Lasioglossum (Dialictus) umbripenne* (Hymenoptera: Halictidae). *Rev. Biol. Trop.* **17**: 199–245.

Wilson, D. S., G. B. Pollock and L. A. Dugatkin. 1992. Can altruism evolve in purely viscous populations? *Evol. Ecol.* **6**: 331–341.

Zuckerkandl, E. and R. Villet. 1988. Concentration-affinity equivalence in gene regulation: convergence of genetic and environmental effects. *Proc. Natl. Acad. Sci. U.S.A.* **85**: 4784–4788.

16 · Intrinsic and extrinsic factors associated with social evolution in allodapine bees

M. P. SCHWARZ, L. X. SILBERBAUER AND P. S. HURST

ABSTRACT

The tribe Allodapini (Family Apidae) contains about 200 described species and is largely restricted to sub-Saharan Africa, the Indo-Oriental region and Australia. Although some species exhibit only solitary (subsocial) behavior, most species show at least transient phases of social organization, ranging from prereproductive assemblages to forms of eusociality involving morphologically distinct and sometimes flightless queens. Sociality varies widely between species within genera, between populations within species, and between colonies within populations, providing a wealth of opportunities for comparative approaches to social evolution. We discuss a variety of intrinsic and extrinsic factors that have the potential to influence social evolution. The allodapine trait of communal, progressive rearing of brood could affect social evolution in a variety of ways. Lack of cell partitions means that immatures are largely dependent on the continuing presence of adults for defense against parasites and predators. This may select for cooperative nesting if brood can be defended by one adult while another forages. Rearing of brood in communal chambers could select for kin association among adult females because of the potential for nestmate parasitism and the inability to restrict parental care to a nestmate's own offspring. Progressive rearing of brood creates the potential for egg production to be decoupled from tasks involved in larval rearing, so that specialization in tasks such as guarding, foraging and nursing could develop after reproduction has been completed. However, the kinds of societies that evolve, and whether they evolve at all, will also depend on a number of extrinsic factors, including the physical characteristics of nesting substrates and predator and parasite pressure. The spatial distribution of nesting substrate could influence selective pressures exerted by enemies at the nest and, combined with the physical durability of the substrate, can also influence how multifemale associations arise and for how long colonies

are able to persist. In the two allodapine species that have been electrophoretically assayed, intracolony relatedness is surprisingly high, given the frequency of plurimatry (multiple queens). High levels of relatedness in these species are due to several factors, including active kin association during nest founding, semisociality in reused nests and, possibly, stochastic effects leading to genetic drift. At least several species show female-biased sex allocation patterns. We argue that such bias has the potential to facilitate eusociality in species with generational overlap during egg production. Finally, we suggest a number of research directions that we believe address the most interesting problems and are, at the same time, tractable.

INTRODUCTION

The apid tribe Allodapini comprises 13 genera, with about 180 described species from sub-Saharan Africa and Australia, and another 20 or so species from the Indo-Oriental region (Michener 1965a, 1971, 1975; Reyes 1991, 1993; Cardale 1993). Most species nest in hollow or pithy plant stems and rear their brood progressively in a communal chamber, which is not divided into cells (Michener 1974). Social organization varies enormously, both within and between species. Although many species frequently show eusocial or semisocial forms of organization, in other species multifemale societies may arise only infrequently or be restricted to brief periods after brood emergence and prior to foundress dispersal (Michener 1990).

In all allodapine species where social behavior has been observed, sociality is facultative. With at least one exception (Houston 1977), the social roles adopted by individual females are not constrained by morphology, but depend rather on proximate factors such as intracolony interactions, the time of the year, and stochastic effects such as adult mortality. This within- and between-species variation in sociality creates unparalleled opportunities for comparative approaches to social evolution, but is a resource that has been largely untapped by researchers. Nevertheless, a variety of factors have been hypothesized to play key roles in allodapine social evolution (Michener 1971, 1974, 1985, 1990; Sakagami 1960; Schwarz 1988a; Wilson 1971). It is therefore surprising that few attempts have been made to empirically assess such hypotheses. This lack of assessment is due, in part, to two factors: (1) we have very inadequate knowledge of the behavioral ecology of all but a few species; (2) there has been no attempt to systematically investigate ecological parameters that are relevant to hypotheses about social evolution.

COMPARATIVE APPROACHES TO SOCIAL EVOLUTION IN ALLODAPINES

Comparative approaches to social evolution can be taken at several levels: (1) the distribution of social traits within a known phylogeny may be used to explore the roles of phylogenetic history and convergent evolution as factors influencing sociality; (2) differences in sociality between conspecific populations may be used to explore the role of the environment in shaping sociality where phylogenetic constraints are minimal; (3) between-colony variation may be used to explore the consequences of different social organizations within a single population; and (4) variation in the social strategies adopted by individuals may be used to examine the inclusive fitnesses associated with these strategies. The current lack of a phylogeny for the allodapine genera and subgenera largely precludes the first of these approaches for investigating variation within the tribe, although useful comparisons can be made between the allodapines and the related tribes Xylocopini, Ceratinini and Manueliini. The most detailed comparative approaches to allodapine sociality have been made at the population and colony levels and much of our chapter will focus on these studies.

In practice, it is often difficult or impossible to separate environmental and genetic factors influencing sociality without recourse to experiments. For example, assume that the ontogenetic development of eusociality depends on nesting substrates lasting for more than one generation, and that two allopatric species nest in substrates that differ in durability; eusociality may be frequent in the species using the more durable substrate, but absent in the other species. Without experimental manipulation it will be difficult to determine whether presence or absence of eusociality can be regarded as a species characteristic, or just a consequence of nesting substrate. This kind of problem is acute in at least a number of Australian allodapine species, and is likely to occur in other regions as well.

Descriptions of allodapine social organization are frequently applied at the species level, but may also be applied to individual colonies (Michener 1971, 1974). A species may be classified as primitively eusocial if most or many surviving colonies ultimately develop reproductively differentiated behavioral castes based on generational overlap, but at the same time individual colonies may be categorized on the basis of their current social constitution. In solitary-founding species, a newly founded nest may be classed as subsocial until the first brood emerge; if these adult daughters then function as workers the colony may become eusocial; finally, if the mother then dies but her daughters become reproductively differentiated the colony may become semisocial. Highly plastic forms of sociality are characteristic of many allodapine species (Michener 1965b, 1971; Schwarz 1986).

The social behavior of allodapines has been comprehensively reviewed by Michener (1974, 1990) and we do not attempt a similar review here. Instead, we explicitly focus on a variety of intrinsic and extrinsic factors that have the potential to influence social evolution in allodapines by facilitating or inhibiting colonial nesting, generational overlap and kin association. By 'intrinsic' factors we mean developmental or life-history traits that are characteristic of the tribe. These include progressive rearing, lack of cells, stem nesting, and extended adult longevity. We use the term 'extrinsic' to refer to environmental factors (both physical and biological) that may influence behavior; these include enemies at the nest, durability and distribution of nesting substrates, climate, and so on. Finally, we also consider the role of genetic relatedness on social evolution. Unlike Evans (1977), we do not treat this as an intrinsic factor since relatedness among interactants may be labile and change with behavioral acts such as nest-switching, kin recognition and sex-ratio bias.

INTRINSIC FACTORS: THE EFFECT OF LIFE-HISTORY TRAITS ON SOCIAL EVOLUTION

Allodapine bees differ from all other bee taxa in that brood are reared progressively and communally in unbranched burrows in hollow or pithy stems (Michener 1974). No species construct cell partitions, and in all genera except *Halterapis* brood are fed more or less continuously throughout their development by adult females. In *Halterapis* each egg is provided with sufficient food (in the form of a pollen ball) to allow complete development, although brood are not separated by cell partitions (Michener 1974). Progressive rearing in a communal burrow has several consequences for social evolution, which we discuss below.

Lack of cell partitions

Because brood are not protected by cell partitions, physical protection from predators and parasites principally involves nest crypsis and adult females guarding the nest (Michener 1971). In single-female nests, brood are unprotected when the adult engages in extranidal activities; if there is strong predator or parasite pressure, this may create the potential for benefits from cooperative brood defense. For example, if two females nest cooperatively, one could defend the nest while the other forages.

Although there are no detailed studies of predators at the nest in allodapines, ants have been implicated as major predators of several species in the Australian genus *Exoneura* (Michener 1974, 1983; Schwarz 1986, 1988a: Schwarz and O'Keefe 1991a). Michener (1971) suggested that predation by ants is a major factor in cooperative nesting in the African allodapine *Allodapula melanopus*, and he found that solitary females suffer higher brood-predation rates than social females. Montane populations of *E. bicolor*, *E. richardsoni* and *E. bicincta* nest in tree-fern fronds, which are also used as nesting substrates by a variety of ant species. Foragers of these species frequently cluster around *Exoneura* entrances. Ants are able to remove brood rapidly if adult bees are not present (M. P. Schwarz, personal observations) and the benefits of cooperative nesting in *E. bicolor* may derive largely or entirely from prevention of brood loss (Schwarz 1994). Lack of cell partitions could further increase vulnerability of brood to predators because diffusion of olfactory signals from pollen and nectar is not retarded by cell partitions or cell lining.

A number of allodapine species show traits suggesting that predation at the nest is significant. *Exoneura* females produce mandibular gland secretions which contain a variety of compounds that may be used in defense and have been shown to be effective in repeling ants (Cane and Michener 1983). Melna and Schwarz (1994) found that in prereproductive colonies of *E. bicolor*, the most clearly distinguished behavioral caste was that of guard. Guards performed very few activities other than guarding. In most multifemale nests of an Oriental species, *Braunsapis hewitti*

(previously reported as *B. sauteriella* and *Allodape marginata*), the queen was the only bee to guard, and this behavior accounted for more than 90% of the time spent in all recorded behaviors (Maeta *et al.* 1992). Mason (1988) examined behavior in two African species, *Allodape exoloma* and *Braunsapis foveata*. She found that in *A. exoloma* females defined as queens did most of the guarding, whereas in *B. foveata* queens spent the least amount of time guarding the nest entrance. The existence of guarding specialization in these genera, even when colonies may contain only two or three individuals, suggests that social evolution has occurred in environments where there is a strong need to defend brood against enemies at the nest.

It is also possible that guarding specialization in allodapines enhances mean fitness of nestmates other than by brood defence. M. P. Schwarz (unpublished data) found that female *E. bicolor* are attacked by an ichneumonid wasp that parasitizes guarding females by inserting its ovipositor between abdominal tergites of guarding bees at the nest entrance. Guarding specialization by *E. bicolor* females may allow guards to act as parasite 'sinks', so that parasitization is restricted to only one behavioral caste within a colony. Silberbauer (1994) has shown that intraspecific pollen robbery may be important in heathland populations of *E. bicolor*, and argues that guards help defend against robbers. Hogendoorn and Leys (1993) have shown that guarding behavior also protects colonies from conspecific pollen robbery in the xylocopine bee *Xylocopa sulcatipes*.

Progressive rearing

Allodapine bees differ strikingly from mass-provisioning bees in that larvae are fed throughout their development by adult females, leading to extended contact between adults and brood. In this section we discuss two consequences that progressive rearing may have for allodapine social evolution: (1) the potential to stockpile eggs and temporally disengage egg production from tasks associated with larval rearing; and (2) selection for adult longevity.

Egg-stockpiling
Because allodapine larvae are fed progressively throughout their development, oviposition does not require the prior construction of a cell or provision of a food mass. In at least *Exoneura*, and probably other genera, egg development from oviposition to larval eclosion may take several weeks. In some species, eggs are laid intermittently over a long period of time, so that nests contain a wide range of

immature instars (see, for example, Michener 1962, 1971). However, in other species such as *E. bicolor* and *E. richardsoni*, eggs are laid rapidly over a short period of time, accumulating as a clutch before larval eclosion begins. This 'stockpiling' of eggs may effectively allow egg production to be temporally disengaged from tasks associated with larval rearing.

In species that lay eggs rapidly over a short period of time, few or no brood-rearing tasks are available for worker-like females during the oviposition period. The lack of cells and burrow linings in allodapine nests, and utilization of pithy or hollow stems, which require little burrowing effort, reduce the tasks that would otherwise be available to workers at this time. Two possible consequences arise from this situation: first, non-reproductive workers may be redundant during the egg-laying phase; second, benefits from non-reproductive task specialization could become decoupled from reproductive activity (Schwarz 1987). In mass-provisioning bees, cell construction, food-provisioning and egg-laying need to be carried out concurrently, or in very rapid succession, creating the potential for increased efficiency in the handling of materials and experience with particular task-specific environments. In mass-provisioners, reproduction by certain individuals could entail trade-offs with their ability to carry out other tasks; for example, gravid females may be inefficient foragers, or the need to track the distribution and state of provisioned cells could make reproductives inefficient as guards. In allodapine species that can stockpile eggs, specialization in foraging, guarding and nursing roles could be still be undertaken, but the onset of such specialization could be delayed until egg production has been completed.

Schwarz (1987) argued that egg-stockpiling may explain why newly founded colonies of *E. bicolor* are quasisocial but colonies in reused nests are semisocial. In newly founded nests, cofoundresses commence laying eggs immediately after nest initiation and egg production is largely completed before larval eclosion begins. This situation differs from reused nests, where eggs are produced by reproductively dominant females over winter: in such nests, larval eclosion commences shortly after spring activity commences, and the requirements of larval rearing create a suite of non-reproductive tasks. Therefore, in reused nests the existence of a range of non-reproductive tasks overlaps with opportunities for reproduction by subordinate females, creating the potential for efficiency trade-offs between reproductive and non-reproductive roles.

Adult longevity

Because allodapine brood are progressively reared, adults must be present in the nest until brood reach at least a post-feeding stage. Along with the need for parental females to protect brood until adult eclosion, this factor could select for generational overlap and therefore opportunities for eusociality. Earlier studies frequently assumed that eusociality in the allodapines was closely associated with progressive rearing, and that the intensity and prolonged nature of adult–brood contact suggested that eusociality had evolved via the 'subsocial route' (Wilson 1971). Indeed, survival of parental females until brood emergence, followed by a period of matrifilial cohabitation before offspring dispersal, is ubiquitous in allodapines.

Adult female longevity is high in at least several allodapine species where sufficient research has allowed longevity to be determined. Maeta *et al.* (1985) reported one female of *B. hewitti* (published as *B. sauteriella*) surviving as an adult in laboratory conditions for approximately 16 months after field collection (after which she was killed). Hurst (1993) found that in *E. bicolor* some females can survive until a second winter, giving them an adult age in excess of 16 months. These extreme forms of adult longevity create opportunities for eusociality, in that older females could produce eggs in a second year of activity after their first brood have emerged as adults.

Although progressive rearing may seem to be associated with adult longevity, evidence from other xylocopine tribes suggests that long life spans in the Allodapini are not be directly attributable to progressive rearing. In the sister tribe xylocopini (which are mass-provisioning bees), adult lifespans as long as two to three years are probable, although the longest definite record is 16 months (Michener 1990). In the Ceratinini, another sister tribe to the Allodapini, *Ceratina japonica* and *C. flavipes* females have been found to live for about three years (Sakagami and Maeta 1977). In ceratinines, brood are initially provided with individual food masses in closed cells, but cell partitions are removed and rebuilt by adults prior to brood maturity, and finally destroyed by the newly emerged adults.

Therefore, the tribes Allodapini, Ceratinini, Xylocopini and probably Manueliini (Daly *et al.* 1987) all contain species showing marked adult longevity as well as primitively social behavior, but only the Allodapini are true progressive provisioners. Consequently, there is no strong evidence that adult longevity depends on progressive rearing. Indeed, it seems likely that extended adult longevity is an ancestral trait for the Xylocopinae, and that progressive provisioning by allodapines evolved later.

Communal rearing

In all allodapine species, the brood of all nestmates are kept in a single undivided chamber where (except in *Halterapis*) they are groomed and fed by adult females. In the following sections we discuss the implications of such communal rearing for parental care, nestmate parasitism and inquilinism.

Implications for parental care

Allodapine larvae show considerable complexity in their morphology, with long setae and frequently with fleshy protuberances from the head and/or body. These features could be adaptations to living in an environment in which there is frequent contact with adults and other immatures (Michener 1974). In *E. bicolor*, and probably other species, larvae are capable of limited movement within the nest (Schwarz 1986). In many species, two or more larvae have been observed feeding from a common food mass and may compete for access to food (Michener 1971, 1990). This larval behavior suggests that it may be difficult or impossible for adults within multifemale nests to restrict food provision and other forms of parental investment to their own offspring. In cooperatively nesting species, this factor could select for kin-association of adult females, since any spill-over of parental care to nestmates' offspring would result in indirect fitness gains (Schwarz 1988a).

Nestmate parasitism

Communal rearing also creates a potential for nestmate parasitism. In multifemale nests, each female has direct access to her nestmates' brood, creating opportunities for brood ejection and oophagy. Oophagy has not been directly observed in non-inquiline allodapines, but in *E. bicolor* egg remnants are occasionally found in the crops of females from multifemale nests (M. P. Schwarz, unpublished data). Oophagy has been recorded in the sister tribes Ceratinini and Xylocopini (Michener 1990). Schwarz (1988a) has argued that nestmate parasitism may increase selection for kin association in social allodapines, since parasitized females would ultimately help rear related offspring. This possibility is discussed later in regard to intra-colony relatedness.

Inquilinism

Finally, communal rearing creates opportunities for inquilinism. Inquilinism has evolved repeatedly within the Allodapini, with parasitic species recorded from seven genera (Michener 1965a, 1970a, 1971; Reyes 1991; Reyes and Michener 1990; Reyes and Sakagami 1990; Batra *et al.* 1993). Allodapine inquilines mostly parasitize hosts with which they appear to be closely related phylogenetically (Michener 1970a, 1983). Communal rearing could facilitate the evolution of inquilinism for at least two reasons: first, host brood are not protected by sealed cells, enabling the inquiline to easily remove them or introduce its own eggs; second, food brought into the nest by the host is freely available to the inquiline. Inquilines are rarely caught on flowers and are thought to subsist on pollen and nectar stored in the host's nest (Michener 1965b, 1970a, 1983). In at least one example, an adult inquiline has been observed to aggressively solicit trophallaxis from host females (Batra *et al.* 1993). If this solicitation was unsuccessful, the inquiline utilized food stored on larvae.

It is possible that inquilinism is linked to the frequency of cooperative nesting among the hosts. For example, if cooperative nesting is frequent in a host species, then tolerance for adult nestmates could be exploited by inquilines if they mimic or adsorb the host's recognition labels; this is thought to be the case in parasitized colonies of *Braunsapis mixta* (Batra *et al.* 1993). Inquilinism involving hosts with larger colonies could also be favored because more resources would be available for exploitation. Conversely, if larger colonies can more effectively repel inquilines, it is possible that cooperative nesting in host species represents, in part, a response to inquiline pressure. Michener (1971) found that in one location 47% of *Braunsapis fascialis* nests were parasitized by the inquiline *Nasutapis straussorum*. About one-third of *B. fascialis* nests collected by Michener contained more than one adult host female. Reyes and Michener (1990) found that the inquiline *Braunsapis breviceps* was found mostly in larger, and probably older, colonies of its host, *B. hewitti*. Two Australian inquiline species, *Inquilina excavata* and *I. schwarzi*, attack *Exoneura* species with marked levels of sociality (Michener 1965b; Cane and Michener 1983; Schwarz 1986). Work is needed to investigate whether cooperative nesting in host species enhances defense against inquilines. We believe that this could be best done by experimentally challenging host colonies of varying sizes with inquilines.

EXTRINSIC FACTORS: THE EFFECTS OF ENVIRONMENT ON SOCIAL EVOLUTION

In the preceding sections we have discussed three allodapine life-history traits that may influence social evolution. However, the effect of any such intrinsic factors will also depend on extrinsic factors deriving from the physical and biological environments. For example, vulnerability of brood to predators will be affected not only by the lack of cell partitions, but also by the number of predators that encounter the nest. In the next two sections we consider two extrinsic factors that may influence allodapine sociality: the characteristics of pithy or hollow stems used as nesting substrates, and climatic factors that may influence seasonal activity and voltinism.

Pithy or hollow stems as nesting substrates

Although a wide variety of bee taxa nest in hollow or pithy stems (Michener 1974), species in the Xylocopinae are the only taxa to exhibit sociality (an exception being the Australian colletid bee *Amphylaeus morosus,* which uses the same nesting substrates as *E. bicolor* (Spessa and Schwarz 1994)). Hollow or pithy stems are likely to entail very different selective environments from those associated with ground- or cavity-nesting. Michener (1970b) has comprehensively reviewed the use of stems and twigs by African allodapines, although he did not investigate the characteristics of such substrates as selective agents affecting social evolution. In the following sections we examine three ways in which stem-nesting may influence sociality.

Limitations on colony size imposed by nesting substrates
The use of hollow or pithy stems as a nesting substrate may limit maximum colony size by setting an upper limit to how many individuals can occupy an unbranched tube and still function efficiently. In the cavity- or shelter-nesting honeybees, colonies frequently contain in excess of 10 000 adult females. In ground-nesting bees, nests can potentially become very large, and colony sizes in excess of 100 adult females have been recorded for both eusocial species (Michener 1974) and presumably communal species (T. F. Houston, quoted in Walker 1986). Although such nests may have only a single entrance, tunnels may become highly branched below the surface. However, in stem-nesting species there is a comparatively small upper limit to the number of brood and workers that can be accommodated

within a single hollow stem. Two constraints may operate to set this limit: (1) stems that have become hollowed for long distances could lose structural resilience and suffer increasing risk of breaking under pressure from wind or rain; and (2) because the nest consists of a single, unbranched burrow, large colony size is likely to lead to 'clogging' of the nest, as adults will need to negotiate passing each other with increasing frequency.

Hollow or pithy stems also deteriorate with age, limiting the number of generations that can utilize a nest before it loses structural integrity, thereby continually resetting colony development back to the founder stage. Maeta *et al.* (1984) found that colonies of *Braunsapis hewitti* (reported as *B. sauteriella*) tended to be larger (and therefore probably older) when nesting in bamboo compared with the herbaceous *Stachytarpheta jamaicensis.* This difference was probably related to the greater durability of bamboo; the authors recommended that the effect of nesting substrates is so important that data from different plants should not be mixed when analyzing colony makeups. Silberbauer and Schwarz (1995) found that heathland colonies of *Exoneura bicolor* nesting in the shrub *Melaleuca squarrosa* were larger (and probably older) than colonies in flower scapes of the grasstree *Xanthorrhoea minor.* Again, it is likely that this reflects a difference in substrate durability, which could affect the potential for colonies to increase over time without the need for dispersal. The largely solitary allodapine *E. (Exoneurella) lawsoni* frequently nests in the narrow stems of herbaceous weeds (Michener 1964), which are unlikely to remain structurally suitable for nesting for much more than one year. Another species in the same subgenus, *E. tridentata,* nests in beetle burrows in the dead branches of the semiarid-zone trees *Halyctryon oleiofolium* and *Acacia papyrocarpa.* Branches of these slow-growing trees may be quite thick (over 3 cm in diameter) and deteriorate very slowly in arid regions, so that nests may be reused for many years. *E. tridentata* is unusual among allodapines in that colonies can become very large (up to 35 females), with morphological variation and effectively flightless queenlike females (referred to as 'majors' by Houston (1977)). It is likely that the evolution of such a social system is contingent upon use of nesting substrates that can be used for long periods of time.

Nest density and behavior of parasites and predators
Michener (1985) has argued that the spatial distribution of nest sites influences social evolution in allodapines and ceratinines, in that both tribes utilize nest sites that are widely

Table 16-1 *Rates of parasitisation by the social parasite* Inquilina schwarzi *in new and established colonies of* Exoneura bicolor *from sites where nesting sites were strongly aggregated (montane population) or dispersed (heathland population)*

Aggregated nest sites (montane population)		Dispersed nest sites (heathland population)	
Re-used nests	new nests	re-used nests	new nests
40%[a]	10.6%[a]	4.7%[c]	1.2%[c]
($n = 25$)	($n = 47$)	($n = 86$)	($n = 1300$)
12.7%[b]	—[b]†	6.3%[d]	0%[d]
($n = 102$)		($n = 32$)	($n = 41$)

Superscripts refer to independent samples: [a], collected from the Grants locality (described in Blows and Schwarz 1991) on 20 Nov. 1993; [b], Grants locality 25 Sep.–11 Oct. 1984; [c], Cobboboonee State Forest (sites described in Silberbauer 1992) Nov. 1990 to Feb. 1993; [d], Wright's Swamp Road, Cobboboonee State Forest, 15 Dec. 1989. † Absence of new nests in this sample arises because the sample was taken prior to foundress dispersal.

dispersed in three-dimensional space. Such a distribution may make nests more difficult targets for wingless predators and parasites compared with ground-nesting bees, such as halictines, in which nest entrances are distributed in two-dimensional space and are frequently aggregated. The nests of Australian allodapines are often widely dispersed (Michener 1965b), but in at least some species, such as *E. bicolor, E. bicincta* and *E. richardsoni*, nests may be highly aggregated around individual tree ferns (Schwarz 1988b). Such aggregation may increase predator and parasite pressure, because these enemies may be able to restrict their search areas to a small, host-rich region near their emergence sites.

The only study to specifically compare parasite pressure between populations that show different levels of nest-site density is that of Schwarz (1994) looking at encyrtid parasitism in separate montane and heathland populations of *Exoneura bicolor*. Levels of cooperative nesting are much greater in the montane habitat where a much higher level of nest aggregation occurs. Encyrtid parasitism rates in the montane populations were approximately twice that of heathland population (10.1% versus 5.7%), suggesting that aggregated nesting sites increase parasite pressure. However, parasitization rates did not significantly vary with colony size in either habitat, suggesting that sociality does not increase protection against this parasite. The same two populations of *E. bicolor* are also attacked by the cuckoo bee *Inquilina schwarzi*, and it seems possible that for this

parasite aggregated nesting could facilitate detection of potential hosts. Table 16-1 summarizes rates of parasitization for two samples from each of the two populations. Although there is variation in parasitization rates between samples within habitats, the montane population, which has strongly aggregated nesting sites, shows a much higher rate of parasitization by *I. schwarzi* than the heathland population. At this time it is not known whether cooperative nesting enhances defense against inquilines, but when inquilines invade host nests they are actively resisted by all nestmates (M. W. Blows, personal communication) suggesting that multifemale nests resist inquilines more effectively.

Nest aggregation and kin-association after foundress dispersal

As noted above, the need for females to disperse from natal nests once a nesting substrate has deteriorated has the potential to continually reset colony development back to a solitary founder stage. This 'resetting' does not occur in montane populations of either *Exoneura bicolor* or *E. richardsoni*, where the large majority of new nests are cooperatively cofounded by groups of close genetic relatives (Fig. 16-1) (Schwarz 1987; M. P. Schwarz *et al.*, unpublished data). It is very likely that such cofounding is linked to the nesting substrate utilised by these populations: montane populations of both species nest in the dead

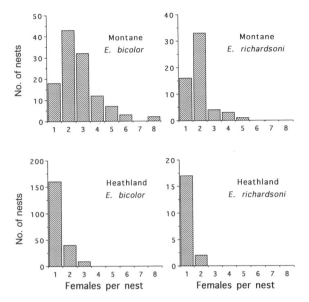

Fig. 16-1. Number of adult female nestmates in newly founded nests of *Exoneura bicolor* and *E. richardsoni* from montane and heathland populations in Victoria, Australia. Data for the montane population of *E. bicolor* is taken from 20 nesting aggregations in Dandenong Ranges National Park (Schwarz 1987). Heathland data are from samples from Cobboboonee State Forest collected on 15 Nov. 1989, 1 Dec. 1989, 7 Dec. 1988 and 14 Dec. 1989. Data for the montane population of *E. richardsoni* were obtained from nests collected from Dandenong Ranges National Park during December 1993. Data for heathland population of *E. richardsoni* were collected from Cobboboonee State Forest during October–December 1989. Substantial samples from the same localities but taken during different years, and not presented here, show the same patterns of cofounding indicated by the figures above.

fronds of tree ferns. Tree-fern fronds are highly aggregated around individual tree ferns, and nesting material is renewed annually when abscission of senescing fronds makes them suitable for nesting. This means that foundresses may only need to disperse very short distances, and this may facilitate kin cofounding by allowing foundresses to locate previous nestmates (Blows and Schwarz 1991).

In contrast to montane populations of *Exoneura*, heathland populations of *E. bicolor* and *E. richardsoni* principally nest in the dead flower scapes of the grasstree *Xanthorrhoea minor*. Unlike tree-fern fronds, this substrate is spatially dispersed and new scapes suitable for nesting do not regularly become available close to existing nest sites.

X. minor flowers profusely after fire, but infrequently at other times. Scapes become suitable for nesting about one year after flowering, and remain useable for about three to six years afterwards. Hence, after fire has moved through an area, there is a sudden increase in nesting substrates, followed by a gradual decrease as scapes deteriorate with age. In contrast to montane populations, both *E. bicolor* and *E. richardsoni* show very low levels of cofounding in the heathland habitat (Schwarz 1994) (Fig. 16-1). When cofounding does occur in *E. bicolor*, cofoundresses do not seem to be related to each other (Silberbauer 1992). Similar habitat-specific differences in cofounding rate have been found for a third species, *E. bicincta*, which also occurs in both areas (Bennett 1990). These results suggest that the distribution of nesting substrates may affect sociality in a variety of species in similar ways.

The effects of climate and voltinism

Social behavior, and particularly eusociality, may be dependent on voltinism. In species with only one generation per year, the existence of eusociality depends on the ability of queens to survive and reproduce for more than one year. In species that produce more than one brood per year, the requirement for adult longevity is not so important. There is some evidence to suggest that brood production in allodapines is related to latitude and altitude, as well as to floral resource availability. *Exoneura* species from northern New South Wales and southern Queensland (which exhibit subtropical climates) characteristically produce more than one brood per year (Michener 1965b). By contrast, montane populations of *E. bicolor* from southern Victoria (which has a cool temperate climate) produce only one brood (Schwarz 1986). Populations of *E. bicolor* from a subcoastal heathland in southern Victoria frequently contain nests that produce a second smaller brood in midsummer (Silberbauer and Schwarz 1995). Although this heathland site occurs at the same latitude as the Victorian montane populations, floral resources are abundant in early spring and summer in the heathland environment, and brood-rearing commences earlier than in montane areas.

Michener (1971) found that most African allodapines produce only one clutch of eggs per year, although egg-laying can occur over an extended period of time and some species seem to lay two clutches of eggs per year. Most of the African allodapine species had brood present throughout the year, although it appears that development slows during periods of low resource availability such as winter in the temperate areas and the dry season in tropical and semitropical areas. Given the potential relationship between eusociality and voltinism, there is a need to conduct studies on species that span a wide range of latitudes to see whether expression of eusociality is linked to latitude.

THE ROLE OF RELATEDNESS IN ALLODAPINE SOCIALITY

Intracolony relatedness is a crucial factor for understanding social traits that may be influenced by kin selection. The allodapines are particularly well suited for investigating the role of relatedness because in all species, with the possible exception of *E. tridentata*, social nesting appears to be a facultative strategy for all adult females. This facultative quality could enable individual strategies to track predictable patterns of relatedness and benefit–cost ratios available for different social behaviors. Thus far, direct measurements of intracolony relatedness, using electrophoretic data, have been made for only two allodapine species. Indirect estimates of relatedness, based on pedigree assumptions, could be unreliable because of the possibilities of multiple insemination and egg-dumping.

Electrophoretic data have been used to measure several aspects of intracolony relatedness in the allodapine bees *E. bicolor* and *E. richardsoni*. Results from these two Australian species provide some insights into the role of relatedness in sociality, particularly since both species show high levels of plurimatry. We will briefly discuss three issues raised by these studies: (1) the implications of well-developed kin recognition during nest initiation for understanding the role of kin selection; (2) the importance of plurimatry (*sensu* Choe 1996) in setting upper limits for intracolony relatedness; and (3) the potential for haplodiploidy and female-biassed sex ratios to facilitate social evolution.

Implications of kin recognition for understanding the role of kin selection

Pleometrosis occurs rarely in bees (Michener 1974) and has the potential to lower relatedness within colonies through association of unrelated gynes. Trap-nesting techniques have shown that the majority of newly founded nests in montane populations of *Exoneura bicolor*, *E. richardsoni* and *E. bicincta* contain more than one foundress. In *E. bicolor*, all cofoundresses lay eggs (Schwarz 1986) and in the other two species cofounded nests containing more than one egg-layer appear to be common (Schwarz 1988c). Electrophoretic data for the two allodapines in which cofoundresses have been assayed indicate moderately high levels of intracolony relatedness (*E. bicolor*, $r = 0.576$ (± 0.056 s.e.), Schwarz 1987; *E. richardsoni* $r = 0.498$ (± 0.152 s.e.), M. P. Schwarz *et al.*, unpublished data). Kin association during nest initiation suggests that kin selection may be important for cooperative nesting in allodapines.

Several studies have indicated that cofounding in *E. bicolor* is not a response to limited nesting sites, and that kin associate even when large numbers of non-kin foundresses are present in a localized area (Schwarz 1986, 1987; Blows and Schwarz 1991). This active association of kin suggests that kin selection may be an important factor in the social behavior of *E. bicolor*. However, another study has suggested that kinship may not be crucial for cooperative nesting; Schwarz and O'Keefe (1991b) carried out an experiment in which female pupae were housed individually in solitary nests prior to adult eclosion in autumn, and were then left in a natural environment for two months before overwintering. They found that most females joined to form multifemale colonies, even though kin association was very unlikely. Although overwintering associations are known for many social and non-social Hymenoptera, they have particular importance in *Exoneura* because reproduction commences prior to spring activity: i.e. in *Exoneura* such association cannot be viewed as being disjunct from reproductive periods. Melna and Schwarz (1994) described an instance of colony fusion involving two small colonies (unrelated to each other) during autumn, and Hurst (1993) found frequent nest-switching among unrelated females during autumn. Given that kin recognition ability is well developed in *E. bicolor*, it is likely that females are able to detect whether or not they are nesting with kin. Consequently, these results *in toto* suggest that kinship is important in cooperative nesting, but not essential.

Cooperative nesting among unrelated *E. bicolor* females could be selected for if cooperation *per se* is obligatory, for example, if predation makes it impossible for solitary nesting females to rear brood successfully to maturation. Schwarz (1988a) argued that if cooperative nesting is obligatory, kin association will ameliorate the consequences of both nestmate parasitism and the inability to restrict parental investment to a female's own direct offspring.

Plurimatry as a factor setting upper limits for intra-colony relatedness

If individuals are not able to discriminate between matri-lines and patrilines within single colonies, then the effects of plurimatry and multiple insemination will tend to lower relatedness within colonies. In *Exoneura bicolor* plurimatry has the potential to significantly limit intracolony related-ness, since most reused nests contain two or more repro-ductives, and in newly founded nests all cofoundresses lay eggs. Surprisingly, mean relatedness among female brood produced in both colony types is moderately high, ranging from $r = 0.499$ (± 0.078 s.e.) in newly founded nests to $r = 0.594$ (± 0.081 s.e.) in reused nests (Schwarz 1987). The mechanisms responsible for these levels of relatedness in the face of frequent plurimatry may therefore be important for understanding social evolution.

We suggest that three factors could help explain the unexpectedly high relatedness among *E. bicolor* brood: (1) even though all cofoundresses are reproductive, any dis-parity in the fecundity of females will decrease genetic var-iance in the brood as a whole, increasing mean intrabrood relatedness above otherwise expected levels; (2) stochasti-city in brood mortality between egg-laying and brood maturation could effectively create genetic drift effects, again leading to greater representation of some matrilines than would be expected under equal sharing of egg-laying among reproductive females; and (3) colonies entering their second or later year of nest reuse become semisocial, such that reproduction is monopolized by only a few adult females in each colony (Schwarz 1986). This monopoliza-tion is reflected by measurements of intrabrood related-ness, in that brood in second- or later-generation nests are more closely related than in the first brood after nest foundation (Schwarz 1987). The first two possibilities have not been explicitly considered in allodapines; given the unusually high levels of intrabrood relatedness in *E. bico-lor*, despite frequent plurimatry, these factors need further examination.

Potential for female-biased sex ratios to facilitate eusociality

Although initial kin selection arguments suggested that haplodiploidy could facilitate eusociality via high sister–sister relatedness (Hamilton 1964; Wilson 1971), low relat-edness of females to their brothers will tend to counteract this effect. It has become apparent that if haplodiploidy

per se is to lower selective thresholds for sib rearing, female-biassed sex allocation is required (Trivers and Hare 1976; Stubblefield and Charnov 1986; Grafen 1986). Stubblefield and Charnov (1986) argued that if female-biased sex alloca-tion ratios are to facilitate eusociality, such bias should be relative to the population mean ratio and occur where potential workers are able to switch from offspring-rearing to sib-rearing. However, Grafen (1986) has argued that female-biased allocation may facilitate eusociality even when the level of female bias in some colonies is below the population mean.

E. bicolor shows very high levels of female-biased sex allocation and this was attributed to local resource enhance-ment (LRE, Charnov 1982; Schwarz 1988c; referred to as 'local fitness enhancement', LFE, by Schwarz 1994). LFE arises when reproductive returns for investment in daugh-ters form non-linear functions of investment. Schwarz (1988a, 1994) suggested that such non-linear returns arise from cooperation between adult sisters, where cooperation enhances mean daughter fitness up to a point where an optimal number of daughters has been produced by a colony. Provided that female bias arises at times when older, adult daughters have opportunities to switch from offspring-rearing to the rearing of siblings, LFE could facilitate eusociality.

Although female-biased sex allocation ratios in *E. bico-lor* could increase relatedness of females to siblings in gen-eral, opportunities to rear siblings are limited by a univoltine life cycle. The absence of such opportunities is not so severe in heathland populations, where a partial second brood may be produced in colonies with advanced brood development (Silberbauer and Schwarz 1995). In such colonies the resulting immatures could be, and prob-ably are, reared by older sisters in the event of parental death. This form of primitive eusociality is common in more northern *Exoneura* populations where more than one brood per year is produced (Michener 1965a) and it also occurs in many African and southeast Asian allodapine species (Michener 1971, 1990). Even in univoltine montane populations of *E. bicolor* where older siblings are not able to rear younger, immature siblings, opportunities for invest-ment in siblings may still arise. Providing food for adult nestmates and guarding behavior in prereproductive autumn colonies constitute well-defined roles and are frequently assumed by newly emerged females. Assump-tion of either behavioral role is likely to incur significant costs but the resulting benefits accrue to all nestmates (Melna and Schwarz 1994). Schwarz (1986) argued that in

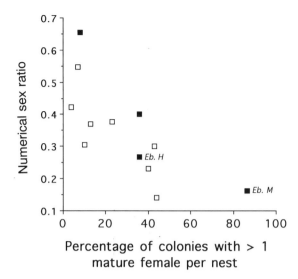

Fig. 16-2. Numerical sex ratios (r) for eight African (open squares) and three Australian (filled squares) allodapine species. The symbols *Eb H* and *Eb M* refer to heathland and montane populations of *Exoneura bicolor* and these data are taken from Schwarz (1994). All other data are taken from Michener (1971, p. 232).

postemergence colonies of *E. bicolor*, these kinds of investments are similar in direction and effect to sib-rearing in eusocial species. Female-biased sex allocation ratios will increase the mean relatedness of females to their siblings as a whole, lowering selective thresholds for group-directed altruism.

To date, LFE or LRE has only been implicated in female-biased sex allocation ratios in *E. bicolor* (Schwarz 1988a, 1994) and in *Xylocopa sulcatipes* (Stark 1992). However, female-biased numerical ratios appear to be common in at least several allodapine species and seem to be positively associated with levels of cooperative nesting (Fig. 16-2). Therefore, the possible facilitative effect of LFE on eusociality needs to be explored further in the Allodapini.

CONCLUSIONS

Despite enormous social variability among colonies, populations, species and genera, most research on allodapines has not involved empirical tests of hypotheses that could explain why sociality is prevalent in some situations but not others. Indeed, for the huge majority of species, the

available data is barely adequate to determine questions such as how long females are able to live, how long colonies are able to persist, what levels of intracolony relatedness occur, or whether disparity in ovary sizes among nestmates truly indicates caste differentiation. With the exception of one or two species, it is virtually impossible to associate social variation with variation in predator pressure, nesting substrates or relatedness among colony members.

In the sections above we have outlined a variety of factors that have the potential to strongly influence social evolution in the allodapines. Current evidence suggests the potential for a complex interplay or covariance between several of these factors in determining social organization. In order to disentangle these factors, they will need to be either statistically or experimentally controlled. Therefore, attempts to assess individual factors should focus on situations where the number of factors that covary with social variation can be held to a minimum.

In order to statistically control covarying factors, it is necessary to obtain large sample sizes, and statistical approaches may therefore only be suitable when colonies can be easily sampled in large numbers. On the other hand, allodapine bees are amenable to experimental manipulation because colonies can be readily maintained in laboratory conditions (Maeta *et al.* 1992; Michener 1972; Schwarz and Overholt 1993) and also subjected to field-based experiments (see, for example, Schwarz and O'Keefe 1991a,b; Schwarz and Blows 1991). Ideally, a combination of the two approaches should be taken, where statistical analyses of natural populations are used to generate hypotheses that are then tested experimentally. In the remaining section we suggest a number of research directions that we feel address the most interesting problems and are, at the same time, tractable.

Areas for future research

Life history and ethology. The only allodapine species that have been intensively studied are *Exoneura bicolor*, *E. variabilis* and *Braunsapis hewitti*. In some cases the findings have been unexpected, with particularly complex forms of social behavior occurring in at least *E. bicolor* (and also likely in *E. tridentata*). Assessment of ecological factors affecting sociality will need to consider the kinds of social behavior that occur within subject species and this requires that detailed ethological studies be carried out. Ethological studies should concentrate on those species that are amenable to ecological or comparative studies.

Social variability among conspecific populations.   Trying to disentangle the various factors that may influence sociality can be most easily approached when differences in phylogenetic histories can be kept to a minimum. This can be most easily done when a study involves conspecific populations. Studies on montane and heathland populations of *E. bicolor* have shown that differences in sociality covary with nesting substrate characteristics and enemies at the nest. Patterns of sex allocation also differ between these two populations, suggesting that at least one life-history trait has responded to selective pressures associated with differences in sociality (Schwarz 1994). Michener (1971) reported interpopulation differences in social nesting in the African allodapine *Braunsapis fascialis* and it is likely that such interpopulation differences also occur in other allodapine species.

Social variability among closely related congeners.   Although comparisons between conspecific populations avoid the confounding effects of differing phylogenetic histories, comparisons between closely related congeners may also be very useful. For example, comparisons of *E. bicolor, E. richardsoni* and *E. bicincta* from the same montane and heathland habitats show that all three species have larger colony sizes and higher rates of cofounding in the montane areas, suggesting that some habitat-specific factors affect all species similarly. The Australian subgenus *Exoneurella* contains four species which differ in their level of cooperative nesting; as discussed above, this may be related to nesting-substrate durability. Comparisons of sociality and factors such as nesting-substrate characteristics, when carried out over several groups of closely related congeners, may throw light on the role of such possible determinants of sociality.

Phylogeny reconstruction.   The two previous research directions require that phylogenetic relationships among subject taxa are known. Cladistic analyses of allodapines based on larval, pupal and adult characteristics have been problematical (Michener 1977). Consequently, there is a need for phylogenetic reconstruction to be further explored using DNA sequence data.

Inquilinism.   Inquilinism has evolved many times in the Allodapini, more than in any other group of bees, and inquilines seem to be phylogenetically related to their hosts (Michener 1970a). As we have discussed above, there are currently no data to determine whether or not social nesting helps resistance to inquilines or whether inquilinism is aided by the nestmate tolerance in hosts that evolves as a consequence of sociality. The first possibility could be assessed using experimental approaches, whereby inquilines are used to challenge host colonies of varying sizes.

Sex allocation bias.   Although sex allocation has only been studied in detail for *Exoneura bicolor*, Maeta *et al.* (1985) reported female-biased ratios in *Braunsapis hewitti*, another allodapine in which sociality is marked. Michener (1971) found female-biased ratios in 17 of 19 African allodapine species for which sex ratio data was obtained, and bias appears to be positively associated with the degree of social nesting (Fig. 16-2). These species provide opportunities to investigate whether local fitness enhancement occur outside *E. bicolor* and, if so, whether it lowers selective thresholds for eusociality.

ACKNOWLEDGMENTS

We thank D. Britton, B. J. Crespi, C. D. Michener, L. Packer, Y. Pamula, M. St. Clair and W. Wcislo for very helpful comments on previous drafts of this manuscript.

LITERATURE CITED

Batra, S. W. T., S. F. Sakagami and Y. Maeta. 1993. Behavior of the Indian allodapine bee *Braunsapis kaliago*, a social parasite in the nests of *B. mixta*. *J. Kansas Entomol. Soc.* **66**: 345–360

Bennett, B. 1990. Comparative sociality and phylogeny in four species of *Exoneura*. B. Sc. (Hons.) Thesis, Department of Zoology, LaTrobe University, Australia.

Blows, M. W. and M. P. Schwarz. 1991. Spatial distribution in a primitively social bee: Does genetic population structure facilitate altruism? *Evolution* **45**: 680–693.

Cane, J. H. and C. D. Michener. 1983. Chemistry and a function of the mandibular gland products of bees of the genus *Exoneura*. *J. Chem. Ecol.* **9**: 1525–1531.

Cardale, J. C. 1993. *Hymenoptera: Apoidea. Zoological Catalogue of Australia.* W. W. K. Houston and G. V. Maynard, eds., vol 10. Canberra: Australian Government Publishing Service.

Charnov, E. L. 1982. *The theory of sex allocation.* (Monographs in Pop. Biol. no. 18.) Princeton: Princeton University Press.

Choe, J. C. 1996. Plurimatry: new terminology for multiple reproductives. *J. Insect Behav.*

Daly, H. V., C. D. Michener, J. S. Moure and S. F. Sakagami. 1987. The relictual bee genus *Manuelia* and its relation to other Xylocopinae. *Pan Pac. Entomol.* **63**: 102–124.

Evans, H. E. 1977. Extrinsic versus intrinsic factors in the evolution of insect sociality. *BioScience* 27: 613–617

Grafen, A. 1986. Split sex ratios and the evolutionary origins of eusociality. *J. Theor. Biol.* 122: 95–121.

Hamilton, W. D. 1964. The genetical evolution of social behavior, Parts I and II. *J. Theor. Biol.* 7: 1–52.

Hogendoorn, K. and R. Leys. 1993. The superseded female's dilemma: ultimate and proximate factors that influence guarding behavior of the carpenter bee *Xylocopa pubescens. Behav. Ecol. Sociobiol.* 33: 371–381

Houston, T. F. 1977. Nesting biology of three allodapine bees in the subgenus *Exoneurella* Michener (Hymenoptera: Anthophoridae). *Trans. R. Soc. S. Austr.* 101: 99–113.

Hurst, P. S. 1993. Reproductive hierarchies in an Australian allodapine bee, *Exoneura bicolor* Smith (Anthophoridae, Xylocopinae). B. Sc.(Hons.) thesis, Department of Zoology, LaTrobe University, Australia.

Maeta, Y., S. F. Sakagami and C. D. Michener. 1985. Laboratory studies on the lifecycle and nesting biology of *Braunsapis sauteriella*, a social xylocopine bee. *Sociobiology* 10: 17–41.

–. 1992. Laboratory studies on the behavior and colony structure of *Braunsapis hewitti*, a Xylocopine bee from Taiwan. *Univ. Kans. Sci. Bull.* 54: 289–333.

Maeta, Y., M. Shiokawa, S. F. Sakagami and C. D. Michener. 1984. Field studies in Taiwan on nesting behavior of a social xylocopine bee, *Braunsapis sauteriella. Kontyû* 52: 266–277.

Mason, C. A. 1988. Division of labour and adult interactions in eusocial colonies of two allodapine bee species. *J. Kansas Entomol. Soc.* 61: 477–491.

Melna, P. A. and M. P. Schwarz. 1994. Behavioral specialization in pre-reproductive colonies of the allodapine bee *Exoneura bicolor* (Hymenoptera: Anthophoridae). *Insectes Soc.* 41: 1–18.

Michener, C. D. 1962. Biological observations on the primitively social bees of the genus 'Allodapula' in the Australian region. *Insectes Soc.* 9: 355–373

–. 1964. The bionomics of *Exoneurella*, a solitary relative of *Exoneura. Pac. Insects* 6: 411–426.

–. 1965a. The lifecycle and social organisation of bees of the genus *Exoneura* and their parasite, *Inquilina. Univ. Kans. Sci. Bull.* 46: 317–358.

–. 1965b. A classification of the bees of the Australian and South Pacific Regions. *Bull. Am. Mus. Nat. Hist.*, vol. 130.

–. 1970a. Social parasites among African allodapine bees. *Zool. J. Linn. Soc.Lond.* 49: 199–215.

–. 1970b. Nest sites of stem and twig inhabiting African bees (Hymenoptera: Apoidea). *J. Entomol. Soc. S. Afr.* 33: 1–22.

–. 1971. Biologies of African allodapine bees (Hymenoptera, Xylocopinae). *Bull. Am. Mus. Nat. Hist.* 145: 219–302.

–. 1972. Activities within artificial nests of an allodapine bee. *J. Kansas Entomol. Soc.* 45: 263–268.

–. 1974. *The Social Behavior of the Bees.* Cambridge, Mass.: Belknap Press of Harvard University Press.

–. 1975. A taxonomic study of African allodapine bees (Hymenoptera, Anthophoridae, Ceratinini). *Bull. Am. Mus. Nat. Hist.* 155: 71–240.

–. 1977. Discordant evolution and classification of allodapine bees. *Syst. Zool.* 26: 32–56.

–. 1983. The parasitic Australian allodapine genus *Inquilina* (Hymenoptera, Anthophoridae). *J. Kansas Entomol. Soc.* 56: 555–559.

–. 1985. From solitary to eusocial: Need there be a series of intervening species? In *Experimental Behavioral Ecology and Sociobiology.* B. Hölldobler and M. Lindauer, eds., pp. 293–305. Stuttgart: Gustav Fischer Verlag.

–. 1990. Caste in xylocopine bees. In *Social Insects: An Evolutionary Approach to Castes and Reproduction.* W. Engels, ed., pp. 120–144. Berlin: Springer-Verlag.

Reyes, S. G. 1991. Revision of the bee genus *Braunsapis* in the Oriental region. *Univ. Kan. Sci. Bull.* 54: 179–207.

–. 1993. Revision of the bee genus *Braunsapis* in the Australian Region. *Univ. Kans. Sci. Bull.* 55: 97–122.

Reyes, S. G. and C. D. Michener. 1990. Observations on a parasitic allodapine bee and its hosts in Java and Malaysia. *Trop. Zool.* 3: 139–149.

Reyes, S. G. and S. F. Sakagami. 1990. A new socially parasitic *Braunsapis* from India with notes on the synonymy of *Braunsapis mixta . J. Kans. Entomol. Soc.* 63: 458–61.

Sakagami, S. F. 1960. Ethological peculiarities of the primitive social bees, *Allodape*, Lepeltier and allied genera. *Insectes Soc.* 7: 231–249.

Sakagami, S. F. and Y. Maeta. 1977. Some presumably presocial habits of Japanese Ceratina bees, with notes on various social types in Hymenoptera. *Insectes Soc.* 24: 319–343.

Schwarz, M. P. 1986. Persistent multi-female nests in an Australian allodapine bee, *Exoneura bicolor. Insectes Soc.* 33: 258–277.

–. 1987. Intra-colony relatedness and sociality in the allodapine bee *Exoneura bicolor. Behav. Ecol. Sociobiol.* 21: 387–392.

–. 1988a. Intra-specific mutualism and kin-association of cofoundresses in allodapine bees. *Monitore Zool. Ital.* 22: 245–254.

–. 1988b. Notes on cofounded nests in three species of social bees in the genus *Exoneura. Victorian Nat.* 105: 212–215.

–. 1988c. Local resource enhancement and sex ratios in a primitively social bee. *Nature (Lond.)* 331: 346–348.

–. 1994. Female biased sex ratios in a facultatively social bee and their implications for social evolution. *Evolution* 48: 1684–1697.

Schwarz, M. P. and M. W. Blows. 1991. Kin association during cofounding in the bee *Exoneura bicolor*: active discrimination, philopatry and familiar landmarks. *Psyche* 98: 241–250.

Schwarz, M. P. and K. J. O'Keefe. 1991a. Order of eclosion and reproductive differentiation in a social allodapine bee. *Ethol. Ecol. Evol.* 3: 233–245.

–. 1991b. Cooperative nesting and ovarian development in females of the predominantly social bee *Exoneura bicolor* Smith after forced solitary eclosion. *J. Austr. Entomol. Soc.* 30: 251–255.

Schwarz, M. P. and L. A. Overholt. 1993. Techniques for rearing allodapine bees in artificial environments. *J. Austr. Entomol. Soc.* **32**: 357–363.

Silberbauer, L. X. 1992. Founding patterns of *Exoneura bicolor* Smith in Cobboboonee State Forest, Southwestern Victoria. *Austr. Zool.* **28**: 67–73.

–. 1994. Intra-specific pollen robbery as a possible factor in selecting for social nesting in *Exoneura bicolor*. In *Les Insectes Sociaux*. A. Lenoir, G. Arnold and M. Lepage, eds., p. 315. Paris: Publications Université Paris Nord.

Silberbauer, L. X. and M. P. Schwarz. 1995. Life cycle and social behavior in a heathland population of an allodapine bee, *Exoneura bicolor*. *Insectes Soc.* **42**: 201–218.

Spessa, A. C. and M. P. Schwarz. 1994. Cooperative nest use in an Australian colletid bee, *Amphylaeus morosus*. In *Les Insectes Sociaux*. A. Lenoir, G. Arnold and M. Lepage, eds., p. 184. Paris: Publications Université Paris Nord.

Stark, R. E. 1992. Sex ratio and maternal investment in the multi-voltine large carpenter bee *Xylocopa sulcatipes*. *Ecol. Entomol.* **17**: 160–166.

Stubblefield, J. W. and E. L. Charnov. 1986. Some conceptual issues in the origin of eusociality. *Heredity* **57**: 181–187.

Trivers, R. L. and H. Hare. 1976. Haplodiploidy and the evolution of the social insects. *Science (Wash., D.C.)* **191**: 249–263.

Walker, K. L. 1986. Revision of the Australian species of the genus *Homalictus* Cockerell. *Mem. Mus. Victoria* **47**: 105–200.

Wilson, E. O. 1971. *The Insect Societies*. Cambridge, Mass.: Belknap Press of Harvard University Press.

17 · Cooperative breeding in wasps and vertebrates: the role of ecological constraints

H. JANE BROCKMANN

ABSTRACT

Theories on the evolution of insect eusociality have developed in some isolation from theories on the evolution of sociality in other organisms. Facultatively eusocial groups are made up of adults who cooperatively rear young that are not direct descendants (alloparental care), with one individual in the group doing most of the breeding. This reproductive suppression is maintained through dominance or other asymmetries. Defined in this way, many species of birds and mammals show facultative eusociality (cooperative breeding). Explanations for the evolution of eusociality are somewhat different from those used to explain cooperative breeding. In particular, relatively little attention has been paid to the Ecological Constraints Model.

Cooperatively breeding vertebrates are thought to be living at the maximum possible population size for the available habitat ('saturated'), with intense competition among conspecifics for breeding opportunities. Among those studying helpers-at-the-nest (birds) or den (mammals), it is generally agreed that delayed breeding occurs when the gain (in inclusive fitness) from helping is greater than the gain from independent breeding. This typically occurs when there is some constraint which either prevents some individuals from attaining breeding status, such as a shortage of territories, or which raises the costs of independent breeding to prohibitive levels, such as when one pair cannot bring in adequate food for their young or when more than two individuals are required to defend resources or ward off predators. When the ecological requirements of a species are specialized to the extent that they cannot find suitable marginal habitats, or if the quality of the environment deteriorates over the season, or if there are marked unpredictable changes in the quality of the environment for initiating breeding, then younger individuals are at a disadvantage relative to established breeders. This can favor helping over dispersal and independent breeding.

Cooperative breeding in primitively eusocial Hymenoptera may be viewed in much the same way. Social insects may also be living at saturated population levels, with few breeding opportunities and intense social competition. What are the constraints on independent nesting? Are there correlations between the degree of difficulty in becoming established as a breeder and the frequency of eusociality? I discuss similarities and differences between hymenopteran and vertebrate cooperative breeding, the applicability of the Ecological Constraints Model and other insights derived from considering these two patterns of behavior to be convergent.

INTRODUCTION

Sociality has evolved repeatedly in distantly related taxa. If the selective pressures favoring sociality in these lineages are found to be similar then we understand an important factor in the evolution of social behavior. Sometimes, however, we are hindered in our ability to develop generalizations by boundaries between taxonomic categories. In this chapter I argue that profound similarities exist in the patterns of sociality found among behaviorally eusocial wasps (Kukuk 1994; also called 'primitively' eusocial by Michener 1974, or facultatively eusocial by Crespi and Yanega 1995) and cooperatively breeding vertebrates. Vertebrate social systems have been well studied from an ecological perspective and considerable agreement exists about the factors favoring the evolution of cooperative breeding. In this chapter I explore whether those same explanations apply to behaviorally eusocial insects. I have selected two groups, sphecid and vespid wasps, because they are two independently evolved cases of behavioral eusociality (bees are covered by Danforth and Eickwort; Yanega; Wcislo; and Schwarz *et al.*, all in this volume). I will focus on one genus of vespids, *Polistes*, because of the data available for this group. Although other genera of Polistinae

Table 17-1. *Terminology used in the social insect and vertebrate literature to describe different types of social behavior*

In all cases these categories refer to animals that are living together on a jointly defended territory or nest and at least some individuals breed on that territory or in that group. The focus is on whether groups originate as matrifilial or parasocial colonies and whether they show reproductive dominance (or delayed breeding) and/or alloparental care (cooperative brood care).

Social insects (joint defense of group nest)	Social vertebrates (joint defense of a group territory or nest)
Parasocial Group[5] (groups formed by adults of the same generation)	
1. Communal[5]	1. Social group
(a) provision same brood cell	(a) use same nest
(i) joint nesting (aggressive)[4]	(i) intraspecific brood parasitism (aggressive)
(ii) nest sharing (amicable)[4]	(ii) joint nesting (amicable)[2]
(b) provision separate brood cells	(b) colonial[2] (separate nesting)
2. Parasocial group with delayed breeding or dominance	2. Delayed breeding or dominance within group of adults[2]
3. Quasisocial[1] (parasocial group with alloparental behavior but no reproductive dominance)	3. Adults, often siblings, live together in group, jointly defend a territory, breed and engage in alloparental behavior[2]; (also called Communal breeding[3])
4. Semisocial[1,2] (parasocial group with reproductive dominance or delayed breeding and alloparental care)	4. Cooperative breeding[3,9]
(a) 1 queen = monogynous[5] = unimatry[6] = (haplometrotic[5] nest initiated by independent founding)	(a) 1 breeder = singular breeding[2]
(b) >1 queen = polygynous[5] = plurimatry[6] (pleometrotic[5] nest initiated by foundress association)	(b) >1 breeder = plural breeding[2]
(i) single nest	(i) separate nesting[2]
(ii) satellite nests (= polydomous nest[1])	(ii) joint nesting[2]
Matrifilial Group[5] (groups formed by overlapping generations)	
5. Intermediate Subsocial I[1]	5. Non-dispersing, non-helping young remain in natal territory or group to breed
6. Matrifilial group with reproductive dominance or delayed breeding	6. Non-dispersing, non-helping young remain in natal group or territory with delayed breeding or reproductive dominance
7. Intermediate Subsocial II[1] (matrifilial group with alloparental behavior and no reproductive dominance)	7. Non-dispersing young remain in natal wterritory or group to breed and engage in alloparental behavior[2] (also called communal breeding[3])
8. Primitively Eusocial[1,5,8] (matrifilial group with reproductive dominance and alloparental care) = behaviorally eusocial[7] = facultatively eusocial[10]	8. Cooperative breeding[3,9]
(a) 1 queen = monogynous[5] = unimatry[6] haplometrosis (nest founded by one queen)	(a) 1 breeder = singular breeding[2]
(i) nest founded by independent founding	
(ii) swarm founding (+ offspring workers)	
(b) >1 queen = polygynous[5] = plurimatry[6] pleometrosis (nest founded by >1 queen)	(b) >1 breeder = plural breeding[2]
(i) single nest	(i) joint nesting[2]
(ii) satellite nests (= polydomous nest[1])	(ii) separate nesting[2]
(iii) foundress association	
(iv) swarm founding (+offspring workers)	

such as *Mischocyttarus* and *Ropalidia* show similar patterns, behavioral eusociality has probably evolved only once in this lineage, so these would not be independent cases. This inquiry may help to provide insight into what has remained a perplexing problem to social insect biologists: the ecological factors favoring the evolution of eusociality.

Terminology and definitions

Crespi and Yanega (1995) argue that many cases of the behavioral pattern known as behavioral or primitive eusociality among insects should be called 'cooperative breeding' because of their similarity to vertebrate social systems (see also Alexander 1974; Wilson 1975; West-Eberhard 1975; Eickwort 1975; Wickler 1978; Koenig and Pitelka 1981; Brockmann 1984, 1990, 1993; Andersson 1984; Emlen and Vehrencamp 1985; Brown 1987; Pollock and Rissing 1988; Strassmann and Queller 1989; Alexander *et al.* 1991; Lacey and Sherman 1991; Seger 1991; Field 1992; Itô 1993; Sherman *et al.* 1995; Keller 1995). Cooperative breeding is defined by the presence of helpers at some or all nests (Brown 1987). A helper is an adult, often a non-breeder, that performs parent-like behavior (alloparental behavior) toward young that are not its own, but in company with at least one of the young's close relatives (thus eliminating brood parasitism, adoption and brood capture; see also Emlen and Vehrencamp 1985). Eusociality is defined as conspecific adults of two generations living together and cooperating to rear young where only one or a few of the adults in the group lay viable eggs (Wilson 1971). In the highly (Michener 1974) or morphologically (Kukuk 1994) eusocial species, most workers are unmated and sterile and the queen is morphologically distinct when she emerges as an adult. In the primitively (Michener 1974) or

facultatively eusocial insects, however, most workers retain their ability to mate and lay eggs and the queen and workers are often difficult to distinguish except by behavior (Strambi 1990; Kukuk 1994; Crespi and Yanega 1995) (Table 17-1).

Among social insects and vertebrates, cooperatively breeding groups are initiated in two ways (Wilson 1971; Vehrencamp 1979; Brockmann 1984) (Table 17-1). First, haplometrotic nests originate from a single queen (or breeding pair in vertebrates) and become eusocial (or a cooperatively breeding group) with the emergence of the first brood who remain at the natal nest and help (referred to as the subsocial route to sociality). Second, social groups may be founded by several adults of the same generation, i.e. by foundress associations (often closely related). These pleometrotic groups become eusocial when the young emerge and remain at the nest and help the parents (referred to as the parasocial route to sociality). Whether one or several of the queens in a foundress association produce offspring depends on the degree of dominance (Pardi 1942, 1948; West 1967; Strassmann 1981d, 1993) and the nature of the social contract between them (Reeve and Nonacs 1992; Reeve and Ratnieks 1993). Among vertebrates, singular breeding occurs when only one female of a group breeds or lays eggs; plural breeding occurs when multiple females breed within a group (Brown 1987). Emlen (1984), Emlen and Vehrencamp (1985) and Crespi and Yanega (1995) distinguish cases in which all individuals in a group produce offspring, calling this communal breeding, as opposed to cases in which some individuals do not breed, which they call cooperative breeding (Table 17-1).

Communal care of young is common in animals (Gittleman 1985; Riedman 1982), but what makes eusociality and cooperative breeding different and interesting from

---

Sources: [1] Wilson 1971; [2] Brown 1987; [3] Emlen 1984; [4] Brockmann 1984; [5] Michener 1974; [6] Choe 1995; [7] Kukuk 1994.

[8] It is common for eusocial and semisocial nests to be be combined, i.e. the nest may start as a semisocial foundress association (of sisters or strangers) but when the daughters emerge the nest becomes eusocial. If the foundresses and cofoundresses remain, then the nest is still called eusocial. It is not uncommon for the newly emerged daughters to kill the queen's cofoundresses, in which case the colony is still called eusocial[c]. One of the queen's daughters may become a secondary queen or take over when the queen dies (also eusocial). Similar patterns occur among vertebrate cooperative breeders. A cooperatively breeding group consists of a breeder and and one or more helpers. The helpers may be the breeder(s)' offspring, relatives of the same generation (often siblings) or unrelated individuals.

[9] Called 'communal' breeding by Brown (1987), but I have not used this term here because of its multiple meanings.

[10] Crespi and Yanega (1995). These authors propose that this form of eusociality be called 'cooperative breeding' in social insects as well vertebrates, leaving the term 'eusocial' for what has been referred to as highly eusocial[1,5] species. They distinguish eusocial species as those with castes, i.e. groups in which some individuals in the group 'become irreversibly behaviorally distinct at some point prior to reproductive maturity'. Their classification differs from the one in this table and in the previous literature because they do not distinguish between subsocial (matrifilial) and parasocial groups.

an evolutionary point of view is that not all individuals in the group breed: there is at least some 'reproductive division of labor' (Wilson 1971). When all adults in a group do not breed, some mechanism must exist that determines who breeds and who does not (Emlen and Vehrencamp 1985). In the obligately eusocial species (Crespi and Yanega 1995), reproductive division of labor is maintained by differential feeding of the larval workers and reproductives and by pheromones produced by the queen. In facultatively eusocial species, reproductive dominance (inhibition) by the queen is usually maintained by behavior and pheromones (Pardi 1948; Wilson 1971; Michener and Brothers 1974; Reeve and Gamboa 1983, 1987) and perhaps by differences in larval nutrition (Turillazzi and Pardi 1977; Hunt 1991, 1993). Similarly, breeders in many cooperatively breeding species maintain dominance behaviorally, through social harassment or chemical communication (Macdonald and Moehlman 1982; Emlen 1984; Rood 1986; Creel and Waser 1991; Reeve and Sherman 1991). An explanation for the evolution of cooperative breeding requires an understanding of the selective pressures favoring reproductive dominance.

Component decisions of cooperative breeding

Cooperative breeding in social insects and vertebrates results from four similar decisions (Brockmann and Dawkins 1979) made by helpers and breeders (Brockmann 1986; Koenig *et al.* 1992; Emlen and Wrege 1994) (Fig. 17-1). First, cooperative breeders differ from other species in that adults remain with their parent(s) or other close relatives rather than dispersing and living on their own, i.e. they are philopatric. In some cases this decision is to stay-and-foray (adults remain on the natal territory and search for breeding opportunities nearby) or depart-and-search (adults disperse to wander in search of breeding opportunities) (Brown 1987; Walters *et al.* 1992). Second, in cooperative breeders, these philopatric adults fail to breed. For some vertebrate cooperative breeders and social insects, these adults never reproduce and remain as workers or helpers their entire lives (Table 17-2). It may be that this failure to breed results from active dominance or suppression by the breeder or it may be that the helper delays her/his own breeding while awaiting a breeding opportunity. Third, when cooperative breeders remain on the natal territory or at the natal nest, they, unlike other species, help rather than just caring for themselves. Helpers or workers provide aid in the form of feeding and caring for the brood, thermoregulation of the

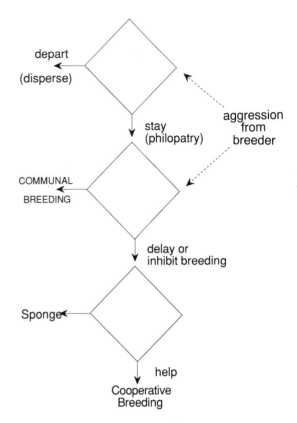

Fig. 17-1. A schematic of the decisions made by individual wasps or other cooperative breeders. According to the model, if the inclusive fitness associated with being philopatric, delaying breeding (reproductive inhibition) and helping is greater than the inclusive fitness expected from alternative routes, then cooperative breeding will be maintained in the population.

brood or nest (incubation and brooding), feeding and grooming the egg-laying or incubating adult, nest-building and nest maintenance and defense against predators, parasites and marauding conspecifics (Litte 1977; Strassmann 1981c; Jeanne 1986). Fourth, the breeder must also decide whether to defend against these non-dispersing individuals or not. Of course, none of these are necessarily conscious calculations; they are evolved rules that influence the choices that animals make (Brockmann and Dawkins 1979). Cooperative breeding and eusociality arise, then, when a similar set of individual decisions are made. The question posed in this chapter is whether these decisions result from similar selective pressures.

When the benefit, measured in inclusive fitness (*rb*), from making a decision is greater than the opportunity

Table 17-2. *The intensity of reproductive dominance or reproductive division of labor*

The table refers to cooperative breeders, i.e. primitively eusocial foundress associations of wasps and cooperatively breeding vertebrates.

Species	Percentage of adults (gynes) in group that never reproduce	Reference
Vertebrate cooperative breeders		
Florida scrub jay	45%	Woolfenden and Fitzpatrick 1990
Splendid fairy wren	17%	Rowley and Russell 1990
Arabian babbler	28%	Zahavi 1990
Green woodhoopoe	79%	Ligon and Ligon 1990
Naked mole rat	99%	Jarvis *et al.* 1994
Damaraland mole rat	92%	Jarvis *et al.* 1994
Polistes wasp foundress associations		
P. annularis	95%	Strassmann 1981a
P. canadensis	79%	West 1967
	96%	West-Eberhard 1969
	99%	Pickering 1980
P. carolinus	95%	Hughes and Strassmann 1988
P. exclamans	58%	Strassmann 1981b, 1985
P. fuscatus	51%	West 1967
	45%	Noonan 1981
	25%	Klahn 1988
P. gallicus	90%	Röseler 1985
P. metricus	54%	Gamboa *et al.* 1978
	91%	Dropkin and Gamboa 1981

costs (c), i.e. what the individual would have gained from the alternative, then a behavior will be favored (i.e. if $rb - c > 0$). This calculation is made on an individual basis: given who this individual is, its size, reproductive condition, the time of year, etc., what are its best options in terms of lifetime reproductive success (both immediate and delayed fitness benefits need to be included; Wiley and Rabenold 1984)? Cooperative breeding will be maintained in a population when a particular combination of decisions (i.e. philopatry, delayed breeding or reproductive inhibition, worker-like behavior and the absence of aggression from the breeder) increases individual inclusive fitness (Koenig and Mumme 1990; Reeve 1991; Koenig *et al.* 1992; Sherman *et al.* 1995). The inclusive fitness gain has both an indirect and a direct component (Brown 1987; Emlen and Wrege 1989). The indirect component is measured in additional collateral kin that the breeder(s) would not have been able to rear without the helper's assistance (Vehrencamp 1979; Grafen 1984; Queller 1989). An individual may also gain direct benefits (personal fitness) from philopatry, delayed breeding and helping (Emlen 1984, 1991; Wiley and Rabenold 1984; Emlen and Vehrencamp 1985). These gains may come from (a) improved experience, (b) greater survivorship, (c) enhanced future breeding opportunities, such as the possibility of inheriting the breeding position of the parent or part of its territory (hopeful reproductive), (d) group dispersal, i.e. the helper may be rearing siblings that will disperse together as a coalition (e.g. a pleometrotic foundress group) (Noonan 1981) or (e) reciprocity, i.e. the helper may be rearing future helpers (Rood 1978) or workers for itself or for the reproductive eggs it has laid. (f) When there is a shortage of mates or in cooperative polyandry, helping may increase an individual's chances of acquiring a mate (Emlen and Wrege 1989). The opportunity costs (c), for each decision are equally important elements in the equation (Fig. 17-1). For example, Koenig *et al.* (1992) have argued that one factor that favors cooperative breeding is the quality of the habitat in which floaters can expect to live while awaiting a breeding opportunity. If that habitat is below the quality of the natal territory, then the individual should delay dispersal and remain on the natal territory if possible.

From the breeder's point of view, cooperative breeding, i.e. reduced aggression and allowing the helper to stay, will be favored when the helper's presence raises the breeder's lifetime reproductive success. Obviously, if helpers do no more than compete with the breeder(s) and its offspring for food, then a breeder will do better by chasing the intruder away (Reyer 1990). Cooperative breeding will evolve only when the helper does something that will increase the breeder's fitness. In most altricial birds, canids and aculeate wasps, for example, there is a fixed nest to which the parents return repeatedly with food which they give or regurgitate to the brood (Andersson 1984). This means that there is much that the helpers can do in building and protecting the nest or den and in supplying the brood with food. Given the many similarities between cooperative breeding and eusociality, it seems appropriate to consider the possibility that the same ecological constraints and selective pressures that are thought to have favored vertebrate cooperative breeding will provide a good model for social insects (Strassmann and Queller 1989; Reeve 1991).

THE ECOLOGICAL CONSTRAINTS MODEL

The consensus in the study of vertebrate social behavior is that cooperative breeding arises under the ecologically peculiar circumstances in which free access to successful independent breeding is strongly limited (Brown 1974; Koenig 1981; Koenig and Pitelka 1981). This has come to be known as the *Ecological Constraints Model* for the evolution of sociality (Emlen 1982a, 1991). Constraints include (1) a saturated habitat, i.e. the spatial localization of some critical resource that is in short supply and thus intense competition over territories that are rarely vacant (Emlen 1991; Walters *et al.* 1992); (2) a lack of suitable mates; or (3) some extreme environmental condition that makes independent reproduction so costly that a pair cannot rear offspring on their own, i.e. high costs of independent reproduction (Emlen 1991). These high costs may arise if some required resource is scare, if cooperation is required to gain access to the resource, if there is intense intraspecific competition, or if group defense is required owing to high predation or parasitism (Emlen and Vehrencamp 1985). Other explanations for the evolution of cooperative

breeding exist (Stacey and Ligon 1987), but these can be subsumed under the more general Ecological Constraints Model. Koenig *et al.* (1992) separate ecological constraints into intrinsic benefits and extrinsic constraints rather than saturated habitats and high costs of independent reproduction. Intrinsic benefits arise when individuals gain by group living, such as through reduced predation or increased foraging efficiency, whereas extrinsic constraints occur when individuals are forced to live in groups to gain access to critical resources for breeding, such as a territory.

Stated in its most general form, the Ecological Constraints Model is little more than a restatement of Hamilton's Rule: cooperative breeding evolves when the benefits in inclusive fitness for being social are greater than the opportunity costs (Fig. 17-1). If the benefits from independent breeding are severely limited for whatever reason, then other routes to reproductive success will be favored. But these models go beyond this simple inequality. In fact, individuals of most species probably face low breeding success, limited habitat availability and difficulties in finding mates and yet only rarely do they evolve cooperative breeding. It is only under a special set of ecological circumstances that this particular pattern is favored.

Ecological constraints on breeding for cooperatively breeding vertebrates

Table 17-3 shows a sampling of known constraints from well-studied, cooperatively breeding vertebrates. For the most part, the suspected cause is habitat saturation or exceptionally high costs of reproduction or both. For example, in Florida scrub jays (*Aphelocoma c. coerulescens*), a mated monogamous pair and their young from the previous year live on well-defined territories (Woolfenden and Fitzpatrick 1984). They occupy a patchily distributed, fire-maintained, oak scrub habitat and intense competition exists among conspecifics for this limited breeding space (Woolfenden and Fitzpatrick 1990). Helpers remain with parents until a territory becomes available; for males, this may involve years of waiting. If a breeder dies or is removed, he or she is immediately replaced by a neighboring helper. Helpers increase the number of fledglings from their parents' nests because they defend the territory, sentinel and mob predators, feed insects to the young, and cache acorns, an important food supply for the winter (Woolfenden and Fitzpatrick 1984; Hailman *et al.* 1994). For many cooperatively breeding birds, the costs of reproduction are

Table 17-3. *Known constraints on breeding for some well-studied cooperatively breeding vertebrates*

Superscript numbers refer to categories on Table 17-1.

Species (taxonomic family)	Ecological constraint	Reference
Singular breeding with offspring helpers[8]		
Florida scrub jay (Corvidae)	quality breeding territory	Woolfenden and Fitzpatrick 1984, 1990
Splendid fairy wren (Malurinae)	quality breeding territory	Rowley and Russell 1990
Superb fairy wren (Malurinae)	mates and quality breeding territory	Pruett-Jones and Lewis 1990
Arabian babbler (Timilinae)	quality breeding territory	Zahavi 1990
Galapagos mockingbird (Mimidae)	quality breeding territory	Curry and Grant 1990
Bushy-crested hornbill (Bucerotidae)	quality breeding territory	Leighton 1986
Red-cockaded woodpecker (Picidae)	nesting cavity	Walters 1990; Walters *et al.* 1992
Green woodhoopoe (Phoeniculidae)	roosting hole	Ligon and Ligon 1988, 1990
Harris' hawk (Accipitridae)	cooperative hunting	Faaborg and Bednarz 1990
Bi-colored wren (Troglodytidae)	predator defense	Rabenold 1990
Stripe-backed wren (Troglodytidae)	predator defense	Rabenold 1990
Seychelles warbler (Sylviidae)	quality breeding territory	Komdeur 1992, 1994
Red fox (Canidae)	breeding opportunity	MacDonald and Moehlman 1982
Silver-backed jackal (Canidae)	predator defense, food supply	Moehlman 1986
Lamprologus brichardi (Cichlidae)	space competitors and egg predators	Taborsky and Limberger 1981; Taborsky 1984
Singular breeding with offspring helpers and other helpers[4,8]		
Pied kingfisher (Alcedinidae)	food supply; females	Reyer 1990
White-fronted bee-eater (Meropidae)	variable food supply	Emlen 1990; Emlen and Wrege 1989, 1991
African wild dog (Canidae)	cooperative hunting; defense; food for pups	Frame *et al.* 1980 Malcolm and Marten 1982
Dwarf mongoose (Viverridae)	predator defense; food for pups energetic costs of reproduction	Rood 1986; Rasa 1986 Creel and Creel 1991
Naked mole rat (Bathyergidae)	predator defense, patchy food cooperative digging	Lacey and Sherman 1991 Sherman *et al.* 1991
Damaraland mole rat (Bathyergidae)	cooperative digging, patchy food	Jarvis *et al.* 1994
	Weak reproductive inhibition	
Plural breeding with offspring helpers and other helpers[4,8]		
Chestnut-bellied starling (Sturnidae)	(unknown)	Wilkinson 1982
Mexican jay (Corvidae)	quality breeding territory quality breeding territory	Brown and Brown 1990
Acorn woodpecker[a,b] (Picidae)	granaries (winter food); quality breeding territory	Koenig and Stacey 1990
Banded mongoose (Viverridae)	predator defense	Rood 1986
Saddle-back tamarin[b] (Callitricidae)	carrying and protecting young	Terborgh and Goldizen 1985; Goldizen 1987, 1989
Black-tailed prairie dog (Sciuridae)	territorial and predator defense	Hoogland 1981
	No reproductive inhibition	
Communal breeding with offspring helpers and other helpers[3,7]		
Spotted hyena[a] (Hyaenidae)	group hunting and territory defense	Mills 1989
African lion (Felidae)	group hunting and defense	Packer 1986
Communal breeding[3]		
Groove-billed ani[a] (Cuculidae)	quality breeding territory	Koford *et al.* 1990; Vehrencamp *et al.* 1988
Pukeko gallinule[a] (Grallidae)	quality breeding territory	Craig and Jamieson 1990
Galapagos hawk[b] (Accipitridae)	quality breeding territory	Faaborg and Bednarz 1990
Dunnock[b] (Prunellidae)	quality breeding territory; food supply; females	Davies 1990

[a] Joint nesting.

[b] Cooperative polyandry.

so high that pairs of adults cannot breed on their own. It takes a team of bee-eaters (*Merops bullockoides*), who are mostly relatives, to supply one brood (Emlen 1990). Fishing in some habitats is so expensive for pied kingfishers (*Ceryle rudis*) that they find it impossible to rear young alone (Reyer 1990). Female stripe-backed wrens (*Campylorhynchus nuchalis*) fight with helpers for vacancies on territories, preferring larger over smaller groups (Zack and Rabenold 1989). When opportunity costs are high and when helping increases fitness, then selection favors group sizes of more than a pair.

Cooperative breeding is not common among mammals, but in most cases intense intraspecific competition for resources and predation have been identified as the most important selective agents (Jennions and Macdonald 1994) (Table 17-3). For example, African wild dogs (*Lycaon pictus*) live in large packs with one breeding female, a few subordinate females (whose estrus is suppressed) who are normally the dominant female's daughters, and a number of males (Malcolm and Marten 1982). The dominant female gives birth to a large litter that is totally dependent on adult food for an entire year. The group hunts together and defends their kills from conspecifics and larger predatory species. Wild dogs are also living at their maximum density in a habitat where they cannot feed their young without assistance from pack members (Macdonald and Moehlman 1982; Malcolm and Marten 1982). Like canids, mongoose species (Viverridae) also vary in sociality from solitary to a social system similar to wild dogs, with one dominant female breeder and suppressed, subordinate daughters (Rood 1986; Creel et al. 1992). The small, diurnal, insectivorous species are extremely vulnerable to predation and if they do not live in a social group they cannot rear young (Rasa 1986, 1987). Naked mole rats (*Heterocephalus glaber*) are highly social and in many ways very similar to eusocial bees, ants and termites (Jarvis 1981; Alexander 1991; Lacey and Sherman 1991). They live in burrows and forage on roots and tubers from tunnels which they dig cooperatively. The colony is made up of close relatives (Reeve et al. 1990) with one large, breeding female, a number of non-working large individuals that serve as soldiers and a number of smaller workers who do most of the digging. Individ-uals retain their ability to breed; when the breed-ing female is removed, one of the larger soldiers quickly assumes the role (Lacey and Sherman 1991). Ovulation by other females is inhibited (Faulkes et al. 1991) probably as a result of the constant shoving and biting

from the breeding female (Reeve 1992; Reeve and Sherman 1991). Ecological constraints favoring cooperative breeding include predation pressure, the patchiness of the animal's food resources and the physical barrier to solitary dispersal owing to the hard soil in their arid environment (Brett 1991).

Helpers are also found in six species of cichlid fishes (*Lamprologus* and *Julidochromis*) from Lake Tanganyika (Taborsky and Limberger 1981). The helpers are young from previous broods who help their monogamous parents guard the cavities that are required as nesting sites. They keep out conspecifics and other space (hole) competitors and they clean the eggs (Taborsky 1987; Siemens 1990). Breeders produce many more young with helpers than without (Taborsky 1984). Naturally or experimentally removed breeders are immediately replaced. For these animals, too, the habitat is densely inhabited by conspecifics and competitors with similar space requirements. These fish are also living in a habitat where breeding alone is less successful than breeding in groups.

Testing the Ecological Constraints Model

The Ecological Constraints Model makes a number of predictions that can be tested (Table 17-4) (Emlen 1991). First, the number of breeders will be stable over long periods, as has been shown in Florida scrub jays. Second, if a breeder is removed, then it will be replaced quickly by a non-breeder, as in Florida scrub jays and cooperatively breeding cichlid fishes. The intensity of the response to removal should be proportional to the quality of the territory from which the breeder was removed (Koenig and Mumme 1987; Stacey and Ligon 1987; Zack and Rabenold 1989). Third, helpers will increase the reproductive success of the breeders and their aid will be directed toward accruing or protecting the resource that is limiting breeder reproductive success, as for example, in the case of cichlids competing for nesting space. Helpers should also increase their own lifetime inclusive fitness by helping. However, there is not necessarily a per capita increase in fitness with each helper because, according to this model, the helpers may be forced into helping because there are no other breeding opportunities (Koenig and Pitelka 1981). Fourth, when helpers are removed, there will be a drop in the reproductive success of the breeders (Emlen 1991; Komdeur 1994). For example, when Florida scrub jay helpers are removed, those groups suffer higher nestling predation and lower rates of fledgling survival than groups in which helpers are allowed to remain

Table 17-4. *Predictions consistent with the Ecological Constraints Model stated in terms of primitively eusocial wasps*

(1) The number of nests in the habitat will be stable over long periods.

(2) If a nesting group is removed then it will be replaced by a non-breeding group. The intensity of the response will correlate with territory quality.

(3) The frequency of non-breeding workers will vary with the degree of difficulty in becoming a breeder or with the degree of environmental harshness, i.e. the frequency of helping will change with the intensity or shortage of resources that are critical to reproduction.

(4) Workers increase the reproductive success of the breeders and their aid will be directed toward accruing or protecting whatever resource is limiting reproductive success.

(5) If a queen is removed, she will be replaced quickly by a non-breeder.

(6) When workers are removed, there will be a drop in the reproductive success of the breeders.

(7) Workers will increase their lifetime inclusive fitness by helping over what they would have achieved had they nested alone.

(8) If the critical resources needed for reproduction are manipulated experimentally, then the frequency of helping will change.

Source: modified from Emlen and Wrege 1989; Emlen 1991.

(Mumme 1992a). Fifth, the frequency of helping should change with the intensity or shortage of resources that are critical to reproduction (Emlen and Vehrencamp 1985). For example, mockingbird pairs are usually intensely territorial, but the Galapagos mockingbirds (*Nesomimus* spp.) have a helper system that is similar to that of Florida scrub jays (Curry and Grant 1990). On islands of the archipelago where the density of birds is higher, the frequency of helping increases compared with islands of lower density, which is good evidence that these birds are living in a 'saturated' habitat with few breeding opportunities (Curry 1989; Kinnaird and Grant 1982). Sixth, if the critical resources needed for reproduction are manipulated experimentally (either increased or decreased), then the frequency of helping should change. Komdeur (1992) and Komdeur *et al.* (1995) moved Seychelles warblers to unoccupied islands where they first occupied high-quality territories. As these areas began to fill, young birds born on these territories remained as helpers rather than dispersing to lower-quality territories, suggesting that the habitat had become saturated. Red-cockaded woodpeckers take 10 months to several years to excavate nesting cavities in living pines. Walters *et al.* (1992) provided additional cavities; these resulted in rapid colonization of previously unoccupied habitat, which is normally very uncommon. This provides good evidence that potential habitat remains unoccupied because of the lack of a critical resource, nesting and roosting cavities.

Most of the evidence collected to date has been correlational; such data are particularly problematic for studies of helper systems. For example, numerous studies purport to

show a correlation between the number of helpers and reproductive success (Emlen 1991). Since these data are generated from naturally occurring family groups (the only data that could be reasonably collected in most species), they amount to a correlation between one year's reproductive success and the next (Leonard *et al.* 1989; Emlen 1991). To avoid this problem, Emlen and Wrege (1991) use a multivariate approach, Koenig and Mumme (1987) use analysis of variance and Woolfenden and Fitzpatrick (1984) use between-year comparisons of the same breeders on the same territories with and without helpers. Experimental manipulations provide a powerful approach. In three of the four studies that have been conducted on birds (Brown *et al.* 1982; Mumme 1992b; Komdeur 1994), and in cooperatively breeding cichlids (Taborsky 1984), a strong effect of helpers has been demonstrated. Powerful as manipulations are, Koenig and Mumme (1990) have criticized the few helper-removal experiments that have been conducted because they do not separate the general benefits of group-living from the benefits of alloparental behavior and reproductive dominance and because they fail to control for the social disruption accompanying the removal of helpers (with the exception of the Komdeur 1994 study).

DOES THE ECOLOGICAL CONSTRAINTS MODEL APPLY TO WASPS?

Is the Ecological Constraints Model a good explanation for cooperative breeding (primitive eusociality) in sphecid and vespid wasps? Are the selective pressures that favor facultative eusociality in wasps the same as those that favor cooperative breeding in vertebrates (Strassmann and Queller 1989; Brockmann 1990; Lacey and Sherman 1991)? Wasps are a particularly good group for this comparison because they are ecologically similar in important respects to cooperatively breeding vertebrates: they are highly mobile, they hunt live prey and they provide intensive parental care for their young. This form of behavior, as well as the fixed nest to which individuals return repeatedly, greatly increases the opportunities as well as the benefits for helping behavior (Alexander 1974; Andersson 1984; Brockmann 1984).

Does the Ecological Constraints Model apply to sphecid wasps?

Despite the fact that nearly all sphecid wasps provide intensive maternal care for a few offspring, only one genus, *Microstigmus*, in the subfamily Pemphredoninae, is eusocial (Matthews 1968a,b; Ross and Matthews 1989a,b) and several other species of closely related genera are suspected of being eusocial (Matthews and Naumann 1988; McCorquodale and Naumann 1988). A few other species of sphecid, scattered through most subfamilies (Fig. 17-2), show various levels of communal nest-sharing (Brockmann and Dawkins 1979; Matthews 1991; Gess and Gess 1989; Wcislo *et al.* 1985), but these are clearly the exception. Most sphecid wasps are solitary and fight when they encounter a conspecific in their nest (Brockmann and Dawkins 1979).

In *Microstigmus* a group of closely related females jointly construct a pendant nest. In *M. comes* the nest is made of pubescence scraped from the underside of a palm frond (Matthews 1968b, 1991). In *M. myersi* females build nests with dirt particles attached to fine rootlet hairs found along an embankment (Demelo and Campos 1993a,b). In both species the nest is given form and suspended with silk, which the adults produce from a gland at the tip of the abdomen. Nest construction is clearly an enormous investment and lone females take far longer to complete the task than groups of females working together (Matthews and Starr 1984). The nest contains only one or a few breeders, and the individuals on a nest are closely related, probably sisters or mother and daughters (Ross and Matthews 1989a,b). Eusocial *Microstigmus* females jointly provision the brood cells with large numbers of collembolans; it may take several hundred of these tiny prey to provision one larva. Workers also defend the nest site from ants, predators, parasites and heterospecific and probably conspecific usurpers. For many reasons then, groups of females are far more successful than lone females (Matthews 1991).

In another Pemphredoninae wasp, *Arpactophilus mimi*, which is closely related to *Microstigmus* and likely eusocial, intense competition exists for the empty mud-dauber cells that they take over for nesting (Matthews 1991). Such competition may be an important factor promoting nest-sharing (or eusociality) in this species (Matthews and Naumann 1988).

Although none are eusocial, nest-sharing is particularly common (about 20 species) among the Cercerini (Elliott *et al.* 1986; Elliott and Elliott 1987; Evans and Hook 1982, 1986; Willmer 1985; Hook 1987; Mueller *et al.* 1992). In *Cerceris antipodes* females reuse natal nests, and take over existing burrows; two females of the same generation may also share nests (McCorquodale 1989a). Females sharing one nest are often highly related and the larger female does much of the foraging whereas the smaller female acts as a guard. The guard expels ants, mutillids and alien females that try to enter the nest; nests with two females have higher success than those with solitary females (McCorquodale 1989c). Despite the different roles by females of different size, eusociality apparently never develops: there is no reproductive inhibition or division of labor (McCorquodale 1988, 1990). These observations are consistent with the interpretation that communal nesting by several females in *Cerceris* is associated with competition for the best nesting sites and nest-sharing allows females to reduce the frequency of usurpation and parasitism (Wcislo *et al.* 1985).

Females of the more social species of *Cerceris* are exceptionally long-lived for a sphecid and overwinter as adults, thus emerging earlier in the spring than animals that must first pupate (McCorquodale 1989b). Among bees there is a strong correlation between sociality and overwintering as adults (Brockmann 1984). Seger (1983) pointed this out in support of his theory that the overwintering life history of most halictid bees predisposes them to associate with their daughters. Although this is certainly true, one cannot help but ask why overwintering as an adult rather than as eggs or larvae has been favored by some but not other species. One possible answer is that overwintering as an adult gives individuals a head start and allows them to begin nesting as soon as conditions are favorable in the spring (Brockmann 1984). If this is correct, it means that competition for favorable overwintering sites (Strassmann 1979), or for nesting space or resources early in the season, may be a critical ingredient

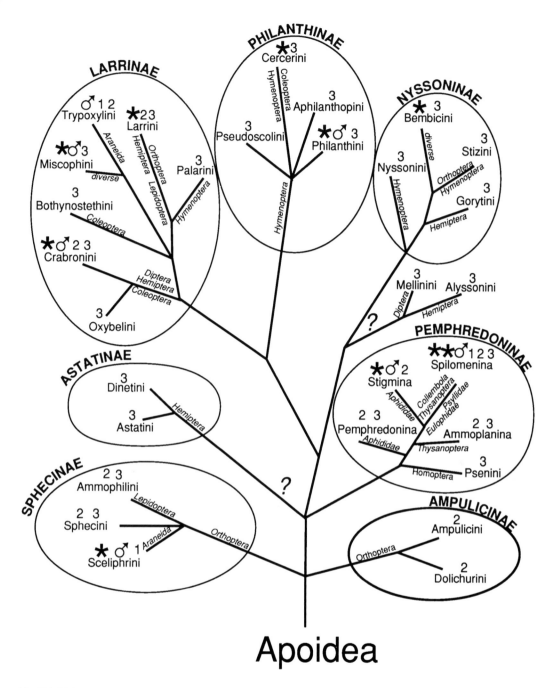

Apoidea

Fig. 17-2. Phylogeny of the Sphecidae showing the prey type (small type on arrows), the occurrence of sociality, the presence of males at the nest, and the type of nest used. The phylogeny is from Bohart and Menke (1976) as modified by Alexander (1992). Prey type is indicated in small letters on the branches. Level of sociality is indicated as '*' for long-term nesting associations among females and '**' for confirmed cases of eusociality (cooperative breeding), i.e. species with reproductive inhibition (from Brockmann and Dawkins 1979; Matthews 1991). The presence of males living on the nest is indicated with '♂' (data from Brockmann and Grafen 1989; Matthews 1991). The nest types are (1) free-standing, independent nests of mud or plant material or both; (2) trap-nesting (nest in existing cavities), (3) ground-nesting (construct nests in soil) (from Bohart and Menke 1976). A '?' on a branch means that the phylogenetic placement of this group is uncertain and/or this lineage is probably not a monophyletic group (Alexander 1992). Within the Pemphredoninae the tribe Pemphredonini has been divided into its four subtribes to show greater detail. Tribes in which the biology is completely unknown are not included.

for reproductive success. So perhaps the tendency to over-winter and the tendency to be social are both favored by intraspecific competition for resources such as suitable nesting sites.

Some level of sociality is found in six other lineages of sphecids (Fig. 17-2) (Brockmann and Dawkins 1979; Gess and Gess 1989; Lin and Michener 1972; Wcislo et al. 1985; Matthews 1991). Wasps from two subfamilies other than *Microstigmus* construct free-standing nests, both from mud. In both cases some level of nest-sharing has evolved in a few species of each group, although apparently without reproductive inhibition (Sphecinae, Eberhard 1972, 1974; Larrinae, Richards 1934; Sakagami et al. 1990; Gobbi et al. 1991). Furthermore, most lineages that evolve sociality are also lineages in which males are often found guarding the nest (Fig. 17-2) (Brockmann and Grafen 1989). Male guards are known to reduce parasitism and predation by ants, thus suggesting that nests of these groups are particularly vulnerable and two adults are needed to rear a successful brood (Hook and Matthews 1980). Finally, among the Sphecidae more trap-nesting (those nesting in pre-existing cavities) than ground-nesting genera have evolved nest-sharing (Brockmann and Dawkins 1979). Suitable locations for nesting are scarce in trap-nesting species and intense competition exists for the few available sites, particularly for the larger species (Krombein 1967). The evolution of group-living in sphecid wasps is associated with lineages suspected of having high nest parasitism or predation and intense competition for nesting sites. Although this conclusion is based on anecdotal information and weak correlations, it is consistent with the Ecological Constraints Model.

Sphecid wasps are now placed in the same superfamily as bees (Brothers and Carpenter 1993), where eusociality has arisen numerous times (Michener 1974). Why is eusociality so much more common among the Apinae than among the Sphecinae (Matthews 1991)? The preconditions for sociality are all present in most of the sphecids as in most bees: extensive parental investment in a few young, a fixed nest to which adults return repeatedly with food for dependent young, high vulnerability to predation and parasitism, intraspecific competition for limited nesting sites and resources, aggregation of nests, philopatry, the opportunity for maternal manipulation, high genetic relatedness, haplodiploidy, and kin or nestmate recognition (Eickwort 1981, 1986; Brockmann 1984; Andersson 1984; Matthews 1991; Pfennig and Reeve 1989, 1993). *Microstigmus*, however, are the only sphecid wasps that use plant fibers, as social

vespids do, to construct a pendant nest which Hansell (1987) has argued is an important factor in the evolution of sociality. They are one of the few sphecids that make a product (silk) that is incorporated into the nest structure, as bees use wax. If silk is expensive to produce then working together may be favored (Matthews 1991). *Microstigmus*, unlike other sphecids, cooperate in nest construction and they provision their nests with the smallest prey, and the largest number of prey, that are used by any sphecid wasp. All of these unusual attributes mean that workers in *Microstigmus*, as in social bees, can contribute much more than just nest protection.

A number of sphecid wasps nest communally, like many vertebrates, which means that all the elements of eusociality are present except that reproductive inhibition has not evolved. Clearly, for many species the decision to stay at the nest does not require a decision to delay breeding as is generally assumed. Both social insect and vertebrate biologists need to focus on the particular conditions that cause reproductive division of labor, dominance, inhibition, suppression or delayed breeding (whatever metaphor seems appropriate) to occur in some group-living species and not others (West-Eberhard 1978; Spradbery 1991; Sherman et al. 1995). This is the essence of the difference between communal and cooperative breeding and its determinants are not understood (Koenig et al. 1992).

Does the Ecological Constraints Model apply to *Polistes* wasps?

Many female paper wasps (gynes) of the genus *Polistes* begin nesting as a member of a group of foundresses (Table 17-5) (Reeve 1991). Although relatedness among cofoundresses is generally high, variation exists within species (Strassmann et al. 1989; Hughes et al. 1993). A foundress association jointly constructs the nest, defends it from predators, parasites and marauding conspecifics, lays the eggs and provisions the first brood of offspring (Reeve 1991; Gamboa et al. 1992). The females maintain a dominance hierarchy with one primary breeder that lays most eggs and remains at the nest and subordinates that lay few eggs and do most of the provisioning (Pardi 1948; West-Eberhard 1969, 1986). The differences among the females in such features as dominance and lifespan are determined by their circumstances (such as time of year), age, or condition (Strassmann and Meyer 1983; Hughes and Strassmann 1988; Mead and Gabouriaut 1993). Like vertebrate cooperative breeders, cofoundresses are similar in size and

Table 17-5. *Demonstrated advantages for foundress associations (pleometrosis) over independent nesting (haplometrosis) early in the season (prior to the emergence of the first daughter brood) in various species of* Polistes *wasps (from field studies)*

Species	Nest with foundress associations (%)	Location	Advantages of foundress associations	Reference
P. annularis	90	Georgia	Higher nest survivorship	Krispyn 1979
	—	Texas	Defense of winter honey caches	Strassmann 1979
	92, 100	Texas	—	Strassmann 1981a,d
	73, 82	Texas	Higher survivorship of adults; greater productivity	Queller and Strassmann 1988
	95–99	Texas	Higher nest survival, higher productivity	Strassmann 1989a,b
			Faster recovery from predation	Strassmann and Queller 1989
	30–31	Texas	—	Strassmann 1991
P. apachus	58	California	Reduced bird predation	Gibo and Metcalf 1978
	91	Texas	—	Lester and Selander 1981
P. bellicosus	78	Texas	Recovery from nest predation	Strassmann et al. 1988
	86	Texas	No effect of bird predation	Strassmann et al. 1987
P. canadensis	98	Colombia	Reduced ant predation	West-Eberhard 1969
	99	Panama	Greater productivity (larger groups make more adults)	Pickering 1980
	98	Panama	—	Itô 1985
P. exclamans	43, 14	Texas	No effect on vulnerability to predation, satellite nest production or ability to recover from predation	Strassmann 1981b
	86	Texas	—	Strassmann 1985
	6	Georgia	—	Hermann et al. 1975
P. fuscatus	43	Ontario	Faster recovery from predation	Gibo 1978
	62	Michigan	Greater productivity and better nest survival	Owen 1962
	—	Michigan	Defense against conspecific intruders	Fishwild and Gamboa 1992
	—	Wisconsin	Larger groups in warmer shelters; faster development	Jeanne and Morgan 1992
	52	Michigan	—	Noonan 1981
	79	Iowa	—	Klahn 1979
	56–61	Michigan	—	West-Eberhard 1969, 1981
P. gallicus	30	Italy	—	Turillazzi et al. 1982
	40	Italy	—	Pardi 1942, 1948
P. metricus	22	Illinois	Reduced predation; increased chance of survival of one queen and thus of nest survivorship	Metcalf and Whitt 1977a,b
	30–35	Kansas	Reduced usurpation by conspecifics	Gamboa 1978; Gamboa et al. 1978
	17	Ohio	Earlier emergence of workers; better adult survivorship	Gamboa 1980
	10–75	Kansas	—	Dropkin and Gamboa 1981
	9	Kansas	—	Bohm and Stockhammer 1977
	8	Kansas	—	Bohm 1972
	41	Ontario	—	Gibo 1978
	42	Kansas	—	Gamboa and Dropkin 1979
P. riparius	1	N. Japan	More likely to rebuild after nest loss	Makino 1989a,b
P. stigma	6	India	—	Suzuki and Ramesh 1992
P. versicolor	83	Panama	—	Itô 1985

Table 17-6. *Experimental and observational evidence regarding the nesting requirements of* Polistes *wasps*

Species	Nesting requirement	Nature of evidence	Reference
Observational evidence			
P. exclamans	Protected nest sites	More visible nests more likely to be knocked down by birds	Strassmann 1981a, 1985
P. fuscatus	Protected nest sites	Unprotected nests destroyed	Gibo 1978
P. gallicus	Protected nest sites	Unprotected nests destroyed	Turillazzi *et al.* 1982
P. metricus	Protected nest sites	Higher predation on unprotected nests	Gibo 1978
Correlation within and among populations			
P. annularis	Protected nest sites	No correlation between nest density and foundress number	Strassmann 1991; Queller and Strassmann 1988
P. exclamans	Less visible nest sites	Correlation between nest density and foundress number	Strassmann 1981c
P. metricus	Protected nest sites	Correlation between nest density and foundress number	Gamboa 1978
Experimental addition of resource			
P. carolinus	Nest boxes	Number of nesting sites does not affect foundress number	Strassmann and Queller 1989
P. fuscatus	Nest boxes	Readily accept nest boxes, most nest boxes used	Klahn and Gamboa 1983
	Nest box temperature	More nests constructed in warm boxes than in cool ones	Jeanne and Morgan 1992
P. metricus	Availability of food	Honey-supplemented colonies emerge earlier and have more fat	Rossi and Hunt 1988
	Nest boxes	High densities have more frequent multiple foundress associations	Gamboa 1980
P. riparius	Pre-emergence nest	Nestless queens readily accept nests	Lorenzi and Cervo 1992

morphology, unlike the queen and her daughters nesting in a matrifilial group later in the season. For this reason I will focus on foundress associations (pleometrotic groups) to examine the predictions of the Ecological Constraints Model (Strassmann and Queller 1989; Reeve 1991).

Is there evidence that habitat saturation favors cooperative breeding in *Polistes* (Table 17-4)? If the wasps are social because of limited availability of nesting sites, one should be able to show that they have specific nesting requirements. Interestingly, little information is available, aside from the general observation that most species seem to locate their nests in protected locations away from wind and rain (Table 17-6). Strassmann (1981c) and Queller and Strassmann (1988) have shown that the more visible nests within a population are more likely to be knocked down by birds. It is also a common observation that wasps will readily take over nest boxes that are provided (Gamboa 1978; Klahn and Gamboa 1983). Jeanne and Morgan (1992) have shown that *Polistes metricus* make

more subtle choices than previously thought, strongly preferring warmer nest boxes over cooler ones. The warmer boxes also contain larger cofoundress groups. Habitat saturation also predicts a positive correlation between wasp density and the frequency of multiple foundress colonies which Gamboa (1978) found in *P. metricus,* a species with high rates of intraspecific nest usurpation. However, in a dense population of *P. annularis* nesting along a cliff, Queller and Strassmann (1988) found no association between nest density and the size of the foundress group. Strassmann and Queller (1989) specifically address the question of habitat saturation by conducting experiments in which they increased the number of nesting sites available to *P. carolinus*. Although the wasps used the nest boxes, Strassmann and Queller found no change in the size of foundress associations and thus they rejected habitat saturation as a likely factor favoring pleometrosis in this species. Taken together, then, the evidence for habitat saturation in *Polistes* is weak.

Is there evidence that high costs of independent reproduction favor sociality? The frequency of non-breeding workers should vary with the degree of difficulty in becoming a breeder (third prediction, Table 17-4). Both Reeve (1991) and Queller (1996) have shown a positive correlation between the size of foundress associations and the probability that nests founded by single females fail. These results suggest that ecological constraints play a role in encouraging some females to join foundress associations rather than nest on their own.

In many species of *Polistes*, females nesting in cofoundress associations are thought to have higher reproductive success than lone females (fourth prediction, Table 17-4; see also Table 17-5), but in each case the evidence is equivocal (Reeve 1991). The factors that have been identified as favoring foundress associations are similar to those that favor cooperative breeding (Table 17-2) and include reduced predation and higher nest survival, faster recovery from predation, higher queen survivorship, reduced usurpation and marauding by conspecifics and higher colony production of reproductives (Reeve 1991; Brockmann 1993; Seger 1993). Experimental manipulations are lacking, but one recent study in a closely related genus of paper wasp showed that queens deprived of their subordinates had greatly reduced success (Clouse 1994). However, like several other studies (Gibo 1974; Klahn 1979; Gadagkar *et al.* 1988; Itô 1993), this one also suggests that females that choose to found nests with a foundress association may be different (in size, condition or behavior) from those that nest alone.

Nestless females also have a third option open to them: usurpation. As the season progresses, females that have lost their nests switch their behavior from rebuilding nests to usurping nests (Klahn 1988; Makino 1989a,b; Nonacs and Reeve 1993). When females take over nests, they systematically destroy the youngest brood, particularly if the nest is nearing the emergence of the first workers (Klahn and Gamboa 1983; Lorenzi and Cervo 1992). This pattern of destruction greatly increases the chance that only workers will emerge and that the reproductive brood from the preceding queen will be eliminated (Nonacs and Reeve 1993). This observation means that the foundress association is of increasing value to the breeder until the workers emerge, as in pleometrotic ants (Choe and Perlman, this volume).

The food that cofoundresses provide for the brood is not normally considered a factor in the evolution of behavioral eusociality, although it is a commonly identified

selective force in studies on the evolution of cooperative breeding in vertebrates (Krebs and Davies 1993). Rossi and Hunt (1988) supplemented *P. metricus* nests with honey and found that young from fed colonies emerged earlier and were fatter than those from unsupplemented controls. Brood cannibalism, apparently resulting from low prey abundance, is common in some species (Kasuya 1983; Hunt 1991). These observations suggest that supplying the young with an adequate and reliable food supply is an important benefit for wasps nesting in groups (Hunt and Nalepa 1993; Wenzel and Pickering 1991).

Taken together, some evidence can be found that high costs of reproduction favor foundress associations in *Polistes* wasps. The arguments differ for different species but defense against conspecifics, parasites and predators and the ability to build up the worker force quickly are the most commonly cited selective pressures (Table 17-5). Despite the ease of working with *Polistes*, careful multivariate analyses of natural variation and experimental manipulations of the sort that have been conducted on cooperatively breeding vertebrates (see, for example, Mumme 1992b; Komdeur 1994) are almost completely lacking. The ecological constraints on cooperative breeding can be established only by coupling correlational studies with experimental manipulations of foundress number (Reeve 1991).

DISCUSSION

Many similarities exist in the pattern of sociality found in behaviorally eusocial wasps and vertebrates. These social systems are so similar that Crespi and Yanega (1995) argue they should be referred to collectively as cooperative breeding. Strassmann *et al.* (1994) come to a similar conclusion for the stenogastrine wasps. Although differences in biology make some comparisons difficult (e.g. the lack of territoriality in social wasps), the essential elements of the two social systems are the same: conspecific adults living together and cooperating in the rearing of non-descendent young where only one or a few adults in the group reproduce. Similar patterns are not necessarily the product of the same selective pressures, however. The Ecological Constraints Model has been generally accepted as the probable explanation for the evolution of vertebrate cooperative breeding. The two most frequently identified selective pressures are the inability of individuals to find suitable breeding opportunities or nesting sites (habitat saturation) and some extreme environmental condition that makes independent reproduction so costly that a pair cannot rear offspring on their own (high costs of independent reproduction). In vertebrates the most frequently identified causes of these high costs are a lack of food for the offspring, exceptionally high predation, intense competition from conspecifics and the necessity for group activities such as cooperative hunting or nest construction (Table 17-3). Among behaviorally eusocial wasps, little evidence exists that a shortage of nesting sites favor group nesting (i.e. habitat saturation seems unlikely) although Strassmann *et al.* (1994) suspect that this may be an important element in the sociality of stenogastrine wasps. For *Microstigmus* and many species of *Polistes*, it seems likely that foundresses band together because of exceptionally high costs of independent reproduction. Together they construct nests, feed young and protect their brood from predation and conspecific usurpation; together they can also recover more quickly once an attack has occurred. Using the weak correlative data that are available, one can say that similar selective pressures have been identified as important contributing factors in the evolution of both vertebrate and wasp cooperative breeding. In this sense, then, sociality in these two groups is not just similar but convergent. These explanations, however, account only for group-living and not for reproductive inhibition, the essential difference between cooperative and communal breeding.

Does reproductive inhibition precede the decision to remain in a group or is it a consequence of that decision (Michener and Brothers 1974)? Is reproductive inhibition externally imposed or self-imposed (Reyer *et al.* 1986; Koenig *et al.* 1992; Emlen and Wrege 1992)? Much has been written about the proximate control and ultimate advantages of reproductive inhibition within colonies of social insects (West-Eberhard 1981; Dew 1983; Röseler 1985, 1991; Engels 1990; Spradbery 1991; Reeve 1991; Keller 1993), but there is little discussion of these topics for vertebrates (Andersson 1984). In some species such as acorn woodpeckers, social competition is an obvious and important element (Koenig and Mumme 1987), but for most vertebrate cooperative breeders establishing dominance is more subtle. Schoech *et al.* (1991) have found that in Florida scrub jays non-breeding adult helpers have reduced gonadal hormones but they also have lower levels of stress hormones (corticosterones) than breeders, suggesting delayed breeding rather than enforced reproductive dominance. Similar results were found for pied kingfishers by Reyer *et al.* (1986). In naked mole rats, on the other hand, non-breeders are constantly being pushed and intimidated by

the queen (Reeve 1992). Non-breeding individuals have both low gonadal hormones and increased levels of corticosterones (Faulkes *et al.* 1991) suggesting that reproductive inhibition is socially induced (Reeve and Sherman 1991) as in some other mammalian cooperative breeders (Jennions and Macdonald 1994). Creel and Creel (1991) argue that in dwarf mongoose the energetic costs of reproduction are so high that individuals without helpers are unable to breed successfully. Reeve and Ratnieks (1993) present a model to explain the intensity of reproductive inhibition (i.e. the proportion of individuals reproducing within a group) in social insects, and their model can be applied to other species of cooperative breeders (Keller and Reeve 1994). According to this model, the degree of reproductive skew within a group should increase (1) when few reproductive options exist for the helper (a measure of the ecological constraints on the subordinate), (2) when the aid given by the helper greatly increases breeder productivity, (3) when the fighting ability of the worker is low and (4) when genetic relatedness is high. Clearly, long-term population data and experimental manipulations are needed to understand the evolution of reproductive inhibition (Reeve 1991).

The presumption in the vast majority of studies is that animals who stay at a natal nest or who join a group will help, but helping is, in fact, a separate decision (Brown 1987; Mumme 1992a). A season of watching wasps (Clouse 1994), mole rats (Reeve 1992) or bee-eaters (Emlen *et al.* 1991) will easily convince one that working is not a necessary consequence of the decision to stay (Poiani 1994). Gadagkar and Joshi (1983) categorize female wasps into fighters, foragers and sitters. Strassmann (1989a) found that *Polistes annularis* females ceased rearing brood far earlier in the season than expected based on weather or what females of other species in the same area were doing. She found that the early loss of the queen resulted in an earlier cessation of brood-rearing than when queens lived longer. Late-emerging females apparently have the option of becoming workers or gynes; when they are more closely related to the brood, they are more likely to become workers. In addition, in drought years females ceased rearing earlier in the season than in wet years, presumably because their chance of increasing the size of the brood was reduced. Together these data suggest considerable flexibility in the willingness of females to help. Similarly, in dwarf mongoose, Creel and Waser (1991) found that when females were more closely related to the dominant female, they were more likely to make substantial investments in the offspring. Clearly, much more research needs to be done to understand the variability in helping that occurs among individuals within groups.

Polyethism, i.e. the development of different roles within a group, is a common feature of social-insect biology and is thought to increase the efficiency of the colony (West-Eberhard 1981; Jeanne 1986, 1991). Even species with no consistent morphological differences between queens and workers show some role specialization based on age, size or time at the nest (Post et al. 1988; Fishwild and Gamboa 1992; Sinha et al. 1993; Clouse 1994). Naked mole rats clearly have some differentiation based on age and size: small non-breeders dig and provision whereas larger individuals guard the nest (Sherman et al. 1991). With this exception, however, studies of vertebrate cooperative breeding seem to have overlooked polyethism. In birds, for example, parents generally incubate and the helpers feed the parents but do not incubate; in a number of species, the helpers do more feeding of fledglings than of young nestlings (which are fed by the parents) and in some species male and female helpers behave differently (Woolfenden and Fitzpatrick 1984; Hailman et al. 1994). More data are needed on behavioral differences among individuals within groups, on ergonomic differences among individuals of different ages or sizes, and even, perhaps, on whether physical changes (or differences) are associated with helping.

If helpers greatly increase the success of breeders, one might expect breeders to produce some offspring that would have little expectation of success on their own (Michener and Brothers 1974; Alexander 1974), as is found in some social insects (Hunt 1991, 1993). Little information of this sort is available for vertebrates (Alexander 1991), but some intriguing observations exist. In Florida scrub jays, for example, some individuals become long-term or perpetual helpers, whereas others breed at a young age (Woolfenden and Fitzpatrick 1984). The chronically male-biased sex ratios of some cooperatively breeding birds suggest that breeders sometimes bias their reproduction toward the sex that is more likely to become helpers (Emlen et al. 1986).

Throughout this discussion one cannot help but be struck by the weak and circumstantial evidence that is used to provide support for the Ecological Constraints Model. Clear predictions can be made (Table 17-4), but few data are available. Most of the evidence is correlative with only a few experimental studies. In most cases the factors that affect reproductive success in cooperative breeders have not been identified. Particularly bothersome is that the factors that favor cooperative breeding are discussed without reference to closely related species that are not breeding cooperatively (Mumme 1992a). Why are Florida scrub jays cooperative breeders when California scrub jays are not (Koenig et al. 1992)? Why are most group-living sphecids communal nesters whereas Microstigmus is a cooperative breeder? Why do some Polistes wasps form large foundress associations when closely related species do not? These remain unanswered questions.

By considering primitive eusociality and cooperative breeding in the same chapter, it is possible to evaluate the generality of our explanations for the evolution of sociality. We are limited in this enterprise by our terminology and by our metaphors. By calling the breeder a 'queen' we conjure up associations that influence our ability to consider alternative hypotheses. By calling a worker a 'helper' or a 'subordinate' and the social system 'cooperative', we fail to see the inherent sources of social competition that are affecting the evolution of social systems (Emlen 1982b, 1991; Brown and Brown 1990; Davies 1990; Sherman et al. 1995). Vertebrate studies tell social insect biologists to consider the role of food and nesting sites in limiting reproductive success, to explore the possible role of ecological constraints and habitat saturation in favoring the evolution of eusociality and to conduct experimental manipulations and multivariate correlational studies to evaluate models. Studies of social insects tell vertebrate biologists to look more carefully at reproductive inhibition and skew, at polyethism and parental manipulation and at the evolutionary history of helping behavior.

ACKNOWLEDGEMENTS

The ideas presented in this chapter have evolved over many years from exchanges with numerous colleagues studying vertebrate social systems and social insects. I thank Walter Koenig, Joan Strassmann, Robert Matthews, Ron Clouse, Bernie Crespi and an anonymous reviewer for reading the manuscript and providing valuable suggestions. This study was initiated while I was supported by a grant from the Visiting Professorships for Women Program of the National Science Foundation and while I was visiting the Department of Biology at Princeton University (1986).

LITERATURE CITED

Alexander, B. A. 1992. An exploratory analysis of cladistic relationships within the superfamily Apoidea, with special reference to sphecid wasps (Hymenoptera). J. Hym. Res. 1: 25–61.

Alexander, R. D. 1974. The evolution of social behavior. *Annu. Rev. Ecol. Syst.* **5**: 323–383.

–. 1991. Some unanswered questions about naked mole-rats. In *The Biology of the Naked Mole-Rat*. P. W. Sherman, J. U. M. Jarvis and R. D. Alexander, eds., pp. 446–465. Princeton: Princeton University Press.

Alexander, R. D., K. M. Noonan and B. J. Crespi. 1991. The evolution of eusociality. In *The Biology of the Naked Mole-Rat*. P. W. Sherman, J. U. M. Jarvis and R. D. Alexander, eds., pp. 3–44. Princeton: Princeton University Press.

Andersson, M. 1984. The evolution of eusociality. *Annu. Rev. Ecol. Syst.* **15**: 165–189.

Bohart, R. M. and A. S. Menke. 1976. *Sphecid Wasps of the World*. Berkeley: University of California Press.

Bohm, M. K. 1972. Reproduction in *Polistes metricus*. Ph. D. dissertation, University of Kansas.

Bohm, M. K. and K. A. Stockhammer. 1977. The nesting cycle of a paper wasp, *Polistes metricus* (Hymenoptera: Vespidae). *J. Kansas Entomol. Soc.* **50**: 275–286.

Brett, R. A. 1991. The ecology of naked mole-rat colonies: burrowing, food, and limiting factors. In *The Biology of the Naked Mole-Rat*. P. W. Sherman, J. U. M. Jarvis and R. D. Alexander, eds., pp. 137–184. Princeton: Princeton University Press.

Brockmann, H. J. 1984. The evolution of social behavior in insects. In *Behavioural Ecology: An Evolutionary Approach*, 2nd edition. J. Krebs and N. B. Davies, eds., pp. 340–361. Oxford: Blackwell Scientific Publications.

–. 1986. Decision making in a variable environment: lessons from insects. In: *Behavioral Ecology and Population Biology*. L. Drickamer, ed., pp. 95–111. Toulouse: Privat Publisher.

–. 1990. Primitive eusociality: comparisons between Hymenoptera and vertebrates. In *Social Insects and the Environment. Proceedings of the 11th International Congress of IUSSI, 1990*. G. K. Veeresh, B. Mallik and C. A. Viraktamath, eds., p. 77. New Delhi: Oxford & IBH Publishing Co. Pvt. Ltd.

–. 1993. Parasitizing conspecifics: comparisons between Hymenoptera and birds. *Trends Ecol. Evol.* **8**: 2–3

Brockmann, H. J. and R. Dawkins. 1979. Joint nesting in a digger wasp as an evolutionarily stable preadaptation to social life. *Behaviour* **71**: 203–245.

Brockmann, H. J. and A. Grafen. 1989. Mate conflict and male behavior in a solitary wasp *Trypoxylon (Trypargilum) politum* (Hymenoptera: Sphecidae). *Anim. Behav.* **37**: 232–255.

Brothers, D. J. and J. M. Carpenter. 1993. Phylogeny of Aculeata: Chrysidoidea and Vespoidea. *J. Hym Res.* **2**: 227–302.

Brown, J. L. 1974. Alternate routes to sociality in jays – with a theory for the evolution of altruism and communal breeding. *Am. Zool.* **14**: 63–89.

–. 1987. *Helping and Communal Breeding in Birds: Ecology and Evolution*. Princeton: Princeton University Press.

Brown, J. L. and E. R. Brown. 1990. Mexican jays: uncooperative breeding. In *Cooperative Breeding in Birds: Long-term Studies of Ecology and Behavior*. P. B. Stacey and W. D. Koenig, eds., pp. 267–288. Cambridge University Press.

Brown, J. L. E. R. Brown, S. D. Brown and D. D. Dow. 1982. Helpers: effects of experimental removal on reproductive success. *Science (Wash., D.C.)* **215**: 421–422.

Choe, J. C. 1995. Plurimatry: new terminology for multiple reproductives. *J. Insect Behav.* **8**: 133–137

Clouse, R. M. 1994. Nesting decisions of the social paper wasp, *Mischocyttarus mexicanus*. M. Sc. thesis, University of Florida.

Craig, J. L. and I. G. Jamieson. 1990. Pukeko: different approaches and some different answers. In *Cooperative Breeding in Birds: Long-term Studies of Ecology and Behavior*. P. B. Stacey and W. D. Koenig, eds., pp. 387–412. Cambridge University Press.

Creel, S. R. and N. M. Creel. 1991. Energetics, reproductive suppression and obligate communal breeding in carnivores. *Behav. Ecol. Sociobiol.* **28**: 263–270.

Creel, S. R., N. Creel, D. E. Wildt and S. L. Monfort. 1992. Behavioural and endocrine mechanisms of reproductive suppression in Serengeti dwarf mongoose. *Anim. Behav.* **43**: 231–245.

Creel, S. R. and P. M. Waser. 1991. Failures of reproductive suppression in dwarf mongooses (*Helogale parvula*): accident or adaptation. *Behav. Ecol.* **2**: 7–15.

Crespi, B. J. and D. Yanega. 1995. The definition of eusociality. *Behav. Ecol.* **6**: 109–115.

Curry, R. L. 1989. Geographic variation in social organization of Galapagos mockingbirds: ecological correlates of group territoriality and cooperative breeding. *Behav. Ecol. Sociobiol.* **24**: 147–160.

Curry, R. L. and P. R. Grant. 1990. Galapagos mockingbirds: territorial cooperative breeding in a climatically variable environment. In *Cooperative Breeding in Birds: Long-term Studies of Ecology and Behavior*. P. B. Stacey and W. D. Koenig, eds., pp. 291–331. Cambridge University Press.

Davies, N. B. 1990. Dunnock: cooperation and conflict among males and females in a variable mating system. In: *Cooperative Breeding in Birds: Long-term Studies of Ecology and Behavior*. P. B. Stacey and W. D. Koenig, eds., pp. 455–485. Cambridge University Press.

Demelo, G. A. R. and L. A. D. O. Campos. 1993a. Nesting biology of *Microstigmus myersi* Turner, a wasp with long-haired larvae (Hymenoptera: Sphecidae, Pemphredoninae). *J. Hym. Res.* **2**: 183–188.

–. 1993b. Trophallaxis in a primitively social sphecid wasp. *Insectes Soc.* **40**: 107–109.

Dew, H. E. 1983. Division of labor and queen influence in laboratory colonies of *Polistes metricus* (Hymenoptera; Vespidae). *Z. Tierpsychol.* **61**: 127–140.

Dropkin, J. A. and G. Gamboa. 1981. Physical comparisons of foundresses of the paper wasp *Polistes metricus* (Hymenoptera: Vespidae). *Can. Entomol.* **113**: 457–461.

Eberhard, W. G. 1972. Altruistic behavior in a sphecid wasp: support for kin-selection theory. *Science (Wash., D.C.)* **175**: 1390–1391.

–. 1974. The natural history and behaviour of the wasp *Trigonopsis cameronii* Kohl (Sphecidae). *Trans. R. Entomol. Soc. Lond.* **125**: 295–328.

Eickwort, G. C. 1975. Gregarious nesting of the mason bee *Hoplitis anthocopoides* and the evolution of parasitism and sociality among megachilid bees. *Evolution* **29**: 142–150.

–. 1981. Presocial insects. In *Social Insects*, vol. 2, H. R. Hermann, ed., pp. 199–280. New York: Academic Press..

–. 1986. First steps into eusociality: the sweat bee *Dialictus lineatulus*. *Fla. Entomol.* **69**: 742–754.

Elliott, N. B. and W. M. Elliott. 1987. Nest usurpation by females of *Cerceris cribrosa* (Hymenoptera: Sphecidae). *J. Kansas Entomol. Soc.* **60**: 397–402.

Elliott, N. B., T. Shlotzhauer and W. M. Elliott. 1986. Nest use by females of the presocial wasp *Cerceris watlingensis* (Hymenoptera: Sphecidae). *Ann. Entomol. Soc. Am.* **79**: 994–998.

Emlen, S. T. 1982a. The evolution of helping I. An ecological constraints model. *Am. Nat.* **119**: 29–39.

–. 1982b. The evolution of helping II. The role of behavioral conflict. *Am. Nat.* **119**: 40–53.

–. 1984. Cooperative breeding in birds and mammals. In *Behavioural Ecology: An Evolutionary Approach*, 2nd edn. J. Krebs and N. B. Davies, eds., pp. 305–339. Oxford: Blackwell Scientific Publications.

–. 1990. White-fronted bee-eaters: helping in a colonially nesting species. In *Cooperative Breeding in Birds: Long-term Studies of Ecology and Behavior*. P. B. Stacey and W. D. Koenig, eds., pp. 489–526. Cambridge University Press.

–. 1991. Evolution of cooperative breeding in birds and mammals. In *Behavioural Ecology: An Evolutionary Approach*, 3rd edn. J. Krebs and N. B. Davies, eds., pp. 301–335. Oxford: Blackwell Scientific Publications.

Emlen, S. T, J. M. Emlen and S. A. Levin. 1986. Sex-ratio selection in species with helpers-at-the-nest. *Am. Nat.* **127**: 1–8.

Emlen, S. T., H. K. Reeve, P. W. Sherman and P. H. Wrege. 1991. Adaptive versus nonadaptive explanations of behavior: the case of alloparental helping. *Am. Nat.* **138**: 259–270.

Emlen, S. T. and S. L. Vehrencamp. 1985. Cooperative breeding strategies among birds. In *Experimental Behavioral Ecology and Sociobiology: In Memoriam Karl von Frisch 1886–1982*. B. Hölldobler and M. Lindauer, eds., pp. 359–374. Sunderland: Sinauer Associates.

Emlen, S. T. and P. H. Wrege. 1989. A test of alternate hypotheses for helping behavior in white-fronted bee-eaters of Kenya. *Behav. Ecol. Sociobiol.* **25**: 303–319.

–. 1991. Breeding biology of white-fronted bee-eaters at Nakuru: the influence of helpers on breeder fitness. *J. Anim. Ecol.* **60**: 309–326.

–. 1992. Parent-offspring conflict and the recruitment of helpers among bee-eaters. *Nature, (Lond.)* **356**: 331–333.

–. 1994. Gender, status and family fortunes in the white-fronted bee-eater. *Nature (Lond.)* **367**: 129–132.

Engels, W. 1990. *Social Insects: An Evolutionary Approach to Castes and Reproduction*. Berlin: Springer-Verlag.

Evans, H. E. and A. W. Hook. 1982. Communal nesting in the digger wasp *Cerceris australis* (Hymenoptera Sphecidae). *Austr. J. Zool.* **30**: 557–568.

–. 1986. Nesting behavior of Australian *Cerceris* digger wasps with special reference to nest reutilization and nest sharing (Hymenoptera, Sphecidae). *Sociobiology* **11**: 275–302.

Faaborg, J. and J. C. Bednarz. 1990. Galapagos and Harris' hawks: divergent causes of sociality in two raptors. In *Cooperative Breeding in Birds: Long-term Studies of Ecology and Behavior*. P. B. Stacey and W. D. Koenig, eds., pp. 357–383. Cambridge University Press.

Faulkes, C. G., D. H. Abbott, C. E. Liddell, L. M. George and J. U. M. Jarvis. 1991. Hormonal and behavioral aspects of reproductive suppression in female naked mole-rats. In *The Biology of the Naked Mole-Rat*. P. W. Sherman, J. U. M. Jarvis and R. D. Alexander, eds., pp. 426–445. Princeton: Princeton University Press.

Field, J. 1992. Intraspecific parasitism as an alternative reproductive tactic in nest-building wasps and bees. *Biol. Rev.* **67**: 79–126.

Fishwild, T. G. and G. J. Gamboa. 1992. Colony defense against conspecifics: caste-specific differences in kin recognition by paper wasps, *Polistes fuscatus*. **43**: 95–102.

Frame, L. H., J. R. Malcolm, G. W. Frame and H. van Lawick. 1980. Social organization of African wild dogs (*Lycaon acrus*) on the Serengeti Plains, Tanzania 1967–1978. *Z. Tierpsychol.* **50**: 225–249.

Gadagkar, R. and N. V. Joshi. 1983. Quantitative ethology of social wasps: time-activity budgets and caste differentiation in *Ropalidia marginata* (Lep.) (Hymenoptera: Vespidae). *Anim. Behav.* **31**: 26–31.

Gadagkar, R., C. Vinutha, A. Shanubhogue and A. P. Gore. 1988. Pre-imaginal biasing of caste in a primitively eusocial insect. *Proc. R. Soc. Lond.* **B233**: 175–189.

Gamboa, G. J. 1978. Intraspecific defense: advantage of social cooperation among paper wasp foundresses. *Science (Wash., D.C.)* **199**: 1463–1465.

–. 1980. Comparative timing of brood development between multiple- and single-foundress colonies of the paper wasp, *Polistes metricus. Ecol. Entomol.* **5**: 221–225.

Gamboa, G. J and J. A. Dropkin. 1979. Comparisons of behaviors in early vs. late foundress associations of the paper wasp, *Polistes metricus* (Hymenoptera: Vespidae). *Can. Entomol.* **111**: 919–926.

Gamboa, G. J., B. D. Heacock and S. L. Wiltjer. 1978. Division of labor and subordinate longevity in foundress associations of the paper wasp *Polistes metricus* (Hymenoptera: Vespidae). *J. Kansas Entomol. Soc.* **51**: 343–352.

Gamboa, G. J., T. L. Wacker, K. G. Duffy, S. W. Dobson and T. G. Fishwild. 1992. Defence against intraspecific usurpation by paper wasp cofoundresses (*Polistes fuscatus*, Hymenoptera, Vespidae). *Can. J. Zool.* **70**: 2369–2372.

Gess, S. K. and F. W. Gess. 1989. Notes on nesting behaviour in *Bembix bubalus* Handlirsch in southern Africa with the emphasis on nest sharing and reaction to nest parasites. *Ann. Cape Prov. Mus. (Nat. Hist.)* **18**: 151–160.

Gibo, D. L. 1974. A laboratory study of the selective advantage of foundress associations in *Polistes fuscatus* (Hymenoptera: Vespidae). *Can. Entomol.* **106**: 101–106.

–. 1978. The selective advantage of foundress associations in *Polistes fuscatus* (Hymenoptera: Vespidae): a field study of the effects of predation on productivity. *Can. Entomol.* **110**: 519–540.

Gibo, D. L. and R. A. Metcalf. 1978. Early survival of *Polistes apachus* (Hymenoptera: Vespidae) colonies in California: a field study of an introduced species. *Can. Entomol.* **110**: 1339–1343.

Gittleman, J. L. 1985. Functions of communal care in mammals. In *Evolution: Essays in Honour of John Maynard Smith.* P. J. Greenwood, P. H. Harvey and M. Slatkin, eds., pp. 187–205. Cambridge University Press.

Gobbi, N., S. F. Sakagami and R. Zucchi. 1991. Nesting biology of a quasisocial sphecid wasp *Trypoxylon fabricator. Jap. J. Entomol.* **59**: 37–51.

Goldizen, A. W. 1987. Facultative polyandry and the role of infant-carrying in wild saddle-back tamarins (*Saguinus fuscicolla*). *Behav. Ecol. Sociobiol.* **20**: 99–109.

–. 1989. Social relationships in a cooperatively polyandrous group of tamarins (*Saguinus fuscicollis*). *Behav. Ecol. Sociobiol.* **24**: 79–90.

Grafen, A. 1984. Natural selection, kin selection and group selection. In *Behavioural Ecology: An Evolutionary Approach*, 2nd edn. J. Krebs and N. B. Davies, eds., pp. 62–86. Oxford: Blackwell Scientific Publications.

Hailman, J. P., K. J. McGowan and G. E. Woolfenden. 1994. Role of helpers in the sentinel behavior of the Florida scrub jay (*Aphelocoma c. coerulescens*). *Ethology* **97**: 119–140.

Hansell, M. 1987. Nest building as a facilitating and limiting factor in the evolution of eusociality in the Hymenoptera. In *Oxford Surveys in Evolutionary Biology*, vol. 4. P. H. Harvey and L. Partridge, eds., pp. 155–181. Oxford University Press.

Hermann, H. R., R. Barron and L. Dalton. 1975. Spring behavior of *Polistes exclamans* (Hymenoptera: Vespidae: Polistinae). *Entomol. News* **86**: 173–178.

Hoogland, J. L. 1981. Nepotism and cooperative breeding in the black-tailed prairie dog (Sciuridae: *Cynomys ludovicianus*). In *Natural Selection and Social Behavior: Recent Research and New Theory.* R. D. Alexander and D. W. Tinkle, eds., pp. 283–310. New York: Chiron Press.

Hook, A. W. 1987. Nesting behavior of Texas *Cerceris* digger wasps with emphasis on nest reutilization and nest sharing (Hymenoptera: Sphecidae). *Sociobiology* **13**: 93–118.

Hook, A. W. and R. W. Matthews. 1980. Nesting biology of *Oxybelus sericeus* with a discussion of nest guarding by male sphecid wasps (Hymenoptera). *Psyche* **87**: 21–37.

Hughes, C. R. and J. E. Strassmann. 1988. Age is more important than size in determining dominance among workers in the primitively eusocial wasp, *Polistes instabilis. Behaviour* **108**: 1–14.

Hughes, C. R., D. C. Queller, J. E. Strassmann and S. K. Davis. 1993. Relatedness and altruism in *Polistes* wasps. *Behav. Ecol.* **4**: 128–137.

Hunt, J. H. 1991. Nourishment and the evolution of the social Vespidae. In *The Social Biology of Wasps.* K. G. Ross and R. W. Matthews, eds., pp. 426–450. Ithaca: Cornell University Press.

–. 1993. Nourishment and social evolution in wasps sensu lato. In *Nourishment and Evolution in Insect Societies.* J. H. Hunt and C. A. Nalepa, eds., pp. 1–34. Boulder: Westview Press.

Hunt, J. H. and C. A. Nalepa. 1993. Nourishment, evolution and insect sociality. In *Nourishment and Evolution in Insect Societies.* J. H. Hunt and C. A. Nalepa, eds., pp. 1–34. Boulder: Westview Press.

Itô, Y. 1985. A comparison of frequency of intra-colony aggressive behavior among five species of polistine wasp (Hymenoptera: Vespidae). *Z. Tierpsychol.* **68**: 152–167.

–. 1993. *Behavior and Social Evolution of Wasps. The Communal Aggregation Hypothesis.* Oxford University Press.

Jarvis, J. U. M. 1981. Eusociality in a mammal: cooperative breeding in naked mole-rat colonies. *Science (Wash., D.C.)* **212**: 571–573.

Jarvis, J. U. M., M. J. O'Rianin, N. C. Bennett and P. Sherman. 1994. Mammalian eusociality: a family affair. *Trends Ecol. Evol.* **9**: 47–51.

Jeanne, R. L. 1986. The evolution of the organization of work in social insects. *Monit. Zool. Ital.* **20**: 119–133.

–. 1991. Polyethism. In *The Social Biology of Wasps.* K. G. Ross and R. W. Matthews, eds., pp. 389–425. Ithaca: Comstock Publishing Associates.

Jeanne, R. L. and R. C. Morgan. 1992. The influence of temperature on nest site choice and reproductive strategy in a temperate zone *Polistes* wasp. *Ecol. Entomol.* **17**: 135–141.

Jennions, M. D. and D. W. Macdonald. 1994. Cooperative breeding in mammals. *Trends Ecol. Evol.* **9**: 89–93.

Kasuya, E. 1983. Behavioral ecology of Japanese paper wasps, *Polistes* spp. (Hymenoptera: Vespidae). II. Ethogram and internidal relationship in *Polistes chinensis antennalis* in the founding stage. *Z. Tierpsychol.* **63**: 303–317.

Keller, L., ed. 1993. *Queen Number and Sociality in Insects.* Oxford University Press.

–. 1995. Social life: the paradox of multiple-queen colonies. *Trends Ecol. Evol.* **10**: 355–360.

Keller, L. and H. K. Reeve. 1994. Partitioning of reproduction in animal societies. *Trends. Ecol. Evol.* **9**: 98–102.

Kinnaird, M. F. and P. R. Grant. 1982. Cooperative breeding by the Galapagos mockingbird *Nesomimus parvulus. Behav. Ecol. Sociobiol.* **10**: 65–73.

Klahn, J. E. 1979. Philopatric and nonphilopatric foundress associations in the social wasp *Polistes fuscatus. Behav. Ecol. Sociobiol.* **5**: 417–424.

–. 1988. Intraspecific comb usurpations in the social wasp *Polistes fuscatus*. *Behav. Ecol. Sociobiol.* **23**: 1–8.

Klahn, J. E. and G. F. Gamboa 1983. Social wasps: discrimination between kin and nonkin brood. *Science (Wash., D.C.)* **221**: 482–484.

Koenig, W. D. 1981. Reproductive success, group size, and the evolution of cooperative breeding in the Acorn Woodpecker. *Am. Nat.* **117**: 421–443.

Koenig, W. D. and R. L. Mumme. 1987. *Population Ecology of the Cooperatively Breeding Acorn Woodpecker.* Princeton: Princeton University Press.

–. 1990. Levels of analysis and the functional significance of helping behavior. In *Interpretation and Explanation in the Study of Animal Behavior: II. Explanation, Evolution and Adaptation.* M. Bekoff and D. Jamieson, eds., pp. 268–303. Boulder: Westview Press.

Koenig, W. D. and F. A. Pitelka. 1981. Ecological factors and kin selection in the evolution of cooperative breeding in birds. In *Natural Selection and Social Behavior: Recent Research and New Theory*, R. D. Alexander and D. W. Tinkle, eds., pp. 261–282. New York: Chiron Press.

Koenig, W. D., F. A. Pitelka, W. J. Carmen, R. L. Mumme and M. T. Stanback. 1992. The evolution of delayed dispersal in cooperative breeders. *Q. Rev. Biol.* **67**: 111–150.

Koenig, W. D. and P. B. Stacey. 1990. Acorn woodpeckers: group-living and food storage under contrasting ecological conditions. In *Cooperative Breeding in Birds: Long-term Studies of Ecology and Behavior.* P. B. Stacey and W. D. Koenig, eds., pp. 415–453. Cambridge University Press.

Koford, R. R., B. S. Bowen and S. L. Vehrencamp. 1990. Groove-billed anis: joint-nesting in a tropical cuckoo. In *Cooperative Breeding in Birds: Long-term Studies of Ecology and Behavior.* P. B. Stacey and W. D. Koenig, eds., pp. 335–355. Cambridge University Press.

Komdeur, J. 1992. Importance of habitat saturation and territory quality for evolution of cooperative breeding in the Seychelles warbler. *Nature (Lond.)* **358**: 493–495.

–. 1994. Experimental evidence for helping and hindering by previous offspring in the cooperative-breeding Seychelles warbler *Acrocephalus sechellensis. Behav. Ecol. Sociobiol.* **34**: 175–186.

Komdeur, J., A. Huffstadt, W. Prast, G. Castle, R. Mileto and J. Wattel. 1995. Transfer experiments of Seychelles warblers to new islands: changes in dispersal and helping behaviour. *Anim. Behav.* **49**: 695–708.

Krebs, J. R. and N. B. Davies. 1993. *An Introduction to Behavioral Ecology.* Oxford: Blackwell Scientific Publications.

Krispyn, J. W. 1979. Colony productivity and survivorship of the paper wasp *Polistes annularis.* Ph. D. dissertation, University of Georgia.

Krombein, K. V. 1967. *Trap-nesting Wasps and Bees: Life Histories, Nests, and Associates.* Washington, D.C.: Smithsonian Institution Press.

Kukuk, P. 1994. Replacing the terms 'primitive' and 'advanced': new modifiers for the term 'eusocial.' *Anim. Behav.* **47**: 1475–1478.

Lacey, E. A. and P. W. Sherman. 1991. Social organization of naked mole-rat colonies: evidence for divisions of labor. In *The Biology of the Naked Mole-Rat.* P. W. Sherman, J. U. M. Jarvis and R. D. Alexander, eds., pp. 275–336. Princeton: Princeton University Press.

Leighton, M. 1986. Hornbill social dispersion: variations on a monogamous theme. In *Ecological Aspects of Social Evolution: Birds and Mammals.* D. I. Rubenstein and R. W. Wrangham, eds., pp. 108–130. Princeton: Princeton University Press.

Leonard, M. L., A. G. Horn and S. F. Eden. 1989. Does juvenile helping enhance breeder reproductive success? A removal experiment on moorhens. *Behav. Ecol. Sociobiol.* **25**: 357–361.

Lester, L. J. and R. K. Selander. 1981. Genetic relatedness and the social organization of *Polistes* colonies. *Am. Nat.* **117**: 147–166.

Ligon, J. D. and S. H. Ligon. 1988. Territory quality: key determinant of fitness in the group-living green wood-hoopoe. In *The Ecology of Social Behavior.* C. N. Slobodchikoff, ed., pp. 229–253. New York: Academic Press.

–. 1990. Green woodhoopoes: life history traits and sociality. In *Cooperative Breeding in Birds: Long-term Studies of Ecology and Behavior.* P. B. Stacey and W. D. Koenig, eds., pp. 31–65. Cambridge University Press.

Lin, N. and C. D. Michener. 1972. Evolution and sociality in insects. *Q. Rev. Biol.* **47**: 131–159.

Litte, M. 1977. Behavioral ecology of the social wasp, *Mischocyttarus mexicanus. Behav. Ecol. Sociobiol.* **2**: 229–246.

Lorenzi, M. C. and R. Cervo. 1992. Behaviour of *Polistes biglumis bimaculatus* (Hymenoptera, Vespidae) foundresses on alien conspecific nests. In *Biology and Evolution of Social Insects.* J. Billen, ed., pp. 273–279. Leuven: Leuven University Press.

Macdonald, D. W. and P. D. Moehlman. 1982. Cooperation, altruism, and restraint in the reproduction of carnivores. In *Perspectives in Ethology*, vol. 5. *Ontogeny.* P. P. G. Bateson and P. H. Klopfer, eds., pp. 433–467. New York: Plenum.

Makino, S. 1989a. Switching of behavioral option from renesting to nest usurpation after nest loss by the foundress of a paper wasp, *Polistes riparius*: a field test. *J. Ethol.* **7**: 62–64.

–. 1989b. Usurpation and nest rebuilding in *Polistes riparius*: two ways to reproduce after the loss of the original nest (Hymenoptera: Vespidae). *Insectes Soc.* **36**: 116–128.

Malcolm, J. R. and K. Marten. 1982. Natural selection and the communal rearing of pups in African wild dogs (*Lycaon pictus*). *Behav. Ecol. Sociobiol.* **10**: 1–13.

Matthews, R. W. 1968a. *Microstigmus comes*: sociality in a sphecid wasp. *Science (Wash., D.C.)* **160**: 787–788.

–. 1968b. Nesting biology of the social wasp *Microstigmus comes* (Hymenoptera: Sphecidae, Pemphredoninae). *Psyche* **75**: 23–45.

–. 1991. Evolution of social behavior in sphecid wasps. In *The Social Biology of Wasps.* K. G. Ross and R. W. Matthews, eds., pp. 570–602. Ithaca: Comstock Publishing Associates.

Matthews, R. W. and I. D. Naumann. 1988. Nesting biology and taxonomy of *Arpactophilus mimi*, a new species of social sphecid (Hymenoptera: Sphecidae) from northern Australia. *Austr. J. Zool.* **36**: 585–597.

Matthews, R. W. and C. K. Starr. 1984. *Microstigmus comes* wasps have a method of nest construction unique among social insects. *Biotropica* **16**: 55–58.

McCorquodale, D. B. 1988. Relatedness among nestmates in a primitively social wasp. *Cerceris antipodes* (Hymenoptera: Sphecidae). *Behav. Ecol. Sociobiol.* **23**: 401–406.

–. 1989a. Nest defense in single- and multi-female nests of *Cerceris antipodes* (Hymenoptera: Sphecidae). *J. Insect Behav.* **2**: 267–276.

–. 1989b. Nest sharing, nest switching, longevity and overlap of generations in *Cerceris antipodes* (Hymenoptera: Sphecidae). *Insectes Soc.* **36**: 42–50.

–. 1989c. Soil softness, nest initiation and nest sharing in the wasp *Cerceris antipodes* (Hymenoptera: Sphecidae). *Ecol. Entomol.* **14**: 191–196.

–. 1990. Oocyte development in the primitively social wasp, *Cerceris antipodes* (Hymenoptera Sphecidae). *Ethol. Ecol. Evol.* **2**: 345–361.

McCorquodale, D. B. and I. D. Naumann. 1988. A new Australian species of communal ground nesting wasp in the genus *Spilomena* (Hymenoptera: Sphecidae: Pemphredoninae). *J. Austr. Entomol. Soc.* **27**: 221–231.

Mead, F. and D. Gabouriaut 1993. Post eclosion sensitivity to social context in *Polistes dominulus* Christ females (Hymenoptera, Vespidae). *Insectes Soc.* **40**: 11–20.

Metcalf, R. A. and G. S. Whitt. 1977a. Intra-nest relatedness in the social wasp *Polistes metricus:* A genetic analysis. *Behav. Ecol. Sociobiol.* **2**: 339–351.

–. 1977b. Relative inclusive fitness in the social wasp *Polistes metricus. Behav. Ecol. Sociobiol.* **2**: 353–360.

Michener, C. D. 1974. *The Social Behavior of the Bees: A Comparative Study.* Cambridge: Harvard University Press.

Michener, C. D. and D. J. Brothers. 1974. Were workers of eusocial Hymenoptera initially altruistic or oppressed? *Proc. Natl. Acad. Sci. U.S.A.* **71**: 671–674.

Mills, M. G. L. 1989. The comparative behavioral ecology of hyenas: the importance of diet and food dispersion. In *Carnivore Behavior, Ecology, and Evolution.* J. L. Gittleman, ed. pp. 125–142. Ithaca: Cornell University Press.

Moehlman, P. 1986. Ecology of cooperation in canids. In *Ecological Aspects of Social Evolution: Birds and Mammals.* D. I. Rubenstein and R. W. Wrangham, eds., pp. 64–86. Princeton: Princeton University Press.

Mueller, U. G., A. F. Warneie, T. U. Grafe and P. R. Ode. 1992. Female size and nest defense in the digger wasp *Cerceris fumipennis* (Hymenoptera, Sphecidae, Philanthinae). *J. Kansas Entomol. Soc.* **65**: 44–52.

Mumme, R. 1992a. Do helpers increase reproductive success? An experimental analysis in the Florida scrub jay. *Behav. Ecol. Sociobiol.* **31**: 319–328.

–. 1992b. Delayed dispersal and cooperative breeding in the Seychelles warbler. *Trends Ecol. Evol.* **7**: 330–331.

Nonacs, P. and H. K. Reeve. 1993. Opportunistic adoption of orphaned nests in paper wasps as an alternative reproductive strategy. *Behav. Proc.* **30**: 47–60.

Noonan, K. M. 1981. Individual strategies of inclusive-fitness-maximizing in *Polistes fuscatus.* In *Natural Selection and Social Behavior: Recent Research and New Theory,* R. D. Alexander and D. W. Tinkle, eds., pp. 18–44. New York: Chiron Press.

Owen, J. 1962. The behavior of a social wasp *Polistes fuscatus* (Vespidae) at the nest, with special reference to differences between individuals. Ph. D. dissertation, University of Michigan.

Packer, C. 1986. The ecology of sociality in felids. In *Ecological Aspects of Social Evolution: Birds and Mammals.* D. I. Rubenstein and R. W. Wrangham, eds., pp. 429–451. Princeton: Princeton University Press.

Pardi. L. 1942. Richerche sui Polistini. V. La poliginia inizial in *Polistes gallicus* L. *Boll. Ist. Entomol. Univ. Bologna* **14**: 1–106.

–. 1948. Dominance order in *Polistes* wasps. *Physiol. Zool.* **21**: 1–13.

Pfennig, D. W. and H. K. Reeve. 1989. Neighbor recognition and context-dependent aggression in a solitary wasp, *Sphecius speciosus* (Hymenoptera: Sphecidae). *Ethology* **80**: 1–18.

–. 1993. Nepotism in a solitary wasp as revealed by DNA fingerprinting. *Evolution* **47**: 700–704.

Pickering, J. 1980. Social biology of *Polistes canadensis*. Ph. D. dissertation, Harvard University.

Poiani, A. 1994. Inter-generational competition and selection for helping behavior. *J. Evol. Biol.* **7**: 419–434.

Pollock, G. B. and S. W. Rissing. 1988. Social competition under mandatory group life. In *The Ecology of Social Behavior.* C. N. Slobodchikoff, ed., pp. 315–334. New York: Academic Press.

Post, D. C., R. L. Jeanne and E. H. Erickson. 1988. Variation in behavior among workers of the primitively social wasp *Polistes fuscatus variatus.* In *Interindividual Behavioral Variability in Social Insects.* R. L. Jeanne, ed., pp. 283–321. Boulder: Westview Press.

Pruett-Jones, S. G. and M. J. Lewis. 1990. Sex ratio and habitat limitation promote delayed dispersal in superb fairy-wrens. *Nature (Lond.)* **348**: 541–542.

Queller, D. C. 1989. The evolution of eusociality: reproductive head starts of workers. *Proc. Natl. Acad. Sci. U.S.A.* **86**: 3224–3226.

–. 1996. The origin and maintenance of eusociality: the advantage of extended parental care. In *Sociality in* Polistes. S. Turillazzi and M. J. West-Eberhard, eds. Oxford University Press, in press.

Queller, D. C. and J. E. Strassmann. 1988. Reproductive success and group nesting in the paper wasp, *Polistes annularis.* In *Reproductive Success: Studies of Individual Variation in Contrasting Breeding Systems.* T. H. Clutton-Brock, ed., pp. 76–98. Chicago: University of Chicago Press.

Rabenold, K. N. 1990. *Campylorhynchus* wrens: the ecology of delayed dispersal and cooperation in the Venezuelan savanna. In *Cooperative Breeding in Birds: Long-term Studies of Ecology and Behavior.* P. B. Stacey and W. D. Koenig, eds., pp. 159–196. Cambridge University Press.

Rasa, O. A. E. 1986. Sociability for survival: Why dwarf mongooses live in groups. In *Behavioral Ecology and Population Biology.* L. Drickamer, ed., pp. 35–39. Toulouse: Privat Publisher.

–. 1987. The dwarf mongoose: a study of behavior and social structure in relation to ecology in a small, social carnivore. *Adv. Stud. Behav.* **17**: 121–163.

Reeve, H. K. 1991. *Polistes.* In *The Social Biology of Wasps.* K. G. Ross and R. W. Matthews, eds., pp. 99–148. Ithaca: Cornell University Press.

–. 1992. Queen activation of lazy workers in colonies of the eusocial naked mole-rat. *Nature (Lond.)* **358**: 147–149.

Reeve, H. K. and G. J. Gamboa. 1983. Colony activity integration in primitively eusocial wasps: the role of the queen (*Polistes fuscatus*, Hymenoptera: Vespidae). *Behav. Ecol. Sociobiol.* **13**: 63–74.

–. 1987. Queen regulation of worker foraging in paper wasps: a social feedback control system (*Polistes fuscatus*, Hymenoptera: Vespidae). *Behaviour* **102**: 147–167.

Reeve, H. K. and P. Nonacs. 1992. Social contracts in wasp societies. *Nature (Lond.)* **359**: 823–825.

Reeve, H. K. and F. L. W. Ratnieks. 1993. Queen–queen conflicts in polygynous societies: mutual tolerance and reproductive skew. In *Queen Number and Sociality in Insects.* L. Keller, ed., pp. 45–85. Oxford University Press.

Reeve, H. K. and P. W. Sherman. 1991. Intracolonial aggression and nepotism by the breeding female naked mole-rat. In *The Biology of the Naked Mole-Rat.* P. W. Sherman, J. U. M. Jarvis and R. D. Alexander, eds., pp. 337–357. Princeton: Princeton University Press.

Reeve, H. K., D. F. Westneat, W. A. Noon, P. W. Sherman and C. F. Aquadro. 1990. DNA "fingerprinting" reveals high levels of inbreeding in colonies of the eusocial naked mole-rat. *Proc. Natl. Acad. Sci. U.S.A.* **87**: 2496–2500.

Reyer, H. U. 1990. Pied kingfishers: ecological causes and reproductive consequences of cooperative breeding. In *Cooperative Breeding in Birds: Long-term Studies of Ecology and Behavior.* P. B. Stacey and W. D. Koenig, eds., pp. 529–557. Cambridge University Press.

Reyer, H. U., J. P. Dittami and M. R. Hall. 1986. Avian helpers at the nest: are they psychologically castrated? *Ethology* **71**: 216–228.

Richards, O. W. 1934. The American species of the genus *Trypoxylon* (Hymenoptera: Sphecoidea). *Trans. R. Entomol. Soc. Lond.* **82**: 173–362.

Riedman, M. L. 1982. The evolution of alloparental care and adoption in mammals and birds. *Q. Rev. Biol.* **57**: 405–435.

Rood, J. P. 1978. Dwarf mongoose helpers at the den. *Z. Tierpsychol.* **48**: 277–287.

–. 1986. Ecology and social evolution in the mongooses. In *Ecological Aspects of Social Evolution: Birds and Mammals.* D. I. Rubenstein and R. W. Wrangham, eds., pp. 131–152. Princeton: Princeton University Press.

Röseler, P. F. 1985. Endocrine basis of dominance and reproduction in polistine paper wasps. In *Experimental Behavioral Ecology and Sociobiology: In Memoriam Karl von Frisch 1886-1982*, B. Hölldobler and M. Lindauer, eds., pp. 259–272. Sunderland: Sinauer Associates.

–. 1991. Reproductive competition during colony establishment. In *The Social Biology of Wasps.* K. G. Ross and R. W. Matthews, eds., pp. 309–335. Ithaca: Comstock Publishing Associates.

Ross, K. G. and R. W. Matthews. 1989a. New evidence for eusociality in the sphecid wasp *Microstigmus comes. Anim. Behav.* **38**: 613–619.

–. 1989b. Population genetic structure and social evolution in the sphecid wasp *Microstigmus comes. Am. Nat.* **134**: 574–598.

Rossi, A. M. and J. H. Hunt 1988. Honey supplementation and its developmental consequences: evidence for food limitation in a paper wasp, *Polistes metricus. Ecol. Entomol.* **13**: 437–442.

Rowley, I. and E. Russell. 1990. Splendid fairy-wrens: demonstrating the importance of longevity. In *Cooperative Breeding in Birds: Long-term Studies of Ecology and Behavior.* P. B. Stacey and W. D. Koenig, eds., pp. 3–30. Cambridge University Press.

Sakagami, S. F., N. Gobbi and R. Zucchi. 1990. Nesting biology of a quasisocial sphecid wasp *Trypoxylon fabricator.* I. Nests and inhabitants. *Jap. J. Entomol.* **58**: 846–862.

Schoech, S. J., R. L. Mumme and M. C. Moore. 1991. Reproductive endocrinology and mechanisms of breeding inhibition in cooperatively breeding Florida scrub jays (*Aphelocoma c. coerulescens*). *Condor* **93**: 354–364.

Seger, J. 1983. Partial bivoltinism may cause alternating sex-ratio biases that favour eusociality. *Nature (Lond.)* **301**: 59–62.

–. 1991. Cooperation and conflict in insects. In *Behavioral Ecology: An Evolutionary Approach*, 3rd edn. J. R. Krebs and N. B. Davies, eds. pp. 338–373. Oxford: Blackwell Scientific Publications.

–. 1993. Opportunities and pitfalls in cooperative reproduction. In *Queen Number and Sociality in Insects*, L. Keller, ed., pp. 1–15. Oxford University Press.

Sherman, P. W., J. U. M Jarvis and R. D. Alexander, eds. 1991. *The Biology of the Naked Mole-Rat.* Princeton: Princeton University Press.

Sherman, P. W., E. A. Lacey, H. K. Reeve and L. Keller. 1995. The eusociality continuum. *Behav. Ecol.* **6**: 102–108.

Siemens, M. 1990. Brood care or egg cannibalism by parents and helpers in *Neolamprologus brichardi* (Poll 1986) (Pisces: Cichlidae): A study on behavioural mechanisms. *Ethology* **84**: 60–80.

Sinha, A., S. Premnath, K. Chandrashekara and R. Gadagkar. 1993. *Ropalidia rufoplagiata*: a polistine wasp society probably lacking permanent reproductive division of labour. *Insectes Soc.* **40**: 69–86.

Spradbery, J. P. 1991. Evolution of queen number and queen control. In *The Social Biology of Wasps*. K. G. Ross and R. W. Matthews, eds., pp. 570–602. Ithaca: Comstock Publishing Associates.

Stacey, P. B. and J. D. Ligon. 1987. Territory quality and dispersal options in the acorn woodpecker, and a challenge to the habitat-saturation model of cooperative breeding. *Am. Nat.* **130**: 654–676.

Strambi, A. 1990. Physiology and reproduction in social wasps. In *Social Insects. An Evolutionary Approach to Castes and Reproduction*. W. Engels, ed., pp. 59–75. Berlin: Springer-Verlag.

Strassmann, J. E. 1979. Honey caches help female paper wasp (*Polistes annularis*) survive Texas winters. *Science (Wash. D.C.)* **204**: 207–209.

–. 1981a. Evolutionary implications of early male and satellite nest production in *Polistes exclamans* colony cycles. *Behav. Ecol. Sociobiol.* **8**: 551–564.

–. 1981b. Kin selection and satellite nests in *Polistes exclamans*. In *Natural Selection and Social Behavior: Recent Research and New Theory*. R. D. Alexander and D. W. Tinkle, eds., pp. 45–60. New York: Chiron Press.

–. 1981c. Parasitoids, predators, and group size in the paper wasp, *Polistes exclamans*. *Ecology* **62**: 1225–1233.

–. 1981d. Wasp reproduction and kin selection: reproductive competition and dominance hierarchies among *Polistes annularis* foundresses. *Fla. Entomol.* **64**: 74–88.

–. 1985. Relatedness of workers to brood in the social wasp *Polistes exclamans* (Hymenoptera: Vespidae). *Z. Tierpsychol.* **69**: 141–148.

–. 1989a. Early termination of brood rearing in the social wasp, *Polistes annularis* (Hymenoptera: Vespidae). *J. Kansas Entomol. Soc.* **62**: 353–362.

–. 1989b. Group colony foundation in *Polistes annularis* (Hymenoptera: Vespidae). *Psyche* **96**: 223–236.

–. 1991. Costs and benefits of colony aggregation in the social wasp *Polistes annularis*. *Behav. Ecol.* **2**: 204–209.

–. 1993. Weak queen or social contract? *Nature (Lond.)* **363**: 502–503.

Strassmann, J. E., C. R. Hughes, D. C. Queller, S. Turillazzi, R. Cervo, S. K. Davis and K. F. Goodnight. 1989. Genetic relatedness in primitively eusocial wasps. *Nature (Lond.)* **342**: 268–270.

Strassmann, J. E., C. R. Hughes, S. Turillazzi, C. R. Solis and D. C. Queller. 1994. Genetic relatedness and incipient eusociality in stenogastrine wasps. *Anim. Behav.* **48**: 813–821.

Strassmann, J. E. and D. C. Meyer. 1983. Gerontocracy in the social wasp *Polistes exclamans*. *Anim. Behav.* **31**: 431–438.

Strassmann, J. E. and D. Queller. 1989. Ecological determinants of social evolution. In *The Genetics of Social Evolution*, M. D. Breed and R. E. Page, eds., pp. 81–101. Boulder: Westview Press.

Strassmann, J. E., D. C. Queller and C. R. Hughes. 1987. Constraints on independent nesting by *Polistes* foundresses in Texas. In *Chemistry and Biology of Social Insects*. J. Eder and H. Rembold, eds., pp. 379–380. Munich: Verlag J. Peperny.

–. 1988. Predation and the evolution of sociality in the paper wasp *Polistes bellicosus*. *Ecology* **69**: 1497–1505.

Suzuki, T. and M. Ramesh. 1992. Colony founding in the social wasp *Polistes stigma* (Hymenoptera Vespidae), in India. *Ethol. Ecol. Evol.* **4**: 333–341.

Taborsky, M. 1984. Broodcare helpers in the cichlid fish *Lamprologus brichardi*: their costs and benefits. *Anim. Behav.* **32**: 1236–1252.

–. 1987. Cooperative behaviour in fish: coalitions, kin groups and reciprocity. In *Animal Societies: Theories and Facts*, Y. Itô, J. L. Brown and J. Kikkawa, eds., pp. 229–237. Tokyo: Japan Scientific Society Press.

Taborsky, M. and D. Limberger. 1981. Helpers in fish. *Behav. Ecol. Sociobiol.* **8**: 143–145.

Terborgh, J. and A. W. Goldizen. 1985. On the mating system of the cooperatively breeding saddle-backed tamarin (*Saguinus fuscicollis*). *Behav. Ecol. Sociobiol.* **16**: 293–299.

Turillazzi, S. and L. Pardi. 1977. Body size and hierarchy in polygynic nests of *Polistes gallicus* (L) (Hymenoptera, Vespidae). *Monit. Zool.* **11**: 101–112.

Turillazzi, S., M. T. Marino Piccioli, L. Hervatin and L. Pardi. 1982. Reproductive capacity of single foundress and associated foundress females of *Polistes gallicus* (L.) (Hymenoptera: Vespidae). *Monit. Zool.* **16**: 75–88.

Vehrencamp, S. L. 1979. The roles of individual, kin, and group selection in the evolution of sociality. In *Handbook of Behavioral Neurobiology*. vol. 3. *Social Behavior and Communication*. P. Marler and J. G. Vandenbergh, eds., pp. 351–394. New York: Plenum Press.

Vehrencamp, S. L., R. R. Koford and B. S. Bowen. 1988. The effect of breeding-unit size on fitness components in groove-billed anis. In *Reproductive Success: Studies of Individual Variation in Contrasting Breeding Systems*. T. H. Clutton-Brock, ed., pp. 291–304. Chicago: University of Chicago Press.

Walters, J. R. 1990. Red-cockaded woodpeckers: a 'primitive' cooperative breeder. In *Cooperative Breeding in Birds: Long-term Studies of Ecology and Behavior*. P. B. Stacey and W. D. Koenig, eds., pp. 67–101. Cambridge University Press.

Walters, J. R., C. K. Copeyon and J. H. Carter. 1992. Test of the ecological basis of cooperative breeding in red-cockaded woodpeckers. *Auk* **109**: 90–97.

Wcislo, W. T., B. S. Low and C. J. Karr. 1985. Parasite pressure and repeated burrow use by different individuals of *Crabro* (Hymenoptera: Sphecidae; Diptera: Sarcophagidae). *Sociobiology* **11**: 115–125.

Wenzel, J. W. and J. Pickering. 1991. Cooperative foraging, productivity, and the central limit theorem. *Proc. Natl. Acad. Sci. U.S.A.* **88**: 36–38.

West, M. J. 1967. Foundress associations in polistine wasps: dominance hierarchies and the evolution of social behavior. *Science (Wash., D.C.)* **157**: 1584–1585.

West-Eberhard, M. J. 1969. The social biology of polistine wasps. *Misc. Publ. Mus. Zool. Univ. Mich.* **140**: 1–101.

−. 1975. The evolution of social behavior by kin selection. *Q. Rev. Biol.* **50**: 1–33.

−. 1978. Polygyny and the evolution of social behavior of wasps. *J. Kansas Entomol. Soc.* **51**: 832–856.

−. 1981. Intragroup selection and the evolution of insect societies. In *Natural Selection and Social Behavior: Recent Research and New Theory*. R. D. Alexander and D. W. Tinkle, eds., pp. 3–17. New York: Chiron Press.

−. 1986. Dominance relations in *Polistes canadensis* (L.), a tropical wasp. *Monit. Zool.* **20**: 263–281.

Wickler, W. 1978. Kin selection and effectiveness in social insect workers and other helpers. *Z. Tierpsychol.* **48**: 100–103.

Wilkinson, R. 1982. Social organization and communal breeding in the chestnut-bellied starling (*Spreo pulcher*). *Anim. Behav.* **30**: 1118–1128.

Wiley, R. H. and K. N. Rabenold. 1984. The evolution of cooperative breeding by delayed reciprocity and queuing for favorable social positions. *Evolution* **18**: 609–621.

Willmer, P. G. 1985. Thermal ecology, size effects and the origins of communal behaviour in *Cerceris* wasps. *Behav. Ecol. Sociobiol.* **17**: 151–160.

Wilson, E. O. 1971. *The Insect Societies*. Cambridge: Harvard University Press.

−. 1975. *Sociobiology*. Cambridge: Harvard University Press.

Woolfenden, G. E. and J. W. Fitzpatrick. 1984. *The Florida Scrub Jay: Demography of a Cooperative-breeding Bird*. Princeton: Princeton University Press.

−. 1990. Florida scrub jays: a synopsis of 18 years of study. In *Cooperative Breeding in Birds: Long-term Studies of Ecology and Behavior*. P. B. Stacey and W. D. Koenig, eds., pp. 239–266. Cambridge University Press.

Zack, S. and K. N. Rabenold. 1989. Assessment, age and proximity in dispersal contests among cooperative wrens: field experiments. *Anim. Behav.* **38**: 235–247.

Zahavi, A. 1990. Arabian babblers: the quest for social status in a cooperative breeder. In *Cooperative Breeding in Birds: Long-term Studies of Ecology and Behavior*. P. B. Stacey and W. D. Koenig, eds., pp. 105–130. Cambridge University Press.

18 · Morphologically 'primitive' ants: comparative review of social characters, and the importance of queen–worker dimorphism

CHRISTIAN PEETERS

ABSTRACT

Ants that exhibit a relatively large proportion of ancestral morphological traits are commonly believed to exhibit less social complexity. This was examined through a comparative review of *Nothomyrmecia*, *Myrmecia*, and the heterogeneous subfamily Ponerinae. A limited dimorphism between winged queens and workers is a common characteristic of morphologically 'primitive' ants. This has two important consequences: (1) queens are not very fecund, and this is reflected in the small size of colonies (a few dozens to several hundreds of workers); (2) solitary foundresses are unable to rear their first generation of workers without hunting outside their nests. In contrast, some of the 'primitive' species having permanently wingless (= ergatoid) queens exhibit considerably greater fecundity, and colonies reach a few thousands. There is also a small number of permanently queenless ponerine ants, in which one or more mated workers reproduce (several dozens to a few hundreds of workers per colony).

Ecological characteristics of 'primitive' ants include their predatory habits (in addition, sweet secretions are collected in various genera), various degrees of diet specialization, and a widespread lack of cooperation among foragers (although several species with ergatoid queens, and larger colonies, exhibit sophisticated recruitment and even group predation). Role specialization among sterile workers (including the influence of age) follows the typical formicid pattern; polymorphic workers occur in only one species. True trophallaxis among nestmates does not exist (two exceptions only); eggs (either reproductive or trophic) are eaten in several species.

Small colony size influences many ecological characteristics, as well as the ability to produce an appropriate number of sexual individuals annually. Independent foundation in 'primitive' taxa may not always be adaptive, because the dealate queens have to take risks while hunting above ground. The widespread occurrence of colony fission may be an evolutionary response to this shortcoming. Once there is no longer independent foundation, natural selection can lead either to the production of ergatoid queens, or to the replacement of the queen caste by gamergates.

INTRODUCTION

All ants live in societies in which the majority of female adults are sterile, with only one or a few mated egg-layers. In ants as well as various other social insects (vespine wasps, some bees, and termites), this reproductive division of labor is characteristically based on the production of two distinct female phenotypes: queens and workers. Morphological caste specialization is arguably a key to the ecological success of the Formicidae. An ever-increasing dimorphism between queens and workers has made possible the elaboration of their social organization (Wheeler 1986).

'Primitive' ants is a colloquial expression used here for a number of unrelated taxa that have retained a relatively large proportion of ancestral (plesiomorphic) morphological characters, e.g. *Nothomyrmecia*, *Myrmecia*, *Apomyrma*, and several amblyoponine genera. The tribe Amblyoponini includes the most generalized species in the large subfamily Ponerinae. This review investigates whether morphologically 'primitive' ants also exhibit primitive social characteristics, as is widely believed. All the Ponerinae are included, for comparative purposes. Special emphasis is placed on the extensive secondary modifications in their reproductive biology. Relevant comparisons with 'higher' ants (detailed in Fig. 18-1) are made. I argue that the degree of queen–worker dimorphism provides a useful comparative framework to investigate the various grades of social complexity.

Colonies of the hypothetical ancestral ants presumably consisted of monomorphic females, and reproductive differentiation occurred among these (this is analogous to what occurs in extant polistine wasps, polybiine wasps, and halictine bees). In ants, such ancestral species are extinct, because

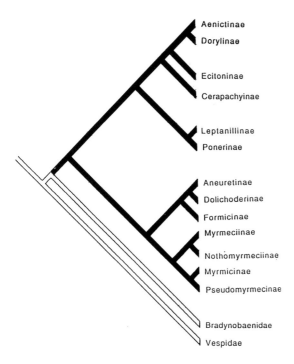

Aenictinae
Dorylinae
Ecitoninae
Cerapachyinae
Leptanillinae
Ponerinae
Aneuretinae
Dolichoderinae
Formicinae
Myrmeciinae
Nothomyrmeciinae
Myrmicinae
Pseudomyrmecinae
Bradynobaenidae
Vespidae

Fig. 18-1. Tentative phylogeny of the Formicidae resulting from a cladistic analysis of the presently recognized ant subfamilies (adapted from Baroni Urbani *et al.* 1992). Two vespoid families were included as outgroups. The four uppermost subfamilies constitute the doryline section. Two minor extant subfamilies have been omitted for clarity.

'primitive' species exhibit morphologically distinct queens and workers. Haskins and Haskins (1950a) and Haskins (1970) previously discussed the archaic social structure of *Myrmecia* and various ponerines, including *Amblyopone*. Caryl Haskins was first to recognize that the study of their social profile might give a picture, however imperfect, of the simple stages through which the 'higher' ants have passed in attaining their present remarkably differentiated social organization. Although *Myrmecia* remains poorly studied, the Ponerinae have in recent years become the focus of many investigations (see review of reproductive biology in Peeters 1993). Almost nothing is known about *Apomyrma* (Brown *et al.* 1970). Following Peeters and Crozier (1988), I use the terms 'queen', 'worker' and 'caste' in a strict morphological sense.

INTERNAL PHYLOGENY OF THE ANTS

Brown (1954) divided the Formicidae into two groups of subfamilies, and this basal dichotomy was recently upheld

by a cladistic analysis based on 68 characters (Baroni Urbani *et al.* 1992) (Fig. 18-1). Subfamilies Apomyrminae and Ponerinae belong to one clade, while Myrmeciinae and Nothomyrmeciinae belong to the other. These four subfamilies exhibit generalized morphological traits within their respective clades (although they probably no longer resemble the ancestors of the other subfamilies). The findings of Baroni Urbani *et al.* (1992) challenge the idea that *Nothomyrmecia* is the least-specialized living ant, but it remains unambiguous that the poneroid complex diverged from the myrmecioid complex early on. Thus two independently derived 'lower' groups are available to compare and illustrate the progressive elaboration of social structure.

The genus *Myrmecia* is the only extant representative of the subfamily Myrmeciinae. It is endemic to the Australian region, with 90 species currently recognized (Ogata 1991). The monospecific *Nothomyrmecia* has been found only in southern Australia, and is currently placed in a subfamily of its own (Taylor 1978). Myrmeciinae and Nothomyrmeciinae appear to be sister groups (Baroni Urbani *et al.* 1992).

The subfamily Ponerinae (1300 species, Bolton 1995) is morphologically very heterogeneous; Bolton (1990b) and Baroni Urbani *et al.* (1992) consider it to be monophyletic, but Hashimoto (1991) and Ward (1994) disagree. Six tribes are currently recognized: Amblyoponini (6 extant genera), Ectatommini (9 genera), Platythyreini (2 genera), Ponerini (23 genera), Thaumatomyrmecini (1 genus) and Typhlomyrmecini (1 genus) (Bolton 1994). The tribe Cerapachyini (e.g. *Cerapachys* and *Sphinctomyrmex*) is no longer included in Ponerinae (Bolton 1990a). A nearly completed revision of tribe Ponerini by William Brown, Jr., will result in the synonymy of several genera under *Pachycondyla* (see Bolton 1994); here, I have used the current traditional names. Ponerine ants flourish world-wide in tropical and subtropical regions, but are poorly represented in the temperate zones. The taxonomic diversity of the Ponerinae is paralleled by very heterogeneous patterns of colony organization, which makes it difficult to distill their essential attributes.

REPRODUCTIVE CHARACTERISTICS OF PONERINAE, *MYRMECIA* AND *NOTHOMYRMECIA*

The existence of phenotypically distinct female castes is a hallmark of ant eusociality. Queen–worker dimorphism can be highly pronounced in the advanced subfamilies, but it is generally small in 'primitive' ants (notwithstanding the absence of wings in workers) (Ponerinae, Brown 1958, 1960, 1976; *Myrmecia*, Haskins and Haskins 1950a; Freeland 1958; Gray 1971c; *Nothomyrmecia*, Taylor 1978). Within the Ponerinae, size dimorphism varies considerably, and reaches a maximum in *Brachyponera lutea* where queens are twice as large as the workers (Haskins and Haskins 1950b). In comparison, myrmicine queens can be up to ten times larger than conspecific workers.

Ovarian specialization

Peeters (1987a) provided the first comparative data on ovarian dimorphism and colony size in Ponerinae, illustrating the importance of specialized queen morphology with respect to egg production. This approach has become common (Villet 1990a; Villet *et al.* 1991; Peeters 1991b; Ito and Ohkawara 1994) and has helped to document considerable interspecific diversity in this subfamily. Many species exhibit little or no difference in the number of ovarioles between the two castes, whereas ovaries are markedly dimorphic in other species (Ponerinae, Peeters 1993; *Nothomyrmecia*, Hölldobler and Taylor 1983; *Myrmecia froggatti*, Ito *et al.* 1994; *M. gulosa*, C. Peeters, unpublished). Caste divergence is even more marked in a few species in which workers lack ovaries (Peeters 1991a; Villet *et al.* 1991; Ito and Okhawara 1994).

A reflection of this limited queen–worker differentiation is the generally low fecundity of ponerine and myrmeciine queens (seldom in excess of 5 eggs per day). In contrast, a monogyne queen of the myrmicine ant *Solenopsis invicta* can lay 150 eggs per hour (Tschinkel 1988).

Associated with limited ovarian dimorphism is the retention of a spermatheca by workers in most ponerine species (Peeters 1991a), as well as in *Nothomyrmecia* (Hölldobler and Taylor 1983) and in at least two species of *Myrmecia* (Crosland *et al.* 1988; Ito *et al.* 1994). Given that ant workers are evolutionarily derived from wasp-like monomorphic females, the presence of a spermatheca in workers can be assumed to be the ancestral condition. The evolution of increasing specialization in the queen caste has often been accompanied by the progressive loss of reproductive attributes in the workers, and thus in 'higher' subfamilies the spermatheca is non-functional or absent in workers. Consequently, they are unable to mate and produce diploid progeny (reviewed in Bourke 1988; Choe 1988). The retention of a spermatheca by workers in many Ponerinae has been highlighted by the occurrence of

worker mating in some species (Table 18-1). Nonetheless, workers lack a spermatheca in other species (Peeters 1991a; Villet *et al.* 1991; Ito and Okhawara 1994). I suggest that the progressive increase in the reproductive specialization of the queen caste eventually eliminated the option of mating by workers, and thus retention of a functional spermatheca was no longer selected for. Ito and Okhawara (1994) offer an alternative interpretation: reduction or loss of worker spermathecae may be a mechanism of 'queen control', but it is not clear how they envisage this.

In 'primitive' taxa, the absence of marked morphological specialization of queens represents the ancestral condition. It also occurs as a secondary modification in representatives of the 'higher' subfamilies (e.g. Formicinae, Formicoxenini (formerly Leptothoracini; Bolton 1994) and *Myrmecina*). In both cases, a limited caste dimorphism imposes constraints on queen behavior during independent colony foundation.

Establishment of new colonies

Colony initiation is a crucial stage in the life history of ants. Independently of workers, a newly mated foundress must find a suitable shelter, lay eggs and rear the first generation of larvae to maturity. In 'higher' taxa, the highly dimorphic queens have large thoraces, and are able to raise their first brood on the metabolic products of wing muscles and fat reserves, as well as storage proteins (Wheeler and Buck 1995). In 'primitive' species, however, the queens' thoraces are not much larger than those of workers, and food must be obtained outside the nest. Thus ponerine and myrmeciine queens establish new colonies in the 'semi-claustral' pattern (Haskins 1970), i.e. they forage above ground at frequent intervals. This exposes them and their brood to predation and accidents. This ancestral condition was first documented in *Myrmecia* and *Amblyopone* in Australia (Wheeler 1933; Haskins and Haskins 1951, 1955). In *Nothomyrmecia*, dealate and inseminated queens have been observed foraging on trees (Hölldobler and Taylor 1983). A similar pattern of colony foundation is exhibited by all the ponerines studied (Kôriba 1963; Colombel 1971; Ward 1981b; Villet 1990d; Villet *et al.* 1989; Dejean and Lachaud 1994). Haskins (1970) reviewed how the 'fully claustral' pattern of colony foundation characteristic of higher ants has been derived through a series of adaptive steps that can still be seen in living ponerine species. In some of these, it appears that the small wing muscles become completely resorbed, and this may partly sustain the founding queen's egg production. A proportion of these eggs can be fed to the first generation of larvae. In *Pachycondyla apicalis*, 95% of eggs laid during colony foundation are eaten, and furthermore young larvae are cannibalized by older ones (Fresneau 1994). Oophagy seems an essential behavior during colony foundation in 'primitive' ants, and is probably associated with the inability to feed larvae by regurgitation (see below), e.g. *Amblyopone australis* (Haskins and Haskins 1951), *Brachyponera senaarensis* (Dejean and Lachaud 1994), *Ectatomma ruidum* (Corbara 1991), *Plectroctena mandibularis* (Villet 1991a) and *Rhytidoponera confusa* (C. Peeters, unpublished). In the laboratory, an *Odontomachus* foundress was prevented from obtaining outside food, but her larvae continued to develop (Haskins and Haskins 1950a), which also suggests the occurrence of egg cannibalism. Fresneau (1994) suggests that the recycling of the founding queen's eggs reduces her need to forage above ground frequently. The only report of a ponerine queen able to start colonies without periodic provisioning of the larvae from outside sources is *Brachyponera lutea* (Haskins and Haskins 1950b), although the manner of larval feeding (oophagy or regurgitation) remains unclear. Not surprisingly, the large difference between queen and worker statures in *B. lutea* is exceptional in the Ponerinae. Caste differentiation is also pronounced in *Paltothyreus tarsatus*, where new queens have large fat reserves when they disperse (Braun *et al.* 1994b).

Fully claustral foundation is impossible in 'primitive' ants owing to the morphological limitations of queens. The conversion of body tissues into food for the first larvae was a vital evolutionary advance in the Formicidae (Hölldobler and Wilson 1990). However, founding queens also forage in various species belonging to two advanced subfamilies (Formicinae: *Cataglyphis bicolor* (Fridman and Avital 1983), *Polyrhachis laboriosa* and *P. militaris* (Lenoir and Dejean 1994); Myrmicinae: *Acromyrmex versicolor* (Rissing *et al.* 1989), *Manica rubida* (Le Masne and Bonavita 1969), tribe Dacetini (Dejean 1987)). Because related species exhibit the fully claustral pattern, it seems clear that this is a secondary modification, presumably as an adaptation to local ecological conditions.

Evolution of flightless queens

Although ant workers are always characteristically wingless, ant queens usually have wings. These enable aerial dispersal and colonization of new habitats, and furthermore the wing muscles constitute a metabolic reserve during colony foundation. Despite these obvious adaptive benefits, a permanently wingless queen caste has evolved in a significant number of ants. Such 'ergatoid' queens (Bolton 1986; Peeters 1991b; Villet 1989) exhibit substantial modifications in their external appearance, because the loss of flight muscles causes their thorax to resemble that of workers (especially in species where queen stature is similar to the workers'). Ergatoid queens exist in all subfamilies, and are the exclusive form in subfamilies belonging to the doryline section (see Fig. 18-1).

Ergatoid queens occur widely in the Ponerinae (12 genera in four tribes; Table 18-1). In *Nothomyrmecia*, queens have very short wings, which appear unsuitable for flying; this brachyptery has not been accompanied by a simplification of the flight sclerites (Taylor 1978). In the genus *Myrmecia*, both brachypterous queens and ergatoid queens occur in addition to normal winged queens (McAreavey 1948; Clark 1951), but their biology and taxonomic occurrence is poorly known. The evolution of ergatoids is often associated with significant modifications in colony organization. Fission occurs instead of independent foundation, i.e. a mother colony becomes divided into two autonomous groups. Ergatoids in various species have much larger ovaries than their workers, despite similar body sizes. Bigger colonies are a consequence of this greater egg-laying specialization (Peeters 1993). The evolution of ergatoids in the Ponerinae has sometimes reached an extreme condition where queens have hypertrophied gasters as an adaptation for the concurrent maturation of large numbers of eggs (for example, in *Simopelta* (Gotwald and Brown 1966)). However, ergatoids retain a relatively low fecundity in other ponerine species (even in some *Leptogenys*), with colonies of fewer than 100 workers (Peeters 1991b).

Comparative data on queen modification in 'primitive' taxa can help understand why wings were lost. *Myrmecia regularis* represents a possible intermediate stage in the evolutionary transition. Queens are winged (although wing development varies between populations), but they dealate *before* dispersal from the parental nest (Haskins and Haskins 1955). Once mated, they found colonies in the partially claustral manner. Wings are thus no longer essential in *M. regularis*, although wing muscles may be. In other species exhibiting a similar behavior, the development of wings may have been selected against. Support for this evolutionary pathway comes from limited laboratory results on the ponerine *Plectroctena mandibularis* (Villet 1991a). An alternative hypothesis (Peeters 1990) is that independent foundation is first replaced by colony fission, and then followed by loss of wings in queens.

Table 18-1. *The occurrence of different morphological types of reproductives in subfamilies Nothomyrmeciinae, Myrmeciinae and Ponerinae*

Only 18 ponerine genera are included; in another 24 genera, only alate queens have been reported. References on ergatoid queen morphology (either taxonomic or biological studies) are listed in Peeters (1991b, 1993); see also Clark (1951); Ito and Ohkawara (1994). AQ, alate queens; AQ*, brachypterous queens; EQ, ergatoid queens; G, gamergates only; Q+G, both dealate queens and gamergates reproduce.

Genus	Reproductives	Key references (gamergates only)
Nothomyrmecia	AQ*	—
Myrmecia	AQ, AQ*, EQ	—
Tribe Amblyoponini (6 genera)		
Amblyopone	AQ, G	Ito 1993a
Onychomyrmex	EQ	—
Tribe Ectatommini (9 genera)		
Discothyrea	AQ, EQ	—
Gnamptogenys	AQ, EQ, G	Gobin *et al.* 1994
Heteroponera	AQ, EQ	—
Proceratium	AQ, EQ	—
Rhytidoponera	AQ, G, Q+G	Haskins & Whelden 1965; Ward 1981b, 1983, 1984; Peeters 1987b
Tribe Platythyreini (2 genera)		
Platythyrea	AQ, EQ, G, Q+G	Villet 1991b,c, 1993; Villet *et al.* 1990; Itô 1995
Tribe Ponerini (23 genera)		
Diacamma	G	Fukumoto *et al.* 1989; Peeters & Higashi 1989; Peeters *et al.* 1992
Dinoponera	G	Haskins & Zahl 1971; Paiva & Brandão 1995; Araujo *et al.* 1990
Harpegnathos	Q+G	Peeters & Hölldobler 1995
Hypoponera	AQ, EQ	—
Leptogenys	AQ, EQ, G	Ito & Ohkawara 1994; Davies *et al.* 1994
Pachycondyla	AQ, G, Q+G	Peeters & Crewe 1986; Peeters *et al.* 1991; Sommer & Hölldobler 1992; Ito 1993c
Hagensia[a]	G	Villet 1992
Megaponera[a]	EQ	—
Ophthalmopone[a]	G	Peeters & Crewe 1985a, 1985b
Plectroctena	AQ, EQ	—
Ponera	AQ, EQ	—
Simopelta	EQ	—
Streblognathus	G	Ware *et al.* 1990

[a] Synonymized under *Pachycondyla* (see Bolton 1994).

Evolutionary loss of the queen caste

In various ponerine tribes and genera, workers reproduce instead of the queens, which are no longer produced (Table 18-1). Several aspects of the biology of queenless ants have been reviewed (Peeters 1991a, 1993). Colonies typically consist of a few dozen workers. 'Gamergates' (mated reproductive workers) are less fecund than queens of related species, and they have shorter lifespans. Fission is the obligatory mode of colony reproduction, because gamergates lack the necessary opportunities for independent foundation. In a few ponerine ants both queens and gamergates reproduce (Table 18-1).

Queenless ponerine species are unique among the ants, because all female inhabitants of a colony are able to mate and produce eggs. Nevertheless, only one or a few workers reproduce. Such absence of specialized repro-ductive traits is analogous to the situation in various polistine wasps and halictine bees, although in Ponerinae it is a secondary modification (Peeters and Crozier 1988). This reproductive organization is akin to the cooperative breeding documented in various vertebrates, such as naked mole rats, wild dogs and dwarf mongooses (Sher-man *et al*. 1991; Jennions and Macdonald 1994). Crespi and Yanega (1995) have advocated that the term 'eusocial-ity' be restricted to societies in which reproductive indivi-duals are 'irreversibly behaviorally distinct prior to reproductive maturity'; I understand this differentiation to be the consequence of morphological specialization, otherwise it cannot be irreversible. This new definition changes the traditional scope of eusociality, and accord-ingly queenless ponerine ants, together with various social wasps and bees, are excluded. Their societies, entirely made up of *totipotent* individuals (i.e. having the potential to express the full behavioral repertoire of the population, and the ability to produce offspring like one-self (Crespi and Yanega 1995)), are better described as 'cooperative breeders'. This contrasts with eusocial ants, which have mutually dependent queens and workers (the latter are not totipotent).

As in cooperatively breeding mammals, dominance hierarchies in queenless ants function to regulate sterility of group members. Physical confrontation among the workers leads to the inhibition of ovarian activity in subor-dinates, and sometimes it regulates mating as well (Heinze *et al*. 1994). In some species there is always only one gamer-gate per colony; other queenless species are polygynous (Peeters 1993).

Gamergates have never been reported outside the Ponerinae. They may be expected to occur in *Myrmecia*, since workers seem morphologically competent to replace queens. Indirect evidence suggests that gamergates should be looked for in *Nothomyrmecia* as well; an electrophoretic analysis of gene–enzyme variation revealed that intracolo-nial relatedness was lower than that predicted with a single, once-mated queen (Ward and Taylor 1981).

Another two novel reproductive modifications have recently been reported in the Ponerinae. In *Pachycondyla obscuricornis*, reproductive 'intercastes' (*sensu* Peeters 1991b, i.e. phenotypic intermediates between queens and work-ers) were found in a queenless colony (Düßmann *et al*. 1996). In *Platythyrea punctata*, virgin workers are able to pro-duce diploid offspring from unfertilized eggs (Heinze and Hölldobler 1995).

ECOLOGICAL CHARACTERISTICS OF PONERINAE, *MYRMECIA* AND *NOTHOMYRMECIA*

Colony size

In contrast to the huge colonies often encountered in the advanced subfamilies (see, for example, Hölldobler and Wilson 1990, Table 3-2), small colonies prevail in 'primi-tive' taxa. In *Nothomyrmecia*, mature nests contain 50–70 workers (Taylor 1978). In *Myrmecia*, colonies often consist of a few hundred workers, but in some species mature colo-nies are much smaller (Haskins and Haskins 1950a; Gray 1974; Itô *et al*. 1994). Colonies of 1000–3000 workers have been found in the larger species of *Myrmecia* (Gray 1974; Higashi and Peeters 1990). In the Ponerinae, colonies have several dozens to hundreds of workers (reviewed in Peeters 1993), with the smallest societies known in ants (9 ± 3 workers) occurring in *Pachycondyla sublaevis* (Peeters *et al*. 1991). Nevertheless, the colonies of a minority of ponerines can contain a few thousand individuals. *Leptogenys distin-guenda* is exceptional, with colonies of up to 50 000 workers (and one ergatoid queen) (Maschwitz *et al*. 1989). This may even be exceeded by *Pachycondyla luteola*, where single colo-nies can be the exclusive inhabitants of individual *Cecropia* trees 30–35 m high (Davidson and McKey 1993; Verhaagh 1994).

Colony size is a reflection of both queen fecundity and average worker longevity. The egg-laying rate of queens determines the upper limit on colony growth, but a further consideration is that enough food must be available for the larvae. Workers in 'primitive' taxa are generally large (length reaches 30 mm in some genera, although in others it is as small as 1.7 mm), and this larger body size may be correlated with greater longevity.

Diet specialization

All 'primitive' ants are armed with a sting and hunt arthro-pods. Although many ponerines are opportunistic in their choice of prey, others are very prey-specific, e.g., *Amblyo-pone pluto* on geophilomorph centipedes (Gotwald and Lévieux 1972), neotropical *Gnamptogenys* spp. on millipedes (Brown 1992), *Proceratium* on spider eggs (Brown 1979), *Leptogenys* sp. 13 (near *kraepelini*) on earwigs (Steghaus-Kovac and Maschwitz 1993) (see also Hölldobler and Wilson 1990, Table 15-1). The genus *Leptogenys* displays several degrees of prey specialization: some species are generalists preying on a wide taxonomic and size range of arthropods, whereas others are narrowly specialized isopod or termite hunters (Steghaus-Kovac and Maschwitz 1993). Scavenging on dead arthropods can be important in a few genera. In *Pachycondyla apicalis*, 45% of items retrieved to the nests were freshly killed prey (with lepidopteran larvae predominating), and as many items were dead insects (Fresneau 1994).

Despite this carnivorous propensity, a few ponerines (belonging mostly to tribe Ectatommini) depend partly on other sources of food. *Paraponera clavata* forage on the foliage of the rainforest understory, in search of arthropod prey as well as sugary secretions from extrafloral nectaries (Young 1977). In *Ectatomma ruidum*, some foragers in a colony are specialized on the nectaries of an orchid, while others hunt (Passera *et al.* 1994). A few generalized ground predators also utilize honeydew from homopterans, and receive secretions from butterfly larvae. *E. ruidum* tend membracids (Weber 1946), as well as myrmecophilous lycaenid and riodinid caterpillars (Robbins 1991; DeVries 1988). *Odontomachus troglodytes* can climb cocoa trees for a short distance to find aphids and coccids, and often construct crude shelters of soil particles over them (Evans and Leston 1971). *O. troglodytes* also attend lycaenid caterpillars (Lamborn 1915). Species in the *Rhytidoponera impressa* group tend homopterans (Ward 1981a); *R. 'metallica'* attend lycaenid larvae (Common and Waterhouse 1981).

In Mexican rainforests, some species of both *Odontomachus* and *Pachycondyla* are highly attracted to the seeds of various herbs; these are retrieved to the nest where the elaiosomes (fleshy appendages containing lipids) are eaten, after which the seeds are discarded (Horvitz and Beattie 1980). *Rhytidoponera* sp. 12 (near *mayri*) has a similarly important role as seed dispersers in the dry sclerophyllous scrub of Australia (Davidson and Morton 1981). In contrast, various other species of *Rhytidoponera* and *Heteroponera* opportunistically consume seeds (Andersen 1991). *Brachyponera senaarensis* is also able to eat seeds. Its diet varies seasonally and geographically: in humid tropical regions of Africa, both seeds and insect prey are collected during the rainy season, whereas the diet consists exclusively of seeds during the dry season, which lasts three months (Dejean and Lachaud 1994). In dry tropical regions, however, foragers react to the absence of seeds in the rainy season by adopting a 100% animal diet (Lévieux 1979). The neotropical *Pachycondyla luteola* feeds mostly on glycogen-rich Müllerian bodies at the base of petioles of young leaves (Davidson and Fisher 1991; Verhaagh 1994).

Myrmecia ants also hunt a variety of insects (including flies and honey bees), but only when larvae are present in the nests. When these are absent, workers forage for nectar almost exclusively (Haskins and Haskins 1950a; Gray 1971b, 1974). *M. desertorum* attend homopterans (Gray 1971b). *Nothomyrmecia* workers forage exclusively at night, climbing low *Eucalyptus* trees where they search for various small insects as well as sweet substances (Taylor 1978).

The predatory habits of 'primitive' ants are shared with their wasp ancestors (e.g. Scoliidae, Tiphiidae). Many Ponerinae hunt a wide range of invertebrates opportunistically, and the extreme prey specialization of some species appears to have evolved secondarily. Indeed it is also exhibited in some myrmicine ants (Hölldobler and Wilson 1990, Table 15-1).

Foraging habits

Many species of ponerine hunt alone (reviewed in Peeters and Crewe 1987). There is no cooperation among foragers, either through the transfer of information about the location of new sources of prey, or through direct assistance during the killing and retrieving of prey. Solitary foraging and the absence of recruitment trails have also been documented in *N. macrops* (Hölldobler and Taylor 1983), as well as in all the species of *Myrmecia* investigated by Haskins and Haskins (1950a) and Gray (1971b). Solitary predation contrasts with the elaborate systems of recruitment and cooperative hunting displayed by other ponerine ants. These include army ant-like swarm raiding in *Onychomyrmex* and some species of *Leptogenys* (Hölldobler *et al.* 1982; Maschwitz *et al.* 1989; Duncan and Crewe 1994), or the invasion of a termite nest by a column of hundreds of workers in *Megaponera foetens*, following its location by a single scout ant who lays a trail back to her nest (Longhurst and Howse 1979).

Peeters and Crewe's (1987) comparative overview of foraging in the Ponerinae emphasized that the occurrence of simple and complex hunting strategies does not reflect phylogenetic relationships. Trail communication has evolved many times independently; and four different trail-pheromone glands have been identified in the Ponerinae (Hölldobler and Wilson 1990). Foraging characteristics seem to result from the unique selective pressures facing each species; for example, in the tribe Ponerini, all species examined have the exocrine glands necessary for recruitment, but only some of them hunt in groups. Together with ecological considerations and the extent of prey specificity, colony size seems an important factor (Beckers *et al.*

1989), i.e. ponerine species with small colonies are not likely to raid in groups. In *Amblyopone*, species that do not recruit have colonies with only one or two dozen workers, but one species from the *reclinata* group, which is able to retrieve centipedes cooperatively, has 100 workers per colony (see Ito 1993b). Similarly, the entire spectrum of foraging strategies is exhibited in *Leptogenys*. In a guild of 12 species studied by Maschwitz *et al.* (1989), five have large colonies and are group- or swarm-raiders, whereas the other species search for prey solitarily (after encountering prey, they either recruit a group of nestmates, or attack and retrieve the prey alone). A cladistic analysis of 22 ant species (Baroni Urbani 1993) also showed that differences in recruitment behavior do not reflect phylogeny, but rather represent species-level adaptations to environmental conditions.

Nesting requirements

All *Myrmecia* spp. nest in the ground except *M. mjobergi*, which constructs its nest in trees (Clark 1951). Founding queens of various species all construct similar simple nests, but as the colonies grow larger, each species exhibit a typical design (Gray 1971a, 1974). Several larger-sized *Myrmecia* have big conspicuous soil mounds, which reach 0.7 m in height in *M. brevinoda* (Higashi and Peeters 1990). Short-lived nests have not been reported in this genus.

The nests of many Ponerinae also occur in the ground; large or small chambers, connected together by galleries, are superficial or extend to considerable depths. Various species invest much labor in building permanent nests; for example, finely shaped chambers in *Centromyrmex sellaris* (Lévieux 1976a), flood-resistant earthen spheres in *Harpegnathos saltator* (Peeters *et al.* 1994), extensive networks of shallow foraging galleries in *Paltothyreus tarsatus* (Braun *et al.* 1994a). In contrast, other species cannot dig their own nests, and are only able to make limited modifications to pre-existing structures; for example, *Amblyopone mutica* reutilizes underground networks of empty cavities (e.g. roots, termite galleries) (Lévieux 1976a). Similarly, *Pachycondyla apicalis* occupies the abandoned galleries made by passalid beetles in rotting wood (Fresneau 1994). Many ponerines inhabit mesic to humid forests, and have ephemeral nests in the leaf-litter stratum, for example, inside or under decaying plant material, under stones, in crevices or in fallen cocoa pods (see Wilson 1959). The dichotomy between permanent and short-lived nests is not related to colony size. *Dinoponera australis* has very

small colonies (13 ± 6 workers), yet its nests often reach one meter in depth (Paiva and Brandão 1995). In contrast, *Leptogenys chinensis* (200–300 workers per colony) nests in any available cavities near the ground (Maschwitz and Schönegge 1983).

Very few ponerine ants make arboreal nests; examples are *Gnamptogenys menadensis* in Indonesia (Gobin *et al.* 1994), *Platythyrea conradti* in Cameroun (Lévieux 1976b), and *Pachycondyla goeldii* in French Guyana (Corbara and Dejean 1996). *Pachycondyla luteola* colonizes exclusively the hollow stems of *Cecropia* trees (Davidson and Fisher 1991; Verhaagh 1994).

A polydomous organization (i.e. a single colony distributed into several distinct nests) has only been reported in *Platythyrea conradti* (Lévieux 1976b) and *Ophthalmopone berthoudi* (Peeters and Crewe 1987).

Frequent nest emigration

Most ants are able to shift their nest site if it becomes unsuitable. Although this behavior has only been reported in one species of *Myrmecia* (Gray 1971b), it is especially ubiquitous in ponerine genera. When nests are ephemeral, or become too small, the ants emigrate elsewhere, e.g. *Rhytidoponera* species in the *impressa* group (Ward 1981a), *Diacamma* sp. from Japan (Fukumoto and Abe 1983). Emigration is efficiently organized in *Megaponera foetens*, where ants and brood move in a distinct column to new sites up to 50 m away (Longhurst and Howse 1979). In *Leptogenys chinensis*, nests are abandoned after about two weeks, and relocation can occur to a new site just a few meters away (Maschwitz and Schönegge 1983). Frequent nest emigrations are often characteristic of species exhibiting colony fission and ergatoid queens. *Pachycondyla marginata* is thus exceptional for a migratory species, because it has winged queens (colonies consist of 500–1500 workers) (Leal and Oliveira 1995).

Army ant (legionary) behavior

Legionary behavior consists of two fundamental features: migration and group predation (Wilson 1958). This characteristic of the doryline and ecitonine ants occurs to a limited extent in a few ponerine genera, i.e. *Onychomyrmex* (Wilson 1958; Hölldobler and Wilson 1990), *Simopelta* (Gotwald and Brown 1966), and some species of *Leptogenys*. Nests are very temporary in aspect, consisting of nothing more than preformed cavities in rotting logs or simple bivouacs in open leaf litter. Legionary behavior is best documented in *L. distinguenda*, where groups of more than 20 000 workers conduct nocturnal raids (Maschwitz *et al.* 1989). Raids are not directed to specific food sources by scouts; nestmates move around in a group and, upon finding prey, immediate cooperation ensures efficient attack. Migrations occur every 1–3 days, usually to an area (5–59 m away) that has just been raided.

Group-raiding ants can feed on large arthropods or the brood of other social insects, which are not normally accessible to solitary foragers. Legionary behavior thus enables great ecological diversification. Indeed, *L. distinguenda* has filled the ecological niche of doryline ants in southeast Asia (Maschwitz *et al.* 1989). All legionary ponerine species have ergatoid queens and large colonies. Unlike many true army ants, the development of brood is not synchronized with the migration events.

INTRACOLONIAL ORGANIZATION

Patterns of food exchange

Trophallaxis
The sharing of food among colony members is an integral part of the specialization of roles associated with sociality (reviewed in Wheeler 1994). After food is brought into the nest by the foragers, it is consumed by those colony members who are not active outside. The most sophisticated form of food sharing is trophallaxis, which is widespread in some groups of advanced ants (Formicinae, Dolichoderinae and Myrmicinae). Following appropriate tactile signals with the antennae and forelegs, a solicited worker regurgitates a drop of liquid from her crop (anterior to the midgut), and this is imbibed by the begging ant. In these advanced groups, larvae are fed exclusively with regurgitated liquid food. In the 'primitive' ants, however, trophallaxis is known in only two species. All colony members, larvae included, feed directly on the arthropod prey retrieved to the nest (larvae have a flexible thorax which allows much movement, and sometimes they are able to crawl within the nest, e.g. *Amblyopone pluto* (Gotwald and Lévieux 1972)). A possible evolutionary precursor of true trophallaxis has been observed in several ponerines, which transport liquids between their mandibles (Fresneau *et al.* 1982; Hölldobler 1985). Foragers exploiting extrafloral nectaries, or aphids and coccids, return to their nest with a droplet of sugary liquid held *outside* their mouths, and this is then shared with nestmates. There is thus no regurgitation from the crop in this 'pseudotrophallaxis',

but the accompanying tactile signals are similar to those observed during regurgitation in 'higher' ants (Hölldobler 1985). Earlier accounts of trophallaxis in *Myrmecia regularis* (Haskins and Whelden 1954) and in various ponerines (Le Masne 1952; Haskins 1970) may have overlooked this distinction. Nevertheless, the regular occurrence of true trophallaxis has recently been documented in *Ponera coarctata* (Liebig *et al.* 1994) and *Hypoponera* sp. (Hashimoto *et al.* 1995).

Egg cannibalism
A different form of intracolonial food exchange is oophagy, which occurs in a wide diversity of ants (reviewed in Crespi 1992; Wheeler 1994). Whether this involves normal reproductive eggs, or special immature oocytes instead, is an important distinction. Typically, the latter ('trophic eggs') lack a chorion and are flaccid, and they differ in shape and histology from reproductive eggs (Passera *et al.* 1968; but see Voss *et al.* 1988). Irrespective of this distinction, oophagy frequently involves worker-laid eggs, which emphasizes that the functional ovaries of workers can play an important social role. This may be especially important in 'primitive' taxa: because the workers do not regurgitate to the larvae or the queen, they can store reserves and utilize them to produce immature oocytes.

Consumption of eggs has been observed in *Nothomyrmecia* (Hölldobler and Taylor 1983) and several *Myrmecia* species (Haskins and Haskins 1950a; Crosland *et al.* 1988). In *M. forceps* and *M. gulosa*, workers lay trophic eggs which are fed to the larvae or the queen (Freeland 1958). However, oophagy seems restricted to a proportion of ponerine genera (albeit distributed in three tribes). In *Amblyopone silvestrii*, reproductive eggs laid by queens are consumed by first and second instar larvae (Masuko 1990). In *Prionopelta amabilis*, the queen is reported to feed almost exclusively on worker-laid eggs (Hölldobler and Wilson 1986). In *Ectatomma*, trophic eggs are laid by workers and differ in color from reproductive ones (Weber 1946; C. Peeters, unpublished data). In *Gnamptogenys menadensis* (with gamergates), virgin workers produce trophic eggs; these are shorter than reproductive eggs and are fed to the larvae (Gobin *et al.* 1994). In various species of the permanently queenless genus *Diacamma*, cannibalism of reproductive eggs laid by virgin workers is part of the behavioral mechanism of hierarchy formation (Peeters *et al.* 1992; Peeters and Tsuji 1993). In *Pachycondyla obscuricornis*, young virgin workers lay eggs in queenright colonies, and these are eaten by the queen and larvae (Fresneau 1984). In orphaned colonies of both *P. obscuricornis* and *P. apicalis*, eggs produced by workers (and dealate virgin queens in *P. obscuricornis*) are destroyed during dominance interactions (Oliveira and Hölldobler 1990, 1991). In the permanently queenless *P. krugeri*, virgin workers lay trophic eggs, which are often fed to the larvae (Villet and Wildman 1991). In *Ponera pennsylvanica*, workers occasionally lay small, sticky eggs, which are subsequently eaten or fed to larvae (Pratt *et al.* 1994).

Egg cannibalism thus occurs in two distinct contexts: trophic eggs are consumed as a strategy of nutrient exchange among nestmates, whereas the destruction of reproductive eggs can be a form of dominance interaction, with incidental trophic benefits for the dominant individuals. Data from various 'primitive' ants do not contradict the general formicid rule that the less frequent the exchange of liquid food by regurgitation, the more frequent the exchange of trophic eggs (Hölldobler and Wilson 1990), although several species seem to lack both behaviors. Eggs may also function in food transfer during the initial stage of colony foundation, but in contrast to the above instances of worker-laid reproductive eggs, the queen's fertilized eggs are involved. Egg consumption by colony-founding queens is not unique to the 'primitive' taxa, however (see, for example, *Solenopsis invicta*, Voss *et al.* 1988; *Atta*, Hölldobler and Wilson 1990; see also Wheeler 1994).

The lack of trophallaxis in 'primitive' ants may contribute to the low fecundity of queens, although this awaits empirical documentation. Reproductives in a colony often have to compete with their nestmates while feeding on prey items. One exception is *Amblyopone silvestrii*, where the queen feeds exclusively on hemolymph from last-instar larvae (which is obtained after biting them) (Masuko 1986). A similar behavior may occur in *Mystrium mysticum* (Wheeler and Wheeler 1988).

Division of labor

Division of colony labor contributes to the ecological success of social insects by enhancing the efficiency and reliability of colony responses to important contingencies (Hölldobler and Wilson 1990). Numerous studies have shown that complicated social organization is attained with a relatively simple repertoire of individual behaviors (reviewed in Robinson 1992). Any empirical description of polyethism proceeds by the following steps: (1) establish that there is role specialization among workers, i.e.

individuals perform only one activity for a sustained period of time; (2) establish whether roles are performed in a predictable temporal sequence (i.e. workers pass through distinct behavioral phases during their lives). Ponerine and myrmeciine ants are ideal subjects to investigate division of labor because their colonies are small, making it possible to study the behavior of all workers as individuals (the workers (often large) are easily marked with paint or numbers). As a consequence, their activities in the colony can be followed with unequaled detail, and the pattern of polyethism may be better understood in the Ponerinae than in any other subfamilies (in contrast, very little is yet known about *Myrmecia*). Polyethism was first demonstrated in ponerines by Bonavita and Poveda (1970).

Behavioral profile of reproductives

All ponerine foundresses initially exhibit a full complement of behaviors, especially since they need to forage outside. Later in colony ontogeny, their repertoire decreases, and important interspecific differences appear. Foundresses of *Pachycondyla apicalis* continue to forage for a while after emergence of the first workers, and they always remain involved in brood care (Fresneau *et al.* 1982). In contrast, queens of *Ectatomma ruidum* are immobile over eggs and larvae, but do not attend them (Corbara 1991). Similarly, the queens of *Ponera pennsylvanica* do not care for brood at all (Pratt *et al.* 1994). There seems to be a comparative trend that queens are less active (oviposition excepted) with increased caste dimorphism, and thus become more behaviorally distinct from workers. The ergatoid queen of *Megaponera foetens* is highly fecund (33 eggs per day); she does little else in the colony, and is surrounded by a distinct retinue of workers (Hölldobler *et al.* 1994). In contrast, queens attract little attention in *E. ruidum* (Corbara *et al.* 1989), *P. apicalis* (Fresneau 1994) and *Nothomyrmecia macrops* (Jaisson *et al.* 1993).

As in some advanced ants (e.g. *Acromyrmex versicolor*, *Linepithema humile*), winged ponerine queens who cannot mate remain in the colonies and participate in maintenance activities (including foraging), e.g. *Harpegnathos saltator* (C. Peeters, unpublished), *Odontomachus affinis* (Brandão 1983), *Pachycondyla apicalis* (Fresneau 1994), *Rhytidoponera confusa* (Ward 1981b).

Not unlike queens, gamergates usually have a behavioral profile distinct from that of sterile workers, although there is much interspecific variation (Villet 1990c, 1991b,c; Ware *et al.* 1990). In contrast to sterile workers, gamergates do not become active outside the nest as they become older.

Role specialization among infertile workers

The investigation of various species belonging to *Ectatomma* and *Pachycondyla* has shown that there is no rigid pattern of division of labor, and this accounts for the occurrence of variations between conspecific colonies. Corbara (1991) documented in *E. ruidum* how the stability of social organization contrasts sharply with the flexibility of role specialization in individuals. There is a distinct chronology in the appearance of functional specializations, but the actual time spent in each functional group seems variable from one individual to another (Corbara *et al.* 1989). A similar pattern is found in *E. tuberculatum*; older workers can revert to brood care if there is a colony need for it (Fénéron 1993). In *P. apicalis*, most workers can be described by three profiles (brood care; non-social activities inside nest; foraging), with remaining individuals having intermediate status (Fresneau and Dupuy 1988). In *Amblyopone silvestrii*, tasks inside the nest are spatiotemporally connected; for example, the dismembering of prey and placing it near larvae, as well as the cleaning of larvae, are often performed by the same workers (Masuko 1996). Thus task specialization is weakly defined, with the exception of foraging. Similarly, in *Nothomyrmecia*, all colony members participate to some extent in all the activities (Jaisson *et al.* 1993). Ethograms of species with gamergates (Villet 1990c, 1991b,c; Villet and Wildman 1991) indicate that conventional role specialization occurs among the infertile workers, and is not based on size differences. Interspecific comparisons are often hindered by methodological variations between researchers.

In newly established colonies, the first generation of workers are remarkably flexible in the performance of the various essential tasks (Lachaud and Fresneau 1987; Corbara 1992; Fresneau 1994). Owing to a pattern of brood cannibalism, new workers do not emerge continuously, but rather in temporal clusters, and they specialize according to the colony needs at the time. Consequently, workers in the same age cohort can follow several different courses of polyethism (Fresneau 1994).

Influence of age on polyethism. Polyethism associated with age is generally observed throughout the social insects (Hölldobler and Wilson 1990, Table 8-3): workers labor in the nest when they are young, and forage outside when they are older. Investigations of advanced ants have generally relied on the color differences among adults (degree of pigmentation of body parts increases with age), and thus

groups of individuals are observed. However, in many studies of ponerine species, the exact age of various numbers of nestmates has been known, and their lifetime profile documented.

Traniello (1978) claimed that groups of age-related task specialists do not exist in *Amblyopone pallipes*. Although this has been widely cited as new evidence of the simple nature of sociality in 'primitive' ants, Traniello's data should be critically examined. His conclusion that there is little role specialization among nestmates was based on a single colony, in which the age of workers was unknown. Indeed, the lack of age polyethism was inferred from separate observations that '5 days-old callows leave the nest to hunt'. Lachaud *et al.* (1988) also investigated polyethism in *A. pallipes* (one colony studied) and documented limited role specialization. However, such behavioral plasticity is similar to what is observed among the first brood of workers during colony foundation in other ponerines. Furthermore, Masuko (1996) studied two colonies of *A. silvestrii* in which the age of most individuals was known, and convincingly demonstrated that role specialization is based on age.

Traniello (1978) explained his singular results by invoking the small size of colonies in *A. pallipes*, together with the occurrence of a single brood per year (apparently an adaptation to survive the long winter in the cool moist forests of the eastern USA and Canada). However, *Ponera pennsylvanica* is another temperate ponerine where a single cohort of workers ecloses nearly simultaneously, and Pratt *et al.* (1994) documented the existence of functional groups of workers, although highly variable in their composition. Pratt *et al.* (1994) also obtained additional data on age polyethism, as have Hölldobler and Wilson (1986) in *Prionopelta amabilis*, Villet (1991c) in *Platythyrea schultzei*, Dejean and Lachaud (1994) in *Brachyponera senaarensis*, and Pratt (1994) in *Gnamptogenys horni*.

Influence of size polymorphism on polyethism.   Only a minority of advanced ants have polymorphic workers (Hölldobler and Wilson 1990, Table 8-2), and this is always associated with role specialization. In almost all ponerine and myrmeciine species, discrete morphological subcastes are absent among workers. Size variation has been recorded, but this is not necessarily associated with polymorphism, which occurs when individuals at the extremes of the size range exhibit distinctly different proportions in body shape ('allometry'). Only in *Megaponera foetens* do workers exhibit both size variation and allometry (Crewe *et al.* 1984); there is an interesting difference in their tempo of activity (majors perform 135 acts per hour, and minors 248 (resting behaviors excluded) (Villet 1990b)), as well as in their energetic cost of retrieving prey (Duncan 1995). In a field study of raiding behavior in *M. foetens*, Longhurst and Howse (1979) showed that minors enter termite galleries and immobilize prey by stinging, and majors carry bundles of termites back to the nest. Furthermore, scout ants are always majors. Workers of *Paraponera clavata* exhibit limited unimodal variation in size, and this is statistically associated with task performance (foraging vs. brood care), although the size distributions of workers performing different tasks overlap almost completely (Breed and Harrison 1988). Claims of weak allometry in *P. clavata* are not supported by the published data (D. Wheeler, personal communication).

Intracolonial variation in the size of workers is especially marked in many of the larger species of *Myrmecia*, although they remain monomorphic (Gray 1971c). In *M. brevinoda*, worker lengths range bimodally from 13 to 36 mm, but this variation is not associated with allometric differences in the shape of body parts (Higashi and Peeters 1990). None the less, such size differences are the basis of polyethism among the workers: hunting, defense and extranidal building were done mainly by large workers of *M. brevinoda*, whereas smaller workers performed intranidal building (Higashi and Peeters 1990).

CONCLUSIONS

The pattern of division of labor in 'primitive' ants seems to differ little from that in 'higher' taxa with monomorphic workers. The behavioral characteristics of individuals are similar to that observed in advanced subfamilies. Individual variability is conspicuous in various Ponerinae, but this may be due to the small worker populations. Age polyethism seems to have been a general phenomenon ever since the inception of sociality in the ants.

The evolution of specialized queen morphology

A tremendous diversity of ecological and reproductive strategies are exhibited in myrmeciine and ponerine ants (partly owing to the wide adaptive radiation within the latter). By excluding some of the species having ergatoid queens, the following core profile of morphologically 'primitive' ants can be distilled out: (1) queens have low fecundity (1–5 eggs per day); (2) colonies consist of dozens or hundreds of workers; (3) predation is the rule; (4) solitary foraging predominates, but occasionally cooperation

and chemical recruitment have evolved. These characters contrast starkly with those of many 'higher' ants, i.e. hundreds of thousands of colony members, dramatic role specialization linked to worker polymorphism, and highly coordinated food-gathering strategies (e.g. *Dorylus*, *Atta*, *Oecophylla* or *Solenopsis*; see Hölldobler and Wilson 1990).

Some derived social traits can be exhibited by 'primitive' ants: (1) colonies can reach a few thousand workers, usually in species with ergatoid queens; (2) trophobiosis with homopterans and butterflies occurs in various ectatommine genera, together with pseudotrophallaxis among nestmates; (3) different degrees of group predation and migratory behavior are exhibited (e.g. in various species with ergatoids, including the amblyoponine *Onychomyrmex*). Nevertheless, other traits that are widespread in 'higher' ants are absent, i.e. no claustral colony foundation, no true trophallaxis (except in *Ponera coarctata* and *Hypoponera* sp.), and no worker polymorphism (except in *Megaponera*).

Limits on queen–worker dimorphism?

A consistent characteristic of morphologically 'primitive' ants is a generally limited divergence between winged queens and workers, and this represents the ancestral condition. Pronounced caste dimorphism occurs only in relatively few species with winged queens (e.g. *Paltothyreus*, *Paraponera*, a few *Pachycondyla*), as well as in several species with ergatoid queens (but this derived condition represents a distinct reproductive strategy, see below). Is there a restriction on the extent of reproductive specialization in 'primitive' queens? There is unlikely to be a structural limit on increasing thorax size, which would allow queens to have bigger wing muscles, and thus found colonies in complete isolation. Tergosternal fusion of abdominal segment IV (the sole apparent apomorphy of the Ponerinae) may be a morphological constraint on gaster enlargement (to accommodate bigger or more active ovaries), although expansion remains possible along the anteroposterior axis, i.e. intersegmental membranes can be stretched as in physogastric ergatoids of *Leptogenys*. Energetic constraints must also be considered: better-specialized queens are more expensive, and small colonies may not be able to produce them in adequate numbers (see below). Furthermore, since caste divergence affects colonial attributes (it is linked to queen fecundity, and thus colony size), the ecological consequences of increased divergence may or may not be selected for.

Small colony size and ecological specialization

Given the possibility of evolving higher queen fecundity, under what conditions is it adaptive for colonies to become more populous? Comparative data indicate that an increase in size alters the ecological profile of colonies markedly. More food needs to be found in the immediate environment to sustain additional larvae, which predatory species may not be able to achieve, even with a larger foraging force (prey density is finite). Furthermore, extra nest space needs to be available to accommodate the additional brood and adults. Wilson (1959) noted that limitation of nest space is an important factor in regulating colony size. For species with distinct requirements in the ground zone of forests (e.g. rotting log, natural cavities), enlargement of an existing nest may not be possible. Indeed, increased colony size means that nesting preferences need to be changed. In contrast, those ponerines that make ephemeral nests anywhere in litter, such as the legionary species, can have colonies with many workers.

Small colony size may thus be adaptive, and many 'primitive' species fill specialized ecological niches. They occur in low densities, and can coexist with ecologically dominant 'higher' ants. Further evidence that simple social traits can be selected for is their sporadic occurrence as secondary modifications in phylogenetically advanced species: (1) limited queen–worker differentiation in some formicines and formicoxenines; (2) small colonies in formicoxenines and various others; (3) solitary foraging and lack of chemical recruitment, e.g. in *Cataglyphis bicolor* (Wehner *et al.* 1983); (4) partly claustral mode of colony foundation, (5) absence of trophallaxis, e.g. two species of *Aphaenogaster* (Delage and Jaisson 1969).

Why is there no trophallaxis in 'primitive' ants? Three factors may act in concert: (1) the retrieval of insect prey makes the exchange of liquid food less appropriate; (2) tergosternal fusion of abdominal segment IV may be a constraint on considerable expansion of the crop; (3) in small colonies, food can be easily exchanged from foragers to the nestmates confined inside, and thus selection for trophallaxis is less strong. The lack of trophallaxis may partly explain why ponerine ants are almost absent from temperate regions. Because the internal storage of nutrients (as fat bodies) is not easily regulated, perennial colonies with carnivorous habits may not survive in regions where hunting is possible during a limited time of the year only. The complete exclusion of army ants (all predatory) from temperate regions supports this idea. Furthermore, ponerine species with ephemeral nests above ground (e.g. rotten logs) would be unable to move deeply into the ground to escape low temperatures.

Evolution of derived reproductive strategies

Colony fission occurs throughout the ants, but is particularly common in 'primitive' taxa. This alternative to independent colony foundation is always exhibited in species having either ergatoid queens or gamergates, as well as in a few species with winged queens (see Peeters 1993).

Independent foundation vs. colony fission

A regular production of gynes is critical to the success of a queen-based breeding system. Colony resources must be allocated to produce a maximum number of gynes with a minimum number of workers (Hölldobler and Wilson 1990). Nuptial flight and mating are periods of high mortality for the young queens of many ants, no matter how highly specialized (Hölldobler and Wilson 1990). In species with large nest populations, individual colonies commonly release hundreds or thousands of winged gynes, and this large reproductive investment appears essential for the success of independent foundation.

Consequent to their limited morphological specialization, the winged queens of 'primitive' ants cannot establish new colonies without foraging. It is intuitive (although awaiting empirical verification) that this requirement leads to increased postdispersal mortality (predators, hostile workers from alien nests, or accidents). Nevertheless, the persistence of semiclaustral foundation in many extant *Myrmecia* spp. and Ponerinae is evidence of its viability under certain ecological conditions. Species having more dimorphic castes may exhibit a higher success rate during colony foundation, or this success may be due to favorable local factors (e.g. assembly of ant competitors). Alternatively, the reproductive investment of some species can be large enough to compensate for high foundress mortality. In *Paltothyreus tarsatus*, several hundreds of specialized big gynes can be reared in colonies having one or more thousands of workers (Braun *et al.* 1994b). Various benefits of dispersal by flight (increased genetic diversity, reduction in local competition, colonization of new habitats) are certainly also involved.

Small colony size restricts the amount of resources that can be invested annually in the production of sexuals. Some species may be unable to produce a sufficiently large number of new gynes, and independent foundation is then selected against (Peeters 1990). Once fission occurs instead, aerial dispersal by queens stops, and mating strategy changes (Peeters 1991a). Thus the continued production of winged female reproductives stops being adaptive (see Tinaut and Heinze 1992). Ergatoid queens may have evolved as a consequence, while in other species gamergates replaced the queen caste.

Ergatoid queens and high fecundity

Because ergatoid queens are accompanied by workers during colony fission, their rate of success is likely to increase greatly (no need to forage, and sometimes they even mate with foreign males inside the nests). Furthermore, since an existing colony can only divide into two (three at the most) viable buds at one time, there will be selection to reduce the number of new queens produced (see Franks and Hölldobler 1987). Thus only a few ergatoids are reared annually, and available colony resources can be used to produce more workers. Indeed, the number of workers determines the viability of daughter colonies, and they represent most of the reproductive investment (Macevicz 1979). This pattern has been well documented in army ants, but empirical data remain scarce in the Ponerinae and *Myrmecia*.

In ant queens generally, morphological specialization serves two distinct functions: (1) dispersal and colony foundation, i.e. large flight muscles, and stored metabolic reserves to produce the first brood of workers; (2) high rate of egg production, i.e. bigger ovaries (more ovarioles, or longer ovarioles) as well as smaller eggs (so that more oocytes can mature simultaneously). The evolution of ergatoid queens clearly reveals that components (1) and (2) can be modified independently. Ergatoids have a worker-like thorax and are incapable of founding colonies independently, but they are highly fecund in various ponerine species where an increase in colony size is adaptive. In other species with ergatoid queens, ovarian specialization remains limited, and colonies are not larger than in species with winged queens. Accordingly, the ecological characteristics of 'primitive' species with ergatoids are extremely varied.

The gamergate strategy is cheap

In queenless species, those workers that do not differentiate as gamergates function as laborers, and consequently reproductive investment (which is equal to worker production, as in species with ergatoid queens) is optimized.

Gamergates have a relatively short lifespan, but they can be replaced in colonies without too many risks to colony survival.

The ecological profiles of species with gamergates are often similar, despite belonging to different tribes and genera. They tend to be solitary foragers without strict prey preferences, and often but not always are found in drier environments. However such combination of traits are also exhibited by various queenright ponerines. More comparative data are needed to derive a complete explanation for the absence of the queen caste in a small number of ants. Future studies of microhabitat preferences, including seasonal shortages in prey availability (trophic shortages may coincide with the period of sexual production, especially since internal storage of nutrients is impossible), may help our understanding of the ecological conditions under which queen reproduction was selected against. Investigation of species in which both queens and gamergates reproduce (Table 18-1) will yield valuable insights into their respective abilities (e.g. *Harpegnathos saltator*, Peeters and Hölldobler 1995).

Future perspectives

Tschinkel (1991) has advocated the systematic collection of descriptive data on the colony attributes of social insects, in order to determine the relationships among these attributes, both at an intraspecific and comparative level. Such sociometric data are badly needed for a large number of morphologically 'primitive' ants (*Myrmecia* in particular), for example, colony size, queen fecundity, queen and worker longevity. Although some of these data are best collected in the laboratory, other data must originate from the field, such as the mortality rate of queens during independent colony foundation, the importance of predators and parasitoids, the number of sexuals produced annually (especially ergatoid queens), and mechanisms of colony fission. Reliable phylogenies are also needed to determine which characters are ancestral or derived.

Specialization of queen morphology is a key to social complexity. The 'primitive' ants exhibit the early stages of the evolutionary divergence between reproductive and helper castes, and thus offer the opportunity to investigate the selective pressures responsible for increased sophistication of colony organization. The occurrence of three distinct reproductive strategies (sometimes in the same genus) is of particular comparative interest.

ACKNOWLEDGEMENTS

I am grateful to Dominique Fresneau (Villetaneuse) for guiding me through the literature on polyethism, and Walter Federle (Würzburg) for references on ant–plant associations. I thank Jae Choe, Bernard Crespi, Robin Crewe, Theo Evans, Konrad Fiedler, Jürgen Liebig, Phil Ward and Diana Wheeler for stimulating comments on various versions of this manuscript.

LITERATURE CITED

Andersen, A. N. 1991. Seed harvesting by ants in Australia. In *Ant–plant Interactions*. C. R. Huxley and D. F. Cutler, eds., pp. 493–503. Oxford: Oxford University Press.

Araujo, C. Z. Dantas de, J.-P. Lachaud and D. Fresneau. 1990. Le système reproductif chez une ponérine sans reine: *Dinoponera quadriceps* Santschi. *Behav. Proc.* 22: 101–111.

Baroni Urbani, C. 1993. The diversity and evolution of recruitment behaviour in ants, with a discussion of the usefulness of parsimony criteria in the reconstruction of evolutionary histories. *Insectes Soc.* 40: 233–260.

Baroni Urbani, C., B. Bolton and P. S. Ward. 1992. The internal phylogeny of ants (Hymenoptera: Formicidae). *Syst. Entomol.* 17: 301–329.

Beckers, R., S. Goss, J. L. Deneubourg and J. M. Pasteels. 1989. Colony size, communication and ant foraging strategy. *Psyche* 96: 239–256.

Bolton, B. 1986. Apterous females and shift of dispersal strategy in the *Monomorium salomonis*-group (Hymenoptera: Formicidae). *J. Nat. Hist.* 20: 267–272.

–. 1990a. Abdominal characters and status of the cerapachyine ants (Hymenoptera: Formicidae). *J. Nat. Hist.* 24: 53–68.

–. 1990b. Army ants reassessed: the phylogeny and classification of the doryline section (Hymenoptera: Formicidae). *J. Nat. Hist.* 24: 1339–1364.

–. 1994. *Identification Guide to the Ant Genera of the World*. Cambridge, Mass.: Harvard University Press.

–. 1995. A taxonomic and zoogeographical census of the extant ant taxa. *J. Nat. Hist.* 29: 1037–1056.

Bonavita, A. and A. Poveda. 1970. Mise en évidence d'une division du travail chez une fourmi primitive. *C.R. Acad. Sci. Paris* 270: 515–518.

Bourke, A. F. 1988. Worker reproduction in the higher eusocial Hymenoptera. *Q. Rev. Biol.* 63: 291–311.

Brando, C. R. 1983. Sequential ethograms along colony development of *Odontomachus affinis* Guérin (Hymenoptera, Formicidae, Ponerinae). *Insectes Soc.* 30: 193–203.

Braun, U., C. Peeters and B. Hölldobler. 1994a. The giant nests of the African Stink Ant, *Paltothyreus tarsatus* (Formicidae, Ponerinae). *Biotropica* 26: 308–311.

Braun, U., B. Hölldobler and C. Peeters. 1994b. Colony life cycle and sex ratio of the ant *Paltothyreus tarsatus* in Ivory Coast. In *Les Insectes Sociaux*. A. Lenoir, G. Arnold and M. Lepage, eds., p. 330. Villetaneuse: Université Paris Nord.

Breed, M. D. and J. M. Harrison. 1988. Worker size, ovary development and division of labor in the giant tropical ant, *Paraponera clavata* (Hymenoptera: Formicidae). *J. Kansas Entomol. Soc.* 61: 285–291.

Brown, W. L. 1954. Remarks on the internal phylogeny and subfamily classification of the family Formicidae. *Insectes Soc.* 1: 21–31.

–. 1958. Contributions towards a reclassification of the Formicidae. II. Tribe Ectatommini (Hymenoptera). *Bull. Mus. Comp. Zool. Harv.* 118: 173–362.

–. 1960. Contributions towards a reclassification of the Formicidae. III. Tribe Amblyoponini (Hymenoptera). *Bull. Mus. Comp. Zool. Harv.* 122: 143–230.

–. 1976. Contributions towards a reclassification of the Formicidae. Part VI. Ponerinae, Tribe Ponerini, Subtribe Odontomachiti. Section A. Introduction, Subtribal characters. Genus *Odontomachus*. *Stud. Entomol.* 19: 67–171.

–. 1979. A remarkable new species of *Proceratium*, with dietary and other notes on the genus (Hymenoptera: Formicidae). *Psyche* 86: 337–346.

–. 1992. Two new species of *Gnamptogenys* and an account of millipede predation by one of them. *Psyche* 99: 275–289.

Brown, W. L., W. H. Gotwald and J. Lévieux. 1970. A new genus of ponerine ants from West Africa (Hymenoptera: Formicidae) with ecological notes. *Psyche* 77: 259–275.

Choe, J. 1988. Worker reproduction and social evolution in ants (Hymenoptera: Formicidae). In *Advances in Myrmecology*. J. C. Trager, ed., pp. 163–187. Leiden: E. J. Brill.

Clark, J. 1951. *The Formicidae of Australia*, vol. 1. *Subfamily Myrmeciinae*. Melbourne: C.S.I.R.O.

Colombel, P. 1971. Recherches sur l'ethologie et la biologie d'*Odontomachus haematodes* (L.) (Hym. Formicoidea, Poneridae). Fondation des colonies par les femelles isolées. *Bull. Soc. Hist. Nat. Toulouse* 107: 442–459.

Common, I. F. and D. F. Waterhouse. 1981. *Butterflies of Australia*, 2nd edn. Sydney: Angus and Robertson.

Corbara, B. 1991. L'organisation sociale et sa genèse chez la fourmi *Ectatomma ruidum* Roger. Ph. D. dissertation, Université Paris XIII, Villetaneuse.

Corbara, B. and A. Dejean. 1996. Arboreal nest building and ant-garden initiation by a ponerine ant. *Naturwissenschaft* 83: 227–230.

Corbara, B., J.-P. Lachaud and D. Fresneau. 1989. Individual variability, social structure and division of labour in the ponerine ant *Ectatomma ruidum* Roger (Hymenoptera: Formicidae). *Ethology* 82: 89–100.

Crespi, B. J. 1992. Cannibalism and trophic eggs in subsocial and eusocial insects. In *Cannibalism – Ecology and Evolution Among Diverse Taxa*. M. A. Elgar and B. J. Crespi, eds., pp. 176–213. Oxford University Press.

Crespi, B. and D. Yanega. 1995. The definition of eusociality. *Behav. Ecol.* **6**: 109–115.

Crewe, R., C. Peeters and M. Villet. 1984. Frequency distribution of worker sizes in *Megaponera foetens* (Fabricius). *S. Afr. J. Zool.* **19**: 247–248.

Crosland, M. W. J., R. H. Crozier and E. Jefferson. 1988. Aspects of the biology of the primitive ant genus *Myrmecia* F. (Hymenoptera: Formicidae). *J. Austr. Entomol. Soc.* **27**: 305–309.

Davidson, D. W. and B. L. Fisher. 1991. Symbiosis of ants with *Cecropia* as a function of light regime. In *Ant–plant Interactions*. C. R. Huxley and D. F. Cutler, eds., pp. 289–309. Oxford University Press.

Davidson, D. W. and D. McKey. 1993. The evolutionary ecology of symbiotic ant-plant relationships. *J. Hym. Res.* **2**: 13–83.

Davidson, D. W. and S. R. Morton. 1981. Myrmecochory in some plants (F. Chenopodiaceae) of the Australian arid zone. *Oecologia (Berl.)* **50**: 357–366.

Davies, S. J., M. H. Villet, T. M. Blomefield and R. M. Crewe. 1994. Reproduction and division of labour in *Leptogenys schwabi* Forel (Hymenoptera: Formicidae), a polygynous, queenless ponerine ant. *Ethol. Ecol. Evol.* **6**: 507–517.

DeVries, P. J. 1988. The larval ant-organs of *Thisbe irenea* (Lepidoptera: Riodinidae) and their effects upon attending ants. *Biol. J. Linn. Soc.* **94**: 379–393.

Dejean, A. 1987. New cases of archaic foundation of societies in Myrmicinae (Formicidae): study of prey capture by queens of Dacetini. *Insectes Soc.* **34**: 211–221.

Dejean, A. and J.-P. Lachaud. 1994. Ecology and behavior of the seed-eating ponerine ant *Brachyponera senaarensis* (Mayr). *Insectes Soc.* **41**: 191–210.

Delage, B. and P. Jaisson. 1969. Etude des relations sociales chez des fourmis du genre *Aphaenogaster. C.R. Acad. Sci. Paris* **268**: 701–703.

Düßmann, O., C. Peeters and B. Hölldobler. 1996. Morphology and reproductive behaviour of intercastes in the ponerine ant *Pachycondyla obscuricornis. Insectes Soc.* **43** (in press).

Duncan, F. D. 1995. A reason for division of labor in ant foraging. *Naturwissenschaften* **82**: 293–296.

Duncan, F. D. and R. M. Crewe. 1994. Group hunting in a ponerine ant, *Leptogenys nitida* Smith. *Oecologia (Berl.)* **97**: 118–123.

Evans, H. C. and D. Leston. 1971. A ponerine ant (Hym., Formicidae) associated with Homoptera on cocoa in Ghana. *Bull. Entomol. Res.* **61**: 357–362.

Fénéron, R. 1993. Ethogenèse et reconnaissance interindividuelle – influence de l'expérience précoce chez une fourmi ponérine (*Ectatomma tuberculatum* Olivier). Ph. D. dissertation, Université Paris XIII, Villetaneuse.

Franks, N. R. and B. Hölldobler. 1987. Sexual competition during colony reproduction in army ants. *Biol. J. Linn. Soc.* **30**: 229–243.

Freeland, J. 1958. Biological and social patterns in the Australian bulldog ants of the genus *Myrmecia. Austr. J. Zool.* **6**: 1–18.

Fresneau, D. 1984. Développement ovarien et statut social chez une fourmi primitive *Neoponera obscuricornis* Emery (Hym. Formicidae, Ponerinae). *Insectes Soc.* **31**: 387–402.

–. 1994. Biologie et comportement social d'une fourmi ponérine néotropicale (*Pachycondyla apicalis*). Ph. D. dissertation, Université Paris XIII, Villetaneuse.

Fresneau, D. and P. Dupuy 1988. A study of polyethism in a ponerine ant: *Neoponera apicalis* (Hymenoptera: Formicidae). *Anim. Behav.* **36**: 1389–1399.

Fresneau, D., J. Garcia Perez and P. Jaisson. 1982. Evolution of polyethism in ants: observational results and theories. In *Social Insects in the Tropics*, vol. 1. P. Jaisson, ed., pp. 129–155. Villetaneuse: Université Paris Nord.

Fridman, S. and E. Avital. 1983. Foraging by queens of *Cataglyphis bicolor nigra* (Hymenoptera: Formicidae): an unusual phenomenon among the Formicinae. *Israel J. Zool.* **32**: 229–230.

Fukumoto, Y. and T. Abe. 1983. Social organization of colony movement in the tropical ponerine ant, *Diacamma rugosum* (Le Guillou). *J. Ethol.* **1**: 101–108.

Fukumoto, Y., T. Abe and A. Taki. 1989. A novel form of colony organization in the 'queenless' ant *Diacamma rugosum. Physiol. Ecol. Jap.* **26**: 55–61.

Gobin, B., C. Peeters and J. Billen. 1994. Dominance interactions and control of reproduction in the queenless ant *Gnamptogenys* sp. from Sulawesi (Ponerinae). In *Les Insectes Sociaux*. A. Lenoir, G. Arnold and M. Lepage, eds., p. 286. Villetaneuse: Université Paris Nord.

Gotwald, W. H. and W. L. Brown. 1966. The ant genus *Simopelta* (Hymenoptera: Formicidae). *Psyche* **73**: 261–277.

Gotwald, W. H. and J. Lévieux. 1972. Taxonomy and biology of a new West African ant belonging to the genus *Amblyopone* (Hymenoptera: Formicidae). *Ann. Entomol. Soc. Am.* **65**: 383–396.

Gray, B. 1971a. Notes on the biology of the ant species *Myrmecia dispar* (Clark) (Hymenoptera: Formicidae). *Insectes Soc.* **18**: 71–80.

–. 1971b. Notes on the field behaviour of two ant species *Myrmecia desertorum* Wheeler and *Myrmecia dispar* (Clark) (Hymenoptera: Formicidae). *Insectes Soc.* **18**: 81–94.

–. 1971c. A morphometric study of the ant species, *Myrmecia dispar* (Clark) (Hymenoptera: Formicidae). *Insectes Soc.* **18**: 95–110.

–. 1974. Nest structure and populations of *Myrmecia* (Hymenoptera: Formicidae), with observations on the capture of prey. *Insectes Soc.* **21**: 107–120.

Hashimoto, Y. 1991. Phylogenetic study of the Family Formicidae based on the sensillum structures on the antennae and labial palpi (Hymenoptera, Aculeata). *Japan J. Entomol.* **59**: 125–140.

Hashimoto, Y., K. Yamauchi and E. Hasegawa. 1995. Unique habits of stomodeal trophallaxis in the ponerine ant *Hypoponera* sp. *Insectes Soc.* **42**: 137–144.

Haskins, C. P. 1970. Researches in the biology and social behavior of primitive ants. In *Development and Evolution of Behavior: Essays in Memory of T. C. Schneirla*. L. Aronson, E. Tobach, D. Lehrman and J. Rosenblatt, eds., pp. 355–388. San Francisco: Freeman.

Haskins, C. P. and E. F. Haskins. 1950a. Notes on the biology and social behavior of the archaic ponerine ants of the genera *Myrmecia* and *Promyrmecia*. *Ann. Entomol. Soc. Am.* **43**: 461–491.

–. 1950b. Note on the method of colony foundation of the ponerine ant *Brachyponera (Euponera) lutea* Mayr. *Psyche* **57**: 1–9.

–. 1951. Note on the method of colony foundation of the ponerine ant *Amblyopone australis* Erichson. *Am. Midl. Nat.* **45**: 432–445.

–. 1955. The pattern of colony foundation in the archaic ant *Myrmecia regularis*. *Insectes Soc.* **2**: 115–126.

Haskins, C. P. and R. M. Whelden. 1954. Note on the exchange of ingluvial food in the genus *Myrmecia*. *Insectes Soc.* **1**: 33–37.

–. 1965. "Queenlessness", worker sibship and colony vs population structure in the Formicid genus *Rhytidoponera*. *Psyche* **72**: 87–112.

Haskins, C. P. and P. A. Zahl. 1971. The reproductive pattern of *Dinoponera grandis* Roger (Hymenoptera, Ponerinae) with notes on the ethology of the species. *Psyche* **78**: 1–11.

Heinze, J. and B. Hölldobler. 1995. Thelytokous parthenogenesis and dominance hierarchies in the ponerine ant, *Platythyrea punctata*. *Naturwissenschaften* **82**: 40–41.

Heinze, J., B. Hölldobler and C. Peeters. 1994. Conflict and cooperation in ant societies. *Naturwissenschaften* **81**: 489–497.

Higashi, S. and C. P. Peeters. 1990. Worker polymorphism and nest structure in *Myrmecia brevinoda* Forel (Hymenoptera: Formicidae). *J. Austr. Entomol. Soc.* **29**: 327–331.

Hölldobler, B. 1985. Liquid food transmission and antennation signals in ponerine ants. *Israel J. Entomol.* **19**: 89–99.

Hölldobler, B., H. Engel and R. W. Taylor. 1982. A new sternal gland in ants and its function in chemical communication. *Naturwissenschaften* **69**: 90.

Hölldobler, B., C. Peeters and M. Obermayer. 1994. Exocrine glands and the attractiveness of the ergatoid queen in the ponerine ant *Megaponera foetens*. *Insectes Soc.* **41**: 63–72.

Hölldobler, B. and R. W. Taylor. 1983. A behavioral study of the primitive ant *Nothomyrmecia macrops* Clark. *Insectes Soc.* **30**: 384–401.

Hölldobler, B. and E. O. Wilson. 1986. Ecology and behavior of the primitive cryptobiotic ant *Prionopelta amabilis* (Hymenoptera: Formicidae). *Insectes Soc.* **33**: 45–58.

–. 1990. *The Ants*. Cambridge, Mass.: Harvard University Press.

Horvitz, C. C. and A. J. Beattie. 1980. Ant dispersal of *Calathea* (Marantaceae) seeds by carnivorous ponerines (Formicidae) in a tropical rain forest. *Am. J. Bot.* **67**: 321–326.

Ito, F. 1993a. Social organization in a primitive ponerine ant: queenless reproduction, dominance hierarchy and functional polygyny in *Amblyopone* sp. (*reclinata* group) (Hymenoptera: Formicidae: Ponerinae). *J. Nat. Hist.* **27**: 1315–1324.

–. 1993b. Observation of group recruitment to prey in a primitive ponerine ant, *Amblyopone* sp. (*reclinata* group) (Hymenoptera: Formicidae). *Insectes Soc.* **40**: 163–167.

–. 1993c. Functional monogyny and dominance hierarchy in the queenless ponerine ant *Pachycondyla* (=*Bothroponera*) sp. in West

Java, Indonesia (Hymenoptera, Formicidae, Ponerinae). *Ethology* **95**: 126–140.

–. 1993d. Queenless reproduction in a primitive ponerine ant *Amblyopone belli* (Hymenoptera: Formicidae) in southern India. *J. N.Y. Entomol. Soc.* **101**: 574–575.

–. 1995. Colony composition of two Malaysian ponerine ants, *Platythyrea tricuspidata* and *P. quadridenta*: sexual reproduction by workers and production of queens (Hymenoptera: Formicidae). *Psyche* **101**: 209–218.

Ito, F. and K. Ohkawara. 1994. Spermatheca size differentiation between queens and workers in primitive ants – relationship with reproductive structure of colonies. *Naturwissenschaften* **81**: 138–140.

Ito, F., N. Sugiura and S. Higashi. 1994. Worker polymorphism in the red-head bulldog ant (Hymenoptera: Formicidae), with description of nest structure and colony composition. *Ann. Entomol. Soc. Am.* **87**: 1–5.

Jaisson, P., D. Fresneau, R. W. Taylor and A. Lenoir. 1993. Social organization in some primitive Australian ants. I. *Nothomyrmecia macrops* Clark. *Insectes Soc.* **39**: 425–438.

Jennions, M. D. and D. W. Macdonald. 1994. Cooperative breeding in mammals. *Trends Ecol. Evol.* **9**: 89–93.

Kôriba, O. 1963. Colony founding of a female of *Brachyponera chinensis* (Emery) in the observation cage (Hymenoptera, Formicidae). *Kontyû* **31**: 285–289 (in Japanese, with English summary).

Lachaud, J.-P. and D. Fresneau. 1987. Social regulation in ponerine ants. In *From Individual to Collective Behavior in Social Insects* (*Experientia* Supplement, vol. 54). J. M. Pasteels and J.-L. Deneubourg, eds., pp. 197–217. Basel: Birkhäuser Verlag.

Lachaud, J.-P., D. Fresneau and B. Corbara. 1988. Mise en évidence de sous-castes comportementales chez *Amblyopone pallipes*. *Actes Coll. Ins. Soc.* **4**: 141–147.

Lamborn, W. A. 1915. On the relationship between certain West African insects, especially ants, Lycaenidae and Homoptera. *Trans. Entomol. Soc. Lond.* **1913**: 436–498.

Leal, I. R. and P. S. Oliveira. 1995. Behavioral ecology of the neotropical termite-hunting ant *Pachycondyla* (= *Termitopone*) *marginata*: colony founding, group-raiding and migratory patterns. *Behav. Ecol. Sociobiol.* **37**: 373–83.

Le Masne, G. 1952. Les échanges alimentaires entre adultes chez la fourmi *Ponera eduardi* Forel. *C.R. Acad. Sci. Paris* **235**: 1549–51.

Le Masne, G. and A. Bonavita. 1969. La fondation des sociétés selon un type archaïque par une fourmi appartenant à une sous-famille évoluée. *C.R. Acad. Sci. Paris* **269**: 2372–76.

Lenoir, A. and A. Dejean. 1994. Semi-claustral colony foundation in the formicine ants of the genus *Polyrhachis* (Hymenoptera: Formicidae). *Insectes Soc.* **41**: 225–234.

Lévieux, J. 1976a. Etude de la structure du nid de quelques espèces terricoles de fourmis tropicales. *Ann. Univ. Abidjan* **C12**: 23–33.

–. 1976b. La nutrition des fourmis tropicales. IV. Cycle d'activité et régime alimentaire de *Platythyrea conradti* (Hymenoptera, Formicidae, Ponerinae). *Ann. Univ. Abidjan* **E9**: 351–365.

–. 1979. La nutrition des fourmis granivores. III. Cycle d'activité et régime alimentaire, en saison des pluies, de *Brachyponera senaarensis* (Hym., Formicidae, Ponerinae). Fluctuations saisonnières. *Insectes Soc.* **26**: 232–239.

Liebig, J., J. Heinze and B. Hölldobler. 1994. Dominance behavior and trophallaxis in the palaearctic ant *Ponera coarctata*. In *Les Insectes Sociaux*. A. Lenoir, G. Arnold and M. Lepage, eds., p. 67. Villetaneuse: Université Paris Nord.

Longhurst, C. and P. E. Howse. 1979. Foraging, recruitment and emigration in *Megaponera foetens* (Fab.) (Hymenoptera: Formicidae) from the Nigerian Guinea savanna. *Insectes Soc.* **26**: 204–215.

Macevicz, S. 1979. Some consequences of Fisher's sex ratio principle for social Hymenoptera that reproduce by colony fission. *Am. Nat.* **113**: 363–371.

Maschwitz, U. and P. Schönegge. 1983. Forage communication, nest moving recruitment and prey specialization in the oriental ponerine *Leptogenys*. *Oecologia (Berl.)* **57**: 175–182.

Maschwitz, U., S. Steghaus-Kovac, R. Gaube and H. Hänel. 1989. A South East Asian ponerine ant of the genus *Leptogenys* (Hym., Form.) with army ant life habits. *Behav. Ecol. Sociobiol.* **24**: 305–316.

Masuko, K. 1986. Larval hemolymph feeding: a nondestructive parental cannibalism in the primitive ant *Amblyopone silvestrii* Wheeler (Hymenoptera: Formicidae). *Behav. Ecol. Sociobiol.* **19**: 249–255.

–. 1990. The instars of the ant *Amblyopone silvestrii* (Hymenoptera: Formicidae). *Sociobiology* **17**: 221–244.

–. 1996. Temporal division of labor among workers in the ponerine ant *Amblyopone silvestrii* (Hymenoptera: Formicidae). *Sociobiology* **28**: 1–21.

McAreavey, J. 1948. Some observations on *Myrmecia tarsata* Smith. *Proc. Linn. Soc. N.S.W.* **73**: 137–141.

Ogata, K. 1991. Ants of the genus *Myrmecia* Fabricius: a review of the species groups and their phylogenetic relationships (Hymenoptera: Formicidae: Myrmeciinae). *Syst. Entomol.* **16**: 353–381.

Oliveira, P. S. and B. Hölldobler. 1990. Dominance orders in the ponerine ant *Pachycondyla apicalis* (Hymenoptera: Formicidae). *Behav. Ecol. Sociobiol.* **27**: 385–393.

–. 1991. Agonistic interactions and reproductive dominance in *Pachycondyla obscuricornis* (Hymenoptera: Formicidae). *Psyche* **98**: 215–225.

Paiva, R. V. and C. R. Brandão. 1995. Nests, worker population and reproductive status of workers, in the giant queenless ponerine ant *Dinoponera* Roger (Hymenoptera Formicidae). *Ethol. Ecol. Evol.* **7**: 297–312.

Passera, L., J. Bitsch and C. Bressac. 1968. Observations histologiques sur la formation des oeufs alimentaires et des oeufs reproducteurs chez les ouvrières de *Plagiolepis pygmaea* Latr. (Hymenoptera: Formicidae). *C.R. Acad. Sci. Paris* **266**: 2270–2272.

Passera, L., J.-P. Lachaud and L. Gomel. 1994. Individual food source fidelity in the neotropical ponerine ant *Ectatomma ruidum* Roger (Hymenoptera: Formicidae). *Ethol. Ecol. Evol.* **6**: 13–21.

Peeters, C. 1987a. The diversity of reproductive systems in ponerine ants. In *Chemistry and Biology of Social Insects*. J. Eder and H. Rembold, eds., pp. 253–254. Munich: Verlag J. Peperny.

–. 1987b. The reproductive division of labour in the queenless ponerine ant *Rhytidoponera* sp.12. *Insectes Soc.* **34**: 75–86.

–. 1990. Reproductive allocation in ponerine ants with or without queens. In *Social Insects and the Environment*. G. K. Veeresh, B. Mallik and C. A. Viraktamath, eds., pp. 357–358. New Delhi: Oxford & IBH Publishers.

–. 1991a. The occurrence of sexual reproduction among ant workers. *Biol. J. Linn. Soc.* **44**: 141–152.

–. 1991b. Ergatoid queens and intercastes in ants: two distinct adult forms which look morphologically intermediate between workers and winged queens. *Insectes Soc.* **38**: 1–15.

–. 1993. Monogyny and polygyny in ponerine ants with or without queens. In *Queen Number and Sociality in Insects*. L. Keller, ed., pp. 234–261. Oxford University Press.

Peeters, C. and R. Crewe. 1985a. Worker reproduction in the ponerine ant *Ophthalmopone berthoudi* – an alternative form of eusocial organization. *Behav. Ecol. Sociobiol.* **18**: 29–37.

–. 1985b. Queenlessness and reproductive differentiation in *Ophthalmopone hottentota*. *S. Afr. J. Zool.* **20**: 268.

–. 1986. Queenright and queenless breeding systems within the genus *Pachycondyla* (Hymenoptera: Formicidae). *J. Entomol. Soc. S. Afr.* **49**: 251–255.

–. 1987. Foraging and recruitment in ponerine ants: solitary hunting in the queenless *Ophthalmopone berthoudi* (Hymenoptera: Formicidae). *Psyche* **94**: 201–214.

Peeters, C. and R. H. Crozier. 1988. Caste and reproduction in ants: not all mated egg-layers are "queens". *Psyche* **95**: 283–288.

Peeters, C. and S. Higashi. 1989. Reproductive dominance controlled by mutilation in the queenless ant *Diacamma australe*. *Naturwissenschaften* **76**: 177–180.

Peeters, C. and B. Hölldobler. 1995. Reproductive cooperation between queens and their mated workers: the complex life history of an ant with a valuable nest. *Proc. Natl Acad. Sci. U.S.A.* **92**: 10 977–10 979.

Peeters, C. and K. Tsuji. 1993. Reproductive conflict among ant workers in *Diacamma* sp. from Japan: dominance and oviposition in the absence of the gamergate. *Insectes Soc.* **40**: 119–136.

Peeters, C., S. Higashi and F. Itô. 1991. Reproduction in ponerine ants without queens: monogyny and exceptionally small colonies in the Australian *Pachycondyla sublaevis*. *Ethol. Ecol. Evol.* **3**: 145–152.

Peeters, C., J. Billen and B. Hölldobler. 1992. Alternative dominance mechanisms regulating monogyny in the queenless ant genus *Diacamma*. *Naturwissenschaften* **79**: 572–573.

Peeters, C., B. Hölldobler, M. Moffett and T. Musthak Ali. 1994. 'Wall-papering' and elaborate nest architecture in the ponerine ant *Harpegnathos saltator*. *Insectes Soc.* **41**: 211–218.

Pratt, S. C. 1994. Ecology and behavior of *Gnamptogenys horni* (Formicidae: Ponerinae). *Insectes Soc.* **41**: 255–262.

Pratt, S. C., N. F. Carlin and P. Calabi. 1994. Division of labor in *Ponera pennsylvannica* (Formicidae: Ponerinae). *Insectes Soc.* **41**: 43–61.

Rissing, S. W., G. B. Pollock, M. R. Higgins, R. H. Hagen and D. R. Smith. 1989. Foraging specialization without relatedness or dominance among co-founding ant queens. *Nature (Lond.)* **338**: 420–422.

Robbins, R. K. 1991. Cost and evolution of a facultative mutualism between ants and lycaenid larvae (Lepidoptera). *Oikos* **62**: 363–369.

Robinson, G. E. 1992. Regulation of division of labor in insect societies. *Annu. Rev. Entomol.* **37**: 637–665.

Sherman, P. W., J. U. M. Jarvis and R. D. Alexander, eds. 1991. *The Biology of the Naked Mole-Rat.* Princeton, N. J.: Princeton University Press.

Sommer, K. and B. Hölldobler. 1992. Coexistence and dominance among queens and mated workers in the ant *Pachycondyla tridentata. Naturwissenschaften* **79**: 470–472.

Steghaus-Kovac, S. and U. Maschwitz. 1993. Predation on earwigs: a novel diet specialization within the genus *Leptogenys* (Formicidae: Ponerinae). *Insectes Soc.* **40**: 337–340.

Taylor, R. W. 1978. *Nothomyrmecia macrops*: a living-fossil ant rediscovered. *Science (Wash., D.C.)* **201**: 979–985.

Tinaut, A. and J. Heinze. 1992. Wing reduction in ant queens from arid habitats. *Naturwissenschaften* **79**: 84–85.

Traniello, J. F. A. 1978. Caste in a primitive **ant**: absence of age polyethism in *Amblyopone. Science (Wash., D.C.)* **202**: 770–772.

Tschinkel, W. R. 1988. Social control of egg-laying rate in queens of the fire ant, *Solenopsis invicta. Physiol. Entomol.* **13**: 327–350.

–. 1991. Insect sociometry, a field in search of data. *Insectes Soc.* **38**: 77–82.

Verhaagh, M. 1994. *Pachycondyla luteola* (Hymenoptera, Formicidae), an inhabitant of *Cecropia* trees in Peru. *Andrias* **13**: 215–224.

Villet, M. 1989. A syndrome leading to ergatoid queens in ponerine ants (Hymenoptera: Formicidae). *J. Nat. Hist.* **23**: 825–832.

–. 1990a. Qualitative relations of egg size, egg production and colony size in some ponerine ants (Hymenoptera: Formicidae). *J. Nat. Hist.* **24**: 1321–1331.

–. 1990b. Division of labour in the Matabele ant *Megaponera foetens* (Fabr.) (Hymenoptera Formicidae). *Ethol. Ecol. Evol.* **2**: 397–417.

–. 1990c. Social organization of *Platythyrea lamellosa* (Roger) (Hymenoptera: Formicidae): II. Division of labour. *S. Afr. J. Zool.* **25**: 250–253.

–. 1990d. Colony foundation in the ponerine ant, *Mesoponera caffraria* (F. Smith) (Hymenoptera: Formicidae). *S. Afr. J. Zool.* **25**: 39–40.

–. 1991a. Colony foundation in *Plectroctena mandibularis* F. Smith and the evolution of ergatoid queens in *Plectroctena* (Hymenoptera: Formicidae). *J. Nat. Hist.* **25**: 979–983.

–. 1991b. Reproduction and division of labour in *Platythyrea* cf. *cribrinodis* (Gerstaecker 1858) (Hymenoptera: Formicidae): comparisons of individuals, colonies and species. *Trop. Zool.* **4**: 209–231.

–. 1991c. Social differentiation and division of labour in the queenless ant *Platythyrea schultzei* Forel 1910 (Hymenoptera: Formicidae). *Trop. Zool.* **4**: 13–29.

–. 1992. The social biology of *Hagensia havilandi* (Forel 1901) (Hymenoptera Formicidae) and the origin of queenlessness in ponerine ants. *Trop. Zool.* **5**: 195–206.

–. 1993. Co-occurrence of mated workers and a mated queen in a colony of *Platythyrea arnoldi* (Hymenoptera: Formicidae). *S. Afr. J. Zool.* **28**: 56–57.

Villet, M., R. Crewe and H. Robertson. 1989. Mating behavior and dispersal in *Paltothyreus tarsatus* Fabr. *J. Insect Behav.* **2**: 413–417.

Villet, M., R. Crewe and F. Duncan. 1991. Evolutionary trends in the reproductive biology of ponerine ants (Hymenoptera: Formicidae). *J. Nat. Hist.* **25**: 1603–1610.

Villet, M., A. Hart and R. Crewe. 1990. Social organization of *Platythyrea lamellosa* (Roger) (Hymenoptera: Formicideae): I. Reproduction. *S. Afr. J. Zool* **25**: 250–253.

Villet, M. and M. Wildman. 1991. Division of labour in the obligately queenless ant *Pachycondyla* (= *Bothroponera*) *krugeri* Forel 1910 (Hymenoptera Formicidae). *Trop. Zool.* **4**: 232–250.

Voss, S. H., J. F. McDonald and C. H. Keith. 1988. Production and abortive development of fire ant trophic eggs. In *Advances in Myrmecology.* J. C. Trager, ed., pp. 517–527. Leiden: E. J. Brill.

Ward, P. S. 1981a. Ecology and life history of the *Rhytidoponera impressa* group (Hymenoptera: Formicidae) I. Habitats, nest sites and foraging behavior. *Psyche* **88**: 89–108.

–. 1981b. Ecology and life history of the *Rhytidoponera impressa* group (Hymenoptera: Formicidae) II. Colony origin, seasonal cycles and reproduction. *Psyche* **88**: 109–126.

–. 1983. Genetic relatedness and colony organization in a species complex of ponerine ants I. Phenotypic and genotypic composition of colonies. *Behav. Ecol. Sociobiol.* **12**: 285–299.

–. 1984. A revision of the ant genus *Rhytidoponera* (Hymenoptera: Formicidae) in New Caledonia. *Austr. J. Zool.* **32**: 131–175.

–. 1994. *Adetomyrma*, an enigmatic new ant genus from Madagascar (Hymenoptera: Formicidae) and its implications for ant phylogeny. *Syst. Entomol.* **19**: 159–175.

Ward, P. S. and R. W. Taylor. 1981. Allozyme variation, colony structure and genetic relatedness in the primitive ant *Nothomyrmecia macrops* Clark (Hymenoptera: Formicidae). *J. Austr. Entomol. Soc.* **20**: 177–183.

Ware, A. B., S. G. Compton and H. G. Robertson. 1990. Gamergate reproduction in the ant *Streblognathus aethiopicus* Smith (Hymenoptera: Formicidae: Ponerinae). *Insectes Soc.* **37**: 189–199.

Weber, N. A. 1946. Two common ponerine ants of possible economic significance, *Ectatomma tuberculatum* (Olivier) and *E. ruidum* Roger. *Proc. Entomol. Soc. Wash.* **48**: 1–16.

Wehner, R., R. D. Harkness and P. Schmid-Hempel. 1983. *Foraging Strategies in Individual Searching Ants* Cataglyphis bicolor *(Hymenoptera: Formicidae). (Information Processing in Animals,* vol. 1.) New York: Gustav Fischer Verlag.

Wheeler, D. E. 1986. Developmental and physiological determinants of caste in social Hymenoptera: evolutionary implications. *Am. Nat.* **128**: 13–34.

–. 1994. Nourishment in ants: patterns in individuals and societies. In *Nourishment and Evolution in Insect Societies.* J. Hunt and C. Nalepa, eds., pp. 245–278. Boulder, Colo.: Westview Press.

Wheeler, D. E. and N. A. Buck. 1995. Storage proteins in ants during development and colony founding. *J. Insect Physiol.* **41**: 885–894.

Wheeler, G. C. and J. Wheeler. 1988. An additional use for ant larvae (Hymenoptera: Formicidae). *Entomol. News* **99**: 23–24.

Wheeler, W. M. 1933. *Colony-founding among Ants, with an Account of some Primitive Australian Species.* Cambridge, Mass.: Harvard University Press.

Wilson, E. O. 1958. The beginnings of nomadic and group-predatory behavior in the ponerine ants. *Evolution* **12**: 24–36.

–. 1959. Some ecological characteristics of ants in New Guinea rain forests. *Ecology* **40**: 437–447.

Young, A. M. 1977. Notes on the foraging of the giant tropical ant *Paraponera clavata* (Formicidae: Ponerinae) on two plants in tropical wet forest. *J. Georgia Entomol. Soc.* **12**: 41–51.

19 · Social conflict and cooperation among founding queens in ants (Hymenoptera: Formicidae)

JAE C. CHOE AND DAN L. PERLMAN

'It is thus that mutual cowardice keeps us in peace.'

Samuel Johnson

ABSTRACT

The colony founding phase is a unique period in the life cycle of social insects in which young reproductives have an option to be either solitary or social. We review the literature on pleometrosis in ants with a special emphasis on the founding processes by *Azteca constructor* and *A. xanthacroa* nesting in *Cecropia* trees. *Azteca* colonies are established in three distinct ways: by single queens, by single-species groups of queens, and by mixed-species groups. Pleometrotic colonies of both single species and mixed species tend to outcompete haplometrotic colonies by (1) producing much larger first worker cohorts; (2) migrating upward within the *Cecropia* tree earlier; and (3) eventually taking exclusive control of the tree. The fact that mixed-species *Azteca* colonies are nearly as successful and cooperative as their single-species counterparts provides a clear-cut case where inclusive fitness is not a prime selective factor for foundress association. Although many studies of pleometrotic ants measure the number of workers produced and the timing of first worker emergence, such measures do not account for reproduction because workers are merely a means by which a queen produces reproductive offspring. Colonies with larger and more rapidly produced worker forces tend to survive better, largely because they are likely to prevail over smaller colonies via interference or exploitation competition. Thus, enhanced worker production plays a role mostly in survival rather than reproduction, although it is possible that colonies with larger worker forces at the start may reach the reproductive stage earlier. We urge future researchers of pleometrotic ants to extend their studies to include the reproductive stage of colony development, and to measure the size and speed of production of reproductive offspring. Compared with bees and wasps, where foundress association may have evolved in response to foraging-related foundress mortality, such predation pressure may have led to the evolution of claustral colony founding in ants and termites. Pleometrosis appears to be a secondary adaptation caused by selection to produce a large worker force in a shorter period of time in those species of ants and termites in which incipient colonies are clumped and destined to experience intense competition. A review of current literature on ant pleometrosis suggests that foundress associations are particularly common in species whose colonies are large and highly territorial.

INTRODUCTION

Social insects establish new colonies in a variety of ways. In species such as honey bees, stingless bees and army ants, newly mated queens found colonies accompanied by workers from the natal colony. In most social insects, however, queens found colonies independently, without the help of workers. These independent foundresses may be claustral, in which case they remain sealed within their nests until the first brood of workers ecloses, or non-claustral, in which case the queens forage outside of the nests while raising their first brood. Although claustral queens may experience reduced predation, they must rely on the histolysis of their wing muscles and fat bodies to supply the energy resources needed to rear the first brood of workers. Therefore, it is imperative for such foundresses to produce workers before their bodily resources are exhausted (Waloff 1957; Keller and Passera 1990).

The number of founding queens also varies in social insects. Although the majority of new colonies are founded by single queens (*haplometrosis*), a significant minority are founded by groups of queens (*pleometrosis*; Wheeler 1933; Hölldobler and Wilson 1977, 1990; Rissing and Pollock 1988). Pleometrotic colony founding is known to occur in a number of vespid wasps (Ross and Matthews 1991; Itô 1993) and various primitively eusocial bees (Michener 1974, 1990a,b; Packer 1993; Schwarz *et al.*, this volume). It is also a common mode of colony foundation for several species of tropical termites (Roisin 1993; Shellman-Reeve, this volume) and over a dozen species of ants.

For the past three decades since Hamilton's (1964, 1972) ingenious theoretical solution to one of Darwin's unsolved mysteries, there has been an explosion of research on the worker sterility problem. Compared with the amount of attention that queen–worker conflicts have received (see, for example, Trivers and Hare 1976; Alexander and Sherman 1977; Nonacs 1986; Bourke 1988; Choe 1988), queen–queen conflicts have not attracted much study (but see Keller 1993). An exception to this trend is an early recognition of the issue by students of pleometrotic social wasps, most notably by West-Eberhard (1967, 1969, 1978). In the life cycle of social insects, the colony founding phase provides a unique window of opportunity to test a variety of models for the origin and maintenance of eusociality, because during this phase young reproductives may have an option to be either solitary or social. By examining the relative costs and benefits of pleometrosis and haplometrosis, one can gain insight by analogy into the evolution of social behavior as vertebrate behavioral ecologists have done through their studies of cooperatively breeding birds and mammals (see, for example, Brown 1987; Creel and Creel 1991).

In most social insects with pleometrotic colony foundation, only one of the multiple foundresses survives to produce reproductive offspring (Hölldobler and Wilson 1977; Rissing and Pollock 1988). Why does a newly mated, potential future queen join others in founding a colony, when she is almost certain to die before making genetic contributions to future sexual generations? Inclusive fitness theory would predict that the foundresses in pleometrotic colonies are relatives and that they may increase their own inclusive fitness by helping the family's total fitness increase. Although this prediction seems to be met in primitively eusocial wasps (Queller *et al.* 1988; Strassmann *et al.* 1989; Hughes et al. 1993; but see Itô 1993), cofounding ant queens have been found to be unrelated (see, for example, Hagen *et al.* 1988; Rissing *et al.* 1989). Thus, the evolution of pleometrosis by non-kin foundresses will most likely be explained by examining ecological costs and benefits of different modes of colony foundation.

We discuss various ecological parameters that may be responsible for the promotion of pleometrotic colony foundation in ants (see Brockmann, this volume, for a similar account on wasps; Schwarz *et al.*, this volume, on allodapine bees; Shellman-Reeve, this volume, on termites; Crespi and Mound, this volume, on thrips). The phylogenetic distribution of pleometrotic ants, and similarities and differences in pleometrosis among bees, wasps, ants, and termites are also analyzed to illustrate the aspects of pleometrosis unique to ants. We describe a previously unexplored form of pleometrosis, mixed-species colony foundation by queens of two distinct *Azteca* species, emphasizing the significance of brood production and intercolony competition. Finally, we attempt to illuminate some of the unanswered questions in the evolution of pleometrosis and suggest directions for future research.

PHYLOGENETIC DISTRIBUTION OF PLEOMETROSIS IN ANTS

Although pleometrotic colonies may occur fortuitously in many subfamilies of ants, most consistently pleometrotic species belong to the three most advanced subfamilies, Myrmicinae, Dolichoderinae, and Formicinae (Fig. 19-1) (see Peeters 1993 for possible exceptions in the Ponerinae). The list of pleometrotic genera we have compiled is by no means comprehensive. We did not include species such as *Nothomyrmecia macrops* (Taylor 1978) in which pleometrosis appears to be the exception rather than the rule. None the less, the current phylogenetic distribution along with ecological characteristics of all known cases of pleometrosis reveals a few important attributes.

First, pleometrosis appears to be an adaptation for intercolony competition in species that are highly territorial and have specific microhabitat requirements for nesting. Most, if not all, species that are included in Fig. 19-1 have large colonies and aggressively defend their territories against other conspecific neighbors. Although pleometrotic *Azteca* species live only inside *Cecropia* trees, ground-nesting species also show strong habitat selection at colony foundation. *Acromyrmex versicolor* queens appear to avoid high temperatures and prefer shaded areas around the bases of large trees (Rissing *et al.* 1986); *Messor pergandei* queens aggregate in ravine bottoms where the soil moisture may be relatively high (Rissing and Pollock 1987); and *Solenopsis invicta* queens seek out elevated microsites (Tschinkel and Howard 1983).

Second, although pleometrotic species occur sporadically in different subfamilies, there exist no marked trends in characteristics of pleometrosis among the three subfamilies. Ants of all three subfamilies have invented remarkably similar sets of modifications and variation on pleometrotic colony development. The only notable deviation is in the occurrence of primary plurimatry (i.e. the retention of multiple foundresses as functional egg-laying

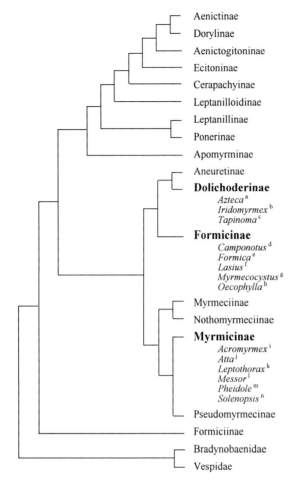

Fig. 19-1. Phylogenetic distribution of ant genera containing one or more pleometrotic species (Ant phylogeny adapted from Baroni Urbani *et al.* 1992; two vespoid families are included as outgroups). References: a, Perlman 1992; b, Hölldobler and Carlin 1985; c, Hanna 1975; d, Mintzer 1979, Fowler and Roberts 1983; e, Deslippe and Savolainen 1995; f, Waloff 1957, Nonacs 1990, Sommer and Hölldobler 1992, 1995; g, Bartz and Hölldobler 1982; h, Peeters and Andersen 1989; i, Rissing *et al.* 1986, 1989; j, Mintzer and Vinson 1985; k, Buschinger 1974; l, Taki 1976, Rissing and Pollock 1986; m, S. P. Cover in Hölldobler and Wilson 1990; n, Tschinkel and Howard 1983.

queens in a mature colony; plurimatry is a new term suggested by Choe (1995) to replace a confusing term, polygyny, for the existence of multiple reproductives in a colony), which has been observed in the Myrmicinae and Dolichoderinae but not in the Formicinae. Incidentally, more obvious cases of aggression and dominance among cofoundresses have been noted in four formicine genera: *Camponotus* (Fowler and Roberts 1983), *Formica* (Deslippe

and Savolainen 1995), *Lasius* (Waloff 1957), and *Myrmeco-cystus* (Bartz and Hölldobler 1982).

Third, pleometrosis appears to be confined principally to claustral species, although *A. versicolor* presents an exception to the rule. Coincidentally, both pleometrosis and claustra-lity tend to be more prevalent among the more advanced taxa. In *A. versicolor*, foraging specialization has apparently evolved among unrelated cofoundresses, although foraging specialists are obviously at a selective disadvantage (Rissing *et al.* 1989). Considering that such foraging specialization accrues fitness benefits in terms of intracolonial comp-etition, non-claustral colony foundation in pleometrotic species appears to be a secondarily derived adaptation.

COLONY FOUNDING AMONG *AZTECA* ANTS

A recent investigation on the colony foundation of two *Cecropia*-nesting *Azteca* species (Perlman 1992) revealed that queens establish colonies in three distinct ways: solita-rily, in single-species groups, and even in mixed-species groups (Fig. 19-2). Such variety of colony-founding behav-ior of these ants provides a unique system to compare the costs and benefits of solitary and social life during a critical period in the life cycle of ants.

In Monteverde, Costa Rica, *Cecropia* saplings (mostly *C. obtusifolia*) are colonized by *A. constructor* and *A. xantha-croa*. We repeatedly monitored young *Azteca* colonies between February 1988 and July 1989 in the field. During the initial census, small holes (approximately 6 mm in dia-meter) were cut into internodes occupied by ants. All *Azteca* queens were removed and uniquely marked on the alitrunk with two or three dots of quick-drying paint. Then the marked queens were returned to their internodes and the holes were plugged with rubber stoppers. The use of rubber stoppers allowed for repeated censuses with little apparent disturbance to either plant or ants. During subse-quent censuses, the plugs were removed one at a time to observe the resident colony, using an otoscope and/or a fiber-optic endoscope. The following sections summarize Perlman's (1992) findings on the founding process and sub-sequent development of *A. constructor*, *A. xanthacroa*, and mixed-species colonies.

Natural history

Cecropia is a moraceous genus of second-growth trees that are common throughout tropical America from Mexico to

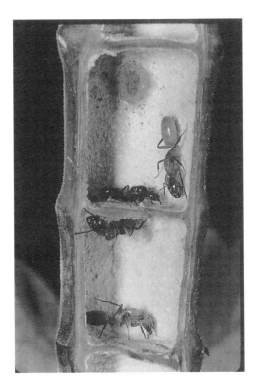

Fig. 19-3. An *Azteca xanthacroa* queen chewing into a *Cecropia* internode.

Fig. 19-2. A longitudinal section of a *Cecropia* sapling, showing two incipient colonies: an *Azteca xanthacroa* colony in the lower internode and mixed-species colony in the upper internode.

Brazil. Several species are myrmecophytes, obligatorily occupied by neotropical dolichoderine ants in the genus *Azteca*. Nesting exclusively in the hollow internodes of the trees, *Azteca* ants maintain a mutualistic relationship with *Cecropia*. The plants provide the ants with shelter and food, glycogen-containing Müllerian bodies produced on hairy pads (trichilia) at the bases of leaf petioles (Rickson 1971). In return, the ants defend the plants against herbivorous insects and encroaching vines (Janzen 1969; Schupp 1986).

Azteca colonies are founded throughout much of the year in Monteverde, with the heaviest activity taking place from February to September. *Cecropia* propagates as either seedlings or stump shoots. Young *Cecropia* trees grow taller by adding new internodes on the top, which are colonized by newly mated *Azteca* queens. Soon after landing on a *Cecropia* sapling, the queen typically explores the upper internodes of the plant for a few minutes, sheds her wings, and chews into a thin, unvascularized area in the wall of one of the uppermost internodes (Fig. 19-3). This process can take from half an hour to nearly two hours. Once inside, she plugs the hole with parenchyma scraped from the inner walls of the internode. Later-arriving queens are able to chew through the entrance plug much more quickly, taking only 5–12 minutes, and are thus exposed to a much lower risk of predation while on the outside of the plant.

Queens begin laying eggs as early as one to three days after colony foundation. By injecting queens in pleometrotic colonies with different oil-soluble vital dyes or feeding them Müllerian bodies soaked in such dyes, we found that all queens lay eggs. Although several potential food sources are available during this stage, the metabolites of the queen's flight muscles and fat bodies appear to be the primary food source. Occasionally, however, *Azteca* queens forage outside the internode, as indicated by the presence of Müllerian bodies in five of 967 colonies (0.52%) observed before they produced workers.

Distribution of queens among internodes

Haplometrotic colonies accounted for 62.9% of all *Azteca* colonies ($n = 739$), while 68.6% of all queens ($n = 1481$) were in pleometrotic colonies. Mixed-species colonies were common: 21.9% of all queens were in mixed colonies. The mean number of queens/internode was 2.00 queens (± 0.08 SE, range $= 1 - 26$, $n = 739$). The distribution patterns of queens per internode were significantly clumped ($p < 0.001$) in single-species colonies of both species (Fig. 19-4).

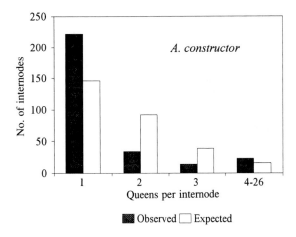

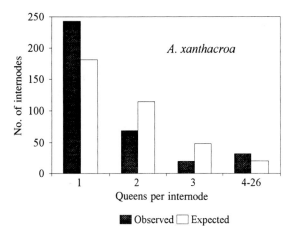

Fig. 19-4. Observed distributions of *Azteca* queens per internode vs. expected distributions.

The distribution patterns of *Azteca* queens appear to be the product of both queen behavior and the architecture and growth patterns of *Cecropia* trees. Although new *Cecropia* seedlings and shoots are present year-round, and thus appropriate nest sites are available for colonization by *Azteca* queens throughout the year, the preference of the queens for colonizing only the terminal three or four internodes strictly limits the number of nest sites available for colonization at any time.

Brood production

Total colony brood production increased with the number of queens in all colony types, i.e., colonies with multiple

queens produced a larger first brood. Although this benefit is measured in terms of the entire colony's first brood production, it is in fact in each queen's best interests to take part in a pleometric group. A haplometrotic foundress cannot raise productivity by increasing the resources she brings to the colony, although she may increase the proportion of her resources allocated to producing workers. However, if she overallocates, she will starve before the colony opens to the outside environment. Therefore, the most effective method for increasing the size of the first brood is to increase the number of foundresses in the colony. Given that claustral colonies do not leave their saplings, and that such colonies depend entirely on queen-produced food, these colonies are participating in a zero-sum game; each queen that a colony acquires is one less that competing colonies in the plant can acquire.

Survival of foundresses

Among pleometric colonies, several *A. constructor* queens frequently survived together months after colony founding, whereas in *A. xanthacroa* and mixed-species colonies only one queen survived beyond the production of the first workers. Not a single *A. xanthacroa* or mixed-species colony contained multiple live queens and more than 50 workers. By contrast, of the 18 *A. constructor* colonies containing at least 100 workers, 15 (83%) held multiple queens. *Azteca constructor* queens in pleometric groups tended to survive longer than those founding alone, whereas *A. constructor* queens in mixed-species colonies survived the shortest time. Neither colony type nor queen number affected *A. xanthacroa* queen survival (Perlman 1992). In other words, *A. xanthacroa* queens survived equally well in both haplometrotic and pleometric as well as mixed-species colonies. In mixed-species colonies, *A. xanthacroa* queens generally survived longer than *A. constructor* queens, although this observation is confounded by the fact that hardly any mixed-species colonies contained a majority of *A. constructor* queens.

Once the first workers eclose and reopen the original queen entrance hole, they begin foraging for Müllerian bodies (Fig. 19-5) and thus change the colony's energy budget drastically. Since the queens are no longer the primary source of food in the colony, interactions among them change drastically as well at this point. Only a single queen survives much past the eclosion of the first workers in most colonies containing *A. xanthacroa* queens. Dingman (1987) observed in the laboratory that *A. xanthacroa* queens are

Fig. 19-5. An *Azteca* worker extracting a Müllerian body from one of a *Cecropia* sapling's trichilia.

Fig. 19-6. *Azteca xanthacroa* queens attacking each other.

considerably more aggressive than *A. constructor* queens (Fig. 19-6). In all *A. constructor* colonies, however, there is no increase in queen–queen conflict when the first workers eclose, and cofoundresses often remain cooperative for a year or more (Fig. 19-7).

Intercolony competition

Intercolony dynamics among founding *Azteca* colonies sharing a *Cecropia* sapling are remarkably rich, complex, and varied. Within a single plant, variables such as the number of queens per colony, the number of workers in the first brood cohort, and internode volume might vary by an order of magnitude or more, and new colonies might be established over a period of up to approximately 17 months. Because no more than one colony per tree reaches sexual maturity (Longino 1989), a queen's probability of reproductive success depends not merely on aspects of her colony; other founding colonies living in her tree may influence her chance of success even more.

Azteca workers do not merely collect Müllerian bodies, they actively stockpile and guard them within the internodes, where they apparently stay mold-free for some time (Fig. 19-8). Such stockpiles not only guard against inclement weather and periods of low production of Müllerian bodies by the plant, but also reduce the food available to competing incipient colonies living in the tree. As part of the competition to stockpile Müllerian bodies, *Azteca* colonies compete to be the first colony in the sapling to migrate to a higher, younger internode. The

Fig. 19-7. Four physogastric *Azteca constructor* queens from a colony that contained several reproductive pupae.

Fig. 19-8. A stockpile of Müllerian bodies inside an internode of a dominant *Azteca* colony. Note the brood intermixed with the hundreds of mold-free Müllerian bodies.

uppermost internodes are much closer to the younger leaves, where the majority of Müllerian bodies are produced, so foraging is much more efficient after migration. A colony's first migration occurs an average of approximately five to ten months after colony foundation, when colonies contain approximately 10–30 workers.

Colonies are at risk during migration, because the queen and brood must reach the new internode via the outside of the tree. Alternatively, some colonies establish 'outposts' in the upper internodes, in which workers can store Müllerian bodies before they transport them to the rest of the colony below. The rest of the colony eventually migrates up to an outpost internode. A colony may create one or more outposts while there are other colonies living in the tree. Outposts mark the beginning of a colony's expansion to multiple internodes. Eventually, one of the colonies in the tree outcompetes neighboring colonies in terms of worker production and resource monopolization; all other *Azteca* colonies in the tree die off before the queens have a chance to produce reproductive offspring. Each mature *Cecropia* tree is exclusively occupied by a single *Azteca* colony.

Differences in reproductive strategies between two *Azteca* species

Azteca constructor and *A. xanthacroa* queens appear to employ different reproductive strategies. Whereas groups of *A. constructor* queens often live together for a year or more, even producing reproductive pupae while in groups, only a single queen survives in pleometrotic *A. xanthacroa* colonies and in most mixed-species colonies. Such differences seem to have profound effects on colony development; pleometrotic *A. constructor* are able to produce worker forces of thousands in little more than a year, whereas *A. xanthacroa* and mixed-species colonies that eliminated all but a single queen at about the time of first worker eclosion appear to be on a slower production schedule. For *A. xanthacroa* queens, cofoundresses are a means to help achieve as large a first cohort of workers as possible. Once the colony becomes an open system, however, the extra reserves provided by cofoundresses during the claustral period are no longer needed. In contrast, although *A. constructor* queens also appear to require exclusive control of the tree before producing reproductive offspring, they do not eliminate cofoundresses as frequently or readily as do *A. xanthacroa* queens (Perlman 1992).

Unique features of pleometrosis in *Azteca*

Mixed-species pleometrosis

Mixed-species colonies of *A. constructor* and *A. xanthacroa* produce first broods similar in size to those of single-species colonies with the same numbers of queens. Furthermore, queens in mixed-species colonies generally do not suffer higher mortality than queens in single-species colonies (Perlman 1992). Selection to produce a large worker force as quickly as possible to outcompete neighboring colonies has led not only to pleometrotic colony founding but also to the development of this remarkable case of interspecific cooperation.

The fact that queens of two different species cooperate in their colony-founding effort and that such colonies are about as successful as single-species pleometrotic colonies essentially eliminates kin selection from a list of potential causes of pleometrosis in this case. This finding further substantiates previous behavioral and genetic observations of low genetic relatedness among cofounding ant queens (Hagen *et al.* 1988; Rissing *et al.* 1989). However, it is still possible that kin selection may play a role when workers eliminate supernumerary queens and/or when queens accept or reject potential joiner queens, particularly in species where cofoundresses remain as a group past the eclosion of the first workers. Genetic relatedness among cofoundresses has not been estimated for such species.

Nature of competition

In most pleometrotic ant species, incipient colonies are clumped (reviewed in Rissing and Pollock 1988, Hölldobler and Wilson 1990). Such an extreme degree of nest-clumping and the territorial nature of mature colonies make competition among incipient colonies inevitable. A common type of competition among pleometrotic ants is interference competition in the form of brood-raiding (see, for example, Bartz and Hölldobler 1982; Tschinkel and Howard 1983; Rissing and Pollock 1986, 1987; Tschinkel 1992). Although Perlman (1992) reports two possible instances of brood raids among *Azteca* colonies based on indirect evidence such as enlarged entrance holes and sudden changes in brood size, brood raiding does not appear to be a common practice among incipient *Azteca* colonies. Instead, *Azteca* colonies are engaged in a form of exploitation competition by controlling access to a limiting resource, Müllerian bodies. Because accumulation of Müllerian bodies within a single *Cecropia* stem is a zero-sum game, it is possible to starve competing colonies by stockpiling and guarding the majority of the Müllerian bodies available. Direct interference competition should be in general much more costly than exploitation competition, because even victorious colonies would also experience a decrease in worker force owing to casualties of fighting.

Nest architecture and defense may be other reasons why brood raiding is not common among incipient *Azteca* colonies. When the first workers of a colony eclose, a single entrance/exit hole is chewed through the scar that grows over the initial queen entrance. These holes are barely large enough for a single *Azteca* worker to enter or exit, thus making it difficult for neighboring colonies to attack in force. The plant tissue surrounding the entrance hole is woody, so it is not a trivial task to enlarge the hole. In addition, *Azteca* workers often stand beside the entrance hole on the inside of an internode, perhaps to defend against intruders. Finally, young colonies often fill the entrance hole with bits of parenchyma, completely sealing the internode's only portal.

Primary plurimatry

Although plurimatry is fairly common in ants, primary plurimatry in which some or all cofoundresses remain as functional queens in a mature colony is rare. Most plurimatrous colonies are formed secondarily by adoption of new queens or fusion with other colonies (Hölldobler and Wilson 1977, 1990). The only known examples of primary plurimatry come from *Atta texana* (Mintzer and Vinson 1985), *Pheidole morrisi* (S. P. Cover in Hölldobler and Wilson 1990), and *Azteca constructor* (Perlman 1992). Pleometrotic colonies are cooperatively breeding societies in which all or most cofoundresses lay eggs and raise brood together (see Crespi and Choe, this volume, for the classification and evolution of social systems). On the other hand, plurimatrous colonies are both cooperatively breeding and eusocial. Queens in a plurimatrous colony form a cooperatively breeding society among themselves and at the same time maintain a eusocial system with their daughters. One reason why the transition from pleometrosis to primary plurimatry is so rare may be the similarity in fighting abilities among cofoundresses (see Reeve and Ratnieks 1993 for discussion).

Then what makes foundresses in *Atta texana*, *P. morrisi* and *Azteca constructor* stay cooperative even after the first workers eclose? One possibility is that, unlike in other pleometrotic species, queens in these species form kin groups at colony foundation. Queens in such a kin group

are selected to be cooperative, because they may gain inclusive fitness benefits in addition to direct personal reproduction.

THE EVOLUTION OF FOUNDRESS ASSOCIATION IN ANTS

Since Hölldobler and Wilson (1977) identified the number of queens in the colony as a critically important trait in the evolution of insect sociality, there have been a number of theoretical models analyzing conditions that favor foundress association over solitary nesting (see, for example, Bartz and Hölldobler 1982; Vehrencamp 1983a,b; Nonacs 1989; Reeve 1991; Perlman 1992; Reeve and Ratnieks 1993). As mentioned above, to date both behavioral and genetic evidence indicate that foundress associations in ants are not kin groups (Taki 1976; Bartz and Hölldobler 1982; Hölldobler and Carlin 1985; Ross and Fletcher 1985; Rissing *et al.* 1986, 1989; Hagen *et al.* 1988). Based on their reproductive skew model, Reeve and Ratnieks (1993; see also Reeve 1991; Keller and Reeve 1994; Reeve and Keller 1995) concluded that relatedness has no net effect on joining decisions. Earlier, a dynamic programming model by Nonacs (1989) made a similar prediction. Therefore, in the absence of the inclusive fitness effects the evolution of foundress association must be examined in terms of ecological costs and benefits.

Survival and reproduction

Do individual cofoundresses survive better than solitary foundresses? To answer this question, we must consider queen mortality due to intracolony competition and intercolony competition separately. Although greater survivorship has often been considered a major advantage for foundress associations, most studies of pleometrotic ants have measured only mortality from intracolony competition during the founding stage; the relationship between the number of foundresses and mortality is not always negative. In *Solenopsis invicta* (Tschinkel and Howard 1983; Tschinkel 1993), *A. constructor* and *A. xanthacroa*, the number of queens in the colony does not affect queen mortality. In *Lasius niger*, queen mortality is lowest in groups of two foundresses and much higher in larger groups and among solitary foundresses (Sommer and Hölldobler 1992). Deslippe and Savolainen (1995) found in *Formica podzolica* that solitary foundresses experienced greater survivorship than pleometrotic foundresses.

Once intercolony mortality caused by interference or exploitation competition are included in the analysis, however, it becomes obvious why foundress associations might be favored over solitary founding. Even though a queen's probability of becoming the sole survivor within a pleometrotic colony decreases as the number of foundresses increases, selection still favors foundress associations because of the decisive advantages of larger worker forces during competition among incipient colonies.

Can a potential joiner queen expect greater personal reproduction by taking part in a pleometrotic colony than by attempting to establish a colony alone? Researchers studying pleometrotic ants typically measure the number of workers produced; several studies have shown positive correlations between the number of foundresses and the number of workers produced early in colony development (see, for example, Waloff 1957; Stumper 1962; Taki 1976; Mintzer 1979, 1987, 1990; Bartz and Hölldobler 1982; Tschinkel and Howard 1983; Hölldobler and Carlin 1985; Rissing and Pollock 1986; Perlman 1992; Sommer and Hölldobler 1995). In at least five species, *Lasius flavus* (Waloff 1957), *L. niger* (Sommer and Hölldobler 1992), *Messor* (= *Veromessor*) *pergandei* (Rissing and Pollock 1986, 1991), *Myrmecosystus mimicus* (Bartz and Hölldobler 1982), and *Solenopsis invicta* (Tschinkel and Howard 1983; Tschinkel 1993), pleometrotic colonies also produce workers faster than haplometrotic colonies.

However, the workers are merely a means by which a queen produces reproductive offspring, not a reproductive goal in their own right. Enhanced production of workers, measured in increased cohort size and/or speed of worker production, can increase a queen's chance of survival and may increase her reproductive output. Colonies with large, rapidly produced worker forces will tend to survive better and/or produce reproductive offspring earlier than colonies with small or slowly produced forces. Although the haplometrotic queen is guaranteed to avoid all *intra*colony competition, *inter*colony competition may be so severe that virtually the only way a queen can survive is to begin a colony with other foundresses. In other words, despite the heavy odds against being the one surviving queen in a pleometrotic colony, a queen may have a better overall chance of surviving to reproduce if she joins other foundresses than if she founds alone. Many of the ant species that exhibit pleometrosis typically have clumped incipient colonies and intense intercolony competition during the early stages of colony development (reviewed in Rissing and Pollock 1988; Hölldobler and Wilson 1990). In such situations,

enhanced early brood production is especially important, given the probability of intercolony brood raids or intense competition for limited food resources.

Survival and reproductive benefits of a large worker force accrue only to surviving queens in the colony; none of these benefits is reaped by queens that die before reproductive offspring are produced. Although students of pleometrosis typically present data on per capita (per queen) production of workers, such data are of limited value. Because the entire worker force takes part in competition with other colonies, it may be more useful to consider the colony's total production of workers. From a pleometrotic queen's perspective, per capita production of workers is a moot point, as long as total production is high enough for the colony to be competitive with neighboring incipient colonies. Granted, a given queen will be best served if each of her cofoundresses produces a large number of workers, but it is unlikely that one queen can increase others' production of workers; a queen must take what she can get and hope for high colony-wide production.

Queen behavior and levels of selection

Group selection has sometimes been implicated for the evolution of foundress association (see, for example, Rissing *et al.* 1989; Wilson 1990; Dugatkin *et al.* 1992; Mesterton-Gibbons and Dugatkin 1992). Although an exceptionally strong form of between-group selection may be at work in *Acromyrmex versicolor*, where cofoundresses are unrelated but specialized in colony tasks (Rissing *et al.* 1989; Wilson 1990), it is unnecessary or even erroneous to invoke such group-selectionist logic for other pleometrotic ant species, as Dugatkin *et al.* (1992) and Mesterton-Gibbons and Dugatkin (1992) have done. Low genetic relatedness and cooperation among cofoundresses alone do not necessarily provide sufficient conditions for group selection to work. It appears that competition for limited nesting sites is the basis for pleometrosis in most of the ant species studied thus far, and thus it is in the interests of individual queens to found colonies in cooperation with other queens if they are to have any chance of reproducing. Cooperative colony foundation in ants is not an example of group selected behavior but of 'by-product mutualism' (*sensu* West-Eberhard 1975).

Rissing and Pollock (1986) found no significant differences among *Messor pergandei* queens in their contributions to the total brood production in the laboratory, and argued that queen–queen aggression should not occur before

Table 19-1. *Comparisons of social attributes among pleometrotic bees, wasps, ants, and termites*

Attributes	Ants	Termites	Bees	Wasps
Mode of founding [.]	claustral[a]	claustral	non-claustral	non-claustral
Genetic relatedness among cofoundresses	low	prob. low	low–high	high
Division of labor among cofoundresses	absent[a]	absent	absent (communal) present (semisocial and eusocial)	present
Competition among incipient colonies	high	high	low–moderate	low–moderate
Level of sociality	highly eusocial	highly eusocial	communal, semisocial and primitively eusocial	primitively eusocial

[a] *A. versicolor* is an exception (Rissing *et al.* 1989).

worker eclosion owing to the energy constraints of claustrality. In contrast, Sommer and Hölldobler (1992) found that foundresses in pleometrotic colonies of *L. niger* differed in the number of eggs they laid and that workers preferentially fed the most fertile foundress in a colony. Overt aggression has been noted among cofoundresses of *Camponotus ferrugineus* (Fowler and Roberts 1983), *F. podzolica* (Deslippe and Savolainen 1995), and *Azteca xanthacroa* (Dingman 1987; Perlman 1992). A somewhat less obvious degree of dominance has also been observed in other species (see, for example, Bartz and Hölldobler 1982; Hölldobler and Carlin 1985). Brood cannibalism (see, for example, Tschinkel 1993) is another indication of potential conflict among cofoundresses. It is in fact often difficult to rule out dominance interactions among cofoundresses, because the behavior may be subtle or may be the result of pheromonal influences (Fletcher and Blum 1983). Social competition theory (West-Eberhard 1979; Vehrencamp 1983a) might have been dismissed too hastily as a possible explanation for the evolution of foundress association in *Acromyrmex versicolor* (Rissing *et al.* 1989).

Interactions among cofoundresses are dynamic processes of conflict and cooperation, far from being blind cooperation. The propensity for cooperation appears to be variable and dependent in part on the intensity of intercolony competition. Individual foundresses must contribute enough to enable their colony to outcompete neighboring colonies, but at the same time they must carefully allocate their metabolic reserves so as to be able to outcompete their cofoundresses after the first workers emerge. Individual foundresses cooperate initially to survive, but when it comes to reproduction, they fiercely compete against one another, typically to the death.

Comparisons with other social insects

Pleometrotic colony founding has been observed among wasps of primitively eusocial Polistinae, Stenogastrinae, and Sphecidae (Keller and Vargo 1993). All known cases of pleometrosis among bees belong to Halictini, Augochlorini, Ceratinini, and Allodapini (Michener 1974, 1990a,b; Packer 1993; Danforth and Eickwort, this volume; Schwarz *et al.*, this volume). Among termites, pleometrosis appears more prevalent in soil-nesting species than in wood-dwelling species (Shellman-Reeve, this volume). Table 19-1 summarizes similarities and differences in some of the major social attributes among pleometrotic bees, wasps, ants, and termites. The generalizations made here are necessarily tentative because of the incomplete knowledge for many taxa, but may provide a conceptual basis for future studies.

At a glance, pleometrotic ants and termites appear to share many of the same attributes, while bees and wasps are more similar to each other than to ants or termites (Table 19-1). Unlike bees and wasps, both ant and termite queens shed their wings before they initiate new colonies. Ant and termite foundresses are also claustral, meaning that they do not forage outside the nest and they rely solely on their own metabolic reserves for rearing the first brood of workers. In other words, most pleometrotic colonies in termites and ants, with the exception of *A. versicolor*, are nutritionally closed systems. Although claustral, founding queens and kings of wood-dwelling termites may feed on wood cambium to successfully rear the first brood (Cook and Scott 1933; Shellman-Reeve 1994). This may explain why foundress association is rare in wood-dwelling termites (J. S. Shellman-Reeve, personal communication).

Convergence in pleometrotic patterns between ants and termites has also been discussed by Roisin (1993).

Although genetic relatedness among cofoundresses tends to be relatively high in wasps (Queller *et al.* 1988; Strassmann *et al.* 1989; Hughes *et al.* 1993), foundress association by kin has yet to be discovered in ants. With direct evidence lacking, Roisin (1993) argues that termite cofoundresses are not related (but see Shellman-Reeve, this volume). Division of labor is absent among termite and ant cofoundresses, *A. versicolor* (Rissing *et al.* 1989) being an exception. On the contrary, it is well developed among cofoundresses in semisocial and primitively eusocial bees (Eickwort 1986; Michener 1974, 1990a,b; Packer 1993; Schwarz *et al.*, this volume) and primitively eusocial polistine wasps (West-Eberhard 1969; Gadagkar 1991; Reeve 1991).

Predation pressure on foundresses may have resulted in different adaptations in different social insects. In wasps, foraging-related mortality of foundresses is considered a major cause of colony failure (see, for example, Strassmann 1981) and may be partly responsible for the evolution of dominance and reproductive division of labor among cofoundresses (Queller 1989). Although data are lacking, the same may be true for bees (G. C. Eickwort, personal communication). In ants and termites, however, the same selection pressure appears to have led to the evolution of claustral colony founding. A trade-off for claustral foundation is the energy constraint, because a founding queen has a fixed amount of resources available for rearing the first brood. Faced with intense competition among incipient colonies, caused by limited nesting sites and highly synchronous breeding, foundress association may have been a secondary adaptation in response to the pressure to produce a large worker force as rapidly as possible. It is intercolony competition that has promoted intracolony cooperation.

UNANSWERED QUESTIONS AND FUTURE STUDIES

Behavior of foundresses

Most theoretical models of foundress association, including the model we present here and others by Nonacs (1989), Reeve (1991) and Reeve and Ratnieks (1993), assume that foundresses are capable of assessing the potential benefits and costs of joining others and selecting an association that will provide a higher fitness return.

Despite these assumptions, little is known about how foundresses make such choices. Krebs and Rissing (1991) demonstrated in the laboratory that *M. pergandei* queens prefer larger foundress associations. Although the foundresses appear to enter the first nest they encounter, they can also abandon nests to join more preferable nests. *Azteca* queens also appear to settle in the first tree upon which they land, but within the tree they appear to examine different internodes (Perlman 1992). They apparently kick their wings off before they choose an internode. It seems reasonable to assume that queens can determine whether an internode is occupied or not after external inspection, because colonizing queens almost always plug their entry hole with parenchyma scraped from the inner walls of the internode. It also seems possible for inspecting queens to determine the species or even colony of origin of queens inside an internode if earlier-arriving queens leave odor traces on the entry plug, but it may be difficult or impossible for later-arriving queens to determine the number of queens in an internode by examining the entry plug. Information on the mechanisms of selecting foundress associations will help us better understand selective agents influencing pleometrosis.

We also need to pay much closer attention to the signs of potential conflicts among cofoundresses and collect more data on relative contributions by individual foundresses. Are cofoundresses engaged in some sort of egg-laying game, as envisioned by Pollock and Rissing (1988)? Should a queen lay many eggs to obtain a large cohort of her own workers, or should she let the other queens exhaust themselves? What determines which queen survives?

Role of kin selection

Estimating genetic relatedness among cofoundresses of primarily plurimatrous species will also add much to our understanding of this dynamic process. It would be interesting to compare the estimates between *A. constructor* and *A. xanthacroa*, since they differ markedly in queen–queen interactions at the time of worker emergence. Other species that exhibit primary plurimatry, e.g. *Atta texana* and *Pheidole morrisi*, can also be compared to samples of haplometrotic congeners. In addition to such between-species comparisons, within-species comparisons may also prove to be useful. In the species mentioned above, not every pleometrotic colony becomes plurimatrous. Some colonies lose all foundresses but one and become unimatrous. One can compare these two types of transition with respect to genetic relatedness among cofoundresses. Although relatives may not be readily available to ant foundresses (Strassmann 1989), queens should still prefer to group with more closely related foundresses whenever they can; genetic analysis of primarily plurimatrous species may provide evidence that inclusive fitness matters.

Alternatively, it is also possible that avoidance of kin may be favored by kin selection. If indeed foundress associations are ultimately competitive, i.e. only one becomes the queen, then a potential queen should avoid competing with relatives since the better she competes, the higher the inclusive fitness cost accrued from defeating a relative. The actual game could become quite complex, if conditional strategies exist. If a group of sisters all have strong competitive ability, they should nest with other non-relatives. If all are small and poorly competitive, kin association may be favored. Occurrence of and conditions for kin association among ant cofoundresses may depend upon the intensity of competition and difference in competitive ability among potential queens.

Long-term comparative studies

Long-term field studies in which both pleometrotic and haplometrotic queens are monitored for the entire colony life cycle are needed to understand the balance between cooperation and conflict among cofoundresses. No studies of pleometrotic ants, with the possible exception of Perlman (1992) on *Azteca*, have followed queens long enough to measure complete lifetime reproductive output. If future researchers find such long-term studies impractical given the lifespans of ant queens and the vagaries of field work, we still urge them to extend their studies at least into the reproductive stage of colony development. In *Azteca*, pleometrotic colonies tend to grow more rapidly and may produce reproductives earlier than neighboring haplometrotic colonies. Thus, surviving queens in pleometrotic colonies may be able to reproduce earlier (see also Tschinkel 1987) and possibly over a longer period than haplometrotic queens (see also Rissing and Pollock 1988; Hölldobler and Wilson 1990). Therefore, data on the speed and quantity of production of reproductive offspring are critically important in determining the ultimate causes of the evolution of foundress association in ants. Such long-term studies seem particularly feasible with *Azteca* ants; *Cecropia* trees can be easily marked and *Azteca* queens can be periodically monitored using the technique developed by Perlman (1992).

Different combinations of *Azteca* species cofound colonies in different geographic regions. Additional studies of pleometrotic colony foundation by different pairs of *Azteca* species should permit a more comprehensive characterization of the ecological parameters responsible for the evolution of foundress association. Comparisons with unrelated, yet ecologically similar species will also provide new insights. For example, Fiala and Maschwitz (1990) found that the southeast Asian ant *Crematogaster borneensis* nests in *Macaranga* plants, which are similar in architecture to *Cecropia*, but founds colonies haplometrotically. It would be interesting to explore differences between these ants and *Cecropia*-nesting *Azteca*.

ACKNOWLEDGEMENTS

We thank Peter Adler, Richard Alexander, Steve Austad, Laura Betzig, Jane Brockmann, Bernie Crespi, the late George Eickwort, Bert Hölldobler, Christian Peeters, Naomi Pierce, Mike Schwarz, Kathrin Sommer and Edward Wilson for commenting on various drafts. JCC was supported by the Richmond Fund of Harvard University, Noyes Fellowship from the Organization for Tropical Studies, Predoctoral Fellowship from the Smithsonian Institution, and Junior Fellowship from the Michigan Society of Fellows. DLP was supported by the Richmond, Kennedy, and deCuevas Funds of Harvard University, Noyes Fellowship from the Organization for Tropical Studies, and Fulbright Scholarship. Sam Wade of Boston Scientific Corporation provided essential technical support.

LITERATURE CITED

Alexander, R. D. and P. W. Sherman. 1977. Local mate competition and parental investment in social insects. *Science (Wash., D.C.)* **196**: 494–500.

Baroni Urbani, C., B. Bolton and P. S. Ward. 1992. The internal phylogeny of ants (Hymenoptera: Formicidae). *Syst. Entomol.* **17**: 301–329.

Bartz, S. H. and B. Hölldobler. 1982. Colony founding in *Myrmecosystus mimicus* Wheeler (Hymenoptera: Formicidae) and the evolution of foundress associations. *Behav. Ecol. Sociobiol.* **10**: 137–147.

Bourke, A. F. G. 1988. Worker reproduction in the higher eusocial Hymenoptera. *Q. Rev. Biol.* **63**: 291–311.

Brown, J. L. 1987. *Helping and Communal Breeding in Birds: Ecology and Evolution.* Princeton: Princeton University Press.

Buschinger, A. 1974. Monogynie und Polygynie in Insektensozietäten. In *Sozialpolymorphismus bei Insekten.* G. H. Schmidt, ed., pp. 862–896. Stuttgart: Wissenschaftliche Verlagsgesellschaft.

Choe, J. C. 1988. Worker reproduction and social evolution in ants (Hymenoptera: Formicidae). In *Advances in Myrmecology*. J. C. Trager, ed., pp. 163–187. New York: E. J. Brill.

–. 1995. Plurimatry: new terminology for multiple reproductives. *J. Insect Behav.* **8**: 133–137.

Cook, S. F. and K. G. Scott. 1933. The nutritional requirements of *Zootermopsis* (*Termopsis*) *angusticollis*. *J. Cell. Comp. Physiol.* **4**: 95–110.

Creel, S. R. and N. M. Creel. 1991. Energetics, reproductive suppression and obligate communal breeding in carnivores. *Behav. Ecol. Sociobiol.* **2**: 263–270.

Deslippe, R. J. and R. Savolainen. 1995. Colony foundation and polygyny in the ant *Formica podzolica*. *Behav. Ecol. Sociobiol.* **37**: 1–6.

Dingman, C. L. 1987. Interspecific pleometrosis in *Azteca constructor* and *Azteca xanthacroa*. Undergraduate honors thesis, Harvard University.

Dugatkin, L. A., M. Mesterton-Gibbons and A. I. Houston. 1992. Beyond the prisoner's dilemma: toward models to discriminate among mechanisms of cooperation in nature. *Trends Ecol. Evol.* **7**: 202–205.

Eickwort, G. C. 1986. First steps into eusociality: the sweat bee *Dialictus lineatulus*. *Fla. Entomol.* **69**: 742–754.

Fiala, B. and U. Maschwitz. 1990. Studies on the south east Asian ant-plant association *Crematogaster borneensis/Macaranga*: adaptations of the ant partner. *Insectes Soc.* **37**: 212–231.

Fletcher, D. J. C. and M. S. Blum. 1983. Regulation of queen number by workers in colonies of social insects. *Science (Wash., D.C.)* **219**: 312–315.

Fowler, H. G. and R. B. Roberts. 1983. Anomalous social dominance among queens of *Camponotus ferrugineus* (Hymenoptera: Formicidae). *J. Nat. Hist.* **17**: 185–187.

Gadagkar, R. 1991. *Belanogaster, Mischocyttarus, Parapolybia*, and independent-founding *Ropalidia*. In *Queen Number and Sociality in Insects*. L. Keller, ed., pp. 149–190. Oxford University Press.

Hagen, R. H., D. R. Smith and S. W. Rissing. 1988. Genetic relatedness among co-foundresses of two desert ants, *Veromessor pergandei* and *Acromyrmex versicolor*. *Psyche* **95**: 191–201.

Hamilton, W. D. 1964. The genetical evolution of social behaviour. I & II. *J. Theor. Biol.* **7**: 1–52.

–. 1972. Altruism and related phenomena, mainly in social insects. *Annu. Rev. Ecol. Syst.* **3**: 193–232.

Hanna, N. H. C. 1975. Contribution à l'étude de la biologie et de la polygynie de la fourmi *Tapinoma simrothi phoenicium* Emery. *C.R. Acad. Sci. Paris* D281: 1003–1005.

Hölldobler, B. and N. F. Carlin. 1985. Colony founding, queen dominance and oligogyny in the Australian meat ant *Iridomyrmex purpureus*. *Behav. Ecol. Sociobiol.* **18**: 45–53.

Hölldobler, B. and E. O. Wilson. 1977. The number of queens: an important trait in ant evolution. *Naturwissenschaften* **64**: 8–15.

–. 1990. *The Ants*. Cambridge, Massachusetts: Harvard University Press.

Hughes, C. R., D. C. Queller, J. E. Strassmann, C. R. Solis, J. A. Negron-Sotomayor and K. R. Gastreich. 1993. The maintenance of high genetic relatedness in multi-queen colonies of social wasps. In *Queen Number and Sociality in Insects*. L. Keller, ed., pp. 153–170. Oxford University Press.

Itô, Y. 1993. *Behaviour and Social Evolution of Wasps: The Communal Aggregation Hypothesis*. Oxford University Press.

Janzen, D. H. 1969. Allelopathy by myrmecophytes: the ant *Azteca* as an allelopathic agent of *Cecropia*. *Ecology* **50**: 147–153.

Keller, L., ed. 1993. *Queen Number and Sociality in Insects*. Oxford University Press.

Keller, L. and L. Passera. 1990. Fecundity of ant queens in relation to their age and the mode of colony founding in ants (Hymenoptera: Formicidae). *Insectes Soc.* **37**: 116–130.

Keller, L. and H. K. Reeve. 1994. Partitioning of reproduction in animal societies. *Trends Ecol. Evol.* **9**: 98–102.

Keller, L. and E. L. Vargo. 1993. Reproductive structure and reproductive roles in colonies of eusocial insects. In *Queen Number and Sociality in Insects*. L. Keller, ed., pp. 16–44. Oxford University Press.

Krebs, R. A. and S. W. Rissing. 1991. Preference for larger foundress associations in the desert ant *Messor pergandei*. *Anim. Behav.* **41**: 361–363.

Longino, J. T. 1989. Geographic variation and community structure in an ant-plant mutualism: *Azteca* and *Cecropia* in Costa Rica. *Biotropica* **21**: 126–132.

Mesterton-Gibbons, M. and L. A. Dugatkin. 1992. Cooperation among unrelated individuals: evolutionary factors. *Q. Rev. Biol.* **67**: 267–281.

Michener, C. D. 1974. *The Social Behavior of Bees: A Comparative Study*. Cambridge, Massachusetts: Harvard University Press.

–. 1990a. Reproduction and castes in social Halictine bees. In *Social Insects: An Evolutionary Approach to Castes and Reproduction*. W. Engels, ed., pp. 77–121. Berlin: Springer Verlag.

–. 1990b. Castes in Xylocopine bees. In *Social Insects: An Evolutionary Approach to Castes and Reproduction*. W. Engels, ed., pp. 123–146. Berlin: Springer Verlag.

Mintzer, A. 1979. Colony foundation and pleometrosis in *Camponotus* (Hymenoptera: Formicidae). *Pan Pac. Entomol.* **55**: 81–89.

–. 1987. Primary polygyny in the ant *Atta texana*: number and weight of females and colony foundation success in the laboratory. *Insectes Soc.* **34**: 108–117.

–. 1990. Colony foundation in leafcutting ants: the perils of polygyny in *Atta laevigata* (Hymenoptera: Formicidae). *Psyche* **98**: 1–5.

Mintzer, A. and S. B. Vinson. 1985. Cooperative colony founding by females of the leaf-cutting ant *Atta texana* in the laboratory. *J. N. Y. Entomol. Soc.* **93**: 1047–1051.

Nonacs, P. 1986. Ant reproductive strategies and sex allocation theory. *Q. Rev. Biol.* **61**: 1–21.

–. 1989. Competition and kin discrimination in colony founding by social Hymenoptera. *Evol. Ecol.* **3**: 221–235.

–. 1990. Size and kinship affect success of co-founding *Lasius pallitarsis* queens. *Psyche* **97**: 217–228.

Packer, L. 1993. Multiple-foundress associations in sweat bees. In *Queen Number and Sociality in Insects*. L. Keller, ed., pp. 215–233. Oxford University Press.

Peeters, C. 1993. Monogyny and polygyny in ponerine ants with or without queens. In *Queen Number and Sociality in Insects*. L. Keller, ed., pp. 234–261. Oxford University Press.

Peeters, C., and A. N. Andersen. 1989. Cooperation between dealate queens during colony foundation in the green tree ant, *Oecophylla smaragdina*. *Psyche* **96**: 39–44.

Perlman, D. L. 1992. Colony founding among *Azteca* ants. Ph. D. dissertation, Harvard University.

Pollock, G. B. and S. W. Rissing. 1988. Social competition under mandatory group life. In *The Ecology of Social Behavior*. C. N. Slobodchikoff, ed., pp. 315–344. New York: Academic Press.

Queller, D. C. 1989. The evolution of eusociality: reproductive head starts of workers. *Proc. Natl. Acad. Sci. U.S.A.* **86**: 3224–3226.

Queller, D. C., J. E. Strassmann and C. R. Hughes. 1988. Genetic relatedness in colonies of tropical wasps with multiple queens. *Science (Wash., D.C.)* **242**: 1155–1157.

Reeve, H. K. 1991. *Polistes*. In *The Social Biology of Wasps*. K. G. Ross and R. W. Matthews, eds., pp. 99–148. Ithaca: Cornell University Press.

Reeve, H. K. and L. Keller. 1995. Partitioning of reproduction in mother–daughter versus sibling associations: a test of optimal skew theory. *Am. Nat.* **145**: 119–132.

Reeve, H. K. and F. L. W. Ratnieks. 1993. Queen–queen conflicts in polygynous societies: mutual tolerance and reproductive skew. In *Queen Number and Sociality in Insects*. L. Keller, ed., pp. 45–85. Oxford University Press.

Rickson, F. R. 1971. Glycogen plastids in Müllerian body cells of *Cecropia peltata* – a higher green plant. *Science (Wash., D.C.)* **173**: 344–347.

Rissing, S. W. and G. B. Pollock. 1986. Social interaction among pleometrotic queens of *Veromessor pergandei* (Hymenoptera: Formicidae) during colony foundation. *Anim. Behav.* **34**: 226–233.

–. 1987. Queen aggression, pleometrotic advantage and brood raiding in the ant *Veromessor pergandei*. *Anim. Behav.* **35**: 975–981.

–. 1988. Pleometrosis and polygyny in ants. In *Interindividual Behavioural Variability in Social Insects*. R. L. Jeanne, ed., pp. 179–221. Boulder: Westview Press.

–. 1991. An experimental analysis of pleometrotic advantage in the desert seed-harvester ant *Messor pergandei* (Hymenoptera: Formicidae). *Insectes Soc.* **38**: 205–211.

Rissing, S. W., R. A. Johnson and G. B. Pollock. 1986. Natal nest distribution and pleometrosis in the desert leaf-cutter ant *Acromyrmex versicolor* (Pergande) (Hymenoptera: Formicidae). *Psyche* **93**: 177–186.

Rissing, S. W., G. B. Pollock, M. R. Higgins, R. H. Hagen, and D. R. Smith. 1989. Foraging specialization without relatedness or dominance among cofounding ant queens. *Nature (Lond.)* **338**: 420–422.

Roisin, Y. 1993. Selective pressures on pleometrosis and secondary polygyny: a comparison of termites and ants. In *Queen Number and Sociality in Insects*. L. Keller, ed., pp. 402–421. Oxford University Press.

Ross, K. G. and D. J. C. Fletcher. 1985. Comparative study of genetic and social structure in two forms of the fire ant *Solenopsis invicta* (Hymenoptera: Formicidae). *Behav. Ecol. Sociobiol.* **17**: 349–356.

Ross, K. G. and R. W. Matthews, eds. 1991. *The Social Biology of Wasps*. Ithaca: Cornell University Press.

Schupp, E. W. 1986. *Azteca* protection of *Cecropia*: ant occupation benefits juvenile trees. *Oecologia (Berl.)* **70**: 379–385.

Shellman-Reeve, J. S. 1994. Limited nutrients in a dampwood termite: nest preference, competition, and cooperative nest defense. *J. Anim. Ecol.* **63**: 921–932.

Sommer, K., and B. Hölldobler. 1992. Pleometrosis in *Lasius niger*. In *Biology and Evolution of Social Insects*. J. Billen, ed., pp. 47–50. Leuven, Belgium: Leuven University Press.

–. 1995. Colony founding by queen association and determinants of reduction in queen number in the ant *Lasius niger*. *Anim. Behav.* **50**: 287–294.

Strassmann, J. E. 1981. Evolutionary implications of early male and satellite nest production in *Polistes exclamans* colony cycles. *Behav. Ecol. Sociobiol.* **8**: 55–64.

–. 1989. Altruism and relatedness at colony foundation in social insects. *Trends Ecol. Evol.* **4**: 371–374.

Strassmann, J. E., C. R. Hughes, D. C. Queller, S. Turillazzi, R. Cervo, S. K. Davis and K. F. Goodnight. 1989. Genetic relatedness in primitively eusocial wasps. *Nature (Lond.)* **342**: 268–270.

Stumper, R. 1962. Sur un effet de groupe chez les femelles de *Camponotus vagus* (Scopoli). *Insectes Soc.* **9**: 329–333.

Taki, A. 1976. Colony founding of *Messor aciculatum* (Fr. Smith) (Hymenoptera: Formicidae) by single and grouped queens. *Physiol. Ecol. Jap.* **17**: 503–512.

Taylor, R. W. 1978. *Nothomyrmecia macrops*: a living-fossil ant rediscovered. *Science (Wash., D.C.)* **201**: 979–985.

Trivers, R. L. and H. Hare. 1976. Haplodiploidy and the evolution of the social insects. *Science (Wash., D.C.)* **191**: 249–263.

Tschinkel, W. R. 1987. The fire ant, *Solenopsis invicta*, as a successful "weed". In *Chemistry and Biology of Social Insects*. J. Eder and H. Rembold, eds., pp. 585–588. München: Verlag J. Peperny.

–. 1992. Brood raiding in the fire ant, *Solenopsis invicta*: field and laboratory studies. *Ann. Entomol. Soc. Am.* **85**: 638–696.

–. 1993. Resource allocation, brood production and cannibalism during colony founding in the fire ant, *Solenopsis invicta*. *Behav. Ecol. Sociobiol.* **33**: 209–223.

Tschinkel, W. R. and D. F. Howard. 1983. Colony founding by pleometrosis in the fire ant, *Solenopsis invicta*. *Behav. Ecol. Sociobiol.* **12**: 103–113.

Vehrencamp, S. L. 1983a. A model for the evolution of despotic versus egalitarian societies. *Anim. Behav.* **31**: 667–682.

–. 1983b. Optimal degree of skew in cooperative societies. *Am. Zool.* **23**: 327–335.

Waloff, N. 1957. The effect of the number of queens of the ant *Lasius flavus* (Fab.) (Hym. Formicidae) on their survival and on the rate of development of the first brood. *Insectes Soc.* **4**: 391–408.

West-Eberhard, M. J. 1967. Foundress associations in polistine wasps: dominance hierarchies and the evolution of social behavior. *Science (Wash., D.C.)* **157**: 1584–1585.

–. 1969. The social biology of polistine wasps. *Misc. Publ. Mus. Zool. Univ. Mich.* **140**: 1–101.

–. 1975. The evolution of social behavior by kin selection. *Q. Rev. Biol.* **50**: 1–33.

–. 1978. Polygyny and the evolution of social behavior in wasps. *J. Kansas Entomol. Soc.* **51**: 832–856.

–. 1979. Sexual selection, social competition, and evolution. *Proc. Am. Phil. Soc.* **123**: 222–234.

Wheeler, W. M. 1933. *Colony-founding among Ants, with an Account of some Primitive Australian Species.* Cambridge, Massachusetts: Harvard University Press.

Wilson, D. S. 1990. Weak altruism, strong group selection. *Oikos* **59**: 135–140.

20 · Social evolution in the Lepidoptera: ecological context and communication in larval societies

JAMES T. COSTA AND NAOMI E. PIERCE

ABSTRACT

We review key ecological and behavioral mechanisms underlying the origin and maintenance of larval sociality in the Lepidoptera. Using communication contexts of group defense, cohesion and recruitment as a framework, we relate social complexity among gregarious caterpillars to three patterns of foraging: patch-restricted, nomadic, and central-place. A review of the incidence of larval gregariousness in the Lepidoptera demonstrates that sociality is widespread in the order, occurring in twenty or more families representing thirteen ditrysian superfamilies, and it is likely to have evolved numerous times in response to different selective pressures. We specifically address the role of sociality in larval defense and resource use, with a focus on (1) signal enhancement in communication systems, (2) differential larval vulnerability, and (3) ant association. Larval Lepidoptera experience the greatest likelihood of mortality in the earliest instars; larval sociality enhances defensive and resource–exploitation signals in these instars, positively influencing survivorship and larval growth. Disease, predation and parasitism, nutrition, and inclusive fitness are discussed in terms of costs and benefits of group living. Finally, we identify two areas where additional research will contribute significantly to an understanding of social evolution in the Lepidoptera: (1) comparative phylogenetic studies, using ecological and communicative characters to trace the origins of caterpillar societies and transitions among them, and (2) larval behavior and ecology, focusing on kin discrimination abilities, assessment of colony genetic structure, and most importantly on the means and contexts of caterpillar communication.

INTRODUCTION

Sociality in the Lepidoptera is characterized by behaviors such as laying eggs in clusters, larval aggregation, and communal roosting by adults. By 'social', we refer to any system in which individuals display reciprocal, cooperative communication (Wilson 1971). Wilson favored this definition partly on the grounds that 'the terms society and social must be defined quite broadly in order to prevent the arbitrary exclusion of many interesting phenomena' (Wilson 1971, p. 5). He stressed that a common denominator in the behavior of all social insects is communication.

While early investigators devised an elaborate classification of insect societies to match the diversity of their subject, an essential distinction was made between 'eusocial' and 'social' species. The term eusocial refers to species exhibiting three social attributes: overlapping generations, cooperative brood care, and reproductive division of labor. The combination of these characteristics is exhibited by complex, integrated societies marked by sophisticated communication systems and caste specializations, represented among the insects by the haplodiploid ants, many bees and wasps, some thrips and beetles, the diplodiploid termites, and certain parthenogenetically reproducing aphids.

The recognition of eusociality was followed by the delineation of a social hierarchy based upon number and complexity of social attributes, with the eusocial species at its apex. This has had the unfortunate effect of infusing studies of social evolution with 'evolutionary ladder' thinking, as reflected in the moniker 'presocial' applied to social Lepidoptera and many other non–eusocial social insects. The term presocial is inappropriate not only because of its implicit teleological progression, but more importantly because it implies that these forms are not yet social, thereby equating the term 'social' with 'eusocial' in a way that underappreciates the complexity of many non–eusocial insect societies, including those of Lepidoptera.

Focussing on the communication criterion of sociality, the common ground that social caterpillars share with other social species is quickly apparent. Social complexity encompasses both signal repertoire (number of signals) and signal specificity (broadcast vs. personal). The simplest signals are non-specific and group-directed, such as those of certain alarm pheromones; the most sophisticated signals are highly specific and often individually or caste-directed, such as the waggle-dance of the honey bee. Communication and cooperation in Lepidoptera are almost entirely confined to the simple, group-directed end of the spectrum. For example, communication for group cohesion in larval societies involves tactile signals or pheromone markers keeping individuals together, whereas insects such as wasps, bees and ants are capable of sophisticated kin recognition and discrimination in addition to simple group cohesion. Nevertheless, certain lepidopteran societies rival the eusocial insects in other respects, such as in the use of pheromonal foraging and recruitment trails.

Fitzgerald and Peterson (1988) and Fitzgerald (1993, 1995) have written recent reviews that discuss many of the ecological and behavioral correlates of lepidopteran sociality. These reviews developed a conceptual framework of understanding sociality in the Lepidoptera in terms of the nature of intra- and interspecific communication exhibited by different groups. We follow this same approach, and explicitly characterize lepidopteran sociality in terms of one or more of three communication contexts: defense, cohesion and foraging. We explicitly discuss the aggregation behavior of ant-associated caterpillars in the Lycaenidae, which have been overlooked in most reviews of lepidopteran sociality (Stamp 1980; Fitzgerald 1993; but see Kitching 1981). Our discussion is confined to sociality among larvae of Lepidoptera rather than adults, largely because sociality in the juvenile stages is far more common and more information is available concerning the behavioral ecology of gregarious juveniles.

We first characterize the levels of complexity observed in lepidopteran sociality, and classify these in terms of the communication contexts exhibited in each case. We then review the distribution of sociality in the Lepidoptera, and present a survey of species described as exhibiting some degree of sociality. The data suggest that gregariousness has evolved repeatedly in the order, and that it is associated in complex ways with such factors as larval host plant, presence of attendant ants, and the relative 'apparency' (visibility or detectability) of the taxa involved. Although considerable advances have been made in particular areas of higher lepidopteran phylogeny (see, for example, Scott 1985; Scott and Wright 1990; Minet 1991; Lee et al. 1992; Martin and Pashley 1992; Weller et al. 1994), the lack of well-resolved phylogenies at many lower taxonomic levels precludes a meaningful comparative study of these traits, and

we use these data to discuss selective pressures that may have been of particular importance in shaping sociality in the Lepidoptera, including signal enhancement, ant-association, and relatedness. Finally, we discuss the evolution of sociality and egg-clustering patterns in the Lepidoptera, and point to areas that require further research.

LEVELS OF COMPLEXITY IN LEPIDOPTERAN SOCIALITY

Adaptations evolved by caterpillars in response to ecological pressures include variability in coloration, such as crypsis, mimicry, and aposematism; acquisition of morphological armature, such as thick cuticles, spines or setae; association with ants; and behavioral modifications such as stem-boring, leaf-mining, leaf-rolling, and leaf-tying. Group-context communication, as the distinguishing feature of sociality in Lepidoptera, is yet another evolutionary response. Three main communication contexts which social Lepidoptera share with other social insects include: (1) alarm or group defense; (2) aggregation or group cohesion; and (3) foraging or resource use (indicating the location and quality of resources).

Fitzgerald and Peterson (1988) and Fitzgerald (1993) identify three levels of sociality in the Lepidoptera, essentially defined by their foraging behavior: patch-restricted, or static foragers, nomadic foragers, and central-place foragers. In each case, concomitant with changes in foraging behavior are changes in alarm and group defense, as well as signals employed in group cohesion. The subsets of social Lepidoptera represented in Fig. 20-1, and discussed below, are defined by the number of these communication characters; these demonstrate the range of social complexity found in the Lepidoptera. Communication signals unique to social species include those involved in promotion of group cohesion and coordinated resource use, both of which are relevant to group contexts only. Such signals are often chemical and serve to define the spatial limits or boundaries of the group and promote group cohesion (Fitzgerald and Costa 1986; Roessingh 1989, 1990) or aid in the location and evaluation of potential food (see, for example, Kalkowski 1966; Masaki and Umeya 1977; Weyh and Maschwitz 1978; Fitzgerald and Peterson 1983; Peterson 1987).

Social lycaenids that are also ant-associated, or myrmecophilous, exhibit a variety of forms of aggregation, although insufficient research has been conducted on either intraspecific communication among these caterpillars, or interspecific communication between caterpillars

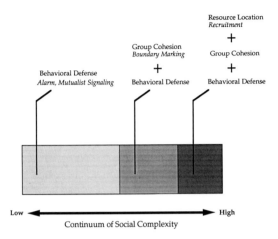

Figure 20-1. Lepidopteran social complexity. Complexity, indicated by intensity of shading, is defined in terms of communication characters: defensive, cohesion, and recruitment. Moving from low to high along the continuum, weakly or facultatively social species exhibit group defense only, more complex social species exhibit group cohesion in addition to group defense, and the most complex lepidopteran societies exhibit recruitment, cohesion, and defense. The continuum is intended to illustrate the range of extant lepidopteran social complexity, and does not represent explicit evolutionary transitions.

and ants, to be able to categorize fully the foraging behavior of many species. Signaling in gregarious ant-associated lycaenids primarily involves the use of ants in defense rather than as a foraging strategy, although the two are closely linked in that attraction of attendant ants may most obviously provide defense against predators and parasitoids, but, thus protected, these larvae are also free to select high-quality foliage and thermally beneficial zones on the host plant.

One question of interest is to what extent attendant ants are directly involved in aggregation of social Lycaenidae. Mathews (1993) demonstrated experimentally that larvae of the Australian lycaenid *Jalmenus evagoras* follow ant trails, and thus ants may play an important indirect role in the aggregation behavior of this species, or in the ability of the larvae to find high-quality food resources by avoiding predator harassment.

Patch-restricted foraging

Lepidoptera whose larvae exhibit patch-restricted foraging represent the most simple form of sociality in the order. Patch-restricted foragers are essentially static, feeding in the same location throughout the larval stage. These species typically construct shelters and feed on leaves incorporated in the structure. As a result, their diet is often a mixture of nutritionally good and poor leaves. As food is exhausted, the shelter may be continually expanded (as in the fall webworm, *Hyphantria cunea*) or the patch occasionally abandoned (as with the palm leaf-skeletonizer, *Homaledra sabalella*). All social Lepidoptera are hypothesized to share the character of group defense, including active group-defensive behaviors such as thrashing and regurgitation (see, for example, Morris 1963; Myers and Smith 1978; Stamp 1984; Peterson 1986; Peterson *et al.* 1987) or attraction of attendant ants in gregarious myrmecophiles (see, for example, Pierce *et al.* 1987; DeVries 1990; Fiedler 1991). These characters are also exhibited by solitary species and, assuming that solitariness represents the plesiomorphic state, they are retained in social species, where their function can become amplified through aggregation. The static lifestyle of patch-restricted species obviates selection for communication beyond alarm signaling and perhaps marking group boundaries. Chemical markers are most commonly responsible for group cohesion, and may occur as a component of the silk deposited by these larvae, or may be applied to silk trails or other substrates by trail-marking glands (Fitzgerald 1993 and references therein).

Nomadic foraging

The next level of complexity in lepidopteran sociality is characterized by species whose larvae engage in nomadic foraging patterns, in which larvae move in groups or bivouacs from patch to patch. Nomads constitute perhaps the greatest number of social lepidopteran species; communication in most of these species appears to be chemically and visually mediated, and it is used in defense, cohesion, and in some cases local orientation to food. Group-cohesion signals are generally chemical in nature, and this use of marker pheromones is likely to be antecedent to resource-use functions such as trail-following and recruitment. Chemical marking and/or trail-following have been demonstrated in such diverse nomadic Lepidoptera as *Malacosoma disstria* (Fitzgerald and Costa 1986), *Euphydryas phaeton* (Stamp 1982), *Asterocampa clyton* (Stamp 1984), *Chlosyne lacinia* (Bush 1969; Stamp 1977), *Hemileuca lucina* (Capinera 1980), *Pieris brassicae* (Long 1955) and some species of *Euselasia* (P. J. DeVries, personal communication). Frequent site abandonment is characteristic of nomadic species, and may represent an evolutionary response to predation (Heinrich 1979, 1993), depletion of local food reserves (Stamp and Bowers 1990a,b) or disease risk.

Larvae of the Australian lycaenid *J. evagoras* appear to follow a nomadic foraging pattern: larvae form loose aggregations composed of individuals of different age classes and presumably different genetic backgrounds, and they forage together diurnally on terminal foliage of their host plants (Common and Waterhouse 1981; Pierce *et al.* 1987). Larvae of this species are also known to produce vibrational calls (DeVries 1991); although these substrate-borne signals have mostly been discussed in terms of signaling to attendant ants (DeVries 1990, 1991) or deterring predators (Downey 1966; Downey and Allyn 1973), it is possible that they also play a role in intraspecific signaling and recruitment behavior.

Central-place foraging

The most sophisticated lepidopteran societies exhibit central-place foraging, and have the ability to communicate the location of food. Central-place foragers often construct shelters but, unlike shelter-building patch-restricted foragers, they feed away from them. Movement between the nest and scattered feeding sites sets the stage for the most complex forms of communication found in the Lepidoptera, recruitment of colonymates to feeding sites. Recruitment involves keeping track of an initial foraging path or food location, usually by means of chemical trails which also convey this information to other larvae. Central-place foraging and trail-marking are also found in some solitary lepidopteran species such as the papilionid *Iphiclides podalirius* (Weyh and Maschwitz 1982) and the charaxine *Polyura pyrrhus* (Tsubaki and Kitching 1986), the larvae of which mark short trails between feeding and resting sites.

Recruitment communication in tent caterpillars improves foraging efficiency by expediting the discovery and use of patchily distributed, high-quality young leaves with relatively little search and exposure time (Fitzgerald and Peterson 1983; Fitzgerald and Costa 1986). Unlike some eusocial insects, eastern tent caterpillars do not carry food to the nest. Like eusocial species, however, they use the tent (nest) as a colony information center for communicating the location of food. Unsuccessful foragers periodically return to the tent; if a successful forager has deposited a recruitment trail, it is detected at the tent and followed to the feeding site (Fitzgerald and Peterson 1983).

Some unsuccessfully foraging larvae may also encounter a recruitment trail before reaching the tent; these trails are also effective in eliciting recruitment.

In a comparative study of trail-marking and trail-following in eastern (*M. americanum*) and forest (*M. disstria*) tent caterpillars, Fitzgerald and Costa (1986) showed that although both species possess trail-marking abilities and prefer trails deposited by fed vs. unfed larvae, only the fixed–base foraging pattern of *M. americanum* leads to recruitment of larvae to food–finds, since larvae return to the tent. By contrast, *M. disstria* often mark trails to a new resting site rather than back to their original site, which 'recruits' colonymates to the new site (promoting colony cohesion), but does not constitute resource-based recruitment. The trail-marking pattern of *M. disstria* may also lead to a higher incidence of colony fragmentation.

Elective recruitment (recruitment based on individual assessment of food quality) has thus far been demonstrated in only two tent caterpillar species, *M. americanum* and *M. neustrium* (Fitzgerald and Peterson 1983; Peterson 1988), but it probably exists in other central-place foraging lasiocampids such as *M. californicum*, *Eriogaster lanestris* and *E. amygdali*, *Gloveria howardi*, and *Eutachyptera psidii* (Fitzgerald 1993, 1995).

Ant-associated lycaenids that exhibit a form of central-place foraging include the Australian species *Paralucia aurifera* (Cushman *et al.* 1994) and *Hypochrysops ignitus* (Common and Waterhouse 1981). These species are housed in earthen or thatch structures (called corrals or byres) constructed by their attendant ants, and are effectively central-place foragers since these structures are static 'nesting' sites. Larvae of these species often lay silken trails, which they follow during their nocturnal foraging bouts. For example, larvae of *Ogyris genoeveva* can use these silken highways to travel extremely rapidly from ant corrals at the base of the host tree to mistletoe feeding sites, sometimes several meters high in the boughs of the tree: relatively great distances from a caterpillar's perspective (Common and Waterhouse 1981). The degree of intra- and interspecific communication between larvae, and between larvae and ants, in these species remains to be investigated.

DISTRIBUTION OF SOCIALITY IN LEPIDOPTERA

We summarize in Table 20-1 key ecological and behavioral data gathered from the literature for social lepidopteran species. Although it is not exhaustive and is likely to contain omissions, this table provides the most comprehensive overview of social Lepidoptera to date.

Inspection of Table 20-1 readily illustrates several points. First, sociality is a widespread phenomenon in the Lepidoptera, occurring in some twenty or more butterfly and moth families representing thirteen ditrysian superfamilies. The taxonomic distribution of social characters suggests that multiple origins of sociality are likely.

Second, social behavior does not correlate in any predictable way with physiological and ecological characteristics such as host specificity or voltinism. The lack of striking patterns of association underscores the point that sociality in the Lepidoptera is likely to have multiple origins, with different species coming to sociality by different paths. However, strong correspondence occurs within some sets of related characters: not surprisingly, shelter construction appears to be more commonly associated with both patch-restricted and central-place foraging, and less commonly with nomadic foraging. This pattern makes sense from a bioenergetic point of view, since silk proteins used in shelter construction are likely to be metabolically expensive, and the nomadic foraging pattern would lead to considerable waste. Central-place foraging, characterizing the most complex lepidopteran societies, is likely to have arisen from both ancestrally nomadic and patch-restricted foraging patterns, a shift reflecting a change in resource use. Such changes were accompanied by changes in the use of communication abilities such as trail-marking.

Certain traits also occur frequently in particular lineages; for example, ant-association is common in the Lycaenidae, and spiny structural defenses are typically observed in the Nymphalidae. Gregariousness in the Lycaenidae is almost entirely confined to ant associated, but otherwise relatively cryptic taxa, with at least two notable exceptions: a social species of Poritiinae has been described to have gregarious, hairy larvae, and larvae of neotropical *Eumaeus* species feed on cycads and sport bright red, aposematic coloration.

Finally, Table 20-1 indicates that the largest gap in our knowledge of social Lepidoptera lies in the feature most essential to their sociality: communication. This has likely arisen because traits involving certain aspects of communication, especially behavior and physiology, are not always obvious and often must be experimentally demonstrated. On the other hand, features such as life cycle, host-plant use, and morphology are more readily measured or observed. In illustrating the taxonomic distribution of

sociality, in terms of which clades are social-rich and social-poor, which have a diversity of social systems, and which have a single system, Table 20-1 serves as a valuable reference point for framing phylogenetic hypotheses and identifying groups most in need of ecological, behavioral and systematic study; several such groups and questions are discussed in detail below.

COMMUNICATION IN THE CONTEXT OF CATERPILLAR ECOLOGY

We next consider selective pressures likely to have been important in the evolution of sociality. In particular, we relate the ecological contexts of communication to modes of larval defense and resource use, two key features of larval biology that mediate growth and adult fecundity, through a discussion of (1) signal enhancement in communication systems, (2) differential larval vulnerability, and (3) ant-association. These observations are integrated into a discussion of the costs and benefits of group living with respect to defense and resource use. Finally, we discuss social evolution in the Lepidoptera, treating the life-history and ecological factors shaping the characteristics of larval societies.

Signal enhancement

The communication-based benefits of sociality take several forms: defense, for example, may be enhanced through improved group-displays (Morris 1976; Shiga 1976; Stamp and Bowers 1988; Lawrence 1990) and shelter construction (Morris 1972a; Fitzgerald and Willer 1983; Damman 1987). Similarly, resource location and assimilation may be improved through a combination of behavioral thermoregulation (Morris and Fulton 1970; Seymour 1974; Capinera *et al.* 1980; Porter 1983; Knapp and Casey 1986; Casey *et al.* 1988; Joos *et al.* 1988) and cooperative or synchronized foraging (Ghent 1960; Fitzgerald 1976; Tsubaki 1981; Tsubaki and Shiotsu 1982; Casey *et al.* 1988; Fitzgerald *et al.* 1988).

The communication modes on which defense and resource use depend may in some cases be facilitated by group expression. A key quality of group contexts which may have favored sociality over solitary life-styles in many lepidopteran species is signal enhancement, a phenomenon wherein the effectiveness or efficiency of signaling improves as the number of individuals sending the signal increases.

Table 20-1. *Sociality in Lepidoptera*

Classification based on Kristensen (1984, 1991), Neilson (1989) and Neilson and Common (1991).

Classification	Species	Defenses[a]			Ant Tended[b]	Foraging pattern[c]	Shelter construc.[d]	Communication[e]				Host specific	Voltinism[g]	References
		Apo.	Struct.	Behav.				Chemical	Visual	Tactile	Acoustic			
TINEOIDEA														
Galacticidae	*Homadaula anisocentra*	?	?	Y	N	PR	None	?	?	?	?	M	?	17
	H. myriospila	?	?	?	N	PR	L,S	?	?	?	?	M	?	11
YPONOMEUTOIDEA														
Heliodinidae	*Heliodines nyctaginella*	?	?	?	N	PR	L,S	?	?	?	?	M	?	11
	H. roesella	?	?	?	N	PR	L,S	?	?	?	?	O	?	11
Plutellidae	*Plutella* spp.	Y	?	Y	N	PR	L,S	?	?	?	?	?	?	68
Yponomeutidae														
Attevinae	*Atteva* spp.	Y	?	Y	N	PR	L,S	P	?	?	?	M/O	?	11; 38; 61
	Prays lambda	?	?	?	N	PR	S	P	?	?	?	M	?B	54
	P. omicron	?	?	?	N	PR	S	P	?	?	?	M	U	54
	Euhyponomeutoides trachydeltus	?	?	P	N	PR	S	P	?	?	?	M	U	54
	Paraswammerdamia lutarea	?	?	P	N	PR	S	P	?	?	?	?	?	22
	Sariodoscelis spp.	?	?	P	N	PR	S	P	?	?	?	M	?B	54
	Yponomeuta cagnagellus	Y	?	Y	N	*PR*	*L,S*	Y	?	?	?	*M*	*U*	*35; 62; 63*
	Y. evonymellus	?	?	P	N	PR	S	P	?	?	?	M	U	54
	Y. intermellus	Y	?	P	N	PR	L,S	P	?	?	?	?	?	11
	Y. kanaiellus	?	?	P	N	PR	S	P	?	?	?	?	U	54
	Y. malinellus	?	?	P	N	PR	S	P	?	?	?	?	U	54
	Y. multipunctellus	Y	?	Y	N	PR	L,S	Y	?	?	?	M	U	38
	Y. padella	?	?	P	N	PR	S	P	?	?	?	?	?	22
	Y. polystictus	?	?	P	N	PR	S	P	?	?	?	M	U	54
	Y. polystigmellus	?	?	P	N	PR	S	P	?	?	?	M	?U	54
	Y. pustulellus	Y	?	P	N	PR	L,S	P	?	?	?	M	?	11
	Y. rorrella	?	?	P	N	PR	S	P	?	?	?	?	?	22
	Y. sociatus	?	?	P	N	PR	S	P	?	?	?	M	U	54
	Y. viginitipunctatus	?	?	P	N	PR	L,S	P	?	?	?	M	M	54
	Xyrosaris lichneuta	?	?	P	N	PR	L,S	P	?	?	?	O	M	54

GELECHIOIDEA													
Coleophoridae	*Homaledra sabalella*	Y	?	?	N	PR	L,S	?	?	?	M	M	38
Ethmiidae	*Ethmia heliomela*	Y	?	?	N	PR	L,S	?	?	?	M	?	11
Oecophoridae													
Xyloryctinae	*Crypsicharis neocosma*	?	?	?	N	PR	L,S	?	?	?	M	?	11
COSSOIDEA													
Cossidae	*Culama* spp.	Y	?	?	N	PR	Under bark	?	?	?	O	?	11
	Macrocyttara expressa	?	?	?	N	PR	Under bark	?	?	?	M	?	11
TORTRICOIDEA													
Tortricidae	*Argyrotaenia pulchellana*	?	?	?	N	PR	?	?	?	?	?	?	92
	Archips cerasivoranus	Y	Setae	?	N	PR	L,S	?	?	?	O	U	26; 38
	A. fervidana	Y	Setae	?	N	PR	L,S	?	?	?	O	?	38
	Cryptoptila australana	Y	?	?	N	PR	L,S	?	?	?	?	?	11
ZYGAENOIDEA													
Limacodidae	*Ctenolita melanosticta*	Y	Sp	?	N	PR	None	?	?	?	O	?	94
	Parasa lepida	?	?	?	N	?PR	?	?	?	?	?	?	40
Zygaenidae	*Artona funeralis*	?	?	?	N	?Nom	None	?	?	?	?	?	36; 53
	Pryeria sinica	Y	?	?	N	Nom	S (1st inst) ?	?	?	?	M	U	81
PYRALOIDEA													
Pyralidae													
Epipaschiinae	*Catamola thyrisalis*	?	?	?	N	PR	L,S	?	?	?	O	?	11
	Macalla ebenina	?	?	?	N	PR	L,S	?	?	?	M	?	11
	M. pyrastris	?	?	?	N	PR	L,S	?	?	?	M	?	11
	Tetralopha robustella	?	?	?	N	CP	L,S	?	?	?	M	U/M	32
Evergestinae	*Evergestis extimalis*	?	?	?	N	PR	S	?	?	?	O	U	31
	E. pallidata	?	?	?	N	PR	S	?	?	?	M	U	31
Gallerinae	*Meyriccia latro*	?	?	?	N	PR	S, flowers	?	?	?	M	?	11
	Omphalocera munroei	?	?	?	N	PR	L	?	?	?	M	M	16
Phycitinae	*Acrobasis consociella*	?	?	?	N	PR	L,S	?	?	?	M	U	31
Pyralinae	*Aglossa pinguinalis*	?	?	?	N	PR	S, debris	?	?	?	P	B	31
	Ocrasa albidalis	Y	?	?	N	PR	L,S	?	?	?	M	?	11
Pyraustinae	*Hyalobathra miniosalis*	?	?	?	N	PR	L,S	?	?	?	M	?	11
	Mutuuraia terrealis	?	?	?	N	PR	S	?	?	?	M	U	31
	Nomophila corticalis	?	?	Y	N	?CP	S	?	?	?	P	?	11
	Paracorsia repandalis	Y	?	?	N	PR	S	?	?	?	M	?	31
	Pyrausta cespitalis	?	?	?	N	PR	S	?	?	?	M	B	31
	P. cingulata	?	?	?	N	PR	S	?	?	?	O	?	31
	Uresiphita reversalis	Y	?	?	N	Nom	L,S	?	?	?	O	M	3

Table 20-1 (*cont.*)

Classification	Species	Defenses[a] Apo.	Struct.	Behav.	Ant Tended[b]	Foraging pattern[c]	Shelter construc.[d]	Communication[e] Chemical	Visual	Tactile	Acoustic	Host specific	Voltinism[g]	References
GEOMETROIDEA														
Geometridae														
Diptychinae	*Venilioides inflammata*	Y	Y	Y	N	PR, Nom	S (1st–2nd inst.)	?	?	?	?	M	?	75
Ennominae	*Mnesampela lenaea*	?	?	?	N	?	L,S	?	?	?	?	?	?	11
	M. privata	?	?	?	N	?	L,S	?	?	?	?	M	?	11
	Zerenopsis leopardina	Y	?	Y	N	PR	S (1st–2nd inst.)	?	?	?	?	M	?	94
Larentiinae	*Hydria prunivorata*	Y	?	Y	N	PR	L,S	?	?	?	?	M	U	66
Oenochrominae	*Naxa seroaria*	?	?	?	N	PR	S	?	?	?	?	?	?	96
URANIOIDEA														
Uraniidae														
Epipleminae	*Epiplema* spp.	N	?	?	N	PR	S	?	?	?	?	?	?	2
HESPERIOIDEA														
Hesperiidae	*Hidari irava*	?	?	?	N	PR	L,S	?	?	?	?	M	?	40
PAPILIONOIDEA														
Lycaenidae														
Polyommatinae	*Anthene emolus*	N	None	?	Y	Nom	None	?	?	?	?	?	?	24
	A. lycaenoides	N	None	?	Y	Nom	None	?	?	?	?	P	?	12
Poritiinae	*Poritia erycinoides*	?Y	Set	?	N	Nom	None	P	?	?	?	?	?	64
	P. sumatrae	N	Set	?	N	Nom	None	?	?	?	?	?	?	95
Riodininae	*Emesis lucinda*	N	None	?	N	?Nom	None	?	?	?	?	O	M	91
	Euselasia cafusa	N	None	?	N	?Nom	None	?	?	?	?	O	M	91
	E. mystica	N	None	?	N	?Nom	None	?	?	?	?	O	M	91
	E. rhodogyne	N	None	?	N	?Nom	None	?	?	?	?	O	M	91
	Hades noctula	Y	None	?	N	?Nom	None	?	?	?	?	O	M	91
	Melanis pixie	Y	None	?	N	?Nom	None	?	?	?	?	O	M	91
Theclinae	*Acrodipsas myrmecophila*	N	None	?	Y	Nom	None	?	?	?	?	Parasites?	?	12
	Arhopala pseudocentaurus	?	?	?	Y	Nom	None	?	?	?	?	?	?	98
	Axiocerses bambana	N	None	?	Y	CP	L,S	?	?	?	?	M	B/M	10
	Crudaria leroma	N	T	?	Y	?Nom	None	?	?	?	?	M	U/M	10
	Drupadia theda	?	?	?	Y	Nom, ?CP	L,S	?	?	?	?	?	?	92

Eumaeus atala	Y	Set	?	N	Nom	None	?	?	?	?	Nom	O	M	35; 69
E. minijas	Y	Set	?	N	Nom	None	?	?	?	?	Nom	O	M	35; 69
Hypochrysops delicia														
delicia	N	None	?	Y	Nom	None	?	?	?	?	Nom	M	?	12
H. ignitus ignitus	N	None	?	Y	CP	Byre	?	?	?	?	CP	P	?	12
H. epicurus	N	None	?	Y	Nom, CP	None	?	?	?	?	Nom, CP	M	?	12
H. cyane	N	None	?	Y	Nom, CP	None	?	?	?	?	Nom, CP	M	?	12
H. miskini	N	None	?	Y	Nom, CP	None	?	?	?	?	Nom, CP	M	?	12
Ogyris amaryllis meridionalis	N	Set	?	Y	Nom	None	?	?	?	?	Nom	M	?	12
O. genoveva	?	None	?	?Y	?Nom	None	?	?	?	?	?Nom	?	?	97
Jalmenus evagoras	N	None	?	Y	Nom	None	?	?	?	?	Nom	O	M	12; 91
J. icilius	N	None	?	Y	Nom	None	?	?	?	?	Nom	O	M	12; 91
J. ictinus	N	None	?	Y	Nom	None	?	?	?	?	Nom	O	M	12; 91
Nymphalidae														
Argynninae														
Phalanta spp.	?	Sp.	?	N	?Nom	?	P	?	?	?	?Nom	P	?	49
Apaturinae														
Asterocampa celtis	?	Ceph. Sp.	?	Y	N	Nom	None	?	?	?	Nom	M	M	48; 69; 72
A. clyton	?	Ceph. Sp.	?	Y	N	Nom	None	?	?	?	Nom	O	M	48; 69; 72
Brassolinae														
Brassolis isthmia	?	?	Y	N	CP	L,S	P	?	?	?	CP	O	B	20; 89
Heliconiinae														
Heliconius doris	Y	Sp	?	N	Nom	None	P	?	?	?	Nom	M	?	19
H. sapho leuce	Y	Sp	?	N	Nom	None	?	?	?	?	Nom	M	?	19
H. hewitsoni	Y	Sp, H	Y	N	?	None	?	?	?	?	None	M	?	19
H. xanthocles	Y	Sp	?	N	Nom	None	?	?	?	?	Nom	M	M	52
Ithomiinae														
Hypothyris euclea valora	Y	?	?	N	?Nom	None	P	?	?	?	?Nom	M	?	19
H. lycaste callispila	Y	?	?	N	?Nom	None	?	?	?	?	?Nom	M	?	19
Mechanitis polymnia isthmia	Y	T	?	N	?Nom		?	?	?	?	?Nom	?	?	19
M. lysimnia doryssus	Y	T	?	N	?Nom		?	?	?	?	?Nom	?	?	19
M. menapis saturata	Y	T	?	N	?Nom		?	?	?	?	?Nom	?	?	19
Melitaeinae														
Anthanassa frisia frisia	Y	Set, Sp	?	N	PR, ?Nom		P	?	?	?	PR, ?Nom	?	?	9
Chlosyne gabbi	Y	Sp	?	N	Nom		P	?	?	?	Nom	O	U	69
C. gorgone	Y	Sp	?	N	Nom		P	?	?	?	Nom	O	U	69
C. hoffmanni	Y	Sp	?	N	PR, Nom	S	P	?	?	?	PR, Nom	O	U	69
C. janais harrisii	Y	Sp	?	N	Nom		P	?	?	?	Nom	?	?	19
C. lacinia	Y	Sp	Y	N	Nom	Y	Y	?	?	?	Nom	O	M	5; 19; 69; 70
C. leanira	Y	Sp	?	N	PR, Nom	S	P	?	?	?	PR, Nom	O	U	69
C. melanarge	Pr	Sp	?	N	Nom		P	?	?	?	Nom	M	?	18

Table 20-1 (cont.)

Classification	Species	Defenses[a]			Ant Tended[b]	Foraging pattern[c]	Shelter construc.[d]	Communication[e]				Host specific	Voltinism[g]	References
		Apo.	Struct.	Behav.				Chemical	Visual	Tactile	Acoustic			
Melitaeinae	C. nyctes	Y	Sp	?	N	Nom	?	?P	?	?	?	O	U	69
	C. palla	Y	Sp	?	N	Nom	?	?P	?	?	?	O	U	69
	Euphydryas aurinia	Y	Sp	Y	N	PR, Nom	S	P	?	?	?	P	?	60; 80
	E. chalcedona	Y	Sp	?	N	PR, Nom	S	P	?	?	?	P	U	69
	E. colon	Y	Sp	?	N	PR, ?Nom	L,S	P	?	?	?	O	?	35; 69
	E. cynthia	Y	Sp	?	N	PR, ?Nom	S	?	?	?	?	?	?	92
	E. debilis	Y	Sp	?	N	PR, ?Nom	S	?	?	?	?	?	?	92
	E. editha	Y	Sp	?	N	Nom	?	P	?	?	?	P	U	69
	E. gilletti	Y	Sp	?	N	PR, Nom	L,S	P	?	?	?	P	U	69
	E. iduna	Y	Sp	?	N	PR, ?Nom	S	?	?	?	?	?	?	92
	E. intermedia	Y	Sp	?	N	PR, ?Nom	S	?	?	?	?	?	?	92
	E. maturna	Y	Sp	?	N	PR, ?Nom	S	?	?	?	?	?	?	92
	E. wolfensbergeri	Y	Sp	?	N	PR, ?Nom	S	?	?	?	?	?	?	92
	E. phaeton	Y	Sp	Y	N	PR, ?Nom	S	P	?	?	?	P	U	71; 69
	Eresia alsima	Y	Sp	Y	N	Nom	?	?	?	?	?	M	?	19
	Melitaea cinxia	Y	Sp	Y	N	PR, Nom	S	?	?	?	?	O	U	8; 80
	M. diamina	Y	Sp	Y	N	PR, Nom	S	?	?	?	?	?	?	92
	M. phoebe	Y	Sp	Y	N	PR, Nom	S	?	?	?	?	?	?	92
	Mellicta asteria	Y	Sp	?	N	PR, Nom	S	?	?	?	?	O	U	92
	M. athalia	Y	Sp	?	N	PR, Nom	S	?	?	?	?	O	U	7; 80
	M. aurelia	Y	Sp	?	N	PR, Nom	S	?	?	?	?	O	U	92
	M. britomartis	Y	Sp	?	N	PR, Nom	S	?	?	?	?	O	U	92
	M. parthenoides	Y	Sp	?	N	PR, Nom	S	?	?	?	?	O	U	92
	M. varia	Y	Sp	?	N	PR, Nom	S	?	?	?	?	O	U	92
	Phyciodes spp.	?	Sp, T	?	N	Most Nom	?	?	?	?	?	M,?O	?	4; 69
	Poladryas minuta	Y	Sp	?	N	?	?	?	?	?	?	O	M	69
Morphinae	Taenaris myops	?	Set, H	?	N	?Nom	?	?	?	?	?	?	?	12
Nymphalinae	Acraea andromacha andromacha	Y	Set, H	?	N	Nom	?	?	?	?	?	O	?	12
	Aglais urticae	Y	Sp	Y	N	PR, Nom, ?CP	L,S	?	?	?	?	M	U	8; 80
	Araschnia levana	Y	Sp	?	N	Nom	?	?	?	?	?	M	B	8
	Cethosia penthesilea paksha	Y	Sp	?	N	Nom	?	?	?	?	?	M	?	12

Taxon														Ref.
C. cydippe chrysippe	Y	Sp	?	N	Nom	?	?	?	?	?	?	M	?	12
Doleschallia bisaltide pratipa	Y	Sp	?	N	?Nom	?	?	?	?	?	?	O	?	14
Hamadryas amphinome	Y	Sp	?	N	Nom	?	?	?	?	?	?	O	M	69
H. fornax	Y	Sp	?	N	Nom	?	?	?	?	?	?	O	M	69
Hypolimnas spp. (most)	Y	Sp	?	N	Nom	S	?	?	?	?	?	P	?	12; 49
Inachis io	Y	Sp	Y	N	PR, Nom	None	?	?	?	?	?	M	U	8; 80
Mynes goeffroyi guerini	Y	Set, Sp	?	N	?Nom	None	?	?	?	?	?	O	?	12
Nymphalis antiopa	Y	Sp	Y	N	Nom	L,S	?	?	?	?	?	O,P	M, B	38; 69
N. californica	Y	Sp	?	N	?Nom	L,S	?	?	?	?	?	O	U, M	69
N. milberti	Y	Sp	?	N	PR, Nom	None	?	?	?	?	?	P	M	35; 69
N. polychloros	Y	Sp	Y	N	?PR, ?CP	None	?	?	?	?	?	P	U	8; 80
N. vau–album	?	Sp	?	N	PR, Nom	None	?	?	?	?	?	P	U	35;69
N. xanthomelas japonica	?	?Sp	?	N	?	None	?	?	?	?	?	?	?	96
Salamis cacta	?	Sp	?	N	?Nom	?	?	?	?	?	?	M	?	49
Symbrenthia lilaea luciana	?	Sp	?	N	?Nom	?	?	?	?	?	?	O	?	14
Satyrinae														
Dioriste spp.	?	H	?	N	?Nom	?	?	?	?	?	?	M	?	19
Megeuptychia antonoe	?	H	?	N	?Nom	?	?	?	?	?	?	M	?	19
Papilionidae														
Papilioninae														
Battus polydamus	Y	O	Y	N	Nom	None	?	?	?	?	?	O	M	69
Papilio anchisiades idaeus	N	O	Y	N	Nom	None	?	?	?	?	?	O	M	69; 90
P. demolion demolion	Y	O	Y	N	Nom	None	?	?	?	?	?	M,O	?	14
P. pelaus	N	O	Y	N	Nom	None	?	?	?	?	?	?	?	4
Parnassius apollo	Y	O	Y	N	Nom	None	?	?	?	?	?	?	?	85
Zerynthia polyxena	Y	O	Y	N	Nom	None	?	?	?	?	?	?	?	67
Pieridae														
Pierinae														
Anaphaeis java	?	?	?	N	?	?	?	?	?	?	?	O	?	93
Ascia monuste	Y	?	?	N	Nom	?	?	?	?	?	?	P	M	69
Aporia crataegi	Y	Set	?	N	CP, Nom	S	?	?	?	?	?	O	?	80
A. hippia japonica	?	?	?	N	?	?	?	?	?	?	?	?	?	96
Catasticta spp.	?	Set	Y	N	?	?	?	?	?	?	?	O	?	19
Colotis amatus	?	?	?	N	Nom	?	?	?	?	?	?	M	?	49
C. phisadia	?	?	?	N	Nom	?	?	?	?	?	?	M	?	49
Delias harpalyce	Y	?	?	N	PR	S	?	?	?	?	?	M	B	12
D. hyparete metarete	Y	Set	?	N	?	?	?	?	?	?	?	O	?	14
D. ninus ninus	Y	Set	?	N	?	?	?	?	?	?	?	O	?	14
Euchetra socialis	Y	None	Y	N	CP	S	P	?	?	?	?	M	U	39; 82

Table 20-1. (cont.)

Classification	Species	Defenses[a]			Ant Tended[b]	Foraging pattern[c]	Shelter construc.[d]	Communication[e]				Host specific	Voltinism[g]	References
		Apo.	Struct.	Behav.				Chemical	Visual	Tactile	Acoustic			
Pierinae	*Eurema blanda*	?	Set	?	N	?	?	?	?	?	?	M,O	?	95
	Hesperocharis crocea	?	Set	?	N	?	?	?	?	?	?	O	?	19
	Neophasia menapia	No	Set, T	?	N	Nom		?	?	?	?	O	U	21; 69
	N. terlooti	No	Set, T	?	N	Nom		?	?	?	?	M	B	21; 69
	Pereute spp.	?	Set	?	N	?		?	?	?	?	?	?	19
	Pieris brassicae	Y	Set	?	N	Nom		?	?	?	?	M	P	51; 80
BOMBYCOIDEA														
Eupterotidae	*Rhabdosia patagiata*	N	?Set	?	N	CP	S	?	?	?	?	O	?	79
	Hyposoides spp.	?	Set	?	N	?CP	S	P	?	?	?	?	?	59
Panacelinae	*Panacela lewinae*	?	Set	Y	N	CP	S	P	?	?	?	P	?	11
	P. nyctopa	?	Set	Y	N	Nom	?	P	?	?	?	P	?	11
	P. pilosa	?	Set	Y	N	CP	S	P	?	?	?	P	?	11
Lasiocampidae														
Lasiocampinae	*Bombycomorpha bifascia*	Y	Set	Y	N	Nom	?	P	?	Y	?	?O	M	78
	Catalebeda cuneilinea	Y	Set	?	N	PR	?	?	?	Y	?	M	?	94
	Eriogaster amygdali	?	?Set	?	N	CP	S	P	?	?	?	O	?	77
	E. arbusculae	N	Set	?	N	CP	S	Y	?	?	?	?	?	65
	E. catax	N	Set	?	N	CP	S, Nom	Y	?	?	?	?	?	57; 92
	E. lanestris	N	Set	?	N	CP	S	Y	?	?	?	P	?	7; 86
	E. philippsi	Y	Set	?	N	CP	S	P	?	?	?	M	?	76
	E. rimicola	?	?	?	N	?Nom	None	P	?	?	?	?	?	57; 92
	Eutachyptera psidii	?	?	?	N	CP	S	P	?	?	?	M	U	13; 29
	Gloveria howardi	?	?	?	N	CP	S	P	?	?	?	M	?	29
	Macrothylacia rubi	Y	Set	?	N	Nom	None	?	?	?	?	?	?	57; 92
	Malacosoma alpicolum	Y	Set	?	N	Nom, ?CP	S	P	?	?	?	P	?	29
	M. americanum	Y	Set	Y	N	CP	S	Y	?	Y	?	O	U	27; 74
	M. californicum	Y	Set	Y	N	CP	S	P	?	Y	?	O	U	74
	M. castrensis	Y	Set	Y	N	Nom, ?CP	S	P	?	?	?	P	U	30
	M. constrictum	Y	Set	Y	N	CP, Nom	S (molting)	P	?	?	?	M	U	74
	M. disstria	Y	Set	Y	N	Nom	None	Y	?	Y	?	P	U	25
	M. franconium	Y	Set	?	N	Nom, ?CP	S	P	?	?	?	P	?	30
	M. incurvum	Y	Set	Y	N	CP	S	P	?	?	?	O	U	74

Family	Species															
	M. luteum	Y	Set	Y	N	?Nom	?	P	?	?	?	?	?	P	?	30
	M. neustria	Y	Set	Y	N	CP	S	Y	?	?	?	?	?	O	U	58
	M. tigris	Y	Set	Y	N	CP, Nom	S (molting)	P	?	?	?	?	?	M	?	74
	Rhinobombyx cuneata	N	?Set	?	N	PR	S	P	?	?	Y	?	?	M	?	94
	Schausinna regia	Y	Set	?	N	PR	S	?	?	?	Y	?	?	O	?	55
Endromidae	*Enromis versicolora*	N	N	?	N	Nom	None	?	?	?	?	?	?	M	U	3a
Lemoniidae	*Sabalia tippelskirchi*	Y	?	Y	N	PR	?	?	?	?	?Y	?	?	?M	?	28; 56
Saturniidae Ceratocampinae	*Amsota senatoria*	Y	T,H	?	N	Nom	?	?	?	?	?	?	?	P	U, B	33; 38
	A. stigma	Y	T,H	?	N	Nom	?	?	?	?	?	?	?	P	U, B	38
	A. virginiensis	Y	T,H	?	N	Nom	?	?	?	?	?	?	?	P	U, B	38
	Dryocampa rubicunda	Y	T,H	?	N	Nom	?	?	?	?	?	?	?	M	U, B	38
	Hemileuca lucina	Y	Spiny T	Y	N	Nom	?	Y	?	?	?	?	?	M	U	15; 73
	H. oliviae	Y	Spiny T	Y	N	Nom	?	Y	?	?	?	?	?	?	?	6
	Hylesia lineata	N	Sp	Y	N	Nom, CP	L,S	P	?	?	?	?	?	P	B	37
	H. acuta	N	Sp	Y	N	CP	L,S	P	?	?	?	?	?	O	U	87
	Samia cynthia	Y	T	Y	N	Nom	?	?	?	?	?	?	?	O	?	2
Ludiinae	*Holocerina smilax*	Y	Sp	?	N	PR	?	?	?	?	?Y	?	?	P	?	94
	Micragone ansorgei	Y	Sp	?	N	PR	?	?	?	?	?Y	?	?	O	?	94
	Vegetia dewitzi	Y	Sp	?	N	PR	?	?	?	?	?Y	?	?	M	?	94
Saturniinae	*Bunaea alcinoe*	Y	Sp	?	N	PR	?	?	?	?	Y	?	?	P	?	42; 94
	B. aslauga	Y	Sp	?	N	PR	?	?	?	?	?	?	?	O	?	94
	Cinabra hyperbius	N	Sp	?	N	PR	?	?	?	?	Y	?	?	M	?	94
	Circula spp	Y	T	?	N	None	None	?	?	?	?	?	?	P	M	54a
	Cirina forda	Y	Sp	?	N	PR	?	?	?	?	Y	?	?	O	?	46; 56
	Copaxa denda	Y	Set, Sp	?	N	Nom	None	?	?	?	?	?	?	O	?	88
	C. escalentei	Y	Set, Sp	?	N	Nom	None	?	?	?	?	?	?	O	?	88
	C. mazaorum	Y	Set, Sp	?	N	Nom	None	?	?	?	Y	?	?	O	?	88
	C. multifenestrata	Y	Set, Sp	?	N	Nom	None	?	?	?	?	?	?	M	?	88
	C. rufinans	Y	Set, Sp	?	N	Nom	None	?	?	?	?	?	?	O	?	88
	Eochroa trimenii	Y	?	Y	N	PR	?	?	?	?	?Y	?	?	M	?	94
	Gynanisa maja	N	Sp	?	N	PR	?	?	?	?	Y	?	?	O	?	44; 94
	Imbrasia belina	Y	Sp	?	N	PR	?	?	?	?	Y	?	?	P	?	94
	I. carnegiei	Y	Sp	?	N	PR	?	?	?	?	Y	?	?	O	?	94
	I. cytherea	Y	Sp	?	N	PR	?	?	?	?	Y	?	?	O	?	94
	I. ertli	Y	?	?	N	PR	?	?	?	?	Y	?	?	O	?	94
	I. gueinzii	Y	Sp	?	N	PR	?	?	?	?	Y	?	?	M	?	94
	I. hecate	Y	Sp	?	N	PR	?	?	?	?	Y	?	?	P	?	41; 94
	I. hochmelii	Y	?	?	N	PR	?	?	?	?	Y	?	?	O	?	47; 94
	I. macrothyris	Y	Sp	?	N	PR	?	?	?	?	Y	?	?	P	?	94

Table 20-1 (*cont.*)

Classification	Species	Defenses[a] Apo.	Struct.	Behav.	Ant Tended[b]	Foraging pattern[c]	Shelter construc.[d]	Communication[e] Chemical	Visual	Tactile	Acoustic	Host specific	Voltinism[g]	References
Saturniinae	*I. petiveri*	Y	Sp	?	N	PR	?	?	?	Y	?	P	?	94
	I. rhodophila	Y	Sp	?	N	PR	?	?	?	Y	?	P	?	45; 94
	I. tyrrhea	Y	?	?	N	PR	?	?	?	Y	?	P	?	94
	I. wahlbergi	Y	Sp	?	N	PR	?	?	?	Y	?	P	?	94
	I. zambesina	Y	Sp	?	N	PR	?	?	?	Y	?	O	?	94
	Lobobunaea acetes	N	?	?	N	PR	?	?	?	?Y	?	?P	?	94
	L. angasana	N	?	?	N	PR	?	?	?	?Y	?	P	?	94
	L. phaedusa	N	?	?	N	PR	?	?	?	?Y	?	?P	?	94
	Melanocera dargei	Y	Sp	?	N	PR	?	?	?	?Y	?	O	?	94
	M. menippe	Y	Sp	?	N	PR	?	?	?	?Y	?	P	?	43; 94
	Opodiphthera engaea	?	?Set	?	N	Nom	?	?	?	?	?	O	?	11
	O. loranthi	Y	Set	?	N	Nom	?	?	?	?	?	O	?	11
	Pavonia pavonia	Y	S	?	N	Nom	None	?	?	?	?	P	U	3a
	Pseudaphelia apollinaris	Y	?	Y	N	PR	?	Y	?	?Y	?	M	?	94
	Pseudobunaea irius	N	?	?	N	PR	?	?	?	Y	?	O	?	94
	P. tyrrhena	N	?	?	N	PR	?	?	?	Y	?	M	?	94
	Rohaniella pygmaea	N	?	?	N	PR	?	?	?	Y	?	M	?	94
	Tagoropsis flavinata	Y	?	Y	N	PR	?	?	?	Y	?	M	?	94
	Ubaena fuelleborniana	Y	Sp	?	N	PR	?	?	?	?Y	?	O	?	94
	Urota sinope	Y	?	?	N	PR	?	?	?	?Y	?	M	?	94
	Usta terpsichore	Y	?	Y	N	PR	?	?	?	?Y	?	O	?	94
	U. wallengrenii	Y	?	Y	N	PR	?	?	?	?Y	?	M	?	94
NOCTUOIDEA														
Arctiidae														
Arctiinae	*Amerila astreus*	Y	Set	?	N	Nom	?	?	?	?	?	M	?	2
	Baroa siamica	Y	Set	?	N	Nom	?	?	?	?	?	M	?	2
	Creatonotus transiens	Y	Set	?	N	Nom	?	?	?	?	?	M	?	2
	Euchaetis egle	N	?	?	N	Nom	?	?	?	?	?	M	?	17
	Halisidota caryae	N	?	?	N	Nom	?	?	?	?	?	O	?	50
	Hyphantria cunea	Y	Set	Y	N	PR	L,S	Y	?	P	?	P	U,M	83; 38
	Lemyra spp.	Y	Set	?	N	Nom	?	?	?	?	?	M	?	2
	Nyctemera spp.	Y	Set	?	N	Nom	?	?	?	?	?	M	?	2
	Tyria jacobaeae	Y	?	?	N	Nom	?	?	?	?	?	M	?	50; 92

Family / Group	Taxon													
	Pericallia galactina trigonalis	2	?	M	?	?	?	Nom	?	?	N	?	Set	Y
	Spilosoma spp.	2	?	M	?	?	?	Nom	?	?	N	?	Set	Y
Crenuchinae	*Syntomeida epilais*	38	T	M	?	?	?	Nom	?	?	N	?	Set	Y
Aganaidae	*Asota* spp.	2; 11	?	M	?	?	?	Nom	?	?	N	?	Set	Y
	Neochera spp.	2	?	M	?	?	?	Nom	?	?	N	?	Set	Y
	Euplocia membliaria	2	?	M	?	?	?	Nom	?	?	N	?	Set	Y
	Anagnia subfascia	2	?	M	?	?	?	Nom	?	?	N	?	Set	Y
Lymantriidae	*Calliteara* spp.	2	?	P	?	?	?	Nom	L/S	?	N	?	Set	Y
	Carriola ecnomoda	2	?	P	?	?	?	Nom	L,S	?	N	?	Set	Y
	Cassidia peninsularis	2	?	P	?	?	?	Nom	L,S	?	N	?	Set	Y
	Cobanilla marginata													
	phaedra	2	?	P	?	?	?	Nom	L,S	?	N	?	Set	Y
	Dura alba	2	?	P	?	?	?	Nom	L,S	?	N	?	Set	Y
	Euproctis spp. (many)	2; 8; 92	U	O,P	?	?	?	Nom, ?CP	L,S	?	N	?	Set	Y
	Ilema vaneeckei	2	?	P	?	?	?	Nom	L,S	?	N	?	Set	Y
	Imaus munda													
	collenettei	2	?	P	?	?	?	Nom	L,S	?	N	?	Set	Y
	Leucoma impressa	2	?	P	?	?	?	Nom	L,S	?	N	?	Set	Y
	Locharna limbata	2	?	P	?	?	?	Nom	L,S	?	N	?	Set	Y
	Lymantria spp.	2	?	P	?	?	?	Nom	L,S	?	N	?	Set	Y
	Numenes contrahens	2	?	P	?	?	?	Nom	L,S	?	N	?	Set	Y
	Redoa micacea	2	?	P	?	?	?	Nom	L,S	?	N	?	Set	Y
	Rhypotoses humida	2	?	P	?	?	?	Nom	L,S	?	N	?	Set	Y
	Scarpona enmomoides	2	?	P	?	?	?	Nom	L,S	?	N	?	Set	Y
	Sircia denudata	2	?	P	?	?	?	Nom	L,S	?	N	?	Set	Y
Noctuidae Chloephorinae	*Camptoloma interiorata*	96	?	?	?	?	?	?	?	?	N	?	?	?
Notodontidae	*Datana major*	38	U,B	O	?	?	?	Nom	None	?	N	?	?	?
	D. ministra	38	U,B	P	?	?	?	Nom	None	?	N	?	?	Y
	D. integerrima	38	U	O	?	?	?	Nom	None	?	N	?	Set	Y
	Ichthyura inclusa	38	U,B	M	?	?	?	CP	L,S	?	N	?	?	N
	Phalera assimilis	96	?	?	?	?	?	?Nom	None	?	N	?	?	?
	P. bucephala	57; 92	?	?	?	?	?	Nom	None	?	N	?	?	Y
	P. bucephaloides	57; 92	?	?	?	?	?	Nom	None	?	N	?	?	Y
	P. sundana	2	?	?	?	?	?	Nom	None	?	N	?	Set	N
	Symmerista canicosta	38	U	M	?	?	?	Nom, CP	None	?	N	?	Set	Y
Oenosandridae	*Oenosandra* spp.	11; 68	?	M	?	?	?	Nom, CP	None	?	N	?	Set	?
Thaumetopoeidae	*Anaphe panda*	59	?	O	?	?	?	?CP	S	P	N	?	Set	?
	A. reticulata	59	?	O	?	?	?	?CP	S	P	N	?	Set	?
	Cynosarga ornata	11	?	M	?	?	?	?	?	?	N	?	?	?

Table 20-1. (cont.)

Classification	Species	Defenses[a] Apo.	Struct.	Behav.	Ant Tended[b]	Foraging pattern[c]	Shelter construc.[d]	Communication[e] Chemical	Visual	Tactile	Acoustic	Host specific[f]	Voltinism[g]	References
Thaumetopoeidae	Discophlebia catocalina	?	?	?	N	?Nom	?	?	?	?	?	M	?	11
	Epanaphe spp.		Set	?	N	?CP	S	P	?	?	?	O	?	59
	Epicoma dispar	?	?	?	N	Nom	?	?	?	?	?	P	?	11
	Ochrogaster spp.		Set	?	N	CP	S	P	?	?	?	M	?	11
	Oenosandra boisduvalii	?	?	?	N	CP	Under bark	P	?	?	?	M	?	11
	Thaumetopoea pinivora	Y	?	?	N	CP,PR	L,S	Y	?	?	?	?	?	92
	T. pityocampa	Y	?	?	N	CP,PR	L,S	P	?	Y	?	M	U	1; 8; 23
	T. processionea	Y	?	?	N	CP,PR	L,S	Y	?	?	?	M	U	8
	Trichiocercus sparshalli	?	?	?	N	CP	Under bark	P	?	?	?	M	?	11
Thyretidae	Metarctia meteus	?	?	?	N	Nom	?	?	?	?	?	O	B	78

[a] Defenses: Apo. = aposematic; Struct. = structures; Behav. = behavioral; Y = yes; N = no; P = probably; Set = setae; Sp = spines; H = horns; T = tubercles; O = osmeterium.

[b] Ant-tended: Y = yes; N = no.

[c] Forging pattern: PR = patch-restricted; Nom = nomadic; CP = central-place.

[d] Shelter construction: L = leaves; S = silk.

[e] Communication: Y = yes; N = no; P = probably.

[f] Host-specificity: M = monophagous; O = oligophagous; Poly = polyphagous.

[g] Voltinism: Uni = univoltine; B = bivoltine; T = trivoltine; Multi = multivoltine.

Sources: (1) Balfour-Browne 1925; (2) Barlow 1982; (3) Bernays and Montllor 1989; (3a) Brookes, 1991; (4) Brown and Heineman 1972; (5) Bush 1969; (6) Capinera 1980; (7) Carlberg 1980; (8) Carter 1982; (9) Chermock and Chermock 1947; (10) Clark and Dickson 1971; (11) Common 1990; (12) Common and Waterhouse 1981; (13) Comstock 1957; (14) Corbet et al. 1992; (15) Cornell et al. 1988; (16) Damman 1987; (17) Dethier 1959a; (18) Dethier 1959b; (19) DeVries 1987; (20) Dunn 1917; (21) Edwards 1897; (22) Emmet 1979; (23) Fabre 1916; (24) Fiedler and Maschwitz 1989; (25) Fitzgerald and Costa 1986; (26) Fitzgerald and Edgerly 1979; (27) Fitzgerald and Peterson 1983; (28) Fontaine 1975; (29) Franclemont 1973; (30) de Freina and Witt 1987; (31) Goater 1986; (32) Hertel and Benjamin 1979; (33) Hitchcock 1961; (34) Hoebeke 1987; (35) Howe 1975; (36) Iwao 1968; (37) Janzen 1984; (38) Johnson and Lyon 1988; (39) Kevan and Bye 1991; (40) Khoo et al. 1991; (41) Lampe 1982a; (42) Lampe 1982b; (43) Lampe 1983a; (44) Lampe 1983b; (45) Lampe 1984; (46) Lampe 1985a; (47) Lampe 1985b; (48) Langlois and Langlois 1964; (49) Larsen 1991; (50) Lawrence 1990; (51) Long 1955; (52) Mallet and Jackson 1980; (53) Mizuta 1968; (54) Moriuti 1977; (54a) Naumann 1995; (55) Ober-prieler 1993; (56) Oberprieler 1995; (57) Patocka 1980; (58) Peterson 1988; (59) Pinhey 1975; (60) Porter 1982; (61) Robinson et al. 1994; (62) Roessingh 1989; (63) Roessingh 1990; (64) Rosier 1951; (65) Rougeout and Viette 1983; (66) Schultz and Allen 1975; (67) Schweizerischer Bund für Naturschutz 1987; (68) Scoble 1992; (69) Scott 1986; (70) Stamp 1977; (71) Stamp 1981a; (72) Stamp 1984; (73) Stamp and Bowers 1988; (74) Stehr and Cook 1968; (75) Staude 1994; (76) Talhouk 1940; (77) Talhouk 1975; (78) Taylor 1949; (79) Taylor 1950; (80) Thomas and Lewington 1991; (81) Tsubaki 1981; (82) Underwood 1994; (83) Warren and Tadic 1970; (85) Weidemann 1988; (86) Weyh and Maschwitz 1978; (87) Wolfe 1988; (88) Wolfe 1993; (89) Young 1985; (90) Young et al. 1986; (91) P. J. DeVries, personal communication; (92) K. Fiedler, personal communication; (93) J. Holloway, personal communication; (94) R. G. Oberprieler, personal communication; (95) M. W. Tan, personal communication; (96) H. Yoshimoto, personal communication; (97) N. E. Pierce, personal observations; (98) K. Fiedler and U. Maschwitz, unpublished observations.

Defensive signals and survivorship

The idea of defensive-signal enhancement through gregariousness has been explored in many theoretical and empirical studies. Guilford (1990) points out that the details of predator–prey interactions are key to understanding the evolution of aposematic coloration. In an early discussion of possible 'predator conditioning' by gregarious larvae, Edmunds (1974) argued that predators adversely affected by ingesting one individual in a group learn to associate their resultant condition with the color patterns exhibited by adjacent larvae. Several theoretical and empirical studies have shown that vertebrate predators can learn to associate distastefulness with conspicuous coloration (see, for example, Brower 1958; Gittleman et al. 1980; Gittleman and Harvey 1980; Harvey et al. 1982; Roper and Redston 1987). Gregariousness could increase the contact rate between predators and aposematic prey, thereby facilitating predator association of warning coloration with unpalatability (Tinbergen et al. 1967; Smith 1974; but see Wiklund and Järvi 1982). Considerable theoretical work has addressed the importance of density-dependence and kin selection in the evolution of aposematism (see, for example, Fisher 1958; Harvey et al. 1982; Guilford 1985; Leimar et al. 1986; Mallet and Singer 1987).

Several studies have explored the effectiveness of group-displayed antipredator behaviors in larvae of Lepidoptera and symphytan Hymenoptera. Group displays generally include defensive regurgitation of noxious compounds and/or vigorous thrashing or flicking of the body (see, for example, Prop 1960; Lyons 1962; Myers and Smith 1978; Stamp 1984; Cornell et al. 1987; Peterson et al. 1987). Although it has often been suggested that aposematic coloration is also more effective in deterring predators in grouped vs. solitary situations (see, for example, Eisner and Kafatkos 1962; Young 1978; Pasteels et al. 1983), Vulinec (1990) points out that there is as yet no experimental evidence directly supporting this claim. Based on studies with gregarious and solitary aposematic caterpillars and bird predators, Sillén-Tullberg (1988, 1990) reports essentially no immediate or 'automatic' benefit accruing to gregarious vs. solitary prey, since gregariousness has initial costs in the form of increased predation risk per capita (cf. Cooper 1992). Insofar as clustering renders aposematic signals more apparent to visually hunting predators capable of such association, it may be expected to increase the efficacy of the aposematic defense. Although myrmecophilous lycaenid species do not appear to exhibit high levels of defensive thrashing (Malicky 1970), the efficacy of chemical and/or acoustic signals in attracting ant attendants that deter predators is improved in a group context, a pattern observed in both larvae and pupae of gregarious species (Pierce and Elgar 1985; Pierce et al. 1987; DeVries 1991).

Other evidence suggesting the importance of defensive-signal enhancement in the evolution of gregariousness and sociality comes from trait-distribution patterns and comparative phylogenetic analysis of some groups. In her review of insect aggregation and its defensive significance, Vulinec (1990) argues that aggregation evolved after other modes of defense such as chemical or structural predator deterrents. Aggregation may thus be seen as an adaptation that increases the effectiveness of signals inherent in warning coloration or structural defenses. This view is supported by the phylogenetic studies of Sillén-Tullberg (1988, 1993) and Sillén-Tullberg and Leimar (1988), who used comparative analyses to show that aposematic coloration in butterfly evolution probably precedes gregariousness. Sillén-Tullberg (1988) states that '... unpalatability is an important predisposing factor for the evolution of ... larval gregariousness in butterflies'. Gregariousness is thus seen to amplify pre-existing antipredator signals in many aposematic butterflies, a conclusion supported by reanalysis (Sillén-Tullberg 1993) for the effects of biassed characters on comparative studies. The presence of many gregarious non-aposematic lepidopteran and hymenopteran larvae (Table 20-1) suggests, however, that aposematism is not a prerequisite for gregariousness, but rather facilitates social evolution.

In summary, the evidence supports the idea that defensive signals are augmented in their expression, and therefore functional effectiveness, in group contexts. The defensive signals of social larvae are not merely directed at predators, but also include a pheromonal, tactile, or visual signal broadcast to the group and acting to coordinate defense.

Foraging signals and resource use

Social facilitation of feeding through foraging-related signals may occur in several ways: (1) by trail-based chemical communication, often exhibited by central-place foragers such as tent caterpillars and other lasiocampids (Fitzgerald 1976; Fitzgerald and Gallagher 1976; Weyh and Maschwitz 1978; Fitzgerald and Edgerly 1979; Carlberg 1980; Fitzgerald and Peterson 1983; Peterson 1987); (2) via synchronization of group feeding schedules (Fitzgerald 1980; Casey et al. 1988; Fitzgerald et al. 1988); (3) through orientation to group feeding sites (Stamp 1981a); and

(4) via group–enhanced establishment of feeding sites (Ghent 1960; Mizuta 1968; Shiga 1976; Tsubaki 1981; Tsubaki and Shiotsu 1982).

Efficiency of central-place foraging is improved as group size increases (see, for example, Fitzgerald and Costa 1986). Since rate of location and communication of resources to colony mates depends upon the number of searching individuals, group foraging efficiency will increase with number of foragers up to a point where density-dependent factors cause it to level off. As a result of information-sharing, central-place foragers capable of recruitment communication reduce average individual search and exposure time, thereby increasing overall survival probability and growth rate of colony members.

Feeding synchronization and group orientation to feeding sites may increase growth rate by raising overall activity levels (Long 1953) or by contributing to the consumption of high-quality food by recruitment to such food patches. Stamp and Bowers (1990a), for example, observed greater survivorship and smaller variance of biomass in larger vs. smaller groups of the saturniid *Hemileuca lucina*. In some species, such as eastern tent caterpillars, group foraging schedules may actually constrain individual feeding frequency, although these caterpillars still grow faster in a social context (Fitzgerald 1993).

As with defensive signals, enhancement of foraging-related signals involved in recruitment increases with the number of potential signalers: the rate of information exchange increases with the number of communicators. However, unlike defense where a signal is simply spreading through the group, recruitment-signal enhancement is manifested as decreased time taken for the average group member to locate food. Enhancement means, in this context, improved foraging through cooperative location of food, improvement stemming from the group-expression of search and recruitment behavior.

Caterpillar castes

Many social insects exhibit morphological or behavioral castes dividing the reproductive and labor effort of the colony. There have been no reports of morphological castes in caterpillar societies; this is not surprising insofar as such castes are generally found in long-lived or clonally reproducing social species. Several authors have, however, explored the possibility of polyethism in various social Lepidoptera. Wellington (1957, 1965) first raised the possibility of behavioral castes in his studies of intracolony foraging variation among *Malacosoma californicum pluviale* larvae. Wellington (1957, 1965) reported that many colonies are composed of relatively 'active' (type I) and 'inactive' (type II) larvae, apparently determined by the amount of yolk deposited in the egg; the type I larvae act as foraging 'leaders' while the type II larvae are 'followers', collectively creating a division of labor. Analyses of other *Malacosoma* species, however, failed to detect any consistent behavioral foraging differences among colony mates (Laux 1962; Franz & Laux 1964; Greenblatt 1974; Greenblatt & Witter 1976; Myers 1978; Shiga 1979; Edgerly & Fitzgerald 1982). Papaj & Rausher (1983) reanalyzed Wellington's (1965) data and found his conclusion of polyethism unsupported.

It thus appears likely that any behavioral variation among tent caterpillars of a given colony is stochastic and does not constitute even a weak division of labor. This conclusion is consistent with Edgerly & Fitzgerald's (1982) observation that *M. americanum* activity levels are normally distributed within colonies. A study of the saturniid *Hemileuca lucina* found that individual levels of activity also vary with age (Cornell *et al.* 1988). Responses to variables such as nutrition and disease are also likely to result in behavioral variation within colonies.

Some authors have treated social facilitation as a weak division of labor in some species (e.g. group-facilitated breaching of plant cuticular defenses by 'biter' larval castes (Ghent 1960; Iwao 1968; Tsubaki 1981)). Insofar as there is no consistent behavioral specialization among larvae, however, facilitation is stochastic and therefore does not reflect behavioral caste differentiation.

Social behavior and larval vulnerability

Non-ant-associated lepidoptera

The larval stage of Lepidoptera is a period of great risk and vulnerability to mortality factors such as predation, desiccation, and starvation. Some instars, however, are likely to be at greater risk than others. We next explore the idea of shifting vulnerability and its relevance to signal enhancement and caterpillar sociality. In what ways do social characters influence larval defense and growth? Several studies on larvae of Symphyta and Lepidoptera have considered the effects of larval size and grouping on survivorship and fecundity. The importance of survivorship is obvious; moreover, fecundity and mating success in the Lepidoptera are often intimately tied to larval size at pupation (Scriber and Slansky 1981; Haukioja and Neuvonen 1985, 1987; Barbosa *et al.* 1986; Boggs 1986; Wickman and Karlsson 1989; Haukioja 1993; Reavey 1993).

Larval size. Relative body size has been used as the basis for determining survival probability in various organisms (Stamp and Bowers 1991). Relative sizes of predators and prey are important determinants of predation levels, as both predator classes and their search modes change as larvae grow (Montllor and Bernays 1993). Early-instar larvae are usually attacked by invertebrate predators such as ants, spiders, stinkbugs and parasitoids (Ayre and Hitchon 1968; Morris 1972a,b; Tilman 1978; Evans 1983; Stamp 1986; Stamp and Bowers 1991), whereas mid– to late instars contend with larger invertebrate predators such as vespid wasps (Rabb and Lawson 1957; Morris 1976; Stamp and Bowers 1988; DeVries 1991) and vertebrate predators such as birds (Dempster 1967; Morris 1972a; Witter and Kulman 1972; Knapp and Casey 1986; Bernays and Montllor 1989; Heinrich 1993).

With respect to larval defense, bigger may be better for a variety of reasons. First, larger caterpillars have a narrower range of predators, since smaller, solitary assailants are often readily rebuffed (Sullivan and Green 1950; Morris 1963; Evans 1982). Second, in shelter-building species, incipient shelters are more easily penetrated by vertebrate or invertebrate predators than are the larger, stouter-walled shelters of older caterpillars. Third, it is often not until later instars that structural and chemical defenses constitute an effective defense. The spines or setae of structurally defended species are proportionately small and poorly developed in newly eclosed larvae, and many chemically defended species require time to accumulate secondary compounds. Such defense phenology may explain why many larvae are cryptic in early instars and only later display aposematic coloration or conspicuous clustering, a pattern observed in many animal species (Booth 1990). These observations suggest that the earliest larval instars are generally more vulnerable than later instars; we describe this early-instar period as a 'vulnerability window'. Early instars are also likely to be more vulnerable to abiotic mortality factors, such as drowning in rainstorms, freezing, or desiccation.

Larval grouping. Group–enhanced growth rates have been reported in a number of symphytan and lepidopteran larvae, including *Neodiprion* spp. (Ghent 1960; Lyons 1962; Henson 1965; Tostowaryk 1972), *Hyphantria cunea* (Watanabe and Umeya 1968; Morris 1976), *Malacosoma* spp. (Shiga 1976; Damman 1987; Peterson 1987), *Pryeria sinica* (Tsubaki 1981; Tsubaki and Shiotsu 1982), and *Halisidota caryae* (Lawrence 1990). It is difficult to disentangle the relative importance of different group-derived factors

influencing growth rate in social species. Since growth is a metabolic process, social characters or their byproducts that affect the location, feeding frequency, and assimilation of resources may be subject to increased efficacy as group size increases. For larvae, socially facilitated feeding may enhance growth rates in several ways: (1) overcoming plant structural defenses (Young and Moffett 1979; Ghent 1960; Tsubaki and Shiotsu 1982; Young 1983); (2) coordinating foraging (Stamp 1981a; Peterson 1987; Casey *et al.* 1988); or (3) constructing group shelters, which may create a favorable microclimate (Fitzgerald 1980; Fitzgerald and Willer 1983; Fitzgerald *et al.* 1988). Recruitment, a form of social facilitation, may improve foraging and growth rate (Peterson 1987) such that larvae more quickly exit the early-instar vulnerability window.

Other observations. Additional observations suggestive of differential vulnerability of larval instars include age-related changes in social behavior of larvae, and 'artificial' increase of egg-batches by adults. The integrity of social groups often erodes over the course of the larval stage, such that the penultimate or ultimate instars abandon the social group and become solitary, or the colony fragments into smaller units (see, for example, Carlberg 1980; Tsubaki and Yamamura 1980; Tsubaki 1981; Porter 1982; Hansen *et al.* 1984a,b; Cornell *et al.* 1987; Pierce *et al.* 1987; Fitzgerald *et al.* 1988; Stamp and Bowers 1988; Lawrence 1990). This phenomenon has been attributed to an easing of selective pressures favoring aggregation in early instars (Chansiguad 1964) or the increased food requirements of older larvae (Dethier 1959a,b; Porter 1982).

Considering that age-related independence is exhibited by many social species even when food appears to be abundant (Tsubaki 1981; Fitzgerald *et al.* 1988), it is likely that early pressures to function as an integrated, cooperative unit are counterbalanced by other factors as the larvae age, such as increased vulnerability to pathogens or predators. Late-instar increases in the cost : benefit ratio stemming from aggregation is consistent with the view that social behaviors are of greatest importance among early instars, which are both most vulnerable to predators and face the greatest hurdles in finding food and establishing feeding sites. For these species, the major benefits of social behavior occur during the early stages of colony growth (Fitzgerald 1993), and the mechanism leading to late-instar abandonment of the colony is likely to vary between species. Hochberg (1991a) points out that there are few instances of solitary early-instar larvae that preferentially associate when older, although one exception appears to be the pine webworm *Tetralopha robustella*, which solitarily mines pine needles in the first few instars and spins small communal tents in later instars (Hertel and Benjamin 1979; Johnson and Lyon 1988).

A second observation concerns oviposition pattern. A positive effect of group size on defense and resource use may make it advantageous for some species to oviposit near existing egg masses, thereby increasing group size at eclosion. Such an oviposition pattern has been observed in several social lepidopteran species (see, for example, Morris 1972b; DeVries 1977; Stamp 1981b; Fitzgerald and Willer 1983; Pierce and Elgar 1985), despite the fact that larger egg clusters sometimes suffer higher rates of parasitism (Stamp 1981b). An interesting mode of increasing batch size is 'social oviposition', in which at least two females simultaneously oviposit eggs in a cluster. This phenomenon has been observed in several species of *Heliconius* butterflies (Turner 1971; Mallet and Jackson 1980), and may be a consequence of resource limitation (i.e. uncommon or ephemeral resources are best exploited by batch laying) or represent a means of increasing group size. Benson *et al.* (1976) report a *Heliconius* cluster of over 800 eggs, a number Mallet and Jackson (1980) attribute to multiple females. To the degree that larval survival or growth rate improves with group size, adjacent or synchronous oviposition suggests that colony family structure is unimportant under some conditions relative to the need for rapid growth or enhanced defense.

Ant-associated Lepidoptera

The twofold advantage of appeasing ants that might themselves be potential predators, and attracting attendant ants that can serve as protective guards against predators and parasitoids, has been essential in the evolution of ant-associated Lepidoptera, especially those conforming to our definition of social Lepidoptera (Hinton 1951; Downey 1962; Ross 1964; Malicky 1970; Pierce and Mead 1981; Pierce and Easteal 1986; Fiedler 1991; Wagner 1993). In particular, Atsatt (1981a) argued that selection for 'enemy-free space' (Askew 1961; Gilbert and Singer 1975; Lawton 1978) has led to the elaborate mutualistic relationships exhibited by many ant-tended lycaenids; this may be especially true of social species. The concept of 'enemy-free space' can likewise be applied to the evolution of other ant-tended gregarious insects such as aphids and membracids in the Homoptera (see, for example, Way 1963; Nault *et al.* 1976; Wood 1977; McEvoy 1979; Bristow 1984).

Differences between lycaenids such as the social Australian species *Jalmenus evagoras* (whose caterpillars are tended by ants from the first instar) and other species such as the North American solitary lycaenid *Glaucopsyche lygdamus* (whose larvae are not strongly attractive to ants until the third instar) suggest that the cost : benefit ratio differs significantly between species for early instars. The dorsal organ, a gland producing sugary secretions, does not develop until the third instar in many lycaenids and riodinids (Clark and Dickson 1956; Ross 1964; DeVries 1988; Fiedler 1991). Production of secretions to appease and reward ants is expensive, and selection should favor the evolution of ant association only when the benefits of tending ants outweigh the costs of their attraction and maintenance. Selection favors early ant-association in *J. evagoras,* and alternative means of larval defense (such as crypsis and burrowing in flower buds) in *G. lygdamus*. Differences between species may be generated by a number of selective forces, including host-plant quality, pressure from predators or pathogens, and availability of alternative means of defense.

In the context of ant-mediated defense of plants, the period of greatest vulnerability to herbivores is thought to occur with the onset of foliar nectar production in ant-protected plants (Tilman 1978; O'Dowd 1979), although larvae of myrmecophilous riodinid butterflies may benefit from both feeding on leaves and drinking from the extrafloral nectaries of their hostplant (DeVries and Baker 1989).

In summary, the probability of mortality in larval Lepidoptera is generally greatest in the earliest instars. Within a given instar, larger groups generally suffer lower per capita mortality rates than smaller groups. Plots of hypothetical, generalized survivorship curves exhibit a general trend from concave-up to concave-down with increasing group size (Fig. 20-2). Group size improves survivorship through concomitant effects on body size and growth rate: early-instar vulnerability to predators and desiccation is inversely related to body size; group size (or simply social context) may increase growth rate, which in turn determines rate of passage to larger, less vulnerable instars.

SOCIAL EVOLUTION IN LEPIDOPTERA: COSTS AND BENEFITS OF LIVING IN LARVAL SOCIETIES

Costs of sociality

Group-living may lead to several types of cost to individuals, including (1) increased conspicuousness to predators, (2) increased transmission rates of pathogens, and

Generalized Survivorship Curve Continuum

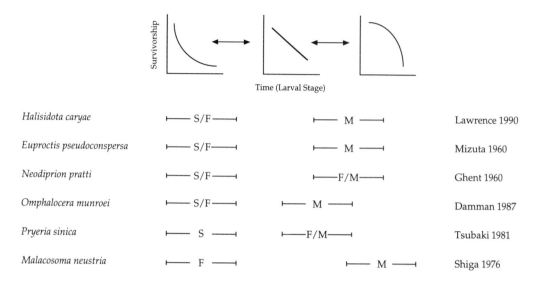

Halisidota caryae	⊢—— S/F ——⊣	⊢—— M ——⊣	Lawrence 1990
Euproctis pseudoconspersa	⊢—— S/F ——⊣	⊢—— M ——⊣	Mizuta 1960
Neodiprion pratti	⊢—— S/F ——⊣	⊢——F/M——⊣	Ghent 1960
Omphalocera munroei	⊢—— S/F ——⊣	⊢—— M ——⊣	Damman 1987
Pryeria sinica	⊢—— S ——⊣	⊢——F/M——⊣	Tsubaki 1981
Malacosoma neustria	⊢—— F ——⊣	⊢—— M ——⊣	Shiga 1976

Figure 20–2. The influence of group size on survivorship in some social Lepidoptera and Symphyta. Three survivorship curves are illustrated as points along a continuum. For purposes of comparison, larval colonies are divided into group-size classes (S = solitary, F = few, M = many); in some cases, 'many' is used to denote intact, natural colonies. Bars for each species and group size are positioned relative to the survivorship curve continuum. Larger groups tend to experience better survivorship as a result of the greater defensive and/or feeding capacity of aggregated individuals. Enhanced group survivorship in some species is in part attributable to accelerated larval growth.

(3) nutritional deficiency under conditions of resource limitation and ensuing competition. Factors leading to fitness trade-offs often interact in a complex fashion; for example, predators and parasites can reduce fitness directly through mortality and indirectly by interfering with feeding and metabolism. Thus, it is most convenient to divide trade-offs into two sections: (1) disease, and (2) predators, parasites, and nutrition.

Disease
Temporal and spatial gregariousness can incur costs through density-dependent controls. Survivorship of social species is reduced, for example, by group–enhanced risk of contracting disease (see, for example, Bucher 1957; Payne *et al.* 1981; Hochberg 1991a,b) or by attracting certain types of predator and parasite (see, for example, Stamp 1981a,b; Knapp and Casey 1986; Pierce *et al.* 1987; Hieber and Uetz 1990; Rosenheim 1990). The role of disease costs in insect social evolution is understudied relative to that of predation, perhaps because the effects of the two are difficult to disentangle (diseased individuals may be less able to

defend themselves against predators). Pathogens are likely to have played a role in shaping sociality in some Lepidoptera, notably in cases where disease risk is influenced by foraging pattern (and hence social complexity).

Another important disease factor is mode of transmission. Pathogens such as certain *Bacillus* bacteria that are transmitted by physical contact pose a greater risk for social groups than those such as polyhedrosis viruses that must be ingested to infect the host. Hochberg (1991a) found that an increase in disease resistance with larval age is more frequently observed in gregarious than in solitary lepidopteran species, which suggests that viruses have historically exerted a selective pressure on social caterpillars.

Predators, parasites and nutrition
Some of the best-studied costs of sociality include conspicuousness to predators and resource competition. Predators that employ foraging strategies involving repeated return to successful foraging sites or intensified searching in the immediate area where prey are encountered place aggregated prey at increased risk relative to solitary prey

(Taylor 1977a,b; Vermeij 1982; Kareiva and Odell 1987; Vulinec 1990). When resources are limiting, aggregation can also incur costs due to increased intraspecific competition and indirect effects on predation. Colonies may deplete their food supply under conditions of high local population density, forcing larvae to abandon the host plant in search of alternative hosts and potentially resulting in significant mortality (Dethier 1959a,b; Chew 1977; Tsubaki and Shiotsu 1982; Stamp 1984).

Trade-offs are often manifested in the foraging pattern of social larvae. Many studies have considered the effects of such factors as predation, food quality, temperature regime, and resource use on caterpillar foraging and developmental patterns, providing measurements of variables such as mortality, developmental rate, mass at pupation, fecundity, etc., in both the laboratory and the field (see reviews by Hassell and Southwood 1978; Scriber and Slansky 1981; Wickman and Karlsson 1989; Montllor and Bernays 1993; Stamp 1993). These studies indicate that larvae exhibit maximum growth rates when feeding on high-quality resources under thermally optimal, 'enemy-free' conditions (Brower 1958; Holloway and Herbert 1979; Price et al. 1980; Schultz 1983). Such conditions are rarely met in the natural world; how and why do larvae deviate from a hypothetical foraging optimum?

Suboptimal conditions include unfavorable climate, low host-plant quality, and predation, often interacting in a complex and interrelated manner (Stamp 1993). Although environmental conditions, plant defenses and leaf quality are recognized as important selective forces, predation risk may have a more immediate effect in influencing the exposure of herbivores to these forces. For example, predation has been hypothesized to play a key role in selecting for caterpillar activity at suboptimal temperatures (such as nighttime foraging to escape diurnal predators) (Heinrich 1979, 1993; Schultz 1983; Fitzgerald et al. 1988), and a large body of evidence indicates that predation can constrain host-plant choice as well as the quality of leaves consumed on a chosen plant. In a series of studies on Hemileuca lucina, Stamp and Bowers (1988, 1990a,b, 1991) demonstrated that harassment by vespid wasp predators reduced caterpillar growth and survivorship by interfering with caterpillar feeding; larvae were frequently induced to move to the shaded host-plant interior where only poor, mature leaves were available. The harassment phenomenon has also been observed in the ecology of other insects, including tent caterpillars (Knapp and Casey 1986), odonates (Heads 1986) and hemipterans (Sih 1980, 1982). In other studies, predation pressure was implicated

in the preference for nutritionally inferior, old, or damaged leaves by the oecophorid leaf-roller Diurnea fagella (Hunter 1987) and the leaf-tying pyralid Omphalocera munroei (Damman 1987) because of the greater defensive potential of these leaves as shelters.

Benefits of sociality

Passive defense and growth effects

It is often difficult to separate communicative and non-communicative factors that affect group defense and resource use. Passive or non-communicative mechanisms contributing to defense include 'group dilution effects' (Hamilton 1971; Turner and Pitcher 1986; Sillén-Tullberg and Leimar 1988; Lawrence 1990; Wrona and Dixon 1991) and early warning against predators (Treherne and Foster 1980, 1981, 1982; Vulinec 1990). Group dilution refers to the 'safety in numbers' concept, whereby a given individual is less likely to be taken by a predator when standing with a group than when alone (Hamilton 1971). In principle, the bigger the group, the more effective the dilution effect. Exceptions to this pattern have been reported, however. For example, Stamp (1981a,b) found that medium-sized groups of larval Euphydryas phaeton experienced lower rates of parasitism than smaller and larger groups. Conversely, Subinprasert and Svensson (1988) observed that the smallest and largest egg clutches of Laspeyresia pomonella had high survivorship compared with medium-sized clutches. In general, however, both survival probability and growth rate tend to be positively correlated with colony size up to a certain point (Evans 1982; Porter 1983; Stamp and Bowers 1988), a phenomenon that is the product of both protective mechanisms and accelerated developmental rates.

Inclusive fitness effects

Genetic mechanisms may play a role in the evolution of cooperation, but there are almost no genetic studies of social lepidopteran species. The selective strength of ecological pressures may be (or historically have been) severe enough to favor cooperation regardless of genetic relatedness between interactants. Indirect fitness will be greater than zero whenever patterns of mating and sperm utilization result in family structure (kinship) within groups. In Lepidoptera, the greatest degree of family structure, full sibships, obtains when colonies are derived from a single batch of eggs (i.e. comprise a single matriline) and the ovipositing female has mated once or uses sperm from one male.

One way to determine the likely importance of kin selection in the evolution of social behavior is to establish whether mechanisms either preserving or undermining group family structure exist, since family structure is integral to the operation of kin selection. Kin discrimination may be the most common mechanism preserving family structure (Fletcher and Michener 1987), whereas structure is undermined by adjacent or synchronous oviposition by mixing family groups (if ovipositing females are unrelated).

In the only study to date addressing kin discrimination in a social caterpillar, Costa and Ross (1993) inferred a lack of kin discrimination among eastern tent caterpillars from their observations of eroding family structure. This erosion occurs through stochastic fusion and fission of unrelated colonies foraging together on the same tree. Despite mixing, however, mating and oviposition in this species set up conditions of both high relatedness within colonies and low colony density on trees, effectively preserving some family structure throughout the larval stage. Thus, insofar as indirect fitness is consistently greater than zero, it is likely to have played some role in *M. americanum* social evolution. The important point is that inclusive fitness effects may result from overt behavioral mechanisms or may be byproducts of behavior and population biology.

Sawfly larvae belong to a group in which inclusive fitness effects are most likely, yet are among the least complex 'caterpillar' societies. In theory, inclusive fitness effects are more readily realized in sawflies because of the relatedness asymmetry of the haplodiploid sex determination system of Hymenoptera; such asymmetries are thought to be key in the evolution of eusocial hymenopteran societies (Wilson 1975; Trivers and Hare 1976). Social communication and interaction in larval sawfly societies is apparently limited to group alarm and defense, and group cohesion. To our knowledge, there are no examples of recruitment communication in sawflies, although a few patch-restricted species construct silken structures. These include species in the pamphiliid genera *Neurotoma*, *Acantholyda* and *Cephalcia*, various members of which are called the 'web-spinning' or 'pine-webbing' sawflies (Peterson 1962; Johnson & Lyon 1988). Many gregarious sawflies are aposematic and exhibit the characteristic sawfly group-defensive behavior of rearing and regurgitating.

Signal enhancement and cooperation

Some authors view predation and parasitism as the major selective force leading to social evolution in insects and other animals (see, for example, Hamilton 1971; Michener 1974; Pulliam and Caraco 1984; Turner and Pitcher 1986; Inman and Krebs 1987; Strassmann *et al.* 1988). Others, focussing primarily on the behavior of larvae in the Symphyta and Lepidoptera, have stressed the importance of social facilitation in feeding (Ghent 1960; Shiga 1976; Tsubaki and Shiotsu 1982; Young 1983). As discussed above, sociality can simultaneously facilitate passive and active defense, thermoregulation, and foraging efficiency in both ant-associated and non-ant-associated contexts. None of these selective factors are mutually exclusive; the most important factor in social evolution in the Lepidoptera is likely to be the enhancement of signals that collectively bear on both group defense and resource use, perhaps providing rapid growth through the vulnerable early larval stages.

DISCUSSION

Lepidopteran social evolution: factors and scenarios

Life history and ecology

No single feature of the ecology, development, genetics, or behavior of social Lepidoptera sets them apart from other social taxa. Ecological factors such as host specificity and voltinism are not consistent predictors of social behavior (Table 20-1), and there appear to be no unusual genetic attributes of Lepidoptera that favor cooperation in the sense that this group lacks the relatedness asymmetry of haplodiploidy and the genetic identity of parthenogenesis. None the less, we identify a suite of life-history and ecological traits collectively shaping and uniquely defining sociality in the order.

Table 20-2 summarizes the key behavioral, life-history, and ecological characteristics of social insects and arachnids. Comparing the social forms of these groups, important similarities and differences are apparent. Virtually all social forms exhibit group, or at least family, defense and cohesion, and communication by tactile or chemical means is nearly universal. Ecological conditions such as predation and resource distribution has resulted in interesting parallels between social Lepidoptera and other social taxa. For example, patch-restricted foragers, found among such diverse taxa as aphids, termites, caterpillars, sawflies and embiids, live in or on their food; recruitment communication, associated with patchy resource distribution, is found in the ants, bees, wasps, caterpillars and termites.

Table 20-2. *Life-history and communication characteristics of social insects and arachnids*

Life-history defines generationally the relationship of social interactants (e.g., parent–parent, parent–offspring, sib–sib). Societies are further shaped by ecological factors influencing defensive and foraging traits. Lepidopteran societies lack parental interaction; communication occurs within larval cohorts, and includes the contexts of foraging, group defense, and group cohesion. See text for discussions of foraging patterns, group defense, and group cohesion. Brood care is broadly defined as defense and/or feeding of immatures by one or both parents.

Taxon	Brood care	Perennial	Foraging Pattern[a]			Group defense	Group cohesion	Nestmate recognition	Modes of communication[b]
			PR	Nom	CP				
Hymenoptera (Eusocial)	Yes	Many	?	Many	Most	Yes	Yes	Yes	V, C, T, A
Hymenoptera (Symphyta)	No[c]	No	Yes	Yes	No	Yes	Yes	?	?V, ?C, ?T
Lepidoptera	No[d]	No	Many	Many	Some	Yes	Yes	?	?V, C, T, A
Coleoptera[e]	Yes	Yes	Yes	No	No	Yes	Yes	?	?C, ?T, A
Thysanoptera	Yes	No	Yes	No	No	Yes	?	?	C
Hemiptera[f]	No	No	Yes	No	No	Yes	?	?	C
Psocoptera	No	Yes	Yes	Yes	No	?	Yes	?	T, ?C, ?A
Zoraptera	No	Yes	Yes	No	No	No	Yes	No	?C, T
Embioptera	Yes	Yes	Yes	No	No	No	?	?	?C, ?T
Dictyoptera[g]	Yes	Yes	Yes	No	Some	Yes	Yes	Yes	C, T
Araneae	Yes	Yes	Yes	No	No	Yes	Yes	?	C, T

[a] PR = patch-restricted; Nom = nomadic; CP = central-place.
[b] V = visual; C = chemical; T = tactile; A = acoustical.
[c] See Dias (1975, 1976, 1982) and Morrow *et al* (1976).
[d] For an exception, see Nafus and Schreiner (1988).
[e] Passalidae.
[f] Aphididae.
[g] Isoptera.

In defensive terms, social Lepidoptera lack soldier castes, but share group-defensive displays with gregarious sawflies and pleometrotic Hymenoptera.

Among the many taxon-specific differences, two general features are apparent. First, the demographic structure of social insect colonies is defined generationally, splitting into those with overlapping generations and those comprising a single-generation cohort. The former colonies are usually perennial or multivoltine; the latter tend to be univoltine. This distinction is important because demographic composition determines the possibility of such social traits as parental care. Second, the communication complexity of social insect colonies is related to foraging pattern. Life-history traits delimit the essential structure and composition of insect societies while ecological factors provide the selective regime favoring particular types of social interactions.

Lepidopteran societies are among the simplest of social insects in terms of demographic composition (typically single-generation) and lifespan (usually annual), while in many cases sharing communication features of more complex social species (such as recruitment). The route of social evolution in many social insects is hypothesized to have begun with a maternal care phase, subsequently elaborated with morphological or behavioral specialization among siblings cooperating in the care and rearing of brood. Lepidopteran adults rarely interact with larval aggregations (but see Nafus and Schreiner 1988), typically abandoning their eggs after perhaps concealing or coating them with accessory-gland secretions or abdominal setae. The absence of adults in most lepidopteran societies also means that they tend to be ephemeral, since eggs are not replenished and the colony exists only as long as the larvae take to mature. The relative simplicity of lepidopteran societies follows from the general lack of parental care or even parental presence, as the parent–offspring communication and reproductive-based cooperation found in many other social taxa are precluded.

Resource use appears to be a factor shaping social complexity in this order. Social interactions beyond alarm and defense are unnecessary for species living in or on abundant resources. For many larvae, seemingly-abundant hostplant leaves are not equally acceptable. Often, larvae can survive on only the youngest foliage (see, for example, Fitzgerald and Peterson 1983; Peterson 1987; Fitzgerald 1993); such leaves are patchily distributed on the host plant, and their exploitation depends on frequent movement or recruitment. The correlation of central-place foraging (and recruitment communication) with patchy resources is quite general among social insects, exhibited by members of such taxonomically widespread social groups as ants, bees, wasps, termites, and butterfly and moth larvae.

The difference in foraging, trail-marking, and trail perception between eastern and forest tent caterpillars, two of the best-studied social lepidopteran species, illustrates how shifts in foraging and communication directly relate to social evolution in this order. As discussed above, these closely related species mark trails before and after feeding. The prefeeding trails are termed 'exploratory trails' (Fitzgerald and Peterson 1983) and may be homologous to the 'personal trails' of trail-marking solitary species (see, for example, Weyh and Maschwitz 1982; Tsubaki and Kitching 1986). Both eastern and forest tent caterpillars deposit postfeeding trails as well. The crucial difference in the social complexity of these species lies in their use of postfeeding trails: the fixed base (tent) of eastern tent caterpillars provides a predictable communication center, setting up conditions for recruitment. By contrast, the postfeeding trails of forest tent caterpillars are as likely to lead to a new resting site as to the site of origin, undermining the use of these trails in recruitment communication.

One scenario for social evolution in Lepidoptera involves the context-elaboration of trail-marking: ancestrally solitary species may have used trails to keep track of food–sites. In groups, trail-marking may initially have played an identical 'personal' function, with group-cohesion simply stemming from mutual marker-recognition. Natural selection may have subsequently favored communication through changes in trail perception (e.g. preference for trails left by colonymates) and behavior (e.g. repeated return to a fixed base). Over evolutionary time, group–enhanced expression of defensive and resource-based signals may have further aided to integrate larvae into a cohesive society. The change from simple webbing to nests reflects elaboration from a purely protective use to a more or less permanent, stable retreat, which simultaneously serves as an information center for foraging-related communication.

Evolution of oviposition patterns

Non-ant-associated caterpillars. Fitzgerald and Costa (1986) and Fitzgerald and Peterson (1988) suggested a general evolutionary pathway for social evolution in the Lepidoptera whereby oviposition patterns facilitating larval encounters were selectively favored as a result of benefits accruing to larvae in chance groupings. In this scenario, the initial benefits of grouping were passive, perhaps involving such factors as predator dilution effects, amplified aposematic signals, and enhanced thermoregulation. This scenario implies that larval success has selectively favored batch oviposition in the Lepidoptera, although eggs may incipiently have been loosely clustered if not specifically batch-laid, as a result of resource limitation.

The inverse pathway was proposed by Hebert (1983), who suggested that the evolution of egg-clustering evolved in response to energetic considerations related to adult feeding habits, and that once egg-clustering evolved, group-favorable behavior and communication could be selected. The crux of Hebert's (1983) argument is the positive correlation of egg-clustering with reduced or absent adult mouthparts. However, this correlation largely occurs along taxonomic lines, and the two characters may be phylogenetically non-independent. In addition, there are many examples of batch oviposition by species capable of feeding as adults.

Courtney (1984) and Stamp (1980) studied batch versus single oviposition in butterflies, and reached different conclusions regarding the evolution of egg clustering. Courtney (1984) argued that the most important benefit to batch-layers is greater fecundity resulting from reduced adult search time, whereas Stamp (1980) argued for protection against desiccation and enhanced defense among other benefits, noting that most species ovipositing in clusters have at least some aposematic larval instars, and many have aposematic eggs. In our view, defensive and larval foraging benefits are probably of greatest importance to the evolution of batch oviposition. It is difficult to evaluate the energetic arguments, since experiments linking fecundity and lifetime reproductive success to oviposition pattern are lacking. The occurrence of aposematic eggs argues for a defensive function, though here, too, experiments evaluating egg predation rates for aposematic vs. non-aposematic batched and singleton eggs are lacking.

Ant-associated caterpillars. Lycaenid gregariousness leads to enhancement of defensive alarm signals just as in non-ant-associated larvae, and it is possible that ant attendance in general permits foraging on high-quality food under thermally beneficial conditions by deterring predators. Because obligately myrmecophilous lycaenids are dependent upon both suitable host plants and attendant ants for survival, resource limitation may have played an important role in the evolution of aggregation behavior in these species. Females of certain myrmecophilous taxa have been shown to use ants and conspecific larvae as cues in oviposition (Atsatt 1981b; Pierce and Elgar 1985; Mathews 1993); in some cases, females deposit larger egg batches in the presence of ants (Atsatt 1981b). Larval vulnerability combined with patchy distribution on limited resources may thus have given rise to active aggregation, in a scenario similar to that proposed by Fitzgerald and Costa (1986). Kitching (1981) pointed out that egg-clustering in lycaenids is often observed in obligate myrmecophiles, especially in Australia, and argued that a causal relationship between the two is likely.

Larval aggregation in ant-tended lycaenids may have played a role in the evolution of species-specificity in lycaenid–ant interactions. Any ant species whose workers are sufficiently good tenders that larvae survive and develop will receive enhanced oviposition by butterflies, because ovipositing females of the aggregating lycaenids are attracted to conspecific larvae. If a particular ant is a consistently strong tender, then selection may favor recognition by ovipositing females of that ant species (by visual or olfactory cues), even in the absence of conspecific larvae (Elgar and Pierce 1988). This may account for the high degree of species-specificity in ant association observed among Australian lycaenids whose larvae aggregate. A curious feature of the Lycaenidae that deserves mention with respect to the evolution of aggregation behavior is that many of the species whose larvae are solitary are also cannibalistic; an important precondition to aggregation behavior in the Lycaenidae is absence of cannibalistic behavior.

Sociality in ant-associated Lepidoptera is, unlike non-ant-associated species, attributable to a particular defensive strategy: employing ant attendants for protection from predators and parasitoids. Because the ants themselves are aggregated, and the lycaenids must rely upon the coincidence of ants and host plants, limitation of both defense and food availability has promoted sociality in these taxa. The rare occurrence of social species amongst the Poritiinae suggests that the trait may have been lost and regained several times.

This 'defensive route' of social evolution is undoubtedly shared by many social Lepidoptera that do not associate with ants, the defenses of which include refuge shelters and chemical and structural deterrents. The ant-associated species are remarkable in employing ants as their primary defense.

Scenarios for lepidopteran social evolution

Batch oviposition is necessary but not sufficient to ensure social interaction. In this sense whether ancestral Lepidoptera laid eggs singly or in batches may not be as important as the selective milieu in which the eggs were laid. Eggs may be deposited in batches owing to adult energetic limitations, resource patchiness, or for unknown historical reasons, but the grouped larvae may disperse, behave antagonistically (e.g. cannibalism) or remain spatially associated upon eclosion. The selective regime experienced by particular species may favor one or the other response; initially 'passive' associations may then experience selective pressures leading to disruption or elaboration of social behaviors. For example, once grouped, larvae are more conspicuous to predators, and increased predation may select for dispersal or socially mediated defenses such as protective webbing or leaf-tying, or coordinated anti-predator behavior.

In attempting to understand the evolution of sociality in Lepidoptera, as well as the transitions between social forms, it is important to note that resource use is intimately connected to foraging pattern. We suggest that nutritional and defensive needs jointly determine the particular pattern of sociality and foraging for a given species. Nomadic (N), patch-restricted (PR) and central-place (CP) foraging hold very different implications for both nutrition and defense: wandering larvae (N and CP foragers) can choose which leaves they eat, seeking profitable patches. Patch-restricted foragers have less choice, feeding on their shelter from within if it is constructed of leaves (e.g. *Hydria prunivorata*) or, if constructed of silk, expanding their shelter to engulf nearby leaves as food becomes exhausted (e.g. *Hyphantria cunea*). In defensive terms, leaf-tying PR foragers rely on their shelter and are often cryptically colored, whereas those inhabiting silken structures

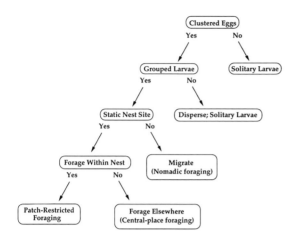

Figure 20–3. Heuristic summary of character states relevant to lepidopteran sociality, presented as a series of either/or options: eggs are deposited singly or in batches, larvae in batched clutches may disperse or group, grouped larvae nest in place or migrate, and larvae nesting in place forage *in situ* (expanding their patch as food is exhausted) or forage elsewhere (returning to the nest site following each foraging bout). This diagram is not intended as an explicit evolutionary scenario.

are often chemically and structurally defended, as N and CP foragers often are (although this needs testing for many groups).

The differences in foraging and defensive ecology of lepidopteran social classes lend themselves to an analysis of social evolution through phylogenetic hypothesis-testing. Fig. 20-3 is a heuristic chart of social states: eggs are deposited either singly or in batches; if in batches, larvae may either disperse or group; grouped larvae either remain in place or migrate; and larvae remaining in place either forage *in situ* (continually expanding the spatial bounds of the patch as food is exhausted) or forage elsewhere (returning to the nest site following each foraging bout). Note that this chart is not intended as an evolutionary scenario; rather, it summarizes the relevant states and their relationships, serving as a starting point for framing such scenarios. Following the framework presented in Fig. 20-3, we divide social-evolutionary hypotheses into two groups: (1) hypotheses concerning the evolution of gregariousness from solitary ancestors; and (2) hypotheses concerning the evolution of particular social states and transitions among these states once gregariousness is achieved. This division is made for convenience, as the two groups are actually part of a continuum.

Social evolution from solitary ancestors.   The only way in which solitary caterpillars systematically differ from social caterpillars is in their solitariness; solitary species, like their social relatives, may be aposematic, sequester and/or regurgitate toxic chemicals, be spiny, hairy, diurnal, nocturnal, uni- or multivoltine, be host-specific or feed broadly. Crypsis is the most apparent lifestyle or character-state consistently differing in frequency between solitary and social caterpillar species, although this depends on how some social species, such as many leaf-tying PR foragers, are scored. Many PR species are certainly not aposematic, if not cryptic *per se*. None the less, clear examples of crypsis and mimicry occur throughout the solitary Lepidoptera and have no counterpart among social species (e.g. twig-mimicking geometrids, leaf–edge-mimicking notodontids, cryptic catocaline noctuids).

Morphological and behavioral defensive and foraging traits such as possession of spines or leaf-tying may be phylogenetically correlated (Table 20-1), suggesting that some ancestral traits may in some cases 'predispose' the evolution of certain social forms over others in a given clade. In other words, have particular solitary lifestyles given rise to particular social lifestyles, or vice versa? For example, are aposematic N foragers consistently derived from ancestrally cryptic or aposematic solitary foragers? Are PR foragers living within webbed leaves derived from solitary leaf-tiers? An important factor leading to different PR strategies may have been ancestral body size. Leaf-tying is observed in a diversity of solitary microlepidoptera, a behavioral trait that was likely to have been preserved and elaborated in a social context. Larger body sizes require active foraging because food is likely to become exhausted locally, leading to either N or CP social systems.

Societies marked by N or PR foraging may be more likely to have evolved from solitary ancestors before CP foraging societies, since the former possess fewer social characters (group cohesion and defense). The solitary to N transition requires simple batch oviposition and group cohesion cues. The transition to PR requires the development of one of two shelter-building classes: leaf structures or silk structures. These 'routes' of social evolution hold different implications for resource use and defense. The question of whether aposematic, chemically defended ancestors more likely to give rise to N foragers could be tested in groups exhibiting the full range of social interactions (solitary plus the three social systems) by mapping foraging or social pattern onto independently generated phylogenies.

Good candidate groups for such analyses, once reliable and largely complete lower-level phylogenies are known, are the Pyraloidea and the Pierinae (see Table 20-1).

Branching order of species typified by different states permits inference of the most likely transitions within a given clade, whether those transitions entail a gain or loss of social characters. The North American pamphiliid sawfly genus *Cephalcia*, for example, has ten species, five of which are gregarious web-spinners; the solitary species construct silken tubes for shelter and the gregarious species are PR foragers (Johnson & Lyon 1988). Knowledge of the phylogenetic branching order of these species could be used to determine whether sociality in this group has arisen from solitary ancestors or vice versa, or whether there has been a more complex pattern of gain and loss of social characters.

Transitions between lepidopteran social forms. Once sociality has arisen, all transitions are possible, though some may be more likely than others. A shift from N to either CP or PR foraging, for example, involves a gain of shelter-building behaviors, but CP foraging also requires the extra step of some form of chemical bookkeeping to relocate the shelter. Trail-marking pheromones are often subsequently used for recruitment in CP foragers. The same analysis described above for exploring the solitary to social transition could also be used to ask whether CP foraging arises from N or PR foraging systems, or vice versa, in particular clades. Groups with a range of social forms (e.g. Saturniide or Thaumetopoeidae, with solitary as well as different social systems represented) hold special promise for this approach.

Focussing on the relationship between nutritional requirements and sociality, we predict that within-host diet breadth will vary with foraging pattern, and thus with social form. CP foragers such as tent caterpillars are often 'leaf specialists' that recruit preferentially to young, newly expanding foliage, whereas PR foragers tend to be leaf generalists in the sense that their feeding is confined to patches of foliage varying in age and nutritional quality. For a given group, host selection and growth experiments, or simple observation, can establish whether member species are leaf specialists or generalists; the co-occurrence of leaf specialization and CP foraging can then be statistically tested by mapping host-use states onto independently derived phylogenies.

Moreover, the evolutionary order of leaf specialization and CP foraging in a given lineage can be useful in inferring whether resource use was more important than defense in the evolution of CP foraging in that lineage,

because the food of leaf specialists is patchily distributed and more efficiently exploited through recruitment communication whereas leaf generalists are presented with an abundance of food. CP foraging in the absence of leaf specialization suggests a defensive role for the nest structure.

There are many other social–evolutionary scenarios that may be tested phylogenetically. For example, we expect shifts in foraging or social pattern to be accompanied by shifts in defense. In other words, are some defenses characteristic of certain foraging or social patterns, such as crypsis with leaf-tying, or aposematism with silk shelter-building or nomadic foraging? There are also patterns worthy of investigation *within* social classes; are there consistent ecological, behavioral or morphological differences between different PR strategies (i.e. leaves vs. silk)?

AVENUES FOR FURTHER INQUIRY

We identify two complementary areas requiring further research. First, comparative phylogenetic approaches will help to assess coincidence of social and ecological characters, as well as patterns of gain and loss of social characters. Several specific questions in need of attention were discussed in the previous section, ranging from the evolution of sociality from solitary ancestors to transitions among social forms and correlated changes in other life-history, defensive or behavioral traits. There are many groups marked by both solitary and social species, often with all three social systems (e.g. Lycaenidae, Nymphalidae, Pierinae, Lasiocampidae, Pyraloidea, Thaumetopoeidae, several saturniid subfamilies); (see Table 20-1). These groups can be used to address the relationship between solitary lifestyles and the social forms to which they are most likely to give rise, notably the importance of solitary defensive and host-use patterns in shaping these parameters in social species. The same taxa can simultaneously serve as focal points for investigations of evolutionary transitions among social forms. Shifts between N, PR and CP foraging are expected to exhibit clade-specific patterns, but may also entail predictable correlated shifts in defense and host use.

Second, a great deal of empirical research is necessary to fill in the gaps in our knowledge of larval ecology and behavior. The characters defining lepidopteran sociality must be better understood before we will be in a position to apply this knowledge to the slowly but steadily accumulating phylogenetic data. The most critical of these characters concern communication mechanisms and the types of larval interactions mediated by communication. Significantly, the communication abilities of the vast majority of social Lepidoptera are altogether unknown, as is the ecological context of such communication (e.g. to what degree do narrow host (or *intra*host) requirements influence the evolution of recruitment communication by creating conditions of patchy resource distribution?). Finally, in terms of behavioral interactions mediated by communication, very little is known about genetic relatedness and kin discrimination abilities of colonymates. Such information is integral to evaluating the potential importance of inclusive fitness and kin selection in the maintenance and evolution of caterpillar societies; this avenue of research is virtually unexplored in the Lepidoptera compared with the wealth of such studies on eusocial Hymenoptera and other social groups.

ACKNOWLEDGEMENTS

We thank Andrew Berry, Jae Choe, Bernie Crespi, Phil DeVries, Terry Fitzgerald, Kathrin Sommer, Nancy Stamp and Diane Wagner for their valuable comments, criticisms, and suggestions. Jeremy Holloway, Kevin Tuck, Hiroshi Yoshimoto, and especially Konrad Fiedler and Rolf Oberpreiler kindly lent their expertise to the improvement of Table 20-1.

LITERATURE CITED

Askew, R. R. 1961. On the biology of the inhabitants of oak galls of Cynipidae (Hymenoptera) in Britain. *Trans. Soc. Br. Entomol.* **14**: 237–268.

Atsatt, P. R. 1981a. Lycaenid butterflies and ants: selection for enemy-free space. *Am. Nat.* **118**: 638–654.

–. 1981b. Ant-dependent food plant selection by the mistletoe butterfly *Ogyris amaryllis* (Lycaenidae). *Oecologia (Berl.)* **48**: 60–63.

Ayre, G. L. and D. E. Hitchon. 1968. The predation of tent caterpillars *Malacosoma americana* [sic] (Lepidoptera: Lasiocampidae) by ants (Hymenoptera: Formicidae). *Can. Entomol.* **100**: 823–826.

Balfour-Browne, F. 1925. The evolution of social life among caterpillars. In *Proceedings of the Third International Congress on Entomology*, Zurich, Switzerland. K. Jordan and W. Horn, eds., pp. 334–339. Weimar: G. Uschmann.

Barbosa, P., P. Martinat and M. Waldvogel. 1986. Development, fecundity and survival of the herbivore *Lymantria dispar* and the number of plant species in its diet. *Ecol. Entomol.* **11**: 1–6.

Barlow, H. S. 1982. *An Introduction to the Moths of South-East Asia.* Kuala Lumpur: H. S. Barlow.

Benson, W. W., K. S. Brown and L. E. Gilbert. 1976. Coevolution of plants and herbivores: passionflower butterflies. *Evolution* **29**: 659–680.

Bernays, E. A. and C. B. Montllor. 1989. Aposematism of *Uresiphita reversalis* larvae (Pyralidae). *J. Lepid. Soc.* **43**: 261–273.

Boggs, C. L. 1986. Reproductive strategies of female butterflies: variation in and constraints on fecundity. *Ecol. Entomol.* **11**: 7–15.

Booth, C. L. 1990. Evolutionary significance of ontogenetic colour change in animals. *Biol. J. Linn. Soc.* **40**: 125–163.

Bristow, C. M. 1984. Differential benefits from ant attendance to two species of Homoptera on New York ironweed. *J. Anim. Ecol.* **53**: 715–727.

Brooks, M. 1991. *A Complete Guide to British Moths.* London: Jonathan Cape.

Brower, L. P. 1958. Bird predation and food plant specificity in closely related procryptic insects. *Am. Nat.* **92**: 183–187.

Brown, F. M. and B. Heineman. 1972. *Jamaica and its Butterflies.* London: E. W. Classey Ltd.

Bucher, G. E. 1957. Disease of the larvae of tent caterpillars caused by a sporeforming bacterium. *Can. J. Microbiol.* **3**: 695–709.

Bush, G. L. 1969. Trail laying by larvae of *Chlosyne lacinia.* *Ann. Entomol. Soc. Am.* **62**: 674–675.

Capinera, J. L. 1980. A trail pheromone from the silk produced by larvae of the range caterpillar *Hemileuca olivae* (Lepidoptera: Saturniidae) and observations on aggregation behavior. *J. Chem. Ecol.* **3**: 655–664.

Capinera, J. L., L. F. Weiner and P. R. Anamosa. 1980. Behavioral thermoregulation by late-instar range caterpillar larvae *Hemileuca oliviae* Cockerell (Lepidoptera: Saturniidae). *J. Kansas Entomol. Soc.* **53**: 631–638.

Carlberg, U. 1980. Larval biology of *Eriogaster lanestris* (Lepidoptera, Lasiocampidae) in SW Finland. *Notulae Entomol.* **60**: 65–72.

Carter, D. 1982. *Butterflies and Moths in Britain and Europe.* London: British Museum of Natural History.

Casey, T. M., B. Joos, T. D. Fitzgerald, M. E. Yurlina and P. A. Young. 1988. Synchronized group foraging, thermoregulation, and growth of eastern tent caterpillars in relation to microclimate. *Physiol. Zool.* **61**: 372–377.

Chansiguad, J. 1964. Observations preliminaires sur les essais d'infestations artificielles avec *Pieris brassicae* L. au stade larvaire dans les conditions naturelles. *Rev. Zool. Agr.* **63**: 55–61.

Chermock, R. L. and O. D. Chermock. 1947. Notes on the life histories of three Floridian butterflies. *Can. Entomol.* **7**: 142–144.

Chew, F. S. 1977. Coevolution of pierid butterflies and their cruciferous foodplants. II. The distribution of eggs on potential foodplants. *Evolution* **31**: 568–579.

Clark, G. C. and C. G. C. Dickson. 1956. The honey gland and tubercles of larvae of the Lycaenidae. *Lepid. News* **10**: 37–43.

–. 1971. *Life Histories of the South African Lycaenid Butterflies.* Cape Town: Purnell.

Common, I. F. B. 1990. *Moths of Australia.* Melbourne: Melbourne University Press.

Common, I. F. B. and D. F. Waterhouse. 1981. *Butterflies of Australia*. London: Angus and Robertson.

Comstock, J. A. 1957. Early stages of *Eutachyptera psidii* (Lasiocampidae), a rare moth from southern Arizona. *Lepid. News* 11: 99–102.

Cooper, W. E. Jr. 1992. Does gregariousness reduce attacks on aposematic prey? Limitations of one experimental test. *Anim. Behav.* 43: 163–164.

Corbet, A. S., H. M. Pendlebury and J. N. Eliot. 1992. *The Butterflies of the Malay Peninsula*, 4th edn. Kuala Lumpur: United Selangor Press Sdn. Bhd.

Cornell, J. C., N. E. Stamp and M. D. Bowers. 1987. Developmental change in aggregation, defense and escape behavior of buckmoth caterpillars, *Hemileuca lucina* (Saturniidae). *Behav. Ecol. Sociobiol.* 20: 383–388.

–. 1988. Variation and developmental change in activity of gregarious caterpillars *Hemileuca lucina* (Saturniidae). *Psyche* 95: 45–58.

Costa, J. T. and K. G. Ross. 1993. Seasonal decline in intracolony genetic relatedness in eastern tent caterpillars: implications for social evolution. *Behav. Ecol. Sociobiol.* 32: 47–54.

Courtney, S. P. 1984. The evolution of egg clustering by butterflies and other insects. *Am. Nat.* 123: 276–281.

Cushman, J. H., V. K. Rashbrook and A. J. Beattie. 1994. Assessing benefits to both participants in a lycaenid–ant association. *Ecology* 75: 1031–1041.

Damman, H. 1987. Leaf quality and enemy avoidance by the larvae of a pyralid moth. *Ecology* 68: 88–97.

Dempster, J. P. 1967. The control of *Pieris rapae* with DDT. I. The natural mortality of the young stages of *Pieris*. *J. Appl. Ecol.* 4: 485–500.

Dethier, V. 1959a. Egg-laying habits of Lepidoptera in relation to available food. *Can. Entomol.* 91: 554–561.

–. 1959b. Food-plant distribution and density and larval dispersal as factors affecting insect populations. *Can. Entomol.* 91: 581–596.

DeVries, P. J. 1977. *Eumaeus minyas*, an aposematic lycaenid butterfly. *Brenesia* 12: 269–270.

–. 1987. *The Butterflies of Costa Rica and their Natural History: Papilionidae, Pieridae, Nymphalidae*. Princeton: Princeton University Press.

–. 1988. The larval ant-organs of *Thisbe irenea* (Lepidoptera: Riodinidae) and their effects upon attending ants. *Zool. J. Linn. Soc.* 94: 379–393.

–. 1990. Enhancement of symbioses between butterfly caterpillars and ants by vibrational communication. *Science (Wash., D.C.)* 24: 1104–1106.

–. 1991. Call production by myrmecophilous riodinid and lycaenid butterfly caterpillars (Lepidoptera): morphological, acoustical, functional, and evolutionary patterns. *Am. Mus. Nov.* no. 3025.

DeVries, P. J. and I. Baker. 1989. Butterfly exploitation of an ant-plant mutualism: adding insult to herbivory. *J. N.Y. Entomol. Soc.* 97: 332–340.

Dias, B. F. 1975. Comportamento presocial de Sinfitas do Brazil Central. I. *Themos olfersii* (Klug) (Hymenoptera: Argidae). *Stud. Entomol.* 18: 401–432.

–. 1976. Comportamento presocial de Sinfitas do Brasil Central II. *Dielocerus diasi* Smith (Hymenoptera: Argidae). *Stud. Entomol.* 19: 461–501.

–. 1982. Maternal care behavior among sawflies. In *The Biology of Social Insects*. M. D. Breed, C. D. Michiner and H. E. Evans, eds., pp. 180–1. Boulder, Colorado: Westview Press.

Downey, J. C. 1962. Myrmecophily in *Plebejus (Icaricia) icarioides* (Lepidoptera: Lycaenidae). *Entomol. News* 73: 57–66.

–. 1966. Sound production in pupae of Lycaenidae. *J. Lepid. Soc.* 20: 129–155.

Downey, J. C. and A. C. Allyn. 1973. Butterfly ultrastructure I. Sound production and associated abdominal structures in pupae of Lycaenidae and Riodinidae. *Bull. Allyn Mus.* 14: 1–47

Dunn, L. H. 1917. The coconut tree caterpillar (*Brassolis isthmia*) of Panama. *J. Econom. Entomol.* 10: 473–488.

Edgerly, J. S. and T. D. Fitzgerald. 1982. An investigation of behavioral variability within colonies of the eastern tent caterpillar *Malacosoma americanum* (Lepidoptera: Lasiocampidae). *J. Kansas Entomol. Soc.* 55: 145–155.

Edmunds, M. 1974. *Defence in Animals: A Survey of Antipredator Defences*. Harlow: Longman.

Edwards, W. H. 1897. *The Butterflies of North America*, vols. 1–3. Boston: Houghton Mifflin and Co.

Eisner, T. and F. C. Kafatkos. 1962. Defense mechanisms of arthropods. X. A pheromone promoting aggregation in an aposematic distasteful insect. *Psyche* 69: 53–61.

Elgar, M. A. and N. E. Pierce. 1988. Mating success and fecundity in an ant-tended lycaenid butterfly. In *Reproductive Success: Studies of Selection and Adaptation in Contrasting Breeding Systems*. T. H. Clutton-Brock, ed., pp. 59–75. Chicago: University of Chicago Press.

Emmet, A. 1979. *A Field Guide to Smaller British Lepidoptera*. London: British Entomol. Nat. Hist. Soc.

Evans, E. W. 1982. Influence of weather on predator/prey relations: stinkbugs and tent caterpillars. *J. N.Y. Entomol. Soc.* 90: 241–246.

–. 1983. Niche relations of predatory stinkbugs (*Podisus* spp., Pentatomidae) attacking tent caterpillars (*Malacosoma americanum*, Lasiocampidae). *Am. Midl. Nat.* 109: 316–323.

Fabre, J. H. 1916. *The Life of the Caterpillar* (reprint). New York: Dodd Mead.

Fiedler, K. 1991. Systematic, evolutionary, and ecological implications of myrmecophily within the Lycaenidae (Insecta: Lepidoptera: Papilionoidea). *Bonn. Zool. Monogr.*, no. 31.

Fiedler, K. and U. Maschwitz. 1989. The symbiosis between the weaver ant, *Oecophylla smaragdina*, and *Anthene emolus*, an obligate myrmecophilous lycaenid butterfly. *J. Nat. Hist.* 23: 833–846.

Fisher, R. A. 1958. *The Genetical Theory of Natural Selection*, 2nd edn. New York: Dover.

Fitzgerald, T. D. 1976. Trail marking by larvae of the eastern tent caterpillar. *Science (Wash., D.C.)* **194**: 961–963.

–. 1980. An analysis of daily foraging patterns of laboratory colonies of the eastern tent caterpillar, *Malacosoma americanum* (Lepidoptera: Lasiocampidae), recorded photoelectronically. *Can. Entomol.* **112**: 731–738.

–. 1993. Sociality in caterpillars. In *Caterpillars: Ecological and Evolutionary Constraints on Foraging.* N. E. Stamp and T. M. Casey, eds., pp. 372–403. New York: Chapman and Hall.

–. 1995. *The Tent Caterpillars.* Ithaca: Cornell University Press.

Fitzgerald, T. D. and J. T. Costa. 1986. Trail-based communication and foraging behavior of young colonies of the forest tent caterpillar *Malacosoma disstria* Hübn. (Lepidoptera: Lasiocampidae). *Ann. Entomol. Soc. Am.* **79**: 999–1007.

Fitzgerald, T. D. and J. S. Edgerly. 1979. Specificity of trail markers of forest and eastern tent caterpillars. *J. Chem. Ecol.* **5**: 564–574.

Fitzgerald, T. D. and E. M. Gallagher. 1976. A chemical trail factor from the silk of the eastern tent caterpillar *Malacosoma americanum* (Lepidoptera: Lasiocampidae). *J. Chem. Ecol.* **2**: 187–193.

Fitzgerald, T. D. and S. C. Peterson. 1983. Elective recruitment communication by the eastern tent caterpillar (*Malacosoma americanum*). *Anim. Behav.* **31**: 417–442.

–. 1988. Cooperative foraging and communication in caterpillars. *BioScience* **38**: 20–25.

Fitzgerald, T. D. and D. E. Willer. 1983. Tent building behavior of the eastern tent caterpillar *Malacosoma americanum* (Lepidoptera: Lasiocampidae). *J. Kansas Entomol. Soc.* **56**: 20–31.

Fitzgerald, T. D., T. M. Casey and B. Joos. 1988. Daily foraging schedule of field colonies of the eastern tent caterpillar *Malacosoma americanum. Oecologia (Berl.)* **76**: 574–578.

Fletcher, D. J. C. and C. D. Michener, eds. 1987. *Kin Recognition in Animals.* Chichester: Wiley.

Fontaine, M. 1975. Un levage ab ovo de *Sabalia tippelskirchi* [sic] (Lép. Eupterotidae – Stryphnopteryginae [sic] selon M. Gaede in Seitz, F. Eth. pp. 301–302). – *Lambillionea* **75** bis (volume jubilaire), pp. 36–39.

Franclemont, J. 1973. Mimallonoidea and Bombycoidea: Apatelodidae, Bombycidae, Lasiocampidae. In *The Moths of America North of Mexico,* R. B. Dominick, D. C. Ferguson, J. G. Franclemont, R. W. Hodges and E. G. Munroe, eds., pp. 1–86. Fasc. 20, Part 1. London: Curwen Press.

Franz, J. M. and W. Laux. 1964. Individual differences in *Malacosoma neustria* (L.). *Proc. XII Int. Congress Entomol.* P. Freeman, ed., pp. 393–394. London: Royal Entomological Society.

Freina, J. J. de and T. J. Witt. 1987. *Die Bombyces und Sphinges der Westpalaearktis,* vol. 1. München: Edition Forschung und Wissenschaft Verlag GmbH.

Ghent, A. W. 1960. A study of the group-feeding behaviour of the jack pine sawfly *Neodiprion pratti banksianae* Roh. *Behaviour* **16**: 110–148.

Gilbert, L. E. and M. C. Singer. 1975. Butterfly ecology. *Annu. Rev. Ecol. Syst.* **6**: 365–397.

Gittleman, J. L. and P. H. Harvey. 1980. Why are distasteful prey not cryptic? *Nature (Lond.)* **286**: 149–150.

Gittleman, J. L., P. H. Harvey, and P. J. Greenwood. 1980. The evolution of aposematic coloration: Some experiments in bad taste. *Anim. Behav.* **28**: 897–899.

Goater, B. 1986. *British Pyralid Moths: A Guide to their Identification.* Colchester: Harley Books.

Greenblatt, J. A. 1974. Behavioral studies on tent caterpillars. M. S. thesis, University of Michigan, Ann Arbor.

Greenblatt, J. A. and J. A. Witter. 1976. Behavioral studies on *Malacosoma disstria* (Lepidoptera: Lasiocampidae). *Can. Entomol.* **108**: 1225–1228.

Guilford, T. 1985. Is kin selection involved in the evolution of aposematic coloration? *Oikos* **45**: 31–36.

–. 1990. The evolution of aposematism. In *Insect Defenses.* D. L. Evans and J. O. Schmidt, eds., pp. 23–62. Albany: State University of New York Press.

Hamilton, W. D. 1971. Geometry for the selfish herd. *J. Theor. Biol.* **31**: 295–311.

Hansen, J. D., J. A. Ludwig, J. C. Owens and E. W. Huddleston. 1984a. Motility, feeding, and molting in larvae of the range caterpillar, *Hemileuca oliviae* (Lepidoptera: Saturniidae). *Environ. Entomol.* **13**: 45–51.

–. 1984b. Larval movement of the range caterpillar, *Hemileuca oliviae* (Lepidoptera: Saturniidae). *Environ. Entomol.* **13**: 415–420.

Harvey, P. H., J. J. Bull, M. Pemberton and R. J. Paxton. 1982. The evolution of aposematic coloration in distasteful prey: a family model. *Am. Nat.* **119**: 710–719.

Hassell, M. P. and T. R. E. Southwood. 1978. Foraging strategies of insects. *Annu. Rev. Ecol. Syst.* **9**: 75–98.

Haukioja, E. 1993. Effects of food and predation on population dynamics. In *Caterpillars: Ecological and Evolutionary Constraints on Foraging.* N. E. Stamp and T. M. Casey, eds., pp. 425–447. New York: Chapman and Hall.

Haukioja, E. and S. Neuvonen. 1985. The relationship between size and reproductive potential in male and female *Epirrita autumnata* (Lep., Geometridae). *Ecol. Entomol.* **10**: 267–270.

–. 1987. Insect population dynamics and induction of plant resistance: the testing of hypotheses. In *Insect Outbreaks.* P. Barbosa and J. Schultz, eds., pp. 411–432. San Diego: Academic Press.

Heads, P. A. 1986. The costs of reduced feeding due to predator avoidance: potential effects on growth and fitness in *Ischnura elegans* larvae (Odonata: Zygoptera). *Ecol. Entomol.* **11**: 369–377.

Hebert, P. D. N. 1983. Egg dispersal patterns and adult feeding behaviour in the Lepidoptera. *Can. Entomol.* **115**: 1477–1481.

Heinrich, B. 1979. Foraging strategies of caterpillars: leaf damage and possible predator avoidance strategies. *Oecologia (Berl.)* **42**: 325–337.

–. 1993. How avian predators constrain caterpillar foraging. In *Caterpillars: Ecological and Evolutionary Constraints on Foraging.* N. E. Stamp and T. M. Casey, eds., pp. 224–247. New York: Chapman and Hall.

Henson, W. R. 1965. Individual rearing of the larvae of *Neodiprion sertifer* (Geoffroy) (Hymenoptera: Diprionidae). *Can. Entomol.* **97**: 773–779.

Hertel, T. and D. M. Benjamin. 1979. Biology of the pine webworm in Florida slash pine plantations. *Ann. Entomol. Soc. Am.* **72**: 816–819.

Hieber, C. S. and G. W. Uetz. 1990. Colony size and parasitoid load in two species of colonial *Metepeira* spiders from Mexico (Araneae: Araneidae). *Oecologia (Berl.)* **82**: 145–150.

Hinton, H. E. 1951. Myrmecophilous Lycaenidae and other Lepidoptera, a summary. *Trans. S. Lond. Entomol. Nat. Hist. Soc.* **1949–1950**: 111–175.

Hitchcock, S. W. 1961. Egg parasitism and larval habits of the orange-striped oakworm. *J. Econ. Entomol.* **54**: 502–503.

Hochberg, M. E. 1991a. Viruses as costs to gregarious feeding behaviour in the Lepidoptera. *Oikos* **61**: 291–296.

–. 1991b. Extra-host interactions between a braconid endoparasitoid, *Apanteles glomerata* L., and a baculovirus for larvae of *Pieris brassicae* L. *J. Anim. Ecol.* **60**: 65–77.

Hoebeke, B. 1987. *Yponomeuta cagnagella* (Lepidoptera: Yponomeutidae): A palearctic ermine moth in the United States, with notes on its recognition, seasonal history, and habits. *Ann. Entomol. Soc. Am.* **80**: 462–467.

Holloway, J. D. and P. D. N. Herbert. 1979. Ecological and taxonomic trends in macrolepidopteran host plant selection. *Biol. J. Linn. Soc.* **11**: 229–251.

Howe, W. H. 1975. *The Butterflies of North America*. New York: Doubleday and Co., Inc.

Hunter, M. D. 1987. Opposing effects of spring defoliation on late season oak caterpillars. *Ecol. Entomol.* **12**: 373–382.

Inman, A. J. and J. Krebs. 1987. Predation and group living. *Trends Ecol. Evol.* **2**: 31–32.

Iwao, S. 1968. Some effects of grouping in lepidopterous insects. In *L'Effet de Groupe chez les Animaux* (Colloq. Internat. Centr. Nat. Rech. Sci. no. 173), pp. 185–212. Paris: Editions du CNRS.

Janzen, D. H. 1984. Natural history of *Hylesia lineata* (Saturniidae: Hemileucinae) in Santa Rosa National Park, Costa Rica. *J. Kansas Entomol. Soc.* **57**: 490–514.

Johnson, W. T. and H. H. Lyon. 1988. *Insects That Feed on Trees and Shrubs*, 2nd edn. Ithaca: Cornell University Press.

Joos, B., T. M. Casey, T. D. Fitzgerald and W. A. Buttemer. 1988. Roles of the tent in behavioral thermoregulation of eastern tent caterpillars. *Ecology* **69**: 2004–2011.

Kalkowski, W. 1966. Feeding orientation of caterpillars during ontogenetic development in *Hyponmeuta evonymellus* L., Lepidoptera, Hyponomeutidae. *Folia Biol. (Cracow)* **14**: 23–46.

Kareiva, P. and G. Odell. 1987. Swarms of predators exhibit "prey-taxis" if individual predators use area-restricted search. *Am. Nat.* **130**: 233–270.

Kevan, P. G. and R. A. Bye. 1991. The natural history, sociobiology, and ethnobiology of *Eucheira socialis* Westwood (Lepidoptera: Pieridae), a unique and little-known butterfly from Mexico. *Entomologist* **110**: 146–165.

Khoo, K. C., P. A. C. Ooi and H. C. Tuck. 1991. *Crop Pests and their Management in Malaysia*. Kuala Lumpur: Tropical Press.

Kitching, R. L. 1981. Egg clustering and the southern hemisphere lycaenids: comments on a paper by N. E. Stamp. *Am. Nat.* **118**: 423–425.

Knapp, R. and T. M. Casey. 1986. Activity patterns, behavior, and growth in gypsy moth and eastern tent caterpillars. *Ecology* **67**: 598–608.

Kristensen, N. P. 1984. Studies on the morphology and systematics of primitive Lepidoptera (Insecta). *Steenstrupia* **10**: 141–191.

–. 1991. Phylogeny of extant hexapods. In *The Insects of Australia*, 2nd edn., vol. 1. I. D. Naumann, P. B. Carne, J. F. Lawrence, E. S. Nielsen, J. P. Spradbury, R. W. Taylor, M. J. Whitten and M. J. Littlejohn, eds., pp. 125–140. Melbourne: Melbourne University Press.

Laux, W. 1962. Individuelle Unterschiede in Verhalted und Leistung des Ringelspinners, *Malacosoma neustria* (L.) *Z. Angew. Zool.* **49**: 465–525.

Lampe, R. E. J. 1982a. Eine Zucht von *Gonimbrasia hecate* Rougeot (Lep.: Saturniidae). *Entomol. Z.* **92**: 222–226.

–. 1982b. Eine Zucht von *Bunaea alcinoe* (Lep.: Saturniidae). *Entomol. Z.* **92**: 329–334.

–. 1983a. Eine Winterzucht von *Melanocera menippe* (Westwood) (Lep.: Saturniidae). *Entomol. Z.* **93**: 41–44.

–. 1983b. Eine Winterzucht von *Gynanisa maja* (Klug) (Lep.: Saturniidae). *Entomol. Z.* **93**: 89–94.

–. 1984. Eine Zucht von *Imbrasia* (*Nudaurelia*) *alopia rhodophila* Walker (Lep.: Saturniidae). *Entomol. Z.* **94**: 121–126.

–. 1985a. Eine Zucht von *Cirina forda* Westwood 1849 (Lep.: Saturniidae). *Entomol. Z.* **95**: 17–20.

–. 1985b. Die Primaginalstadien von *Gonimbrasia tyrrhea hoehneli* Rogenhofer 1891 (Lep.: Saturniidae). *Entomol. Z.* **95**: 330–331.

Langlois, T. H. and M. H. Langlois. 1964. Notes on the life-history of the hackberry butterfly, *Asterocampa celtis*, on South Bass Island, Lake Erie (Lepidoptera: Nymphalidae). *Ohio J. Sci.* **64**: 1–11.

Larsen, T. B. 1991. *The Butterflies of Kenya and their Natural History*. Oxford University Press.

Lawrence, W. S. 1990. The effects of group size and host species on development and survivorship of a gregarious caterpillar *Halisidota caryae* (Lepidoptera: Arctiidae). *Ecol. Entomol.* **15**: 53–62.

Lawton, J. H. 1978. Host-plant influences on insect diversity: The effects of space and time. *Symp. R. Entomol. Soc. Lond.* **9**: 105–125.

Lee, C. S., B. A. McCool, J. L. Moore, D. M. Hillis, and L. E. Gilbert. 1992. Phylogenetic study of heliconiine butterflies based on morphology and restriction analysis of ribosomal RNA genes. *Zool. J. Linn. Soc.* **106**: 17–31.

Leimar, O., M. Enquist and B. Sillén-Tullberg. 1986. Evolutionary stability of aposematic coloration and prey unprofitability: a theoretical analysis. *Am. Nat.* **128**: 469–490.

Long, D. B. 1953. Effects of population density on larvae of Lepidoptera. *Trans. R. Entomol. Soc. Lond.* **104**: 533–585.

—. 1955. Observations on subsocial behaviour in two species of lepidopterous larvae, *Pieris brassicae* L. and *Plusia gamma* L. *Trans. R. Entomol. Soc. Lond.* **106**: 421–437.

Lyons, L. A. 1962. The effect of aggregation on egg and larval survival in *Neodiprion swainei* Midd. (Hymenoptera: Diprionidae). *Can. Entomol.* **94**: 49–58.

Malicky, H. 1970. New aspects on the association of between lycaenid larvae (Lycaenidae) and ants (Formicidae, Hymenoptera). *J. Lepid. Soc.* **24**: 190–202.

Mallet, J. L. and D. A. Jackson. 1980. The ecology and social behaviour of the Neotropical butterfly *Heliconius xanthocles* Bates in Colombia. *Zool. J. Linn. Soc.* **70**: 1–13.

Mallet, J. and M. S. Singer. 1987. Individual selection, kin selection, and the shifting balance in the evolution of warning colours: the evidence from butterflies. *Biol. J. Linn. Soc.* **32**: 337–350.

Martin, J. A. and D. P. Pashley. 1992. Molecular systematic analysis of butterfly family and some subfamily relationships (Lepidoptera: Papilionoidea). *Ann. Entomol. Soc. Am.* **85**: 127–139.

Masaki, S. and K. Umeya. 1977. Larval life. In *Adaptation and Speciation in the Fall Webworm*. T. Hidaka, ed., pp. 13–29. Japan: Kodansha.

Mathews, J. N. A. 1993. Aggregation and mutualism in insect herbivores. Ph. D. dissertation, University of Oxford.

McEvoy, P. B. 1979. Advantages and disadvantages of group living in treehoppers (Homoptera: Membracidae). *Misc. Publ. Entomol. Soc. Am.* **11**: 1–13.

Michener, C. D. 1974. *The Social Behavior of Bees: A Comparative Study.* Cambridge, Mass.: Belknap Press of Harvard University Press.

Minet, J. 1991. Tentative reconstruction of the ditrysian phylogeny (Lepidoptera: Glossata). *Entomol. Scand.* **22**: 69–95.

Mizuta, K. 1960. Effects of individual number on the development and survival of the larvae of two lymantriid species living in aggregation and in scattering. *Jap. J. Appl. Entomol. Zool.* **4**: 146–152.

—. 1968. The effect of larval aggregation upon survival, development, adult longevity and fecundity of a zygaenid moth *Artona funeralis* Butler. *Bull. Hiroshima Agr. Coll.* **3**: 97–107.

Montllor, C. B. and E. A. Bernays. 1993. Invertebrate predators and caterpillar foraging. In *Caterpillars: Ecological and Evolutionary Constraints on Foraging*. N. E. Stamp and T. M. Casey, eds., pp. 170–202. New York: Chapman and Hall.

Moriuti, S. 1977. *Fauna Japonica: Yponomeutidae S. Lat. (Insecta: Lepidoptera)*. Tokyo: Keigaku Publishing Co.

Morris, R. F. 1963. The effect of predator age and prey defense on the functional response of *Podisus maculiventris* Say to the density of *Hyphantria cunea* Drury. *Can. Entomol.* **95**: 1009–1020.

—. 1972a. Predation by wasps, birds, and mammals on *Hyphantria cunea. Can. Entomol.* **104**: 1581–1591.

—. 1972b. Fecundity and colony size in natural populations of *Hyphantria cunea. Can. Entomol.* **104**: 399–409.

—. 1976. Relation of parasite attack to the colonial habit of *Hyphantria cunea. Can. Entomol.* **108**: 833–836.

Morris, R. F. and W. C. Fulton. 1970. Models for the development and survival of *Hyphantria cunea* in relation to temperature and humidity. *Mem. Entomol. Soc. Can.*, no. 70.

Morrow, P. A., T. E. Bellas and T. Eisner. 1976. *Eucalytus* oils in the defensive oral discharge of Austrian sawfly larvae (Hymenoptera: Pergidae). *Oecologia.* **24**: 193–206.

Myers, J. H. 1978. A search for behavioural variation in first laid eggs of the western tent caterpillar and an attempt to prevent a population decline. *Can. J. Zool.* **56**: 2359–2363.

Myers, J. H. and J. N. M. Smith. 1978. Head flicking by tent caterpillars: a defensive response to parasite sounds. *Can. J. Zool.* **56**: 1628–1631.

Nafus, D. M. and I. H. Schreiner. 1988. Parental care in a tropical nymphalid butterfly *Hypolimnas anomala. Anim. Behav.* **36**: 1425–1431.

Nault, L. R., M. E. Montgomery and W. S. Bowers. 1976. Antaphid association: role of aphid alarm pheromone. *Science (Wash., D.C.)* **192**: 1349–1351.

Naumann, Z. S. 1995. *Die Saturniiden-Fauna von Sulawesi, Indonesia.* Ph. D. Dissertation, Free University of Berlin.

Nielson, E. S. 1989. Phylogeny of major lepidopteran groups. In *The Hierarchy of Life*. B. Ferholm, K. Bremer and H. Jörnvall, eds., pp. 281–294. Cambridge: Elsevier.

Nielson, E. S. and I. F. B. Common. 1991. Lepidoptera (Moths and Butterflies). In *The Insects of Australia*, 2nd edn., vol. 2. I. D. Naumann, P. B. Carne, J. F. Lawrence, E. S. Nielsen, J. P. Spradbury, R. W. Taylor, M. J. Whitten, and M. J. Littlejohn, eds., pp. 817–915. Melbourne: Melbourne University Press.

Oberprieler, R. G. 1993. Biological notes on the eggar moth *Schausinna regia* (Grünberg) (Lepidoptera: Lasiocampidae). *Metamorphosis* **4**: 73–78.

—. 1995. *The Emperor Moths of Namibia.* Pretoria: Ecoguild Publishers.

O'Dowd, D. 1979. Foliar nectar production and ant activity on a neotropical tree, *Ochroma pyramidale. Oecologia (Berl.)* **43**: 233–248.

Papaj, D. R. and M. D. Rausher. 1983. Individual variation in host location by phytophagous insects. In *Herbivorous Insects: Host-Seeking Behavior and Mechanisms*. S. Ahmad, ed., pp. 77–124. New York: Academic Press.

Pasteels, J. M., J. C. Gregoire and M. Rowell-Rahier. 1983. The chemical ecology of defense in arthropods. *Annu. Rev. Entomol.* **28**: 263–289.

Patocka, J. 1980. Die Raupen und Puppen der Eichenschmetterlinge Mitteleuropas. *Monogr. Angew. Entomol.* **23**: 1–188.

Payne, C. C., G. M. Tatchell, and C. F. Williams. 1981. The comparative susceptibilities of *Pieris brassicae* and *P. rapae* to a granulosis virus from *P. brassicae. J. Invert. Pathol.* **38**: 273–280.

Peterson, A. 1962. *Larvae of Insects: An Introduction to Nearctic Species*, part 1. *Lepidoptera and Plant-Infesting Hymenoptera*. Columbus, Ohio: A. Peterson.

Peterson, S. C. 1986. Breakdown products of cyanogenesis: repellency and toxicity to predatory ants. *Naturwissenschaften* **73**: 627–628.

–. 1987. Communication of leaf suitability by gregarious eastern tent caterpillars (*Malacosoma americanum*). *Ecol. Entomol.* **12**: 283–289.

–. 1988. Chemical trail marking and following by caterpillars of *Malacosoma neustria*. *J. Chem. Ecol.* **14**: 815–823.

Peterson, S. C., N. D. Johnson and J. L. LeGuyader. 1987. Defensive regurgitation of allelochemicals derived from host cyanogenesis by eastern tent caterpillars. *Ecology* **68**: 1268–1272.

Pierce, N. E. and S. Easteal. 1986. The selective advantage of attendant ants for the larvae of a lycaenid butterfly, *Glaucopsyche lygdamus*. *J. Anim. Ecol.* **55**: 451–462.

Pierce, N. E. and M. A. Elgar. 1985. The influence of ants on host-plant selection by *Jalmenus evagoras*, a myrmecophilous lycaenid butterfly. *Behav. Ecol. Sociobiol.* **16**: 209–222.

Pierce, N. E. and P. S. Mead. 1981. Parasitoids as selective agents in the symbiosis between lycaenid butterfly larvae and ants. *Science (Wash., D.C.)* **211**: 1185–1187.

Pierce, N. E., R. L. Kitching, R. C. Buckley, M. F. J. Taylor, and K. F. Benbow. 1987. The costs and benefits of cooperation between the Australian lycaenid butterfly, *Jalmenus evagoras*, and its attendant ants. *Behav. Ecol. Sociobiol.* **21**: 237–248.

Pinhey, E. C. G. 1975. *Moths of Southern Africa*. Cape Town: Tafelberg Pub. Ltd.

Porter, K. 1982. Basking behaviour in larvae of the butterfly *Euphydryas aurina*. *Oikos* **38**: 308–312.

–. 1983. Multivoltinism in *Apanteles bignellii* and the influence of weather on synchronization with its host *Euphydryas aurina*. *Entomol. Exp. Appl.* **34**: 155–162.

Price, P. W., C. E. Bouton, P. Gross, B. A. McPheron, J. N. Thompson and A. E. Weis. 1980. Interactions among three trophic levels: influence of plants on interactions between insect herbivores and natural enemies. *Annu. Rev. Ecol. Syst.* **11**: 41–65.

Prop, N. 1960. Protection against birds and parasites in some species of tenthredinid larvae. *Arch. Neerl. Zool.* **13**: 380–447.

Pulliam, R. H. and T. Caraco. 1984. Living in groups: is there an optimal group size? In *Behavioral Ecology: An Evolutionary Approach*, 2nd edn. J. R. Krebs and N. B. Davies, eds., pp. 122–147. London: Blackwell Scientific Publications.

Rabb, R. L. and F. R. Lawson. 1957. Some factors influencing the predation of *Polistes* wasps on the tobacco hornworm. *J. Econ. Entomol.* **50**: 778–784.

Reavey, D. 1993. Why body size matters to caterpillars. In *Caterpillars: Ecological and Evolutionary Constraints on Foraging*. N. E. Stamp and T. M. Casey, eds., pp. 248–279. New York: Chapman and Hall.

Robinson, G. S., K. R. Tuck and M. Schaffer. 1994. *A Field Guide to the Smaller Moths of South-East Asia*. Kuala Lumpur: Malaysian Nature Society and the British Museum of Natural History.

Roessingh, P. 1989. The trail-following behaviour of *Yponomeuta cagnagellus*. *Entomol. Exp. Appl.* **51**: 49–57.

–. 1990. Chemical trail marker from silk of *Yponomeuta cagnagellus*. *J. Chem. Ecol.* **16**: 2203–2216.

Roper, T. J. and S. Redston. 1987. Conspicuousness of distasteful prey affects the strength and durability of one-trial avoidance learning. *Anim. Behav.* **35**: 739–747.

Rosenheim, J. A. 1990. Density-dependent parasitism and the evolution of aggregated nesting in the solitary Hymenoptera. *Ann. Entomol. Soc. Am.* **83**: 277–286.

Rosier, J. P. 1951. Notes on Lepidoptera. II. Metamorphosis of some Javanese butterflies. *Idea* **9**: 26.

Ross, G. N. 1964. Life history studies on Mexican butterflies. III. Nine Rhopalocera (Papilionidae, Nymphalidae, Lycaenidae) from Ocotol Chico, Vera Cruz. *J. Res. Lepid.* **3**: 207–229.

Rougeout, P. C. and P. Viette. 1983. *Die Nachtfalter Europas und Nordwestafrikas*, vol. 1. Schwärmer and Spinner. Keltern: E. Bauer.

Schultz, J. C. 1983. Habitat selection and foraging tactics of caterpillars in heterogeneous trees. In *Variable Plants and Herbivores in Natural and Managed Systems*. R. F. Denno and M. S. McClure, eds., pp. 61–90. New York: Academic Press.

Schultz, D. E. and D. C. Allen. 1975. Biology and descriptions of the cherry scallop moth *Hydria prunivorata* (Lepidoptera: Geometridae). *Can. Entomol.* **107**: 99–106.

Schweizerischer Bund für Naturschutz. 1987. *Tagfalter und ihre Lebensrüme.*. Basel: Schweizerische Burd für Naturschutz.

Scoble, M. J. 1992. *The Lepidoptera*. Oxford University Press.

Scott, J. A. 1985. The phylogeny of butterflies (Papiliodoidea and Hesperioidea). *J. Res. Lepid.* **23**: 241–281.

–. 1986. *The Butterflies of North America: A Natural History and Field Guide*. Stanford: Stanford University Press.

Scott, J. A. and D. M. Wright. 1990. Butterfly phylogeny and fossils. In *Butterflies of Europe*, vol. 2. *Introduction to Lepidopterology*. O. Kudrna, ed., pp. 152–208. Wiesbaden: AULA-Verlag.

Scriber, J. M. and F. Slansky, Jr. 1981. The nutritional ecology of immature insects. *Annu. Rev. Entomol.* **26**: 183–211.

Seymour, R. 1974. Convective and evaporative cooling in sawfly larvae. *J. Insect Physiol.* **20**: 2447–2457.

Shiga, M. 1976. Effect of group size on the survival and development of young larvae of *Malacosoma neustria testacea* Motschlshy (Lepidoptera: Lasiocampidae) and its role in the natural population. *Konty.* **44**: 537–553.

–. 1979. Population dynamics of *Malacosoma neustria testacea* (Lepidoptera, Lasiocampidae). *Bull. Fruit Tree Res. Stn*, A6: 59–168.

Sih, A. 1980. Optimal behavior: can foragers balance two conflicting demands? *Science (Wash., D.C.)* **210**: 1041–1043.

–. 1982. Foraging strategies and avoidance of predation by an aquatic insect, *Notonecta hoffmanni*. *Ecology* **63**: 786–796.

Sillén-Tullberg, B. 1988. Evolution of gregariousness in aposematic butterfly larvae: a phylogenetic analysis. *Evolution* **42**: 293–305.

–. 1990. Do predators avoid groups of aposematic prey? An experimental test. *Anim. Behav.* **40**: 856–860.

–. 1993. The effect of biased inclusion of taxa on the correlation between discrete characters in phylogenetic trees. *Evolution* **47**: 1182–1191.

Sillén-Tullberg, B. and O. Leimar. 1988. The evolution of gregariousness in distasteful insects as a defense against predators. *Am. Nat.* **132**: 723–734.

Smith, J. N. M. 1974. The food searching behaviour of two European thrushes II: the adaptiveness of the search patterns. *Behaviour* **49**: 1–61.

Stamp, N. E. 1977. Aggregation behavior of *Chlosyne lacinia* (Nymphalidae). *J. Lepid. Soc.* **31**: 35–40.

–. 1980. Egg deposition patterns in butterflies: why do some species cluster their eggs rather than deposit them singly? *Am. Nat.* **115**: 367–380.

–. 1981a. Effect of group size on parasitism in a natural population of the Baltimore checkerspot *Euphydryas phaeton*. *Oecologia (Berl.)* **49**: 201–206.

–. 1981b. Parasitism of single and multiple egg clusters of *Euphydryas phaeton* (Nymphalidae). *J. N. Y. Entomol. Soc.* **89**: 89–97.

–. 1982. Behavioral interactions of parasitoids and Baltimore checkerspot caterpillars *Euphydryas phaeton*. *Environ. Entomol.* **11**: 100–104.

–. 1984. Foraging behavior of tawny emperor caterpillars (Nymphalidae: *Asterocampa clyton*). *J. Lepid. Soc.* **38**: 186–191.

–. 1986. Physical constraints of defense and response to invertebrate predators by pipevine caterpillars (*Battus philenor*: Papilionidae). *J. Lepid. Soc.* **40**: 191–205.

–. 1993. A temperate region view of the interaction of temperature, food quality, and predators on caterpillar foraging. In *Caterpillars: Ecological and Evolutionary Constraints on Foraging*. N. E. Stamp and T. M. Casey, eds., pp. 478–505. New York: Chapman and Hall.

Stamp, N. E. and M. D. Bowers. 1988. Direct and indirect effects of predatory wasps (*Polistes* **sp.**: Vespidae) on gregarious caterpillars (*Hemileuca lucina*: Saturniidae). *Oecologia (Berl.)* **75**: 619–624.

–. 1990a. Variation in food quality and temperature constrain foraging of gregarious caterpillars. *Ecology* **71**: 1031–1039.

–. 1990b. Phenology of nutritional differences between new and mature leaves and its effect on caterpillar growth. *Ecol. Entomol.* **15**: 447–454.

–. 1991. Indirect effect on survivorship of caterpillars due to presence of invertebrate predators. *Oecologia (Berl.)* **88**: 325–330.

Staude, H. S. 1994. Notes on the life history of the inflamed tigerlet, *Veniliodes inflammata* Warren, 1894 (Lepidoptera: Geometridae). *Metamorphosis* **5**: 122–127.

Stehr, F. W. and E. F. Cook. 1968. A revision of the genus *Malacosoma* Hübner in North America (Lepidoptera: Lasiocampidae): systematics, biology, immatures, and parasites. *U. S. Natl. Mus. Bull.*, no. 276.

Strassmann, J. E., D. C. Queller, and C. R. Hughes. 1988. Predation and the evolution of sociality in the paper wasp *Polistes bellicosus*. *Ecology* **69**: 1497–1505.

Subinprasert, S. and B. W. Svensson. 1988. Effects of predation on clutch size and egg dispersion in the codling moth *Laspeyresia pomonella*. *Ecol. Entomol.* **13**: 87–94.

Sullivan, C. R. and G. W. Green. 1950. Reactions of larvae of the eastern tent caterpillar and of the spotless fall webworm to pentatomid predators. *Can. Entomol.* **82**: 52.

Talhouk, A. S. 1940. The oak tree tent caterpillar, *Eriogaster philippsi*, Bartel; its life history, habits and parasites in Lebanon. *Entomol. Rec.* **52**: 87–89.

–. 1975. Contribution to the knowledge of almond pests in East Mediterranean countries. I. Notes on *Eriogaster amygdali* Wilts. (Lepid., Lasiocampidae) with a description of a new subspecies by E. P. Wiltshire. *Z. Angew. Entomol.* **78**: 306–312.

Taylor, J. S. 1949. Notes on Lepidoptera in the eastern Cape province (Part 1). *J. Entomol. Soc. S. Afr.* **12**: 78–95.

–. 1950. Notes on *Phiala* [= *Rhabdosia*] *patagiata* Aur., the Karoo tent caterpillar. *Entomol. Mem. Union S. Afr.* **2**: 219–229.

Taylor, R. J. 1977a. The value of clumping to prey. *Oecologia (Berl.)* **30**: 285–294.

–. 1977b. The value of clumping to prey when detectability increases with group size. *Am. Nat.* **111**: 229–301.

Thomas, J. and R. Lewington. 1991. *The Butterflies of Britain and Ireland*. London: Dorling Kindersley.

Tilman, D. 1978. Cherries, ants and tent caterpillars: timing of nectar production in relation to susceptibility of caterpillars to ant predation. *Ecology* **59**: 686–692.

Tinbergen, N., M. Impekoven, and D. Franck. 1967. An experiment on spacing-out as a defence against predation. *Behaviour* **28**: 307–327.

Tostowaryk, W. 1972. The effect of prey defense on the functional response of *Podisus modestus* (Hemiptera: Pentatomidae) to densities of the sawflies *Neodiprion swainei* and *N. pratti banksianae* (Hymenoptera: Neodiprionidae). *Can. Entomol.* **104**: 61–69.

Treherne, J. E. and W. A. Foster. 1980. The effects of group size on predator avoidance in a marine insect. *Anim. Behav.* **28**: 1119–1122.

–. 1981. Group transmission of predator avoidance in a marine insect: the Trafalgar effect. *Anim. Behav.* **29**: 911–917.

–. 1982. Group size and anti-predator strategies in a marine insect. *Anim. Behav.* **30**: 536–542.

Trivers, R. L. and H. Hare. 1976. Haplodiploidy and the evolution of social insects. *Science (Wash., D.C.)* **191**: 249–263.

Tsubaki, Y. 1981. Some beneficial effects of aggregation in young larvae of *Pryeria sinica* Moore (Lepidoptera: Zygaenidae). *Res. Pop. Ecol.* **23**: 156–167.

Tsubaki, Y. and R. L. Kitching. 1986. Central-place foraging in larvae of the charaxine butterfly, *Polyura pyrrhus* (L.): a case study in a herbivore. *J. Ethol.* **4**: 59–68.

Tsubaki, Y. and Y. Shiotsu. 1982. Group feeding as a strategy for exploiting food resources in the burnet moth *Pryeria sinica*. *Oecologia (Berl.)* **55**: 12–20.

Tsubaki, Y. and N. Yamamura. 1980. A model descriptive of gregariousness of colonial insect larvae. *Res. Pop. Ecol.* **21**: 332–344.

Turner, G. F. and T. J. Pitcher. 1986. Attack abatement: a model for group protection by combined avoidance and dilution. *Am. Nat.* **128**: 228–240.

Turner, J. R. G. 1971. Studies of Müllerian mimicry and its evolution in burnet moths and Heliconiid butterflies. In *Ecological Genetics and Evolution*. E. R. Creed, ed., pp. 224–260. Oxford: Blackwell Scientific Publications.

Underwood, D. L. A. 1994. Intraspecific variability in host plant quality and ovipositional preferences in *Eucheira socialis* (Pieridae: Lepidoptera). *Ecol. Entomol.* **19**: 245–256.

Vermeij, G. J. 1982. Unsuccessful predation and evolution. *Am. Nat.* **120**: 701–720.

Vulinec, K. 1990. Collective security: aggregation by insects in defense. In *Insect Defenses*. D. L. Evans and J. O. Schmidt, eds., pp. 251–288. Albany: State University of New York Press.

Wagner, D. 1993. Species-specific effects of tending ants on the development of lycaenid butterfly larvae. *Oecologia (Berl.)* **96**: 276–281.

Warren, L. O. and M. Tadic. 1970. The fall webworm, *Hyphantria cunea* (Drury). *Univ. Arkans. Agric. Expt. Stn. Bull.*, no. 795.

Watanabe, N. and K. Umeya. 1968. Biology of *Hyphantria cunea* Drury (Lepidoptera; Arctiidae) in Japan. IV. Effects of group size on survival and growth of larvae. *Jap. Plant Prot. Serv. Res. Bull.* **6**: 1–6.

Way, M. J. 1963. Mutualism between ants and honeydew-producing Homoptera. *Annu. Rev. Ecol. Syst.* **8**: 307–344.

Weidemann, H.-J. 1988. *Tagfalter*, vol. 2. Melsungen: Neumann-Neudamm.

Weller, S. J., D. P. Pashley, J. A. Martin, and J. L. Constable. 1994. Phylogeny of noctuoid moths and the utility of combining independent nuclear and mitochondrial genes. *Syst. Biol.* **43**: 194–211.

Wellington, W. G. 1957. Individual differences as a factor in population dynamics: The development of a problem. *Can. J. Zool.* **35**: 293–323.

–. 1965. Some maternal influences on progeny quality in the western tent caterpillar, *Malacosoma pluviale* (Dyar). *Can. Entomol.* **97**: 1–14.

Weyh, R. and U. Maschwitz. 1978. Trail substance in larvae of *Eriogaster lanestris* L. *Naturwissenschaften* **65**: 64.

–. 1982. Individual trail-marking by larvae of the scarce swallowtail *Iphiclides podalirius* L. (Lepidoptera; Papilionidae). *Oecologia (Berl.)* **52**: 415–416.

Wickman, P. -O., and B. Karlsson. 1989. Abdomen size, body size, and the reproductive effort of insects. *Oikos* **56**: 209–214.

Wiklund, C. and T. Jrvi. 1982. Survival of distasteful insects after being attacked by naive birds: a reappraisal of the theory of aposematic coloration evolving through individual selection. *Evolution* **36**: 998–1002.

Wilson, E. O. 1971. *The Insect Societies*. Cambridge, Mass.: Belknap Press of Harvard University Press.

–. 1975. *Sociobiology: The New Synthesis*. Cambridge, Mass.: Belknap Press of Harvard University Press.

Witter, J. A. and H. M. Kulman. 1972. A review of the parasites and predators of tent caterpillars (*Malacosoma* spp.) in North America. *University of Minnesota Agric. Expt. Stn. Bull.*, no. 289.

Wolfe, K. L. 1988. *Hylesia acuta* (Saturniidae) and its aggregate larval and pupal pouch. *J. Lepid. Soc.* **42**: 132–137.

–. 1993. The *Copaxa* of Mexico and their immature stages (Leipdoptera: Saturniidae). *Trop. Lepidopt.* **4**: 1–26.

Wood, T. K. 1977. Role of parent females and attendant ants in maturation of the treehopper, *Entylia bactriana* (Homoptera: Membracidae). *Sociobiology* **2**: 257–272.

Wrona, F. J. and R. W. Jamieson Dixon. 1991. Group size and predation risk: a field analysis of encounter and dilution effects. *Am. Nat.* **137**:186–201.

Young, A. M. 1978. A communal roost of the butterfly *Heliconius charitonius* L. in Costa Rican premontane tropical wet forest (Lepidoptera: Nymphalidae). *Entomol. News* **89**: 235–243.

–. 1983. On the evolution of egg placement and gregariousness of caterpillars in the Lepidoptera. *Acta Biother.* **32**: 43–60.

–. 1985. Natural history notes on *Brassolis isthmia* Bates (Lepidoptera: Nymphalidae) in northeastern Costa Rica. *J. Res. Lepid.* **24**: 385–392.

Young, A. M. and M. W. Moffett. 1979. Studies on the population biology of the tropical butterfly *Mechanitis isthmia* in Costa Rica. *Am. Midl. Nat.* **101**: 309–319.

Young, A. M., M. S. Blum, H. H. Fales and Z. Bian. 1986. Natural history and ecological chemistry of the neotropical butterfly *Papilio anchisiades* (Papilionidae). *J. Lepid. Soc.* **40**: 36–53.

21 · Sociality and kin selection in Acari

YUTAKA SAITO

ABSTRACT

Social organization in the subclass Acari involves parental care and various forms of communal behavior. I first describe subsocial behavior in two spider mite species and then discuss three examples of complex subsociality in Prostigmata and two in Mesostigmata. Aggregation within a narrow space is a common trait observed in all subsocial species. The deposition of feces at certain places (or other nest sanitation behavior) is also common to these species. Subsociality in the Acari is always associated with male-haploid or thelytokous genetic systems, and kin selection under male-haploid genetic system may be important in the evolution of sociality. I next discuss why there is no evidence of eusociality in the Acari, although highly developed subsociality is observed and high relatedness of individuals is expected. One of the plausible explanations is that their simple neural systems decrease the probability of kin recognition; another possibility is that continuous inbreeding decreases the likelihood of the origin of eusociality.

INTRODUCTION

Mites and ticks (Acari) are small organisms with low mobility, and such low mobility may increase the likelihood of interaction among kin. If kin selection (Hamilton 1964a,b) is an important prime mover of social evolution in animals, then we might expect to find examples of sociality in the Acari. However, evidence of social behavior in this group has been scarce; for example, Buskirk (1981) included only 40 lines describing mite behavior in a review of arachnid sociality. During the past decade, however, several fascinating examples of mite society have come to light. In this chapter, I first describe the patterns of subsocial lives of two spider mite species, in order to demonstrate what kinds of social interactions between mite individuals have been observed. Second, evidence of kin selection in a subsocial species of mite is shown, in order to assess the importance of such selection for the evolution of sociality.

Next, I describe the overall patterns of social organization in the Acari, including parent–offspring and intersexual interactions. Finally, I assess why there is no evidence of eusociality in Acari.

WHAT IS THE ACARI?

There are two hypotheses concerning the cladistic status of the subclass Acari, polyphyly (Hammen 1970) and monophyly (Norton et al. 1993). By both hypotheses, mites have been placed close to the subclass Araneae (spiders) and include very diversified species that inhabit terrestrial, aquatic, and subterranean conditions, and show phytophagous, predaceous, parasitic and saprophagous food habits (Krantz 1978).

According to Norton et al. (1993), there are two orders in Acari, Parasitiformes and Acariformes, the former including four suborders and the latter three. Ixodida (ticks) and Mesostigmata are well-known suborders in the Parasitiformes, because they include many ectoparasites of vertebrates and many predaceous species that are effective control agents of pest arthropods.

Three suborders in the Acariformes are also well-known groups: Prostigmata includes chiggers and spider mites, which are deleterious to human beings or plants; Oribatida are known as important decomposers of litter in forests; and Astigmata includes many species injurious to stored products, which are also important as allergens for human health. Because of their minute size and of the scarcity of taxonomists, only 30 000 species have been described so far; however, it has been estimated that there are at least a half-million species of mite in the world (Krantz 1978).

SUBSOCIAL SPIDER MITES

Spider mites (Prostigmata, Tetranychidae) exhibit two examples of complex subsociality (intermediate subsocial II, as defined by Wilson 1975) in the spider mite genus

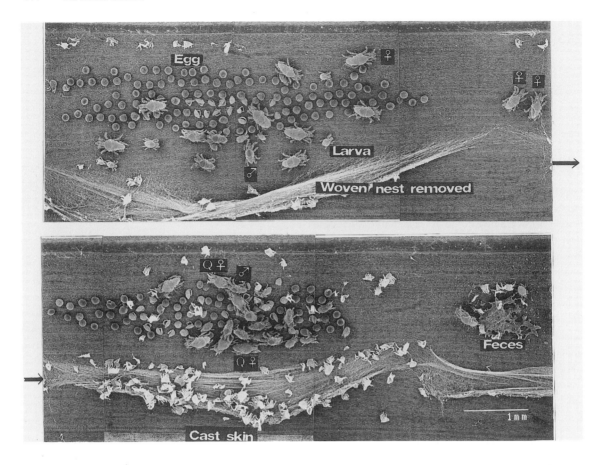

Fig. 21-1. Nesting pattern of *Schizotetranychus longus* Saito on *Sasa* bamboo, photographed under a scanning electron microscope (SEM). Q quiescent stage; arrows indicate continuity between the parts of the photograph.

Schizotetranychus. Schizotetranychus longus Saito and *S. miscanthi* Saito, two closely related species (Saito 1990a; Osakabe *et al.* 1993), exhibit the same nesting habit: they live in a nest constructed by weaving silk threads over depressions along the midrib or curled edges of leaves of host plants. They feed on leaf substances using piercing–sucking stylets. Their sociality is characterized by four traits (Saito 1986a, 1990b): (1) spatiotemporal overlap of generations (two or three generations living gregariously in woven nests); (2) cooperative nest building, repairing and enlarging; (3) cooperative use of nest space and resources (nest sanitation); and (4) cooperative brood defense by females and a male or males against predators.

The nesting pattern of *S. longus* (Fig. 21-1) shows that multiple females and males live together with many of their sons and daughters, sometimes with their grandchildren, in a silken nest. One or two specific areas near nest entrances are used for deposition of frass (Fig. 21-1), such that colony members show coordination in nest sanitation. A nest is sometimes initiated by multiple females (Fig. 21-2); nests are completed faster by multiple females than by single females (Y. Saito, unpublished data). The nests observed in nature often consist of several cells (chambers), which are used simultaneously by nest members (Fig. 21-3). Because new nest cells are added before the previously established cells have become unavailable as resources, this behavior may represent a form of cooperation in nest enlarging and repairing.

Furthermore, in both *S. longus* and *S. miscanthi* (Saito 1986a,b, 1990b; Yamamura 1987) females defend their nest and progeny against predators, such as phytoseiid and stigmaeid mites; the defense is more effective if there are many females per nest (Fig. 21-4). A more impressive and important feature of these societies is defensive behavior by

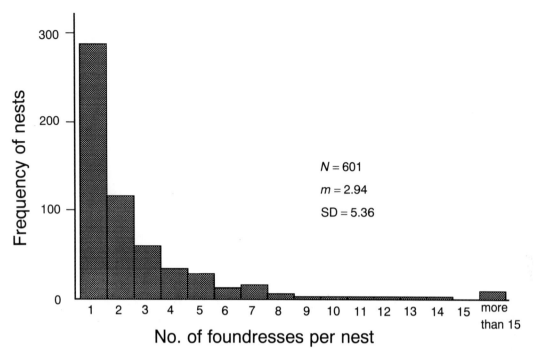

Fig. 21-2. Frequency of number of foundresses per spring nest of *Schizotetranychus longus* Saito in late May to early June 1982 in Sapporo. *N*, number of nests observed; *m*, average number of foundresses per nest; SD, standard deviation (after Saito 1987; reproduced by permission of the Society of Population Ecology).

males. Under the theory of kin selection, males of male-haploid species are not expected to evolve parental care behavior (Hamilton 1972). In these mite species, however, males living in a nest defend their offspring from predators (Figs. 21-5 and 21-6). As shown in Fig. 21-5, a single male of *S. longus* can defend its nest against phytoseiid predator larvae with the probability of about 0.4; in a nest including two males, predator larvae were killed with a probability of about 0.8. By contrast, *S. miscanthi* shows no such cooperative male defense, but males fight mortally with each other to establish a harem. However, a male that has established a harem strongly defends the nest and offspring from predators as well as from conspecific males.

The relationship between females and males can also be considered to be cooperative with respect to nest defense in *S. longus* (Fig. 21-7), and possibly in *S. miscanthi*. A nest containing two females and one or more males is more effectively defended from predators than a nest containing individuals of only one sex. Behavioral observations have shown that females first recognize the predator intrusion, and communicate the emergency to cohabiting males (and vice versa); in a case recorded on film (Saito 1986b), a female

first found a predatory intruder and became excited (running around the nest space with repeated jostling behavior). Then, as soon as the female contacted a male, which had never recognized the intruder directly, the latter also became excited and began searching within the nest. They continued searching and attacked the intruder. Thus, Saito (1986a) referred to the behavior of *S. longus* as biparental defense.

Why is there biparental care in both *S. longus* and *S. miscanthi*, but a large difference in male mating system between the two species? Saito (1995b) reported that there is latitudinal variation in male aggressiveness of *S. miscanthi*: aggressiveness, measured experimentally as the frequency of male mortal fighting within nests, is positively correlated with the winter temperature of the areas where populations were sampled (Fig. 21-8). In spider mites that occur in the temperate zone, where females overwinter in diapause, the proportion of overwintering females that are inseminated and the probability of overwintering by non-diapause males are critical factors that determine kin relationships within a spring breeding group (Saito 1987, 1990b, 1995b). Because spider mite females can produce males by

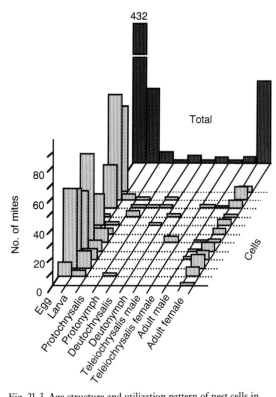

Fig. 21-3. Age structure and utilization pattern of nest cells in *Schizotetranychus longus* Saito in late May to early June 1982 in Sapporo (after Saito 1986a; © 1986 Springer-Verlag, reproduced with permission).

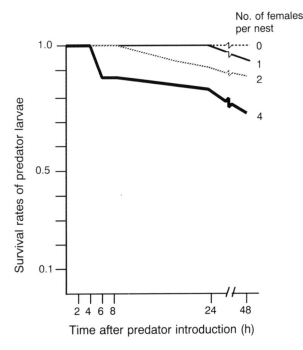

Fig. 21-4. Effect of various densities of *Schizotetranychus longus* Saito females (mothers) on success of counter-attack against predator larvae (after Saito 1986b).

arrhenotoky, the occurrence of mother–son mating is highly probable within nests founded by a small number of females (Saito 1987). Furthermore, the absence of overwintering males in spring nests enhances sib-mating in the succeeding generation produced by the overwintered females. In such cases, a significant difference in the relatedness of nest members (especially between males) would be expected among populations from various localities, where the probability of the overwintering of males and uninseminated females changes according to the severity of winter conditions. Because there are differences in the frequency of male overwintering among local populations (Saito 1995b), a shift in antagonism between males, following the change in the relatedness between interacting males in a nest, may be explicable by kin selection.

Males of *S. miscanthi* have two conflicting interests in their nest: to increase their personal fitness by monopolization of females, and to gain in inclusive fitness by defending cooperatively (with other males) their offspring against

predators. Because males cannot distinguish their kin from strangers (Saito 1994b), conflicts between males may be influenced by the mean relatedness between interacting males. A male that lives in cooler regions with his 'brothers' within a nest gains no advantage in combat; consequently he tends to increase his inclusive fitness (via his brothers and other relatives) through cooperative defense. If, however, a male often meets strange males (non-kin) who, in warmer regions, occasionally intrude from the other nests, he should keep his priority of mating with females through eliminating his competitors by mortal fighting. The cooperative brood defense by males of *S. longus* (Saito 1986b), which inhabits cooler regions of Japan, thus becomes understandable from the expected high relatedness between cohabiting males, which is probably caused by mother–son inbreeding (Saito 1987). This hypothesis could be tested by measurement of relatedness using genetic markers, and by experimental analysis of the effects of relatedness on male fighting and parental care.

The observations mentioned above show that these two spider mite species have well-developed subsociality and that their behavior has evolved in the context of kin

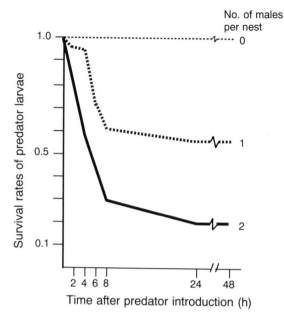

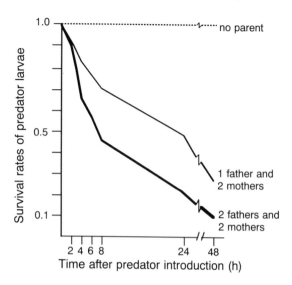

Fig. 21-5. Effect of densities of *Schizotetranychus longus* Saito males (fathers) on success of counter-attack against predator larvae (after Saito 1986b).

Fig. 21-7. Effect of various combinations of *Schizotetranychus longus* Saito parents on success of counter-attack against predator larvae (after Saito 1986b).

selection. To analyze the prime movers of the evolution of the subsociality and the reasons why the subsociality has not evolved to eusociality, we must trace traits common to the Acari and peculiar to the Tetranychidae.

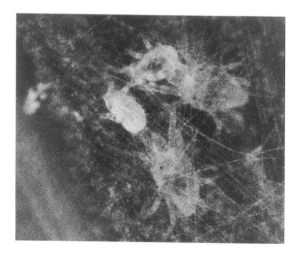

Fig. 21-6. Photograph of counter-attack behavior by males of *Schizotetranychus longus* Saito against larvae of the predator *Typhlodromus bambusae* Ehara.

OVERVIEW OF SOCIALITY IN ACARI

Prostigmata

The suborder Prostigmata includes many families that show various kinds of social interaction. The family Tetranychidae, which includes many species that are serious pests of plants, has been relatively well studied behaviorally. For example, precopulatory mate guarding and male aggressiveness (Potter *et al.* 1976), family colonizing episodes and their adaptive significance (Mitchell 1973), sex-ratio selection (Wrensch 1993), nest-making behavior (Yokoyama 1932; Saito 1979) and life-type (weaving habit) classification (Saito 1983, 1985) have been discussed for this group. Such fragmentary knowledge did not attract attention from the viewpoint of social organization until Saito (1986a, 1990b) reported that there are two spider mite species that, as described above, exhibit highly developed subsociality.

The silk-producing (webbing) habit and other peculiarities of the Tetranychidae, i.e. male-haploid genetic system, gregarious habits and life history, may be related to the evolution of their sociality (cf. Fig. 21-10). Most species of this family make webs over their eggs, protecting them against poor climatic conditions and/or predators. For example, the species *Aponychus corpuzae* Rimando and *Eurytetranychus japonicus* Ehara weave fine and dense webs over their eggs, thus investing energy in their offspring after

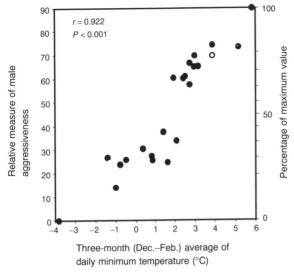

Fig. 21-8. Relative measure of male aggressiveness of
Schizotetranychus miscanthi Saito, positively correlated with winter
coldness. The relative measure of aggressiveness was estimated as
the relative death rate of one of the paired males introduced in a
nest for 5 days. Each experiment included more than 25
replications with the same number of control experiments. The
death rate of each population observed was transformed as a
relative value when that of the control was converted to 1.0.
Furthermore, the value of each population was arcsin transformed,
after it had been recalculated as a percentage of the maximum
value. The open circle represents the data for the control strain
(maintained for 2–3 years) (after Saito 1990b), and black circles
represent local populations observed (after Saito 1995b).

egg-laying. In five genera in Japan, *Tetranychus*, *Oligonychus*,
Sasanychus, *Eotetranychus* and *Schizotetranychus*, much variation exists in the utilization of webbing. In the last two
genera in particular, several species constructing very
sophisticated nest webs are known (Fig. 21-9). For example, females of *Eotetranychus suginamensis* Yokoyama, *E. shii*
Ehara and *Schizotetranychus brevisetosus* Ehara make dense
nest webs over depressions on leaf undersurfaces and live
gregariously within them. They always deposit their feces
on the web-nest roof (when depositing, climbing onto the
top of the nest), perhaps to reinforce their nests (Fig. 21-9)
(Yokoyama 1932). In these species, the nest appears to serve
as a shelter for offspring, such that parental care extends
into the nymphal stages. Table 21-1 shows the range of
nest utilization patterns observed in spider mites. Most
mite species with nest-weaving continue their cooperative
living in a nest for a relatively short period, probably

because of resource or habitat deterioration as well as
their short lifespans. If the leaf resource enclosed by nests
lasts sufficiently for mites to live longer, then the nest may
come to include many individuals, and the cooperative
interaction between individuals may become enhanced
(Table 21-1). In *S. longus* and *S. miscanthi*, the nest is maintained during at least two generations partly because of the
nest-enlarging habit (cf. Fig. 21-1) as well as the presence of
suitable host quality and quantity.

A species of Cheyletidae, *Hemicheyletia morii* Ehara,
which is thelytokous and predaceous, has a somewhat different kind of sociality (Mori 1989). Mites of this species
live gregariously on a woven mat (Mori 1989 called it a
nest) along the leaf edge or midrib of tropical evergreen
plants (*Ficus* and *Bambusa* in Malaysia). Individuals stand
in a circle surrounding their web mat, opening their powerful chelicerae towards the outside. Once a victim is trapped
by the chelicerae of a single mite, most of the nest members
simultaneously aggregate to the victim, beginning to feed
on it without aggression between them. Furthermore, one
of the nest members usually tries to throw the victim's
dead body from their nest after predation, which may function as nest sanitation as well as cleaning the trapping site
(Mori 1989). Eggs are laid in a web shelter made on the nest
web mat, and immatures live together with their mothers.
Group hunting and feeding with immatures strongly
suggest that females feed and protect their offspring;
the society of this species is a highly developed state of
subsociality.

Highly aggregative herbivorous species are found in
Eriophyidae and Tarsonemidae, both of which exhibit
male-haploidy. In some species of eriophyids in the genus
Eriophyes, which make galls in leaf tissue (Krantz 1978),
aggregation should have led to social interactions between
individuals, if there is selection for division of labor as in
gall aphids (Stern and Foster, this volume) and thrips
(Crespi and Mound, this volume). Indeed, eriophyids are
capable of complex behavior; Michaelska and Aoxiang
(1991) observed in eriophyid mites that males guarded an
area, in which their spermatophores were placed, against
conspecific males. However, no sufficient behavioral studies have so far been conducted on the species in this
family.

Extremely prolonged viviparity has been reported in
Pygmephoridae (a closely related family of Tarsonemidae) (Kaliszewski and Wrensch 1993). Both sexes of *Siter-
optes graminum* (Reuter) eclose and mate within their
mother's body (Baker and Wharton 1952; Kaliszewski and

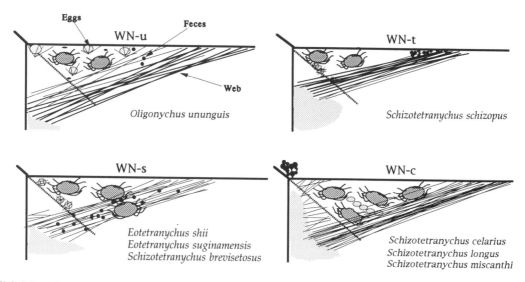

Fig. 21-9. Schematic expression of the nesting patterns of spider mites (WN, woven nest life types (after Saito 1983)).

Wrensch 1993). Although such viviparity is thought to be a form of parental investment, these species should not be considered to exhibit parental care.

Finally, males of the freshwater mite species *Unionicola formosa* (Dana et Whelpley), which is a symbiont of a freshwater mussel, establish harems and engage in mortal male–male combat (Dimock 1983). The genetic system is unknown for this species, although a species of the same genus is reported to be male-diploid (Solokov 1954).

Mesostigmata

The suborder Mesostigmata includes predaceous, omnivorous and parasitic species of mites. Usher and Davis (1983) stated that cannibalism in a predaceous soil mite, *Hypoaspis aculeifer* (Can.) (Laelapidae), which has a male-haploid genetic system, is not observed except under extreme conditions of hunger. In addition, they claim that this predator does not 'overkill' its collembolan prey, and

Table 21-1. *Social organization of spider mites*

No. of nests made per female life	Web density	Staying period in a nest	Overlapping generations	Cooperation between individuals	Life type[a]
Several	Sparse	Short	No (mother and larvae)	None	WN-u
Several	Medium to dense	Medium	At most 2 generations	In nest sanitation (depositing frass on nest web)	WN-s, WN-t
Single	Extremely dense	Long	2–3 generations	In nest sanitation (depositing frass at fixed places), cooperative nest repairing and enlargement	WN-c

[a] WN, 'woven nest' type; the letters after the hyphens are the initials of species representing each life type (see also Fig. 21-9).

that predation efficiency increases with predator density, especially for large prey (suggesting a kind of cooperative hunting). Therefore, they concluded that *H. aculeifer* has a kind of subsocial life. Although I hesitate to define mites with such behavior as subsocial, their study suggests consideration of the presence of kin selection for social adaptations in predaceous species. Indeed, similar phenomena have been also observed in spiders. Buskirk (1981) pointed out that cessation of intraspecific predatory behavior in spiders sometimes occurs at times when other sources of food are plentiful and aggregation is ecologically advantageous (see also Uetz and Hieber, this volume). Such a tendency may thus represent a precondition for the evolution of sociality (Luczak 1971).

The Phytoseiidae, which includes many important species of mite used for pest management, may be one of the important groups of mites in which to seek examples of subsociality. Although there are many studies of their life histories (Takafuji and Chant 1976), behavior (Sabelis and Dick 1985), sex ratios (Sabelis and Nagelkerke 1993) and genetic systems (e.g. paternal genome loss, a kind of male-haploidy, Shulten 1985), there have been few reports on social interactions in this family.

Low rates of cannibalism have been reported in some phytoseiid species, such as *Phytoseiulus persimilis* Athias-Henriot and *Amblyseius longispinosus* Evans (Mori *et al.* 1990). Furthermore, these two species have another peculiarity common to *H. aculeifer*, i.e. they have non-feeding larvae (Takafuji and Chant 1976; Saito and Mori 1981; Murphy and Sardar 1991). Non-feeding larvae, as observed in these phytoseiidae and *H. aculeifer*, may be an adaptation to avoid cannibalism in kin groups (Y. Saito and M. Toyama, unpublished).

To examine the hypothesis that non-feeding larvae evolved by kin selection, I will compare several phytoseiid species having different ecology. *Amblyseius eharai* Amitai et Swirski and *Iphiseius degenerans* Berlese must feed before molting (Takafuji and Chant 1976; Saito and Mori 1981), but there are no data on the rate of cannibalism in these species. They have relatively slow development, feed on sparsely distributed prey and have an ability to eat plant substances such as pollen (Saito and Mori 1975; Takafuji and Chant 1976). If there is little chance of sib-cannibalism, then larvae would be selected to eat as soon as possible after hatching.

By contrast, adults of *P. persimilis* and *A. longispinosus*, whose larvae are non-feeding, are efficient predators that feed mainly on living spider mites, and their developmen-

tal periods are quite short (Takafuji and Chant 1976; Saito and Mori 1981). Females of these species lay many eggs in a colony of spider mites in the genus *Tetranychus* in a relatively short time. Thus, if larvae of these predators must eat before molting, their food inevitably includes sibling eggs as well as hatching larvae (in fact, starving nymphal mites will eat conspecific eggs and larvae). In such a situation, non-feeding larvae may have evolved under kin selection, because prevention of sib-cannibalism is advantageous. In a similar context, it is important to note that first-instar larvae of spiders do not feed (Savory 1964). In addition, spider eggs deposited as a large egg mass hatch simultaneously, so that all larvae from the same egg mass probably have little chance to feed on their sibling eggs.

Thus, some phytoseiids aggregate and interact between sibling individuals without cannibalism. There is an interesting report of predatory behavior in a phytoseiid species: Cloutier and Johnson (1993) observed that young nymphs of *Amblyseius cucumeris* Oudemans have difficulty preying on thrips larvae by themselves, such that they can develop normally only when there are conspecific gravid females who hunt the prey for them. Although Cloutier and Johnson (1993) did not consider this phenomenon as social organization, I suggest that this phytoseiid species performs a kind of parental care, feeding of young. Such tendencies may provide an important background to the evolution of developed sociality in phytoseiids, as in spiders (Seibt and Wickler 1987).

There is an interesting example of a finely organized subsociality in moth-ear mites. Sociality in the Acari was first discovered by Treat (1975). A moth-ear mite, *Dicrocheles phalaenodectes* Treat, which parasitizes the ear chambers of noctuid moths, utilizes different chambers of the ear for different purposes: the first and second chambers are used for feeding and oviposition, and the third chamber is used for mating and deposition of feces. Thus, there is a regulation mechanism for public sanitation. Furthermore, adults of this mite defend their chambers against intruders. Because the chambers include many of their offspring, this defense behavior may represent a kind of parental care (similar to subsocial spider mites), even if they give no special care directly to offspring (Treat 1975). The genetic system of this mite species is assumed to be male-haploidy (Norton *et al.* 1993).

The most noteworthy aspect of the colonies of moth-ear mites is its unilaterality, i.e. most adult mites parasitize only one of the moth's ears, even if they reach the host at different times. This habit may be an adaptation for increasing the survival of their host, and thereby their own survival. If so, then this behavior has evolved by group selection (Maynard Smith 1976), because such cooperation should only evolve in the context of reducing the probability of extinction of the group (i.e. a moth with mite colonies in both ears).

Donzé and Guerin (1994) reported a case of subsociality in male-haploid mites parasitizing honey-bee brood. In *Varroa jacobsoni* Oudemans, they observed that there is a fecal accumulation on the wall of the honey bee's brood cell that serves as a rendezvous site for mating and as a deposition place for cell sanitation. Furthermore, the foundress only feeds at a particular single site on the bee larva, probably increasing offspring survival through avoidance of drowning owing to hemorrhaging of the host.

However, there appear to be fewer cases of cooperation or kin selection in Mesostigmata than in Prostigmata. One reason for this difference may be the lack of philopatry in most species of Mesostigmata, as a result of their omnivorous and predaceous habits. If there are species that live gregariously in a restricted habitat for a long period, and there are habitats that permit high longevity, some kinds of social organization through kin selection may be expected. More studies are necessary on mites of this suborder, especially moth-ear mites.

Ixodida (Mesostigmata)

This suborder is divided into two subgroups, soft-shell ticks (Argasidae) and hard-shell ticks (Ixodidae), most of which are parasites of endothermic animals (Krantz 1978). Most taxa in this suborder exhibit male-diploidy or thelytoky, and sometimes these two systems are simultaneously present in a single species. The male-diploid genetic system, and parasitic habits, of members of this taxon may have resulted in their apparent lack of sociality. However, they do exhibit some complex and interesting mating behavior (Feldman-Musham 1986).

Oribatei and Astigmata

Most families of Sarcoptiformes (Oribatei + Astigmata) have a male-diploid genetic system (Norton *et al.* 1993). Although from a consideration of genetic system there seems to be a relatively low probability of social organization in these groups, the aggregative nature of some taxa in this group may lead to behavioral interaction among kin.

For example, pheromonal communication systems for aggregation and for avoiding predators' attacks (alarm pheromones) have been reported in many Astigmata species (Kuwahara *et al.* 1979; Kuwahara 1991). If a group consists of kin, as is probable, the mechanism of social organization at group level may be kin selection. Some life-historical studies on mites associated with bees (Eickwort 1993) suggest gregarious life in a narrow space, such as bee nests, and such association of individuals may also result in close interactions between kin.

GENETIC, MORPHOLOGICAL, DEVELOPMENTAL AND BEHAVIORAL FACTORS IN THE EVOLUTION OF MITE SOCIAL BEHAVIOR

In Table 21-2, I depict factors related to the evolution of spider mite subsociality. First, overlap of generations in time and space, which is one of the important traits for social organization (Wilson 1975), has been enhanced in some spider mite species under stable environmental conditions. A temporally stable habitat may help spider mites to survive longer in the same place. However, the longer a colony remains in the same location, the larger the risks of accumulating natural enemies (Southwood 1977) (see also Edgerly, this volume). Therefore, mites living in such habitats should have evolved adaptations to avoid or repel such enemies (e.g. parental defense). Comparisons between the life histories of several spider mite species (Fig. 21-10) show that there appears to be a positive relationship between the extent of generation overlap and the relative stability of their habitats. The extent of generation overlap in *S. longus* is the largest (it reaches about three generations in Fig. 21-10) and appears to be closely related to the organization of sociality, in particular the presence of defense and nest sanitation. One of the ecological prerequisites of social organization of spider mites is thus a stable habitat with sufficient resources to enable mites to overlap generations (Saito 1986a).

Second, the degree of aggregation of individuals is also important for social organization. If organisms lay their eggs in masses and are semelparous, then aggregation of individuals should be inevitable, at least during their early developmental stages if there is no selection for dispersal (e.g. danger of cannibalism in predaceous species, such as spiders). However, most spider mite species exhibit continuous iteroparity, such that there is less tendency for aggregation of individuals. In fact, some spider mite species

Table 21-2. *Conditions and factors affecting social organization of Acari*

Level of factors	Items	Enhancing factors	Elements of sociality
Genetics	Genetic system	Male-haploidy and thelytoky	High relatedness (high tolerance to inbreeding)
Morphology	Movement ability	Low mobility	High relatedness (by inbreeding)
Life history	Development and oviposition	Continuous iteroparity and rapid development	Overlapping generations
Behavior	Distribution pattern of individuals	Aggregation	High probability of sib interaction
Ecology	Shareable labor among individuals	Prevention of predation; nest-building and hunting	Cooperative nest building, hunting and defense
Environment	Resources and habitat conditions	Spatiotemporally stable	Longer life and aggregation, producing generation overlap

making no web do not form aggregations (e.g. *A. corpuzae*). Thus, aggregation for ecological reasons (such as selection to create a woven nest) appears to be the second important ecological prerequisite for social organization. Nest formation by webs probably evolved as a refuge from predators and shelter from adverse climates in a stable habitat, and it has simultaneously enhanced aggregation of individuals in a narrow space. Moreover, prolongation of the useful period of a single nest may be a function of the density of individuals and the amount of resources enclosed.

Another factor that appears to be responsible for the extent of mite social evolution is whether there is labor that can be shared between individuals. In spider mites, there is little labor that can be engaged in by different individuals, so that there appears to be little opportunity for selection of division of labor. Nest building and repairing of webs, observed in *S. longus* and *S. miscanthi* may thus be completed by nest members in a communal way.

Among mites, inbreeding and a female-biased sex ratio are both thought to be peculiar to male-haploid

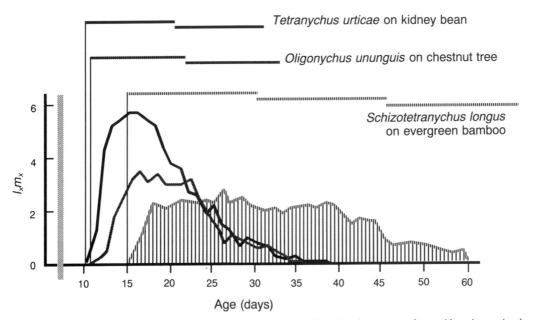

Fig. 21-10. Three generations potentially overlap in *Schizotetranychus longus* Saito. The $l_x m_x$ curves of two spider mite species. l_x = age-specific survival rate, m_x = (age-specific oviposition) × (proportion of females) × (egg hatchability).

species (Hamilton 1967). These phenomena are important with respect to kin selection, because they increase the relatedness between interacting individuals. However, as I describe in the next section, 'inbreeding' may become a double-edged sword for the evolution of sociality.

INBREEDING, OUTBREEDING, AND THE PRESENCE AND ABSENCE OF EUSOCIALITY

Hamilton (1987) stated that the 3/4-relatedness hypothesis does not explain why societies of highly related individuals are not more cooperative. This difficulty is exemplified by the weak interactions between individuals of many clonal species and also by the lack of eusociality in many male-haploid species in which individuals live gregariously with their kin, e.g. subsocial spider mites. The absence of castes in these organisms has been due, for example, to constraints of morphology (simple neural systems and a lack of kin recognition ability) or ecology (a lack of shareable labor). However, if such species have been inbreeding for a

long time, because of their low mobility (Mitchell 1973) and male-haploidy (the latter making inbreeding more likely (Matsuda 1987; Atmer 1991)), the lack of a shift from outbreeding to inbreeding in their evolutionary process suggests another explanation for the phenomenon.

Saito (1994a) proposed a hypothesis to explain three basic questions in eusocial evolution, i.e. (1) why the evolution of eusociality has occurred so often in male-haploids; (2) why the worker and soldier castes are restricted to adult females in male-haploids; and (3) why societies with highly related individuals are not more cooperative from the expected nature of mutations affecting the evolution of non-reproductive castes (i.e. altruism). By this hypothesis, unevenness in the reproductive ability of bisexual organisms between sibs, which may facilitate the origin of caste differentiation, is caused by mildly deleterious recessive genes, which cause homozygous females to be sterile under inbreeding conditions (Fig. 21-11). Saito (1994a) focussed on the similarity of equilibrium frequencies of female-limited deleterious recessives between male-haploidy and male-diploidy, and on the lower frequency of

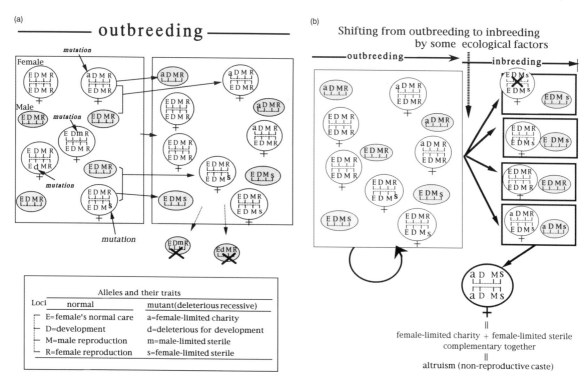

Fig. 21-11(a). A schema of the possible process of linking up the sterile and altruistic genes in male-haploidy under outbreeding.
(b). A schema (continued from a) of the possible expression process of the sterile and altruistic genes in male-haploidy when the mating structure shifts from outbreeding to inbreeding.

Table 21-3. *Asymmetry in equilibrium frequency of a deleterious recessive gene*

Variables: u, mutation rate; s selection coefficient.

Genetic systems	Common to both sexes	Sex-limited	
		female	male
Male-haploidy	$3u/s$	$\sqrt{3u/2s}$	$3u/s$
Male-diploidy	$\sqrt{u/s}$ [a]	$\sqrt{2u/s}$	$\sqrt{2u/s}$
Supposed depression	Decrease in immature survival (egg lethality and abnormal morphs)	Decrease in fertility (sterility) and in adult longevity and abnormal morphs	
Supposed peculiarities of male-haploidy	High tolerance of inbreeding conditions	Depression occurs selectively in adult females (Helle and Overmeer 1973; Bruckner 1978)	

[a] This approximation is only accurate when s is relatively small. If s is large, the frequency is much lower than this value (Werren 1993).

non-sex-limited genes in male-haploidy than in male-diploidy under outbreeding conditions (Crozier 1985) (Table 21-3). Such a deleterious effect of female-limited recessives occurs only in adult females under male-haploidy, where both females and males have a high tolerance to inbreeding. Assume that two recessive mutations occur by chance at different loci in a gene pool, one governing female sterility and the other affecting females' care behavior towards non-offspring (i.e. erroneous care; that is, charity). If these genes become complementary when linking up and becoming homozygous in an individual, then such a gene set may be regarded as an 'altruistic gene' (Saito 1994a, 1995a). The frequencies of such deleterious genes tend to reach their own equilibria depending upon mutation and selection rates. If matings occur at random (outbreeding) and the population is sufficiently large, these genes on different loci become linked with each other by chance in the genome of heterozygous females (Fig. 21-11A). By contrast, under inbreeding conditions these genes are separately exposed in homozygous females and are easily removed by selection without any chance of linking up together (Fig. 21-12). Therefore, the likelihood of retaining such genes and of their linking up are higher under outbreeding than under inbreeding.

If the above scenario is correct, then the conditions under which the linked genes are expressed as altruism become important. The shift from outbreeding to inbreeding is an important prerequisite for the evolution of altruism in male-haploid organisms, as shown in Fig. 21-11B.

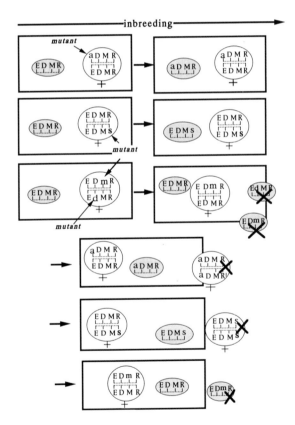

Fig. 21-12. A schema of the selection process of sterile and altruistic genes in male-haploidy under consistently inbreeding conditions.

Organisms of low mobility, such as mites, cannot satisfy the conditions of Saito's (1994a) model, because a single gene for sterility or erroneous care arising by mutation is rapidly selected against without any chance of complementary expression (Fig. 21-12). In other words, mites have never undergone shifts from outbreeding to inbreeding, and I believe that this is one of the reasons why we have no eusociality in the subclass Acari.

On the other hand, it is very difficult to evaluate the effect of the shift from outbreeding to inbreeding at the original state of eusocial evolution in the Hymenoptera, aphids, thrips, termites, and the naked mole rat. As Michod (1993) recently stated, whether inbreeding can have different roles to play in the initial stages of social evolution than it does in the maintenance of established social structures remains an open problem.

Outbreeding is also important to avoid inbreeding depression, such that organisms may first evolve kin-recognition ability to breed with partners other than kin (Bateson 1983; Spiess 1987). Such recognition mechanisms could be a preadaptation for the evolution of eusociality, whereby individuals should discriminate kin and non-kin for increasing their inclusive fitness through altruism and detecting non-cooperative cheaters (Hamilton 1987). This hypothesis may also explain why there is little evidence of eusociality in organisms having kin aggregation and low mobility, as they have no kin-recognition ability (Saito 1994b). However, this reasoning does not explain why there is more evidence of social evolution under male-haploidy than male-diploidy, which is one of the major questions of eusocial evolution. If kin-recognition mechanisms are preadaptations to eusocial life, then they should evolve more easily under male-diploidy than under male-haploidy, because of the former's extremely low tolerance to inbreeding (Werren 1993; see also Table 21-3).

It is currently very difficult to judge the plausibility of the arguments described above. The first hypothesis basically concerns the question of what gene set (if it really exists) governs caste differentiation in eusocial species; the second is concerned with the mechanical or neural background of kin selection. Thus, the two explanations need not be mutually exclusive.

SUGGESTIONS FOR FUTURE RESEARCH

From the point of view of a further discovery of sociality in Acari, as stated repeatedly, the Prostigmata (especially the families Tetranychidae, Eriophyidae and Cheyletidae) is a very promising group, because it involves male-haploid and thelytokous genetic systems and has a very aggregative nature. However, I wonder whether we will discover eusocial species of Acari in the future, because there is little shareable labor in their lives. If my hypothesis (Saito 1994a) is correct, the breeding structure (i.e. the high probability of inbreeding) has perhaps also suppressed such an evolutionary change.

On the other hand, the genetic background of the origin of sociality should be studied from the view of population genetics: What kind of deleterious genes are retained in a gene pool and how frequent are they? I suggest that most species of Acari appear to be very good material for such genetic examinations, because they have short developmental periods and are very easy to rear experimentally. Diversity in the genetic systems of a family is also a great advantage for such studies. Therefore, I believe that we can obtain much information about the effects of kin selection and the constraints of genetics on eusocial evolution from this subclass; such information may be of great help in reconstructing the original states or backgrounds of eusocial evolution, as suggested in Saito (1994a).

ACKNOWLEDGEMENTS

I thank the editors of this book, B. Crespi and J. C. Choe, who kindly invited me to contribute to this book and suggested much important information. Thanks are also due to D. L. Wrensch, who recommended me as a participant of this book to the editors. The manuscript benefitted greatly from the comments of Y. Itô, N. Yamamura and S. Kudo. I also thank L. Keller and G. Donzé, who gave me valuable comments.

LITERATURE CITED

Atmer, W. 1991. On the role of males. *Anim. Behav.* **41**: 195–205.

Baker, E. W. and G. W. Wharton. 1952. *An Introduction to Acarology.* New York: The Macmillan Company.

Bateson, P. 1983. Optimal outbreeding. In *Mate Choice.* P. Bateson ed., pp. 255–259. Cambridge University Press.

Buskirk, R. E. 1981. Sociality in the Arachnida. In *Social Insects.* vol. 2. H. R. Hermann, ed., pp. 282–367. New York: Academic Press.

Bruckner, D. 1978. Why are there inbreeding effects in haplo-diploid systems? *Evolution* **32**: 456–458.

Cloutier, C. and S. Johnson. 1993. Interaction between life stages in a phytoseiid predator: western flower thrips prey killed by adults as food for protonymphs of *Amblyseius cucumeris. Exp. Appl. Acarol.* **17**: 441–449.

Crozier, R. H. 1985. Adaptive consequences of male-haploidy. In *Spider Mites. Their Biology, Natural Enemies and Control*, vol. 1A. W. Helle and M. W. Sabelis, eds., pp. 201–222. Amsterdam: Elsevier.

Dimock, R. V. Jr. 1983. In defense of the harem: interaspecific aggression by male water mites (Acari: Unionicolidae). *Ann. Entomol. Soc. Am.* 76: 463–465.

Donzé, G. and P. M. Guerin. 1994. Behavioral attributes and parental care of *Varroa* mites parasitizing honeybee brood. *Behav. Ecol. Sociobiol.* 34: 305–319.

Eickwort, G. C. 1993. Evolution and life-history patterns of mites accociated with bees. In *Mites, Ecological and Evolutionary Analyses of Life-History Patterns.* M. A. Houck, ed., pp. 218–251. New York: Chapman and Hall.

Feldman-Musham, B. 1986. Observations on the mating behavior of ticks. In *Morphology, Physiology, and Behavioral Biology of Ticks.* J. R. Sauer and J. A. Hair, eds., pp. 217–232. Chichester: Ellis Horwood Limited.

Hamilton, W. D. 1964a. The genetical evolution of social behaviour I. *J. Theor. Biol.* 7, 1–16.

–. 1964b. The genetical evolution of social behaviour II. *J. Theor. Biol.* 7: 17–52.

–. 1967. Extraordinary sex ratio. *Science (Wash., D.C.)* 16: 477–488.

–. 1972. Altruism and related phenomena, mainly in social insects. *Annu. Rev. Ecol. Syst.* 3: 193–232.

–. 1987. Kinship, recognition, disease and intelligence: constraints of social evolution. In *Animal Societies. Theories and Facts.* Y. Itô, J. L. Brown and J. Kikkawa, eds., pp. 81–102. Tokyo: Japan Scientific Societies Press.

Hammen, L. van der. 1970. La phylogénese des Opilioacarides, et leurs affinites avec les autres acariens. *Acarologia* 12: 465–473.

Helle, W. and W. P. J. Overmeer. 1973. Variability in tetranychid mites. *Annu. Rev. Entomol.* 18: 97–120.

Kaliszewski, M. and D. L. Wrensch. 1993. Evolution of sex determination and sex ratio within the mite cohort tarsonemina (Acari: Heterostigmata). In *Evolution and Diversity of Sex Ratio in Insects and Mites.* D. L. Wrensch and M. A. Ebbert, eds., pp. 192–213. London: Chapman and Hall.

Krantz, G. W. 1978. *A Manual of Acarology*, 2nd edn. Corvallis: Oregon State University Book Stores.

Kuwahara, Y. 1991. Pheromonal communication of mites and ticks. In *Modern Acarology*, vol. 1. F. Dubabek and V. Bukva, eds., pp. 43–52. The Hague: Academia, Prague and SPB Academic Publishing bv.

Kuwahara, Y., H. Fukami, S. Ishii, K. Matsumoto and Y. Wada. 1979. Pheromone study on acarid mites. II. Presence of alarm pheromone in the mold mite, *Tyrophagous putrescentiae* Schrank (Acarina: Acaridae) and the site of its production. *Jap. J. Sanit. Zool.* 30: 309–314.

Luczak, J. 1971. Skupienia srodowiskowe i socjalne pajakow. *Wiad. Ekol.* 17: 404–412.

Matsuda, H. 1987. Conditions for the evolution of altruism. In *Animal Societies. Theories and Facts .* Y. Itô, J. L. Brown and J. Kikkawa, eds., pp. 67–80. Tokyo: Japan Scientific Societies Press.

Maynard Smith, J. 1976. Group selection. *Q. Rev. Biol.* 51: 277–283.

Michaelska, K. and S. Aoxiang. 1991. Spermatophore deposition and competition between males of an eriophyid mite *Vasates robiniae* (Acari: Eriophyidea) in the presence of female quiescent deutonymphs. *Abstract of the 22nd International Ethological Conference*, Kyoto, 22–29 August 1991, no. P-8-57. (Unpublished.)

Michod, R. E. 1993. Inbreeding and the evolution of social behavior. In *The Natural History of Inbreeding and Outbreeding.* N. W. Thornhill, ed., pp. 74–96. Chicago: The University of Chicago Press.

Mitchell, R. 1973. Growth and population dynamics of a spider mite (*Tetranychus urticae* K., Acarina: Tetranychidae). *Ecology* 54: 1349–1355.

Mori, H. 1989. The social predaceous mite, *Hemicheyletia morii* Ehara, and its behavior. *Abstract of International Symposium on Biological Control: A Century of Success*, Riverside, 27–30 March 1989, no. 51. (Unpublished.)

Mori, H., Y. Saito and H. Natao. 1990. Use of predatory mites to control spider mites (Acari, Tetranychidae) in Japan. In *The Use of Parasitoids and Predators to Control Agricultural Pests.* J. Bay-Petersen, ed., pp. 142–156. (FFTC Book Ser. no. 40.) Taipei: FFTC.

Murphy, P. W. and M. A. Sardar. 1991. Resource allocation and utilization contrasts in *Hypoaspis aculeifer* (Can.) and *Alliphis halleri* (G & R. Can.) (Mesostigmata) with emphasis on food source. In *The Acari. Reproduction, Development and Life-History Strategies.* R. Schuster & P. W. Murphy, eds., pp. 302–311. London: Chapman and Hall.

Norton, R. A., J. B. Kethley, D. E. Johnston and B. M. O'Connor. 1993. Phylogenetic perspectives on genetic systems and reproductive modes of mites. In *Evolution and Diversity of Sex Ratio in Insects and Mites.* D. L. Wrensch and M. A. Ebbert, eds., pp. 8–99. New York: Chapman and Hall.

Osakabe, Mh., Y. Saito and Y. Sakagami. 1993. Protein differences detected by two-dimensional electrophoresis in the species complex of *Schizotetranychus celarius* (Banks). *Exp. Appl. Acarol.* 17: 757–764.

Potter, D. A., D. L. Wrensch and D. E. Johnston. 1976. Guarding, aggressive behavior, and mating success in male two-spotted spider mites. *Ann. Entomol. Soc. Am.* 69: 707–711.

Sabelis, M. W. and M. Dick. 1985. Long-range dispersal and searching behaviour. In *Spider Mites. Their Biology, Natural Enemies and Control*, vol. 1B. W. Helle and M. W. Sabelis, eds., pp. 141–160. Amsterdam: Elsevier.

Sabelis, M. W. and K. Nagelkerke. 1993. Sex allocation and pseudoarrhenotoky in phytoseiid mites. In *Evolution and Diversity of Sex Ratio in Insects and Mites.* D. L. Wrensch and M. A. Ebbert, eds., pp. 477–511. New York: Chapman and Hall.

Saito, Y. 1979. Study on spinning behavior of spider mites III. Responses of mites to webbing residues and their preferences for particular physical conditions of leaf surfaces (Acarina: Tetranychidae). *Jap. J. Appl. Entmol. Zool.* **23**: 82–91 (in Japanese with English summary).

–. 1983. The concept of "life types" in Tetranychidae. An attempt to classify the spinning behaviour of Tetranychinae. *Acarologia* **24**: 377–391.

–. 1985. Life types of spider mites. In *Spider Mites. Their Biology, Natural Enemies and Control,* vol. 1A. W. Helle and M. W. Sabelis, eds., pp. 253–264. Amsterdam: Elsevier.

–. 1986a. Biparental defense in a spider mite (Acari, Tetranychidae) infesting *Sasa* bamboo. *Behav. Ecol. Sociobiol.* **18**: 377–386.

–. 1986b. Prey kills predator: counterattack success of a spider mite against its specific phytoseiid predator. *Exp. Appl. Acarol.* **2**: 47–62.

–. 1987. Extraordinary effects of fertilization status on the reproduction of an arrhenotokous and sub-social spider mite (Acari: Tetranychidae). *Res. Pop. Ecol.* **29**: 57–71.

–. 1990a. Two new spider mite species of the *Schizotetranychus celarius* complex (Acari: Tetranychidae). *Appl. Entomol. Zool.* **25**: 389–396.

–. 1990b. 'Harem' and 'non-harem' type mating systems in two species of subsocial spider mites (Acari, Tetranychidae). *Res. Pop. Ecol.* **39**: 263–278.

–. 1994a. Is sterility by deleterious recessives an origin of inequalities in the evolution of eusociality? *J. Theor. Biol.* **166**: 113–115.

–. 1994b. Do males of *Schizotetranychus miscanthi* (Acari, Tetranychidae) recognize kin in male competition? *J. Ethol.* **12**: 15–18.

–. 1995a. "Altruistic gene" *a priori* involves two different deleterious gene effects. *J. Theor. Biol.* **174**: 471–472.

–. 1995b. Clinal variation in male-to-male antagonism and weaponry in a subsocial mite. *Evolution* **49**: 413–417.

Saito, Y. and H. Mori. 1975. The effects of pollen as an alternative food for three species of phytoseiid mites (Acarina: Phytoseiidae). *Mem. Fac. Agr. Hokkaido Univ.* **9**: 236–246 (in Japanese with English summary).

–. 1981. Parameters related to potential rate of population increase of three predaceous mites in Japan (Acarina: Phytoseiidae). *Appl. Entomol. Zool.* **16**: 45–47.

Savory, T. 1964. *Arachnidae.* London: Academic Press.

Seibt, U., and W. Wickler. 1987. Gerontophagy versus cannibalism in the social spiders *Stegodyphus mimosarum* Pavesi and *Stegodyphus dumicola* Pocock. *Anim. Behav.* **35**: 1903–1905.

Shulten, G. G. M. 1985. Pseudo-arrhenotoky. In *Spider Mites. Their Biology, Natural Enemies and Control,* vol. 1B. W. Helle and M. W. Sabelis, eds., pp. 67–71. Amsterdam: Elsevier.

Southwood, T. R. E. 1977. Habitat, the templet for ecological strategies? *J. Anim. Ecol.* **46**: 337–365.

Spiess, E. B. 1987. Discrimination among prospective mates in *Drosophila.* In *Kin Recognition in Animals.* D. J. C. Fletcher and C. D. Michener, eds., pp. 73–119. Chichester: John Wiley and Sons.

Solokov, I. I. 1954. The chromosome complex of mites and its importance for systematics and phylogeny. *Trud. Leningr. Obshch. Estestvois. Otd. Zool.* **72**: 124–159.

Takafuji, A. and D. A. Chant. 1976. Comparative studies on two species of predaceous phytoseiid mites (Acarina: Phytoseiidae), with special reference to their responses to the density of their prey. *Res. Pop. Ecol.* **17**: 255–310.

Treat, A. E. 1975. *Mites of Moths and Butterflies.* Ithaca: Comstock Publishing Associates.

Usher, M. B. and P. R. Davis. 1983. The biology of *Hypoaspis aculeifer* (Canestrini) (Mesostigmata): Is there a tendency towards social behaviour? *Acarologia* **24**: 243–250.

Werren, J. H. 1993. The evolution of inbreeding in haplodiploid organisms. In *The Natural History of Inbreeding and Outbreeding. Theoretical and Empirical Perspectives.* N. W. Thornhill, ed., pp. 42–59. Chicago: The University of Chicago Press.

Wilson, E. O. 1975. *Sociobiology. The New Synthesis.* Cambridge: Belknap Press of Harvard University Press.

Wrensch, D. L. 1993. Evolutionary flexibility through haploid males or how chance favors the prepared genome. In *Evolution and Diversity of Sex Ratio in Insects and Mites.* D. L. Wrensch and M. A. Ebbert, eds., pp. 118–149. New York: Chapman and Hall.

Yamamura, N. 1987. Biparental defense in a subsocial spider mite. *Trends Ecol. Evol.* **2**: 261–262.

Yokoyama, K. 1932. New tetranychid mites attacking the mulberry leaf. Contribution I. Bionomics and external structures of *Tetranychus suginamensis* n. sp. *Bull. Imp. Sericult. Exp. Sta.* **8**: 229–282, 2 pls. (in Japanese with English summary).

22 · Colonial web-building spiders: balancing the costs and benefits of group-living

GEORGE W. UETZ AND CRAIG S. HIEBER

ABSTRACT

Colonial, or territorial permanent-social, web-building spiders in the orb-weaving families Araneidae and Ulobor-idae exhibit a combination of solitary and social behavior. These spiders maintain individual territories within the aggregation or colony and do not disperse before reaching reproductive maturity. Although all colony members contribute to the communal silk framework of the colony to some degree, each individual builds, occupies and defends its own orb web, and prey are usually not shared. In this chapter, we focus on the ecological conditions and behavioral characteristics that influence the evolution of sociality. We describe and compare the web structure and group-living arrangements seen in a variety of orb-weaving species, as well as apparent behavioral transitions from simpler to more complex forms of social behavior within taxa.

Long-term ecological research on a single taxon (*Metepeira*) has identified the primary benefits (enhanced prey capture and reduced silk costs) and costs (increased rates of predation and egg-sac parasitism) of group-living. Here, we present new data from our research on one species in this genus (*M. incrassata*) and examine how these benefits and costs vary with group size and affect individual fitness.

Extensive research on a variety of spider taxa suggests that differences in life histories and the internal organiza-tion of colonies (regarding web territories and communal retreats) reflect divergent subsocial and parasocial path-ways to the evolution of group-living. However, there remain a number of unanswered questions concerning the genetic structure of colonial web-building spider popula-tions. The degree of relatedness within and between colo-nies has not yet been studied, and the potential role of kin selection in the evolution of colonial behavior in spiders remains unclear.

INTRODUCTION

Most of the world's 34 057 described spider species (N. Platnick, personal communication) are solitary, aggres-sive, territorial and cannibalistic predators. However, approximately 50 species of spider exhibit some form of group-living, and this behavior has apparently evolved independently several times (Shear 1970; Kullman 1972; Brach 1977; Burgess 1978; Buskirk 1981; Krafft 1982; Uetz 1986a; D'Andrea 1987). Social spiders have not been stu-died as comprehensively as the social insects or vertebrates (Wilson 1971, 1975; Brown 1975; Oster and Wilson 1978; Wittenberger 1981; Hermann 1982); this is unfortunate, as the 'social' spiders are so clearly atypical for their taxon. These spiders are particularly interesting because they lack both the haplodiploid sex-determining mechanism suggested as one important force in the evolution of most eusocial insects (Hamilton 1964; Wilson 1971, 1975; Oster and Wilson 1978) and the major features of insect social-ity: reproductive division of labor and morphological castes. In these regards, spider sociality more closely resembles that of some vertebrates and other insect taxa than the social insects (see Crespi and Choe, Chapter 24, this volume). These attributes make spiders a unique alter-native model system for understanding the evolution of social behavior and the various factors that have directed that evolution.

Social behavior in the spiders is a continuum ranging from temporary aggregations of a few webs to permanent web colonies containing thousands of individuals. Social-ity in spiders appears to take two basic forms; spiders are often referred to as either 'colonial' or 'social' based on their spatial organization and level of cooperative behavior (Burgess and Uetz 1982). In general, 'colonial' spiders join individual webs together within a communal web frame-work, but individuals build and occupy their own webs within a colony (Burgess and Witt 1976; Uetz and Burgess

1979; Burgess and Uetz 1982). In contrast, 'social' spiders live together in complex web-nests, cooperate in web construction and prey capture, and engage in communal feeding and sometimes indiscriminate brood care (Shear 1970; Kullman 1972; Burgess 1978; Buskirk 1981; Krafft 1982; Burgess and Uetz 1982; Smith 1983; Rypstra 1983). As terminology varies (see Crespi and Choe, Chapter 24, this volume; Wcislo, Chapter 1, this volume), we will seek to be consistent wherever possible with the other chapter on spider sociality by Avilés (this volume), who uses the terminology of D'Andrea (1987).

Among the colonial spiders, there is considerable variation in living arrangements, despite similarities in overall social structure. As a consequence, there have been difficulties in developing precise definitions of social structure, and in reconciling the levels of social behavior seen in spiders with the terminology previously established for the social insects. The term most commonly encountered is 'colonial' (Lubin 1974; Buskirk 1975a; Rypstra 1979; Burgess and Uetz 1982), although 'communal' (Cloudsley-Thompson 1981; Smith 1985), 'communal/territorial' (Jackson 1978) and 'territorial permanent-social' (D'Andrea 1987) have also been employed to describe this type of social organization. There is, however, general agreement that this form of group-living is distinctly different from the solitary foraging behavior demonstrated by most spiders and the cooperative behavior exibited by the 'social' or 'non-territorial permanent-social' spiders (Avilés, this volume).

REVIEW: ATTRIBUTES OF SELECTED TAXA

All of the colonial web-building spiders studied to date are from the orb-weaving families Araneidae, Tetragnathidae and Uloboridae, which have recently been shown to form a monophyletic group (Coddington and Levi 1991). However, phylogenetic relationships among and within orb-weaving taxa are currently undergoing re-examination and are still unclear. Moreover, the degree to which the social behavior of species within these taxa has been studied varies greatly. Nonetheless, evolutionary insights may be gained from a comparison of various behavioral and ecological characteristics of selected genera and species.

Common features of colonial or territorial permanent-social behavior include individual web-building within a common silk framework, individual prey capture and aggressive defense of the web. Cooperation in these species is limited in most cases to the construction of the silken framework of the colony, although communal prey capture and feeding are seen in some species (Fowler and Gobbi 1988; Breitwisch 1989; Binford and Rypstra 1992).

One way to examine differences among colonial orb-weaving species incorporates the relationship between the common web framework, the individual prey capture web and the retreat. In some genera (*Cyrtophora*, *Metepeira*), the individual web may be seen as a multipurpose 'territory' (Davies 1978), in that it may include a foraging site (the web), a habitation (the retreat), a mating site (web, retreat or both) and a breeding or egg-laying site (the retreat). In other genera (*Parawixia* (= *Eriophora*), *Metabus*, *Philoponella*), spiders forage on individual webs built within the communal scaffolding, but occupy a communal retreat off the web when they are not foraging. In these groups, the communal retreat may be used by both juvenile and adult spiders, or as a common defended area for egg-sac deposition.

This classification based on web structure, while appearing superficial, may reflect a more fundamental difference between taxa. Species exhibiting territorial defense of individual retreats within aggregations clearly contrast with those sharing a communal retreat separate from the foraging web (i.e. tolerance of conspecifics at distances of less than one centimeter, or closer than one body length). The structure of colonial webs is the product of the behavior of individuals whose social organization may also reflect significantly different behavioral origins on divergent pathways to sociality. Within the evolution of sociality in insects, as well as in spiders, sociality may have two distinct origins: the subsocial and parasocial pathways (Shear 1970; Buskirk 1981; Uetz 1986b). The subsocial route originates with extended parental care or prolonged association of siblings from the same egg sac, and may be influenced by kin selection. The alternative parasocial route originates from aggregation of possibly unrelated individuals, because of favorable ecological conditions, and may or may not involve kin selection. Evidence for both these origins of group-living may be seen in several orb-weaving taxa. In the absence of established phylogenies, comparisons based on the relationship between colonial web structure and social structure are appropriate.

To better understand the breadth of colonial living arrangements in various taxa, and to gain insights regarding the evolution of colonial behavior, we will briefly review a number of selected examples that illustrate the range of colonial behaviors seen among spiders. In this review, we will first compare colonial species within several genera of the family Araneidae (*Cyrtophora*, *Metepeira*, *Metabus*, *Parawixia* (= *Eriophora*)) in order to identify similarities and differences in social structure. Then, we will examine evolutionary trends among several well-studied species within a single genus (*Philoponella*) in the family Uloboridae. Finally, we will illustrate the ecological costs and benefits of group-living for a single species, *Metepeira incrassata* (Araneidae), upon which our field research is based.

Fortuitous aggregations of individuals of solitary species: evidence for parasocial origins

Many orb-weavers, and many other web-building spiders, have been observed to build webs together as the result of favorable concentrations of web attachment points, or favorable concentrations of prey. Web clumping is commonly seen in many otherwise solitary spiders, including *Dictyna folicola* (Dictynidae) (Honjo 1977), *Holocnemus pluchei* (Pholcidae) (Jakob 1991), *Nuctenea sclopetaria* (Araneidae) (McCook 1889; Burgess and Uetz 1982), *Metepeira daytona* (Araneidae) (Schoener and Toft 1983), *Zygiella x-notata* (Araneidae) (LeBorgne and Pasquet 1987a,b, 1994), *Nephila clavipes* (Tetragnathidae) (Brown *et al.* 1985; Rypstra 1983, 1985; Uetz and Hodge 1990) and *Tetragnatha elongata* (Tetragnathidae) (Gillespie 1987), and in both intra- and interspecific groups of orb-weavers in field enclosures fed at high levels of prey (Rypstra 1983).

The most frequently cited factor influencing aggregative behavior among solitary spiders is prey availability. For example, *Nephila clavipes* (Tetragnathidae) from tropical and subtropical habitats in the Americas build solitary webs, but can also be found in aggregations (2–20 individuals) where there is sufficient space to locate the webs, climatic factors are favorable, and prey density is high (Brown *et al.* 1985; Rypstra 1985; Uetz and Hodge 1990). Rypstra (1985) found both aggression and aggregation size in *Nephila* to be influenced by food availability. Experimental food deprivation resulted in increased aggression and dispersion of web aggregations; supplementation of prey increased aggregation size and maintained low aggression (Rypstra 1985). Gillespie (1987) found similar results when she manipulated prey densities on experimental groups of *Tetragnatha elongata* (Tetragnathidae). Excess prey, at high spider densities, allowed individual spiders to significantly reduce web building or give it up altogether. In addition, spiders indiscriminately used the silk of other spiders to capture prey. At low prey densities, spiders maintained individual orb-webs for prey capture.

Another mechanism responsible for aggregative behavior in orb-weaving spiders is an attraction to silk, or 'sericophily' (McCook 1889; Enders 1977; Gillespie 1987). Many solitary and facultatively colonial spider species exhibit a tendency to locate webs in places where they encounter silk, which may serve as an indicator of suitable web sites (Riechert and Gillespie 1986). Experimental studies with *Z. x-notata*, which aggregate around windows, has shown a tendency of spiders to build webs near or attached to the silk structures of conspecifics (LeBorgne and Pasquet 1987 a,b). Several authors have suggested that the kinds of behaviors seen in these fortuitous aggregations around extremely high prey densities, namely tolerance of shared silk, use of common web attachment points, and reduced aggression, constitute preadaptations for colonial web-building (Shear 1970; Kullman 1972; Burgess and Witt 1976; Burgess and Uetz 1982).

Territorial permanent-social spiders with individual defended retreats

Orb-weavers in the genera *Cyrtophora* and *Metepeira*, although phylogenetically distant (Levi and Coddington 1983), share a superficial similarity in that they both use a three-dimensional web design. Webs of individuals are complex, and include a prey-catching orb and a retreat within a barrier web consisting of irregular silk lines. Spiders occupy retreats above a horizontal orb (*Cyrtophora*), or to one side of a vertical orb (*Metepeira*), and actively defend them.

Cyrtophora
Spiders of the genus *Cyrtophora* are common orb-weavers from southeast Asia, Africa and the Australo-Pacific region. They are found in open areas, grasslands, forest clearings and forest edges, and can occur solitarily (*C. monulfi*, *C. cicatrosa*, *C. moluccensis*: Lubin 1973, 1974), in loose aggregations or small groups (*C. monulfi*, *C. moluccensis*: Lubin 1974), or in colonies of over 1000 individuals (*C. cicatrosa*, *C. moluccensis*, *C. citricola*: Tikader 1966; Clyne 1969; Lubin 1973, 1974; Sabath *et al.* 1974; Rypstra 1979; Buskirk 1981; Subrahmanyam 1986). All of the spiders in this genus use a similar web consisting of a horizontal, slightly domed, non-sticky orb-web with a network of tangled barrier threads above and below the orb. Individual webs tend to be quite durable and can be relatively large. In aggregations and colonies, the webs are connected to one another by shared barrier webs above and below the individual orb-webs.

A number of authors have suggested that the evolution of coloniality in the *Cyrtophora* has been caused primarily by the exploitation of new habitat; that is, aggregation facilitated invasion of prey-rich microhabitats, such as tree-fall gaps or the spaces between treetops, which are too large to be spanned by the webs of solitary spiders (Lubin 1974; Brach 1977; Rypstra 1979; Buskirk 1981). Once grouped, other benefits such as increased foraging efficiency, reduced web maintenance, and predator protection may have reinforced the advantages of grouping.

Dispersal may also play a role in the evolution of colonial behavior. Dispersal rates in *Cyrtophora* vary (Kullman 1958; Blanke 1972; Lubin 1974), but in some species (e.g. *C. moluccensis* and *C. citricola*) colonies have few dispersing individuals (Kullman 1958; Lubin 1974). Low dispersal rates coupled with maternal tolerance of spiderlings in the colony or web framework (*C. moluccensis* and *C. citricola*: Lubin 1974) would result in rapid colony growth and high relatedness among colony members. Because of this high relatedness, Lubin (1974) suggests that it is advantageous for individuals in colonies to remain together and share resources, because the benefits outweigh the potential costs (competition for prey and increased aggression).

Metepeira

Members of the genus *Metepeira* build three-dimensional web complexes with retreats, although in contrast to *Cyrtophora*, the orb is oriented vertically with a signal line connecting the web hub to the retreat. The existence of temporary aggregations in some solitary species (e.g., *M. labyrinthea*, McCook 1889; *M. daytona*, Schoener and Toft 1983) in which three-dimensional webs of unrelated individuals are joined together, as well as permanent colonies in other species (*M. spinipes*, *M. incrassata*) with individuals building webs within a common silk framework, suggest an evolutionary transition between the two arrangements. If so, this may be evidence supporting a parasocial origin for coloniality in this group.

Colonies of *Metepeira* are found in a variety of habitats (Uetz *et al.* 1982; Uetz and Hodge 1990), but all have in common the requirement of rigid structures for web support (Uetz and Burgess 1979; Schoener and Toft 1983). For example, *M. spinipes* occurs in the agricultural valleys surrounding Mexico City where the habitat is primarily corn fields, agave plantations, and open range grazing with scattered cattle yards. Webs are located in relatively permanent structures including *Agave* spp., dead branches of trees, shrubs and man-made objects. Colony size and spacing

within colonies of *Metepeira spinipes* varies with the availability of prey (Uetz *et al.* 1982). Where prey are scarce, spiders live in small, widely spaced groups, but where prey are abundant, colonies are large and densely packed. Within colonies, spiders defend webs and retreats, as in *Cyrtophora*. As the behavior and ecology of *Metepeira* will be discussed in much greater detail below, this brief description will suffice for this review of taxa.

Territorial periodic- or permanent-social spiders with shared central retreats

Spiders in the genera *Metabus* and *Parawixia* (=*Eriophora*) build two-dimensional, vertically oriented orb-webs within a communal silk framework. In these species, there is also a communal retreat located off to the side of the colony, where individuals aggregate when not foraging on their orbs. Although individuals may defend their orbs against intrusion by other colony members (similar to *Cyrtophora* and *Metepeira*), they also tolerate conspecifics at extremely close quarters while in the communal retreat.

Metabus

Buskirk (1975a,b) studied *Metabus gravidus* (Araneidae) in Costa Rica, and found colonies of five to 75 individuals building webs within a communal framework attached to various vegetation points over running water. These spiders use a communal retreat, emerging during the day to build individually defended orb-webs in the pre-existing communal framework of silk. Prey captured by individuals are not shared among colony members. Both spacing within the colony and web defense rely on agonistic behavior, but surprisingly, this behavior is suspended when individual spiders take down their orb-webs and move into the communal retreat to spend the night. Buskirk (1975a) suggests that habitat selection, a preference for spinning within a group and maintenance of year-round populations result in permanent *Metabus* colonies. Based on an analysis of habitat parameters, web construction behaviors and prey capture rates (in locations in and around the edges of streams where these spiders occur), Buskirk concluded that the primary advantage gained by colonial web-building is the ability to exploit habitats that are unusable by solitary individuals (Buskirk 1975a).

Parawixia (=*Eriophora*)

Parawixia (=*Eriophora*) *bistriata* (Araneidae), an orb-weaver from semiarid habitats in Paraguay, Argentina, and Brazil, shows a number of behavioral characteristics similar to *Metabus*. Spiders share a communal retreat during the day, and each night individuals construct their own orb-webs within a communally generated scaffold of silk lines that radiate out from the retreat to nearby vegetation (Fowler and Diehl 1978; Sandoval 1987; Fowler and Gobbi 1988). However, this spider differs from *Metabus* in a number of important respects. First, this spider exhibits facultative communal prey capture and feeding; operating solitarily when prey is of equal size or smaller than the captor, but operating in groups of two or more when the prey is larger than the spiders involved (Fowler and Diehl 1978; Fowler and Gobbi 1988). This variability in capture behavior may be facilitated by the fact that spiders can move quickly from one orb to the next, because the orbs are placed within a single plane, forming extensive prey capture 'sheets' (Sandoval 1987, translated by L. Avilés, personal communication). Furthermore, colonies of *P. bistriata* are territorial periodic-social, composed of synchronously developing immature individuals from the same egg sac (Fowler and Diehl 1978). Upon maturing, the males and females within a colony disperse, lay eggs away from the colony site, and die. Fowler and Diehl (1978) suggest that the synchronous development of group-living siblings, facultative communal prey capture and feeding, and the dispersal of adults to lay eggs (and thus start new colonies) seen in *P. bistriata* is a suite of adaptations for colonizing suitable microclimates in a semiarid environment where the prey are patchily distributed in space and time.

Evolutionary transitions within a single genus: *Philoponella*

A number of species in the family Uloboridae are reported to be colonial, particularly members of the genus *Philoponella* (Muma and Gertsch 1964; Opell 1979). These species show a range of colonial behaviors that may have evolved along a number of different pathways.

Facultative coloniality: a parasocial origin?

One important evolutionary development seen among *Philoponella* spp. is facultative grouping in response to environmental factors. Both *Philoponella semiplumosa* (Spiller 1992) in the Bahamas and *P. oweni* (Smith 1982, 1983) in the American southwest and Mexican Sonoran deserts (Muma and Gertsch 1964) show solitary spiders and communal groups coexisting in the same habitat. For both of these spiders, solitary individuals build horizontal orbs

with an attached irregular barrier web and the spider sits at the hub of the orb. In colonies, individual webs are connected to one another by common barrier webs.

Surprisingly, web-site attachments seem to be more important than prey abundance or variance in explaining social variation in facultatively colonial *Philoponella*. Smith (1982) found that both solitary individuals and communal groups of 2–40 *P. oweni* individuals coexisted in the same habitat. The shift between these two states was mediated by the abundance of web-site attachments: when attachment points were scarce most spiders were colonial; when they were more abundant spiders were solitary. She did find, however, that colonies were located in microhabitats with high prey density (Smith 1982, 1983).

Similar results were seen in two studies of another species. In a short-term study of facultative colonial behavior within a single habitat, Spiller (1992) found no relationship between prey abundance and colony size in *P. semiplumosa*, and no effect of prey supplementation on colony size. Lahmann and Eberhard (1979) found considerable between-habitat variation in the frequency of solitaries and colonial groups of *P. semiplumosa* in Costa Rica, but within a single habitat found no difference in prey capture rates of solitary and colonial individuals. Evidence from these studies and those of Smith suggest that the tendency to aggregate is highly plastic, and that colonial behavior may have evolved in some *Philoponella* spp. as a response to variation in a number of habitat factors.

Communal retreats: subsocial behavior?

Other interesting developments seen within the genus *Philoponella* are the use of a communal retreat and communal prey capture behavior. In *P. oweni*, individual females in a colony maintain their own orb-webs and actively defend them, but use one or more centrally located retreats in the colony where they deposit and stay with their egg sacs (Smith 1982, 1983). Prey capture in this species, however, is always an individual effort. *Philoponella republicana*, on the other hand, is a highly communal, retreat-using species that constructs colonies in areas of complex rainforest understory or small tree-fall gaps (sites with large numbers of colony attachment points) with high prey abundance in Panama, Trinidad, and northern South America (Simon 1891; Hingston 1932; Lubin 1980; Smith 1985; Binford and Rypstra 1992). Colonies consist of attachment lines, individual orb-webs, and a centrally located retreat of tangled non-sticky silk. Individuals move off their orb-webs into the retreats in the evenings or when they are disturbed, and adult males and females guarding egg-sacs spend much of their time here as well. Spacing within the colony is mediated by relative prey abundance; individuals increase their distance from others as prey declines (Smith 1985). If prey declines to extremely low levels, then the whole colony may abandon the web site and move to a new location (Smith 1985). New colonies may be started by groups of immatures dispersing *en masse* (Lubin 1980). This species is distinctly different from *P. oweni*, however, in that it shows occasional cooperative prey capture (Binford and Rypstra 1992). This behavior is more frequent when the prey being captured is larger than any spider. Solitary spiders usually capture prey no larger than themselves.

Communal webs and cooperative prey capture: a transition to higher levels of sociality?

Within some species of *Philoponella*, a complete loss of individual territories (orb-webs) and a switch to a communal web (with retreat use maintained) and cooperative prey capture is seen. Struhsaker (1969) found that individuals of *Uloborus mundior* (= *P. republicana*) in colonies (14–51 individuals) were not restricted in movement over any part of the colonial web. However, he found no evidence of cooperation in the wrapping, killing, or eating of prey. Breitwisch (1989), working on an undescribed *Philoponella* species in wet lowland forest in southern Cameroon, found a single colony composed of an 'immense sheet web of a tangled mass of strands with no discernable individual orb-webs, although there were many distinguishable, more or less horizontal, layers'. Several thousand individuals (mostly immatures) occupied this colony, positioned primarily in the lower layers. Instead of a central retreat, the juveniles and adults made use of the many fallen leaves trapped within the layers as retreats. In contrast to *P. republicana*, Breitwisch (1989) found cooperative prey capture in this undescribed species. He found no correlation between the size of captured prey and the number of individual spiders participating in the attack, but feeding group size was positively correlated with prey size.

How did the use of communal webs and cooperative prey capture evolve in this group? Binford and Rypstra (1992) suggest that cooperative prey capture within the genus *Philoponella* is strongly connected to communal retreat use and the breakdown of territoriality. Presumably, cooperation evolved through low dispersal rates and the resulting high kinship of individuals within the colony. Two other aspects of uloborid biology may also have played a part. Uloborids are cribellate spiders: they use cribellate silk that is sticky like 'velcro' (rather than ecribellate silk, which is sticky because of an applied glue) and they have no poison glands. The loss of individual orb-webs as prey-capture devices and the development of communal prey-capturing traps may have been facilitated by the fact that cribellate silk can be fashioned into sheets that are still effective in holding prey. Once communal web-building evolved, the size range of prey available would likely increase. Since these spiders lack poison glands, individuals could not easily subdue the larger items, and cooperative prey capture would be advantageous. Measurements of prey capture efficiency as a function of prey size at the individual level, comparing solitary spiders and those in groups, could address this question.

METEPEIRA: INSIGHTS FROM A MODEL SYSTEM

Several orb-weaving spiders in the genus *Metepeira* (Araneidae) exhibit colonial social organization (Burgess and Witt 1976; Uetz and Burgess 1979; Burgess and Uetz 1982; Schoener and Toft 1983). Studies of these species have shown considerable geographic variation in group size, spacing, life history, and reproductive output among related species within a single taxon (Burgess and Uetz 1982; Uetz *et al.* 1982, 1987; Uetz 1983, 1985, 1986b, 1988a,b; Benton and Uetz 1986). Along with environmental factors (e.g. food availability), there may also be genetic differences between populations that influence spacing and tolerance of other individuals (Uetz and Cangialosi 1986; Uetz *et al.* 1987; Cangialosi and Uetz 1987). As *Metepeira* is probably the best studied of all the colonial spiders, these studies provide insight about the nature of selection for colonial web-building.

The two species of *Metepeira* studied most intensively, *M.* sp. 'a' (tentatively named *atascadero*) and *M. incrassata*, represent the extremes in the range of social behavior for the members of this genus that have been studied (Uetz *et al.* 1982, 1987; Uetz and Cangialosi 1986; Cangialosi and Uetz 1987; Uetz and Hodge 1990). *Metepeira atascadero* occurs in desert–mesquite grassland habitat near San Miguel de Allende, Guanajuato, Mexico. Conditions are typically severe and fluctuating in this habitat, with low seasonal rainfall ($<60 \, \mathrm{cm \, yr^{-1}}$), low relative humidity (45–65%) and wide daily temperature range (approx. 8–33 °C). *M. atascadero* live solitarily or in small groups of two to ten individuals in patches of cactus or mesquite, maintaining considerable distance between individuals. Variation in climatic conditions and their subsequent effect on prey have a dramatic influence on colonial behavior in this species. In years when prey availability is low, approximately 65–80% of the spiders forage solitarily; the remaining 20–35% forage in small groups of two to five individuals. In years of high prey abundance (during or following El Niño events) approximately 30–40% of the spiders forage solitarily; the remaining 60–70% forage primarily in groups of two to ten individuals (Hieber and Uetz 1990). *Metepeira incrassata* occurs in the East-Central mountains above the Gulf of Mexico, in Fortin de las Flores, Veracruz. Here the habitat is primarily moist tropical (second-growth) forest vegetation with banana and coffee plantations, high rainfall (170–220 cm yr^{-1}), high humidity (68–99%), and a moderate daily temperature range (20–32 °C). These spiders live in dense groups ranging in size from fewer than ten individuals to several thousand individuals, spanning large spaces between trees along the forest edge in coffee and banana plantations, and in power lines along roadsides (Uetz 1985). The contrasting environments of these two species have impacts on many aspects of the ecology of these spiders, and have served as a basis for numerous ecological comparisons (see Uetz and Hodge 1990).

Ecological benefits and costs of colonial web-building

Foraging

Previous studies of several colonial orb-weaving species have shown that spiders in colonies capture more prey than solitary individuals, and that prey captured per spider increases with colony size (Rypstra 1979, 1989; Spiller 1992; Uetz 1985, 1986b, 1988a,b). In *Metepeira*, this is due to the 'ricochet effect' (Uetz 1989), which results from prey escaping from one web being caught by another spider in the colony.

Foraging is not only affected by the average amount of food obtained, but also by the variance, which influences the risk of starvation (Caraco 1981; Pulliam and Milikan 1982; Caraco and Pulliam 1984; Pulliam and Caraco 1984; Real and Caraco 1986). Group-foraging reduces variance in prey intake in spiders (Uetz 1988a). Risk-sensitivity models predict that solitary foraging (the risk-prone strategy) should occur where prey availability is less than or equal to an individual's maintenance requirements. In contrast, groups (the risk-averse strategy) should arise only

where prey availability is greater than needed (Uetz 1988a,b).

The divergent foraging strategies of *M. atascadero* and *M. incrassata* illustrate risk-sensitivity in foraging (Uetz 1988a,b; Uetz and Hodge 1990). In the desert grassland habitat, where average available daily prey biomass is less than or equal to individual needs and of relatively high variability, most *M. atascadero* forage solitarily. It is better in this habitat to forage alone and risk the possibility of starving for the chance of encountering a site with above-average prey availability, rather than to forage in a group and reduce variance in the already inadequate level of food, thereby ensuring starvation. During El Niño years, with high levels of rain and abundant prey, and in localized sites where prey availability exceeds individual daily needs, *M. atascadero* are more likely to forage in groups. In contrast, conditions in the tropical montane rain-forest or agriculture sites are more moderate, and as a result prey availability is well above (two to three times) the threshold required for daily energetic needs; variability in prey is much lower than in the desert site. In this habitat, *M. incrassata* forage in a risk-averse manner in large groups, thereby ensuring a considerable margin of safety from starvation.

The relationship between habitat and prey availability, and the foraging decisions imposed by risk-sensitivity considerations, suggest why sociality is so rare in spiders (Gillespie 1987; Uetz 1988a,b; Rypstra 1989). In most habitats, solitary foraging is the appropriate strategy for spiders because the availability of prey is probably variable, and because the amount that can be obtained by each spider is likely to be close to or less than maintenance requirements. In contrast, in certain habitats with abundant prey (e.g. tropical forest sites), spiders link their webs because the subsequent reduction in prey variance makes it advantageous to do so, and as a consequence gain other benefits of foraging socially, such as increased prey capture rates or increased prey size (Uetz 1988a, 1989).

Tolerance and aggressive behavior

Habitat and prey availability have a profound effect on the frequency and size of groups in these two species of spiders; they also affect other factors that may play a role in the tendency for grouping. Several researchers have suggested that tolerance of conspecifics is an essential step towards the evolution of sociality in spiders (Shear 1970; Buskirk 1981; Burgess and Uetz 1982; Rypstra 1986), and it has been demonstrated that tolerance varies with food availability (Rypstra 1989).

Metepeira atascadero are actively foraging on webs from June to November; because of the seasonal availability of prey, they produce only one generation per year. Spiderlings overwinter in their egg sacs (December to May) and emerge into an environment bereft of existing colonies, with patches of vegetation suitable for web location scattered throughout the habitat. Web sites are critical for survival, and web defense is highly aggressive, often escalating to dangerous levels (Uetz and Hodge 1990; Hodge and Uetz 1995).

In contrast to *Metepeira atascadero*, *M. incrassata* occur year-round in a habitat with abundant prey, and higher degrees of tolerance are expected (Rypstra 1986). Reproduction is continuous, generations overlap, and although there is no parental care, the young emerge from egg sacs and stay within the colony. Although agonistic encounters are frequent, they are usually resolved with minimal levels of aggression (Uetz and Hodge 1990; Hodge and Uetz 1995).

In these two *Metepeira* species, differences in levels of aggressive behavior persist under identical controlled conditions in the laboratory, with abundant food (Uetz and Cangialosi 1986; Cangialosi and Uetz 1987). These findings suggest a genetic basis for differences in aggressive behavior, and support earlier predictions about environmental severity and the evolution of tolerance (Uetz and Hodge 1990; Hodge and Uetz 1995). For *M. atascadero*, coloniality is facultative, because climatic factors and prey availability fluctuate between years. Under these conditions, any short-term advantage of social grouping for spiders in good years would be counterbalanced by selection against grouping in subsequent (more common) years with low prey availability. However, *M. incrassata* exhibits overlapping generations, long-term colony persistence and reduced dispersal (G. W. Uetz and C. S. Hieber, personal observations). These factors may increase the potential for inbreeding, leading to increased relatedness among colony members, which may temper the inherent cannibalistic tendency of these animals and the aggressive behavior with which they defend territories.

Predators and parasites

Coloniality has a cost – increased vulnerability to predators and parasites with increasing colony size – because the sedentary nature, high density, and conspicuous silk of colonies make them easier for predators to locate (Lubin 1974; Rypstra 1979; Buskirk 1981; Smith 1982; Hieber and Uetz 1990; Uetz and Hieber 1994). However,

colonial behavior also has advantages in this regard. Grouping facilitates detection of parasitoids and improves egg-sac defense (Lubin 1974), and increases levels of protection from predators for adult spiders and spiderlings (Shear 1970). Within the genus *Metepeira*, it appears that colonial behavior results in both disadvantages and benefits with regard to predators and parasites.

Hieber and Uetz (1990) compared levels of egg-sac predation across the range of naturally occurring colony sizes in *M. atascadero* and *M. incrassata*. For *M. atascadero*, in typical years with low prey, egg-sac production is sparse, and the overall rates of parasitism are low (3.4–6.9%). Increased rates of parasitism (as high as 40%) are seen in years when prey are abundant and (consequently) spider egg production is higher. However, even though there are more egg-sacs per female in groups during these years, there is no relationship between group size and individual risk of egg-sac predation. For *M. incrassata*, rates of egg-sac predation are relatively constant from year to year, but there is a significant positive relationship between colony size and frequency of egg-sac predation. This represents an increased cost to individual spiders in terms of egg loss.

Although increased attacks from predators are a consequence of sociality for spiders (Spiller and Schoener 1989), the costs incurred by colonial spiders may be balanced by gains from a number of antipredator mechanisms. For example, Rayor and Uetz (1990, 1993) found that predator attack and capture rates varied with the position of the spider within the colony. Rate of attack by wasps was higher for spiders on the periphery, but lower in the core of the colony. These findings suggest that there is a 'selfish herd effect' (Hamilton 1971) operating in colonial webs: animals in the center of a group may decrease their chances of predation by surrounding themselves with others.

Simply being in a colony may also confer antipredator benefits on an individual in two ways: through the 'encounter effect' and the 'dilution effect' (Turner and Pitcher 1986; Inman and Krebs 1987; Wrona and Dixon 1991). Both of these attack-abatement mechanisms appear to be operating in *M. incrassata* colonies attacked by predatory wasps (Uetz and Hieber 1994), but they operate in different ways and at different times in the predator's attack. The encounter effect reduces the probability of a predator's location of a colony; the dilution effect reduces individual risk once a colony is encountered.

There is a general pattern of increasing rate of encounter with predators with increasing colony size; however, predator encounter rate is lower than expected in the

(a)

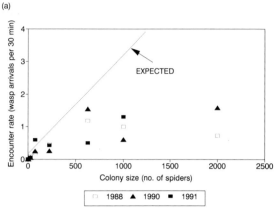

(b)

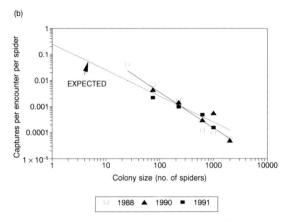

Fig. 22-1. Relationship between group size and wasp predation rates in *Metepeira incrassata*. (a) Encounter rates with wasp predators at the colony level increase with group size, but at a rate lower than that predicted; (b) rates of capture per encounter suggest a dilution effect, but the decrease in risk with group size is steeper than predicted by simple numerical dilution. (From Uetz and Hieber (1994), © 1994 Oxford University Press, used with permission.).

smallest and largest colonies (Fig. 22-1a). Solitary individuals or small colonies are never encountered by wasp predators, suggesting they are below the threshold size for detection. In addition, the rate of encounter does not increase beyond the level seen for groups of 500. An explanation for this observed non-linear relationship may be that visual apparency does not increase linearly with colony size (i.e. much of the expansion takes place 'behind' the three-dimensional colony and its true size is masked from view). In addition, interindividual spacing of spiders within a colony generally decreases as the colony size increases (Uetz *et al.* 1982; Uetz and Hodge 1990). As a result, spiders in large groups enjoy increased protection from predator encounter because they are no more likely to be located than those individuals in smaller groups. Equally important, these results suggest that the encounter effect operating by itself is enough to confer fitness advantages on spiders living in the smallest and largest colonies (Inman and Krebs 1987).

Once a colony has been located by predators, the risk of attack and capture for an individual are predicted by the dilution effect to be an inverse function of colony size (Hamilton 1971; Foster and Treherne 1981; Inman and Krebs 1987; Wrona and Dixon 1991). Our data support this prediction (Fig. 22-1b), but our field observations of the attack process indicate that the relationship is more complex than simple dilution. This complexity is due to two different factors operating simultaneously. First, wasps may attack more than one spider in a colony after it is located, which may offset any gain in fitness the spiders

may realize because of a dilution effect. Second, in spite of multiple attacks, there is an overall decrease in risk for individual spiders living in groups, because the capture success of wasp predators decreases with group size. The network of interconnected silk lines provides an 'early warning effect', informing spiders about the approach of predators (Lubin 1974; Buskirk 1975b; Uetz 1986b; Hodge and Uetz 1992; G. W. Uetz, J. Boyle and C. S. Hieber, unpublished). This is analogous to the 'many eyes' vigilance system seen in flocks of birds (Bertram 1978; Kenward 1978), or the 'Trafalgar' effect demonstrated by ocean skaters (Treherne and Foster 1982). These observations suggest that although increased rates of predator encounter are a consequence of group-living in colonial web-building spiders (Spiller and Schoener 1989), complex attack-abatement mechanisms provide protection as well.

Group size variation in *Metepeira incrassata*: field studies of benefits, costs, and fitness

In *M. incrassata*, colony size varies considerably, from occasional solitary individuals and small groups of 10–20 individuals to enormous colonies with thousands of individuals. Given this range in colony size, we expect that the benefits and costs associated with group-living change with group size, and affect the individual spider in different ways. To assess the differences in various ecological benefits and costs with group size in this species, we undertook a set of field studies from 1988 to 1991

(Hieber and Uetz 1990; Uetz and Hieber 1994). The study was conducted at our field site in Fortin de las Flores (described above and in Benton and Uetz 1986; Uetz and Hodge 1990). We present here a synopsis of previous and current studies.

Methods

We censused spider colonies each year (1988, 1990, 1991) on two roadside transects (15 m × 4 km) along the edges of coffee and banana plantations. From the censused populations, we chose three to five colonies for observation in each of the following size classes: solitary, 2–9, 10–49, 50–149, 150–499, 500–999, 1000–1999 and >2000 spiders per colony; these reflect the grouping and uneven distribution apparent within the range of naturally occurring colonies (Fig. 22-2).

We used a combination of all-occurrence sampling and focal individual sampling to assess the frequency of behaviors and events (Altmann 1974; Martin and Bateson 1993). Colonies or selected individuals were observed for 30 minutes at time, during which all occurrences of prey capture (groups) and all behaviors (individuals) were recorded. To ensure that these observations were independent of one another (did not follow immediately at the same colony) and covered the entire daily activity period equally, both colony size-class and time period for observation were assigned at random in advance. From these data and published values we were able to estimate the potential energetic gain from prey (Rogers *et al.* 1976; Tanaka 1991; S. E. Riechert, personal communication), the energetic costs of silk and web-building (Lubin 1973; Peakall and

Witt 1976; Tanaka 1989) and the metabolic costs of behavioral activities (Anderson 1970; J. F. Anderson, personal communication). From these values, we developed estimates of the daily energy budget for spiders in colonies of various sizes. We estimated reproductive success from collections of females and egg-sacs (a minimum of five to ten individuals from each of three colonies in each colony size class). Female body size was recorded, and spiderlings and unhatched eggs were counted for determination of fecundity.

Group size and energetic costs and benefits

The transition from solitary to group-living is accompanied by a major shift in energy budget for the individual spider (Fig. 22-3). The cost of silk and web maintenance are relatively high for solitary spiders, and constitute the majority of the spiders' energy budget, leaving a negative net energy gain. This may explain why solitary *M. incrassata* are found so rarely, and why experimental solitaries do not survive (G. W. Uetz and C. S. Hieber, personal observations). However, the shift from one to a few spiders reverses this prey energy gain vs. silk cost imbalance, presumably because of the energy savings that accrue from shared construction and maintenance of the communal silk framework in which the prey-capture web is hung. As colony size increases, the per capita net energy gain increases as well. Even though spiders in larger colonies spend more time in active behaviors and therefore may incur more energetic expenses, these expenses are apparently offset by increases in energy from higher rates of prey capture.

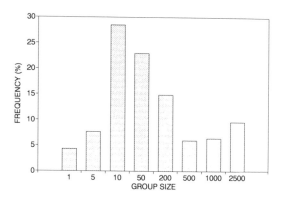

Fig. 22-2. Frequency distribution of group size (categorized as median of each size class) of *Metepeira incrassata* during the censuses taken from 1988 to 1991.

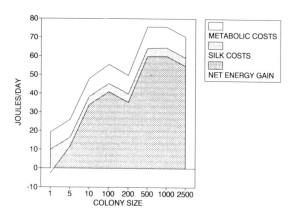

Fig. 22-3. Estimated individual daily energy budget for *M. incrassata* for groups of varying size (see text for details).

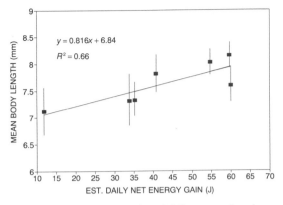

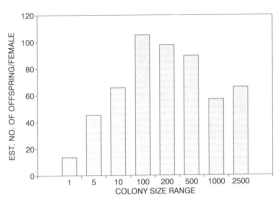

Fig. 22-4. Relationship between estimated daily energy gain and mean body size for adult female *M. incrassata*. Female body size varies significantly with estimates of individual net energy gain among colonies of varying size (One-way ANOVA: $F_{5,58} = 3.73$, $p < 0.01$; Pearson's $r = 0.812$, $p < 0.05$).

Fig. 22-6. Estimates of individual reproductive success (based on egg sac collections and hatching success data) in *M. incrassata* colonies of varying size. Fecundity varies significantly among colony size categories (One-way ANOVA, $F_{6,33} = 2.91$, $p < 0.05$).

Group size and fitness

The rise in net energy gain with increasing colony size suggests that growth rates should be higher and adult body size greater in the larger colonies (Vollrath 1987a; Jakob and Dingle 1990; Uetz 1992). A comparison of body size of adult females in colonies of varying size supports this prediction (Fig. 22-4). This variation may have important fitness considerations, as fecundity is significantly correlated with female body size (Fig. 22-5). Estimated reproductive success (based on mean fecundity and probability of hatching success for each colony size-class) shows an increase with colony size followed by a decline in the largest colonies (Fig. 22-6). Solitary individuals have the lowest mean reproductive success, which is probably due to low rates of

growth and maturation or the inability to obtain a mate (we only collected one egg sac from nine solitary females). Spiders in groups of 50–500 have the highest levels of reproductive success. Surprisingly, those at both ends of the range have equivalent levels of reproductive success, but for different reasons. Spiders in small colonies have a lower level of net energy gain, are smaller, and consequently have reduced fecundity. Spiders in larger colonies, however, suffer a higher rate of egg-sac predation (Hieber and Uetz 1990); as many as 25% of egg-sacs in colonies of over 1000 spiders are parasitized.

Balancing costs and benefits

The relationship between group size and reproductive success suggests there is an optimal colony size (100–200 spiders). However, colonies of optimal size (i.e. in which individuals have maximal reproductive success) would become much larger as a consequence of successful reproduction, resulting in a decrease in individual reproductive success. Spiders would therefore appear to be caught in a 'cruel bind', as leaving the group and going off on one's own would result in an even more drastic reduction in fitness. Hence, spiders in the largest colonies are faced with a complex set of trade-offs between energetic gains, and costs due to predation and parasitism.

The 'selfish herding' behavior of female spiders in large colonies demonstrated by Rayor and Uetz (1990, 1993) reflects the trade-offs between these benefits and costs. Spiders on the periphery of colonies get more prey, but

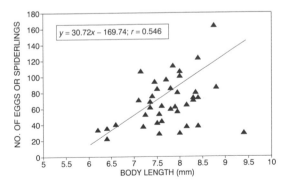

Fig. 22-5. Relationship between fecundity (expressed as number of eggs or spiderlings per female) and adult female body size for *M. incrassata*. (Pearson's $r = 0.546$, $p < 0.05$).

are subject to a higher risk of predation; consequently, fitness is higher in the central core of the colony (Rayor and Uetz 1990). Trade-offs vary with spider age and size, as predators appear to preferentially attack the largest spiders, and smaller spiders gain disproportionately more prey if they build webs on the periphery (Rayor and Uetz 1993). As a consequence, there are ontogenetic shifts in position within the colony, which creates a dynamic size hierarchy over time. This size hierarchy, with larger spiders defending space within the central core, and smaller spiders on the periphery, is the result of size- and owner-biassed outcomes of aggressive encounters (Hodge and Uetz 1995) as well as pre-emptive web-building by the largest spiders, who build their webs first (L. S. Rayor and G. W. Uetz, in preparation). Current research is centered on developing and testing a dynamic optimization model (E. M. Jakob, M. Mangel and G. W. Uetz, unpublished data) to predict fitness outcomes of web location decisions by spiders of different size classes.

Metepeira incrassata has provided an in-depth look at the ecological costs and benefits of group-living for colonial orb-weaving spiders. One unique aspect of this species is that so many of the mechanisms proposed as benefits of group-living for bird flocks and ungulate herds (enhancement of foraging, reduction in variance of prey capture, predator encounter and dilution effects, early warning of predator approach, selfish herding) appear to be operating simultaneously. Group-living in *Metepeira* clearly involves a complex trade-off between these benefits and the costs (increased aggression, increased rates of predation, increased rates of egg-sac parasitism) associated with living in groups, and consequently individual fitness may vary with group size and position within the colony.

SYNTHESIS: COMPARISONS, CONSTRAINTS AND COMMON THREADS

The social behavior of spiders inevitably invites comparison with insect sociality. However, it is important to remember that although spiders and insects are both arthropods, the relatively ancient evolutionary divergence and subsequent radiation of their respective subphyla (Chelicerata and Uniramia) creates an asymmetry: spiders are a single Order within the Class Arachnida, whose diversity in species (34 057), morphology and lifestyle pales by comparison to that of the 26 Orders and perhaps one to ten million species of the Class Insecta. Perhaps it is a result of this asymmetry alone that sociality is so rare and limited in diversity among the spiders. However, there are two significant differences between these taxa that relate to sociality and its evolution that also deserve mention: (1) spiders are all predatory (and potentially cannibalistic), whereas insects occupy a myriad of trophic niches; and (2) the haplodiploid sex-determining mechanism frequently cited as promoting kin selection and social evolution in the Hymenoptera is lacking in spiders. As a consequence of these differences, spiders would seem far less likely to evolve social behavior beyond its most primitive forms: resource-based aggregation and parental care. As such, the sociality of spiders appears similar to that seen in a number of vertebrate species, as well as numerous insect and arachnid orders discussed in other chapters of this volume (Embioptera, Edgerly; some Isoptera, Shellman-Reeve; Lepidoptera, Costa and Pierce; Acari, Saito).

Despite the rarity of social behavior in spiders, the fact that it exists at all provides a unique opportunity to examine the evolution of 'exceptions to the rule'. Research on colonial spiders to date has provided evidence for both parasocial and subsocial pathways in the evolution of group-living for a variety of orb-weaving species. Evolution of more complex sociality (i.e. cooperation in prey capture, brood care) may be constrained by the territorial social structure inherent in orb-weaving spiders imposed by the nature of the web (Lubin 1974) and the negative impact of predators and parasitoids which accrues from grouping (Smith 1982; Spiller and Schoener 1989; Hieber and Uetz 1990; Uetz and Hieber 1994). Alternatively, the evolution of some forms of cooperation may be fostered by the high degree of relatedness within colonies and the potential for kin or group selection (Avilés, this volume).

Are there constraints on the evolution of social behavior in colonial orb-weavers?

Lubin (1974) implies that there is a limit to the evolution of more complex levels of sociality (i.e. non-territorial webs or cooperative prey capture) in colonies such as those of *Cyrtophora* and *Metepeira*, and suggests that this limit is set by the nature of orb-webs themselves. Orb-webs can only be constructed and used by one individual at a time, and they must be spaced in a way so as to maximize prey capture. As such, she argues, the nature of orb-webs precludes grouping webs together to form a continuous sheet (as seen in the social, or cooperative periodic- or permanent-social species). This would limit the evolution of cooperative prey capture, even if aggression within the colony is reduced. However, there are a number of observations that argue against this view. For example, the vertically oriented individual orb-webs within *Parawixia* colonies are arranged side by side along a single plane, resulting in extensive 'sheets' across which individuals can travel quickly and capture prey (Sandoval 1987). Even though the orb-webs of *Metepeira* are not arranged within a single plane, and do not form sheets, individuals often move rapidly from one web to another on the communal scaffold in response to very large prey items (G. W. Uetz and C. S. Hieber, personal observations). Given the advantages of facultative cooperation in subduing large prey items in *P. bistriata* (Fowler and Diehl 1978; Fowler and Gobbi 1988), and the potential for high genetic relatedness within colonies, the possibility of reduced territorial aggression and evolution of cooperation are not precluded, even among orb-weavers (Shear 1970).

A common thread: three-dimensional web design

Evidence for the parasocial pathway, and the origin of coloniality as a consequence of ecological factors, is substantial. Despite the fact that colonial web-building has evolved several times in phylogenetically-distant genera, there is one feature common among all the colonial orb weavers and the solitary, but facultatively aggregating species: their webs are three-dimensional aggregates, composed of a planar or slightly domed prey-capture orb and varying amounts of three-dimensional barrier webbing on one or both sides of the orb. This composite structure allows individual spiders to attach webs together without any change in behaviors related to construction of the prey-capture orb, or its use in prey capture. The scenario initially proposed by Shear (1970) and elaborated on by Buskirk (1981), Uetz (1988b) and Rypstra (1989) appears credible: spiders aggregate in areas of high prey density, attaching webs together as a consequence of silk attraction ('sericophily'), reduced aggression and risk-aversion, all of which are facilitated by the three-dimensionsional organization of individual webs. As a consequence, spiders gain the benefits of group-living, which include increased prey, reduced variance in prey, energetic savings from reduction in silk costs and increased tolerance.

For example, the communal web framework that supports the individual webs of *Metepeira* may have evolved from a modular system of individual spider 'territories' attached at their periphery in an aggregation of solitary individuals. The subsequent communal framework may contribute to increased prey knockdown or ricochet, or transmit vibrations allowing detection of prey escaping from other webs, or the approach of predators, all of which may allow energetic gains and survival benefits for individual colony members (Uetz 1986b, 1989). At the same time, the increased stability of the common framework (reinforced by the web-building of multiple individuals) reduces individual silk costs, thereby resulting in energetic savings for all colony members.

The scenario posed above depends on environmental conditions that foster high prey abundance and prolonged aggregation (Uetz 1988a). Once spiders live in colonies throughout their lives, selection pressures for aggression would be reduced, and more permanent forms of sociality could evolve. As seen in *M. incrassata*, although spiders still defend their webs, the boundaries of individual space within this system have effectively dissolved, and silk lines elaborated upon by cooperative effort serve as a common scaffolding for movement throughout the colony. Under these circumstances, the genetic structure of colonies would be expected to change to permanent groups with overlap of generations, reduced dispersal, and higher degrees of relatedness among colony members.

Evidence for subsocial behavior among orb-weavers?

We have presented evidence for a subsocial pathway to the evolution of cooperation in the behavior of several orb-weaving species (*Metabus*, *Parawixia* and *Philoponella*). Presumably, the use of communal retreats by *M. gravidus* and *P. bistriata* provides protection against predators or parasites (Buskirk 1981). The stepwise mechanisms underlying the evolution of this behavior are less clear. Communal retreat use requires reduction of the aggressive tendencies between individuals within the colony. How this may have arisen in *M. gravidus* is unclear, although the potential evolutionary pathway to retreat use shown by *P. bistriata* is conceptually clearer. It is likely that all the individuals in a *P. bistriata* colony are immature siblings, and have all emerged from a single egg sac (Fowler and Diehl 1978; Smith 1982). Colonies may then represent extended family groups, resulting in kin selection for tolerance. The high levels of relatedness among colony members would foster cooperation, and could explain the use of retreats and the reduction in territoriality with regard to web-defense.

Evidence for a subsocial pathway, and the evolution of cooperative social behavior, in orb-weavers is, however, circumstantial. The evolution of cooperation among orb-weavers represents the most critical unanswered question, which perhaps holds the key to understanding the evolution of sociality in spiders. Avilés (this volume) discusses research on the highly social spiders that live communally and cooperate in prey capture. Is an evolutionary transition between colonial behavior and these higher forms of sociality possible? Do colonial spiders represent a evolutionarily intermediate stage, or an alternate form of social behavior? Evidence presented here for *Philoponella*, i.e. use of communal retreats as egg sites, reduced use of orbs for prey capture, and cooperation in prey capture, might suggest that colonial orb-weavers could evolve more complex and cooperative forms of sociality. However, as far as we know, no species has been found within the orb-weaving Araneidae or Uloboridae that exhibits anything other than prey size-dependent facultative cooperation (Fowler and Gobbi 1988). On the other hand, higher sociality (cooperative prey capture, brood care) is seen in several species in other taxa, such as the Theridiidae, Dictynidae, and Agelenidae (which all build tangle or sheet webs). Given that the Uloboridae, Araneidae and Theridiidae are more closely related (within the same major clade, the Orbiculariae (Coddington and Levi 1991)), the possibility of an as yet undiscovered, 'missing link' species is open.

UNANSWERED QUESTIONS AND FUTURE RESEARCH

Given the likelihood of inbreeding and relatedness within colonies, and the potential importance of kin selection (and possibly interdemic group selection) in the evolution of social behavior, little is known about the genetic structure of populations of colonial spiders. Indirect evidence can be found in the varying degrees of heterozygosity seen in *M. incrassata* and *M. atascadero* from a species-level electrophoretic study (Uetz *et al.* 1987); lower heterozygosity values in *M. incrassata* suggest higher inbreeding. However, this study involved collections of individuals from colonies of varying size in separate geographic populations, and did not take into account the within- and between-colony component of genetic variability within the population. In contrast, a number of studies using a variety of molecular techniques have documented high degrees of genetic relatedness within and among colonies of social species such as *Achaearanea wau* (Theridiidae) (Lubin and Crozier 1985), *Agelena consociata* (Agelenidae)

(Roeloffs and Riechert 1988), *Stegodyphus sarasinorum* (Eresidae) (Smith and Engel 1994) and *Anelosimus* sp. (Theridiidae) (Smith 1986, 1987; Smith and Hagen 1996). The widespread availability of molecular techniques should make these unanswered research questions accessible to any who wish to pursue them.

In addition to a need for studies of genetic structure of colonial spiders, there is a paucity of information about sex ratio in these species. A number of studies have shown highly female-biased sex ratios in the non-territorial permanent-social spiders (Lubin and Crozier 1985; Lubin 1991; Vollrath 1986, 1987b; Frank 1987; Avilés 1986; Avilés and Maddison 1991; Avilés, this volume). These extraordinary sex ratios are apparently biased at the egg stage (Avilés and Maddison 1991; Vollrath 1986, 1987b), at least in the theridiid genera *Anelosimus* and *Achaearanea*, although information on other species is lacking. Lubin (1991) found that the sex ratio of *A. wau* colonies was female-biased, but changed over development (adult sex ratios were less biased than those of juveniles), and varied with colony size (larger colonies were more likely to be female-biased). Avilés (1986) has hypothesized that female-biased sex ratios in the social spiders may evolve as a consequence of their unique population structure and dynamics. Given the variation in dispersal rates within and among species of colonial orb-weavers, sex ratio might be expected to vary as well. Our own observations of colonial *Metepeira* species suggest an initial 1 : 1 male : female ratio at the egg stage, but a progressive change towards a female-biased sex ratio as a consequence of male mortality (during the reproductive period) and female longevity. A comparison with other colonial species with different social structure would clearly be valuable.

Another obvious question about colonial spiders that has not been addressed in any depth concerns the reproductive strategies of male spiders in the colony. Although a number of studies have examined the relationships between male combat, mate-guarding, mating success and fitness (see reviews in Christenson 1984; Watson 1993), there have been few studies of the behavior of male spiders in the social context (Lubin 1986). Given the trade-off for males inherent in colonial living – ease of mate location versus increased competition for mates (Rubenstein 1987; Lubin 1986) – studies of the relationship between male size, fighting ability and mating success in colonial web-builders should provide valuable insights about the role of sexual selection in an alternative social system.

Most of the species of colonial and social spiders that have been studied to date are found in tropical habitats, and are often conveniently located at or near established research stations (Riechert 1984; Riechert *et al.* 1986). Although the occurrence of social spiders in tropical habitats is not surprising, as moderate environments and abundant insect faunas create conditions favorable for grouping, there may be biases created by the location of research stations. We have often lamented that colonial *Metepeira* spp. have never been found on any of the numerous research stations in Mexico, where research facilities would make our work easier! Considering the potential importance of new findings to unanswered evolutionary questions, one is left to wonder what else might be out there if only we looked hard enough (for example, note the recent findings of Evans and Main (1993) and Avilés (1993)). Given how rapidly these habitats are disappearing, it may be wise to speed up the search.

ACKNOWLEDGMENTS

This research was supported by National Science Foundation grants BSR-8615060 and BSR-9109970, National Geographic Society Grants 3095-85 and 4428-90, the Exline-Frizzel Arachnological Research Fund and a Summer Reseach Grant from Saint Anselm College. We thank the Mexican Government, Direccion General de Conservacion y Ecologia de los Recursos Naturales, Direccion de Flora y Fauna Silvestres, for permission to work in Mexico. We especially thank Ana Valiente de Carmona and family, Becky Cotera, Blanca Alvarez and Loli Alvarez-Garcia for the use of their properties as research sites. We are grateful for field assistance, manuscript review, and other forms of collaboration in research by Maggie Hodge, Sam Marshall, Linda Rayor, Beth Jakob, Stim Wilcox, Jon Coddington, Brent Opell, Andrea McCrate, David Kroeger, Jay Boyle, Steve Leonhardt, Frank Light, Veronica Casebolt, Rebecca Forkner and Will McClintock. We also thank Bernie Crespi, Leticia Avilés and an anonymous reviewer, who made valuable comments on the manuscript. Most importantly, we are grateful to our wives and families for tolerating our absence during periods of fieldwork.

LITERATURE CITED

Altmann, J. 1974. Observational study of behavior: sampling methods. *Behaviour* **49**: 227–265.

Anderson, J. F. 1970. Metabolic rates of spiders. *Comp. Biochem. Physiol.* **33**: 51–72.

Avilés, L. 1986. Sex ratio bias and possible group selection in the social spider *Anelosimus eximius*. *Am. Nat.* **128**: 1–12.

—. 1993. Newly-discovered sociality in the neotropical spider *Aebutina binotata* Simon (Dictynidae?). *J. Arachnol.* **21**: 184–193.

Avilés, L. and W. Maddison. 1991. When is the sex ratio biased in social spiders?: chromosome studies of embryos and male meiosis in *Anelosimus* species (Araneae: Theridiidae). *J. Arachnol.* **19**: 126–135.

Benton, M. J. and G. W. Uetz. 1986. Variation in life-history characteristics over a clinal gradient in three populations of a communal orb-weaving spider. *Oecologia (Berl.)* **68**: 395–399.

Bertram, B. C. R. 1978. Living in groups: predators and prey. In *Behavioral Ecology: An Evolutionary Approach*. J. R. Krebs and N. B. Davies, eds., pp. 64–96. Oxford: Blackwell Scientific Publications.

Binford, G. J., and A. L. Rypstra. 1992. Foraging behavior of the communal spider, *Philoponella republicana*. *J. Insect Behav.* **5**: 321–335.

Blanke, R. 1972. Untersuchengen zür Oekophysiologie und Oekethologie von *Cyrtophora citricola* Forskal (Araneae: Araneidae) in Andalusien. *Forma Functio* **5**: 125–206.

Brach, V. 1977. *Anelosimus studiosus* (Araneae: Theridiidae) and the evolution of quasisociality in theridiid spiders. *Evolution* **31**: 154–161.

Breitwisch, R. 1989. Prey capture by a west African social spider (Uloboridae: *Philoponella* sp.). *Biotropica* **21**: 359–363.

Brown, J. L. 1975. *The Evolution of Social Behavior*. New York: Norton.

Brown, S. G., E. M. Hill, K. E. Goist, P. A. Wenzl and T. E. Christenson. 1985. Ecological and seasonal variations in a free-moving population of the golden-web spider, *Nephila clavipes*. *Bull. Bri. Arachnol. Soc.* **6**: 313–319.

Burgess, J. W. 1978. Social behavior in group-living species. *Symp. Zool. Soc. Lond.* **42**: 69–78.

Burgess, J. W. and G. W. Uetz. 1982. Social spacing strategies in spiders. In *Spider Communication: Mechanisms and Ecological Significance*. P. N. Witt and J. S. Rovner, eds., pp. 317–351. Princeton: Princeton University Press.

Burgess, J. W. and P. N. Witt. 1976. Spider webs, design, and engineering. *Interdisc. Sci. Rev.* **1**: 322–355.

Buskirk, R. E. 1975a. Coloniality, activity patterns and feeding in a tropical orb-weaving spider. *Ecology* **56**: 1314–1328.

—. 1975b. Aggressive display and orb defense in a colonial spider, *Metabus gravidus*. *Anim. Behav.* **23**: 560–567.

—. 1981. Sociality in the Arachnida. In *Social Insects*, vol. 2. H. R. Hermann, ed., pp. 281–367. New York: Academic Press.

Cangialosi, K. R. and G. W. Uetz. 1987. Spacing in colonial spiders: effects of environment and experience. *Ethology* **76**: 236–246.

Caraco, T. 1981. Risk sensitivity and foraging groups. *Ecology* **62**: 527–531.

Caraco, T. and H. R. Pulliam. 1984. Sociality and survivorship in animals exposed to predation. In *A New Ecology: Novel Approaches to Interactive Systems*. P. W. Price, C. N. Slobodchickoff and W. S. Gaud, eds., pp. 279–309. New York: Wiley Interscience.

Christenson, T. 1984. Alternative reproductive tactics in spiders. *Am. Zool.* **24**: 321–332.

Cloudsley-Thompson, J. L. 1981. On communal and social spiders. *Br. Arachnol. Soc., Secretary's Newsl.*, no. 32, November 1981.

Clyne, D. 1969. *A Guide to Australian Spiders.* Melbourne: Nelson.

Coddington, J. A. and H. W. Levi.1991. Systematics and evolution of spiders (Araneae). *Annu. Rev. Ecol. Syst.* **22**: 565–592.

Davies, N. B. 1978. Ecological questions about territorial behavior. In *Behavioral Ecology: An Evolutionary Approach.* J. R. Krebs and N. B. Davies, eds., pp. 317–350. Sunderland: Sinauer.

D'Andrea, M. 1987. Social behaviour in spiders (Arachnida: Araneae). *Monit. Zool. (N.S.) Monogr.* **3**: 1–156.

Enders, F. 1977. Web-site selection by orb-web spiders, particularly *Argiope aurantia* Lucas. *Anim. Behav.* **25**: 694–712.

Evans, T. A. and B. Y. Main. 1993. Attraction between social crab spiders: silk pheromones in *Diaea socialis. Behav. Ecol.* **4**: 99–105.

Foster, W. A. and J. E. Treherne. 1981. Evidence for the dilution effect in the selfish herd from fish predation on a marine insect. *Nature (Lond.)* **293**: 466–467.

Fowler, H. G. and J. Diehl. 1978. Biology of a Paraguayan colonial orb-weaver *Eriophora bistriata* (Rengger) (Araneae, Araneidae). *Bull. Br. Arachnol. Soc.* **4**: 241–250.

Fowler, H. G. and N. Gobbi. 1988. Cooperative prey capture by an orb-web spider. *Naturwissenschaften* **75**: 208–209.

Frank, S. 1987. Demography and sex ratio in social spiders. *Evolution.* **41**: 1267–1281.

Gillespie, R. G. 1987. The role of prey availability in aggregative behavior of the orb weaving spider *Tetragnatha elongata. Anim. Behav.* **35**: 675–681.

Hamilton, W. D. 1964. The genetical evolution of social behavior, Parts 1 and 2. *J. Theor. Biol.* **7**: 1–52.

–. 1971. Geometry for the selfish herd. *J. Theor. Biol.* **31**: 295–311.

Hermann, H. R. 1982. *Social Insects,* vol. 1. New York: Academic Press.

Hieber, C. S. and G. W. Uetz. 1990. Colony size and parasitoid load in two species of colonial *Metepeira* spiders from Mexico (Araneae: Araneidae). *Oecologia (Berl.)* **82**: 145–150.

Hingston, H. W. G. 1932. *A Naturalist in the Guiana Forest.* New York: Longmans, Green.

Hodge, M. A. and G. W. Uetz. 1992. Anti-predator benefits of single and mixed-species grouping by *Nephila clavipes* (L.) (Araneae: Tetragnathidae). *J. Arachnol.* **20**: 212–216.

–. 1995. A comparison of agonistic behaviour of colonial web-building spiders from desert and tropical areas. *Anim. Behav.* **50**: 963–972.

Honjo, S. 1977. Social behavior of *Dictyna follicola* Bos et Str. (Araneae: Dictynidae). *Acta Arachnol.* **27**: 213–219.

Inman, A. J. and J. Krebs. 1987. Predation and group living. *Trends Ecol. Evol.* **2**: 31–32.

Jackson, R. R. 1978. Comparative studies of *Dictyna* and *Mallos* (Araneae: Dictynidae). I. Social organization and web characteristics. *Rev. Arachnol.* **1**: 133–164.

Jakob, E. M. 1991. Costs and benefits of group living for pholcid spiderlings: losing food, saving silk. *Anim. Behav.* **41**: 711–722.

Jakob, E. M. and H. Dingle. 1990. Food level and life history characteristics in a pholcid spider (*Holocnemus pluchei*). *Psyche* **97**: 96–109.

Kenward, R. E. 1978. Hawks and doves: factors affecting success and selection in goshawk attacks on wood pigeons. *J. Anim. Ecol.* **47**: 449–460.

Krafft, B. 1982. The significance and complexity of communication in spiders. In *Spider Communication: Mechanisms and Ecological Significance.* P. N. Witt and J. S. Rovner, eds., pp. 15–66. Princeton: Princeton University Press.

Kullman, E. 1958. Beobachtungen des Netzbaues und Betrage zür Biologie von *Cytophora citricola* Forskal (Araneae: Araneidae). *Zool. Jb. (Abt. Syst. Oekol. Geogr. Tiere).* **86**: 181–216.

–. 1972. Evolution of social behavior in spiders (Araneae; Eresidae and Theridiidae). *Am. Zool.* **12**: 419–426.

LeBorgne, R. and A. Pasquet. 1987a. Influences of aggregative behaviour on space occupation in the spider *Zygiella x-notata* (Clerck). *Behav. Ecol. Sociobiol.* **20**: 203–208.

–. 1987b. Influence of conspecific silk-structures on the choice of a web-site by the spider *Zygiella x-notata* (Clerck). *Rev. Arachnol.* **7**: 85–90.

–. 1994. Presence of potential prey affects web-buiding in an orb-weaving spider *Zygiella x-notata. Anim. Behav.* **47**: 477–480.

Levi, H. W. and J. Coddington. 1983. Progress report on the phylogeny of the orb-weaving family Araneidae and the superfamily Araneoidea (Arachnida:Araneae) (Abstract). *Verh. Naturwiss. Ver. Hamburg.* **26**: 151–154.

Lahmann, E. J. and W. G. Eberhard. 1979. Factors selectivos que afectan la tendencia a agruparse en la arana colonial *Philoponella semiplumosa* (Araneae: Uloboridae). *Rev. Biol. Trop.* **27**: 231–240.

Lubin, Y. D. 1973. Web structure and function: the non-adhesive orb-web of *Cytophora moluccensis* (Doleshall) (Araneae: Araneidae). *Forma Funct.* **6**: 337–358.

–. 1974. Adaptive advantages and the evolution of colony formation in *Cyrtophora* (Araneae: Araneidae). *Zool. J. Linn. Soc.* **54**: 321–339.

–. 1980. Population studies of two colonial orb-weaving spiders. *J. Zool. Soc.* **70**: 265–287.

–. 1986. Courtship and alternative mating tactics in a social spider. *J. Arachnol.* **14**: 239–258.

–. 1991. Patterns of variation in female-biased colony sex ratios in a social spider. *Biol. J. Linn. Soc.* **43**: 297–311.

Lubin, Y. D., and R. H. Crozier. 1985. Electrophoretic evidence for population differentiation in a social spider *Achaearanea wau* (Theridiidae). *Insectes Soc.* **32**: 297–304.

Martin, P. and P. Bateson. 1993. *Measuring Behaviour:An Introductory Guide.* Cambridge University Press. 222pp.

McCook, H. C. 1889. *American Spiders and their Spinning Work,* vol. 1. Philadelphia: publ. by author and Academy of Natural Sciences of Philadelphia.

Muma, M. H. and W. J. Gertsch. 1964. The spider family Uloboridae in North America north of Mexico. *Am. Mus. Nov.* **2196**:
1–43.

Opell, B. D. 1979. Revision of the genera and tropical American
species of the spider family Uloboridae. *Bull. Mus. Comp. Zool.*
148: 443–547.

Oster, G. F., and E. O. Wilson. 1978. *Caste and Ecology in the Social
Insects.* (*Monogr. Pop. Biol.*, No. 12.) Princeton: Princeton University Press.

Peakall, D. B. and P. N. Witt. 1976. The energy budget of an orb-
web building spider. *Comp. Biochem. Physiol.* A**54**: 187–190.

Pulliam, H . R. and T. Caraco. 1984. Living in groups: is there an
optimal group size? In *Behavioral Ecology: An Evolutionary
Approach.* J. R. Krebs and N. B. Davies, eds., pp. 122–147.
Oxford: Blackwell Scientific Publications.

Pulliam, H. R. and G. C. Milikan. 1982. Social organization in
the non-reproductive season. In *Avian Biology*, vol. 6. D. S.
Farner and J. R. King , eds., pp. 45–87. New York: Academic
Press.

Rayor, L. S. and G. W. Uetz. 1990. Trade-offs in foraging success
and predation risk with spatial position in colonial spiders.
Behav. Ecol. Sociobiol. **27**: 77–85.

–. 1993. Ontogenetic shifts within the selfish herd: predation risk
and foraging trade-offs with age in colonial web-building
spiders. *Oecologia (Berl.)* **95**: 1–8.

Real, L. and T. Caraco. 1986. Risk and foraging in stochastic environments. *Annu. Rev. Ecol. Syst.* **17**: 371–390.

Riechert, S. E. 1984. Why do some spiders cooperate? *Agelena consociata*, a case study. *Fla. Entomol.* **68**: 105–116.

Riechert, S. E., R. Roeloffs and A. C. Echterncht. 1986. The ecology of the cooperative spider *Agelena consociata*. *J. Arachnol.* **14**:
175–192.

Riechert, S. E. and R. G. Gillespie. 1986. Habitat choice and utilization in web-building spiders. In *Spiders, Webs, Behavior, and
Evolution.* W. A. Shear, ed., pp. 23–48. Stanford: Stanford University Press.

Roeloffs, R. and S. E. Riechert. 1988. Dispersal and population-
genetic structure of the cooperative spider, *Agelena consociata*,
in west African rainforest. *Evolution* **42**: 173–183.

Rogers, L. E., W. T. Hinds and R. L. Buschbom. 1976. A general
weight vs. length relationship for insects. *Ann. Entomol. Soc. Am.*
69: 387–389.

Rubenstein, D. I. 1987. Alternative reproductive tactics in the
spider *Meta segmentata*. *Behav. Ecol. Sociobiol.* **20**: 229–237.

Rypstra, A. L. 1979. Foraging flocks of spiders: a study of aggregate
behavior in *Cytophora citricola* Forskal (Araneae: Araneidae) in
west Africa. *Behav. Ecol. Sociobiol.* **5**: 291–300.

–. 1983. The importance of food and space in limiting web-spider
densities; a test using field enclosures. *Oecologia (Berl.)* **59**:
312–316.

–. 1985. Aggregations of *Nephila clavipes* (L.) (Araneae, Araneidae) in
relation to prey availability. *J. Arachnol.* **13**: 71–78.

–. 1986. High prey abundance and a reduction in canabalism: the
first steps to sociality in spiders (Arachnida). *J. Arachnol.* **14**:
193–200.

–. 1989. Foraging success of solitary and aggregated spiders:
insights into flock formation. *Anim. Behav.* **37**: 274–281.

Sabath, M. D, L. E. Sabath, and A. M. Moore. 1974. Web, reproduction, and commensals of the semisocial spider *Cyrtophora
moluccensis* (Araneae: Araneidae) on Guam, Mariana Islands.
Micronesica **10**: 51–55.

Sandoval, C. P. 1987. Aspectos da ecologia e socialidade de uma
aranha colonial: *Eriophora bistriata* (Rengger 1936) (Aranaeidae).
Ph. D. thesis, Univ. Estadual de Campinas, Brazil.

Schoener, T. W. and C. A. Toft. 1983. Dispersion of a small-island
population of the spider *Metepeira datona* (Araneae: Araneidae) in
relation to web-site availability. *Behav. Ecol. Sociobiol.* **12**: 121–128.

Shear, W. A. 1970. The evolution of social phenomena in spiders.
Bull. Br. Arachnol. Soc. **1**: 65–76.

Simon, E. 1891. Observations biologiques sur les Arachnides. *Ann.
Soc. Entomol. Fr.* **60**: 5–14.

Smith, D. R. R. 1982. Reproductive success of solitary and communal *Philoponella oweni* (Araneae: Uloboridae). *Behav. Ecol. Sociobiol.* **11**: 149–154.

–. 1983. Ecological costs and benefits of communal behavior in a
presocial spider. *Behav. Ecol. Sociobiol.* **13**: 107–114.

–. 1985. Habitat use by colonies of *Philoponella republicana* (Araneae:
Uloboridae). *J. Arachnol.* **13**: 363–373.

–. 1986. Population genetics of *Anelosimus eximius* (Araneae: Theridiidae). *J. Arachnol.* **14**: 201–218.

–. 1987. Genetic variation in solitary and cooperative spiders of the
genus *Anelosimus* (Araneae: Theridiidae). In *Chemistry and Biology of Social Insects.* J. Eder and H. Rembold, eds., pp. 347–348.
Munich: Verlag J. Peperny.

Smith, D. R. and M. S. Engel. 1994. Population structure in an
Indian cooperative spider. *J. Arachnol.* **22**: 108–113.

Smith, D. R. and X. Hagen. 1996. Population structure and interdemic selection in the cooperative spider *Anelosimus eximius*
(Araneae: Theridiidae). *J. Evol. Biol.*, in press.

Spiller, D. A. 1992. Relationship between prey consumption and
colony size in an orb spider. *Oecologia (Berl.)* **90**: 457–466.

Spiller, D. A. and T. S. Schoener. 1989. Effect of a major predator
on grouping of an orb-weaving spider. *J. Anim. Ecol.* **58**: 509–523.

Struhsaker, T. T. 1969. Notes on the spiders *Uloborus mudiar*
(Chamberlin and Ivie) and *Nephila clavipes* (Linneaus) in
Panama. *Am. Midl. Nat.* **82**: 611–613.

Subrahmanyam, T. V. 1986. An introduction to the study of Indian
spiders. *J. Bombay Nat. Hist. Soc.* **65**: 726–743.

Tanaka, K. 1989. Energetic cost of web construction and its effect
on web relocation in the web-building spider *Agelena limbata*.
Oecologia (Berl.) **81**: 459–464.

–. 1991. Food consumption and diet composition of the web-building spider, *Agelena limbata* in two habitats. *Oecologia (Berl.)* **86**:
8–15.

Tikader, B. K. 1966. Studies on some biology of Indian spiders. *J. Bengal Nat. Hist. Soc.* **35**: 6–11.

Treherne, J. E. and W. A. Foster. 1982. Group size and anti-predator strategies in a marine insect. *Anim. Behav.* **32**: 536–542.

Turner, G. F. and T. J. Pitcher. 1986. Attack abatement: a model for group protection by combined avoidance and dilution. *Am. Nat.* **128**: 228–240.

Uetz, G. W. 1983. Sociable spiders. *Nat. Hist.* **92**: 62–79.

–. 1985. Ecology and behavior of *Metepeira spinipes* (Araneae: Araneidae), a colonial web-building spider from Mexico. *Natl. Geogr. Res. Rep.* **1985**: 597–609.

–. 1986a. Symposium: Social Behavior in Spiders. *J. Arachnol.* **14**: 145–281.

–. 1986b. Web building and prey capture in communal orb weavers. In *Spiders, Webs, Behavior, and Evolution.* W. A. Shear, ed., pp. 207–231. Stanford: Stanford University Press.

–. 1988a. Risk-sensitivity and foraging in colonial spiders. In *Ecology of Social Behavior.* C. A. Slobodchikoff, ed., pp. 353–377. San Diego: Academic Press.

–. 1988b. Group foraging in colonial web-building spiders: evidence for risk sensitivity. *Behav. Ecol. Sociobiol.* **22**: 265–270.

–. 1989. The "ricochet effect" and prey capture in colonial spiders. *Oecologia (Berl.)* **81**: 154–159.

–. 1992. Foraging strategies of spiders. *Trends Ecol. Evol.* **7**: 155–159.

Uetz, G. W. and J. W. Burgess. 1979. Habitat structure and colonial behavior in *Metepeira spinipes* (Araneae: Araneidae), an orb-weaving spider from Mexico. *Psyche* **86**: 79–89.

Uetz, G. W. and K. R. Cangialosi. 1986. Genetic differences in social behavior and spacing in populations of *Metepeira spinipes*, a communal-territorial orb weaver (Araneae: Araneidae). *J. Arachnol.* **14**: 159–173.

Uetz, G. W. and C. S. Hieber. 1994. Group size and predation risk in colonial web-building spiders: analysis of attack abatement mechanisms. *Behav. Ecol.* **5**: 326–333.

Uetz, G. W. and M. A. Hodge. 1990. Influence of habitat and prey availibility on spatial organization and behavior of colonial web-building spiders. *Natl. Geogr. Res.* **6**: 22–40.

Uetz, G. W., T. C. Kane and G. E. Stratton. 1982. Variation in the social grouping tendency of a communal web-building spider. *Science (Wash., D.C.)* **217**: 547–549.

Uetz, G. W., T. C. Kane, G. E. Stratton and M. J. Benton. 1987. Environmental and Genetic influences on the social grouping tendency of a communal spider. In *Evolutionary Genetics of Invertebrate Behavior.* M. D. Huettel, ed., pp. 43–53. New York: Plenum.

Vollrath, F. 1986. Environment, reproduction, and the sex ratio of the social spider *Anelosimus eximius. J. Arachnol.* **14**: 266–281.

–. 1987a. Growth, foraging, and reproductive success. In *Ecophysiology of Spiders.* W. Nentwig, ed., pp. 357–370. New York, Berlin, Heidelberg: Springer.

–. 1987b. Eusociality and extraordinary sex ratios in the spider *Anelosimus eximius* (Araneae: Theridiidae). *Behav. Ecol. Sociobiol.* **18**: 283–287.

Watson, P. 1993. Foraging advantage of polyandry for female sierra dome spiders (*Linyphia litigiosa:* Linyphiidae) and assessment of alternative direct benefit hypotheses. *Am. Nat.* **141**: 440–465.

Wilson, E. O. 1971. *The Insect Societies.* Cambridge: Belknap Press.

–. 1975. *Sociobiology, the New Synthesis.* Cambridge: Harvard University Press.

Wittenberger, J. F. 1981. *Animal Social Behavior.* Boston: Duxbury Press.

Wrona, F. J. and R. W. Dixon. 1991. Group size and predation risk: a field analysis of encounter and dilution effects. *Am. Nat.* **137**: 186–201.

23 · Causes and consequences of cooperation and permanent-sociality in spiders

LETICIA AVILÉS

ABSTRACT

Social behavior that involves cooperation in nest building, prey capture, feeding and brood care has arisen independently several times among spiders. In most cases, this form of cooperative behavior has led to the subdivision of the social spider populations into collections of perpetually inbreeding colony lineages that give rise to daughter colonies or become extinct without mixing with one another. The possible consequences of such population structure include intercolony selection leading to female-biased sex ratios, complex population dynamics leading to frequent colony extinction, and low genetic variability resulting from the isolation of the colony lineages, their small size at foundation, and their high rate of turnover. When trying to explain the phylogenetic distribution of cooperative behavior in spiders, therefore, we need not only to consider the preadaptations that may have facilitated the origin of their sociality (e.g. an irregular web and extended maternal care in the ancestral species), but also the consequences that such highly subdivided population structure might have on the patterns of extinction and speciation of the phylogenetic lineages of social spiders.

INTRODUCTION

When trying to understand the phylogenetic distribution of social behavior in a given group of organisms, we can look backwards in time from the origin of sociality and ask the question: what predisposed those particular lineages to the evolution of social behavior? However, in order to explain the presence of social species in some branches of phylogenetic trees and not in others we must also consider how sociality might have affected the subsequent evolution, extinction, and speciation of the species lineages in which it arose. In this paper, I approach the question of the phylogenetic distribution of cooperative behavior in spiders by looking both backwards and forwards in time from the

origin of their sociality. I will limit my argument to the species that engage in cooperative prey capture and feeding, remain in association through at least part of the adult period of their lives, and, for the most part, have communal brood care. As I will argue here, this form of cooperative behavior has had a major impact on the population biology of those species, leading, in most instances, to the subdivision of their populations into multiple colony lineages perpetually isolated from one another. I will start out by briefly reviewing the biology of these species and what is known about their phylogenetic distribution. I will then consider, one at a time, the issues of the preadaptations that might have facilitated the origins of their sociality and the consequences that such origins might have had on the biology and further evolution of those species.

SOCIAL BEHAVIOR IN SPIDERS

Social behavior, ranging from aggregations of individual webs to cooperative care of the brood, is known in a few dozen spider species out of the 34 000 described (Buskirk 1981; D'Andrea 1987). Two criteria have been used to classify the types of sociality represented by these species (D'Andrea 1987): (1) whether or not the aggregations or colonies last throughout the life cycle of their members, giving rise to either permanent or temporary societies, and (2) whether or not the spiders maintain individual territories (i.e. webs) within the nests, giving rise to either loose aggregations with limited cooperation or tight societies with a wide range of cooperative behaviors. According to these criteria, most of the spider species displaying some form of social behavior can be assigned to four categories:

(1) non-territorial permanent-social (also known as 'quasisocial', Wilson 1971, or simply, 'cooperative', Riechert 1985), examples of which include the well-known *Anelosimus eximius* (Theridiidae), *Agelena consociata* (Agelenidae) and *Stegodyphus mimosarum* (Eresidae);

(2) territorial permanent-social (also known as 'communal territorial', e.g. Uetz and Cangialosi 1986, or 'colonial', e.g. Lubin 1980), including *Metabus gravidus* (Araneidae), *Philoponella republicana* (Uloboridae), *Dictyna calcarata* (Dictynidae) and *Scytodes fusca* (Scytodidae) (Buskirk 1981; D'Andrea 1987; Bowden 1991; Uetz and Hieber this volume);

(3) non-territorial periodic-social (also known as 'subsocial'; Wilson 1971; Krafft 1979), which include *Anelosimus studiosus* (Theridiidae), *Stegodyphus lineatus* (Eresidae), *Theridion saxatile* (Theridiidae) and *Phryganoporus candidus* (=*Badumna candida*) (Desidae) (D'Andrea 1987; Downes 1993); and

(4) territorial periodic-social, such as *Parawixia* (=*Eriophora*) *bistriata* (Araneidae) from Uruguay (see, for example, Fowler and Diehl 1978).

According to more general classification schemes (see, for example, Wilson 1971; Crespi and Yanega 1995), the species in category (1) would be labeled as 'quasisocial' (Wilson 1971) or 'cooperative breeders' (subcategory quasisocial), those in category (2) would be labeled as 'communal', and those in categories (3) and (4) would be considered 'subsocial'. The species on which this chapter will focus include all the non-territorial permanent-social species, which are the only ones that exhibit alloparental care, plus three related non-territorial periodic-social species in which siblings continue to cooperate past the onset of reproductive maturity. Species that could potentially be included in this review, but on which not enough information is yet available, include a cooperative *Philoponella* sp. from Cameroon (Breitwisch 1989) and a recently discovered social *Argirodes* sp. from Australia (M. Whitehouse and R. Jackson, personal communication).

THE NON-TERRITORIAL PERMANENT-SOCIAL OR 'COOPERATIVE' SPIDERS

Spider species in which adult members of a generation share a single communal nest and engage in cooperative prey capture and feeding are distributed in ten genera and seven families and occur in habitats ranging from thornbush savanna to tropical rainforest, mostly in tropical regions of the world (Table 23-1). The nests of most of these species, which are built and maintained cooperatively, consist of a silk mesh built around or supported by vegetation and include a relatively protected area serving as a refuge and a prey-capture snare (Fig. 23-1). Species in only two genera, *Diaea* and *Delena*, lack the prey-capture snare because they hunt prey without the aid of a web (see species descriptions below). Several individuals may participate in prey capture and in bringing the prey to the more protected areas of the nests where communal feeding takes place (Buskirk 1981; D'Andrea 1987). With the exception of the species in the genus *Diaea* in which the mothers disperse before egg-laying, care of the brood is communal and may involve simply the passive protection of egg sacs and young by locating them in the more protected areas of the nests (see, for example, Uetz 1983), through to assisted egg-sac building (see, for example, Kraus 1988) or active brood-guarding and regurgitation-feeding (see, for example, Kullmann *et al.* 1971/72; Christenson 1984; but see Lubin 1982) in some species.

A feature that critically differentiates these species from the eusocial insects is that all spiders in a colony are totipotent, i.e. they can perform all the tasks in the colonies, including reproduction (see, for example,. Darchen and Delange-Darchen 1986; Lubin 1995). The only division of labor present in the colonies is based on sex and age: males normally do not participate in the activities of the colonies and juveniles gradually accrue new tasks to their behavioral repertoire (see, for example, Avilés 1993a; Lubin 1995). Further, because individuals in a nest do not live long enough to overlap with their adult offspring, the colonies consist of more or less separate maternal and offspring generations. However, competition is not entirely absent among contemporaneous members of a generation, and in certain species some individuals may fail to acquire sufficient resources to reproduce (Vollrath and Rohde-Arndt 1983; Rypstra 1993). There is evidence to suggest that this uneven reproductive success is a result of increased competition for resources within the growing colonies (see, for example, Vollrath 1986a; Seibt and Wickler 1988a; Rypstra 1993; J. Henschel, personal communication), in clear parallel with intraspecific competition in animal populations where not all individuals reproduce. Some authors, however, have argued that the lack of universal reproduction warrants the label 'eusocial' for at least one of the species (Vollrath 1986a; Rypstra 1993), on the grounds that the individuals that fail to reproduce are still active members of the colonies. That these individuals do not disperse, however, may simply reflect the risks of dispersal, the difficulty of finding a mate outside the colony, and the impossibility of knowing ahead of time (before the opportunity to reproduce has passed) how well a given individual will fare in the struggle for resources. The term 'eusociality', in my opinion, should be restricted to those

Table 23-1. *Habitat, phenology, and periodic-social relatives of the cooperative spiders*

Family	Genus and species	Distribution	Habitat[a]	Phenology	Periodic–social species in same or related genus[b]	Source
Agelenidae	*Agelena consociata*	Gabon and Central African Republic	Understory of tropical rainforest, 500–600 m	Aseasonal	*Coelotes terrestris*, *C. atropos*	1, 2, 3, 4
	Agelena republicana	Gabon	Forest edge of tropical rainforest, 500–600 m	?	As *Ag. consociata*	5
Dictynidae	*Aebutina binotata*	Amazonian Brazil, Ecuador and Peru	Understory of tropical rainforest, 210 m	Aseasonal	Relatives unknown	6, 7, 8
	Mallos gregalis	Central Mexico	Mountainous areas with distinct rainy and dry seasons, 2500 m	Annual, seasonal	None, territorial permanent-social spp. in *Mallos-*, *Mexitlia* and *Dictyna*	9, 10, 11
Eresidae	*Stegodyphus dumicola*	Southern third of Africa	Dry thornbush savanna	Annual, seasonal	*S. lineatus* and other *Stegodyphus* spp.	12, 13, 14
	Stegodyphus mimosarum	Eastern Africa, south of the Sahara	Dry thornbush savanna	Annual, seasonal	*S. lineatus* and other *Stegodyphus* spp.	12, 13
	Stegodyphus sarasinorum	India, Sri Lanka and eastern Afghanistan	Cultivated ares, dry scrub jungle, along river courses	Annual, seasonal	*S. pacificus*	12, 15, 16
Oxyopidae	*Tapinillus* sp.	Amazonian Ecuador	Tropical rainforest	Aseasonal	Other *Tapinillus* spp.	17
Sparassidae	*Delena cancerides*	Widespread in Australia and assoc. islands	Under exfoliating bark of dead trees	?	Species in *Isopoda*, *Pediana* and *Olios*	18, 19
Theridiidae	*Achaearanea disparata*	Gabon and Ivory Coast	Understory and forest edge of tropical rainforest	?	*Ac. mundula* (=*tesselata*), *Ac. tepidariorum*	20, 21
	Achaearanea vervortii	Papua New Guinea	Above 2000 m elevation	?	*Ac. kaindi*, *Ac. mundula*	22, 23
	Achaearanea wau	Papua New Guinea	Forest edge of montane rainforest, 1200–1700 m	Aseasonal	As *Ac. vervortii*	22, 23, 24
	Anelosimus domingo	Surinam, Ecuador and Peru	Understory of tropical rainforest	Aseasonal	*An. studiosus*, *An. jucundus*	25, 26, 27, 28
	Anelosimus eximius	Panama to southern Brazil	Understory and forest edge of tropical rainforest, 30–880 m	Aseasonal	As *An. domingo*	25, 29, 30
	Anelosimus lorenzo	Paraguay	Recorded from citrus orchards	?	As *An. domingo*	31
	Anelosimus rupununi	Northern South America	Clearings and forest edge of tropical rainforest	?	As *An. domingo*	25, 32, 27

				T. pinctum, varians, lunatum, ovatum, saxatile, impressum, sisyphium	
	Theridion nigroannulatum	Amazonian Ecuador, Peru and Venezuela	Understory of tropical rainforest, 200–640 m	Aseasonal	33, 34
Thomisidae	*Diaea ergandros*	SE Australia and Tasmania	Dry, closed-canopy eucalypt forests	Annual, seasonal	Other Australian *Diaea* spp.	35
	Diaea megagyna	E Australia, N New South Wales, S Queensland	Dry, closed-canopy eucalypt forests	Annual, seasonal	Other Australian *Diaea* spp.	35
	Diaea socialis	SW Australia	Jarrah and Karri eucalypt forests	Biennial, seasonal	Other Australian *Diaea* spp.	35, 36

[a] Elevation data reflects only areas where the species have been studied, as given in the respective sources.

[b] List of periodic-social relatives for species in the genera *Agelena, Mallos, Stegodyphus, Anelosimus,* and *Theridion* taken from D'Andrea (1987).

Sources: 1, Krafft 1970; 2, Darchen 1980; 3, Riechert *et al.* 1986; 4, Pain 1964; 5, Darchen 1967a,1976; 6, Simon 1892; 7, Avilés 1993a; 8, L. Rayor, personal communication; 9, Diguet 1909; 10, Burgess 1976; 11, Jackson and Smith 1978; 12, Kraus and Kraus 1988; 12, Seibt and Wickler 1988b; 14, J. Henschel, personal communication; 15, Kullmann *et al.* 1971/72; 16, Bradoo 1972; 17, Avilés 1994; 18, Rowell 1987; 19, Rowell and Avilés 1995; 20, Darchen 1968; 21, Darchen and Ledoux 1978; 22, Levi *et al.* 1982; 23, Lubin 1982, 1991; 24, Lubin and Robinson 1982; 25, Levi 1963; 26, Levi and Smith 1982; 27, Rypstra and Tirey 1989; 28, L. Avilés unpublished; 29, Vollrath 1982, 1986b; 30, Avilés 1992; 31, Fowler and Levi 1979; 32, Levi 1972; 33, Levi 1983, H. Levi, personal communication; 34, L. Avilés and W. Maddison, unpublished; 35, Evans 1996; 36, Main 1988.

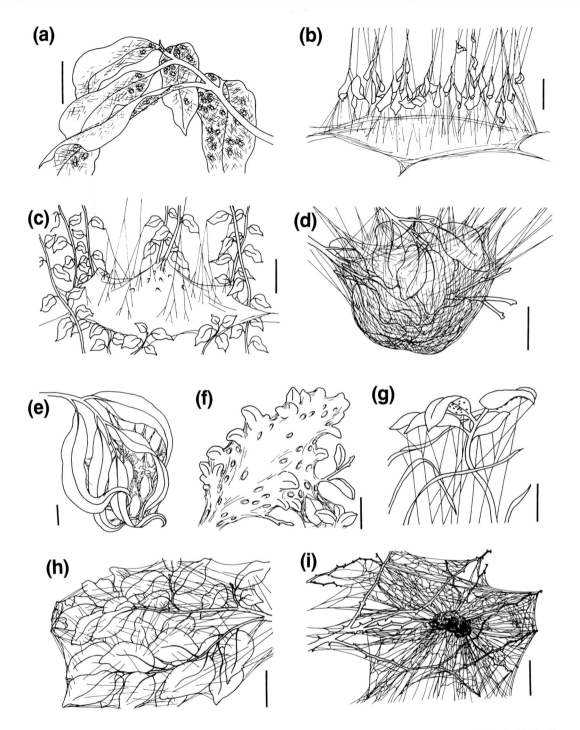

Fig. 23–1. Representative nests of species in nine of the ten genera where cooperative spiders have been described: (a) *Aebutina binotata*, drawn from photograph in Avilés (1993a); (b) *Achaearanea wau*, drawn from photograph in Lubin (1986); (c) *Agelena consociata*, original drawing based on D'Andrea (1987) and Krafft (1970); (d) *Anelosimus eximius*, redrawn from Simon (1891); (e) *Diaea socialis*, original drawing based on Main (1988); (f) *Mallos gregalis*, drawn from photograph in Uetz (1983); (g) *Theridion* cf. *nigroannulatum*, drawn from nest in the field (L. Avilés and W. Maddison, unpublished); (h) *Tapinillus* sp., drawn from photograph in Avilés (1994); (i) *Stegodyphus dumicola*, drawn from photograph in Seibt and Wickler (1988b). Drawings by Kris Sonderegger, 1994. Scale bars (approximately): (a) 5 cm; (b) 20 cm; (c) 10 cm; (d) 10 cm; (e) 2 cm; (f) 5 cm; (g) 20 cm; (h) 5 cm; (i) 10 cm.

cases in which sterility or subfertility, rather than being a side product of competition, has been selected for as a socially adaptive trait (see also Crespi and Choe, Chapter 24, this volume). There is as yet no evidence that such has been the case in any of the cooperative spiders.

Another feature that critically differentiates the cooperative spiders from the eusocial insects, in particular from the eusocial Hymenoptera and termites, is that, with the exception of species in the genera *Delena*, *Tapinillus* and *Diaea*, mating occurs among nestmates (reviewed in Riechert and Roeloffs 1993; Smith and Engel 1994). This inbreeding process, which results in separate generations of spiders succeeding each other within a nest, is responsible for colony growth. Colonies that reach a relatively large size proliferate by means of budding, fission, or the production of small propagules (see species descriptions below). Colonies may also go extinct frequently and for reasons that are yet unknown. Colonies, therefore, appear as inbred population lineages perpetually isolated from one another. The lack of a regular outbreeding phase makes the cooperative spiders unlike most other social insects or arachnids. Only one ant (Itow *et al.* 1984; Tsuji 1988) and some species of social psocid (Mockford 1957; New 1973) appear to have a comparable degree of population subdivision.

Here, I provide a brief description of each of the cooperative spiders so far described, focusing on the demographic and population-level aspects of their biology, which are fundamental to the main arguments of this paper and which have only recently become the subject of study. The sociality of seven of the twenty species discussed below has only been reported since the publication of the last review of social spider biology (D'Andrea 1987). The sociality of an eighth species, *Theridion nigroannulatum*, is reported here for the first time. Three points will emerge from the descriptions that follow: (1) that a strongly subdivided population structure, that comes hand in hand with a highly female-biased sex ratio, is a common theme among the cooperative spiders (Fig. 23-2); (2) that beyond their basic similarities, each species is unique in the structure of its nests and colonies, phenology, and population dynamics; and (3) that there is a great need for quantitative estimates of the population parameters that define each species (Table 23-2) in order to develop and test hypotheses of why and how the similarities and differences arose. My hope is that the discussion that follows, where I make a first attempt at explaining, if not the differences, the similarities, will make evident the wealth of opportunities that these species provide not only for the study of social

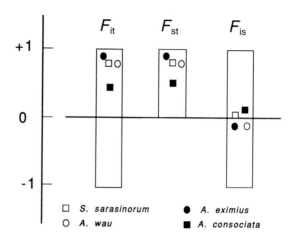

Fig. 23–2. Three-level *F*-statistics obtained from allozyme electrophoresis studies of four of the cooperative spiders, reprinted with permission from Smith and Engel (1994). The length of the boxes on the y-axis indicates the range of possible values for the three statistics, the symbols represent the estimated values. The allozyme results show that the large overall departure from panmixia (F_{it}) found in these species is due to subvision of their populations in colony lineages (F_{st}), rather than to non-random mating within the colony lineages (F_{is}).

evolution, but also for the fields of population ecology and evolutionary biology in general.

Species descriptions

In the following descriptions I will use the term 'colony' for the group of individuals that occupy a continuous and nuclear web structure that I will call a 'nest' (retreat + prey-capture snare). Except in extremely large nests, individuals in a colony would tend to interact with each other at least a few times during their lifetimes. Often, several nests may occur near each other (within one to a few dozen meters) in what I call a 'colony cluster' or 'colony complex' (or 'colony patch' (Seibt and Wickler 1988b)). Exchange of individuals may occur among nests in a complex that are connected by silklines. However, because such nests would most likely be recently derived from a common ancestral colony, such exchange of individuals would not imply mixing among different colony lineages. This usage is similar to that of Vollrath (1982), Lubin and Robinson (1982) or Seibt and Wickler (1988b), but unlike that of Pasquet and Krafft (1989) or Riechert and Roeloffs (1993) who define a 'colony' as the ensemble of individuals that occupy all of the nests in a complex.

Table 23–2. *Population structure and dynamics of the cooperative spiders*

Species	Colony size[1]	Intercolony migration[2]	Proliferation threshold	Propagule size[3]	Turnover rate[4]	Sex ratio (% males)[5]	Eggs per sac[6]	Source
Aebutina binotata	14–104 (800)	unlikely	1 generation?	14–104; $m = 40$	100%?[d]	8%[j]	11–17 (14.8)	1, 2
Achaearanea disparata	up to 100s?	?	?	?	?	?	20–32	3
Ac. vervortii	100s?	?	?	?	?	0.4, 14.5%[a]	?	4
Ac. wau	few–1800	no	>100	1–44, $\bar{x} = 19.1$	47%[e]	11%[j]	0–40 (15.8)	4, 5, 6
Agelena consociata	0–2213, $\bar{x} = 16$	no	?	?	?	8, 13%[p,a]	2–21	7, 8, 9
Ag. republicana	dozens	?	?	?	?	19%[p,a]	1–11	10
Anelosimus domingo	1–>1000	unlikely	?	?	?	8%[e]	13–16 (15.2)	11, 12
An. eximius	1–>50 000	no	>1000	1–dozens	21%[e]	9%[e]	25–57	1, 11, 12
An. lorenzo	>200	?	?	?	?	biased[a]	?	13
An. rupununi	up to 1000	?	?	?	?	?	?	13
Delena cancerides	1–300	yes	?	?	?	50%[e]	?	14
Diaea ergandros	1 + offspr.	yes	1 generation	1	100%[d]	50%[a]	15–80 (45)	15
D. megagyna	1 + offspr.	unlikely	1 generation	1	100%[d]	20%[a]	?	15
D. socialis	1 + offspr. (56)	no	1 generation	1	100%[d]	28%[e]	20–86 (48)	15, 16, 17
Mallos gregalis	up to 20 000	?	?	?	?	13%[a]	10–20	18, 19
Stegodyphus dumicola	1–261 (647)	unlikely	?	1–45, $m = 1$	67%[e]	17%[e]	61–104	20, 21, 22
S. mimosarum	1–371	unlikely	?	1–?	?	9.7%[a]	15–48 (26.3)	20
S. sarasinorum	1–120 (900)	no	?	1–?	?	22%[a]	60–115	23, 24
Tapinillus sp.	7–100 (700)	likely	?	?	?	50%[e]	11–24	25
Theridion nigroannulatum	1–4000	?	?	1–5, $m = 1$	?	13%[p]	?	26

[1] Min–max no. of mature individuals (max. colony size, if juveniles are included).

[2] Among nests lacking silk-line connections.

[3] range: m = median or $\bar{x}$ = mean.

[4] [d] = % of colonies dividing or dispersing; [e] = % of established colonies going extinct.

[5] Percentage of males among: [e] = embryos; [j] = juveniles; [p] = preadults; [a] = adults.

[6] Range (mean, if available).

Source: 1, Avilés 1992; 2, Avilés 1993a; 3, Darchen 1968; 4, Lubin 1991; 5, Lubin and Robinson 1982; 6, Lubin and Crozier 1985; 7, Riechert *et al.* 1986, 8, Riechert and Roeloffs 1993; 9, Pain 1964; 10, Darchen 1967a; 11, L. Avilés, unpublished.; 12, Avilés and Maddison 1991; 13, Fowler and Levi 1979; 14, Rowell and Avilés 1995; 15, Evans 1996; 16, Main 1988; 17, Rowell and Main 1992; 18, Jackson and Smith 1978; 19, Burgess 1976; 20, Seibt and Wickler 1988b; 21, J. Henschel, personal communication; 22, C. Varas and L. Avilés, unpublished; 23, Bradoo 1972, 1973, 1975; 24, Jacson and Joseph 1973; 25, Avilés 1994; 26, L. Avilés and W. Maddison, unpublished.

Aebutina binotata Simon (Dictynidae?)

This spider lives in the understory of Amazonian rainforests (Table 23-1) where it forms colonies that contain up to a few dozen adult females and at most 800 developing offspring (Avilés 1993a). The nests consist of one or a few contiguous leaves covered on both surfaces by a fine layer of silk (Fig. 23-1a). The spiders sit underneath the leaves, with adult females lined up along the edges and the offspring concentrated towards the center. Two features appear unique to this species: (1) the colonies do not remain in the same location for more than one generation; and (2) the regular method of colony proliferation appears to be fission (Avilés 1992). Prior to mating and egg laying, preadult males and females move as a group to locations that may be dozens of meters away. The males die or emigrate soon after mating; the mothers remain with their offspring until they near reproductive maturity (Avilés 1992). The development of the offspring, which lasts about six months from egg to adult, is highly synchronized within colonies, but asynchronous among colonies (Avilés 1992). The sex ratio among preadult individuals is highly female-biased (Table 23-2). Colony fission apparently takes place when parts of expanding colonies move to nearby leaves. It is not known whether loss of entire colonies occurs in this species.

Achaearanea disparata Denis (Theridiidae)

This tropical rainforest species from western Africa (Table 23-1) forms nests that are collections of up to hundreds of subunits, each unit consisting of a horizontal sheet web 30–40 cm in diameter and one or a few curled-leaf retreats placed about 30 cm above it (Darchen 1968; Darchen and Ledoux 1978). Forest-edge nests can extend up to 10 m in height and width; nests inside the forest are smaller and mostly lack the refugia and the clear subunit structure (Darchen and Ledoux 1978). Each of the subunits may house from one to seven adults, up to 50 spiderlings, and a handful of egg sacs (Darchen 1968). The nests may expand by accretion of new leaf retreats sought out by older spiders when the growing spiderlings crowd their old retreat (Darchen 1968). There are no data on whether alloparental care occurs, how independent nests arise, or whether and how often colonies die. Sex-ratio data are also lacking.

Achaearanea vervortii Chrysanthus (Theridiidae)

Ac. vervortii lives in Papua New Guinea, at higher elevations than its congeneric *Ac. wau* (Table 23-1). Its nests consist of a basal horizontal sheet web and a vertical barrier holding numerous curled-leaf retreats within it (Levi *et al.* 1982). The adult sex ratio is highly female-biased (Table 23-2).

Achaearanea wau Levi (Theridiidae)

This species occurs in tree-fall clearings and forest edges of montane rainforest in Papua New Guinea (Table 23-1). The nests, which may contain over a thousand individuals (Lubin 1991), consist of a basal sheet web and a superior tangled maze where one to several curled-leaf retreats are suspended (Fig. 23-1b). The generations, which are clearly distinct because of a synchronization in mating, egg-laying, and offspring development, last for 7–8 months with an overlap of about three months between mother and offspring (Lubin and Robinson 1982). The sex ratio is highly female-biased (Table 23-2).

Daughter colonies of this species form by the concerted emigration of groups of up to 44 females that leave their natal nest after having been inseminated (Lubin and Robinson 1982; Lubin 1991). Only colonies in their second or third generation and containing over 100 individuals proliferate. Such proliferation events are apparently responsible for the formation of colony clusters that may occupy areas up to $0.01 \, km^2$ and contain up to several dozen nests (Lubin and Crozier 1985). The colonies in a cluster tend to be synchronized with each other in the stage of their life cycle, but they may be out of synchrony with colonies in neighboring populations. Males have occasionally been seen to join nearby daughter colonies; otherwise no internest migration appears likely. Six out of seven local populations were fixed for one or the other allele of the one polymorphic locus found among 22 scorable allozyme loci (Lubin and Crozier 1985). Colony extinction, which may be widespread and sudden (Lubin and Crozier 1985), may involve up to 50% of the colonies per generation (Lubin 1991).

Agelena consociata Denis (Agelenidae)

This spider inhabits the understory of tropical rainforests of Gabon and the Central African Republic (Table 23-1). The nests, which may occur in clusters of up to 27 units (Riechert *et al.* 1986), consist of one or more silk galleries serving as retreats, a basal horizontal sheet web, and a scaffolding of vertical and oblique threads connecting the nest to the vegetation above (Fig. 23-1c) (Darchen 1965, 1979). The nests contain from a few to over two thousand individuals (Darchen 1979; Riechert *et al.* 1986), with a female-biased sex ratio among adults and preadults (Table 23-2).

New colonies may arise passively when expanding nests become fragmented as a result of heavy rain or falling branches (Krafft 1970; Darchen 1976, 1978; Roeloffs and Riechert 1988) or, as inferred experimentally, when groups of spiders are transported by animal carriers that walk or fly through the colonies (Roeloffs and Riechert 1988). Active dispersal has not been observed (Roeloffs and Riechert 1988), although its occurrence has been inferred from the presence of colonies that contain mostly adult and preadult females (Darchen 1978). Allozyme electrophoresis showed that although over one-third of twenty-eight loci scored were polymorphic, heterozygosity was extremely low and colonies only 30 m apart could be fixed for different alleles (Roeloffs and Riechert 1988; Riechert and Roeloffs 1993). Thirty-seven out of 323 nests studied by Riechert et al. (1986) went extinct during a two-year study period, most of them during the two wet seasons.

Agelena republicana Darchen (Agelenidae)

This African species occurs in the same geographic area as *Ag. consociata* (Table 23-1), but on the forest edge (usually along water courses) and at higher distances from the ground (3–20 m) (Darchen 1967a). The nests, which are smaller (up to 120 cm in length) and lighter than those of *Ag. consociata* (Table 23-2), occur on the distal parts of branches, often enveloping them. They consist of a horizontal sheet, a central refuge, and a few vertical and oblique threads (Darchen 1967a, 1976). Hundreds of nests, either clustered or in isolation, can be seen along dozens of meters of continuous vegetation (Darchen 1967a, 1976). There are no direct observations of colony proliferation, although Darchen (1976) suggests a method of budding similar to that of *Ag. consociata*. No data on colony extinction are available. The preadult and adult sex ratio is highly female-biased (Table 23-2).

Anelosimus domingo Levi (=*An. saramacca* Levi) (Theridiidae)

This spider occurs in the understory of tropical rainforests of northern South America (Table 23-1). The nests are basket-shaped and consist of an external base-sheet, an internal mesh with parts of live vegetation serving as refugia, and a scaffolding of vertical and oblique threads connecting this structure to the vegetation above. The nests are on average much smaller (Table 23-2) and lighter than those of *An. eximius* (the largest, reported by Rypstra and Tirey (1989), measured 110 cm × 140 cm × 35 cm) and may contain any number of spiders from a single female plus

her offspring to just over a thousand individuals (L. Avilés, personal observations). Embryo sex ratios are highly female-biased (Table 23-2). The colonies appear to be much less abundant than those of *An. eximius* and tend not to occur in clusters (L. Avilés, personal observations). There are no data on colony proliferation or extinction.

Anelosimus eximius Keyserling (Theridiidae)

This neotropical spider (Table 23-1) builds nests that may measure up to a few meters in width and length, the largest of any of the cooperative spiders (Table 23-2). They may contain from one female plus her offspring to several tens of thousands of spiders (Vollrath 1982; Pasquet and Krafft 1989; Avilés 1992; Venticinque et al. 1993). The nests, which may occur in clusters of up to a few dozen, are similar to those of *An. domingo* (Fig. 23-1d), except that very large ones are no longer hemispherical but follow the contours of the vegetation. Nests have been found both in the understory and forest edge (Vollrath 1986b; Avilés 1992; but see Pasquet and Krafft 1989), with nests in the forest edge being on average significantly larger than those in the understory (Avilés 1992).

The generations last for about eight months, with a three-month overlap between mother and offspring (Avilés 1986). Different colonies in a locality, or even within a cluster, may occur at different stages of their life cycle (L. Avilés, personal observations), although this is not as obvious as in *Ae. binotata* or *Ac. wau* where the generations within a nest are more clearly distinct (see above). The sex ratio, measured from embryos through adults, is highly female-biased (Table 23-2).

Large colonies (>1000 individuals) give rise to new ones by budding or dispersal of gravid females that build new nests either in groups or individually (Overal and Ferreira 1982; Vollrath 1982; Avilés 1992; Venticinque et al. 1993). The survivorship of newly founded colonies is very low, although it improves as the number of foundresses increases (Vollrath 1982; Christenson 1984; Avilés 1992; Venticinque et al. 1993; Leborgne et al. 1994). Extinction of established colonies can be widespread and sudden (Vollrath 1982) and may involve an average of 21% of the colonies per generation (as measured at an understory site in Eastern Ecuador; Avilés 1992). Allozyme electrophoresis uncovered extremely low levels of polymorphism, with most of the variation being distributed at the among-colony level (Smith 1986; D. Smith and R. Hagen, personal communication).

Anelosimus lorenzo Levi (Theridiidae)

An. lorenzo has been recorded from citrus orchards in Paraguay (Fowler and Levi 1979). Its webs are similar to those of other social *Anelosimus* (Fig. 23-1d) and can reach volumes close to one and a half cubic meters. Fowler and Levi (1979) suggest that colony foundation may be carried out by small groups of spiders because 'small colonies were always found in trees adjacent to large ones'. There are no data on colony persistence, although it is clear that the colonies may last for several generations (Fowler and Levi 1979). The adult sex ratio is female-biased (Table 23-2).

Anelosimus rupununi Levi (Theridiidae)

This species occurs in cultivated areas, clearings, and edges of tropical rainforests of northern South America (Table 23-1). A photograph in Levi (1972) shows a mesh-like structure covering a large portion of the crown of a tree, much like an *An. eximius* nest. The colonies may contain up to 1000 individuals (Fowler and Levi 1979).

Delena cancerides Walckenaer (Sparassidae)

This Australian huntsman spider (Table 23-1) forms colonies of up to 300 individuals that live under the bark of dead *Acacia*, *Casuarina* and other trees with exfoliating bark (Rowell 1987; Rowell and Avilés 1995). The spiders in a nest live in tight physical contact with each other and when kept in the laboratory they may participate in joint prey capture and share their food. Unlike all the other species, in which tolerance extends to conspecifics from even distant nests, *D. cancerides* are extremely aggressive towards members of foreign colonies. In addition, the mating system is outbred, as suggested by allozyme data and a lack of sex-ratio bias (Rowell and Avilés 1995). Finally, unlike all but the species in the genus *Diaea*, prey capture does not involve the use of a web snare. It is not known whether alloparental care is present, how new colonies arise, for how long they last, or how outbreeding is accomplished.

Diaea socialis Main, D. megagyna Evans and D. ergandros Evans (Thomisidae)

These social crab spiders from Australia and Tasmania (Table 23-1) lack alloparental care as they form colonies consisting of a single adult female and her offspring (Main 1988; Evans 1996). Although these species should technically be considered as non-territorial periodic-social, they have been included in this review because

siblings cooperate past the onset of reproductive maturity. In *D. socialis* and *D. megagyna*, mating takes place among siblings; in *D. ergandros*, males migrate among colonies prior to mating (Main 1988; Evans 1996). In all three cases, mated females disperse before egg-laying. As would be expected from their respective breeding structures, *D. socialis* and *D. megagyna* have female-biased sex ratios, whereas *D. ergandros* has an even sex ratio (Table 23-2). Two of the social *Diaea* spp., therefore, are genetically similar to the highly inbred cooperative spiders and appear to be at the transition point between periodic- and permanent-sociality.

The *Diaea* species form nests that consist of bundles of eucalyptus leaves bound together with silk threads (Fig. 23-1e). The spiders occupy the spaces in the interior and forage through portholes on the surface of the nest. From a core built by the founding female, the nests are enlarged with the help of the offspring, who continue to build after their mother's death (Main 1988; Evans 1996). Mature nests of *D. socialis* may contain up to 200 leaves and 56 mature offspring (Main 1988). *D. socialis* has a two-year life cycle that tracks the phenology of the *Eucalyptus* species with which it is associated. Its colonies may occur in two developmental stages at a given locality, depending on the year and the seasonal conditions (Main 1988; B. Y. Main, personal communication). The other two species have annual life cycles, with the young in *D. ergandros* being produced in the summer (Evans 1996). In *D. socialis*, only one in five incipient nests produces adult offspring (Main 1988).

Mallos gregalis Simon (Dictynidae)

This species occurs in mountainous areas of central Mexico (Table 23-1), where it appears to have a seasonal life cycle, with young being produced in November, after the end of the rainy season (Buskirk 1981, based on Diguet 1909). The nests, which are occupied and enlarged by successive generations (Burgess 1976), may reach several square meters (Diguet 1909; Burgess 1976) and contain up to 20 000 spiders (Jackson and Smith 1978). They are built surrounding the branches of trees or bushes (Fig. 23-1f) and consist of an external envelope of cribellar silk and an interior mass of silken galleries (Diguet 1909; Burgess 1976; Uetz 1983). The adult sex ratio in laboratory colonies was highly female-biased (Table 23-2). There are no field data on colony proliferation or extinction, although in the laboratory webs with just a few or single individuals appeared occasionally (Jackson 1978).

Stegodyphus dumicola Pocock (Eresidae)

S. dumicola occurs in southern Africa (Table 23-1) where it can be found in isolated local patches containing dozens to hundreds of nests, with interpatch distances of 0.1–10 km (Seibt and Wickler 1988b). The nests are built around the twigs or branches of spiny bushes or trees, closer to the ground than those of *S. mimosarum* (Seibt and Wickler 1988b). They consist of a central, sponge-like, mass of silk serving as a refuge, and one or more vertical sheet webs where insects become entangled (Fig. 23-1i). The nests contain from a single female up to a few hundred individuals (Table 23-2), although the effective colony size may be larger because contiguous nests may be interconnected by common sheet-webs (Seibt and Wickler 1988b). The adult sex ratio, which is preponderantly female, reflects a primary sex ratio bias (C. Varas and L. Avilés, unpublished; Table 23-2).

The nests are occupied and enlarged by successive generations, which are annual and clearly distinct (Seibt and Wickler 1988b). After laying their eggs at the end of the summer, mothers tend their offspring until their death in May. The offspring overwinter on their own, enlarge their nest after the rains in October, and reach reproductive maturity by the following summer (Seibt and Wickler 1988b). New nests may be formed by the budding of expanding nests (Wickler and Seibt 1993) or when gravid females, either by themselves (82% of the cases), or in groups of up to 45 individuals, disperse at the end of the mating season (J. Henschel, personal communication). At a site in Namibia, only 6% of 159 solitary nests survived to the next season, as opposed to 33% of 9 social units (J. Henschel, personal communication). The nests are attacked by numerous predators, in particular ants (J. Henschel, personal communication). Allozyme electrophoresis revealed a marked degree of population subdivision (Wickler and Seibt 1993).

Stegodyphus mimosarum Pavesi (Eresidae)

This east African species is widespread in thornbush country (Table 23-1) where it builds nests on terminal branches of shrubs and trees (Seibt and Wickler 1988b). The nests are similar to those of *S. dumicola*, except that the sheet webs are less clearly developed, more numerous, and highly irregular. As in *S. dumicola*, the nests contain from a single female to a few hundred individuals (Table 23-2) and are occupied and enlarged by successive generations (Seibt and Wickler 1988b). The generations are discrete and have an annual life cycle similar to that of

S. dumicola. The adult sex ratio is highly female-biased (Table 23-2).

As in *S. dumicola*, the colonies occur in discrete patches that may contain hundreds of nests (Seibt and Wickler 1988b). There are no direct observations of colony proliferation, though nests containing single females and their egg sac can be seen surrounding old damaged nests (Wickler 1973). Nests that had been seen flourishing for years could deteriorate from one year to the next. A patch of 286 nests, seen healthy in 1982, had only 113, 50, and 0 surviving nests one, two, and five years later, respectively (Seibt and Wickler 1988b). In this species, allozyme electrophoresis revealed a similar degree of population subdivision as in *S. dumicola* (Wickler and Seibt 1993).

Stegodyphus sarasinorum Karsh (Eresidae)

This spider from the Indian subcontinent (Table 23-1) builds nests similar to those of the previous two species, in aggregations that may occasionally cover whole trees (Bradoo 1972; Jacson and Joseph 1973). The nests contain from a single female up to 120 adults (Jacson and Joseph 1973) or up to several hundred juveniles (Table 23-2). As in the previous two species, the nests are occupied and enlarged by successive generations, which have an annual life cycle similar to that of *S. dumicola* (Bradoo 1972, 1975; Jacson and Joseph 1973). The adult sex ratio is female-biased (Table 23-2).

New nests may be formed by: (1) groups of spiderlings that move short distances along silk bridges to establish nests often connected to the maternal colony (Bradoo 1972, 1975; Jacson and Joseph 1973); (2) single inseminated females that disperse greater distances along silk lines; or (3), less commonly, ballooning of spiderlings or adults (Jacson and Joseph 1973; Jambunathan 1905; but see Henschel *et al.* 1995). There are no data on colony extinction, although Bradoo (1972) notes that 'old nests do not last long in this species'. Only two out of 25 scorable allozyme loci examined from 29 colonies were polymorphic, with most colonies consisting of identical homozygotes (Smith and Engel 1994).

Tapinillus sp. (Oxyopidae)

This species from the tropical rainforests of northeastern Ecuador (Table 23-1) is unusual among the lynx spiders because of both its sociality and its web-building (Avilés 1994). Its nests, which may occur in clusters of up to seven units, consist of a light, three-dimensional irregular mesh that envelops the distal parts of branches (Fig. 23-1h).

They contain from a few to a few dozen spiders, although one unusually large colony contained 100 individuals of the larger size-classes and 600 spiderlings (Avilés 1994). The age structure is highly variable, with colonies that contain only members of one age class to colonies that have all instars represented. Unlike the other species, the sex ratio is 1 : 1, from embryos to preadults (Table 23-2).

A decrease in the proportion of males among adults, as well as allozyme electrophoresis data (L. Avilés, unpublished), suggest that there is migration of males among colonies. It is not clear how colonies originate or for how long they last. Many colonies, however, remain in place and grow from one year to the next, some apparently giving rise to new colonies by budding (Avilés 1994). Other colonies disappear, although it is not known whether they move, disperse, or go extinct (Avilés 1994).

Theridion cf. *nigroannulatum* Keyserling (Theridiidae)
Colonies of this species have been observed in the understory of tropical rainforests of eastern Ecuador (L. Avilés and W. Maddison, unpublished), although the species has also been recorded from Peru and Venezuela (Table 23-1). The nests consist of numerous silk lines that drop to the ground from one to a few contiguous low-vegetation leaves underneath which the spiders sit (Fig. 23-1g). In colonies of this species, adult females usually occur alongside their egg sacs or spiderlings of a fairly homogenous stage of development. Fifty percent of the nests observed at a site in eastern Ecuador contained a single female; most others contained between 2 and 32 adult females. One intermediate-sized nest contained about 240 spiders; an extraordinarily large nest of 3.5 m × 2.5 m × 4.0 m was estimated to contain 4000 spiders (L. Avilés and W. Maddison, unpublished). It is not known whether the colonies undergo several generations of growth before dispersing, although the size of this unusually large colony suggests such to be the case. Dispersal of this large colony gave rise to numerous propagules containing one to five adult females (L. Avilés and W. Maddison, unpublished). The sex ratio among preadults is female-biased (Table 23-2).

PHYLOGENETIC RELATIONSHIPS

Although our knowledge of spider systematics is still fragmentary (see, for example, Forster *et al.* 1990; Coddington and Levi 1991), some general statements can be made regarding the phylogenetic relationships of these species. First, the species belonging to different spider families represent independent derivations of cooperation and permanent-sociality: the families to which they belong are well spread over the phylogenetic tree of spiders (as compiled by Coddington and Levi 1991), most of the species they contain are solitary, and no social representatives are known from related families (Fig 23-3). Thus, there have been at least seven independent derivations of cooperative behavior in spiders. The two social species in the family Dictynidae, *M. gregalis* and *Ae. binotata*, are also certainly independently derived: the rest of the known *Mallos* species are solitary and *Aebutina* may not be a true dictynid (Lehtinen 1967). Within the family Theridiidae there are three genera that contain permanent-social species, *Anelosimus*, *Achaearanea*, and *Theridion*. Here again, these probably represent independent derivations of cooperation since it is unlikely that these three genera will be found to be monophyletic (Forster *et al.* 1990), each genus contains mostly solitary species, and solitary life is most likely the primitive condition in the Theridiidae since no social species are known from the hypothesized sister families Nesticidae and Synotaxidae (Forster *et al.* 1990) or from other genera in the family. If the presumption of three independent derivations in the Theridiidae is correct, we would have at least 10 independent derivations of cooperative behavior in spiders, one for each genus in which it occurs. Finally, within the genus *Stegodyphus*, the work of Kraus and Kraus (1988, 1990) suggests three independent derivations of sociality since each of the permanent-social representatives belongs to one of the three species groups in this genus, which otherwise contains solitary or periodic-social representatives.

Thus, we have at least 12 independent derivations of cooperative behavior in spiders, only four of which correspond to genera that contain more than one species: the genus *Agelena* with two African species, the genus *Achaearanea* with two New Guinean and one African species, the genus *Anelosimus* with four Neotropical species, and the genus *Diaea* with three Australian species. Whether any of these genera represent single or multiple derivations of sociality and whether sociality may have been reversed in some cases remains to be determined. The similarity of some species pairs is suggestive of a common origin in some cases. The two *Achaearanea* species from New Guinea, for instance, apparently belong to the same species group (Levi *et al.* 1982); *An. lorenzo* and *An. rupununi* differ only 'by the shape of the conductor in the male palpus' (Fowler and Levi 1979); the three Australian thomisids appear 'morphologically very similar' (Evans 1996);

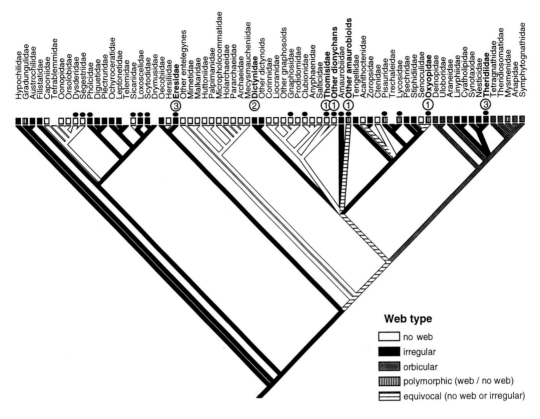

Fig. 23–3. Distribution of web type, extended maternal care and cooperation in spider families, after the phylogeny compiled by Coddington and Levi (1991). Families in which cooperative species have evolved are bold faced and the minimum number of independent derivations of cooperation within them are circled underneath the respective family names. The families for which instances of maternal care of the spiderlings have been reported are marked with a black oval (sources: Forster and Forster 1973; Buskirk 1981; Levi 1982; Eberhard 1986; D'Andrea 1987). The type of web characteristic of each family is coded in white, gray, or black squares just below these ovals. A hypothetical tracing of the evolution of web type is shown on the phylogeny, as reconstructed by MacClade v. 3.0 (Maddison and Maddison 1992), with the character 'web type' nested within the state 'web present'. Among others, 'Other dionychans' contains the Family Sparassidae and 'Other amauroboids' contains the Family Agelenidae.

and the occurrence of the two *Agelena* species in the same geographic area is worth noting. Because of the larger number of cooperative species it contains, *Anelosimus* appears the most promising genus for future systematic work, as we do not yet know how *An. eximius* and *An. domingo* fit with respect to each other, with respect to the probable species pair of *An. lorenzo* and *An. rupununi*, and with respect to their other non-social relatives.

PREADAPTATIONS TO COOPERATION IN SPIDERS

The occurrence of twelve independent derivations of cooperative behavior may seem extraordinary in a group of

organisms known for their aggressive and cannibalistic instincts. As noted by Krafft (1979), however, a predisposition to sociality might be present in all spiders given that they spend their first instar after eclosion confined within the common sac where their mother laid her eggs. This initial phase of tolerance among siblings may have served as the basis from which longer-lasting associations could have arisen (Shear 1970; Kullmann 1972; Burgess 1978; Krafft 1979).

It is clear, however, that cooperation is not evenly distributed across spider clades. With only one exception (Darchen 1967b), all cases of cooperative or communal behavior in spiders have taken place within the Infraorder Araneomorphae ('true' spiders (Coddington and Levi

1991)). No argument has been made as to why this might the case, although it may simply be related to the many more opportunities there are for the origin of a trait in a clade of 32 000 species vs. one of only 2 000 species (in the Infraorder Mygalomorphae, the sister clade to the Araneomorphae (Coddington and Levi 1991)).

Within the araneomorph spiders, cooperative or communal behavior occurs more often within families whose lifestyle depends on the use of a web (irregular or orbicular) (Shear 1970; Krafft 1979; Buskirk 1981; D'Andrea 1987). Among permanent-social spiders, for instance, ten of the postulated twelve independent derivations of cooperation involve web-building species, whereas about half the families consist of wandering, web-less spiders (Fig. 23-3). The importance of the web may depend on: (1) its capacity to provide a physical link among individuals, thus providing a means for communication (Shear 1970) and cohesiveness to the group (Luczar 1971, cited in Buskirk 1981; Krafft 1979); and (2) its being an expensive structure, and, thus providing an incentive for nest-sharing and communal web-building (Riechert 1985). Among the cooperative spiders, the social lynx spider is of particular interest because, although it belongs to a family of mostly wandering spiders, the genus it belongs to, *Tapinillus*, is apparently exceptional in building three-dimensional irregular webs (Griswold 1983; Mora 1986). The two instances of cooperation outside of a web-bound life style, *Diaea* and *Delena*, are cases in which silk is used as a means to tie together or secure their nests, but in which prey capture is entirely independent of a web. In *Diaea*, it can be argued that its nest structure does provide a substrate for the transmission of vibrations and is complex enough to favor nest sharing and cooperative building. *Delena*, which has an under-bark life style, on the other hand, appears entirely different from the other cooperative spiders and probably had different forces behind its social evolution (Rowell 1987; Rowell and Avilés 1995).

Another trend that is evident from examining the distribution of sociality on the phylogenetic tree of spiders is that all instances of cooperation involving prey capture and feeding have occurred in clades other than those containing geometric orb-weavers (Fig. 23-3). This has been the case even though aggregative behavior occurs repeatedly in the orb-weaver families Araneidae and Uloboridae (Uetz and Hieber, this volume). As argued by several authors (Krafft 1979; Buskirk 1981; D'Andrea 1987), the primary reason for this phenomenon might be that the orb-web cannot be simultaneously built or utilized by more than one individual, and, thus, it imposes a barrier on

more complex forms of sociality. There is only one notable exception to this apparent rule, the territorial periodic-social *P. bistriata* from Paraguay, in which preadult siblings cooperate in prey capture and share their food (Fowler and Diehl 1978; Penteado Sandoval 1987). Their ability to do so may depend on the arrangement of their orbs along a single plane, forming what amounts to a common sheet web along which the spiders can run freely from one orb to the next. The only other case of cooperation in a species belonging to an orb-weaving family, *Philoponella* sp. from Cameroon, has taken place in a species that has altogether lost its orb (Breitwisch 1989).

Although there are probably thousands of species with irregular webs, only 16 of those are known to be cooperative and permanent-social. What else is needed? It has been argued that an extension of the initial period of tolerance can be facilitated in clades in which maternal care places siblings in a situation of cohabitation beyond their period of development within the egg sac (Shear 1970; Kullmann 1972; Burgess 1978; Krafft 1979). If this period of tolerance is then extended through adulthood (see below), a permanent-social species would arise. We would expect, therefore, that cooperative permanent-social species would occur more frequently within clades in which maternal or periodic-social behavior tends to be present. As shown in Fig. 23-3, such has apparently been the case. The only exception appears to be *Mallos* (Table 23-1, Fig. 23-3), as extended maternal care is not characteristic of this genus (Jackson 1978). *M. gregalis* could have been derived from a territorial permanent-social ancestor via the parasocial route (Jackson 1978), although this hypothesis is not supported by a recent phylogenetic reconstruction of the genus *Mallos* (now split into *Mallos* and *Mexitlia* (Lehtinen 1967; J. Bond, personal communication)). In general, a formal test of the preadaptive value of extended maternal care awaits a phylogenetic analysis at the species-within-genera level, given that maternal care is relatively rare among spiders (Buskirk 1981; D'Andrea 1987) and has probably been derived within families or genera.

Some authors have suggested that paedomorphosis might be the mechanism by which adult individuals have retained the tolerant phase (Burgess 1978; Kraus and Kraus 1988, 1990; Wickler and Seibt 1993), pointing to the trend towards smaller body size (which suggests maturation at a younger instar) that characterizes some of the permanent-social species relative to their periodic-social or solitary relatives (Kraus and Kraus 1988, 1990; Wickler and Seibt 1993). By contrast, other authors have stressed the importance of the food supply in determining the length of the tolerant phase, pointing out that periodic-social spiderlings can be made to remain together for significantly longer periods of time if provided with a plentiful food supply (Krafft *et al.* 1986; Ruttan 1990; Gundermann *et al.* 1993; J. Schneider, personal communication; see also Rypstra 1986). This latter reasoning points to the need to address the environmental conditions which, in addition to the phylogenetic potential, would have been required for the evolution of cooperation and permanent-sociality in spiders.

ECOLOGICAL CORRELATES OF COOPERATION AND PERMANENT-SOCIALITY IN SPIDERS

An indication of the environmental conditions favoring group-living in spiders may come from the mostly tropical distribution of the cooperative, permanent-social species (Table 23-1). Several arguments can be made to explain this distribution: (1) the more plentiful year-round food supply in the tropics may facilitate a delay in the timing of dispersal in periodic-social species (Ruttan 1990; Uetz 1992; but see Riechert *et al.* 1986); (2) only in the tropics are there prey of a large enough size to make cooperation profitable (Ward 1986; Rypstra and Tirey 1990; G. Uetz, personal communication); (3) competition in the tropics is more intense, thus creating the pressure for a species to escape to more open ecological niches (e.g. exploitation of large prey, which are inaccessible to solitary spiders); (4) predation in the tropics is more intense, thus selecting for longer and more elaborate forms of maternal care and, perhaps, group defense (J. Henschel, personal communication); (5) frequent rains in the tropical rainforests would tend to destroy webs more frequently, thus favoring nest-sharing (Riechert *et al.* 1986); (6) the lack of marked seasons in the tropics should allow continuity of generations, thus, making permanent-sociality possible (Riechert *et al.* 1986); (7) living in an aseasonal or mildly seasonal environment should facilitate the persistence of populations that consist of subunits (i.e. colonies) that are subject to high rates of extinction (since they will not be synchronized in the stage of their life cycle and, thus, will be less likely to go extinct simultaneously); and (8) because of the vastly larger number of spider species living in the tropics (Coddington and Levi 1991), there are correspondingly many more opportunities for the steps towards permanent-sociality to have taken place. A combination of some or most of these arguments probably applies in the different cases.

Although mostly tropical, there are some interesting contrasts in the habitats occupied by the different species that require further comment, in particular the contrasts between wet vs. dry habitats and seasonal vs. aseasonal environments (Table 23-1). In addition to year-round prey availability and large prey size, which would make group-living possible and advantageous, the amount of precipitation may be a critical factor selecting for nest-sharing in wet habitats, whereas defense against predators may be the predominant factor in dry habitats. Riechert *et al.* (1986), for instance, showed that smaller colonies of *Ag. consociata* were significantly more likely to go extinct in the rainforests of Gabon; J. Henschel (personal communication) found that larger colonies of *S. dumicola* in Namibia are better able to withstand attacks by ants. Regarding seasonal vs. aseasonal environments, let me note two interesting facts about the species in the genera *Stegodyphus* and *Diaea* that might explain in part how they can inhabit seasonal environments: the spiders in both of these genera use their mother's body as food supply at the start of the low-food season (Seibt and Wickler 1987; Evans *et al.* 1995) and they build the most compact and sealable nests of any of the species (Fig. 23-1). Juveniles of the African and Indian *Stegodyphus*, for instance, are able to close the entrance to their nest from June to September in order to isolate themselves during the winter season in southern Africa (J. Henschel, personal communication) or the monsoon season in India (Bradoo 1972).

Although the reasons for spider sociality in the tropics appear compelling, several studies have shown that social living is not without costs. Although spiders living in groups are able to catch larger prey (Nentwig 1985; Ward 1986; Rypstra 1990; Rypstra and Tirey 1990; Pasquet and Krafft 1992), produce less silk per individual (Riechert 1985; Tietjen 1986; Jakob 1991), have better survivorship (Vollrath 1982; Riechert 1985; Avilés 1992; J. Henschel, personal communication), enjoy better access to mates (Krafft 1979) and have surrogate care-givers for their offspring (Christenson 1984; Avilés 1993a), several recent studies have shown that social groups may attract more parasites (see, for example, Griswold and Meikle-Griswold 1987; Wickler and Seibt 1988; but see Cangialosi 1990) and that feeding efficiency (Ward and Enders 1985; Jakob 1991; but see Rypstra and Tirey 1990; Uetz 1992) and fertility of individual spiders may decrease with increasing group size (Riechert 1985; Vollrath 1986a; Seibt and Wickler 1988a,b; Wickler and Seibt 1993). Presumably in the temperate zones, in absence of the facilitating conditions above discussed, the benefits of communal living would not be large enough to compensate for its costs, and thus cooperative spider societies would not have arisen. But even in the tropics, there must be a colony size at which communal living is no longer profitable. At that point, the dispersal of individual spiders seeking to improve their own reproductive potential would bring about the population-level processes by which new colonies are born.

CONSEQUENCES OF COOPERATION AND PERMANENT-SOCIALITY

Despite the occurrence of twelve independent derivations of cooperation, the presence of only 17 non-territorial permanent-social species (excluding the three periodic-social *Diaea*) appears surprising for a group of organisms comprising over 34 000 species. Granted that new social species are still being discovered, it is unlikely that this figure will go up enough to alter significantly the proportion that these species represent among the hundreds or thousands of tropical species that could have developed this level of sociality. So, another question to address is: why so few species? This brings me to considering what I believe have been the population-level consequences of the evolution of cooperation and permanent-sociality in spiders that may have made it difficult for these species to become established and persist in evolutionary time.

The first most immediate consequence of the evolution of cooperation in the ancestral phylogenetic lineage of most of these species was probably a switch from an outbred to an inbred breeding system, and thus from periodic- to permanent-sociality. This transition would have taken place if the colony size yielding the maximum advantage to social living could only be attained after several generations of colony growth. If the benefits of communal living and the costs of dispersal were comparable for both sexes, then males and females would have remained in their natal colonies to mate. This might explain why in most of the non-territorial permanent-social species mating takes place among nestmates (see species descriptions), whereas their periodic-social relatives typically disperse before attaining sexual maturity (see, for example, Brach 1977; Nentwig and Christenson 1986). Because of inbreeding depression, the switch from one breeding system to another would have constituted a barrier for the successful establishment of the newly arisen social isolates, so that even though the steps towards the evolution of permanent-sociality may have taken place many times, only few species would have been successful in undergoing this transformation. Those that were successful would have seen their gene pool subdivided into more or less permanently isolated colony lineages, with the possibility that selection among those lineages could become an important factor in their further evolutionary history.

Theory developed since the late 1960s (Williams 1966; Maynard Smith 1964, 1976; Leigh 1983) has emphasized the stringent conditions necessary for selection among groups to override counteracting individual selection within them. In the case of interdemic selection, these conditions would include: (1) the presence of a relatively large number of small population units or demes, (2) little or no migration among them, (3) relatively small sizes of the demes at foundation, and (4) a relatively high rate of deme turnover, comparable to the rate of replacement of the individuals within them. As seen from the species descriptions above, these conditions are almost perfect descriptions of the population structure and dynamics of these spiders. In even relatively small geographic areas, the populations of these spiders may comprise hundreds or thousands of colonies. The colonies are relatively small, averaging in size only in the dozens, hundreds, or low thousands; only *An. eximius* and *M. gregalis* have colonies that reach into the tens of thousands (Table 23-2). Barring accidental mixing when individuals dispersing to found new colonies happen to encounter each other (Smith 1986), there is no regular mixing of colony lineages in any of the species so far studied, with the exception of *D. cancerides*, *Tapinillus* sp., and one of the social *Diaea* spp. (Table 23-2). The size of the groups at foundation is often a single individual, and usually no more than a few dozen (Table 23-2). Finally, colony extinction appears to be quite common, on the order of 20–70% of the colonies per generation (Table 23-2).

If intercolony selection has been a factor in the evolution of these species, is there any evidence of its occurrence? That is, are there any traits whose evolution cannot be explained except by the presence of a higher level of selection? The highly female-biased sex ratios that characterize most of the non-territorial permanent-social spiders (Table 23-2) might be one such trait (Avilés 1986, 1993b). Because of their rarity, we would expect that within colonies males would have an inordinate reproductive success. Consequently, and according to the usual Fisherian argument (Fisher 1930), selection within colonies should tend to restore a 1:1 sex ratio. By means of computer simulations, Avilés (1993b) showed that selection for less biased sex ratios should operate within the colonies despite levels of inbreeding and population subdivision comparable to those present in the cooperative spiders, and that an overall sex-ratio bias can only be maintained if a preponderance of females increases the proliferation success of the colony lineages in which it occurs (see also Frank 1987). Higher proliferation success in species such as *D. socialis*, which disperse after a fixed number of generations of colony growth (Table 23-2), could result from the production of more or bigger propagules, while in species such as *An. eximius* or *Ac. wau*, which disperse when the colonies reach a particular size, it could result from reaching earlier the size at which proliferation takes place. Testing this hypothesis, both with respect to the adaptive significance of biased sex ratios and with respect to the match between estimated sex-ratio values and those expected under particular conditions of population subdivision and group turnover (Avilés 1993b), constitutes one of the challenges for the next few years. For now, it is clear that, at least in the broad sense, the species shown to lack a sex-ratio bias, *D. cancerides*, *Tapinillus* sp. and *D. ergandros*, apparently also lack the population subdivision necessary for intercolony selection to take place (Table 23-2).

Why most species turned to local mating, while a few maintained outbreeding, is an issue that will need to be addressed in the future. In addition, it has yet to be determined whether traits other than sex ratio, perhaps acting on colony survivorship rather than fertility, might have also evolved in response to intercolony selection. It is important to emphasize, however, that the occurrence of intercolony selection as a novel evolutionary force would have only become possible once permanent-sociality, and the resulting population subdivision, was well established. Consequently, intercolony selection *cannot* be used as an argument to explain the origin of sociality in these spiders.

As we have seen, the transition from periodic- to permanent-sociality would have transformed each colony into a self-contained population unit which we might expect to be subject to the usual ecological processes governing the growth of populations. As I will argue next, taking into account this status of social spider colonies should be critical to our understanding of their patterns of growth, proliferation and extinction. Small colonies, for instance, should tend to grow more or less exponentially to reach sooner the colony size at which social living is most advantageous. The lack of dispersal from smaller colonies would reflect the advantages of cooperation at small colony sizes, while the dispersal events leading to the production of new colonies probably reflect the point at which the costs of living in large, overcrowded colonies outweigh the risks of emigration. Are there similar arguments that might explain the frequent colony extinction events observed in these species?

Critical to the argument that follows is the observation that the colonies of the non-territorial permanent-social spiders grow in discrete generations (i.e. adults of the parental generation die before their offspring reach reproductive maturity) and that their per capita growth rate is affected by colony size (see, for example, Riechert 1985; Vollrath 1986a; Seibt and Wickler 1988b; Rypstra 1993; J. Henschel, personal communication). In the 1970s, May (1974) and May and Oster (1976) showed that populations growing under these two circumstances may overshoot their equilibrium population size and then crash as a result of the ensuing overcrowding. These population-size oscillations would have a self-perpetuating nature and, at relatively large values of the growth parameter, would occur with no apparent periodicity, i.e. chaotically. It has been noted by several authors (e.g. Thomas *et al.* 1980) that if these oscillations occur in combination with a certain amount of environmental or demographic stochasticity, they can reduce the mean time to extinction of the populations. The significance of this effect would increase as the net reproductive output of the populations increases.

Although many factors are probably responsible for the destruction of social spider colonies in the field (e.g. predation, J. Henschel, personal communication; disease, Riechert and Roeloffs 1993; or natural disasters, Main 1988; Seibt and Wickler 1990), extinction frequently comes about without any apparent external cause and involves quite large and apparently healthy colonies (see, for example, Lubin and Crozier 1985; Avilés 1992; L. Avilés, unpublished data). This observation, in combination with the theoretical considerations discussed above, leads me to suggest that the high frequency of colony extinction events observed in some of the cooperative spiders (Table 23-2) is due, at least in part, to unstable colony dynamics resulting from the combination of size-dependent colony growth, discrete generations, and a relatively large reproductive output. Although the brood sizes of the cooperative permanent-social species are generally smaller that those of their less social or asocial relatives (Buskirk 1981), the proportion of those offspring surviving to maturity probably increases as a result of social living. Additionally, their characteristic sex-ratio bias in favor of the egg-laying sex should lead to a many-fold increase of the net reproductive output of the colonies relative to a similarly sized colony with equal numbers of males and females. Both cooperation and female-biassed sex ratios, therefore, by magnifying the reproductive potential of the colonies, may lead to colony-size oscillations, perhaps of a chaotic nature (L Avilés, in preparation). Although dispersal could prevent a crash, owing to constraints on the stage of a colony's life cycle in which dispersal can occur (e.g. once individuals have matured and mated) and imperfect information about the future level of resources in a fluctuating environment, timely dispersal events may not always occur. Extinction, just like dispersal, may thus be related to demographic processes internal to the colonies.

It is important to emphasize that this hypothesis can only explain the frequency with which already-established colonies go extinct. It does not explain why any particular colony goes extinct or why newly founded colonies have a difficult time becoming established. This hypothesis does not apply, therefore, to species, such as *Ae. binotata* or *D. socialis*, that tend to produce daughter colonies after a single generation of colony growth. In addition, this hypothesis does not exclude other sources of extinction that might be important in some species. Besides explaining in part the high frequency of colony extinction, this hypothesis might also explain the brood-size reduction, relative to other species in their genera, that has been documented in some of the non-territorial permanent-social spiders (see, for example, Kullmann 1972; Enders 1976; Krafft 1979; Buskirk 1981, fig. 7, p. 344): had the spiders maintained the brood sizes of their ancestral species, the amplitude of the oscillations and, therefore, the rate of extinction of their colonies might have been too large to be countered by the rate of production of new colonies. Thus, other experiments in spider sociality might have failed because the species did not strike the right balance and went extinct.

The following scenario describes the sequence of events that I am suggesting occurred following the onset of cooperation and permanent-sociality in spiders. As mating became confined to within the colony lineages and the colonies became growing populations subject to the effects of density-dependence, oscillations of their colony sizes probably caused a fraction of the colonies to become extinct every generation while new colonies were being produced by the proliferation of other colonies that dispersed successfully. These processes gave rise to a relatively fast rate of group turnover that would have created the conditions under which intercolony selection could become a

significant evolutionary force. Selection for a reduced brood size, tending to stabilize the oscillations, would then have taken place more or less simultaneously with selection for more female-biased sex ratios, tending to increase the rate of colony proliferation; making more females, rather than increasing the brood size, would have been a better way to maintain or increase growth rate since the additional females could help take care of any additional offspring produced. These two variables, brood size and sex ratio, would have equilibrated at values that would depend on the specific features of each of the species, such as (1) the spider's body size; (2) the maximum size that their colonies could reach (probably a function of the nest architecture, the characteristics of the vegetation used as substratum, and the availability of prey in their particular environments); (3) the minimum viable size of colonies at foundation, and (4) the rate of intercolony migration, if any. All of these processes combined would certainly have had an impact on the genetic structure of these species, as is evidenced by several studies that show extremely low levels of allozyme variability in these species, both between and, especially, within colonies (reviewed in Riechert and Roeloffs 1993; Smith and Engel 1994; see Fig. 23-2).

RIPPLE WAVES UP THE PHYLOGENETIC TREE?

The onset of cooperation in a species lineage may appear to be of significance mostly at the level of the interacting individuals, but, as I have argued above for the cooperative spiders, its consequences to processes at the population level may be far-reaching. Could these population-level effects also affect the evolutionary fate of the phylogenetic lineages in which sociality arose? That is, in the case of the spiders, are speciation or extinction of the cooperative permanent-social species more likely given their population structure and dynamics? At the moment, we can only speculate about the answers.

One could argue, for instance, for a faster rate of speciation because the isolated colony lineages could diverge from each other and give rise to more than one species (see, for example, Smith 1986). However, even though population subdivision has certainly reduced the amount of genetic variation within the colony lineages, it has apparently not significantly increased the variation among them (see allozyme studies cited above). The reason for this pattern might be that, because of intercolony selection and fast turnover rate of the colony lineages, vast geographic areas may be taken over by the descendant colonies of just one or a few successful lineages (Avilés 1993b). Thus, there would not be enough intercolony variation or sufficient time for species divergence.

One could also argue for greater species vulnerability to extinction because of the potential risk of global extinction brought about by the high rate of local colony extinction. However, if, as argued above, chaos in the growth patterns of the colonies is responsible for the high rate of colony extinction, this very feature may also be the species' best protection against global extinction. A recent theoretical study by Allen *et al.* (1993) suggests that chaos tends to increase the persistence of species with subdivided populations because it causes the growth trajectories of the different subpopulations to become uncorrelated with each other, so that at no time are all likely to crash simultaneously.

A higher rate of species extinction may also occur in permanent-social species because the low overall genetic variability would tend to increase colony vulnerability to disease and parasites and lead to poor adaptability to other aspects of changing environments (see, for example, Riechert and Roeloffs 1993). However, the widespread distribution and local abundance of species such as *An. eximius* or the two African *Stegodyphus* spp. suggest that at least some of these species are extremely successful in their habitats.

Finally, what really matters in terms of long-term evolutionary success is the balance between rates of speciation and extinction. We know that social spiders are rare, but do they occur at the tips of phylogenetic trees, as would be the case if extinction was high and speciation low? Are the social species each others' closest relatives in the genera that contain several representatives, as would be the case if speciation, rather than extinction, was high? Can sociality be reversed once it arises? These are questions that can only be addressed by reconstructing the phylogenetic relationships of the genera that contain the non-territorial permanent-social species. Expanded exploration of the tropical regions is also needed to uncover new species to add to our data set. In the meantime, we can use the species known so far to address what I consider are fascinating problems in evolutionary biology and population ecology: group selection, sex ratio evolution, chaos, and extinction. By discovering cooperation and permanent-sociality, spiders have given us an unexpected gift: twenty species (so far) and at least twelve independent derivations of a grand-scale natural experiment in population biology.

ACKNOWLEDGMENTS

Special thanks to G. Uetz for pointing out the importance of large prey items in the evolution of cooperative prey capture; G. Binford for much-needed help in compiling web type and spider size information; J. Henschel, Y. Lubin and D. Rowell for sharing data, specimens, and experience with their social spiders; M. Engel, T. Evans, J. Henschel, Y. Lubin, J. Schneider, D. Smith, G. Uetz, and M. Whitehouse for sharing their unpublished manuscripts; and B. Crespi, J. Choe, J. Henschel, Y. Lubin, W. Maddison, B. Main, D. Rowell, J. Schneider, W. Wickler and two anonymous reviewers for comments on the manuscript. The field studies during which I became acquainted with several of the species covered here were financed by The National Geographic Society, Sigma Xi, Harvard University, the University of California at Berkeley, and the NSF-funded Analysis of Biological Diversification RTG from the University of Arizona. I also had the logistic support of the Pontificia Universidad Católica del Ecuador, Nuevo Mundo Travel, and the City Ecuadorian Production Company.

LITERATURE CITED

Allen, J. C., W. M. Schaffer and D. Rosko. 1993. Chaos reduces species extinction by amplifying local population noise. *Nature (Lond.)* **364**: 229–232.

Avilés, L. 1986. Sex ratio bias and possible group selection in the social spider *Anelosimus eximius. Am. Nat.* **128**: 1–12.

–. 1992. Metapopulation biology, levels of selection and sex ratio evolution in social spiders. Ph. D. dissertation, Harvard University.

–. 1993a. Newly-discovered sociality in the neotropical spider *Aebutina binotata* Simon (Dictynidae?). *J. Arachnol.* **21**: 184–193.

–. 1993b. Interdemic selection and the sex ratio: A social spider perspective. *Am. Nat.* **142**: 320–345.

–. 1994. Social behaviour in a web building lynx spider, *Tapinillus* sp. (Araneae, Oxyopidae). *Biol. J. Linn Soc.* **51**: 163–176.

Avilés, L. and W. Maddison. 1991. When is the sex ratio biased in social spiders?: Chromosome studies of embryos and male meiosis in *Anelosimus* species (Araneae, Theridiidae). *J. Arachnol.* **19**: 126–135.

Bowden, K. 1991. The evolution of sociality in the spitting spider, *Scytodes fusca* (Araneae: Scytodidae) – evidence from observations of intraspecific interactions. *J. Zool. (Lond.)* **223**: 161–172.

Brach, V. 1977. *Anelosimus studiosus* (Araneae: Theridiidae) and the evolution of quasisociality in Theridiid spiders. *Evolution* **31**: 154–161.

Bradoo, B. L. 1972. Some observations on the ecology of social spider *Stegodyphus sarasinorum* Karsch (Araneae: Eresidae) from India. *Orient. Ins.* **6**: 193–204.

–. 1973. The cocoon spinning behaviour and fecundity of *Stegodyphus sarasinorum* Karsch (Araneae: Eresidae) from India. *J. Bombay. Nat. Hist. Soc. J.*: **72**: 392–400.

–. 1975. The sexual biology and morphology of the reproductive organs of *Stegodyphus sarasinorum* Karsch (Araneae: Eresidae). *Entomol. Mon. Mag.* **111**: 239–247.

Breitwisch, R. 1989. Prey capture by a West African social spider (Uloboridae: *Philoponella* sp.). *Biotropica* **21**: 359–363.

Burgess, J. W. 1976. Social spiders. *Scient. Am.* **234**: 101–106.

–. 1978. Social behavior in group-living spider species. *Symp. Zool. Soc. Lond.* **42**: 69–78.

Buskirk, R. 1981. Sociality in the Arachnida. In *Social Insects*, vol. 4. H. R. Hermann, ed., pp. 282–367. New York: Academic Press.

Cangialosi, K. 1990. Social spider defense against kleptoparasitism. *Behav. Ecol. Sociobiol.* **27**: 49–54.

Christenson, T. 1984. Behaviour of colonial and solitary spiders of the Theridiid species *Anelosimus eximius*. *Anim. Behav.* **32**: 725–734.

Coddington, J. and H. W. Levi. 1991. Systematics and evolution of spiders (Araneae). *Annu. Rev. Ecol. Syst.* **22**: 565–592.

Crespi, B. J. and D. Yanega. 1995. The definition of eusociality. *Behav. Ecol.* **6**: 109–115.

D'Andrea, M. 1987. Social behaviour in spiders (Arachnida, Araneae). *Ital. J. Zool.* (N.S.), Monogr. 3.

Darchen, R. 1965. Ethologie d'une araignée sociale, *Agelena consociata* Denis. *Biol. Gabonica* **1**: 117–146.

–. 1967a. Une nouvelle araignée sociale du Gabon *Agelena republicana* Darchen. *Biol. Gabonica* **3**: 31–42.

–. 1967b. Biologie d'une mygale gabonaise nouvelle *Macrothele darcheni* Benôit (Araneidae: Dipluridae). *Biol. Gabonica* **3**: 253–257.

–. 1968. Ethologie d'*Achaearanea disparata* Denis, Araneae, Theridiidae, araignée sociale du Gabon. *Biol. Gabonica* **4**: 5–25.

–. 1976. La fondation de nouvelles colonies d'*Agelena consociata* et d'*Agelena republicana* araignées sociales du Gabon. Problemes Ecoethologiques. In *Comptes Rendues Ilme Colloquium d'Archnologie* Les Eyzies, France, pp. 20–39.

–. 1978. Les essaimages de l'Araignée Sociale, *Agelena consociata* Denis, dans la forét gabonaise (III). *C.R. Acad. Sci. Paris* **287**: 1035–1037.

–. 1979. Relations entre colonies d'agélénides sociaux du Gabon. Précisions sur les essaimages. II. *Bull. Biol. Fr. Belg.* **113**: 3–29.

–. 1980. Les populations d'*Agelena consociata* Denis Araignée sociale, dans la forét primaire gabonaise. Leur répartition et leur densité. *Ann. Sci. Nat. Zool. Paris* **14**: 19–26.

Darchen, R. and B. Delange-Darchen. 1986. Societies of spiders compared to the societies of insects. *J. Arachnol.* **14**: 227–238.

Darchen, R. and J. C. Ledoux. 1978. *Achaearanea disparata*, araignée sociale du Gabon, synonyme ou espece jumelle d'*A. tessellata*, solitaire. *Rev Arachnol.* **1**: 121–132.

Diguet, L. 1909. Sur l'araignée mosquero. *C.R. Acad. Sci. Paris* **148**: 735–736.

Downes, M. 1993. The life history of *Badumna candida* (Araneae: Amaurobioidea). *Austr. J. Zool.* **41**: 441–66.

Eberhard, W. 1986. Maternal behavior in *Apollophanes punctipes* (O.Pickard-Cambridge)(Araneae,Thomisidae). *J.Arachnol.* **14**: 398.

Enders, F. 1976. Clutch size related to hunting manner of spider species. *Ann. Entomol. Soc. Am.* **69**: 991–998.

Evans, T. A. 1996. Two new species of social crab spiders (Genus *Diaea*) from Eastern Australia, their natural history and distribution. In *Australasian Spiders and their Relatives: Papers in Honour of Barbara York Main*. M. S. Harvey, ed. *Suppl. Rec. West. Austral. Mus.*, Perth, in press.

Evans, T. A., R. J. Wallis and M. A. Elgar. 1995. Making a meal of mother. *Nature (Lond.)* **376**: 299.

Fisher, R. A. 1930. *The Genetical Theory of Natural Selection*. New York: Dover.

Forster, R. R. and L. M. Forster. 1973. *New Zealand Spiders: An Introduction*. Auckland and London: Collins Press.

Forster, R. R., N. I. Platnick and J. Coddington. 1990. A proposal and review of the spider family Synotaxidae (Araneae, Araneoidea), with notes on Theridiid interrelationships. *Bull. Am. Mus. Nat. Hist.* **193**: 1–116.

Fowler, H. G. and J. Diehl. 1978. Biology of the Paraguayan colonial orb-weaver, *Eriophora bistriata* (Rengger) (Araneae, Araneidae). *Bull. Br. Arachnol. Soc.* **4**: 241–250.

Fowler, H. G. and H. W. Levi. 1979. A new quasisocial *Anelosimus* spider (Araneae, Theridiidae) from Paraguay. *Psyche* **86**: 11–18.

Frank, S. A. 1987. Demography and sex ratio in social spiders. *Evolution* **41**: 1267–1281.

Griswold, C. E. 1983. *Tapinillus longipes* (Taczanowski), a web-building lynx spider from the American tropics (Araneae: Oxyopidae). *J. Nat. Hist.* **17**: 979–985.

Griswold, C. E. and T. Meikle-Griswold. 1987. *Archaeodictyna ulova*, new species (Araneae: Dictynidae), a remarkable kleptoparasite of group-living eresid spiders (*Stegodyphus* spp., Araneae: Eresidae). *Am. Mus. Nov.* **2897**: 1–11.

Gundermann, J. L., A. Horel and B. Krafft. 1993. Experimental manipulations of social tendencies in the subsocial spider *Coelotes terrestris*. *Insectes Soc.* **40**: 229–229.

Henschel, J. R., J. Schneider and Y. D. Lubin. 1995. Dispersal mechanisms of *Stegodyphus* (Eresidae): Do they balloon? *J. Arachnol.* **23**: 202–204.

Itow, T., K. Kobayashi, M. Kubota, K. Ogata, H. T. Imai and R. H. Crozier. 1984. The reproductive cycle of the queenless ant *Pristomyrmex pungens*. *Insectes Soc.* **31**: 87–102.

Jackson, R. R. 1978. Comparative studies of *Dictyna* and *Mallos* (Araneae, Dictynidae): I. Social organization and web characteristics. *Rev Arachnol.* **1**: 133–164.

Jackson, R. R. and S. E. Smith. 1978. Aggregations of *Mallos* and *Dictyna* (Araneae, Dictynidae): population characteristics. *Psyche* **85**: 65–80.

Jacson, C. C. and J. Joseph. 1973. Life-history, bionomics and beha-viour of the social spider *Stegodyphus sarasinorum* Karsch. *Insectes Soc.* **20**: 189–204.

Jakob, E. M. 1991. Costs and benefits of group living for pholcid spiderlings: losing food, saving silk. *Anim. Behav.* **41**: 711–722.

Jambunathan, N. S. 1905. The habits and life-history of a social spider (*Stegodyphus sarasinorum* Karsch). *Smithsonian. Misc. Coll.* **47**: 365–372.

Krafft, B. 1970. Contribution à la biologie et à l'éthologie d'*Agelena consociata* Denis (Araignée sociale du Gabon). I partie. *Biol. Gabonica* **6**: 197–301.

–. 1979. Organisations des sociétés d'araignées. *J. Psychol.* **1**: 23–51.

Krafft, B., A. Horel and J. M. Julita. 1986. Influence of food supply on the duration of the gregarious phase of a maternal-social spider, *Coelotes terrestris* (Araneae, Agelenidae). *J. Arachnol.* **14**: 219–226.

Kraus, M. 1988. Cocoon-spinning behavior in the social spider *Stegodyphus dumicola* (Arachnida, Araneae): cooperating females as "helpers". *Verh. Naturwiss. Ver. Hamb.* **30**: 305–309.

Kraus, O. and M. Kraus. 1988. The genus *Stegodyphus* (Arachnida, Araneae). Sibling species, species groups and parallel origin of social living. *Verh. Naturwiss. Ver. Hamb.* **30**: 151–254.

–. 1990. The genus *Stegodyphus*: systematics, biogeography and sociality (Araneidae, Eresidae). *Acta Zool. Fenn.* **190**: 223–228.

Kullmann, E. 1972. Evolution of social behavior in spiders (Araneae; Eresidae and Theridiidae). *Am. Zool.* **12**: 419–426.

Kullmann, E., St. Nawabi and W. Zimmermann. 1971/72. Neue Ergebnisse zur Brutbiologie cribellater Spinnen aus Afghanistan und der Serengeti. *Z. Kölner Zoo* **14**: 87–108.

Leborgne, R., B. Krafft and A. Pasquet. 1994. Experimental study of foundation and development of *Anelosimus eximius* colonies in the tropical forest of French Guiana. *Insectes Soc.* **41**: 179–189.

Lehtinen, P. 1967. Classification of the Cribellate spiders and some allied families, with notes on the evolution of the suborder Araneomorpha. *Ann. Zool. Fenn.* **4**: 199–468.

Leigh, E. G. 1983. When does the good of the group override the advantage of the individual? *Proc. Natl. Acad. Sci. U.S.A.* **80**: 2985–2989.

Levi, H. 1963. The American spiders of the genus *Anelosimus* (Araneae, Theridiidae). *Trans. Am. Microsc. Soc.* **82**: 30–48.

–. 1972. Taxonomic-nomenclatural notes on misplaced Theridiid spiders (Araneae: Theridiidae), with observations on *Anelosimus*. *Trans. Am. Microsc. Soc.* **91**: 533–538.

–. 1982. Arthropoda. In *Synopsis and Classification of Living Organisms*. Parker, S. P., ed., vol. 2. New York: McGraw-Hill.

–. 1983. American *Theridion*. *Bull. Mus. Comp. Zool.* **129**: 539–540.

Levi, H. W., Y. D. Lubin and M. H. Robinson. 1982. Two new *Achaearanea* species from Papua New Guinea with notes on other Theridiid spiders (Araneae: Theridiidae). *Pac. Insects* **24**: 105–113.

Levi, H. W. and D. R. Smith. 1982. A new colonial *Anelosimus* spider from Suriname (Araneae: Theridiidae). *Psyche* **89**: 275–278.

Lubin, Y. D. 1980. Population studies of two colonial orb-weaving spiders. *Zool. J. Linn. Soc.* **70**: 265–287.

–. 1982. Does the social spider *Achaearanea wau* (Theridiidae) feed its young? *Z. Tierpsychol.* **60**: 127–134.

–. 1986. Courtship and alternative mating tactics in a social spider. *J. Arachnol.* **14**: 239–257.

–. 1991. Patterns of variation in female-biased colony sex ratios in a social spider. *Biol. J. Linn. Soc.* **43**: 297–311.

–. 1995. Is there division of labour in the social spider *Achaearanea wau* (Theridiidae)? *Anim. Behav.* **49**: 1315–1323.

Lubin, Y. D. and R. H. Crozier. 1985. Electrophoretic evidence for population differentiation in a social spider *Achaearanea wau* (Theridiidae). *Insectes Soc.* **32**: 297–304.

Lubin, Y. D. and M. H. Robinson. 1982. Dispersal by swarming in a social spider. *Science (Wash., D.C.)* **216**: 319–321.

Maddison, W. and D. Maddison. 1992. *MacClade*, Version 3.0. Sunderland: Sinauer Associates.

Main, B. Y. 1988. The biology of a social thomisid spider. *Austr. Arachnol.* **5**: 55–73.

May, R. M. 1974. Biological populations with non-overlapping generations: stable points, stable cycles and chaos. *Science (Wash., D.C.)* **186**: 645–647.

May, R. M. and G. F. Oster. 1976. Bifurcations and dynamic complexity in simple ecological models. *Am. Nat.* **974**: 573–599.

Maynard Smith, J. 1964. Group selection and kin selection. *Nature (Lond.)* **201**: 1145–1147.

–. 1976. Group selection. *Q. Rev. Biol.* **51**: 277–283.

Mockford, E. L. 1957. Life history studies on some Florida insects of the genus *Archipsocus* (Psocoptera). *Bull. Fla. St. Mus. Biol. Sci.* **1**: 253–254.

Mora, G. 1986. Use of web by *Tapinillus longipes* (Araneae: Oxyopidae). In *Proc. 9th Int. Cong. Arach., Panama 1983*. W. Eberhard, Y. Lubin and B. Robinson, eds., pp. 173–175. Washington: Smithsonian.

Nentwig, W. 1985. Social spiders catch larger prey: a study of *Anelosimus eximius* (Araneae: Theridiidae). *Behav. Ecol. Sociobiol.* **17**: 79–85.

Nentwig, W. and T. E. Christenson. 1986. Natural history of the non-solitary sheetweaving spider *Anelosimus jucundus* (Araneae: Theridiidae). *Zool. J. Linn. Soc.* **87**: 27–35.

New, T. R. 1973. The Archipsocidae of South America (Psocoptera). *Trans. R. Entomol. Soc. Lond.* **125**: 57–105.

Overal, W. L. and P. R. Ferreira da Silva. 1982. Population dynamics of the quasisocial spider *Anelosimus eximius* (Araneae: Theridiidae). In *The Biology of Social Insects*. M. D. Breed, C. D. Michener and H. E. Evans, eds., pp. 181–182. Boulder: Westview Press.

Pain, J. 1964. Premières observations sur une espèce nouvelle d'araignée sociales, *Agelena consociata* Denis. *Biol. Gabonica* **1**: 47–58.

Penteado Sandoval, C. 1987. Aspectos da ecologia e socialidade de uma aranha colonial: *Eriophora bistriata* (Rengger, 1936) (Araneidae). M.S. thesis, Instituto de Biologia Universidade Estadual de Campinas, Campinas, Brazil.

Pasquet, A. and B. Krafft. 1989. Colony distribution of the social spider *Anelosimus eximius* (Araneae, Theridiidae) in French Guiana. *Insectes Soc.* 36: 173–182.

–. 1992. Cooperation and prey capture efficiency in a social spider, *Anelosimus eximius*. *Ethology* 90: 121–133.

Riechert, S. E. 1985. Why do some spiders cooperate? *Agelena consociata*, a case study. *Fla. Entomol.* 68: 105–116.

Riechert, S. E. and R. M. Roeloffs. 1993. Evidence for and consequences of inbreeding in the cooperative spiders. In *The Natural History of Inbreeding and Outbreeding*. N. Thornhill, ed., pp. 283–303. Chicago: The University of Chicago Press.

Riechert, S. E., R. M. Roeloffs and A. C. Echternacht. 1986. The ecology of the cooperative spider *Agelena consociata* in equatorial Africa. *J. Arachnol.* 14: 175–191.

Roeloffs, R. and S. E. Riechert. 1988. Dispersal and population-genetic structure of the cooperative spider, *Agelena consociata*, in West African rainforest. *Evolution* 42: 173–183.

Rowell, D. 1987. The population genetics of the huntsman spider *Delena cancerides* (Sparassidae: Arachnida). Doctoral dissertation, Australian National University.

Rowell, D. and L. Avilés 1995. Sociality in a bark-dwelling hunstman spider from Australia, *Delena cancerides* Walckenaer (Araneae: Sparassidae). *Insectes Soc.* 42: 287–302.

Rowell, D. and B. Y. Main. 1992. Sex ratio in the social spider *Diaea socialis* (Araneae: Thomisidae). *J. Arachnol.* 20: 200–206.

Ruttan, L. M. 1990. Experimental manipulations of dispersal in the subsocial spider, *Theridion pictum*. *Behav. Ecol. Sociobiol.* 27: 169–173.

Rypstra, A. L. 1986. High prey abundance and a reduction in cannibalism: the first step to sociality in spiders (Arachnida). *J. Arachnol.* 14: 193–200.

–. 1990. Prey capture and feeding efficiency of social and solitary spiders: a comparison. *Acta. Zool. Fenn.* 190: 339–343.

–. 1993. Prey size, social competition and the development of reproductive division of labor in social spider groups. *Am. Nat.* 142: 868–880.

Rypstra, A. L. and R. Tirey. 1989. Observations on the social spider, *Anelosimus domingo* (Araneae, Theridiidae), in southwestern Peru. *J. Arachnol.* 17: 368.

–. 1990. Prey size, prey perishability and group foraging in a social spider. *Oecologia (Berl.)* 86: 25–30.

Seibt, U. and W. Wickler. 1987. Gerontophagy versus cannibalism in the social spiders *Stegodyphus mimosarum* Pavesi and *Stegodyphus dumicola* Pocock. *Anim. Behav.* 35: 1903–1904.

–. 1988a. Why do "Family spiders", *Stegodyphus* (Eresidae), live in colonies? *J. Arachnol.* 16: 193–198.

–. 1988b. Bionomics and social structure of "Family Spiders" of the genus *Stegodyphus*, with special reference to the African species *S. dumicola* and *S. mimosarum* (Araneidae, Eresidae). *Verh. Naturwiss. Ver. Hamb.* 30: 255–303.

–. 1990. The protective function of the compact silk nest of social *Stegodyphus* spiders (Araneae, Eresidae). *Oecologia (Berl.)* 82: 317–321.

Shear, W. A. 1970. The evolution of social phenomena in spiders. *Bull. Br. Arachnol. Soc.* 1: 65–76.

Simon, E. 1891. Observations biologiques sur les arachnides. I. Araignées sociales. *Ann. Soc. Entomol. France,* 60: 5–14.

–. 1892. *Histoire Naturelle des Araignées,* pp. 221–222. Paris: Roret.

Smith, D. R. 1986. Population genetics of *Anelosimus eximius* (Araneae, Theridiidae). *J. Arachnol.* 14: 201–217.

Smith, D. R. and M. S. Engel. 1994. Population structure in an Indian cooperative spider, *Stegodyphus sarasinorum* Karsch (Eresidae). *J. Arachnol.* 22: 108–113.

Thomas, W. R., M. J. Pomerantz and M. E. Gilpin. 1980. Chaos, asymmetric growth and group selection for dynamical stability. *Ecology* 61: 1312–1320.

Tietjen, W. J. 1986. Effects of colony size on web structure and behavior of the social spider *Mallos gregalis* (Araneae, Dictynidae). *J. Arachnol.* 14: 145–157.

Tsuji, K. 1988. Obligate parthenogenesis and reproductive division of labor in the Japanese queenless ant, *Pristomyrmex pungens*: comparison of intranidal and extranidal workers. *Behav. Ecol. Sociobiol.* 23: 247–255.

Uetz, G. W. 1983. Sociable spiders. *Nat. Hist.* 92: 62–79.

–. 1992. Foraging strategies of spiders. *Trends Ecol. Evol.* 7: 155–159.

Uetz, G. and K. Cangialosi. 1986. Genetic differences in social behavior and spacing in populations of *Metepeira spinipes*, a communal-territorial orb weaver (Araneae, Araneidae). *J. Arachnol.* 14: 159–173.

Venticinque, E. M., H. G. Fowler and C. A. Silva. 1993. Modes and frequencies of colonization and its relation to extinctions, habitat and seasonality in the social spider *Anelosimus eximius* in the Amazon (Araneidae: Theridiidae). *Psyche* 100: 35–41.

Vollrath, F. 1982. Colony foundation in a social spider. *Z. Tierpsychol.* 60: 313–324.

–. 1986a. Eusociality and extraordinary sex ratios in the spider *Anelosimus eximius* (Araneae: Theridiidae). *Behav. Ecol. Sociobiol.* 18: 283–287.

–. 1986b. Environment, reproduction and the sex ratio of the social spider *Anelosimus eximius* (Araneae, Theridiidae). *J. Arachnol.* 14: 267–281.

Vollrath, F. and D. Rohde-Arndt. 1983. Prey capture and feeding in the social spider *Anelosimus eximius*. *Z. Tierpsychol.* 61: 334–340.

Ward, P. I. 1986. Prey availability increases less quickly than nest size in the social spider *Stegodyphus mimosarum*. *Behaviour* 97: 213–225.

Ward, P. I. and M. M. Enders. 1985. Conflict and cooperation in the group feeding of the social spider *Stegodyphus mimosarum*. *Behaviour* 94: 167–182.

Wickler, W. 1973. Ueber Koloniegründung und soziale Bindung von *Stegodyphus mimosarum* Pavesi und anderen sozialen Spinnen. *Z. Tierpsychol.* 32: 522–531.

Wickler, W. and U. Seibt. 1988. Two species of *Stegodyphus* spiders as solitary parasites in social *S. dumicola* colonies (Araneida, Eresidae). *Verh. Naturwiss. Ver. Hamb.* **30**: 311–317.

–. 1993. Pedogenetic sociogenesis via the "sibling-route" and some consequences for *Stegodyphus* spiders. *Ethology* **95**: 1–18.

Williams, G. C. 1966. *Adaptation and Natural Selection*. Princeton, N. J.: Princeton University Press.

Wilson, E. O. 1971. *The Insect Societies*. Cambridge, Mass.: Belknap Press.

24 · Explanation and evolution of social systems

BERNARD J. CRESPI AND JAE C. CHOE

ABSTRACT

We review the causes of the evolution of social systems and
the methods used in their analysis. First, we discuss the
roles of genetics, phenotypic traits, ecology (basic necessary resources and natural enemies) and demography in
the origin and evolution of sociality, and synthesize the
effects of these conditions in a comparative assessment of
the predictions of optimal skew models. The models provide a useful framework to explaining and predicting
social systems, but would benefit from expansion in the
range of their assumptions and more explicit connection
to ecological and demographic selective pressures.
Second, we review the purposes and usefulness of alternative social system lexicons. We conclude that the trade-off
between universality and taxon-specific precision of terms
can usefully be addressed by explanation of social terms for
each comparative test coupled with striving for recognition
of convergence across the broadest possible taxonomic
range. Finally, we provide an overview of current adaptationist methods used for analyzing social systems, focussing on approaches that utilize phylogenetic information.
Integration of comparative with behavioral–ecological
methods, especially experimentation, promises to lead to
the next series of insights and critical data for tests of
theory.

INTRODUCTION

This volume has had three main objectives. First, we have
tried to bring together the widest possible diversity of
social insects and arachnids, to elucidate necessary and sufficient conditions for the origin and maintenance of different social systems. In this chapter, we first review the
evidence for associations between social systems and
genetic, phenotypic, ecological and demographic variables. This review involves searching for common selective
pressures among diverse taxa, to assess what sets of conditions appear important in producing and maintaining
particular social systems.

Second, we have encouraged the authors to critically
appraise the meaning of the terms 'eusocial', 'cooperatively
breeding' and 'communal', to more clearly outline the
nature of the phenomena that we are seeking to understand. In this chapter, we will examine the usefulness of
alternative terminology systems (Wilson 1971; Gadagkar
1994; Kukuk 1994; Crespi and Yanega 1995; Sherman et al.
1995; Wcislo, this volume) for recognition of convergence
and divergence, and for testing of social-behavior models.

Third, we have asked the authors to bring phylogenetically based comparative approaches to bear on the problems of understanding the evolution of sociality within
their taxonomic group. Here, we review the results of
these efforts, and assess the efficacy of different approaches
to studying sociality. We focus mainly on phylogenetic
methods, which have recently become much more accessible, and experimental approaches, which have been
remarkably little used. From the data and hypotheses presented in this volume, we hope for both a critical overview
and appraisal of the past 30 years of research since the
modern era of social behavior studies began, and a
renewed look to the future, through recognizing the most
informative taxa, the most important theory and questions, and the most productive methods.

THE SOCIAL BESTIARY

The taxa in this volume represent a bewildering diversity
of forms. Their genetic, phenotypic, ecological and demographic idiosyncrasies can, however, be organized under
the rubric of a set of factors presumed to encapsulate the
main predictors and correlates of social systems. We will
first review the evidence for the effects of these factors individually for the taxa discussed in this book. This exercise
allows us to take stock of our understanding of the role of
each of these putatively causal variables as agents of social
evolution. Next, we will analyze the assumptions and predictions of recent 'optimal skew' models that integrate
these variables into a single scheme (reviewed in Keller

and Reeve 1994). The rationale for these models is that neither relatedness, nor ecological constraints, nor any other factors operate in isolation from others, and that many social systems share common elements of dominant–subordinate interaction. We will address two questions: (1) how well the assumptions of the model are met in different taxa; and (2) how well these models can explain variation in social systems across a broad taxonomic scale.

Genetics and relatedness

Genetics and relatedness refer to aspects of differences in genetic systems among taxa that affect the evolution of sociality (Hamilton 1964; Reeve 1993; Yamamura 1993; Saito, this volume). Testing for the effects of genetics and relatedness on social systems requires showing that variation in relatedness leads to variation in social behavior, either within or among species. Thus far, several patterns have emerged from the taxonomically sporadic pattern of relatedness measurements: (1) relatedness is high, often above one-half, in taxa with facultative eusociality and high reproductive skew, such as eusocial *Lasioglossum* spp., *Microstigmus comes*, *Polistes* spp., *Bombus* spp. and *Exoneura* spp. (reviewed in Gadagkar 1991a; Ross and Carpenter 1991); (2) relatedness varies from high to very low in many obligately eusocial species, and reaches zero in some multiple-foundress associations (Hölldobler and Wilson 1990; Ross and Carpenter 1991; Choe and Perlman, this volume); and (3) relatedness is low in many, though not all, communal bees and wasps (reviewed in Kukuk and Sage 1994).

These three patterns are suggestive of relatedness effects on the origin and maintenance of eusociality, but robust tests require all else being equal or similar for other causal factors. Such controls can be performed or approximated within species either through experimental manipulation of within-colony relatedness in socially polymorphic species such as halictine bees (Wcislo, this volume), allodapine bees (see, for example, Schwarz et al., this volume) or *Polistes* foundress associations (Reeve 1991; Brockmann, this volume), or through analysis of among-population differences where the effects of the main confounding variables can be assessed (see, for example, Schwarz et al., this volume). Among species, effects of relatedness and other genetic variables can be analyzed through comparisons of sister taxa that differ in their social systems, or using other phylogenetically based tests. For

example, the hypothesis of Seger (1983a) and Sherman (1979), that elevated chromosome numbers predispose to eusociality, cannot be evaluated with available data (*contra* Andersson 1984) because of the confounding effects of shared ancestry (Felsenstein 1985). This hypothesis should be analyzed by comparing sister taxa between eusocial and non-eusocial forms, or through use of empirical null distributions or independent contrasts (Garland et al. 1992, 1993). Given that recently developed molecular methods allow efficient generation of phylogenies (reviewed in Simon et al. 1994) and precise determination of relatedness, including estimates for colonies and pairs of interactants (see, for example, Hughes and Queller 1993; Queller et al. 1993; Avise 1994), experimental and phylogenetically based tests of relatedness effects, such as Hamilton's (1964) original hypothesis, should be among the priorities of students of sociality. However, for such tests to be meaningful, they must be combined with data on benefits and costs of alternative social behaviors, perhaps focussing on the taxa whose descendents bracket the transition in social system. Moreover, within-species variation in relatedness and other parameters should be analyzed to assess the expected accuracy of character-optimization inferences.

Overall, the genetic systems of recent additions to the social bestiary described here do not bode well for a significant comparative association of haplodiploidy with eusociality: these include an apparently diploid eusocial beetle (Kent and Simpson 1992; Kirkendall et al., this volume), two diploid eusocial rodents (Jarvis and Bennet 1993; Jarvis et al. 1994), and a lack of eusociality in group-living, haplodiploid mites (Saito, this volume). However, one benefit of having a second haplodiploid group, the Thysanoptera, showing eusocial, subsocial and communal forms, is that it allows wider tests for convergence of relatedness among social systems, and for convergence of relatedness effects such as those due to relatedness asymmetry. Thus, to the extent that any of the various effects of haplodiploidy, including high relatedness, asymmetric relatedness, other population-genetic influences (Reeve 1993; Yamamura 1993), sex allocation adjustment (reviewed in Nonacs 1986; Boomsma and Grafen 1990), or male-production by virgins (reviewed in Bourke 1988; Choe 1988) are shown to be important in predicting social systems in both Thysanoptera and Hymenoptera, the large differences in ecology between these taxa would suggest that unobserved, confounding third variables are not the cause of the association.

Phenotype

The presence and state of particular phenotypic traits, such as weaponry or defensive behavior (Starr 1985, 1989; Kukuk et al. 1989), inability of workers to store sperm (Peeters 1991), voltinism and longevity (see, for example, Seger 1983b), flight polymorphism (Taylor 1978), ability to manipulate objects and build (Hamilton 1972; Evans 1977), other elements of parental care (Nalepa 1994), hemimetaboly (Alexander et al. 1991), polylecty (Danforth and Eickwort, this volume), neural and information-processing capacity (Bernays and Wcislo 1994) and fecundity variation (West-Eberhard 1978; Brown and Pimm 1985; Strassmann and Queller 1989) may alter the likelihood that a particular social system evolves. For example, the presence of some traits, such as univoltinism in some bees (Wcislo et al. 1993) or cannibalism in some spiders (Uetz and Hieber, this volume), may prevent a social system from evolving. Among thrips, soldiers have evolved only in lineages with conspecific fighting by adults during gall formation (Crespi and Mound, this volume); among aphids, soldiers are found in taxonomic groups that also exhibit fighting among first-instar larvae over gall initiation sites (Stern and Foster, this volume).

Other effects of phenotypic variation may be more subtle and only alter probabilities of evolutionary change rather than preventing transitions. For example, hemimetaboly may favor the evolution of sociality in termites, cockroaches and aphids by allowing useful helping by juveniles and flexibility in juvenile developmental trajectories (Alexander et al. 1991; Benton and Foster 1992; Nalepa 1994; Roisin 1994), and holometaboly may engender high variation in adult fertility, if newly eclosed adults can be viewed as similar-sized shells variably filled with nutrients (Huxley 1932, pp. 60–61; Gadagkar et al. 1988; Wheeler 1994).

The essence of such effects of phenotype is whether a species would evolve a different social system were suitable variation available for selection. Of course, oak trees and earthworms are not social for any number of reasons, so examining the role of particular traits in the evolution of social systems is meaningful only among social species and their close relatives. In this context, one method of evaluating the role of a given trait would be to use tests for comparing whether a social system is more or less likely to evolve under one or the other states of a trait of interest (such as the presence of cannibalism or weaponry) (Maddison 1990, 1994; Ridley and Grafen 1996). Phylogenetically

based tests allow inference of evolutionary trajectories, and thus, in principle, offer clues for disentangling causes from effects. Another possible method for evaluating the importance of phenotype would be experimental manipulation of traits, such as body size or fecundity, to create a wider range of variation than is normally observed (see, for example, Anholt 1991), and determining whether social interactions change in response. One difficulty with testing for effects of phenotype is that we may actually be evaluating the role of a selective factor that both produced a phenotypic trait and favored a social system (H. K. Reeve, personal communication); this type of problem can best be addressed using experiments or multivariate comparative tests.

Ecology

Ecological factors influence the survival and fecundity of individuals alone and in different-sized groups, the costs of dispersal, and the costs and benefits of division of labor (Emlen 1982; Vehrencamp 1983a,b). Although the importance of ecological factors in the evolution of sociality has been stressed for cooperatively breeding vertebrates (see, for example, Woolfenden and Fitzpatrick 1978; Emlen 1991), and, more recently, for invertebrates (Brockmann 1984, this volume; Strassmann and Queller 1989; Alexander et al. 1991), specification of the different qualities of nest sites, food, and natural enemies that favor or disfavor different social systems has seldom been attempted in a comparative context. For analysis of what Alexander et al. (1991) call the 'basic necessary resource' for sociality, the following qualities should be critical: expansibility, improvability, divisibility, mean and variance of duration of nests with respect to generation time, ability of one individual to dominate, spatial and temporal patchiness of nest sites, food distribution in time and space, mean and variance of food size, food quality, safety, defensibility, and the nature of enemies (host specific, attacking individuals or whole colonies, adults or juveniles). Which of these variables, alone or in combination, show convergences among the disparate taxa and social systems described in this volume? Are there ecological correlates of social systems and the variation within them, and, if so, what are the selective pressures involved?

To assess the role of different ecological factors in group formation, dispersal costs, division of labor, and other social dynamics, we will focus first on the nature of nest sites and food sources (the 'basic necessary resource' of Alexander *et al.* 1991) and seek to identify convergences and causes of differences between convergent taxa. Subsocial behavior was reviewed with this type of approach by Tallamy and Wood (1986) and Tallamy (1994), so we consider mainly communal, cooperatively breeding, and eusocial societies. Next, we will examine in detail the role of natural enemies in the evolution of social systems and their components, with respect to the type of attackers and the intensity of the selective pressure. As we shall see, resources and attackers are intimately enmeshed, although the nature of their interaction over evolutionary time remains to be determined.

Basic necessary resources

Galls. One of the more striking convergences found in this volume is the presence of soldiers in both gall aphids and gall thrips (Crespi and Mound, this volume; Stern and Foster, this volume). Galls are defensible (but with variable levels of actual safety), have some but limited expansibility, and allow for feeding on high-quality nutrients in the nursery; they have durations that seldom span more than one or two generations, and their ease of formation probably varies considerably among species, but may often be low (Crespi 1996; Foster and Northcott 1994). In aphids and thrips, eusociality in the form of soldiers probably originated in the selective context of high rates of attack by predators and parasites and moderate ability to defend against them, and a duration of the resource long enough for soldiers to repay the investment in their production. Other selective factors may include low success rates of independent founders due to high intraspecific competition (Crespi and Mound, this volume) or, perhaps, intrinsic difficulties involved in the timing or mechanics of gall formation (see, for example, Whitham 1992).

One of the main differences between aphids and thrips is the life-history stage of the soldiers: young larvae in aphids and adults in thrips. This difference may be due to the efficiency of weaponry with respect to nature of attackers. Thus, gall-forming thrips are attacked mainly by highly sclerotized adult *Koptothrips*, which can be repelled only by sclerotized adults; by contrast, many important aphid enemies can be repelled by masses of early-instar larvae, whose stylet or foreleg weaponry should be nearly as effective as that of later instars (Stern and Foster 1996, this volume). For both taxa, attackers can destroy entire colonies, such that the inclusive fitness of defenders can be high in spite of little or no reproduction. However, the compatibility of defense and reproduction or growth,

owing to the local presence of food, may select against total sterility, at least in thrips and the gall aphids in which soldiers can molt to adulthood.

Galls would appear to provide suitable arenas for the evolution of reproductive dominance, since they are small, enclosed and easily patrolled. Whereas aphids have not evolved dominance behavior, owing to an absence of genetic conflict within colonies (Stern and Foster, this volume), gall thrips may lack such dominance because: (1) soldiers probably have higher fighting ability than the foundress; (2) defense is more physiologically compatible with reproduction than in social forms that forage; and (3) soldiers gain directly by defense of the gall, because they can lay eggs and kleptoparasitic invaders would, eventually, attempt to kill them.

A second, less obvious convergence is between gall aphids and some termites: in these taxa, food is found within the relative safety of the nest, soldiers have evolved, and juveniles help, permanently or temporarily (see also Abe 1991; Crespi 1994). Shellman-Reeve (this volume) shows that termite species with temporary helping exhibit lower nest-site stability than species with permanent helping (worker castes), and Stern and Foster (this volume) describe a remarkably similar difference between gall aphid species with permanent and facultative soldiers.

Silk and similar secretions. A wide variety of the species discussed in this volume use silk or a similar material to form their 'basic necessary resource', and have converged on social systems that are subsocial, communal or facultatively communal, but not, except in *Microstigmus* wasps, cooperatively breeding or eusocial. This menagerie includes spiders, embiids, psocids, mites, some lepidopteran larvae, and some thrips on *Acacia*. Silk, or secretions that similarly serve to catch prey or enclose a space with a built membrane, engenders a number of distinct ecological and social consequences. Silk is expensive to produce (Uetz and Hieber, this volume), continuously expansible, readily divisible into discrete but contiguous sections (Ward and Enders 1985), and can be eaten as a form of recycling (spiders); silk also forms a shelter that can be used to catch food (spiders), enclose food (mites, thrips, psocids, embiids), provide only shelter and protection (lepidopterans), or create a home and nursery (*Microstigmus*).

Why are silk and similar materials associated with communal or subsocial life, but not, except in *Microstigmus* wasps, with social systems involving reproductive dominance? Subsociality and group-living may be favored because the resource is localized, long-lived and expensive to produce, such that its costs can be shared and spread over time, and its benefits are gained by offspring, other relatives, or unrelated individuals (Riechert 1985; Uetz and Hieber, this volume). Several hypotheses may explain the absence of reproductive division of labor in silken habitats. First, division of labor may be of little benefit to silk inhabitants in terms of increased group survival or reproduction. No silk producers except *Microstigmus* forage away from their domicile, and in most of them, food is either readily available on the substrate (lepidoptera, embiids, psocids, thrips) or arrives unpredictably and must be subdued (spiders). As a result, the only tasks to be performed are prey capture (spiders), defense, and silk production. Except when prey are dangerous to subdue, prey capture is unlikely to involve altruism because the attacking spider will gain from the behavior through increased feeding opportunities (Ward and Enders 1985). Silken habitats may be more difficult to effectively defend than wood, soil, or galls owing to their relative permeability and diffuse spatial structure (such that there is normally no single site to defend), and the unpredictable dispersion of food in the structure may make resource monopolization difficult (H. K. Reeve, personal communication). Alternatively, silk-inhabiting creatures may have few of the types of enemies that can be expelled by a dangerous altruistic act, or a phenotype that does not involve structures modifiable by selection into effective weapons. Thus, psocids and embiids escape notice or represent unprofitable prey owing to their silken packaging (Edgerly, this volume), and lepidopteran larvae join together in group displays to advertise unprofitability (Costa and Pierce, this volume). Similarly, silk may be difficult to defend against conspecific intruders (e.g. psocids and spiders) (Edgerly, this volume). All of these considerations would make the evolution of division of labor less likely.

Second, reproductive dominance may not have evolved in silk because of weak ecological constraints. Thus, the success of dispersing individuals may be high in silk producers relative to gall-formers or wood or soil-inhabiters, because their food (foliage, algae, or flying insects) is relatively easy to locate and obtain, and small silken domiciles are relatively easy to create. This hypothesis is, however, unlikely to apply to embiids (Edgerly, this volume) or some types of social spider in which success in independent colony initiation appears low (see, for example, Christianson 1984; Avilés, this volume).

Third, in some taxa, such as spiders, embiids, psocids, and perhaps some lepidoptera, the spatial diffuseness of webs may make it physically difficult and costly for one individual to monitor and dominate others. This consideration does not apply in *Microstigmus* wasps, the one exception to the silken rule, which forms small, enclosed hanging nests of silk. Indeed, in this species silk production may have facilitated the evolution of eusociality by increasing the difficulties of breeding independently (Matthews 1991).

Finally, group selection for rapid colony growth may select against any reduction in fecundity that division of labor would entail, in some social spiders (Lubin 1995).

Testing these ideas concerning the selective pressures for subsocial and communal behavior in silk-producing species requires analysis of the importance and efficacy of defense, the costs and benefits of silk production and group-living in silk, and the success of dispersers. Such analyses are likely to be important in discerning the selective pressures that lead to reproductive dominance and division of labor.

Social symbioses in wood and dung. Wood and dung share a number of ecological features, foremost among them being low nutrient value (Nalepa 1994) and high defensibility once the resource is in use. The main difference among them is resource size and duration; burrows and food in wood are larger relative to their exploiters, more expansible and longer-lived. The low nutrient value of wood and dung has convergently selected for symbioses and exploitation of microorganisms in passalid beetles, some cockroaches and their termite descendants (Nalepa 1994), dung beetles (Halffter, this volume), and ambrosia beetles (Kirkendall *et al.*, this volume). Such symbioses may have selected for parental behavior because offspring gain symbionts or inoculate, or processed food, from parents or alloparents (termites, passalids, cockroaches) or rely upon them to tend the farm or culture (ambrosia beetles, dung beetles). Nalepa (1994) explains how proctodeal transfer of symbionts to offspring was important in facilitating the origin of subsociality (rather than favoring the origin of obligate group-living because of the need to exchange flagellates) in prototermites, because it established the behavioral basis of trophallactic exchange and engendered an avenue of nepotistic reproduction that favored the evolution of reproductive division of labor (see also Honigberg 1970).

Apparent social convergences of wood and dung resources include eusociality in some termites and

Austroplatypus incompertus ambrosia beetles, and subsociality in some bark beetles, cockroaches, and passalid beetles (Tallamy 1994). Among some termites and *A. incompertus*, success at independent colony formation is very low (Shellman-Reeve, this volume, for termites), relatedness is probably near one-half (Reilly 1987), and expanding the nest system involves specialized, long-term work. However, at least in beetles, natural enemies may often be a relatively weak selective pressure because the colony is easily defended in a wooden fortress with no or few entrances and little traffic at any entrance.

Soil. Compared with all other resources, soil is the most permanent, and provides the best opportunities for expansion, inheritance, and effective defense (Alexander *et al.* 1991). Moreover, digging and protecting the entrance provide predictably important, specialized tasks, and soil nests are usually small enough that physical dominance by one individual is economically feasible. The importance of difficulties in initiating soil nests has been stressed for mole rats and many Hymenoptera (see, for example, Sherman *et al.* 1991; McCorquodale 1989a), such that the costs of colony initiation may often be outweighed by the benefits of remaining within the home burrow.

Demonstrating that diverse soil-inhabiting creatures have converged on eusociality, or evolved it in parallel, as a result of its unusual characteristics involves identifying particular aspects of soil burrows as critical for the origin and maintenance of eusociality. Such an analysis is difficult in ants, owing to a lack of suitable variation in social structure and habitat. Among halictine bees, one evolutionary transition from soil to wood was accompanied by a shift from eusocial to solitary life (in the lineage leading to *Augochlora pura*; see Michener 1990; Danforth and Eickwort, this volume). Additional cases of such within-clade habitat variation are necessary to test rigorously for effects of nest site on social systems.

Enemies

Although natural enemies have long been considered as one of the most important selective pressures for the evolution of social systems (Lin 1964; Michener 1985), few studies have addressed the questions of: (1) how different types of enemies, with qualitatively and quantitatively different effects, select for different social behavior (see, for example, Starr 1985); and (2) what patterns of covariation are expected in nature between social systems and enemy pressure and success. Natural enemies can be categorized

into those that can wipe out entire colonies, those that prey upon adults or juveniles, and those that parasitize adults or juveniles. Any of these types can be either host-specific, and therefore possibly more liable to coevolve with their hosts, or generalized, and therefore possibly a more diffuse selective pressure (reviewed in Thompson 1994). Moreover, attackers can be either conspecifics (usually usurpers of the nest and perhaps work force), or heterospecifics. How should these various types of enemies affect the evolution of parental care, communal life, cooperative breeding, and eusociality?

Destruction of whole colonies. Heterospecific or conspecific enemies that can wipe out the genetic success of entire colonies in a short time should represent a strong selective force for the evolution and maintenance of division of labor, in species capable of defense against such threats (e.g. aphids, thrips, paper wasps, and some bees) (West-Eberhard 1975; Starr 1985, 1989; Kukuk *et al.* 1989; Alexander *et al.* 1991). Social behavior mitigates against such attacks by both defense with highly effective weapons (stings, termite mandibles and chemicals), and more rapid colony recovery after attack and loss or damage of the resource (Gibo 1978; Starr 1990; Strassmann and Queller 1989; but see Reeve 1991). Behavioral and morphological specialization may lead to increased efficiency in repelling invaders and concomitantly involve a decrease in fecundity, via either reduced ability to feed or a trade-off between investment in reproduction and defensive morphology.

Predation and parasitism of adults and juveniles. Predation of adult individuals has been demonstrated as one of the main selective pressures for group-living in many vertebrates (see, for example, Alexander 1974; Hoogland and Sherman 1976). However, advantages to group-living without division of labor predicated on this selective pressure have seldom been considered for social insects or arachnids (but see Christianson 1984). Selection by predation and parasitism of single adults could affect the benefits of division of labor through assured fitness returns from helping when young, especially under high mortality rates of helping or dispersing adults (Queller 1989, 1994; Gadagkar 1990). Moreover, in *Polistes* the best explanation for multiple-female association appears to be 'survivorship insurance', whereby the presence of more females reduces the odds that the work force will be reduced to zero by predation on individuals (Reeve 1991). High rates of parasitism of adults, by enemies such as conopid flies that make individuals non-reproductive but need not reduce ability to help (Knerer and Atwood 1967; Strassmann and Queller 1989), may also select for helping by making independent breeding by first-brood offspring a less viable option. It is more difficult to imagine that the enemies of adults could favor communal life; indeed, nests could be easier to detect if they have multiple inhabitants.

Predation and parasitism of juveniles, singly or en masse, clearly selects for parental care (Tallamy and Wood 1986; Tallamy and Schaefer, this volume), but do they also select for communal life, cooperative breeding, or eusociality? Having a guard at the nest benefits colony members through protection against attacks on juveniles in many Hymenoptera with different social systems (see, for example, Lin 1964; Wcislo et al. 1988; McCorquodale 1989b; Matthews 1991; Garófalo et al. 1992). However, such selection for guarding should not lead to cooperative breeding unless division of labor increases the presence of guards and their success, nor should it favor eusociality unless castes, and perhaps morphological specialization, lead to increased efficacy of protection (Hamilton 1972). Some wasp species with communal nests defend them effectively without a division of labor (see, for example, Wcislo et al. 1988, but see also Sakagami et al. 1990), although defense by multiple females, perhaps more likely under cooperative breeding or eusociality, may help against persistent threats (see, for example, Lin 1964). Could natural enemies select for cooperative breeding or division of labor only when defending is dangerous and kin are helped? Because natural enemies are considered as being such important selective pressures for sociality (Lin and Michener 1972), studies of pressure from enemies and costs and benefits of defense in species with different social systems, especially eusociality, are long overdue.

Host specificity. High host specificity of attackers could affect the evolution of sociality by leading to social 'arms races', manifested in the adaptations and counter-adaptations found in gall aphids (Moffett 1989; Stern and Foster 1996) or subsocial Hemiptera (Eberhard 1975). This speculative hypothesis can be tested by measuring rates of attack and success of natural enemies in a phylogenetic context, and by assessing the ecological success (e.g. rarity, range and species diversity) of related solitary and social forms (Tallamy and Schaefer, this volume). For example, gall thrips may sometimes cospeciate with their kleptoparasitic *Koptothrips* enemies (P. Abbot, B. J. Crespi and D. Carmean, unpublished data), and hosts involved in cospeciation may have relatively high kleptoparasitism rates.

Conspecifics. Attack by individuals of one's own species may be an especially effective factor favoring multiple-female groups, because such attacks involve usurpation of entire nests (e.g. in *Polistes* (Gamboa 1978), *Xylocopa* (Hogendoorn and Velthuis 1993), and halictine bees (Kaitala et al. 1990); reviewed in Field (1992)), and thus represent an especially strong selective pressure. Moreover, attackers and defenders are likely to be more evenly matched when conspecific than when heterospecific, because parties have the same weaponry and neither is likely to 'win' or 'lose' the battle over evolutionary time. Thus, selection for defense against conspecifics may remain more unrelenting in the long term. Similarly strong selection pressure is expected from attack by parasites closely related to their hosts that usurp nests or nest and work forces (e.g. socially parasitic allodapines and polistines). Testing these hypotheses requires quantification of rates of parasitism by different types of enemy. The evolutionary effects of selection by conspecific and phylogenetically related attackers depends on whether social behavior facilitates defense; if not, successful social parasites could select for solitary or communal life (Field 1992). This idea could be tested in allodapine bees, which exhibit both a high incidence of social parasitism and high levels of intraspecific plasticity in social behavior (Schwarz et al., this volume).

Evolutionary dynamics of defenders and enemies. How should conspecific and heterospecific enemy pressure and success covary with sociality in nature, if enemies are strong selective forces? Species more strongly subject to selection by attacks of predators and parasites should have adaptations against them, some of which will involve group living (e.g. selfish herd effects, Hamilton 1971; Wcislo 1984), behavioral specialization as guards or soldiers (bees, some aphids), or both behavioral and morphological specialization (e.g. in aphids, thrips and termites). Thus, at least at the origin of the social adaptation against enemies, the incipiently social species should be subject to higher rates of successful attack from enemies. Whether or not such higher rates of parasite success persist depends upon the efficacy of defenders, but, if sociality is maintained by attackers, then we expect that parasite pressure should remain higher in taxa with sociality than without it. Parasitism rates are no higher in one-female than in two-female nests of a halictine bee (Wcislo et al. 1993) and a xylocopine bee (Hogendoorn and Velthuis 1993). By contrast, lower rates of parasitism in communal than solitary nests have been reported for *Cerceris* wasps (McCorquodale 1989b), a megachilid bee (Garófalo *et al.* 1992) and a halictine bee (Abrams and Eickwort 1981). *Polistes* wasps show higher vulnerability of single-female nests to conspecific usurpation (Gamboa 1978; Suzuki and Murai 1980; Klahn 1988), and per capita heterospecific brood parasitism increases with colony size (Reeve 1991). In thrips, hemipterans and cockroaches, predator attack and success against social forms appear high in comparison with related less social taxa (Crespi and Mound, this volume; Tallamy and Schaefer, this volume; Nalepa and Bell, this volume). Indeed, some such species may become trapped by their suite of enemies in evolutionary cul-de-sacs from which the only escape is extinction. By contrast, in taxa that show high intraspecific variation in social behavior, the frequencies of nest-sharing, guarding, intranest competition for provisions, and perhaps even reproductive division of labor may fluctuate over a short evolutionary time scale (Michener 1985; Wcislo *et al.* 1985).

Over a long evolutionary time scale, taxa exhibiting different social systems should exhibit different diversities of natural enemies, such as kleptoparasites, if: (1) different social systems are differentially likely to give rise to kleptoparasitic forms, perhaps via intraspecific parasitism; (2) some social systems are more vulnerable to infiltration; or (3) some third variable selects for joint variation in social system and kleptoparasite success. For example, kleptoparasitic bee species appear to be more common in lineages with eusocial forms, such as Halictinae and Xylocopinae, than in those with predominantly communal and solitary forms, such as Nomiinae, Rophtinae and Andrenidae (Hamilton 1972; Wcislo and Cane 1996; W. T. Wcislo, personal communication).

One of the main difficulties in understanding the role of natural enemies in the evolution of sociality is distinguishing cause from consequence (Starr 1985). For example, natural enemies might be expected to become a stronger selective pressure *after* a species has evolved social life, because sociality engenders a valuable resource (e.g. a nest or gall) or concentrated food source (the nursery) that more readily attracts enemies (Lin 1964; Smith 1982). This problem can be avoided or mitigated in several ways: (1) by analyzing among-population covariation between pressure from enemies and sociality (e.g. in allodapines, Schwarz *et al.*, this volume); (2) by experimentally manipulating parasite pressure, by excluding, simulating, or adding attackers, to see if social systems change in response; (3) by quantifying attack and success rates of enemies for related species that differ in sociality but are similar for other variables relevant to the attackers of interest (e.g. in gall aphids or thrips); or (4) by quantifying attacker pressure and success in a phylogenetic context, and inferring ancestral rates of attack in relation to transitions between social systems.

Demography

Survivorship

Effects of demography on the evolution of social systems are probably pervasive yet subtle and difficult to measure. Survivorship, relative to generation time and duration of the nest, is important with respect to the presence and extent of generation overlap, and the retention of reproductive options, such as inheritance of the nest (Myles 1988) or dispersal some time after reproductive maturity. First, lives spanning several generations, or longer than the duration of nests, should select for the maintenance of behavioral flexibility, because the inclusive fitness benefits from alternative social strategies should change substantially over time. For example, anthophorid bees, such as *Xylocopa* and *Exoneura bicolor* (Schwarz *et al.*, this volume) exhibit long lifespans and high levels of behavioral plasticity, expressed in their absence of castes. Similar flexibility and long lifespans are shown by all cooperatively breeding birds and mammals (Alexander *et al.* 1991), with the possible exception of naked mole rats (Sherman *et al.* 1991).

Second, high mortality of adults should favor helping when individuals can increase their inclusive fitness soon after adulthood, by helping the offspring of relatives (Queller 1989, 1994; Gadagkar 1990, 1991b). This hypothesis may help explain helping in both some wasps (Gadagkar 1990) and at least one cooperatively breeding bird (Clarke 1984). In other cooperatively breeding birds, however, relatively *low* extrinsic adult mortality has been hypothesized as an important condition favoring cooperative breeding (Woolfenden and Fitzpatrick 1984)! Could low extrinsic mortality rates that evolved *after* the origin of cooperative breeding have led to delayed senescence?

Third, if helpers are engaging in risky behavior, then they should undergo accelerated senescence, while the lifespan of reproductives evolves to become longer (Alexander *et al.* 1991). Such an effect is expected to promote caste divergence, and it could be evaluated by testing for an interspecific association between high extrinsic mortality rates (due, for example, to foraging), and some measure of senescence (see, for example, Promislow 1991).

Fourth, possibilities for inheritance of the nest should select against risky helping behavior, and for a long lifespan (Shellman-Reeve, this volume). The 'laziness' of some individuals in social groups, such as *Polistes* and *Ropalidia* wasps (Reeve and Gamboa 1983, 1987; Chandrashekara and Gadagkar 1992) and naked mole rats (Reeve 1992), may be explained by such selective pressures, but further data are needed to demonstrate that the lazy indeed inherit the nest, and that lifespans are longer in species for which inheritance is more of a possibility.

Fifth, Yanega (1988) suggests that among-year variation in mortality rates during the second yearly generation of halictine bees selects for diapause by some first-brood females that would otherwise become workers. By this hypothesis, direct diapause should be common in species of bivoltine bees and wasps for which mortality rates are highly variable among years (owing, for example, due to epidemics or weather variation), and exclusively solitary life could evolve when second-generation mortality is sufficiently high, or fully social life histories could evolve when mortality is consistently low.

Fecundity
Variation in fecundity may affect social evolution in several ways (West-Eberhard 1975; Brown and Pimm 1985). First, higher levels of intraspecific variation in fecundity, or age at reproductive maturity, could select for helping if the resultant 'losers' in independent reproduction options benefit more from helping than from dispersing (West-Eberhard 1975; Gadagkar 1991b; Roisin 1994). Wing polymorphism may be a preadaptation for such fecundity differences (Taylor 1978) as well as a template for the evolution of division of labor. Moreover, some taxa may be unlikely to exhibit high intraspecific variation in fecundity because food availability is relatively constant spatially and temporally (for example, in passalids, embiids, psocids, and some plant-feeding taxa such as thrips and aphids). By contrast, in Hymenoptera, resources may vary considerably in time and space (Hunt 1994; Nalepa 1994); this variability could select for social foraging in species with progressive provisioning (Wenzel and Pickering 1991).

Second, some forms of helping, such as defense, may be favored more strongly in species with relatively low fecundity, because slower growth rates make groups more vulnerable to predation and select for soldiers (Stern and Foster 1996). This idea may apply at least to aphids and thrips, in which galls of some of the species with soldiers persist for multiple generations (Crespi and Mound, this volume).

Is soldier production in termites related to colony growth rate? Does the r–K selection continuum apply to other social insects, such as ants or bees (Iwata and Sakagami 1966), and if so, might some taxa, such as passalids, dung beetles, many cockroaches, and xylocopine bees be ensconced in K-selected life histories from which the evolution of large colonies is prohibitive? In passalids, inability to shunt nitrogen efficiently among individuals (as do termites; Nalepa 1994) could represent a proximate mechanism for such a constraint.

MODELS FOR ANALYZING
SOCIAL SYSTEMS

Aspects of genetics, phenotype, ecology and demography interact in their influences on social systems. Thus, to explain the phylogenetic distribution of social systems, the effects of these variables should be considered jointly. Vehrencamp (1983a,b) developed one of the first models that integrates the effects of ecological costs and benefits of sociality with effects of genetic relatedness. Vehrencamp's model assumes that: (1) each group has a single manipulator (dominant), who monitors and controls resource allocation and breeding within the group; (2) subordinates can only avoid manipulation by leaving to attempt breeding solitarily or joining another group; and (3) mean per capita fitness of the group depends on group size. Given these assumptions, the model predicts the type of societies (degree of reproductive bias and relatedness of group members) that result from maximization of the dominant's inclusive fitness, given the dominant's option of allowing some reproduction by the subordinates (a 'staying' incentive) (Reeve and Ratnieks 1993), and subordinates' options of staying or leaving. Four variables are used to predict social systems: (1) relatedness; (2) probability of successful dispersal; (3) fitness effects of grouping, defined by the per capita reproduction in groups relative to the per capita reproduction of individuals breeding alone; and (4) an efficiency factor, which quantifies the ecological benefits of reproductive division of labor and the fitness costs of within-group competition for reproduction.

Reeve and Ratnieks (1993) refine and extend this model to include effects of relative fighting ability of individuals within groups, and classify social systems involving plurimatrous (multiple-reproductive; see Choe 1995) associations. As in the basic model, a dominant individual controls the reproduction of others (subordinates) in such a way as to maximize its fitness; but now the subordinate has the additional option of challenging the dominant to a lethal fight. Dominants thus have the new option of offering 'peace incentives', reproduction of subordinates that dominants allow to make lethal fighting a less viable option (see Kennedy 1992 for a discussion of possible difficulties arising from such anthopomorphic terminology).

The models seek to explain: (1) group formation; (2) the amount of variation in reproduction in a group (the skew); and (3) the extent of fighting within the group. The models are useful in that they quantify the arguments of Lin and Michener (1972), Alexander (1974) and West-Eberhard (1975) regarding the interactions of relatedness, mutualism and parental manipulation effects, and they make a number of falsifiable predictions, most intuitive but some novel (summarized in Reeve and Ratnieks 1993; Reeve and Keller 1995). The main predictions of the models, with respect to which values of the above parameters are expected in differing social systems, are as follows.

Communal societies are expected when imposition of bias does not maximize the inclusive fitness of dominants, because group-living is similarly beneficial to individuals (in terms of individual offspring production) as breeding alone, and the benefits of division of labor are low. Under communal breeding, per capita reproduction should remain constant, or should increase, with varying group size, and ecological constraints (*sensu* Emlen 1991) should be relatively weak. Alternatively, dominants may prefer to bias reproduction, but may be incapable of doing so as a result of homogeneity of fighting abilities, lack of spatial proximity, or other natural history details that lead to violation of the model's assumptions. To the extent that individuals benefit from grouping, they should prefer to cooperate with relatives rather than non-relatives. However, high relatedness is more compatible with cooperative breeding (discussed below) than communal life whenever dominance can be expressed, because subordinate individuals thereby garnish more inclusive fitness from helping.

Cooperatively breeding societies, which involve some degree of reproductive division of labor, may exhibit any degree of skew, from marginally above zero to unity. The associations of variables predicted from the model (Reeve and Ratnieks 1993; Keller and Reeve 1994) include: (1) smaller staying and peace incentives with larger group productivity and stronger ecological constraints; (2) larger peace incentives with higher fighting ability of the subordinate; and (3) smaller staying and peace incentives with higher relatedness. Higher skew should normally be associated with higher relatedness, higher levels of aggressive

'testing' of dominants by subordinates, and stronger division of labor and willingness of subordinates to engage in risky tasks (Keller and Reeve 1994). However, the presence of mildly risky tasks that greatly increase per capita reproduction can give rise to high skew and counterintuitive performance of the risky tasks by dominants. Indeed, skew models predict that, if work performed is mildly risky and greatly increases per capita reproduction, then a 'reversed' division of labor, with dominant individuals specializing in the more risky work, can result (Reeve and Ratnieks 1993). This type of 'role-reversed' social system could be stable because subordinates have little incentive to engage in risky work when skew is high. In such situations, high ecological constraints and large benefits of grouping keep the subordinate with its hard-working but dominant nestmate, despite its low reproduction (see Reeve and Ratnieks 1993, pp. 78–80, for details and more predictions).

The transition from cooperative breeding to eusociality (*sensu* Crespi and Yanega 1995) is not encompassed within these models, since they assume that individuals are totipotent (Keller and Reeve 1994). In the models, the greatest degree of skew is expected under strong ecological constraints and low fighting ability of subordinates, and higher relatedness makes complete skew more likely as relative fighting abilities converge. Because eusociality involves an evolutionary loss of ability to become a dominant reproductive, it should evolve in situations where ability to become a dominant is reliably very low when skew is high; this may occur when benefits of helping are high and involve low compatibility with reproduction, or when individual lifespans are short relative to generation time and duration of the nest. The higher the ecological benefits of helping, the lower relatedness may be and make a lifetime of helping worthwhile. These ideas can be tested by searching for differences between related taxa with cooperative breeding and eusociality, in such taxa as ponerine ants, allodapine and halictine bees, aphids, and termites with temporary vs. permanent workers.

Findings that would contradict the assumptions or predictions of model include: (1) showing that dominant individuals cannot or do not adjust skew optimally for their own interests; or (2) discovery of any society with an unexpected pattern of parameter covariation, such as low relatedness and high skew when ecological constraints are weak. Data thus far available to test the predictions of the skew models are of two main types. First, data from experimental manipulations of *Polistes* wasps are consistent with the existence of peace incentives (Reeve and Nonacs 1992), although other interpretations have been proposed (Strassmann 1993; Reeve and Nonacs 1993). Further experimental tests of this type, coupled with analysis of maternity, are required to determine if the presence and extent of reproduction by subordinates fits the predictions of the models. Second, the expected patterns of covariation of the model variables can be tested within and between species (Vehrencamp 1983a; Reeve 1991; Keller and Reeve 1994; Reeve and Keller 1995; Bourke and Heinze 1994). Such tests have thus far supported the predictions of the models, but further tests are needed to: (1) discern the scope of the models by testing their critical assumption of dominant-regulated skew; (2) analyze their risky, counterintuitive predictions; and (3) measure their parameters with sufficient accuracy for quantitative tests and assignment of differences in skew between and within taxa to particular model variables.

In the next section of this chapter we will assess the assumptions and predictions of skew models, using the taxa discussed in this volume. The authors in this volume have provided us with hypothesized values and ranges, for their taxonomic group, of four variables relevant to the models: (1) relatedness; (2) costs of dispersal; (3) effects of group size on per capita reproduction; and (4) the fitness of biassed groups relative to unbiassed groups, which encapsulates benefits of division of labor and costs of social strife, and two additional variables, (5) relative fighting abilities of interactants; and (6) temporal variability, over lifespans, in expected inclusive fitness from helping vs. dispersing. We discuss only those variables that are relevant or appear important, for the taxon and social systems under consideration. The purposes of this exercise are to: (1) assess how well the assumptions of the model apply on a broad taxonomic scale; (2) determine whether the patterns of covariation of the variables predicted by the skew models are observed; and (3) help point out which variables, in which taxa, are most in need of further study.

Embiids. Antipaluria urichi, the only embiid studied in enough detail to provide estimates of the important parameters, is facultatively communal and solitary (Edgerly, this volume). It probably exhibits low to moderate relatedness (except perhaps where wingless males inbreed and females seldom immigrate) and reasonably high success of dispersers, if not in independent founding then in joining an established colony. Communal life may be favored because of the high failure rate of solitary females starting galleries, and the benefits of using existing silk; indeed,

females in communal groups produce more eggs per egg mass than solitary females. However, these extra eggs are lost to parasitism, such that average egg hatch rates are the same for solitary and communal females. The observation that newly adult females usually disperse suggests that breeding alone is a better option than staying (if a good breeding site is found), and that females have lower per capita reproduction in communal groups (but see Choe 1994). However, any such cost of group-living may vary with the age of a colony: females in relatively recently formed communal groups should enjoy high fecundity and low parasitism rates (since parasites have not yet built up locally), whereas female fitness in older communal groups should be lower than that of solitary females (J. S. Edgerly, personal communication). Thus, temporally increasing selection by parasites, and colony-founding behavior dependent on success in finding a good site, may maintain the social polymorphism.

The lack of reproductive dominance in embiids could be due to low relatedness, similarity of abilities to fight and reproduce, relatively weak ecological constraints, or a lack of suitable work to be performed (see discussion of silk-inhabitants above). But could helping be favored in embiids that provision rather than graze (Edgerly, this volume)?

Termites. In termites that inhabit a non-expansible wood resource, cooperative breeding is associated with high relatedness and high ecological constraints due to low availability of high-quality food and high predation on dispersers (Shellman-Reeve, this volume). The lack of termites without helping by juveniles should be due to demographic benefits of colony growth over immediate reproduction, but this idea has not yet been tested. In some taxa, workers aggressively prevent their sibs from becoming reproductive (see also Roisin 1994), but in no species are there dominant–subordinate relationships between primary reproductives and their offspring, possibly because workers are equally related to sibs and offspring and reproductives are much more efficient at offspring production (Alexander *et al.* 1991; see also Shellman-Reeve, this volume, for discussion of complications due to inbreeding and sibling supplemental reproductives).

The main difference between termites that are cooperatively breeding (with temporary helping by workers) and those that are eusocial (with permanent workers) is the predictably long duration of the basic necessary resource, which is continuously expansible in eusocial forms

(Abe 1991, Shellman-Reeve, this volume). Thus, the longer the duration of the food and nest resource, the greater the helper's opportunity for cumulative inclusive fitness benefits (Shellman-Reeve, this volume). This is one of the clearest cases of an ecological difference between cooperatively breeding and eusocial forms.

Aphids. Aphids with soldiers do not fit well within the skew framework because the models implicitly assume the presence of genetic conflict. Stern and Foster (1996; this volume) discuss the evolution of soldiers from the viewpoint of resource allocation by the clone, and show how soldiers can be advantageous under low colony growth rates, high predation rates, and high soldier effectiveness. Ecological constraints could affect the evolution of aphid soldiers in two ways. First, if individuals are constrained to stay on their natal plant (owing to a lack of adaptation to host alternation), then investment in soldiers can be favored (Stern and Foster, this volume). Second, in situations where individuals have a choice between helping and leaving (i.e. dispersing as a nymph or developing to the alate stage), then individuals should stay and defend when this decision would lead, on average, to higher inclusive fitness than breeding elsewhere. The decision to leave or stay would depend upon success of dispersers, and thus is influenced by ecological constraints.

Thrips. In gall thrips with soldiers, relatedness is probably high, the success of dispersers is probably low, and survival of individuals in a gall is expected to increase with the number of soldiers present. Soldiers lay eggs in at least some species, but there does not appear to be a dominant-subordinate relationship between foundresses and soldiers (Crespi 1992a; Crespi and Mound, this volume; B. J. Crespi and L. A. Mound, personal observation), for reasons discussed above.

Bark and ambrosia beetles. In the eusocial *Austroplatypus incompertus*, relatedness among a foundress's offspring is unknown but may be one-half, ecological constraints on independent breeding are very high owing to a lack of male involvement in burrow foundation and the difficulty of establishing and maintaining a burrow in a living tree, and the relationship between per capita fitness and group size is unknown (Kirkendall *et al.*, this volume). Whether or not foundresses exhibit dominance behavior is unknown; if helpers cannot mate in the burrow and reproduce (as suggested by their lack of sperm), then their repro-

duction is precluded. Moreover, because helpers lose tarsal segments and cannot leave the colony, they have no options but to help extend the burrow system and maintain and defend the burrow entrance. These considerations suggest that dominance behavior is not expected in *A. incompertus*, despite the conditions of high skew, high relatedness, and strong ecological constraints that otherwise favor it.

In other ambrosia beetles, Kirkendall *et al.* (this volume) suggest that cooperative breeding may exist in taxa within which the wood resource is sufficiently long-lived relative to generation time. Because some of these taxa exhibit haplodiploidy (e.g. the Xyleborini) and others are diploid (e.g. Corthylini), and ecological constraints may be high, such ambrosia beetles offer excellent prospects for study of the origin and evolution of cooperative breeding.

Burying beetles. Burying beetles are restricted to one generation of breeding by the duration of their food source. However, multiple adults of each sex will sometimes breed together on large carcasses (Trumbo 1992, 1995; Scott 1994; Eggert and Müller, this volume). These individuals are unrelated, and ecological constraints, measured as the likelihood that beetles do not find carcasses, are probably high. Scott (1994) analyzed communal breeding in *Nicrophorus tomentosus* using the skew framework, and she shows that both females gain in per capita reproduction by breeding communally whenever carcasses are large and flies are present, because neither female can lay sufficient eggs to use the whole carcass and they both remove fly eggs and larvae (see also Trumbo 1992, 1995; Eggert and Müller, this volume). By contrast, Eggert and Müller (this volume) suggest that larger female *Nicrophorus* would always gain by excluding small ones from the carcass, but are unable to do so effectively on larger carcasses. *Nicrophorus* behavior on carcasses of different sizes provides a rare opportunity to study the ecological basis of the evolution of communal sociality vs. interactions involving dominance (Robertson *et al.* 1997).

Allodapine bees. *Exoneura bicolor*, the one allodapine bee studied in enough detail for meaningful discussion, exhibits social variability associated with montane vs. heathland habitats (Schwarz *et al.*, this volume). In montane populations, relatedness is high (0.5–0.6) and ecological constraints are weak. In heathland populations, relatedness is high (about 0.5) in reused nests but effectively zero in new nests (because relatives are not available); moreover, costs of dispersal are higher than in montane populations because nest sites are more dispersed. In both habitats, the extent of reproductive skew differs between reused nests, with high skew, and new nests, with very low skew. Moreover, per capita brood production increases with group size in both habitats, and it is higher among bees in reused nests, perhaps because they initiate reproduction earlier. Heathland populations exhibit higher levels of solitary nesting, probably because kin are less available for cofounding.

The coincidence of weak ecological constraints and low skew in new nests in montane habitats is consistent with predictions of the skew model, and the higher skew in reused nests in both habitats may be associated with their advantage in per capita brood production. However, the absence of dominance behavior in this species is inconsistent with the high levels of skew in reused nests unless colonies are strictly matrifilial with singly mated foundresses (Reeve and Keller 1995); are foundresses so long-lived, relative to nest duration, that supersedure is exceedingly rare and aggressive testing of reproductives is not selected for?

Xylocopine bees. In some *Xylocopa* bees, the prediction of role reversal from skew models is fulfilled: nests often contain two females, a dominant forager and a subordinate, nest-guarding female (Velthuis 1987; Michener 1990; Stark 1992). Moreover, some of the assumptions of the model appear to be met, in that nest productivity is greatly increased by the presence of a subordinate guard bee (because the dominant forages longer), foraging may be risky, but not excessively so, given the usual long lifespan of *Xylocopa* females (Hogendoorn and Velthuis 1993), and subordinates sometimes inherit the nest. The situation in these bees is more complicated in several ways, in that: (1) a clear division of labor does not always exist in multi-female nests (see, for example, Stark *et al.* 1990); (2) the degree of skew probably differs considerably both between *Xylocopa* species and between multifemale nests within species (Stark *et al.* 1990; Hogendoorn and Velthuis 1993). However, analysis of this variation, especially in comparison to other bees, such as *Ceratina* spp. and other allodapines (see, for example, Maeta *et al.* 1985; Mason 1989), which inhabit similar types of nest but do not show such strong role-reversal (Sakagami and Maeta 1977, 1985, 1987), should allow rigorous tests of skew models.

Halictine bees. We apply the skew model to two of the better-known taxa in halictines: the bivoltine *Lasioglossum*

zephyrum and similar bivoltine species with eusociality (Wcislo, this volume; Yanega, this volume), and the communal species *L. hemichalceum,* studied by Kukuk and Sage (1994). In bivoltine species with eusociality, relatedness is usually high (see, for example, Crozier *et al.* 1987; Kukuk 1989; Packer and Owen 1994), constraints on independent nest-founding in summer are high (see, for example, Sakagami 1977; Yanega, this volume), aggression is common among nestmates, and productivity per capita declines with increasing group size (Michener 1964). However, the degree to which foundresses can optimize skew in their own interests is variable (Richards *et al.* 1995, *Halictus ligatus*) and depends on queen size and vigor (Kukuk and May 1991, *L. zephyrum*), so the assumptions of the model may not be closely met. In contrast to eusocial halictines, the communal *L. hemichalceum* exhibits low relatedness, probable low constraints on nest-founding, and per capita productivity remaining constant with group size (Kukuk and Sage 1994). Moreover, females often engage in trophallaxis, which is not found in eusocial halictines, and they coexist without aggression. These differences between eusocial and communal societies are consistent with predictions of skew models, but there are sufficiently many differences between the bees exhibiting different societies that it is difficult, as yet, to separate and identify their causes.

Ants. The ecological and genetic causes of different social systems in plurimatrous ant species have been analyzed using skew models (Reeve and Ratnieks 1993; Bourke and Heinze 1994) with considerable success. Multiple-foundress associations in ants are characterized by zero relatedness and high ecological constraints, as single foundresses almost never survive (Hölldobler and Wilson 1990; Choe and Perlman, this volume).

Another important form of variation in ant societies is cooperative breeding, found in some ponerine ants without queens (Peeters 1991, this volume; Ito and Higashi 1991; Ito 1993a,b). Peeters (1991, this volume) describes a suite of traits, common in ponerines, that may influence the evolution of their social systems, including limited queen–worker dimorphism, low fecundity and small colonies, solitary predation on arthropods sometimes involving prey specialization, spermathecae in workers, seasonal trophic stress which prevents high investment in alates, and nest fission and frequent nest movement. These traits may favor cooperative breeding, rather than eusociality, because ecological specialization keeps colonies small, and the presence of small colonies and fission means that any given individual has a non-trivial opportunity to become a reproductive, through either inheritance or colony fission. This hypothesis could be tested by comparing colony sizes and prey specialization between ponerines with and without queens, once sufficient phylogenetic information is available. One set of observations consistent with the high-skew, high-aggression prediction of skew models is high levels of aggression in ponerines without queens (Ito 1993a,b; Peeters and Tsuji 1993; Heinze *et al.* 1994); such levels should vary with the frequency of supersedure, except in species where becoming a gamergate is more a matter of timing than prowess or vigor (see, for example, Peeters and Higashi 1989; Ito and Higashi 1991).

Lepidoptera. In social groups of larval lepidoptera, as in aphids, the fitness criterion for helping vs. dispersing decisions is successful molting to adulthood (Costa and Pierce, this volume). Relatedness has been estimated in only one species, *Malacosoma americanum,* as about 0.5, but other species are likely to exhibit similarly high relatedness depending on mating, sperm utilization, and the likelihood of colony fusion (Costa and Ross 1993; Costa and Pierce, this volume). In terms of foraging and pupal size, there are strong benefits to group-living (Costa and Pierce, this volume). The lack of dominant–subordinate relations in caterpillars, despite high relatedness and high costs of solitary development, could be due to the lack of possible benefits to dominance in a situation where all individuals must forage and effective defense involves signal enhancement through coordinated group behavior (Costa and Pierce, this volume). But are there caterpillars that 'cheat', taking advantage of scouting and silk production by others?

Mites. Relatedness is probably quite high in group-living mites, the success of dispersers is probably low, and success in fighting off attackers increases with numbers of defenders (Saito, this volume). Given that mites have clear capabilities to evolve fighter morphs (Timms *et al.* 1981), why have soldiers not evolved? Several possibilities include attacks by natural enemies being rare, web duration being short in relation to generation time, or an insufficiently large improvement in defensive ability through specialized morphology and behavior.

Spiders. In colonial web-building spiders in the genus *Metepeira* (Uetz and Hieber, this volume), the costs of dispersal are probably fairly high, relatedness varies from

low to high, and per capita fitness is highest in medium-sized groups. Individuals compete within the group for good positions, but 'losers' normally do not leave, probably because the costs of building and maintaining a solitary web are so high. Cooperative spiders (Avilés, this volume) exhibit high relatedness and high costs of dispersal, and probable low variation in fighting ability. Group-living is apparently advantageous because, although per capita fertility may decrease as colonies increase beyond a certain size, adult and offspring survival is enhanced in social groups. Spiders are unusual in that they cooperate, and (sometimes) show high relatedness, but exhibit no reproductive dominance behavior; are the benefits of division of labor very low, as discussed above? Is cheating not favored because relatedness is high, and intercolony selection is strong?

Rypstra (1993) and Vollrath (1986) discuss the possibility that *Anelosimus eximius* should be considered eusocial, because it exhibits generation overlap, cooperative brood care and web building, and only some females reproduce while many remain uninseminated. It is not yet clear, however, if this species has a division of labor that involves a trade-off between reproduction and helping, which is required under the eusociality definitions of Michener (1974), Wilson (1971) and Crespi and Yanega (1995). If all females are striving to feed and reproduce, and some fail owing to lack of ability to compete successfully for food (Rypstra 1993), then variation among individuals in reproduction and alloparental care cannot be considered socially adaptive unless it can be demonstrated that helpers reproduce less because they help, and reproductive individuals help less because they reproduce. For example, if all spiders reproduce when food is abundant then it is difficult to consider colonies eusocial when food is scarce; but if non-reproductive females feed reproductive ones, or preferentially allow them to feed, then a division of labor would be demonstrated. *Zethus miniatus,* a eumenid wasp with nest-sharing, offers a parallel case: females less successful in social competition for resources are more worker-like and less fecund (West-Eberhard 1988).

Comparative assessment of skew models

How well do the skew models fare in this broad review? First, the limited evidence available thus far suggests that communal societies differ from cooperatively breeding and eusocial ones in several of the predicted ways. Many communal societies exhibit low relatedness and an increase or constancy in per capita reproduction with increased group size (Kukuk and Sage 1994 for Hymenoptera; Eggert and Müller, this volume, for burying beetles). Moreover, most species with communal nests also show a high incidence of solitary nesting (e.g. bees, spiders, thrips, embiids), which suggests that ecological constraints are weak. Some species with communal nests, however, exhibit fairly high relatedness among nestmates (the wasp *Cerceris antipodes* (McCorquodale 1988) and some spiders (Avilés, this volume)), or probably do (the bee *Microthurge corumbae* (Garófalo et al. 1992)), owing to preferential association with kin or nest reuse among sibs. Is success at independent nest initiation especially low in these species, and if so, why do they not exhibit reproductive dominance and cooperative breeding?

Second, in contrast to the patterns for species with communal nests, nestmate relatedness appears to be high in the vast majority of species that show cooperative breeding (bees, wasps, and vertebrates; but see also Ligon 1983 for cases of helping by unrelated individuals). Moreover, in most species per capita reproduction decreases with increased group size (Michener 1964; Smith 1990; Reeve 1991; but see Wcislo et al. 1993). These data are clearly in accord with the main expectation from skew models. However, there are apparently no species showing reproductive division of labor and very low but positive relatedness, as might sometimes be expected under high ecological constraints. Moreover, although success of single dispersing females is low in many taxa (e.g. *Polistes* wasps, social spiders, embiids, many halictines, *Austroplatypus incompertus* beetles), we do not yet know if it is *lower* in cooperative breeders than in related communal or solitary forms (see, for example, Raw 1977). A similar lack of critical data exists in the study of cooperatively breeding birds (Smith 1990).

Third, some cooperatively breeding xylocopine bees exhibit a type of role-reversed social system predicted by the model, although the presence in nature of the model parameters yielding this system remains to be tested.

Fourth, some taxa with high ecological constraints, probable high relatedness, and differences in fighting ability, including some spiders, mites, and thrips with soldiers, do not exhibit reproductive dominance, although some of them show forms of division of labor. These observations suggest that taxon-specific or other factors mitigate against the evolution of dominance; we have discussed possible causes for the lack of dominance in these taxa above. Analysis of the evolution of reproductive dominance

requires experimental study of species showing intraspecific and intranest variation, or comparison of closely related species, some peaceful and others not.

The assumptions and predictions of skew models appear to fit best for some Hymenoptera and cooperatively breeding vertebrates, perhaps because the nature of their nests and territories usually allows effective control. Further development of skew models, leading to more broad taxonomic application, should incorporate: (1) the behavioral option of intraspecific parasitism; (2) variable ability to dominate; (3) frequency-dependence of social dynamics between dominants and subordinates; and (4) interactions among subordinates. Moreover, robust testing of skew models requires separation of skew due to social causes from skew due to other aspects of life history (Crespi and Yanega 1995). Despite these suggestions and caveats, the models help substantially in organizing the integration of genetic, phenotypic, ecological and demographic causes of social evolution described in detail above, and serve as a solid base for taxon-specific analyses.

A final observation emerging from this review is that numerous group-living taxa (*Lasioglossum hemichalceum*, burying beetles, some allodapines, lepidopteran larvae, cooperative spiders such as *Anelosimus eximius*, and some communal thrips) exhibit communal social systems (see Table I-1) with forms of mutualistic cooperation, including trophallaxis and building, that in some cases are not kin-selected and appear vulnerable to cheating. What ecological factors affect the presence and magnitude of cooperation in egalitarian societies? Demonstrating the presence, and ecological causes, of mutualism (Vehrencamp 1979), group selection (Avilés, this volume), or other forms of cooperation (Connor 1995a,b) in such taxa would greatly advance our understanding of the diversity of social systems.

WHAT'S IN A NAME

Our second main goal in this volume has been exploring the usefulness of different social lexicons. In particular, to what extent will the categorization schemes and philosophies of Gadagkar (1994), Crespi and Yanega (1995), Sherman et al. (1995), or Wcislo (this volume) lead to new insights about the causes of social behavior (see also Downes 1995)? The traditional categorization of insect societies (Michener 1969, 1974; Wilson 1971; Danforth and Eickwort, this volume) has several important advantages: (1) facility of communication between students of sociality;

(2) division of social systems into three hierarchically structured components, whose presence and causes can be analyzed separately; (3) clear and accurate application to the characterization of colonies of halictines (and other bees), for which it was originally developed (Batra 1996; Michener 1969). However, this lexicon also faces two problems: (1) uncertainty as to the precise meaning of crucial terms, such as reproductive division of labor; and (2) lack of unambiguous applicability to various taxa outside the Hymenoptera and Isoptera, such as social vertebrates. Thus, even when terms are applied to groups at particular times, rather than to species (Eickwort 1981), comparisons among diverse taxa, especially vertebrates and invertebrates, may be inadvertently avoided or thwarted. Moreover, as Kukuk (1994) points out, the traditional division of eusocial forms into 'primitive' and 'advanced' creates phylogenetic and non-adaptationist biasses, and categorization of eusocial taxa based on morphological vs. behavioral differentiation and cyclical vs. permanent colonial life is more objective and precise.

The main novel effect of the approach of Crespi and Yanega (1995) is to view many social insects, including some ponerine ants, *Ropalidia* wasps, *Xylocopa* bees, allodapine bees, augochlorine bees, some stenogastrine wasps (Strassmann *et al.* 1994) and 'lower' termites (with respect to helpers, not soldiers), as cooperative breeders, because individuals retain totipotency throughout life, as do most or all social vertebrates. This idea can be analyzed by searching for systematic differences in putative causal factors between these taxa and related forms with castes (i.e., permanent differences in life-history trajectory after reproductive maturity). We predict that two factors, the temporal variability and predictability of helping vs. dispersing success over lifetimes (Koenig *et al.* 1992), and individual lifespan relative to generation time, may be sufficient to differentiate between cooperatively breeding and eusocial societies.

The approach of Gadagkar (1994) and Sherman *et al.* (1995) focusses at points along a spectrum of degree of alloparental care or reproductive skew, and views the selective pressures producing this variation as differing quantitatively along the axis, with no thresholds or qualitative shifts. The usefulness of such an approach should come from observing similarities in causal factors between species, populations or colonies with similar degrees of skew; for example, ecological constraints, demographic variables, or relatedness, covary with skew in systematic ways (see above). However, societies with the same value for skew may differ considerably in the nature of their behavioral interactions (Crespi and Yanega 1995; Keller and Perrin 1995) and the selective pressures that have shaped them, so focussing on one dimension may also blind us to the causes of among-taxon diversity.

Wcislo (this volume) points out the arbitrariness of all categorizations not based on phylogenies, and he recommends that social definitions be tailored to specific taxa and questions. He stresses the importance of recognizing homology and convergence, and the role of phylogenies in this goal. But given the phylogenetic knowledge that the traits of two or more taxa are not homologous, judging whether they exhibit convergence, rather than being sufficiently different to have resulted from different selective pressures, is still problematic (Edwards and Naeem 1994): traits and definitions will vary considerably in scale and scope. The lexicons described here have as their goal the broadest possible social scope, which they envision at a risk of recognizing false convergences but a hope of drawing conclusions that are as general as possible. By contrast, Wcislo's view encourages treating the myriad components of social systems more separately, as mosaics of traits with some nested in others. Thus, instead of asking 'when does eusociality evolve', we can ask under what circumstances we predict group-living, cooperative group-living, generation overlap, reproductive dominance, or division of labor with or without totipotency (see Brockmann, this volume). As these questions become more specific they reach into particular clades and species (such as Lepidoptera; see Costa and Pierce, this volume), and predictive ability increases but generality may be lost (Crespi 1996). Such flexibility of social definitions is also useful because it allows for incorporation of factors about which we are presently unaware. Moreover, the lack of typology in this approach avoids the blinding of biologists to substantial observed variability in social behavior within taxa.

For analysis of the broadest questions in social system evolution, we believe that the approaches of Gadagkar (1994), Sherman *et al.* (1995) and Crespi and Yanega (1995) (see also Keller and Perrin 1995) are more compatible than they appear: skew due to social causes is a fundamental effect and cause of social dynamics, and its correlates and magnitude in communal and cooperatively breeding taxa should help to indicate common causes (Keller and Reeve 1994). However, once permanent differences in life history have evolved, we believe that the rules of social games have changed in such a way as to make interpretations of skew values non-comparable to those in forms without castes.

In such cases, one of the critical assumptions of skew models, that dominant individuals' interests are met at the stable outcome of social contract negotiation, becomes irrelevant because the parties are mutually dependent. As a result, any observed skew values are probably largely a function of sex allocation ratio disputes, interworker squabbles, and occasional revolts once the queen has become less valuable to her minions (Trivers and Hare 1976; Crespi 1992b; Bourke 1994).

The utility and compatibility of the continuum and discrete social system ideas will ultimately be determined by two sources of evidence: data on skew values, plotted to detect multimodality which would indicate that a simple continuum does not exist, and data on whether or not the social categories described by Crespi and Yanega (1995) differ with respect to the variables that produce them. Wcislo's (this volume) approach of defining social attributes for each comparative test should spur test of specific social attributes, but we must still determine to what extent the aggregate of traits in a given nest or population are analogous in diverse taxa (Pagel 1994), to search for convergence and common causes at the broadest scale. Finally, the main usefulness of the traditional Batra–Michener–Wilson categorization lies in its universality and cohesive influence on social evolution literature to date. Its utility for some taxa, however, comes at a cost of difficulty in comparison of diverse social taxa, which hinders the unification of social theory. There is a fine line to walk between the dangers of typology and misinterpretation. A good start is to describe behavioral variability in detail before attempting classification, and to define usages of terms explicitly in every comparative context.

METHODS FOR ANALYZING SOCIAL ADAPTATION

Three types of method for the analysis of ultimate (evolutionary) questions are available for the study of adaptation: functional design (Williams 1966; Thornhill 1990), measurement of selection (Lande and Arnold 1983; Crespi 1990), and comparative analysis (reviewed in Harvey and Pagel 1991; Ridley and Grafen 1996). How are these methods used in the study of social behavior? Functional design analyses encompass three types of study: (1) descriptive, correlative data collection within species; (2) modeling and testing fits to predictions; and (3) manipulative experiments. In the analysis of sociality (and in contrast to other fields of behavioral ecology, such as

foraging studies) the former two methods have been used almost to the exclusion of the third. Brockmann (this volume) discusses the several experimental investigations of the ecological constraints model conducted for vertebrates, and concludes that such methods can usefully be applied to insects. Indeed, the few field experimental studies of social insect behavioral dynamics undertaken thus far (see, for example, Makino 1989; Mueller 1991; Reeve and Nonacs 1992; Rypstra 1993) have been remarkably successful and illuminating. Laboratory experiments have been similarly rare but productive (see, for example, Greenberg 1979; Kukuk and Crozier 1990; Kukuk and May 1991). Further such field and laboratory manipulations are critical to separating cause from effect, especially given the multitude of causal variables affecting social interactions.

The second method for analyzing sociality, measurement of selection, involves quantification of phenotypes and fitness components (see, for example, Lande and Arnold 1983; Crespi 1990). For social insects, the phenotype is usually a behavioral decision, or a social trait such as group size, and fitness is the fitness consequences of the alternatives (Queller and Strassmann 1989; Keller 1993). The main difficulty of this approach is that costs and benefits in the inclusive fitness equation $rb - c > 0$ refer to differences between the two alternatives, both of which a single individual cannot perform. Queller and Strassmann (1988, 1989) describe a way out of this dilemma: experimental manipulations to change an individual's option, followed by inclusive fitness measurement. Such studies can help to demonstrate whether individuals gain from helping or whether they are making the best of a bad job, and whether individuals are altruistic, manipulated, or mutualistic (Vehrencamp 1979). Moreover, inclusive fitness studies compel recognition and quantification of the benefits and costs of alternative social behaviors. Given the increasing ease of precise relatedness measurements, manipulative studies followed by fitness measurement (see, for example, Mueller et al. 1994) deserve renewed interest.

The third method for analysis of sociality is comparisons among species and higher taxa. Use of comparative methods and phylogenetics for the study of sociality has thus far been focussed mainly on determining numbers and locations of origins and losses of various forms of sociality (Packer 1991; Cameron 1993; Richards 1994; Stern 1994: Crespi 1996), and which transitions, of all possible ones, have occurred between social systems (Gittleman

1981; Gross and Sargent 1985; Carpenter 1989, 1991; Crespi 1996; Székely and Reynolds 1995; Danforth and Eickwort, this volume). These studies have shown, or suggested, that eusociality is ancient in some clades and is lost in others, that multiple origins of eusociality often occur in ecologically homogeneous groups, and that the observed transitions between social systems in a large taxonomic group often represent only a small subset of the possible transitions.

All of the chapters in this volume discuss the taxonomic or phylogenetic distribution of sociality and explore comparisons between related forms, sister taxa when possible, that differ socially. A few chapters go further, and conduct or discuss cladistic analyses and mapping of social traits onto their trees (Tallamy and Schaefer, Smith, Danforth and Eickwort, Crespi and Mound, this volume). However, in only a few cases, in this volume and elsewhere, is enough information available to investigate concurrent inferred change in both sociality and one of more variables that may affect its evolution (Hunt 1994; Crespi 1996; Crespi and Mound, this volume; Danforth and Eickwort, this volume).

Herein lies one possible future for social behavior studies: inferring phylogenies for all clades in which eusociality or some other social system has been gained or lost, and choosing for detailed behavioral–ecological study those non-social and social species that bracket the transition. Once such data are gathered, focusing on the causes of benefits and costs of alternative strategies, and relatedness, we ask in what traits these pairs of social and non-social sister taxa differ. Such traits should be those that facilitated or precipitated the transition, whereas the traits that are common to a clade containing both social and non-social forms may represent necessary conditions for origin or loss (Crespi 1996). Moreover, to the extent that these 'facilitating' traits differ between origins and losses of sociality in a reasonably homogeneous clade, we can test hypotheses of how traits evolving *after* the origin of eusociality might restrain its loss, or search for other differences in selection with respect to origin vs. maintenance.

Arguments against the accuracy of inferring ancestral states (see, for example, Frumhoff and Reeve 1994; Ridley and Grafen 1996) are, at this early stage, irrelevant to such a research strategy. If sociality and its causes are highly labile phylogenetically (for example, in halictine bees (Michener 1985)), then such analyses should yield noise and no discernible patterns unless parallelism is rampant. However, we have few ways of knowing the labi-

lity of social traits, aside from the observation of within-population variability, until such analyses are performed. Indeed, de Queiroz and Wimberger (1993) have shown that behavioral traits exhibit no more homoplasy than morphological ones; this evidence argues that rates of character change are sufficiently low, relative to speciation rates, for substantial information about changes in behavior to persist in phylogenetically structured data. Phylogenetic conservatism should not, however, be interpreted as 'phylogenetic constraint' or 'inertia' (Reeve and Sherman 1993), because phylogeny, due only to speciation and extinction, has no effect on evolutionary constancy (Pagel 1994). Instead, causes of low levels of homoplasy should be investigated using hypotheses of stabilizing selection, genetic correlation or lack of variation, or other causes of stasis (Williams 1992).

The main uses of comparative data in a phylogenetic framework, in addition to sister-taxon comparisons, include: (1) test of specific, directional hypotheses, such as the conjecture of Tallamy and Schaefer (this volume) that subsociality in Hemiptera is an ancient trait with dubious adaptive value relative to other methods of reproduction; (2) analyses of the degree to which a trait is conserved phylogenetically or subject to convergence (see, for example, Edwards and Naeem 1993; Baroni Urbani 1993); (3) test for life-history differences between taxa with and without cooperative breeding (Poiani and Jermiin 1994); (4) analyses of scenarios for assembly of character complexes through time, such as Smith's (this volume) ideas concerning the evolution of paternal care in Belostomatidae; (5) test of hypotheses of constraint, genetic or phenotypic traits that prevent otherwise adaptive transitions due to selection (for example, in *Kladothrips* without soldiers (Crespi 1996)); (6) separation of cause from effects through the inference of temporal sequences (see, for example, Coddington 1988; Baum and Larson 1991), which may allow us to partition the causes of the origins of social systems from the effects of their maintenance; (7) tests of West-Eberhard's (1986, 1987, 1988) idea that within-species behavioral variation serves as a wellspring of evolutionary change and novelty, by measuring the extent of behavioral plasticity in a phylogenetic context (for example, does facultative usurpation give rise to social parasitism; Wcislo 1987) (see also Crozier 1992); and (8) statistical analysis of the effects of social systems on speciation and extinction rates (Slowinski and Guyer 1993), such as the hypothesis of Avilés (this volume) that some spider social systems lead to demographic chaos and high rates of colony extinction (possibly affecting species persistence), and Tallamy and Schaefer's (this volume) idea that subsociality engenders ecological failure and low diversification rates. Indeed, aphids and thrips with soldiers might also be considered as desperate ecological holdouts against barbarian invasion, given their high rates of group destruction (Moffett 1989; Foster 1990; B. J. Crespi, unpublished data), and, for some species, extreme rarity (Kurosu and Aoki 1991).

Some authors may despair that the database for statistical tests of the causes of the origin or loss of rarely-evolved social systems, such as eusociality, is impossibly small. However, recently developed tests, involving comparison of the magnitude of independent contrasts and the creation of clade-specific empirical null distributions (Garland *et al.* 1992, 1993), allow statistics with as few as one species showing a trait of interest.

Fine-scale sister-taxon comparisons, and statistical tests of specific hypotheses, can be used to infer the necessary and sufficient conditions for the evolution of sociality in any one taxonomic group. To the extent that these conditions can be shown to concur among diverse taxa, through an ability to predict social systems from some set of traits, our search for convergences will be rewarded. Few of the chapters in this volume have come so far as to attempt prediction of phylogenetic distributions of social systems, but we hope that the description of natural history and taxonomy, and the formulation of hypotheses, in all of the chapters will expedite future integration of phylogenetics into the study of adaptation.

The above three approaches all address evolutionary questions, rather than proximate mechanisms (Sherman 1988). For the analysis of sociality, physiological, neurological, and modeling studies designed to uncover the mechanisms responsible for individual behaviors provide another avenue of research (Bernays and Wcislo 1994), one that has developed largely independently from evolutionary inquiries except, to some extent, in the field of kin recognition. The advantage of knowing proximate mechanisms is that they provide a window into the sensory and motivational world of the insect or arachnid. To what aspects of its environment is a bee or spider attuned, and how do behavioral stimuli translate into physiological changes? How are reaction norms for social behaviors environmentally cued (Wcislo, this volume)? The 'how' questions can inform 'why' analyses by delineating the causes of behavioral variation; for example, hormone levels may mediate dominance and reproduction (Röseler 1991), and coincident tracking of behavior and hormone titres may help demonstrate who

controls reproduction in a colony, and to what extent. Ultimately, knowledge of mechanisms can also guide in the construction of realistic strategy sets for optimization studies, and show just how well-adapted, or constrained, is the behavior of our six- and eight-legged friends.

PROSPECTUS

Where do we go from here? We believe that recent extensions and intimations of the range of taxa available for study, newly developed theoretical approaches that integrate the roles of relatedness, manipulation and mutualism, and new molecular and phylogenetic methods, will lead to more rapid progress in the study of sociality than ever before.

First, the diversity of taxa analyzed in this volume attests to the recently widened range of social creatures, but also underscores the desperate need for depth in analysis of little-studied taxa, such as beetles, embiids, mites, psocids, stenogastrine wasps, and the solitary and communal Hymenoptera related to cooperative breeding and eusocial forms (Sakagami 1980; Smith 1990; Koenig *et al.* 1992; Crespi 1996). We hope that this volume will serve as a guide to which taxa will best reward future studies of the natural-history core of social-behavior research, especially for workers in the tropics and subtropics where this core is rapidly being eroded by development.

Second, the models of Vehrencamp, Reeve, Keller and Ratnieks (Vehrencamp 1983a,b; Reeve and Ratnieks 1993; Keller and Reeve 1994; Reeve and Keller 1995) provide a useful unifying framework for the analysis of sociality, as they mediate between the vastly diverse genetics, phenotypes, ecology and demography of the social bestiary and the set of social systems actually observed. The models provide new insights into cause–effect relations (for example, skew causing aggression, rather than the reverse) and predict how sets of variables should be interrelated; such prediction structures the search for convergences. Currently, however, the models are applicable to only a subset of social creatures, their assumption of optimal skew adjustment requires rigorous tests, and additional parameters and behaviors, such as mortality rates and inheritance possibilities, need to be incorporated. However, the broad-scale analysis of the models' predictions conducted here suggests that they provide a realistic simplification of nature for many taxa, and capture the main causes of social variability.

One of the main uses of models is in telling us what to measure in nature; consideration of the parameters of skew models highlights our current ignorance of the costs of dispersal and the effects of group size and division of labor on individual and colony reproduction. The study of arthropod sociality has, except for a few clear exceptions (see, for example, Bourke and Heinze 1994) developed rather independently from the behavioral–ecological adaptationist program, so useful for understanding mating systems and foraging, of explaining behavior from variation in resource and enemy distributions in space and time. Is sociality (rather than solitary life) more common in bees that nest in disturbed areas (Danforth and Eickwort, this volume) because nests are aggregated, and parasites become common (Knerer and Atwood 1967)? How do wasps with aggressive and peaceful societies (Keller and Reeve 1994) differ ecologically? To use skew models effectively, the ecological variables affecting each parameter must be identified and quantified.

Third, we believe that molecular methods are transforming social behavior studies. Within the next few years, we should have methods with sufficient statistical power to infer relatedness accurately and reasonably easily between any groups or pairs of individuals in any taxon (Queller *et al.* 1993; Avise 1994; Schierwater *et al.* 1994). Coupling such measurements with data on individual behavior will allow much more robust tests of the assumptions and predictions of skew (and other) models than have previously been possible. Similarly, recent developments in ability to infer phylogenies should lead to easier integration of phylogenetic with behavioral–ecological data. Indeed, given the rarity of origins and losses of cooperative breeding and eusociality for most forms, phylogenetic hypotheses should be available for most of the main clades within the next five years.

Integration of phylogenetic with behavioral–ecological data should help in addressing some of the most fundamental questions in the evolution of social behavior. To the extent that results from interspecific and intraspecific studies concur, variation in social systems among and within all animals will ultimately be explicable from some set of preconditions and genetic and demographic variables. This set of variables should, in turn, be predictable from myriad aspects of genetics, phenotype and ecology more or less unique to each lineage. By contrast, if vagaries of history such as multiple adaptive peaks or irreversible transitions thwart among-species prediction, then answers will be idiosyncratic and clade-specific, though, we hope, no less satisfying.

Brown (1994) describes the history of the study of avian social behavior as involving cycles of fission, leading to research programs narrowly focused by paradigm and methods, and fusion, whereby formerly disparate disciplines converge on the same problems. We hope that this volume will initiate a new round of convergence among students of sociality, by drawing diverse taxa, terminology, and approaches together in the analysis of some of the most spectacular products of evolutionary history.

ACKNOWLEDGEMENTS

We are very grateful to Leticia Avilés, Jane Brockmann, Jim Costa, Bryan Danforth, Janice Edgerly, William Foster, Larry Kirkendall, Chris Nalepa, Christian Peeters, Kern Reeve, Mike Schwarz, Jan Shellman-Reeve, David Stern, Bill Wcislo and Mark Winston for helpful comments.

LITERATURE CITED

Abe, T. 1991. Ecological factors associated with the evolution of worker and soldier castes in termites. *Ann. Entomol.* **9**: 101–107.

Abrams, J. and G. C. Eickwort. 1981. Nest switching and guarding by the communal sweat bee *Agapostemon virescens* (Hymenoptera, Halictidae). *Insectes Soc.* **28**: 105–116.

Alexander, R. D. 1974. The evolution of social behavior. *Annu. Rev. Ecol. Syst.* **5**: 325–383.

Alexander, R. D., K. M. Noonan and B. J. Crespi. 1991. The evolution of eusociality. In *The Biology of the Naked Mole Rat.* P. W. Sherman, J. U. M. Jarvis and R. D. Alexander, eds., pp. 3–44. Princeton: Princeton University Press.

Andersson, M. 1984 The evolution of eusociality. *Annu. Rev. Ecol. Syst.* **15**: 165–189.

Anholt, B. R. 1991. Measuring selection on a population of damselflies with a manipulated phenotype. *Evolution* **45**: 1091–1106.

Avise, J. 1994. *Molecular Markers.* New York: Chapman and Hall.

Baroni Urbani, C. 1993. The diversity and evolution of recruitment behaviour in ants, with a discussion of the usefulness of parsimony criteria in the reconstruction of evolutionary histories. *Insectes Soc.* **40**: 233–260.

Batra, S. W. T. 1996. Nests and social behavior of halictine bees of India. *Indian J. Entmol.* **28**: 375–393.

Baum, D. A. and Larson, A. 1991. Adaptation reviewed: a phylogenetic methodology for studying character macroevolution. *Syst. Zool.* **40**: 1–18.

Benton, T. G. and W. A. Foster 1992. Altruistic housekeeping in an aphid. *Proc. R. Soc. Lond.* B247: 199–202.

Bernays, E. A. and W. T. Wcislo. 1994. Sensory capabilities, information processing, and resource specialization. *Q. Rev. Biol.* **69**: 187–204.

Boomsma, J. J. and A. Grafen. 1990. Intraspecific variation in ant sex ratios and the Trivers-Hare hypothesis. *Evolution* **44**: 1026–1034.

Bourke, A. F. G. 1988. Worker reproduction in the higher eusocial Hymenoptera. *Q. Rev. Biol.* **63**: 291–311.

–. 1994. Worker matricide in social bees and wasps. *J. Theor. Biol.* **167**: 283–292.

Bourke, A. F. G. and J. Heinze. 1994. The ecology of communal breeding: the case of multiple-queen leptothoracine ants. *Phil. Trans. R. Soc. Lond.* B345: 359–372.

Brockmann, J. 1984. The evolution of social behaviour in insects. In *Behavioural Ecology. An Evolutionary Approach*, 2nd edn. J. R. Krebs and N. B. Davies, eds., pp. 340–361. Oxford: Blackwell Scientific Publications.

Brown, J. L. 1994. Historical patterns in the study of avian social behavior. *Condor* **96**: 232–243.

Brown, J. L. and S. L. Pimm. 1985. The origin of helping: the role of variability in reproductive potential. *J. Theor. Biol.* **112**: 465–477.

Cameron, S. 1993. Multiple origins of advanced eusociality in bees inferred from mitochondrial DNA sequences. *Proc. Natl. Acad. Sci. U.S.A.* **90**: 8687–8691.

Carpenter, J. M. 1989. Testing scenarios: wasp social behavior. *Cladistics* **5**: 131–144.

–. 1991. Phylogenetic relationships and the origin of social behavior in the Vespidae. In *The Social Biology of Wasps*. K. G. Ross and R. W. Matthews, eds., pp. 1–32. Ithaca: Cornell University Press.

Chandrashekara, K. and R. Gadagkar. 1992. Queen succession in the primitively eusocial tropical wasp *Ropalidia marginata* (Lep.) (Hymenoptera: Vespidae). *J. Insect. Behav.* **5**: 193–209.

Choe, J. C. 1988. Worker reporoduction and social evolution in ants (Hymenoptera: Formicidae). In *Advances in Myrmecology*. J. C. Trager, ed., pp. 163–187. New York: E. J. Brill.

–. 1994. Communal nesting and subsociality in a webspinner, *Anisembia texana* (Insecta: Embiidina: Anisembiidae). *Anim. Behav.* **47**: 971–973.

–. 1995. Plurimatry: new terminology for multiple reproductives. *Insectes Soc.* **8**: 133–137.

Christianson, T. E. 1984. Behaviour of colonial and solitary spiders of the theridiid species *Anelosimus eximius*. *Anim. Behav.* **32**: 725–734.

Clarke, M. F. 1984. Co-operative breeding by the Australian bell miner *Manorina melanophrys* Latham: a test of kin selection theory. *Behav. Ecol. Sociobiol.* **14**: 137–146.

Coddington, J. A. 1988. Cladistic tests of adaptational hypotheses. *Cladistics* **4**: 3–22.

Connor, R. C. 1995a. Altruism among non-relatives: alternatives to the "Prisoner's Dilemma". *Trends Ecol. Evol.* **10**: 84–86.

–. 1995b. The benefits of mutualism: a conceptual framework. *Biol. Rev.* **70**: 427–457.

Costa, J. T. III and K. G. Ross. 1993. Seasonal decline in intracolony genetic relatedness in eastern tent caterpillars: implications for social evolution. *Behav. Ecol. Sociobiol.* **32**: 47–54.

Crespi, B. J. 1990. The measurement of selection on phenotypic interaction systems. *Am. Nat.* **135**: 32–47.

–. 1992a. Eusociality in Australian gall thrips. *Nature (Lond.)* **359**: 724–726.

–. 1992b. Cannibalism and trophic eggs in subsocial and eusocial insects. In *Cannibalism – Ecology and Evolution in Diverse Taxa*, M. Elgar and B. Crespi, eds., pp. 176–213. Oxford University Press.

–. 1994. Three conditions for the evolution of eusociality: are they sufficient? *Insectes Soc.* **41**: 395–400.

–. 1996. Comparative analysis of the origins and losses of eusociality: causal mosaics and historical uniqueness. In *Phylogenies and the Comparative Method in Animal Behavior*. E. Martins, ed., pp. 253–287. New York: Oxford University Press.

Crespi, B. J. and D. Yanega. 1995. The definition of eusociality. *Behav. Ecol.* **6**: 109–115.

Crozier, R. H. 1992. The genetic evolution of flexible strategies. *Am. Nat.* **139**: 218–223.

Crozier, R. H., B. H. Smith and Y. C. Crozier. 1987. Relatedness and population structure of the primitively eusocial bee, *Lasioglossum zephyrum* (Hymenoptera: Halictidae) in Kansas. *Evolution* **41**: 902–911.

de Queiroz, A. and P. H. Wimberger. 1993. The usefulness of behavior for phylogeny estimation: levels of homplasy in behavioral and morphological characters. *Evolution* **47**: 46–60.

Downes, M. F. 1995. Australasian social spiders: what is meant by 'social'? *Rec. W. Austr. Mus. Suppl.* **52**: 25–32.

Eberhard, W. G. 1975. The ecology and behavior of a subsocial pentatomid bug and two scelionid wasps: strategy and counterstrategy in a host and its parasites. *Smithson. Contrib. Zool.* **205**: 1–39.

Edwards, S. V. and S. Naeem. 1993. The phylogenetic component of cooperative breeding in perching birds. *Am. Nat.* **141**: 754–789.

–. 1994. Homology and comparative methods in the study of avian cooperative breeding. *Am. Nat.* **143**: 723–733.

Eickwort, G. C. 1981. Presocial insects. In *Social Insects*, vol. 2. H. R. Hermann, ed., pp. 199–280. New York: Academic Press.

Emlen, S. 1982. The evolution of helping. I. An ecological constraints model. *Am. Nat.* **119**: 29–39.

–. 1991. Evolution of cooperative breeding in birds and mammals. In *Behavioural Ecology. An Evolutionary Approach*, 3rd edn. J. R. Krebs and N. B. Davies, eds., pp. 301–337. Oxford: Blackwell Scientific Publications.

Evans, H. E. 1977. Extrinsic versus intrinsic factors in the evolution of insect sociality. *BioScience* **27**: 613–617.

Felsenstein, J. 1985. Phylogenies and the comparative method. *Am. Nat.* **125**: 1–15.

Field, J. 1992. Intraspecific parasitism as an alternative reproductive tactic in nest-building wasps and bees. *Biol. Rev.* **67**: 79–126.

Foster, W. A. 1990. Experimental evidence for effective and altruistic colony defense against natural predators by soldiers of the gall-forming aphid *Pemphigus spyrothecae* (Hemiptera: Pemphigidae). *Behav. Ecol. Sociobiol.* **27**: 421–430.

Foster, W. A. and P. A. Northcott. 1994. Galls and the evolution of social behaviour in aphids. In *Plant Galls: Organisms, Interactions, Populations.* M. A. J. Williams, ed., pp. 161–182. Oxford: Claredon Press.

Frumhoff, P. C. and H. K. Reeve. 1994. Using phylogenies to test hypotheses of adaptation: a critique of some current proposals. *Evolution* **48**: 172–180.

Gadagkar, R. 1990. Evolution of eusociality: the advantage of assured fitness returns. *Phil. Trans. R. Soc. Lond.* B329: 17–25.

–. 1991a. On testing the role of genetic asymmetries created by haplodiploidy in the evolution of eusociality in the Hymenoptera. *J. Genet.* **70**: 1–31.

–. 1991b. Demographic predisposition to the evolution of eusociality: a hierarchy of models. *Proc. Natl. Acad. Sci. U.S.A.* **88**: 10 993–10 997.

–. 1994. Why the definition of eusociality is not helpful to understand its evolution and what we should do about it. *Oikos* **70**: 485–487.

Gadagkar, R, C. Vinutha, A. Shanubhogue and A. P. Gore. 1988. Pre-imaginal biasing of caste in a primitively eusocial insect. *Proc. R. Soc. Lond.* B233: 175–189.

Gamboa, G. J. 1978. Intraspecific defense: advantage of social cooperation amoung paper wasp foundresses. *Science (Wash., D.C.)* **199**: 1463–1465.

Garland, T., P. H. Harvey and A. R. Ives. 1992. Procedures for the analysis of comparative data using phylogenetically independent contrasts. *Syst. Biol.* **41**: 18–32.

Garland, T., A. W. Dickerman, C. M. Janis, and J. A. Jones. 1993. Phylogenetic analysis of covariance by computer simulation. *Syst. Biol.* **42**: 265–292.

Garófalo, C. A. E. Camillo, M. J. O. Campos and J. C. Serrano. 1992. Nest re-use and communal nesting in *Microthurge corumbae* (Hymenoptera, Megachilidae), with special reference to nest defense. *Insectes Soc.* **39**: 301–311.

Gibo, D. L. 1978. The selective advantage of foundress associations in *Polistes fuscatus* (Hymenoptera: Vespidae): a field study of the effects of predation on productivity. *Can. Entomol.* **110**: 519–540.

Gittleman, J. L. 1981. The phylogeny of parental care in fishes. *Anim. Behav.* **29**: 936–941.

Greenberg, L. 1979. Genetic component of bee odor in kin recognition. *Science (Wash., D. C.)* **206**: 1095–1097.

Gross, M. R. and R. C. Sargent. 1985. The evolution of male and female parental care in fishes. *Am. Zool.* **25**: 807–822.

Hamilton, W. D. 1964. The genetical evolution of social behaviour. *J. Theor. Biol.* **7**: 1–52.

–. 1971. Geometry for the selfish herd. *J. Theor. Biol.* **31**: 295–311.

–. 1972. Altruism and related phenomena, mainly in social insects. *Annu. Rev. Ecol. Syst.* **3**: 193–232.

Harvey, P. H. and M. Pagel. 1991. *The Comparative Method in Evolutionary Biology.* Oxford University Press.

Heinze, J., B. Hölldobler and C. Peeters. 1994. Conflict and cooperation in ant societies. *Naturwissenschaften* **81**: 489–4967.

Hogendoorn, K. and H. H. W. Velthuis. 1993. The sociality of *Xylocopa pubescens*: does a helper really help? *Behav. Ecol. Sociobiol.* **32**: 247–257.

Hölldobler, B. and Wilson, E. O. 1990. *The Ants.* Cambridge: Harvard University Press.

Honigberg, B. 1970. Protozoa associated with termites and their role in digestion. In *Biology of Termites.* vol. 2. K. M. Krishna and F. M. Weesner, eds., pp. 1–36. New York: Academic Press.

Hoogland, J. and P. W. Sherman. 1976. Advantages and disadvantages of bank swallow (*Riparia riparia*) coloniality. *Ecol. Monogr.* **46**: 33–58.

Hughes, C. R. and D. C. Queller. 1993. Detection of highly polymorphic microsatellite loci in a species with little allozyme polymorphism. *Molec. Ecol.* **2**: 131–138.

Hunt, J. H. 1994. Nourishment and evolution in wasps sensu lato. In *Nourishment and Evolution in Insect Societies.* J. H. Hunt and C. A. Nalepa, eds., pp. 211–244. Boulder, **Co.**: Westview Press.

Huxley, J. 1932. *Problems of Relative Growth.* London: Methuen.

Ito, F. 1993a. Social organization in a primitive ponerine ant: queenless reproduction, dominance hierarchy and functional polygyny in *Amblyopone sp.* (*reclinata* group) (Hymenoptera: Formicidae: Ponerinae). *J. Nat. Hist.* **27**: 1315–1324.

–. 1993b. Functional monogyny and dominance hierarchy in the queenless ponerine ant *Pachycondyla* (= *Bothroponera*) sp. in West Java, Indonesia (Hymenoptera, Formicidae, Ponerinae). *Ethology* **95**: 126–140.

Ito, F. and S. Higashi. 1991. A linear dominance hierarchy regulating reproduction and polyethism of the queenless ant *Pachycondyla sublaevis*. *Naturwissenschaften* **78**: 80–82.

Iwata, K. and S. F. Sakagami. 1966. Gigantism and dwarfism in bee eggs in relation to the modes of life, with notes on the number of ovarioles. *Jap. J. Ecol.* **16**: 4–16.

Jarvis J. U. M. and N. C. Bennett. 1993. Eusociality has evolved independently in two genera of bathyergid mole-rats but occurs in no other subterranean mammal. *Behav. Ecol. Sociobiol.* **33**: 253–260.

Jarvis, J. U. M., M. J. O'Riain, N. C. Bennett and P. W. Sherman. 1994. Mammalian eusociality: a family affair. *Trends Ecol. Evol.* **9**: 47–51.

Kaitala, V., B. H. Smith and W. M. Getz. 1990. Nesting strategies of primitively eusocial bees: a model of nest usurpation during the solitary stage of the nesting cycle. *J. Theor. Biol.* **144**: 445–471.

Keller, L. 1993. The assessment of reproductive success of queens in ants and other social insects. *Oikos* **67**: 177–180.

Keller, L. and N. Perrin. 1995. Quantifying the degree of eusociality. *Proc. R. Soc. Lond.* B260: 311–315.

Keller, L. and H. K. Reeve. 1994. Partitioning of reproduction in animal societies. *Trends Ecol. Evol.* **9**: 98–102.

Kennedy, J. S. 1992. *The New Anthropomorphism*. New York: Cambridge University Press.

Kent, D. S. and J. A. Simpson. 1992. Eusociality in the beetle *Austroplatypus incompertus* (Coleoptera: Curculionidae). *Naturwissenschaften* **79**: 86–87.

Koenig, W. D., F. A. Pitelka, W. J. Carmen, R. L. Mumme and M. T. Stanback. 1992. The evolution of delayed dispersal in cooperative breeders. *Q. Rev. Biol.* **67**: 111–150.

Klahn, J. 1988. Intraspecific comb usurpation in the social wasp *Polistes fuscatus*. *Behav. Ecol. Sociobiol.* **23**: 1–8.

Knerer, G. and C. E. Atwood. 1967. Parasitization of social halictine bees in southern Ontario. *Proc. Entomol. Soc. Ont.* **97**: 103–110.

Kukuk, P. F. 1989. Evolutionary genetics of a primitively eusocial bee, *Dialictus zephyrus*. In *Evolutionary Genetics of Social Insects*. R. E. Page, Jr. and M. D. Breed, eds., pp. 183–202. Boulder, Co.: Westview Press.

–. 1994. Replacing the terms "primitive" and "advanced": new modifiers for the term eusocial. *Anim. Behav.* **47**: 1475–1478.

Kukuk, P. and R. H. Crozier. 1990. Trophallaxis in a communal halictine bee *Lasioglossum (Chilalictus) erythrurum*. *Proc. Natl. Acad. Sci. U.S.A.* **87**: 5402–5404.

Kukuk, P. K., G. C. Eickwort, M. Raveret-Richter, B. Alexander, R. Gibson, R. A. Morse and F. Ratnieks. 1989. Importance of the sting in the evolution of sociality in the Hymenoptera. *Ann. Entomol. Soc. Am.* **82**: 1–5.

Kukuk, P. K. and B. P. May. 1988. Dominance hierarchy in the primitively eusocial bee *Lasioglossum (Dialictus) zephyrum*: is genealogical relationship important? *Anim. Behav.* **36**: 1848–1850.

–. 1991. Colony dynamics in a primitively eusocial bee *Lasioglossum (Dialictus) zephyrum* (Hymenoptera: Halictidae). *Insectes Soc.* **38**: 171–189.

Kukuk, P. F. and G. K. Sage. 1994. Reproductivity and relatedness in a communal halictine bee *Lasioglossum (Chilalictus) hemichalceum*. *Insectes Soc.* **41**: 443–455.

Kurosu, U. and S. Aoki. 1991. Why are aphid galls so rare? *Evol. Theor.* **10**: 85–99.

Lande, R. and S. Arnold. 1983. The measurement of selection on correlated characters. *Evolution* **37**: 1210–1226.

Ligon, D. J. 1983. Cooperation and reciprocity in avian social systems. *Am. Nat.* **121**: 336–384.

Lin, N. 1964. Increased parasite pressure as a major factor in the evolution of social behavior in halictine bees. *Insectes Soc.* **11**: 187–192.

Lin, N. and C. D. Michener. 1972. Evolution of sociality in insects. *Q. Rev. Biol.* **47**: 131–159.

Lubin, Y. 1995. Is there division of labour in the social spider *Achaearanea wau* (Theridiidae)? *Anim. Behav.* **49**: 1315–1323.

Maddison, W. P. 1990. A method for testing the correlated evolution of two binary characters: are gains or losses concentrated on certain branches on a phylogenetic tree? *Evolution* **44**: 539–557.

–. 1994. Phylogenetic methods for inferring the evolutionary history and processes of change in discretely valued characters. *Annu. Rev. Entomol.* **39**: 267–292.

Maeta, Y., S. F. Sakagami and C. D. Michener. 1985. Laboratory studies on the life cycle and nesting biology of *Braunsapis sauteriella*, a social xylocopine bee (Hymenoptera: Apidae). *Sociobiology* **10**: 17–41.

Makino, S. 1989. Switching of behavioral option from renesting to nest usurpation after nest loss by the foundress of a paper wasp, *Polistes riparius*: a field test. *J. Ethol.* **7**: 62–64.

Mason, C. A. 1989. Division of labor and adult interactions in eusocial colonies of two allodapine bee species (Hymenoptera: Anthophoridae) *J. Kansas Entomol. Soc.* **61**: 477–491.

Matthews, R. 1991. Evolution of social behavior in sphecid wasps. In *The Social Biology of Wasps*. K. G. Ross and R. W. Matthews, eds., pp. 570–602. Ithaca: Cornell University Press.

McCorquodale, D. B. 1988. Relatedness among nestmates in a primitively social wasp, *Cerceris antipodes* (Hymenoptera: Sphecidae). *Behav. Ecol. Sociobiol.* **23**: 401–406.

–. 1989a. Soil softness, nest initiation, and nest sharing in the wasp, *Cerceris antipodes* (Hymenoptera: Sphecidae). *Ecol. Entomol.* **14**: 191–196.

–. 1989b. Nest defense in single- and multifemale nests of *Cerceris antipodes* (Hymenoptera: Sphecidae). *J. Insect Behav.* **2**: 267–276.

Michener, C. D. 1964. Reproductive efficiency in relation to colony size in hymenopterous societies. *Insectes Soc.* **11**: 317–342.

–. 1969. Comparative social behavior of bees. *Annu. Rev. Entomol.* **14**: 299–342.

–. 1974. *The Social Behavior of the Bees*. Cambridge: Harvard University Press.

–. 1985. From solitary to eusocial: need there be a series of intervening species? In *Experimental Behavioral Ecology and Sociobiology*. B. Hölldobler and M. Lindauer, eds., pp. 293–305. Sunderland Mass.: Sinaeur Associates.

–. 1990. Caste in Xylocopine bees. In *Social Insects: An Evolutionary Approach to Castes and Reproduction*. W. Engels, ed., pp. 120–144. Berlin: Springer-Verlag.

Moffett, M. W. 1989. Samurai aphids, survival under siege. *Nat. Geogr.* September, pp. 406–422.

Mueller, U. G. 1991. Haplodiploidy and the evolution of facultative sex ratios in a primitively eusocial bee. *Science (Wash., D.C.)* **254**: 442–444.

Mueller, U. G., G. C. Eickwort and C. F. Aquadro. 1994. DNA fingerprinting analysis of parent-offspring conflict in a bee. *Proc. Natl. Acad. Sci. U.S.A.* **91**: 5143–5147.

Myles, T. G. 1988. Resource inheritance in social evolution from termies to man. In *The Ecology of Social Behavior*. C. N. Slobodchikoff, ed., pp. 379–423. New York: Academic Press.

Nalepa, C. A. 1994. Nourishment and the origin of termite eusociality. In *Nourishment and Evolution in Insect Societies*. J. H. Hunt and C. A. Nalepa, eds., pp. 57–104. Boulder, Co.: Westview Press.

Nonacs, P. 1986. Ant reproductive strategies and sex allocation theory. *Q. Rev. Biol.* **61**: 1–21.

Packer, L. 1991. The evolution of social behavior and nest achitecture in sweat bees of the subgenus *Evylaeus* (Hymenoptera: Halictidae): a phylogenetic approach. *Behav. Ecol. Sociobiol.* **29**: 153–160.

Packer, L. and R. E. Owen. 1994. Relatedness and sex ratio in a primitively eusocial halictine bee. *Behav. Ecol. Sociobiol.* **34**: 1–10.

Pagel, M. 1994. The adaptationist wager. In *Phylogenetics and Ecology*. P. Eggleton and R. I. Vane-Wright, eds., pp. 29–51. New York: Academic Press.

Peeters, C. 1991. The occurrence of sexual reproduction among ant workers. *Biol. J. Linn. Soc.* **44**: 141–152.

Peeters, C. and S. Higashi. 1989. Reproductive dominance controlled by mutilation in the queenless ant *Diacamma australe*. *Naturwissenschaften* **76**: 177–180.

Peeters, C. and K. Tsuji. 1993. Reproductive conflict among ant workers in *Diacamma* sp. from Japan: dominance and oviposition in the absence of the gamergate. *Insectes Soc.* **40**: 119–136.

Poiani, A. and L. S. Jermiin. 1994. A comparative analysis of some life-history traits between cooperatively and non-cooperatively breeding Australian passerines. *Evol. Ecol.* **8**: 471–488.

Promislow, D. E. L. 1991. Senescence in natural populations of mammals: a comparative study. *Evolution* **45**: 1869–1883.

Queller, D. C. 1989. The evolution of eusociality: reproductive head starts of workers. *Proc. Natl. Acad. Sci. U.S.A.* **86**: 3224–3226.

–. 1994. Extended parental care and the origin of eusociality. *Proc. R. Soc. Lond.* B**256**: 105–111.

Queller, D. C. and J. E. Strassmann. 1988. Reproductive success and group nesting in the paper wasp, *Polistes annularis*. In *Reproductive Success – Studies of Individual Variation in Contrasting Breeding Systems*. T. Clutton-Brock, ed., pp. 76–96. Chicago: University of Chicago Press.

–. 1989. Measuring inclusive fitness in social wasps. In *The Genetics of Social Evolution*. M. D. Breed and R. E. Page, Jr., eds., pp. 103–121. Boulder, Co.: Westview Press.

Queller, D. C., J. E. Strassmann and C. R. Hughes. 1993. Microsatellites and kinship. *Trends Ecol. Evol.* **8**: 285–288.

Raw, A. 1977. The biology of two *Exomalopsis* species (Hymenoptera: Anthophoridae) with remarks on sociality in bees. *Rev. Trop. Biol.* **25**: 1–11.

Reeve, H. K. 1991. Polistes. In *The Social Biology of Wasps*. K. G. Ross and R. W. Matthews, eds., pp. 99–148. Ithaca: Cornell University Press.

–. 1992. Queen activation of lazy workers in colonies of the eusocial naked mole-rat. *Nature (Lond.)* **358**: 147–149.

–. 1993. Haplodiploidy, eusociality and the absence of male parental and alloparental care in Hymenoptera: a unifying genetic hypothesis distinct from kin selection theory. *Phil. Trans. R. Soc. Lond.* B**342**: 335–352.

Reeve, H. K. and G. J. Gamboa. 1983. Colony activity integration in primitively eusocial wasps: the role of the queen (*Polistes fuscatus*, Hymenoptera: Vespidae). *Behav. Ecol. Sociobiol.* **13**: 63–74.

–. 1987. Queen regulation of worker foraging in paper wasps: a social feedback control system (*Polistes fuscatus*, Hymenoptera: Vespidae). *Behaviour* **102**: 147–167.

Reeve, H. K. and L. Keller. 1995. Partitioning of reproduction in mother-daughter versus sibling associations: a test of optimal skew theory. *Am. Nat.* **145**: 119–132.

Reeve, H. K. and P. Nonacs. 1992. Social contracts in wasp societies. *Nature (Lond.)* **359**: 823–825.

–. 1993. Weak queen or social contract? Reply. *Nature (Lond.)* **363**: 503.

Reeve, H. K. and F. L. W. Ratnieks. 1993. Resolutions of conflicts in polygynous societies: mutual tolerance and reproductive skew. In *Queen Number and Sociality in Insects*. L. Keller, ed. pp. 45–85. Oxford University Press.

Reeve, H. K. and P. W. Sherman. 1993. Adaptation and the goals of evolutionary research. *Q. Rev. Biol.* **68**: 1–32.

Reilly, L. M. 1987. Measurements of inbreeding and average relatedness in a termite population. *Am. Nat.* **130**: 339–349.

Richards, M. H. 1994. Social evolution in the genus *Halictus*: a phylogenetic approach. *Insectes Soc.* **41**: 315–325.

Richards, M. H., L. Packer and J. Seger. 1995. Unexpected patterns of parentage and relatedness in a primitively eusocial bee. *Nature (Lond.)* **373**: 239–241.

Ridley, M. and A. Grafen. 1996. How to study discrete comparative methods. In *Phylogenies and the Comparative Method in Animal Behavior*. E. Martins, ed., pp. 76–103. Oxford University Press.

Riechert, S. E. 1985. Why do some spiders cooperate? *Agelena consociata*, a case study? *Fla. Entomol.* **68**: 105–116.

Robertson, I. C., W. G. Robertson and B. Roitberg. 1997. Mutual tolerance and the origin of communal associations in female burying beetles. *Behav. Ecol. Sociobiol.*, submitted.

Roisin, Y. 1994. Intragroup conflicts and the evolution of sterile castes in termites. *Am. Nat.* **143**: 751–765.

Röseler, P.-F. 1991. Reproductive competition during colony establishment. In *The Social Biology of Wasps*. K. G. Ross and R. W. Matthews, eds., pp. 309–335. Ithaca: Cornell University Press.

Ross, K. G. and J. Carpenter. 1991. Population genetic structure, relatedness, and breeding systems. In *The Social Biology of Wasps*. K. G. Ross and R. W. Matthews, eds., pp. 451–479. Ithaca: Cornell University Press.

Rypstra, A. L. 1993. Prey size, social competition, and the development of reproductive division of labor in social spider groups. *Am. Nat.* **142**: 868–880.

Sakagami, S. F. 1977. Seasonal change of nest survival and related aspects in an aggregation of *Lasioglossum duplex* (Dalla Torre), a eusocial halictine bee (Hymenoptera: Halictidae). *Res. Pop. Ecol.* **19**: 69–86.

–. 1980. Bionomics of halictine bees in northern Japan. I. *Halictus (Halictus) tsingtonensis* (Hymenoptera, Halictida) with notes on the number of origins of eusociality. *Kontyû* **48**: 526–536.

Sakagami, S. F. and Y. Maeta. 1977. Some presumably presocial habits of Japanese *Ceratina* bees, with notes on various social types in Hymenoptera. *Insectes Soc.* 24: 319–343.

–. 1985. Multifemale nests and rudimentary castes in the normally solitary bee *Ceratina japonica* (Hymenoptera: Xylocopinae). *J. Kansas Entomol. Soc.* 57: 639–656.

–. 1987. Multifemale nests and rudimentary castes of an "almost" solitary bee *Ceratina flavipes*, with additional observations on multifemale nests of *Ceratina japonica* (Hymenoptera, Apoidea). *Kontyû* 55: 391–409.

Sakagami, S. F., N. Gobbi and R. Zucchi. 1990. Nesting biology of the quasisocial sphecid wasp *Trypoxylon fabricator*. I. Nests and inhabitants. *Jap. J. Entomol.* 58: 846–862.

Schierwater, B., B. Streit, G. P. Wagner and R. DeSalle. 1994. *Molecular Ecology and Evolution: Approaches and Applications.* Basel: Springer-Verlag.

Scott, M. P. 1994. Competition with flies promotes communal breeding in the burying beetle, *Nicrophorus tomentosus*. *Behav. Ecol. Sociobiol.* 34: 367–373.

Seger, J. 1983a. Conditional relatedness, recombination, and the chromosome numbers of insects. In *Advances in Herpetology and Evolutionary Biology. Essays in Honor of Ernest E. Williams.* A. G. J. Rhodin and K. Miyata, eds., pp. 596–612. Cambridge: Harvard University Press.

–. 1983b. Partial bivoltinism may cause alternating sex-ratio biases that favour eusociality. *Nature (Lond.)* 301: 59–62.

Sherman, P. W. 1979. Insect chromosome numbers and eusociality. *Am. Nat.* 113: 925–935.

–. 1988. The levels of analysis. *Anim. Behav.* 36: 616–619.

Sherman, P. W., J. U. M. Jarvis and R. D. Alexander, eds. 1991. *The Biology of the Naked Mole Rat.* Princeton: Princeton University Press.

Sherman, P. W., E. A. Lacey, H. K. Reeve and L. Keller. 1995. The eusociality continuum. *Behav. Ecol.* 6: 102–108.

Simon, C., F. Frati, P. Flook, A. Beckenbach, B. J. Crespi and H. Liu. 1994. Evolution, weighting, and phylogenetic utility of mitochondrial gene sequences and a compilation of conserved polymerase chain reaction primers. *Ann. Entomol. Soc. Am.* 87: 651–701.

Slowinski, J. B. and C. Guyer. 1993. Testing whether certain traits have caused amplified diversification: an improved method based on a model of random speciation and extinction. *Am. Nat.* 142: 1019–1024.

Smith, B. H. 1987. Effects of genealogical relationship and colony age on the dominance hierarchy in the primitively eusocial bee *Lasioglossum zephyrum. Anim. Behav.* 35: 211–217.

Smith, D. T. 1982. Reproductive success of solitary and communal *Philoponella oweni* (Araneae: Uloboridae). *Behav. Ecol. Sociobiol.* 11: 249–256.

Smith, J. N. M. 1990. Summary. In *Cooperative Breeding in Birds: Long Term Studies of Ecology and Behavior.* P. B. Stacey and W. D. Koenig, eds., pp. 593–611. Cambridge University Press.

Stark, R. E. 1992. Cooperative nesting in the multivoltine large carpenter bee *Xylocopa sulcatipes* Maa (Apoidea: Anthophoridae): do helpers gain or lose to solitary females? *Ethology* 91: 301–310.

Stark, R. E., A. Hefetz, D. Gerling and H. H. W. Velthuis. 1990. Reproductive competition involving oophagy in the socially nesting bee *Xylocopa sulcatipes. Naturwissenschaften* 77: 38–40.

Starr, C. K. 1985. Enabling mechanisms in the origin of sociality in the Hymenoptera – the sting's the thing. *Ann. Entomol. Soc. Am.* 78: 836–840.

–. 1989. In reply, is the sting the thing? *Ann. Entomol. Soc. Am.* 82: 6–8.

–. 1990. Holding the fort: colony defense in some primitively social wasps. In *Insect Defenses. Adaptive Mechanisms and Strategies of Prey and Predators.* D. L. Evans and J. O. Schmidt, eds., pp. 421–463. Stony Brook: State University of New York Press.

Stern, D. L. 1994. A phylogenetic analysis of soldier evolution in the aphid family Hormaphididae. *Proc. R. Soc. Lond.* B256: 203–209.

Stern, D. L. and W. A. Foster. 1996. The evolution of soldiers in aphids. *Biol. Rev.* 71: 27–79.

Strassmann, J. E. 1993. Weak queen or social contract? *Nature (Lond.)* 363: 502–503.

Strassmann, J. E. and D. C. Queller. 1989. Ecological determinants of social evolution. In *The Genetics of Social Evolution.* M. D. Breed and R. E. Page, Jr., eds., pp. 81–101. Boulder, Co.: Westview Press.

Strassmann, J. E., C. R. Hughes, S. Turillazzi, C. Sólis and D. C. Queller. 1994. Genetic relatedness and incipient eusociality in stenogastrine wasps. *Anim. Behav.* 48: 813–821.

Suzuki, H. and M. Murai. 1980. Ecological studies of *Roparidia* (sic) *fasciata* in Okinawa island. I. Distribution of single- and multiple-foundress colonies. *Res. Pop. Ecol.* 22: 184–195.

Székely, T. and J. D. Reynolds. 1995. Evolutionary transitions in parental care in shorebirds. *Proc. R. Soc. Lond.* B262: 57–64.

Tallamy, D. W. 1994. Nourishment and the evolution of paternal investment in subsocial arthropods. In *Nourishment and Evolution in Insect Societies.* J. H. Hunt and C. A. Nalepa, eds., pp. 21–55. Boulder, Co.: Westview Press.

Tallamy, D. W. and T. K. Wood. 1986. Convergence patterns in subsocial insects. *Annu. Rev. Entomol.* 31: 369–390.

Taylor, V. A. 1978. A winged élite in a subcortical beetle as a model for a prototermite. *Nature (Lond.)* 276: 73–75.

Thompson, J. N. 1994. *The Coevolutionary Process.* Chicago: University of Chicago Press.

Thornhill, R. 1990. The study of adaptation. In *Interpretation and Explanation in the Study of Animal Behavior.* M. Bekoff and D. Jamieson, eds., pp. 31–62. Boulder, Co.: Westview Press.

Timms, S., D. N. Ferro and R. M. Emberson. 1981. Andropolymorphism and its heritability in *Sancassania berlesei* (Michael) (Acari: Acaridae). *Acarologia* 22: 391–398.

Trivers, R. L. and H. Hare. 1976. Haplodiploidy and the evolution of the social insects. *Science (Wash., D.C.)* 191: 249–263.

Trumbo, S. T. 1992. Monogamy to communal breeding: exploitation of a broad resource base by burying beetles (*Nicrophorus*). *Ecol. Entomol.* **17**: 289–298.

–. 1995. Nesting failure in burying beetles and the origin of communal associations. *Evol. Ecol.* **9**: 125–130.

Vehrencamp, S. L. 1979. The roles of individual, kin and group selection in the evolution of sociality. In *Handbook of Behavioral Neurobiology*, vol 3. *Social Behavior and Communication*. P. Marler and J. G. Vandenbergh, eds., pp. 351–394. New York: Plenum Press.

–. 1983a. A model for the evolution of despotic versus egalitarian societies. *Anim. Behav.* **31**: 667–682.

–. 1983b. Optimal degree of skew in cooperative societies. *Am. Zool.* **23**: 327–335.

Velthuis, H. H. W. 1987. The evolution of sociality: ultimate and proximate factors leading to primitive social behaviour in carpenter bees. In *From Individual to Collective Behaviour in Social Insects*. J. M. Pasteels and J. L. Deneubourg, eds., pp. 405–434. Basel: Birkhäuser.

Vollrath, F. 1986. Eusociality and extraordinary sex ratios in the spider *Anelosimus eximius* (Araneae: Theriidae). *Behav. Ecol. Sociobiol.* **18**: 283–287.

Ward, P. I. and M. M. Enders. 1985. Conflict and cooperation in the group feeding of the social spider *Stegodyphus mimosarum*. *Behaviour* **94**: 167–182.

Wcislo, W. T. 1984. Gregarious nesting of a digger wasp as a "selfish herd" response to a parasitic fly (Hymenoptera: Sphecidae; Diptera: Sarcophagidae). *Behav. Ecol. Sociobiol.* **15**: 157–160.

–. 1987. The roles of seasonality, host synchrony, and behavior in the evolutions and distributions of nest parasites in Hymenoptera (Insecta), with special reference to bees (Apoidea). *Biol. Rev.* **62**: 515–543.

Wcislo, W. C. and J. H. Cane. 1996. Floral resource utilization by solitary bees (Hymenoptera: Apoidea) and exploitation of their stored foods by natural enemies. *Annu. Rev. Entomol.* **41**: 257–286.

Wcislo, W. T., B. S. Low and C. J. Karr. 1985. Parasite pressure and repeated burrow use by different individuals of *Crabro* (Hymenoptera: Sphecidae; Diptera: Sarcophagidae). *Sociobiology* **11**: 115–125.

Wcislo, W. T., M. J. West-Eberhard and W. G. Eberhard. 1988. Natural history and behavior of a primitively social wasp, *Auplopus semialatus*, and its parasite, *Irenangelus eberhardi* (Hymenoptera: Pompilidae). *J. Insect Behav.* **1**: 247–260.

Wcislo, W. T., A. Wille and E. Orozco. 1993. Nesting biology of tropical solitary and social sweat bees, *Lasioglossum (Dialictus) figueresi* Wcislo and *L. (D.) aeneiventre* (Friese) (Hymenoptera: Halictidae). *Insectes Soc.* **40**: 21–40.

Wenzel, J. W. and J. Pickering. 1991. Cooperative foraging, productivity, and the central limit theorem. *Proc. Natl. Acad. Sci. U.S.A.* **88**: 36–38.

West-Eberhard, M. J. 1975. The evolution of social behavior by kin selection. *Q. Rev. Biol.* **50**: 1–33.

–. 1978. Polygyny and the evolution of social behavior of wasps. *J. Kansas Entomol. Soc.* **51**: 832–856.

–. 1986. Alternative adaptations, speciation, and phylogeny. *Proc. Natl. Acad. Sci. U.S.A.* **83**: 1388–1392.

–. 1987. Flexible strategy and social evolution. In *Animal Societies: Theories and Facts*. Y. Itô, J. L. Brown and J. Kikkawa, eds., pp. 35–51. Tokyo: Japan Science Society Press.

–. 1988. Phenotypic plasticity and "genetic" theories of insect sociality. In *Evolution of Social Behavior and Integrative Levels. The T. C. Schnierla Conference Series*, vol. 3. G. Greenberg and E. Tobach, eds., pp. 123–133. New Jersey: Lawrence Erlbaum Associates.

Wheeler, D. 1994. Nourishment in ants: patterns in individuals and societies. In *Nourishment and Evolution in Insect Societies*. J. H. Hunt and C. A. Nalepa, eds., pp. 245–278. Boulder, Co.: Westview Press.

Whitham, T. G. 1992. Ecology of *Pemphigus* gall aphids. In *Biology of Insect-Induced Galls*. J. O. Shorthouse and O. Rohfritsch, eds., pp. 225–237, New York: Oxford University Press.

Williams, G. C. 1966. *Adaptation and Natural Selection*. Princeton: Princeton University Press.

–. 1992. *Natural Selection. Domains, Levels and Challenges*. Oxford University Press.

Wilson, E. O. 1971. *The Insect Societies*. Cambridge: Belknap Press of Harvard University Press.

Woolfenden, G. E. and J. W. Fitzpatrick. 1978. The inheritance of territory in group-living birds. *BioScience* **28**: 104–108.

–. 1984. *The Florida Scrub Jay: Demography of a Cooperatively-Breeding Bird*. Princeton: Princeton University Press.

Yamamura, N. 1993. Different evolutionary conditions for worker and soldier castes: genetic systems explaining caste distribution among eusocial insects. *J. Theor. Biol.* **161**: 111–117.

Yanega, D. 1988. Social plasticity and early-diapausing females in a primitively social bee. *Proc. Natl. Acad. Sci. U.S.A.* **85**: 4374–4377.

Organism index

Subject index